Thermal Performance of the Exterior Envelopes of Whole Buildings XIII International Conference

Clearwater, Florida, USA
4-8 December 2016

ISBN: 978-1-5108-6001-8

Printed from e-media with permission by:

Curran Associates, Inc.
57 Morehouse Lane
Red Hook, NY 12571

Some format issues inherent in the e-media version may also appear in this print version.

Printed by Curran Associates, Inc. (2018)

For permission requests, please contact Oak Ridge National Laboratory at the address below.

Oak Ridge National Laboratory
P.O. Box 2008
Oak Ridge, TN 37831-6070

Phone: (865) 576-7658
Fax: (865) 576-5812

Additional copies of this publication are available from:

Curran Associates, Inc.
57 Morehouse Lane
Red Hook, NY 12571 USA
Phone: 845-758-0400
Fax: 845-758-2633
Email: curran@proceedings.com
Web: www.proceedings.com

Thermal Performance of the Exterior Envelopes of Whole Buildings XIII International Conference

December 4–8, 2016
Clearwater, FL

Session 1

Practices: Attics

Field Measurements of Moisture in Cold Ventilated Attics with Different Types of Insulation Materials and Vapor Barrier 1
Thor Hansen, Eva B. Møller

Monitoring of Two Unvented Roofs with Air-Permeable Insulation in Climate Zone 2A 14
Kohta Ueno, Joseph W. Lstiburek

Cost-Effective High-Performance Unvented Attics 34
Francis Babineau, Jr

Principles: Insulation Performance And Material Properties

Wind Washing Effects on Mineral Wool Insulated Sheathings 46
Randy Van Straaten, John Straube, Trevor Trainor, Antoine Habellion

Improving Polyiso Thermal Performance at Low Temperature 59
John B. Letts, Jennifer Yao, Michael J. Hubbard

Determination of Linear Thermal Transmittance of Vacuum Insulation Panels (VIPs) by Measurement in a Guarded Hot-Plate (GHP) Apparatus or a Heat-Flow Meter (HFM) Apparatus 68
Christoph Sprengard, Andreas H. Holm, Christine Maderspacher

Session 2

Practices: Heritage and Retrofit Considerations for Buildings

Refurbishing Heritage and Historic Buildings—Key Motivation, Benefits and Challenges 75
Torben Valdbjørn Rasmussen

Treatment of Rising Damp in Historic Buildings Using a Wall Base Hygro-Regulated Ventilation System—Case Study 84
V.P. de Freitas, M.F.C. Barbosa, A.S. Guimarães, J.M.P.Q. Delgado

Study of Energy Intensity and WWRs of Existing Office Buildings 95
Viktoriya Mykytyak, Hitesh Doshi, Ramani Ramakrishnan

Principles: Walls—Moisture and Durability, Part 1

Using Fiber Insulation as a Means of Drying Internally Insulated Walls 102
Ekaterina Tzekova, Kim Pressnail

A Public Domain, Transient, Experimental Database on the Hygrothermal Performance of Durable, Energy-Efficient Full Basement Foundation Walls in a Cold Climate 113
Anna C. Harmon, Louise F. Goldberg

Session 3

Practices: Taking Stock

Field Demonstration of an Attic Retrofit Strategy Using Cellulose Insulation and Phase-Change Material 124
Kaushik Biswas, Som Shrestha, Andre O. Desjarlais, Nicholas M. Josefik, Andrew J. Nelson, Larry D. Stephenson, Jeffrey L. Lattimore, Robert De Vries

Evaluating Summertime Overheating in Multi-Unit Residential Buildings Using Surveys and In-Suite Monitoring 135
Marianne F. Touchie, Ekaterina S. Tzekova, Jeffrey A. Siegel, Bryan Purcell, Jonathan Morier

Online Airtightness Savings Calculator for Commercial Buildings in the United States, Canada, and China 152
Som Shrestha, Diana Hun, Lisa Ng, Andre Desjarlais, Steven Emmerich, Laverne Dalgleish

Principles: Material Durability

Critical Property Contrasts of Fluid-Applied Air and Water Barrier Membranes Used for Envelope Chemistries, Performance, and Durability 161
Patrick H. Young, Ben Meyer

Critical Freeze/Thaw Saturation Measurement of In-Service Masonry 177
Randy Van Straaten, Trevor Trainor, Chris Schumacher

Thermal and Moisture Properties of Extruded Polystyrene from Inverted Roofs: A Pilot Study 187
Leslie B.B. Peer, Fitsum Tariku, Wendy Ying Simpson

Session 4

Practices: Airtightness in Buildings

Commercial Building Airtightness Testing—Lessons Learned from the Red River College Airtightness Testing Program 197
Gary Proskiw, Rob Spewak, Kevin Knight, Cory Carson

Impact of Large Building Airtightness Requirements 206
Lorne Ricketts

Air Sealing Tight Commercial and Institutional Buildings 214
David Bohac, James Fitzgerald, Martha Hewett, Joshua Novacheck, Andrew Lutz

Principles: Moisture and Durability, Part 2

Protocol to Evaluate the Moisture Durability of Energy-Efficient Walls 223
Philip Boudreaux, Simon Pallin, Diana Hun, Manfred Kehrer, Roderick Jackson, Andre Desjarlais

A Systematic Method of Assessing the Durability of Wood-Frame Wall Assemblies: Towards the Limit-States Design Approach 235
Michael A. Lacasse, Martin Morelli

Risk of Condensation and Mold Growth in Highly Insulated Wood Frame Walls 246
Hamed H. Saber, Michael A. Lacasse, G. Ganapathy, Silvio Plescia, Anil Parekh

Session 5

Practices: Ventilation and Air Movement

Practical Applications and Case Study of Temperature-Based Smart Ventilation Controls 265
Michael Lubliner, Paul Francisco, Eric Martin, Iain Walker, Brennan Less, Robin Vieira, Rick Kunkle, Zachary Merrin

Smart Ventilation Control of Indoor Humidity in High-Performance Homes in Humid U.S. Climates 272
Brennan Less, Iain Walker

Measured Commercial and Institutional Building Pressures 281
David Bohac, Joshua Quinnell, Martha Hewett, James Fitzgerald, Joshua Novacheck, Andrew Lutz

Principles: Attic and Roof Performance

Sealed Attics Exposed to Two Years of Weathering in a Hot and Humid Climate 290
William Miller, Sudhir Railkar, Ming Shiao, Andre Desjarlais

Field Study on the Thermal and Hygrothermal Performance of Insulated Ventilated Nail Base System 305
Ming L. Shiao, Sudhir Railkar, William A. Miller, Andre Desjarlais

Annual Energy and Heat Flows in Vented and Sealed Attics—Parametric Study: Climate Zone 2A 317
Mikael Salonvaara, Achilles Karagiozis, William Miller

Session 6

Practices: Horizontal Envelope Assemblies

Innovative Roof Design for Passive Low-Income Housing in Western India 327
Madeline Gradillas, Qin Zhang, Emma Nelson, Bradley Tran, Leon Glicksman, Tejas Kotak

Vegetative Roof Performance During Summer—Critical Analysis of the Irrigation Effect 335
S. de Freitas, N.N.M. Ramos, V.P. de Freitas, H.M. Künzel, D. Zirkelbach

Effective Floor Cavity and Knee Wall Construction Techniques in Two-Story Homes in Hot Climates 348
Charles R. Withers, Jr.

Principles: Indoor Environmental and Thermal Comfort

Performance Factor for Floor Heating Systems Using New Analytical Formula 356
Henrik Karlsson, Carl-Eric Hagentoft

Measured Differences of Ground and Space Temperatures for Side-by-Side Slab-on-Grade Residences with and without Carpet 364
Robin K. Vieira, Danny S. Parker, Jamie Kono, Eric Martin, John Sherwin

Characterizing Indoor Humidity for Comparison Studies: The Moisture Balance Approach 372
William B. Rose, Stacy L. Gloss

Session 7

Practices: High-Performance Insulation Systems

Method to Evaluate and Develop Next-Generation Vacuum Insulation Panels for Implementation in the Retrofit of Existing Building Envelopes 380
Kyle Vansice, Ulrike Passe

Development of High-Performance Composite Foam Insulation with Vacuum Insulation Cores 389
Kaushik Biswas, Andre Desjarlais, Douglas Smith, John Letts, Jennifer Yao

Field Demonstration of Superior Performance of Advanced Building Envelope Technologies 396
Nitin Shukla, Alliston Watts, Jan Kosny

Principles: Rain

Analysis of Wind-Driven Rain Exposure Based on Long-Term Monitoring 407
Fitsum Tariku, Wendy Ying Simpson, Elsa Ngudjiharto

Laboratory Study of Rates of Inward Leakage in Seven Different Gaps in a Facade Exposed to Driving Rain or Water Splash 415
Lars Olsson

Session 8

Practices: Insulated Assemblies

Thermal and Moisture Performance of Buried Ducts 423
Mikael Salonvaara, Katrina Keeley, Achilles Karagiozis

Field Test on Two Interior Insulation Systems with Large Thickness—Influence of Orientation and Airtightness 432
M. Stelzmann, B. Nusser, U. Möller

Approach to Assessing Continuous Insulation Products 439
Jonathan Smegal, John Straube, Theresa Weston, Aaron Grin

Principles: Intentional and Unintentional Airflow

Duct Leakage Repeatability Testing 452
Iain Walker, Max Sherman, Paul W. Francisco

Airflow through Lightweight Wall Assemblies—Influence of Size and Location of Leakages 459
Philipp Kölsch, Daniel Zirkelbach, Bernd Nusser, Ralf Wagner, Andreas Zegowitz, Hartwig Künzel

Simulating Air Leakage in Walls and Roofs Using Indoor and Outdoor Boundary Conditions 485
Simon Pallin, Diana Hun, Philip Boudreaux

Session 9

Practices: Hygrothermal Performance and Durability

Comparison of Measured Hygrothermal Performance of Wood-Frame Walls Built with Continuous Exterior Insulation versus Walls Built with Housewrap and Oriented Strand Board Sheathing in Single-Family Homes in a Cold Climate 493
Gary D. Parsons, Brian Lieburn, Greg Stewart, Peter Eastman

Measured and Predicted Moisture Performance of High-R Wall Assemblies in Cold Climates 504
Trevor M. Trainor, Jonathan Smegal, John Straube, Anil Parekh

Field Evaluation of Thermal and Moisture Response of Highly Insulated Wood-Frame Walls 515
Michael A. Lacasse, Hamed H. Saber, Wahid Maref, G. Ganapathy, Silvio Plescia, Anil Parekh

Principles: Thermal Performance and Thermal Bridging

Potential and Limitations of Infrared Thermography on Unventilated Walls 527
Katrien Maroy, Nathan Van Den Bossche, Marijke Steeman

Thermal Analysis of Curtain Wall Systems—A Parametric Study 540
Nathan Van Den Bossche, Wahid Maref

Session 10

Practices: Performance of Wall and Glazing Systems

Window Condensation—Theory into Practice 554
Alex McGowan

Thermal Performance Evaluation of Walls with Gas-Filled Panel Insulation 561
Som Shrestha, Andre Desjarlais, Dahtzen Chu, Axy Pagan-Vazquez

Mind The Gap: Summary of Window Residential Retrofit Solutions 567
Joseph Petersen, Katherine Cort, Sarah Widder, Thomas Culp, Greg Sullivan

Principles: Innovative Approaches

Application of Glass Bubbles for Vacuum Insulation 576
Boyce Chang, Mufit Akinc

Big Area Additive Manufacturing Applied to Buildings 583
Kaushik Biswas, Randall Lind, Brian Post, Roderick Jackson, Lonnie Love, Johney Green, Jr., Lucas Tryggestad, Andrew Obendorf, Leif Eikevik, Maged Guerguis

Economic Efficiency of Thermal Insulation Aimed at Saving Energy: A Critical Assessment 591
Andreas Holm, Christoph Sprengard

Session 11

Practices: Assessment of Heat and Moisture Flow

Evaluation of a Modified Co-Heating Test for In-Situ Measurements of Thermal Transmittance of Single-Family Houses 598
Angela Sasic Kalagasidis, Emma Brycke, Jannicke Nilssen, Pär Johansson

A Simplified Energy Savings Calculation Tool for the Lifetime Energy and Environmental Impact Assessment of Insulation Materials 609
Mahabir Bhandari, Som Shrestha, Kaushik Biswas, Andre Desjarlais

Moisture-Related Durability of In-Service High-R Wall Assemblies in Pacific Northwest Climates 617
Jonathan Smegal, Robert Lepage, Chris Schumacher

Principles: Whole-Building Performance

Using the Thermal Mass of a Building to Reduce the Magnitude of the Peak Power Demand of the Primary Heating System—A Whole-Building Simulation with Parametric Analysis 628
Victor Fransson, Dennis Johansson, Hans Bagge

Implementation and Validation of a Long-Wave Heat Exchange Model 636
Matthias Winkler, Florian Antretter, Jan Radon

A New Approach for Analysis of Complex Building Envelopes in Whole-Building Energy Simulations 645
Jan Kosny, Charlie Curcija, Anthony D. Fonanini, Handan Liu, Elizabeth Kossecka

Session 12 and Session 13

Practices: Low-Energy Buildings

Evaluation of Energy Efficiency of U.S. Army Hard Shelters 661
Som Shrestha, Axy Pagan-Vazquez, Dahtzen Chu, Megan Kreiger, Kaushik Biswas

Building Energy and Envelope Performance of a Near Net Zero Energy Building 668
Stanley D. Gatland II, Julie M. Dinh-Ngoc, Remi Goulet, Parhum Delgoshaei, David Riley, Diana Elliott

Principles: Building Envelope Performance Studies

High-Rise Wood Building Enclosures 679
Graham Finch

Guiding Design Teams by Hygrothermal, Energy, and Thermal Comfort Analysis While Managing Uncertainty 693
Ivan Lee, Patrick Roppel, Randy Van Straaten, Jamie McKay

Overcladding for Thermal Performance and Building Resiliency 706
Daniel McKelvey, David Copley, Kendal Foster, Allison Wilson

Thermal Performance of the Exterior Envelopes of Whole Buildings XIII International Conference

Sheraton Sand Key Resort
Clearwater Beach, FL
December 4-8, 2016

U.S. DEPARTMENT OF ENERGY

ASHRAE

National Institute of BUILDING SCIENCES

OAK RIDGE National Laboratory

BUILDINGS XIII CONFERENCE

Organizing Committee

Karma Sawyer, Conference Chairperson
U.S. Department of Energy
Email: karma.sawyer@ee.doe.gov

André Desjarlais, Conference Director
Oak Ridge National Laboratory
Email: desjarlaisa@ornl.gov

Steve Comstock, Representative
ASHRAE
Email: comstock@ashrae.org

Peter Adams, Chair, Practices Track
Morrison Hershfield Limited
Email: padams@morrisonhershfield.com

Theresa Weston, Chair, Practices Track
DuPont
Email: theresa.a.weston@usa.dupont.com

Florian Antretter, Chair, Principles Track
Fraunhofer Institute for Building Physics
Email: florian.antretter@ibp.fraunhofer.de

Sam Glass, Chair, Principles Track
U.S. Dept. of Agriculture
Email: svglass@fs.fed.us

Hartwig Kuenzel, Chair, Principles Track
Fraunhofer Institute of Building Physics
Email: hartwig.kuenzel@ibp.fraunhofer.de

Wahid Maref, Chair, Workshops
National Research Council Canada
Email: wahid.maref@nrc-cnrc.gc.ca

Simon Pallin, Chair, Workshops
Oak Ridge National Laboratory
Email: pallinsb@ornl.gov

Laverne Dalgleish, Committee Member
Building Professionals Consortium
Email: ldalgleish@buildingprofessional.ca

Michael Ennis, Committee Member
SPRI
Email: m.ennis@mac.com

Diana Fisler, Committee Member
Johns Manville
Email: diana.fisler@jm.com

Stanley Gatland, III, Committee Member
CertainTeed Corporation
Email: stanley.d.gatland@saint-gobain.com

Carl-Eric Hagentoft, Committee Member
Chalmers University of Technology
Email: carl-eric.hagentoft@chalmers.se

Hugo Hens, Committee Member
University of Leuven (KU Leuven)
Email: hugo.hens@bwk.kuleuven.be

Andreas Holm, Committee Member
FIW Munich
Email: holm@fiw-muenchen.de

Roderick Jackson, Committee Member
Oak Ridge National Laboratory
Email: jacksonrk@ornl.gov

Achilles Karagiozis, Committee Member
Owens Corning
Email: achilles.karagiozis@owenscorning.com

Jan Ko[ny, Committee Member
Fraunhofer CSE
Email: jkosny@fraunhofer.com

Alex McGowan, Committee Member
Levelton Consultants
Email: amcgowan@levelton.com

Carsten Rode, Committee Member
DTU Civil Engineering, Tech. Univ. of Denmark
Email: car@byg.dtu.dk

David Roodvoets, Committee Member
DLR Consultants
Email: dlrconsult@charter.net

Mikael Salonvaara, Committee Member
Owens Corning
Email: harri.m.salonvaara@owenscorning.com

Paul Shipp, Committee Member
USG Corporation
Email: pshipp@usg.com

Fitsum Tariku, Committee Member
British Columbia Institute of Technology
Email: fitsum.tariku@bcit.ca

Sam Taylor, Committee Member
Consultant
Email: samuel.taylor.sr@gmail.com

Karen M. Triplett, Conference Coordinator
Oak Ridge National Laboratory
Email: triplettkm@ornl.gov

Tammy Moore, Conference Coordinator
Oak Ridge National Laboratory
Email: mooretm1@ornl.gov

Adrienne Weber, Conference Communications
CSRA / Oak Ridge National Laboratory
Email: weberam@ornl.gov

Program

Sunday, December 4th				
7:00 a.m.– 8:00 a.m.	**Continental Breakfast, Lobby II**			
8:00 a.m. – 12:00 p.m.	**Workshop 1** Performance of Air Barrier Systems, Part 1 Ballroom – Beach	**Workshop 2** Thermal Mass, Part 1 Ballroom – Gulf	**Workshop 3** Heat and Moisture in Buildings Ballroom – Palm	**Workshop 5** How to Evaluate the Risk of Mold Using the Mold Growth Index Ballroom – Bay
12:00 p.m. – 1:00 p.m.	**Lunch (on your own)**			
1:00 p.m. – 5:00 p.m.	**Workshop 1** Performance of Air Barrier Systems, Part 2 Ballroom – Beach	**Workshop 2** Thermal Mass, Part 2 Ballroom – Gulf	**Workshop 4** Insulation Materials and Systems Ballroom – Palm	**Workshop 6** Probability Assessment of Performance and Cost (RAP–RETRO) Ballroom – Bay
6:00 p.m. – 8:00 p.m.	**Reception, Poolside**			

<table>
<tr><th colspan="3">Monday, December 5th</th></tr>
<tr><td>7:00 a.m. – 8:00 a.m.</td><td colspan="2">Continental Breakfast, Lobby II</td></tr>
<tr><td>8:00 a.m. – 10:00 a.m.</td><td colspan="2">Plenary Session, Grand Ballroom

William Rose
A building science to-do list

Carl-Eric Hagentoft
Trends and challenges in building physics during the last 30 years: A Swedish perspective</td></tr>
<tr><td>10:00 a.m. – 10:30 a.m.</td><td colspan="2">Break, Lobby II</td></tr>
<tr><td rowspan="2">10:30 a.m. – 12:00 p.m.</td><td colspan="2">Session 1</td></tr>
<tr><td>Practices:
Attics
Chair: David Roodvoets

Field measurements of moisture in cold ventilated attics with different types of insulation and vapor barriers
Thor Hansen – Paper #45

Monitoring of two unvented roofs with air-permeable insulation in Climate Zone 2A
Kohta Ueno – Paper #99

Cost-effective, high-performance unvented attics
Francis Babineau – Paper #103</td><td>Principles:
Insulation Performance and Material Properties
Chairs: Andreas Holm and Manfred Kehrer

Wind-washing effects on mineral wool insulated sheathings
Randy Van Straaten – Paper #67

Improving polyiso thermal performance at low temperature
John Letts – Paper #159

Determination of linear thermal transmittance of vacuum-insulation panels by measurement in a guarded hot-plate apparatus or a heat-flow meter apparatus
Christoph Sprengard – Paper #17</td></tr>
<tr><td>12:00 p.m. – 1:30 p.m.</td><td colspan="2">Lunch (on your own)</td></tr>
</table>

Time	Practices	Principles
	Session 2	
1:30 p.m. – 3:00 p.m.	***Practices:*** **Heritage & Retrofit Considerations for Buildings** *Chair: Kim Pressnell* Refurbishing heritage and historic buildings: Key motivations, benefits and challenges ***Torben Valdbjørn Rasmussen – Paper #4*** Treatment of rising damp in historic buildings using a wall based hygro-regulated ventilation system: Case studies ***Vasco Peixoto de Freitas – Paper #34*** The impact of window-to-wall ratio on energy intensity of existing commercial buildings in Ontario and Quebec ***Viktoriya Mykytyak – Paper #56***	***Principles:*** **Walls – Moisture and Durability, Part 1** *Chair: Diana Hun* Using fiber insulation as a means of drying internally insulated walls ***Kim Pressnail – Paper #1*** A public domain, transient, experimental database on the hygrothermal performance of durable, energy-efficient, full-basement foundation walls in a cold climate ***Louise Goldberg – Paper #13***
3:00 p.m. – 3:30 p.m.	**Break, Lobby II**	
	Session 3	
3:30 p.m. – 5:00 p.m.	***Practices:*** **Taking Stock** *Chair: Kaushik Biswas* Field demonstration of an attic retrofit strategy using cellulose insulation and phase change material ***Kaushik Biswas – Paper #26*** Evaluating summertime overheating in multi-unit residential buildings using surveys and in-suite monitoring ***Ekaterina Tzekova – Paper #19*** Online airtightness savings calculator for commercial buildings in the US, Canada and China ***Diana Hun – Paper #39***	***Principles:*** **Material Durability** *Chair: Mikael Salonvaara* Critical property contrasts of fluid-applied air- and water-barrier membranes used for envelope: Chemistries, performance and durability ***Patrick Young – Paper #46*** Critical freeze-thaw saturation measurement of in-service masonry ***Randy Van Straaten – Paper #97*** Field study of moisture storage and thermal properties of buried extruded polystyrene ***Leslie Peer – Paper #149***
5:00 p.m. – 7:00 p.m.	**Reception, Island Ballroom: Sponsored by the Air Barrier Association of America (ABAA)**	

<table>
<tr><th colspan="3">Tuesday, December 6th</th></tr>
<tr><td>7:30 a.m. – 8:30 a.m.</td><td colspan="2">Continental Breakfast, Lobby II</td></tr>
<tr><td>8:30 a.m. – 10:00 a.m.</td><td colspan="2">Special Presentation, Grand Ballroom

Karma Sawyer
Innovation Required for the Next Generation of Building Envelopes</td></tr>
<tr><td>10:00 a.m. – 10:30 a.m.</td><td colspan="2">Break, Lobby II</td></tr>
<tr><td rowspan="2">10:30 a.m. – 12:00 p.m.</td><td colspan="2">Session 4</td></tr>
<tr><td>Practices:
Air Tightness in Buildings
Chair: Laverne Dalgleish

Commercial building airtightness testing: Lessons learned from the Red River College airtightness testing program
Gary Proskiw – Paper #24

Impact of large building airtightness requirements
Lorne Ricketts – Paper #16

Air sealing tight commercial and institutional buildings
Dave Bohac – Paper #129</td><td>Principles:
Walls – Moisture and Durability, Part 2
Chair: Fitsum Tariku

Protocol to evaluate the moisture durability of energy-efficient walls
Philip Boudreaux – Paper #38

A systematic method of assessing the durability of wall assemblies: Towards the limit-states design approach
Martin Morelli – Paper #91

Risk of condensation and mold growth in highly insulated wood-frame walls
Hamed Saber – Paper #109</td></tr>
<tr><td>12:00 p.m. – 1:30 p.m.</td><td colspan="2">Lunch (on your own)</td></tr>
</table>

<table>
<tr><td></td><td colspan="2">Session 5</td></tr>
<tr><td>1:30 p.m. – 3:00 p.m.</td><td>Practices:
Ventilation and Air Movement
Chair: Alex McGowan

Practical applications and case study of temperature smart ventilation controls
Michael Lubliner – Paper #82

Smart ventilation control of indoor humidity in high performance homes in humid U.S. climates
Brennan Less – Paper #87

Measured commercial and institutional building pressures
Dave Bohac – Paper #130</td><td>Principles:
Attic and Roof Performance
Chair: Bill Rose

Sealed attics exposed to two years of weathering in a hot and humid climate
William Miller – Paper #11

Field study on the thermal and hygrothermal performance of insulated, ventilated nail base system
Ming Shiao – Paper #12

Annual energy and heat flows in vented and sealed attics: Parametric study – climate zone 2A
Mikael Salonvaara – Paper #137</td></tr>
<tr><td>3:00 p.m. – 3:30 p.m.</td><td colspan="2">Break, Lobby II</td></tr>
<tr><td></td><td colspan="2">Session 6</td></tr>
<tr><td>3:30 p.m. – 5:00 p.m.</td><td>Practices:
Horizontal Envelope Assemblies
Chair: Mike Ennis

Innovative roof design for passive low-income housing in western India
Leon Glicksman – Paper #5

Vegetative roof performance during summer: Critical analysis of the irrigation effect
S. de Freitas – Paper #35

Effective floor cavity and knee wall construction techniques in two-story homes in hot climates
Chuck Withers – Paper #70</td><td>Principles:
Indoor Environmental and Thermal Comfort
Chair: Stanley Gatland

Performance factor for floor heating systems using new analytical formula
Carl-Eric Hagentoft – Paper #30

Measured differences of ground and space temperatures for side-by-side slab on grade residences with and without carpet
Robin Vieira – Paper #71

Characterizing indoor humidity for comparison studies: The moisture balance approach
William Rose – Paper #120</td></tr>
<tr><td>5:00 p.m. – 6:30 p.m.</td><td colspan="2">Workshop 7
Building Science Education Update
Ballroom, Beach</td></tr>
<tr><td>6:30 p.m. – 8:30 p.m.</td><td colspan="2">Reception, Island Ballroom
Sponsored by Armatherm™</td></tr>
</table>

Wednesday, December 7th		
7:30 a.m. – 8:30 a.m.	**Continental Breakfast, Lobby II**	
	Session 7	
8:30 a.m. – 10:00 a.m.	***Practices:*** **High-Performance Insulation Systems** *Chair: Diana Fisler* Method to evaluate and develop next generation vacuum insulation panels for implementation in the retrofit of existing building envelopes ***Kyle Vansice – Paper #52*** Development of high performance composite foam insulation with vacuum insulation cores ***Kaushik Biswas – Paper #64*** Field demonstration of superior performance of advanced building technologies ***Nitin Shukla – Paper #153***	***Principles:*** **Rain** *Chair: Hugo Hens* Analysis of wind-driven rain exposure based on long-term monitoring ***Fitsum Tariku – Paper #78*** Laboratory study of rates of inward leakage in seven different gaps in a façade exposed to driving rain or water splash ***Lars Olsson – Paper #126***
10:00 a.m. – 10:30 a.m.	**Break, Lobby II**	
	Session 8	
10:30 a.m. – 12:00 p.m.	***Practices:*** **Insulated Assemblies** *Chair: Patrick Roppel* Thermal and moisture performance of buried ducts ***Mikael Salonvaara – Paper #136*** Field test on two interior insulation systems with large thickness: Influence of orientation and airtightness ***Mario Stelzmann – Paper #60*** Determining the performance of an exterior continuous insulation wall assembly ***Jonathan Smegal – Paper #145***	***Principles:*** **Intentional and Unintentional Airflow** *Chair: Hugo Hens* Duct-leakage repeatability testing ***Iain Walker – Paper #27*** Airflow through lightweight wall assemblies: Influence of size and location leakages ***Philipp Kolsch – Paper #58*** Simulating air leakage in walls and roofs using indoor and outdoor boundary conditions ***Simon Pallin – Paper #115***

12:00 p.m. – 1:30 p.m.	**Lunch (on your own)**	
	Session 9	
1:30 a.m. – 3:00 p.m.	***Practices:*** **Hygrothermal Performance and Durability** *Chair: David Yarbrough* Comparison of measured hygrothermal performance of wood frame walls built with continuous exterior insulation… ***Gary Parsons – Paper #23*** Measured and predicted moisture durability performance of high-*R* wall assemblies in cold climates ***Trevor Trainor – Paper #84*** Field evaluation of thermal and moisture response of highly insulated wood-frame walls ***Michael Lacasse – Paper #96***	***Principles:*** **Thermal Performance and Thermal Bridging** *Chair: Jan Kośny* Potential and limitations of infrared thermography on unventilated walls ***Katrien Maroy – Paper #158*** Thermal analysis of curtain wall systems: A parametric study ***Nathan Van Den Bossche – Paper #55***
3:00 p.m. – 3:30 p.m.	**Break, Lobby II**	
	Session 10	
3:30 p.m. – 5:00 p.m.	***Practices:*** **Performance of Wall and Glazing Systems** *Chair: David Kayll* Window condensation: Theory into practice ***Alex McGowan – Paper #8*** Thermal performance evaluation of walls with gas-filled panel insulation ***André Desjarlais – Paper #92*** Demonstration and evaluation of low-e storm windows and advanced window coverings in the PNNL lab homes ***Joseph Petersen – Paper #142***	***Principles:*** **Innovative Approaches** *Chair: Carl-Eric Hagentoft* Alternative core materials for vacuum insulation ***Boyce Chang – Paper #54*** Big- area additive manufacturing applied to buildings ***Kaushik Biswas – Paper #83*** Economic efficiency of thermal insulation aimed at saving energy: A critical assessment ***Andreas Holm – Paper #124***
6:00 p.m. – 8:00 p.m.	**Conference Dinner, Grand Ballroom**	

Thursday, December 8th		
7:30 a.m. – 8:30 a.m.	**Continental Breakfast, Lobby II**	
	Session 11	
8:30 a.m. – 10:00 a.m.	***Practices:*** **Assessment of Heat and Moisture Flow** *Chair: Brian Stoik* Evaluation of a modified co-heating test for in-situ measurements of thermal transmittance of single family houses ***Angela Sasic Kalagasidis – Paper #57*** A simplified energy savings calculation aid in tool for the lifetime energy and environmental impact assessment of insulation materials ***Mahabir Bhandari – Paper #95*** Moisture-related durability of in-service high-R wall assemblies in Pacific Northwest climates ***Jonathan Smegal – Paper #132***	***Principles:*** **Whole Building Performance** *Chair: Katrin Klingenberg* Using the thermal mass of a building to reduce the magnitude of the peak power demand of the primary heating system: A whole-building simulation with parametric analysis ***Victor Fransson – Paper #107*** Implementation and validation of a longwave heat exchange model ***Florian Antretter- Paper #147*** A new approach for analysis of complex building envelopes in whole-building energy simulations ***Jan Kośny – Paper #151***
10:00 a.m. – 10:30 a.m.	**Break, Lobby II**	
	Session 12 and Session 13	
10:30 a.m. – 12:00 p.m.	***Practices:*** **Low-Energy Buildings** *Chair: Chris Mathis* Evaluation of energy efficiency of U.S. Army hard shelters: B-huts and SIP-hut ***Som Shrestha – Paper #93*** Building energy and envelope performance of a near net-zero energy building ***Stanley Gatland – Paper #157***	***Practices:*** **Building Envelope Performance Studies** *Chair: Benjamin Meyer* High-Rise Wood Building Enclosures ***Graham Finch – Paper #7*** Guiding design teams by hygrothermal, energy and thermal comfort analysis while managing uncertainty ***Ivan Lee – Paper #88*** Case study: Over-cladding for thermal performance and building resiliency ***Dan McKelvey – Paper #128***

12:00 p.m. – 1:00 p.m.	**Lunch (on your own)**	
1:00 p.m. – 5:00 p.m.	**Workshop 8** US DOE and IEA Building Envelope Road Map Ballroom, Beach	**Workshop 9** (Title TBD) Ballroom, Gulf

Workshop Overview

Workshop 1

Performance of Air Barrier Systems, Part 1
Sunday, December 4, 2016
Ballroom – Beach
8:00 a.m. – 12 noon

Workshop Host:
Laverne Dalgleish

Abstract and Session information not available at time of printing.

Workshop 2

Thermal Mass, Part 1
Sunday, December 4, 2016
Ballroom – Gulf
8:00 a.m. – 12 noon

Workshop Host:
Jan Ko[ny, Ph.D.
Fraunhofer CSE, Director, Building Enclosures and Materials,
Fraunhofer Center for Sustainable Energy Systems CSE, Boston, MA.

Since 2001, a group of building scientists, architects, and researchers has been meeting during the Thermal Performance of Exterior Envelopes Conference to discuss the current research and new application trends in the areas of thermal storage and thermal stability of buildings.

A modern "massive building" is one, which can utilize thermal excitations coming from the exterior environment to improve overall energy efficiency, manage the building peak energy demand, and allow a coordination of the building energy consumption and solar energy generation with electric power grid dynamics. Significant energy-savings are possible by blending thermally massive building envelopes with natural elements such as sunlight, temperature fluctuations, breezes, and appropriate landscaping. When properly combined, these elements produce a building with less reliance on purchased energy, less environmental impact, and improved thermal comfort.

Today, building designs utilizing either conventional thermal mass or phase change materials are receiving wider acclaim due to both energy cost concerns and the increasing interest in the environmental impact of buildings.

Workshop 3

Heat and Moisture in Buildings
Sunday, December 4, 2016
Ballroom – Palm
8:00 a.m. – 12 noon

Chaired by:
Dr. Wahid Maref
National Research Council Canada

Abstract: The workshop will cover Heat, Air and Moisture in Buildings. The speakers will go from fundamentals to practice by presenting case studies or/and by modelling and field experiment in order to assess the hygrothermal performance of the building envelopes.

Since the building envelopes are becoming more and more insulated because of the new energy requirements, around the world, on new construction, or by adding insulation to the existing building either from the interior or from the exterior, problem of interstitial condensation, thermal bridges and other physical phenomena are becoming a real issue for the energy performance of building and its durability. In-situ thermography techniques, guidelines and calculation techniques for European Standard are presented in this workshop. This will help to assess the impact of thermal bridges on the overall heat loss for poor design, standard practice, etc.

New sensors technology for roofing application is presented in this workshop. Practical applications, installation details of the sensors added to installations during various stages of the building life cycle, including new construction, existing buildings, site condition assessments, and live monitoring results shall be presented.

- Thermal bridges — N. Van Den Bossche
- Structure monitoring — making the invisible visible — J. Teetaert
- Impact Of Fenestration On The Overall Thermal Performance of The Building Envelope— by Mario D. Gonçalves

Workshop 5

How to Evaluate the Risk of Mold Using the Mold Growth Index
Sunday, December 4, 2016
Ballroom – Bay
8:00 a.m. – 12 noon

Workshop Host:
T. Ojanen

This workshop will present the Mold Growth Index (MGI) model, its application and how it has been used for decades in Europe. It will also explain:

- How it was developed and which factors that will influence the risk of mold.
- Evaluation of critical mold growth levels in structures.
- The assessment criteria and how these relate to MGI.
- The parameters used in the numerical model and what they represent.
- Examples of how MGI is used and how the result will vary for different materials.

In addition, the type of materials tested to define a numerical model will be presented, and under which conditions. Finally, ongoing research will be discussed and if further development if the numerical model is expected.

Workshop 1 (cont'd from AM session)

Performance of Air Barrier Systems, Part 2
Sunday, December 4, 2016
Grand Ballroom – Beach
1:00 p.m. – 5:00 p.m.

Workshop Host:
Laverne Dalgleish

Abstract and Session information not available at time of printing.

Workshop 2 (cont'd from AM session)

Thermal Mass, Part 2

Sunday, December 4, 2016
Ballroom – Gulf
1:00 p.m. – 5:00 p.m.

Workshop Host:
Jan Ko[ny, Ph.D.
Fraunhofer CSE, Director, Building Enclosures and Materials,
Fraunhofer Center for Sustainable Energy Systems CSE, Boston, MA.

Since 2001, a group of building scientists, architects, and researchers has been meeting during the Thermal Performance of Exterior Envelopes Conference to discuss the current research and new application trends in the areas of thermal storage and thermal stability of buildings.

A modern "massive building" is one, which can utilize thermal excitations coming from the exterior environment to improve overall energy efficiency, manage the building peak energy demand, and allow a coordination of the building energy consumption and solar energy generation with electric power grid dynamics. Significant energy-savings are possible by blending thermally massive building envelopes with natural elements such as sunlight, temperature fluctuations, breezes, and appropriate landscaping. When properly combined, these elements produce a building with less reliance on purchased energy, less environmental impact, and improved thermal comfort.

Today, building designs utilizing either conventional thermal mass or phase change materials are receiving wider acclaim due to both energy cost concerns and the increasing interest in the environmental impact of buildings.

Workshop 4

Insulation Materials and Systems
Sunday, December 4, 2016
Ballroom – Palm
1:00 p.m. – 5:00 p.m.

Workshop Hosts:
Hartwig Kunzel, Ph.D., Fraunhofer IBP, Andreas Holm, Ph.D., FIW Munich, and André Desjarlais, ORNL

What are the trends in building insulation? How do new systems perform and how durable are they? Where are the limits? Appreciate the dependence of thermal performance on temperature, humidity and air flow. What are the pitfalls of testing and evaluating the thermal properties of insulation systems? We will discuss appropriate protection and maintenance measures to ensure long service life of insulated building assemblies. Learn about the recycling options for insulation materials and systems.

Workshop 6

Probability Assessment of Performance and Cost (RAP–RETRO)
Sunday, December 4, 2016
Ballroom – Bay
1:00 p.m. – 5:00 p.m.

Workshop Host:
Prof. Carl-Eric Hagentoft – Chalmers University of Technology, Sweden

During this year the IEA-EBC annex on Reliability of Energy Efficient Building Retrofitting - Probability Assessment of Performance and Cost (RAP-RETRO) has been finalized.
Retrofit measures are of the utmost importance for upgrading the existing building stock. We need to design and realize robust retrofitting with low energy demand and life cycle costs, while controlling risk levels for performance failure. But, many building owners are only interested in the initial capital cost. Looking at actual risks in performance and the costs incurred highlights the need for life cycle thinking.

In short, the annex provides decision support data and tools for energy retrofitting measures. The tools are based on probabilistic methodologies for prediction of energy use, life cycle cost and functional performance. A risk assessment framework has been developed and applied on case studies. The process of risk management and how to include probabilistic assessment tools during the building process has been outlined as well as exemplifications of how to write guidelines.

Some results from the annex will be presented, where after group discussions will be on; How to target and facilitate such analyses in the common practice both in projects related to new and existing buildings.

Plenary Session

Monday, December 5, 2016
8:00 a.m. to 10:00 a.m.
Grand Ballroom

Welcoming Remarks

8:00 a.m.

Karma Sawyer
Conference Chairperson
U.S. Department of Energy, Building Technologies Office (DOE-BTO)

8:10 a.m.
Patricia Graef
ASHRAE

8:20 a.m.
Henry Green
NIBS

Keynote Speakers

Mr. William Rose
A building science to-do list

Mr. Carl-Eric Hagentoft
Trends and challenges in building physics during the last 30-years: A Swedish perspective

Practices Session 1: Attics

Monday, December 5, 2016
10:30 a.m. to 12 p.m.
Ballroom – Beach/Gulf

Session Chair:

David Roodvoets, Committee Member
DLR Consultants
Email: dlrconsult@charter.net

Introductions
David Roodvoets

10:30 a.m. *"Field Measurements of Moisture in Cold Ventilated Attics with Different Types of Insulation Materials and Vapor Barrier"*
Thor Hansen

11:00 a.m. *"Monitoring of Two Unvented Roofs with Air-Permeable Insulation in Climate Zone 2A"*
Kohta Ueno

11:30 a.m. *"Cost-Effective High Performance Unvented Attics"*
Francis Babineau

Principles Session 1: Insulation Performance and Material Properties

Monday, December 5, 2016
10:30 a.m. to 12 noon
Ballroom – Palm/Bay

Session Chairs:

Andreas Holm
FIW Munich
Email: holm@fiw-muenchen.de

Manfred Kehrer
Email:

Introductions
Andreas Holm and Manfred Kehrer

10:30 a.m. — *"Wind washing effects on mineral wool insulated sheathings"*
Randy Van Straaten

11:00 a.m. — *"Improving polyiso thermal performance at low temperature"*
John Letts

11:30 a.m. — *"Determination of linear thermal transmittance of vacuum-insulation panels by measurement in a guarded hot-plate apparatus or a heat-flow meter apparatus"*
Christoph Sprengard

Practices Session 2: Heritage & Retrofit Considerations for Buildings

Monday, December 5, 2016
1:30 p.m. to 3:00 p.m.
Ballroom – Beach/Gulf

Session Chair:

Kim Pressnail
Toronto Atmospheric Fund
E-mail: pressna@ecf.utoronto.ca

Introductions
Kim Pressnail

1:30 p.m. *"Refurbishing heritage and historic buildings - Key motivations, benefits and challenges"*
Torben Valdbjørn Rasmussen

2:00 p.m. *"Treatment of Rising Damp in Historic Buildings Using a Wall Base Hygro-Regulated Ventilation System – Case Study"*
Vasco Peixoto de Freitas

2:30 p.m. *"The impact of window-to-wall ratio on energy intensity of existing commercial buildings in Ontario and Quebec"*
Viktoriya Mykytyak

Principles Session 2: Walls – Moisture and Durability, Part 1

Monday, December 5, 2016
1:30 p.m. to 3:00 p.m.
Ballroom – Palm/Bay

Session Chair:

Diana Hun
Oak Ridge National Laboratory
Email: hunde@ornl.gov

Introductions
Diana Hun

1:30 p.m. *"Using fiber insulation as a means of drying internally insulated walls"*
Ekaterina Tzekova

2:15 p.m. *"A public domain, transient, experimental database on the hygrothermal performance of durable, energy-efficient, full-basement foundation walls in a cold climate"*
Louise Goldberg

Practices Session 3: Taking Stock

Monday, December 5, 2016
3:30 p.m. to 5:00 p.m.
Ballroom – Beach/Gulf

Session Chair:

Kaushik Biswas
Oak Ridge National Laboratory
E-mail: biswask@ornl.gov

Introductions
Kaushik Biswas

3:30 p.m. *"Field demonstration of an attic retrofit strategy using cellulose insulation and phase change material"*
Kaushik Biswas

4:00 p.m. *"Evaluating summertime overheating in multi-unit residential buildings using surveys and in-suite monitoring"*
Marianne Touchie

4:30 p.m. *"Online airtightness savings calculator for commercial buildings in the US, Canada and China"*
Diana Hun

Principles Session 3: Material Durability

Monday, December 5, 2016
3:30 p.m. to 5:00 p.m.
Ballroom – Palm/Bay

Session Chair:

Mikael Salonvaara
Owens Corning
E-mail: harri.m.salonvaara@owenscorning.com

Introductions
Mikael Salonvaara

3:30 p.m. *"Critical property contrasts of fluid-applied air- and water-barrier membranes used for envelope: Chemistries, performance and durability"*
Patrick Young

4:00 p.m. *"Critical freeze-thaw saturation measurement of in-service masonry"*
Randy Van Straaten

4:30 p.m. *"Field study of moisture storage and thermal properties of buried extruded polystyrene"*
Leslie Peer

Special Presentation

Tuesday, December 6, 2016
8:30 a.m. to 10:00 a.m.
Grand Ballroom

Karma R. Sawyer, PhD
Acting Program Manager
Emerging Technologies Program
Building Technologies Office, Department of Energy

"Innovation Required for the Next Generation of Building Envelopes"

Practices Session 4: Air Tightness in Buildings

Tuesday, December 6, 2016
10:30 a.m. to 12 noon
Ballroom – Beach/Gulf

Session Chair:

Laverne Dalgleish
Building Professionals Consortium
Email: ldalgleish@buildingprofessional.ca

Introductions
Laverne Dalgleish

10:30 a.m. *"Commercial Building Airtightness Testing – Lessons Learned from the Red River College Airtightness Testing Program"*
Gary Proskiw

11:00 a.m. *"Impact of Large Building Airtightness Requirements"*
Lorne Ricketts

11:30 a.m. *"Air Sealing Tight Commercial and Institutional Buildings"*
Dave Bohac

Principles Session 4: Walls – Moisture and Durability, Part 2

Tuesday, December 6, 2016
10:30 a.m. to 12 noon
Ballroom – Palm/Bay

Session Chair:

Fitsum Tariku
British Columbia Institute of Technology
Email: fitsum.tariku@bcit.ca

Introductions
Fitsum Tariku

10:30 a.m. *"Protocol to Evaluate the Moisture Durability of Energy Efficient Walls"*
Philip Boudreaux

11:00 a.m. *"A Systematic Method of Assessing the Durability of Wall Assemblies: Towards the Limit-States Design Approach"*
Martin Morelli

11:30 a.m. *"Risk of Condensation and Mold Growth in Highly Insulated Wood-Frame Walls"*
Hamed Saber

Practices Session 5: Ventilation and Air Movement

Tuesday, December 6, 2016
1:30 p.m. to 3:00 p.m.
Ballroom – Beach/Gulf

Session Chair:

Alex McGowan
Levelton Consultants
Email: amcgowan@levelton.com

Introductions
Alex McGowan

1:30 p.m. *"Practical Applications and Case Study of Temperature Smart Ventilation Controls"*
Michael Lubliner

2:00 p.m. *"Smart Ventilation Control of Indoor Humidity in High Performance Homes in Humid U.S. Climates"*
Brennan Less

2:30 p.m. *"Measured Commercial and Institutional Building Pressures"*
Dave Bohac

Principles Session 5: Attic and Roof Performance

Tuesday, December 6, 2016
1:30 p.m. to 3:00 p.m.
Ballroom – Palm/Bay

Session Chair:

Bill Rose
Illinois University
E-mail: wrose@illinois.edu

Introductions
Bill Rose

1:30 p.m. *"Sealed Attics Exposed to Two Years of Weathering in a Hot and Humid Climate"*
William Miller

2:00 p.m. *"Field Study on the Thermal and Hygrothermal Performance of Insulated Ventilated Nail Base System"*
Ming Shiao

2:30 p.m. *"Annual Energy and Heat Flows in Vented and Sealed Attics – Parametric Study - Climate Zone 2A"*
Mikael Salonvaara

Practices Session 6: Horizontal Envelope Assemblies

Tuesday, December 6, 2016
3:30 p.m. to 5:00 p.m.
Ballroom – Beach/Gulf

Session Chair:

Mike Ennis
SPRI
Email: m.ennis@mac.com

Introductions
Mike Ennis

3:30 p.m. *"Innovative roof design for passive low income housing in Western India"*
Leon Glicksman

4:00 p.m. *"Vegetative roof performance during summer: Critical analysis of the irrigation effect"*
S. de Freitas

4:30 p.m. *"Effective Floor Cavity and Knee Wall Construction Techniques in Two-Story Homes in Hot Climates"*
Charles R. Withers, Jr.

Principles Session 6: Indoor Environmental and Thermal Comfort

Tuesday, December 6, 2016
3:30 p.m. to 5:00 p.m.
Ballroom – Palm/Bay

Session Chair:

Stanley Gatland, II
CertainTeed Corporation
Email: Stanley.d.gatland@saint-gobain.com

Introductions
Stanley Gatland, II

3:30 p.m. *"Performance factor for floor heating systems using new analytical formula"*
Carl-Eric Hagentoft

4:00 p.m. *"Measured differences of ground and space temperatures for side-by-side slab on grade residences with and without carpet"*
Robin Vieira

4:30 p.m. *"Characterizing indoor humidity for comparison studies: The moisture balance approach"*
William Rose

Practices Session 7: High-Performance insulation Systems

Wednesday, December 7, 2016
8:30 a.m. to 10:00 a.m.
Ballroom – Beach/Gulf

Session Chair:

Diana Fisler
Johns Manville
Email: diana.fisler@jm.com

Introductions
Diana Fisler

8:30 a.m. *"Method to evaluate and develop next generation vacuum insulation panels for implementation in the retrofit of existing building envelopes"*
Kyle Vansice

9:00 a.m. *"Development of high performance composite foam insulation with vacuum insulation cores"*
Kaushik Biswas

9:30 a.m. *"Field demonstration of superior performance of advanced building technologies"*
Nitin Shukla

Principles Session 7: Rain

Wednesday, December 7, 2016
8:30 a.m. to 10:00 a.m.
Ballroom – Palm/Bay

Session Chair:

Hugo Hens
University of Leuven (KU Leuven)
Email: hugo.hens@bwk.kuleuven.be

Introductions
Hugo Hens

8:30 a.m. *"Analysis of wind-driven rain exposure based on long-term monitoring"*
Fitsum Tariku

9:15 a.m. *"Laboratory study of rates of inward leakage in seven different gaps in a façade exposed to driving rain or water splash"*
Lars Olsson

Practices Session 8: Insulated Assemblies

Wednesday, December 7, 2016
10:30 a.m. to 12 noon
Ballroom – Beach/Gulf

Session Chair:

Patrick Roppel
Morrison Hershfield
Email: proppel@morrisonhershfield.com

Introductions
Patrick Roppel

10:30 a.m. *"Thermal and moisture performance of buried ducts"*
Mikael Salonvaara

11:00 a.m. *"Field test on two interior insulation systems with large thickness: Influence of orientation and airtightness"*
Mario Stelzmann

11:30 a.m. *"Determining the performance of an exterior continuous insulation wall assembly"*
Jonathan Smegal

Principles Session 8: Intentional and Unintentional Airflow

Wednesday, December 7, 2016
10:30 a.m. to 12 noon
Ballroom – Palm/Bay

Session Chair:

Hugo Hens
University of Leuven (KU Leuven)
Email: hugo.hens@bwk.kuleuven.be

Introductions
Hugo Hens

10:30 a.m. *"Duct-leakage repeatability testing"*
Iain Walker

11:00 a.m. *"Airflow through lightweight wall assemblies: Influence of size and location leakages"*
Philipp Kolsch

11:30 a.m. *"Simulating air leakage in walls and roofs using indoor and outdoor boundary conditions"*
Simon Pallin

Practices Session 9: Hygrothermal Performance and Durability

Wednesday, December 7, 2016
1:30 p.m. to 3:00 p.m.
Ballroom – Beach/Gulf

Session Chair:

David Yarbrough
R&D Services
Email: dave@rdservices.com

Introductions
David Yarbrough

1:30 p.m. *"Comparison of Measured Hygothermal Performance of Wood Frame Walls Built w/Continuous Exterior Insulation vs Walls Built w/Housewrap & Oriented Strand Board Sheathing in Single Family Homes in a Cold Climate"*
Gary Parsons

2:00 p.m. *"Measured and predicted moisture durability performance of high-R wall assemblies in cold climates"*
Trevor Trainor

2:30 p.m. *"Field evaluation of thermal and moisture response of highly insulated wood-frame walls"*
Michael Lacasse

Principles Session 9: Thermal Performance and Thermal Bridging

Wednesday, December 7, 2016
1:30 p.m. to 3:00 p.m.
Ballroom – Palm/Bay

Session Chair:

Jan Ko[ny
Fraunhofer CSE
Email: jkosny@fraunhofer.org

Introductions
Jan Ko[ny

1:30 p.m. *"Potential and limitations of infrared thermography on unventilated walls"*
Katrien Maroy

2:15 p.m. *"Thermal analysis of curtain wall systems: A parametric study"*
Nathan Van Den Bossche

Practices Session 10: Performance of Wall and Glazing Systems

Wednesday, December 7, 2016
3:30 p.m. to 5:00 p.m.
Ballroom – Beach/Gulf

Session Chair:

David Kayll
Morrison Hershfield
Email: DKayll@morrisonhershfield.com

Introductions
David Kayll

3:30 p.m. *"Window condensation: Theory into practice"*
Alex McGowan

4:00 p.m. *"Thermal performance evaluation of walls with gas-filled panel insulation"*
André Desjarlais

4:30 p.m. *"Demonstration and evaluation of low-e storm windows and advanced window coverings in the PNNL lab homes"*
Joseph Petersen

Principles Session 10: Innovative Approaches

Wednesday, December 7, 2016
3:30 p.m. to 5:00 p.m.
Ballroom – Palm/Bay

Session Chair:

Carl-Eric Hagentoft
Chalmers University of Technology
Email: carl-eric.hagentoft@chalmers.se

Introductions
Carl-Eric Hagentoft

3:30 p.m. *"Alternative core materials for vacuum insulation"*
Boyce Chang

4:00 p.m. *"Big-Area Additive Manufacturing applied to buildings"*
Kaushik Biswas

4:30 p.m. *"Economic efficiency of thermal insulation aimed at saving energy: A critical assessment"*
Andreas Holm

Practices Session 11: Assessment of Heat and Moisture Flow

Thursday, December 8, 2016
8:30 a.m. to 10:00 a.m.
Ballroom – Beach/Gulf

Session Chair:

Brian Stroik
Tremco Incorporated
Email: BStroik@tremcoinc.com

Introductions
Brian Stroik

8:30 a.m. *"Evaluation of a modified co-heating test for in-situ measurements of thermal transmittance of single family houses"*
Angela Sasic Kalagasidis

9:00 a.m. *"A simplified energy savings calculation aid in tool for the lifetime energy and environmental impact assessment of insulation materials"*
Mahabir Bhandari

9:30 a.m. *"Moisture-related durability of in-service high-R wall assemblies in Pacific Northwest climates"*
Jonathan Smegal

Principles Session 11: Whole Building Performance

Thursday, December 8, 2016
8:30 a.m. to 10:00 a.m.
Ballroom – Palm/Bay

Session Chair:

Katrin Klingenberg
Passive House California (PHCA)
Email: katrin@passivehouse.us

Introductions
Katrin Klingenberg

10:30 a.m. *"Using the thermal mass of a building to reduce the magnitude of the peak power demand of the primary heating system: A whole-building simulation with parametric analysis"*
Victor Fransson

11:00 a.m. *"Implementation and validation of a longwave heat exchange model"*
Florian Antretter

11:30 a.m. *"A new approach for analysis of complex building envelopes in whole-building energy simulations"*
Jan Ko[ny

Practices Session 12: Low-Energy Buildings

Thursday, December 8, 2016
10:30 a.m. to 12 noon
Ballroom – Beach/Gulf

Session Chair:

Chris Mathis
Mathis Consulting Company
Email: Chris@mathisconsulting.com

Introductions
Chris Mathis

10:30 a.m. *"Evaluation of energy efficiency of U.S. Army hard shelters: B-huts and SIP-hut"*
Som Shrestha

11:15 a.m. *"Building energy and envelope performance of a near net-zero energy building"*
Stanley Gatland

Practices Session 13: Building Envelope Performance Studies

Thursday, December 8, 2016
10:30 a.m. to 12 noon
Ballroom – Palm/Bay

Session Chair:

Ben Meyer
DuPont
Email: benjamin.meyer@dupont.com

Introductions
Ben Meyer

10:30 a.m. *"High-Rise Wood Building Enclosures"*
Graham Finch

11:00 a.m. *"Guiding design teams by hygrothermal, energy and thermal comfort analysis while managing uncertainty"*
Ivan Lee

11:30 a.m. *"Case study: Over-cladding for thermal performance and building resiliency"*
Dan McKelvey

Workshop 8

US DOE and IEA Building Envelope Road Map
Thursday, December 8, 2016
Ballroom – Beach
1:00 p.m. – 5:00 p.m.

***Moderator/Presenter*:**
Marc LaFrance, US Department of Energy

The workshop will include the detailed presentation of technical goals and recommended actions that have been published in the US Department of Energy's 2016 Multiyear Plan along with the International Energy Agencies Energy Efficient Building Envelope Technology Roadmap (2013). The workshop is intended to engage participates to provide suggestions on areas for high priority activities by a full range of US entities. Furthermore, areas that participants believe that the Federal government should provide leadership and support on will be discussed in more detail. Building sector market players will be encouraged to share their future activities, as well as lessons learned in different countries since the conference attracts a diverse audiences from many countries with varying building practices, climates, challenges, and technological solutions.

Workshop 9

Title TBD

Thursday, December 8, 2016
Ballroom – Gulf
1:00 p.m. – 5:00 p.m.

Workshop Host:
TBD

Abstract and Session information not available at time of printing.

Field Measurements of Moisture in Cold Ventilated Attics with Different Types of Insulation Materials and Vapor Barrier

Thor Hansen, Msc **Eva B. Møller, PhD**

ABSTRACT

To save energy, large amounts of insulation material are currently installed in traditional cold naturally ventilated attics. This is the case in new buildings as well as in existing buildings. Consequently the temperature in the ventilated attic decreases which reduces the capability to remove humid air by ventilation. This can lead to moisture-related damage in the construction. Therefore, the general recommendation in Denmark is to install an air and vapor tight barrier if the insulation material has a thickness of more than 150 mm (6 inch). This very strict recommendation is on the safe side based on non-documented experience.

As part of an investigation to verify or potentially refine the recommendation, measurements have been executed in ten different Danish dwellings for the evaluation of the moisture penetration into the cold ventilated attics. In the test dwellings, the type and amount of insulation material vary and so does the vapor barrier, some test buildings have no vapor barrier.

Unfortunately there have been some complications in collecting data from the original planed thirty dwellings, based on the remaining ten dwellings it is not possible to make a solid conclusion which can verify or refine the current recommendation. However, from the limited number of dwellings there is a trend that the hygroscopic properties of the insulation material might have an essential impact on the moisture level in the attic, compared to the effect of either the vapor barrier or the amount of insulation material.

INTRODUCTION

In recent years, the amount of insulation in buildings has increased, both in new buildings and in renovation of the existing building stock. This has led to a significant reduction of heat loss to cold, naturally ventilated attics. This reduction implies that the temperature in the attic is decreased, which causes a declined capability to remove moist air penetrating from below into the attic (Nik et al. 2012). The lacking capability to remove moist air has led to an increased risk of moisture and mold growth damages to the attic. This is often the result of a combination of high insulation thickness, lack or limited ventilation rate of the attic, and leakage in the air barrier and/or vapor barrier against the attic.

In Table 1, the overall heat transfer coefficient (U-factor) and an approximated thickness of the insulation material on the ceiling/roof constructions are stated according to the different versions of the Danish Building Regulation (BR) (Energistyrelsen 2015a, 2015b). The Danish Building Regulation from 2015 (BR2015) stipulates no specified U-factor, but an energy performance of the building, which has been estimated to correspond to the U-factor and thickness stated in Table 1. Insulation thickness is based on a λ-value of 0.04 W/m·K (0.28 Btu·in./h·ft^{2}·°F).

In the Danish Building Regulations from 1966, it was stated for the first time that a vapor barrier should be established on the warm side of the insulation (Energistyrelsen 1966). Before that time, the description in the regulations was limited to outlining that all constructions should be erected in a way that prevented condensation of moisture. Since 1998, a vapor barrier is no longer mandatory, but condensation in the attic must be avoided and a vapor barrier is suggested. This implies that in buildings erected before 1966 there is normally

Thor Hansen is a doctoral student and **Eva B. Møller** is a senior researcher at the Danish Building Research Institute/Aalborg University, Copenhagen, Denmark.

Table 1. U-Factors for Ceiling and Roof Constructions and the Corresponding Approximated Insulation Thickness According to the Different Danish Building Regulations (BR) and Commonly Used Vapor Barriers Over Time

Year	U-Factor, $W/m^2 \cdot K$	U-Factor, $Btu/h \cdot ft^2 \cdot °F$]	Insulation Thickness, mm	Insulation Thickness, in.	Vapor Barrier
1961	0.45	0.08	80	3	—
1966	0.45	0.08	80	3	Alu-paper-foil
1972	0.45	0.08	80	3	Alu-paper-foil
1977	0.20	0.04	200	8	Alu-paper-foil/PE-foil
1982	0.20	0.04	200	8	Alu-paper-foil/PE-foil
1985	0.20	0.04	200	8	Alu-paper-foil/PE-foil
1998	0.15	0.03	250	10	PE-foil
2008	0.15	0.03	250	10	PE-foil
2010	0.10	0.02	350	14	PE-foil
2015	0.07	0.01	500	20	PE-foil

no vapor barrier. The first kind of vapor barrier used in Denmark was a thin paper with a coat of bitumen. Later a paper foil with one surface of a thin aluminum coat (alu-paper-foil) was introduced. Today, the most used type of vapor barrier is a polyethylene foil (PE foil). In the period from the mid-1970s to the 1990s, both types of foils were used. None of the BRs prescribe the material of the vapor barrier; however, Table 1 shows the most common materials used over time.

The U.S. Department of Energy (1978) recommended that a vapor barrier should be used in a climate zone corresponding approximately to today's U.S. Climate Zones 6–8 (DOE 2013) for the IECC Climate Zone map, and it was recommended in climate zones corresponding approximately to today's Climate Zones 3–5.

Moisture can penetrate the ceiling by convection through cracks or by diffusion through the materials. Moisture transport by convection is normally much higher than by diffusion. In Denmark prior to 2007 it was considered that if the ceiling was airtight, e.g., a plastered ceiling, it would prevent moisture transport by convection and would be sufficient to prevent damaging moisture penetration since the moisture penetrating by diffusion and minor cracks would be limited by ventilating the attic. However, because the thickness of the insulation material has increased for both renovation and new constructions, the heat loss to the ventilated attic is reduced, which leads to a temperature decrease in the attics. Consequently, the relative humidity (RH) in the attic increases, which means a higher risk of moisture-related problems in the attic such as mold growth (Harderup and Arfvidsson 2013; Brandt et al. 2007). Therefore, the Danish Building Research Institute (SBi) has written in their Guideline 224 and Guideline 240 (Brandt et al. 2013; Møller 2012) that plastered ceilings with no cracks can no longer be considered sufficiently tight to ensure an acceptable humidity level in the attic if the insulation thickness is above 150 mm (6 in.). If a higher insulation thickness is applied, the guideline calls for a vapor barrier to be used, regardless of what kind of insulation material is used.

Unfortunately, establishing a tight vapor barrier in an existing attic is very difficult and the need for a vapor barrier might therefore mean that more than 150 mm (6 in.) insulation material in existing attics is not cost-efficient. If a larger amount of insulation material does not cause moisture problems, the recommendations should be changed and thereby make it possible to reduce the energy loss through the ceiling.

The aim of the work related to this paper is to introduce results from moisture measurements in a collection of different Danish dwellings with a varying ceiling construction regarding the amount and type of insulation material and the use of different types of vapor barriers. All of the different buildings are designed with a cold-ventilated attic. The results should contribute to an overall recommendation with focus on defining the optimal combination of type and the total amount of insulation material when insulating ceilings beneath cold-ventilated attics without vapor barrier.

THEORY

The moisture content in attics is a combination of humidity in the external air and humidity from the internal air coming into the attic, either by penetrating through leaks in the ceiling (convection) or by diffusion through the material. In the following subsection, the general equation for estimating the amount of moisture migrating into the attic by convection and diffusion is shown. Furthermore, an equation for calculating the internal moisture excess is shown, together with different humidity classes from the European Standard—EN/ISO 13788 (2012). On the basis of the given equations, some examples are give an indication of the amount of moisture moving into the attic either through an airtight ceiling with or without

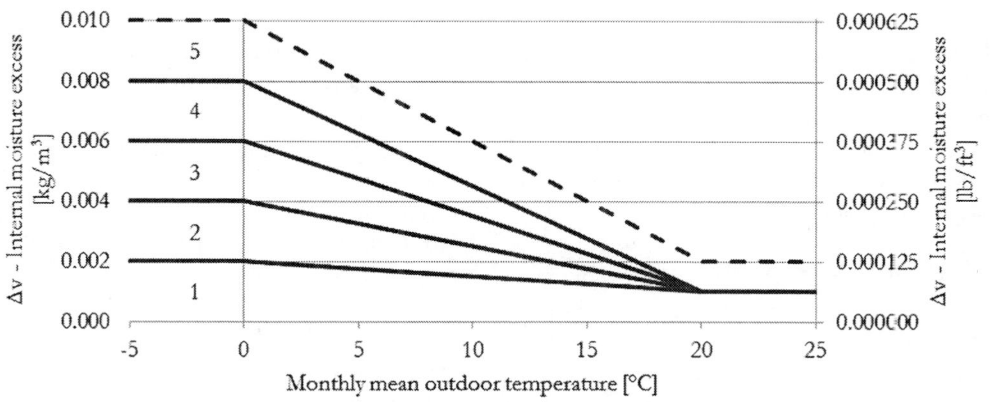

Figure 1 *The moisture excess of the indoor climate for the different humidity classes based on EN/ISO 13788 (2012).*

a vapor barrier or a ceiling with cracks. A short description of the moisture storage capability for different types of insulation materials and the chosen thresholds for this research are also listed.

Convection and Diffusion

By convection, the transport of moist air occurs along with air movements, which occur when there is a difference in the air pressure over a given volume of air. This could be due to a difference in the temperature, wind pressure or air pressure caused by a ventilation system. When looking at moisture transport due to convection through a crack in a thin layer, the production rate is given by the following (Brandt et al. 2013):

$$G_c = c \cdot A_{crack} \cdot \Delta p_a^{0.5} \cdot v_i \tag{1}$$

where

G_c = the internal moisture production rate due to convection, kg/s

c = the constant, 0.8 [m/(s · Pa0.5)]

A_{crack} = the area of the crack, m^2

Δp_a = the air pressure difference over a material, Pa

v_i = the humidity of internal air by volume, kg/m^3

Diffusion is driven by the difference in the water vapor pressure/humidity of air by volume, and will neutralize from the highest amount. The moisture transfer by pure water vapor diffusion is given by the following (EN/ISO 13788 2012):

$$g = \delta_0 \cdot \frac{\Delta p}{s_d} \tag{2}$$

where

g = the density of water vapor flow rate, kg/(m^2 · s)

δ_0 = the water vapor permeability of air with respect to partial vapor pressure, $2 \cdot 10^{-10}$ kg/(m·s·Pa)

Δp = the internal vapor pressure excess, Pa

s_d = the water vapor diffusion-equivalent air layer thickness, m

The total moisture excess due to diffusion can be described as:

$$G_d = g \cdot A_{sur} \tag{3}$$

where

G_d = the internal moisture production rate due to diffusion, kg/s

A_{sur} = the surface area of the ceiling, m^2

Humidity Classes

The amount of heat and moisture transported through the ceiling will vary with the indoor and outdoor climate. The moisture content of the indoors is given by the humidity of external air by volume plus an internal moisture contribution that depends on the use of the building. This is expressed by the following equation (EN/ISO 13788 2012):

$$\Delta_v = v_i - v_e = \frac{G}{n \;\; V} \tag{4}$$

where

Δ_v = the internal moisture excess, kg/m^3

v_i = humidity of internal air by volume, kg/m^3

v_e = humidity of external air by volume, kg/m^3

G = the total internal moisture production rate due to convection and diffusion, kg/h

n = is the air change rate, h^{-1}

V = is the internal volume, m^3

The internal moisture excess, Δ_v, is significant for the moisture transport as it determines the moisture content between indoor and outdoor; an increased internal moisture excess means increased moisture transport through the ceiling. According to EN/ISO 13788 (2012), the internal humidity load can be described by five humidity classes. The internal moisture excess for each humidity class is shown in Figure 1.

Dwellings in Denmark are normally either in humidity class 2 or 3. Danish single family houses are generally in the humidity class 2 while apartments and terrace houses are normally humidity class 3 (Brandt et al. 2013). Figure 1 shows that the internal moisture excess is highest in the winter due to less ventilation of the indoor climate.

The RH in the attic can be simulated by using Equation 3, where the moisture excess is given by the moisture penetrating into the attic by convection and diffusion through the ceiling.

Relative Humidity in Ventilated Attics—Examples

To illustrate the difference between moisture penetration through a ceiling by convection and diffusion, a series of examples is shown. The parameters for the examples listed in Table 2 are used to calculate the amount of moisture that moves into the attic. Both the indoor and the outdoor climate is based on normal Danish winter conditions. The approximated air pressure difference over a ceiling is based on the work by Brandt et al. (2013), and the air change rate in the attic is varied for comparison. Harderup and Arfvidsson (2013) used the values 0.2 and 2.0 h^{-1} in traditional attics.

To show the change in the attic's RH, the calculations were performed with different insulation thicknesses and air change rates of the attic. The equations shown previously were used to make the calculations in Tables 3 and 4. The temperature used in the attic was estimated based on a simple heat loss calculation (no radiation).

From Tables 3 and 4 it can be seen that moisture diffusion through an airtight vapor barrier for an indoor humidity level at 50% rh is $4.5 \cdot 10^{-6}$ kg/h ($0.99 \cdot 10^{-5}$ lb/h) per square metre (square foot) ceiling, or 0.11 g ($0.24 \cdot 10^{-3}$ lb) each day per square metre (square foot). However, when looking at convection through a small crack (1 cm^2 [0.155 in^2]) in the ceiling, the amount is 96.39 g (0.21 lb) each day per square metre (square foot). When calculating the internal moisture excess to a ventilated attic space with an air change rate of 0.5 h^{-1}, the internal moisture excess is $8.9 \cdot 10^{-6}$ and $8.0 \cdot 10^{-3}$ kg/m^3 ($0.056 \cdot 10^{-5}$ and $50 \cdot 10^{-5}$ lb/ft^3) for diffusion and convection, respectively, which means that the moisture transport is around a thousand times higher by convection than diffusion.

When looking at the impact of the RH in the attic, the internal moisture excess in the example with diffusion will not increase the RH significantly, but when considering the moisture transport due to convection, the RH will be 100% and condensation of moisture will occur.

Figure 2 shows the impact of the air change rate. From this it can be concluded that the total amount of moisture excess due to diffusion is in such order that the contribution alone will almost reach the threshold for condensation. Therefore any supply from convection will lead to condensation.

The figure also shows that it is not possible to raise the air change rate to such a degree that the contribution caused by convection can be removed. The amount of insulation material changes the temperature in the attics and consequently the threshold for condensation. This means that diffusion contribution is closer to the threshold when the amount of insulation material is high.

If the air change rate in the attic is changed from 0.5 h^{-1} to 2 h^{-1}, the RH will still be 100%, but the amount of condensed moisture will have decreased significantly.

Similar calculations have been performed at different indoor RH levels with and without vapor barrier, and are illustrated in Figure 3.

Figure 3 shows the influence of the indoor air humidity for attics without vapor barrier and different amounts of insulation material as well as the indoor climate and air change rate in the attics. It can be seen that the air change rate has a larger effect than the indoor climate. With an air change rate of 2 and an indoor air RH of 40% there will be no condensation in the attic if there is no convection, whereas at 60% rh the diffusion alone will be at the threshold for condensation (solid line), independent on the amount of insulation material. The amount of insulation material has an influence on how close the moisture excess caused by diffusion is to the threshold for condensation. For example, if the insulation thickness is low, the construction is more robust when it comes to condensation, although in all cases the moisture excess due to diffusion alone is above the threshold for mold growth (dotted line).

If the temperature in the attic is increased because of solar radiation, e.g., an increase of 10°C (50°F), the threshold for condensation will be increased to approximately 0.018 kg/m^3 (0.0011 lb/ft^3), which means that it is possible to dry out previously condensed moisture.

Table 2. Parameters Used in the Examples

Parameter	Value	Unit	Value	Unit
Indoor temperature	20	°C	68	°F
Indoor RH	50	% rh	50	% rh
Outdoor temperature	0	°C	32	°F
Outdoor RH	90	% rh	90	% rh
Sd-value, vapor barrier	100	m	328	ft
Sd-value, 150 insulation	0.15	m	0.5	ft
Sd-value, 400 insulation	0.4	m	1.3	ft
Sd-value, plasterboard	0.1	m	0.3	ft
Crack in ceiling	0.0001	m^2	0.0011	ft^2
Approximated air pressure different over a ceiling	2.5	Pa	0.052	lbf/ft^2
Ceiling area	1	m^2	10.8	ft^2
Attic volume	1	m^3	35.3	ft^3

Different Insulation Materials—Different Moisture Storage

The insulation material primarily used in Denmark is mineral wool, which is either produced from glass or stone. This makes the insulation material non-hygroscopic. This is the opposite of cellulose fiber material, which is hygroscopic because its fundamental element is raw wood or recycled

Table 3. Relative Humidity in a Ventilated Attic with a Variation in Insulation Thickness and Air Change Rate for 50% Indoor RH (SI)

Insulation Thickness, t (m)	Diffusion, d, or Convection, c	Air Change Rate, n (h^{-1})	Internal Moisture Production Rate, G (kg/h)	Internal Moisture Excess (kg/m^3)	Attic Air Humidity by Volume due to Diffusion (kg/m^3)	Attic Air Humidity by Volume due to Convection (kg/m^3)	Total Attic Air Humidity by Volume (kg/m^3)	Attic Temperature (°C)	Attic RH (%)
0.15	d	0.5	$4.5 \cdot 10^{-06}$	$8.9 \cdot 10^{-06}$	0.004	—	0.004	1.5	81.3
0.15	c	0.5	$4.0 \cdot 10^{-03}$	$8.0 \cdot 10^{-03}$	0.004	0.012	0.016	1.5	100
0.4	d	0.5	$4.4 \cdot 10^{-06}$	$8.9 \cdot 10^{-06}$	0.004	—	0.004	0.5	87.1
0.4	c	0.5	$4.0 \cdot 10^{-03}$	$8.0 \cdot 10^{-03}$	0.004	0.012	0.016	0.5	100
0.15	d	2	$4.5 \cdot 10^{-06}$	$2.2 \cdot 10^{-06}$	0.004	—	0.004	1.5	81.2
0.15	c	2	$4.0 \cdot 10^{-03}$	$2.0 \cdot 10^{-03}$	0.004	0.006	0.010	1.5	100
0.4	d	2	$4.4 \cdot 10^{-06}$	$2.2 \cdot 10^{-06}$	0.004	—	0.004	0.5	86.9
0.4	c	2	$4.0 \cdot 10^{-03}$	$2.0 \cdot 10^{-03}$	0.004	0.006	0.010	0.5	100
0.15	d	10	$4.5 \cdot 10^{-06}$	$4.5 \cdot 10^{-07}$	0.004	—	0.004	1.5	81.1
0.15	c	10	$4.0 \cdot 10^{-03}$	$4.0 \cdot 10^{-04}$	0.004	0.005	0.009	1.5	100
0.4	d	10	$4.4 \cdot 10^{-06}$	$4.4 \cdot 10^{-07}$	0.004	—	0.004	0.5	86.9
0.4	c	10	$4.0 \cdot 10^{-03}$	$4.0 \cdot 10^{-04}$	0.004	0.005	0.009	0.5	100

Table 4. Relative Humidity in a Ventilated Attic with a Variation in Insulation Thickness and Air Change Rate for 50% Indoor RH (I-P)

Insulation Thickness, t (in.)	Diffusion, d, or Convection, c	Air Change Rate, n (h^{-1})	Internal Moisture Production Rate, G (lb/h)	Internal Moisture Excess (lb/ft^3)	Attic Air Humidity by Volume due to Diffusion (lb/ft^3)	Attic Air Humidity by Volume due to Convection (lb/ft^3)	Total Attic Air Humidity by Volume (lb/ft^3)	Attic Temperature (°F)	Attic RH (%)
6	d	0.5	$0.99 \cdot 10^{-5}$	$0.056 \cdot 10^{-5}$	$25 \cdot 10^{-5}$	—	$25 \cdot 10^{-5}$	34.7	81.3
6	c	0.5	$882 \cdot 10^{-5}$	$50 \cdot 10^{-5}$	$25 \cdot 10^{-5}$	$75 \cdot 10^{-5}$	$100 \cdot 10^{-5}$	34.7	100
16	d	0.5	$0.97 \cdot 10^{-5}$	$0.056 \cdot 10^{-5}$	$25 \cdot 10^{-5}$	—	$25 \cdot 10^{-5}$	32.9	87.1
16	c	0.5	$882 \cdot 10^{-5}$	$50 \cdot 10^{-5}$	$25 \cdot 10^{-5}$	$75 \cdot 10^{-5}$	$100 \cdot 10^{-5}$	32.9	100
6	d	2	$0.99 \cdot 10^{-5}$	$0.014 \cdot 10^{-5}$	$25 \cdot 10^{-5}$	—	$25 \cdot 10^{-5}$	34.7	81.2
6	c	2	$882 \cdot 10^{-5}$	$12.5 \cdot 10^{-5}$	$25 \cdot 10^{-5}$	$37.5 \cdot 10^{-5}$	$62.5 \cdot 10^{-5}$	34.7	100
16	d	2	$0.97 \cdot 10^{-5}$	$0.014 \cdot 10^{-5}$	$25 \cdot 10^{-5}$	—	$25 \cdot 10^{-5}$	32.9	86.9
16	c	2	$882 \cdot 10^{-5}$	$12.5 \cdot 10^{-5}$	$25 \cdot 10^{-5}$	37.5 · 10-5	$62.5 \cdot 10^{-5}$	32.9	100
6	d	10	$0.99 \cdot 10^{-5}$	$0.003 \cdot 10^{-5}$	$25 \cdot 10^{-5}$	—	$25 \cdot 10^{-5}$	34.7	81.1
6	c	10	$882 \cdot 10^{-5}$	$25 \cdot 10^{-5}$	$25 \cdot 10^{-5}$	$31.3 \cdot 10^{-5}$	$56.3 \cdot 10^{-5}$	34.7	100
16	d	10	$0.97 \cdot 10^{-5}$	$0.003 \cdot 10^{-5}$	$25 \cdot 10^{-5}$	—	$25 \cdot 10^{-5}$	32.9	86.9
16	c	10	$882 \cdot 10^{-5}$	$25 \cdot 10^{-5}$	$25 \cdot 10^{-5}$	$31.3 \cdot 10^{-5}$	$56.3 \cdot 10^{-5}$	32.9	100

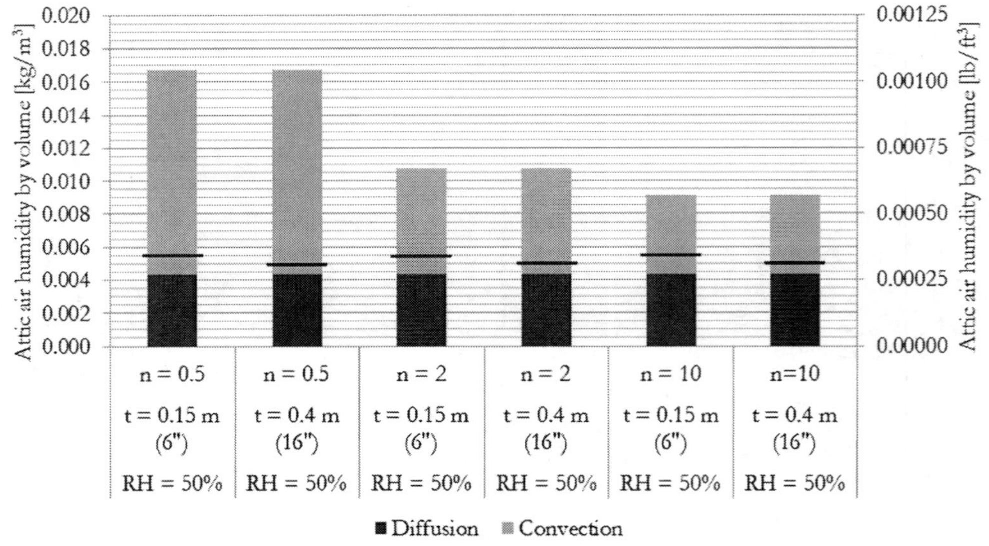

Figure 2 *The moisture contribution caused by diffusion and convection resulting in air humidity by volume in attics with a vapor barrier, showing different amounts of insulation material,* t, *and air change rate,* n, *(h^{-1}). The black horizontal line marks the threshold for condensation at different temperatures in the attics.*

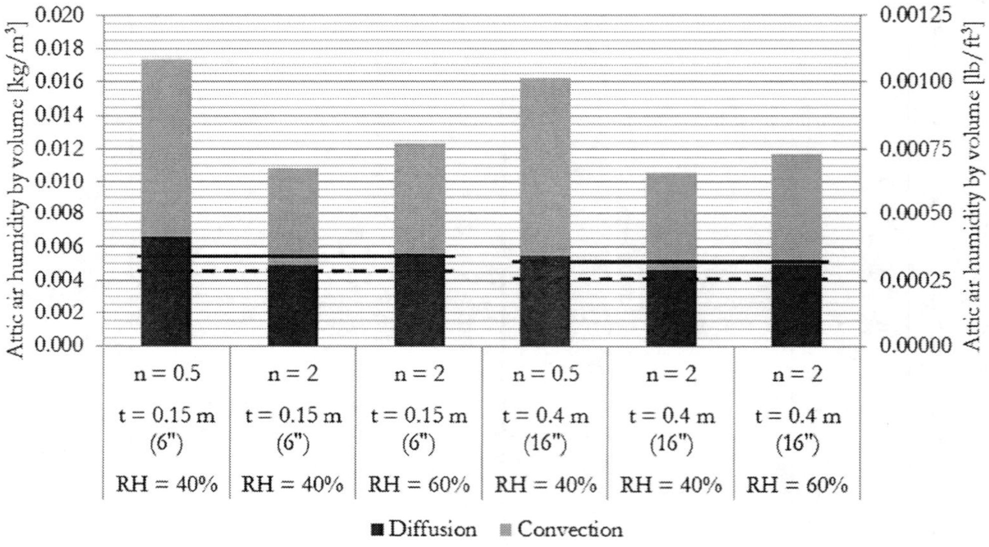

Figure 3 *The moisture contribution caused by diffusion and convection resulting in air humidity by volume in attics without a vapor barrier, showing different air change rates,* n, *(h^{-1}); amounts of insulation material,* t; *and different indoor air humidity. The black horizontal solid line marks the threshold for condensation and the dotted line is the threshold for mold growth at different temperatures (because of different insulation thickness) in the attics.*

newspapers, cardboard, and etc. Cellulose fiber insulation based on paper has recently become more popular.

The main difference between these two insulation materials is the hygroscopic effect, and the resulting capability for moisture storage which can be described in a sorption isotherm as shown in Figure 4.

From Figure 4 it can be seen that insulation material based on cellulose fiber can contain 10 times more moisture than glass fiber at a RH of 90%.

Thresholds

The timber in building constructions can be degraded by two kinds of biologic attacks—mold growth and rot. Mold growth does not decay timber, but it may have a negative impact on the indoor environment. Conversely, rot decays timber, but it does not influence the indoor environment (Singh 1999). Mold grows on the surface of the material, while rot grows inside the timber.

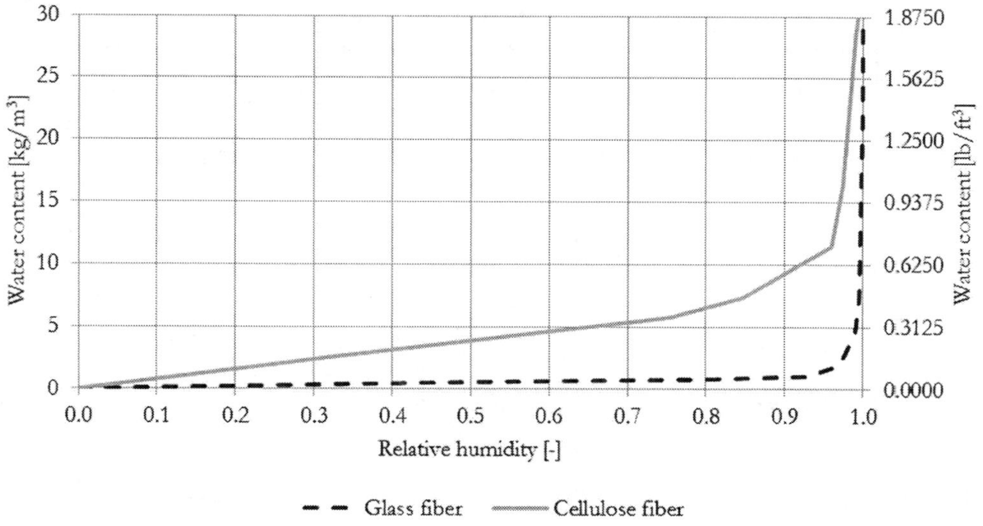

Figure 4 *Sorption isotherm for glass fiber and cellulose fiber insulation. The data are retrieved from Fraunhofer (2015) databases; respectively, Fraunhofer-IBP - Holzkirchen, Germany—ISOVER GW Integra ZKF 032 (glass fiber) and MASEA Database, Germany (cellulose fiber).*

The different mold species have varying favorable growth conditions, but the general "starting point" is 75% rh (which corresponds to a wood-moisture content of 15% by weight at 20°C (68°F). The stated humidity is on the surface of the material or in the pores in the surface. For rot attack in the timber, the moisture content must be above 20% by weight, which corresponds to 85% rh at 20°C (68°F).

Besides humidity, mold growth also depends on temperature. The optimal temperature is 20°C to 30 °C (68°F to 86 °F) for most of species.

The threshold value in this research, defining a moisture problem, is set to be 85% rh and a wood-moisture content of 18% by weight. The reason for these criteria is based on the work by Sedlbauer (2001) and that the temperature in the examined attics is expected to be around 5°C to 12°C (41°F to 54°F) in the period where the moisture problems will most likely occur. Although mold growth may stop for periods of time where the temperature/RH is unfavorable, damage is done because there is a danger that particles from dead mold in the attic might affect the indoor environment below the ceiling (Valbjoern 2003).

MATERIALS AND METHODS

To evaluate the impact of the moisture in the attic when the insulation thickness is increased without establishing a new vapor barrier, a field survey of a selected collection of different dwellings in Denmark was conducted. The following parameters were considered for the selection:

- Type of vapor barrier
- Orientation of the roof surface
- Amount of insulation
- Type of insulation

Low ventilation rate is expected to have a high influence on the moisture level in attics, possibly much more than the other parameters; in attics with low ventilation rate other effects may not have been detectable. Furthermore, it is normally not difficult to establish sufficient ventilation openings. Consequently, ventilation rate has not been further investigated.

Thirty owners were willing to participate in the research; unfortunately, only 10 dwellings are represented in this paper because of complications in collecting data from the installed sensors. The 10 dwellings are listed in Table 5, where the different selected parameters are stated as well.

Each dwelling was inspected in summer and in winter. At each inspection the wood-moisture content of the wooden part in the attic was measured. In connection with the first inspection, a series of sensors was placed to measure the temperature and the RH every hour at different locations in the attic and construction. The location of the measuring points is as follows:

- Attic space (ridge)
- In attic just below roof surface at both orientations
- Just above the insulation material
- On top of the vapor barrier (if no vapor barrier, then on top of the ceiling)
- Inside the building
- Outside the building (only if there is not another test dwelling nearby)

Data from sensors positioned just above the insulation material and on top of the vapor barrier/ceiling will later be used for evaluating the risk of moisture forced into the insulation material by high temperatures in the attic during summertime. In this paper, the data collection has only been during wintertime; therefore, data from these sensors is omitted.

Table 5. Description of Each Test Building

Test Type	Test Number	Original Insulation, mm (in.)	Building Year	Vapor Barrier	Roof Orientation	Insulation Type*	Additional Insulation, mm (in.)	Insulation Type*	Total Insulation, mm (in.)
A	1	—	2015	None	E/W	—	600 (24)	CL	600 (24)
A	2	150 (6)	1964	None	N/S	MW	200 (8)	CL	350 (14)
B	5	250 (10)	2004	PE-foil	N/S	MW	—	—	250 (10)
B	6	200 (8)	1979	PE-foil	N/S	MW	—	—	200 (8)
B	7	150 (6)	1996	PE-foil	N/S	MW	—	—	150 (6)
B	8	250 (10)	1998	PE-foil	NW/SE	MW	—	—	250 (10)
C	10	100 (14)	1969	Alu-paper- foil	N/S	MW	200 (8)	CL	300 (12)
C	11	100 (4)	1969	Alu-paper- foil	N/S	MW	350 (14)	CL	450 (18)
C	12	100 (4)	1971	Alu-paper- foil	N/S	MW	200 (8)	CL	300 (12)
D	23	100 (4)	1971	Alu-paper- foil	N/S	MW	100 (4)	MW	200 (8)

* CL = cellulose, MW = mineral wool

To make a comparison between the indoor climate and the climate in the attic, the water vapor content in kg/m^3 is calculated from the measured temperature and RH.

RESULTS

When the sensors were installed, the ventilation openings were visually assessed and found to comply with the guidelines, i.e., ventilated through air gaps at eaves and valves or openings at the ridge, which means a ventilation rate of approx. 2 h^{-1} on average. Furthermore, a visual inspection of mold growth did not show any remarkable mold growth.

Basis of Evaluation

Because of the selection to register the temperature and RH every hour, the calculation of air humidity by volume has a high fluctuation. Therefore, it was chosen to evaluate the raw data with a calculation of the moving average for a period of one week. An example of this is illustrated in Figure 5.

Only evaluations of data from sensors at the ridge are presented in this paper because there have been no significant differences in the measurements from sensors positioned in the ridge and roof underlay.

Outdoor Climate

Collection of the outdoor climate data was performed by five test buildings within 40–70 km (25–40 mi); the variation is shown in Figure 6. An average value of these five measurements is used for the calculation of the moisture excess to the indoor climate and in the attics.

Internal Moisture Excess

The internal moisture excess is calculated on basis of the differences in the indoor and outdoor climate. The variation over the period is shown in Figure 7. The two black curves represent buildings with no vapor barrier in the ceiling. Two dwellings (test numbers 5 and 6) have a sizable internal moisture excess compared to the other test buildings.

Moisture Level in Attics

On the basis of the different parameters, the moisture level in attics is shown in Figures 8 to 11. See the figure labels for descriptions of the parameters.

Moisture Excess to Attic

For evaluation of the moisture excess to the attic for all ten dwellings two figures are shown. Figure 12 illustrates the differences with and without a vapor barrier and Figure 13 illustrates the impact of the thickness of the total amount of insulation material.

DISCUSSION

The results show minor differences in the moisture level in the attics. However, the tendency is the opposite of what is expected:

- The moisture level is lowest without vapor barrier
- High insulation amount means low moisture level
- Internal moisture level has a minor influence on the moisture level in the attic

Only the hypothesis that hygroscopic insulation material can have some effect on reducing the moisture level in the attic seems to be supported. Unfortunately, test buildings without vapor barriers are also those with high insulation amount and hygroscopic insulation material. The explanation could be that the effect of hygroscopic material is higher than the effect of insulation thickness and vapor barrier. Having said that, it must

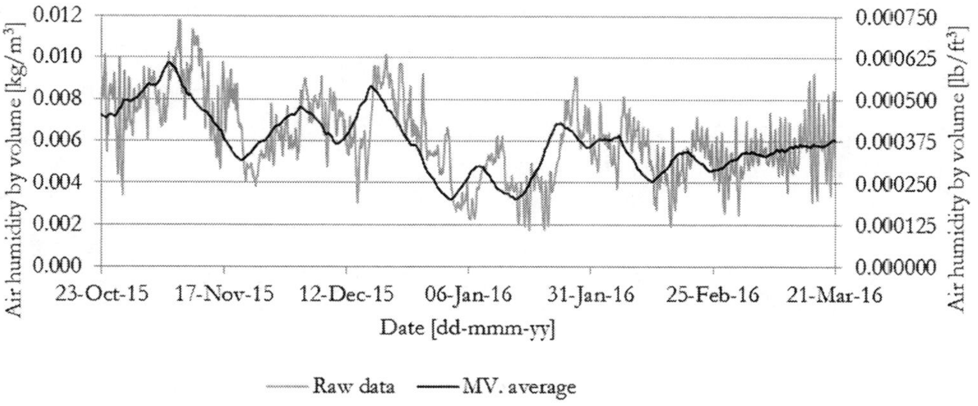

Figure 5 *Difference between the measured data (raw data) and the use of moving average (MV. average) for illustrating the variation of air humidity by volume. The moving average period is one week.*

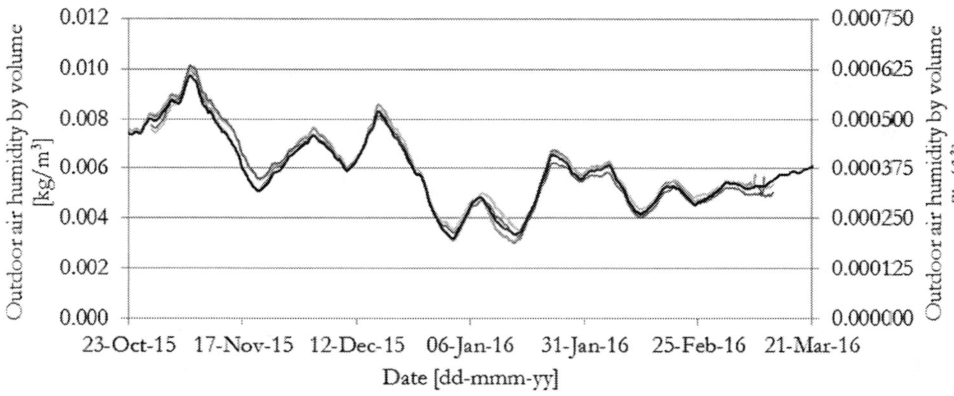

Figure 6 *Measured outdoor climate at five different locations within 40–70 km (25–40 mi).*

be emphasized that the differences in moisture levels in the attics are small, and the number of test buildings without vapor barrier and high amount of insulation material is limited to two.

Figures 2 and 3 are based on calculations using an outdoor temperature of 0°C (32°F). This level of temperature was detected in the beginning of January and mid-February to mid-March. The measurements in these periods show a moisture level close to the calculated moisture excess casued by diffusion only. Consequently, the moisture excess caused by convection is limited and, therefore, the ceilings must be considered to be almost airtight.

According to Figure 3, even ceilings without convection must have a ventilation rate at least above 0.5 h^{-1} to prevent the risk of condensation. This corresponds to the measured moisture levels in the attics and no visual signs of condensation. Nevertheless, it can be seen from the collected data that the moisture level in all the tested attics is near the threshold value for the risk of mold growth, which means that insulated ceilings against ventilated attics is a fragile construction and small failures can lead to moisture-related problems. However, when sensors have been installed in the attics there has been no visual signs of mold growth. If the lowest threshold value, shown in Figure 11 (black broken line with dot), was the actual threshold, mold growth would have been expected. Apparently the threshold value is higher; a more likely threshold could be an RH of 95% at temperatures above 10°C (50°F). The reason for this higher threshold could be that instationary conditions hamper mold growth, which therefore start at a higher moisture level.

Because of complications with the owners of the test buildings, is has only been possible to show results from 10 of the dwellings, which means some parameters are missing to have a comprehensive basis for a final evaluation of the moisture levels in ventilated attics. The missing parameters are buildings with a high amount of insulation material and buildings with no vapor barrier and different types of insulation material. Measurements from these types of buildings are currently being

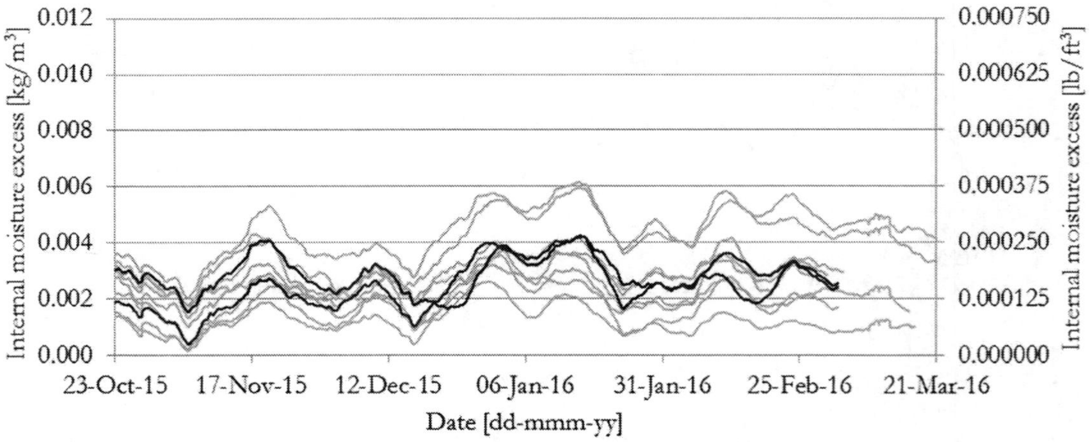

Figure 7 *Internal moisture excess for all 10 locations. The black lines represent buildings where there is no vapor barrier in the ceiling (Type A).*

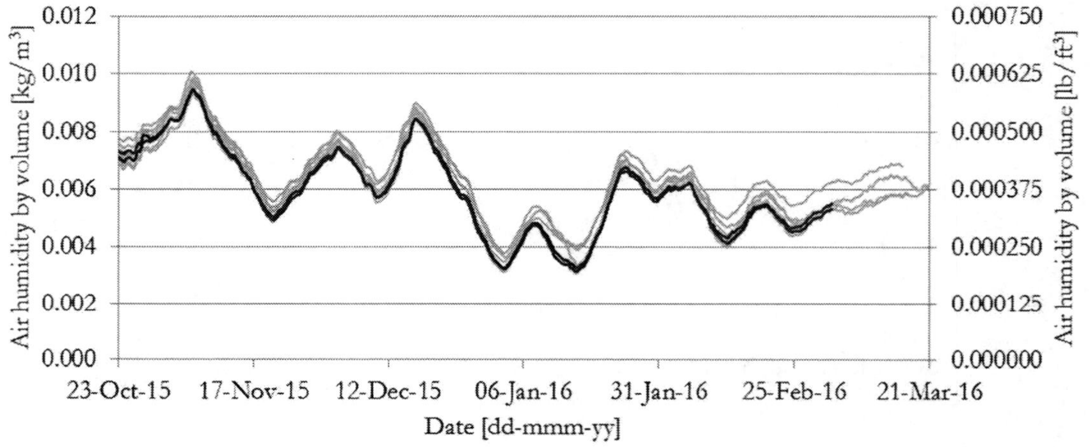

Figure 8 *Moisture level at the ridge in attics. The black lines represent buildings where there is no vapor barrier in the ceiling (Type A).*

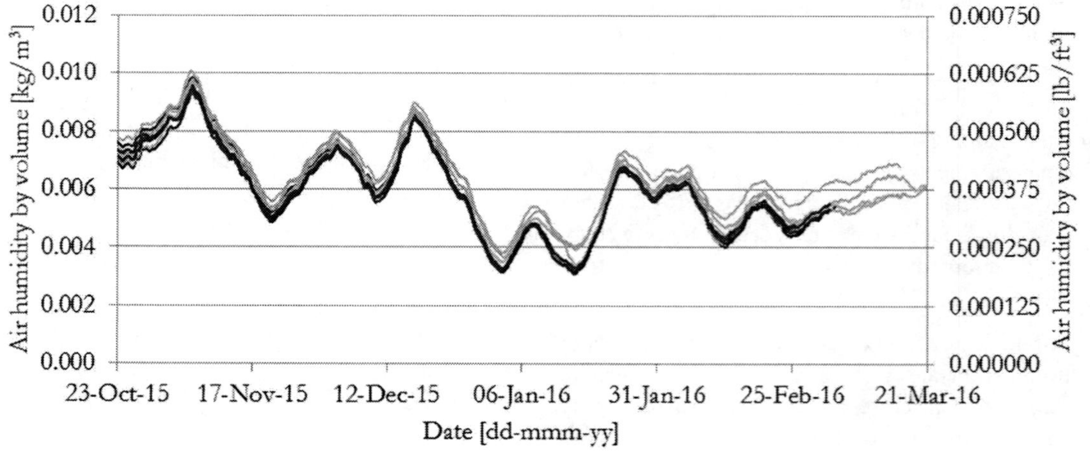

Figure 9 *Moisture level at the ridge in attics. The black lines represent buildings where the total amount of insulation material is 300 mm (12 in.) or more.*

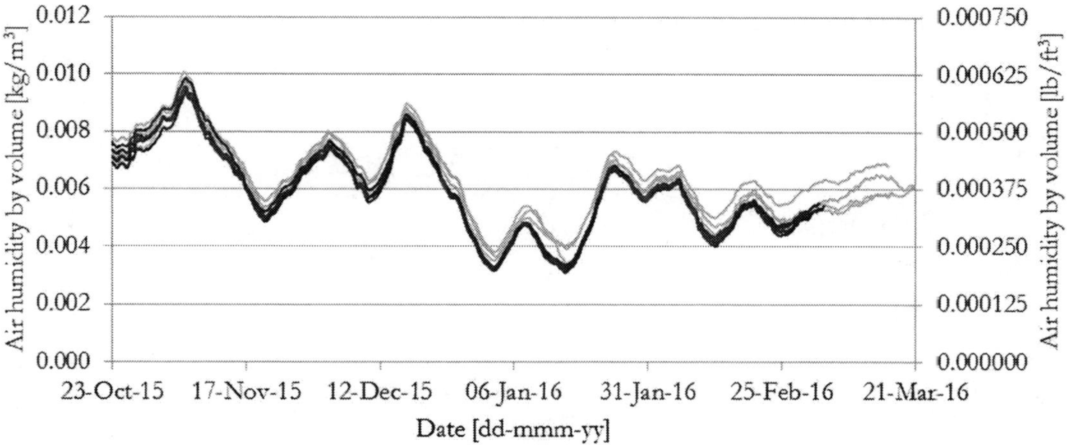

Figure 10 *Moisture level at the ridge in attics. The black lines represent buildings where the insulation material is cellulose based and grey lines represent mineral wool based.*

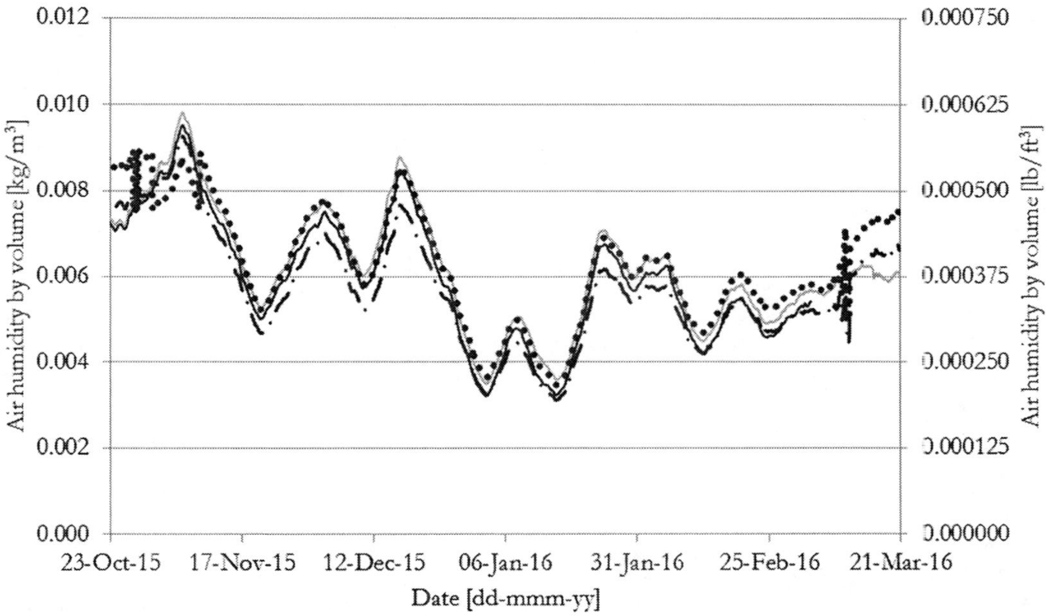

Figure 11 *Average moisture level at ridge in attics compared with threshold values for mold growth. Solid black line is the average for buildings with 300 mm (12 in.) or larger amount of insulation material, while the solid grey line represents buildings with a smaller amount of insulation material. Two different threshold values are shown, calculated with, respectively 15°C (59°F) and 85% rh (black broken line with dot) and where the temperature is 10°C (50°F) and 95% rh (black dotted line).*

collected together with results from a full-size test building with different ceiling constructions and controlled indoor environments.

Before the current conservative recommendation is changed, the differences in existing buildings and the significance of vapor tightness, hygroscopic properties of insulation material, and amount of insulation material must be evaluated. This must be done by combining field measurements and tests under controlled conditions with well-chosen parameter variations. Therefore, these measurements are only a part of a larger project that also includes a full-size test building with six different ceiling constructions. To valuate other possible constructions, the project will also include simulations. Nevertheless, models for the simulation will be validated with measurements before more parameter variations will be explored, as simulations with convection can be problematic.

CONCLUSION

On the basis of the collected data, it is not possible to conclude that the current guidelines for vapor barriers in

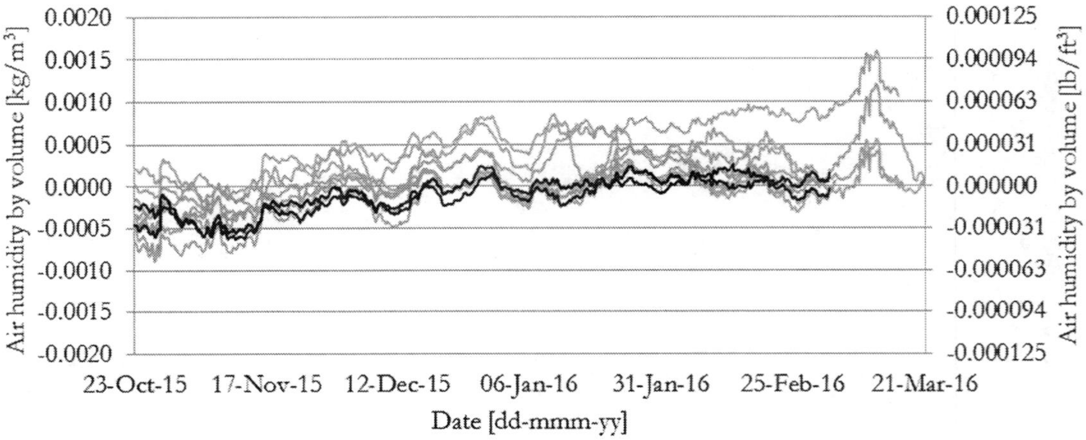

Figure 12 *Moisture excess to the attic. Black lines represent buildings where there is no vapor barrier in the ceiling (Type A).*

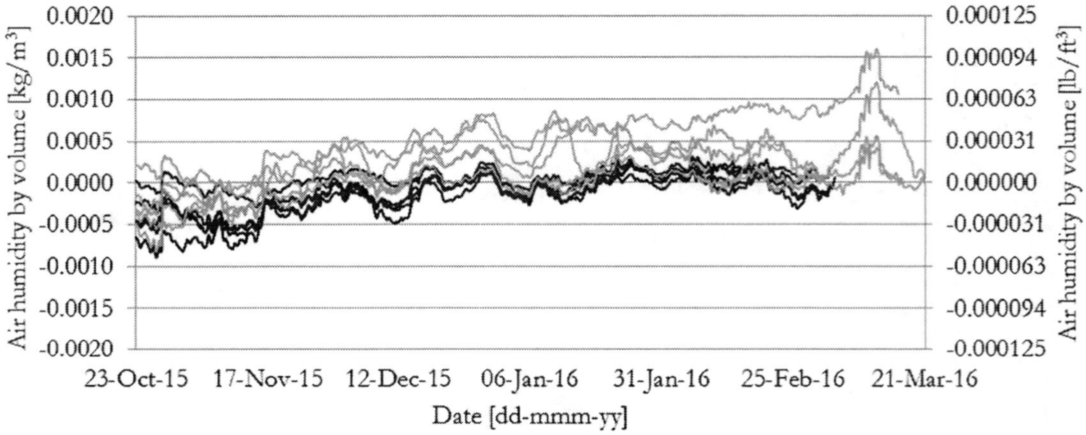

Figure 13 *Moisture excess to the attic. Black lines represent buildings where the total amount of insulation on the ceiling is 300 mm (12 in.) or more.*

highly insulated attics can be changed, because of the limited amount of test buildings and lack in the parameter variation. However, it seems that there is a trend that neither vapor barriers nor the amount of insulation material, or the level of indoor moisture level have essential impact on the moisture level in the attic, whereas the hygroscopic properties of the insulation material may have an influence. On the other hand, the measured dwellings have too little variation to draw solid conclusions; the test buildings with hygroscopic insulation materials are generally well insulated, and two of these are the only ones without vapor barrier. Calculations show that the ventilation rate of the attics may have a high impact, but sufficient ventilation is normally easy to establish and was partly therefore not a selected parameter in this study; all attics were ventilated according to recommendations.

ACKNOWLEDGMENTS

The authors would like to thank the National Building Fund (Landsbyggefonden), The Landowners' Investment Foundation (Grundejernes Investeringsfond), Association of Danish Insulation Manufacturers (VarmeIsoleringsForeningen), and the Danish Construction Association (Dansk Byggeri) for financial support of the research project DaLo of which this project is a part.

REFERENCES

Brandt, E., et al. 2007. (39) 07 10 29 Dampspærrer i klimaskærmen - fugttransport og materialer.: Byg-Erfa.

Brandt, E., et al. 2013. SBi-anvisning 224 - Fugt i bygninger 2. udgave. København SV: Statens Byggeforskningsinstitut, Aalborg Universitet.

DOE. 1978. Material criteria and installation practices for the retrofit application of insulation and other weatherization materials. Washington, D.C.: U.S. Department of Energy.

DOE. 2013. Building science-based climate maps—Building america top innovation. Washington, D.C.: U.S. Department of Energy.

EN/ISO 13788. 2012. *Hygrothermal performance of building components and building elements—Internal surface temperature to avoid critical surface humidity and interstitial condensation—Calculation methods*. Brussels: European Committee for Standardization.

Energistyrelsen. 1966. *The building regulations*. (Bygningsreglementet). http://w2l.dk/file/502084 /br_seksogtres.pdf.

Energistyrelsen. 2015a. *The building regulations* (Bygningsreglementet). http://bygningsreglementet.dk/.

Energistyrelsen. 2015b. *Consolidation act of the handbook for energy consultants* (HB2014). www.retsinformation.dk/Forms/R0710.aspx?id=161845.

Fraunhofer. 2015. *WUFI*. www.wufi.de.

Harderup L.E., and J. Arfvidsson. 2013. Moisture safety in cold attics with thick thermal insulation. *Journal Of Architectural Engineering* December:265–78.

Møller, E.B. 2012. SBi-anvisning 240—Efterisolering af småhuse—byggetekniske løsninger 1. udgave. Hørsholm: Statens Byggeforskningsinstitut, Aalborg Universitet.

Nik, V.M., A.S. Kalagasidis, and E. Kjellström. 2012. Assessment of hygrothermal performance and mould growth risk in ventilated attics in respect to possible climate changes in Sweden. *Building and Environment* 55:96–109.

Sedlbauer, K. 2001. *Prediction of mould fungus formation on the surface of and inside building components*. Stuttgart: Fraunhofer Institute for Building Physics.

Singh, J. 1999. Dry Rot and other wood-destroying fungi: Their occurrence, biology, pathology and control. *Indoor Built Environment*:3–20.

Valbjoern, O. 2003. Undersøgelse og vurdering af fugt og skimmelsvampe i bygninger.:By og Byg Anvisning 204.

Monitoring of Two Unvented Roofs with Air-Permeable Insulation in Climate Zone 2A

Kohta Ueno
Associate Member ASHRAE

Joseph W. Lstiburek, PhD, PEng
Fellow ASHRAE

ABSTRACT

Unvented roof assemblies can bring attic mechanical systems into conditioned space, negating ductwork losses. However, in previous work, unvented roofs with air-permeable (fibrous) insulation, instead of air-impermeable insulation (spray foams) have shown localized moisture accumulation at the ridge. This research is a test implementation of two unvented roof assemblies insulated with air-permeable insulation (netted and blown fiberglass or adhered fiberglass) in hot/humid climates. One test roof is located in Houston, TX and has asphalt shingles; the other is in Orlando, FL, with concrete tile; both are in Climate Zone 2A. Given that localized moisture accumulation and failures occurred at the ridge in previous unvented roofs, a diffusion vent (open to water vapor but closed to airflow) was installed at the highest points in the roof assembly to allow for the wintertime release of moisture. The diffusion vent is an opening at the ridge and hips covered with a water-resistant but vapor-open membrane. As a control comparison, portions of the roof were constructed as a typical unvented roof (self-adhered membrane at ridge).

Collected data indicate that the diffusion vent roof shows greater moisture safety and less wintertime moisture accumulation than the conventional, unvented roof design. The unvented roof had winter periods of 95%–100% rh, with other sensors indicating possible condensation; high moisture levels were concentrated at the roof ridge. In contrast, the diffusion vent roofs had drier conditions. In the spring, as outdoor temperatures warmed, all roofs dried well into the safe range (10% MC or less).

INTRODUCTION

Unvented roofs (also known as *cathedralized attics*) are an established concept in residential building; such roofs are typically implemented with spray polyurethane foam (SPF) at the underside of the roof deck (Rudd and Lstiburek 1998; Lstiburek 2006; Lstiburek 2011). Primary advantages to this construction are that (a) the attic ductwork and heating, ventilating, and air-conditioning (HVAC) system are brought into the thermal and air barrier of the conditioned space, and (b) air leakage at the ceiling plane (often complicated by interior detailing) is eliminated, which typically results in improved overall airtightness. Advantages in hurricane-prone areas include reduced wind-driven rain entry, by eliminating attic vents, and potentially reduced hurricane blow-off issues, due to differential pressurization. Disadvantages include higher first cost and greater surface area (conductive losses); the latter factor is often negated by eliminating duct leakage to the exterior.

However, unvented roof assemblies have a higher risk of wintertime interstitial (hidden) condensation issues than vented roof assemblies. These condensation risks occur when the roof sheathing is colder than interior dew point; in contrast, vented roof assemblies remove interior-sourced moisture before it can condense on the roof sheathing. These risks are addressed in current building code with the requirement for a minimum amount of air-impermeable insulation—such as rigid continuous insulation or spray foam—on the exterior side of the roof assembly. This exterior insulation raises the temperature of the condensing surface, or the underside of the air-impermeable insulation. The air-impermeable insulation requirements are presented (by climate zone) in

Kohta Ueno is a senior associate and **Joseph W. Lstiburek** is a principal at Building Science Corporation, Westford, MA.

Section R806.4, "Unvented Attic Assemblies," of the International Residential Code (ICC 2009). For Climate Zones (CZs) 1A and 2A, the requirement is a minimum of R-5 (RSI 0.9). For reference, requirements for ventilated roof assemblies are covered in Sections R806.1 through R806.3 (ICC 2009).

This research covers two test implementations of unvented roofs in hot/humid climates (CZ 2A): one in Houston, TX, with asphalt shingle roofing, and the other in Orlando, FL, with concrete tile roofing. The roofs were insulated with air-permeable fiberglass insulation, with a diffusion vent detail at the ridge to allow for wintertime removal of accumulated moisture. One goal of the research was to determine the assembly's resistance to condensation, or the design's moisture safety. This research has an additional benefit of potentially reducing the first cost of unvented roofs by eliminating SPF, which is a higher-cost material than fibrous insulations. This research is covered in detail in Ueno and Lstiburek (2015a; 2015b).

PREVIOUS FIELD WORK

The authors made discoveries on the hygrothermal behavior of unvented roofs in hot/humid climates during field implementation; several of these roofs were built in Houston, TX circa 2001 (CZ 2A). The typical roof assembly was asphalt shingle, roofing felt, oriented strand board (OSB) sheathing, and netted and blown cellulose insulation. During the first fall season of operation, a homeowner reported seeing moisture condensation and dripping from the ridge of the unvented attic, from the interior. Field investigation (Figure 1, left) revealed elevated wood moisture contents (MCs) at the ridge and evidence of previous condensation (i.e., rusted staples). The cellulose insulation at the peak had a noticeably "packy" or compactable texture as opposed to the fluffy texture of dry cellulose, indicating previous wetting (Rose and McCaa 1998). This failure was localized at the peak/ridge of the roof. These moisture issues were remediated by removing the insulation at the roof ridge, which solved the problem, but with an associated thermal penalty.

A similar issue arose in two production homes in Jacksonville, Florida (CZ 2A) circa 2000. Again, the roof assembly was asphalt shingles, roofing felt, OSB sheathing, and netted and blown cellulose insulation. During a site visit, the cellulose and roof sheathing were wet to the touch, and the steel truss plate was corroded; however, the wood had not decayed (Figure 1, right). In response, the insulation was removed near the ridge (roughly 2 ft [610 mm]) to increase surface temperatures of the roof sheathing. This roof has been rechecked in 2003 and 2014; roof sheathing conditions remain dry.

Similar localized ridge problems (colloquially referred to as *ridge rot*) have been documented in unvented compact cellulose roof assemblies in northern California (CZ 3C). Lower portions of the roof were dry and undamaged.

BACKGROUND LITERATURE

An overview of the use of dense-pack cellulose insulation (Lstiburek 2010) shows examples of the problems that can occur when using the material in compact roof assemblies. A comprehensive analysis of the use of dense-pack insulation of compact roofs (Schumacher and LePage 2012) noted the potential for interior air-leakage-sourced condensation in cold and mixed climates. Their hygrothermal analysis suggests that failures are not endemic in houses retrofitted with dense-packed roofs, because only moderate airtightness is achieved. Therefore, interior moisture generation is sufficiently diluted by wintertime air leakage to result in low wintertime interior relative humidity (RH) levels.

(a) (b)

Figure 1 *(a) Unvented cellulose roof ridge moisture content measurements in Houston, TX and (b) corrosion of ridge truss plate in Jacksonville, FL.*

Salonvaara et al. (2013) ran hygrothermal simulations models of vented and unvented roof assemblies in various configurations in CZs 1 through 4. Interior RH levels were set at design conditions; the roof assemblies consisted of asphalt shingles, felt paper, OSB sheathing, and insulation (open cell SPF [ocSPF] or fiberglass). Cases were run for no water intrusion and for a rainwater leak at the roof sheathing. The authors concluded that a vented attic assembly has better drying ability than unvented assemblies in all climates. Most unvented ocSPF assemblies had safe moisture contents (MCs) without rainwater leakage, but in colder climates (CZ 4) with very vapor-open ocSPF (54 perm in. or 78 ng/[Pa·s·m]), peak MCs reached risky levels. With the addition of rainwater leakage, MCs were substantially higher, resulting in failure in most CZs. The addition of a 1 perm (57 ng/[Pa·s·m^2]) vapor retarder coating on the interior side of the ocSPF resulted in lower MCs without rain leakage, due to reduction of interior-sourced vapor flow). However, the addition of a vapor retarder resulted in higher MCs with rainwater leakage, due to reduced inward drying. Two final cases used blown fiberglass insulation (100 perm in. [145 ng/(Pa·s·m]) instead of ocSPF: when vapor-open interior netting was used for installation, acceptable MCs were found in CZs 1 and 2, but riskier conditions were seen in CZs 3 and 4. Adding netting with a vapor permeance of 10 perms (570 ng/[Pa·s·m^2])] reduced risks to reasonable levels in CZs 1 through 3.

Boudreaux et al. (2013) presented data collected at eight houses in CZs 3A and 4A (four with vented attics and four with sealed attics) to understand their moisture performance. Their data showed that attic absolute moisture levels (vapor pressures) were consistently higher in the sealed attics than in the vented attics. Similarly, the vapor pressures in the main spaces were higher in the sealed attic houses. When the individual house data were plotted, the two groups overlapped. The worst-performing unvented attic had the highest air leakage (10 ach_{50}) of the sample; the authors described other factors that influenced interior moisture levels, including occupant density and cooling setpoint. Plots of vapor pressure at various sections of the roof assembly indicated that the sealed attic roof typically had patterns of daytime drying (desorption into the attic, driven by solar gain) and nighttime wetting (adsorption from the attic air). Several sources of moisture for the sealed attic were discussed, including roof sheathing storage, solar-driven moisture from moisture stored in asphalt shingles (judged unlikely by the authors), and exterior air leakage (likely, given the downward stack effect during the summer).

Lstiburek (2014) discussed unvented attics, or an alternate term of *conditioned attic*. In earlier construction, attic ductwork was typically leaky and supply leakage would slightly pressurize the attic. This would cause air leakage (i.e., air communication) from the attic to the main space, thus diluting attic moisture with drier air.

The moisture source in the attic was primarily from interior, rather than exterior sources. Moisture would concentrate in the attic because the moisture-laden air was more buoyant (less dense) than dry air. But as airtight ductwork became more common, incidental space conditioning provided to unvented attics due to communication between the attic and the main space decreased. The resulting problem has been higher humidity levels in unvented and sealed attics, especially with vapor-open insulation materials. Interior moisture is adsorbed and desorbed from the roof sheathing with daily warm and cold cycles. When air communication is eliminated or reduced, moisture accumulates in the attic due to moisture buoyancy. The author also mentions Houston-area field research about the effect of vapor-permeable and vapor-impermeable shingle underlayments on inward vapor drives; the study found no measurable differences between these roof test bays. The author concludes that shingle inward vapor drive issues are not a significant moisture source.

Pallin et al. (2014) built on Boudreaux et al.'s (2013) field studies, running whole-house energy, moisture, and airflow simulations of unvented roof houses, followed by one-dimensional hygrothermal models of air leakage into the unvented roof assembly. The authors ran over 400 simulations, changing variables such as leakage areas, duct leakage, and interior moisture generation rates. The simulations indicated that low interior moisture generation rates present no significant risks, moderate generation rates result in some risky cases, and high generation rates result in some noticeable failure cases. Lower enclosure leakage was associated with higher risks (which retain interior moisture generation in winter), as was midrange attic-to-exterior air leakage. These simulations showed no effect of duct leakage on moisture risks; these failures occurred in winter, when cooling system dehumidification would not reduce interior moisture loads. Based on these simulations, the authors recommended that unvented attics should be built to be as airtight as possible, and interior moisture generation rates should be controlled. They noted that higher air attic-to-exterior air leakage rates—as occur in practice—have moisture risks.

SITE DESCRIPTIONS, TEST ASSEMBLIES, AND INSTRUMENTATION

Two unvented test roofs were built and instrumented (Figure 2). One was in Friendswood, TX (Houston area, CZ 2A); it is a two-story, 4028 ft^2 (374 m^2), slab-on-grade model house that was unoccupied during testing. The other was a two-story, 3600 ft^2 (334 m^2) slab-on-grade house in Orlando, FL, and was occupied after completion of construction.

Both test houses had unvented/sealed attics that were insulated at the roofline. An experimental diffusion vent detail was installed at the ridge, which allows for vapor diffusion drying at the ridge and tops of hips, while preventing airflow/air leakage; it is discussed in detail below. As a control comparison, one portion of the roof was constructed as a typical unvented roof, with vapor-impermeable, self-adhered membrane at the ridge (a Class I vapor barrier). The team's hypothesis is that moisture accumulation would occur in the control roof, and the experimental diffusion vent roofs would show sufficient drying to avoid failure. Both roofs were insulated with fiberglass insulation, which is air permeable and vapor permeable (in the range of 120 perm in. [170 ng/(Pa·s·m)]).

The Houston roof (Figure 3, left) was insulated with spray-applied/adhered fiberglass (1.8 lb/ft^3 [29 kg/m^3] density) that was left exposed, with no interior finish. The roof assembly consists of asphalt shingles, underlayment (#15 asphalt saturated felt), 7/16 in. (11 mm) oriented strand board (OSB) sheathing, and spray-applied fiberglass at R-38 (RSI 6.7), encapsulating the 2 × 8 (typical) rafters.

The Orlando roof (Figure 3, right) roof assembly consists of concrete barrel tile screwed down to the sheathing, self-adhered modified asphalt 60 mil roofing underlayment with a woven polyester surface, ½ in. (12 mm) plywood sheathing, and blown fiberglass insulation (R-38/RSI 6.7) supported by netting stapled to the roof truss top chords (Owens Corning 2015).

No intentional space conditioning was provided in the unvented attic spaces; the space "floats" at conditions between the interior setpoint and exterior conditions. In the Houston attic, humidification was provided in the final (third) winter, to substitute for the lack of moisture generation at the unoccupied model house.

(a) **(b)**

Figure 2 *Front views of (a) Houston, north elevation, and (b) Orlando (west elevation) test roof houses.*

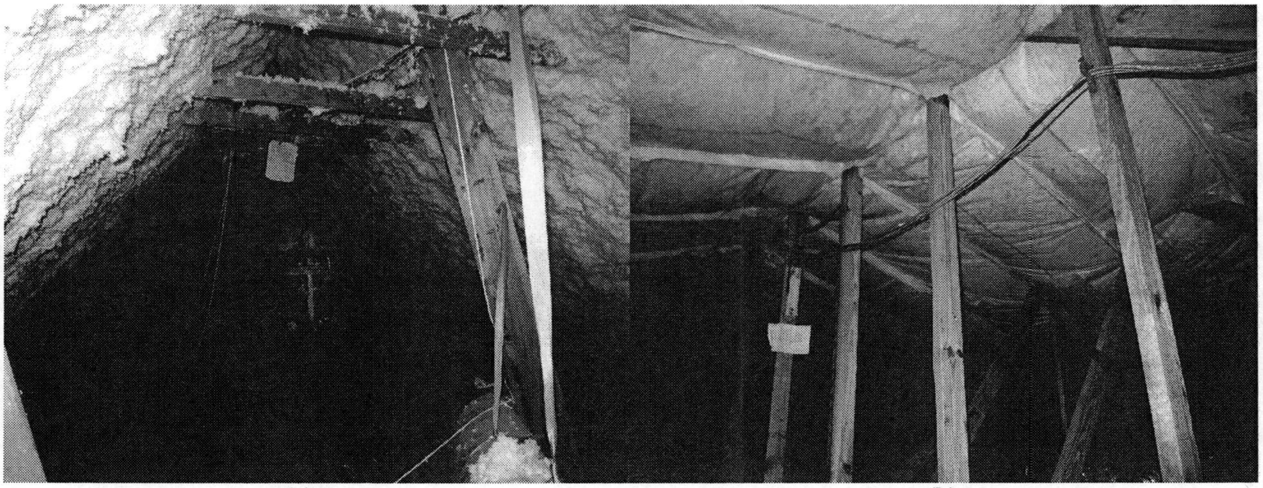

(a) **(b)**

Figure 3 *Attic interiors: (a) spray-applied fiberglass in Houston and (b) netted/blown fiberglass in Orlando.*

Ridge-Top Diffusion Port Design

The experimental roof assembly at this site has a diffusion vent detail, which allows for vapor diffusion drying at the ridge and tops of hips, while still functioning as an air barrier. The detail is used at the peak of the roof rafter bay cavities (at the ridge and hip conditions); it was not used at the roof valleys. The construction details for Houston and Orlando are shown in Figure 4 and Figure 5.

In both roofs, a portion of the roof sheathing (±3 in. [76 mm]) near the ridge was omitted, similar to typical a slot vent roof ridge detail. This opening was covered with a strip of a highly vapor-permeable roof membrane, composed of a tear-resistant polyethylene terephthalate fabric with a diffusive waterproof dispersion coating. The manufacturer's specified vapor permeance is 214 perms (12,200 ng/[Pa·s·m^2]) dry cup and 550 perms (31,500 ng/[Pa·s·m^2]) wet cup. The membrane was taped to the roof sheathing to create an air barrier and detailed to shed water. At the roof hips of the Houston house, a series of 2 in. (51 mm) diameter holes were drilled in the roof sheathing at the top 2–3 ft (610–915 mm) of the hip. The structural engineer did not accept the omitted sheathing detail at the hips but accepted the drilled hole detail. The slot vent detail was accepted at the roof hips in Orlando. The diffusion vent was covered with typical ridge vent materials, per local practice.

Test Roof Designations

Test roofs were instrumented to capture multiple roof orientations, roof geometries, and ridge materials (diffusion vent versus nondiffusion vent). Instrumentation was typically concentrated the ridge, because moisture typically accumulates at roof peaks. This is referred to as a *ridge package* composed of sensors, shown in Figure 6.

The measurements at the Houston roof (Table 1) include comparisons between roof ridges with diffusion vents and

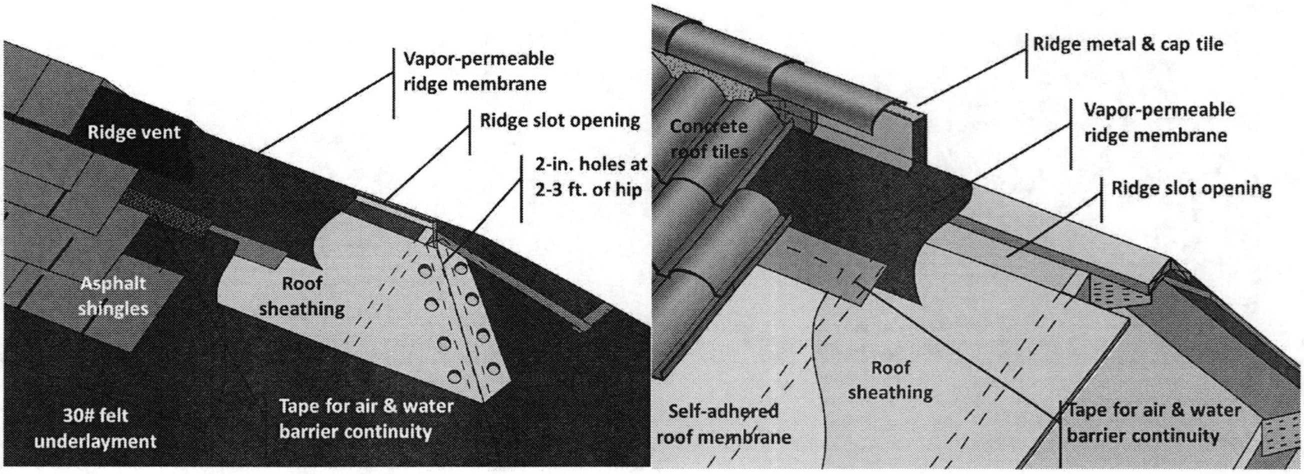

Figure 4 *Ridge (and hip) diffusion vent design: (a) with asphalt shingle in Houston and (b) concrete tile and self-adhered membrane in Orlando.*

Figure 5 *Diffusion vent details (left to right): Houston asphalt shingle diffusion vent ridge detail; Houston hip detail with ridge vent; Orlando tile roof ridge detail, with ridge metal tile support.*

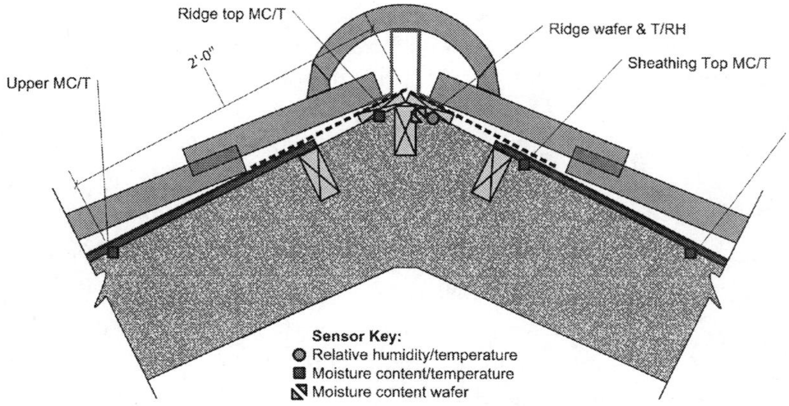

(a)

(b)

Figure 6 *(a) Diagram of ridge package of instrumentation for Orlando (tile) roof and (b) photo of ridge package.*

unvented (self-adhered membrane) at the ridge. Hip conditions were also measured. The attic geometry includes a larger attic on the east side and smaller attic on the west side. The two attics are connected via a small passage.

The measurements at the Orlando roof (Table 2) had a similar comparison of diffusion vent and unvented ridge conditions and hips. In addition, two roof-wall (RW) conditions (without diffusion vents) were monitored; one with a long run of downhill roof, another with a short run. The house has a smaller attic over the second floor and a larger attic over the first floor that is not covered by the upper story.

Sensor Setup and Data Acquisition System

Measurements in the roof assemblies included temperature and relative humidity (RH), wood sheathing moisture content, and a wood wafer sensor intended as an RH surrogate and condensation indicator. The latter is a wood-based relative humidity surrogate sensor, described in detail by Ueno and Straube (2008); the moisture content of the wood (eastern white pine) is measured via electrical resistance. A typical ridge package is shown below in Figure 6.

Roof sheathing MCs on both sides of the ridge were measured roughly 24 in. (610 mm) from the ridge to capture the extent of the wetting. In addition, at each rafter/truss bay with a ridge package, sheathing MC and temperature are measured lower (or downhill) on the roof sheathing; the number of measurements depended on the roof geometry. These "downhill" MCs measure the moisture accumulation in the rafter bays with the same condition but lower on the roof.

In addition, conditioned attic (i.e., interior) temperature and RH were measured at three pairs of locations; measurements were taken high and low (roughly the 1/3 and 2/3 vertical points) to capture the effects of thermal and/or moisture stratification in the space. Interior main space temperature and RH were measured via a sensor installed in the return duct of the space-conditioning system near the interior grille. Exterior temperature and RH were recorded on site.

Table 1. Houston (Asphalt Roof) Monitoring Locations

Designation	Description	Attic
DVR1	Diffusion vent ridge	East (larger) attic
UVR1	Unvented (control) ridge	East (larger) attic
DVR2	Diffusion vent ridge	West (smaller) attic
UVR2	Unvented (control) ridge	West (smaller) attic
DVR3	Diffusion vent ridge	East (larger) attic
DVH4	Diffusion vent hip	East (larger) attic
DVH5	Diffusion vent hip	East (larger) attic
DVR6	Diffusion vent ridge	West (smaller) attic

Table 2. Orlando (Tile Roof) Monitoring Locations

Designation	Description	Attic
UVR1	Unvented (control) ridge	Upper (over 2nd)
DVR1	Diffusion vent ridge	Upper (over 2nd)
DVR2	Diffusion vent ridge	Upper (over 2nd)
DVR3	Diffusion vent ridge	Lower (over 1st; front of building)
DVH1	Diffusion vent hip (lower)	Lower (over 1st)
DVH2	Diffusion vent hip (upper)	Lower (over 1st)
RW1	Roof wall, no diffusion vent (long roof downhill)	Lower (over 1st)
RW2	Roof wall, no diffusion vent (short roof downhill)	Lower (over 1st)

Data were collected by a measurement and control system installed in the attic. Measurements were taken at five-minute intervals and average values recorded hourly; data were downloaded remotely. No battery backup for the data logger was provided; however, the unit has nonvolatile memory, and resumes data collection after a power failure.

Further information on these monitoring methods is provided in Straube et al. (2002).

Houston Attic Humidification

Given the unoccupied condition of the model house in the Houston roof house, no moisture generation occurred, resulting in lower than expected interior RH levels. Therefore, between the second and third winter, a humidification system was installed in the attic to simulate occupant moisture generation. A heated reservoir-style humidifier was commissioned; by mid-October 2015, it was generating a steady 22.6 lb/day (10.3 L/day). Based on ASHRAE Standard 160 (ASHRAE 2009), this is slightly less than the design moisture generation rate for a two-bedroom residence. Although this is a five-bedroom house, this rate was considered a reasonable moisture generation rate for a source directly introduced to the attic.

AIR LEAKAGE TESTING

As part of the commissioning process, both experimental houses and attics were tested for air leakage. Energy Conservatory instruments (Minneapolis Blower Door Model 3 and DG-700 Pressure and Flow Gauge; ± 3% flow accuracy) were used for multipoint depressurization testing. Full descriptions of the air leakage testing regimen and results are provided in Ueno and Lstiburek (2015a; 2015b).

The Houston house was tested with the attic hatch either open or closed; the air leakage measurements are normalized based on an area/volume takeoff that includes the entire attic. Results are shown normalized by volume (air changes per hour at 50 Pa [ach_{50}]) and surface area (cfm_{50}/ft^2 [$L/s{\cdot}m^2$] enclosure area) and are stated in terms of equivalent leakage area (EqLA) in Table 3. EqLA provides a physical indication of the relative leakage areas of successive tests (Energy Conservatory 2012). Opening the hatch caused the air leakage to increase by only 14%. In both cases, air leakage was less than 0.25 cfm_{50}/ft^2 (1.2 $L/s{\cdot}m^2$) and fewer than 2.5 ach at 50 Pa (half or less than the 5 ach_{50} requirement in the 2012 IECC (ICC 2012) for CZs 1 and 2). The pressures across closed attic hatches were measured with the main space at –50 Pa to demonstrate the relative leakage connection to the interior versus the exterior. The attic-to-exterior pressure drop was 76% to 86% of the total, which is consistent with the hatch open/closed results, or a relatively airtight attic.

Table 3. Houston (Asphalt Shingle) Air Leakage Measurements

	cfm_{50} (L/s)	ach_{50} (h^{-1})	cfm_{50}/ft^2 ($L/s{\cdot}m^2$)	EqLA, in^2 (m^2)
Hatch Closed	2494 (1177)	2.2	0.22 (1.1)	257 (0.166)
Hatch Open	2846 (1343)	2.5	0.25 (1.2)	293 (0.189)

Overall, the roof plane was relatively airtight; using an infrared camera during depressurization, few thermal anomalies were seen at the roof plane. Indoor/outdoor temperature differences were 21°F to 31°F (11.7°C to 17.5°C) during these observations; although the roof was heated by morning sun, induced air leakage was clearly visible. Infrared observation of the roof peaks showed no indications of air leakage at the diffusion vent details. However, leakage often occurred at transitions and connections between roofs and walls, especially at complicated details with multiple intersecting planes. For instance, leakage was evident at dormer and gable end intersecting planes at the front of the house and roof-wall connections at the rear of the house (Figure 7).

The Orlando roof geometry included an upper attic, over the second floor, and a lower attic, at the first floor where it was not covered by the second floor. The house was tested with the two attic hatches open and closed in various combinations, as shown in Table 4. The overall house airtightness with both attic hatches open was 7.6 ach_{50} or 0.58 cfm_{50}/ft^2 (2.9 $L/s{\cdot}m^2$) enclosure area. This result exceeds the 5 ach_{50} target for CZ 2. This was followed by multizone/multifan air leakage depressurization guarded testing to quantify zone leakage to exterior versus adjoining zones, as described by Feustel (1989), and Energy Conservatory (2015). The tests revealed significant leakage from the lower attic to the exterior (196 $in.^2$ EqLA [0.13 m^2]). In comparison, the upper attic had 35 $in.^2$ EqLA to the exterior, and the main conditioned space (first and second floors) had 311 $in.^2$ EqLA (0.20 m^2) to the exterior (albeit without door gaskets or finish floors).

Both of these air leakage results indicate that complex details such as roof-to-wall intersections, and conditioned-to-unconditioned attic walls were significant sources of air leakage. However, the field (main body) of the roof and the diffusion vent ridge detail were not significant sources of air leakage. Spray foam unvented roofs are typically very airtight, due to the monolithic air barrier created with the insulation material. In contrast, unvented roofs built with fibrous insulation will likely require greater attention at connection details to achieve equivalent airtightness levels. In hot/humid climates, air leakage may increase attic moisture loads in summer and moderate them in winter.

MONITORING RESULTS: HOUSTON

Overview and Boundary Conditions

Data have been collected in Houston from early February 2014 through June 2016, providing over two years of results and three winters of data.

Exterior climate (based on Houston Intercontinental/IAH airport data) is summarized in Table 5: the first (partial) winter was colder than normal, but the second and third winters were warmer.

Interior conditions were measured at six conditioned attic locations (three high/low pairs) and the interior main space. Attic temperatures ran parallel to interior temperatures, albeit slightly warmer in summer and cooler in winter. Maximum attic-to-interior differences were in the 5°F to 7°F (3°C to 4°C) range. There was a clear diurnal temperature cycle due to solar gain, and there was evidence of vertical temperature stratification. However, a stronger signal was that some portions of the attic were warmer at times of day. This was ascribed to direct solar gain at the west-facing roof plane, and/or the ratio of enclosure gains/losses to incidental conditioning from attic duct leakage.

Conditioned attic dew points were typically between interior and exterior conditions in the summer: a two-week excerpt from July 2014 is shown in Figure 8 (peak summer conditions). The attic runs at a consistently higher dew point than interior, as the space was not directly conditioned or dehumidified. Dew point had a strong diurnal cycle, with peaks occurring between 3 and 6 p.m., which is consistent with moisture adsorbed in sheathing being driven out by solar gain and dry-bulb temperature difference. After sunset, the sheathing tended to adsorb moisture from the air and pull down the dew point as it cooled.

The measurements show moisture stratification in the attic. The highest (south-high) sensor shows the highest peaks; north low and south low sensors showed the lowest peaks.

In the first winter, interior dew points were relatively high, due to completion of construction activities (painting and drywall). In the second winter, dew points were low (40°F–60°F [5°C–15°C]) due to a lack of interior moisture generation. In the third winter, attic dew points were higher, due the addition of humidification at 22.6 lb/day (10.3 L/day), resulting in dew points typically in the 50°F to 63°F/10°C to 17°C range. The attic dew point was higher than interior main space conditions.

(a) (b)

Figure 7 *Infrared image of Houston attic during air leakage depressurization testing, showing roof-wall air leak.*

Table 4. Orlando (Concrete Tile) Air Leakage Measurements

	cfm_{50} (L/s)	Δ cfm_{50} (L/s)	ach_{50} (h^{-1})	cfm_{50}/ft^2 ($L/s{\cdot}m^2$)	EqLA, in^2 (m^2)
Both Hatches Closed	4429 (2090)		6.7	0.51 (2.6)	456 (0.294)
Lower Hatch Open	5012 (2366)	583 (275)	7.6	0.58 (2.9)	516 (0.333)
Upper Hatch Open	4473 (2111)	44 (21)	6.8	0.52 (2.6)	461 (0.297)
Both Hatches Open	5009 (2364)	580 (274)	7.6	0.58 (2.9)	516 (0.333)

Roof Assembly Conditions

Moisture measurements, including RH at the roof peak (moisture accumulation point) and wood MCs throughout the roof could be compared among the experimental assemblies.

Results for a control unvented roof (UVR1) are shown in Figure 9. This and the following plots show roof sheathing and wafer sensor MCs on the left-hand axis. The roof peak RH is plotted on the right hand axis, with a ten-day rolling average superimposed. Roof peak RHs reached high levels (90%+) in the first winter, but as exterior conditions warmed, RHs quickly fell to 40% to 50% due to solar heating driving moisture inward, out of the roof assembly. In the second winter (2014–2015), interior RH levels were low and outdoor conditions mild; RH maxima were much lower (60%–80%), followed by summertime drying. The missing data from April 2015 to August 2015 was due to an instrumentation failure for all RH and MC channels. In the third winter (2015–2016), interior humidification was added, and peak RHs again rose to 90%+ levels. As exterior conditions warmed, RH levels again dropped.

MC measurements were consistent with RH measurements; the first winter MCs were higher than recommended, peaking over 20%, but fell to safe levels (under 10%) in the summer. Similarly, the second winter saw a rise in MCs, but drier than the first winter. The third winter saw a rise in MCs similar to the first winter, with some MC maxima in the 25% to 30% range. The MCs show a spatial pattern, with the highest MCs typically at or near the roof ridge, and higher on the north side than the south side.

A similar plot is shown for an experimental diffusion vent roof (DVR1) in Figure 10. These diffusion vent roofs had a markedly different pattern from the unvented roofs: instead of a seasonal swing, the RH and MC measurements were relatively stable over most of the measurement period. In these roofs, the first winter MCs and RHs were much drier and safer than the unvented roofs. Peak MCs remained lower than 20%. Summertime RHs and MCs are higher than in the unvented roofs, but well within safe conditions. In the second winter (2014–2015), there was a slight rise in RHs

Table 5. Houston (IAH) Exterior Climate HDD/CDD Summary

	HDD, 65°F (18°C)	CDD, 65°F (18°C)	HDD, % of Normal	CDD, % of Normal
Normal (IAH Airport)	1525 (847)	2893 (1607)	—	—
Winter 2013–2014	1739 (966)	—	114%	—
Summer 2014	—	3037 (1687)	—	105%
Winter 2014–2015	1461 (812)	—	96%	—
Summer 2015	—	3344 (1858)	—	116%
Winter 2015–2016	1012 (562)	—	66%	—

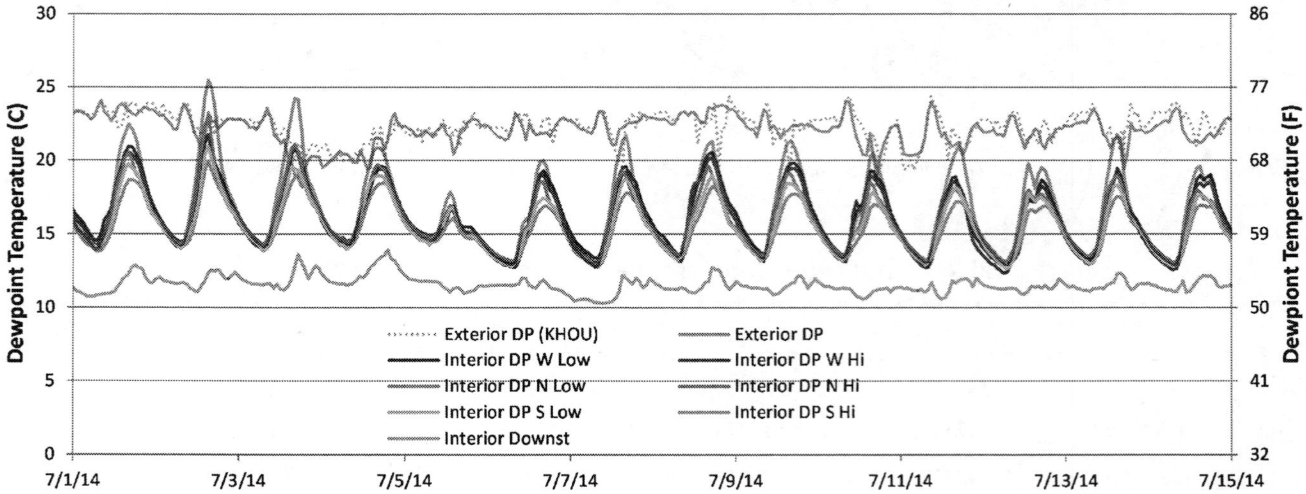

Figure 8 *Houston dew-point temperatures in the summertime (July 2014) detail.*

and wood MCs, but well within the safe range—typically lower than 15%. The third, humidified winter, however, saw a sharp rise in wood MCs. The ridge wafer MC remained low (peaking at 15%), but two north-side sheathing MCs peaked in the 20% to 25% range. This indicates that the diffusion vent provides localized drying at the ridge, but areas away from the diffusion vent may accumulate moisture at high interior RH conditions. As outdoor temperatures warmed, all MC fell back to the safe range.

Roof ridge RH measurements cycled on a diurnal basis; most readings are 50% to 80%. The diurnal RH cycling in the diffusion vent roofs shows a greater range than that in the unvented roofs. When plotted, the roof peak RH shows a consistent sawtooth pattern with a slow rise and a sharp daily fall. The fall in RH coincides with morning solar warming; in the late afternoons (sunset) the RH started to rise, continuing overnight into the morning. The MCs had a similar type of sawtooth behavior but over a much smaller range.

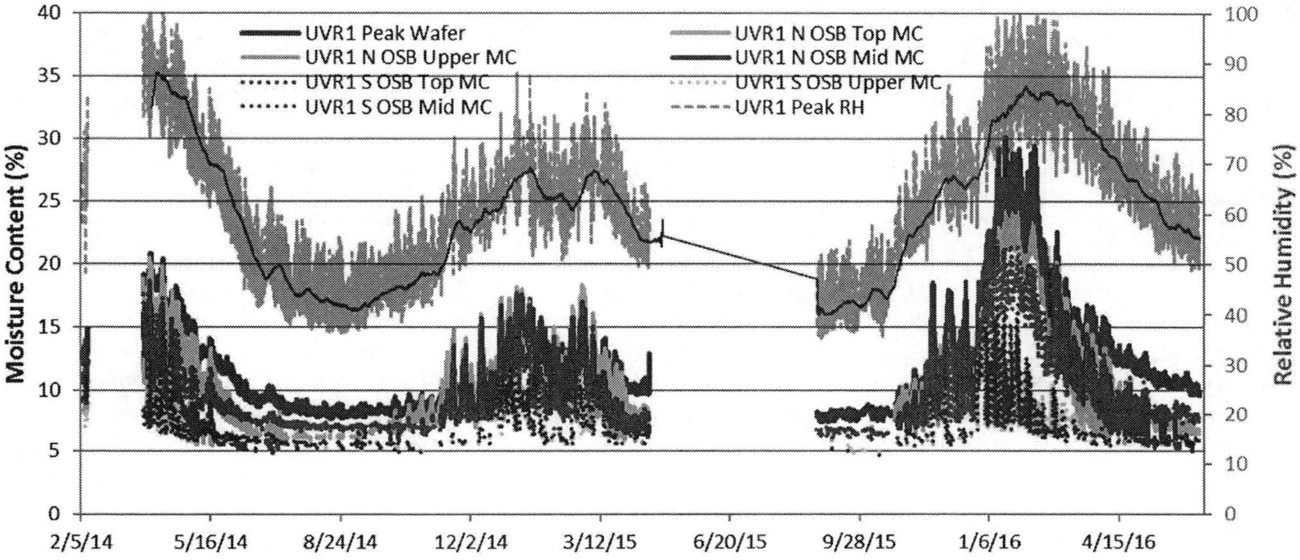

Figure 9 *Houston unvented roof (UVR1) roof MC and RH measurements.*

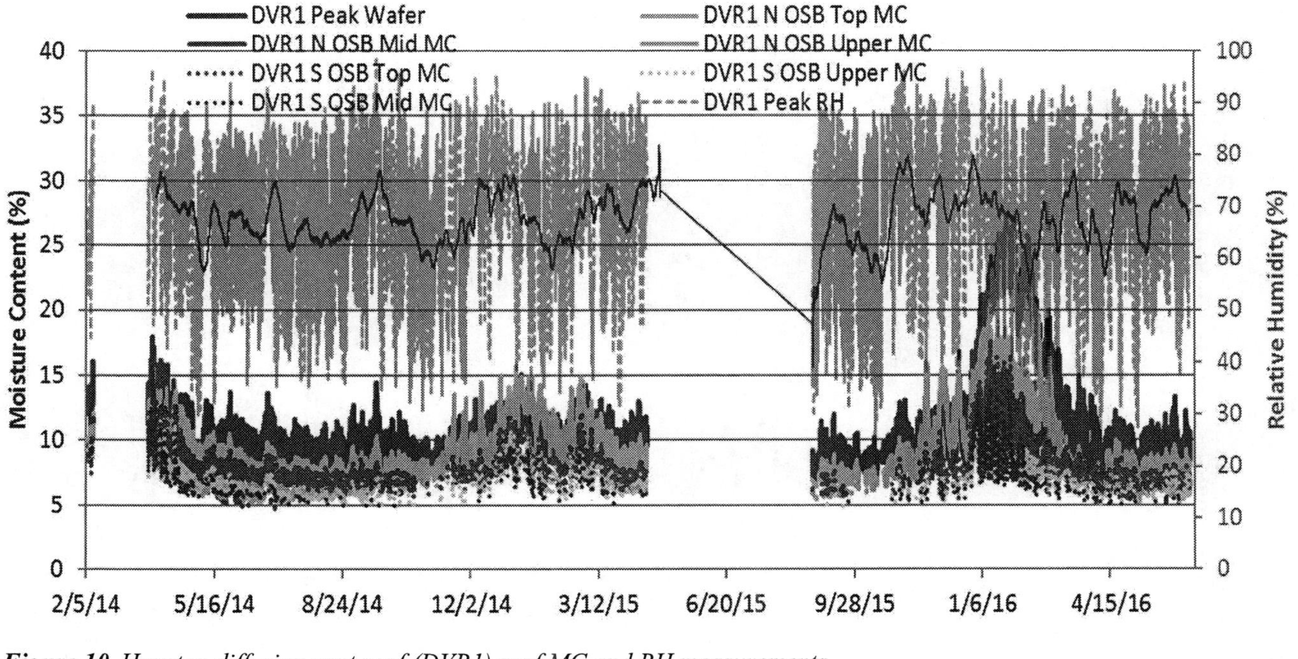

Figure 10 *Houston diffusion vent roof (DVR1) roof MC and RH measurements.*

Two monitored locations were hip geometries (DVH4 and DVH5); the hips are different from the ridge diffusion vents because the diffusion vent area is smaller (2 in. [51 mm] diameter drilled holes near the peak). These roofs exhibit behavior partway between the unvented and diffusion vent roofs, consistent with the reduction of exposed diffusion drying area.

The two roof systems (UVR1 and DVR1) are plotted together in Figure 11, isolating ridge RH and the wafer sensor at the ridge. A comparison of RH patterns shows that the unvented roof has higher RHs in winter during high dew-point conditions and lower in summer, while the diffusion vent roof has high variability but lower wintertime peaks. A similar pattern is seen at the ridge wafers: unvented roof is wetter in winter and drier in summer. The wafer sensor is useful in that it maps to long-term moisture accumulation patterns and, hence, the potential for moisture damage.

These patterns are shown Figure 12, which shows five-day details of summer behavior (August 2014, Figure 12a) and winter behavior under humidified conditions (February 2016, Figure 12b).

During the summer, unvented RHs are lower (35%–55%) than diffusion vent RHs (40%–90%). The greater diurnal cycling of the diffusion vent roof is evident, due to the coupling to exterior conditions at the vapor-open membrane. It results in greater drying during daytime heating and moisture uptake at night. However, the ridge wafers show that overall accumulation at the ridge is negligible in both roofs, with MCs in the 8% to 11% range.

In contrast, in winter, the unvented roof shows greater long-term moisture accumulation at the ridge wafer (20%–30% MC), while the diffusion vent ridge remains in the safe 10% to 14% range. Unvented roof RHs show peak levels hitting 100% rh at midday conditions, likely due to trapped moisture being desorbed from the sheathing.

The field monitoring work covered eight roof locations. To summarize the data, results are presented as box-and-whisker plots that show the overall range and distribution of measurements. The box-and-whisker plots show median (center of box), lower and upper quartile of data (extent of box), and maximum/minimum (extension lines). The ridge RHs are shown in Figure 13a and wafer MCs in Figure 13b. Given the varying boundary conditions, the final season (August 2015–June 2016) is plotted, which reflects humidified winter conditions.

There is substantial overlap between the unvented and diffusion vent roofs. The median and quartile RHs are higher in the diffusion vent roofs; this is consistent with the lack of seasonal wet-dry swings (i.e., DVR RHs remain higher in summer). However, DVR1, 2, 3, and 6 have lower maxima than UVR1 and 2.

The wafer MCs provide an indication of long-term moisture accumulation, indicating higher outliers in the unvented roof assemblies; the diffusion vent roofs have consistently lower maxima, except for DVH5. During decommissioning, DVH5 was found to have been assembled incorrectly, with the diffusion vent ports constricted by vapor-impermeable layers.

ASHRAE Standard 160 (ASHRAE 2009, 2011) provides guidance on moisture analysis for building envelope design, including the moisture performance evaluation criteria. Moisture failure criteria are defined as hours with a "30-day running average surface RH over 80% when the 30-day running average surface temperature is between 5°C and 40°C (41°F and 104°F)" (ASHRAE 2011). Failure criteria for the assemblies were calculated using 30-day running averages of the ridge RH and temperature; hourly pass/fail results (i.e., hours with likely mold growth) were tabulated. This analysis showed that both unvented roof assemblies (UVR1/UVR2) had hours failing ASHRAE Standard

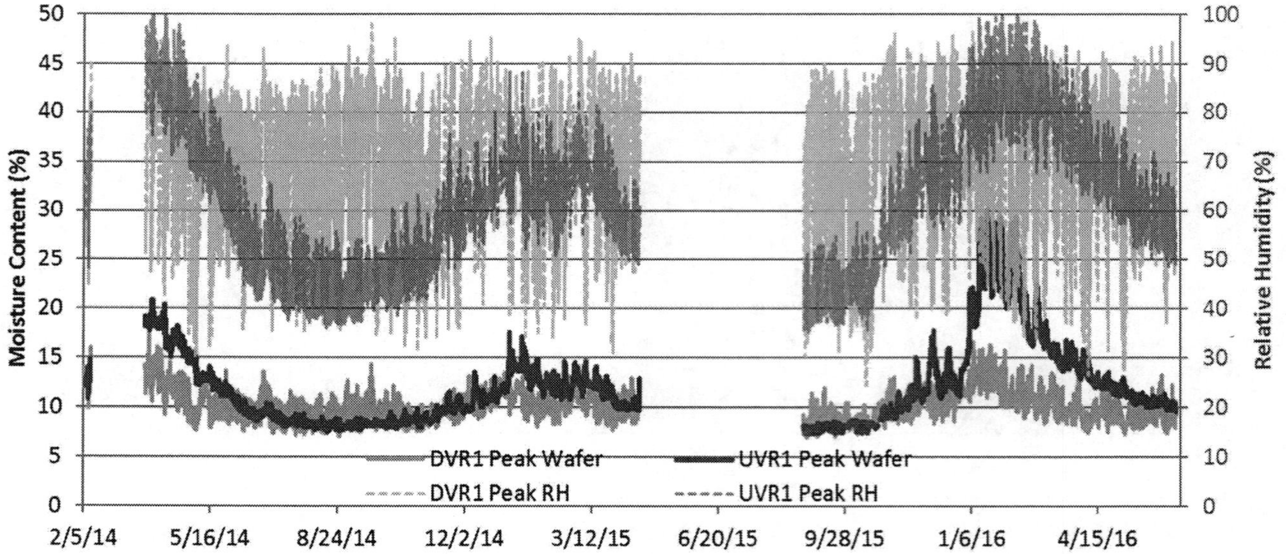

Figure 11 *Houston unvented (UVR1) and diffusion vent (DVR1) roof wafer MC and RH comparison.*

160 criteria in the first partial winter (427 and 238 hours, respectively); no diffusion vent roofs had failing hours. Admittedly, ASHRAE Standard 160 is known to provide overly stringent failure criteria (i.e., provides false positives of failure).

Overall, these results indicate that the diffusion vent roof had less wintertime moisture accumulation than the unvented roof and the potential for avoiding the ridge rot failures seen in the field in earlier work.

Inward Drive Experiment

Capillary water accumulation in shingle laps and inward solar moisture vapor drives have been suggested as contributors to unvented attic humidity problems. However, this mechanism was refuted by hygrothermal modeling by Boudreaux et al. (2013) and by a field comparison of roof underlayments by Lstiburek (2014). However, given this opportunity, the issue was studied further in Houston: a 2 ft (610 mm) square section of south-facing roof was isolated using 1 in. (25 mm) foam blocking covered with construction tape to reduce its vapor permeance (Figure 14a). After insulation, the interior side was covered with a layer of 0.093 in. (2.4 mm) clear acrylic plastic and air sealed at the perimeter with flashing tape (Figure 14 b).

Instrumentation included sheathing MC/T and interior-side RH/T and wafer sensors (Figure 14 right). The intent was to capture moisture that was driven through the exterior side of the assembly (shingles, underlayment, and sheathing) into the insulated, sealed cavity. Inward-driven moisture would accumulate at the insulation-to-acrylic plastic interface and was measured by the T/RH and wafer sensors. Sheathing MC was measured to determine whether moisture within the closed system increased over time as moisture was driven from one side to the other seasonally. This experiment was not intended to capture any diurnal or seasonal increase of MCs due to accumulated interior-sourced moisture in the sheathing. Instead, it

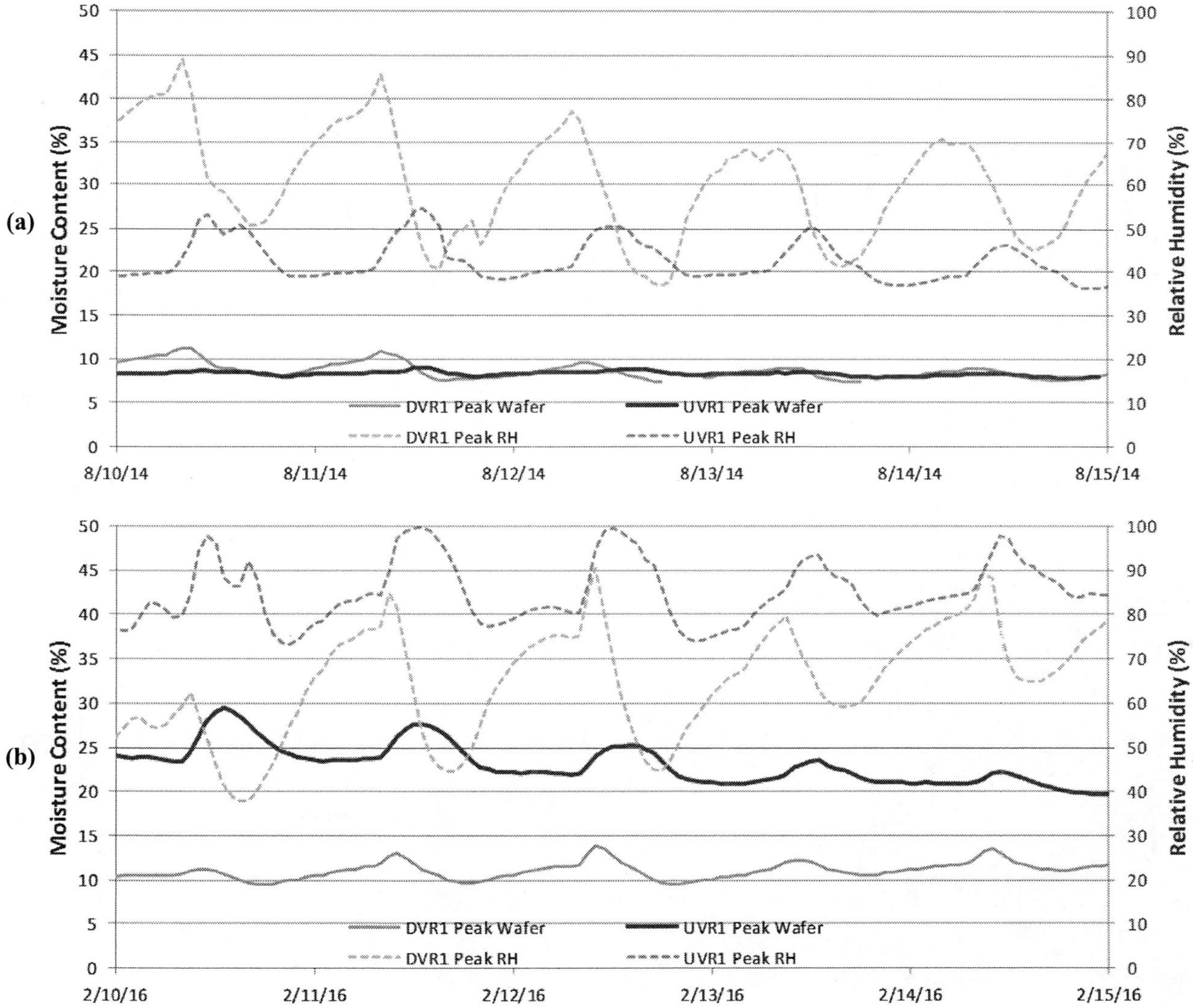

Figure 12 *Houston unvented and diffusion vent wafer MC and RH in (a) summer and (b) winter, five-day details.*

was intended to only capture inward-driven moisture from the exterior source. The initial sheathing MC was measured with a handheld meter at 8% to 9% MC.

Selected data are shown in Figure 15; the interior-side RH cycles seasonally (higher in summer) and daily, due to solar gain. Summertime peaks over 90% occur, but with no evidence of condensation (100% rh). Comparable data were measured by the interior wafer sensor. The wafer results were examined for correlations with exterior weather events: interior and exterior dew point had weak relationships. Precipitation was also plotted for correlation, given that wetted shingles are a suspected moisture source for inward drive. Although some wafer MC spikes correlated with precipitation, many other events were uncorrelated. The tightest correlation of boundary conditions to wafer response was the 24-hour average of the exterior sheathing temperature; there is excellent correlation to the MC rise and fall patterns. This is consistent with solar-heated sheathing desorbing moisture that is driven to the interior side. The wafer MCs rise and fall within a fixed range, indicating that moisture is not being added to the closed" system. The exterior sheathing MC was compared with other sensors in the same roof; similar MC levels were seen, also indicating a lack of moisture gain.

Overall, the lack of high moisture levels (condensation/100% rh) and the stable interior and exterior MC ranges (i.e., no ratcheting increase of moisture) point away from inward moisture vapor drives as a mechanism that adds humidity to the attic. This is consistent with Künzel and Sedlbauer's (2001) measurements of compact (cathedral cellulose) roof assemblies with clay and concrete barrel tile near Munich, Germany (roughly CZ5). A strong inward vapor drive effect was seen in summertime in the south-facing clay tile roof (30%+ MC at the rafters), but no effect was seen in the concrete tile roof. This effect was ascribed to the difference in water uptake or water absorption coefficient; concrete tile is much less absorptive than clay tile. The minimal storage and absorption in asphalt shingle roofing is consistent with the field results.

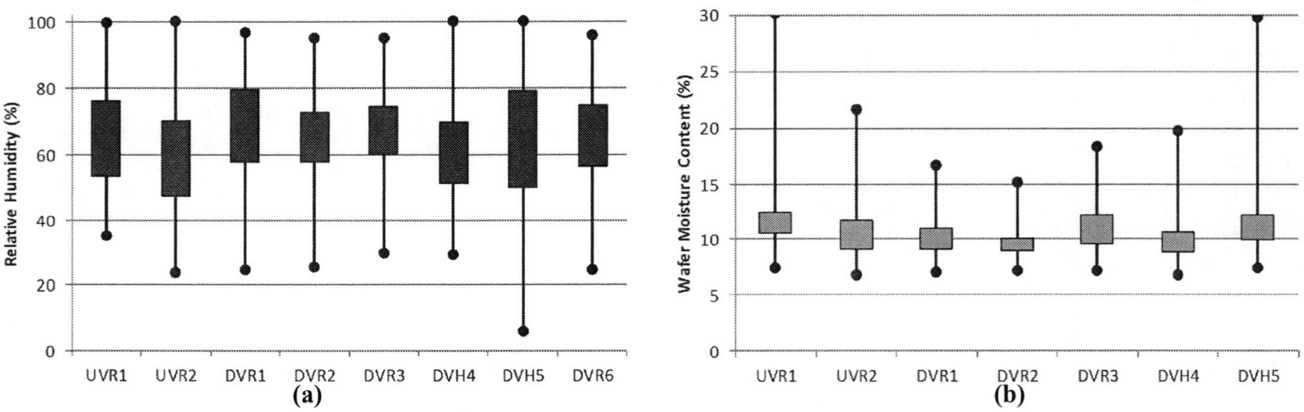

Figure 13 *Houston box-and-whisker plots for roof ridge moisture conditions: (a) RH and (b) wafer MC August 2015 to June 2016.*

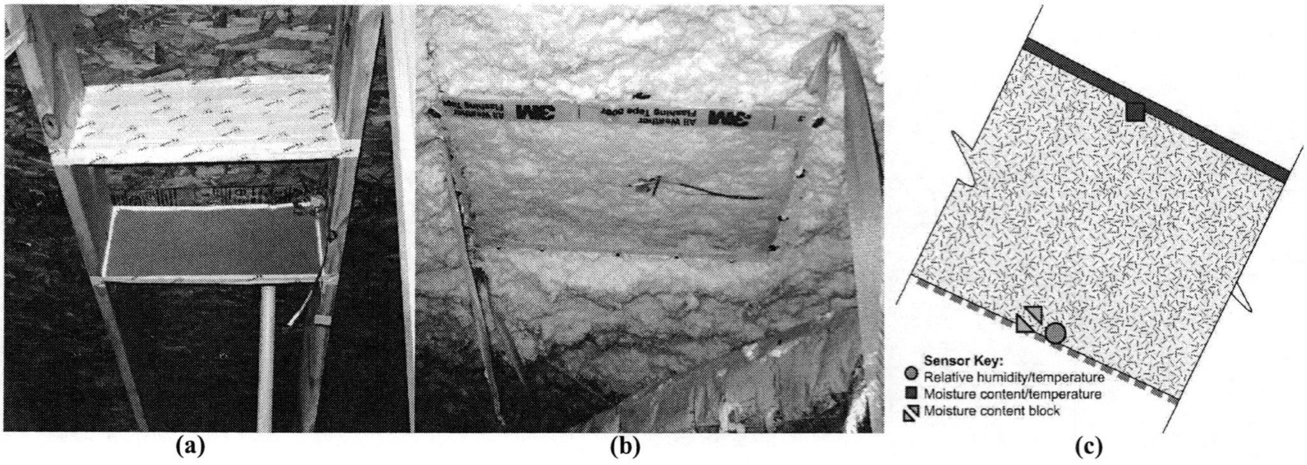

Figure 14 *Inward drive box (a) before and (b) after insulation and clear plastic cover and (c) sensor package.*

MONITORING RESULTS: ORLANDO

Overview and Boundary Conditions

Data have been collected in Orlando from early November 2014 through June 2016, providing roughly 20 months of results and two winters of data, with one partial winter. Exterior climate taken from Orlando airport data is summarized in Table 6. Both winters had lower than normal heating degree days and summers with higher than normal CDD. Although both locations are in CZ 2A, the heating load in Orlando is much smaller than Houston (580 versus 1525 HDD 65°F [322 versus 847 HDD 18.3°C], respectively). The difference is also demonstrated by wintertime 99.6% design heating temperatures (37.7°F [3.2°C] Orlando versus 29.1°F [–1.6°C] Houston), per climate data in ASHRAE (2013). Of course, with milder/warmer winters, interstitial condensation problems are reduced. Humidity loads between the two locations are more comparable; 0.4% design dew-point temperatures are 77.6°F [25.3°C] Orlando and 77.9°F [25.5°C] Houston (ASHRAE 2013).

Interior temperature and RH in the conditioned attics were measured at six points; one high/low pair in the attic above the second floor and two high/low pairs in the attic above the first floor. Patterns were similar to the Houston monitoring: attic temperatures ran parallel to interior (main space) conditions. However, attic temperatures were warmer in summer, and cooler in winter, with maximum excursions in summer of 7°F [4°C]. In summer, the second floor attic tracked warmer than others; in winter the second floor attic and the exposed front first floor attic tracked colder than interior. Paired sensors indicated temperature stratification in the attics.

Dew points in the attic (Figure 16) were substantially higher than the interior main space in the summer, with peak dew-point differences/excursions in the 27°F (15.2°C range). Attic dew points showed a strong diurnal swing, consistent with moisture being adsorbed and desorbed from the roof sheathing. The second floor attic consistently had the highest summertime dew points; air leakage testing indicated that this attic was much tighter than the first floor attic. This result is consistent with moisture stratification within the house driv-

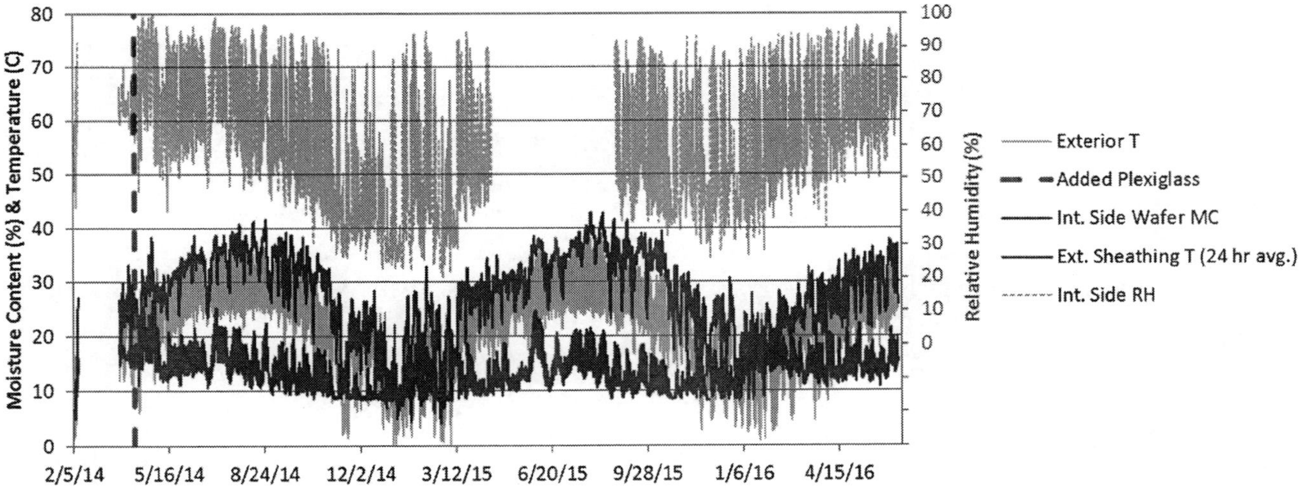

Figure 15 *Inward drive data, including interior-side RH and wafer MC, exterior T, and exterior sheathing T (24-hour rolling average).*

Table 6. Orlando (MCO) Climate HDD/CDD Summary

	HDD, 65°F (18.3°C)	CDD, 65°F (18.3°C)	HDD, % of Normal	CDD, % of Normal
Normal	580 (322)	3428 (1904)	—	—
Winter 2013–2014	410 (228)	—	71%	—
Summer 2014	—	3573 (1985)	—	104%
Winter 2014–2015	497 (276)	—	86%	—
Summer 2015	—	4266 (2370)	—	124%
Winter 2015–2016	378 (210)	—	65%	—

ing attic humidity levels, rather than exterior air leakage as the primary factor. Attic dew-point measurements again showed strong signs of vertical stratification.

Roof Assembly Conditions

RH measurements at the roof peak and wood MCs throughout the roof were plotted in a similar manner to the Houston roof; Orlando results for the unvented roof and diffusion vent roof are shown in Figure 17 and Figure 18, respectively. The gap in data (winter 2014–2015) was due to site power issues; moisture-generating construction activities (gypsum board finishing and painting) occurred during this period.

In the unvented roof, similar to Houston work, ridge RHs in the first winter were high (90%–100%); the ridge wood wafer sensor peaked over 30% (indicating possible condensation). However, RH and MCs dropped rapidly in summer to safe levels, due to the inward temperature gradient. In the second winter, RHs and MCs rose to similar levels, with RH peaks near 100% for extended periods.

Similar to Houston, the diffusion vent roof shows less of a seasonal RH swing (with only brief excursions over 90% RH) and lower peak wood MCs. Wood MCs hit a maximum of 18% at the ridge sheathing, and 15% in other roof locations.

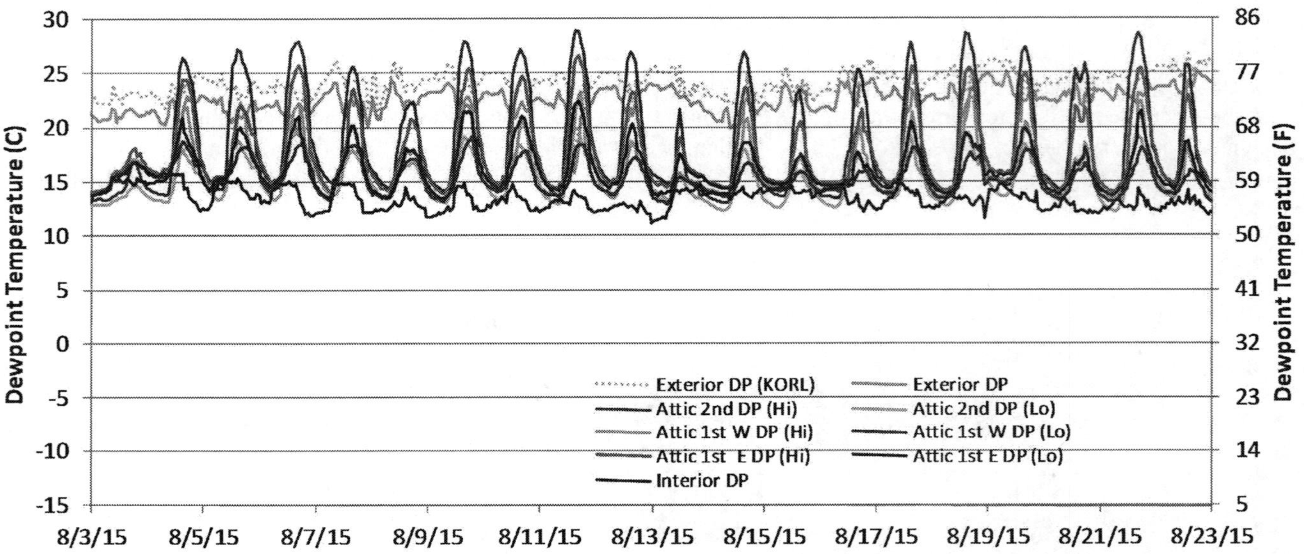

Figure 16 Orlando unvented attic summertime (August 2015) dew points, with interior and exterior dew point.

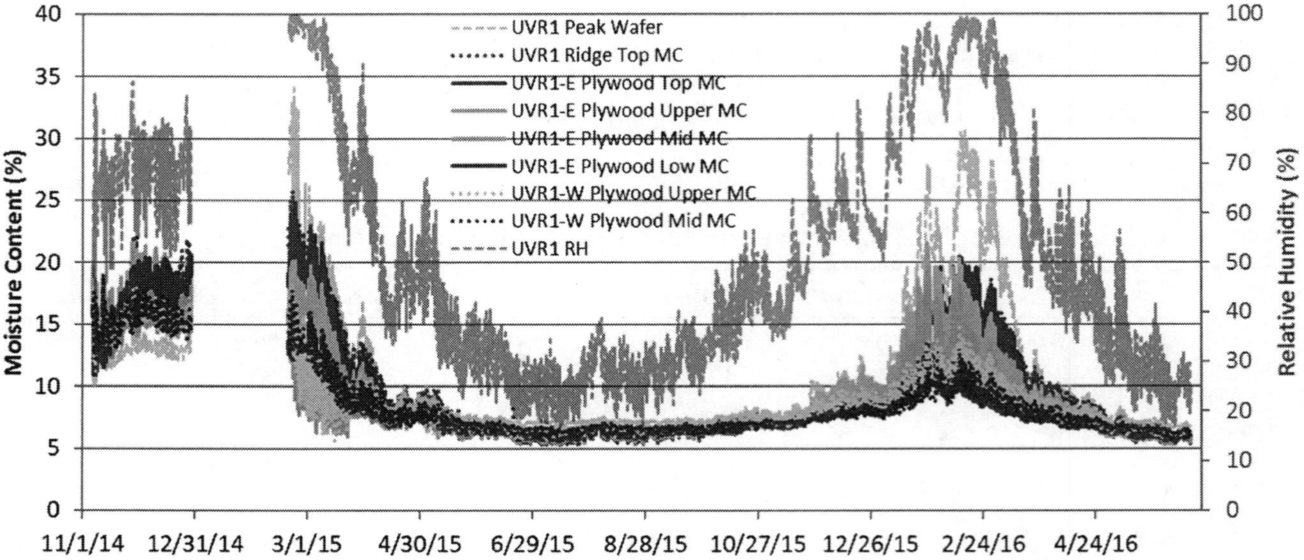

Figure 17 Orlando unvented roof (UVR1) roof MC and RH measurements.

Again, the two roofs were plotted together (RH and wafer MC) in Figure 19, allowing direct comparison; the higher RHs and MCs in the unvented roof ridge during the first winter are evident.

The eight experimental roofs are compared with box-and-whisker plots, showing ridge RH (Figure 20a) and wafer MC (Figure 20b) for the entire monitoring period.

There was only one unvented control roof at this test site; there were diffusion vent ridges and hips (DVR/DVH) and two roof-wall (RW) intersections (no diffusion vent). Similar to the Houston roof, the diffusion vent roofs showed a higher median and distribution of RHs than the unvented roof, but the unvented roof had the highest wintertime excursions. The wafer MC data shows the strong effect of the winter outliers in the unvented roof.

One hip measurement (DVH2) and two roof-wall measurements (RW1/RW2) have higher RH and wafer MC distributions than the remaining diffusion vent roofs. This is likely due to the shading patterns created by the building, as shown in Figure 21. The lower diffusion vent hip measurement (DVH1) is fully exposed to the sun, while the upper measurement (DVH2) is shaded by the adjacent second story and roof. The roof-wall interfaces (RW1 and RW2) lie directly north of the wall in shadow.

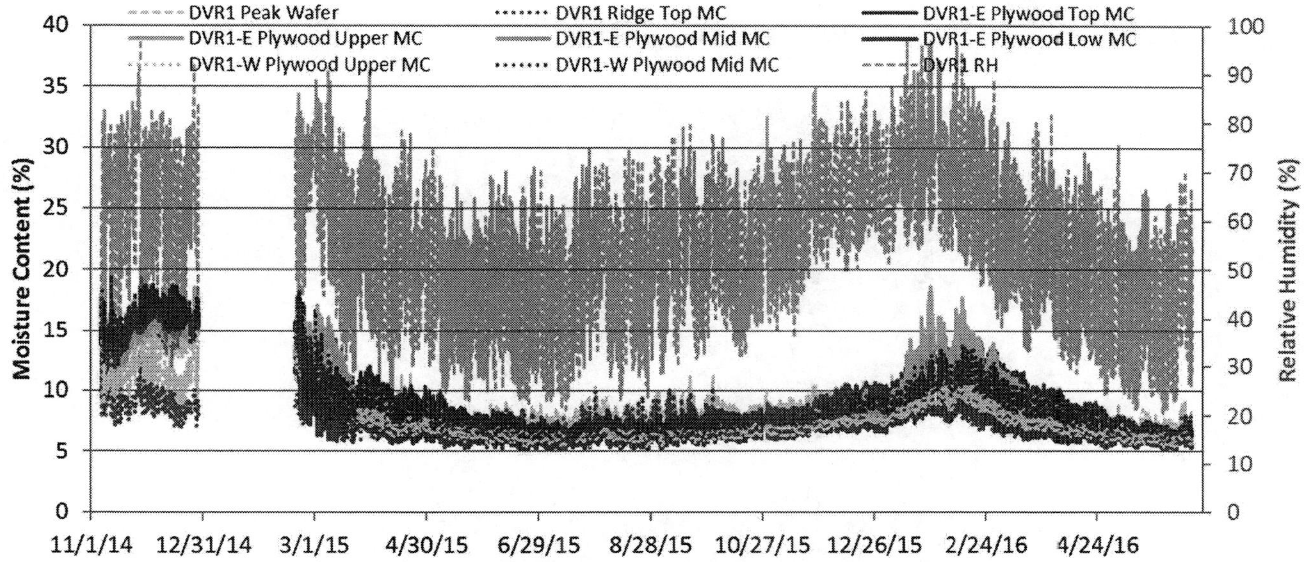

Figure 18 *Orlando diffusion vent roof (DVR1) roof MC and RH measurements.*

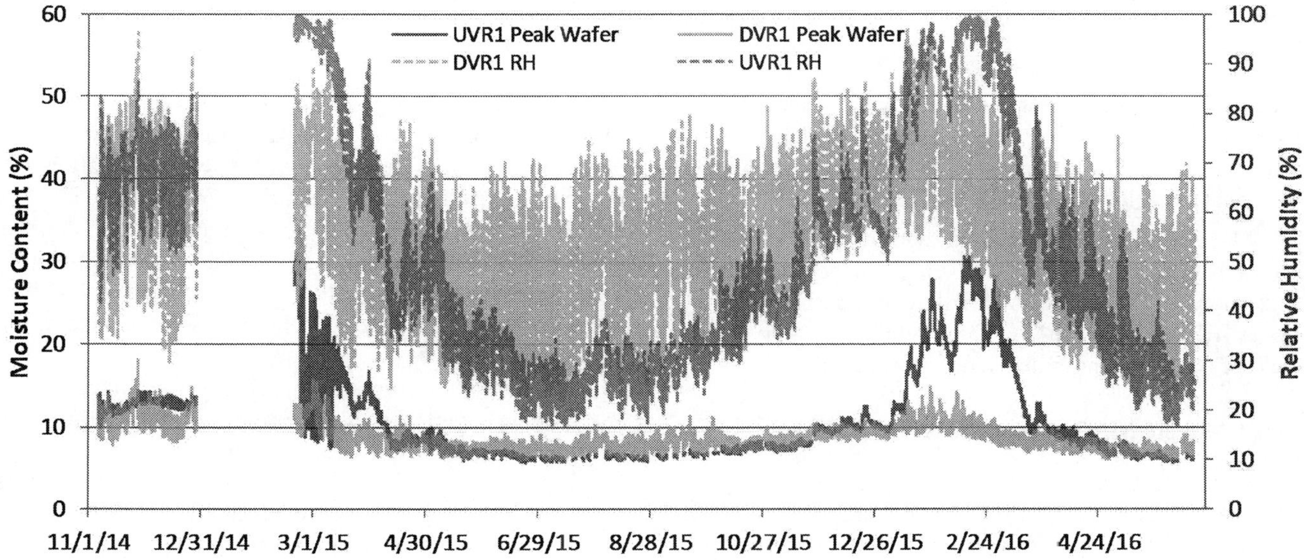

Figure 19 *Orlando unvented (UVR1) and diffusion vent (DVR1) roof wafer MC and RH comparison.*

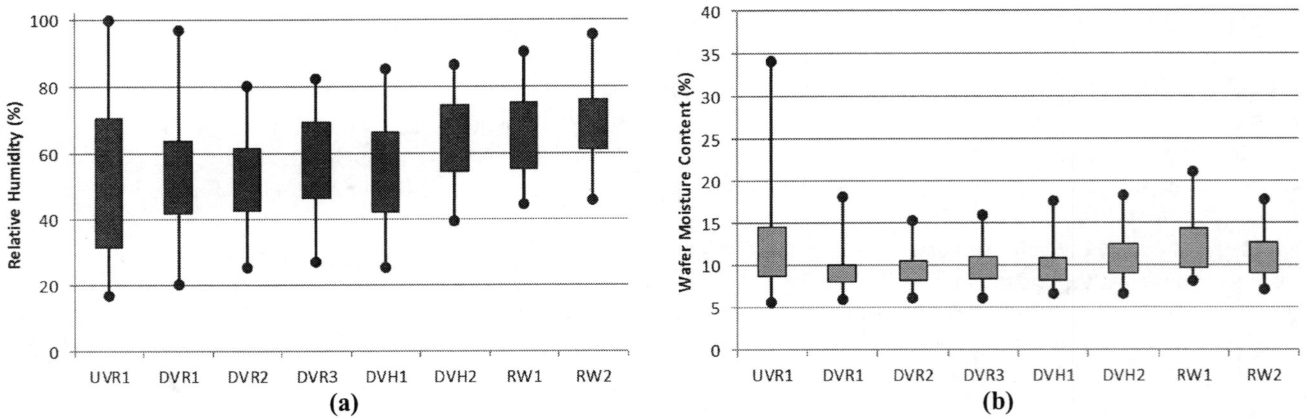

***Figure 20** Orlando box-and-whisker plots for roof ridge moisture conditions: (a) RH and (b) wafer MC.*

***Figure 21** Overhead view of roof shading with key sensor locations.*

This was confirmed by plotting roof sheathing temperatures at these locations; this shading resulted in cooler sheathing temperatures. In addition, these temperatures did not show the same diurnal cycling as in the exposed roofs.

Most of the roofs have hours with ridge RH measurements that exceed 80% rh. However, given the discontinuity (missing portion) of data, an ASHRAE Standard 160 analysis (ASHRAE 2009, 2011), which relies on 30-day rolling averages, does not provide useful information.

DECOMISSIONING AND DISASSEMBLY

At the conclusion of the experiment, the data logging equipment was removed, and the experimental unvented (self-adhered membrane) portions were retrofitted with the diffusion vent detail. Interior and interstitial conditions were examined during this final visit.

In the Houston (asphalt shingle) house, no visible signs of moisture damage were seen from the interior or during partial exterior roof disassembly. However, insulation was not removed from roof bays, limiting surveys of the framing and sheathing conditions. DVH5 was found to have shingles overlapping the diffusion vent port (Figure 22), explaining measurements similar to the unvented roof assemblies. Shingles were cut away to open the diffusion vents to the exterior.

In the Orlando (tile roof) house, spotting (likely mold) was seen on the interior insulation netting at the peak/ridge, tapering in intensity coming down the roof, extending roughly 4 ft (1.2 m) downslope (Figure 23). This issue was only seen at the highest ridge (upper attic, or UVR1/DVR1). Spotting was also seen at the top end of the wood truss vertical member, typically where covered by insulation netting (Figure 23a). The truss spotting was only seen at the

Figure 22 *Houston (a) roof hip diffusion vent holes, (b) DVH-5 blocked by shingles, and (c) retrofit.*

Figure 23 *Orlando tile roof interior conditions showing spotting on insulation netting and truss vertical member.*

unvented roofs, and not at the diffusion vent roofs. This growth is entirely consistent with measured moisture accumulation in the upper attic during the first winter, due to construction moisture. This moisture then became vertically stratified in the attic, resulting in condensation on upper netting and truss surfaces. Truss spotting occurred only in the unvented attic, due to greater local accumulation than the diffusion vent roof. Infrared observations showed that truss spotting is consistent with thermal bridging. The monitored data show the problems occurred in the first winter but did not recur in the second winter. These observations are entirely consistent with previous field observations of unvented attics with fibrous insulation.

Temperature and moisture (dew point) stratification were also directly measured in attics at both sites with a handheld meter. Measurements showed stratification consistent with the long-term monitoring patterns.

ANALYSIS AND CONCLUSIONS

Two unvented or conditioned attic houses were monitored in CZ 2A; key points from the data include the following.

Both unvented attics ran at temperature conditions near interior setpoint; daily temperature excursions ran high in summer and cool in winter. Attic conditions result from the balance of enclosure losses/gains (conductive, solar gain, and air leakage) versus conditioning losses/gains from duct leakage, air leakage from interior space, and conduction.

Attic dew points also showed strong daily cycling that was consistent with desorption (during the day) and adsorption (at night) of moisture from the roof sheathing. Sheathing adsorption/desorption is likely greater in roofs than walls due to color (often dark) and orientation (assembly experiences direct solar gain and night-sky cooling). These effects are intensified with the use of vapor- open insulation materials (open-cell polyurethane spray foam or fibrous insulation) and when an interior air barrier and vapor retarder are not used (per these test roofs).

Interior dew-point peaks were at times higher than outdoor dew-point conditions; Lstiburek (2014) recommends intentional space conditioning of the attic space to provide dehumidification. Both temperature and dew point showed vertical stratification, which is consistent with the concentrated moisture issues seen at the roof ridge/highest point.

The unvented roof assemblies had a strip of vapor-impermeable (Class I vapor barrier), self-adhered membrane at the ridge; the resulting assembly is moisture-coupled only to interior conditions. During the summer, the thermal gradient drives stored moisture inward, out of the assembly, resulting in dry conditions. In winter, when interior dew points are sufficiently elevated, moisture accumulation occurs at the ridge; wintertime peaks were much higher than 20% MC. For reference, decay fungi are inhibited when MC is lower than 20% (FPL 2010); optimum growth occurs in the 25% to 30% MC range. However, at drier interior conditions, moisture accumulation was minimal and well within the safe range. In the following spring, with warmer weather, the roof dried to safe levels. Moisture accumulation issues were seen primarily at the high points of the roof assembly (e.g., ridge); these issues were not seen in lower portions. Wood MCs were consistent with RH measurements.

In contrast, the diffusion vent roofs had a strip of vapor-permeable membrane at the opening; ridge conditions are coupled to both interior and exterior conditions, creating a flow-through assembly. Moisture can adsorb from and desorb to both the interior and exterior. The resulting ridge RH conditions showed strong diurnal oscillations. Higher RHs were seen in the diffusion vent roof in summertime (than the unvented roof) but within the safe range. Drier conditions were seen in the diffusion vent roof in wintertime, due to the outward drying of the moisture accumulating at the ridge.

The results collected to date show moisture accumulation at the unvented roof ridges, but the levels did not create catastrophic failures, based on decommissioning observations. However, the accumulation is consistent with the failures seen in previous field work on unvented attics with fibrous insulation. High interstitial moisture levels were seen in the first winter of operation, which involved drying of construction moisture; this is consistent with the timing of the failures seen historically in the field. With interior humidification, the Houston diffusion vent roof showed greater drying and less moisture accumulation than the unvented roof. Even if actual failures did not occur in the test roofs, results indicate that the diffusion vent assembly has a greater margin of safety than an unvented assembly with vapor-impermeable materials at the ridge.

A study of the solar inward-driven moisture through asphalt roof shingles in Houston indicated that no moisture accumulation was occurring due to inward drives. Instead, the fixed amount of moisture within the closed test assembly was driven to the interior or exterior due to the thermal gradient; the moisture levels seen at the interior side closely tracked exterior roof sheathing temperature.

FURTHER WORK

Overall, the use of a diffusion vent at the roof ridge appears to be a promising technology, especially in milder climate zones. The test climates (Houston and Orlando; CZ 2A) are not highly challenging for wintertime condensation but were selected because problems were previously observed in the field using these assemblies in these climates.

Of course, in colder climate zones, interstitial condensation risks increase. These risks might be counteracted by interior airflow and vapor flow control; in these test roofs, no attempt was made to control air or moisture vapor at the interior side of the assembly. In addition, as seen in the collected data, interior dew point affects the level of moisture risk. Unvented roof assemblies using fibrous insulation and interior air/vapor control will be studied in future research.

ACKNOWLEDGMENTS

The authors would like to thank the U.S. Department of Energy's Building America Program for funding the research underlying this paper. This research was also made possible by generous donations of materials, access to test buildings, and support by Cosella Dörken, David Weekley Homes, DuPont Building Innovations, Johns Manville, and Owens Corning.

REFERENCES

ASHRAE. 2009. ANSI/ASHRAE Standard 160-2009, *Criteria for moisture-control design analysis in buildings*. Atlanta: ASHRAE.

ASHRAE. 2011. ANSI/ASHRAE Standard 160-2009, *Criteria for moisture-control design analysis in buildings*, Addendum a. Atlanta: ASHRAE.

ASHRAE. 2013. *ASHRAE Handbook—Fundamentals*. Atlanta: ASHRAE.

Boudreaux, P., S. Pallin, and R. Jackson. 2013. *Moisture Performance of Sealed Attics in the Mixed-Humid Climate*. Technical Report ORNL/TM-2013/525. Oak Ridge, TN: Oak Ridge National Laboratory, Energy and Transportation Science Division.

Energy Conservatory. 2012. Minneapolis Blower Door™ Operation Manual for Model 3 and Model 4 Systems. http://www.energyconservatory.com/sites/default/files/documents/mod_3-4_dg700_-_new_flow_rings_-_cr_-_tpt_-_no_fr_switch_manual_ce_0.pdf.

Energy Conservatory. 2015. General Guarded Blower Door Testing Guidance. http://energyconservatory.com/wp-content/uploads/2015/01/General-Guarded-Blower-Door-Testing-Guidance.pdf.

Feustel, H.E. 1989. Measurements of air permeability in multizone buildings. *Energy and Buildings* 14:103–16.

FPL. 2010. *Wood Handbook—Wood As an Engineering Material*. Technical Report, FPL-GTR-190. Madison, WI: U.S. Department of Agriculture, Forest Service, Forest Products Laboratory.

ICC. 2009. *2009 International Residential Code for One- and Two-family Dwellings*. Country Club Hills, IL: International Code Council.

ICC. 2012. *2012 International Energy Conservation Code*. Country Club Hills, IL: International Code Council.

Künzel, H., and K. Sedlbauer. 2001. Steildächer-feuchte- und wärmetechnische Ausbildung. Bauphysik-Kalendar, Ernst & Sohn, 13086 Berlin, S. 459–82.

Lstiburek, J. 2006. Understanding attic ventilation. *ASHRAE Journal* 48(4):36–38, 40, 42–45. Atlanta: ASHRAE. https://buildingscience.com/documents/digests/bsd-102-understanding-attic-ventilation.

Lstiburek, J. 2010. Building sciences: Don't be dense with insulation. *ASHRAE Journal* August, 54–57. Atlanta: ASHRAE. https://buildingscience.com /documents/insights/bsi-043-dont-be-dense.

Lstiburek, J. 2011. A crash course in roof venting. *Fine Homebuilding Magazine* August/September, 68–72. Newtown, CT: Taunton Press. www.buildingscience. com/documents/published-articles/pa-crash-course-in -roof-venting/view.

Lstiburek, J. 2014. Building sciences: Cool Hand Luke meets attics. *ASHRAE Journal* April, 52–57, Atlanta: ASHRAE. www.buildingscience.com/documents /insights/bsi-077-cool-hand-luke-meets-attics.

Owens Corning Insulating Systems, LLC. 2015. Owens Corning PROPINK® High Performance Conditioned Attic System Installation Guidelines, Pub No. 10019449-B. October 2015. http://www2.owenscorning.com/literature /pdfs/HPCA%20Installation%20Instructions.pdf. Accessed January 2016.

Pallin, S., P. Boudreaux, and R. Jackson. 2014. Indoor climate and moisture durability performances of houses with unvented attic roof constructions in a mixed-humid climate. Technical Report ORNL/TM-2014/549. Oak Ridge, TN: Oak Ridge National Laboratory, Energy and Transportation Science Division.

Rose, W., and D. McCaa. 1998. Temperature and moisture performance of wall assemblies with fiberglass and cellulose insulation. *Proceedings: Thermal Performance of the Exterior Envelopes of Buildings VII*, 133–44. Atlanta: ASHRAE. ISBN 1-883413-70-2.

Rudd, A.F., and J.W. Lstiburek. 1998. Vented and sealed attics in hot climates. *ASHRAE Transactions*. TO-98-20-3.

Salonvaara, M., A. Karagiozis, and A. Desjarlais. 2013. Moisture performance of sealed attics in Climate Zones 1 to 4. *Proceedings: Thermal Performance of the Exterior Envelopes of Buildings XII*. Atlanta: ASHRAE.

Schumacher, C.J., and R. LePage, R. 2012. BA-1308: Moisture control for dense-packed roof assemblies in cold climates: Final measure guideline. www.buildingscience.com /documents/bareports/ba-1308-moisture-control-dense -packed-roof-assemblies-cold-climates/view.

Straube, J., D. Onysko, and C. Schumacher. 2002. Methodology and design of field experiments for monitoring the hygrothermal performance of wood frame enclosures. *Journal of Thermal Envelope and Building Science* 26(2).

Ueno, K., and J. Straube. 2008. Laboratory calibration and field results of wood resistance humidity sensors. *Proceedings of BEST 1 Conference*, Minneapolis, June 10–12.

Ueno, K., and J.W. Lstiburek. 2015a. Field testing unvented roofs with asphalt shingles in cold and hot-humid climates. U.S. Department of Energy/Building America Report DOE/GO-102015-4705. http://www.nrel.gov /docs/ fy15osti/64543.pdf.

Ueno, K., and J.W. Lstiburek. 2015b. Field testing of an unvented roof with fibrous insulation, tiles, and vapor diffusion venting. U.S. Department of Energy/Building America Report DOE/GO-102015-4764. http:// apps1.eere.energy.gov/buildings/publications/pdfs /building_america/64999.pdf.

Cost-Effective High-Performance Unvented Attics

Francis Babineau, Jr, PE
Member ASHRAE

ABSTRACT

For buildings with heating and cooling equipment and ducts in the attic, the benefits of shifting the thermal boundary from the ceiling to the roof are well documented. However, the practical aspects of installing effective thermal control at the roof can be technically challenging and/or costly. The most common solution in the U.S. is spray polyurethane foam, which can be considerably more expensive than other insulation materials. For warm and dry climates, fiberglass is a cost-effective material option, but some installation methods can become complex. This paper describes methods for installing fiberglass under a roof that are relatively simple for installers and use off-the-shelf products. Modeling and testing has been completed to demonstrate thermal and moisture performance and long-term physical durability. An example project from California is highlighted as a case study. The project demonstrates installation on a whole-roof scale and includes instrumentation to measure thermal and moisture performance in the field. Changes to the California energy code are driving the industry towards more creative solutions for unvented attics.

INTRODUCTION

For buildings with heating and cooling equipment and ducts in the attic, the benefits of shifting the thermal boundary from the ceiling to the roof are well documented (Rudd et al. 2005; Hendron et al. 2002; DOE 2013). Energy code requirements in the U.S. are becoming increasingly stringent for ductwork located in such a hostile environment (ICC 2015). In California, the upcoming 2016 version of the *Building Energy Efficiency Standards for Residential and Nonresidential Buildings*, more commonly referred to as Title 24, requires a high level of performance for conventional vented attics with ductwork that includes insulation either above or below the roof deck along with the typical practice of insulating the attic floor. Rather than insulating both the roof deck and the attic floor, a more cost-effective option may be to create an unvented attic and insulate at the roof deck, thereby bringing the ducts and HVAC equipment into semi-conditioned space. However, the practical aspects of installing effective thermal control at the roof can be technically challenging and/or costly. For warm and dry climates, fiberglass is a cost-effective insulation material option, but installation methods can become complex. This paper presents a method for installing fiberglass under a pitched roof deck that is relatively simple for installers and utilizes off-the-shelf products. Calculations were completed to provide an initial assessment of thermal performance; simulations were performed to assess moisture performance; and laboratory evaluations were used to check long-term physical durability. Finally, the proposed system was installed in a home in a warm, dry climate. We also describe the lessons learned during installation of this project. It is our belief that the insulation system explored in this paper can be a cost-effective option for builders who wish to implement unvented sealed attics with fiberglass insulation.

CURRENT OPTIONS

The most common method to insulate and air seal an unvented attic is through the use of spray polyurethane foam (SPF). This practice has been used in the insulation industry for many years with considerable success (SPFA 2013).

Francis Babineau Jr. is a research manager in Corporate R&D at Johns Manville, Littleton, CO.

Depending on the climate zone where the building is located, open-cell and/or closed-cell SPF materials are often appropriate options (Schumacher 2007). However, some homebuilders would prefer to not use SPF for a variety of reasons, including cost, impact on construction schedule, or safety/environmental impact. Other alternatives include insulating with foam boards (installed either above or below the roof deck) or with air-permeable insulation such as fiberglass, mineral wool, or cellulose. Foam boards may also present cost and installation challenges associated with additional material layers creating a thicker roof assembly, which could lead to challenges with finishing the roof edge and connections with other building elements. Air permeable insulations offer the potential for a lower-cost option, so long as the following three main items can be addressed:

- **Cost-effectively air seal the roof and eaves.** Air leakage between an unvented attic and the exterior climate can seriously degrade the energy performance of the system, especially if HVAC ducts remain leaky and if air is allowed to move freely between the attic and conditioned space. Air leakage around or through roof-deck insulation can also be a main contributor to moisture durability issues (Hendron et al. 2002).
- **Effectively and practically install the insulation up against the roof deck.** Many wood-framed attics are constructed using wood trusses. In this scenario, the top chords of the trusses are often 2 × 4 or 2 × 6 lumber, which result in framing cavities not deep enough to accommodate most code-level insulation amounts. Air-permeable insulation, such as fiberglass, should ideally be installed in complete contact with the roof deck and with minimal compression. Additionally, any preferred methods would also cover the roof framing, to reduce thermal bridging.
- **Manage attic moisture.** A primary concern with the use of unvented attics is the potential for moisture to accumulate at the underside of the roof sheathing during cold weather. Research has shown that the good long-term performance of ventilated attics is due to the removal of moisture that passes into the attic from the interior. Unvented attic systems must either include one or more means to manage and/or remove attic moisture, or the mechanical systems in the house must include means to limit the interior humidity levels.

Both the International Residential Code (IRC) and the California Building Code (CBC) allow unvented attics to be insulated with air-permeable insulation only, in appropriately dry climates and in conjunction with vented roof claddings (IRC 2015, CBC 2013). Many air-permeable insulation options end up leaving the roof framing exposed, reducing the thermal effectiveness of the system. Other systems may require complex, multistep installations or multiple layers of material.

PROPOSED SYSTEM

The unvented attic insulation system evaluated in this study uses fiberglass batts installed such that they are held securely against the bottom of the roof deck with minimal compression. Air sealing, which is described in more detail later, was achieved by the use of canned foam and caulk to seal all roof-deck seams, eave blocking, and gable walls. After air sealing was completed, the batts were mechanically secured up against the bottom of the roof deck. Thermal performance was maintained by using batts that were wider than the space between the framing and allowing the batts to expand laterally to cover adjacent roof framing. Moisture management in the test case was addressed by installing the system in a relatively warm, dry climate (DOE climate zone 3B) and below vented tile roof cladding.

Materials and Installation. R-30 and R-38 fiberglass batts have nominal thicknesses of approximately 10.25 and 13 in. (260.3 and 330.2 mm), respectively. When typically choosing batt insulation to fit between framing, the insulation comes pre-cut to fit between framing with just enough lateral compression for a friction fit. For this unvented attic system, the specified batts were selected to be slightly oversized, to extend more completely over the roof framing. Installation of the batts is best achieved by mechanical fastening, with a focus on maintaining good contact with the roof deck with minimal compression or gaps. A photo of an end-view of batts installed into the framing is shown in Figure 1. Adhesives were evaluated for this application, but the required amount of adhesive was not considered as cost-effective as mechanical fastening, and potentially not as durable. Another option could also have been to insulate the roof similar to a metal building, with batts between the roof framing and a separate layer of continuous insulation installed across the surface of the roof framing. This option was not pursued due to complexity and installation cost concerns. Details of the installation process are presented later in the paper.

Thermal Performance. Thermal performance for the attic insulation system was estimated based on the parallel-path method described in the Chapters 25 and 27 of *ASHRAE Handbook—Fundamentals* (ASHRAE 2013) to calculate the overall thermal transmittance (U-factor) of the roof assembly. Performance of wood-frame assemblies tend to be reasonably described by this approach. To demonstrate energy code compliance, the California Building Energy Efficiency Standards, Joint Appendix JA4 (CEC 2016) also prescribes this calculation method when estimating the performance of nonstandard building assemblies not otherwise listed in the appendix. Appendix JA4 also includes assumptions to be used for the thermal resistance of noninsulation building materials, air films, and framing fraction for various assemblies including roofs. The tabulated calculations are shown below in Table 1, along with source references for values used. These estimates

Figure 1 *Example photo of attic insulation system with fiberglass batts. Note that the batts compress slightly at the roof framing but are able to expand to sideways to form a layer of continuous insulation over the framing.*

Table 1a. Thermal Calculations for Fiberglass Unvented Attic (I-P)

Assembly Component	Component R-values, h·ft²·°F/Btu		Source and Notes
	Framing Path	Cavity Path	
Ext. air film	0.17	0.17	CA T24, JA4
Roof cladding	0.44	0.44	CA T24, JA4
Roof underlayment	0.06	0.06	CA T24, JA4
OSB roof deck	0.48	0.48	CA T24, JA4
2 × 4 truss top chord	4.53	—	*ASHRAE Handbook—Fundamentals,* Ch. 26
Cavity insulation	—	10.23	Cavity portion of R-38 batt
Continuous insulation	27.77	27.77	CI portion of R-38 batt
Attic air space	0.80	0.80	CA T24, JA4
Interior ceiling gypsum	0.45	0.45	CA T24, JA4
Int. air film	0.61	0.61	CA T24, JA4
Total R-value for each path	**35.31**	**41.01**	
Area fraction, %	7%	93%	CA T24, JA4
Overall assembly U-factor, Btu/h·ft²·°F	**0.025**		

Table 1b. Thermal Calculations for Fiberglass Unvented Attic (SI)

Assembly Component	Component RSI, m²·K/W		Source and Notes
	Framing Path	Cavity Path	
Ext. air film	0.03	0.03	
Roof cladding	0.08	0.08	
Roof underlayment	0.01	0.01	
OSB roof deck	0.08	0.08	
2 × 4 truss top chord	0.80	—	
Cavity insulation	—	1.80	
Continuous insulation	4.89	4.89	All values converted from Table 1a
Attic air space	0.14	0.14	
Interior ceiling gypsum	0.08	0.08	
Int. air film	0.11	0.11	
Total RSI for each path	**6.21**	**7.22**	
Area fraction, %	7%	93%	
Overall assembly U-factor, W/m²·K	**0.142**		

indicate the proposed system should meet the thermal performance requirements for the State of California. System-scale laboratory thermal performance is planned for a future phase of the project, and *in situ* measurements will also be made on the test home described later in this paper.

Moisture Performance. An unvented attic with air-permeable insulation is allowed under both the IRC and the CBC, provided it is installed under a vented tile roof and in a warm, dry climate (IRC CZ 2B and 3B, CBC CZ6-15). To better understand any potential moisture risks, a series of moisture simulations were performed, using WUFI Pro v5.3 (Kuenzel and Kiessl 1997). The one dimensional nature of the simulations does not account for heat or moisture transfer into framing or due to convection in the large attic air space. Ventilation under the roof tiles and air exchange between the attic and the interior of the home were included in the simulations. The simulation models were based on and validated against previous work done for CEC for attics in California climates (Lstiburek and Schumacher 2011). The material layers in the modeled assembly are shown in Figure 2. The exterior climate used for the simulations was Fresno, CA, since that was the location of the test home. The prediction for interior conditions was EN 15026 (BSI 2007), which is included as an interior climate function in the WUFI software. This model considers the outside temperature and adjusts the interior relative humidity (RH) and allows for normal or high moisture levels to account for different levels of interior moisture production and/or ventilation. The baseline interior climate was the EN 15026 function with the normal moisture level setting.

The initial performance metric used to assess moisture durability was the moisture content of the inner surface (inner 1/16 in.) of the oriented strand board (OSB) roof decking. Previous work (Straube et al. 2010; Black and Straube 2007) has suggested that wood moisture content above 16% for more than 28 days (672 hours) may pose a risk of microbial growth on wood and sustained moisture content above 28% may result in wood decay beginning. Graphs of OSB roof-deck moisture content and relative humidity next to the roof-deck inner surface, based on the initial simulation, are presented in Figure 3. To further explore the effect of interior relative humidity on the system, additional simulations were performed, adjusting the interior humidity downwards by 5% and 10% from the EN 15026 baseline, only during the winter. To explore elevated interior moisture, a simulation was also run using the EN15026 interior climate with high moisture loading. The roof-deck moisture contents from all four simulations are shown in Figure 4, and the interior relative humidities through the simulation year are shown in Figure 5. Comparison of the simulated roof-deck moisture content versus the thresholds defined above are presented in Table 2. The simulations suggest that across a broad range of likely interior relative humidities, this attic system could experience elevated moisture, which could pose a risk to the wood roof deck. Both the normal and high EN 15026 interior climates result in roof-deck moisture content that exceeds the criteria for mold risk described by Straube (Straube et al. 2010). Interior relative humidity in the winter needs to reach 40% or lower to prevent the roof deck from going above 16% moisture content for over the 672 hour threshold. Relatively small changes in interior RH result in large changes in peak roof-

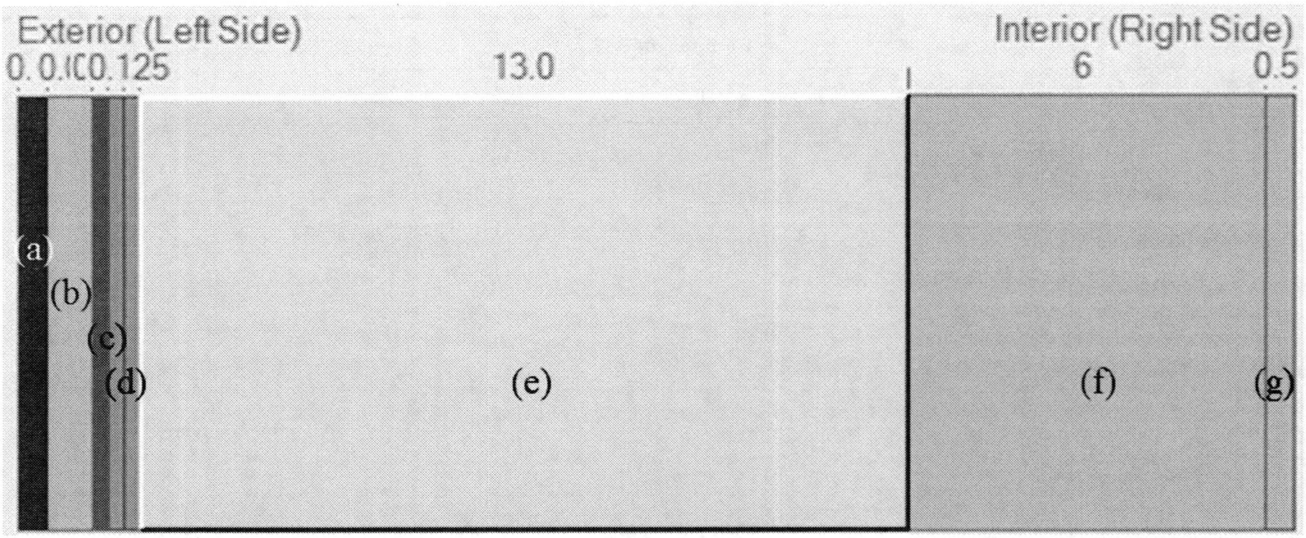

Figure 2 *Attic system material layers for WUFI simulations: (a) clay roof tile, (b) ventilated air space below roof tiles, (c) roof underlayment, (d) OSB roof decking, (e) R-38 fiberglass, (f) attic air space, and (g) gypsum board ceiling.*

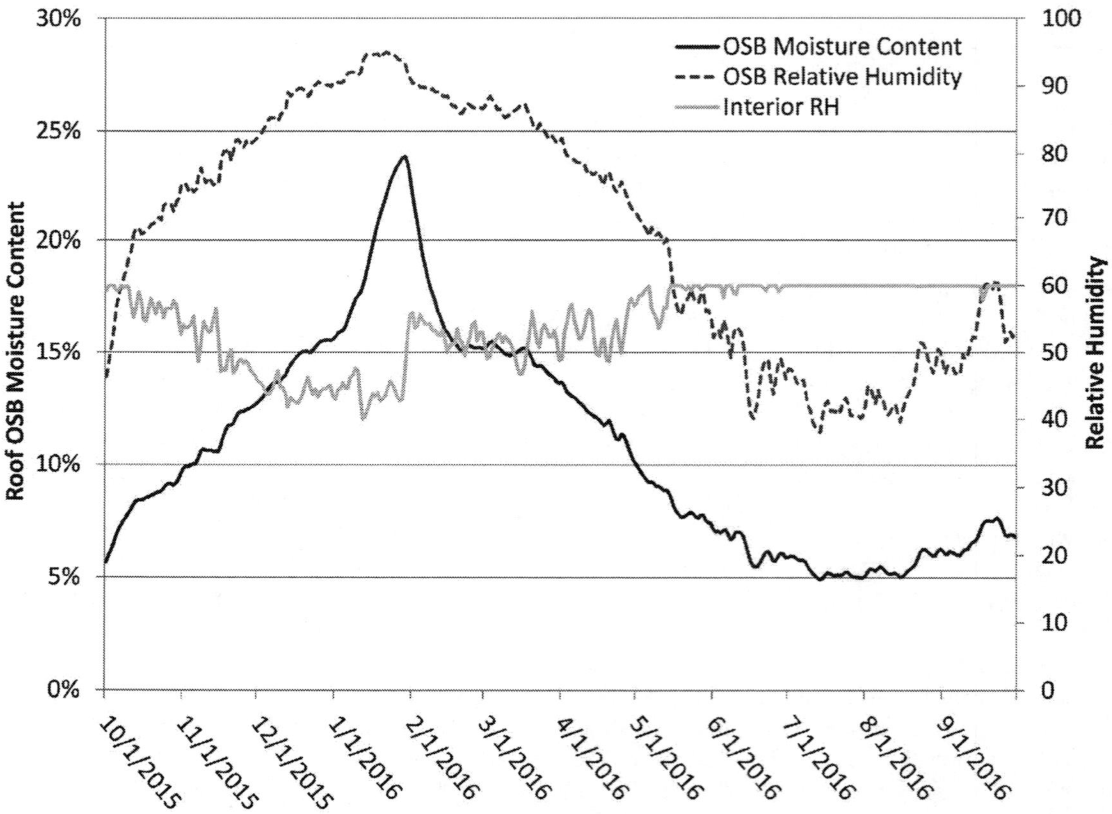

Figure 3 *Simulated roof-deck moisture content, relative humidity next to the roof deck, and interior relative humidity from WUFI simulations of test attic in Fresno, CA for one year. The curves are based on average daily values from the hourly simulation data. Even though the roof does see elevated moisture content in the winter, it dries completely during the warmer summer months.*

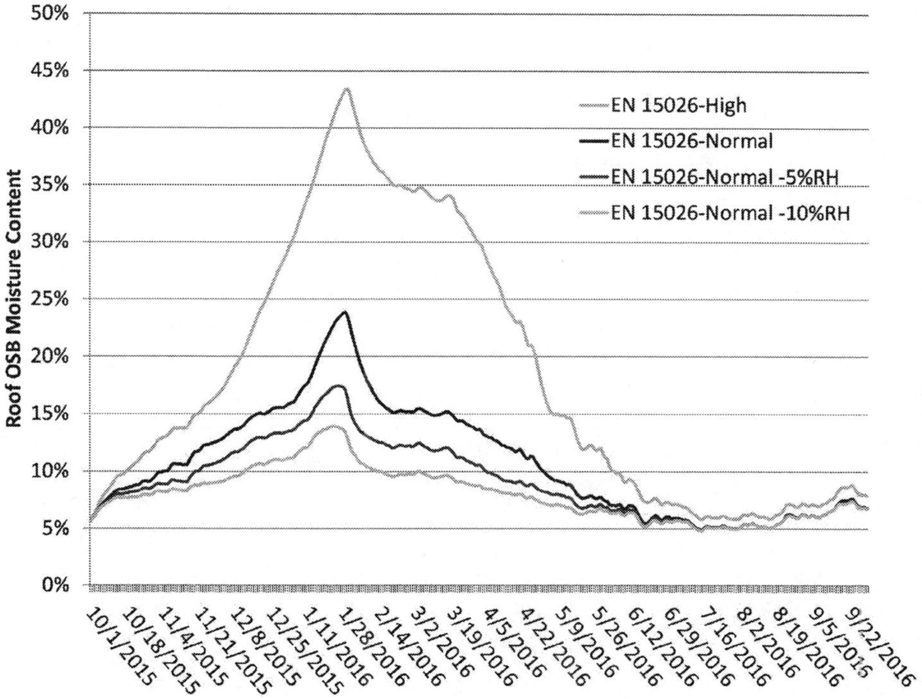

Figure 4 *WUFI-simulated roof-deck moisture content, due to different interior relative humidity conditions. Exterior climate was Fresno, CA for one year. The curves are based on average daily values from the hourly simulation outputs. Peak moisture content varies widely with interior RH but does dry completely during the warmer summer months.*

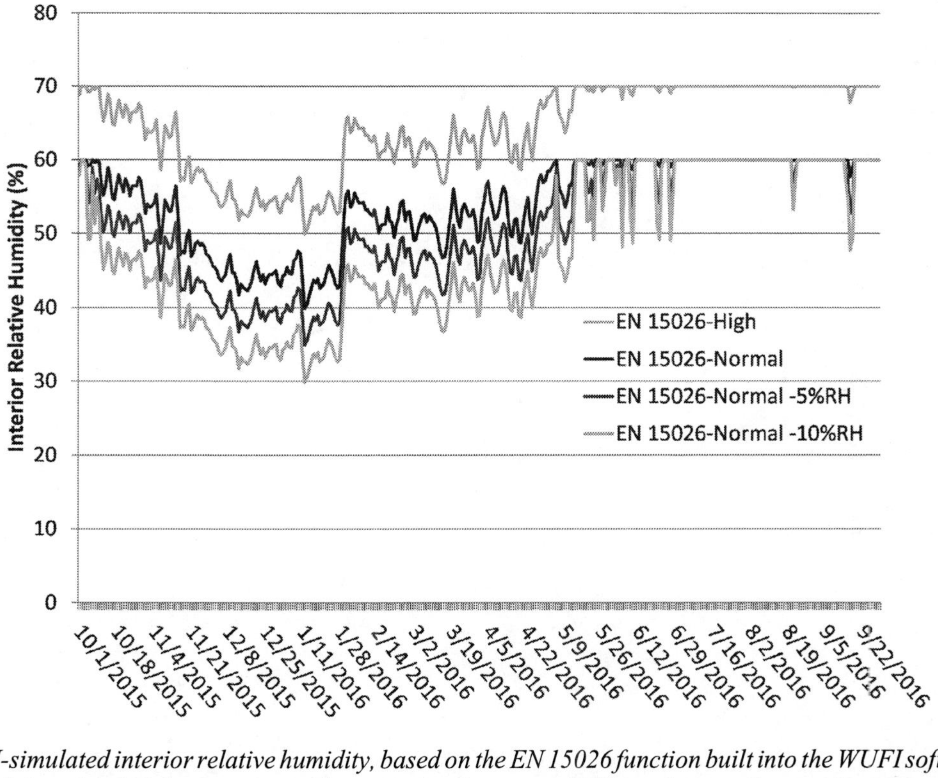

Figure 5 *WUFI-simulated interior relative humidity, based on the EN 15026 function built into the WUFI software. The curves at –5% rh and –10% rh were created by exporting the original RH conditions, altering the selected data, and creating a new climate file for WUFI to use.*

deck moisture, as shown in the comparison between the moisture content graphs with normal and high interior climate moisture levels. An increase of interior RH by about 10% results in peak roof-deck moisture content increasing by 20%. The simulations suggest that the moisture performance of the unvented attic is highly dependent on interior relative humidity.

One aspect the moisture risk criteria by Straube does not account for is drying, and that moisture risk can decrease if conditions are unfavorable for mold or microbial growth. As an alternate criteria for moisture durability, output from the simulations were also used to calculate an hourly mold index, as defined by Ojanen (Ojanen et al. 2010). Hourly mold index was calculated based on a one-year simulation, using the EN 15026 normal interior climate. The definition of *mold index* by Ojanen includes classification descriptions of mold index ranging from 0 to 6, where a mold index of 3 is the first instance of visible mold growth with the naked eye. Mold index <3 was used as an alternate criterion for acceptable moisture risk to the wood roof deck. The hourly mold index from the one-year simulation is shown in Figure 6. The trend in mold index from the 1-year simulation suggests there could be a long-term mold risk, if the mold index continued to incrementally increase year over year. To further explore this trend, additional simulations were performed, and mold index calculated, for five- and ten-year periods. Figure 7 shows the five-year mold index calculations, along with moisture content and relative humidity at the interior surface of the roof deck. Even though the roof moisture content and RH show a cyclic behavior, with no year-over-year trend, the mold index continues to increase each year, though it does not reach the threshold value of 3. Extending the analysis out to ten years in Figure 8, we can begin to see the asymptotic nature of the mold index trend. Extending the trend line for mold index shown on the graph in Figure 8 out to a fifty-year lifetime for a house (438,000 hours) indicates the mold index ends up fluctuating about a mean of approximately 1.6. This level of expected microbial activity would still be below the risk threshold (mold index >3).

Table 2. Simulated Moisture Content Analysis of OSB Roof-Deck Inner Surface

Interior Condition	Hours > 16% Moisture Content	Hours > 28% Moisture Content	Max Moisture Content (mass%)	Min Moisture Content (mass%)
EN 15026, high	3806	2344	43%	6%
EN 15026, normal	983	0	24%	5%
EN 15026, normal, –5% rh	344	0	17%	5%
EN 15026, normal –10% rh	0	0	14%	5%

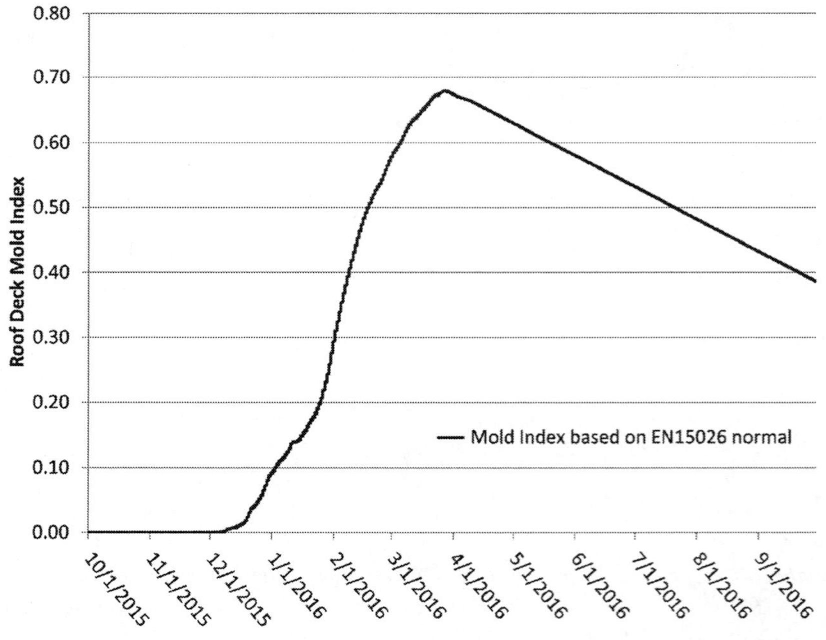

Figure 6 *Roof-deck mold index, based on simulations with interior climate using EN 15026, normal moisture loading.*

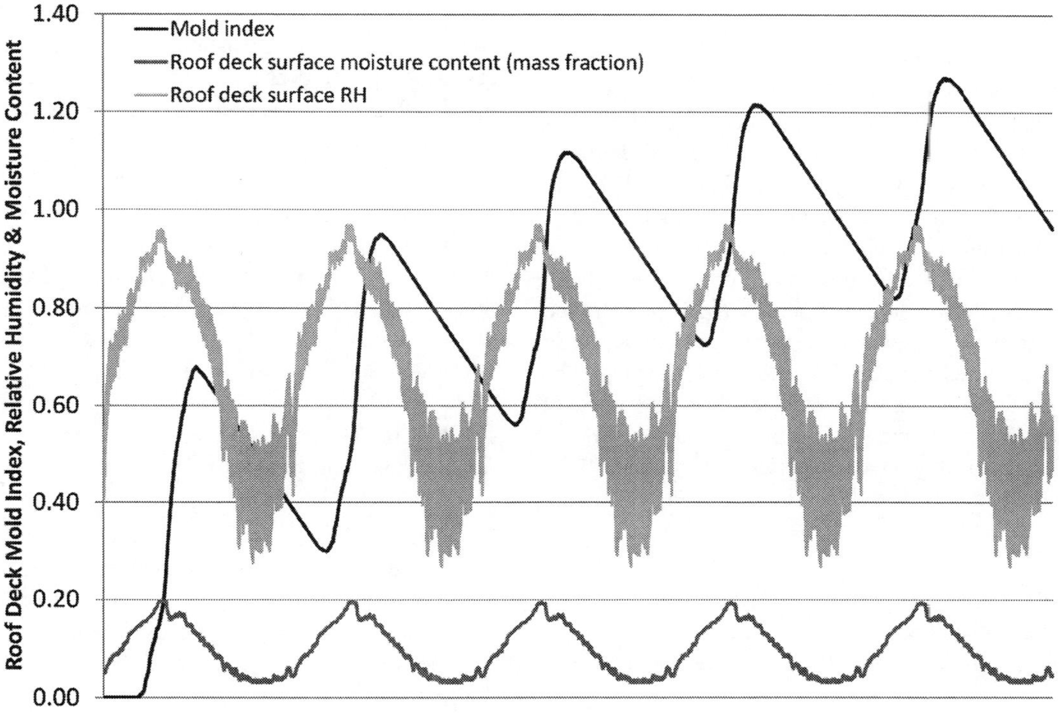

Figure 7 *Roof mold index, based on a five-year simulation with interior climate using EN 15026, normal moisture loading. Graph includes relative humidity and moisture content at the roof-deck inner surface. The mold index is incrementally higher each year, even though roof-deck moisture content and RH return to minimum values each year.*

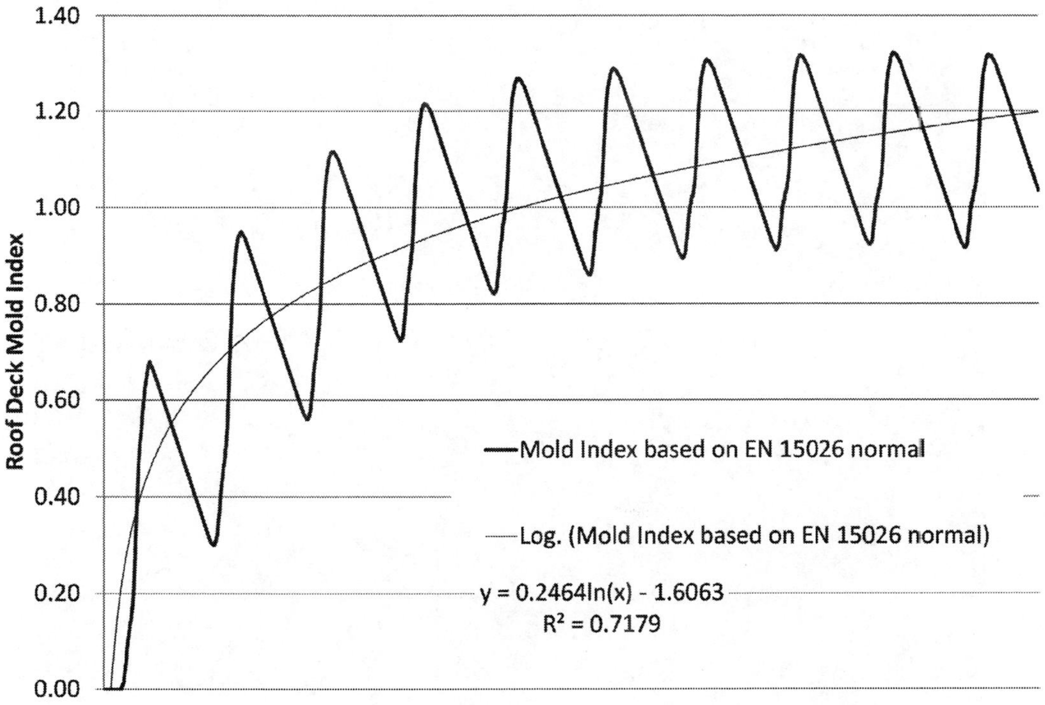

Figure 8 *Roof mold index, based on a ten-year simulation with interior climate using EN 15026, normal moisture loading. In this graph, the asymptotic trend of the mold index curve can be observed. The trend line is a best-fit log function. For a fifty-year building service life, the mold index trend approaches an average annual value of 1.6.*

TEST HOUSE

A builder in California expressed interest in trialling this attic insulation system on a home in the Central Valley, near Fresno. The home was 3609 ft^2, two-stories, with a 6:12 roof pitch. The roof deck was approximately 3200 ft^2. The project team included Lawrence Berkley National Lab (LBNL), who assisted with insulation system design and is leading the project to install instrumentation in the house to monitor moisture and thermal and energy performance. Other project team members included DeYoung Properties (builder), Broken Drum Insulation (insulation contractor), and BIRAenergy (home energy modeling and CA code compliance). The purpose of the test home project was to evaluate the practical aspects of installing fiberglass batts up against a pitched roof deck and to verify the thermal and moisture performance of the system in a home that is intended to be zero-energy ready.

Installation

Air Sealing. Air-sealing of the roof consisted of applying one-part canned polyurethane foam to all roof-deck seams and installing and sealing blocking at all the eaves, as shown in Figure 9. Both air sealing and insulation under the roof deck was completed prior to installation of any plumbing, electrical, HVAC, or ceiling gypsum board to maximize attic access. One result of this sequencing decision is that the insulation contractor must still go back to the home and perform additional air sealing and insulation repair to penetrations through the roof and attic.

Insulation Install. The insulation contractor for this project opted to install the batts using medium gauge wire (baling wire). Each 24 × 48 in. batt was secured to the roof deck with three sections of wire, each secured up to the truss top chord on each side of the batt. An illustration and photos of the installation steps are described in Figure 10 and the list below:

1. Starting at the roof ridge, staple each section of wire to the side of the truss top chord. Wires are spaced to evenly support the batts.
2. Position the first batt in place, tucking it fully up into the corner where the truss meets the roof deck. Tilt the batt into place and tuck the other edge in place. Compression from the roof framing creates enough friction to support the batt while the wiring is secured.

(a)

(b)

Figure 9 *Roof-deck air sealing: (a) roof-deck seams and ridge and (b) blocking and sealing of eaves.*

Figure 10 *Attic insulation system install steps. Numbers correspond to the install steps under the section "Insulation Install."*

3. Bend the first section of wire to form a 90° corner at the outer edge of the batt. Bring the wire across and bend to create a second corner on the other side. Bring the wire back up to the next top chord and secure with staples. Double the wire back out and twist it back onto itself to form a more rigid side support. Repeat with the other two sections of wire.
4. Steps 2 to 4 are repeated with the next batt, until the sections of wire run out, and then repeat all steps, working along the ridges and then down the roof slope for the subsequent rows.

The installers for this project were already familiar with this installation method, as it is frequently used in California under wood-framed low-slope commercial roofs. Air sealing the roof deck took approximately 4 to 5 man-hours. The installation of the batts was completed in approximately 54 man-hours. This included two particularly challenging areas, where an intermediate layer of roof sheathing slowed the install process. It is expected that future projects will take less time. Figure 11 shows a photograph of a section of roof with all batts installed. With a skilled installer, batts can be secured to the roof deck with minimal compression and virtually no gaps or voids.

Instrumentation

As part of a separate project, LBNL has installed a large suite of sensors in the test house to monitor thermal and moisture performance of the roof-attic system, along with monitoring comfort and energy usage in the home while it is occupied. For the roof-attic system, the following sensors were installed:

- Temperature sensors below the roof tile, at the roof decking, in the attic air, at the ceiling gypsum, and in the occupied space
- Heat flux transducers at the roof decking
- Relative humidity sensors in the attic and in the occupied space
- Condensation sensors at the roof decking

The data acquisition system will feed into a web-accessible monitoring system. Team members will have online data access, and triggers are in place to alert the team if moisture performance of the system is more severe than expected.

CONCLUSION

This project is exploring the development of a relatively simple means for insulating an unvented attic with fiberglass batts. This initial study evaluated the following aspects of the system:

- Predicted thermal performance
- Predicted moisture performance
- Practical aspects of product installation
- Estimates of cost effectiveness

***Figure 11** A section of the roof with insulation installation completed.*

Thermal performance was estimated based on parallel-path calculations of the U-factor of the roof-attic assembly, using assumptions required by the California energy code (Title 24). These calculations suggest the system should provide acceptable thermal performance. However, system-scale testing in a calibrated hot box is planned. Moisture performance of the system was assessed via hourly hygrothermal simulations (WUFI), with the simulation outputs analyzed against two different sets of moisture durability criteria. Simulated moisture content of the wood roof deck indicated acceptable, but marginal, moisture performance. Analysis of roof-deck moisture content highlighted that moisture durability of unvented attics is strongly dependent on interior humidity levels in winter. Analysis using the mold index metric initially suggested a trend towards an eventual moisture problem (high mold index value). Further analysis, calculating the mold index for ten years, demonstrated that this metric can approach a long-term limit that suggests the roof-attic system may not encounter moisture problems. The findings from the mold index analysis suggest that applications of this performance metric may need to extend out many years to properly assess moisture performance and durability of assemblies.

Working closely with an insulation contractor who is already familiar with this method of installation greatly simplified this aspect of the project. Demonstrations of the install technique to the builder, construction superintendent, rater, and third-party inspector made it possible to address concerns prior to construction of the test house. Air sealing the roof deck was relatively straightforward, as was applying the process of wired batts to a pitched roof rather than a flat roof. With only one test house completed, enough data is not available to answer the question of cost effectiveness. However, it is expected that future projects will provide a more data to evaluate cost effectiveness, as well as provide opportunities for optimizing the installation procedure.

CONTINUING WORK

Development of this insulation approach will continue. Thermal performance will be verified, by testing several variations in a calibrated hot-box (CHB) apparatus. The CHB testing will investigate the thermal effects of small gaps and voids that could occur with this method, as well as potential solutions. Moisture and thermal performance will be monitored via the instrumentation in the test house in Fresno, with future WUFI models adjusted to match collected field data. Installation methods will be optimized through continued collaboration with builders and insulation contractors in California to reduce the install labor and better compare against other unvented attic solutions.

ACKNOWLEDGMENTS

The authors would like to acknowledge DeYoung Properties and Broken Drum Insulation. Without your interest, active participation, and valuable input this project would not have been possible.

REFERENCES

ASHRAE. 2013. *2013 ASHRAE Handbook—Fundamentals*. Atlanta: ASHRAE.

Black, C.D., and J.F. Straube. 2007. Mould resistance of full scale wood frame wall assemblies. *Proceedings of Eleventh Canadian Building Science and Technology Conference*. Alberta: Banff.

BSI. 2007. BS EN 15026:2007, *Hygrothermal performance of building components and building elements. Assessment of moisture transfer by numerical simulation*. London: British Standards Institution.

CBC. 2013. *2013 California building standards code*. Sacramento: State of California.

CEC. 2016. *2016 Building energy efficiency standards for residential and nonresidential buildings*. Sacramento: State of California.

DOE. 2013. Building America Top Innovations Hall of Fame Profile: Unvented, Conditioned Attics. http://energy.gov/sites/prod/files/2014/01/f6/1_1c_ba_innov_unventedconditionedattics_011713.pdf.

Hendron, R., R. Anderson, P. Reeves, and E. Hancock. 2002. *Thermal performance of unvented attics in hot-dry climates*. Report NREL/TP-550-30839. Golden, CO: National Renewable Energy Laboratory.

IRC. 2015. *2015 International residential code*. Country Club Hills, IL: International Code Council.

Künzel, H.M. 2006. *WUFI® PC-Program for calculating the coupled heat and moisture transfer in buildings*. Holzkirchen, Germany: Fraunhofer Institute for Building Physics.

Künzel, H.M., and K. Kießl. 1997. Calculation of heat and moisture transfer in exposed building components. *International Journal of Heat and Mass Transfer* 40(1):159–67.

Lstiburek, J., and C. Schumacher. 2011. *Hygrothermal analysis of California attics*. Westford, MA: Building Science Press.

Miller, W., A. Desjarlais, and M. LaFrance. 2010. Roof and attic design guidelines for new and retrofit construction of homes in hot and hold climates. *ASHRAE Transactions* 116(2).

Ojanen, T., H. Viitanen, R. Peuhkuri, K. Lähdesmäki, J. Vinha, and K. Salminen. 2010. Mold growth modeling of building structures using sensitivity classes of materials. *Proceedings of Thermal Performance of the Exterior Envelopes of Whole Buildings XI.*

Rudd, A. 2005. Field performance of unvented cathedralized (UC) attics in the USA. *Journal of Building Physics* 29(2):145–69.

Schumacher, C. 2007. Unvented roof assemblies for all climates. *Building Science Digest* 149. Westford, MA: Building Science Press.

SPFA. 2013. *SPFA Builder's reference handbook*. Fairfax, VA: Spray Polyurethane Foam Alliance.

Straube, J., Smegal, J., and J.B. Smith. 2010. Moisture-safe unvented wood roof systems. Presented at the Second Building Enclosure Science & Technology conference, Portland, OR, April 12–14.

Wolters Kluwer Law & Business. 2006. *Energy Design Update* Febuary 2006. New York: Aspen Publishing

Wind Washing Effects on Mineral Wool Insulated Sheathings

Randy Van Straaten, MASc, PEng
Student Member ASHRAE

John Straube, PhD, PEng
Associate Member ASHRAE

Trevor Trainor, MASc

Antoine Habellion, MEng, MAS

ABSTRACT

Wind washing *is wind-driven air movement through or behind thermal insulation within enclosures. This bulk movement of air increases heat loss, resulting in increased energy consumption, risk of condensation on cooled surfaces, and increased space-heating loads. Previous studies have investigated airflow bypassing the insulation through gaps, around insulation boards, and through low-density fibrous insulation products. The impact of wind washing on well-installed mineral wool board products used outside the air barrier as continuous insulation has not been previously studied in detail. This paper presents a methodology for predicting wind washing impacts and laboratory measurement of heat loss impacts for insulation products exposed to cavity airflows.*

It is recommended that average winter wind speeds (or ASHRAE mean coincident wind speed) be used for wind washing analysis. A method is presented for adjusting wind speeds for sheltering and height impacts based on approaches used in structural wind load calculations. Hourly average air speeds as high as 0.1 and 0.7 m/s (0.4 and 2.4 ft/s) were predicted using simple airflow network analysis for well-ventilated brick cladding and slot panels, respectively. Analysis of full-scale wind tunnel pressure tap data for a small house was used to predict vinyl siding cavity air speeds as high as 0.5 m/s (1.5 ft/s). Heat flow measurements were taken for a number of insulation products in a purpose-built apparatus. Of the mineral wool board samples tested, only the 25 and 50 mm (1 and 2 in.) thick 70 kg/m^3 (4.4 pcf) samples showed measurable wind washing impact. This impact was as much as 0.03 RSI (R-0.2) reduction of thermal performance. Hence, wind washing impacts are expected to be small for well-installed mineral wool board continuous insulation.

INTRODUCTION

Wind washing is wind-driven air movement through or behind thermal insulation within enclosures. This bulk movement of air increases heat loss, resulting in increased energy consumption, risk of condensation on cooled surfaces, and increased space-heating loads. Previous research has shown that wind washing of air-permeable insulations is most significant for installations with

- higher air velocities along the insulation surface due to higher wind exposure and/or more open cladding systems (Bankvall 1978; Henning 1983; Taylor and Phillips 1983; Uvsløkk 1996);
- higher insulation air permeability, typically correlating with lower density and/or less thickness (Bankvall 1978; Henning 1983; Taylor and Phillips 1983; Yarbrough and Tooe 1983; Tanner and Ghazi 1996);
- installation imperfections resulting in air gaps around and behind the insulation (Bankvall 1978; Hens et al. 2001); and
- greater scouring of insulation due to eddies in the cavity airflow, and/or greater air momentum from direct wind exposure (Bankvall 1978; Berlad 1979; Silberstein 1991; Deseyve and Bednar 2005; Janssen and Hens 2007).

Figure 1 illustrates the process of wind washing, and Figure 2 illustrates two measures to limit it: air cavity compart-

Randy Van Straaten is a senior research engineer at RDH Building Science Laboratories, Waterloo, Ontario, and a doctoral candidate in the Department of Civil and Environmental Engineering at Western University, London, Ontario. **John Straube** is a principal at RDH Building Science Laboratories and a professor in the Department of Civil Engineering, University of Waterloo, Waterloo, Ontario. **Trevor Trainor** is a building science researcher at RDH Building Science Laboratories. **Antoine Habellion** is a building science specialist at ROXUL Inc., Milton, Ontario.

mentalization and the application of a so-called "wind barrier" on the exterior of the insulation. Uvsløkk (1996) recommended a wind barrier air permeance no greater than 1.4E-5 $m^3/m^2 \cdot s \cdot Pa$, based on laboratory tests and field monitoring of a small test house. Mineral wool insulation boards used as sheathing have air permeances greater than Uvsløkk's recommendation, hence some designers choose to place wind barriers over top of them. However, Uvsløkk's recommendation was for a membrane installed over low-density fibreglass batt insulation without an air barrier. Therefore, his finding may not apply to unprotected insulation installed outboard of solid surfaces (often air barrier membranes), as is becoming common in North America. More air-permeable exterior insulation products may act as effective wind barriers themselves in such modern wall systems.

This paper presents a methodology for predicting wind washing impacts. The wind exposures are illustrated using a number of cold North American cities. The paper also presents laboratory measurement of heat loss impacts for insulation products exposed to cavity airflows. Based on these and the methodology, the impact of wind washing on mineral wool board products is predicted.

WIND SPEEDS FOR WIND WASHING ANALYSIS

Prevailing wind speed (WS) and direction vary depending on geographical location. Wind speeds for structural wind load requirements are published in building codes, but design speeds do not occur often and are too high for use in predicting wind washing impacts. Average annual wind speeds may also have limited suitability because higher wind speeds may occur in summer when wind washing is less of a concern. For heating degree day (HDD) energy analysis, one would ideally use a single HDD-weighted wind speed. However, such values are not readily available and the value would have a dependence on the balance temperature for the spaces being analyzed, complicating its derivation.

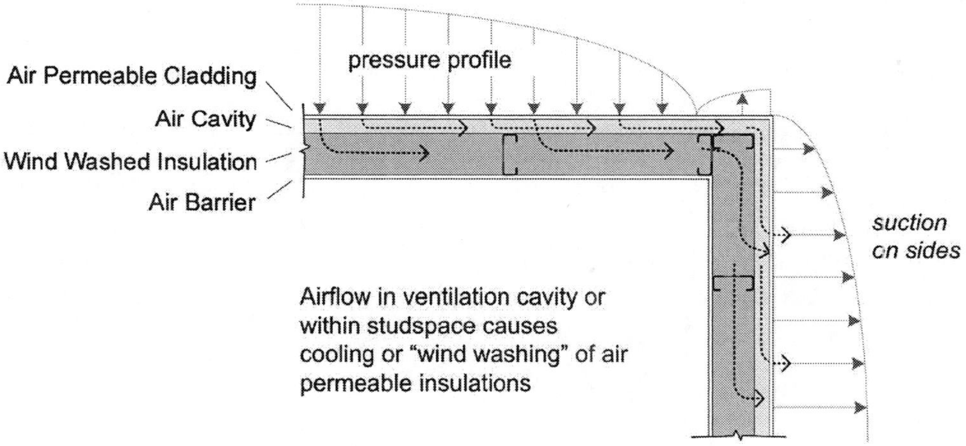

Figure 1 *Illustration of wind washing effects on a wall—plan view (adapted from Straube 2001).*

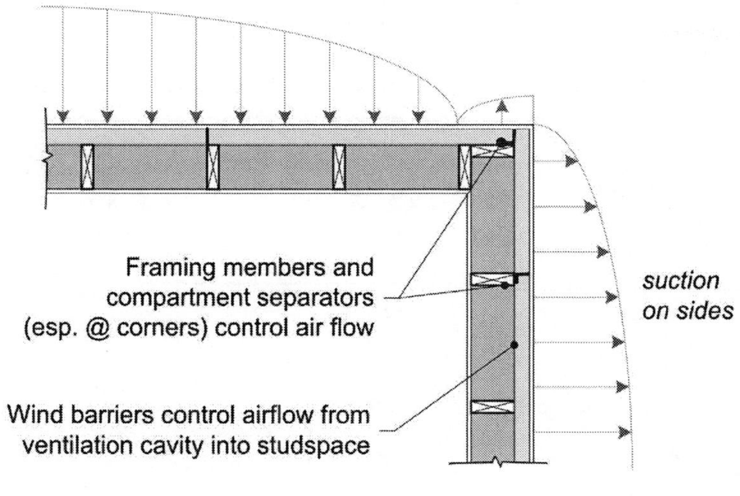

Figure 2 *Measures to address wind washing—plan view (adapted from Straube 2001).*

Three different approaches using ASHRAE climatic data (ASHRAE 2013) were considered for determining appropriate wind speeds for wind washing analysis. The first approach considered was to use mean January wind speeds and dry-bulb temperatures. This method is the simplest approach but does not account for the coincidence of both cold temperatures and high wind speeds. The second approach is to use the highest wind speed (0.04%) for the coldest month of the year and the mean coincident dry-bulb (MCDB) temperatures associated with that wind speed. This approach accounts for both cold temperatures and high wind speeds but applies more weight to wind speed than to temperature. A third approach is to choose the heating design temperature (99.6%) and the mean coincident wind speed (MCWS) associated with that temperature. This approach is similar but applies more weight to temperature than to wind speed.

Table 1 lists these values for some cold North American cities. The coldest month with 0.04% wind speeds (3.5 h/yr) are 6 to 9 m/s (13 to 20 mph) greater than January averages. However, the 99.6% dry-bulb temperature is as much as 20°C (36°F) colder than the January average and as much as 30°C (54°F) colder than both the coldest month average temperature and the temperature coincident with 0.04% wind speeds. It is assumed that the significant temperature difference would result in the 99.6% dry-bulb condition being a worst-case peak heating load and condensation condition for most locations. Furthermore, the use of the MCWS has the advantage of correlating with the temperature typically used for heating load calculations. MCWS values are used in the analysis presented in this paper.

The wind speeds given in Table 1 are based on the data from anemometers placed at 10 m (32.8 ft) heights in open fields. Straube (1998) used wind speed adjustments from structural building code calculations to adjust the wind speeds in driving rain analysis to adjust for local sheltering by surrounding buildings and landscapes (hills, cliffs, etc.) as well as height. It is proposed to use a similar approach for wind washing analysis. The results of this approach are given in Table 2 for a small house in the suburbs, a small house in the countryside, and a tall building in a city centre, where height, h, and sheltering (captured by logarithmic exponent, α) affect wind exposure. These values are used to estimate height-dependent mean wind speed as follows:

$$U = WS(h/10\ \mathrm{m})^{\alpha} \qquad (1)$$

Table 1. Cold Weather Wind Speeds in ASHRAE Climatic Design Data Tables

Location	Average January		Coldest Month Wind Speed		Coldest Month Wind Speed		Mean Coincident Wind Speed	
	DB, °C (°F)	WS, m/s (mph)	MCDB, °C (°F)	0.04%WS, m/s (mph)	MCDB, °C (°F)	0.1% WS, m/s (mph)	99.6% DB, °C (°F)	MCWS, m/s (mph)
Toronto	–4.7 (24)	4.9 (11)	–4.7 (24)	14.2 (32)	–4.1 (25)	12.7 (28)	–18.1 (–1)	4.8 (11)
Chicago	–3.9 (25)	4.9 (11)	–0.5 (31)	12.1 (27)	–0.6 (31)	11.1 (25)	–18.6 (–3)	4.9 (11)
New York City	0.8 (33)	5.6 (13)	–1.3 (30)	14.2 (32)	–0.9 (30)	12.8 (29)	–10.1 (14)	7.4 (17)
Edmonton	–12 (10)	3.3 (8)	–2.9 (27)	11.6 (26)	–6.7 (20)	10.1 (23)	–32.6 (–27)	2.1 (5)
Denver	–0.3 (31)	3.8 (9)	4.0 (39)	13.4 (30)	2.3 (36)	11.9 (27)	–17.5 (1)	3.5 (8)

Table 2. Adjusted Wind Speeds for Sample Building Types

Location	MCWS Open Field h = 10 m (32.8 ft) m/s (mph)	Small House in Suburbs h = 2 m (6. 6ft) α = 0.25 m/s (mph)	Small House in Countryside h = 2 m (6.6 ft) α = 0.13 m/s (mph)	Large Building in City Centre h = 30 m (98 ft) α = 0.25 m/s (mph)
Toronto	4.8 (11)	3.2 (9)	3.8 (7)	9.1 (20)
Chicago	4.9 (11)	3.3 (9)	3.9 (7)	9.3 (21)
New York City	7.4 (17)	4.9 (13)	5.9 (11)	14.1 (32)
Edmonton	2.1 (5)	1.4 (4)	1.7 (3)	4.0 (9)
Denver	3.5 (8)	2.3 (6)	2.8 (5)	6.7 (15)

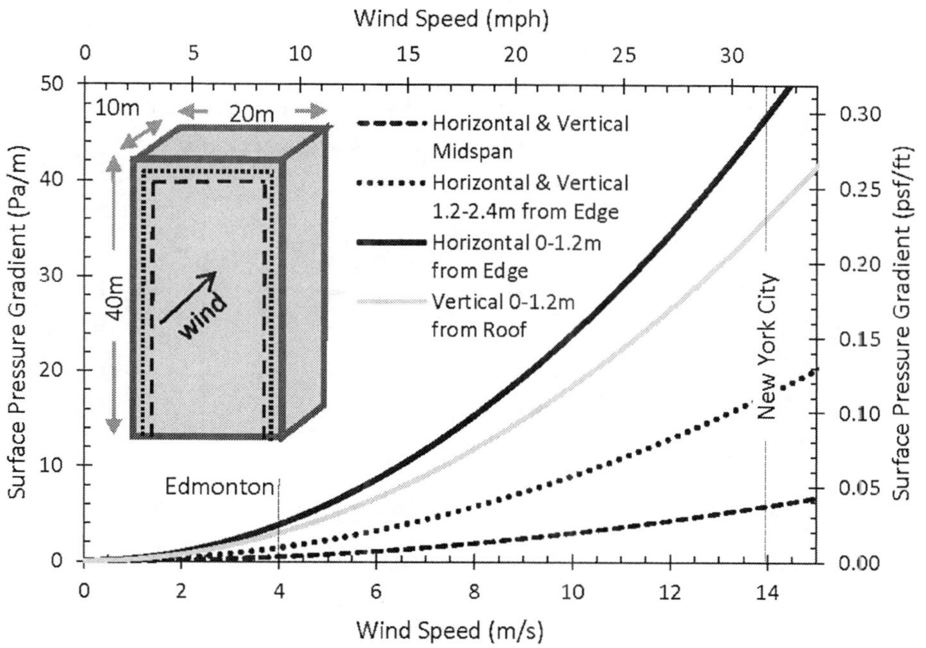

Figure 3 *Vertical and horizontal surface pressure gradients for a tall building.*

Table 3. Calculated Tall-Building Surface Pressure Gradients, Pa/m (psf/ft)

Location	Midspan	Tall Building in City Centre		
		1.2–2.4 m (4–8 ft) from Edge	Within 1.2 m (4 ft) of Edge	Within 1.2 m (4 ft) of Edge
	Horizontal/Vertical	Horizontal/Vertical	Horizontal	Vertical
Toronto	2.5 (0.016)	7.5 (0.048)	20.1 (0.128)	15.1 (0.096)
Chicago	2.6 (0.017)	7.9 (0.050)	20.9 (0.133)	15.7 (0.100)
New York City	6.0 (0.038)	17.9 (0.114)	47.7 (0.304)	35.8 (0.228)
Edmonton	0.5 (0.003)	1.4 (0.009)	3.8 (0.024)	2.9 (0.018)
Denver	1.3 (0.008)	4.0 (0.025)	10.7 (0.068)	8.0 (0.051)

WIND PRESSURES DRIVING CAVITY AIRFLOW

Winds generate temporally and spatially varying pressures on building surfaces. These pressures are reported from field measurements, wind tunnel testing, and modeling, typically as nondimensional pressure coefficients. Straube (1998) used the differences in mean pressure coefficients measured on the surface of a small test building to estimate ventilation rates for a series of brick vent combinations. Inculet et al. (1997) measured pressure gradients, ΔP, on a building surface in a wind tunnel and divided the mean values by distance, L, and dynamic pressure, $0.5\rho U^2$(where ρ is the air density), to calculate gradient coefficients, as follows:

$$G = \Delta P / 0.5 L \rho U^2 \quad (2)$$

Predictions of surface pressure gradients using MCWS and Inculet's normalization approach are plotted in Figure 3 for a range of wind speeds. These results show a dramatic rise within 2.4 m (8 ft) of the wall edges. The edge of a tall building is predicted to have a 3.8 Pa/m (0.024 psf/ft) gradient in Edmonton and a 47.7 Pa/m (0.304 psf/ft) gradient in New York City. Midspan gradients are 0.5 and 6 Pa/m (0.003 and 0.038 psf/ft) for these cities. Values are tabulated in Table 3.

Cope et al. (2012) measured surface pressures on a small test house exposed to wind in a full-scale wind tunnel. The horizontal and vertical gradient results for winds perpendicular to the wall surface are plotted in Figure 4 for windward and leeward wall exposures.

Horizontal gradients range from 0.16/m to 0.43/m (0.05/ft to 0.13/ft) and vertical gradients range from 0.06/m to 0.27/m

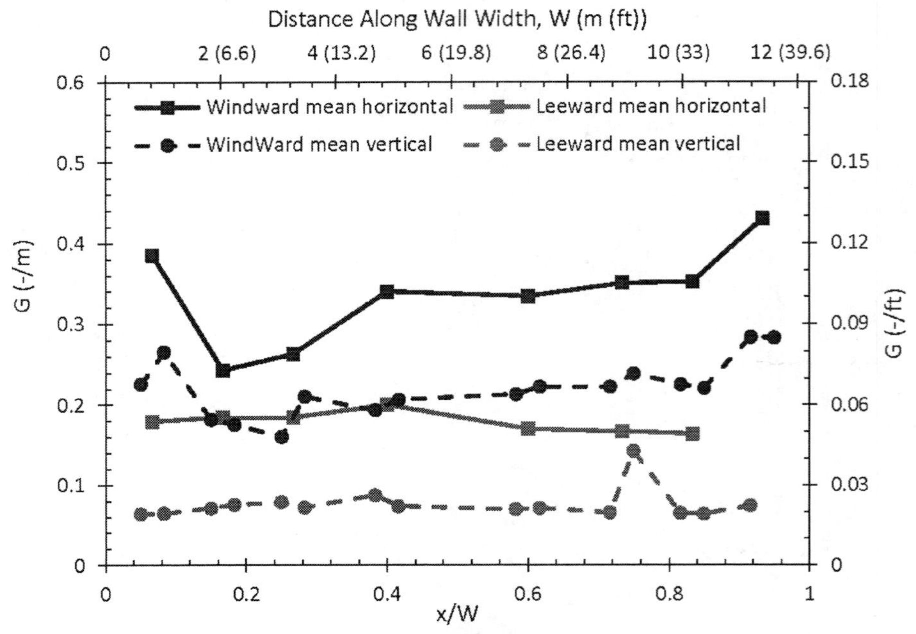

Figure 4 *Mean surface pressure gradients along width of wall for perpendicular wind.*

(0.02/ft to 0.08/ft). Using these results as "low" and "high" values, surface pressure gradients are plotted for a range of wind speeds in Figure 5 and tabulated in Table 4. Horizontal gradient values for a suburban house in Edmonton and a house in the countryside in New York City show as < 1 and 3–9 Pa/m (<0.006 and 0.019–0.057 psf/ft), respectively.

Cavity Pressure Gradients

Uvsløkk (1996) measured cavity pressure gradients of 0.1–0.5 Pa/m (0.001–0.003 pcf/ft) for mean wind speeds of 3 m/s (6.7 mph) in a 23 mm (~1 in.) clear air gap behind wood siding of a small test house. To determine cavity pressure coefficients for vinyl siding, the cavity pressure data, which were also available in Cope et al.'s (2012) data set, were analyzed. Horizontal and vertical cavity pressure gradients from this data set are shown in Figure 6. Average windward mean horizontal and vertical gradients across the width are 0.2/m and 0.1/m (0.06/ft and 0.03/ft). Both averages are significantly lower than the surface gradients.

CAVITY AIRFLOWS

Cavity pressure gradients can be used to calculate airspeeds using the Darcy-Weisbach equation and assuming the cavity flow is equivalent to fully developed laminar flow between parallel plates as follows:

$$\Delta P_c = \frac{96}{\mathrm{Re}} \cdot \gamma \frac{\rho V_c^2}{2} \cdot \frac{L}{D_h} \tag{3}$$

where

ΔP_c = cavity pressure gradient

V_c = mean cavity velocity

γ = cavity blockage factor

Re = Reynolds number based on hydraulic diameter

D_h = hydraulic diameter

H = cavity depth

L = cavity length

Equivalent vertical and horizontal vinyl siding cavity depths of 1 and 4 mm (1/32 and 5/32 in.) from Van Straaten (2004) were used to estimate cavity velocities for a range of wind speeds, as shown in Figure 7. Wind speeds between 1.4 and 5.9 m/s (3 and 13 mph) correlate to horizontal cavity air speeds of 0.03 and 0.5 m/s (0.1 and 1.6 ft/s) and vertical air speeds of 0.004 and 0.06 m/s (0.01 and 0.2 ft/s).

Vinyl siding cavity flow rates based on measured cavity pressure coefficients are tabulated for five cold cities in Table 5. The vertical air speeds are quite low, whereas the horizontal air speeds are as high as 0.3 m/s (1 ft/s).

Field-measured hourly average cavity air velocity values for various cladding systems as high as 0.6 m/s (2 ft/s) have been published in previous studies, as shown in Table 6. The reported values are similar to those calculated for vinyl siding.

A complex model would be needed to estimate vinyl siding airflow rates given its complex vent opening geometry. Straube (1998) used simple pipe flow models to estimate cavity airflow rates for cladding systems with discrete vents in series with a cavity as follows:

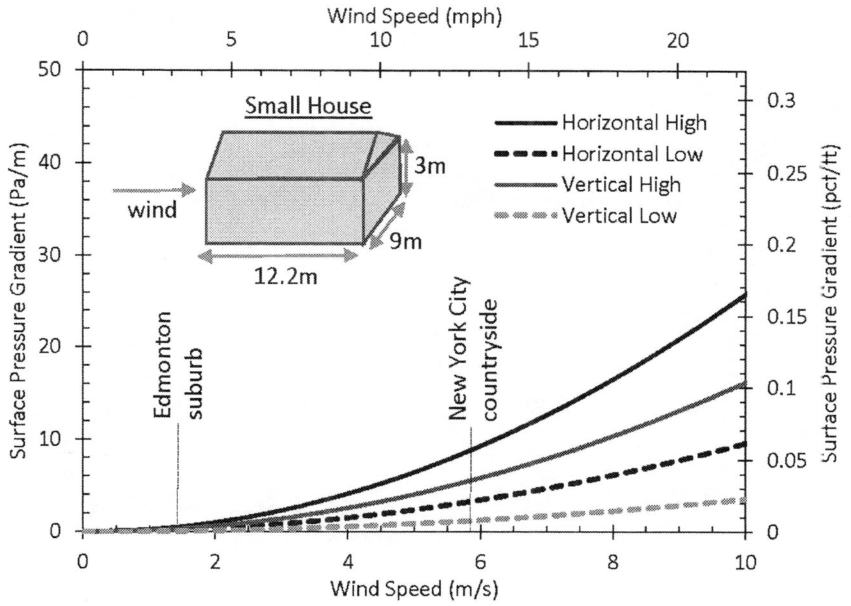

Figure 5 *Surface pressure gradients for range of measured small-house values.*

Table 4. Calculated Small-House Surface Pressure Gradients, Pa/m (psf/ft)

Location	Small House in Countryside				Small House in Suburbs			
	Low		High		Low		High	
	Horizontal	Vertical	Horizontal	Vertical	Horizontal	Vertical	Horizontal	Vertical
Toronto	1.4 (0.009)	0.5 (0.003)	3.8 (0.024)	2.4 (0.015)	1.0 (0.006)	0.4 (0.002)	2.7 (0.017)	1.7 (0.011)
Chicago	1.5 (0.009)	0.6 (0.004)	3.9 (0.025)	2.5 (0.016)	1.0 (0.007)	0.4 (0.002)	2.8 (0.018)	1.7 (0.011)
New York City	3.3 (0.021)	1.3 (0.008)	9.0 (0.057)	5.7 (0.036)	2.4 (0.015)	0.9 (0.006)	6.3 (0.040)	4.0 (0.025)
Edmonton	0.3 (0.002)	0.1 (0.001)	0.7 (0.005)	0.5 (0.003)	0.2 (0.001)	0.1 (0.0005)	0.5 (0.003)	0.3 (0.002)
Denver	0.7 (0.005)	0.3 (0.002)	2.0 (0.013)	1.3 (0.008)	0.5 (0.003)	0.2 (0.001)	1.4 (0.009)	0.9 (0.006)

$$\Delta P = C_{en} \cdot \frac{\rho V_{en}^2}{2} + \frac{96}{\mathrm{Re}} \cdot \gamma \frac{\rho V_c^2}{2} \cdot \frac{L}{D_h} + C_{ex} \cdot \frac{\rho V_{ex}^2}{2} \quad (4)$$

Entrance and exit loss coefficients, C_{en} and C_{ex}, were taken from duct design manuals and measured data. A number of assumptions have been made in the cavity flow calculations. Negligible inertial and compression effects are assumed. The use of absolute pressure gradients ignores air mixing within the cavity and assumes that all airflow completely clears the cavity. Davidovic et al. (2012) published a review of Straube's (1998) model observing that transition losses would exist between the vents and the cavity and would therefore resist the airflow. Using these simplifications should result in conservative (high) cavity velocities in this study. The model was found by Basset and McNeil (2009) to correlate well with cladding ventilation over the course of a day but not for shorter increments.

Analyses using this simple model have been replicated both for brick with limited ventilation consisting of a 25 mm (1 in.) cavity with 50% cavity blockage connecting 10 by 60 mm (3/8 by 2 1/2 in.) top and bottom vents and with good ventilation consisting of a 25 mm (1 in.) clear cavity with similar bottom vents and open top slot. A smooth panel over a 50 mm (2 in.) cavity with 12 mm (1/2 in.) slot opening has also been analyzed. The results are shown in Figure 8. It is noted that the model for the smooth panel is dominated by the pressure drop at the slot openings and hence is relatively independent of cavity height.

The minimally ventilated brick cavity (common in North America) is predicted to have low air speeds in most exposures. The cavity air velocity for the other claddings is tabulated for the same five cold cities in Table 7. The well-

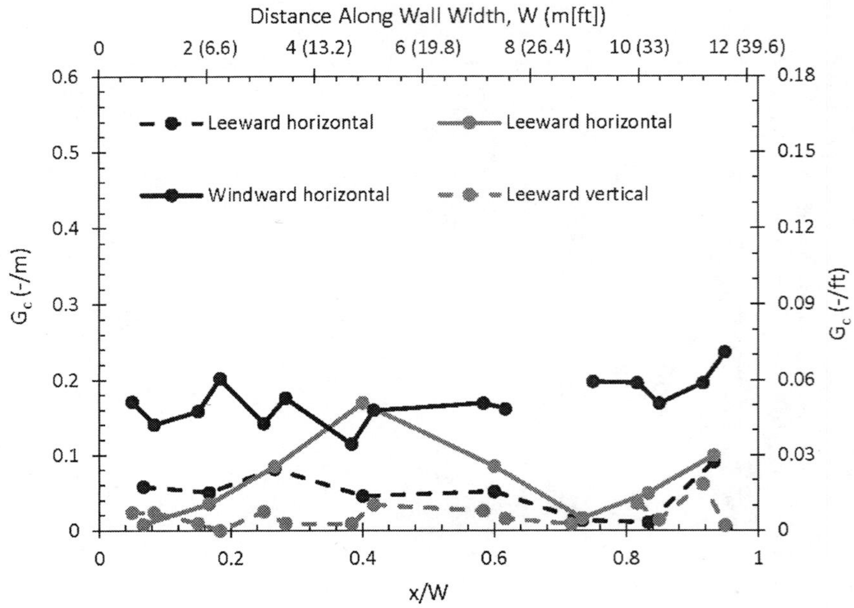

Figure 6 *Mean vinyl siding cavity pressure gradients along width of wall for perpendicular wind.*

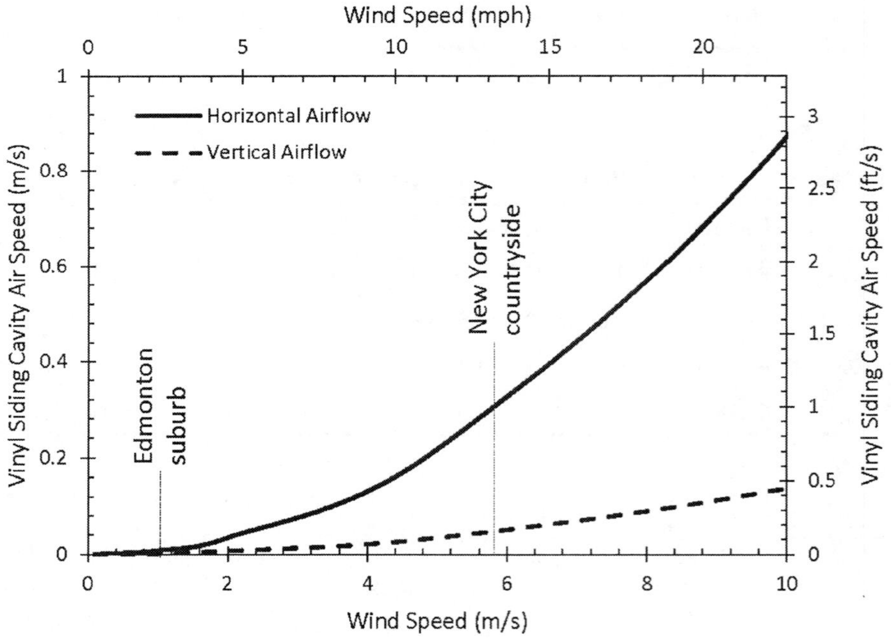

Figure 7 *Vinyl siding cavity airflow predicted from measurement of cavity pressure gradients.*

ventilated brick cases range up to 0.11 m/s (0.36 ft/s) cavity speed. The smooth panel cases with horizontal or vertical slots range up to 0.74 m/s (2.4 ft/s). In summary, the flow behind many types of cladding, when used on tall and exposed buildings or on a small suburban house, will fall with the lower end in the range of 0.1 to 1.0 m/s (0.33 to 3.3 ft/s). This range of cavity air velocities was used for the results presented in the next section.

EXPERIMENTALLY MEASURED WIND WASHING HEAT LOSS

Yarbrough and Tooe (1983) constructed a test rig to measure the impact of airflows over 152 mm (6 in.) of loose fill insulation. The tests included mineral wool with densities between 25 and 34 kg/m^3 (1.6 and 2.2 pcf). The results showed an increase in wind-washing-induced heat loss with decreas-

Table 5. Predicted Cavity Air Speeds for Vinyl Siding, m/s (ft/s)

Location	Small House in Countryside		Small House in Suburbs	
	Horizontal	Vertical	Horizontal	Vertical
Toronto	0.13 (0.43)	0.02 (0.07)	0.09 (0.30)	0.01 (0.03)
Chicago	0.13 (0.43)	0.02 (0.07)	0.09 (0.30)	0.02 (0.07)
New York City	0.30 (0.98)	0.05 (0.16)	0.21 (0.69)	0.03 (0.10)
Edmonton	0.02 (0.07)	0.004 (0.01)	0.02 (0.07)	0.003 (0.01)
Denver	0.07 (0.23)	0.01 (0.03)	0.05 (0.16)	0.01 (0.03)

Table 6. Measured Cavity Velocities Found in Literature, m/s (ft/s)

Reference	Cladding and Exposure	Mean Wind Speed	Mean Cavity Velocity
Schwartz (1973)	Open slots to 40 mm (1.5 in.) cavity behind smooth panels in 18-story building	0–8 m/s (0–26 ft/s)	0.2–0.6 m/s (0.7–2 ft/s)
Künzel and Mayer (1983)	Open slots to 20 mm (3/4 in.) cavity behind smooth panels in three-story building	3 m/s (10 ft/s)	0.06–0.16 m/s (0.2–0.5 ft/s)
Gudum (2003)	Top and bottom slot to 25 mm (1 in.) clear cavity in small single-story test building	0.7–2.1 m/s (2–7 ft/s)	0.12–0.27 m/s (0.4–0.9 ft/s)
Falk and Sandin (2013)	Open slots to 25 mm (1 in.) cavity behind smooth panels in single-story test building	2 m/s (7 ft/s)	0.05m/s (0.2 ft/s)

ing density. Extrapolating the results for densities of mineral wool insulation boards of 70 kg/m^3 (4.4 pcf) and greater would suggest negligible wind washing impacts at cavity airflows of 1 m/s (3 ft/s).

An apparatus was developed to measure the impact of cavity airflow on heat flow for a number of mineral wool insulation products as shown in Figure 9. The test approach and apparatus was similar to that of Yarbrough and Tooe (1983) but included several improvements, as follows:

- The meter section was surrounded with guard heaters maintained within ±0.02°C (±0.05°F) of the plate temperature to limit flanking losses.
- A thicker 6.4 mm (1/4 in.) aluminum plate was used for the meter section to encourage more even temperature distribution.
- A custom electric heating system using Omega-N160 nickel chromium heating wire was installed at to the bottom of the meter section's aluminum plate with fine spacing to further ensure temperature uniformity.
- A long 1625 mm (64 in.) guard length was installed upstream of the meter area to allow airflow disturbances and uneven air velocities at the air cavity entrance to dissipate before reaching the meter section (see Figure 9).
- For simplicity, a thick layer of insulation was used in lieu of a lower guard heater to limit heat loss in this direction.

To calculate the thermal conductance, U, of the test sample, the Fourier heat transfer equation is used:

$$U = [A \cdot (T_m - T_s)]/q \qquad (5)$$

where

A = sample area
T_m = temperature of the meter plate below the sample
T_s = temperature of the top surface of the sample
q = heat flux

Multiple surface temperatures and the electrical heating system voltage and current were recorded at 5 min intervals with a Campbell Scientific CR1000 data acquisition system. The meter plate and the four side guard plates were constantly maintained at 60°C±0.02°C (140°F±0.05°F) during testing using a relay that was controlled by the CR1000. The heat flux through the sample was calculated as the electrical power provided to the meter plate minus flanking losses. The flanking losses were through the passive lower guard and were determined by a calibration procedure. Airflow velocity was measured by setting a Dwyer 641RM-12-LED air velocity transmitter with an accuracy of ±0.15 m/s (±0.5 ft/s) in the middle air cavity depth over the center of the meter plate. To factor out the effects of sample thermal resistance temperature dependence, a correction factor was applied to the airflow R-values based on Lasercomp FOX 314 heat flow

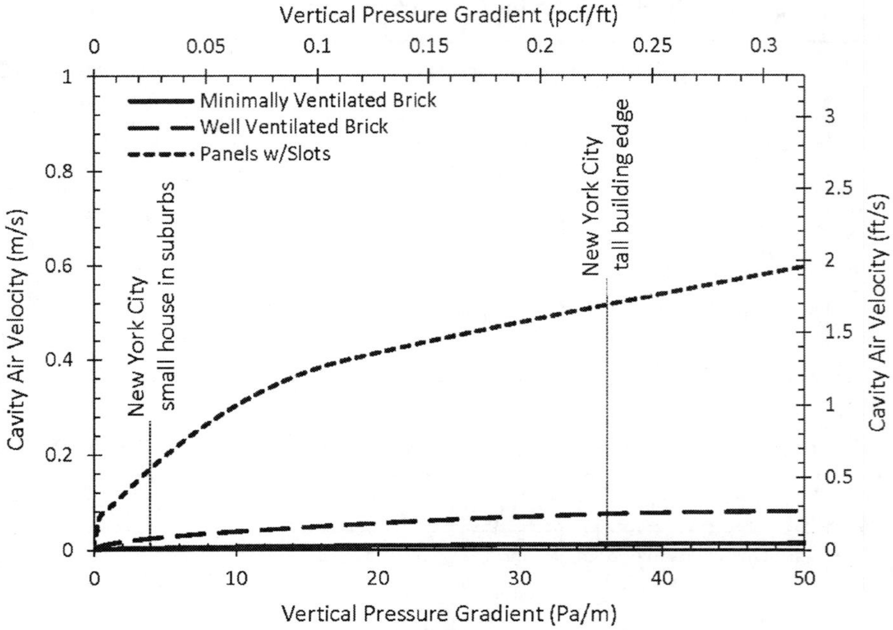

Figure 8 *Cavity air velocities for simple cladding systems.*

Table 7. Calculated Air Velocities behind Cladding, m/s (ft/s)

Location	Tall Building in City Centre			Small House in Countryside		Small House in Suburbs	
	Midspan	1–2 m	1 m Edge	Low	High	Low	High
			Well-Ventilated Brick (Vertical Flow)				
Toronto	0.03 (0.10)	0.05 (0.17)	0.07 (0.24)	0.01 (0.04)	0.03 (0.09)	0.01 (0.04)	0.02 (0.08)
Chicago	0.03 (0.10)	0.05 (0.17)	0.07 (0.24)	0.01 (0.05)	0.03 (0.10)	0.01 (0.04)	0.02 (0.08)
New York City	0.05 (0.15)	0.08 (0.26)	0.11 (0.37)	0.02 (0.07)	0.04 (0.15)	0.02 (0.06)	0.04 (0.12)
Edmonton	0.01 (0.04)	0.02 (0.07)	0.03 (0.10)	0.01 (0.02)	0.01 (0.04)	0.005 (0.02)	0.01 (0.03)
Denver	0.02 (0.07)	0.04 (0.12)	0.05 (0.17)	0.01 (0.03)	0.02 (0.07)	0.01 (0.03)	0.02 (0.06)
			Horizontal Slots—1 m (3.3 ft) Vent Separation				
Toronto	0.16 (0.52)	0.28 (0.92)	0.47 (1.54)	0.12 (0.39)	0.20 (0.65)	0.10 (0.32)	0.16 (0.54)
Chicago	0.16 (053)	0.29 (0.94)	0.48 (1.57)	0.12 (0.39)	0.20 (0.66)	0.10 (0.33)	0.17 (0.55)
New York City	0.25 (0.82)	0.44 (1.45)	0.74 (2.42)	0.18 (0.61)	0.31 (1.01)	0.15 (0.50)	0.26 (0.84)
Edmonton	0.07 (0.22)	0.12 (0.39)	0.20 (0.65)	0.05 (0.16)	0.08 (0.27)	0.04 (0.14)	0.07 (0.23)
Denver	0.11 (0.37)	0.20 (0.66)	0.34 (1.11)	0.08 (0.28)	0.14 (0.46)	0.07 (0.23)	0.12 (0.39)
			Vertical Slots—1 m (3.3 ft) Vent Separation				
Toronto	0.16 (0.52)	0.28 (0.92)	0.40 (1.33)	0.07 (0.23)	0.15 (0.51)	0.06 (0.19)	0.13 (0.42)
Chicago	0.16 (0.53)	0.29 (0.94)	0.41 (1.36)	0.07 (0.24)	0.16 (0.52)	0.06 (0.20)	0.13 (0.43)
New York City	0.25 (0.82)	0.44 (1.45)	0.64 (2.08)	0.11 (0.36)	0.24 (0.80)	0.09 (0.30)	0.20 (0.66)
Edmonton	0.07 (0.22)	0.12 (0.39)	0.17 (0.56)	0.03 (0.10)	0.07 (0.21)	0.02 (0.08)	0.05 (0.18)
Denver	0.11 (0.37)	0.20 (0.66)	0.29 (0.95)	0.05 (0.17)	0.11 (0.36)	0.04 (0.14)	0.09 (0.30)

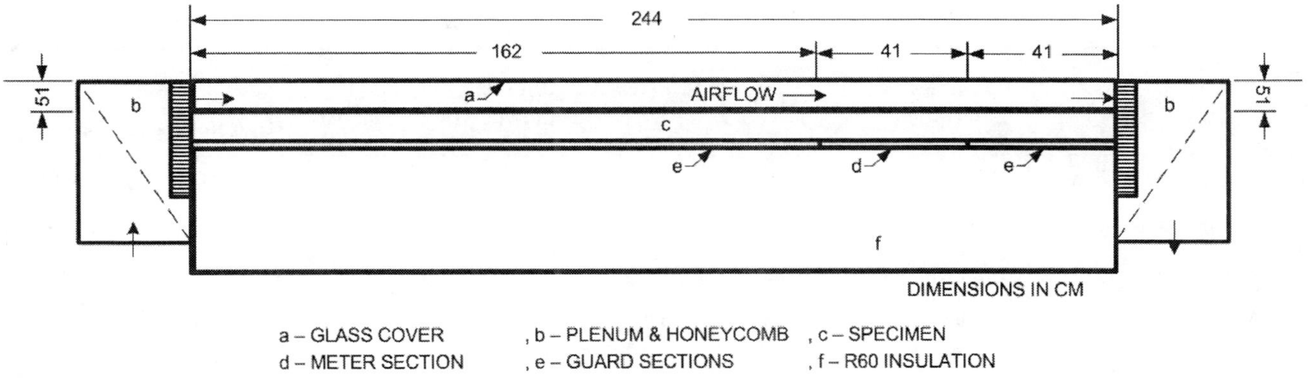

Figure 9 *Diagram of test apparatus.*

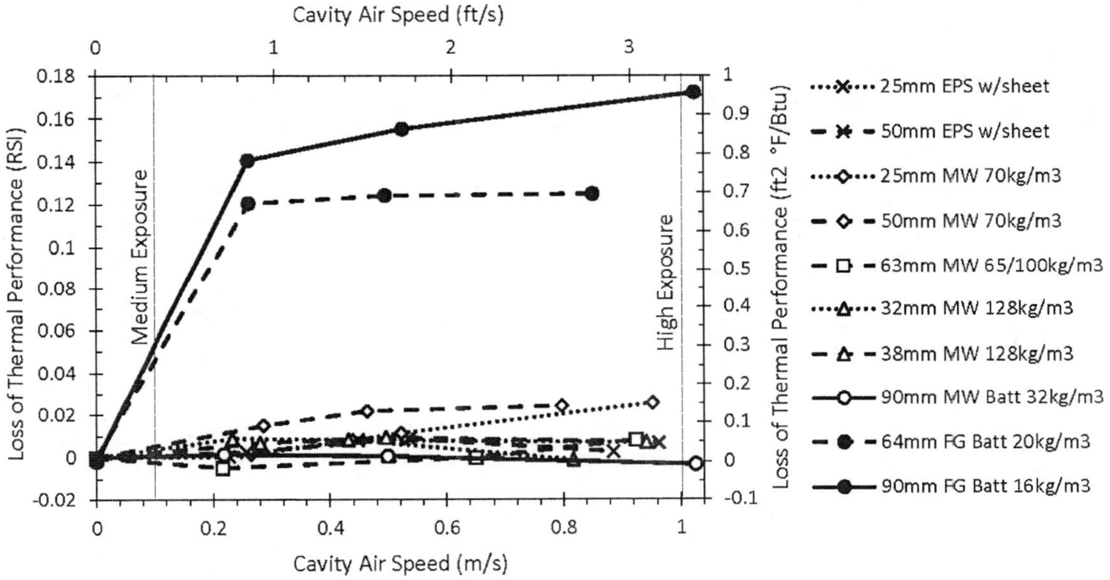

Figure 10 *Cavity air speed impact on absolute mineral wool board thermal performance.*

meter measurements. For a temperature difference of 20°C (68°F), the heat flow uncertainty was calculated to be ±0.02 W/m^{2}·°C (±10.0035 Btu/h·ft^{2}·°F). For insulations with RSI values from 0.65 to 2.47 (R-3.7 to R-14), this uncertainty is 0.02 to 0.08 RSI (R-0.1 to R-0.5).

The test results are plotted in Figure 10, showing measured loss of thermal performance for a range of cavity air speeds. The results for the various insulation materials tested are further tabulated in SI units in Table 8 and in I-P units in Table 9. The tables include rated, ASTM C518 (ASTM 2015) sample test results, test results from the wind washing test rig but with no applied airflow (still air), and estimated loss of thermal performance based on test results for 0.1 and 1 m/s (0.3 and 3 ft/s) cavity airflow. These air cavity speeds were selected as approximate medium and high wind exposures.

Plastic-film-faced expanded polystyrene (EPS) boards were used to verify the measurement process. These samples should not be affected by cavity airflow, as they have an air-impermeable facer and were installed with carefully taped joints. The maximum performance variation over the test range of air speeds was less than 0.01 RSI (R-0.05), and still air values were within 0.01 RSI (R-0.05) of ASTM C518 measurements.

Two fibreglass batt samples (low density and high air permeability) were used as products with known susceptibility to wind washing to confirm the ability of the apparatus to measure wind washing effects. The R-8 and R-13 rated fiberglass samples showed 0.12 and 0.17 RSI (R-0.7 and R-1) losses of thermal resistance, respectively, for 0.1 and 1 m/s (0.3 and 3 ft/s) cavity air speeds.

The 70 kg/m^{3} (4.4 pcf) samples were the only mineral wool insulation boards tested showing measurable wind washing impact. The maximum 0.03 RSI (R-0.2) impact is within the range of estimated test uncertainty, which is assumed to

Table 8. Summary of Wind Washing Testing Results (SI)

Material	Product Description	Thickness	RSI (20°C) Rated	RSI (40°C) ASTM C518	RSI (45°C) Still Air	Loss of RSI 0.1 m/s	Loss of RSI 1 m/s
Plastic-faced foamboard	Type 1 EPS	25 mm	0.65	0.61	0.60	<0.01	<0.01
Plastic-faced foamboard	Type 1 EPS	50 mm	1.30	1.12	1.13	<0.01	<0.01
Mineral wool board	70 kg/m^3	25 mm	0.74	0.70	0.72	<0.01	0.03
Mineral wool board	70 kg/m^3	50 mm	1.11	1.35	1.41	0.01	0.03
Mineral wool board	128 kg/m^3	32 mm	1.89	1.64	1.65	<0.01	<0.01
Mineral wool board	128 kg/m^3	38 mm	0.90	0.92	0.93	<0.01	<0.01
Mineral wool board	65/100 kg/m^3	64 mm	1.08	1.04	1.06	<0.01	<0.01
Mineral wool batt	32 kg/m^3	90 mm	2.47	2.27	2.17	<0.01	<0.01
Fibreglass batt	R-8 rated	64 mm	1.41	1.12	1.29	0.12	0.13
Fibreglass batt	R-13 rated	90 mm	2.29	1.92	1.91	0.12	0.17

Table 9. Summary of Wind Washing Testing Results (I-P)

Material	Product Description	Thickness	R- (68°F) Rated	R- (104°F) ASTM C518	R- (113°F) Still Air	Loss of R-Value 0.3 ft/s	Loss of R-Value 3 ft/s
Plastic-faced foamboard	Type 1 EPS	1 in.	R-3.7	R-3.5	R-3.4	<0.05	<0.05
Plastic-faced foamboard	Type 1 EPS	2 in.	R-7.4	R-6.4	R-6.4	<0.05	<0.05
Mineral wool board	4.4 pcf	1 in.	R-4.2	R-4.0	R-4.1	<0.05	R-0.2
Mineral wool board	4.4 pcf	2 in.	R-6.3	R-7.7	R-8.0	R-0.05	R-0.2
Mineral wool board	8 pcf	1 3/8 in.	R-10.7	R-9.3	R-9.4	<0.05	<0.05
Mineral wool board	8 pcf	1 1/2 in.	R-5.1	R-5.2	R-5.3	<0.05	<0.05
Mineral wool board	4/6.2 pcf	2 1/2 in.	R-6.1	R-5.9	R-6.0	<0.05	<0.05
Mineral wool batt	2 pcf	3 1/2 in.	R-14.0	R-12.9	R-12.3	<0.05	<0.05
Fibreglass batt	R-8 rated	2 1/2 in.	R-8.0	R-6.4	R-7.3	R-0.7	R-0.7
Fibreglass batt	R-13 rated	3 1/2 in.	R-13.0	R-10.9	R-10.8	R-0.7	R-1.0

explain why the impacts on the 50 mm (2 in.) sample appear greater than those for the thinner 25 mm (1 in.) sample.

The test method was designed to provide a simple repeatable cavity flow scenario of eddy-free flow parallel to the insulation surface. It was also found during the study that installation care significantly affected the testing results for all samples. This observation suggests a significant impact of airflow paths through and beneath poorly installed insulated board insulations. It is noted that samples were installed with more care and against a flatter, more consistent surface than would be typical of field applications. Differences in flexibility and roughness of insulated sheathings could affect the vulnerability of such products to installation deficiencies. The effects of gaps between and behind samples is recommend for future studies.

It is assumed that openings are relatively small and that eddies within the wind are not carried into the cavity and the momentum of the wind is not impacting the material directly. Hypothetically, for large openings (such as slots larger than 50 mm [2 in.]) these phenomena could have significant scouring effects on fibrous insulations, increasing heat loss. These cases should be explored through further analysis and field measurements.

Finally, components causing partial blockages within the cavity may accelerate, redirect, and/or create eddy structures within the cavity airflow. It is assumed that such affects are limited to small, narrow areas, limiting heat loss and interior surface temperature impacts.

CONCLUSION

A methodology for predicting wind washing impacts has been presented in this paper. It is proposed to use average winter wind speeds or MCWS for wind washing analysis. The use of wind speed adjustments for sheltering and height-effect-based building code structural wind load calculation have also been proposed. Air speeds as high as 0.74 m/s (1.7 ft/s) were predicted using simple airflow network analysis for well-ventilated brick cladding and slot panels, respectively. Analysis of full-scale wind cavity tunnel pressure tap data for a small house was used to predict vinyl siding cavity air speeds as high as 0.3 m/s (0.7 ft/s).

Heat flow measurements were taken for a number of mineral wool board products. The 25 and 50 mm (1 and 2 in.) thick 70 kg/m^3 (4.4 pcf) mineral wool board samples tested showed wind washing impacts to reduce thermal performance as much as 0.03 RSI (R-0.2). These values are small and practically negligible for design considerations.

ACKNOWLEDGMENTS

We would like to thank ROXUL Inc. for supporting and providing valuable input to this research. We would also like to thank Anne Cope and Murray Morrison from the Institute for Business and Home Safety for providing time series data from their full-scale wind tunnel tests.

NOMENCLATURE

α	=	exponent based on surface roughness
γ	=	blockage factor for the cavity
ρ	=	density of air
A	=	sample area
C	=	loss coefficients for the entrance and exit
D	=	hydraulic diameter of the cavity
DB	=	dry-bulb temperature
G	=	pressure gradient coefficient
h	=	height
L	=	cavity length
MCDB	=	mean coincident dry-bulb temperature (with 0.04% and 0.1% wind speed)
MCWS	=	mean coincident wind speed (with 99.6% design dry-bulb temperature)
ΔP	=	pressure difference
q	=	heat flux through sample
T	=	temperature
U	=	adjusted wind speed or thermal conductance of sample
V	=	air cavity air speed
W	=	wall width
WS	=	wind speed
x	=	horizontal distance from edge of wall

Subscripts

c	=	cavity
en	=	entrance
ex	=	exit
h	=	hydraulic
H	=	horizontal
m	=	meter
S	=	surface
V	=	vertical

REFERENCES

ASHRAE. 2013. *ASHRAE Handbook—Fundamentals*. Atlanta: ASHRAE.

ASTM. 2015. ASTM C518-15, *Standard test method for steady-state thermal transmission properties by means of the heat flow meter apparatus*. West Conshohocken, PA: ASTM International.

Bankvall, C. 1978. Forced convection: Practical thermal conductivity in an insulated structure under the influence of workmanship and wind. In *Thermal transmission measurements of insulations* (ASTM STP660), pp. 409–25. West Conshohocken, PA: ASTM International.

Bassett, M., and S. McNeil. 2009. Ventilation measured in the wall cavities of high moisture risk buildings. *Journal of Building Physics* 32(4):291–303.

Berlad, A., Tutu, R., Jaung, and Y. Yeh. 1979. *Interim progress report on an investigation of energy transport in porous insulator systems, ORNL/SUB-7551/I.* Oak Ridge, TN: Oak Ridge National Laboratory.

Cope, A., J. Crandell, D. Johnston, V. Kochkin, Z. Liu, L. Stevig, and T. Reinhold. 2012. Wind loads on components of multi-layer wall systems with air-permeable exterior cladding. *Proceedings of the ATC-SEI Advances in Hurricane Engineering Conference.*

Davidovic, D., J. Piñon, E. Burnett, and J. Srebric. 2012. Analytical procedures for estimating airflow rates in ventilated, screened wall systems (VSWS). *Building and Environment* 47:126–37.

Deseyve, C., and T. Bednar. 2005. Increased thermal losses caused by ventilation through compact pitched roof constructions—In situ measurements. *Proceedings of the 7th Symposium on Building Physics in the Nordic Countries* 2:881–87.

Falk, J., and K. Sandin. 2013. Ventilated rainscreen cladding: Measurements of cavity air velocities, estimation of air exchange rates and evaluation of driving forces. *Building and Environment* 59:164–76.

Gudum, C. 2003. Moisture transport and convection in building envelopes—Ventilation in light weight outer walls. Doctoral thesis, Technical University of Denmark, Lyngby, Denmark.

Henning, G.N. 1983. Energy conservation with air infiltration barriers. In *Thermal insulation, materials, and systems for energy conservation in the '80s* (ASTM

STP789), pp. 551–58. West Conshohocken, PA: ASTM International.

Hens, H., A. Janssens, and W. Depraetere. 2001. Hygrothermal performance of masonry cavity walls with very low U-factor: A test house evaluation. *Proceedings of Thermal Performance of the Exterior Envelopes of Whole Buildings VIII International Conference.*

Inculet, D., D. Surry, and A. Davenport. 1997. Unsteady pressure gradients and their implications for pressure equalized rainscreens. *ICBEST'97 Conference Proceedings.*

Janssens, A., and H. Hens. 2007. Effects of wind on the transmission heat loss in duo-pitched insulated roofs: A field study. *Energy and Buildings* 39(9):1047–54.

Künzel, H., and E. Mayer. 1983. Untersuchung über die notwendige Hinterlüftung an Außenwandbekeidung aus großformatigen Bauteilen. Schriftenreihe Bundesminister für Raumordnung, Bauwesen, und Städtebau; 3.

Schwarz, B. 1973. Witterungsbeansphruchung von Hochhausfassaden, HLH Bd. 24, 12:376–84.

Silberstein, A., E. Auguis, and D. McCaa. 1991. Forced convection effects in fibrous thermal insulation. In *Insulation materials: Testing and applications*, vol. 2 (ASTM STP1116), pp. 292–309. West Conshohocken, PA: ASTM International.

Straube, J. 1998. Moisture control and enclosure wall systems. Doctoral thesis, University of Waterloo, Waterloo, Ontario, Canada.

Straube, J. 2001. Air flow control in building enclosures; More than just air barriers. *Proceeding of 8th Buildingg Science & Technology Conference*, pp. 282–302.

Tanner, C., and K. Ghazi. 1996. Warmebrucken von hinterlufteten Fassaden. EMPA Bericht 158 740, Dubendorf.

Taylor, B., and A. Philips. 1983. Thermal transmittance and conductance of roof construction incorporating fibrous insulation. In *Thermal insulation, materials, and systems for energy conservation in the '80s* (ASTM STP789), pp. 479–501. West Conshohocken, PA: ASTM International.

Uvsløkk, S. 1996. The importance of wind barriers for insulated timber frame constructions. *Journal of Building Physicss* 20(1):40–62.

Van Straaten, R. 2004. Ventilation of cladding systems. Master's thesis, University of Waterloo, Waterloo, Ontario, Canada.

Yarbrough, D., and L. Tooe. 1983. Effects of air movement on thermal resistance of loose-fill insulation. In *Thermal insulation, materials, and systems for energy conservation in the '80s* (ASTM STP789), pp. 529–41. West Conshohocken, PA: ASTM International.

Improving Polyiso Thermal Performance at Low Temperature

John B. Letts, PhD **Jennifer Yao, PhD** **Michael J. Hubbard**

ABSTRACT

Polyiso (polyisocyanurate) is an important insulation material that has successfully overcome environmental and performance challenges for over 30 years. Recently, concerns about its low mean temperature performance relative to a mean temperature of 75°F have been raised. R-Values at mean temperatures lower than 75°F, such as such as 40°F, have historically increased. However, polyiso blown with pentanes today have reduced R-value performance at 40°F and lower mean temperatures, contrary to expectations. This paper will address these concerns, review appropriate mean temperatures for North America, and offer several innovative solutions.

INTRODUCTION

Polyisocyanurate (polyiso), a closed-cell thermoset polmer with low conductivity gases, has a long history of high performance and addressing environmental and performance challenges. The industry switched blowing agents from trichloroflouromethane (CFC-11) to 1,1-Dichloro-1-flouroethane (HCFC-141b) and then to pentanes to address environmental concerns (ozone depletion). Pentanes have very low GWP numbers. Likewise, this industry is a leading advocate and user of long-term thermal resistivity (LTTR) that was developed in cooperation with industry partners including a national laboratory, which is a more realistic performance estimate over as-manufactured and six-month R-values. Recently, it has come to our attention that R-values measured at lower mean temperatures, such as 40°F, yield R-values lower than at the industry standard measured at a mean temperature of 75°F contrary to expectations (Building Science Corporation 2013; Graham 2010a, 2010b, 2014, 2015).

This paper will discuss the current situation, the appropriate mean temperature to determine the R-value for most of North America, and solutions to mean temperature R-values below 75°F.

A review of product standards in ASTM Book 04.06 (ASTM 2016) show the percent difference increase in between the R-value per inch at 75°F mean temperature and at a mean temperature (actually integrated over a temperature range of about 40°F from 20°F to 60°F) of 40°F is between 8.0% and 10.5% for expanded polystyrene (EPS), extruded polystyrene (XPS), and polyiso. For polyiso it is 10%. R-value will be used henceforth to mean R-value per inch ($ft^2 \cdot h \cdot °F/Btu \cdot in.$) As might be expected, there is variability in the percentage. Since the R-value of polyiso is not routinely measured at 40°F, a review of low mean temperature R-values along with historical data on older polyiso boards blown with CFC-11 and HCFC-141b is useful. Low mean temperature R-values of other insulations (EPS and XPS) was also measured.

One method to measure the low temperature R-value performance of insulation materials is to divide their R-value at a mean temperature of 40°F by their R-value at a mean temperature of 75°F, which yields a low mean temperature ratio (henceforth to be designated as the ratio). A ratio above 1.0 means that the R-value at 40°F is higher than the R-value at 75°F and vice versa. Or more specifically, the ratio of 1.05 means the 40°F R-value is 5% higher

John B. Letts is technical director, Insulations, **Jennifer Yao** is a scientist III, and **Michael J. Hubbard** is a research manager at Firestone Building Products, Inc., Indianapolis, Indiana.

than the R-value at a mean temperature of 75°F. Historically, polyiso R-value as a function of mean temperatures has increased at mean temperatures below 75°F to at least 40°F, or this ratio is greater than 1.0.

A review of published R-values at mean temperatures of 75°F and 40°F for polyiso expanded with CFC-11 or HCFC-141b gave a ratio percentage of 1.042 to 1.170, with an average of 1.084 for CFC-11 and 1.032 to 1.083 with an average of 1.058 for HCFC-141b (Singh 2015). Likewise, random recent XPS and EPS samples averaged 1.052 for XPS and 1.062 for EPS. These results are shown in Table 1.

It is known that as the mean temperature decreases, the R-value of most insulation increases. The two most important factors that influence the R-value as the mean temperature decreases is the radiation component and if any condensation occurs with the blowing agent. Condensation with blowing agents like isopentane and n-pentane typically will decrease the R-value. Keeping the isopentane/n-pentane from condensing will increase R-value at low mean temperatures.

Recent work in our lab and in our industry partner's (a major partner to the polyiso industry) lab has begun to determine the parameters that contribute to this issue, and separate solutions have been found. In addition, the polyiso industry has studied and begun the discussion on what is the appropriate mean temperature for R-value measurements for polyiso and other insulation types in North America.

EXPERIMENT

Foam was prepared three ways: one, in the laboratory using small amounts of material and the components mixed at low, ambient pressure with a modified drill press; two, at high

Table 1. R-Value Ratio Measured at 40°F and 75°F Mean Temperatures for Selected Foam Insulations

Insulation	Expanding Agent	R-Value at 75°F, $ft^2 \cdot h \cdot °F/Btu \cdot in.$ (m·K/W)	R-Value at 40°F, $ft^2 \cdot h \cdot °F/Btu \cdot in.$ (m·K/W)	Ratio
Polyiso	CFC-11	7.73 (53.6)	8.55 (59.3)	1.106
Polyiso	CFC-11	7.55 (52.3)	8.83 (61.2)	1.170
Polyiso	CFC-11	6.74 (46.7)	7.06 (48.9)	1.048
Polyiso	CFC-11	5.96 (41.3)	6.29 (43.6)	1.055
Polyiso	CFC-11	7.58 (52.5)	7.91 (54.8)	1.043
Polyiso	HCFC-141b	6.58 (45.6)	7.13 (49.4)	1.083
Polyiso	HCFC-141b	5.96 (41.3)	6.21 (43.0)	1.042
Polyiso	HCFC-141b	7.50 (1.08)	7.91 (54.8)	1.055
Polyiso	HCFC-141b	6.98 (48.4)	7.42 (51.4)	1.063
Polyiso	HCFC-141b	5.83 (40.4)	6.27 (43.5)	1.075
Polyiso	HCFC-141b	5.67 (39.3)	5.85 (40.6)	1.032
XPS	NA	4.94 (34.2)	5.19 (36.0)	1.051
XPS	NA	5.06 (35.1)	5.40 (37.4)	1.067
XPS	NA	4.76 (33.0)	4.98 (34.5)	1.047
XPS	NA	5.14 (35.6)	5.92 (41.0)	1.044
EPS	NA	3.95 (27.4)	4.24 (29.4)	1.072
EPS	NA	3.84 (26.6)	4.12 (28.6)	1.075
EPS	NA	3.83 (26.5)	4.01 (27.8)	1.046
EPS	NA	3.70 (25.6)	3.90 (27.0)	1.055

pressure (2000 psi) with modest amounts of material for a 42 by 13.25 by 4 in. mold; and three, on a high speed, high pressure (2000 psi) full commercial laminator.

In the small scale laboratory, approximately 325 grams of total material were mixed at 2000 rpm for 8 seconds and then deposited in a 165 oz paper container. Samples for R-Value, density, compressive strength, dimensional stability were obtained from this container. A generic formulation is shown in Table 2.

The scale-up at a high pressure apparatus was designed to produce foam at nearly the same processing conditions seen in full-scale production, except smaller amounts. A movable mold with the reinforced paper facers was utilized to mimic actual laydown conditions seen in commercial lines. Below are the parameters of this high-pressure machine:

- Machine: Cannon A-40 high pressure foaming machine
- Mold size: 42 × 13.25 × 4 in.
- Chemical throughput: 150 g/s
- Typical shot size: 1000 g targeting a 1.55 pcf free-rise density in the formulation
- Mix pressure: 2000 psi
- Chemical temperature: 75°F–80°F
- Mold temperature: 140°F
- Demold time: 8 minutes

A commercial laminator that injects the two streams at 2000 psi at high speed was used to prepare the third set of samples.

APPROPRIATE MEAN TEMPERATURES

Historically, R-values have been measured at a mean temperature of 75°F, typically with at least a 40°F temperature difference between the high and low temperature plates. ASTM standards for insulations all have mandatory R-values at a mean temperature of 75°F, as does the Federal Trade Commission (FTC). The FTC mandated data be reported at 75°F, because it realized that testing at different temperatures would yield different results and by mandating a test temperature they could eliminate some gamesmanship with R-value reporting. But why report at a mean temperature of 75°F? If the summer and winter data for all climate zone is averaged, it yields 63°F. Then, if solar radiation to the surface is added, the temperature increases further. So, a mean temperature of 75°F is not bad as a national average. Nonetheless, it seems that the most appropriate mean temperature to measure R-values in south Florida would be different from that for northern Minnesota. The NRCA (National Roofing Contractors Association) and other have noted that R-values measured at mean temperatures lower than 75°F, such as 40°F, are lower than expected.

Table 2. Standard Laboratory Formulation

Component	Parts by Weight
Polyol	100
Flame retardant	10
Catalyst 1	2.7
Catalyst 2	0.3
Surfactant	2
Water	0.25
Iso/n-pentane	14.0–25.5
Co-blowing agent	2.5–10.0
Polymeric MDI	185.9

However, what is the appropriate mean temperature for buildings from Duluth, MN to Miami, FL? In a report put together by Dr. James Hoff of Tegnos Research for Polyisocyanurate Insulation Manufacturers Association (PIMA) titled “Thermal Resistance and Mean Temperature: A Report for Building Design Professionals,” which can be found on the PIMA website (www.polyiso.org) (Hoff 2016), Hoff made several points: one, mean reference temperature is not the same as average outdoor temperatures; two, the average mean reference temperature data varies in the winter months from 45°F to 70°F when going from Climate Zone 7 to Climate Zone 1, respectively, and in the summer months from 66°F to 76°F again from Climate Zone 7 to Climate Zone 1, which covers the vast majority of the United States; three, the R-value data from the 2010 NRCA study were close to PIMA member-published R-values. Figures 1 and 2 show the mean reference temperatures (calculated by averaging the actual mean outdoor ambient temperatures for a particular climate zone and the intended indoor design temperature) for Zones 1 through 7 in the winter and summer months, respectively. The mean temperatures in these figures are a good approximation, but, as noted above, there are other things that can influence them, such as the location of the insulation material in the roof/wall and solar/sky radiation. Exterior surface temperatures can be greater than the ambient air temperature on a sunny day and less than ambient air temperature at night.

In this report, Hoff took the indoor temperature—set at 68°F—and averaged it with the monthly average outdoor temperature for cities ranging from Duluth, MN to Miami, FL. Based on this work, there are not many situations where measuring the mean temperature less than 40°F needs to be considered.

Calculations from the report on three cities (Duluth, MN; Indianapolis, IN; and Baltimore, MD) from temperature Zones, 7, 5, and 4, respectively, show the raw numbers (Table 3).

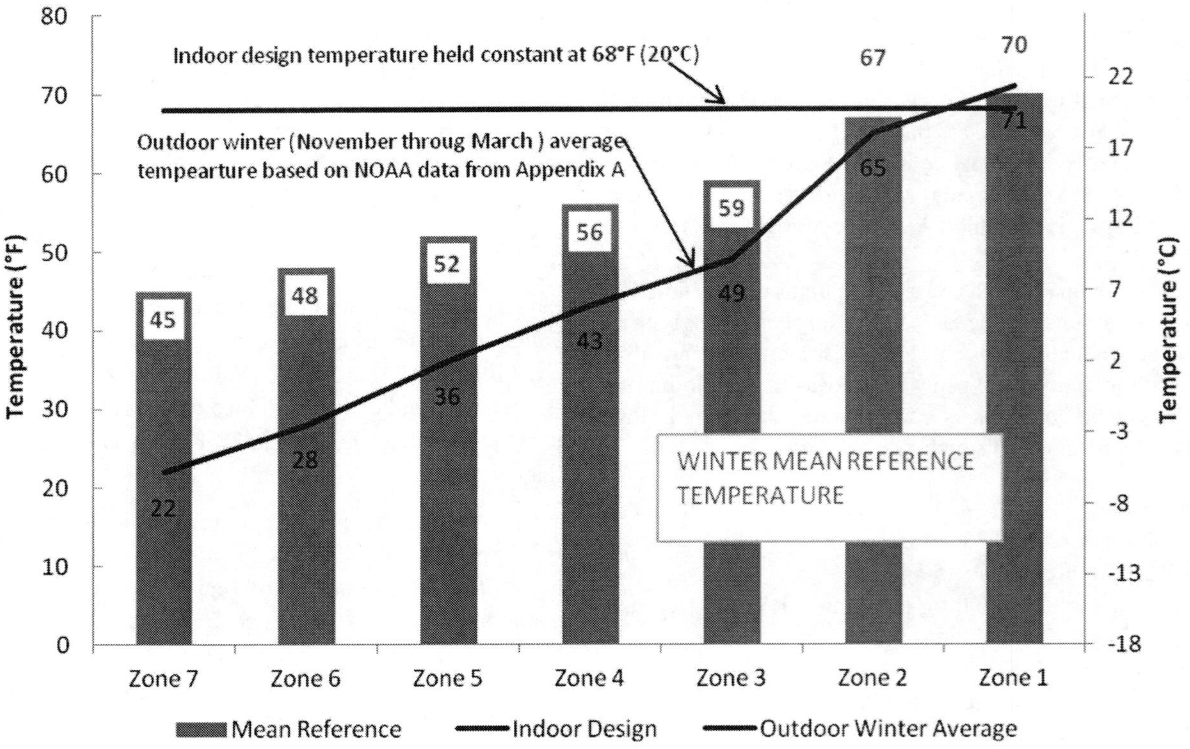

Figure 1 *Mean reference temperatures by Climate Zone for winter conditions.*

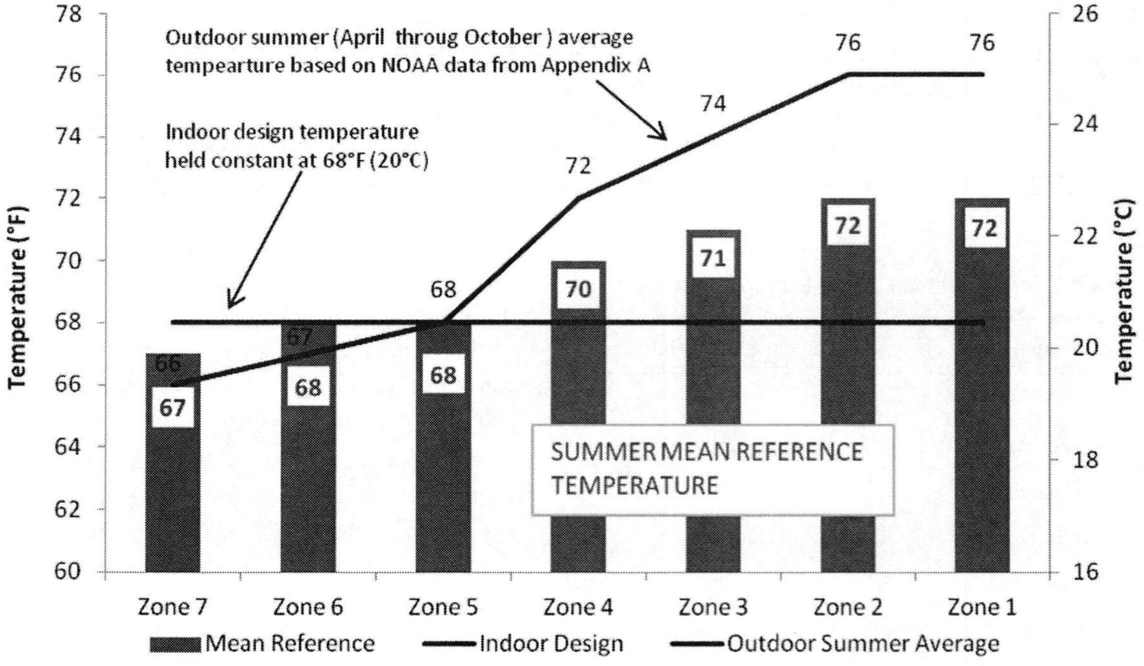

Figure 2 *Mean reference temperatures by Climate Zone for summer conditions.*

Table 3. Mean Outdoor and Reference Temperatures for Selected North American Cities

Duluth, MN, °F (°C)
ASHRAE Zone 7 (Winter Mean Reference = 45°F, Summer Mean Reference = 66°F)

	Jan	Feb	Mar	Apr	May	Jun	Jul	Aug	Sep	Oct	Nov	Dec	Winter Ave.	Summer Ave.
Mean outdoor ambient	11 (–12)	17 (–8.3)	27 (–2.7)	39 (3.8)	48 (8.8)	58 (14)	66 (19)	66 (19)	58 (14)	46 (7.7)	31 (–0.56)	17 (–8.3)	21 (–6.0)	63 (17)
Inside design	68 (20)	68 (20)	68 (20)	68 (20)	68 (20)	68 (20)	68 (20)	68 (20)	68 (20)	68 (20)	68 (20)	68 (20)	68 (20)	68 (20)
Mean reference (midpoint)	40 (4.4)	43 (6.1)	48 (8.9)	54 (12)	58 (14)	63 (17)	67 (19)	67 (19)	68 (20)	57 (14)	50 (10)	43 (6.1)	45 (8.2)	66 (19)

Indianapolis, IN, °F (°C)
ASHRAE Zone 5 (Winter Mean Reference = 52°F, Summer Mean Reference = 68°F)

	Jan	Feb	Mar	Apr	May	Jun	Jul	Aug	Sep	Oct	Nov	Dec	Winter Ave.	Summer Ave.
Mean outdoor ambient	26 (3.3)	30 (–1.1)	40 (4.4)	51 (11)	62 (17)	71 (22)	75 (24)	73 (23)	66 (19)	54 (12)	42 (5.6)	31 (–5.6)	34 (1.1)	73 (23)
Inside design	68 (20)	68 (20)	68 (20)	68 (20)	68 (20)	68 (20)	68 (20)	68 (20)	68 (20)	68 (20)	68 (20)	68 (20)	68 (20)	68 (20)
Mean reference (midpoint)	47 (8.3)	49 (9.4)	54 (12)	60 (16)	65 (18)	70 (21)	72 (22)	71 (22)	67 (19)	61 (16)	55 (13)	50 (10)	51 (11)	71 (22)

Baltimore, MD, °F (°C)
ASHRAE Zone 4 (Winter Mean Reference = 54°F, Summer Mean Reference = 71°F)

	Jan	Feb	Mar	Apr	May	Jun	Jul	Aug	Sep	Oct	Nov	Dec	Winter Ave.	Summer Ave.
Mean outdoor ambient	35 (1.7)	36 (2.2)	46 (7.7)	56 (7.8)	66 (19)	76 (24)	80 (27)	78 (26)	71 (22)	60 (16)	50 (10)	39 (3.9)	41 (5.0)	78 (26)
Inside design	68 (20)	68 (20)	68 (20)	68 (20)	68 (20)	68 (20)	68 (20)	68 (20)	68 (20)	68 (20)	68 (20)	68 (20)	68 (20)	68 (20)
Mean reference (midpoint)	51 (11)	52 (11)	57 (14)	62 (17)	67 (19)	72 (22)	74 (23)	73 (23)	70 (21)	64 (18)	59 (15)	54 (12)	55 (13)	73 (23)

ENHANCE LOW MEAN TEMPERATURE R-VALUES RESULTS

In the course of investigating this low temperature phenomenon, our business partner has found a proprietary solution to this problem, which is designated as Solution A. The R-value ratio (40°F/75°F) is approximately 1.007 with a standard deviation of 2.2% based on extended plant trials.

In Firestone's studies, work was started with co-blowing agents. A list of molecular weights, boiling points, the total Hansen solubility parameters, R-values at mean temperature of 75°F and 40°F, and their ratios are given in Table 4. In all cases except the last row, 25 mole % of these co-blowing agents were used; the balance of the blowing agent (75 mole %) was a proprietary blend of isopentane/*n*-pentane. All these foams were made in the laboratory and their R-values measured soon after preparation.

A concern with present day polyiso blown with iso/n-pentane is condensation of the pentanes as the temperature decreases. Hence, addition of 15% by weight of either *n*-butane (boiling point 32°F) or isobutane (boiling point 10.9°F) to the pentane mixture in the laboratory was explored to determine if the ratio 40°F to 75°F could be increased above 1.00. The vapor pressure at 40°F, for example, should be higher than for the iso/*n*-pentane, especially for the isobutane blend. When we used the same formulation in the previous work except the iso/n-pentane was modified with n-butane or isobutane, the R-values at 75°F were similar, but the 40°F R-values were better than the control. However, the R-value at 40°F were still below the R-value at a mean temperature of 75°F. Their ratios were 0.967 and 0.976 for *n*-butane and isobutane, respectively (Table 5). Although they were better, they were still 3.4% and 2.2% lower than R-values at 75°F.

Most of the work done with these co-blowing agents has been done in the lab at atmospheric pressure and the R-values measured at different mean temperatures have been initial R-values or soon after preparation. Additional R-values were measured after a month of aging to determine if the ratio of

Table 4. Co-Blowing Agent Physical Properties Including R-Values at 40°F and 75°F Mean Temperatures and Their Ratios

Co-Blowing Agent				Main Blowing Agent	R-Value @ 75°F, ft²·h·°F/Btu·in. (m·K/W)	R-Value @ 40°F, ft²·h·°F/Btu·in. (m·K/W)	Ratio (R-Value @ 40°F/ R-Value @ 75°F)
25 Mole %	Molecular Weight	Boiling Point, °C (°F)	Total Hansen Parameters				
Acetone	58	56 (13)	20	Iso-/*n*-pentane	6.66 (46.2)	7.11 (49.3)	1.07
Methyl ethyl ketone	72	80 (27)	19	Iso-/*n*-pentane	6.59 (45.7)	7.03 (48.7)	1.07
Methyl formate	60	32 (0.0)	18.7	Iso-/*n*-pentane	6.57 (45.5)	6.94 (48.1)	1.06
Chloroform	119	61 (48)	19	Iso-/*n*-pentane	6.73 (46.7)	6.97 (48.3)	1.04
Methylal	76	42 (5.6)	17.5	Iso-/*n*-pentane	6.61 (45.8)	6.84 (47.4)	1.03
Hexane	86	69 (21)	14.9	Iso-/*n*-pentane	6.06 (42.0)	5.42 (37.6)	0.89
Toluene	92	111 (44)	18.2	Iso-/*n*-pentane	6.55 (45.4)	7.07 (49.0)	1.08
Ethyl acetate	88	77 (25)	18.1	Iso-/*n*-pentane	6.54 (45.3)	7.06 (48.9)	1.08
Tetrahdyrofuran (THF)	72	66 (19)	19.4	Iso-/*n*-pentane	6.61 (45.8)	7.02 (48.7)	1.06
None	NA	NA	NA	Iso-/*n*-pentane	6.39 (44.3)	5.62 (39.0)	0.88

Table 5. R-Value Ratio with Addition of 15 wt.% Low Boiling Hydrocarbons

	Standard	Isobutane	*n*-butane
Isopentane/*n*-pentane	24		
Iso/n-pentane + 15% isobutane		23	
Iso/n-pentane + 15% *n*-butane			24.4
Density, psi (kg/m³)	1.65 (26.4)	1.62 (25.9)	1.56 (25.0)
Compressive stren (X), psi (kPa)	31.1 (214)	30.2 (208)	28.3 (195)
Compressive stren (Y), psi (kPa)	14.3 (98.6)	14.4 (99.3)	13.0 (89.6)
R-Value, ft²·h·°F/Btu·in. (m·K/W)			
75°F	6.42 (44.5)	6.45 (44.7)	6.46 (44.8)
40°F	5.58 (38.7)	6.21 (43.0)	6.30 (43.7)
Ratio (40°F/75°F)	0.869	0.963	0.975

40°F to 75°F mean temperature R-values would change with time. While it is expected that the absolute R-values measurements at 40°F and 75°F will decrease, the ratio should remain approximately the same. Table 6 shows the R-value ratios for methyl formate initially and after a month of aging. The initial ratio was 1.056 and the ratio after one month remained essentially the same at 1.063.

Most of our work has been focused on R-value performance at mean temperatures of 75°F and 40°F and to a lesser extent between these two temperatures. The basis for this is ASTM standards for insulation materials are measured at a mean temperature of 75°F, and the nonmandatory sections R-values measured between 40°F and 75°F have a lot of merit. Based on all of this information, a study of R-value as a function mean temperature from 75°F to 25°F for one co-blowing agent, methyl formate, was conducted. Figure 3 is a graph of the initial and aged R-values for 25 mole % methyl formate and for the aged samples R-values measured between 75°F and 25°F. It is important to note the inflection point in Figure 3 is between 40°F and 50°F. The R-value at 75°F after 153 days was 5.96, at 40°F it was 6.34, and at 25°F it was 6.07.

Table 6. The Ratio (R-Value @ 40°F/R-Value @ 75°F) as a Function of Time for Methyl Formate

Mole% Co-Blowing Agent	25% Methyl Formate
Initial R-value, ft^2·h·°F/Btu·in. (m·K/W)	
75°F	6.57 (45.5)
40°F	6.94 (48.1)
Ratio (R-value @ 40°F/R-value @ 75°F)	1.06
Month aged R-value, ft^2·h·°F/Btu·in. (m·K/W)	
75°F	6.19 (42.9)
40°F	6.58 (45.6)
Ratio (R-value @ 40°F/R-value @ 75°F)	1.06

Investigation into this new technology was conducted mostly in the lab at room temperatures and with relatively small amounts of material. However, recent work with the co-blowing agent, methyl formate, used high pressure (2000 psi) injection of the polyol blend with the polymeric isocyanate and enough material made to quickly fill an approximate 4 by 14 by 4 in. mold. Table 7 outlines a number of experiments that were conducted with this equipment. We looked at two different polyols, different amounts of methyl formate, and with and without Solution A.

The ratio increases as the amount of methyl formate increases from 5.0% for 20 mole % methyl formate to 8.1% for 25 mole % methyl formate and 12.1% for 30 mole % methyl formate, albeit there was a change in polyols and two of the samples had Solution A.

DISCUSSION

As mentioned before, several organizations have reported that the polyiso of today (blown with pentanes) have R-values that are frequently—but not always—lower at 40°F mean temperature versus 75°F mean temperature. It has been noted that the R-value versus mean temperature curve is odd for polyiso, especially around mean temperatures of 40°F and lower. Whereas some condensation of the blowing agent would be expected since *n*-pentane's boiling point is 36°C and isopentane's boiling point is 28°C, HCFC-141b's boiling point is 32°C at 1 atm, which is right in between these two pentanes but it does not have this oddity in its R-value versus mean temperature curve.

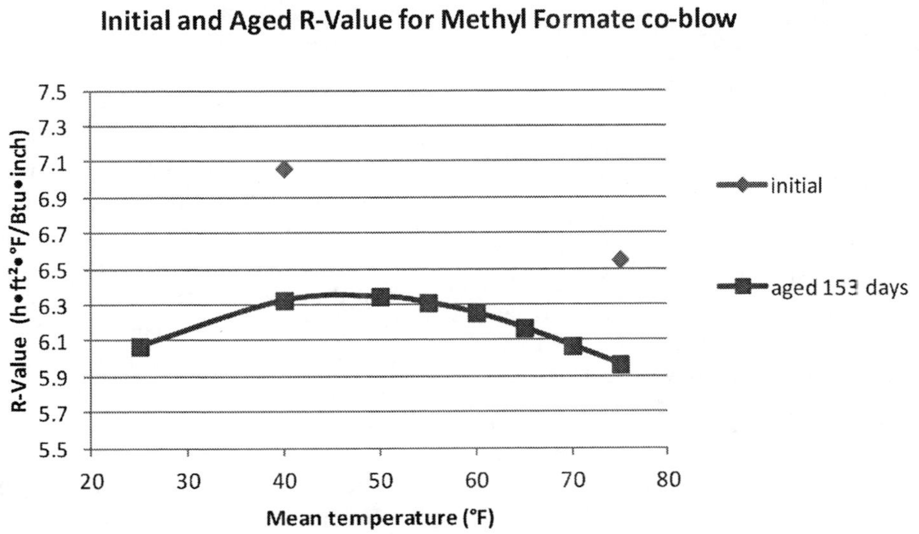

Figure 3 *Aged methyl formate co-blown R-values as a function of mean temperature.*

Table 7. High-Pressure Scale-Up Trials with Methyl Formate

Mole% Methyl Formate	25	20	30	0	0
Polyol 1	Yes	Yes			Yes
Polyol 2			Yes	Yes	
Solution A	Yes	No	Yes	Yes	Yes
R-value, $ft^2 \cdot h \cdot °F/Btu \cdot in.$ (m·K/W)					
75°F	6.61 (45.8)	6.57 (45.5)	6.44 (44.6)	6.55 (45.4)	6.64 (46.0)
40°F	7.15 (49.6)	6.90 (47.8)	7.22 (50.0)	6.55 (45.4)	6.58 (45.6)
Ratio	1.081	1.050	1.121	1.000	0.991
Percent change	8.1%	5.0%	12.1%	0.0%	–1.1%
Average compressive strength (X, Y, Z), psi (kPa)	15.9 (109)	15.6 (108)	21.8 (150)	22 (152)	17.5 (121)

Our research focused on not only finding solutions but why did this phenomenon exists. This oddity associated with R-values versus lower mean temperatures seems to be centered on high index polyiso with pentane blowing agents. The working theory we used is that pentanes are soluble in at least some domain(s) of the polymer. The concentration of the low conductivity gas is reduced, and, therefore, the R-value decreases. The partial pressure of high conductivity gases (nitrogen and oxygen) increases and this percentage may be augmented by gases trapped in the cell walls. It seems reasonable that as the pentanes condense as temperature decreases, the pentanes are being absorbed into the polymer network, which will lead to more pentanes condensing at the same temperature, ultimately decreasing the gaseous pentane concentration until equilibrium is attained.

During our examination of the low mean temperature R-value issue, we tried to think of the problem at the molecular polymer level. If the pentanes are absorbing into the polymer network in the struts and cell walls of the foam much like moisture condenses on glass on a high humidity day, and if these absorption sites or surface areas are less available for the pentanes to interact with the polymer domain, then the pentanes may not condense as readily. A variety of solvents have proved surprisingly very effective in increasing the R-value at a mean temperature of 40°F relative to the R-value at a mean temperature of 75°F. Most of these solvents have boiling points that are higher than the pentanes, but, still, their 40°F R-values are better than the control without them.

These solvents tie up the sites for either absorption or adsorption, because they are better solvents than the pentanes (isopentane and *n*-pentane), which free these pentanes to remain in the gaseous phase. Table 4 shows the results of a study where good solvents yield excellent 40°F results relative to R-values at 75°F.

It appears the solubility determined by the total Hansen solubility parameter needs to be about 17.0 and above. n-pentane has a total Hansen solubility parameter of 14.5. Hexane with a total Hansen solubility parameter of 14.9 did not yield good R-value results at a mean temperature of 40°F (relative to 75°F), as expected based on its similarity to *n*-pentane.

With one exception (methyl formate), all the co-blowing agents have higher boiling points than the isopentane/n-pentane mixture (and methyl formate is about the same) with corresponding lower vapor pressures. Although at least in some cases (for example acetone with a low thermal conductivity at 75°F) higher R-values at a mean temperature of 75°F might be anticipated, poorer results would be expected at a 40°F mean temperature due to its lower vapor pressure. However, this is not the case with acetone or the other co-blowing agents. If these co-blowing agents are soluble in the polymer network, then the isopentane and n-pentane is not as free to interact with the polymer network, and it remains in the gaseous phase where its can contribute to R-value at these lower temperatures.

There is not much of a drop-off past 40°F with this technology, as illustrated in Figure 3. The variability seen in other studies (Building Science Corporation 2013; Graham 2010a, 2010b, 2014, 2015) (and also in this study, especially for Solution A) when comparing R-Values at mean temperatures of 40°F versus 75°F, is probably related to the relatively steeper part of the R-Value versus mean temperature curve around 40°F. As the inflection point temperature for a particular formulation increases, as seen with pure pentane blends, the steep portion of the curve is more centered around 40°F. Hence, any small changes in the polymer composition could have an effect on the inflection point and therefore the ratio of R-values at mean temperatures of 40°F and 75°F.

In general, the results in our small scale laboratory have been verified in these large scale experiments (Table 7). Additionally, we confirmed that this new technology with co-blowing has a synergistic effect with Solution A. It appears that as the amount of co-blowing agent increases the ratio of R-values at 40°F relative to R-values at 75°F increases as well. The R-value at a mean temperature of 75°F is unaffected, although there may be a slight increase depending on the solvent used. Other work has also shown this effect.

In the last two columns of Table 7 we have demonstrated the efficacy of Solution A; R-value performance at a mean temperature of 40°F is nearly equal to that at a mean temperature of 75°F. It is interesting to note that even with 30 mole % methyl formate, the addition of Solution A with the polyol 2 yielded the highest ratio (12%), and the compressive strength was still very good. However, much more work needs to be done to optimize this technology.

CONCLUSION

Below are our conclusions for this paper:

- The surprisingly poor low mean temperature (40°F) R-value of polyiso produced with pentanes has been confirmed in our studies.
- Work done for PIMA by Dr. James Hoff demonstrates that based on actual U.S. and Canadian climate data, the most suitable mean winter heating reference temperatures for testing the R-value of polyiso and any other thermal insulation should be no less than 45°F for the coldest climates and up to 72°F for the warmest climates in North America.
- There is an R-value inflection point between 40°F and 75°F in this work that corresponds to the most useful mean temperatures for most of the United States.
- Solution A by a business partner shows R-values that are roughly equivalent at mean temperatures of 40°F and 75°F.
- A second solution has focused on the solubility of blowing agents and the use of co-blowing agents. Surprisingly, even high boiling point solvents perform well with pentanes to achieve excellent low temperature R-value performance.
- Our second solution has been shown to work with multiple polyols and multiple additives. There appears to be a positive synergistic effect when Solution A is combined with the second solution.

ACKNOWLEDGMENTS

The authors would like to thank Doug Toomey for assistance in the laboratory for many of these experiments. Additionally, we appreciate the assistance of Huntsman Polyurethanes for the high pressure work and Sachi Singh for many useful discussions.

REFERENCES

ASTM. 2015a. ASTM C578-15, *Standard specification for rigid, cellular polystyrene thermal insulation.* West Conshokocken, PA: ASTM International.

ASTM. 2015b. ASTM C1289-15, *Standard specification for faced rigid cellular polyisocyanurate thermal insulation board.* West Conshokocken, PA: ASTM International.

ASTM. 2016. *ASTM Volume 04.06, Thermal insulation; building and environmental acoustics.* West Conshokocken, PA: ASTM International.

BSC. 2013. BSC Information Sheet 502. Understanding the temperature dependence of R-Values for polyisocyanurate roof insulation. Westford, MA: Building Science Corporation. Avalaible: www.buildingscience.com

Graham. 2010a. R-value concerns. *Professional Roofing* 24.

Graham. 2010b. Revised R-values. *Professional Roofing* 20.

Graham. 2014. A discrepancy in polyiso's R-values. *Professional Roofing* 16.

Graham. 2015. Testing R-values. *Professional Roofing* 14.

Hoff, James L., PhD with TEGNOS Research, Inc, 2016. Thermal resistance and mean temperature: A report for building design professionals. Available: www.polyiso.org

Singh, Sachi. 2015. Email May 7, 2015.

Determination of Linear Thermal Transmittance of Vacuum Insulation Panels (VIPs) by Measurement in a Guarded Hot-Plate (GHP) Apparatus or a Heat-Flow Meter (HFM) Apparatus

Christoph Sprengard **Dr. Andreas H. Holm** **Christine Maderspacher**

ABSTRACT

The determination of the linear thermal transmittance, ψ, at the edges of vacuum insulation panels (VIPs) is usually carried out by numerical simulations according to ISO 10211 to determine the overall heat transfer through these high-performance insulating material. But it can be carried out by measurement in a guarded hot-plate apparatus (GHP) or in a heat-flow meter apparatus (HFM) as well, if additional measures take the nonuniform temperature distribution within the surfaces of the VIP assembly with joints into account. Therefore VIPs have to be assembled in a way that joints come to lay within the metering area. Temperature sensors have to be placed directly on the joints (area strongly influenced by thermal bridge), in the area slightly influenced by the edge effect, and in the undisturbed area (center of panel [COP]) as well. The width of the influenced areas depend strongly on the cross conduction within the barrier layers of the envelope and possible cover layers on the surfaces of the VIPs. Further influence comes from the heating and cooling plates of the measurement apparatus itself that tend to even out the temperature gradient at the edges of the specimens. The measured temperature differences have to be area weighted and averaged before calculating the equivalent thermal conductivity result for the VIP.

INTRODUCTION

Vacuum insulation panels (VIP) are insulating elements with very low thermal transmittance that are used for insulation in buildings or household appliances, like refrigerators. VIPs contain a pressure-resistant support core, commonly fumed silica or mineral wool, wrapped in a barrier foil, evacuated, and sealed (Simmler and Brunner 2005). VIPs show the lowest thermal conductivity of all insulating materials. Fresh values in the center of panel (COP) can be as low as 0.0018 W(m•K) for core materials based on mineral fibers (1/10th of fresh PU-foam) and 0.0030 W/(m•K) for fumed-silica cores (1/10th of polystyrene) (Sprengard and Spitzner 2011a). The barrier foils are either aluminum laminated films (ultra barrier laminates), metallized plastic films (high barrier laminates), or a combination of both (hybrid envelopes). The laminates differ in their permeability values for water vapor and air. As both laminate types make use of aluminum as inorganic barrier material (thermal conductivity aluminum $\lambda = 160$ W/m•K), the envelopes and sealing layers are an inevitable thermal bridge at the edges of the panels. The severity of the thermal bridging effect is strongly connected to the thickness and number of the aluminum layer(s) in the laminates, the edge design of the panels (single or multilayer edge design), the gap width between two panels installed next to each other, and the filling material of that joint (Sprengard and Spitzner 2011a, 2011b). Further influence comes from the use of cover layers on the panel surfaces mostly used to protect the panels during installation and use. Layers made of wood, massive plastics, rubber, metal sheets, and similar materials lead to an increase of the thermal bridging effects at the panel edges due to cross conduction in the cover layers towards the panel edge (Sprengard and Holm 2014). A slight decrease of thermal bridging effect can be achieved by use of insulating material cover layers such as PU-foam, polystyrene, or similar materials of comparatively low thermal conductivity. These materials function as additional thermal resistance and are able to suppress the cross conduction (Sprengard and Holm 2013).

Christoph Sprengard is a civil engineer and head of the Department of Research and Development, and **Christine Maderspacher** is research fellow in the Department of Research and Development at FIW, München, Germany. **Andreas H. Holm** is a professor of building physics at the University of Applied Sciences in Munich and managing director, at FIW, München, Germany.

Aluminum laminated films (ultra barrier properties) have aluminum layers from 6 to 12 μm, where metallized plastic films have two to three layers of evaporated aluminum on PET substrates in the thickness range of 30 nm to 100 nm each. The thermal bridging effect (linear thermal transmittance ψ) is therefore much higher for aluminum laminated films compared to metallized plastic films. The differences of single and overlapping edge designs for VIP with varying thicknesses are shown in Figure 1 (Sprengard and Holm 2014).

The mean pore size of the core material is the decisive property for the limit for the internal pressure of VIPs where gaseous thermal conduction is still eliminated. With larger mean pore size of the core material, the VIPs become sensitive to internal pressure increase. Therefore, the comparatively large pore sizes of mineral fiber cores (mostly micro glass fibers with a mean pore size of about 200 to 300 μm) allow only a very small pressure increase inside the panels. The internal pressure has to be kept below 1 to 2 mbar to ensure the suppression of the gaseous conduction. Derived from the projected service life for building applications of 25 to 50 years, the allowed permeation rate for air and water vapor has to be very low to ensure a pressure increase of less than 0.05 mbar/year. To date, this is only possible with aluminum laminated films, leading to large thermal bridging effects at the edges of the fiber-core VIPs. Additional precautions such as getters and drying agents inside the VIPs become necessary. Subsequently, almost all VIPs produced for building applications in Europe (projected service lifetime more than 25 years) have fumed silica cores allowing the use of multilayer metallized plastic films with much lower thermal bridging effects but still very good permeation rates for air and water vapor.

To determine the overall thermal performance of VIPs for building application, it is necessary to take the thermal bridging effects of the panel edges into account. The magnitude of the thermal bridging effect at the panel edge is decisive for the total performance of the panel. The total performance also depends on the perimeter to area ratio, where small panels perform significantly worse than big panels. Additional thermal bridging effects result from the application and from mounting and fixing (Quenard and Sallé 2005). These effects influence the thermal transmittance of constructions containing VIPs. To date, the scientific community distinguishes between various effects that have to be added to the panel's performance ("as placed on the market") and effects that result

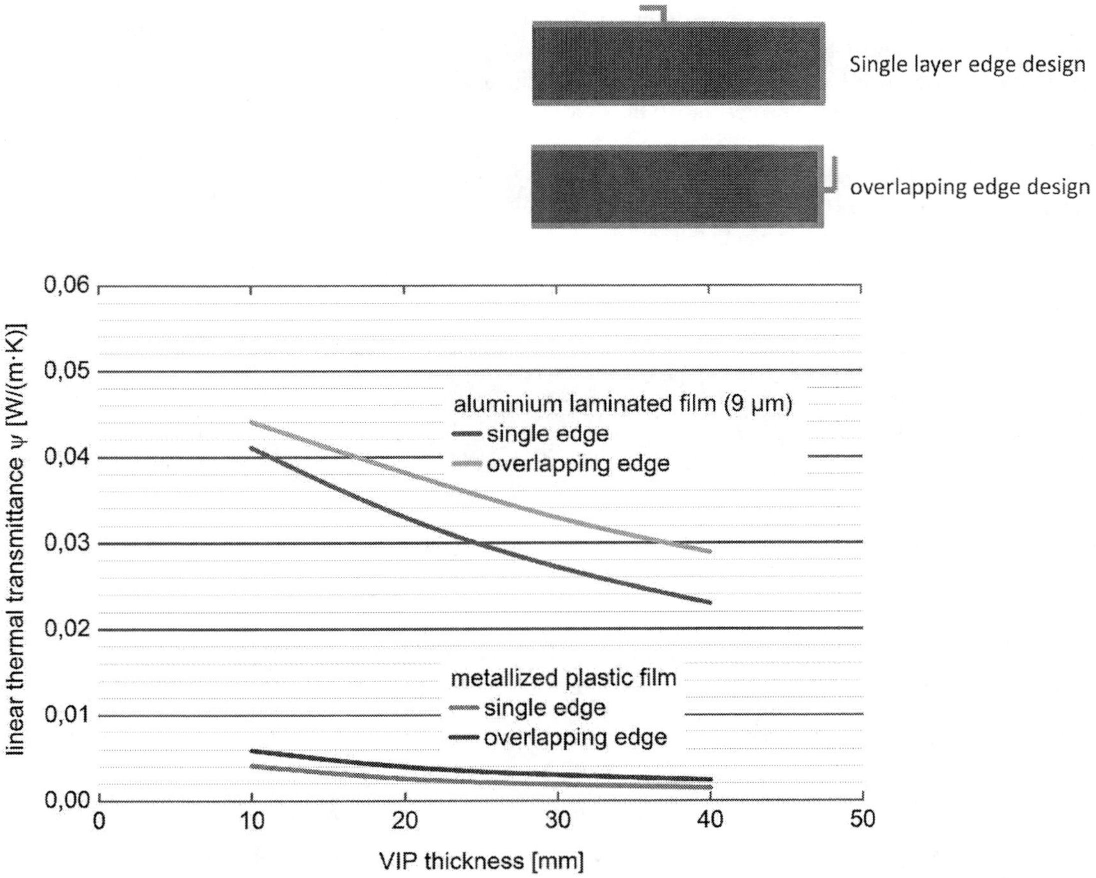

Figure 1 *Comparison of linear thermal transmittance for aluminum laminated films (9 μm Al) and metallized plastic films (3 × 60 nm Al) for two edge designs.*

from the application where the panel is used. Thermal bridging effects are usually calculated using numerical simulations according to ISO 10211, but they can be measured as well using guarded-hot-plate (GHP) or heat-flow-meter (HFM) technique on the joint of two panels (according to EN 12667 with adjustments).

APPROACH

This study compares measurements of the thermal bridging effect on the edges of panels with fumed silica core material and metallized plastic films in two panel thicknesses (20 mm and 50 mm) with effects on glass fiber-core VIPs with various envelope concepts. The fiber-core VIPs are enwrapped in: (a) aluminum laminated film, (b) metallized plastic film and, (c) a combination of both (hybrid solution, where one side of the foil bag is made from a metallized plastic film and the other side is made from an aluminum laminated film; see Figure 2). Table 1 gives an overview on the tested panels.

Heat Transfer and Background

Thermal performance calculations of VIP and constructions containing VIP make use of the U-factor (thermal transmittance) for a single element or component. Results depend on the size of the panel or on the size of the construction and the thermal bridging effects at the edges and the construction. For performance declarations at least, all two-dimensional effects have to be taken into account (Ghazi-Wakili et al. 2004).

The first step is to measure the thermal conductivity in the center of panel λ_{COP} in a GHP or HFM apparatus and calculate the thermal transmittance in the undisturbed area U_0. Then the thermal bridging effects for the edges of the panels have to be added by their linear thermal transmittance ψ and their length. If one wants to calculate the overall thermal performance of a construction, the 3D effects of mounting and fixing need to be taken into account as well by their point thermal transmittance χ and by their number n, especially if insulation layer penetrating mechanical fasteners are used. Also, three-dimensional effects cannot be ignored, as their effect easily exceeds the 3% criterion in ISO 6946.

The heat transfer coefficient is defined in ISO 13789 with following formula:

$$H_D = \sum_i A_i U_i + \sum_k l_k \psi_k + \sum_j \chi_j n_j \tag{1}$$

where

H_D = direct heat transfer coefficient between the heated or cooled space and the exterior through the building envelope, W/K

A_i = the area of element i of the building envelope, in m^2

U_i = the thermal transmittance of element i of the building envelope, $W/(m^2 \cdot K)$

l_k = the length of linear thermal bridge k, m

ψ_k = linear thermal transmittance of thermal bridge k, taken from tables or catalogues prepared in accordance with ISO 14683 or calculated according to ISO 10211, W/(m·K)

χ_j = point thermal transmittance of point thermal bridge j, calculated according to ISO 10211, W/K

When complementing this formaula with the temperature difference, the total heat flow (in W) through an element can be derived. The effective U-factor U_{eff} can then be derived from panel size and (hypothetic) temperature difference (this can be set to 1 K in Q_{eff} and U_{eff} calculations).

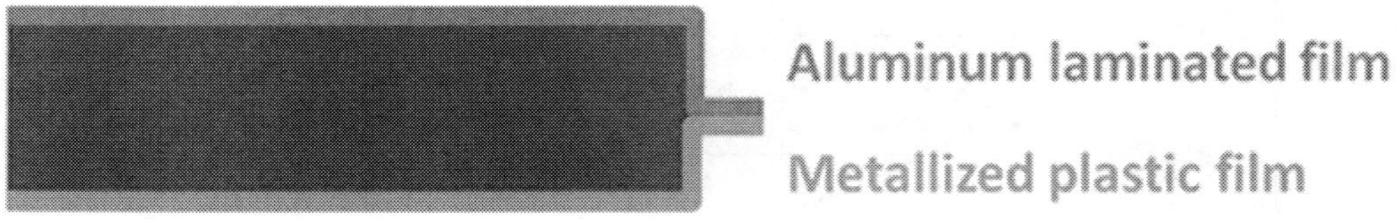

Figure 2 *Design of hybrid envelope.*

Table 1. Overview of Investigated Panels

Panel	Core	Barrier	Dimensions, cm
FIB-AL	Glass fiber	Aluminum laminated	80 × 80 × 1,5
FIB-ME	Glass fiber	Metallized film	80 × 80 × 1,5
FIB-HY	Glass fiber	Hybrid solution	80 × 80 × 1,5
SIL-ME.2	Fumed silica	Metallized film	60 × 60 × 2
SIL-ME.5	Fumed silica	Metallized film	60 × 60 × 5

$$Q_{eff} = \Delta\vartheta \cdot (U_0 \cdot A_{panel} + \Sigma\psi_i \cdot l_i + \Sigma\chi_i \cdot n_i) \quad (2)$$

$$U_{eff} = \frac{Q_{eff}}{A_{panel} \cdot \Delta\vartheta} \quad (3)$$

where

U_{eff} = effective U-factor of whole element, W/(m²•K)

Q_{eff} = total heat flux through element, W

A_{panel} = area of element, m²

$\Delta\vartheta$ = temperature difference, K

U_0 = U-factor of element without thermal bridges (center of element—one-dimensional), W/(m²•K)

ψ_i = linear thermal transmittance (e.g., for edges of panels), W/(m•K)

χ_i = point thermal transmittance (e.g., for mechanical fasteners), W/K

l, n = Length of the rims m; number of anchors, dimensionless

Conventional insulating materials are sold by their thermal resistance or their thermal conductivity stated in product data sheets or technical approvals. To be comparable with these materials, VIP manufacturers need to state a single thermal performance value for their products. Therefore, equivalent thermal resistance for the elements or equivalent thermal conductivity for the insulating material is derived from thermal transmittance and the thickness of the panels using various inputs. λ_{eff} can be derived from U_{eff} obtained by two- or three-dimensional numerical simulations of all thermal bridging effects, including additional calculations for various panel sizes. Direct determination is possible by measurement of U_{eff} in a hot box or by measuring R_{eff} in a GHP or HFM apparatus. The differences are obvious: when numerical simulations are used, all effects can be determined separately (e.g., different edge designs, gasket strips, or adhesive tapes). When using hot-box measurements, 2D and 3D effects can be considered but only in whole. It is not possible to determine single effects, but the performance can be measured directly. When using GHP or HFM measurements, the same limitations as for hot-box measurements apply. Additionally these measurements can only be used to measure two-dimensional effects.

$$\lambda_{eff} = \frac{d_{panel}}{\frac{1}{U_{eff}} - R_s} \quad (4)$$

or

$$\lambda_{eff} = \frac{d_{panel}}{R_{eff}} \quad (5)$$

where

λ_{eff} = equivalent thermal conductivity for a VIP including two- and three-dimensional thermal bridging effects, W/(m·K)

d_{panel} = thickness of the VIP, m

R_s = sum of internal and external thermal resistances at the surfaces of the panel, m²·K/W

R_{eff} = thermal resistance of the VIP including thermal bridges—taken from a measurement of a joint assembly, m²·K/W

If U_{eff} is determined by measurement of a joint assembly in a GHP or a HFM apparatus from R_{eff}, the cold-side and warm-side surface resistances are negligible. Then the term R_s in Equation 4 can be omitted (Ghazi-Wakili et al. 2011).

Determination of ψ by Measurement in HFM Apparatus

The determination of the linear thermal transmittance, ψ, can be carried out in a guarded heat-flow meter (HFM) apparatus, if additional measures take the nonuniform temperature distribution within the surfaces of the VIP assembly with joints into account. For measurement, two VIPs are assembled within the apparatus with joint arranged within the metering area (see Figure 3). Temperature sensors are placed directly on the joint (area strongly influenced by thermal bridge) in the area slightly influenced by the edge effect and in the undisturbed area (COP) as well.

The measured temperature differences on the different areas have to be area weighted and averaged before calculating the equivalent thermal conductivity result for the VIP. Figure 4 shows the defined influence areas for the weighting of the

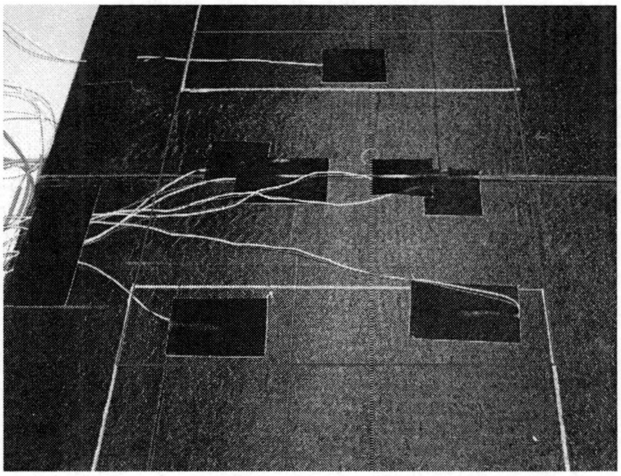

Figure 3 *Picture of a vacuum-insulation-panel joint assembly in between two VIPs with two thermocouples attached directly to the joint, two near the joint (the slightly influenced area), and three in the center of panel (COP). The white marks show the metering areas of earlier λ_{COP}-measurements. The red mark shows the metering area for the edge assembly.*

temperature differences and the positioning of the temperature sensors. The width of the influence areas is derived from the temperature gradient obtained from finite difference simulations of the edge area (Sprengard and Holm 2015). In addition to the barrier properties, the results are influenced by the same criteria that influence the linear thermal transmittance (e.g. joint filler material and its dimensions, edge design, etc.) The width for this study is shown in Table 2. The values are given separately for each panel. The following equation is used to calculate the area weighted temperature difference:

$$\Delta\theta_m = \frac{A_{COP} \cdot \Delta\theta_{COP} + A_{SA} \cdot \Delta\theta_{SA} + A_{joint} \cdot \Delta\theta_{joint}}{A_{COP} + A_{SA} + A_{joint}} \quad (6)$$

where

$\Delta\theta_m$ = area weighted temperature difference for joint assembly, K
A_{COP} = center of panel area, m^2
A_{SA} = area slightly affected, m^2
A_{joint} = joint area (strongly affected), m^2
$\Delta\theta_{COP}$ = temperature difference for COP area, K
$\Delta\theta_{SA}$ = temperature difference for SA area, K
$\Delta\theta_{joint}$ = temperature difference for joint area, K

With this area weighted temperature difference, the equivalent thermal resistance of the assembly, including thermal bridge $R_{eq\,ja}$ is derived from the GHP or HFM measurement according to the respective standards, considering steady state criteria and uncertainty. Equivalent thermal conductivity for the assembly, including 2D thermal bridging effect at the panel edges within the metering area $\lambda_{eq\,ja}$, can be calculated.

$$\lambda_{eq\,ja} = \frac{\Phi \cdot d_{panel}}{A \cdot \Delta\vartheta_m} \quad (7)$$

The linear thermal transmittance, ψ, can be determined by comparison of a GHP or HFM measurement in the center of a panel and the GHP or HFM measurement of the joint assembly by applying the following formula. The surface resistances can be omitted for this calculation of the GHP and HFM setups, as it is assumed that the heating and cooling plates are in perfect thermal contact with the specimens:

$$\psi = \frac{A}{d \cdot l_\psi} \cdot (\lambda_{eq\,ja} - \lambda_{COP}) \quad (8)$$

where

ψ = linear thermal transmittance for the joints in the metering area, W/(m·K)
A = metering area of the GHP or HFM apparatus used for the measurement, m^2
d = thickness of the specimens (equal thickness of joint assembly and COP specimens required for this method), m
l_ψ = length of the joints within the metering area, m
$\lambda_{eq\,ja}$ = equivalent thermal conductivity including edge effects for the specific joint assembly, W/(m·K)
λ_{COP} = thermal conductivity for center of panel, W/(m·K)

Table 2. Influence Width on Each Panel Side

Panel	Joint Area	Slightly Affected Area	Center of Panel
FIB-AL	20 mm	30 mm	200 mm
FIB-MET	2 mm	15 mm	233 mm
FIB-HY	20 mm AL/ 2 mm MET	30 mm AL/ 15 mm MET	200 mm AL/ 233 mm MET
SIL-MET.20	2 mm	11 mm	237 mm
SIL-MET.50	2 mm	13 mm	235 mm

EXPERIMENTAL PROCEDURE

A guarded heat-flow-meter apparatus is used for the measurement with calibrated heat-flow-meters on the warm and cold-side of the specimens. The heating and cooling plates are 800 × 800 mm, with a metering section of 500 × 500 mm. The measurements are carried out with temperatures of the cool and hot plate of 2,5°C/17,5°C (mean temperature 10°C) and 15,5°C/30,5°C (mean temperature 23°C). The temperature difference for both measurements is 15 K. The heat-flow meters are made of 8 mm thick black rubber. To ensure even contact between panels and hot/cold plates, a thin silicone layer was placed between the plates and the test specimen. The stop criterion of EN 12667 is applied for the measurement, but the measuring time is at least 24 h. All temperature sensors and thermopile signals are monitored for the complete duration of the measurement.

Special precautions are needed to prepare the joint assembly. Loose seams of multilayer edge design (Figure 5[b]) are fixed tightly with adhesive tape to the surfaces of the VIPs, ensuring that all foldings are towards the same side. The joint area is filled with a preferably thin precompressed gasket strip (vinyl rubber), not exceeding 5 mm (see Figure 5 for details). For this study, only multilayer edge designs are investigated. The measurement uncertainty for the λ_{COP} measurement is in between 3% and 4%, depending directly on the accuracy of the thickness measurement (e.g., fumed silica core VIPs have more parallel and even surfaces). The uncertainty for the $\lambda_{eq\,ja}$ measurement is around 5% to 8% (decisive properties: influence area determination and width of the joint), leading to an uncertainty of more than 8% for the determination of ψ. The given λ values are for the investigated joint of two panels. To calculate the effective thermal conductivity of a panel, the ψ value has to be divided by 2 before multiplying it with the full perimeter length of the panel.

RESULTS

The results for the thermal conductivity in the center of panel λ_{COP}, the equivalent thermal conductivity of the joint assembly $\lambda_{eq\ ja}$, and the resulting linear thermal transmittance ψ for the multilayered joint of two panels with precompressed gasket strip are shown in Table 3. For the measured ψ values, the effective thermal conductivity λ_{eff} of a panel of the size 1.0 × 0.5 m is calculated, and the increase in thermal conductivity including edge effects is stated in percent, based on the initial λ_{COP} values. Numerical simulations of the thermal bridging effect of the two fumed silica core VIPs SIL MET 20 and SIL MET 50 have been carried out to compare the results with the measurements.

DISCUSSION AND FURTHER RESEARCH NEEDS

Good agreement between numerical simulation and measurement were obtained for the 20 mm VIP with fumed silica core. The result for the 50 mm fumed silica VIP is significantly higher from measurement than ψ from numerical simulation, even with consideration of an increased uncertainty of 8%. These deviations can partly be traced to an increased λ_{COP} for one of the two adjacent VIPs in the measurement of the 50 mm VIPs joint assembly caused by a creeping leakage. An additional reason could be the larger width of the joint in the measurement compared to the calculation, caused by the higher force needed to compress the gasket strip of the 50 mm VIP compared to the 20 mm VIP.

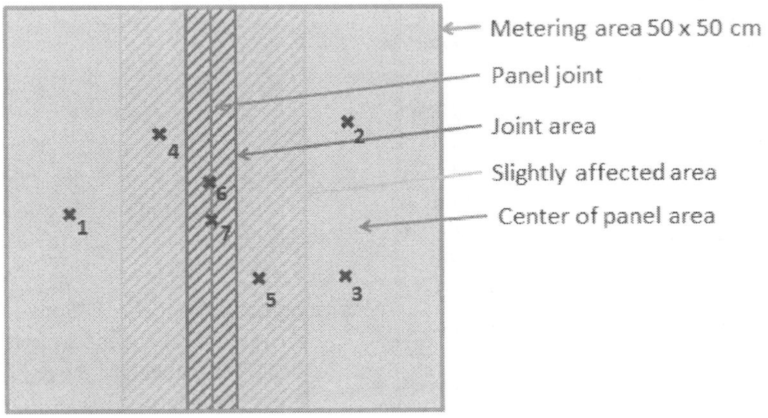

Figure 4 *Sketch of influence areas and positioning the thermocouples.*

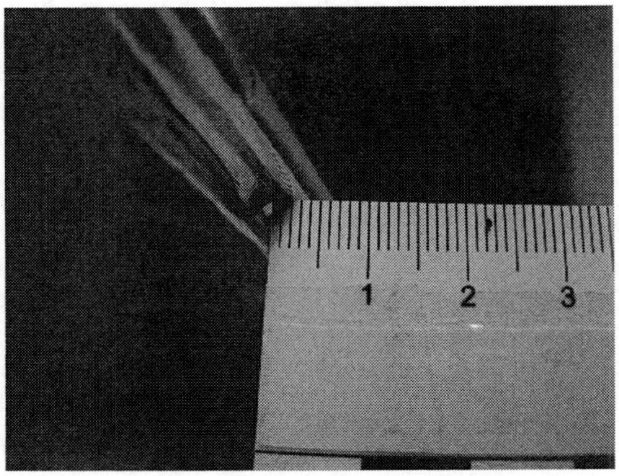

(a)

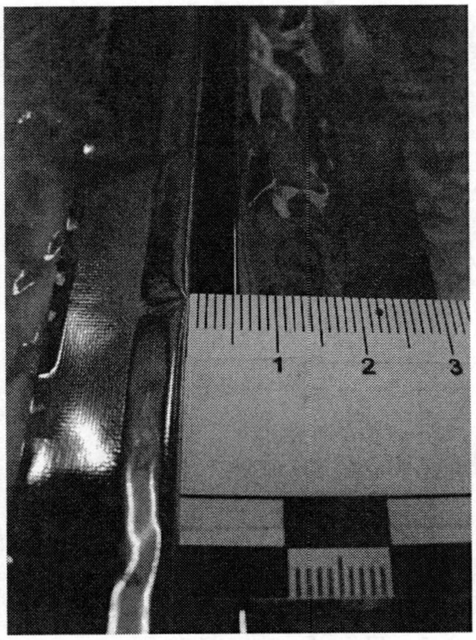

(b)

Figure 5 *(a) 3 to 4 mm gap on one panel edge, resulting from outstanding flanges of the seams. (b) Loose seams of both adjacent VIPs folded towards the same side. Joint filled with precompressed gasket strip (vinyl rubber).*

Table 3. Multilayer Edge Design—Results

	FIB-AL	FIB-MET	FIB-HY	SIL-MET.20	SIL-MET.50
λ_{COP}, W/(m•K)	0.0020 ± 0.00008	0.0022 ± 0.000088	0.0021 ± 0.000084	0.00435 ± 0.00014	0.0050 ± 0.00015
$\lambda_{eq\ ja}$, W/(m•K)	0.0056 ± 0.000448	0.0027 ± 0.000216	0.0032 ± 0.000256	0.00498 ± 0.00035	0.0062 ± 0.00043
$\psi_{measured}$, W/(m•K)	0.117 ± 0.00937	0.016 ± 0.00128	0.037 ± 0.00296	0.015 ± 0.0012	0.013 ± 0.00104
$\psi_{calculated}$, W/(m•K)	—	—	—	0.0144	0.0073
λ_{eff} panel size, 1.0 × 0.5 m, W/(m•K)	0.0079	0.0030	0.0039	0.0054	0.0070
λ increase from λ_{COP} to λ_{eff}, %	293	35	85	23	39

The strip decompressed during the measurement, leading to an increase of the joint dimension. Additionally, the 50 mm panels were made of two core slabs stacked on each other (20 and 30 mm), but not fully congruent, leading to small steps at the edges (approximately 2 mm), resulting in significantly wider joints and partly expanded compressed gasket strips.

The measurement of linear thermal transmittance ψ depends on a large variety of influencing factors that need to be considered for highest level of accuracy. Even small deviations in the measurement setup can lead to big uncertainties for the results. In order to use measurements for the determination of ψ in future, the definition of exact boundary conditions for comparable results is inevitable. More tests are needed in the future to find out about the practicability of measurements on VIP with other edge designs, other barrier materials, and for VIP with cover layers. More experiences will be gathered during the round-robin tests within the work for IEA Annex 65, Superinsulating Materials.

The first results show how demanding the measurement of the thermal bridging effect of VIPs in a HFM or GHP apparatus is. For the determination of the thermal performance of VIPs according to the product standard (currently under development), the recommendation is to use numerical simulations —preferably finite-difference method—instead of measurements, as it seems to be easier to define the boundary conditions.

REFERENCES

Ghazi-Wakili, K., R. Bundi, and B. Binder. 2004. Effective thermal conductivity of vacuum insulation panels. *Building Research & Information* 32(4):293.

Ghazi-Wakili, K., T. Stahl, and S. Brunner. 2011. Effective thermal conductivity of a staggered double layer of vacuum insulation panels. *Energy and Buildings* 43(6):1241.

Quenard, D., and H. Sallée. 2005. From VIP's to building facades: Three levels of thermal bridges, in: *Proceedings to the International Vacuum Insulation Symposium IVIS*.

Simmler, H., and S. Brunner. 2005. Vacuum insulation panels for building application—Basic properties, aging mechanisms and service life. *Energy and Buildings* 37(2005):1122–31.

Sprengard, C., and M.H. Spitzner. 2011a. Optimierung der energetischen Eigenschaften und der Wirtschaftlichkeit von VIP-Paneelen durch die optimale Kombination vonKieselsäure-, Mineralfaser- und EPS-Dämmstoff (Optimization of thermal and economic properties of VIP by the optimized combination of fumed silica, mineral fibers and expanded Polystyrene). Research Report, Fraunhofer IRB Verlag,Stuttgart, 143 S, 2011 (in German only—Abstract in English).

Sprengard, C., and M.H. Spitzner. 2011b. Investigations into ageing and thermal bridging in vacuum insulation panels (VIP) for construction applications. *Bauphysik* 33(4)(2011) (German Journal for Building Physics, in German only).

Sprengard, C., and A.H. Holm. 2013. Numerical examination of thermal bridging effects at the edges of vacuum-insulation-panels (VIP) in various constructions. *Proceedings of the 11th International Vacuum Insulation Symposium (IVIS)*, Zurich 2013, P. 7.

Sprengard, C., and A.H. Holm. 2014. Numerical examination of thermal bridging effects at the edges of vacuum-insulation-panels (VIP) in various constructions. *Energy and Buildings*. Elsevier (2014), http://dx.doi.org/10.1016/j.enbuild.2014.03.027.

Sprengard, C., and A.H. Holm. 2015. Determination of linear thermal transmittance of vacuum insulation panels by measurement in a guarded hot plate (GHP) or a heat-flow-meter (HFM) apparatus. *Proceedings of the 12th Vacuum Insulation Symposium*, Nanjing, 2015, P. 292–95.

Refurbishing Heritage and Historic Buildings—Key Motivation, Benefits and Challenges

Torben Valdbjørn Rasmussen, MSC (Civeng), PhD

ABSTRACT

Heritage and historic buildings are known for having very high energy consumption as well as an unacceptable indoor climate standard compared to modern requirements. Therefore, there is an intensified interest in refurbishing existing buildings including heritage and historic buildings. As many of the heritage and historic buildings are privately owned, refurbishment of these buildings is needed to ensure that they remain part of the attractive building stock. Otherwise, these buildings will turn into unused monuments or deteriorate, because there is no financial incentive to maintain them. Preserving recognized heritage values gives rise to challenges when refurbishing heritage buildings as well as many historic buildings. This might raise issues that can have a high impact on the refurbishing project of such buildings. From a number of case studies, the paper identifies key parameters that, in spite of recognized challenges, may motivate an owner, including end users, to undertake refurbishment of a heritage or historic building. Furthermore the paper describes recognized benefits as well as the challenges that need to be taken into account for future maintenance and use of a refurbished building.

INTRODUCTION

Over the past decade, indoor climate and energy consumption requirements of new buildings have gradually been tightened in many countries. Tightened requirements for the energy consumption of new buildings were initiated by a global priority to develop a climate-change policy that creates measures to mitigate climate-change effects. This process has been going on for more than a decade. On a global priority scale, efforts to mitigate the impacts of climate change have focused on reducing greenhouse gas emissions within the framework of the Kyoto Protocol (United Nations 1998), which has been implemented partly on an international level and partly through individual national initiatives. Efforts to mitigate climate change focus on reducing greenhouse gas emissions with emphasis on reducing the CO2 emissions. In many countries, buildings using energy for heating and comfort are one of the principal energy consumers and therefore there is a great interest in reducing buildings' energy consumption and in providing a CO2-neutral energy supply. Reducing the amount of energy used for heating and comfort provides an opportunity to produce less energy or to use the saved energy for other purposes. Older buildings, especially heritage and historic buildings, are categorized as having very high energy consumption as well as a very low indoor climate standard compared with modern requirements (Kragh and Rose 2011). As many of the heritage and historic buildings are privately owned, refurbishment of these buildings is needed to ensure that they remain part of the attractive building stock. Otherwise, these buildings will turn into unused monuments or deteriorate if there is no financial incentive for maintaining them (Sächsishes Staatsministerium des Inneren 2011). Only if an owner has a particular interest in the building, e.g., for cultural or sentimental reasons, then they might be interested in maintaining a loss-making building. Therefore, there is an intensified interest in refurbishing existing buildings including heritage and historic buildings.

However, refurbishment of heritage and many historic buildings is a complicated process where the individual's

Torben Valdbjørn Rasmussen is a senior researcher at the Danish Building Research Institute, Aalborg University, København, Denmark.

ideas are often neglected in favor of architectural, material, contractual, or layout issues. Because heritage buildings are an important part of the local or national cultural identity, they do not have an end-use date. Thus, these buildings need to be preserved in perpetuity and their corresponding impact on climate change needs to be taken into account. The extent to which governments support refurbishment of heritage buildings depends on the global priority and the ability of leaders to reach agreements that can mitigate the impacts of climate change (IPCC 2014; Rasmussen and Hansen 2011).

This paper identifies the key motivations and benefits that inspire an owner to refurbish heritage and historic buildings. This paper also describes the challenges that an owner needs to address both during the process of identifying feasible refurbishment measures as well as adaptation to climate change and a different risk assessment level. Risk assessment includes complying with the requirements and control levels for the indoor climate and assessing the changes in the building physics from the original construction. For example, changing the air infiltration rate, moisture load, and moisture content over the year can potentially increase the risk of degradation of organic materials and mold growth. This paper presents three case studies involving the refurbishment of privately owned heritage and historic buildings. Apart from restoration and modernization measures, this paper focuses on feasible indoor climate and energy performance measures and evaluates the obtained results.

HERITAGE AND HISTORIC BUILDINGS

The refurbishment of the buildings *Elmehuset* (Munch 2012), *Kavalergården* (Tommerup and Laustsen 2012), and *Fæstningens Materialgård* (Funch 2011) will be examined here. The refurbishment includes restoration, energy upgrading and renovation.

Elmehuset is a historic building rated in the highest category of buildings worthy of preservation, while Kavalergården and Fæstningens Materialgård are listed, and as such all three of the studied cases are assessed to be among the best or most characteristic of their kind and period. A total of 13% of all buildings in Denmark are assessed to be heritage buildings, and given a status as listed or worthy of preservation. Approximately 9000 buildings are listed and approximately 350,000 buildings have been assessed as being worthy of preservation, divided into several categories.

Buildings considered worthy of preservation are designated as such by using the Survey of Architectural Values in the Environment (SAVE) method (InterSAVE 1995), and protected on a regional level by local authorities. Listed buildings are protected by the National Heritage Agency. This listing is based on the Danish Act on Protecting Buildings and the Conservation of Buildings and the Built Environment (Government of Denmark 2005), which means that all building work, beyond routine maintenance, requires a permit from the National Heritage Agency.

Case A—Elmehuset

The building Elmehuset (Munch 2012), shown in Figure 1, was constructed in 1887 as part of the Poor Relief Hospital for elderly people and is located in Copenhagen. Elmehuset is part of a complex called The Old People's Town. Today the building hosts a comfortable home and care center for autistic children. The building is a 3-story rectangular masonry building of red brick with a slate roof and white-painted windows. At some locations on the ground floor there are large shutters intended to cover the windows—a legacy from World War II, when the rooms behind them were used as shelters. The shutters are also subject to a strict preservation order.

Case B—Kavalergården

The listed complex Kavalergården (Tommerup and Laustsen 2012) was constructed in 1877 as part of Bernstorff Palace located north of Copenhagen, and shown Figure 2. The complex consists of three buildings: a residential building with a half-timber construction on the first floor, a building housing

Figure 1 *Elmehuset.*

Figure 2 *Kavalergården.*

the infirmary with a half-timber construction on the first floor, and a building with stables. The buildings form a single complex surrounding a courtyard and consist of 1- to 3-story brick buildings with yellow lime-washed ground-floor facades, red-painted doors and gates, and white-painted windows.

The complex was run-down at the time the refurbishment was initiated. It had been unused for a long period of time and was in need of repair due to lack of maintenance.

Case C—Fæstningens Materialgård

The history of the listed complex, Fæstningens Materialgård (Funch 2011), shown in Figure 3, goes back to the 17th and 18th centuries when the old royal materials yard had to be replaced by a new one. Construction of the complex began in 1740 with a new, very distinguished residence for the supervisor. The complex consists of four individual listed buildings: the supervisor's residence, with a mix of historic styles and joinery details; a half-timbered warehouse with a longitudinal beam structure and rough, simple detailing; an office building with longitudinal partition walls and a characteristic room layout and joinery details and with a monopitch roof; and the original timber structure with very large dimensions. The buildings form a single complex surrounding a courtyard and consist of 1- to 3-story brick buildings with red tile roofs, yellow lime-washed facades, green-painted doors and gates, and white-painted windows. Only the original warehouse building of the complex is built as a half-timbered structure.

Figure 3 *Office building at the complex Fæstningens Materialgård.*

The use of the individual buildings of the complex has changed several times in the course of their history. Recently, the buildings were used for different office-type functions.

REFURBISHMENT

Each of the three owners wanted to carry out refurbishment that combined renovation, restoration, and an overall upgrading of the individual buildings and complexes to a standard that would make them attractive for accommodation and offices as well as conference and meeting facilities.

Motivation

Todays' building requirements must be met to use and preserve our national heritage and thus make our heritage and historic buildings an integrated part of everyday life today. Because of the common identity in a region, a city, or in a country, it is very important to protect heritage and historic buildings to preserve the individual characteristics and the attractiveness and competitiveness of a region, a city, or a country. Furthermore heritage is very important evidence of cultural background which has immense economic effects on, for example, tourism, attractiveness, etc. On the other hand, today's requirements must be met for heritage buildings to be an attractive part of the building stock.

Besides restoration and energy upgrading, heritage buildings need to meet modern standards, which is why the studied buildings were renovated. Renovations included a changed layout, new bathroom facilities, new kitchen facilities, new stairwells, improved acoustic insulation, new wiring, and the implementation of sprinkler and alarm systems.

Restoration was carried out to protect and preserve all remaining original elements, including exterior walls, most windows (Figure 4), decorations, the layout of the building, and the main construction, as well as the original loadbearing elements.

Several case studies have shown that there might not be conflicting interests between preserving heritage and refurbishing heritage buildings modern standards of indoor climate, energy consumption, climate protection measures, and goals (3encult 2013).

Benefits

The refurbishment of the privately owned heritage buildings for commercial use was studied. The cases offer a variety of attractive facilities like accommodation, offices, and conference and meeting facilities, as shown in Figures 5 and 6. Owners provide modern, up-to-date facilities integrated in a

Figure 4 *Original window frame in the stable building at Fæstningens Materialgård.*

heritage context that creates a unique atmosphere, which, for some businesses, supports the brand, and the image benefits products and creates increased value. The refurbishment has the potential of uncovering heritage qualities of the facility to increase its value and show the use of quality material, joints, and handicraft rarely seen today. The uses of a heritage building usually change from its original use, creating the opportunities to decorate the facilities with an inspiring atmosphere between the original and the new use. In this transformation, "the lamp of memory and tales" (Ruskin 1984), told through the original building, its components, and signs of use, is very important.

Challenges

It is a challenge by itself to identify feasible refurbishment measures that will not compromise the recognized heritage values both of individual buildings and of a complex of several individual buildings. Measures need to be agreed on through a process of selecting feasible refurbishment measures including renovation, energy upgrading and restoration measures for the individual heritage buildings and for the complex as a whole (Figures 7 and 8). The process depends on the cooperation between authorities and the owner of a heritage building. For the investigated heritage buildings it led to the identification and implementation of feasible measures for restoration and renovation, including energy upgrading, which at the same time allows for the preservation of identified heritage values inherent in the studied cases. The process of identifying feasible measures used in these buildings is outlined and described by Rasmussen (2014). In the end, the individual measures as well as the entire project must comply with requirements in terms of the following:

- Conservation and heritage
- Cost-benefit, including rental opportunities, operation and maintenance

Figure 5 *Cell offices at the old stables at Fæstningens Materialgård.*

Figure 6 *The entrance is placed in one of the old gateways for horse wagons at Kavalergården.*

Figure 7 *Foundations of brick on granite blocks at Kavalergården.*

Figure 8 *Roof construction with slate on batten on rafts, and a cold attic room at the pitch, at Elmehuset.*

- Architecture, including appearance, functionality and interior design
- Structural design, including risk assessment and building physics, energy consumption, CO2 emission, air infiltration, and indoor climate (draught, heat and light control)
- future weather loads from wind, rain, moisture, temperature, UV-rays and from extreme snowfall.

Table 1. Feasible Refurbishment Measures for the Three Case Studies

Measure	A	B	C
Low-energy glazing	✓	✓	✓
Reduced building envelope air permeability (0.16 to 0.5 h^{-1})			✓
Ventilation via opening of windows			✓
Balanced ventilation with a standard exchange including cooling	✓		✓
Centralized domestic hot water supply	✓		✓
Additional insulation of external walls, i.e. in the kitchen area and in utility rooms	✓	✓	✓
New insulated ground slab		✓	✓
Radiator heating	✓		✓
Underfloor heating on the ground floor			✓
Improved roof thermal insulation	✓	✓	✓
External solar shading with a reduction factor of 0.5			✓
Energy-saving light sources	✓	✓	✓
Daylight-controlled lighting		✓	✓
Centralized control of electrical components			✓
Cooling via a centrally placed unit where excess heat is transferred to the outside air	✓		✓
Shared canteen, meeting and conference facilities	✓	✓	✓

A is the case A study Elmehuset.
B is the case B study Kavalergården.
C is the case C study Fæstningens Materialgård.

FEASIBLE MEASURES

Individual feasible refurbishment measures for each case study, with influence on the indoor climate and energy performance, are shown in Table 1. The identified feasible refurbishment measures were found through a process involving the presence of relevant assessment authorities each playing their specific role. The relevant assessment authorities involved were the following:

- The National Heritage Agency, which had to evaluate each measure specifically for each building from a conservation and heritage point of view. Assessments were only for guidance. The final, official approval was only to be given once the entire project was submitted. For buildings that are not classified as listed, local authorities play this role.

- The Danish Working Environment Authority, which had to evaluate each measure specifically in terms of the acceptability of the workspace.
- Owner/Developer, who had to evaluate each measure specifically for each building from the owner's perspective. The impact of each measure was evaluated in relation to cost-benefit, rental opportunities, and operation and maintenance.
- Architects, who had to evaluate each measure specifically for each building from an architectural point of view. Among other issues, the form, appearance, functionality, and interior design were evaluated. These evaluations were further supplemented wherever possible with more general assessments and views.
- Structural engineers, who had to evaluate each measure specifically for each building from a structural design perspective. The impact of each measure on the existing construction was risk assessed, especially with respect to building physics, e.g., moisture levels. Wherever possible, a more general assessment of the individual measures was given with references to relevant literature.
- HVAC engineers, who had to assess each measure specifically for each building in terms of energy and the indoor climate. The impact of the measure in terms of energy-saving effect and room temperature was critically evaluated. Wherever possible, a more general assessment of the measure was given, with references to relevant literature.

The process identified feasible refurbishment measures that fulfill the requirement of creating synergy between the interests of preserving heritage values and developing reasonable cost-benefit solutions that meet the requirements for the future use of the building through a number of meetings. Basically, a general list including all potential measures needs to be considered in the process, narrowing down measures to the feasible refurbishment measures for the individual buildings. As a starting point, the list should be elaborated to include all energy-upgrading measures regardless of the architectural and heritage value of the buildings. The proposals should not take into account the location, actual building geometry, costs, use and function of the building or similar features that could immediately mean that the proposed measures were not to be implemented. The process narrowing down proposed measures to feasible measures is in detail described by Rasmussen (2014). Identifying feasible refurbishment measures for the three studied cases, Elmehuset (Munch 2012), Kavalergården (Tommerup and Laustsen 2012), and Fæstningens Materialgård (Funch 2011) was carried out using the process described by Rasmussen (2014).

DISCUSSION

The outcome of the refurbishment of the individual buildings was to implement feasible measures including renovation and restoration, and meeting climate protection measures and goals composed of upgrading the energy performance and creating attractive facilities for accommodation, offices, conferences, and meetings, including having an acceptable indoor climate.

Feasible refurbishment measures are characterized by fulfilling the requirement of creating synergy between the interests of preserving the identified heritage values and developing reasonable cost-benefit solutions that meet the requirements for the future use of the building.

Feasible refurbishment measures were found for refurbishment, including restoration and energy upgrading of each of the individual buildings in the three case studies. Measures were found using the process for selecting feasible comprehensive refurbishment measures for heritage buildings described by Rasmussen (2014). The process includes the cooperation between authorities and the owner of a historic or heritage building. In these cases, public authorities were represented by the National Heritage Agency and the Danish Working Environment Authority and, in the case of Elmehuset, local authorities. The owner was advised by architects, structural engineers, and HVAC engineers. Through a number of meetings between the parties, feasible refurbishment measures were identified. The identification of feasible refurbishment measures were based on a design brief of the individual measures and synergy between the individual measures including building history, current condition of the construction and individual building materials and identified heritage values, as well as energy and indoor evaluation calculations. As a starting point, measures should include all the potential energy-saving solutions regardless of the architectural and heritage value of the buildings. The proposals should not take into account the location, actual building geometry, costs, use, and function of the building or similar features that would immediately mean that the proposed measures were not to be implemented. This means that all available measures should be considered. Measures such as windows, double glazing, solar screening, solar shading, added insulation of outer walls, added insulation at ceilings and ground slab, airtightness, heat pump, light bulbs, daylight control, central heating and cooling system, power supply, ground heating source, central technical system controlling electricity, air exchange, domestic hot-water supply, rainwater collection, solar panels, photovoltaics, heat emission, ventilation, use of room, and so on, should be considered.

In all three case studies, measures were implemented with due respect for the core heritage values of the individual buildings and complexes without compromising identified heritage values.

On this basis the lessons learned regarding the energy consumption of refurbished historic and heritage buildings (Munch 2012; Tommerup and Laustsen 2012; Funch 2011;

Rasmussen et al. 2015), special attention must be drawn to the initial planning of their new use. Implementing shared kitchen and canteen facilities within the original buildings a lesson was learned that such decisions changed the overall demand for energy as these facilities do have a high energy consumption if provided with an acceptable indoor climate (Rasmussen et al. 2015). Indoor climate considerations showed that it was necessary to add significant cooling capacity to cool the canteen, kitchen and the meeting facilities, a result of the initial planning where a whole heritage building or large parts of it was given a service function. In all buildings of the three case studies the initial planning showed the need for a significant cooling load which resulted in a negative influence on the energy consumption related to the climate change goals.

Before refurbishment, Fæstningens Materialgård had an unacceptable indoor climate and its energy consumption was 470.92 MWh per year for operating offices in the building complex. After refurbishment, energy consumption was 399.01 MWh per year operating offices and a shared canteen with an acceptable indoor climate, a Category C level (ISO 2005). The needed cooling load comprises 10.3% of the energy consumption per year for the refurbished building complex with an acceptable indoor climate (Funck 2011). The energy needed to raise the thermal environment from an unacceptable level to an acceptable level, Category C, increases the energy consumption and reduces the climate change goals. A layout where energy-consuming facilities such as a kitchen or a canteen are located elsewhere than within a heritage building, e.g., nearby or in a building of its own, should be considered when needed.

Although these case studies show synergy between the interests in preserving heritage values and energy-saving measures, it is only the case for some of the available measures. In these case studies implementation of balanced ventilation with a standard exchange system including cooling and radiator heating was implemented by taking advantage of larger cavities located behind cavity walls in front of spandrels under the windows. Unfortunately, many measures involve a conflict between the heritage values and energy savings (Sächsishes Staatsministerium des Inneren 2011). Airtightening the buildings is one of the feasible measures in this study; since the early 1970s, it has been done to many existing buildings, by replacing old windows with new, more airtight windows, for example. This resulted in reduced energy consumption and contributed to reaching the climate change goals, but unfortunately the humidity in many buildings also increased, creating moisture-related indoor climate problems. The reason was a biased focus on energy savings. Airtightening buildings should only be done in combination with the possibility to ventilate, either automatically and preferably with heat recovery or manually where users must be aware of the necessity of actively ensuring ventilation, e.g., by opening windows regularly. Most heritage buildings have wooden beams in the horizontal structures. Ultimately, high moisture content in wooden beam ends can result in a collapse of the beam. Mold growth is less severe at the beam end. Both failure modes need to be avoided. Rasmussen (2011) recommended the installation of monitoring equipment connected to an automatic control system of the indoor air to reduce the humidity as needed. Morelli and Svendsen (2013) suggested leaving a 200 mm gap just above the floor, thus creating a thermal bridge that will ensure higher temperatures at the beam end compared with a fully additional insulated wall. In some cases, a combination of the two measures might be necessary.

Previously, Krebs and Collet (1981) concluded that the most important factors for the moisture conditions at the beam ends are as follows:

- Geometry of the structure
- Interior temperature and relative humidity
- Production of interior moisture
- Outdoor climate (e.g., temperature, relative humidity, driving rain intensity, and wind velocity)
- Material properties for wood and masonry
- Air changes around the beam end

As listed buildings—including when upgraded to save energy—are expected to preserve our cultural heritage for posterity, the list of feasible measures for energy upgrading (Table 1) should be compared with the climate changes that may be expected. In the EU region these are as follows (Rasmussen 2013):

- More frequent and heavier rainfall
- More extreme temperatures in summer and heat waves of long duration, including more hours of sun
- Warmer and more humid winters
- More frequent and heavier storms
- Extreme snowfall

Some of the expected changes will worsen moisture conditions at especially beam ends, e.g., more frequent and heavier rainfall and warmer and more humid winters. On the other hand, more hours of sun mean higher indoor temperatures, which decrease the relative humidity. Unfortunately, it also means that more cooling is needed and cooling increases the energy consumption significantly.

As a result, the most sustainable solution might be to limit the use of historic and heritage buildings; to accept that some buildings are not suitable for all occupation and may never become suitable for all purposes. In some cases, it is not feasible to use a building for canteen facilities, which are in humidity Class 4 according to ISO 13788 (ISO 2013), but it may be possible to use it for offices (humidity Class 2). The difference is that at an outdoor temperature of 0°C, the indoor moisture supply is 4 g/m^3 in humidity Class 2 but 8 g/m^3 in humidity Class 4.

Finally, measures that improve the energy performance of a heritage building should be reversible; it must be possible to reinstall measures and restore the building to its former state and to change the use of the building. In this way, measures that seemed necessary at the time in order to make the building attractive can be removed later if the building needs to fulfill another purpose or at a time when energy performance and climate goals change.

CONCLUSION

The refurbishment of each of the buildings investigated in the three case studies showed a successful implementation of feasible measures including renovation, restoration, and compliance with climate protection requirements and goals including upgrading of the energy performance, thus creating attractive facilities for accommodation, offices, and conference and meeting rooms providing an acceptable indoor climate. However, facilities with a high energy consumption such as a kitchen and a canteen located within a heritage building were discovered to increase the overall energy consumption substantially if provided with an acceptable indoor climate. This concludes that the most sustainable solution includes limitations in the use of heritage buildings and to accept that some buildings are not suitable for all occupations and may never become suitable for all purposes.

The study demonstrates that it is possible to refurbish heritage buildings, including improving the energy performance and indoor climate while not compromising identified heritage values, and in this way ensuring that they remain part of the attractive building stock.

A main issue concerning the refurbishment of heritage buildings is to establish an airtight building. Apart from resulting in a large heat loss, air infiltration also causes indoor climate discomfort in the form of draughts and asymmetrical temperature distribution. Furthermore, the layout and use of a heritage building should be taken into consideration, as some measures result in higher moisture levels at critical parts of the building. The consequence might be mold growth or even deterioration of structural parts. As heritage per definition is expected to be preserved for posterity, expected changes in weather loads from climate changes must also be taken into consideration.

ACKNOWLEDGMENT

The work on which this paper is based was supported by Priority 1: Fostering Innovations of the Baltic Sea Region Programme 2007–2013. Climate Change, Cultural Heritage & Energy Efficient Monuments, Co2olBricks.

REFERENCES

3encult. 2013. Efficient energy for EU cultural heritage. Baltic Sea Region Programme 2007–2013. Bolzano, Italy: EURAC Research. www.3encult.eu/en/project/welcome/default.html

Climate Change, Cultural Heritage & Energy Efficient Monuments. 2013. Co2olBricks. Baltic Sea Region Programme 2007–2013. Hamburg: Co2olBricks. www.co2olbricks.eu

Funch, M. 2011. Energy refurbishment of Fæstningens Materialgård. Hamburg: Co2olBricks. www.co2olbricks.eu/fileadmin/Redaktion/Dokumente/Publications/Refurbishment_Faestningens_Materialgard.pdf

Government of Denmark. 2005. Act No. 1088. The consolidated listed buildings and preservation of buildings and urban environments. www.kulturarv.dk/fileadmin/user_upload/kulturarv/english/The_Consolidated_Act_No._1088_on_Listed_Buildings_and_Preservation_of_Buildings_and_Urban_Environments.pdf. Copenhagen: Agency for Culture and Palaces.

InterSAVE. 1995. International Survey of Architectural Values in the Environment. Copenhagen: Ministry of the Environment and Energy, the National Forest and Nature Agency, Division of Town Preservation. www.sns.dk/byer-byg/Netpub/INTRSAVE/TEKST/CONTENTS.HTM

IPCC. 2014. Climate change 2014: Impacts, adaptation, and vulnerability. IPCC Working Group II Contribution to AR5. New York: Cambridge University Press. http://ipcc-wg2.gov/AR5/

ISO. 2005. EN ISO 7730. Ergonomics of the thermal environment—Analytical determination and interpretation of thermal comfort using calculation of the PMV and PPD indices and local thermal comfort criteria. Geneva: International Organization for Standardization.

ISO. 2013. ISO 13788. Hygrothermal performance of building components and building elements—Internal surface temperature to avoid critical surface humidity and interstitial condensation—Calculation methods. Geneva: International Organization for Standardization.

Kragh, J., and J. Rose. 2011. Energy renovation of single-family houses in Denmark utilising long-term financing based on equity. Elsevier Ltd. 88(6)2245–53.

Krebs, H.J., and P.F. Collet. 1981. Indvendig Efterisolering: Indmurede Bjælkeenders Fugt—og Temperaturforhold i Etagekryds. [in Danish]. 1st ed. Taastrup: Danish Technological Institute.

Morelli, M., and S. Svendsen. 2013. Investigation of interior post-insulated masonry walls with wooden beam ends. *Journal of Building Physics* 36:265–93.

Munch, R. 2012. Upgrading the energy performance of Elmehuset in the old people's town. Hamburg: Co2olBricks. http://www.co2olbricks.eu/fileadmin/Redaktion/Dokumente/Publications/Elmehuset_Copenhagen_Danmark.pdf

Rasmussen, T.V. 2011. Retrofitting listed buildings: Measures, savings and requirements. *The Open Construction and Technology Journal*. Bentham Open. 5:174–81.

Rasmussen, T.V., and E.J.D.P. Hansen. 2011. Climate change: Mitigating effects vs. adapting to climate change. Proceedings of the World Sustainable Building Conference: SB11. pp. 15. RIL—Finnish Association of Civil Engineers.

Rasmussen, T.V. 2013. *Strategy for climate change adaptation.* In proceedings of SB13 Southern Africa: Creating a Resilient and Regenerative Build Environment. red. / Fidelis A. Emuze. pp. 44–53. Department of Built Environment. Central University of Technology. Free State.

Rasmussen, T.V. 2014. Refurbishing Fæstningens Materialgård: A heritage complex. *Journal of Civil Engineering* and Architecture 8(7):888–97.

Rasmussen, T. V., E.B. Møller, and T.C. Buch-Hansen. 2015. Extensive renovation of heritage buildings—Reduced energy consumption and CO2 emissions. *The Open Construction & Building Technology Journal.*

Ruskin, J. 1984. *The seven lamps of architecture.* New York: Farrar. Straus and Giroux..

Sächsishes Staatsministerium des Inneren. 2011. Energetische Sanierung von Baudenkmalen. Handlungsanleitung für Behörden. Denkmaleigentürmer. Architekten und Ingenieure [in German]. Dresden, Germany.

Tommerup, H.M., and J.B. Laustsen. 2012. Upgrading the energy performance at Kavalergården. Hamburg: Co2ol-Bricks. www.co2olbricks.eu/fileadmin/Redaktion/Dokumente/Publications/Kavalergarden_Copenhagen_Danmark.pdf

United Nations. 1998. *Kyoto protocol to the United Nations framework convention on climate change.* New York: United Nations. http://unfccc.int/resource/docs/convkp/kpeng.pdf.

Treatment of Rising Damp in Historic Buildings Using a Wall Base Hygro-Regulated Ventilation System—Case Study

V.P. de Freitas
FEUP-Construct-LFC

M.F.C. Barbosa
FEUP-Construct-LFC

A.S. Guimarães
FEUP-Construct-LFC

J.M.P.Q. Delgado
FEUP-Construct-LFC

ABSTRACT

The treatment of rising damp is a problem difficult to solve and of the utmost importance in historic buildings. In addition, the traditional techniques used to deal with this kind of problem are sometimes not effective enough, in particular, when dealing with walls of considerable thickness and heterogeneous materials. We consider the use of air with controlled relative humidity the most suitable principle, and we have been developing studies that use wall base ventilation systems with a hygro-regulated device.

The efficiency of the ventilation system was confirmed by experimental and numerical studies and, also, the geometry of the ventilation channel was studied to define the best configuration. The system was implemented in different buildings and particularly in a church with the goal of developing a prototype of the hygro-regulated device. Based on the acquired knowledge, the HUMIVENT system was patented. This device receives information of the relative humidity and temperature probes at the air inlet and air outlet of the system and compares the vapor pressure in those points controlling the wall base ventilation process.

Having in mind the application of scientific and technological knowledge from building physics to practice, this paper presents a simplified model to design the hygro-regulated system (geometry and ventilation rate) and presents the specification of its different components and information about the implementation in a real case study.

INTRODUCTION

The treatment of rising damp is a problem difficult to solve and of the utmost importance in historic buildings. Several treatment techniques were developed, not always with the required efficiency. As examples, the following techniques can be mentioned (Franzoni 2014): creating a physical or chemical barrier (Barberot 1927; Downing 1850; Mariz Ferreira 2009; Massari 1971), creating a potential against the capillarity (Henriques 1993), applying atmospheric drainage (CSTC 1985; Henriques 1993), applying a coating with controlled porosity (Groot et al. 2009), and concealing the anomalies (Henriques 1993). Several systems were also patented (James 2003; Robert 2004; Baldszuhn 1990).

Bearing in mind all the developed techniques, we consider the use of air with controlled relative humidity the most suitable principle; that is why we have been developing studies that use the wall base ventilation system with a hygro-regulated device.

The research started in 2000 with experimental laboratory tests to evaluate the effect of a wall base ventilation system in the level achieved by the damp front. It was concluded that the damp front is clearly lower when using the wall with a base ventilation system (Torres and de Freitas 2007) (Figure 1[a]).

Subsequently, numerical studies with a 2-D advanced hygrothermal software (WUFI 2D) were developed and showed similar results compared with the experimental ones previously mentioned, even without considering the interface between materials in the moisture transfer process (Figure 1[b]) (Torres and de Freitas 2002). So, the efficiency of the ventilation system was confirmed by both experimental and numerical results (Figure 1).

The system geometry has a great importance in the wall base ventilation system behavior and therefore some experimental studies were developed considering three configurations presented in Figure 2. The results indicate that

V.P. de Freitas is a full professor, **M.F.C. Barbosa** is a doctoral student, **A.S. Guimarães** is an assistant professor, and **J.M.P.Q. Delgado** is post-doctoral researcher in the Department of Civil Engineering, Faculty of Engineering of University of Porto, Porto, Portugal.

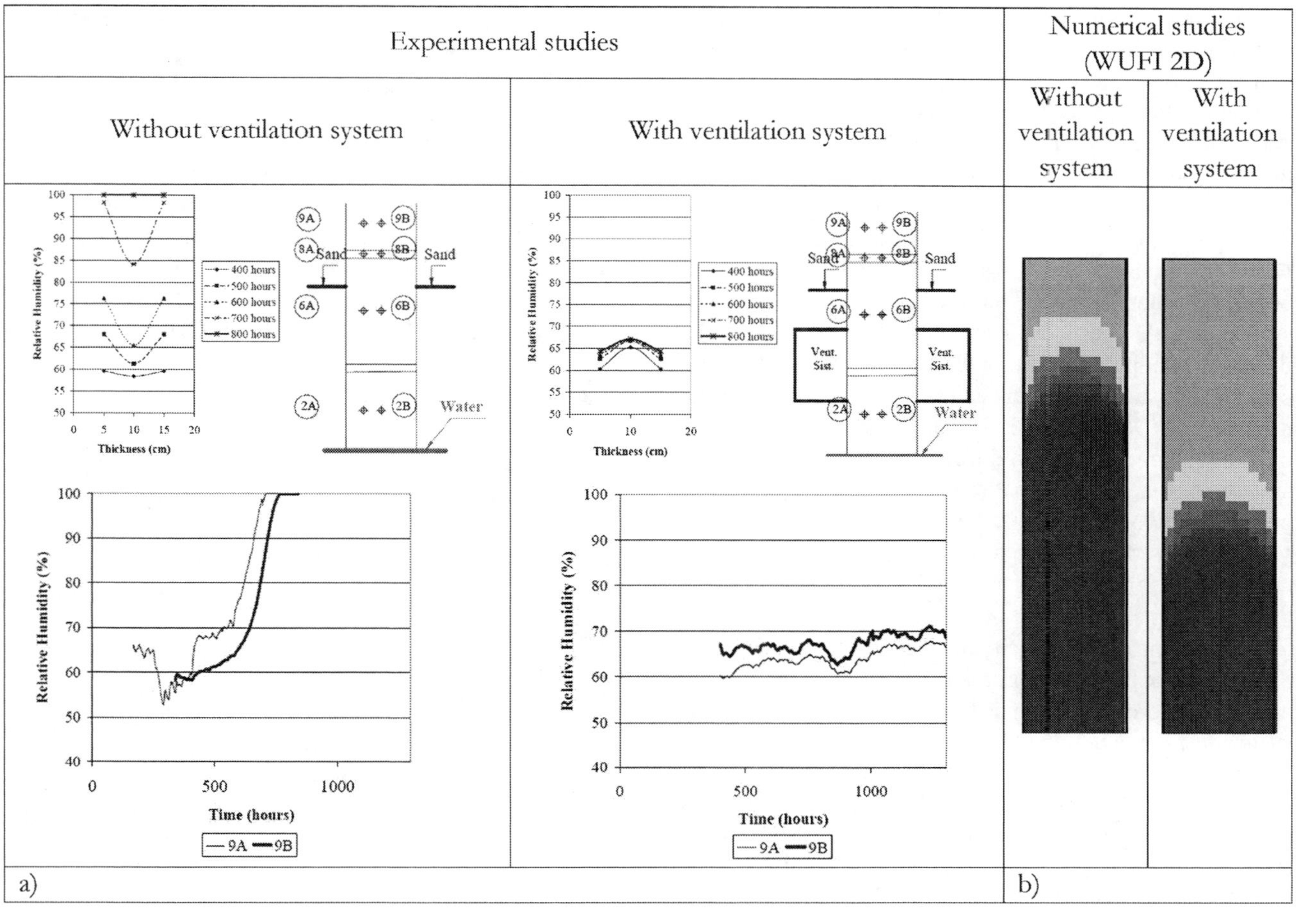

Figure 1 (a) Experimental evaluation of the relative humidity at a representative level (9A and 9B), indicating that, with the ventilation system, the damp front did not achieve this level (Torres and de Freitas 2006). (b) Relative humidity profiles obtained by a 2-D advanced hygrothermal model (Torres and de Freitas 2002).

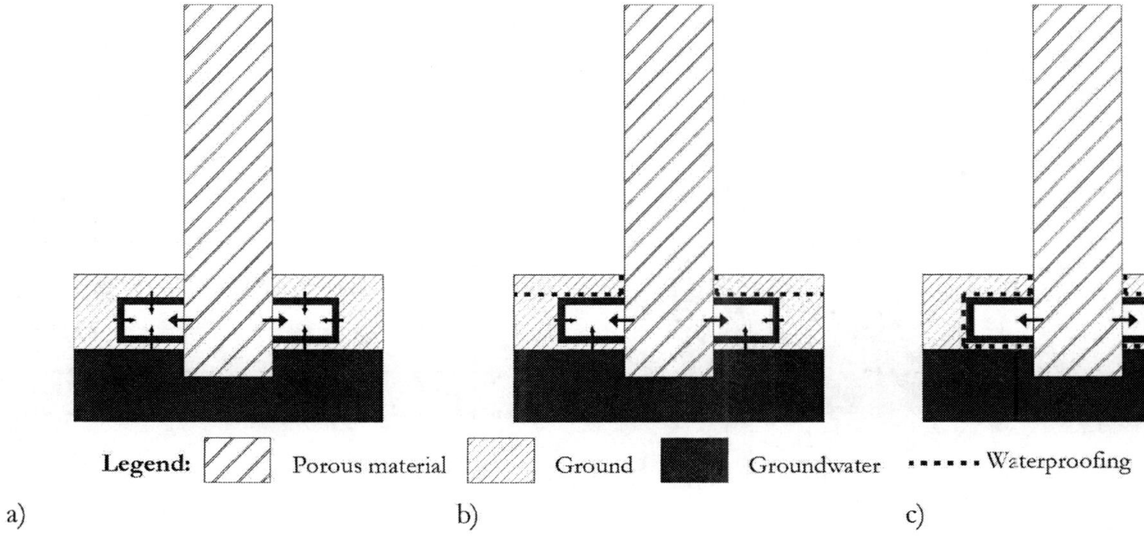

Figure 2 System geometry: (a) Configuration 1—without waterproofing, (b) Configuration 2—ground floor waterproofing, (c) Configuration 3—ventilation system waterproofing.

Configuration 3 is the best for the maximization of the system efficiency (de Freitas et al. 2007). However, the implementation of Configuration 3 is not always possible. In fact, Configuration 2 is the most common in practice and presents a good performance. All the research work has been done with porous materials such as stone or brick masonry.

The system was implemented in a church with the goal of developing the prototype for the hygro-regulated device. Based on this study it was possible to determine that the vapor pressure gradient, between the air outlet and the air inlet of the system, was the most adequate criterion to define the system operation conditions, ensuring that, with a positive difference of vapor pressure, the system can work without condensation occurrences (Delgado et al. 2013). Figure 3 illustrates how much moisture was removed from this wall base ventilation system and also the seasonal influence in the moisture transport. When ΔP value is above the dashed line (Figure 3[a]) the ventilation works, and when it is below the dashed line the system stops to control the condensation risk (Guimaraes et al. 2013).

Based on the acquired knowledge, the HUMIVENT device that controls the ventilation was patented (de Freitas et al. 2011). In order to optimize the overall performance of the system, the extraction is now controlled by a hygro-regulated fan that comes into operation when the water vapor pressure at the entrance is lower than the water vapor pressure at the exit (that results from the combination of the temperature and the relative humidity) and when the relative humidity of the entrance air is higher than a certain predefined value to guarantee no salt crystallization/dissolution problems inside the system.

The control module receives information from two probes (temperature and relative humidity at the entrance and at the exit), calculates the air vapor pressure, and evaluates the positive or negative sign of the pressure differential, switching on or off the fan.

When the system is turned off the air velocity inside the system is null, but when the system is in an active mode the air velocity takes a constant value, which could also be controlled.

Having in mind the application of scientific and technological knowledge from building physics to practice, this paper presents a simplified model to design the hygro-regulated system (geometry and ventilation rate) and presents the specification of its different components and information about the implementation in a real case study.

SYSTEM DESIGN—SIMPLIFIED MODEL

The Principle

The hygro-regulated ventilation system requires the installation of a channel in the base of the wall in order to control the drying process and the rising damp. A ventilated channel should contour the wall with one side in contact with it (see Figure 2). The ventilation system is located underground and the air admission and exhaustion is performed in the beginning and the end of the channel.

In practice, it is necessary to pre-size the channel geometry. The simplified model pretends to be a tool to size the most important parameters of the wall base ventilation system (Guimarães, de Freitas, and Delgado 2012). The three most important design parameters are the wall length (L), the system height (h_s), and the air velocity inside the system (u).

Design Procedure

The designer engineer must define the maximum acceptable level of the damp front in the wall, h_{max}, which should be below the floor level. Figure 4 illustrates the involved parameters such as h_a (ground height adjacent to the wall), h_s (system height), b (wall thickness), and L (wall length).

In this proposed model there is a balance between the absorption and evaporation (Equation 1).

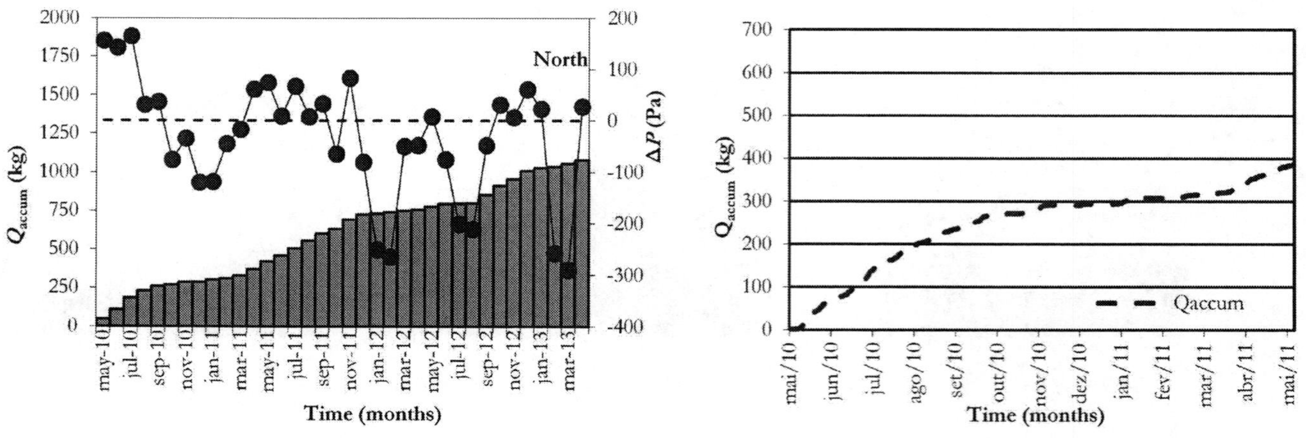

Figure 3 *(a) Accumulated amount of water removed (Q_{accum}) by the system and gradient of vapor pressure (ΔP) between air inlet and outlet. (b) Seasonal influence is clearly proved by the reduced water removal during winter (November to March) (Guimarães, Delgado, and de Freitas 2012).*

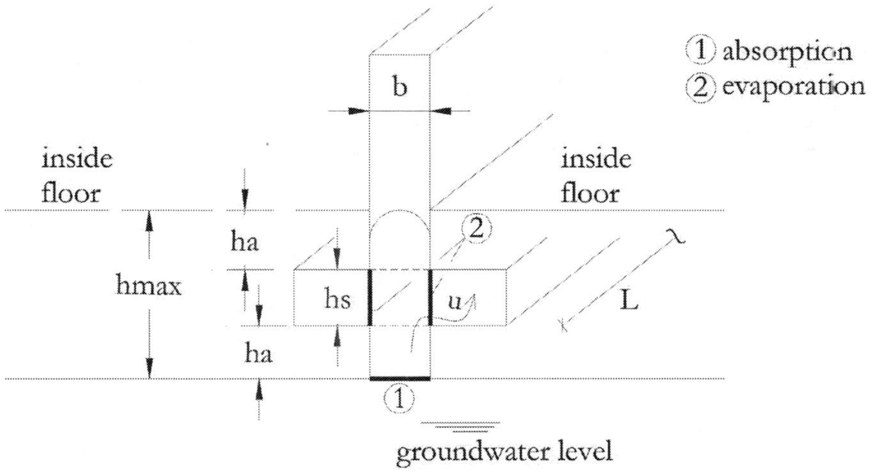

Figure 4 *Geometric design parameters—interior wall.*

Equation 1 represents the simplified model and the equilibrium between absorption (by the surface 1) and evaporation (by the surface 2) (Guimarães, Delgado, and de Freitas 2010). In the nomenclature the significance of the parameters is described:

$$\underbrace{\frac{b \cdot S^2}{2w \cdot h_{max}}}_{\text{absorption}} = \underbrace{e_s \cdot h_s}_{\text{evaporation}} <=>$$

$$\underbrace{\frac{b \cdot S^2}{2w \cdot h_{max}}}_{\text{absorption}} = \underbrace{\frac{(c^* - c_0)}{\rho_w}\sqrt{\frac{4D_m}{\pi\frac{L}{u}}} \cdot h_s}_{\text{evaporation}} \tag{1}$$

Input Parameters

The application of the simplified model requires some input parameters, namely, geometrical data, wall material properties, and environmental conditions. Table 1 summarizes those parameters.

Simplifications

The wall base hygro-regulated ventilation system model takes into account some simplifications that facilitate the model application. It should be mentioned that these simplifications are acceptable according to previously published studies (Guimarães, Delgado, and de Freitas 2012).

The model simplifications are the following:

Simplification 1. This is a steady-state model (average conditions), and the values defined for each variable are unchangeable over time.

Simplification 2. It is assumed that the air velocity inside the system, u (m/s), is constant throughout the channel, but it is possible to change this velocity.

Simplification 3. The system height, h_s (m), should be quantified through the inner side of the channel. The thickness of the channel has no relevant influence on the final outcomes.

Simplification 4. When it is not possible to know the maximum moisture ratio by volume of the material, it is considered that the amount of water in the open pores is 85% of the total pore volume in a porous material, corresponding to all the open pores.

Simplification 5. It is considered that the inner side of the ventilation channel is isothermal. However, the temperature next to the wall is different from the air temperature inside the system, due to the latent heat resulting from the changing process of liquid water to water vapor and due to the sensible heat resulting from the diffusion and convection process. Nevertheless, a small change in the temperature does not significantly affect the results. It is suggested the same value of T inside the system be used to calculate the c^*, c_0, and D_m parameters. If there is no available T data inside the system, the average temperature can be used.

Simplification 6. The interface effect between the layers of the material is not considered in this model. The model is applicable to monolithic materials without joints. However, according to the experimental work developed before (de Freitas 1992), the interface effect between layers delays the humidification process (de Freitas et al. 1996). So, this is a conservative model.

Simplification 7. It is assumed that the absorption of water occurs exclusively at the base wall level. There is no absorption through the wall in contact with the ground. In this model, the wall surface is considered waterproofed down to the groundwater level, as shown in Figure 5.

Table 1. Input Parameters

	Symbol	Units	Definition	Data Source	Description
Geometry	b	m	Wall thickness	Project data	Should be measured in the building plan.
Geometry	h_a	m	Ground height	Project data	Should be measured in the wall cross section. Represents the ground height adjacent to the wall, excluding the system height.
Geometry	L (*)	m	Wall length	Project data	Should be measured in the building plan.
Properties	S	m/s$^{1/2}$	Sorptivity	Equation	$S \cong A / \rho_w$ (Hanzic and Ilic 2003; Hall and Hoff 2002) where A is the water absorption coefficient (kg/m^2·s$^{1/2}$) and ρ_w is the water density (kg/m^3).
Properties	θ_w or w	m^3/m^3	Moisture ratio by volume	Equation	When the value is not available, the following equation can be used instead: $w = 0,85 \times \varepsilon$ (Hall and Hoff 2002), where ε is the porosity (m^3/m^3).
Properties	c^*	kg/m^3	Vapor concentration at the wall surface	Equation	$c^* = 0,002167\dfrac{p}{273,15 + T}$ where $p = 611 e^{\frac{17,08T}{234,18 + T}}\Phi$ and p is the water vapor partial pressure (Pa), T is the air temperature (°C) and ϕ is the relative humidity at the wall surface, which is 100%.
Properties	ρ_w	kg/m^3	Water density	—	1000 kg/m^3
Environment	D_m	m^2/s	Water vapor diffusion coefficient in the air	Equation	$D_m = 2,23E^{-5}\left[\dfrac{T + 273,15}{273,15}\right]^{1,5}$ where T is the air temperature (°C).
Environment	c_0	kg/m^3	Vapor concentration at air inlet	Equation	$c_0 = 0,002167\dfrac{p}{273,15 + T}$ where $p = 611 e^{\frac{17,08T}{234,18 + T}}\Phi_0$ and p is the water vapor partial pressure (Pa), T is the air temperature (°C) and ϕ_0 is the air inlet relative humidity.
Environment	T	°C	Inlet air temperature—average	Measurement	*In situ* measurements
Environment	ϕ_0	—	Inlet relative humidity—average	Measurement	*In situ* measurements

(*) – Iterative input

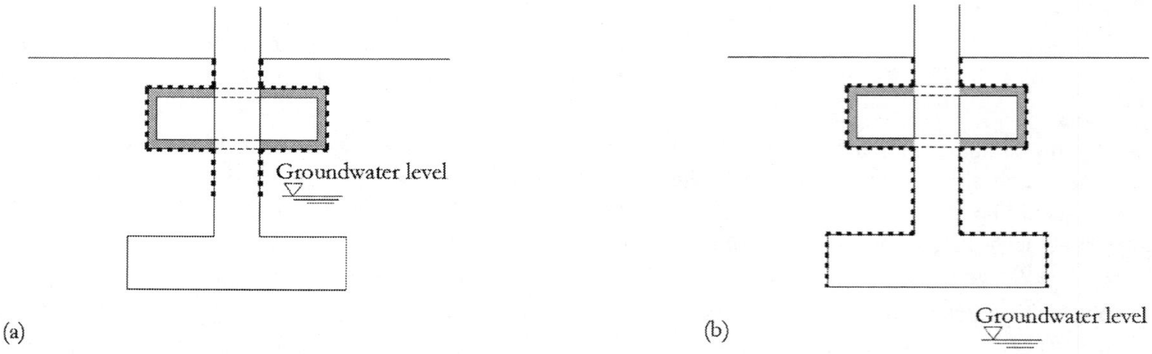

Figure 5 *Waterproofed wall until the groundwater level: (a) groundwater level above the foundation structure or (b) groundwater level below the foundation structure.*

Model Limitations

It is of the utmost importance to define the limitations of this model:

Limitation 1. It is assumed that the wall and the ventilation system have to be symmetric.

Limitation 2. The lower level of the system must be necessarily above the groundwater level. The groundwater level should be at least 10 cm below the system (in order to ensure that the water transport inside the channel is exclusively performed by diffusion).

Limitation 3. This model is applicable to walls with less than 1 meter ($b \leq 1{,}00$ m). It is known that the amount of water removed by the ventilation system (kg/m^3) decreases with increasing thicknesses. This means that the ventilation system has a decreasing efficiency with increasing wall thickness (Torres and López 2015a; Torres and de Freitas 2010; Torres and López 2015b).

MODEL APPLICATION

For this practical application, we choose an historic building with an inner room, *B* (Figure 6[a]). It was our purpose to install a wall base ventilation system in both sides of the *B* room walls, with the configuration shown in Figure 6(b). The two parts of the ventilation system (symmetrical) have a similar development, $L_{B1} = L_{B2}$.

The next steps must be followed:

Step 0. Verify the model applicability, namely following the points 1 to 3 that represent the model limitations:

- Point 1. The environmental conditions are symmetric because it is an inner room.
- Point 2. The lower level of the ventilation system is 10 cm above the groundwater level.
- Point 3. The walls are 0.5 m thick.

All conditions are adequate for the simplified model application.

Step 1. Define h_{max}.

In this example it was considered that the maximum level of the wet front stays at 10 cm below the floor level, to prevent moisture appearance in the base of the walls ($h_{max} = h_a + h_s - 0{,}10 = 0{,}90$ m) (as shown in Figure 7). The evaporation should be carried out only by the system, taking into account that there is no evaporation through the wall in contact with the ground, because a waterproof layer will be applied in both sides of the wall.

Step 2. Parameters determination: b, S, w, h_a, c^*, c_0, and D_m, according to Table 1.

Table 2 presents the values used in this example.

Step 3. Sizing the parameters: L, h_s, and u, which define the system geometry.

The definition of a wall base ventilation system for room *B* requires the value of $L_{wall} = L_{B1} = L_{B2}$ and the verification of the resulting u and h_s values.

In Step 3, the designer engineer defines the system length (in this case, equal to the wall length, 40m) and the variables u and h_s that can assume several values but are dependent on one another. It should be mentioned that in some cases the system height (h_s) is limited by constructive reasons.

To help the designer engineer, a software called "HUMIVENT pré-dim" was developed (Guimarães, Delgado, and de Freitas 2012). This software asks for the input parameters presented in Table 1 (whose values for this application example are shown in Table 2) and, based on the Equation 1, stretches different solutions, allowing the selection of the most suitable one.

In Table 3, different acceptable solutions are presented. The final solution for this specific case selected by the designer engineer is as follows: L = 40m, h_s = 0,30 m, and u = 0,38m/s.

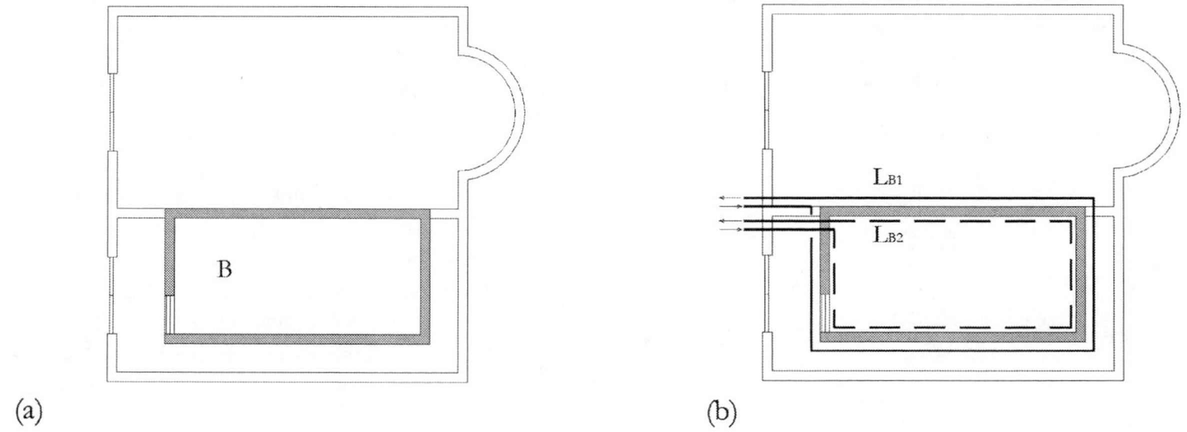

Figure 6 *(a) Room plant. (b) Ventilating system configuration B1 and B2, installed on both sides of the room walls.*

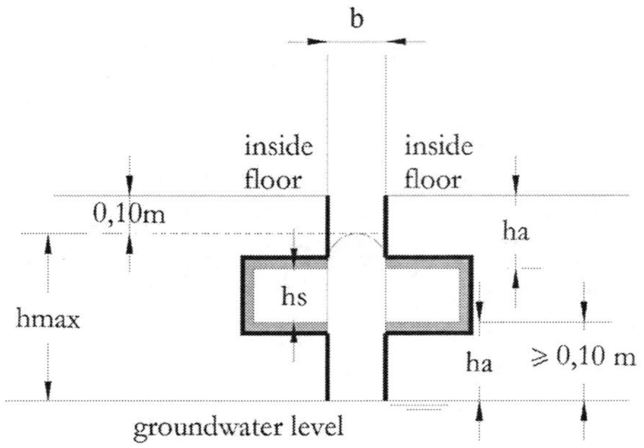

Figure 7 *Definition of* h_{max}.

Table 2. Input Parameters—Quantification

Geometry	$b = 0{,}50$ m
Geometry	$h_a = h_{max} - h_s + 0{,}10$
Geometry	$L = 40$ m (Development of the wall)
Properties	For stone walls: $S \cong \frac{0{,}025}{1000} = 0{,}000025\ \text{m/s}^{1/2}$
Properties	For stone walls: $w = 0{,}85 \times 0{,}20 = 0{,}17\ \text{m}^3/\text{m}^3$
Properties	For an air inlet temperature of 18°C and relative humidity of 100%: $p = 611 e^{\frac{17{,}08 \times 18}{234{,}18 + 18}} \cdot 1 = 2067{,}8$ Pa; $c^* = 0{,}002167 \frac{2067{,}8}{18 + 273{,}15} = 0{,}01539\ \text{kg/m}^3$
Properties	$\rho_w = 1000\ \text{kg/m}^3$
Environment	For an air temperature of 18°C: $D_m = 2{,}23E^{-5}\left[\frac{(18 + 273{,}15)}{273{,}15}\right]^{1{,}5} = 0{,}000025\ \text{m}^2/\text{s}$
Environment	For an inlet temperature of 18°C and relative humidity of 60%: $p = 611 e^{\frac{17{,}08 \times 18}{234{,}18 + 18}} \cdot 0{,}60 = 1240{,}7$ Pa; $c_0 = 0{,}002167 \frac{1240{,}7}{18 + 273{,}15} = 0{,}009234\ \text{kg/m}^3$
Environment	$T = 18$°C
Environment	$\phi_0 = 60\%$

Table 3. Different Acceptable Solutions

h_{max} = 0,9 m (Figure 7)	
h_s, m	**u, m/s**
0,50	0,14
0,40	0,22
0,30	**0,38**
0,26	0,50
0,19	1,00

CASE STUDY—IMPLEMENTATION OF A PROTOTYPE

General Application in Historic Buildings

The system and the HUMIVENT device was installed in several Portuguese historic buildings. Figure 8 illustrates the backside of the church (Figure 8[a]), the intervention in the ground floor including a waterproof layer (Figure 8[b] and 8[c]), and the implementation of the channel *in situ* (Figure 8[d]). The inlet air is carried out by an opening on the facade, so the air comes from the outside. The exhaustion is controlled by a hygro-regulated ventilation system.

The HUMIVENT (Figure 8[f]) device is a control module linked with two relative humidity and temperature probes. The data acquisition system collects the information from the probes and the control module orders the fan to turn on or to turn off according to the functioning criterion.

Case Study—Implementation and Details

The case study here presented is a church where a concrete channel was placed with $h_s = 200$ mm, immediately below the floor and in the inner face of the walls (Figure 9[a]).

Figure 9 represents the plan of the church (Figure 9[a]), a detail of the channel (Figure 9[c]), the configuration and position of the exterior probes (Figure 9[b]), a general scheme of the system (Figure 9[d]), and the position of the fan (Figure 9[e]).

CONCLUSIONS

The main conclusions of this paper are the following:

1. The treatment of rising damp in historic buildings is a problem difficult to solve.
2. Despite the existence of many treatment solutions and techniques, we consider the use of air with controlled relative humidity the most suitable treatment principle.
3. A hygro-regulated wall base ventilation device is essential to optimize the overall efficiency of the system and to control condensation risks inside.
4. The experimental study carried out in field showed that the most suitable criterion is to turn off the ventilation if

Figure 8 *(a) The backside of the church. (b) Removal of the pavement stones. (c) Ground floor waterproofing. (d) Installation of the ventilation channel. (e) Air inlet. (f) HUMIVENT device to control the fan taking into account the information from the two probes installed.*

the vapor pressure at the air outlet is lower than the vapor pressure at the air inlet.

5. A simplified methodology was developed for designing the wall base ventilation system to treat the rising damp.
6. The application of this simplified methodology was presented.
7. All the important parameters and system geometry were detailed, studied, and evaluated.
8. An example of an *in situ* application of the system was performed.
9. The wall base ventilation hygro-regulated system is a simple and economic technique for treating rising damp in historic buildings.
10. Figure 10 shows the interior of the church several years after the wall base ventilation system has been installed. It is possible to observe that the pathology associated with rising damp phenomenon has disappeared.

ACKNOWLEDGMENTS

M.F.C. Barbosa would like to thank FCT for financial support through the grant PD/BD/52658/2014 and J.M.P.Q. Delgado would like to thank FCT for financial support through the grant SFRH/ BPD/84377/2012. The authors would like to thank engineer Pedro Gonçalves and professor engineer Vasco Peixoto de Freitas for all the contributions to the design of the case study.

NOMENCLATURE

A	=	water absorption coefficient, kg/m^2·s$^{1/2}$
B	=	room
B1, B2	=	ventilation system inside and outside the wall
b	=	thickness of the wall, m
c_0	=	vapor concentration at the inlet of the system, kg/m^3
c^*	=	vapor concentration at the wall surface, kg/m^3
D_m	=	water vapor diffusion coefficient in the air, m^2/s
e_s	=	system evaporation, m/s
ε	=	porosity, m^3/m^3
ϕ	=	relative humidity, dimensionless
h_a	=	ground height adjacent to the wall, m
h_s	=	system height, m
h_{max}	=	maximum level of the damp front in the wall, m
L	=	length of the wall/system, m
p	=	water vapor partial pressure, Pa
ρ_w	=	water density, kg/m^3
S	=	sorptivity, m/s$^{1/2}$
T	=	air temperature, °C or K
θ_w or w	=	moisture ratio by volume, m^3/m^3
u	=	air velocity inside the system, m/s

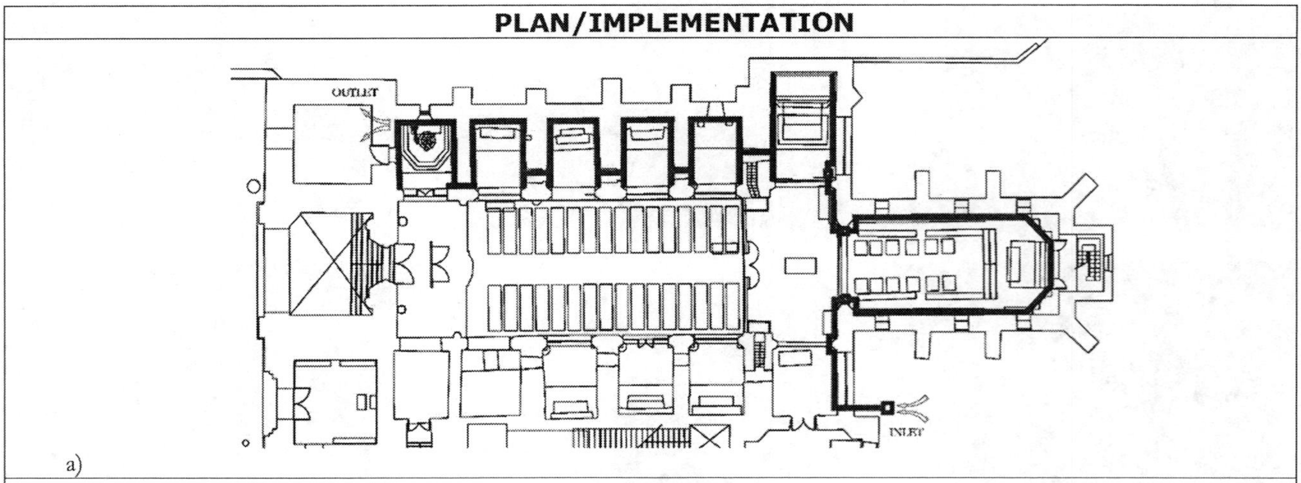

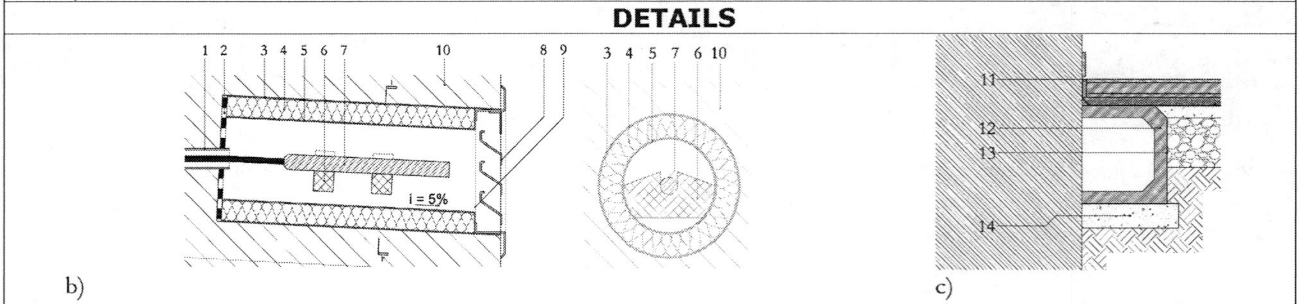

Legend: 1-VD pipe, 2-vapor barrier, 3-aluminium, 4-mineral fiber, 5-aluminium, 6-support, 7-outdoor temperature and relative humidity probe, 8-grid of outdoor probe, 9-anti-insect stainless steel mesh, 10-masonry wall, 11-waterproofing/vapor barrier, 12-precasted concrete channel, 13-vapor barrier, 14-concrete

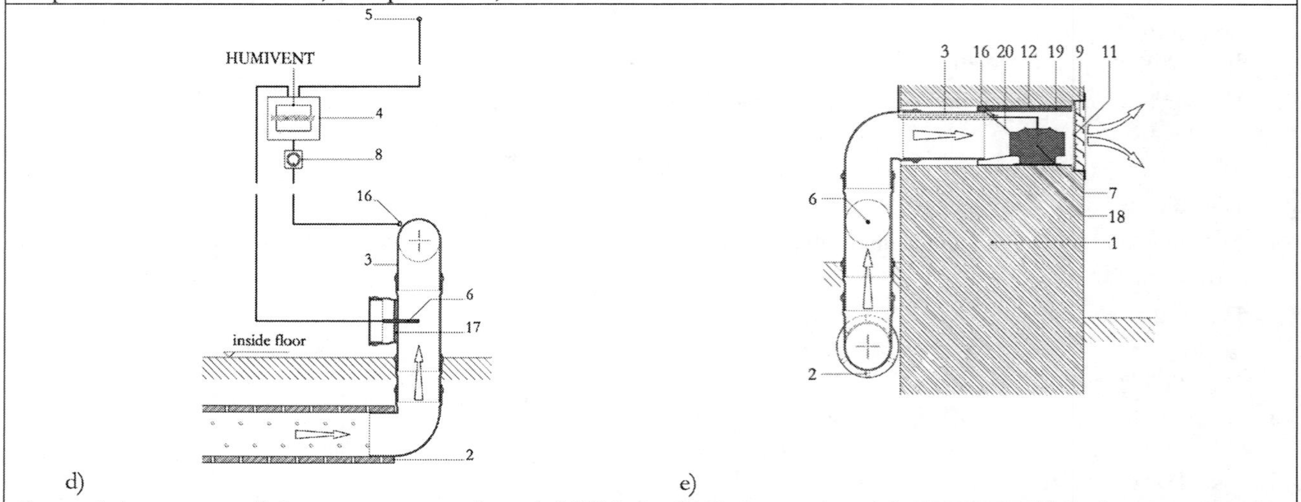

Legend: 1-masonry wall, 2-precast concrete channel, 3-PVC pipe ϕ200, 4-control module HUMIVENT, 5-admission air probe, 6-exhaustion air probe, 7-fan, 8-variable speed adjuster, 9-exhaustion grid of the interior ventilation system, 10-grid, 11-anti-insect stainless steel mesh, 12-aluminium, 13-mineral fiber, 14-aluminium, 15-bitumen concrete, 16-fan cable, 17-extruded polystyrene, 18-flexible foam rubber, 19-flexible polyurethane foam, 20-galvanized steel transition element

Figure 9 *(a) Plan of the church/implementation of the system—unscaled. (b) Probe (longitudinal and cross section)—unscaled. (c) Principle of the ventilation system—unscaled. (d) HUMIVENT control—unscaled. (e) Ventilator—unscaled.*

(before) (after)

Figure 10 *Performance before and after treatment.*

REFERENCES

Barberot, E. 1927. *Tratado práctico de edificación*. 2ª ed. ed. Barcelona.

Baldszuhn, B. 1990. Device for ventilating and venting, in particular, the foundation of a structure, Patent DE 39 02 065 A1. In https://google.com/patents/DE3902065A1?cl=en (14/10/2016). DE.

CSTC. 1985. Les procèdes de traitement des maçonneries contre l'humidité ascensionnelle, Note d'information technique 162. Centre Scientifique et Technique de la Construction. Bruxelles.

de Freitas, V.P. 1992. Transferência de Humidade em Paredes de Edifícios—Análise do fenómeno de interface. PhD, FEUP.

de Freitas, V.P., V. Abrantes, and P. Crausse. 1996. Moisture migration in building walls—Analysis of the interface phenomena. *Building and Environment* 31(2):99–108. doi: 10.1016/0360-1323(95)00027-5.

de Freitas, V.P., A.S. Guimaraes, and J.M.P.Q. Delgado. 2011. The "Humivent" device for rising damp treatment. *Recent Patents on Engineering* 5(3):233-240. doi: 10.2174/187221211797636863.

de Freitas, V.P., A.S. Guimarães, and M.I.M. Torres. 2007. The experimental characterization of a wall base ventilation system for the treatment of rising damp in historical buildings. 12th Symposium for Building Physics, Dresden, Germany, 29 to 31 of March, 2007.

Delgado, J.M.P.Q., A.S. Guimaraes, and V.P. de Freitas. 2013. Analysis and Monitoring of the Drying Process of a Hygro-Regulated Wall Base Ventilation System Implemented in a Historical Church to Control Rising Damp. *Drying Technology* 31(4):385-392. doi: 10.1080/07373937.2012.737395.

Downing, A.J. 1850. The architecture in country houses, including designs for cottages, farm-houses and villas with remarks on interiors, furniture's and the best modes of warming and ventilating. New York.

Franzoni, Elisa. 2014. Rising damp removal from historical masonries: A still open challenge. *Construction and Building Materials* 54:123–36. doi: 10.1016/j.conbuildmat.2013.12.054.

Groot, G., R. Van HeesS, and T. Wijffels. 2009. Selection of plasters and renders for salt laden masonry substrates. *Construction and Building Materials* 23.

Guimaraes, A.S., J.M.P.Q. Delgado, and V.P. de Freitas. 2013. Rising damp in walls: Evaluation of the level achieved by the damp front. *Journal of Building Physics* 37(1):6–27. doi: 10.1177/1744259112453822.

Guimarães, A.S., J.M.P.Q. Delgado, and V.P. de Freitas. 2010. Mathematical analysis of the evaporative process of a new technological treatment of rising damp in historic buildings. *Building and Environment* 45 (11):2414-2420. doi: http://dx.doi.org/10.1016/j.buildenv.2010.05.007.

Guimarães, A.S., V.P. de Freitas, and J.M.P.Q. Delgado. 2012. Wall base ventilation system to treat rising damp. 5th International Building Physics Conference - IBPC 2012, Kyoto, Japan, 28 to 31 May, 2012.

Guimarães, A.S., J.M.P.Q. Delgado, and V.P. de Freitas. 2012. *Wall base ventilation systems design to the treatment of rising damp*. Edited by LAP Lambert Academic Publishing GmbH & Co. KG.

Hall, C., and W.D. Hoff. 2002. Water transport in brick, stone and concrete. London, UK; New York, NY: Taylor & Francis.

Hanzic, L., and R. Ilic. 2003. Relationship between liquid sorptivity and capillarity in concrete. *Cement and Concrete Research* 33 (9):1385–88. doi: 10.1016/S0008-8846(03)00070-X.

Henriques, F.M.A. 1993. Acção da Humidade em Paredes - Formas de Manifestação, Critérios de Quantificação e Análise de Soluções de Reparação. PhD, Universidade Técnica de Lisboa - UTL.

James, M.P. 2003. Basement water drainage conduit and methods of use thereof, Patent US 6 598 360 B1. In https://www.google.com/patents/US6598360 (14/10/2016). U.S.

Mariz Ferreira, L. 2009. El Azulejo en la arquitectura de la ciudad de Oporto [1850-1920] caracterización e intervención. PhD, Universidad del País Vasco, Facultad de Bellas Artes.

Massari, G. 1971. Bâtiment humides et insalubres. Pratique de leur assainissement. Paris, Eyrolles.

Robert, M.T. 2004. Drying system for structural waterproofing, Patent U.S. 2004 / 0139688 A1. In http://ip.com/patapp/US20040139688 (5/09/2010). U.S.

Torres, M.I.M., and V.P. de Freitas. 2002. Modeling versus experimental study of rising damp in historical buildings. XXX IAHS-World Congress on Housing 2002, Coimbra, Portugal, September 9-13, 2002.

Torres, M.I.M., and V.P. de Freitas. 2006. Rising damp in historical buildings: A serious building pathology. CIB W086, Rome.

Torres, M.I.M., and V. P. de Freitas. 2007. Treatment of rising damp in historical buildings: wall base ventilation. *Building and Environment* 42(1):424–35. doi: 10.1016/j.buildenv.2005.07.034.

Torres, M.I.M., and V. P. de Freitas. 2010. The influence of the thickness of the walls and their properties on the treatment of rising damp in historic buildings. *Construction & Building Materials* 24(8):1331-9. doi: 10.1016/j.conbuildmat.2010.01.004.

Torres, M.I.M., and Óscar López. 2015a. Drying out walls after a flood: influence of the presence of rising damp. Rehab Structures 2015, Azores, Portugal.

Torres, M.I.M., and Óscar López. 2015b. Ventilation system for drying out walls after a flood: influence of the thickness of the walls. PANACM 2015-MECOM 2015, Buenos Aires, Argentine.

Study of Energy Intensity and WWRs of Existing Office Buildings

Viktoriya Mykytyak, MBSc **Hitesh Doshi, MASc, PEng**
Ramani Ramakrishnan, DSc, PEng

ABSTRACT

Thermal performance of the building envelope is a major factor to be considered in the design of new buildings. Because the relative thermal resistance of glass is low, the amount of fenestration in the building envelope is usually controlled. For example, SB-10 of the Ontario Building Code considers a window-to-wall ratio (WWR) of 40% as the prescriptive threshold for the thermal resistance of the enclosure. However, WWR is not a parameter that is included in the reporting of energy intensity and the energy performance of buildings. Studies on the energy influence of WWR and building envelope have focused on the influence of building geometry based on modeling. The current study investigated the relationship of the energy intensity of the existing commercial and institutional building to the WWR, through recorded energy use. Fifteen office buildings in Ontario and Quebec, Canada, were included in the study. Electricity and natural gas consumption were obtained through the energy tracking service, which monitors billing data using the EnergyCAP software. An overall picture of the building's monthly and annual energy consumption was thus determined. The investigation showed that WWR directly influenced the energy intensity of the building. However the impact was not consistent. The energy intensity varied measurably even for buildings with similar WWR. These results, based on the actual measured data, raise the question about appropriateness of using WWR as a threshold for prescriptive requirements. The results of the study are described in this paper.

INTRODUCTION

Energy conservation and sustainability are the main focus for the construction industry, in particular to the commercial and institutional sector, due to their very high demand for energy. According to National Resources Canada (2008), commercial and institutional (C&I) establishments accounts for two thirds of the building industry. Among the commercial and institutional establishments, Ontario accounts for the highest amount of energy consumption (approximately 30%), and Quebec stands third with 23% of overall energy consumption. Therefore, the C&I sector in Ontario and Quebec requires close attention in terms of energy efficiency. Typically, efforts to reduce energy consumption in the C&I sector are geared towards research and development of energy-efficient building technologies such as equipment, heating and air conditioning (HVAC), lighting, and ventilation systems. It is not very common to proceed with passive technologies such as window-to-wall ratio (WWR) to achieve energy efficiency in the existing building. Despite common strategies, WWR solution has been promoted on a large scale through different mechanisms, including building codes and incentives. Supplementary standards such as SB-10 of the Ontario Building Code address WWR specifications and compliances as to their implementation. The Ontario Building Code requirements in this regard are modeled after ASHRAE Standard 90.1. For design of new buildings, a WWR of 40% is considered a threshold in Ontario for prescriptive solutions to the thermal resistance of the enclosure.

A number of Canadian agencies as the Conference Board of Canada and the Natural Resources Canada's Office of Energy Efficiency gather data on the energy intensity of the construction industry (CBC 2013; OEE 2013). The Confer-

Viktoriya Mykytyak is a graduate of the Master of Building Science program and **Hitesh Doshi** and **Ramani Ramakrishnan** are professors at Ryerson University, Toronto, Ontario.

ence Board of Canada evaluates energy intensity in terms of the GDP, but OEE determines the energy consumption in gigajoules per square meter (GJ/m^2). In addition, BOMA assessed a large data set of buildings and evaluated their energy consumption (BOMA 2014). The result of quantifying the energy intensity data gathered by these agencies showed that the average energy intensity is around 1.22 GJ/yr/m^2. However, it is to be noted that the energy intensity is expressed solely in terms of the total floor area of the building without identifying individual influencing parameters such as WWR.

Harvey, in his study of the essential principles of low-energy design, has quantified the necessary steps such as building orientation (this includes also surface area-to-volume ratio and thermal mass), high performance building envelope, passive systems (heating, cooling, ventilation, and daylighting), individual energy devices (to ensure that energy devices can be controlled individually based on necessity of users), and proper commissioning of the building (Harvey 2013). One of the determining factors of the building's energy load, according to Harvey, is WWR. Proper selection of WWR can help in reducing energy intensity.

Building energy audits are usually conducted in the C&I sector to analyze the energy consumption of the building and to make a preliminary list of energy efficiency measures. Some of its goals are to reduce wastage at the source, use of high-efficiency products, optimize use of facilities to ensure the best possible comfort for occupants, and, of course, minimize energy consumption. However, energy audits very rarely include any discussion on building envelope assemblies, and, hence, WWR is not considered during energy audit either.

Finally, substantial research is available in the literature that has studied the impact of energy efficiency and fenestration (ASHRAE 2012; Ko 2009; Public Works and Government Services Canada 2002). The main findings of these studies are that WWR is a very important factor that impacts energy use compared to other components of building envelope, due to the relatively lower thermal resistance of the glazing portion of the building envelope.

It can be seen, therefore, that WWR is an important factor that should be considered in the design of buildings. Dogrusoy and Tureyen (2007) and Ross (2009) in their studies showed that WWR, has slightly overpowering results on energy use in commercial buildings than other building geometry parameter. In addition, Ferdous's (2012) research concluded that WWR influences the energy intensity of existing office buildings. The current OBC requirements put a threshold of WWR of 40% up to which the thermal characteristics of the building envelope may be chosen based on prescriptive requirements. However, the rationale for the 40% threshold is not very well understood.

All of the studies that have been examined were carried out on buildings using energy modeling and typical shapes. There is hardly any information on the energy consumption in existing buildings and its relationship to the WWR. One of the reasons for study of existing buildings is that the shapes assumed in these studies are ideals shapes and much different from the architectural form assumed in these studies. The current study explored the possibility of using actual energy consumption data along with actual architectural forms and building envelope characteristics to examine the type of relationship that exists between the WWR and the energy consumption. Only a summary of the study is presented in this report. Complete details of the full study are contained in Mykytyak (2015).

THE STUDY

The focus of the study was to determine and quantify the extent of WWR influences on energy usage of existing office buildings based on actual energy usage information. The investigation consisted of reviewing and evaluating utility bills and calculations of energy consumption of each building. The required data was provided by the Operational and Maintenance department of SNC Lavalin (2015). This allowed the study to rely on the actual energy consumption of office buildings located in South Central Canada. Fifteen buildings with different architectural characteristics were selected for the study. These fifteen buildings were selected from a portfolio of over forty buildings managed by a single property manager for different clients. Energy audits including details about the energy consumption over a period of years were available for these forty buildings. Upon examination of these forty buildings, fifteen were chosen for the study based on a similar nature of occupancy and, therefore, operational characteristics. The fifteen buildings, six in Toronto area, five in Ottawa, and four in Montreal area, are located in three weather regions. These different regions required normalizing of the energy use data for differences in weather in order for them to be compared for the purpose of this article. The architectural aspects varied for these buildings, and there was no further shortlisting done on the basis of architectural features. Two of these architectural features, along with the age being the third, were used as variables in the study. The choice of these fifteen buildings cannot be considered random and therefore it is not the intention of this study to draw inferences based on statistical significance. The results of this study, therefore, have to be viewed with the above limitation in mind and be considered as providing a case-study-based response to the main questions asked: (a) How does energy intensity of existing case study office buildings compare with averages for similar buildings? (b) How does the energy intensity of existing case-study buildings vary with the WWR of that building? The investigation followed a simple three-step process: (a) building description such as shape, age, and height; (b) calculation of energy consumption and WWR; and (c) analysis of energy as a function of WWR. Details of this framework are included in Mykytyak (2015).

Selected Buildings

Fifteen office buildings were selected for this study. Eleven of the case study buildings are located in Ontario and four are located in Quebec. Buildings were categorized as to their shape, height, and age. The building categories are summarized in Table 1 below. The following properties of the buildings can be surmised from Table 1:

- Shape: Three are square, four are rectangular, two are complex, and six are complex-rectangular buildings.
- Height: Five are low-rise (1 to 3 stories), five are mid-rise (4 to 5 stories), and five are high-rise buildings (6 and above).
- Age: One was constructed in 1873, three in 1970s, ten in 1980s, and one in 1990.

Energy Use of Buildings

The energy consumption information from these Level 1 audits were available for all buildings. Utility bill information (electricity and natural gas consumption) for a range of years was obtained through energy tracking service, which monitors billing data using the EnergyCAP software application. This data was considered in the analysis to obtain an overall picture of the building's monthly and annual energy consumption. Details of these calculations and data are reported by Mykytyak (2015). Because weather varies from day to day and year to year, normalization measures provide the best solution to represent the weather impacts on energy consumption of the building. Despite normalization for weather, some variations might occur resulting from the use and occupancy of the building. These factors have not been recorded and are difficult to take into account. Energy used for heating and cooling is directly dependent on how cold or hot it was during a specific year or season. Weather normalization was applied to the year with the lowest energy usage so it could be compared to the year with highest energy usage. Comparing normalized data showed actual energy consumption not dependent on weather during such periods. Data for heating and cooling degree days was taken from the weatherstats.ca web page. A trend line was developed from a scatter plot of data with the x axis representing degree days and y axis representing energy consumption. An example of this analysis is shown for a building in Figure 1 and represented by the trend line below:

$$y = 3.1724\,x + 1340.5 \tag{1}$$

From the above equation, normalized average energy index was calculated as 1.14 GJ/yr/m^2. This calculated value is used for further comparisons. Figure 1 also shows linear progression for normalized data with equations.

WWR Determination

To calculate WWR for each case study building, architectural drawings were utilized for four case-study buildings: #1, #2, #3, and #4. From elevations as well as cross section detail drawings WWR was calculated for each building. For the rest

Table 1. Buildings Description Summary

Building Number	Location	Shape	Height (Number of Stories)	Construction Year	Gross Floor Area, m^2
1	Toronto	Square	12	1989	25,424
2	Mississauga	Complex	6	1988	8225
3	Toronto	Rectangular	8	1970	9240
4	Mississauga	Complex	10	1988	15,712
5	Mississauga	Complex	1	1988	6332
6	Toronto	Rectangular	2	1980	2130
7	Ottawa	Rectangular	4	1993	8180
8	Ottawa	Complex	5	1989	5020
9	Ottawa	Square	3	1989	4854
10	Ottawa	Square	5	1989	8270
11	Ottawa	Square	4	1988	6162
12	Montreal	Complex	7	1873	3518
13	Dorion	Rectangular	2	1977	8022
14	Cote-St-Luc	Rectangular	5	1976	7943
15	Gatineau	Complex	3	1989	5518

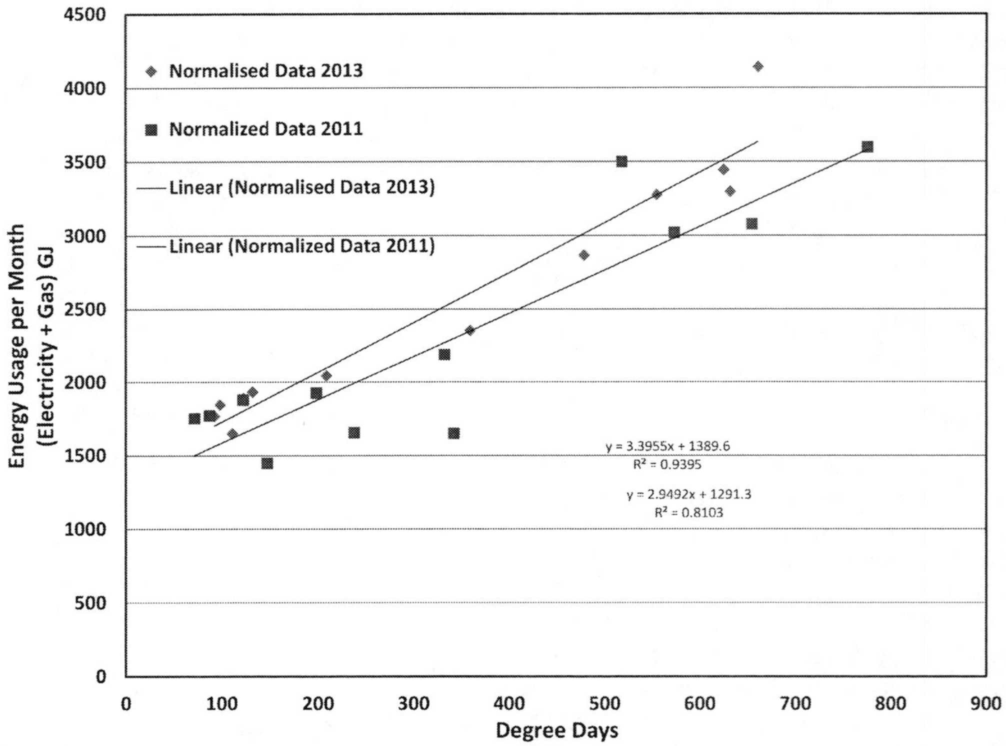

Figure 1 *Example building—normalized energy intensity calculations.*

of the buildings without architectural drawings, the WWR was determined based on site visit and photographs of elevations. The WWR is summarized in Table 2. In Table 2, the reference to Google Maps indicates that the shape of the building was determined from the image acquired from Google Maps and confirmed through field observations.

ANALYSIS

Analysis of Energy Audits

All of the audit reports contain the following information: overview and general information on the building, analysis of the utility bills, energy index calculations and load factor analysis, the initial selection of energy conservation measures (ECM), and final energy conservation measures. The conservation measures section includes existing conditions and proposed solutions for various building systems and equipment, such as lighting fixtures replacement, upgrade of thermostat control, retrofit of heating/cooling equipment, and others. Each report provides an executive conservation measure summary with a list of high level cost estimate for all upgrades. Reports also provide the next step section, where anticipated implementations are discussed and a high level plan of actions is provided.

After analyzing all 15 energy audit reports, it was observed that the building envelope was only briefly discussed in four reports. Building 3 report proposes to replace single-pane outdated windows with new, double-

Table 2. WWRs of Buildings

Building Number	Shape	WWR	Notes
1	Square	44%	Drawings provided
2	Complex	27%	Drawings provided
3	Rectangular	39%	Drawings provided
4	Complex	42%	Drawings provided
5	Complex	40%	Site visit
6	Rectangular	40%	Site visit
7	Rectangular	32%	Google maps
8	Complex	43%	Google maps
9	Square	42%	Google maps
10	Square	42%	Google maps
11	Square	42%	Google maps
12	Complex	19%	Google maps
13	Rectangular	25%	Google maps
14	Rectangular	23%	Google maps
15	Complex	38%	Google maps

pane, thermally broken, solar reflective windows. The Building 5 report proposes to upgrade the existing roof due to ponding and signs of deterioration. The Building 9 and 11 reports propose to upgrade existing windows by adding supplemental window glazing/shading to help reduce the heat penetration during the cooling season. One can surmise then that building envelope upgrade is considered only in a limited manner as a part of the potential energy and cost savings measure during early energy audits analysis. However, most of the reports claim that the energy audit is based on an established process to identify measures for reducing energy consumption (perform renovations to eliminate thermal bridges), measures for improving occupant comfort, and measures for reducing greenhouse emissions. However, a very small percentage of them discuss building envelope concerns. It is possible that this may be the case because envelope retrofits are often done in conjunction with other performance issues and often have higher payback than other measures.

Energy Intensity Analysis

Below is the summary of all 15 case-study buildings. Table 3 provides the year range for which energy bills were available, electricity consumption and cost, gas consumption and cost, and energy intensity result for each building. All case study buildings are office facilities with typical office hours and, in most cases, with typical occupancies. There were some anomalies that are discussed in more detail in Mykytyak (2015).

As it was specified previously, energy intensity benchmark value is used from the BOMA report, where average energy use intensity for certified Office Buildings, is 27.1 ekWh/ft^2/yr (or 1.05 GJ/yr/m^2). The comparison of energy intensity of each case study building to BOMA's average value was also shown in Table 3. Two buildings, #8 and #10, show significantly higher value from the benchmark value, and Building #5 shows very low value in comparison to the rest of the buildings. More than half of the case study buildings have energy index values around the BOMA benchmark, as in Table 3.

The impact of building shape on the energy intensity did not appear to show any particular trend. Energy intensity of buildings with complex shapes and square shapes showed a larger variation compared to buildings with rectangular shapes. There is no definitive statistically valid conclusion that can be drawn with respect to the impact of the shape. In terms of the height, the low-rise, mid-rise, and high-rise buildings on average have an energy intensity of 1.00, 1.54, and 1.13 GJ/m^2. respectively. The mid-rise buildings appear to have a higher intensity. Further explanation of these variations can be found in Mykytyak (2015).

Table 3. Summary of Annual Average Energy Use

Building Number	Utility Bill Range	Electricity Usage, kWh	Gas Usage, m^3	Average Energy Intensity, GJ/m^2	Normalized Average Energy Intensity, GJ/m^2	BOMA Datum, GJ/m^2
1	2007–2014	5,356,500	256,540	1.13	1.14	1.05
2	2007–2014	2,121,843	82,820	1.20	1.21	1.05
3	2007–2014	1,912,410	158,362	1.38	1.36	1.05
4	2007–2014	3,421,950	58,556	1.08	1.09	1.05
5	2007–2014	454,070	17,995	0.36	0.36	1.05
6	2007–2014	237,525	26,523	0.86	0.93	1.05
7	2007–2014	2,358,440	31,962	1.18	1.21	1.05
8	2007–2014	1,829,845	81,765	1.92	1.93	1.05
9	2007–2014	1,358,050	4,622	1.04	1.07	1.05
10	2008–2014	6,121,642	n/a	2.66	2.34	1.05
11	2007–2014	1,827,60	n/a	1.07	1.1	1.05
12	1998–2014	739,530	14,663	0.91	0.94	1.05
13	2003–2014	1,692,200	8,189	0.91	0.92	1.05
14	2000–2014	3,060,854	n/a	1.39	1.47	1.05
15	2007–2014	1,907,417	64,139	1.68	1.69	1.05

The impact of age of the building on energy intensity can also be surmised from Table 3. None of the buildings that were studied have undergone any substantially different energy upgrades. Most of the buildings were constructed between 1970 and 1993; only one building was constructed in 1873. Building #12, which was built in the 19th century, has relatively low energy consumption in comparison to the newer buildings. This building has a WWR of 19%, and its impact is reflected in the low-energy intensity. On the other hand, two buildings with the highest energy consumption were both constructed in 1989 and had WWR of just over 40%. There were other buildings with similar WWR but with lesser energy intensity. The next section discusses the impact of WWR in greater detail.

WWR Analysis

Different researchers have identified different impacts of the WWR on the energy consumption in buildings. In this approach (Mykytyak 2015), contributions to the energy load as a result of different architectural aspects such as orientation, height, WWR, and thermal resistance values were determined based on reported values for existing buildings. From these contributions, the WWR contributions were predicted for case study buildings. The results are presented in Table 4. An energy balance approach using information about the performance of a typical office building was used to predict the amount by which the existing WWR contributed to the overall energy consumption.

Table 4. WWR and Energy Consumption

Building Number	Average Energy Intensity, $GJ/yr/m^2$	WWR, %	WWR and Energy Consumption, %
1	1.1	44	13%
2	1.2	27	15%
3	1.4	39	11%
4	1.1	42	13%
5	0.5	40	11%
6	1.0	40	8%
7	1.2	32	16%
8	1.9	43	13%
9	1.1	42	17%
10	1.8	42	27%
11	1.1	42	18%
12	0.9	19	15%
13	0.9	25	15%
14	1.5	23	17%
15	1.7	38	14%

Table 4 summarized the impact in percentage that WWR has on overall building energy consumption, based on heating, cooling, and lighting energy balance. The same loads percentage distribution was applied for all buildings but based on normalized energy intensity value, electricity, and gas consumption—percentages vary; some buildings do not use natural gas for building operations, and this also influenced calculations. As per these results, WWR impacts significant percentage of energy consumption for these case study buildings. As can be seen, the WWR can have an impact varying from 8% to 27%, depending on the various aspects of each building.

The relationship between energy intensity and the WWR of the 15 case-study buildings can be surmised from Table 4. The WWR range of the case study buildings is generally between 20% to 45%, with one building that has a WWR lower than 20%. In addition, WWR was analyzed against the age of the buildings to see if newer buildings have higher fenestration areas. It was determined that newer building do have higher WWR, and Building #12, which was constructed in 1873 has the lowest WWR.

Analysis Summary

Energy intensity of existing building is directly impacted by the WWR; however, the actual data from 15 case study buildings showed that the impact was inconsistent. Most of the buildings have energy intensity around 1–1.25 $GJ/yr/m^2$, meaning that they are relatively energy-efficient buildings in relationship to the BOMA benchmark. The energy intensity varies widely in case study buildings with more complex shape and ones that are in the mid-rise height category. It is also seen that although WWR has an impact on energy intensity, the relationship is not simple. The energy intensity varied widely even for buildings with relatively similar WWR.

CONCLUSIONS

WWR is directly related to energy intensity of the existing building. Fifteen case-study buildings were analyzed to calculate how energy intensity of existing case study office buildings compares with averages for similar buildings and those that would be resulting from the application of current standards. BOMA average energy use intensity for certified office buildings (1.05 $GJ/yr/m^2$) was selected as the energy intensity benchmark value. Calculations for case study buildings showed that the building shape did not directly influence energy intensity; however, analysis based on the age of the building showed that the newer buildings (except for one with very low WWR) are more energy efficient, and analysis on buildings' height showed that mid-rise study buildings have higher energy intensity. More importantly, the study shows that there are buildings with lower WWR that have higher energy intensity than buildings with higher WWR. Currently the prescriptive requirements of the OBC, SB-10, apply to buildings with WWR below 40%. The results of this study, which examined the buildings with similar envelope characteristics, indicate factors other than the WWR are important in the actual energy use of the buildings.

REFERENCES

ASHRAE. 2012. *ASHRAE Handbook—Fundamentals*. Atlanta: ASHRAE.

BOMA. 2014. *BOMA BEST Energy and Environment Report*. Retrieved from http://www.bomabest.com/news-publications/publications/.

CBC. 2013. How Canada performs: A report card on Canada, energy intensity. The Conference Board of Canada. Retrieved March 16, 2015, from http://www.conferenceboard.ca.

Dogrusoy, I., and M. Tureyen. 2007. A field study on determination of preferences for windows in office environments. *Building and Environment* 42(10):3660–68.

Ferdous, T. 2012. *Determining the effect of building geometry on energy use patterns of office developments*. Toronto, Ontario: Major Research Project presented to Ryerson University.

Harvey, L.D. 2013. Recent advances in sustainable buildings: Review of the energy and cost performance of the state-of-the-art best practices from around the world. *Annual Review of Environment and Resources* 38:281–309.

Ko, D.H. 2009. *Fenestration guideline for energy and daylight efficiency: Evaluation and prediction of performance in office buildings*. Chicago: Illinois Institute of Technology. ProQuest, UMI Dissertations Publishing.

Mykytyak, V. 2015. The impact of window-to-wall ratio on energy intensity of existing office buildings. MBSc Research paper, Department of Architectural Science, Ryerson University, Toronto.

National Resources Canada. 2008. 2008 Commercial & Institutional Consumption of Energy Survey. http://oee.nrcan.gc.ca/publications/statistics/cices08/pdf/cices08.pdf

OEE. 2013. *Survey of commercial and institutional energy use: Buildings 2009*, Summary Report. Ottawa: Natural Resources Canada, Office of Energy Efficiency.

Public Works and Government Services Canada. 2002. Daylighting guide for Canadian commercial buildings. Public Works and Government Services Canada.

Ross, B.M. 2009. Design with energy in mind: Toward a low-load and high-satisfaction civic architecture in the Great Lakes Basin. Master of Architecture thesis, University of Waterloo.

SB. 2011. Supplementary Standard SB-10, *Energy efficiency supplement*. Toronto: Ministry of Municipal Affairs and Housing.

SNC Lavalin. 2015. Personal Communication. SNC Lavalin Construction (Ontario) Inc., Toronto.

Using Fiber Insulation as a Means of Drying Internally Insulated Walls

Ekaterina Tzekova, PhD
Associate Member ASHRAE

Kim Pressnail, PhD

ABSTRACT

Previous field studies in Toronto, Canada, have demonstrated the drying ability of a dedicated vented air space when historic solid masonry walls are internally insulated using spray-foam insulation. These studies found that when masonry walls with vented air spaces were exposed to sun-driven moisture, the drying potential of the walls could be improved. During field tests, a multi-component envelope system with an air- and vapor-permeable polypropylene waffle material was used to create the air space. The work reported here, however, investigates an alternative way of creating the vented air space using air-permeable rock wool insulation in an effort to reduce construction time and costs. The venting performance of air-permeable rock wool insulation was tested in a climate simulator. Two different insulation densities were subjected to high inward vapor pressure gradients using a simulated solar gain. In addition, various inlet and outlet vent areas were tested. The performance of the rock wool insulation when it was used as part of the venting system was compared to the performance of clear air spaces with varying thicknesses when subjected to similar vapor pressure gradients and ventilation areas. Laboratory tests revealed that walls with a clear 19 mm (3/4 in.) air space removed between 59% and 95% of the total moisture entering the walls. In comparison, similar walls that incorporated rock wool insulation and no clear air space removed between 52% and 90% of moisture. These tests have revealed the potential of using air-permeable insulation to remove moisture from walls in lieu of a clear air space.

BACKGROUND

One of the main challenges of improving the thermal performance of older buildings is the preservation of the historic facade. Improving thermal performance without changing the facade requires insulating from the interior. However, in cold climates, this approach reduces the average temperature of the masonry, leaving it more vulnerable to salt spalling and freeze-thaw damage (Groot and Gunneweg 2012). In one study by Van Der Klugt (1988), laboratory and field investigations revealed that masonry drying rates after installing interior insulation were reduced by approximately 25%. Reducing the drying rate can negatively affect brick durability, particularly where bricks have a moderate-to-high absorbency and a high exposure to exterior moisture. Such bricks have been found to be at an increased risk of freeze-thaw damage (Van Der Klugt 1998; Lstiburek 2010).

A further complication is the variability of brick properties. Predicting the freeze-thaw durability of old masonry buildings is difficult because brick properties vary, even within the same wall. Properties vary because the bricks used in historic buildings were not all created equal. Due to the temperature variation within beehive kilns, the constituents in some older bricks are more poorly fused than others (Laefer et al. 2004; Karagiozis and Kuenzel 2009). This means that historical bricks will not perform equally—with some bricks being more durable than others. Ultimately, varying brick properties coupled with cooler and more damp wall conditions means that predicting brick durability following insulation retrofits can be challenging and can lead to uncertainties.

Many studies have evaluated the effects of ventilating various cladding materials. A comprehensive study funded by ASHRAE examined the moisture performance of different

Ekaterina Tzekova is the building research manager at the The Atmospheric Fund, Toronto, Ontario. **Kim Pressnail** is an associate professor in the Department of Civil Engineering, University of Toronto, Toronto, Ontario.

sheathing membranes within ventilated walls featuring a variety of cladding materials, including brick veneer. The study found that ventilation can be used to reduce the amount of moisture entering a variety of wall systems and should be incorporated wherever possible (Burnett et al. 2004).

A study in Germany (Popp et al. 1983), as reported by Stovall and Karagiozis (2004), revealed that even small vent openings with a clear air space improved drying rates of asbestos cement and wood siding. To reach this conclusion, the researchers measured the air velocity and the number of air changes behind the siding. A similar study found that drying rates could also be improved with increased levels of solar radiation (Straube and Burnett 1998). The authors found that solar heating increased the thermal buoyancy and moisture carrying capacity of the ventilation air in the air space. A third study found that providing a 25 mm (1 in.) vented air space was able to effectively reduce the amount of sun-driven moisture entering a series of walls in a laboratory setting (Pressnail et al. 2003).

Stovall and Karagiozis (2004) modeled a vented air space using computational fluid dynamics (CFD) to best approximate air movement in a wood frame with brick veneer cladding. The wall was modeled with 89 mm (3.5 in.) glass fiber insulation, 166 mm (6.5 in.) brick veneer, and two different air space thicknesses: 19 and 50 mm (3/4 and 2 in.). Different models were examined with a variety of wind velocities and exterior temperatures and compared to empirical data. This study found that there was no observable mass flow rate difference between the 19 and 50 mm (3/4 and 2 in.) air spaces. However, the authors found that by doubling the distance between vent holes, the airflow rate also doubled. Furthermore, the researchers concluded that the airflow within the air space was driven by buoyancy due to differences between the outdoor air temperature and the temperature within the air space.

A similar study used WUFI 4.1 to model vented air spaces (Straube and Finch 2009). Airflow and air leakage were represented as moisture sources, while the effects of drainage and ventilation were modeled as moisture sinks. The vented air space was shown to reduce sun-driven moisture as well as increase drying rates in walls with absorptive claddings such as brick veneer. With high ventilation rates within the vented air space, the walls were able to dry faster.

After considering these and other studies, the question arises: will the provision of a vented air space improve the drying potential of internally insulated solid masonry walls in a drying climate? To answer this question, two field studies were carried out at the University of Toronto (Tzekova et al. 2011, 2012; Tzekova 2015). Results from single-family and commercial building field trials showed that where a vented air space was incorporated, net drying occurred on the solar-exposed faces. The air space constructed in the field was created using a polymer mesh normally used to protect air spaces from mortar droppings during brick veneer construction. The polymer mesh was applied to the inboard face of the masonry wall and then a spun-bonded polyolefin, air-resistant sheathing was installed over the polymer mesh to ensure that the insulating foam did not penetrate the polymer mesh. Next, a layer of polyurethane spray foam was applied over the protected air space system. While this envelope system ultimately functioned well, this method of construction was labor-intensive and expensive given the number of materials involved.

To reduce construction time and material expenses, the laboratory investigation presented here researched alternative ways of creating a vented space. In lieu of the polymer mesh and spray-foam system, the performance of air-permeable rock wool insulation was evaluated through a number of laboratory experiments.

LABORATORY TESTS

This section describes the two sets of tests that were conducted in a laboratory setting. The first tests were conducted as initial, exploratory tests of the walls and are referred to as *preliminary tests*. After these tests were carried out, the experimental procedures were improved and a second series of tests were conducted. These tests are referred to as *modified tests*.

Test Apparatus

Various walls were tested using the climate simulator in the Building Science Laboratory at the University of Toronto. Figure 1 shows a plan view of the laboratory where the various wall sections were tested. The climate simulator consisted of two rooms: a warm room and a cold room. The temperatures within the warm and cold rooms were controlled by means of heaters and a chiller. To reduce the temperature variation within these rooms, particularly during defrost cycles, two guard rooms were constructed. The air temperature within these guard rooms fluctuated by less than ±1.5°C (±2.7°F). As the temperature was kept relatively constant throughout the tests, and the guard rooms were isolated from the rest of the laboratory during the tests. Prior to the start of each test, the guard rooms were allowed to reach equilibrium, minimizing any relative humidity fluctuations during testing.

The air temperature within each guard room was regulated using type T thermocouples, and readings were stored in a data acquisition (DAQ) unit. Once the temperature within a particular space was beyond a preset range, the DAQ unit sent a signal to a series of relays. These relays were connected to electric resistance heaters used to control the ambient air temperatures in both guard rooms. Circulation fans ensured mixing of the air within in guard room 2, where a light array was used. Temperature fluctuations larger than ±1.5°C (±2.7°F) occurred only during the 30 min chiller defrost cycle periods, which took place every 6 h. During the tests, the air temperatures in guard rooms 1 and 2 were generally maintained between 19°C and 21°C (66°F and 70°F). Guard room 2 was kept at the same air temperature as guard room 1 to ensure that the vapor pressures were the same.

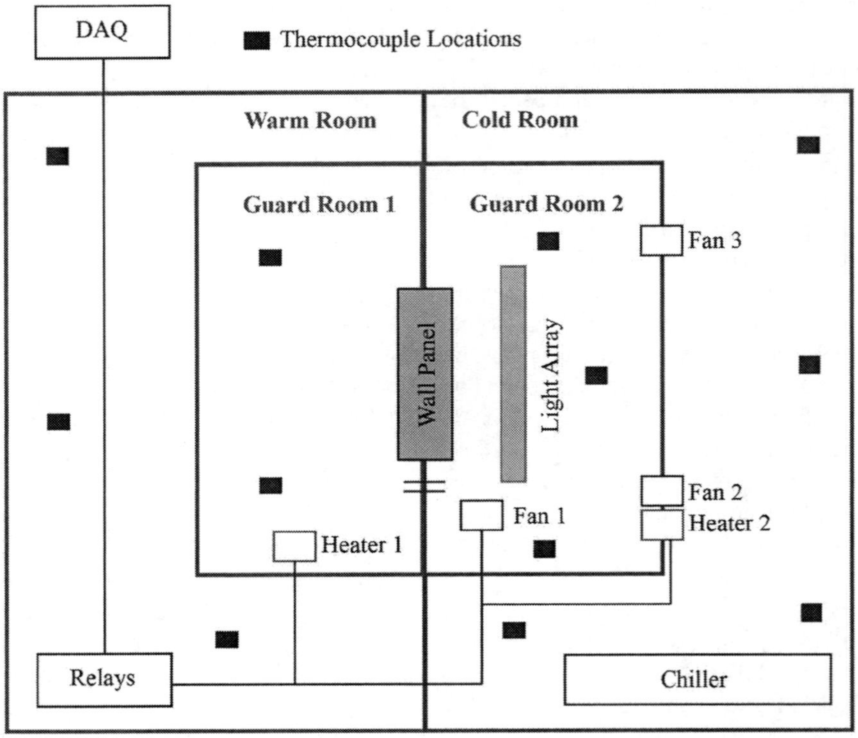

Figure 1 *Plan view of the climate simulator showing the temperature control systems.*

Air and surface temperature readings were measured using thermocouples. Readings were taken every 15 s and stored in the DAQ unit. Based on predefined channel groupings, a computer program (HTBasic) calculated the average temperature for each zone. The program was set to automatically control the heaters and fans for guard room 1, guard room 2, and the warm room. The cold room chillers were controlled by an independent thermostat, located on the side of the climate simulator control unit. Multiple temperature readings were taken from each zone to ensure an accurate representation. The thermocouples were mounted approximately 30 cm (1 ft) from the ceiling and the floor.

Thermocouples were also placed on and within the test wall samples. Thermocouples were mounted at the top, middle, and bottom portions of the pine siding as well as between the polyethylene and rock wool insulation. Wireless data loggers (SMT A2) with internal temperature and relative humidity sensors were mounted at the rock wool and pine siding interface as well as near the top and bottom of each test wall. When walls with a clear air space were tested, these sensors recorded the conditions within the air space. The operating conditions of the data loggers are 0°C to 40°C (32°F to 104°F), 5% to 100% rh (noncondensing), while the accuracy is ±1°C (2°F) and ±5% rh (SMT 2014). Measurements were taken every 15 s and downloaded at the end of each test.

A light array was used to increase the wall surface temperatures in order to create a vapor pressure gradient that would drive moisture into the sample test walls. The nominal power used by the light array was similar to the power used by previous researchers (L. Kan 1999; V. Kan 2002). For the preliminary tests a nominal power of 1228 W/m^2 (115 W/ft^2) was supplied to the light array by using a total of four 150 W and five 250 W bulbs. Although a nominal power of 1228 W/m^2 was supplied by the array, only a portion of the power supplied is actually absorbed as radiation by the walls. The light bulbs were evenly spaced along six rows, alternating between high and low wattage, to reduce the likelihood of local hot areas developing on the radiated surface. After some trials in which the distance between the wall and the light array was varied, the array was positioned approximately 30 cm (1 ft) from the wall for the preliminary tests. This distance was found to provide a reasonably constant wall surface temperature. The goal in the preliminary series of tests was to provide an irradiated wall surface temperature that was approximately 10°C (18°F) warmer than the air temperature in guard room 2.

Preliminary Tests: Sample Preparation

All of the test walls were constructed using 38 × 140 mm (2 × 6 in.) spruce framing. Each wall, including the studs, was approximately 0.40 m (1.3 ft) wide by 2.05 m (6.7 ft) high and 160 mm (0.5 ft) thick without an air space or 180 mm (0.6 ft) thick with an air space. Studs were spaced 406 mm (16 in.) on center. Figure 2 shows a plan view of the two wall types used during the preliminary testing. Wall A was constructed without any vent holes and without a clear air space. With this type of construction, the rock wool insulation was placed directly

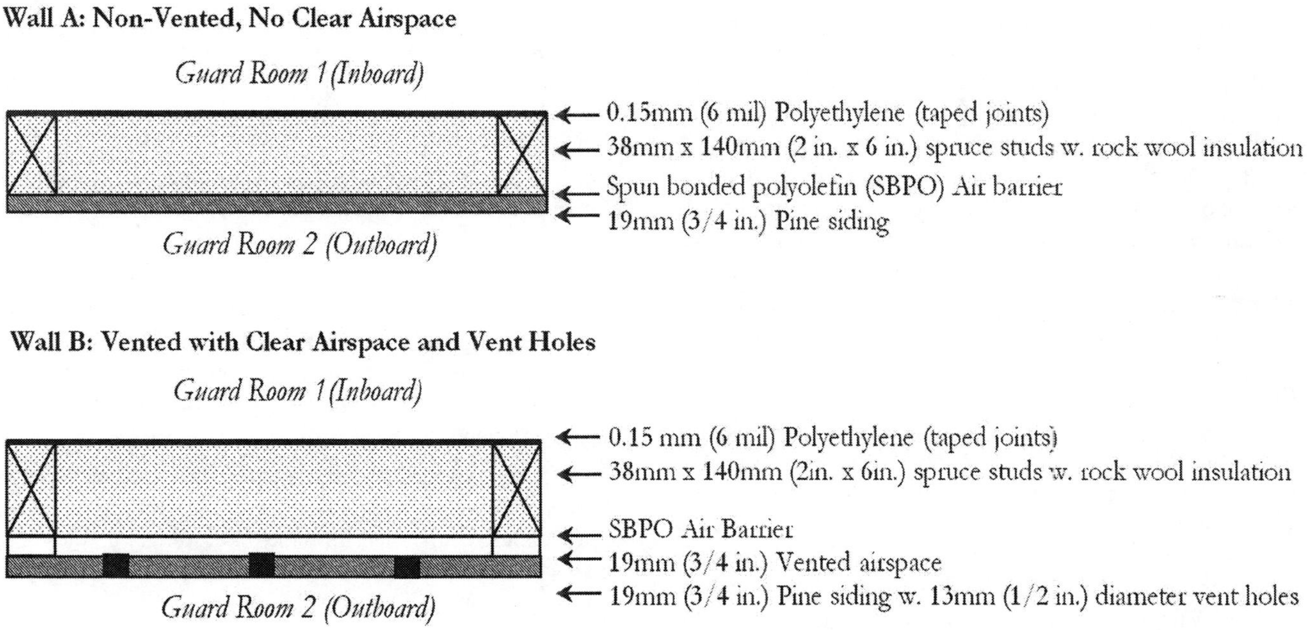

Figure 2 *Plan view of preliminary test walls: wall A = nonvented, no clear air space and wall B = vented with clear air space and vent holes.*

against the pine siding. Wall B was constructed with vent holes and with a 19 mm (3/4 in.) clear air space inboard of the pine sheathing and 13 mm (1/2 in.) diameter vent holes.

The vent holes were installed at the bottom and top of the test wall to facilitate the passage of air from guard room 2 into the clear air space. This air space was vented by means of six 13 mm (1/2 in.) diameter holes that were drilled in the pine siding. These holes were uniformly spaced at the tops and bottoms of the walls, resulting in a vent area ratio of 1000 mm^2 for every square meter (0.15 $in.^2/ft^2$) of wall surface area. The air space was created with various materials such as spruce strapping, extruded polystyrene strips, and circular pucks. The latter pucks were 5 cm (2 in.) in diameter and were installed in a checkerboard pattern on the insulation.

The inboard face consisted of a 0.15 mm (6 mil) polyethylene vapor retarding membrane. This membrane was taped and sealed so that air leakage was minimized across the assembly. The stud framing was also taped with vapor-impermeable tape so that the studs would not absorb any condensed moisture during the testing period. The stud cavity was filled with rock wool insulation; two types were examined—3.87 and 4.23 $m^2 \cdot K/W$ (R-22 and R-24) (ROXUL 2014b). Next the outboard face of the stud cavity was covered with a spunbonded polyolefin (SBPO) air-retarding membrane that was air-sealed to the framing with vapor-impermeable tape.

To allow air to move more freely through the air-permeable cavity insulation, an air-retarding membrane and plywood sheathing were omitted between the insulation and pine siding. Finally, 19 mm (3/4 in.) Eastern White Pine siding was affixed to the test wall. Each piece of siding was shiplapped and measured 400 × 140 mm (1.3 × 0.5 ft) wide. This pine siding served as an exterior cladding that was the source of moisture for the tests. Prior to assembly, the pine siding was immersed in cold water for up to 48 h. While the soaking period varied slightly, the purpose of the soaking was to provide a moisture source for the test walls. Following soaking, the pine siding was simply mounted on each test wall. However, when the last sample was tested in the preliminary round of tests, the pine siding was altered slightly. In order to reduce differential shrinkage and the resulting distortion of the wood siding during testing, the end grain of the pine siding was sealed with tape.

Preliminary Tests: Procedure

Prior to assembly of each test wall, all of the building materials were weighed. In the case of the pine siding, each piece was individually weighed. Following wall assembly, each test wall was exposed to the light array for a period of 24 h. Previous studies have shown that a 24 h period is a sufficient amount of time for moisture to be evaporated from the pine siding (L. Kan 1999; V. Kan 2002).

At the end of the test period, each wall was disassembled and all of the components were weighed again. The amount of moisture deposited within the insulation was measured by weighing each individual insulation batt. Paper towels were used to absorb excess moisture that had condensed on the polyethylene, the air barrier layers, and the surface of the sealed studs.

The moisture lost between setup and tear down was estimated by gravimetric measurement. The first pine board was weighed twice—once at the start of the wall assembly and once just before the polyethylene sheet was installed. This pine board dried the most amount of moisture into the ambient air during the test setup. The last pine board was assumed to have no moisture loss since the polyethylene sheet was installed immediately after the board was adhered to the wall. A linear function of moisture loss was used to estimate the amount of water evaporating from the remaining pine boards. The time increment between the installation of each board was generally the same. The tear-down procedure was similar, where the last pine board removed dried the most amount of moisture into the ambient air.

Preliminary Tests: Results

Table 1 summarizes the wall types and variables tested in the preliminary series of tests. The vent area ratio was calculated by dividing the vent hole area by the wall surface area. The average temperature differences between the surface temperature of the pine (T_s) and ambient air temperature in guard room 2 (T_a) are presented in Table 1. In addition, the average vapor pressure difference between the saturated siding and the outboard face of the polyethylene is shown.

The wall types referenced in Table 1 are described in Figure 2. The uncertainty associated with the temperature and relative humidity values recorded by the sensors was propagated through the vapor pressure difference and ranged between 75 and 140 Pa. Following soaking, the average moisture content of the pine siding defined as the weight of the water divided by the dry weight of the wood was found to range between 24% and 35%. At the end of the tests, the pine siding had an average moisture content that ranged between 14% and 19%.

Table 1 shows that there was some variability in the temperature difference created between the surface temperature of the pine siding and the ambient air in guard room 2. This variability resulted in vapor pressure differences that varied as well. The average air temperatures in guard rooms 1 and 2 ranged between 17.4°C and 19.4°C (63°F and 67°F) and 18.5°C and 20.0°C (65°F and 68°F), respectively. The pine surface temperatures showed larger variations, between 23.2°C and 30.4°C (74°F and 87°F). This variability may be attributed to small changes in the positioning of the thermocouples on the pine siding. If the light array bulbs were shining directly on the temperature sensor, that sensor recorded higher values. Furthermore, positioning of the fans, which helped to circulate the air in guard room 2, may also have affected the surface temperatures observed.

Vapor pressures were calculated using the temperature readings from the thermocouples and relative humidity sensors. Throughout all of the tests, there was a vapor pressure gradient driving moisture into the wall. This occurred because the vapor pressure at the inboard surface of the pine was always higher than the vapor pressure at the outboard face of the polyethylene.

Figure 3 shows the interstitial condensation that was measured gravimetrically and expressed as a percentage of the total moisture evaporated from the pine siding. Samples 3, 5, and 6 were tested twice, and the mean of the two condensation results is shown in Figure 3.

Figure 3 shows that the vented (wall B) and nonvented (wall A) samples behaved very differently. Tests of wall A showed that between 2.9% and 3.8% of the moisture supplied from the pine condensed within the walls. The propagation of error based on the laboratory scale measurements of moisture was approximately ±0.2%. Approximately 30% to 55% of the interstitial condensation was found within the insulation and between 25% and 42% was deposited on the polyethylene. The remaining condensed moisture was found on the SBPO layer. In comparison, wall B tests that included a vented air space had no measurable interstitial condensation.

Overall, less than 4% of the moisture supplied by the pine siding became interstitial condensation in the case of the unvented walls. This amount is rather small, and it is likely that most of the moisture supplied by the pine siding evaporated from the outboard exposed face of the pine. Due to the small amount of moisture gain in the preliminary test walls, no correlations between variables such as the amount of moisture gained, vapor pressure, or total moisture dried from the wood siding were

Table 1. Preliminary Tests Summary

Sample Number	Number of Tests	Wall Type	Insulation Type, m^2·K/W (Btu/h·°F·ft^2)	Clear Air Space	Vent Area Ratio, mm^2/m^2 (in.^{2}/ft^2)	Average $T_s - T_a$	Average Vapor Pressure Differential, Pa
1	1	Wall A	4.23 (R-24)	No	None	6.0	2450
2	1	Wall A	3.87 (R-22)	No	None	7.5	1810
3	2	Wall B (strap)	4.23 (R-24)	19 mm (3/4 in.)	1000 (0.15)	8.1	1560
4	1	Wall B (strips)	4.23 (R-24)	19 mm (3/4 in.)	1000 (0.15)	7.5	2280
5	2	Wall B (pucks)	4.23 (R-24)	19 mm (3/4 in.)	1000 (0.15)	8.4	2380
6	2	Wall A (end)	4.23 (R-24)	No	None	12.1	3710

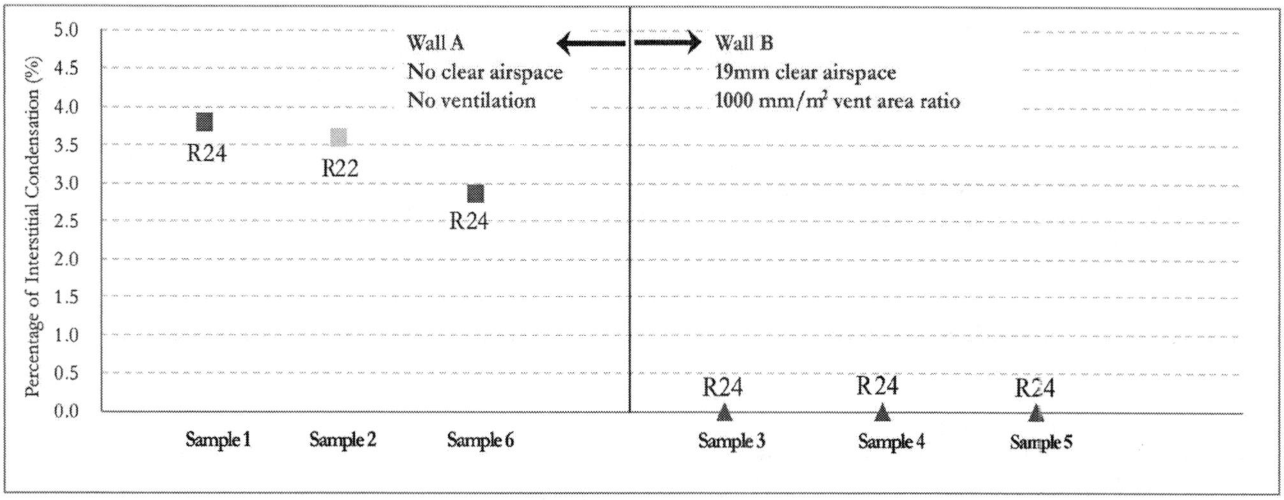

Figure 3 *Percentages of interstitial condensation in the preliminary tests.*

found. Lastly, no significant differences were observed between the performance of the 3.87 and 4.23 m^2·K/W (R-22 and R-24) walls. This is, once again, likely due to the relatively small amount of interstitial condensation deposited in the walls.

As the pine boards dried during the tests, shrinkage gaps formed between boards, creating numerous drying pathways. Consequently, even wall A, which was constructed to be non-vented, actually experienced some natural venting. Air that moved behind the pine siding likely removed a portion of the moisture that might otherwise have been deposited within the walls. In the case of the wall B tests, because no interstitial condensation was observed, it is likely that the venting capacity of the walls was not approached.

Modified Tests: Sample Preparation and Changes in Test Procedures

During the preliminary tests, the majority of the moisture supplied by the pine siding likely dried into the ambient air of guard room 2. Therefore, in order to better test the ventilation drying of the walls, the experimental apparatus was modified. The purpose of the modifications was to increase the moisture loading of the walls in order to observe the venting capabilities of the various walls. The construction of the samples was modified to reduce the moisture evaporation from the outboard pine surface as well as to maximize the amount of moisture entering the walls. To reduce the moisture drying into guard room 2, a 0.15 mm (6 mil) polyethylene layer was placed over the outboard face of the siding for the modified series of tests. This layer was secured with tape to ensure moisture did not escape into the ambient air of guard room 2. The polyethylene layer forced the vast majority of moisture within the pine siding to be driven into the wall instead of being removed by outboard surface evaporation and ventilation pathways between the boards. An additional modification was also made: the SBPO air barrier layer was removed for the modified tests so that air could move more freely through the air-permeable cavity insulation.

All of the test procedures for the modified tests were kept the same as those of the preliminary tests with two exceptions: the positioning of the light array and the test duration. To increase the vapor pressure gradient across the wall, the light array was moved 50 mm (2 in.) closer to the wall. This led to an increase in the temperature difference between the pine siding and the inboard face of the rock wool insulation. Consequently, the vapor pressure gradient through the wall also increased. To eliminate the effects, if any, of the climate chamber temperature fluctuations during defrost cycles that occurred every 6 h, the duration of the test was also shortened to 6 h.

Three basic wall configurations were tested in the modified series, as shown in Figure 4. Wall C was constructed without any vent holes and without a clear air space. Wall D was constructed with vent holes and with a 19 mm (3/4 in.) clear air space. Finally, wall E was constructed with vent holes but without a clear air space. This wall type was introduced to determine whether ventilation drying could occur through the air-permeable rock wool insulation. With this type of construction, the rock wool insulation was placed directly against the pine siding.

In addition to the 4.23 m^2·K/W (R-24) rock wool insulation tested previously, double-density rock wool was also included in the modified tests. This type of insulation has two densities measuring 65 and 100 kg/m^3 (0.0023 and 0.0036 lb/in.^{3}) (ROXUL 2014a). The insulation was installed such that the lower-density portion of the insulation was in contact with the pine siding or clear air space, depending on the wall type. The double-density insulation installed was 125 mm (5 in.) thick with a thermal resistance of 3.78 m^2·K/W (R-21.5) (ROXUL 2014a).

Wall C: Non-Vented, No Clear Airspace

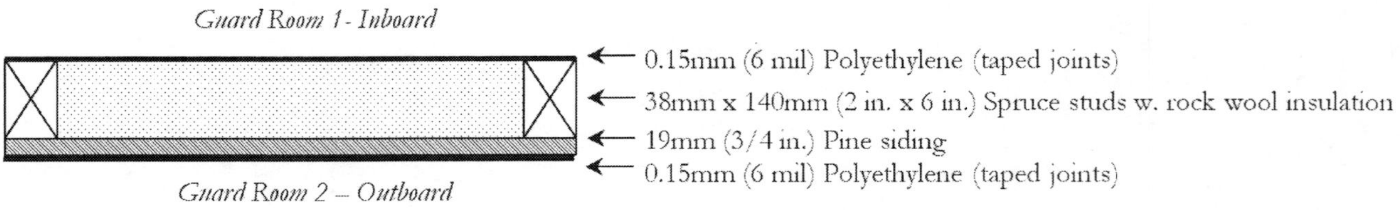

Wall D: Vented with Clear Airspace

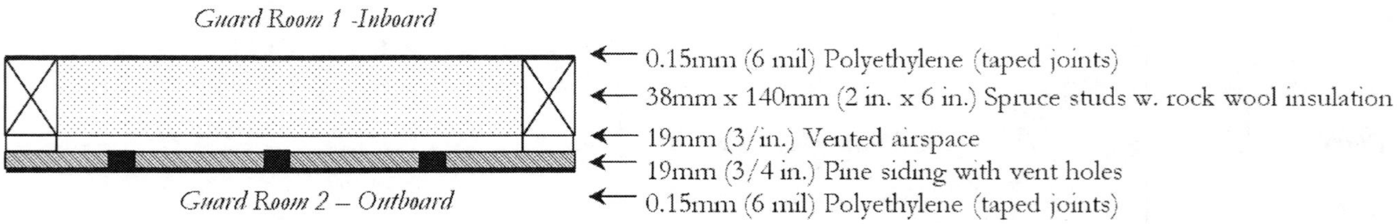

Figure 4 *Plan view of modified test walls: wall C = nonvented, no clear air space, wall D = vented with clear air space, and wall E = vented, no clear air space.*

Table 2. Modified Tests Summary

Sample Number	Number of Tests	Wall Type	Insulation Type	Clear Air Space	Vent Area Ratio, mm^2/m^2 ($in.^2/ft^2$)	Average $T_s - T_a$	Average Vapor Pressure Differential, Pa
7	1	Wall C	3.87 $m^2 \cdot K/W$ (R-21.5) double density	No	None	29.3	2720
8	1	Wall D	3.87 $m^2 \cdot K/W$ (R-21.5) double density	19 mm (3/4 in.)	1000 (0.15)	16.0	1581
9	1	Wall D	4.23 (R-24)	19 mm (3/4 in.)	1000 (0.15)	31.2	2680
10	1	Wall D	3.87 $m^2 \cdot K/W$ (R-21.5) double density	19 mm (3/4 in.)	2000 (0.29)	26.6	1710
11	1	Wall E	3.87 $m^2 \cdot K/W$ (R-21.5) double density	No	1000 (0.15)	31.0	3840
12	1	Wall E	4.23 (R-24)	No	1000 (0.15)	20.1	4280
13	1	Wall E	3.87 $m^2 \cdot K/W$ (R-21.5) double density	No	2000 (0.29)	33.8	2770
14	1	Wall E	3.87 $m^2 \cdot K/W$ (R-21.5) double density	No	3000 (0.43)	31.8	2880
15	1	Wall E	3.87 $m^2 \cdot K/W$ (R-21.5) double density	No	10,000 (1.44)	16.6	3530

A variety of vent area ratios (defined as the vent hole area over the wall surface area) were also explored to determine the effect of the vent area on the amount of moisture removed. The vent area ratios tested were: 1000, 2000, 3000, and 10,000 mm^2/m^2 (0.15, 0.29, 0.43, and 1.44 $in.^2/ft^2$). The vent holes allowed air from guard room 2 to enter and leave the wall system. It is important to note that whenever a test wall incorporated vent holes in the siding, corresponding holes were also installed in the adjacent polyethylene layer. At each hole location, the polyethylene was secured by stapling and taping the polyethylene to the pine siding, thereby ensuring that drying did not occur from the outboard face of the siding.

MODIFIED TESTS: RESULTS AND DISCUSSION

Table 2 shows a summary of the samples tested with the modified test procedure. The walls tested featured two different insulation types and a variety of vent area ratios. Furthermore, some of the walls had a 19 mm (3/4 in.) clear air space, while others had the pine siding installed so that it was directly in contact with the insulation. The uncertainty associated with the temperature and relative humidity values recorded by the sensors was propagated through the vapor pressure difference and ranged between 80 and 280 Pa.

At the beginning of the testing, the average pine moisture content ranged from 23% to 37% for all three wall types. This initial moisture content was similar to the initial moisture content used in the preliminary tests. However, at the end of the modified tests, the moisture content ranged between 18% and 28%, compared to 14% and 19% when the preliminary tests were carried out. Because the outboard surface was covered with polyethylene, this higher moisture content was expected because the pine could only dry from the inboard face. In addition, less moisture dried from the pine siding because the test duration was shortened from 24 to 6 h.

The temperature differences between the siding and the ambient air in guard room 2 were generally higher than in the preliminary tests. This was because the light array was situated 5 cm (2 in.) closer to the wall. The placement was modified to increase the pine surface temperatures and consequently to increase the vapor pressure gradient across the wall.

In accounting for the moisture supplied by the pine siding when it was dried for 6 h, the moisture lost during setup (after weighing but before the pine siding and polyethylene were installed) and tear down (before weighing but after the polyethylene and pine siding were removed) was estimated by measurement. Thus, the net moisture that dried once the wall was assembled could be determined.

Sample 7 was used as a base case for evaluating the amount of moisture supplied during testing. The amount of interstitial condensation found within sample 7 was expected to be the maximum amount of moisture one would expect to find within any wall, given that this sample was essentially a sealed system without any ventilation. Interestingly, despite attempts to completely seal the wall system, when sample 7 was tested, 84% of the net moisture that dried from the pine siding could be found within the wall. Therefore, approximately 16% of the net moisture supplied to the wall was unaccounted for when a moisture balance was carried out. It is likely that this unaccounted-for moisture was lost by uncontrolled air leakage through small penetrations in the tape around the polyethylene perimeter and framing members. Because the other sample walls tested had some means of venting, the question arose, how much water was unaccounted for when these other walls were tested?

To address this question, estimates of the upper and lower bounds of moisture removed by ventilation were made. A lower bound estimate was made by assuming that 16% of the total net moisture supplied by the pine was lost by discontinuities during testing of all samples. Prior to conducting subsequent sample wall tests, sample 7 was retested a number of times, consistently achieving 16% moisture loss. Although subsequent tests included vent holes, the vapor pressure gradient caused by the light array is driving the moisture inwards; therefore, moisture lost through the vent holes is likely minimal. Thus, the amount of moisture unaccounted for when sample 7 was tested was used as a base case. This lower-bound estimate was calculated with the assistance of Equation 1 and applied to every test. This is a conservative estimate and likely underestimates how much moisture was actually removed by ventilation. To estimate the maximum amount of water that could have been removed by ventilation, it was assumed any liquid water not found within the wall system was removed by ventilation. This upper bound estimate was calculated using Equation 2 and applied to each test. The actual amount of moisture removed by ventilation is likely between the lower bound and upper bound estimates.

$$\text{Lower Bound:}\quad \text{Moisture Removed by Ventilation} = X - Y - w \tag{1}$$

$$\text{Upper Bound:}\quad \text{Moisture Removed by Ventilation} = X - Y \tag{2}$$

Once the upper and lower bounds were calculated, the findings were plotted. Figure 5 shows the amount of moisture removed by ventilation for wall D (vented with clear air space) and wall E (vented without a clear air space). In Figure 5, when considering walls with a vent area ratio of 1000 mm^2/m^2 (0.15 $in.^2/ft^2$), the amount of moisture removed by ventilation in the 3.87 $m^2 \cdot K/W$ (R-21.5) walls without a clear air space (sample 11 = 58% to 71%) was only slightly less than amount of moisture removed by ventilation of the 3.87 $m^2 \cdot K/W$ (R-21.5) walls with a clear air space (sample 8 = 59% to 72%). The difference in ventilation drying between these two samples is small enough that it is within the bounds of experimental error. These tests indicate that even if a clear air space is not provided, so long as vent holes are in place a similar amount of moisture can be removed by ventilation. This is an interesting finding because it shows that some ventilation drying is occurring through the air-permeable rock wool insulation in walls without a clear air space. However, it is likely that the observed venting is occurring through the insulation as well as along the interface between the insulation and moist siding. Although steps were taken to ensure contact between the insulation and the siding, given the irregular surface of the insulation, it is likely that some venting is occurring along the interface. Future research work will examine the relative contribution of venting along the interface and through the insulation.

Slightly larger differences in the moisture removed by ventilation were observed between samples with a clear air space and samples without a clear air space when 4.23 $m^2 \cdot K/W$ (R-24) cavity insulation was tested. Walls with clear air spaces removed more moisture than walls without clear air spaces. The absolute differences between the lower and upper bound

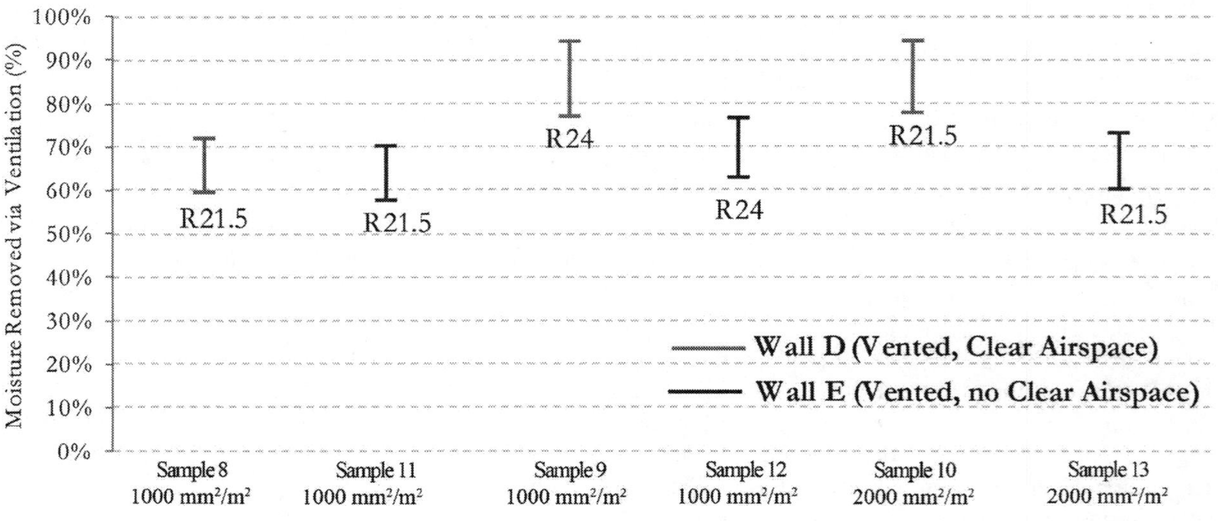

Figure 5 *Comparison between the moisture removed by ventilation for the vented clear air space samples (wall D) and vented air space samples without a clear air space (wall E).*

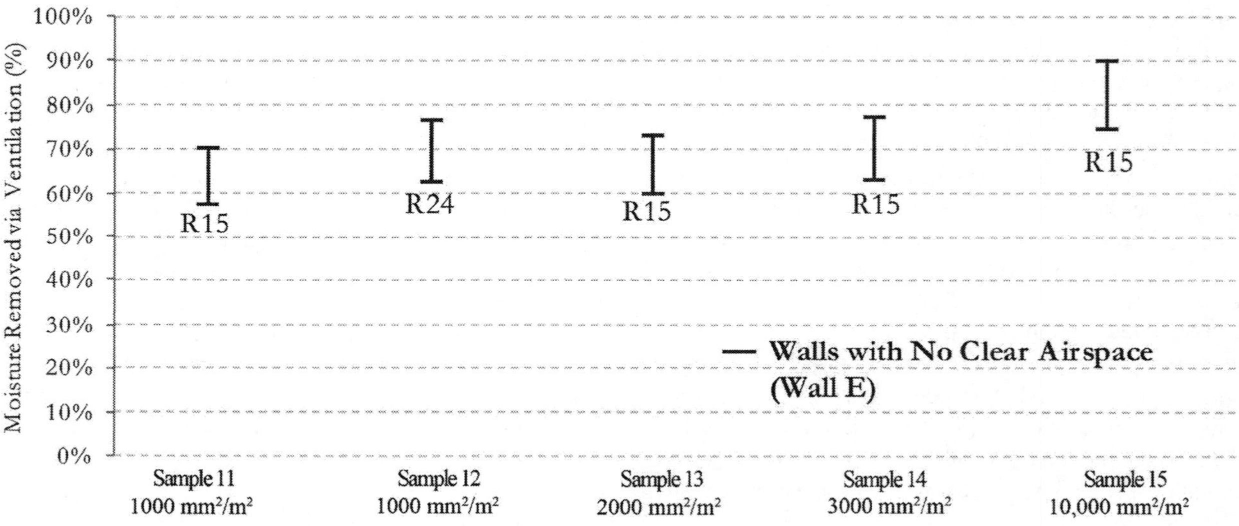

Figure 6 *Walls without a clear air space: percentage of moisture removed by ventilation for various ventilation ratios.*

values ranged between 14% and 18% when samples 9 and 12 were compared. When the vent area ratio was increased from 1000 to 2000 mm^2/m^2 (0.15 to 0.29 $in.^2/ft^2$) of wall area, walls with clear air spaces removed more moisture than walls without clear air spaces. However, these differences are small, which indicates that there is an opportunity to use air-permeable insulation in lieu of a clear air space for ventilation drying of walls.

Next, Figure 6 shows the effects of increasing the vent area ratio on the percentage of moisture removed by ventilation. All of the samples shown in Figure 6 are constructed based on wall E, without a clear air space. As shown in the figure, walls with a vent area ratio of 1000 mm^2/m^2 (0.15 $in.^2/ft^2$) removed approximately 58% to 71% of the moisture entering the test walls. When the vent area ratio was doubled to 2000 mm^2/m^2 (0.29 $in.^2/ft^2$), the amount of moisture removed by ventilation increased to 60% to 74%. Similarly, increasing the vent area ratio by a factor of 10 in sample 15 removed the largest amount of moisture, between 74% and 90%.

These experiments also examined the influence of two types of insulation. Sample 12 used 4.23 $m^2 \cdot K/W$ (R-24) rock wool insulation with a density of 32 kg/m^3 (0.0012 $lb/in.^3$), while the remaining tests incorporated 3.87 $m^2 \cdot K/W$ (R-21.5) insulation with dual density between 65 and 100 kg/m^3 (0.0023 and 0.0036 $lb/in.^3$). Samples 11 and 12 showed that there appears to be a small difference between the moisture removed in walls featuring the two types of insulation when all other wall characteristics remain the same. With the less dense

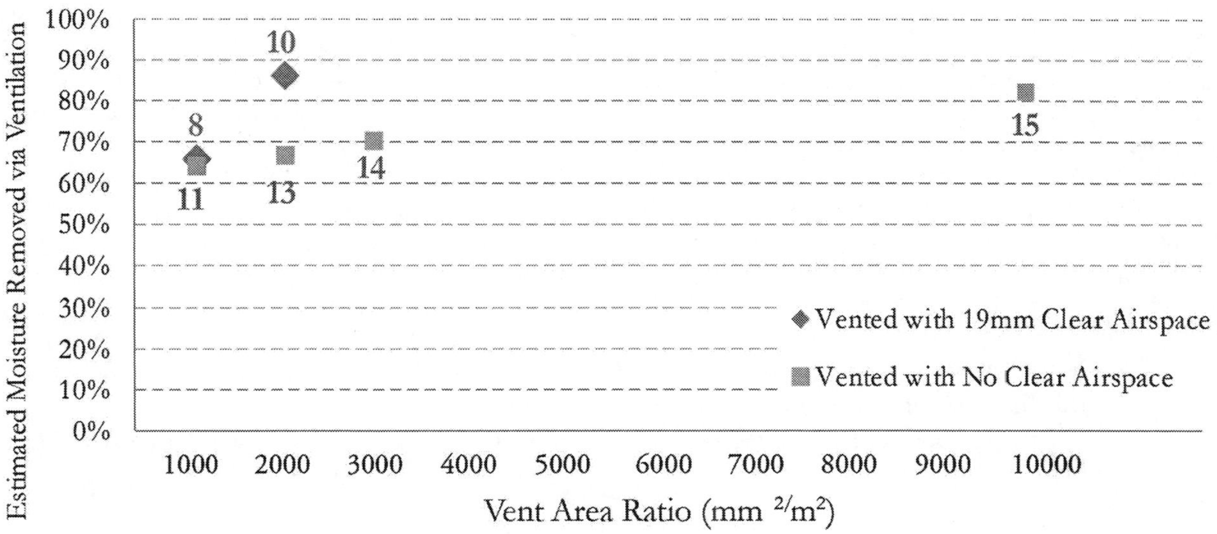

Figure 7 *Moisture removed by ventilation versus vent area ratio for walls with and without a clear air space.*

4.23 $m^2·K/W$ (R-24) insulation, a higher percentage of moisture was removed. These two samples indicated that insulation density could affect how easily ventilation air flows through the insulation, ultimately affecting the amount of moisture removed from the walls. Additional testing is recommended to confirm this observation.

Figure 7 plots the estimated amount of moisture removed from the walls against the vent area ratio. Samples 8 and 10 are shown in orange, featuring wall D with a clear air space. Samples 11 and 13 through 15 are shown in blue, featuring wall E without a clear air space. All of the walls in Figure 7 have 3.87 $m^2·K/W$ (R-21.5) double-density insulation.

Figure 7 shows that as the vent area ratio increased, so did the percentage of moisture removed. This was true for walls featuring a 19 mm (3/4 in.) clear air space as well as for walls without a clear air space. Furthermore, Figure 7 shows that sample 10, constructed with a 19 mm (3/4 in.) clear air space and 2000 mm^2/m^2 (0.29 $in.^2/ft^2$) vent area ratio, performed similarly to sample 15, which had a much higher vent area ratio but no clear air space. These tests indicate that in order to maximize the drying potential of the walls, larger vent area ratios can be used as an alternative to the provision of a clear air space. Figure 7 also shows that the difference of moisture removed between walls with and without a clear air space is small at lower vent area ratios. This difference increases when higher vent area ratios are introduced, such as in samples 10 and 13.

Although the temperature conditions in guard room 1 were relatively constant for all tests, differences in the ambient air temperature in guard room 2 and the pine surface temperatures resulted in non-identical boundary conditions during the modified tests. To assess the effects, if any, of these variable boundary conditions, the relationship between the moisture removed and the vapor pressure gradient was reviewed for consistency. As expected, the percentage of moisture removed by ventilation decreased slightly when the vapor pressure difference increased. However, when linear regression was carried out between the moisture removed and the vapor pressure gradient, a coefficient of determination (R^2) of 0.25 was found, indicating a weak linear fit. The coefficient of determination was low because the amount of moisture removed was influenced by a variety of other variables such as the vent area ratio, insulation density, and the construction of a clear air space. Therefore, future tests involving a parametric study of the vapor pressure as it affects moisture removal are recommended.

CONCLUSIONS

This research examined the performance of a vented air space that was constructed using air-permeable rock wool insulation. A preliminary series of tests examined the influence of insulation density, type of air space offsets, and vent area ratios. In these preliminary tests, gaps formed between the shiplap siding, which allowed air to flow behind the siding. This likely resulted in the majority of moisture drying by evaporation from the exposed surface and by natural ventilation. To maximize the amount of moisture being driven into the walls, the construction and testing procedures were modified. A sheet of polyethylene was installed over the pine siding in the modified tests to reduce surface evaporation and natural venting between the siding boards.

The modified series of tests compared the performance of walls with a clear 19 mm (3/4 in.) air space and vent holes to walls featuring only vent holes without a clear air space. Walls with vent holes but without a clear air space removed 52% to 90% of the total moisture entering the walls. The difference of moisture removed was related to increasing the vent area ratio from 1000 to 10,000 mm^2/m^2 (0.15 to 1.44 $in.^2/ft^2$). In comparison, walls featuring a clear air space removed 59% to

95% of moisture. The small differences between these tests indicate that a clear air space is not the only means of removing moisture from test walls. Air-permeable insulation along with adequately sized vent holes can also promote ventilation drying. Thus, mineral fiber insulation can be used to provide thermal insulation while promoting drying of internally insulated masonry walls. This means that energy use and durability can both be improved when historic masonry buildings are internally insulated with vapor-permeable insulation.

NOMENCLATURE

T_a = ambient air temperature in guard room 2

T_s = surface temperature of the pine siding

w = 16% unaccounted-for moisture based on sample 7 (nonvented, sealed wall)

X = net moisture dried, %

Y = interstitial condensation, %

ACKNOWLEDGMENTS

The authors gratefully acknowledge funding provided by the Natural Sciences and Engineering Research Council and the Neil B. Hutcheon Bequest. The authors also acknowledge ROXUL Inc. for donating materials for the laboratory testing.

REFERENCES

Burnett, E., J. Straube, and A. Karagiozis. 2004. *ASHRAE 1091 synthesis report and guidelines.* Technical report. Atlanta: ASHRAE.

Groot, C., and J. Gunneweg. 2012. The influence of material characteristics and workmanship on rain penetration in historic fired clay brick masonry. *HERON* 55(2):141–53.

Kan, L. 1999. An investigation of measures to prevent dummer condensation in walls. Master's thesis, University of Toronto.

Kan, V. 2002. Controlling sun driven moisture: a laboratory investigation. Master's thesis, University of Toronto.

Karagiozis, A., and H. Kuenzel. 2009. The effect of air cavity convection on the wetting and drying behaviour of wood frame walls using a multi-physics approach. *Journal of ASTM International* 6(10):1–15.

Laefer, D., J. Boggs, and N. Cooper. 2004. Engineering properties of historic brick: Variability considerations as a function of stationary vs. non-stationary kiln types. *Journal of the American Institute for Conservation* 43(3):255–72.

Lstiburek, J. 2010. BSI-001: The perfect wall. *Buidling Science Insights.* Westford, MA: Building Science Corporation. https://buildingscience.com/documents/insights/bsi-001-the-perfect-wall.

Popp, W., H. Kuenzel, and E. Mayer. 1983. *Untersuchugen uber die notwendige Hin terluftung an Auzenwandbekeidung aus gors\zformatigen Bauteilen.* Technical Report [in German]. Fraun- hofer Institut fur Bauphysik, Forschungsbericht B Ho.

Pressnail, K., J. Timusk, L. Kan, B. Dong, and V. Kan. 2003. In search of a wall for all seasons: Controlling sun-driven moisture. *Proceedings of the 9th Canadian Conference on Building Science and Techology*, pp. 157–70.

ROXUL. 2014a. ROXUL CavityRock® exterior cavity wall insulation. Milton, Ontario, Canada: ROXUL, Inc. http://www.roxul.com/products/commercial/products/roxul+cavityrock.

ROXUL. 2014b. ROXUL ComfortBatt® Commercial thermal insulation product. Milton, Ontario, Canada: ROXUL, Inc. http://www.roxul.com/products/commercial/products/roxul+comfortbatt.

SMT. 2014. SMT-A2—Wireless data acquisition unit. SMT-A2 datasheet. Vancouver, British Columbia, Canada: SMT Research Ltd. http://www.smtresearch.ca/resources_files/RS-1100%20SMT-A2%20Datasheet%20Web_0.pdf.

Stovall, T., and A. Karagiozis. 2004. *Airflow in the ventilation space behind a rain screen wall.* Technical report. Atlanta: ASHRAE.

Straube, J., and E. Burnett. 1998. Vents, ventilation and masonry veneer wall systems. *Proceedings of the Eighth Canadian Masonry Symposium*, pp. 194–207.

Straube, J., and G. Finch. 2009. Ventilated wall claddings: Review, field performance, and hygrothermal modeling. Tech. Rep. 0906. Westford, MA: Building Science Corporation.

Tzekova, E.S. 2015. Improving the thermal performance and durability of historic masonry buildings. Doctoral thesis, University of Toronto.

Tzekova, E.S., K.D. Pressnail, C. Binkley, N.T. Pearson, P. Pasqualini, and C. Aikin. 2011. Integrating a vented airspace into a spray-foam insulated solid masonry historic building in a cold climate: A case study. *Proceedings of the13th Canadian Conference on Building Science and Technology.*

Tzekova, E.S., K.D. Pressnail, M.F. Touchie, N. Pearson, and P. Pasqualini. 2012. Evaluating the wetting and drying of a vented airspace in a spray-foam insulated solid masonry historic building. *Proceedings of the 3rd International Conference on Heritage and Sustainable Development.*

Van Der Klugt, L. 1988. Frost testing by uni-directional freezing. *British Ceramic* 87(1):8–12.

A Public Domain, Transient, Experimental Database on the Hygrothermal Performance of Durable, Energy-Efficient Full Basement Foundation Walls in a Cold Climate

Anna C. Harmon, PE **Louise F. Goldberg, PhD**

ABSTRACT

A database, spanning two years in a cold, Zone 7 climate, that describes the measured transient energy and hygrothermal performance of insulated hollow CMU basement walls, has been published in the public domain. The database includes experimental data from walls with IECC 2012-compliant full-wall interior and half-wall exterior insulation and data from vapor-sealed rim joist cavities. The database has been used to test the temperature and heat flux predictions of a three-dimensional thermal simulation program yielding an average root mean square error of 1.22°C for soil temperatures and 0.42 W/m²·K for CMU wall interior heat fluxes for the particular experimental wall system simulated. The data allowed insight to the heat transport effects of hollow CMU wall buoyant cavity flow loops and their impact on the wall heat fluxes and soil temperatures and showed that these loops cannot be ignored in basement thermal simulations. If it exists, the importance of a sub-slab water table and its effect on the soil temperatures adjacent to CMU walls with interior insulation was demonstrated. Heat transport from the sub-slab water table in a cold climate was the dominant heat source affecting the observed soil temperatures adjacent to interior-insulated foundation walls while heat loss through the footing and wall below the interior insulation had no measured impact. Including ambient and ground interface phase change in a snow model is necessary to accurately simulate the observed ground temperatures within 12 in. of grade. A solid CMU wall bond beam with a permeance of 0.1 to 1 U.S. perms (Class II) was sufficient to isolate a vapor-sealed rim joist cavity from vapor saturated CMU cores so that condensation on the rim board interior face was avoided with and without exterior insulation.

INTRODUCTION

Research was undertaken to experimentally evaluate the energy and hygrothermal performance of retrofit foundation wall insulation systems compliant with the insulation requirements of the 2012 IECC and the hygrothermal performance standards of the 2009 MN Energy Code over a continuous two-year monitoring period. The research yielded a transient experimental database that has proved effective in testing the accuracy and validity[1] of building foundation system thermal simulations. In addition, numerous experimental insights into the heat and moisture transport physics were observed during the experiment. A summary of the more interesting observations and an example of the utility of the database are discussed.

Experimental Setup

The experiment was located in Cloquet, Minnesota (IECC climate zone 7) at the University of Minnesota's Cold Climate Research Facility (CRRF). Existing concrete masonry unit (CMU) basement walls were retrofitted with insulation, an interior water separation plane (WSP), and a 2 × 3 open cavity stud wall covered with ½ in. gypsum board. The WSP was either an adhered peel-and-stick rubberized asphalt membrane or a nonadhered ¼ in. thick dimpled high-density polyethylene (HDPE) sheet. There were a total of four

1. *Validity* is defined here as "replication of the heat transport continuum mechanics applicable to a particular experiment."

Anna C. Harmon is a senior associate at Avant Energy and, formerly, a graduate student at the University of Minnesota, Minneapolis, MN. **Louise F. Goldberg** is an adjunct professor at the University of Minnesota, St. Paul, MN, and principal of Lofrango Engineering, Schererville, IN.

test bays with wall insulation and one additional bay that was used for a rim joist experiment. Each bay had north and south exposures. Table 1 shows the construction details of the four full-wall test bays.

Instrumentation was installed to collect the following data: temperature, absolute humidity, heat flux, soil temperature and moisture content, CMU moisture content, and interior and exterior boundary conditions. The details of the sensors and instrumentation are reported in Goldberg and Harmon (2015a). The data measurements were distributed across the height and width of a plane perpendicular to the wall at the test bay centerline with measurements in the soil, at condensing surfaces, in the hollow CMU cores, and at the boundaries. Figure 1 shows the instrumentation layout for Bays 1 and 2.

Database

The experimental database includes transient data from all five experimental wall systems and spans the period from 11/10/2012 to 11/24/2014. It is in the public domain and is accessible through a persistent URL linking to the University of Minnesota's Data Repository (Goldberg and Harmon 2015b). The database is in a standard flat file, comma-separated value format with one file per day and is readily accessible via standard spreadsheet and database software. Each time record includes all the data recorded at that time step with a maximum of 607 entries. Accompanying documentation includes a copy of the peer-reviewed final project report and a complete description of the nomenclature, database access details, and instrumentation layout.

DISCUSSION OF SOME SALIENT EXPERIMENTAL RESULTS FROM THE DATABASE

CMU Buoyant Cavity Flow

If the CMU hollow cores were filled with still air, the horizontal heat transfer would be by conduction and the wall temperatures would decrease from the interior to the exterior

Table 1. Wall System Construction Details

Bay	Water Separation Plane (WSP)	Insulation	Soil Classification (USDA Soil Triangle/ USC)
1	Adhered	Interior, full-height, 3 in. XPS	Loam/sandy-silt (ML)
2	Adhered	Interior, full-height, 3 in. XPS	Sand/sand (SP)
3	Nonadhered	Interior, full-height, 3 in. XPS	Sand/sand with silt (SP-SM)
4	Adhered	Exterior, half-height, 3 in. XPS	Sand/sand with silt (SP-SM)

during the heating season (after November 1 and before May 1). This temperature profile was not observed for Bay 2N at 69 and 40 in. above the slab (Figure 2). At these locations, the hollow core was warmer than the interior face shell and the exterior face shell was colder than the soil 5 in. from the face of the wall (Harmon 2015).

Figure 2 also shows that during the heating season, the interior face shell was warmer than the exterior face shell as expected, indicating a net interior to exterior heat transfer across the CMU. Figure 3 shows that during the heating season, the Bay 2N hollow core temperatures decreased from the bottom to the top of the wall. The observed temperature profiles likely are indicative of buoyant cavity flow that yields a cold air plume dropping down the exterior face of the cavity and a warm air plume rising up the interior face of the cavity. The airflow depresses the exterior face shell temperature below the soil temperature and elevates the hollow core temperature above the interior face shell temperature. This demonstrates that for below-grade walls insulated on the interior face, buoyant cavity flow is a significant horizontal heat transfer mechanism that is larger than still air thermal conduction (Figure 11) during the heating season.

Impact of Exterior/Interior Insulation Placement on Heat Transfer

A comparison of the interior CMU face shell surface heat fluxes for the northern wall exposure is shown in Figure 4. Bays 1, 2, and 3 were fitted with full-height interior RUS-15 extruded polystyrene (XPS) insulation compliant with the 2012 IECC for the entire experiment. Bay 4 was uninsulated during the first heating season and thus served as an uninsulated control for the first heating season. During the second heating season, Bay 4 was insulated with exterior RUS-15 XPS insulation extending from the top of the wall to 38 in. below grade. The balance of the wall system was the same for Bays 1 to 4. In these circumstances, Figure 4 shows that during the first heating season the three bays with interior insulation experienced a heat loss of about 5 $W/m^2 \cdot K$ at a location 12 in. below grade. At the same height, the uninsulated Bay 4 experienced a heat loss of about 15 $W/m^2 \cdot K$, showing a three-fold increase in the heat loss compared to the insulated wall. During the second heating season, the Bay 4 upper half-wall exterior insulation yielded a heat loss of about 7 $W/m^2 \cdot K$ while the heat loss for the walls with interior insulation was slightly reduced compared with the first heating season at about 4.5 $W/m^2 \cdot K$. Hence, RUS-15 exterior insulation over the upper half of the wall yielded about a 55% larger heat loss than full-wall interior insulation. Thus, as a foundation wall retrofit strategy, half-wall exterior insulation does not offer any energy performance advantages compared with full-wall interior insulation.

Rim Joist Cavity/CMU Core Vapor Coupling

One of the key questions pertaining to retrofitting CMU walls with interior insulation is the impact of the insulation on the rim joist cavity hygrothermal performance, particularly when the cavity is vapor isolated from the interior space. An experiment

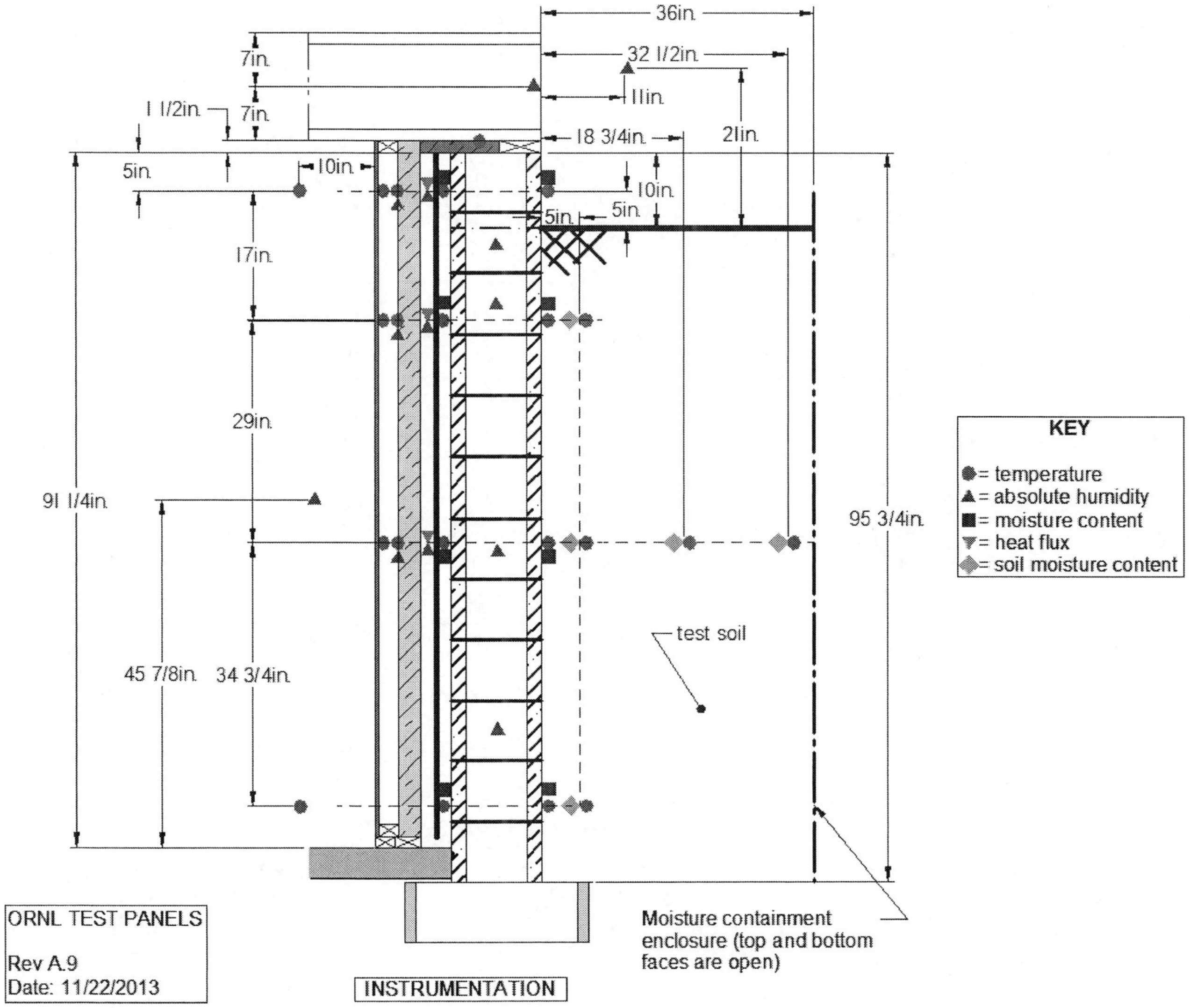

Figure 1 *Instrumentation layout for Bays 1 and 2.*

was conducted to investigate the moisture coupling between a very well-sealed rim joist cavity and the cores of a CMU wall fitted with an interior WSP (6-mil. polyethylene). Cases with and without exterior rim board insulation were investigated and the exterior insulation case is shown in Figure 5. The sensor and instrumentation details are reported in Goldberg and Harmon (2015a). In this case, the solid bond beam block provides the only vapor retarder between the CMU cores and the rim joist cavity.

A summary of the results is shown in Figure 6 for the hygrothermally more severe northern exposure. In both the insulated and uninsulated cases, the CMU cores maintained 100% relative humidity (RH) throughout the monitoring period, while the rim joist cavity midpoint RH was generally lower in the insulated case but did not exceed about 75% in either case. Also shown are the interior and exterior bond beam CMU face shell moisture contents that were less than 6% by volume over the monitoring period. No condensation occurred on the rim board interior surface (dry-bulb temperature greater than dew-point temperature) over the monitoring period in both the insulated and uninsulated cases (Goldberg and Harmon 2015a, Figures 46 and 47). Thus, the data demonstrate that a solid CMU bond beam provides adequate vapor separation (by calculation, effectively a Class II vapor retarder[2]) between saturated CMU cores and a vapor-sealed rim joist cavity.

Influence of Sub-Slab Soil Domain on the Soil Heat Transfer

It was expected that the temperatures of the CMU wall face shells and adjacent soil during the heating season would be influenced primarily by heat lost through the uninsulated slab, the wall beneath the retrofitted insulation, and the foot-

2. Vapor permeance in the range of 0.1 to 1 U.S. perm.

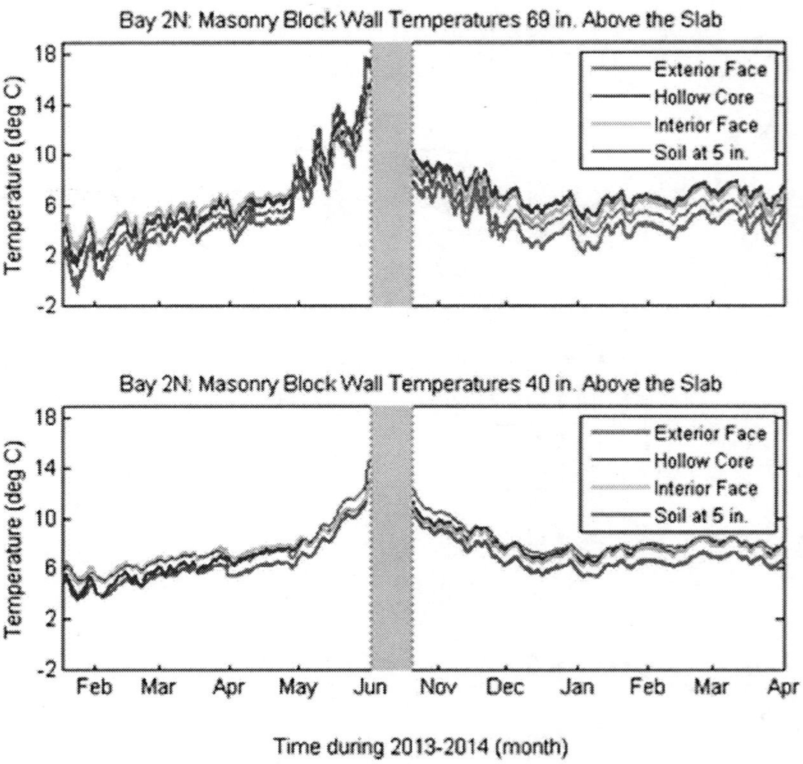

Figure 2 *Heating season wall temperatures for Bay 2N at 69 and 40 in. above the slab.*

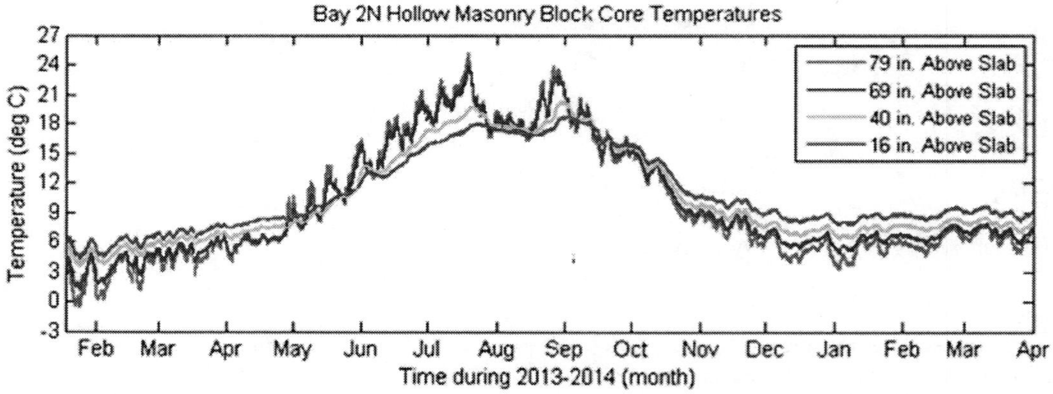

Figure 3 *Hollow core temperatures for Bay 2N.*

ing. However, initial simulation/experiment soil temperature comparisons 5 in. away from the walls revealed experimental temperatures significantly higher than those simulated throughout the year (approx. 5°C during the cooling season and 3°C during the heating season) (Goldberg and Harmon 2015a). To investigate the cause of this discrepancy, a 5 by 8 ft sheet of 3 in. thick extruded polystyrene (XPS) insulation was placed on top of the concrete floor slab flush with the wall framing (Figure 7) in Bay 3 (south and north). The insulation was centered on the line of wall and soil sensors and in place from 12/13/2013 until 1/22/2014.

It was expected that the floor insulation would decrease the heat loss through the base of the wall and, in turn, lower the wall and soil temperatures. However, there was no change in the temporal temperature gradients when the insulation was installed or removed, as shown for the north wall in Figure 8 (Harmon 2015). Further, it was expected that the corresponding adjacent Bay 2 temperature profiles (same wall insulation and similar soil) without the temporary floor insulation would have been different from those of Bay 3. However, the Figure 8 temperature profiles were very similar to the corresponding profiles for Bay 2N, as was the case for the south wall. Even

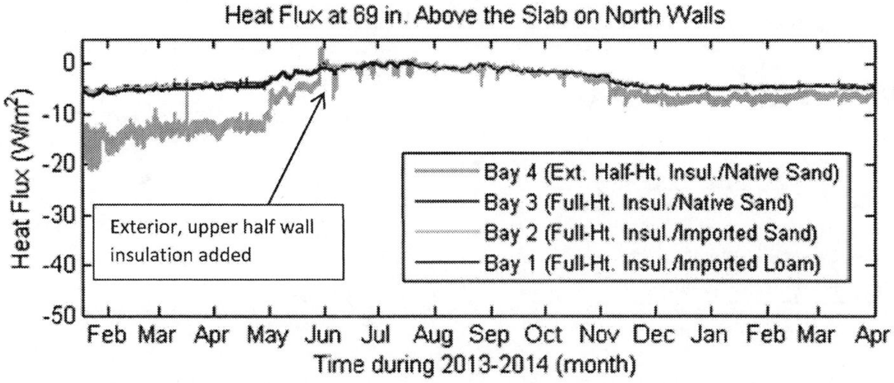

Figure 4 *North exposure wall heat flux comparison.*

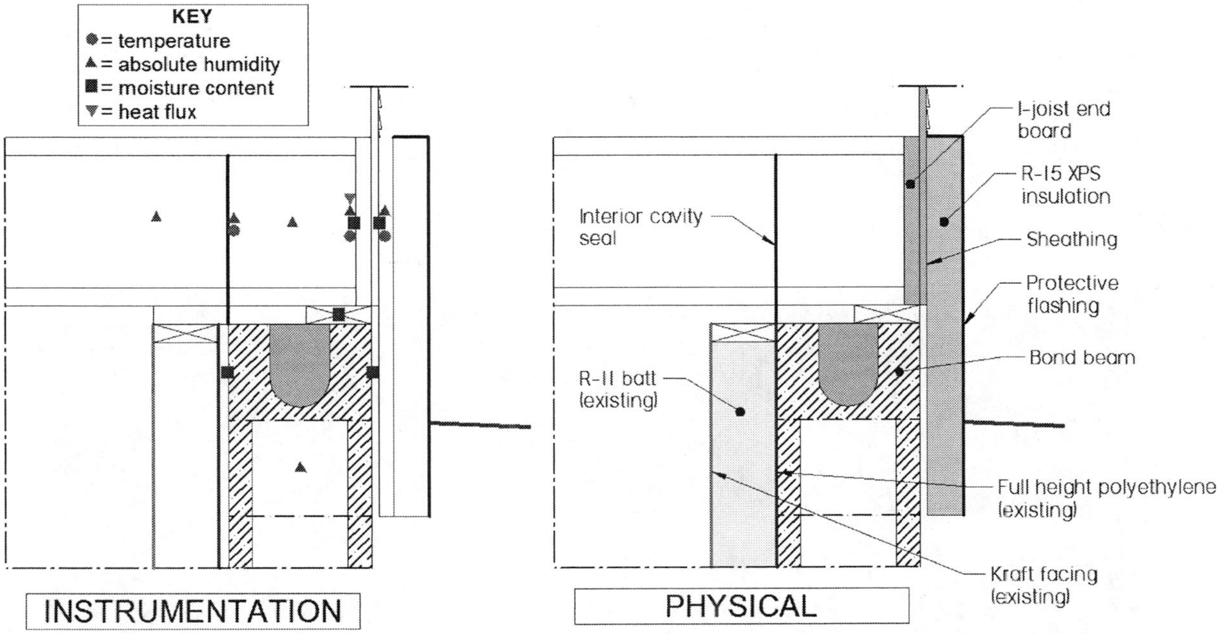

Figure 5 *Insulated rim joist cavity experimental configuration.*

though the center of the basement floor was not covered with temporary insulation, the 5 ft strip of perimeter insulation was sufficient at least to yield observable changes in the wall and soil temperature profiles 5 in. above the slab if heat loss through the base of the wall was the primary heat source elevating the experiment temperatures. This led to the conclusion that heat loss through the base of the uninsulated wall was not the primary soil heat source.

It was then hypothesized that a high ground water table beneath the CRRF had the potential to act as a heat source during the heating season. It was later verified that the water table beneath the CRRF was approximately 13 in. below the bottom of the footings (Harmon 2015). To evaluate this hypothesis, basic heat transfer calculations were performed to estimate the following heat transfers (Harmon 2015):

1. Between the basement and the soil below the slab (Q_{base})
2. Between the outdoor ambient environment and the soil below the slab (Q_{crib})
3. Between the basement and the soil adjacent to the wall (Q_{horz})

Two-dimensional calculations were performed on a 1 ft wide foundation strip that included half the basement width (10 ft) and a soil domain that extended 4 ft beyond the wall[3] and 8 ft below the bottom of the footing. A schematic of the

3. In this interior insulation configuration, the bulk of the heat transfer affecting Domain A occurs within the masonry block wall and the soil within 1 ft or so of the wall. Including more than 4 ft is not justified for the granularity of the calculation described.

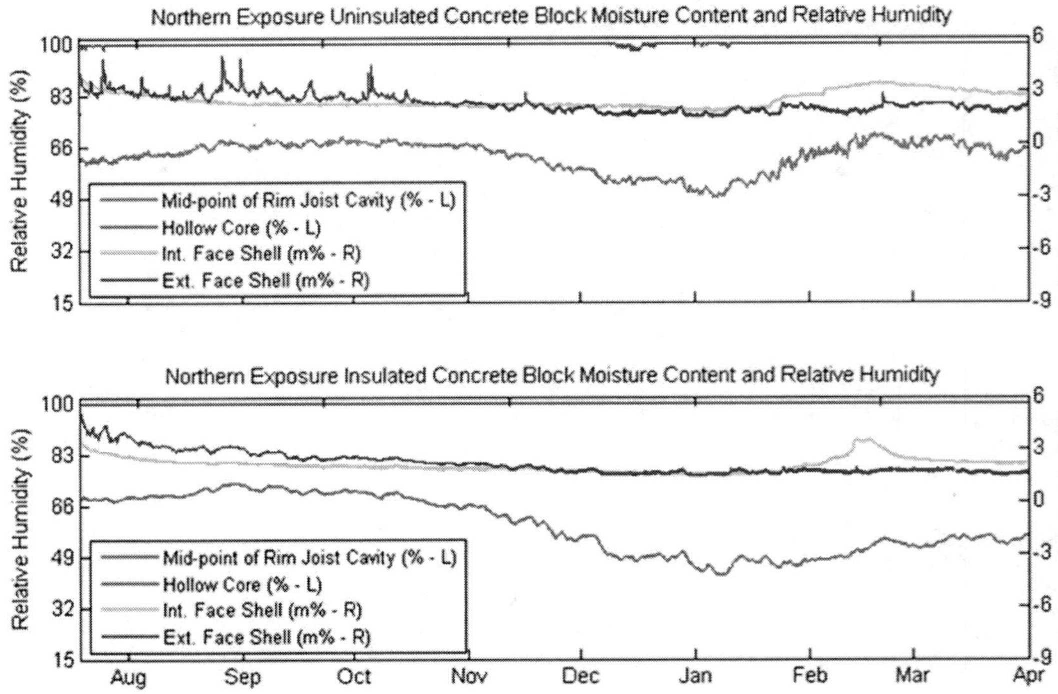

Figure 6 *CMU core/rim joist cavity moisture coupling.*

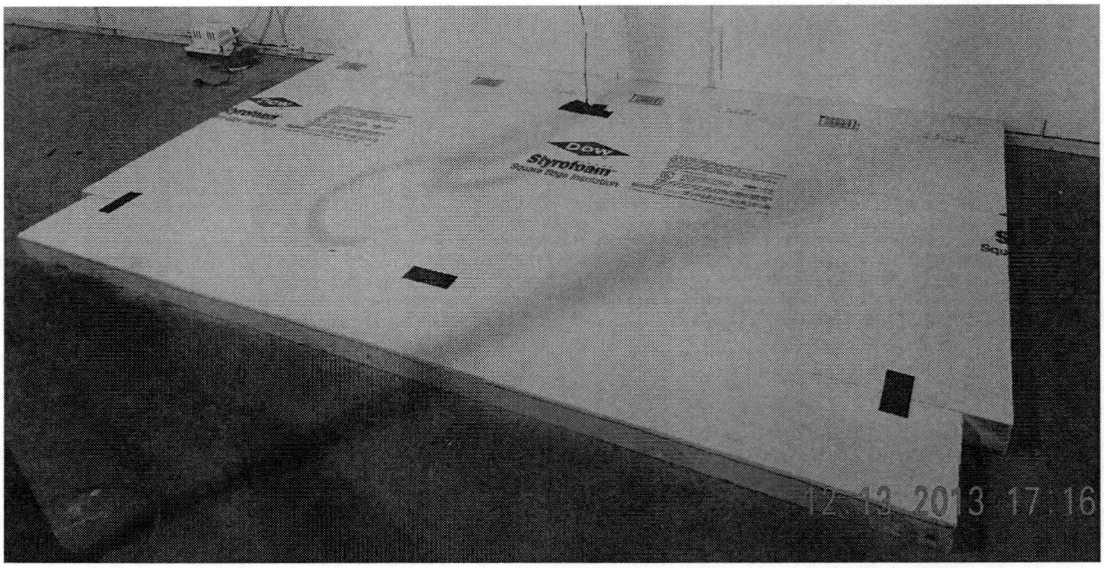

Figure 7 *Temporary floor insulation in Bay 3S.*

soil domain and heat flows used in the calculations is shown in Figure 9. The calculations included a time-varying temperature variable, T_m, at 4 ft below the bottom of the footing. For the calculations, the deep ground temperature, T_{dg}, was held constant at 7.22 °C. It was assumed that all heat transferred between the basement and Domain A and the soil surface and Domain A was stored in or lost from Domain A.[4] This simplification was based on the floor insulation experiment that showed no change to the soil temperatures adjacent to the wall, thus indicating that heat lost from the basement through the wall beneath the insulation could be ignored, at least to a first order of approximation.

4. A more sophisticated approach allowing Q_{horz} to be isolated from Q_{crib} entails treating domain E as a separate control volume with an independently evaluated mean temperature.

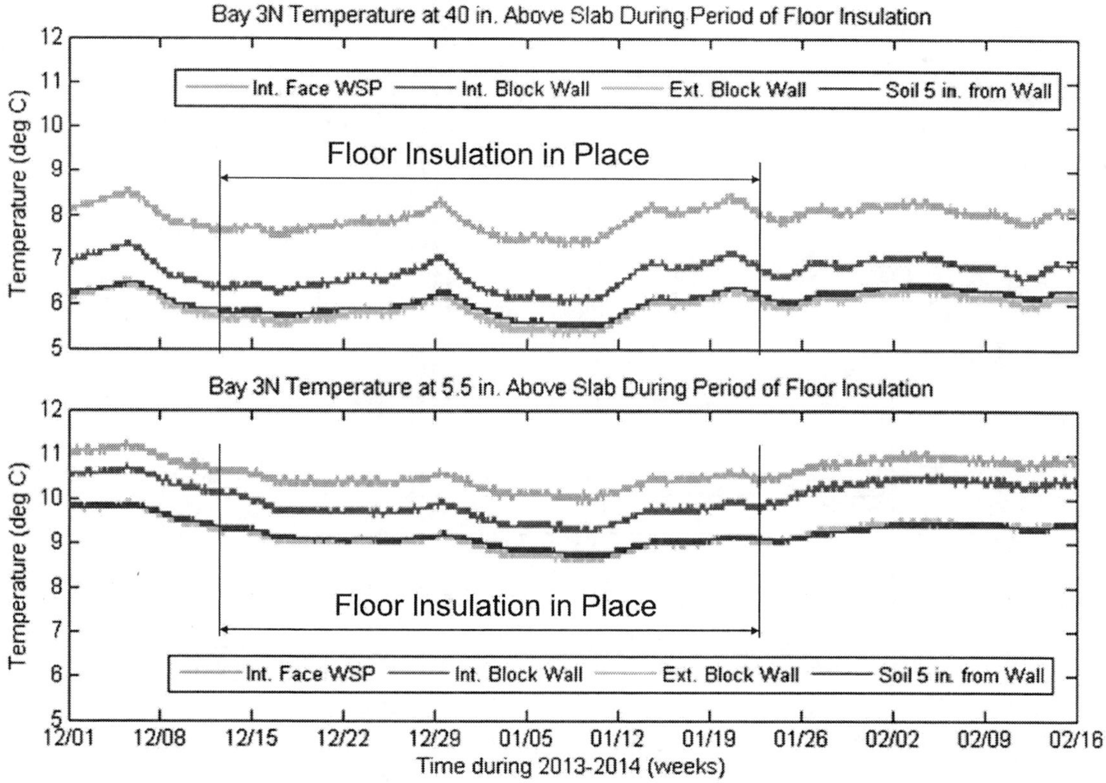

Figure 8 *Wall temperature profile with temporary floor insulation.*

The annual heat transferred from the basement, deep ground, and soil surface was calculated using time-discrete trapezoidal integration of the heat transfers as a function of T_m and the basement and outdoor ambient experimental boundary temperatures. A discrete analytic solution for T_m was derived by solving the transient energy balance for the sub-slab soil domain. Under these conditions, the horizontal heat loss below the insulation (Q_{horz}) could be estimated (derivation in Harmon 2015).

A number of heat transfer cases were evaluated, and the calculation results for the two most pertinent cases are shown in Figure 10. Case B had the top of the water table located at the bottom of the footing and Case C had the top of the water table located at the bottom of the soil domain (8 ft below the bottom of the footing). Positive heat transfer values are shown by the arrows in Figure 9. If there was heat gain and loss over the year, the gains (+) and losses (–) are provided in parentheses.

Q_{base} was greater for Case B than it was for Case C owing to the amount of water in the soil below the slab. More water in the soil domain (Case B) meant the soil domain had a higher thermal conductance and a significantly higher heat capacity, thus increasing the heat loss from the basement and allowing the transferred heat to be stored before being lost to the deep ground and ambient surroundings. Hence, Q_{crib} (heat loss to ambient air) increased with increasing Q_{base}.

The net Q_{crib} loss was larger in magnitude than Q_{horz} (horizontal heat loss beneath insulation) by a factor of about 1.9 for Case B and 4 for Case C. The factors were larger for the heating season only. This confirmed numerically that heat stored in the soil domain below the slab and lost to the soil surface was the primary driver of the soil temperatures adjacent to the wall. It also accounted in principle for the substantial discrepancy (5°C) initially observed between the simulated and experimental soil temperatures at the base of the wall during the cooling season (the discrepancy was lower [3°C] in the heating season). Specifically, the elevated temperature of the high heat capacity sink beneath the slab (relative to the deep ground boundary temperature) yielded higher soil temperatures adjacent to the wall during the cooling season. Thus, this simple analysis elucidated the underlying heat transfer physics that explained the experimental observations and that were confirmed by the more rigorous simulation calculations discussed below.

Simulation/Experiment Result Comparisons

One of the key objectives of the research was the generation of a transient experimental database enabling the testing of the accuracy and physical validity of foundation earth contact simulation programs. The effectiveness of the database for this purpose was investigated using the BUilding Foundation Energy Transport Simulation (BUFETS) program

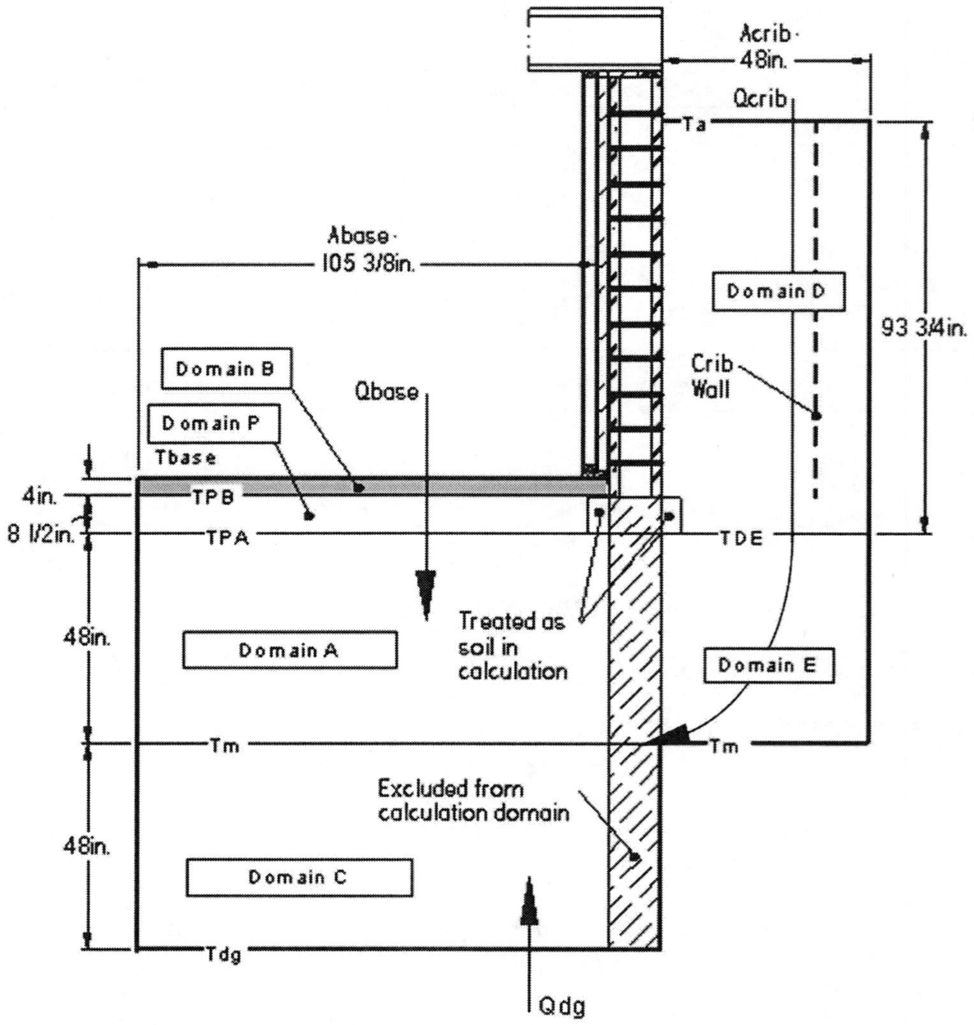

Figure 9 *Soil domain and heat flows for heat transfer calculations.*

(see Goldberg and Harmon 2015a for a full description). BUFETS is based on a rigorous application of the continuum mechanics energy conservation balance that includes a jump energy balance across the frost front for describing solid/liquid phase change. Air and moisture (vapor and liquid) transport are not included in the simulation but their energy impacts may be represented by temperature and moisture dependent models or by measured data. BUFETS has been successfully tested (Goldberg and Mosiman 2015) against the analytic IEA BESTEST case GC10a (Neymark and Judkoff 2008).

Figure 11 depicts the correlation between the hourly averaged transient simulation and experiment soil temperatures measured 5 in. away from the foundation wall in Bay 2N over a 1.5-year period. The root mean square errors over three time periods corresponding approximately to the first and second heating seasons separated by a cooling season are reported in Table 2. The best correlation was achieved at 40 in. above the slab with a 1°C or less root mean square (RMS) error. Five inches above the slab, the RMS error varied between 0.9°C and 2°C, while at 69 in. above the slab (12 in. below grade), the error similarly ranged between 0.7°C and 2.1°C. Thus, an overall RMS error of about 2°C or less was considerably improved over the initial poor correlation (discussed above) as a result of the following:

- **Inclusion of a sub-soil water table model corresponding to Case B in Figure 10.** This reduced the heating season simulation/experiment discrepancy at the base of the wall from 3°C and 5°C to RMS errors of 1.24°C (or less) and 1.91°C in the heating and cooling seasons, respectively. Thus, the simple analysis described above was supported by the simulation results.
- **Invocation of a phase change model to describe the thermal impact of the snow cover in the heating season based on measured snow depth, diffuse, direct and long-wave surface radiation.** This model approximates the effect of sublimation and melting at the snow surface

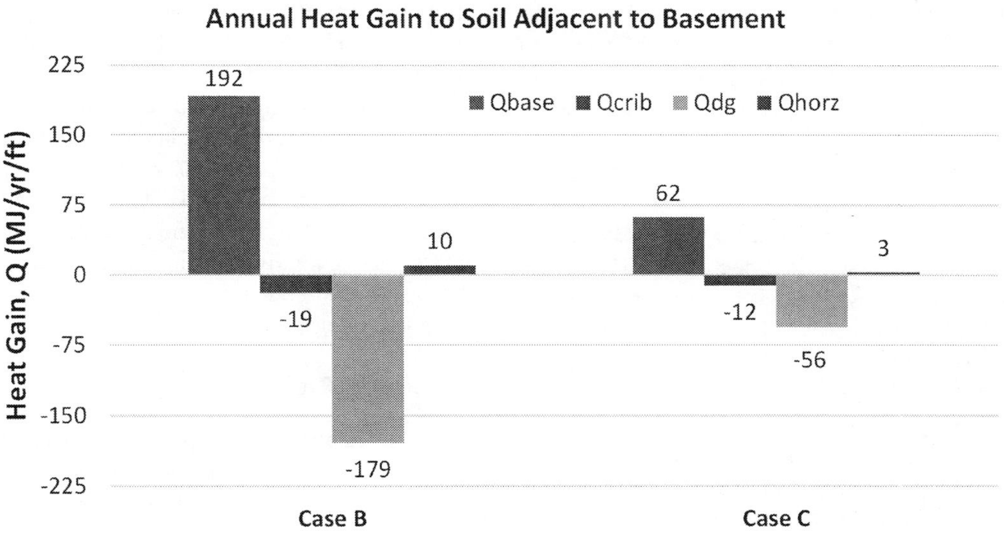

Figure 10 *Annual heat gain to the soil adjacent to the basement.*

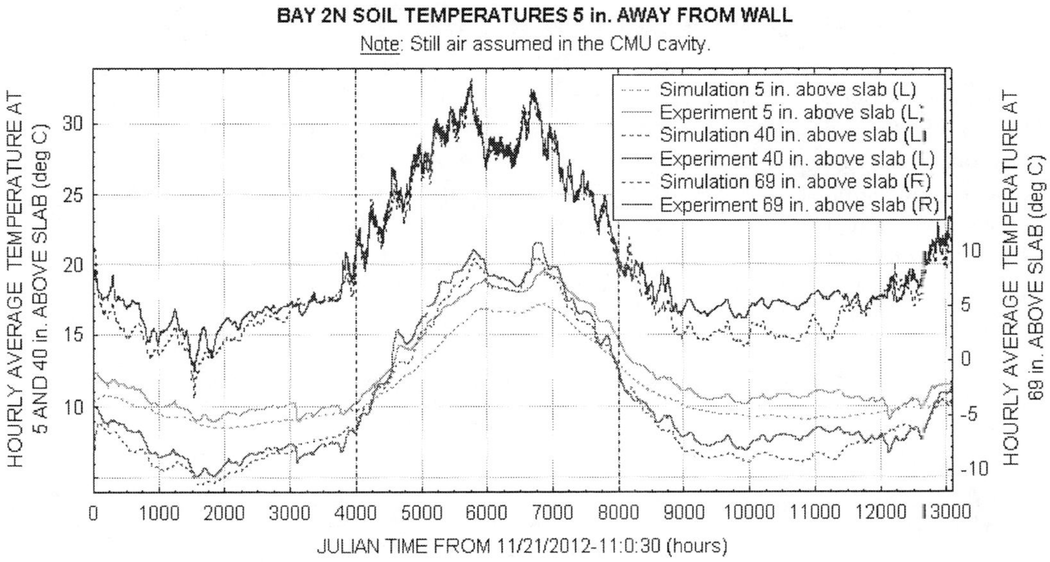

Figure 11 *Bay 2N simulation and experiment soil temperature profiles.*

and melting at grade. During the first and second heating seasons, without the phase-change model, the RMS errors 12 in. below grade were 4.2°C and 4.5°C, respectively. With the phase-change model in place, the RMS errors were reduced to 1.5°C and 2.1°C, respectively. Thus, the impact of the snow phase change model was substantial. During the cooling season (with the snow model inactive), the RMS error amounted to 0.7°C 12 in. below grade. Thus, it is apparent that, at least, the snow model adopted needs to be improved to achieve lower RMS errors during the heating season relative to the cooling season errors with an inactive phase-change model.

Table 2. Soil Temperature Simulation/Experiment Profile Root Mean Square Errors

Time Period, h	5 in. above the Slab, K	40 in. above the Slab, K	69 in. above the Slab, K
0–4000 (heating season)	0.85	0.79	1.46
4000–8000 (cooling season)	1.91	0.89	0.74
8000 –13091 (heating season)	1.24	1.03	2.11

Figure 12 examines the correlation between the measured and simulated heat fluxes on the CMU interior face/insulation interface. In this case, the heat flux RMS errors range from 0.4 to 0.5 $W/m^2 \cdot K$ that, relative to the average experimental heat flux magnitude over the monitoring period of 3.26 $W/m^2 \cdot K$, corresponds to an approximate absolute error of 15%. This is a result of the CMU air cores being simulated as still air with an experimentally derived equivalent thermal conductance (see Huelman et al. 2015) rather than as cavities with buoyant cavity flows as discussed above. During the heating seasons, the simulated CMU interior face shell temperatures are colder than those measured with RMS errors of 2.7°C and 3.2°C because the simulation ignores the impact of the rising warm air plume alongside the interior CMU face shell. The absence of buoyant cavity flow may also contribute to the soil temperature heating season RMS errors 69 in. above the slab shown in Figure 11. The cooling season RMS error of 1.3°C is half or less of the heating season RMS error and is of the same order of magnitude as the adjacent soil temperature RMS error, possibly owing to the absence of or significant reduction in buoyant cavity flow at the top of the wall during the cooling season (Harmon 2015).

Hence, inadequate accounting of the effects of CMU buoyant cavity flow results in inaccurate simulation of the wall surface heat flux. The complexities arising from this are evident at the season transitions (3000–4000 and 12000–13000 hours) where unaccounted for buoyant cavity flow dynamics yield counter-intuitive, relatively low temperature discrepancies combined with relatively high heat flux discrepancies. Because the installed RUS-15 interior insulation yields small heat fluxes, buoyant cavity flow has a relatively small impact on the soil temperatures that are dominated by the grade surface boundary conditions and the heat transfer from the sub-slab water table as observed and simulated. The experiment/simulation comparison demonstrates that the experimental database is useful for evaluating the accuracy of thermal foundation simulation codes and for gaining a quantitative understanding of their physical validity in terms of the principal heat and moisture transport phenomenology occurring in the experiment.

CONCLUSIONS

- The database is useful as it provides experimental data for assessing the accuracy of thermal simulation programs, and it enables an insightful understanding of the dominant thermal transport mechanics observed experimentally.
- The experimental data demonstrate that buoyant cavity flow exists in a CMU wall with interior insulation and also show that this cavity flow yields higher horizontal heat flow through the wall compared with conduction heat transport through still air. Therefore, this flow needs to be adequately accounted for in thermal simulations of CMU walls.
- IECC 2012 compliant RUS-15 interior insulation reduces the heat flow through the upper half of a hollow CMU foundation wall by a factor of 3 compared with the same wall when uninsulated in a cold climate.

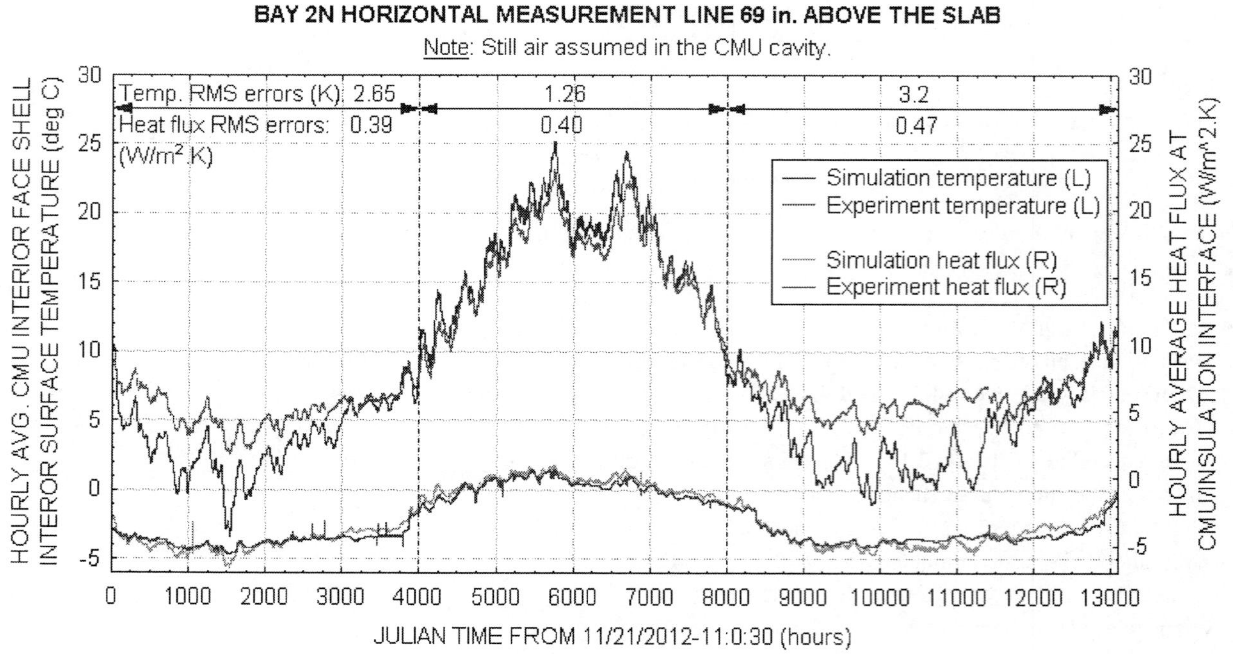

Figure 12 *Bay 2N CMU wall interior surface heat flux and temperature.*

- On a hollow CMU wall, upper half-wall IECC 2012 compliant RUS-15 exterior insulation yields a 55% larger heat flow 12 in. below grade than full-wall interior RUS-15 insulation in a cold climate. Thus, in this climate, half-wall exterior insulation does not offer any thermal performance advantages over full-wall interior insulation.
- In a vapor-sealed rim joist cavity, a solid CMU wall bond beam (vapor permeance of 0.1 to 1 U.S. perms—Class II) provides sufficient vapor isolation from saturated hollow CMU cores so that condensation on the rim board interior face does not occur with or without exterior insulation.
- On a hollow CMU wall with RUS-15 full-height, interior insulation, the heat transport through the footing and wall below the insulation relative to the heat transport from the liquid-saturated soil beneath the basement has a negligible measured impact on the soil temperatures adjacent to the wall in a cold climate under pseudo-steady-state conditions.
- If a water table exists beneath the slab, its depth and the proximity of its surface to the slab are critical for accurately calculating the basement foundation thermal transport in a cold climate.
- A simple heat transport model that captures the essence of experimentally observed thermal phenomena in a full basement can offer the same insight as a detailed, three-dimensional thermal transient simulation.
- A snow model that includes phase change at the ambient and ground interfaces is necessary for accurately simulating basement heat transport during the heating season in a cold climate.
- Heating season simulation/experiment CMU wall heat flux errors of about 15% arise from ignoring buoyant cavity flows in the hollow cores of CMU foundation walls with full-wall interior insulation.

ACKNOWLEDGMENTS

The research reported was carried out with funding provided by the National Renewable Energy and Oak Ridge National Laboratories of the U.S. Department of Energy. While the authors gratefully acknowledge this support, they assume complete responsibility for the contents herein.

REFERENCES

Goldberg, L.F., and A.C. Harmon. 2015a. Cold climate foundation retrofit experimental hygrothermal performance: Cloquet Residential Research Facility Laboratory Results, National Renewable Energy Laboratory, U.S. Dept. of Energy, prepared under subcontract no. KNDJ-0-40338-04. http://www.nrel.gov/docs/fy15osti/63319.pdf.

Goldberg, L.F., and A.C. Harmon. 2015b. Two years of transient cold-climate building foundation hygrothermal experimental data [dataset]. Retrieved from the Data Repository for the University of Minnesota, http://dx.doi.org/10.13020/D65P4C.

Goldberg, L.F., and G. Mosiman. 2015. The energy savings potential of optimized slab-on-grade foundation insulation retrofits. *J. Green Buildings* 10(3):116–36.

Harmon, A.C. 2015. The hygrothermal performance of cold climate basement walls retrofitted with insulation and a water separation plane. Master's thesis, University of Minnesota, http://hdl.handle.net/11299/174774.

Huelman, P., L. Goldberg, and R. Jacobson. 2015. Innovative retrofit insulation strategies for concrete masonry foundations, National Renewable Energy Laboratory, U.S. Dept. of Energy, prepared under subcontract no. KNDJ-0-40338-03, http://apps1.eere.energy.gov/buildings/publications/pdfs/building_america/insulation-strategies-concrete-masonry.pdf.

Neymark, J. and R. Judkoff. 2008. International Energy Agency Building Energy Simulation Test and Diagnostic Method (IEA BESTEST), In-Depth Diagnostic Cases for Ground Coupled Heat Transfer Related to Slab-On-Grade Construction, NREL Technical Report no. NREL/TP-550-43388.

Field Demonstration of an Attic Retrofit Strategy Using Cellulose Insulation and Phase-Change Material

Kaushik Biswas, PhD **Som Shrestha, PhD** *Member ASHRAE* **Andre O. Desjarlais**

Nicholas M. Josefik **Andrew J. Nelson**

Larry D. Stephenson, PhD **Jeffrey L. Lattimore** **Robert De Vries**

ABSTRACT

The U.S. Department of Defense (DOD) has set aggressive building energy performance goals to comply with the requirements of the 2007 Energy Independence and Security Act (EISA 2007), and is introducing state-of-the-art technologies into its new building design and building upgrade programs. A key part of this strategy is the identification, proof-testing, and introduction of promising new technologies into the building stock that will be needed to achieve mandated energy reduction goals. This article describes an attic retrofit strategy using batt and loose-fill cellulose insulation combined with phase-change material (PCM) applied in an existing DOD building. The subject building is located at Fort Bragg, North Carolina, and is a one-story building with ducts in the attic. Originally, the attic contained ceiling insulation and was unconditioned with soffit and gable vents. The project consists of three stages. First, the original attic was instrumented with temperature, heat flux and humidity sensors, and monitored over the heating and cooling seasons. In stage two, the attic vents were sealed and the roof decks, soffits, and gable ends were insulated with cellulose insulation. Finally, in stage three, low-cost bio-based PCM mats were installed on the interior surfaces of the cellulose insulation at roof decks and gables. Pre- and post-retrofit data are being collected over at least 9-month periods for each stage of analysis.

INTRODUCTION

This project was designed to measure and demonstrate the effectiveness of phase-change material (PCM) combined with cellulose insulation in an existing U.S. Army building at Fort Bragg, NC. This project was undertaken in support of the building energy performance goals of the U.S. Department of Defense (DOD) to comply with the requirements of the Energy Independence and Security Act of 2007 (EISA 2007). The DOD is introducing state-of-the-art designs and technologies into its new building design and building upgrade programs. A key part of this strategy is the identification, proof-testing, and introduction of promising new technologies into the building stock that will be needed to achieve the mandated energy reduction goals. Loose-fill or blown insulation enhanced with distributed PCM has recently been tested by Oak Ridge National Laboratory (ORNL) (Kosny et al. 2010; Shrestha et al. 2011). Biswas et al. (2011) and Kosny et al. (2012) also demonstrated the efficacy of bio-based PCM mats in reducing the heat fluxes through a test roof, when incorporated with other energy-saving technologies such as above-sheathing-ventilation, low-emissivity surface, and rigid insulation. The effectiveness of PCM insulation technologies need to be proven in the DOD building stock to enable its integration into DOD planning and programs and widespread implementation into the building portfolio. Further, insulating the roof and sealing the attic, especially if HVAC equipment and ducts are present in the attic (as was the case in Fort Bragg), can result in significant energy savings (Rudd et al., 1999, Morshead, 2007).

Oak Ridge National Laboratory (ORNL), in collaboration with the U.S. Army Construction Engineering Research Laboratory (CERL), installed building data (temperature, heat flows, etc.) monitoring instrumentation in a selected test building in Fort Bragg. The building originally contained an

Kaushik Biswas and **Som Shrestha** are research and development staff and **Andre Desjarlais** is the program manager for building envelope research at Oak Ridge National Laboratory, Oak Ridge, TN. **Larry D. Stephenson** (retired), **Jeffrey L. Lattimore**, and **Nicholas Josefik** are researchers and **Andrew J. Nelson** is the chief of the Energy Branch at the Construction Engineering Research Laboratory, Champaign, IL. **Robert De Vries** is a member of technical services at Nu-Wool Co., Inc., Jenison, MI.

unconditioned and vented attic, which contains the space conditioning ducts, with ceiling insulation. This article describes the conversion of the vented attic into a sealed and insulated attic. The project is being done in three phases. First, the original attic was monitored for a period of over 12 months. In phase two, the roof decks, eaves and gable ends were sealed and insulated with batt and blown-in cellulose insulation, while the ceiling insulation was removed. Finally, in phase three, PCM mats were added adjacent to the interior (facing the attic) surfaces of the roof deck and gable insulation. Pre- and post-retrofit performance data are planned to be collected over at least 9- to 12-month periods for each stage to cover the summer, winter, and spring/autumn seasons, and analyzed to evaluate the effectiveness of the retrofit measures.

TEST BUILDING AND DATA ACQUISITION

Figure 1 shows the classroom building at Fort Bragg, North Carolina that has been chosen for testing the PCM insulation technology. It is a single-story building with a footprint of about 308 m^2. It has a gable roof, with the roof decks facing north and south, and east and west gable ends. The baseline building contained an unconditioned attic, which was vented at the gables and soffits. Ducts for air distribution are located within the attic.

Figure 2 shows the original attic of the test building, with uninsulated roof deck and fiberglass ceiling insulation. Figure 3 shows a schematic of the building instrumentation. Sensor arrays were installed under the roof decks and inside the knee walls of the attic. The insets show the instrumentation configuration at the roof deck (top left) and knee walls (bottom left) and the weather station (top right). At each roof location in the original attic (during phase 1) a heat flux transducer (HFT) was installed at the interior roof deck surface using a 30.5 × 30.5 cm^2 sheet of oriented strand board (shown in Figure 3). During phase two retrofit (addition of cellulose), the HFTs were moved so that they were located close to the interior (attic-facing) surface of the insulation, as seen in Figure 3.

Figure 1 *Test building (south face).*

Thermistors (*T*) were installed on the exterior shingle surface and in line with the bottom of the existing rafter at the roof deck; the latter were moved to the inside surface on the cellulose insulation after the phase 2 retrofit. While retrofitting with PCM, additional thermistors were installed on inside surfaces of the PCM mats. Combination temperature and relative humidity (T/RH) sensors were located at the interior roof deck surfaces and at mid-depth of the insulation to be added (at the interface of the batt and blown cellulose insulation). Finally, two sets of thermistors and RH sensors were added to the interior walls of the classroom. On one gable end, a weather station was installed. There are pyranometers and a pyrgeometer to measure the incident solar and infrared radiation on the roof decks. The weather station also contains an instrument suite to measure the outdoor temperature and relative humidity, precipitation, pressure, and wind velocity.

In addition to the ongoing data monitoring, air leakage rates of the building envelope and the space-conditioning ducts were measured. The leakage tests were done under baseline (pre-retrofit) building conditions and after retrofitting the attic with cellulose insulation. Duct leakage tests were also performed. The building air leakage rate was measured according to ASTM E779 (ASTM 2010), and the duct leakage was measured using ASTM E1554 Test Method B (ASTM 2007). The results of the leakage tests, done under pre- and post-retrofit conditions, will be used in numerical simulations. The simulations will be performed to determine the actual energy-saving potential of the retrofit measures for the test building, and to extrapolate the energy savings to different climate zones.

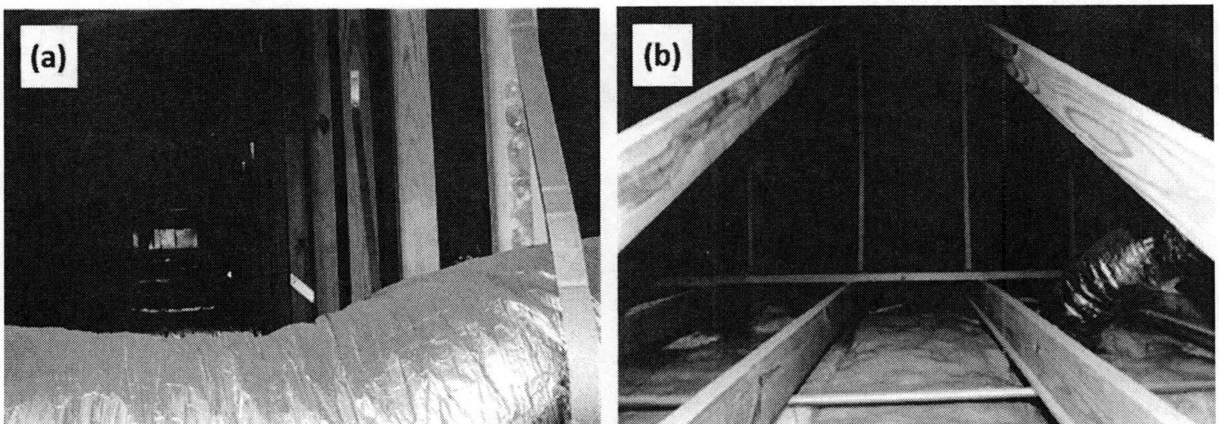

Figure 2 *Original unconditioned and vented attic with (a) ducts and (b) ceiling insulation.*

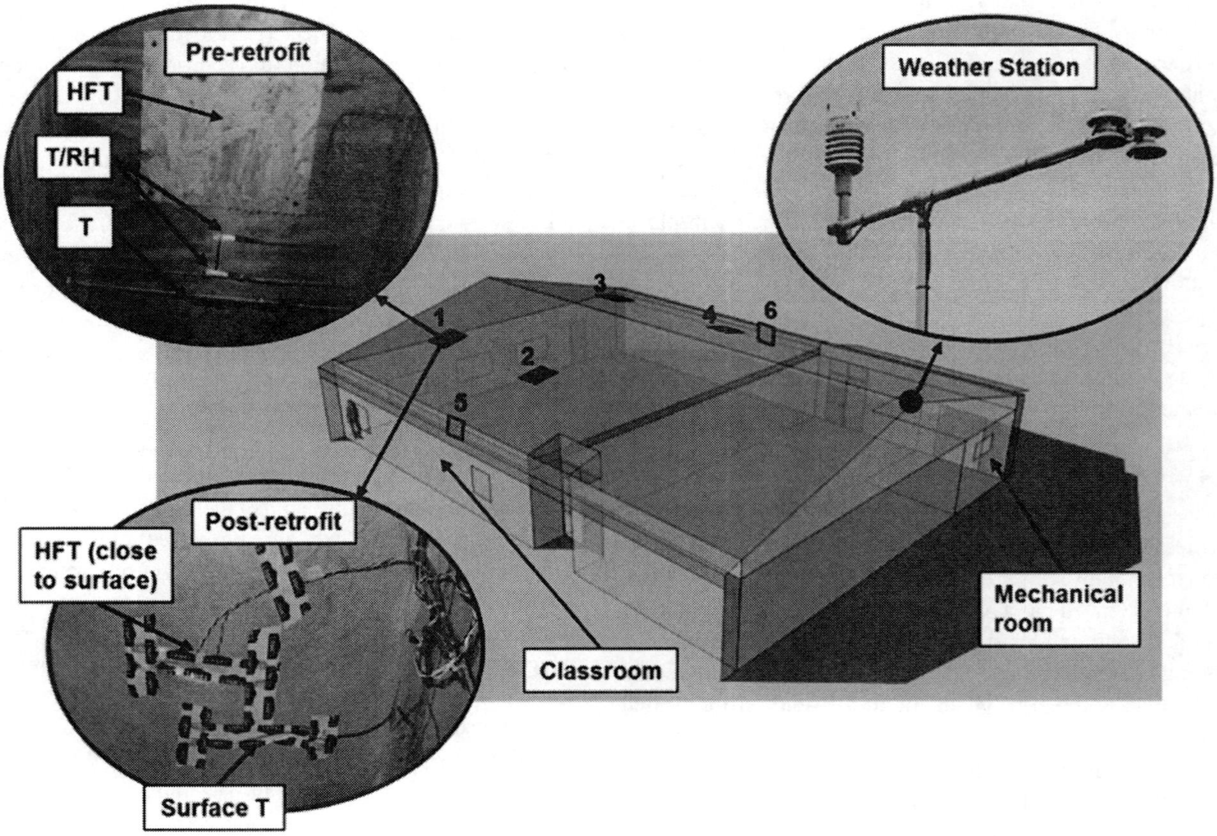

Figure 3 *Building data acquisition systems.*

BUILDING RETROFIT AND INSULATION DETAILS

Originally, the attic contained fiberglass insulation above the ceiling. The fiberglass insulation was nominally 3.3 $m^2 \cdot K/W$. The attic underwent a two-stage retrofit. During June–July, 2014, the first retrofit took place. During this time, the gable and soffit vents were closed. The target insulation thermal resistance (R-value) under the roof deck and the gable ends was about 5.28 $m^2 \cdot K/W$, which required about 9 in. (22.9 cm) of cellulose insulation. The attic contains trusses constructed from 2 × 6 and 2 × 4 wood studs. The original rafters under the roof decks consisted of 2 × 6 studs yielding a depth of 14 cm. To get the additional depth, 2 × 4 wood studs (of 8.9 cm depth) were attached to the underside of the existing rafters. Additional wood studs were also added at the gable ends to achieve the required depth. Figure 4 shows the original and modified framework within the attic. The retrofit work was confined to the attic interior; the roof components above the deck (weather barrier, shingles, etc.) were not modified during the retrofit activities.

The insulation was added in two layers. The first, outer layer consisted of 14 cm thick cellulose batts followed by an 8.9 cm layer of dense-packed blown cellulose. Nets were added to the bottom edges of the extended rafters and the inner edges of the gable end frames to support the blown cellulose. The nominal density of the blown cellulose was 56.0 kg/m^3. A single layer of batt insulation was added at the knee walls, which comprise a small fraction of the overall surface area within the attic. During this retrofit, the original fiberglass insulation on the ceiling was removed.

Figure 5 shows the added cellulose batt and blown insulation. Table 1 lists the measured thermophysical properties of the insulation materials. The thermal conductivities were measured using a heat flow meter apparatus (HFMA), following ASTM C518 (ASTM 2015) guidelines. Based on HFMA measurements, the total R-value of the added cellulose insulation was 5.06 $m^2 \cdot K/W$.

The second retrofit took place during June 2015. Boo-based PCM mats were added to the underside of and inside the blown cellulose insulation added under the roof deck and at the gable ends, respectively. The PCM was encapsulated within plastic pouches and covered with a metalized polymer film. The effect of the thermal emittance of the metalized film facing the attic will be investigated and evaluated to isolate the impact of the PCM. Table 2 lists the relevant physical properties of the PCM mat, as provided by the manufacturer. M-value is a measure of the latent heat of the PCM per unit area. The melting point stated in Table 2 is the temperature at the peak heat input during the melting test using differential scanning calorimetry (DSC) (Gmelin and Sarge 2000); the freezing point is the temperature at the peak heat release during the

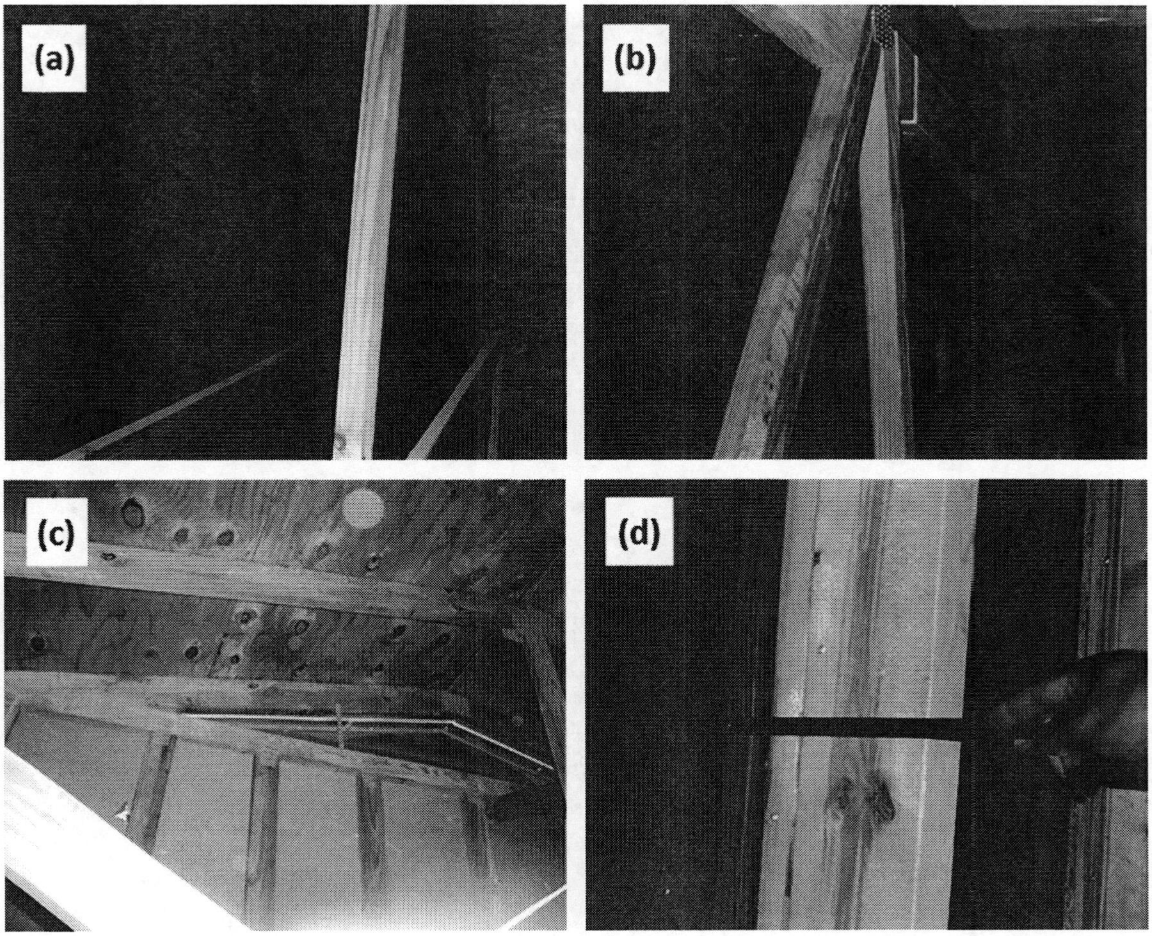

Figure 4 *Original and modified attic framework: (a) original 2 × 6 wooden rafters, (b) extended rafters with 2 × 4 wood studs added to achieve a depth of 22.9 cm, (c) original framing at the gable ends, and (d) extended framing at the gable ends achieving a depth of 22.9 cm.*

freezing test. Table 2 also lists the melting and freezing initiation and end temperatures (listed within brackets), which were estimated based on the DSC data provided by the manufacturer. By comparing these data with the measured PCM surface temperatures, the initiation and end temperatures would indicate whether the PCM underwent partial or full melting/freezing, or none at all, following the diurnal temperature variations within the attic. Figure 6 shows the PCM mat and installation of the mats under the roof deck insulation and on the gable ends. PCM mats were not added to the knee walls.

RESULTS AND DISCUSSION

Temperature and Heat Flux Data

In this section some comparative temperature and heat flux data are shown, to evaluate the efficacy of the two retrofit measures. Figure 7 shows the temperature trends across the insulated roof deck following the phase 2 (cellulose) retrofit. Data from four summer and four winter days are shown. As expected, the shingle and roof deck temperatures exhibit the highest diurnal fluctuations, followed by the mid-insulation (*Mid-ins.*) location. The insulation surface (*Ins. Surf.*) and attic temperatures showed very little fluctuations (~2°C–3°C). In the absence of any ceiling insulation, the attic temperatures were very close to the conditioned space temperatures.

In order to realistically compare the energy-saving potential of the retrofit measures, whole-building numerical simulations are needed and efforts to do so are underway. In addition, the models can be extrapolated to evaluate the energy savings under different climate conditions, which is useful because the DOD portfolio contains similar buildings in a wide range of climate zones. Here, comparative attic temperatures and heat fluxes at the deck/insulation surface from the different phases are shown. For comparison, three summer days with similar outdoor temperatures and solar irradiance on the roofs were chosen, as shown in Figure 8. The days chosen for comparison were: Pre-retrofit—7/14–7/16/2012, Phase 2—7/12–7/14/2014, and Phase 3—7/26–7/28/2015. Figure 9 shows the comparative temperature and heat flux data from the three phases.

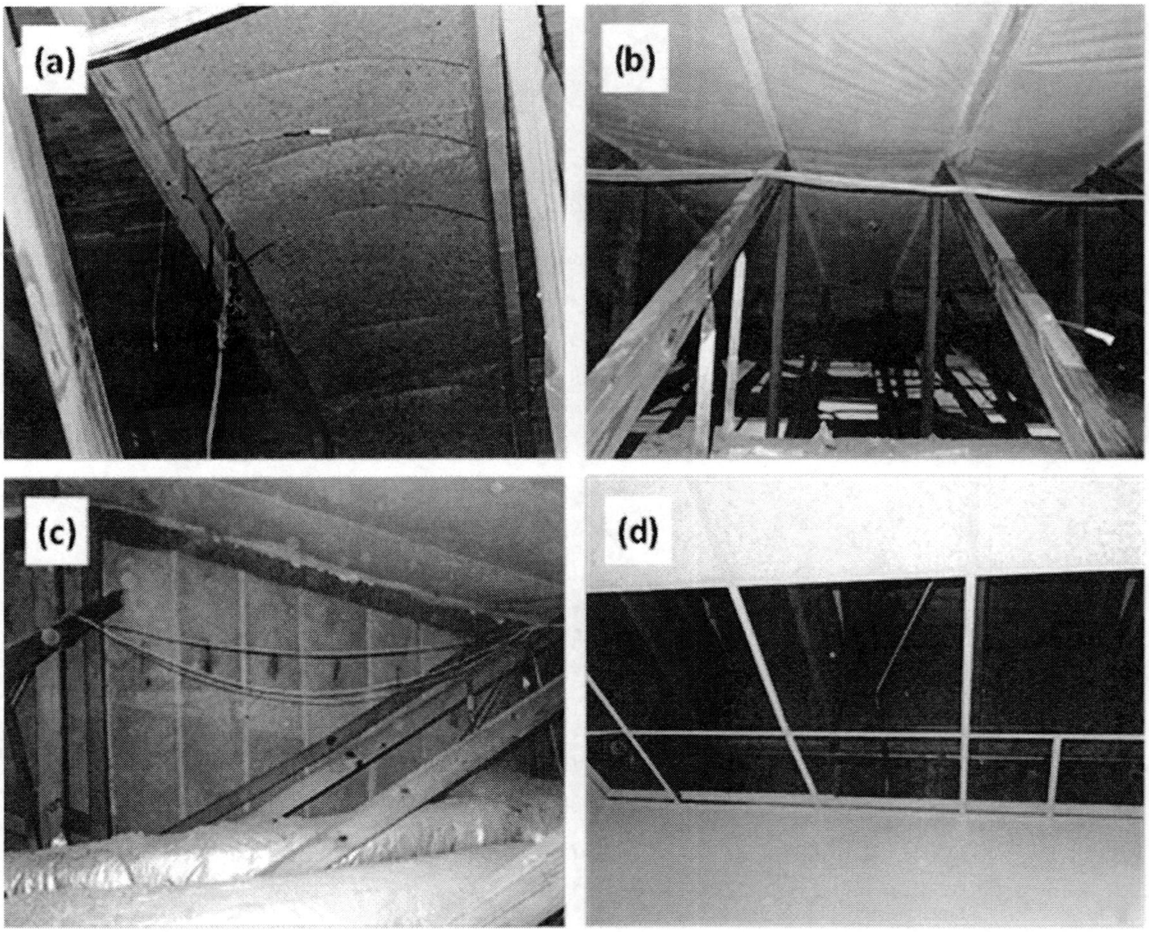

Figure 5 *Attic retrofit with cellulose insulation: (a) outer layer of batt insulation under the roof deck, (b) blown cellulose added under the roof deck, (c) batt and blown cellulose under the roof deck and at a gable end, and (d) batt insulation at a knee wall.*

Table 1. Measured Properties of Cellulose Insulation Materials

	Installed Thickness, cm	Density, kg/m^3	Conductivity, W/m·K	Thermal Resistance, m^2·K/W
Cellulose batt	14	38.9	0.048	2.91
Blown cellulose	8.9	56.0*	0.041	2.15

* Nominal installed density, per insulation installer

Between 20°C and 25°C reductions in peak attic temperatures and 75%–80% reduction in peak heat fluxes were observed after the attic was retrofitted. The results are similar to expectations, because an uninsulated roof deck of an unvented attic will experience very high temperature gradients (due to solar irradiance) and fluctuations, compared to the attic-facing surface of an insulated roof in a sealed attic.

Based on current data, it is difficult to gauge the effect (or lack thereof) of adding the PCM layer on the energy consumption of the building. However, comparing the measured surface temperatures across the PCM mats against the phase transition temperatures can indicate whether the PCM is undergoing phase change or not. Figure 10 shows the top and bottom PCM surface temperatures at one roof location. For comparison, the melting and freezing start and end temperatures are also shown. It is evident that during summer and early autumn the PCM was predominantly molten. Around the end of September, the PCM started freezing partially, but never completely froze during the period of monitoring. Further analysis and modeling are ongoing to estimate the energy-saving impact of the PCM layer.

Table 2. PCM Properties

Property	Value
Composition	Fatty acid and ester blend
M-value	579.23 kJ/m^2
Melting point	24.9°C [16.1°C–29.8°C]
Freezing point	22.0°C [22.7°C–14.1°C]
Latent heat	200 kJ/kg
Mat dimensions	1.2 × 0.6 m
Mat weight	3.57 kg/m^2
PCM density	2.74 kg/m^2

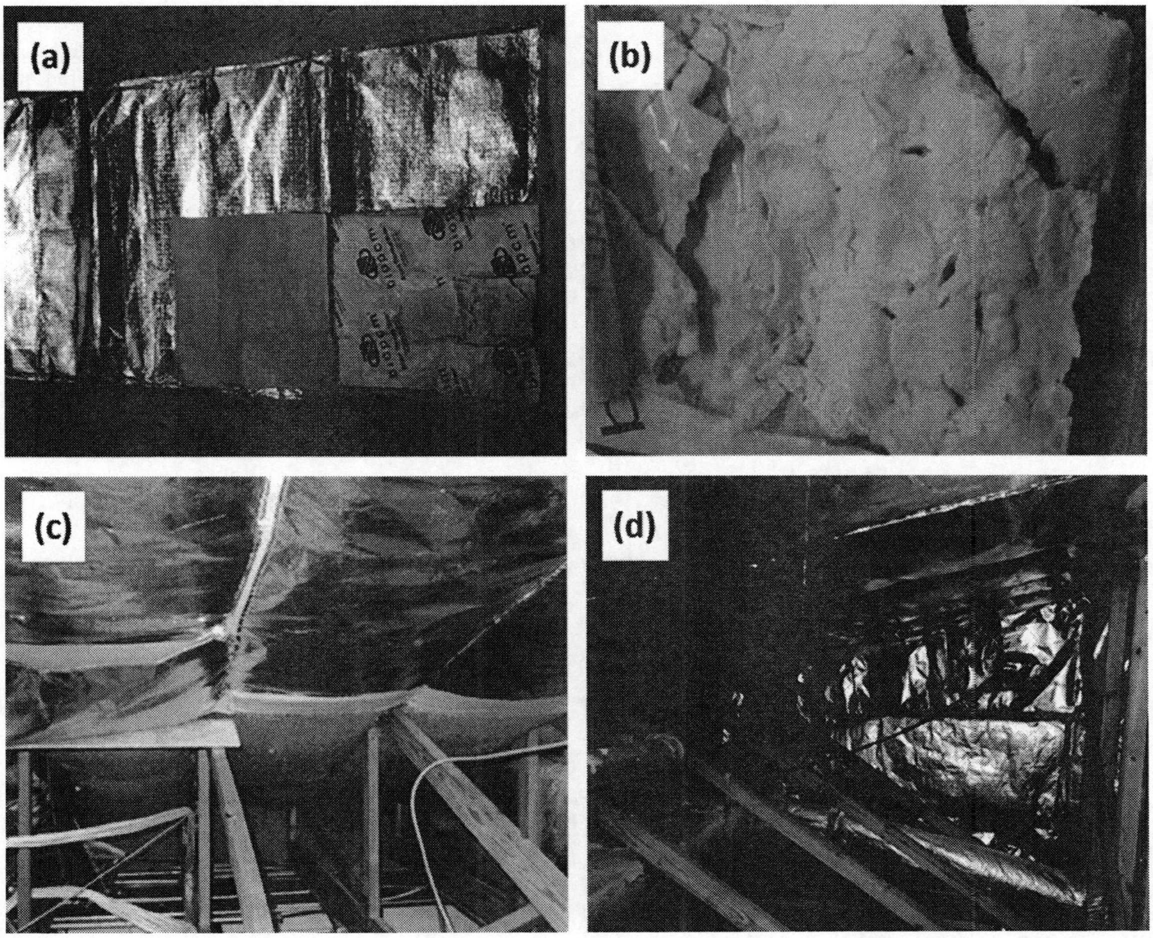

Figure 6 *(a) PCM mat with metalized film and plastic pouches containing PCM, (b) PCM, (c) PCM mats being added under the roof insulation, and (d) roof deck and gable end retrofitted with insulation and PCM mats.*

Building Power Consumption Data

The U.S Army created the Meter Data Management System (MDMS) to monitor and track the energy consumption of building on the fixed installations. Fort Bragg management is currently monitoring the subject building and uploading the data to the MDMS. These data will be analyzed to identify energy reduction trends attributed to the installation of the insulation and PCM. By comparing 15-minute energy consumption data for each phase, the energy performance characteristics of each technology will be analyzed. Initial

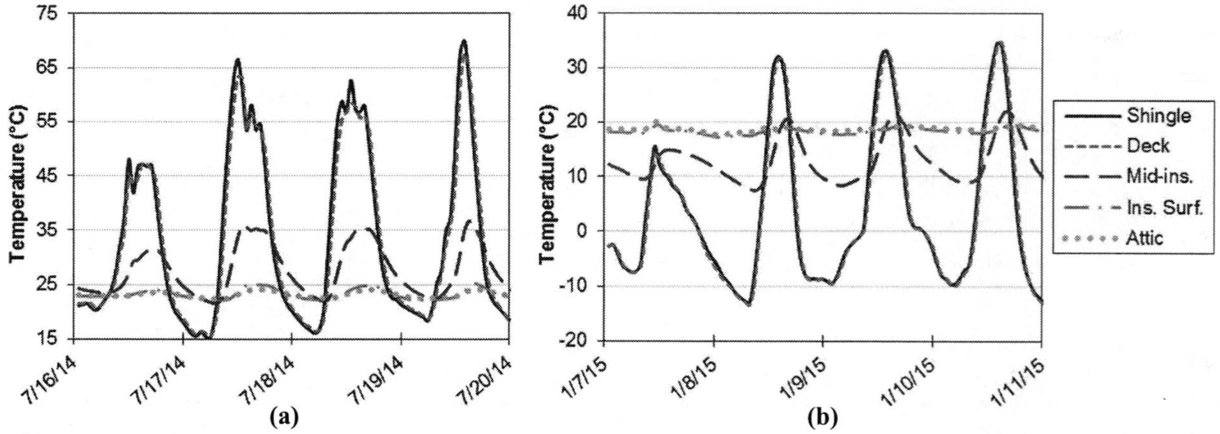

Figure 7 *Measured temperatures across the insulated roof during (a) summer and (b) winter of phase 2.*

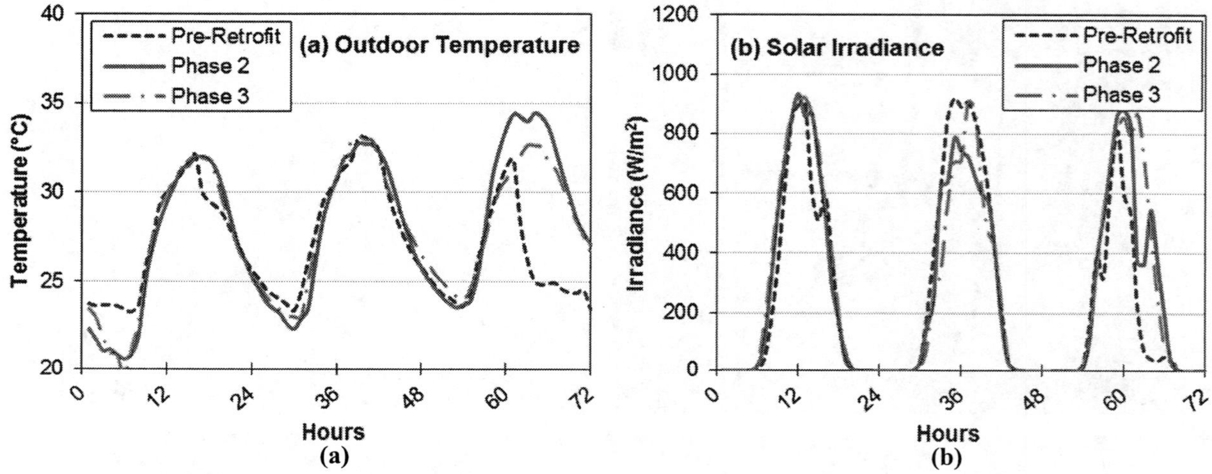

Figure 8 *Weather data during three days of similar conditions from the three project phases: (a) outdoor temperatures, and (b) solar irradiance on the south roof.*

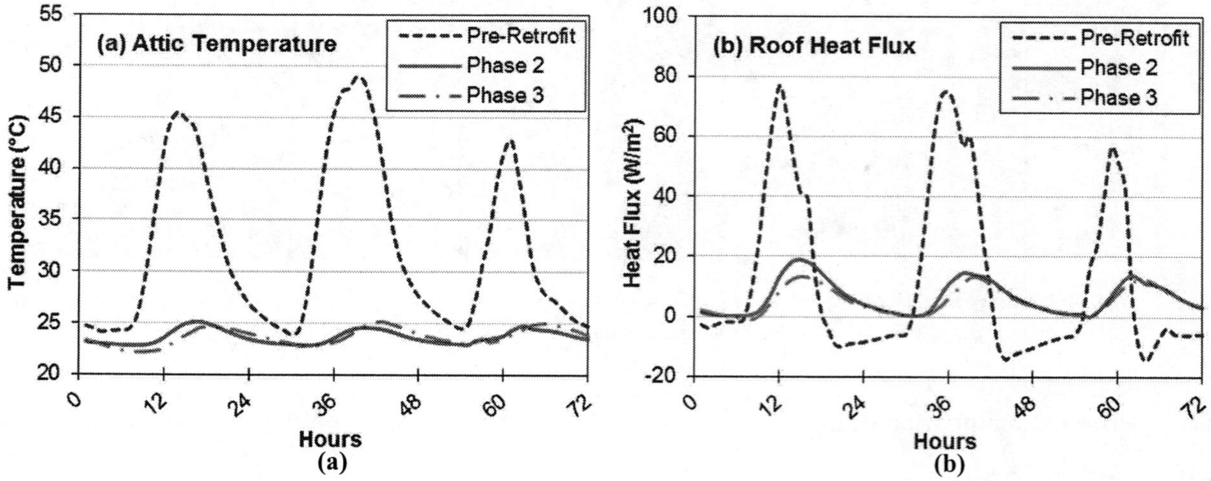

Figure 9 *Comparison of measured (a) attic temperatures and (b) roof heat fluxes over three days of similar weather conditions during the three project phases.*

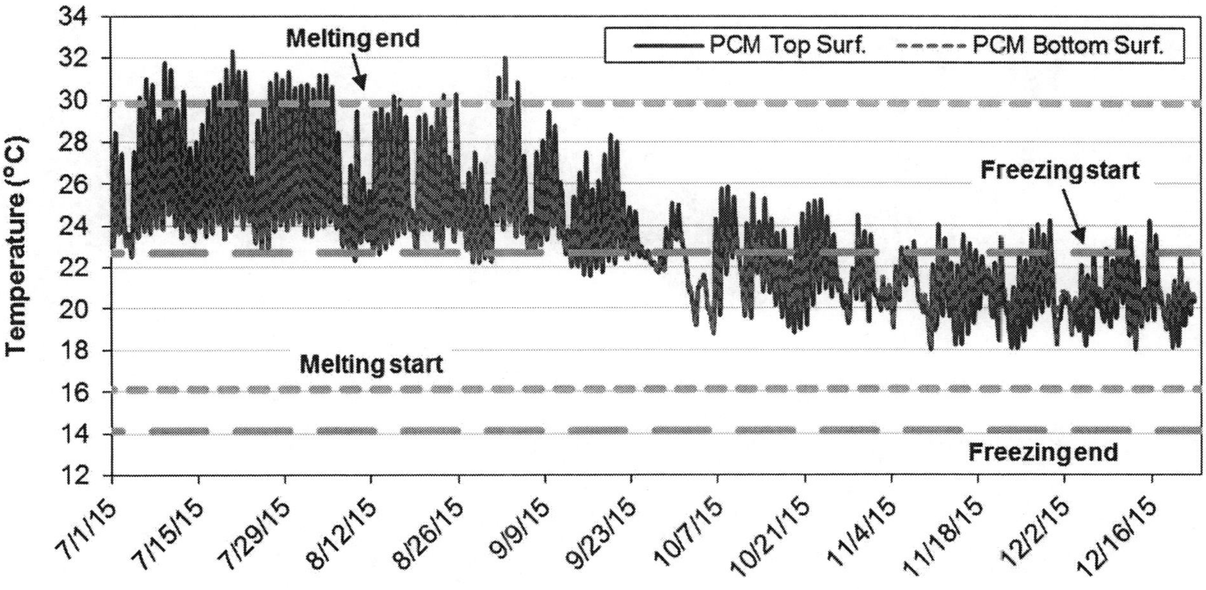

Figure 10 *Measured PCM top and bottom surface temperatures compared to phase transition initiation temperatures.*

analysis of the pre-retrofit and phase 2 energy consumption data showed a reduction in both total energy and average power consumption. Analyzing 12 months of electricity consumption during the pre-retrofit (July 2013–June 2014) and phase 2 (July 2014–June 2015) periods showed a decrease from 63,764 to 54,604 kWh, a 14% reduction caused by addition of cellulose to the roof deck and gables and converting the attic to a semiconditioned space (consequently, bringing the HVAC ducts within the thermal envelope). The average power consumption also decreased from 7.3 kW during the pre-retrofit time period to 6.2 kW during phase 2. Heating and cooling degree days were within 1% of each other over these time periods. Similar analysis will be performed comparing the phase 3 energy consumption to the first two phases, to isolate the impact of the PCM.

Humidity-Related Issues

One concern regarding sealing and insulating attics is the rise of humidity-related issues. In fact, once the phase 2 (cellulose) retrofit was complete, the occupants complained of high humidity levels within the conditioned space (classroom). This was also corroborated by measurements from the two RH sensors within the classroom, as shown in Figure 11. The higher humidity, especially during the peak summer weeks of July and early August, was caused by the substantial reduction in the runtimes of the HVAC equipment. When the issue was identified, two dehumidifiers were placed inside the conditioned space to minimize the effects of the increased humidity. After an investigation, it was determined that the duty cycle of the building heat pumps were not optimized to the new building configuration. The duty cycle was adjusted, which lowered the humidity in the classroom and the dehumidifiers were able to be removed.

An additional potential issue is the inability of the roof deck to dry following any water intrusion (due to wind-driven rain, for example) because of the addition of insulation and the PCM mats with near-impermeable metalized barrier films. As a mitigation measure, the PCM mats were installed with ~2.5 cm gaps between adjacent mats, to allow water vapor movement. Figure 12 and Figure 13 show the measured RH near one roof location, at the roof deck, mid-insulation and at a nearby location in the attic, as well as the 30-day running average RH at the roof deck during phases 2 and 3, respectively. This roof location exhibited the highest 30-day average RH at the roof deck. Per ANSI/ASHRAE Standard 160, Addendum A (ASHRAE 2009), a 30-day running average RH of less than 80% is needed to minimize problems associated with mold growth on the surfaces of components of building envelope assemblies.

During phase 2, the 30-day average RH peaked at 79.6% at the northwest section of the roof, as seen in Figure 12. During phase 3, after the addition of the PCM, the worst-case location within the roof experienced a 30-day average RH exceeding 80% during a few weeks in January and February and with a peak value of 88% in early February. On the south facing roofs, the 30-day average RH remained below 80% all the time during both phases 2 and 3. The initial analysis indicates that the addition of PCM mats has potential for adverse moisture-related impacts in the north-facing roofs. Monitoring of the RH levels within the attic and the roof locations is being continued. Mitigating action will be considered based on the current results.

SUMMARY, CONCLUSIONS AND FUTURE WORK

The article reports a two-phase attic retrofit of a classroom building on an army base located in Fort Bragg, North Carolina. The phase 1 retrofit involved addition of combina-

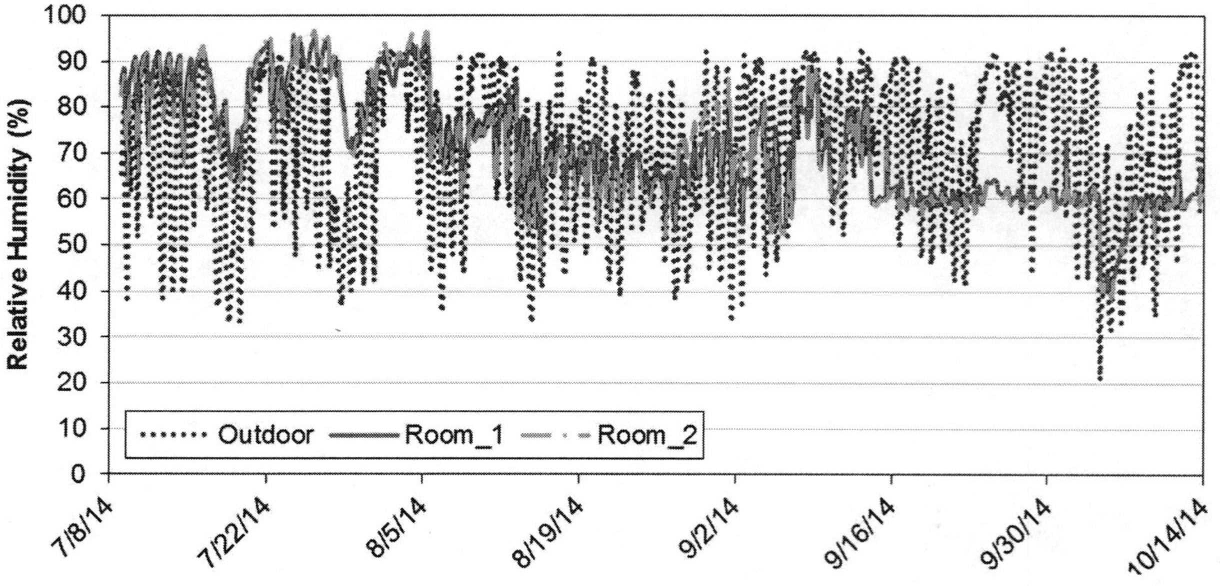

Figure 11 *Measured RH in the conditioned space (classroom) following phase 2 (cellulose) retrofit.*

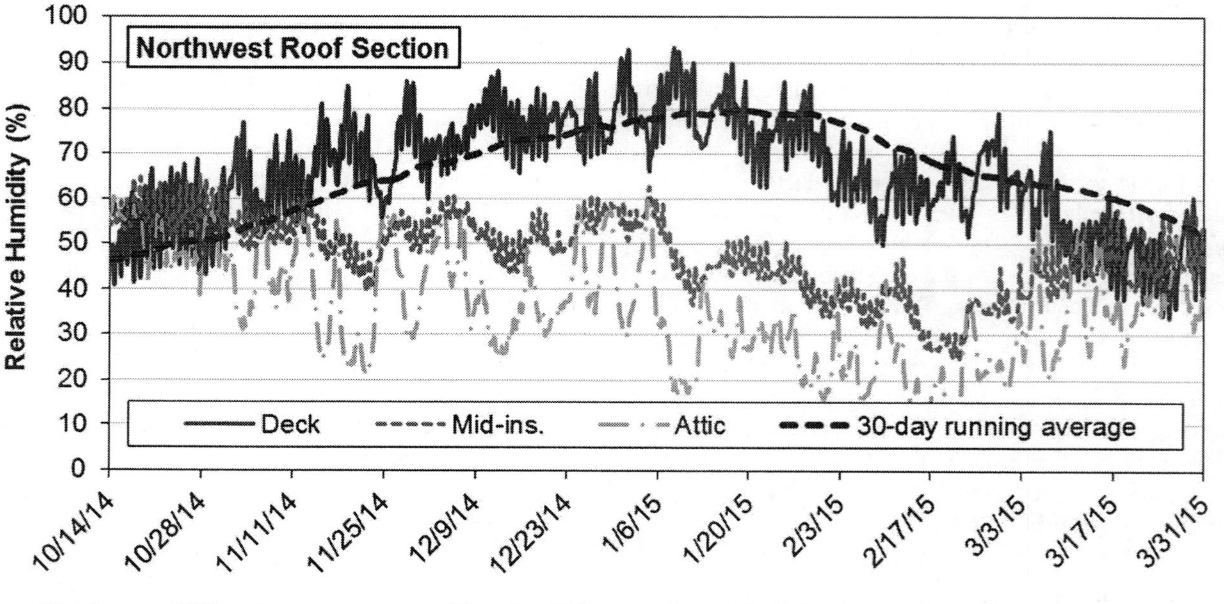

Figure 12 *Measured RH at the worst-case roof location following phase 2 (cellulose) retrofit.*

tion batt and blown cellulose insulation, yielding 5.06 $m^2 \cdot K/W$ of thermal resistance, under the roof deck and the gable walls. During phase 2 retrofit, bio-based PCM mats were added to the interior (attic-facing) surfaces of the roof and gable insulation. The building and attic are instrumented with a weather station and temperature, humidity, and heat flux sensors. The data collection periods extend over 9–12 months or more for the pre-retrofit and the two post-retrofit phases.

The temperature and heat flow data were analyzed to determine the thermal performance of the roof and gable insulation and PCM mats. The summertime peak heat fluxes through the roof were reduced by 75% to 80% by adding the insulation and PCM, compared to the pre-retrofit condition. Most of the reductions were caused by the insulation; the benefits of the PCM layer with this configuration and with reduced amounts of insulation are topics for ongoing and future analyses. In addition to reduction in the heat fluxes, bringing the mechanical system (supply and return ducts) inside the conditioned building envelope reduces the adverse energy impact of duct leakage. Thus, insulating the roofs and gables and sealing the attic resulted in substantially reduced HVAC runtimes. The data from the humidity sensors were

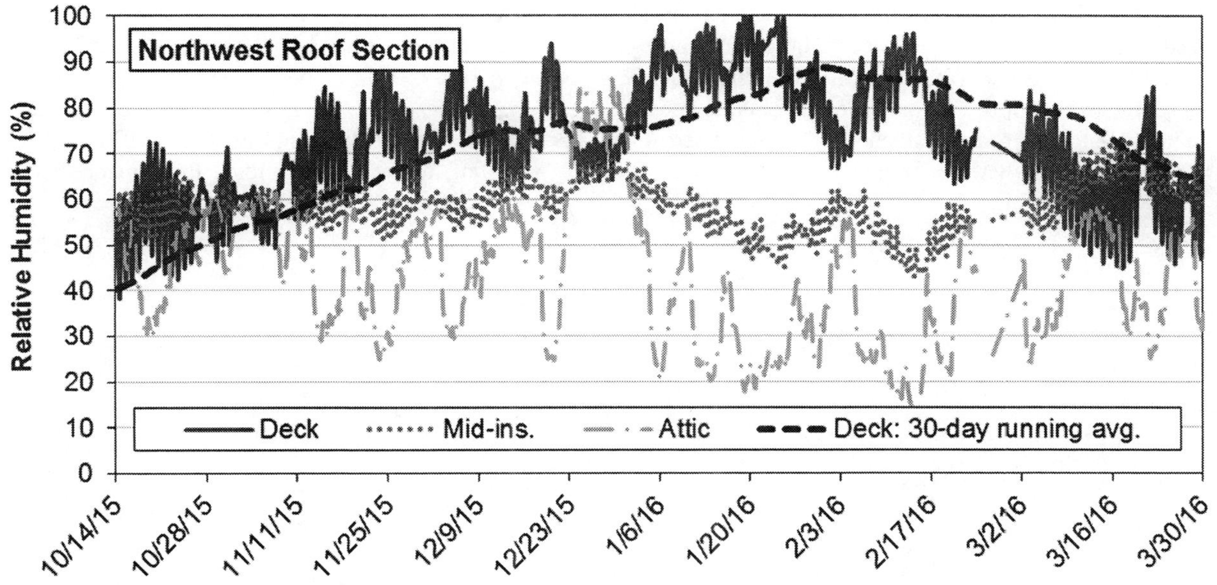

Figure 13 *Measured RH at the worst-case roof location following phase 3 (cellulose + PCM) retrofit.*

analyzed, which indicated potential for some moisture-related impacts on the roof decks due addition of insulation and the PCM mats, and mitigating action will be considered.

In the future, building power consumption data from the MDMS will be analyzed and reported. Further, whole-building EnergyPlus (DOE 2015) simulations are planned. EnergyPlus model of the subject building will be created and validated against measured sensor data, and then utilized to estimate the energy savings attributable to the retrofit measures.

ACKNOWLEDGMENTS

The funding for this project has been provided by the Department of Defense's Environmental Security Technology Certification Program (ESTCP). Oak Ridge National Laboratory (ORNL) is supported by the Building Technologies Office of the U. S. Department of Energy under contract No. DE-AC05-00OR22725 with UT-Battelle, LLC. The contributions of Phillip Childs and Jerald Atchley of ORNL in instrumenting the test building, data gathering and troubleshooting are gratefully acknowledged. The pre-retrofit building and duct leakage tests were performed by Jonathan Coulter and Christine Maurer of Advanced Energy (Raleigh, NC). The building retrofits were performed by Henry Venema and his crew from Nu-Wool Company, Inc.

REFERENCES

ASHRAE. 2009. ANSI/ASHRAE *Standard 160—Criteria for moisture-control design analysis in buildings*. Addendum A. Atlanta: ASHRAE.

ASTM. 2015. ASTM C518-15. *Standard test method for steady-state thermal transmission properties by means of the heat flow meter apparatus*. Philadelphia: ASTM International.

Biswas, K., W. Miller, P. Childs, J. Kosny, and S. Kriner. 2011. *Performance evaluation of a sustainable and energy efficient re-roofing technology using field-test data*. Proceedings of the 2011 NRCA International Roofing Symposium, National Roofing Contractors Association, Washington D.C.

DOE. 2015. EnergyPlus. http://apps1.eere.energy.gov/buildings/energyplus/. Washington, D.C.: U.S. Department of Energy.

EISA. 2007. *Energy independence and security act*. Washington, D.C.: U.S. Government.

Gmelin, E., and S.M. Sarge. 2000. Temperature, heat and heat flow rate calibration of differential scanning calorimeters. *Thermochimica Acta* 347:9–13.

Morshead J. 2007. Insulating unvented attics with spray foam: Closed-cell polyurethane foam provides the insulation, air barrier, and vapor retarder necessary for an unvented attic assembly. *The Journal of Light Construction*. March. www.jlconline.com/how-to/insulation/insulating-unvented-attics-with-spray-foam_o.

Kosny, J., D. Yarbrough, P. Childs, S. Shrestha, W. Miller, J. Atchley, M. Bianchi, J. Smith, T. Fellinger, E. Kossecka and E. Lee. 2010. *Theoretical and experimental thermal performance analysis of building shell components containing blown fiberglass insulation enhanced with phase-change material (PCM)*. Proceedings of the Thermal Performance of the Exterior Envelopes of Whole Buildings XI, Clearwater Beach, FL.

Kosny, J., K. Biswas., W. Miller, and S. Kriner. 2012. Field thermal performance of naturally ventilated solar roof with PCM heat sink. *Solar Energy* 86:2504–14.

Rudd, A., J. Lstiburek, K. Ueno. 1999. Unvented-cathedralized attics: Where we've been and where we're going. Research Report 990. Westford, MA: Building Science Corporation. http://buildingscience.com/documents/reports/rr-9904-unvented-cathedralized-attics-where-we-ve-been-and-where-we-re-going/view.

Shrestha, S., Miller, W., Stovall, T., Desjarlais, A., Childs, K., Porter, W., Bhandari, M., and S. Coley. 2011. *Modeling PCM-enhanced insulation system and benchmarking EnergyPlus against controlled field data*. Proceedings of Building Simulation 2011: 12th Conference of International Building Performance Simulation Association, p 800–7, November 2011.

Evaluating Summertime Overheating in Multi-Unit Residential Buildings Using Surveys and In-Suite Monitoring

Marianne F. Touchie, PhD
Associate Member ASHRAE

Ekaterina S. Tzekova, PhD
Associate Member ASHRAE

Jeffrey A. Siegel, PhD
Member ASHRAE

Bryan Purcell

Jonathan Morier

ABSTRACT

Multi-unit residential buildings (MURBs) are an important source of housing in many urban regions including Toronto, Canada, where MURBs provide over half of the city's housing. Many of these MURBs were constructed following the second world war (Touchie et al. 2013) and are in need of renewal to address energy performance and indoor environmental quality (IEQ) concerns. The Toronto Atmospheric Fund (TAF), a nonprofit corporation whose mandate is to reduce greenhouse gas emissions in the City of Toronto, and a local social housing provider have partnered to improve the performance of seven post-war MURBs through retrofits. To help prioritize retrofit options and assess IEQ, a monitoring program including pre-retrofit resident surveys and in-suite monitoring was implemented. The resident surveys included questions about basic demographics, resident behavior, and perceptions about IEQ. To supplement the surveys, temperature, relative humidity, and mean radiant temperature were monitored for one year in 65 suites. This paper examines the findings from both the in-suite monitoring and the resident surveys to determine the IEQ issues that should be addressed during the retrofit process. The analysis presented focuses on the summertime condition, which is of particular concern given the vulnerable populations housed in these buildings and the lack of central cooling facilities. The findings suggest that an air temperature threshold may be insufficient to ensure thermal comfort given the range of survey responses associated with the observed suite temperatures. Based on the data collected, it was not possible to explain the differences between the survey responses related to thermal comfort and the monitored conditions. Factors such as air speed, exposure to solar gains, and resident perceptions may contribute to these differences. The paper concludes with a discussion of some insights from the comparison between the survey and monitoring data and how they best can be used to inform the design of future retrofits and monitoring programs.

BACKGROUND

Southern Ontario contains almost 2000 post-war multi-unit residential buildings (MURBs) built between 1945 and 1984 that house nearly one million people (ERA Architects et al. 2010). This region contains more post-war MURBs than high-rise buildings of any type (12 stories or greater) in any other city in North America except New York City (ERA Architects 2010). Therefore, the environmental impact of these buildings is significant: in the City of Toronto alone, the entire MURB sector, including buildings constructed since 1984, is responsible for over 2.5 million tonnes eCO_2 of emissions every year (Touchie et al. 2013). These MURBs exhibit relatively high average energy use intensities between 290–330 ekWh/m^2, while some of the worst performers consume over 400 ekWh/m^2 (ERA Architects 2010; Touchie et al. 2013). As many of these towers were constructed in the post-war era, they are in need of renewal across a variety of building components including energy-related systems, creating an opportunity for major performance improvements.

In addition to having considerable energy and carbon emission implications, many of these high-rise apartment towers experience thermal comfort (IEQ) issues. In particular, comfort issues include chronic overheating, lack of temperature control at the suite level, and localized areas of high

Marianne F. Touchie is an assistant professor in the Department of Civil Engineering and Department of Mechanical Engineering, University of Toronto, Toronto, ON, Canada. **Ekaterina S. Tzekova** is a building research manager, **Bryan Purcell** is director of policy and programs, and **Jonathan Morier** is energy and environmental quality coordinator at The Atmospheric Fund, Toronto, ON, Canada. **Jeffrey A Siegel** is a professor in the Department of Civil Engineering, University of Toronto, Toronto, ON, Canada.

humidity in the kitchens and bathrooms. Although there has been extensive research on thermal comfort in commercial buildings and other workplace environments, less is known about the residential indoor environment, particularly in the multi-unit residential sector. This work aims to address this gap.

As North Americans spend more than 65% of their time in residential buildings (Klepeis et al. 2001; Leech et al. 2002), this building type can have a significant impact on occupants. A strong correlation has been found between areas of high social need in Southern Ontario and multi-residential tower neighborhoods (ERA Architects 2010). Many social housing MURB residents experience mobility and health issues and are particularly vulnerable to the quality of the indoor environment, including thermal comfort. Through energy retrofits there is an opportunity to enhance occupant comfort and health, in addition to reducing operating costs and energy consumption. However, prior to designing and implementing retrofits, it is important to understand the existing thermal comfort issues and the associated causes so they can be directly addressed by the retrofit. To gather this information, resident surveys and in-suite monitoring were used.

Occupant surveys have been used extensively to assess employee satisfaction with workplace indoor environmental quality (CBE 2004); however, use of surveys in the residential market has been less common. One survey of residential buildings in Hong Kong found that thermal comfort was considered the most important IEQ attribute by the vast majority of respondents (Lai and Yik 2009).

Previous work has shown that residents value and exercise control over their interior environment. A Toronto-based study using surveys and submetered energy data found that the majority of the variability of energy use for heating (57%) and cooling (84%) in buildings with in-suite controls was linked to occupant behavior and demographics, not the physical characteristics of the building (Brown et al. 2015). A Danish study, with a survey of over 2400 households in a variety of residential building types, found that residents highly valued the ability to open windows, even if they had mechanical ventilation (Frontczak 2012). And an Israeli study that collected short-term measurements of air and globe temperatures, relative humidity, and air speed while conducting resident surveys found that survey responses indicated that the conditions associated with satisfaction with the thermal environment varied depending on the residents' ability to control those conditions (Becker and Paciuk 2009). For example, the air temperatures considered comfortable in suites without active cooling were higher than those with cooling. At present, the residents in the seven study buildings do not have control over their heating system and lack access to central cooling, although about half of the monitored suites contain A/C units.

There have been a number of studies monitoring the indoor environment of schools and office buildings. However, this work has largely focused on evaluating ventilation levels and a variety of air quality indicators such as volatile organic compounds and particulate matter, to name a few, rather than thermal comfort and extreme temperatures. A school in Munich that was monitored to evaluate the thermal comfort found that night ventilation was able to reduce maximum daytime temperatures in the classrooms and improve thermal comfort (Wang et al. 2015).

Other studies have combined occupant survey data with monitoring data. A three-year study of 12 low-energy office buildings in Germany evaluated thermal comfort through occupant surveys and indoor temperature monitoring and found that 41% of occupants were dissatisfied with temperature in the summer compared to only 24% in the winter, even though indoor temperature and humidity levels were similar between the two seasons (Pfafferott et al. 2007). The authors concluded that lower occupant satisfaction cannot be explained by room temperature only. The study also found that extreme summer temperatures in 2003 made it difficult for buildings without active cooling controls to maintain a comfortable environment. A similar study in Germany found that occupant thermal comfort in the workplace correlates poorly with indoor temperature (Wagner and Gossauer 2007). One Californian study of 16 apartments in three buildings, which served a low-income population, monitored a number of IEQ parameters, including temperature and relative humidity. The monitoring was conducted over two one-week periods before and after energy retrofits that were also intended to improve IEQ (Noris et al. 2013). The researchers found improved post-retrofit thermal comfort based on a calculated comfort zone. They also conducted a survey but did not indicate how the thermal comfort responses compared to the measured data.

To the authors' knowledge there are no studies that combine the use of continuous long-term in-suite monitoring of thermal comfort parameters with resident comfort surveys in the MURB sector. This study addresses this gap to explore the correlation between actual comfort data and comfort perceived by the occupants, among other objectives.

OBJECTIVES

Data collected through surveys and in-suite monitoring were used to assess resident comfort, among other things, and inform the retrofit design process of seven multi-residential buildings. By using these data to determine the most significant IEQ issues, retrofit measures can be prioritized. Overall, these data provide new information about the quality of the indoor environment in post-war MURBs used for social housing and establish the need for IEQ improvements as part of building energy retrofits.

A round table led by Toronto's Public Health department in March 2015 identified the health impacts of extreme heat in MURBs and the building operational challenges that influence the residents' ability to stay cool during summer months (Campbell 2015). In a recent report, Toronto Public Health indicated that by 2049, the city will experience an additional 46 days with exterior temperatures exceeding 30°C (Meaney and Jackson 2015). A similar trend is expected in urban regions across North America in the coming decades (Kahn 2015). With a projected increase in days where the exterior temperature will exceed 30°C, an assessment of current comfort conditions is necessary to develop strategies for dealing with extreme temperatures in the future. Of particular concern are populations most vulnerable to extreme heat, such as residents of social housing, many of whom live in post-war MURBs. As such, this paper presents an analysis of the survey and monitoring data to assess summertime thermal comfort in 65 suites across seven post-war MURBs. The following objectives, presented in order of importance, were used to guide the analysis:

1. Determine when overheating occurs and how widespread this issue is across the seven buildings and, if possible, what factors drive the prevalence of these conditions.
2. Determine if the resident survey responses correlate well to the monitored data and whether they can provide sufficient indication of the existence of summertime thermal comfort issues (i.e., can surveys be used in lieu of in-suite monitoring to reliably identify overheating?)
3. Use data to prioritize the thermal comfort issues that need to be addressed during the building retrofits.

METHODOLOGY

The study involved collection of data from resident surveys and in-suite monitoring prior to the retrofit of seven social housing MURBs owned by the local social housing provider. Table 1 summarizes the basic characteristics of the seven buildings in the study as well as the number of suites surveyed and monitored in each building.

All seven buildings are heated through perimeter hydronic baseboards without any in-suite thermostats or controls. There is no central cooling or ceiling-mounted fans in the suites, but 47% of the monitored suites have free-standing or window-mounted air conditioning (A/C) units. Some residents also have free-standing fans. Fresh air is provided through a pressurized corridor ventilation system. Fresh air supply rates measured at the rooftop intake were 45%–50% below the ASHRAE Standard 62.1 requirements of 14,000–27,000 cfm, depending on the building size (ASHRAE 2013b). Buildings A–E have central bathroom exhaust fans, while Buildings F and G exhaust bathroom air direct to the exterior through the wall of each suite. Only Building D has kitchen exhaust to the exterior. Units at the other buildings have either a recirculating range hood or no form of kitchen exhaust. Rooftop exhaust flow rates measured in Buildings A and B were 25% below the 9200 cfm code requirements. In-suite exhaust flow rates measured in 31 suites across Buildings A–E were 25%–65% below the 25–50 cfm code requirements (ASHRAE 2013b).

In Buildings A–B and F–G, each glazing unit consists of a fixed single-glazed unit over two layers of single-glazed double sliders, as shown in Figure 1. Buildings C–E have the same configuration with an additional set of single-glazed double sliders below, also shown in Figure 1. Apartments on the fourth floor at Buildings A–B have fixed double-glazed units over the sliders. The window-to-wall ratios are presented in Table 1 and representative photos of one building on each site are shown in Figure 2. Many of the suites had interior blinds or curtains, but the use of these shading devices was not monitored. The as-built drawings do not indicate a low-emissivity coating on the double-glazed units. This was verified through site observations.

There were two data collection methods used in this study: surveys of building residents and monitoring of IEQ parameters in suites using sensors. First, the survey design, suite selection, and administration of the surveys are described, followed by an outline of the monitoring program design, suite selection, and data collection.

Resident Surveys

Survey Design. The survey was designed to capture basic details about the residents, including number and age of suite occupants and behaviors such as smoking, cooking, bathing,

Table 1. Summary of Building Parameters

Bldg.	Const. Year	No. of Stories	Floor Area, ft^2	No. of Suites	No. of Suites Surveyed (% of Total)	No. of Suites Monitored (% of Total)	Type of Occupancy	Window-to-Wall Area Ratio
A	1972	4	113,580	213	30 (15%)	10 (5%)	Seniors	0.17
B	1972	4	97,920	184	25 (15%)	8 (4%)	Seniors	0.15
C	1965	7	126,540	217	33 (15%)	13 (6%)	Bachelor	0.26
D	1965	7	35,960	58	6 (10%)	3 (5%)	Seniors	0.39
E	1965	11	116,820	196	31 (16%)	9 (5%)	Bachelor	0.25
F	1974	19	148,485	165	25 (15%)	10 (7%)	Family	0.24
G	1974	18	206,928	204	30 (15%)	12 (6%)	Family	0.16

Figure 1 *Typical window configurations at (a) Buildings A–B, (b) Buildings C–E, and (c) Buildings F–G.*

Figure 2 *Photos of one representative building from each site (a) Site 1: Buildings A–B, (b) Site 2: Buildings C–E, and (c) Site 3: Buildings F–G.*

and cleaning. Residents were asked to rate thermal comfort during summer and winter as well as indicate the existence of odors or drafts and excessive moisture or dryness in the air. They were also asked about any measures they had taken to alter their interior environment such as use of heaters, air conditioners, fans, odor control, humidifiers/dehumidifiers, operation of doors and windows, and use of weatherstripping. A number of previous studies were consulted in the development of the survey questions (Bohac and Hewett 2004; Engvall et al. 2000; Frontczak 2012; CCOHS 2014; Reijula 2004). This paper is based on the results of the pre-retrofit survey. An identical survey will be conducted following completion of the retrofits in each building (estimated winter 2017), and the results will be compared to the pre-retrofit survey.

Suite Selection. The resident surveys were completed prior to installation of the monitoring equipment, so a greater number of suites were surveyed than what was required for the monitoring program. The survey was conducted in 180 suites and monitoring was done in 65 of those suites. This larger sample of surveyed suites served two purposes: to allow an additional pool of surveyed suites to draw from in the event of attrition and to capture a larger sample size than what was possible with the monitored suites. Furthermore, participants were excluded from the monitoring phase of the study if they intended to move out of their suite in the next 12–18 months.

Excluding the top and bottom floors, the highest and lowest floors in each building were targeted first, moving closer to the neutral pressure plane as the surveying proceeded. The intent with this approach was to capture survey responses in the suites likely to experience the most extreme thermal comfort conditions in summer and winter. Suites with same layout and exterior exposure were also targeted, so that they could be directly compared. This excluded corner and end suites in addition to the top and bottom floors, as mentioned. Thus, random sampling was not employed in participant recruitment.

Administration of Surveys. Prior to the start of the project, an invitation to an information session was delivered in writing to each of the suites and fliers were posted throughout the building. One information session was held on each of the three sites (each site contained two or three buildings). The survey was administered in person by a third-party research and analytics firm in January and February 2015. Administrators went door-to-door and conducted the surveys electronically using tablets that required a response to each question before moving on to the next. A script was used to deliver the survey in the same way to each resident. Compensation for any inconvenience and time associated with study participation was provided in the form of gift cards from the grocery store located closest to each building. Participants were able to withdraw from the study at any time and could keep any gift cards received to date.

Monitoring

The in-suite monitoring program was implemented to evaluate resident comfort, provide context to the survey results, and help establish the pre-retrofit baseline indoor environment.

Program Design. Sensor packages were deployed in each suite to assess various IEQ parameters. The sensor package, shown in Figure 3, included a HOBO U12 data logger with integrated air temperature and relative humidity sensors. The mean radiant temperature (MRT) sensor consisted of an external temperature probe shielded from the room by a hollow, black plastic hemisphere and was connected to the data logger through an external channel. Table 2 includes details of the sensors used. Prior to installation, all sensors were collocated for a six-hour period and the resulting data were checked to ensure the difference between the recorded values and the mean and median of all recorded values were within the accuracy of each sensor.

Figure 4 shows the approximate location of the sensor packages at Buildings F and G. Each sensor package was installed in a similar location in each suite approximately 30 cm from the ceiling on an interior wall to ensure the sensor was out of direct solar radiation for most of the day. Since the MRT sensors were not positioned in the middle of the suites, according to ASHRAE Standard 55 (ASHRAE 2013a), they are only used to compare the relative differences in MRT at the sensor location between suites rather than provide an absolute value of the MRT.

Suite Selection and Data Collection. From the group of 180 survey participants, summertime data was collected from 65 suites that were instrumented with IEQ monitoring equipment. The suites chosen for monitoring were selected based on orientation and location in each building, not random sampling methods. Ideally, groups of suites horizontally or vertically adjacent to one another were chosen to facilitate isolation of certain explanatory variables such as exposure to solar radiation

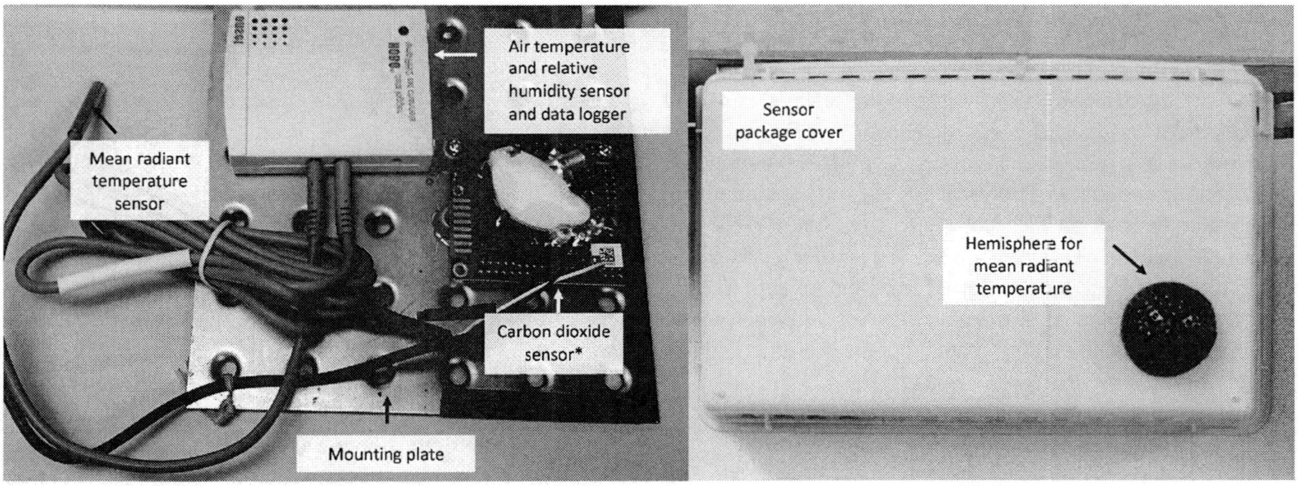

(a) (b)

Figure 3 *(a) Sensor package components and (b) sensor package cover.*
** Data from the carbon dioxide sensors is excluded from the analysis presented in this paper.*

Table 2. Summary of Sensor Specifications

Measurement	Sensor Type	Accuracy	Accuracy Threshold
Air temperature	Onset HOBO U12-13	±0.35°C	0°C–50°C
Mean radiant temperature	Onset TMCx-HD	±0.25°C	0°C–50°C
Relative humidity	Onset HOBO U12-13	±2.5% rh	10%–90%

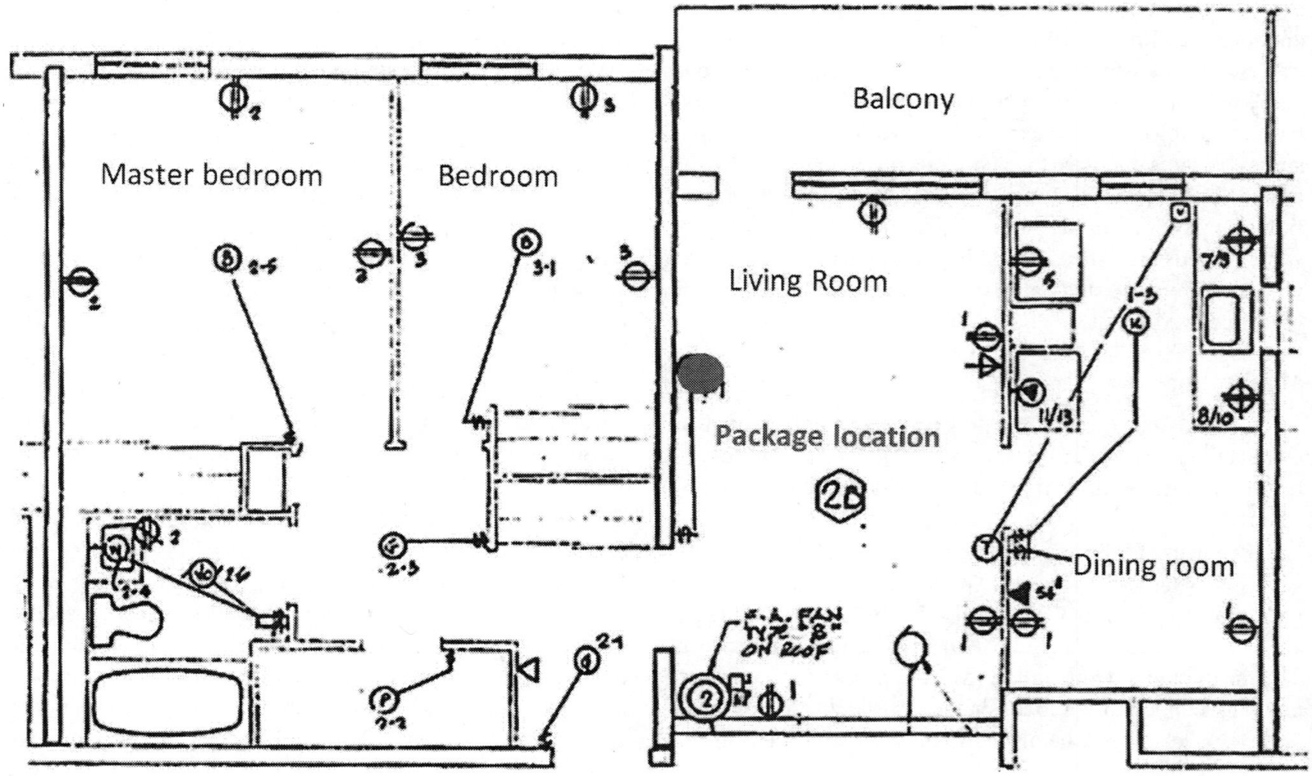

Figure 4 *Typical sensor package location—Buildings F and G.*

and the effects of stack action.The resulting selected units and a summary of window orientations are shown in Figures 5 and 6, respectively. The IEQ monitoring program is ongoing and will continue until one year after the retrofit work is completed. This paper focuses on data collected from June 1, 2015 to August 31, 2015. Measurements were recorded once every 15 minutes and data were collected from the loggers in each suite every three months. During the site visits, the presence of air conditioners and fans were noted, but the presence and use of interior blinds and curtains was not.

RESULTS AND DISCUSSION

This section begins with an overview of the survey and monitoring data separately. Then the relationship between the two data sources is investigated and the reasons for different comfort conditions are explored.

Resident Surveys

The surveys were conducted over a period of ten days between January and February 2015. Respondents were asked how comfortable they were in their apartment during the summer months and how often they used of air conditioners and fans. In addition to the surveys, the presence of air conditioners was visually verified during the monitoring period. Table 3 shows the percentage of respondents in each building who reported feeling "Too cold," "Just right," "Too warm," or "Don't know."

Just over half of the residents across all seven buildings reported feeling too warm during the summer months. However, there were large differences between the buildings. Only 32% of the respondents in Building B reported feeling "Too warm," while 77% of respondents reported feeling "Too warm" in Building G. To explore the physical causes of this difference, the percentage of residents reporting "Too warm" conditions were plotted against building height in Figure 7.

The strong correlation here suggests a relationship between building height and summertime discomfort. However, when the responses were distributed along the height of the building, there was no clear trend of greater thermal discomfort near the top of the building, as perhaps expected due to stack effect. Instead, the "Too warm" responses are distributed throughout the building. This suggests that other factors are impacting perceived thermal comfort such as solar radiation, air movement, and/or resident behaviors and preferences.

A detailed shadow study is beyond the scope of this paper, but examination of the building orientations, shown in Figure 5, and site observations during the summer were used to explain some of the differences between the buildings. In Figure 5, the buildings sizes and the separation between buildings on Sites 1 and 2 are to scale relative to one another, while the buildings on Site 3 are further apart than they appear. The number of suites monitored and their approximate location in each building are shown on the floor plans. Where the

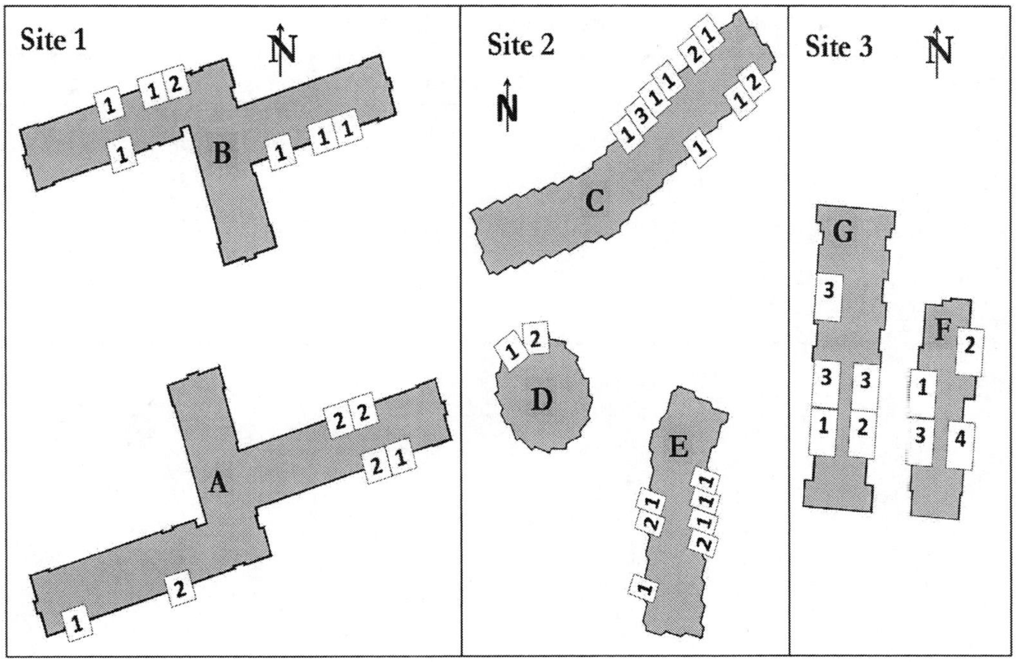

Figure 5 *Location and number of suites monitored at each site.*

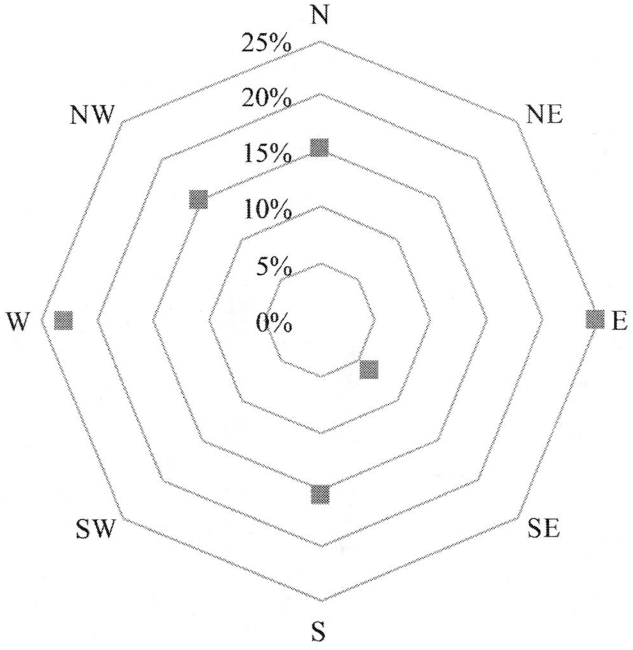

Figure 6 *Window orientations by percentage of total number of monitored suites.*

numbers are greater than 1, the monitored suites are located directly above and below one another. Figure 6 shows the glazing orientation, expressed as a percentage of the total number of monitored units.

When comparing buildings on the same site, the differences in satisfaction with thermal comfort may relate to the impact of shading from nearby trees and buildings. For example, greater levels of perceived comfort in Buildings A and B could be a result of trees shading the South side of each building and the fact that no suites facing East or West were monitored. Similarly, Building G shades Building F in the late afternoon and early evening, which might contribute to the slightly higher comfort reported by the residents in Building F. Buildings C and E are interesting in that Building C has North and South exposure while Building E has East and West exposure, which means different parts of each building will see solar radiation at different times of the day. This will be explored further using the monitored temperature data below. Another explanation for varied levels of thermal comfort could be the use of interior blinds and curtains. However, resident use of interior shading devices was not monitored or assessed through the surveys, so it is not possible to test this theory.

The survey data also showed how often residents used a fan and/or air conditioner in their suites. Figure 8 shows how this equipment use varies with the perceived summertime comfort. Residents who said they were "Too warm" during the summer months and did not have access to a fan and/or A/C represented 13% and 14% of total respondents, respectively. Fans and/or A/C units were used daily by 18% (fans) and 19% (A/C) of respondents who considered conditions "Too warm". Only 5% (fans) and 7% (A/C) of respondents who considered conditions "Just right" used this equipment daily.

Table 3. Survey Responses Related to Thermal Comfort

Building	A	B	C	D	E	F	G	Total
Too cold	0%	0%	3%	0%	3%	4%	0%	2%
Just right	53%	68%	45%	67%	39%	20%	10%	40%
Too warm	47%	32%	48%	33%	52%	72%	77%	54%
Don't know	0%	0%	3%	0%	6%	4%	13%	4%
Number of respondents	30	25	33	6	31	25	30	180

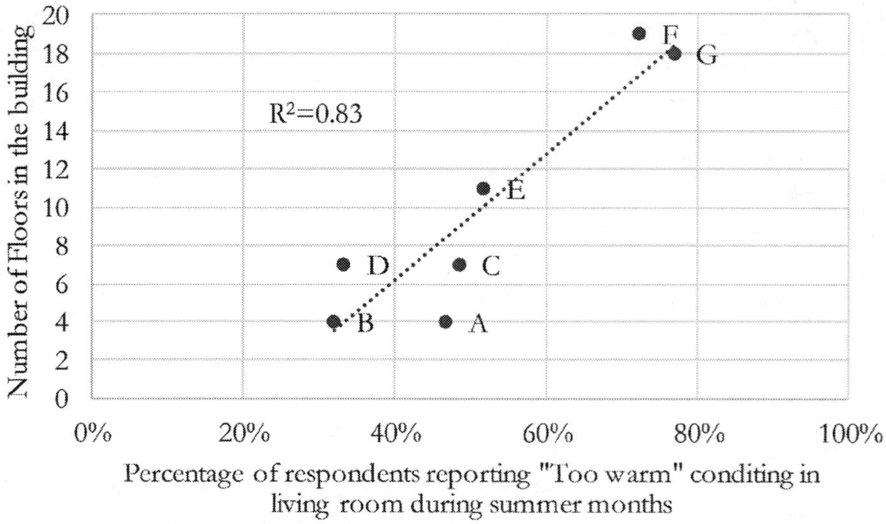

***Figure 7** Influence of building height on percentage of respondents reporting "Too warm" conditions.*

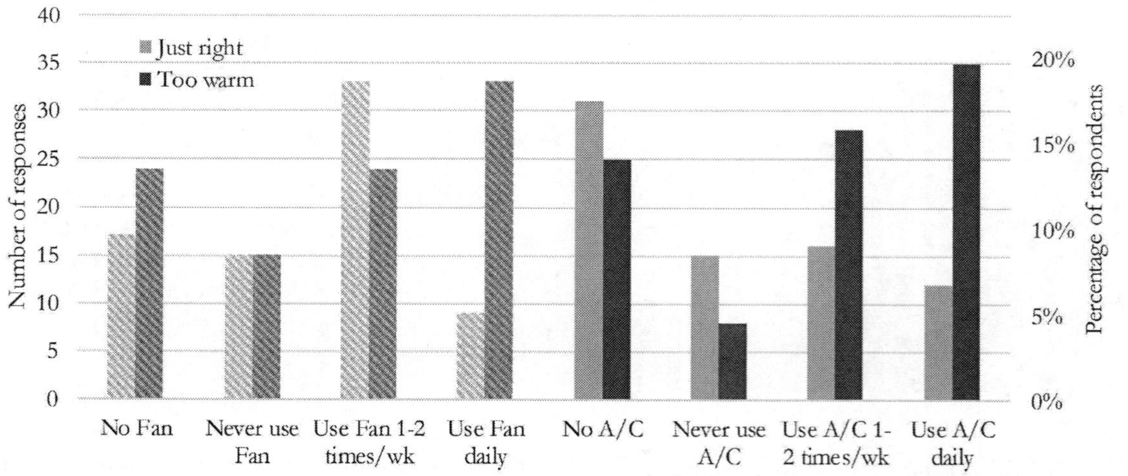

Note: 170 responses appear here in both the fan and A/C responses. "Don't know" (n=8) and "Too cool" (n=2) responses were omitted from the bar chart but "Percentage of respondents" is based on the total sample size of 180.

***Figure 8** Thermal comfort survey response and reported use of cooling controls.*

The survey did not include questions about whether residents chose to use fans and/or A/Cs to maintain "Just right" conditions or whether they did not need to use this equipment because the conditions were "Just right". However, the greater daily use of both fans and A/Cs by respondents who were "Too warm" suggests that, despite these temperature control measures, they were unable to achieve comfortable conditions in their suites. Fan and A/C use was also examined by building in an attempt to explain the differences in perceived comfort between the buildings, as shown in Figure 9.

Figure 9 shows the greater prevalence of fan use in Buildings A and B. This would contribute to improved thermal comfort due to greater air movement in the suite and may be one of the reasons for fewer complaints of "Too warm" conditions. A/C use was most prevalent in Buildings F and G, which also had the greatest numbers of residents reporting conditions were "Too warm." This suggests that A/C as a temperature control mechanism was relatively ineffective in these buildings despite reported widespread use. Air leakage paths around the air-conditioner units driven by pressure differentials can potentially allow the cool, conditioned air to escape towards the exterior. Site observations have confirmed that many A/C units are not adequately installed, which could result in reducing the A/C effectiveness in cooling the units.

Monitoring

The monitoring included collection of air temperature, approximate mean radiant temperature, and relative humidity data in 15-minute intervals. The air temperature data is presented in this section and the MRT and relative humidity data are incorporated into the development of comfort zones when the monitoring and survey data are compared.

Analysis of the indoor air temperatures in the seven buildings revealed chronic overheating throughout the summer. Based on all of the monitored suites, the average air temperature was 27.7°C, while maximum recorded temperatures exceeded 34°C. During heat alerts issued by the City, approximately 80% of the suites surpassed 30°C during the day. Figures 10 through 12 show the percent of time suites exceeded 26°C or 28°C during the summer, by building. Suites with air conditioners are excluded from these graphs. Building D was not plotted because only three suites were monitored. A 26°C threshold was examined initially because this temperature is being explored as a maximum upper limit for rental MURBs. However, because the suites on Site 1 and Site 3 spent so much time above this 26°C threshold, the percentage of time above 28°C was used for a more meaningful comparison.

A sensitivity analysis revealed that the all suites exceeded 24°C for 100% of the time during the summer. Buildings A and B exceeded a 30°C threshold for approximately 20% of the time. In comparison, the rest of the buildings in the study rarely reached 30°C.

With the exception of a few outliers, nearly 100% of the suites in Buildings A and B were above 26°C the entire summer. Figure 10 shows that suites were over 28°C between 70% –90% of the time in June and July. The greater variation in Building A may be a result of the localized impact of shading from the surrounding trees. Despite these high air temperatures, the residents in Buildings A and B were also the most satisfied with 53% and 68%, respectively, reporting conditions were "Just right" during summer months, as shown in Table 3.

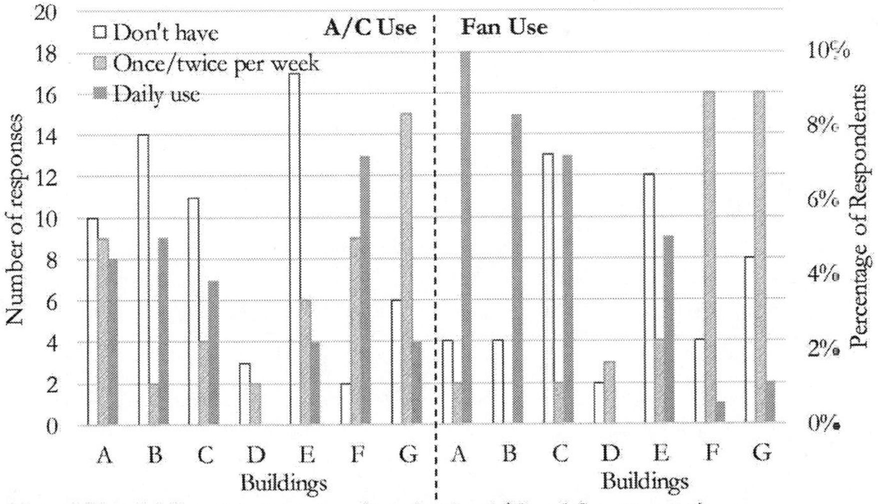

Note: 155 and 148 responses appear here for the A/C and fan use questions, respectively. "Never use" (n=25 and 31, respectively) and "Don't know" (n=0 for both) responses were omitted from the chart. "Percentage of respondents" is based on the total sample size of 180 respondents.

Figure 9 *Reported A/C and fan use.*

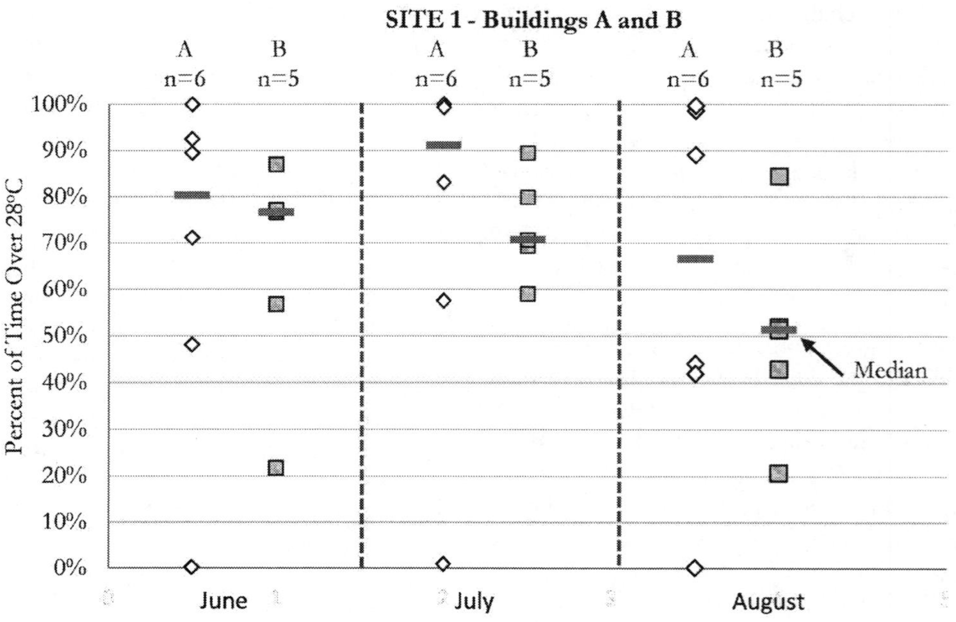

Figure 10 *Percentage of time suites at Site 1 are above 28°C.*

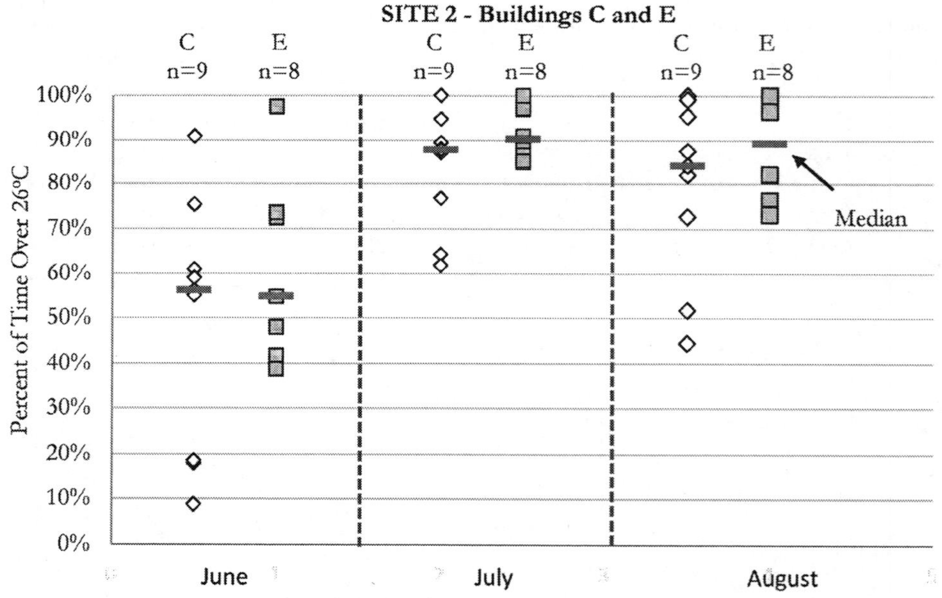

Figure 11 *Percentage of time suites at Site 2 are above 26°C.*

This may be due to the prevalence of daily fan use, shown in Figure 9, or other reasons that could not be verified with the data collected in this study, including less exposure to solar radiation or differences in resident preferences. It is interesting to note that Buildings A and B had the smallest operable glazing areas, compared to the rest of the buildings in the study. One review paper reported that seniors are less sensitive and take more time to respond to extreme heat exposure, despite being severely affected by it (Kenny et al. 2010). A telephone survey in New York City found that, while almost 80% of the seniors who participated had A/Cs, the majority did not use them for various reasons (Lane et al. 2014). Some also noted a preference for fans over A/C, did not feel the heat, or said they were used to the heat. Research has also indicated that seniors often underreport pain because they believe it is part of the normal aging process or in many cases deny it

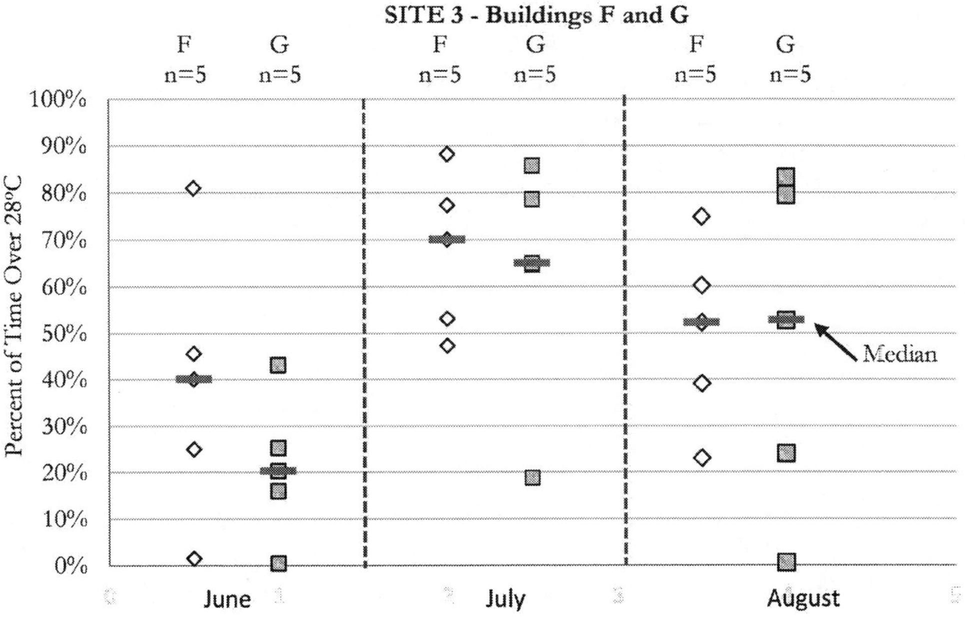

Figure 12 *Percentage of time suites at Site 3 are above 28°C.*

because they do not want to be a nuisance (Kaye et al. 2010; Ferrell et al. 2010; Herr and Mobily 1997). These studies indicate that discomfort due to overheating could similarly be underreported by senior residents.

Buildings C and E spent the least amount of time above 26°C. The temperature difference between these buildings, as shown in Figure 11, could be related to their orientation. Building E has suites facing East/West while Building C is oriented with suites facing North/South. However, it is unclear why these buildings are cooler than the other two sites, especially given the limited use of A/Cs as shown in Figure 9.

Larger temperature differences were observed between Buildings F and G, particularly in June when the suites in Building F were, on average, over 28°C for 40% of the time while Building G suites were over the same threshold for 20% of the time, as shown in Figure 12. It is interesting to note that despite the possible afternoon shading provided by Building G and greater daily A/C use, Building F was considerably warmer than the near-identical Building G at the start of the summer. Towards August, the temperature differences between these two buildings decreased. It is unclear what is driving this trend or the higher temperature differences at the start of the summer.

To explore the effects of orientation on indoor temperatures, North- and South-facing suites in Buildings A and B were compared to one another as shown in Figure 13. If more than one suite is available at a particular location, a suite that was representative of the average was chosen for the plot. Two days are shown: a typical summer day, when the average exterior temperature was 21°C, and a hot summer day when the average exterior temperature was 28°C. Suites with air conditioners are excluded from these comparisons. In addition, the monitored suites were all on the second floor (with the exception of Building B, South, which was on the third floor), minimizing any potential differences in temperature due to building height. Based on the t-test and f-tests, the mean and variance from each sample of data were found to be significantly different.

The hot day comparisons, shown in Figure 13, indicate that the South-facing suites in both buildings are considerably warmer than the North-facing suites. Temperatures on the South side also increase several degrees midmorning; these high temperatures are then maintained for the remainder of the day. It is also interesting to note that high indoor temperatures between 28°C and 31°C are maintained during nighttime at both elevations, while the exterior temperature drops to 22°C. Given that the previous day (July 28) had an average temperature of 28°C, these data indicate that suites do not experience a cooling benefit from the diurnal temperature swing.

The average summer day in June also exhibited a large difference between the indoor and outdoor temperatures. However, temperature differences between the orientations were a lot closer for both buildings. In fact, average conditions in Building B show a slightly higher temperature at the north than at the south. large deciduous trees are located along the South end of the site, which can help shade units located at that elevation. The data here could be influenced by differences relating to resident behavior in the individual units.

The effects of East and West orientation on temperature were also explored. Throughout the day, the temperature increases above 30°C were observed at 7 a.m. on the East side and 5 p.m. on the West side. However, the East-facing suite was consistently hotter than the West-facing suite, which could have been partially shaded by Building D, which is situated in close proximity.

Next, the influence of building height on indoor temperature was explored as shown in Figure 14 for a hot summer day. Building E was used instead of the taller Buildings F and G because a larger number of suites without air conditioning were available for the comparisons. The two units included in this analysis face East to eliminate the effects of orientation on temperature differences.

The typical summer day is not shown because the difference between suites on the third and ninth floors is, on average, within the sensor uncertainty. The t-test and f-tests for this comparison also indicated that the mean and variance from each sample of data were not significantly different because the suite temperatures were so similar. On the other hand, there was nearly a 2°C difference between the top and bottom floors

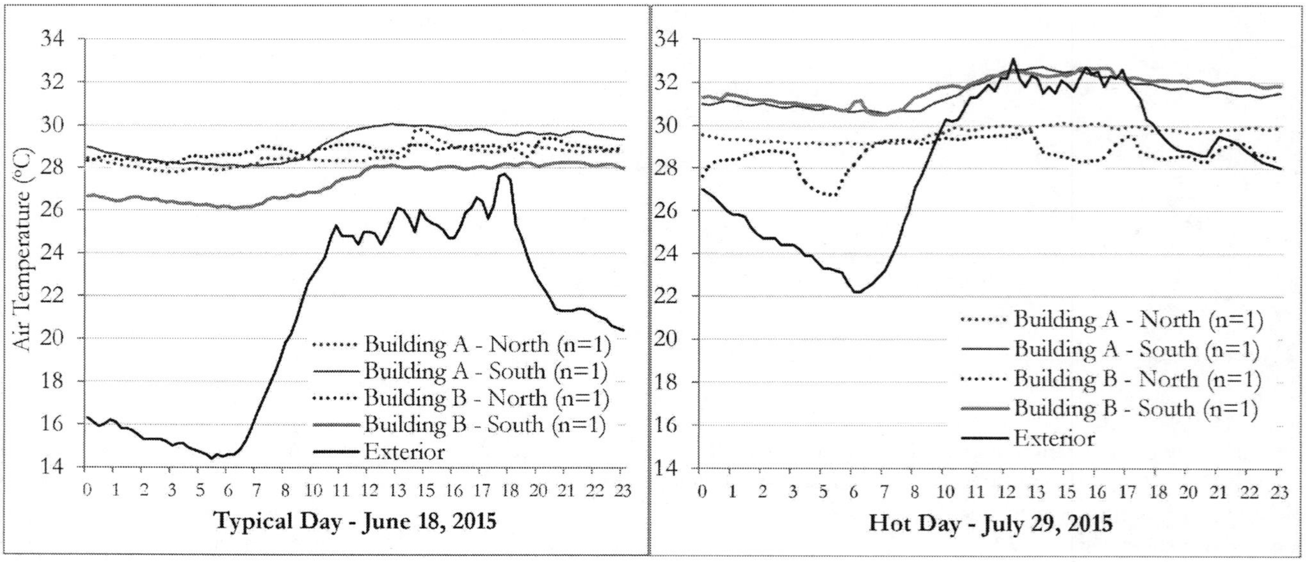

Figure 13 *Influence of suite orientation on air temperature in Buildings A and B.*

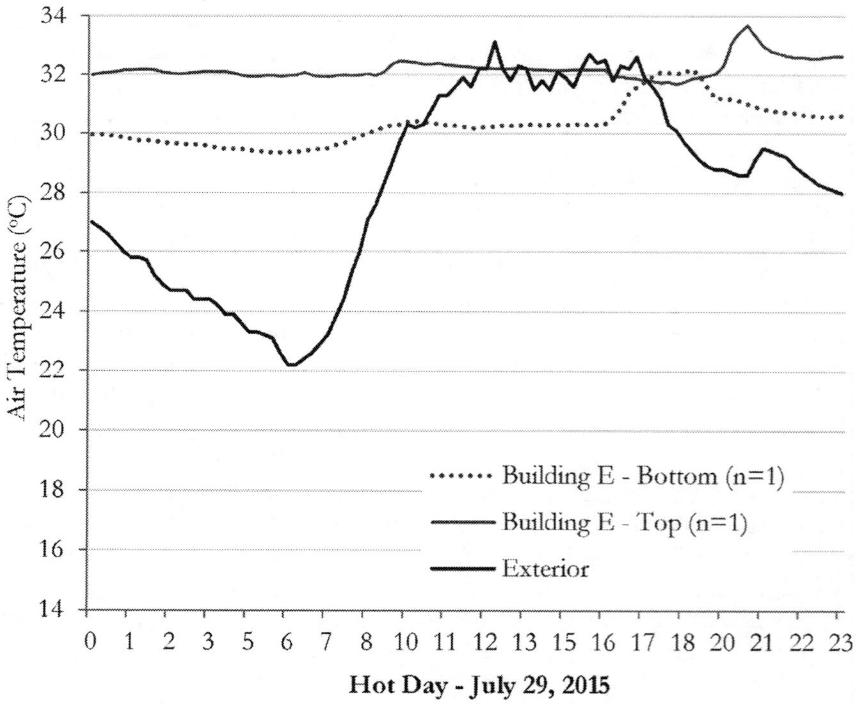

Figure 14 *Influence of suite height on air temperature in Building E.*

for the hot summer day. As shown in the other buildings, diurnal changes in the outdoor temperature do not translate to similar changes indoors.

COMPARISON AND AGREEMENT

The purpose of the comparison between the survey responses and monitoring data was to determine if the surveys present a consistent indication of thermal comfort issues during the summer months, thereby eliminating the need for costly long-term in-suite monitoring.

First, the survey responses related to thermal comfort satisfaction were compared with the suite air temperature on a site basis in Figure 15. Each data point used to construct the box plot is the average temperature in each suite over the entire summer period inclusive of June, July, and August. Only suites with at least 99% of the possible data collected this three month period were included here.

For Sites 1 and 3, the "Too warm" responses are associated with higher interior temperatures when compared to the "Just right" responses. However, the median temperatures associated with the "Too warm" responses are not consistent between sites. This is problematic if these data are used to identify a threshold at which most residents would be comfortable as there is no agreement between sites on which temperatures are "Just right" and which are "Too warm." Similarly, an Israeli study also found poor correlations between operative temperature and perceived thermal comfort (Becker and Paciuk 2009).

As air temperature alone does not yield a consistent correlation to survey responses about perceived thermal comfort, the air temperature, MRT, and relative humidity data were used to determine the percentage of time the suite was outside of the comfort zone. To determine this percentage, the conditions in the suite for each 15-minute increment were compared to a zone where an estimated 80% of the population is likely to be comfortable. This comfort zone was generated by the adaptive method in the Centre for the Built Environment's Thermal Comfort Tool based on ASHRAE Standard 55-2013 using the observed air temperature, MRT, and relative humidity data (Hoyt et al. 2013). It was assumed that resident's clothing level and metabolic rate were 0.5 clo and 1.0 met, respectively. These values are indicative of an average person in a seated and relaxed position. It was also assumed the air speed in the suite was 0.15 m/s. No data were collected on these three parameters but a sensitivity analysis was conducted on the air speed. Decreasing the airspeed to 0.1 m/s resulted in increasing the percent of time outside the comfort zone by an average of 2%, because reducing air circulation will have increased discomfort in units where overheating is an issue. Increasing the airspeed to 0.2 m/s resulted in decreasing the percent of time outside the comfort zone by an average of 6%.

A significant limitation of this approach is the inability to verify the assumptions made in order to generate the comfort zone. For example, by assuming a single air speed in all suites, the time spent outside the comfort zone could be overestimated if air speeds generated by a fan or breeze through an open window are greater than 0.15 m/s. Similarly, they could be underestimated in stagnant air conditions, as illustrated by the sensitivity analysis. Nevertheless, the

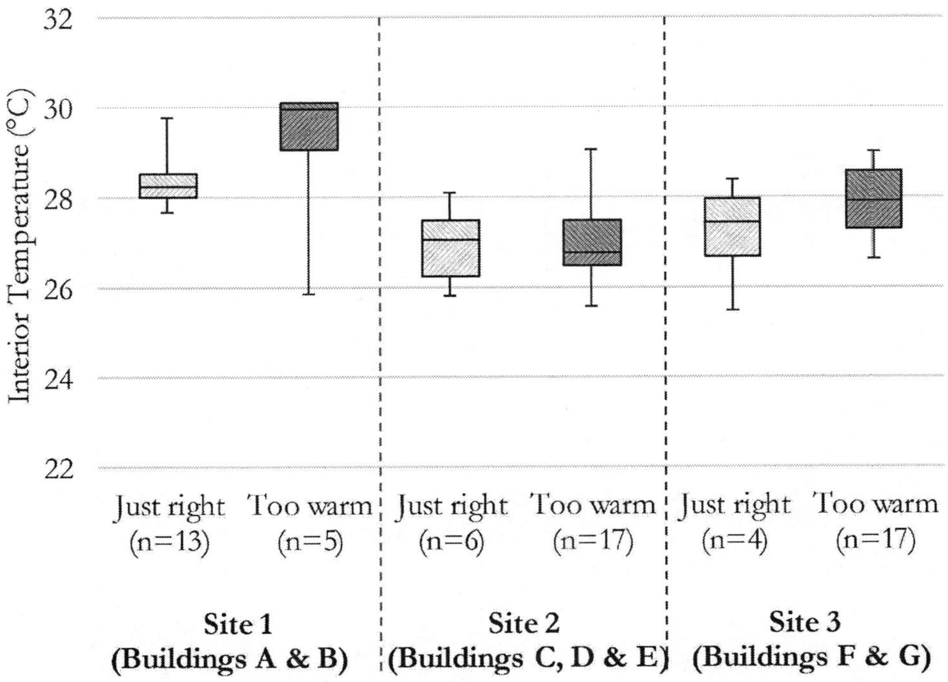

Figure 15 *Influence of interior air temperature on thermal comfort survey responses.*

comfort evaluations provide a way to evaluate the indoor conditions and allow for comparisons between the buildings based on a standard set of assumptions. Figure 16 shows how the percentage of time outside the assumed comfort zone relates to the respective survey responses from each suite. All suites were outside the comfort zone due to the high temperatures.

Similar to the air temperature analysis, there was no consistent percentage of time outside the comfort zone associated with the "Too warm" responses. Furthermore, there was a greater variation in the range of percentages that corresponded to a particular survey response when compared to the air temperature ranges in Figure 14. This increased variability was likely due to the uniform assumptions made about the comfort zone for all suites. In reality, there would be variations in air movement, clothing level, and metabolic rate impacting the actual thermal comfort perceptions of the residents.

Despite the lack of consistent conditions associated with each survey response, the differences between survey responses within each site and between the sites themselves were investigated further. A number of factors related to the occupants and the building could contribute to these differences. Although the conditions in Buildings A and B were the warmest and were outside the comfort zone for the greatest amount of time, residents were also the most satisfied with the thermal comfort in their suites. This paradoxical finding could also be because of incorrect assumptions in the comfort zone model or because the senior residents have fundamentally different behaviors or preferences compared to the residents in other buildings, as discussed earlier (e.g., Kenny et al. 2010). Given the prevalence of daily fan use identified in the survey responses for these two buildings, as shown in Figure 9, it is possible that the air speed in the comfort zone model is underestimated. A higher air speed can potentially reduce the time spent outside the comfort zone.

The influence of A/Cs was also investigated as shown in Figure 17. While A/Cs clearly reduce the median temperature in the suite compared to suites without A/C, the time these suites are above 28°C is still 65% and 55% for Buildings A and B, respectively. Therefore, either A/C units are inadequate for cooling these spaces or the residents are using them infrequently.

The different metabolic rates associated with age and activity level of residents may also be a contributing factor. Buildings A and B house seniors who may be less mobile than residents in the other buildings, contributing to a lower metabolic rate and perhaps a tolerance of higher temperatures despite a greater vulnerability. Through anecdotal evidence, some of the residents of Buildings A and B have difficulty opening and closing their windows which, if left closed, could contribute to higher interior temperatures. The New York City study found that some seniors thought it was incorrect to use fans with the window open and so kept windows closed during fan use (Lane et al. 2014). Another factor contributing to the comfort of the residents in Buildings A and B is the presence of a cooling center in the common area where many residents spend their time. Despite high in-suite temperatures, perhaps the residents perceive greater comfort in their buildings compared to the other sites because of access to a cooled space.

Sites 2 and 3 had greater numbers of residents reporting conditions were "Too warm." To explore the differences in conditions associated with the thermal comfort survey responses for the remainder of the sites, the responses were plotted by building as shown in Figure 18.

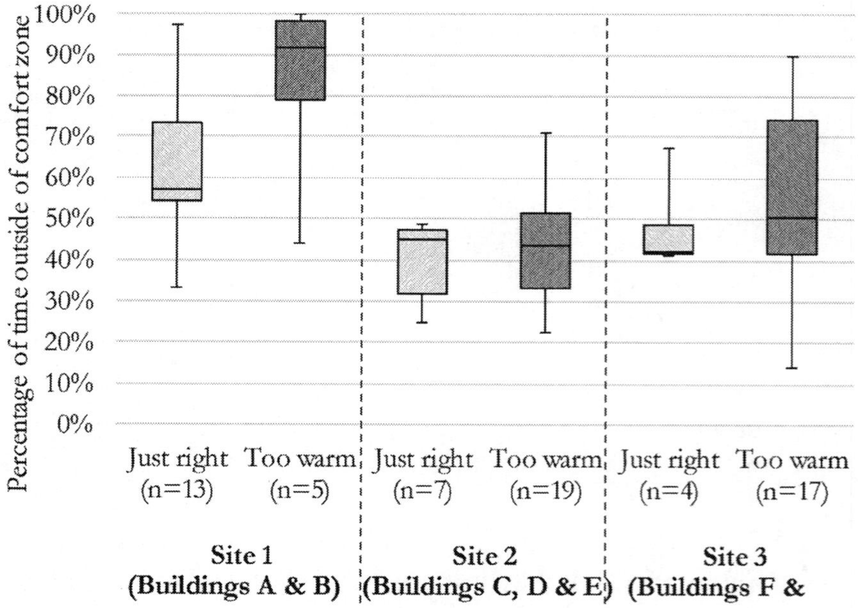

***Figure 16** Influence of time outside the comfort zone on thermal comfort survey responses, by site.*

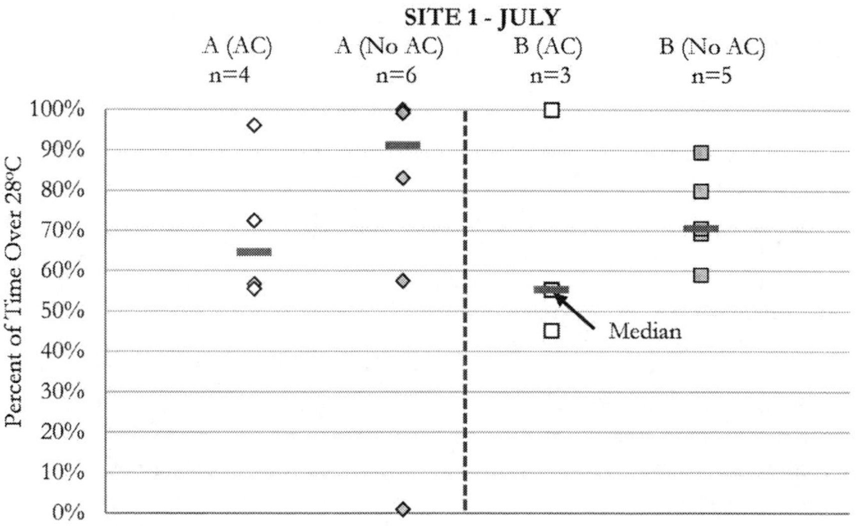

Figure 17 *Influence of interior air temperature on thermal comfort survey responses.*

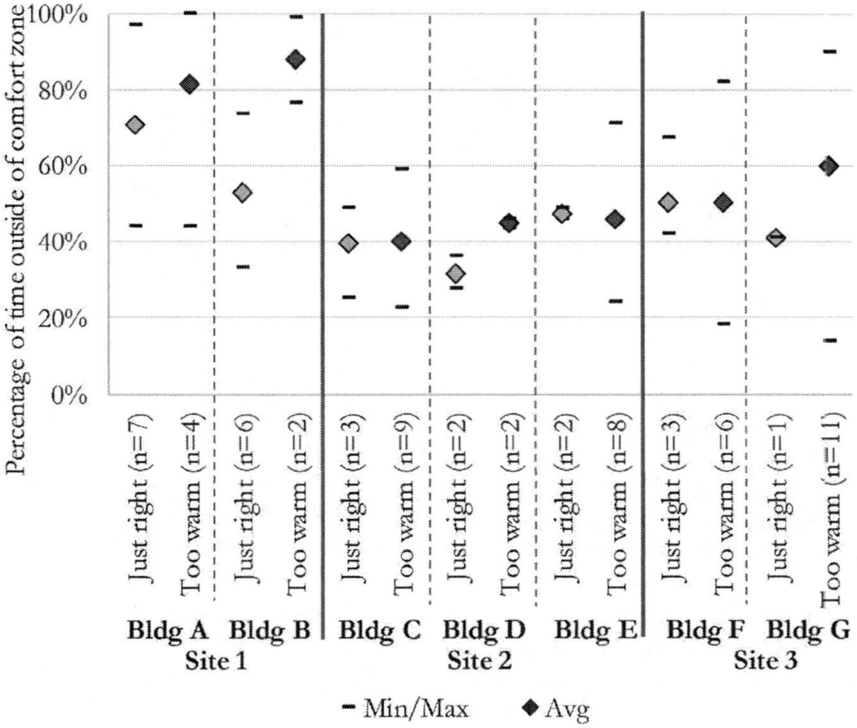

Figure 18 *Influence of time outside the comfort zone on thermal comfort survey responses, by site.*

As observed at the site-level analysis, there is little difference between the average time spent outside the thermal comfort zone and the resident thermal comfort survey response, though the variation of the "Too warm" conditions is generally greater than the "Just right" conditions on Sites 2 and 3. This could indicate that resident perceptions significantly affect what conditions are considered to be "Just right" or "Too warm," with some residents being more accepting of the "Too warm" conditions.

Buildings C–G are taller (7–19 stories) than Buildings A and B (4 stories each); thus, the majority of the monitored suites in Buildings C–G cannot benefit from shading from trees. However, given their height, they may benefit from more natural ventilation due to greater wind speeds at these heights, provided the windows and/or balcony doors are kept open. However, no data were gathered on air speed, so this hypothesis cannot be confirmed.

CONCLUSIONS AND IMPLICATIONS FOR RETROFIT PROCESS

The monitoring and survey data has shown that significant overheating occurs during the summer months across the seven subject buildings. More than half of survey respondents reported conditions were "Too warm" during summer months. Thermal dissatisfaction appears to increase with height of building; however, this trend is not related to stack effect. Instead, it is suspected that exposure to solar radiation as well as resident behaviors and preferences are influencing thermal dissatisfaction. The monitoring data collected showed extremely high interior air temperatures throughout the summer, where all buildings saw interior temperatures exceeding 26°C more than half the time. These high temperatures meant that suites fell outside the thermal comfort zone as evaluated using the ASHRAE Standard 55-2013 adaptive method with a set of uniform assumptions for clothing level, metabolic rate, and air speed.

Differences were observed between the resident survey responses and the monitoring data. For example, Buildings A and B had the highest recorded temperatures and the highest percentage of time spent outside the comfort zone. However, the residents in these buildings were also the most satisfied with the thermal comfort in their suites compared to Buildings C–G. Based on the data presented here, it is not appropriate to use surveys in lieu of in-suite monitoring to reliably identify overheating, as the temperatures and time spent outside of the comfort zone do not correspond to the same thermal comfort response across all buildings and sites. Therefore, it may not be appropriate to use an air temperature threshold to ensure comfort, unless the temperature is sufficiently low, which may be impractical to maintain and could result in conditions that are considered too cool for some residents.

Additional types of in-suite data are required to generate a more accurate thermal comfort zone. In particular, air speed measurements would be a valuable addition to this analysis. This work also suggests that solar exposure is a contributor to thermal discomfort. A more targeted study of how the use of blinds, curtains, or other solar shading can impact thermal comfort is also needed. Light meters located near windows and/or a more representative measure of MRT would provide an indication of exposure to solar gains and whether the residents used shading devices.

Overheating in suites can have several implications for the building retrofits. Survey respondents who considered conditions "Too warm" also reported frequent use of A/Cs and fans which, particularly in Buildings F and G, may not be the most effective way of combating overheating. With extreme temperatures persisting throughout the summer, this frequent use of A/Cs and fans can result in high electricity consumption across the seven buildings. To improve the thermal comfort, passive measures that reduce solar heat gain as well as cooling strategies are strongly recommended. The installation of exterior blinds and/or replacing windows with those that have a low solar heat gain coefficient can help reduce solar heat gain during the summer. These types of measures should be considered in particular for Buildings A–B, where the majority of the windows are not shaded by balcony overhangs. Resident education about the impact of blinds, shades, and curtains on thermal comfort is also essential for these passive measures to be effective. In addition, adding conditioned spaces on site or providing a dedicated outdoor air supply or cooling to each suite with air sealing measures can help to reduce extreme temperatures. However, these active measures should only be contemplated after the passive measures have been adopted.

ACKNOWLEDGMENTS

The authors would like to acknowledge Natural Resources Canada, the Federation of Canadian Municipalities, Enbridge Gas Distribution, and the Toronto Atmospheric Fund for financing of the survey data collection and monitoring program. The authors would also like to thank the local social housing provider, residents, and building managers for their cooperation in this study.

REFERENCES

ASHRAE. 2013a. ASHRAE Standard 55-2013, *Thermal environmental conditions for human occupancy*. https:// Atlanta: ASHRAE. www.ashrae.org/resources--public ations/bookstore/standard-55.

ASHRAE. 2013b. ASHRAE Standard 62.1, *Ventilation for acceptable indoor air quality*. Atlanta: ASHRAE. https:// www.ashrae.org/resources--publications/bookstore /standards-62-1--62-2.

Becker, R., and M. Paciuk. 2009 Thermal comfort in residential buildings—Failure to predict by standard model. *Building and Environment* 44:948–60.

Bohac, David L., and Martha J. Hewett. 2004. Reduction in environmental tobacco smoke transfer in Minnesota multifamily buildings using air sealing and ventilation treatments. *Center for Energy and Environment* 104.

Brown, C., M. Gorgolewski, and A. Goodwill. 2015. Using physical, behavioral, and demographic variables to explain suite-level energy use in multi residential buildings. *Building and Environment* 89:308–17.

Campbell, M. 2015. Addressing Extreme Heat and Health in Vulnerable Populations. Presentation.

CBE. 2004. Occupant Indoor Environmental Quality (IEQ) Survey™. Center for the Built Environment. http:// www.cbe.berkeley.edu/research/survey.htm.

CCOHS. 2014. Indoor Air Quality—General: OSH Answers. Government of Canada. Canadian Centre for Occupational Health and Safety. http://www.ccohs.ca /oshanswers/chemicals/iaq_intro.html.

Engvall, K., C. Norrby, J. Bandel, M. Hult, and D. Norback. 2000. Development of a multiple regression model to identify multi-family residential buildings with a high prevalence of sick building syndrome (SBS). *Indoor Air* 10:101–10.

ERA Architects. 2010. ERA Architects; Planning Alliance; Cities Center at the University of Toronto Tower Neighbourhood Renewal in the Greater Golden Horseshoe. Toronto: ERA Architects.

Ferrell, B.A., B.R. Ferrell, and D. Osterweil. 1990. Pain in the nursing home. *Journal of the American Geriatric Society* 38:409.

Frontczak, M. 2012. Questionnaire survey on factors influencing comfort with indoor environmental quality in Danish housing. *Building and Environment* 50:56–64.

Hoyt, T., S. Stefano, A. Piccioli, D. Moon, and K. Steinfeld. 2013. CBE Thermal Comfort Tool. Center for the Built Environment.Berkeley: University of California Berkeley. http://cbe.berkeley.edu/comforttool.

Herr, K.A., and P.R. Mobily. 1997. Chronic pain in the elderly. *Gerontological Nursing: Chronic Illness and the Older Adult*. New York, NY: Springer Publishing Co.

Kahn, B. 2015. A Surge in Heat Wave "Danger Days" is expected in coming decades. *Scientific American*.

Kaye, A.D. 2010. Pain management in the elderly population: A review. *The Ochsner Journal* 10(3):179–87.

Kenny, G.P., J. Yardley, C. Brown, R.J. Sigal, and O. Jay. 2010. Heat stress in older individuals and patients with common chronic diseases. *CMAJ* 182:1053–60.

Klepeis, N.E., W.C. Nelson, J.P. Robinson, A.M. Tsang, P. Switzerm, J.V. Behar, S.C. Hern, and W.H. Englemann, 2001. The National Human Activity Pattern Survey (NHAPS): A resource for assessing exposure to environmental pollutants. *Journal of Exposure Analysis and Environmental Epidemiology* 11:231–52.

Lai, J.H.K., and F.W.H. Yik. 2009. Perception of importance and performance of the indoor environmental quality of high-rise residential buildings. *Building and Environment* 44:352–60.

Lane, K., K. Wheeler, K. Charles-Guzman, M. Ahmed, M. Blum, K. Gregory, N. Graber, N. Clark, and T. Matte. 2014. Extreme heat awareness and protective behaviors in New York City. *Journal of Urban Health* 91:403–14.

Leech, J.A., W.C. Nelson, R.T. Burnett, S. Aaron, and M.E. Raizenne. 2002. It's about time: A comparison of Canadian and American time-activity patterns. *Journal of Exposure Analysis and Environmental Epidemiology* 12: 427–32.

Meaney, M., and E. Jackson. 2015. *Options for reducing risk from extreme heat to vulnerable populations living in multi-residential buildings*. Toronto: ICLEI Canada for Toronto Public Health.

Noris, F., G. Adamkiewicz, W.W. Delp, T. Hotchi, M. Russell, B.C. Singer, M. Spears, K. Vermeer, and W.J. Fisk. 2013. Indoor environmental quality benefits of apartment energy retrofits. *Building and Environment* 170–78.

Pfafferott, J., S. Herkel, D.E. Kalz, and A. Zueschner. 2007. Comparison of low-energy office buildings in summer using different thermal comfort criteria. *Energy and Building* 39:750–57.

Reijula, K. 2004. Assessment of indoor air problems at work with a questionnaire. *Occupational and Environmental Medicine* 61:33–38.

Touchie, M.F., C. Binkley, and K.D. Pressnail. 2013. Correlating energy consumption with multi-unit residential building characteristics in the city of Toronto. *Energy and Buildings* 66:648–56.

Wagner, A., and E. Gossauer. 2007. Thermal comfort and workplace occupant satisfaction—Results of field studies in German low-energy office buildings. *Energy and Buildings* 39(7):758–69.

Wang, Y., J. Kuckelkorn, F.U. Zhao, D. Lui, A. Kirschbaum, J.L. Zhang. 2015. Evaluation on classroom thermal comfort and energy performance of passive school building by optimizing HVAC control systems. *Building and Environment* 89:86–106.

Online Airtightness Savings Calculator for Commercial Buildings in the United States, Canada, and China

Som Shrestha, PhD, BEMP
Member ASHRAE

Diana Hun, PhD

Lisa Ng, PhD
Member ASHRAE

Andre Desjarlais

Steven Emmerich
Fellow ASHRAE

Laverne Dalgleish
Associate Member ASHRAE

ABSTRACT

The relative contribution of air leakage to heating and cooling loads has been increasing as the thermal resistance of commercial building envelopes continues to improve. Easy-to-access data are needed to convince building owners and contractors that enhancing the airtightness of new and existing buildings is the next logical step to achieve a high-performance building envelope. To this end, Oak Ridge National Laboratory, the National Institute of Standards and Technology, the Air Barrier Association of America, and the U.S.-China Clean Energy Research Center for Building Energy Efficiency partnered to develop an online calculator that estimates the potential energy and cost savings in major U.S., Canadian, and Chinese cities from improvements in airtightness. This tool will have a user-friendly graphical interface that accesses a database of CONTAM and EnergyPlus pre-run simulation results and will be available to the public at no cost. Baseline leakage rates are either user specified or selected by the user from a list of supplied leakage rates. Users will then enter the expected airtightness after the installation of an air barrier system. Energy costs are estimated based on the building location and other user inputs. This paper provides an overview of the methodology implemented in this calculator as well as example results. The deployment of this calculator could influence construction practices, contributing to significant reductions in energy use and greenhouse gas emissions from the United States, Canada, and China.

INTRODUCTION

The U.S. Department of Energy's (DOE) *Windows and Building Envelope Research and Development: Roadmap for Emerging Technologies* (DOE 2014) indicates that improving airtightness is among the most cost-effective strategies to decrease energy loads due to the building envelope. This conclusion is based on the fact that air leakage (i.e., infiltration and exfiltration) is responsible for about 6% of total energy used by commercial buildings in the United States, or about 15% of primary energy consumption in commercial buildings that is attributable to fenestration and building envelope components in 2010 was due to air leakage (DOE 2014). Nevertheless, improving airtightness is not always recognized by owners of commercial buildings, as they have been slow in acknowledging and diminishing the detrimental effects of air leakage on energy use and other aspects of building performance. The construction industry needs a credible, easy-to-use tool that estimates potential energy and financial savings in a standardized manner so designers and contractors can give building owners compelling reasons to invest in reducing air leakage.

Although air leakage has long been recognized as a key contributor to heating and cooling loads, methods that estimate its effects on energy consumption vary due to the complexity of this task (Crawley et al. 2008; Goel et al. 2014; Gowri et al. 2009; Ng et al. 2012). Comprehensive building design and energy simulations should take into account the fact that air leakage rates vary due to the operation of heating, ventilation, and air-conditioning (HVAC) systems, occupancy, and weather (i.e., indoor-to-outdoor temperature and wind). However, typical energy simulations tend to take shortcuts to expedite the analysis, such as assuming constant leak-

Som Shrestha and **Diana Hun** are research and development staff members and **Andre Desjarlais** is the program manager of the Building Envelope Systems Research Group at Oak Ridge National Laboratory, Oak Ridge, TN. **Lisa Ng** is a mechanical engineer and **Steven Emmerich** is the group leader of the Indoor Air Quality and Ventilation Group at the National Institute of Standards and Technology, Gaithersburg, MD. **Laverne Dalgleish** is the executive director of the Air Barrier Association of America, Walpole, MA.

age rates and/or using simplified algorithms, which can lead to under- or over-estimated energy usage. Also, use of macro climate data versus micro climate data may impact the pressure distribution on buildings, and hence the infiltration rate. However, quantifying the impact of using macro climate data is beyond the scope of this paper.

Oak Ridge National Laboratory (ORNL), the National Institute of Standards and Technology (NIST), the Air Barrier Association of America (ABAA), and the U.S.-China Clean Energy Research Center for Building Energy Efficiency (CERC BEE) are collaborating to develop an online calculator that will be free to the public, user friendly, and powerful enough to address the previously mentioned variables when estimating energy savings due to improvements in airtightness. Figure 1 describes the general steps to achieve this goal. The tool will use a database of EnergyPlus pre-run simulation results for DOE commercial prototype buildings. The main difference between the online calculator and the procedure followed in the DOE prototypes is that the calculator utilizes CONTAM-calculated air changes per hour (ACH) or air leakage rates as inputs, whereas the prototypes make simplified assumptions that are described in the following sections of this paper. CONTAM (Dols and Polidoro 2015) is a multizone airflow and contaminant transport analysis software developed at NIST and validated by multiple studies, such as Haghighat and Megri (1996), Chung (1996), Emmerich (2001), and Emmerich et al. (2004). This software takes into account multiple variables, such as weather conditions, envelope airtightness, and HVAC system operation, to calculate air leakage rates through the building enclosure. The CONTAM-calculated hourly air leakage rates are imported into DOE's whole-building energy simulation software EnergyPlus (DOE 2016a) with the CONTAM Results Export Tool (Polidoro et al. 2016). EnergyPlus is then used to calculate the effect of air leakage on energy consumption.

In addition to CONTAM, the Airflow Network module in EnergyPlus could have been used to calculate the air leakage rates through the building envelope. However, comparing results from CONTAM and the Airflow Network were beyond the scope of this project. Future efforts may cover this assessment.

The ultimate objective of the tool is for users to be able to estimate expected energy and financial savings for different airtightness levels in commercial buildings that are located in the United States, Canada, and China. This paper presents an overview of this calculator and results for a stand-alone retail building prototype in Chicago, Winnipeg, and Shanghai.

BUILDING MODELS

In order to cover a large percentage of the common building types in the United States, the calculator uses DOE

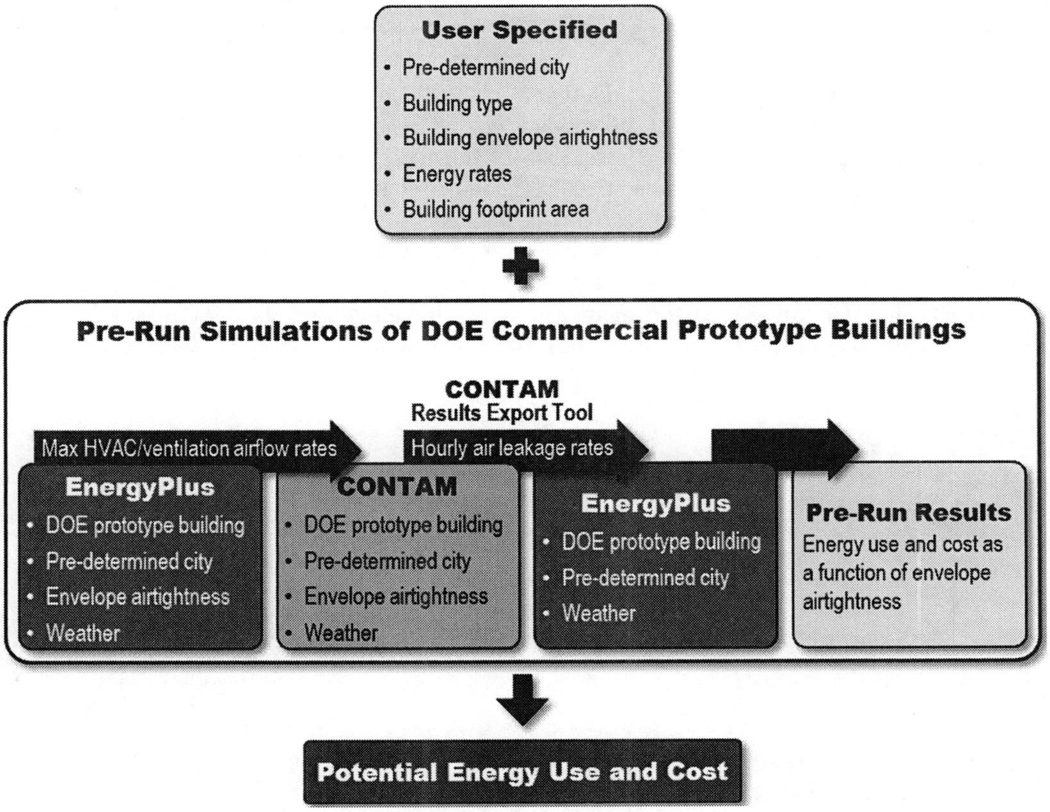

Figure 1 *General procedure to estimate potential energy costs for different levels of envelope airtightness in DOE commercial prototype buildings.*

commercial prototype building models (DOE 2016b). These prototypes were derived from DOE commercial reference building models (DOE 2016c) and represent about 80% of new construction. Moreover, these prototypes cover 16 commercial building types, including mid- to high-rise residential buildings in 17 climate locations defined in ASHRAE/IES Standard 90.1-2013 (ASHRAE 2013b). The variables that are prescribed in these models include building envelope components, HVAC equipment types and efficiency, and occupancy schedules. As Standard 90.1 evolves, Pacific Northwest National Laboratory (PNNL) modifies these models with input from ASHRAE Standing Standard Project Committee 90.1 members and building industry experts. Features of the building models and a detailed description of their development are provided by Goel et al. (2014) and the Building Energy Codes Program website (DOE 2016b).

The first phase in the development of the calculator will cover three prototype building models (stand-alone retail, medium office, and mid-rise apartment) in 45 cities in the United States, 5 cities in Canada, and 5 cities in China. Models that represent typical commercial buildings in Canada and China are not available in the public domain; therefore, the DOE prototypes will also be used in these two countries.

EXAMPLE CALCULATIONS

The example in this paper uses the DOE prototype building model for a stand-alone retail building (Figure 2). The main characteristics of this prototype are based on ASHRAE/IES Standard 90.1-2013 and listed in Table 1. Note that Table 1 describes the simplified method used with prototype buildings to take into account the effects of HVAC operation on air leakage rates. This method assumes that the air leakage rate is 1 L/s·m^2 at 75 Pa when the HVAC is off and that the leakage rate decreases by 75% when the HVAC is on (Gowri et al. 2009). This approach is followed because EnergyPlus does not consider the effects of HVAC operation and wind direction on air leakage unless the Airflow Network module is used, which is not typically done because it is not a trivial task. In contrast, the online calculator utilizes CONTAM to estimate air leakage rates. A complete description of the prototype building is provided by DOE (DOE 2016d).

Ng et al. (2012) developed CONTAM models using EnergyPlus models of the ASHRAE/IESNA Standard 90.1-2004 prototype buildings as a baseline, and they were updated for this effort based on the ASHRAE/IES Standard 90.1-2013 models. The EnergyPlus and CONTAM models shared the same building geometry, occupancy, heating and cooling set points, and outdoor air ventilation requirements. However, the building zoning was modified in the CONTAM models in instances where additional zones were needed to support realistic airflow analyses (e.g., elevator shafts and restrooms). Modeling these additional zones is important to properly capture pressure relationships and airflow patterns in buildings. The present work utilizes the CONTAM model generated for the stand-alone retail building that includes a restroom that is not present in the prototype building model.

In order to determine the HVAC supply flow rates that would be used in CONTAM, a preliminary comparison was made of the maximum values that are calculated by EnergyPlus for the prototype stand-alone retail building in different cities by Ng et al. (2012). EnergyPlus results varied by less than 10% on average among the evaluated cities. Since the HVAC system modeled in CONTAM would retain approximately 10% more supply air than return air, the differences found in the maximum supply rates did not warrant changing their values in the CONTAM models for each city. Thus, the supply flow rates that were obtained for Chicago were applied to Winnipeg and Shanghai.

The stand-alone retail building has five thermal zones as shown in Figure 2. All zones, except the front entry, are conditioned in the summer and winter according to the set points listed in Table 1. These temperatures were scheduled in CONTAM because CONTAM does not perform thermal calculations. In the prototype building models, the front entry had a cooling set point of 38°C in the summer. However, it was assumed in the CONTAM model that the temperature in this

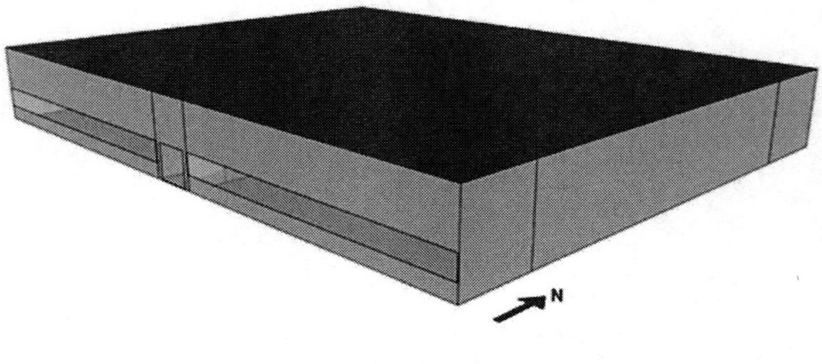

(a)

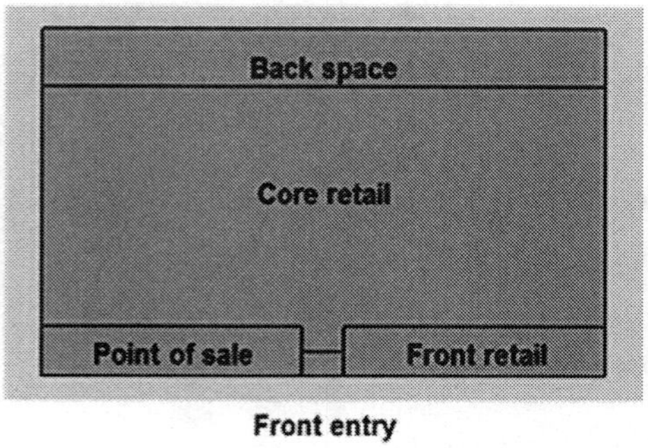

(b)

Figure 2 *Stand-alone retail building prototype: (a) building shape and orientation and (b) layout of five thermal zones (DOE 2016d).*

Table 1. Modeling Specifications of Stand-Alone Retail Building Prototype (DOE 2016d)

Characteristic	Description
Floor area, m^2	2300 (length 54.3 m × width 42.4 m)
Number of floors	1
Floor to ceiling height, m	6.1
Window-to-wall ratio, % Windows on south-facing façade	25.4
Building Envelope	
Walls	20.3 cm concrete masonry block + insulation per ASHRAE/IES Standard 90.1-2013 + 1.3 cm drywall
Roof	Roof membrane + insulation per ASHRAE/IES Standard 90.1-2013 + metal decking
Window U-factor and solar heat gain coefficient (SHGC)	Per ASHRAE/IES Standard 90.1-2013
Foundation	15.2 cm concrete slab-on-grade + insulation per ASHRAE/IES Standard 90.1-2013
Air leakage rates for prototype buildings (not used in the present study)	HVAC off = 1 L/s·m^2 at 75 Pa HVAC on = 25% of HVAC off rate = 0.25 L/s·m^2 at 75 Pa
HVAC	
Heating type	Gas furnace inside the packaged air-conditioning unit
Cooling type	Packaged air-conditioning unit
Size	Autosized to design day
Efficiency	Based on climate location and design cooling/heating capacity and ASHRAE/IES Standard 90.1-2013 requirements
Thermostat set point, °C	23.9 cooling/21.1 heating
Thermostat setback, °C	29.4 cooling/15.6 heating
Ventilation	Per ASHRAE Standard 62.1-2013 (ASHRAE 2013a)

zone was equal to the outdoor temperature in the cooling months. Also, in the EnergyPlus model the front entry had scheduled air leakage with a maximum flow rate of 0.94 m^3/s that varied between 0% and 100%, corresponding to unoccupied and occupied periods, respectively. This large air leakage was not modeled in CONTAM because its high flow rates would artificially increase the leakage of the entire building envelope in the whole-building air leakage rate data that would be exported to EnergyPlus. The outdoor air economizers and night cooling options in the EnergyPlus models were not implemented in the CONTAM models because CONTAM does not perform thermal calculations and would not be able to predict when economizers or night cooling options would be activated. Indoor temperatures in the CONTAM model were scheduled according to the set points in the EnergyPlus model.

Table 2 lists the four levels of airtightness that were assumed in the simulations. These include the slab and below-grade envelope area in the normalization of the air leakage rate, which is why they are referred to as *six-sided envelopes*, as well as the assumption that the air leakage is equally distributed over all exterior surfaces. The six-sided value is used in many building codes and standards; however, the CONTAM and EnergyPlus models assume no air leakage through the exterior envelope that is not exposed to ambient air. The baseline value in Table 2 was calculated using the average leakage rate for commercial buildings reported by Emmerich et al. (2005) of 9 L/s·m^2 at 75 Pa for a five-sided envelope. The baseline of 5.4 L/s·m^2 at 75 Pa was obtained by multiplying the average leakage rate by the five-sided to six-sided envelope area ratio of the stand-alone retail building prototype. Table 2 also lists three target levels for improved airtightness at 75 Pa: 2 L/s·m^2 is the most stringent of three options in the *2015 International Energy Conservation Code* (ICC 2015) because it involves a blower door test, whereas the other two options are based on laboratory tests per ASTM E2357 (ASTM 2011) and ASTM E2178 (ASTM 2013); 1.25 L/s·m^2 is the airtightness required by the U.S. Army Corps of Engineers (USACE

2012); and 0.25 L/s·m^2 is the leakage rate targeted by the DOE buildings envelope roadmap (DOE 2014). Emmerich and Persily (2014) analyzed the NIST U.S. commercial building air leakage database and found that the 79 buildings categorized as having an air barrier had an average six-sided leakage of 1.39 L/s·m^2 at 75 Pa, which was 70% below the average leakage of the 290 buildings without an air barrier (i.e., 4.33 L/s·m^2 at 75 Pa) and is similar to the second target level above. Zhivov et al. (2014) reported the average six-sided leakage for a set of 285 new and retrofitted military buildings constructed to the USACE specifications to be 0.9 L/s·m^2.

The three cities that were evaluated are Chicago, IL, USA; Winnipeg, Canada; and Shanghai, China. Table 3 shows their DOE climate zones and the locations of the corresponding prototype building models that were used in the simulations. CONTAM was used to calculate the hourly air leakage rates for the prototype building for each of these cities. Table 4 lists the ACH results for when the HVAC system is on ($ACH_{HVAC\ on}$), when the HVAC system is off ($ACH_{HVAC\ off}$), and the annual average (ACH_{avg}). Results indicate that ACH_{avg} for Winnipeg is the highest, followed by Chicago and Shanghai. This is mainly due to differences in weather among the cities; for example, the annual average wind speed for these cities is 4.78, 4.56, and 3.25 m/s, respectively. Results suggest that reducing the air leakage rate from 5.4 to 2 L/s·m^2 at 75 Pa led to a decrease in ACH_{avg} of about 75% across the three locations. By further lowering the leakage rate to 1.25 and 0.25 L/s·m^2 at 75 Pa, ACH_{avg} was reduced by about 86% and 98%, respectively, compared to the baseline.

Table 2. Assumed Building Envelope Airtightness Levels for a Six-Sided Envelope

Case	Air Leakage Rate at 75 Pa, L/s·m^2	Source
Baseline	5.4	Emmerich et al. (2005)
1	2.0	ICC (2015)
2	1.25	USACE (2012)
3	0.25	DOE (2014)

Table 3. Evaluated Cities

City	Climate Zone	Prototype Building Model Used in Calculator
Shanghai, China	3A (warm, humid)	Memphis, TN, USA
Chicago, IL, USA	5A (cold, humid)	Chicago, IL, USA
Winnipeg, Canada	7 (very cold)	Duluth, MN, USA

Table 4. Predicted Air Changes per Hour

Leakage Rate at 75 Pa, L/s·m^2	Air Changes per Hour, 1/h			$\frac{ACH_{HVAC\ on}}{ACH_{HVAC\ off}}$, %	Decrease in ACH_{avg} from Baseline ACH, %
	HVAC On	HVAC Off	Annual Average		
Chicago					
5.4	0.2077	0.2861	0.2389	73	
2.0	0.0366	0.1061	0.0642	35	73
1.25	0.0117	0.0664	0.0334	18	86
0.25	0.0009	0.0134	0.0059	7	98
Winnipeg					
5.4	0.2804	0.3684	0.3154	76	
2.0	0.0571	0.1366	0.0887	42	72
1.25	0.0201	0.0855	0.0461	23	85
0.25	0.0012	0.0172	0.0076	7	98
Shanghai					
5.4	0.1021	0.1823	0.1340	56	
2.0	0.0118	0.0675	0.0340	18	78
1.25	0.0037	0.0422	0.0190	9	88
0.25	0.0006	0.0085	0.0037	7	98

As previously stated, in order to estimate the hourly $ACH_{HVAC\,on}$, the DOE commercial prototype building models assume that this number is 25% of $ACH_{HVAC\,off}$. However, Table 4 shows that using multizone airflow simulations, this percentage is closely linked to the airtightness of the envelope. For example, when the building enclosure leakage rate was 5.4 L/s•m^2 at 75 Pa, $ACH_{HVAC\,on}$ was 56% to 76% of $ACH_{HVAC\,off}$. In contrast, this ratio decreased to 7% when the envelope airtightness was 0.25 L/s•m^2 at 75 Pa. This implies that the approach followed by users of prototype building models significantly underestimates the contribution of air leakage to energy consumption when the HVAC system is on in buildings with leaky enclosures, while the opposite occurs in buildings with very tight envelopes.

Figure 3 illustrates the HVAC energy use as a function of the building envelope airtightness level in Winnipeg. Results indicate that improving airtightness from 5.4 to 2 L/s•m^2 at 75 Pa led to an 18% and 55% decrease in electricity and natural gas use, respectively.

Energy costs were calculated using the annual energy outputs from EnergyPlus and the annual average prices of

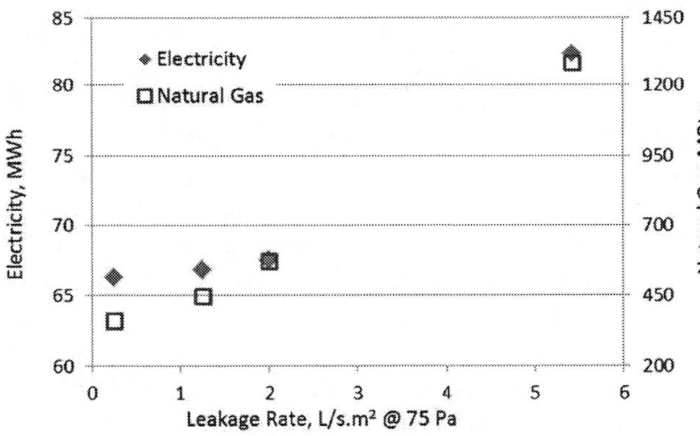

Figure 3 *Annual HVAC energy use for a prototype stand-alone retail building in Winnipeg based on the CONTAM+EnergyPlus approach.*

Table 5. Energy Prices

Location	Electricity Price	Natural Gas Price
Chicago	\$0.0933/kWh [a]	\$8.86/1000 ft^3 [b]
Winnipeg	C\$0.14/kWh [c] (≈\$0.10/kWh)	C\$0.1605 m^3 [d] (≈\$3.4/1000 ft^3)
Shanghai	¥0.781/kWh [e] (≈\$0.12/kWh)	¥3.65/m^3 [f] (≈\$15.9/1000 ft^3)

[a] http://www.eia.gov/electricity/sales_revenue_price/
[b] http://www.eia.gov/dnav/ng/ng_sum_lsum_a_EPG0_PCS_DMcf_a.htm
[c] https://www.ovoenergy.com/guides/energy-guides/average-electricity-prices-kwh.html
[d] http://www.economicdevelopmentwinnipeg.com/uploads/document_file/natural_gas_rates.pdf?t=1433529826
[e] http://news.asean168.com/a/20150413/5318.html
[f] http://gas.gold600.com/

electricity and natural gas listed in Table 5. Figures 4, 5, and 6 show the annual HVAC energy cost in Chicago, Winnipeg, and Shanghai, respectively, as a function of building envelope leakage rate. The figures also present quadratic regression equations. The high coefficients of determination (i.e., R^2 >

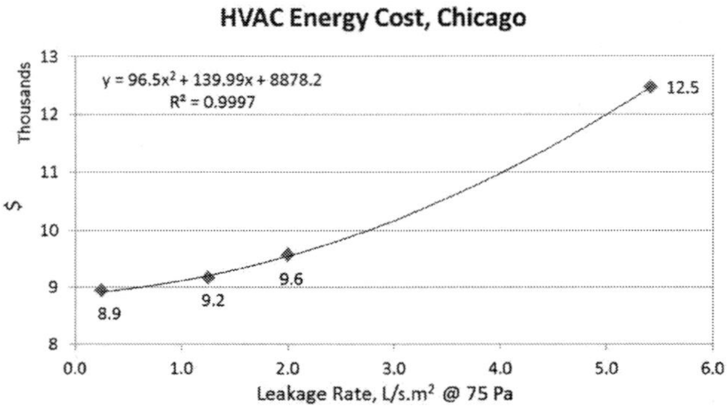

Figure 4 *Annual HVAC energy cost for a prototype stand-alone retail building in Chicago based on the CONTAM+EnergyPlus approach.*

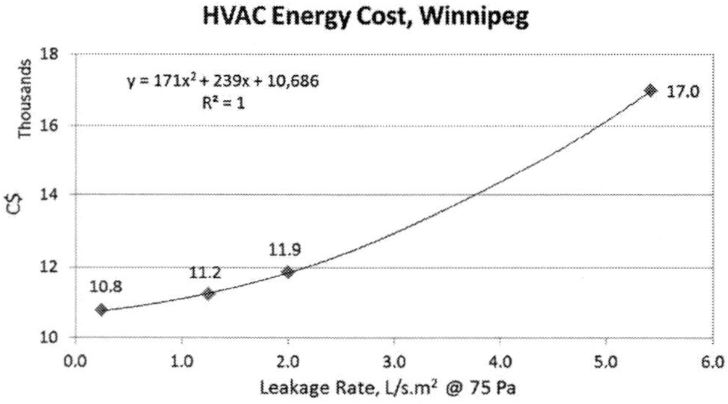

Figure 5 *Annual HVAC energy cost for a prototype stand-alone retail building in Winnipeg based on the CONTAM+EnergyPlus approach.*

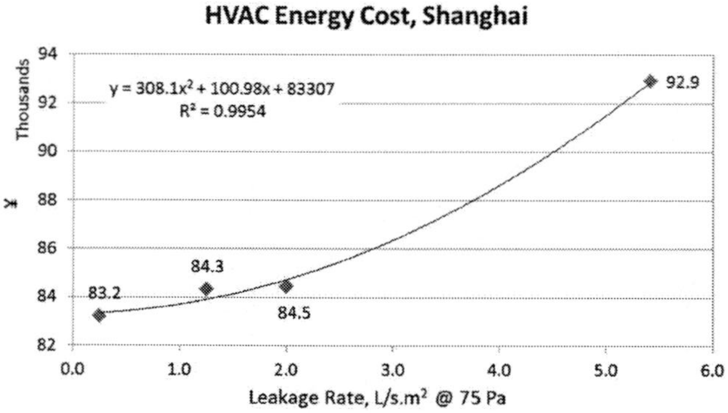

Figure 6 *Annual HVAC energy cost for a prototype stand-alone retail building in Shanghai based on the CONTAM+EnergyPlus approach.*

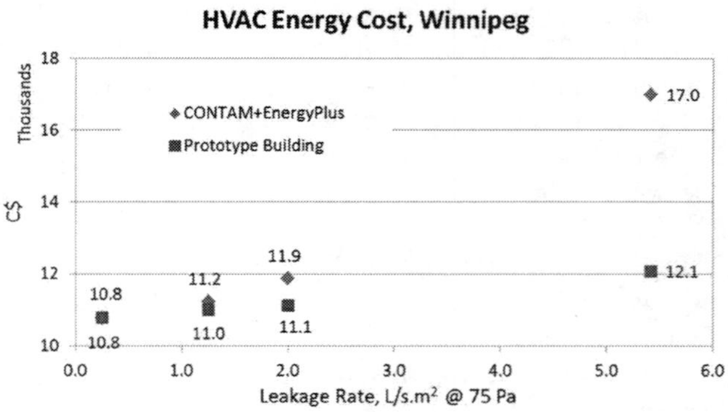

Figure 7 *Annual HVAC energy costs in Winnipeg from the CONTAM+EnergyPlus and the prototype building models.*

0.995) suggest that the calculator may be able to use quadratic equations to estimate energy costs for any given airtightness level. Similar equations could be derived for the heating and cooling costs, as well as for energy usage.

Further evaluations were performed to get a better understanding of the improvements that the CONTAM+EnergyPlus approach offers. To serve as a reference, simulations were conducted for Winnipeg using only EnergyPlus; that is, hourly air leakage rates from CONTAM were not imported into EnergyPlus and the prototype building assumption that $ACH_{HVAC\,on}$ equals 25% of $ACH_{HVAC\,off}$ was made Table 6 shows the results from these simulations, and Table 7 compares the ACH values that were obtained through these two approaches. These data illustrate how the simplified method used in the prototype building models underestimates the ACH and its corresponding impact on energy use. The effects are more noteworthy in leaky buildings, where the two approaches had ACH differences of 70% when the HVAC system is on and an annual average discrepancy in ACH of 49%. These differences decrease as the envelope becomes tighter, although they remained significant even when the leakage rate of the enclosure was 2 $L/s{\cdot}m^2$ at 75 Pa.

Figure 7 compares the annual HVAC energy costs in Winnipeg that were calculated with the CONTAM+EnergyPlus and the prototype building methods. As previously stated, differences are greater in buildings with leakier envelopes: the discrepancy in buildings with a leakage rate of 5.4 $L/s{\cdot}m^2$ at 75 Pa amounted to nearly $5000 per year. Ongoing projects at ORNL will help validate these estimated energy savings.

CONCLUSION

ORNL and NIST combined their expertise to develop a procedure that will be used in an online airtightness calculator. This procedure is different from other common methods used in energy analysis in that it uses hourly air leakage rates that are estimated by taking into account key variables such as building leakage rate, weather conditions, and HVAC operation. The calculator will provide energy cost estimates as a

Table 6. Predicted Air Changes per Hour in Winnipeg Using the Prototype Building Leakage Rate Reduction Method

Leakage Rate at 75 Pa, $L/s{\cdot}m^2$	Air Changes per Hour, 1/h			$\frac{ACH_{HVAC\ on}}{ACH_{HVAC\ off}}$, %	Decrease from Baseline Annual Average ACH, %
	HVAC On	HVAC Off	Annual Average		
5.4	0.0841	0.2765	0.1605	30	
2.0	0.0310	0.1024	0.0594	30	63%
1.25	0.0194	0.0641	0.0372	30	77%
0.25	0.0039	0.0128	0.0074	30	95%

Table 7. Comparison of Air Changes per Hour from CONTAM+EnergyPlus and the Prototype Building Simulation Approaches

Leakage Rate at 75 Pa, $L/s{\cdot}m^2$	Difference Between CONTAM+EnergyPlus and the Prototype Building Simulation ACH values, %		
	HVAC On	HVAC Off	Annual Average
5.4	70	25	49
2.0	46	25	33
1.25	3	25	19
0.25	-225	26	3

function of building envelope airtightness for DOE commercial prototype buildings in cities in the United States, Canada, and China. To demonstrate the CONTAM+EnergyPlus procedure, the paper presents an example where a prototype stand-alone retail building is simulated in Chicago, Winnipeg, and Shanghai. Results demonstrate that methods using simplified assumptions, such as $ACH_{HVAC\,on}$ equals 25% of $ACH_{HVAC\,off}$, underestimate the air leakage rates and the effects of building envelope airtightness on energy use. In the stand-alone retail building prototype example in Winnipeg, this discrepancy amounted to nearly $5000 per year for a building with a leakage rate of 5.4 $L/s{\cdot}m^2$ at 75 Pa; ongoing projects at ORNL will help validate these estimated energy savings. The calculator that is under development will be a powerful, credible, and easy-to-use tool that designers and contractors can utilize to estimate the energy and financial savings that building owners could achieve by reducing the air leakage.

ACKNOWLEDGMENTS

The authors would like to thank the U.S. Department of Energy, the Air Barrier Association of America, and the U.S.-China Clean Energy Research Center for Building Energy Efficiency for funding this research.

REFERENCES

ASHRAE. 2013a. ANSI/ASHRAE Standard 62.1-2013, *Ventilation for acceptable indoor air quality*. Atlanta: ASHRAE.

ASHRAE. 2013b. ANSI/ASHRAE/IES Standard 90.1-2013, *Energy standard for buildings except low-rise residential buildings*. Atlanta: ASHRAE.

ASTM. 2011. ASTM E2357-11, *Standard test method for determining air leakage of air barrier assemblies. West Conshohocken, PA: ASTM International.*

ASTM. 2013. ASTM E2178-13, *Standard test method for air permeance of building materials. West Conshohocken, PA: ASTM International.*

Chung, K.C. 1996. Development and validation of a multi-zone model for overall indoor air environment prediction. *HVAC&R Research* 2:376–85.

Crawley, D.B., J.W. Hand, M. Kummert, and B.T. Griffith. 2008. Contrasting the capabilities of building energy performance simulation programs. *Building and Environment* 43(4):661–73.

DOE. 2014. Windows and building envelope research and development: Roadmap for emerging technologies. Washington, DC: U.S. Department of Energy. http://energy.gov/sites/prod/files/2014/02/f8/BTO_windows_and_envelope_report_3.pdf

DOE. 2016a. *EnergyPlus*. Retrieved February 2016 from https://energyplus.net. Washington, DC: U.S. Department of Energy.

DOE. 2016b. Commercial prototype building models. Retrieved February 2016 from https://www.energycodes.gov/commercial-prototype-building-models. Washington, DC: U.S. Department of Energy.

DOE. 2016c. Commercial reference buildings. Retrieved February 2016 from http://energy.gov/eere/buildings/commercial-reference-buildings. Washington, DC: U.S. Department of Energy.

DOE. 2016d. 90.1 prototype building models sand alone retail. Retrieved February 2016 from https://www.energycodes.gov/901-prototype-building-models-stand-alone-retail. Washington, DC: U.S. Department of Energy.

Dols, W.S., and B. Polidoro. 2015. *CONTAM user guide and program documentation*. Technical Note 1887. Gaithersburg, MD: National Institute of Standards and Technology.

Emmerich, S.J. 2001. Validation of multizone IAQ modeling of residential-scale buildings: A review. *ASHRAE Transactions*, vol. 107(2).

Emmerich, S.J., S.J. Nabinger, A. Gupte, and C. Howard-Reed. 2004. Validation of multizone IAQ model predictions for tracer gas in a townhouse. *Building Services Engineering Research and Technology* 25:305–16.

Emmerich, S.J., T.P. McDowell, and W. Anis. 2005. Investigation of the impact of commercial building envelope airtightness on HVAC energy use. NISTIR 7238. Gaithersburg, MD: National Institute of Standards and Technology.

Emmerich, S.J., and A.K. Persily. 2014. Analysis of U.S commercial building envelope air leakage database to support sustainable building design. *The International Journal of Ventilation*, vol. 12.

Goel, S., R. Athalye, W. Wang, J. Zhang, M. Rosenberg, Y. Xie, R. Hart, and V. Mendon. 2014. *Enhancements to ASHRAE Standard 90.1 prototype building models*. PNNL-23269. Richland, WA: Pacific Northwest National Laboratory.

Gowri, K., D. Winiarski, and R. Jarnagin. 2009. *Infiltration modeling guidelines for commercial building energy analysis*. PNNL-18898. Richland, WA: Pacific Northwest National Laboratory.

Haghighat, F., and A.C. Megri. 1996. A comprehensive validation of two airflow models—COMIS and CONTAM. *Indoor Air* 6:278–88.

ICC. 2015. *2015 International Energy Conservation Code*. Washington, DC: International Code Council.

Ng, L.C., A. Musser, S.J. Emmerich, and A.K. Persily. 2012. *Airflow and indoor air quality models of DOE reference commercial buildings*. Technical Note 1734. Gaithersburg, MD: National Institute of Standards and Technology.

Polidoro, B., L.C. Ng, and W.S. Dols. 2016. *CONTAM results export tool*. Technical Note 1912. Gaithersburg, MD: National Institute of Standards and Technology.

USACE. 2012. *Air leakage test protocol for building envelopes*. Washington, DC: U.S. Army Corps of Engineers.

Zhivov, A., D. Herron, J.L. Durston, M. Heron, and G. Lea. 2014. Airtightness in new and retrofitted US army buildings. *International Journal of Ventilation*, vol. 12.

Critical Property Contrasts of Fluid-Applied Air and Water Barrier Membranes Used for Envelope: Chemistries, Performance, and Durability

Patrick H. Young, PhD **Ben Meyer, RA, LEED AP**

ABSTRACT

Two primary elements needed to achieve an effective, energy-efficient commercial envelope are the ability to control air movement and prevent water infiltration. A product enjoying increased usage to mitigate enclosure air and water leakage is a fluid-applied membrane. Furthermore, buildings are expected to move with lateral loads and expand and contract with temperature, so a third critical element to address would be durability by using fluid-applied membranes with elastomeric character. Elastomeric character of membranes is determined by looking at several properties (e.g,. ultimate elongation, dynamic recovery, Young's modulus), as well as retaining these characteristics under normal exposure conditions like heat, water, and application variations.

Contrary to common practice, ultimate elongation alone does not accurately define the total elastomeric character of the membrane since the number is obtained at catastrophic failure. Higher ultimate elongation will not necessarily translate to improved elastomeric behavior over a wider dynamic range. In fact, the opposite is frequently true. Beneficial elastomeric character should be defined by both the elongation of the membrane and its ability to recover.

The results confirm that polymers with a high ultimate elongation will not necessarily translate to improved elastomeric character over a broader dynamic range. However, materials displaying a high elongation with good elastic recovery will exhibit those same qualities over a broad dynamic range. Hence, elongation to break is only one small fragment of the property puzzle when evaluating and comparing fluid-applied membranes. Elastic recovery at a specific elongation is also a key component.

This paper will discuss property comparisons between different fluid-applied chemistries and relate them to specifications required in the commercial building market.

INTRODUCTION

The heavy commercial building market continues to mandate that industrial structures are more durable, energy efficient, and environmentally friendly while managing a solution that will always be cost effective. Two primary elements needed to achieve a durable and energy efficient commercial structure are the ability to control air movement and the elimination of water leakage (Shohet and Paciuk 2006). Additionally, buildings are known to expand and contract with temperature, so a third critical element to address durability would be to specify materials with elastomeric character (Harris 2001; Straube and Burnett 2005; Warsek 1986). The most common solution to influence air and water leakage is the installation of an elastomeric barrier membrane around the opaque areas of the building envelope (Amstock 2001).

Product datasheets describing the properties of these barrier membranes will frequently associate ultimate elongation with elastomeric properties per ASTM Standard D412 (ASTM 2013). However, ultimate percent elongation alone does not systemically correlate with the total elastomeric character of the membrane because the property is obtained at catastrophic failure (Argon and Bessonov 1977; Arruda and Boyce 1993; Berstrom and Boyce 1999; Bloomberg et al. 1992;

Patrick H. Young is a senior research associate and **Ben Meyer** is a building science architect with the DuPont Protection Systems Building Envelope Group in Richmond, VA.

Brock 2006; Gordon et al. 1975). Furthermore, higher ultimate elongation as measured by ASTM D412 will not necessarily translate to improved elastomeric behavior over a broader dynamic range. In fact, the opposite is frequently true. Beneficial elastomeric character should be defined by both the percent elongation of the membrane and its ability to recover (Lawson et al. 2008; Luo and Ghosh 2003; Billmeyer 1984). The ASTM Standard 412 method does allow for reporting of a tensile set number, which begins to address the recovery aspect of the elastomeric membrane, but there is little guidance on how to create a valid number for fluid-applied membranes. The minimum criterion for elastomeric behavior in coatings appears to have converged on between 95% and 100% recovery at 100% elongation (Brock 2006; Rawlins and Storey 2012).

An additional parameter to benchmark the efficacy of a fluid-applied weather barrier is the overall coverage rate consistency on porous substrates such as concrete masonry unit (CMU) (Litvan 1995). A critical component of coverage rate is in the statistical distribution of basis weights required to achieve the manufacturer's recommended thickness for the entire job. All fluid-applied weather barriers identify a coverage rate that is a statistical average, and it is this parameter that a contractor, architect, etc. will use to estimate the cost of a job. An allowance of 10% contingency is customary for cost budgeting during the contract document development and bidding phase (Phelan 2014).

Significance of Temperature Cycling Events

A National Ready Mixed Concrete Association study measured frequency and number of freeze/thaw cycles per year and the presence of moisture in the substrate across cities and climate zones. For example, a sidewalk at Penn State in State College, PA measured 120 cycles/year noted as a hard, wet freeze (NRMCA 2004). Multiplying the ground freezing rate of 120 cycles per year by a 50- and 100-year expected design life of buildings shows 600 to 1200 expansion contraction cycles for the design life of a building. This is actually a conservative estimate since surfaces not buffered by the ground, such as a building, will freeze more regularly. Furthermore, extreme summer heat resulting in high-expansion events on building surfaces are not accounted for in the study. It is critical for fully bonded membranes on building walls to withstand hundreds of cycling events and not yield in strength and elastomeric recovery while continuing to maintain air and water performance properties.

There are several sources of normal expansion and contraction in exterior walls such as moisture/relative humidity changes, structural creep, chamber and live floor load movements, as well as seasonal temperature variations. A simple, single variable scenario shown in Figure 1 illustrates a joint movement of 100% caused by the thermal linear expansion of steel studs sheathed with exterior gypsum using tight 1/32 in. joints and coated with a fluid-applied membrane. In this example, the 4 ft steel stud between horizontal gypsum

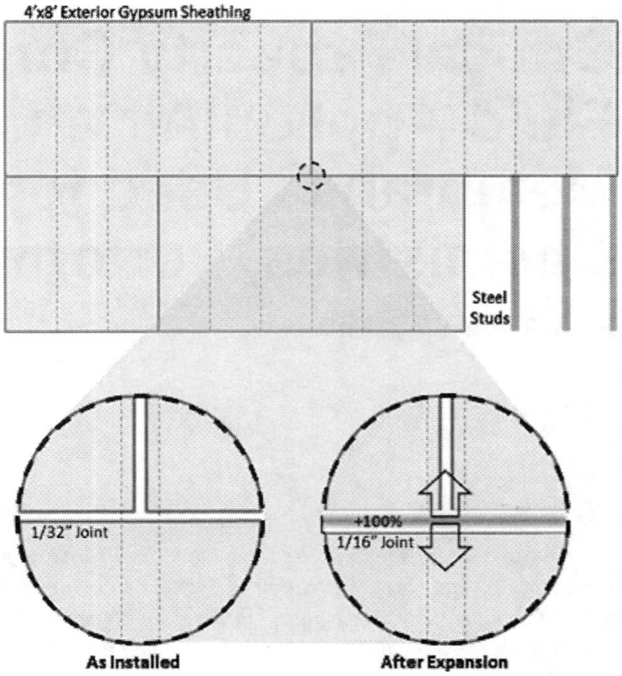

Figure 1 *100% expansion joint scenario on gypsum board with steel studs.*

sheathing courses experiences expansion from 0°F to 100°F. The expansion coefficient of steel is 8.0×10^{-6} (in./in.°F). Plugging these numbers into Equation 1 below results in a calculated expansion of the 4 ft stud between sheathing joints to be ~1/32 in.

$$dL = L_o\ \alpha\ (t_1 - t_o) \tag{1}$$

where

dL = change in length

L_o = initial length

α = coefficient of thermal expansion

t_1 = initial temperature

t_o = final temperature

A second common scenario is illustrated by a CMU masonry wall, which is expected to allow for ½ in. of movement over a 20 ft horizontal wall section (Ballast 2007). This translates, per a typical 16 in. wide CMU, to roughly be 1/32 in. of expected movement in the individual mortar head joint as the movement is transferred over to the designed control joints. This scenario can result in much more than 100% movement, as the initial starting position of the joint is much closer to zero in the CMU scenario. A safety, or design factor of 2× to 3× can be applied to the 100% movement scenario to help ensure the amount of change can be accommodated for with normal construction tolerances and that the fluid-applied material can recover to its original state.

Significance of Permeability

In the case of vapor-impermeable fluid applied, the membrane plays a dual role: air and vapor control. While air barriers could be installed anywhere in the building envelope, a vapor retarder location and its use is both climate and design specific. For example, vapor retarders are required by code only in cold climates and must be installed at the warm-in-winter side of the envelope. The impact of an air barriers' vapor permeance on moisture management is beyond the scope of this paper, but building physics must always be considered when an unintended vapor retarder is used in a wall assembly. Vapor permeability could impact moisture management in the building envelope. However, there are no clear requirements for the vapor permeance of an air barrier, leaving this decision to the building envelope designer.

The expected service life for the building enclosure is longer than for other building systems, which makes the building envelope durability even more critical. Moisture is one of the major factors affecting durability. Building assemblies may periodically get wet, but if allowed to dry they can still exhibit acceptable performance. Moisture problems only occur when buildings get wet and stay wet long enough under adverse conditions for materials to deteriorate (Spinu 2012). Good enclosure design must consider both sides of the moisture balance: minimize the risk of wetting and maximize drying potential, as shown in Figure 2. If a wall assembly is able to dry, it may experience some wetting without long-term durability risks. Vapor diffusion as a wetting source is highly overestimated, while vapor diffusion as a drying mechanism is poorly understood and often neglected (Spinu 2012). The optimum vapor permeance will depend on the building envelope design and climate zone and must be assessed for each project.

Chemistry

Fluid-applied polymeric membrane chemistries can be generalized into four categories: silyl terminated, silicone, rubber based, and acrylic based.

Silyl-terminated polymers may have polyether or polyurethane backbones and are typically greater than 90% solids. Equation 2 is an example of a silyl-terminated polyether backbone (STPE). Equation 3 is an example of a silyl-terminated polyurethane backbone added (STPU).

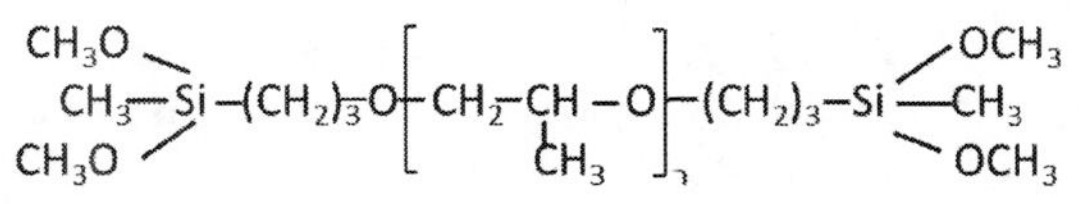

(2)

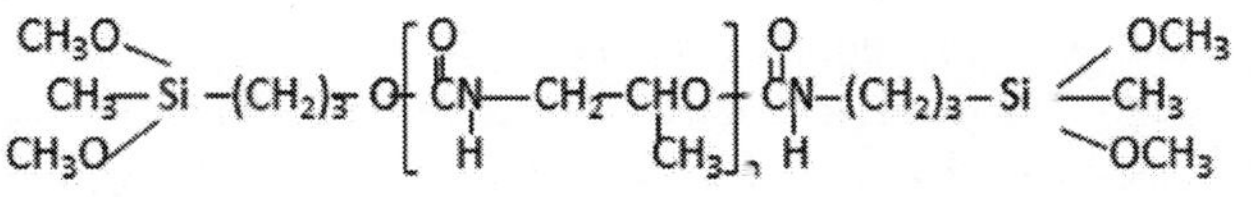

(3)

Acrylics are usually water based. Rubber-based chemistries may be solvent or water-based and all can vary from 30% to 80% solids. Equation 4 is a generalized formula for polymethacrylate and methacrylic acid, respectively.

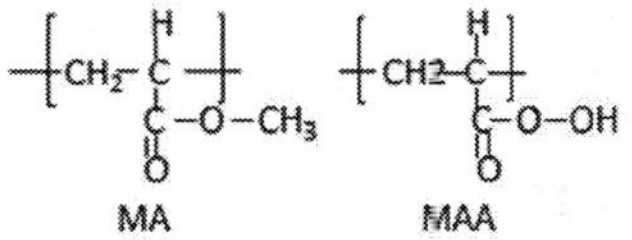

(4)

Equation 5 is a generalized formula for butyl rubber, which is a copolymer of isobutylene and isoprene.

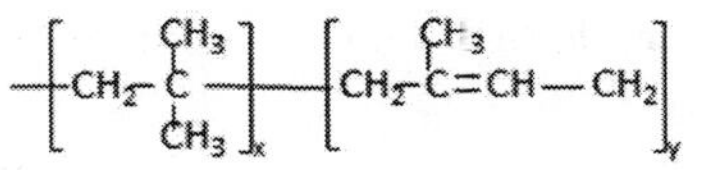

(5)

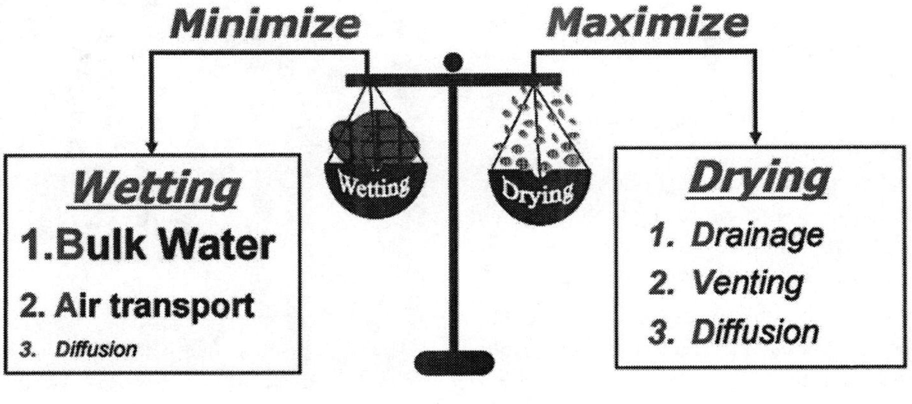

Figure 2 *Importance of moisture balance in a building enclosure design.*

The amide, ester, and carboxylic acid linkages of STPUs and acrylics indicate a strong propensity for hydrogen bonding (Mattia and Painter 2007; Hao et al. 2010). The ether linkages of the STPE polymers can form hydrogen bonds but they are much weaker than STPUs and acrylics. Electrostatic forces in rubbers are primarily composed of weak van der Waals interactions (Florys 1946). The presence of stronger electrostatic forces such as hydrogen bonding can result in the enhancement of some physical properties, while others might be negatively affected (Elias 1987; Craver 1983; Seymour et al. 1984). These relationships will be discussed in more detail later.

The purpose of this study was to provide a performance baseline comparison for common commercial-grade construction membranes that have been third-party evaluated by Air Barrier Association of America (ABAA). Properties assessed were consistency of coverage rate, percent expansion after heating/cooling without mechanical stress, percent elongation (breakage after the material is held for 5 seconds), and percent recovery after being held to a specific elongation for five seconds. These properties were tested at the manufacturer's recommended thickness for each product. The study was also expanded to test the same materials at 180°F (relates to AAMA 711/714 Type III and AAMA 501.5 Thermal Cycling). A relationship of elastic recovery to the electrostatic forces of certain polymer chemistries will also be discussed.

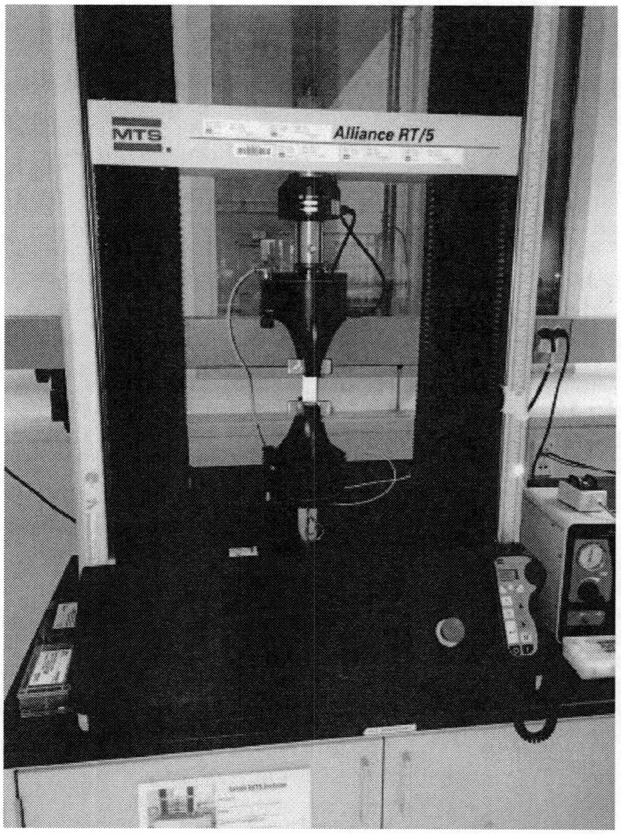

Figure 3 Instron device used in testing.

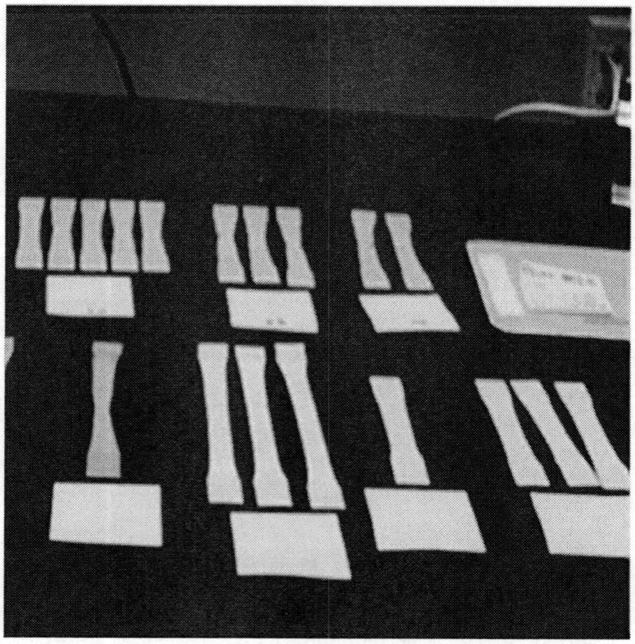

Figure 4 Membrane recovery in process.

EXPERIMENTAL

Room Temperature Instron

Samples were prepared by creating a draw down of each product using a draw-down device. The draw down was laid at the manufacturer's recommended wet thickness for each product. After curing for a minimum of two weeks, five samples were cut into 1× 5 in. samples for testing.

An Instron was used for elongation testing and is shown in Figure 3. A one square inch of Teflon was attached to the inside face of each clamp to prevent deformation of the membrane while removing samples. Each sample was then loaded into the Instron and taken to a specified percent elongation prior to failure. After reaching a specified elongation, the sample was held for five seconds before the Instron jaws were released back to their original position.

After elongation, the sample was placed on the table and the length was measured at 0, 1, 2, and 24 hours using a ruler. Depending on the tackiness of the product, some samples were placed on a rougher surface to prevent fusing to the table and restricting recovery. Representative examples of samples being allowed to relax are shown in Figure 4.

A series of five 1 × 5 in. samples were measured to obtain a statistical representation at each percent elongation evaluated. Testing covered the range of elongation from 100% to 1000%. Data was collected until greater than 50% of the samples broke. The elongation when this event occurred was identified as elongation at break after five seconds.

Elevated Temperature Instron

Draw-down samples were made as before for each brand of liquid barrier. The draw downs were allowed to cure for at least two weeks and samples were cut into 1 × 5 in. rectangles. The cut samples were allowed to condition at 180°F in an environmental chamber designed to enclose the Instron grippers for a few minutes and then loaded into the Instron machine to test elongation (Figure 5). After conditioning, the samples were loaded vertically with a two-inch gap between grips and with pieces of Teflon covering the clamp faces to prevent the membrane from adhering to the apparatus.

The elevated temperature chamber was then moved over the grip assembly and sample. Five samples of each brand were stretched to 100%, 200%, and 300% elongation at a temperature of 180°F (82.2°C). The sample was then held at length for five seconds. The sample was brought back to its original position, the hot box rolled off the grips, and the sample unclamped. The length of each sample was measured at 1 hour, 2 hours, and 24 hours from the initial stretch to determine recovery of each brand. The measurement at 24 hours was selected as the final length and used to determine percent recovery. This was then compared to the recovery at room temperature.

When more than half of the samples broke at a specific elongation, it was identified as the elongation to break and no recovery measurement was recorded.

Thermal Mechanical Analysis (TMA)

A 0.0787 in. (2 mm) wide strip of each membrane material was cut and mounted between TMA film clamps. Care must be taken, because some of the membrane materials are tacky, rendering the sample mounting very challenging. The clamped sample was set up in the film and fiber probe of a TA Instruments Q400 TMA in tensile mode (Figure 6). A preload force of 0.005N (~0.5 grams) was applied and the initial length of the sample was measured. The sample was then heated and cooled in two cycles from –20°F to 180°F at 9°F/min (–30°C to 82°C at 5°C/min) with five-minute isothermal holds between cycles. The dimensional change was measured after each isothermal hold for five minutes.

Coverage Rate

Ten samples were made from each brand of competitive liquid weather barrier to evaluate coverage rates and mass variation on 12 × 12 in. CMU when applied at the manufacturer's recommended thickness. The final calculated mass of each product was determined by weighing the amount of liquid weather barrier left in a sampling cup before and after application. The difference in measurement is identified as the mass applied. The weather barrier material was applied with a foam paint brush to minimize material deposited on the brush itself. The thickness was periodically checked with a mil gauge to obtain the manufacturer's thickness.

DISCUSSION

Table 1 shows all the competitive samples used in this study along with their chemistries and the specified wet thickness per the manufacturer's datasheet, accessed in November of 2015.

Figure 5 *Instron with elevated temperature chamber.*

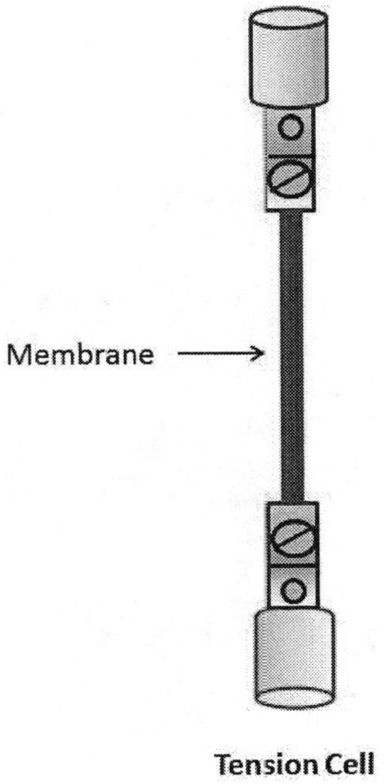

Figure 6 *Tension mode sample cell designed with a thermal mechanical analyzer.*

Table 1. Recommended Thickness and Chemistry for Each Competitive Sample

Sample ID	Wet Thickness, Mils	Polymer Chemistry
STPE A	26	Silyl-terminated polyether
STPE B	12	Silyl-terminated polyether
STPE C	25	Silyl-terminated polyether
Acrylic A	60	Acrylic
Acrylic B	70	Acrylic
Acrylic C	60	Acrylic
Acrylic D	10	Acrylic
Acrylic E	68	Acrylic
Acrylic F	80	Acrylic
Silicone	26	Silicone
Rubber	10	Rubber
STPU	12	Silyl-terminated polyurethane

Room Temperature Instron

Figure 7 is a bar chart comparing the percent recovery of all the competitive fluid-applied membranes tested at 300% elongation after being held for five seconds. STPE C exhibited the best overall elongation and recovery numbers of all the other competitors (>99% recovery at 300% elongation for five seconds). Several of the competitive materials broke at or below 300% elongation and are shown with red Xs in Figure 7. Every competitor to STPE C with STPE chemistry broke below the 300% elongation threshold. One membrane (Acrylic D) broke below 100% elongation. All the remaining competitive membranes (primarily acrylic chemistry) showed significant levels of inelastic recovery to varying degrees when compared to STPE C.

Figures 8 and 9 corroborate each membrane's capacity to recover over the complete dynamic range of elongation values. These comparisons were also made on membranes at the manufacturer's recommended thickness. Considering the chemical components in each material, they were separated into two groups. Figure 8 is the membrane grouping that compares all the products evaluated using acrylic chemistry. Figure 9 is the membrane grouping that compares all the products evaluated using STPE, STPU, silicone, and rubber chemistries. Each graph identifies the 300% elongation comparison criterion and how all membranes behaved above and below this threshold.

The acrylic-chemistry-based elastomeric materials, shown in Figure 8, displayed the widest dynamic range of elongation. However, over most of that range, they also exhibited significant inelastic recovery, which is most likely due to intermolecular bond cleavage.

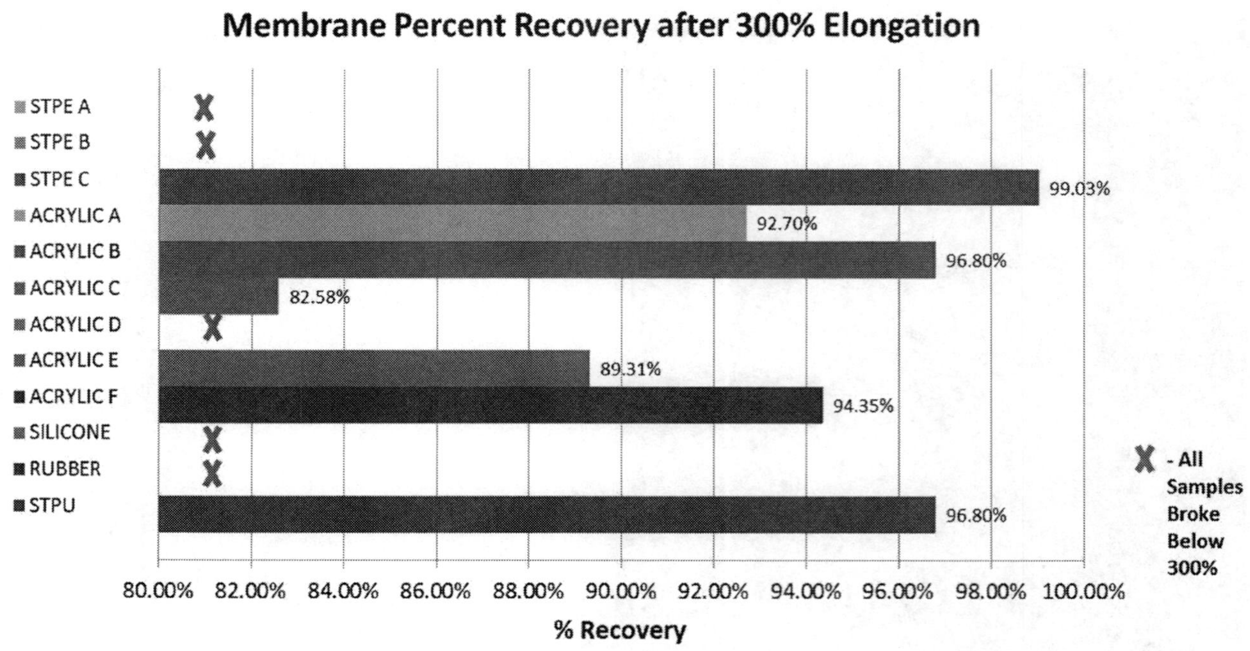

Figure 7 *Membrane percent recovery after 300% elongation (manufacturer's recommended thickness).*

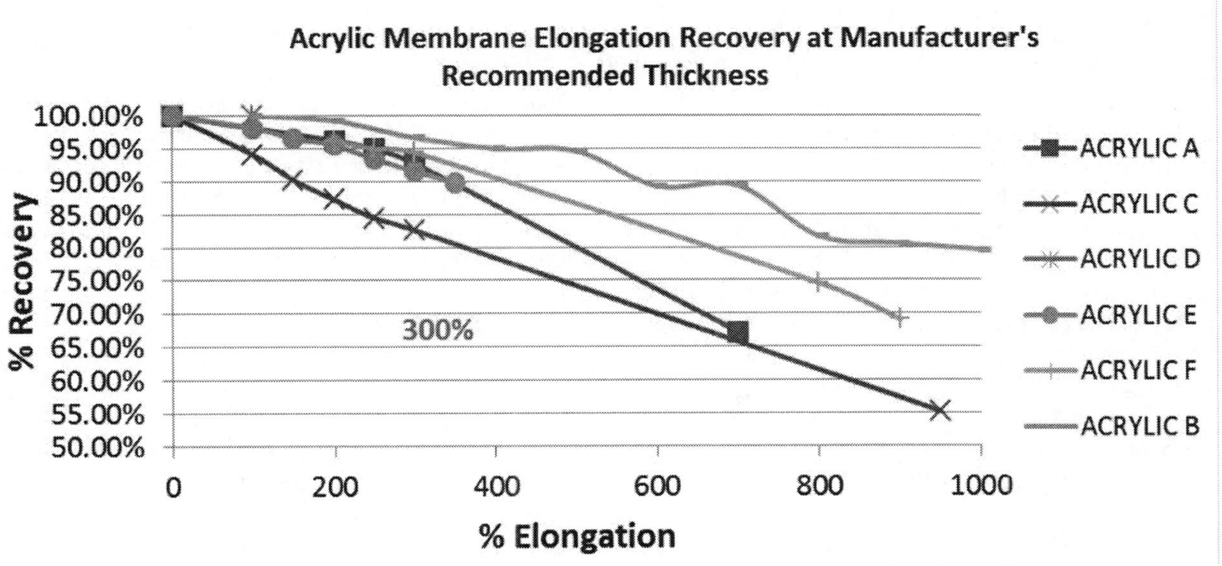

Figure 8 *Acrylic membrane elongation recovery at manufacturer's recommended thickness.*

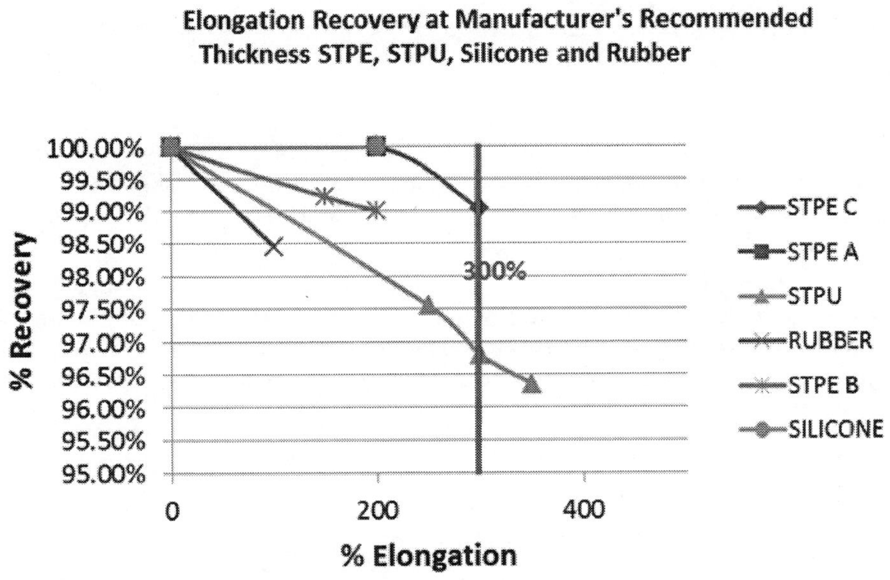

Figure 9 *Elongation recovery at manufacturer's recommended thickness for STPE, STPU, silicone, and rubber.*

All the STPE chemistries in Figure 9 exhibited a narrower dynamic range of elongation when compared to the acrylics but displayed recoveries that were almost completely elastic over their specified range. The STPU competitor exhibited a moderately wide dynamic range of elongation but demonstrated inelastic recovery early in the testing, most likely due to intermolecular bond cleavage. The silicone competitor showed slight inelastic recovery beginning at 100% elongation and then broke at all elongations above 100%. STPE C had the best recovery percentage overall within its dynamic range of elongation (>99%).

The direct result of stretching an elastomer is a decrease in entropy because the degrees of freedom of molecular motion become increasingly restricted (Billmeyer 1984). Conversely, complete elastic recovery results in positive entropy through sufficient freedom of molecular motion, which is achieved via high localized chain mobility (Mishra and Yagei 2008). While localized chain mobility in the polymer matrix needs to be high, macro scale chain mobility must be limited if excellent recovery characteristics are desired (Elias 1987). Macro scale chain mobility is frequently minimized by creating a complex network of primary bond cross links (Seymour et al. 1984).

Secondary bond forces, such as hydrogen bonding, need to be minimized due to an ability to restrict the localized chain mobility (Billmeyer 1984). Hence, the presence of hydrogen bonding can play a key role in elastic recovery dynamics of an elastomeric polymer. As the material is stretched, the intermolecular hydrogen bonds can rupture due to increased stress, resulting in slippage between the polymer chains. Therefore, while higher elongation may be attainable (as shown in Figures 8 and 9), the localized chain slippage is not completely reversible, resulting in inelastic recovery early in the stretching process. Hence, while the elongation at break may initially appear impressive, the membrane will typically experience inelastic recovery commencing at much lower elongations due to an irreversible cleavage of the hydrogen bonds which are restricting the localized chain mobility. Intermolecular hydrogen bonding is significant in both the acrylic and STPU polymers.

Elevated Temperature Instron

Since heating causes elastomeric polymer chains to contract (Brown 1963), stretching of fluid-applied membranes under a heat load should result in a lower elongation to break (Roundy and Rogers 2013). Table 2 shows each product with their elongation at break (five second duration) measured at the manufacturer's recommended wet thickness. The dry film thickness will differ for several of these membranes due to varying percent solids in each competitor's mixture. The elongation to break was measured for each membrane at 70°F and 180°F respectively. The percent loss of elongation at 180°F was then calculated.

Several membranes (Table 2) displayed significant percent loss in elongation due to temperature. Acrylic A exhibited the worst with a percent loss of 95%, Acrylic C and rubber deteriorated 78% and 80% respectively, and Acrylic F deteriorated 74%. The elongation to break of the silicone competitor had the lowest measurable percent loss at 0%, while STPE C was second with a percent loss of 34%. Since the environmental chamber only allowed for elongations up to 300%, two membranes (STPU and Acrylic B) never broke at elevated temperature so the percent loss when comparing the temperature difference was not quantifiable.

Comparisons of room temperature elongation recovery to those at elevated temperatures (180°F) are shown in Table 3. Overall, each brand had lower elongation at break and worse percent recovery under elevated temperatures compared to their properties at room temperature.

It should be noted that several of the membranes using acrylic chemistry softened and became tacky when heat was applied. STPE B and STPU also became soft and tacky under heat, a behavior suggesting the polymer underwent a phase transition or a continuation of the curing stage. Once heat was removed from any of the acrylic based samples, the product once again hardened and behaved as a pseudothermoplastic polymer. The silicone competitor appeared to shrink upon cooling from the initial heat cycle.

All the STPEs were able to resist the 180°F temperature without significant visible stress. STPE Cs elongation to break under heat was ~230% and without heat was approximately 350%. Furthermore, the STPE C product was able to achieve

Table 2. Measured Elongation to Break at Room Temperature and Heated in the Environmental Chamber

Sample ID	Thickness Wet, mils	Elongation to Break,* 70°F	Elongation to Break,* 180°F	Percent Loss of Elongation, 180°F
STPE A	26	300%	150%	50%
STPE B	12	250%	135%	46%
STPE C	26	350%	230%	34%
Acrylic A	55	750%	40%	95%
Acrylic B	65	1000%	>300%	N/A**
Acrylic C	60	950%	210%	78%
Acrylic D	14	76%	N/A	N/A
Acrylic E	60	450%	250%	44%
Acrylic F	70	950%	250%	74%
Silicone	24	250%	250%	0%
Rubber	10	150%	30%	80%
STPU	12	400%	>300%	N/A**

* Method per discussed in the experimental section.
** Environmental chamber only allows for a 300% elongation to break.

Table 3. Summary of Elongation and Recovery Data at Room Temperature and 180°F

Brand	% Elongation	% Recovery at 70°F	% Recovery at 180°F	% Loss of Recovery at 180°F	Comments at 180°F in the Environmental Chamber
STPE A	100%	100%	N/A	100%	Not enough sample to run hot Instrons
STPE B	100%	100%	100.00%	0.00%	Sample broke at 135% elongation
	200%	100%		100%	
STPE C	100%	100%	100.00%	0.00%	Sample broke at 230% elongation
	200%	100%	100.00%	0.00%	
Acrylic A	100%	100.00%	N/A	100%	Sample broke at 40% elongation
Acrylic B	100%	100.00%	96.85%	3.15%	
	200%	99.26%	96.15%	3.13%	Elongation at break greater than 300%
	300%	96.62%	86.96%	10.00%	
Acrylic C	100%	94.07%	93.46%	0.65%	Sample broke at 210% elongation
	200%	87.23%	85.47%	2.00%	
Acrylic D	100%	N/A	N/A	N/A	Sample broke at less than 100%
Acrylic E	100%	98.15%	95.01%	3.20%	Sample broke at 250%
	200%	93.66%	90.40%	3.50%	
Acrylic F	100%	100.00%	95.01%	5.00%	Sample broke at 250%
	200%	96.21%	91.12%	5.30%	
Rubber	100%	98.45%	N/A	100%	Sample broke at 30% elongation
Silicone	100%	100.00%	101.14%	N/A	Sample broke at 250% elongation
	200%	100.00%	100.50%	N/A	Shrinkage occurred after heat
STPU	100%	100.00%	100.00%	0.00%	
	200%	100.00%	99.10%	0.90%	Elongation at break greater than 300%
	300%	96.80%	94.86%	2.00%	

100% recovery at 180°F and 200% elongation. The silicone competitor was the only other membrane that was able to attain 100% recovery at 180°F and 200% elongation.

Acrylic D failed below 100% elongation in both the room temperature and 180°F temperature testing. Acrylic A and the rubber membrane were not elastic enough to achieve a 100% elongation under heat load. Both of their elongation break points deteriorated to under 50%. Acrylic C had low elastic recovery percentages at room temperature and 180°F when stretched at both 100% and 200%. All data is shown in Table 3.

Figure 10 compares the elastic recovery in a heated environment to that of room temperature and normalizing all membranes to an elongation of 200%. All the Acrylic A, rubber, and STPE B membranes broke at or before reaching 200% elongation in a heated environment. All products had a lower elongation at break under heat than at room temperature. STPE C and silicone membranes demonstrated 100% recovery at 200% elongation at room temperature and at 180°F. When comparing the behavior of the other membranes at 200% elongation at room temperature and then 180°F, Acrylic B, C, E, F and STPU displayed the most significant deterioration in percent recovery (refer to Table 3).

Figures 11 and 12 represent an individual graphical comparison of the elongation recovery behavior of each competitive product at room temperature and 180°F heat load. Figure 11 displays the acrylic chemistry products and Figure 12 displays the STPE, STPU, silicone, and rubber chemistries.

Previous room temperature recovery experiments confirmed that most of the acrylic materials displayed a broad dynamic range of elongation percentages but exhibited inelas-

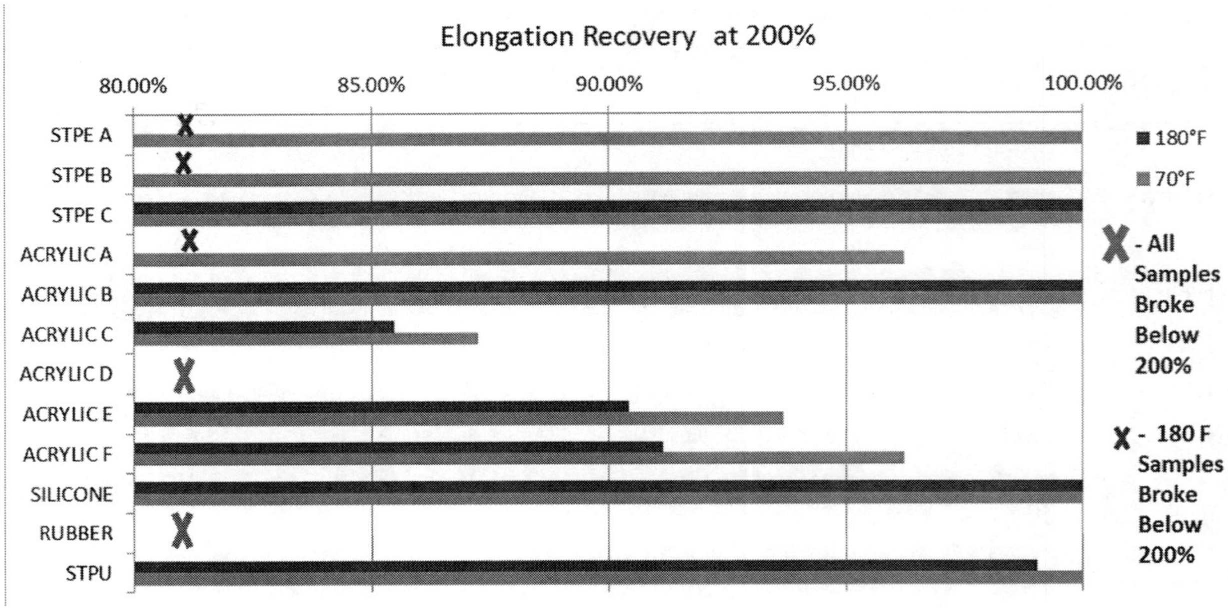

Figure 10 *Elongation recovery at 200%. Room temperature and heated.*

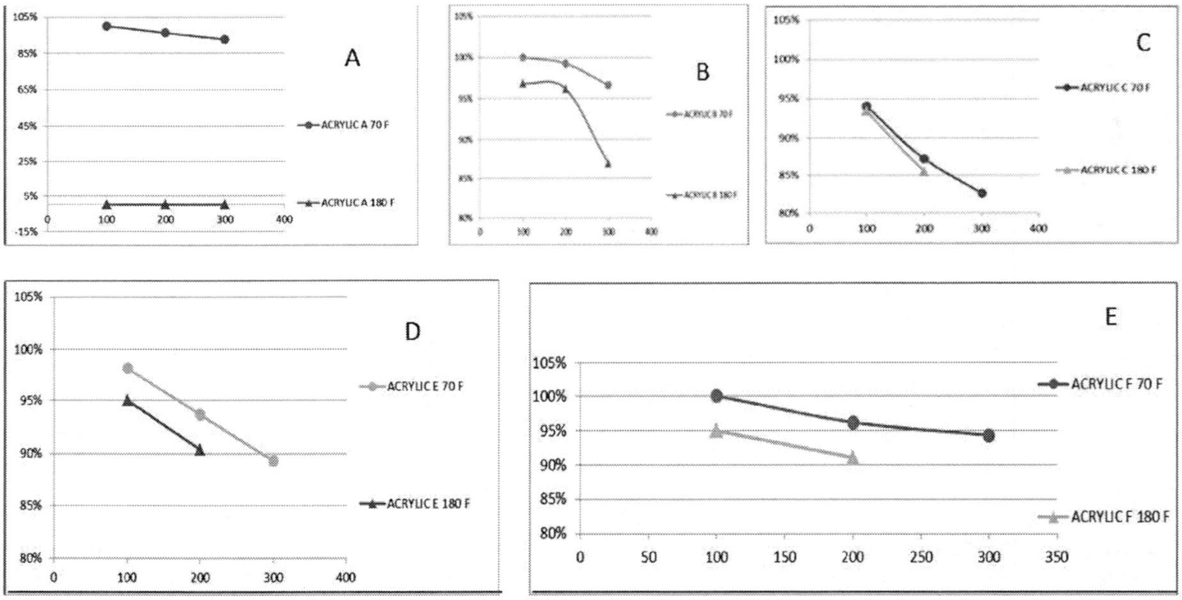

Figure 11 *Acrylic recovery from 100% to 300% elongation. Room temperature and heated.*

tic recovery over most of that measured range. Acrylic D was not plotted because the membrane was not capable of achieving a minimum of 100% elongation at room temperature or 180°F. Acrylic A was not able to achieve a minimum of 100% elongation at 180°F (Figure 11a). Acrylic B reached 300% elongation at both room temperature and 180°F but inelastic recovery was significant (Figure 11b). Acrylic C, E, and F exhibited inelastic recovery in both the room temperature and 180°F testing (Figure 11c, 11d and 11e).

While the percent elongations were not as high for the STPE polymers, their recovery at room temperature and elevated temperature was >99%. The STPE B samples broke at 200% elongation when heated to 180°F (Figure 12a). STPE C had good recovery at 200% elongation at room temperature and at 180°F (Figure 12b). The rubber competitor broke prior to achieving an elongation of 100% (Figure 12c). The silicone competitor showed good recovery at 200% elongation at both room temperature and

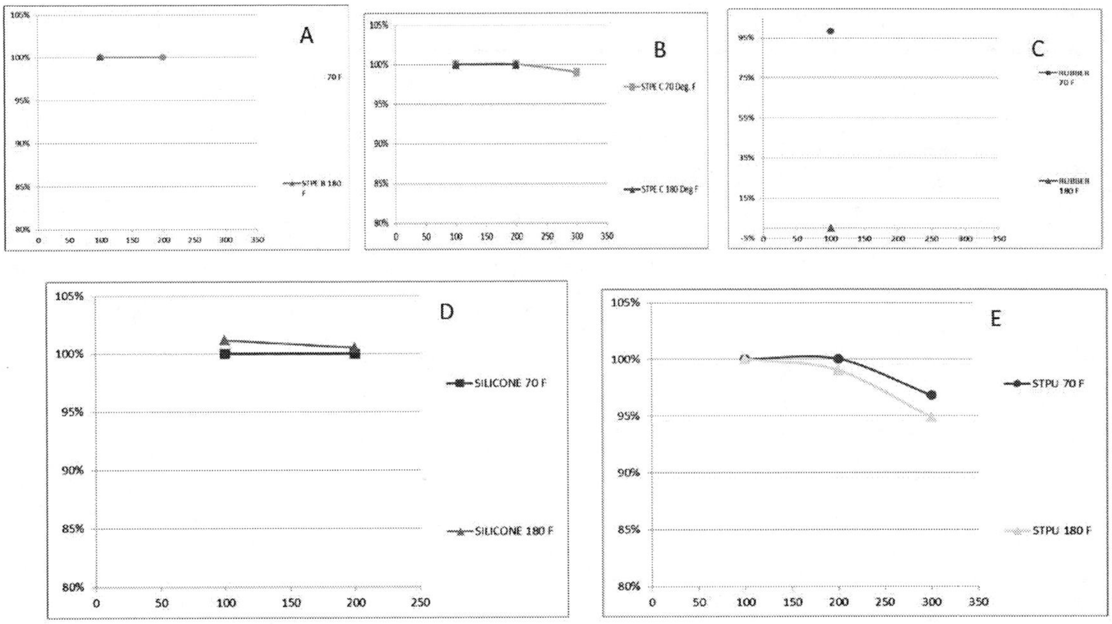

***Figure 12** STPE, STPU, silicone, and rubber recovery from 100% to 300% elongation. Room temperature and heated.*

180°F (Figure 12d). The STPU polymer began demonstrating inelastic recovery at a lower elongation when heated at 180°F (Figure 12e).

Thermal Mechanical Analysis

A thermal mechanical analyzer (TMA) can provide insight into a fluid-applied membrane's dimensional behavior during heating/cooling while under minimal mechanical stress. Table 4 shows all the samples run on the TMA with their resulting dimensional changes.

Sequence 1 is the initial measurement at room temperature. Sequence 2 is the first heating cycle to 180°F and held for five minutes. Sequence 3 is the first cooling cycle to –20°F and held for five minutes. Sequences 4 and 5 are the second heating and cooling cycles. Sequence 6 returns the sample to room temperature for the final measurement after five minutes.

Figure 13 is a graphical representation of the dimensional change of each membrane during every sequence of the heating/cooling (Sequences 1–6). The most significant dimensional change occurs in the first heating cycle (Sequence 2). Membranes with the highest first cycle expansions were acrylic chemistries (~11%). STPEs/STPU expanded only ~1%–3% during the first heating cycle.

A zone of acceptable dimensional stability was defined in Figure 13 to range from ±2%. The membranes falling within this acceptable zone also displayed less than 1% permanent distortion after cycling. The maximum dimensional change of STPE C, STPE A, Acrylic C, silicone, rubber, and STPU all fell within the acceptable range after cycling back to room temperature.

Acrylic A and Acrylic B displayed the highest expansions after cycling. STPE B had the highest overall expansion of the STPE chemistry products at 7%. Permanent distortions of this magnitude will place an unwarranted stress on the interface between the fluid applied and the wall including the expansion joint. The result can be premature failure of the weather barrier.

To visually demonstrate how the effect of stress placed on an exterior surface due to thermal expansion of the fluid-applied membrane with heat could result in premature failure, draw downs were done on 75 gauge polyester film and allowed to cure at the manufacturer's recommended thickness. Prior to the draw downs, the polyester film was annealed at 200°F to minimize any residual dimensional stability effect of the film itself. Two samples of the each membrane on polyester film were then placed in an oven at 180°F for ten minutes and allowed to cool.

The results are shown in Figure 14. The unheated membranes on polyester film exhibited little to no residual distortion after cure. However, after heating, Acrylic A revealed some buckling due to expansion of the membrane at 180°F, while STPE C continued to remain flat. The buckling that was observed with the Acrylic A/PET film composite is a confirmation of the expansion that was observed in the TMA analysis shown in Figure 13.

Coverage Rate

Different fluid-applied competitors were evaluated for variations in coverage rate on CMU. 12 × 12 in. squares of CMU were used to evaluate how the basis weight fluctuated while attempting to preserve a consistent membrane thick-

Table 4. Thermal Mechanical Analysis Cycling of Competitive Fluid-Applied Membranes

Sample ID	Initial Measurement	180°F First Cycle	−20°F First Cycle	180°F Second Cycle	−20°F Second Cycle	75°F Second Cycle
	Sequence 1	Sequence 2	Sequence 3	Sequence 4	Sequence 5	Sequence 6
	% Expansion/ Shrinkage	% Expansion/ Shrinkage	% Expansion/ Shrinkage	% Expansion/ Shrinkage	% Expansion/ Shrinkage	% Expansion/ Shrinkage
STPU	0	1.2	0.7	1.6	1	1.2
STPE C	0	1.3	−1	1.5	−0.9	0.4
STPE A	0	1.3	−0.7	1.5	−0.6	0.5
Acrylic A	0	11.1	12.7	15.5	13.9	15.2
Silicone	0	1.9	−1.2	1.9	−1.1	0.5
Acrylic C	0	2.1	0.7	2.1	0.7	1.3
Rubber	0	1.8	0.3	1.7	0.3	1
Acrylic F	0	3.9	2.3	4.4	2.4	3.4
Acrylic B	0	9.0	8.8	10.7	9.1	10.1
STPE B	0	2.8	5.8	5.7	7.6	7

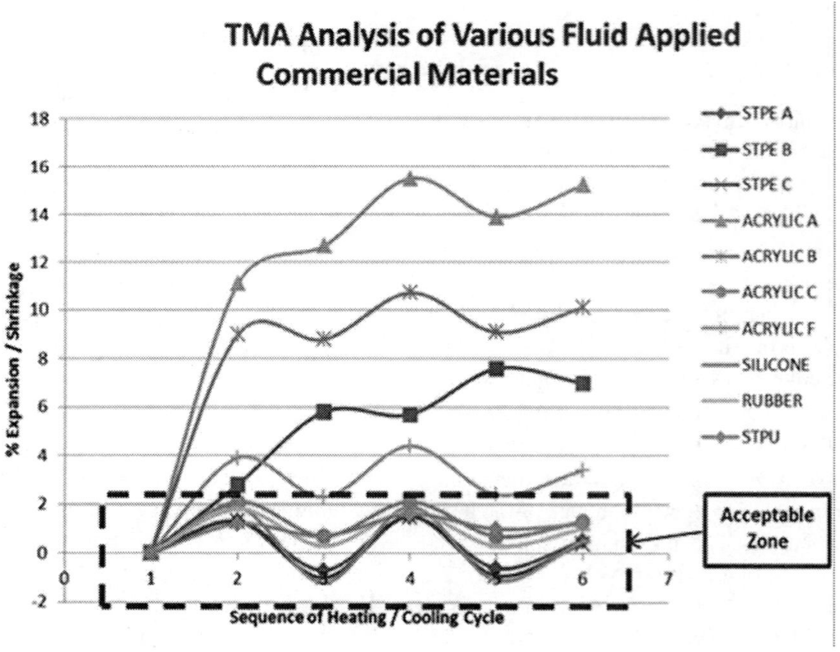

Figure 13 *TMA analysis of various fluid-applied commercial material.*

ness. The coverage rate data was evaluated using sum of the squares error calculations. Sum of the squares attempts to weight the distribution of the data from the statistically correct answer (in this case, the mean of the basis weight). Statistical samples were made using the recommended coverage rate from each competitor. The goal was to determine the numerical deviation in the basis weight of material used to reproduce the manufacturer's recommended thickness.

Figure 15 is a series of box plots showing the standard deviation of basis weights to achieve a constant thickness. The standard deviation was always higher when using the acrylic based fluid-applied competitors. All the other chemistries, except for STPE B, exhibited a more highly restricted distribution, indicating an area where the variation between coverage rate and thickness is considered acceptable. A region of acceptable variation in coverage rate when comparing all manufacturers is defined for this paper as ~10% or less.

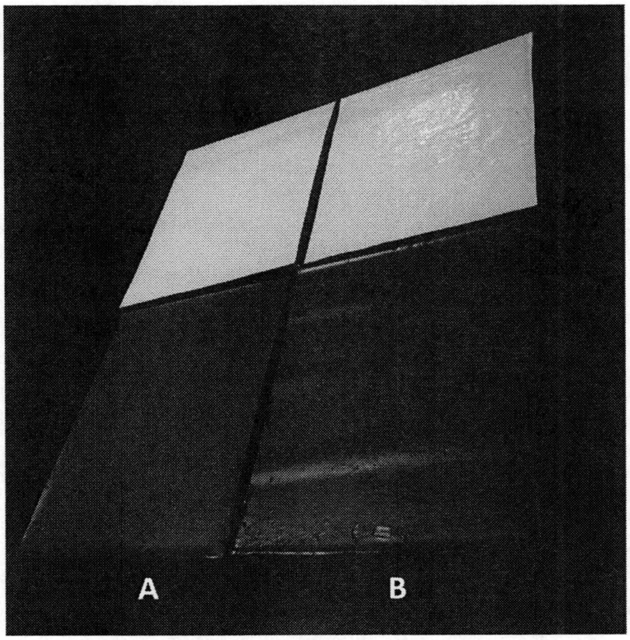

Figure 14 *Fluid-applied draw downs at manufacturer's recommended thickness onto 75 gauge polyethylene terephthalate (PET) film. Blue is Acrylic A and white is STPE C. A is room temperature and B is heated to 180°F for ten minutes.*

Figure 16 is a regression plot of sum of the squares versus final dry membrane thickness. The R^2 for this plot was 77.2%, indicating a good correlation between measured and predicted values.

Figure 17 is a regression plot of percent membrane solids versus sum of the squares. While the correlation of Figure 17 is less predictable than Figure 16, a relationship between higher percent solids and sum of the squares error is easily observed. Hence, the statistical relationship between thickness and basis weight is more compelling when the membrane is applied at high percent solids.

CONCLUSIONS

Table 5 is a summation of the property contrasts measured on each competitor's membrane. Every membrane was judged against a minimum acceptable criteria identified for each property test. STPE C was the only fluid-applied membrane to pass all of the criteria proposed in Table 5. STPE C also had the second lowest percent loss of elongation to break when relating room temperature and 180°F temperature (34%). Furthermore, STPE C has a perm rating of 25.

The silicone membrane had the lowest percent loss of elongation to break when comparing room temperature and 180°F (0%). However, the silicone material failed criteria put forth in the 300% recovery test and has a very low perm rating of 6.

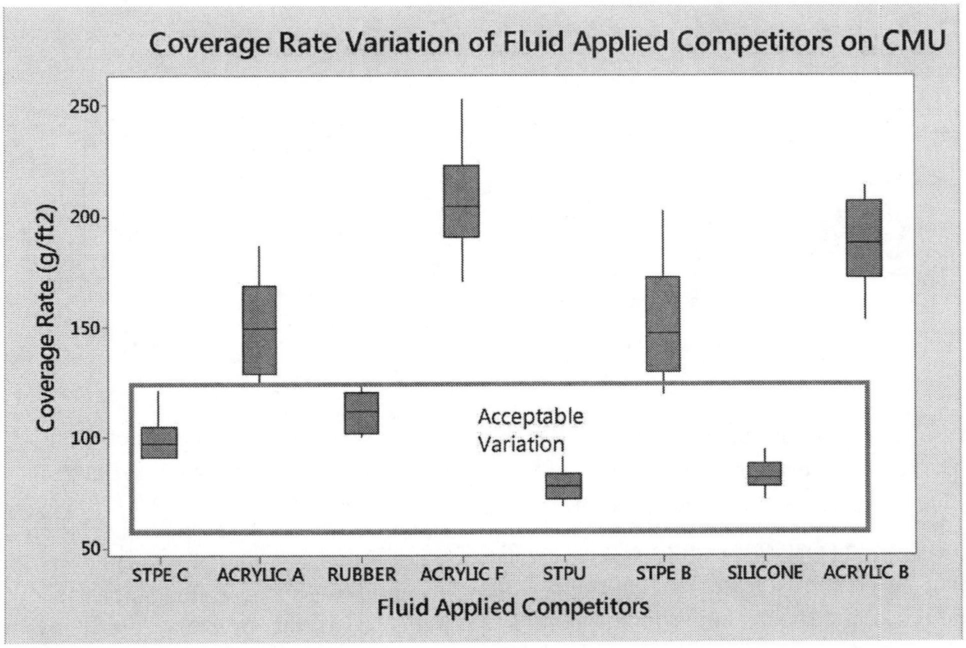

Figure 15 *Coverage rate variation of fluid-applied competitors on CMU.*

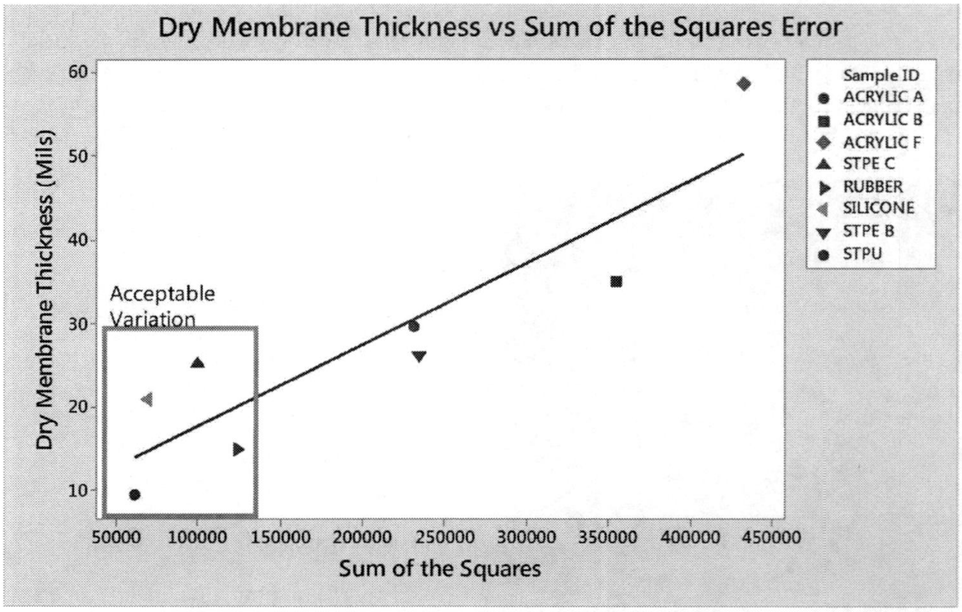

Figure 16 *Dry membrane thickness versus sum of the squares error.*

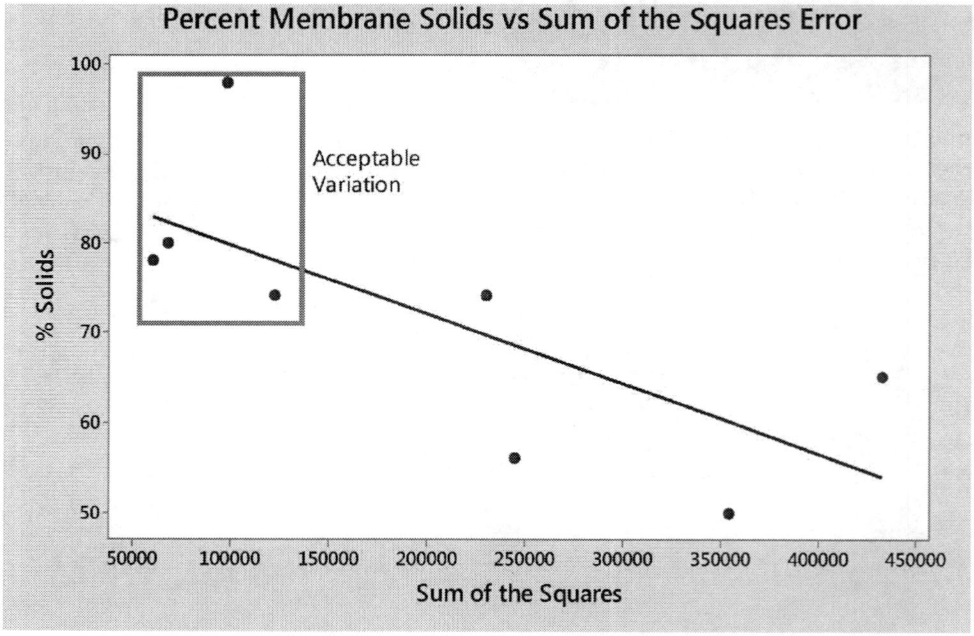

Figure 17 *Percent membrane solids versus sum of the squares error.*

Acrylic C was the only acrylic chemistry to pass one of the property tests (TMA analysis). The other acrylic membranes failed every property test listed in Table 5.

Coverage rate variation on CMU quantified via sum of the squares statistics is greatest at lower percent solids and higher membrane thickness. Hence, high solids and lower dry weight thickness fluid-applied materials should exhibit the most statistically reproducible coverage rate on CMU.

The results confirm that polymers with a high ultimate elongation will not necessarily translate to improved elastomeric character over a broader dynamic range. However, materials displaying a high elongation with good elastic recovery will exhibit those same qualities over a broad dynamic range. Hence elongation to break is only one small fragment of the property puzzle when evaluating and comparing fluid-applied membranes. Elastic recovery at a specific elongation is also a key component.

Table 5. Summation of Property Contrasts When Comparing All Competitive Fluid-Applied Products

Sample ID	Wet Thickness Mils*	Perms E96 Method B*	Polymer Chemistry	TMA Analysis**	Sum of the Squares***	% Loss of Elongation to Break from Room Temperature to 180°F	>99% Recovery at 300% Elongation at Room Temperature	>99% Recovery at 200% Elongation at 180°F
STPE A	26	32	Silyl-terminated polyether	Pass	N/A	50%	Fail	N/A
STPE B	12	18	Silyl-terminated polyether	Fail	Fail	46%	Fail	Fail
STPE C	25	25	Silyl-terminated polyether	Pass	Pass	34%	Pass	Pass
Acrylic A	60	14	Acrylic	Fail	Fail	95%	Fail	Fail
Acrylic B	70	12	Acrylic	Fail	Fail	N/A	Fail	Fail
Acrylic C	60	12	Acrylic	Pass	N/A	78%	Fail	Fail
Acrylic D	10	10	Acrylic	N/A	N/A	N/A	Fail	Fail
Acrylic E	68	15	Acrylic	N/A	N/A	44%	Fail	Fail
Acrylic F	90	21	Acrylic	Fail	Fail	74%	Fail	Fail
Silicone	26	6	Silicone	Pass	Pass	0%	Fail	Pass
Rubber	10	18	Rubber	Pass	Pass	80%	Fail	Fail
STPU	12	13	Silyl-terminated polyurethane	Pass	Pass	N/A	Fail	Fail

* From manufacturer's datasheet accessed in November 2015.
** A zone of acceptable dimensional stability was defined to range from ±2%.
*** Membrane groupings that had sum of the squares error variations of less than 150,000 were defined as acceptable.

REFERENCES

Amstock, J. 2001. Chapter 19, *Handbook of adhesives and sealants in construction*. New York: McGraw Hill.

Argon, A.S., and M.I. Bessonov. 1977. Plastic deformation in polyimides with new implication on the theory of plastic deformation of glassy polymers. *Philosophical Magazine* 35:917–33.

Arruda, E.M., and M.C. Boyce. 1993. A three dimensional constitutive model for the large stretch behavior in elastomers. *Journal of the Mechanics and Physics of Solids* 41:389–412.

ASTM. 2013. ASTM Standard D412, *Standard test methods for vulcanized rubber and thermoplastic elastomers—Tension*. West Conshohocken, PA: ASTM International.

Ballast, D.K. Chapter 19, Documenting and Enforcing Tolerances, *2007 Handbook of Construction Tolerances*, 2nd Ed. Hoboken, NJ: John Wiley and Sons.

Bergstrom, J., and M.C. Boyce. 1999. Mechanical behavior of particle filled elastomers. Rubber Chemistry and-Technology 72:633–56.

Billmeyer, F.W. 1984. Chapter 19, *Textbook of polymer science*. New York: John Wiley & Sons.

Bloomberg, P.A., S. Mahmood, R.J. Smales, and O.F. Makinson. 1992. Comparative elasticity tests for elastomeric (non-putty) impression materials. *Australian Dental Journal* 37(5):346–52.

Brock, M. 2006. Performance requirements of wall coatings: The facts and the fiction. *Concrete Repair Bulletin* July/August, 16–18.

Brown, J.B. 1963. *Thermodynamics of a rubber band. American Journal of Physics* 31:397.

Craver, C.D., ed. 1983. Chapter 18, *Polymer characterization Spectroscopic, chromatographic and physical instrumental methods*. Washington, DC: American Chemical Society.

Elias, H.G. 1987. Chapter 12, *Mega molecules*. New York: Springer-Verlag.

Florys, P. 1946. Effects of molecular structure on physical properties of butyl rubber. *Journal of Industrial & Engineering Chemistry* 38(4):417–36.

Gordon, G., L. Lower, and L. Carbary. 1975. Using rheology test methods to assess durability of cured elastomers undergoing cyclic deformation. *Journal of ASTM International* 6(3):313–27.

Harris, S. 2001. Chapter 3, *Building pathology: Deterioration, diagnostics and intervention*. New York: John Wiley & Sons.

Hao, J., G. Yuan, H. Weidong, H. Cheng, C. Han, and C. Wu. 2010. Interchain hydrogen-bonding-induced association of polyacrylic acid)-graft-poly(ethylene oxide) in water. *Macromolecules* 43:2002–08.

Lawson, N.C., N.O. Burgess, and M.S. Litaker. 2008. Tensile elastic recovery of elastomeric impression materials. *Journal of Prosthetic Dentistry* 100(1):29–33.

Litvan G.G. 1995. Waterproofing of parking garage structures with sealers and membranes: The Canadian experience. *Construction and Building Materials* 10(1):95–100.

Luo, L., and A. Ghosh. 2003. Elastic and inelastic recovery after plastic deformation of DQSK steel sheet. *Journal of Engineering Materials and Technology* 125(3):237–46.

Mishra, M., and Y. Yagei. 2008. Part IV, *Handbook of Vinyl polymers: Radical polymerization, process and technology*. Boca Raton, FL: CRC Press.

Mattia, J., and P. Painter. 2007. A comparison of hydrogen bonding and order in polyurethane and poly (urethane urea) and their blends with poly (ethylene glycol). *Macromolecules* 40(5):1546–54.

NRMCA. 2004. Freeze thaw resistance of pervious concrete. www.rmc-foundation.org/images/PCRC%20Files/Durability%20&%20Maintenance/Freeze-Thaw%20Resistance%20of%20Pervious%20Concrete.pdf.

Phelan, M, ed. 2014. *RS Means Assemblies Cost Data*, 39th Annual Edition. RK1010. Norwell, MA: Reed Construction Data.

Rawlins, J.W., and R.F. Storey, eds. 2012. The Waterborne symposium. *Proceedings of the Thirty-Ninth Annual International Waterborne, High-Solids, and Powder Coatings Symposium*. DEStech Publications, Inc. 252–55.

Roundy, D., and M. Rogers. 2013. Exploring the thermodynamics of a rubber band. *American Journal of Physics* 81(1):20–23.

Shohet, I., and M. Paciuk. 2006. Service life prediction of exterior cladding components under failure conditions. *Construction Management and Economics* 24:131–48.

Seymour, R.B., and C.E. Carraher. 1984. Chapter 1, *Structure property relationships in polymers*. New York: Plenum Press.

Spinu, M. 2012. Design without compromise: Balancing durable and energy efficient buildings. *The Construction Specifier* 65(11):36–53.

Straube, J. E., and E.F.B. Burnett. 2005. *Building science for building enclosures*. Westford, MA: Building Science Press.

Warsek, K. 1986. Why Sealant Joints Fail. http://www.buildingdiagnostics.com/articles/sealantjoints.html.

Critical Freeze/Thaw Saturation Measurement of In-Service Masonry

Randy Van Straaten, PEng
Student Member ASHRAE

Trevor Trainor, MASc

Chris Schumacher, MASc
Member ASHRAE

ABSTRACT

When planning an insulation retrofit for an existing masonry building, it is important to analyze the freeze/thaw durability of the masonry units. The critical freeze/thaw saturation value is an important measure of durability, and frost dilatometry is an effective method of determining this value. This study reports developments in the field sampling of masonry units and critical saturation testing of in-service masonry.

The challenge when sampling masonry is to ensure that selected units are representative of the entire building or capture the range of material properties of units on the building. Two approaches to sampling are explored. Bulk sampling involves the removal of a large number of samples with the goal of randomly capturing the range of units of interest. Alternatively, field testing of a relevant material property can be performed to identify units that capture the range of material on the facade. A nondestructive field drying rate measurement technique is presented along with pilot study results suggesting it could completely replace the need for bulk sampling.

Significant variation in the permanent strain resulting from the frost dilatometry method has been found in previous testing. It is thought that this variation is due to lack of material homogeneity within the brick, resulting in variations between sample slices. It appears from the current study that use of the mean strain for as little as four samples at each saturation level will allow accurate determination of critical saturation. It further appears that use of a 95th percentile strain value for each saturation level could alternatively be used as a conservative method. Overall, the proposed new methodology was found to have a high level of reproducibility for critical saturation measurement.

INTRODUCTION

Older, load-bearing masonry buildings are common throughout North America. They are often in desirable urban locations, are aesthetically pleasing with architectural significance, and have solid structures and useful window areas and floor plans. Hence, they are good candidates for renovation and conversion. Given modern demands for low energy usage, thermal comfort, and carbon emission reductions, thermal insulation retrofits are a highly desirable part of renovation for this type of building stock.

From a building science point of view, adding insulation to the exterior of existing buildings is always the preferred approach for retrofits. This practically eliminates thermal bridging while simultaneously protecting the load-bearing structure from all exterior climatic elements. However, for aesthetic and historic reasons, the exterior of many load-bearing buildings must remain exposed, and interior retrofits must be considered. Unfortunately, the addition of interior insulation will change the thermal and moisture balance of any wall assembly and, in some cases, can initiate moisture problems such as freeze/thaw damage in masonry units by decreasing the drying capacity while simultaneously reducing the temperature of the inner wythes (Straube and Schumacher 2006). The increased risk of freeze/thaw damage has caused many designers and owners to avoid the addition of insulation altogether, thereby missing an opportunity that is rarely presented in many buildings.

Randy Van Straaten is a senior research engineer at RDH Building Science Inc. in Toronto, ON. **Trevor Trainor** is a research engineer and **Chris Schumacher** is a principal at RDH Building Science Laboratories in Waterloo, ON.

One method used to assess the risk of freeze/thaw degradation is frost dilatometry. The typical frost dilatometry process involves a site review and material sample collection, testing of project masonry units, hygrothermal analysis, and communication to inform project teams of the risks associated with various thermal insulation and other related retrofit measures. Case studies of such projects have been published (Ueno et al. 2013a) and the methodology described in detail in previous publications (Ueno et al. 2013b; Van Straaten 2014). A critical component of the methodology is to determine the performance threshold for the masonry at risk. Within the frost dilatometry methodology, a critical freeze/thaw saturation is determined and utilized in analysis. This paper describes critical saturation measurements along with recent developments in material sampling.

CRITICAL SATURATION BACKGROUND

Fagerlund (1977) found that porous, brittle materials below a certain moisture content level can be freeze/thaw cycled repeatedly without any measurable damage. Freeze/thaw damage can be quantified as irreversible growth in specimen volume, whose measurement is termed *dilatometry*. Freeze/thaw strain, Δ_{ft}, is calculated within the method from initial and post freeze/thaw length measurements, l_0 and l_{ft}, as follows:

$$\Delta_{ft} = \frac{l_{ft} - l_0}{l_0} \quad (1)$$

Fagerlund further found that freeze/thaw damage occurring at moisture levels above this critical saturation level is linearly proportional to the saturation level. This is illustrated in Figure 1, which shows a hypothetical critical saturation of 66% determined from the *x*-axis intercept of the strain of the damaged specimens (having strain greater than zero).

Freeze/thaw degradation of masonry is understood to be due to a number of mechanisms (Mensinga et al. 2010). A review of testing results from a number of projects did not reveal clear correlation of critical saturation with visible indicators (Van Straaten 2014). It was further shown that there can even be significant variation in sample lots where historical information and visible indicators suggest units were from the same manufactured lot. As a result of these factors, projects utilizing frost dilatometry to assess freeze/thaw degradation risk need to empirically determine the critical saturation of in-service masonry through lab testing. This study reports developments in field sampling and critical saturation testing of in-service masonry. Project brick testing results reported in the paper are all from testing completed by the authors.

FIELD SAMPLING OF MASONRY UNITS

Two general approaches to sampling are explored in this study. The challenge when sampling masonry is to ensure that selected units capture the range of material properties of units on the building. Bulk sampling involves removal of a large number of samples with the goal of randomly capturing the range of units. Alternatively, nondestructive field testing of a relevant material property can be performed to identify units that capture the range of material on the facade. Each method is presented separately in this section, examining current approaches, results of recent field applications, and a discussion of the practicality of each approach.

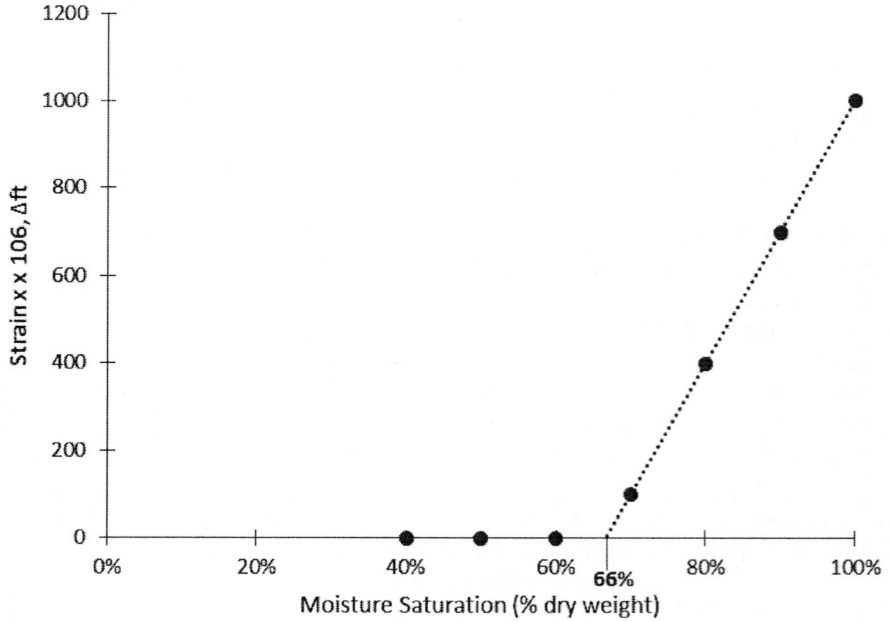

Figure 1 *Illustration of strain measurements after a number of freeze/thaw cycles.*

Bulk Sampling

Bulk sampling involves removal of a large number of masonry units from a building for testing. Assuming the bulk samples will be removed as part of a condition assessment for the building, the major limitation of this approach is the cost and timing required for full testing of a large sample set.

The sample size required for critical saturation testing has been reduced by characterizing and prescreening the population on the basis of water uptake testing. Water uptake testing is used for prescreening samples for three reasons: (1) it is used to derive liquid transport coefficients, which characterize capillary liquid transport (the dominant moisture transport mechanism in moist materials) (Wilson et. al. 1999); (2) the values reflect pore size distributions, which relate to a range of other properties; and (3) the measurements are done on whole sample units, avoiding the need for cutting and limiting the time required to obtain useful properties of the material.

Figure 2 shows water uptake results for a sample set of 47 bricks from an industrial to residential conversion project in Ottawa. The bricks are labeled in the plot with a number

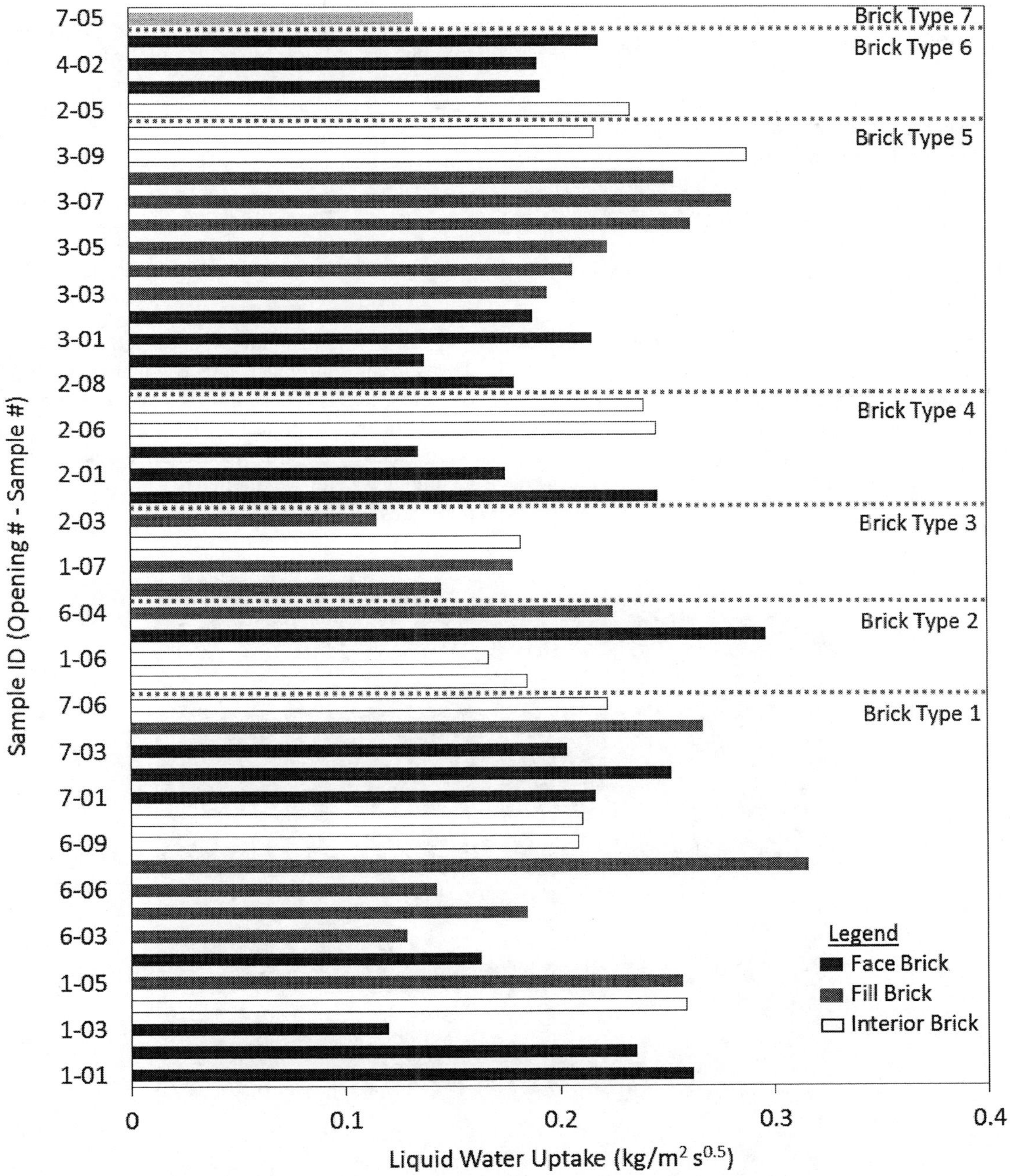

Figure 2 *Measured water uptake results for a large project sample set of 47 bricks.*

indicating their extraction location followed by a sample number (location#-sample#). The bars in the plot are black for exterior wythe bricks, gray for fill bricks, and white for interior wythe bricks. The bars are grouped into the seven brick types identified by different brick frog patterns. The variety of brick types was only revealed after they were removed from the building. The plot shows no correlations of visual characteristics with water uptake values; however, the water uptake values for this project were all similar. Within other projects, water uptake rates often vary by one or two orders of magnitude (Van Straaten 2014). For this project, the samples chosen for detailed testing primarily captured the different brick types and then secondarily the range of water uptake results.

The removal of a large number of bricks was feasible in this project due to the availability of samples from intrusive inspections performed to assess structural deterioration. A limitation to sampling using this approach is that the samples are not randomly selected. In many cases, openings will be targeted at areas of interest, typically with existing deterioration issues. Ideally the selection of intrusive inspection locations should be done in coordination with capturing masonry samples from areas showing the greatest vulnerability to freeze/thaw degradation.

In other projects, removal of a significant number of samples from a building can be both impractical and cost prohibitive. Challenges are most significant for projects with large stone masonry units or masonry units that are difficult to ship and replace. For these projects, nondestructive field tests are typically considered and will be discussed in the next subsection.

Nondestructive Field Testing

A common nondestructive field test for masonry is the RILEM tube test described by Driscoll and Gates (1993). The test results generally correlate well with water uptake. However, it is difficult to utilize RILEM tube testing to prescreen a large number of masonry samples due to the limited number of samples an operator can test in one day. It is also difficult, if not impossible, to test masonry with rough surfaces and/or significant cracks due to water leakage.

An alternative nondestructive screening test was developed by the authors using a Tramex Moisture Plus electrical impedance-based moisture content meter. This drying technique involves spraying 15 ml of water from a spray bottle held 20 cm from the target bricks at the start of the test. Moisture readings are taken before wetting and then two and ten minutes after wetting. The moisture content readings after wetting are expected to decrease (dry) as the water first redistributes to the interior of the brick and dries back towards the exterior. Drying (or drying/redistribution) rate is calculated as follows:

$$\dot{M} = \frac{M_{2min} - M_{10min}}{8min} \tag{2}$$

The drying technique was piloted on a Chicago project. The various buildings were originally constructed in two phases, which is reflected by the cladding type (either red or beige bricks, see Figure 3). As a minimum, sampling from both brick types for the project was required and the drying technique was further applied to capture the range of material properties for both brick types.

Figure 3 *Photos showing the red brick and beige brick facades on the test building.*

Three locations were evaluated on the buildings with red bricks and five locations on the buildings with beige bricks. Eighty bricks were tested in approximately four hours. The operator only needed a step ladder, a spray bottle, the Tramex meter, and an audio recorder for recording results.

The initial moisture meter readings and drying rates are plotted in Figure 4 for these bricks. The red bricks tested at Location 3 were initially damp and had a low drying rate. However, the beige bricks at Locations 4 and 5 were also initially somewhat damp but had a wide range of drying rates. The remainder of the bricks had a range of drying rates and were all initially dry. The one exception was one of the beige bricks at Location 3, which was damp. This brick had a high drying rate but other dry bricks at the same location showed a similar drying rate. Based on these measurements, six red and six beige sample bricks were selected on the basis of capturing the range of drying rates and different initial moisture contents within the same wall areas.

A comparison of the lab water uptake measurements and field drying rates for sample bricks from the Chicago project selected for full testing is shown in Figure 5. A clear linear correlation was found for the red bricks. A nonlinear correlation is apparent for the beige bricks but is less clear.

The initial moisture content may vary based on material properties of the brick but will also vary based on exposure and assembly. It may be possible in future projects to conduct hygrothermal analysis for these specific locations using recent weather data to estimate masonry material properties and/or local exposure factors based on such initial moisture readings. It may also be possible to conduct the analysis immediately after the collection of the reading and use the result to assist with sample selection.

The correlation between the dry/redistribution rates suggests the technique could completely replace the need for bulk sampling. This would be quicker, allow evaluation of more samples, and impose less restrictions on sampling locations.

CRITICAL FREEZE/THAW SATURATION MEASUREMENT

Modern critical saturation measurement of bricks is described in detail by Van Straaten (2014). The process basically involves the following steps:

1. cut masonry into slices (typically 12 × 20 × 100 mm);
2. dry in oven, then cool to room temperature;
3. measure dry weights;
4. saturate with water by vacuum saturation;
5. measure saturated weights;
6. dry slices to a range of saturation levels;
7. measure initial lengths;
8. seal slices in plastic wrap;
9. freeze/thaw slices in chilled bath;
10. allow slices to warm to room temperature; and then
11. measure final lengths to determine strain.

Van Straaten (2014) also provided suggestions for a number of improvements aimed at reducing the time and cost requirements, including the following:

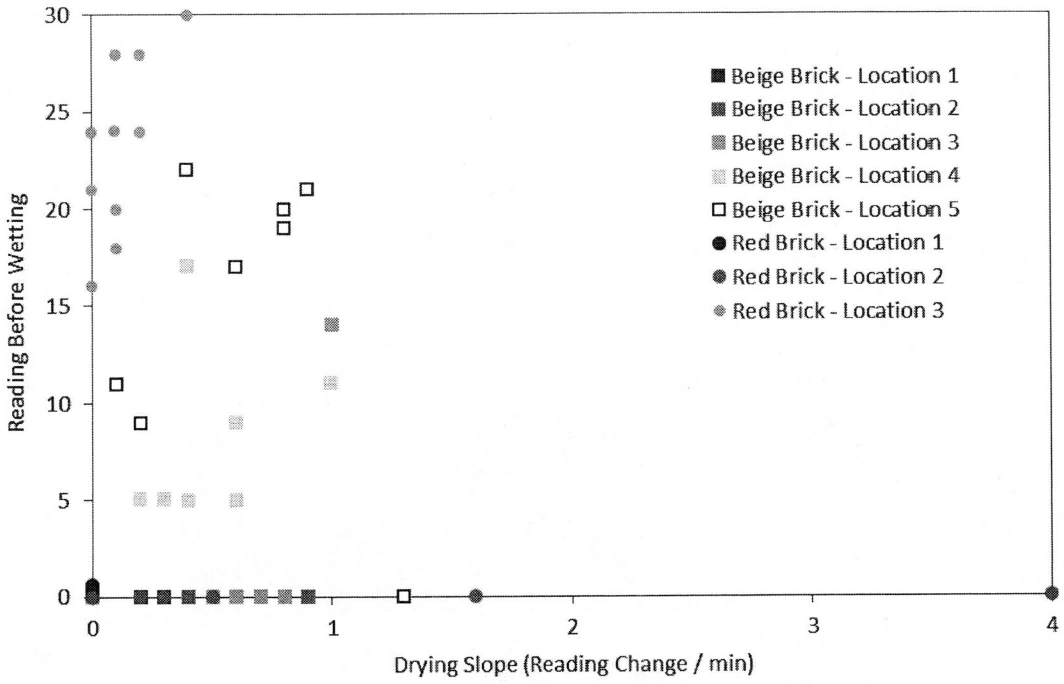

Figure 4 *Initial moisture meter readings and drying slope measured on project wall locations.*

1. direct measurement of the slice surfaces in lieu of set pins,
2. freeze/thaw cycle routine optimized to minimize runtimes, and
3. simplification of plastic sealing method.

A concern with the methodology is that for some materials it was found that length measurements were not repeatable. This issue was explored through a repeatability study of the slice length measurement and is reported in this section. Furthermore, high strain variance has been found for highly saturated samples in some cases after freeze/thaw cycling. A further study is also reported investigating this issue and evaluating the within-lab reproducibility of critical saturation measurements.

Sample Slice Length Measurements Repeatability

Limestone sample slices were cut from a single block on a table wet saw with guiding brackets to achieve square cuts. Limestone was used because it has a hard cut surface to limit mechanical wear from repeat testing. Three operators used a micrometer mounted within a measurement jig (see Ueno et al. 2013b) within the same lab to measure the length of limestone slices. The results of eight consecutive measurements of individual samples are shown in Figure 6 for six sample slices. Note that these measurements have all been done prior to freeze/thaw cycling in the lab.

A slight decline in the lengths of some samples can be seen and was suspected to be due to mechanical wear. The rate of wear, *m*, was calculated using a method of least-squares linear curve fit to each consecutive length measurement for individual samples with resultant values given in Figure 6.

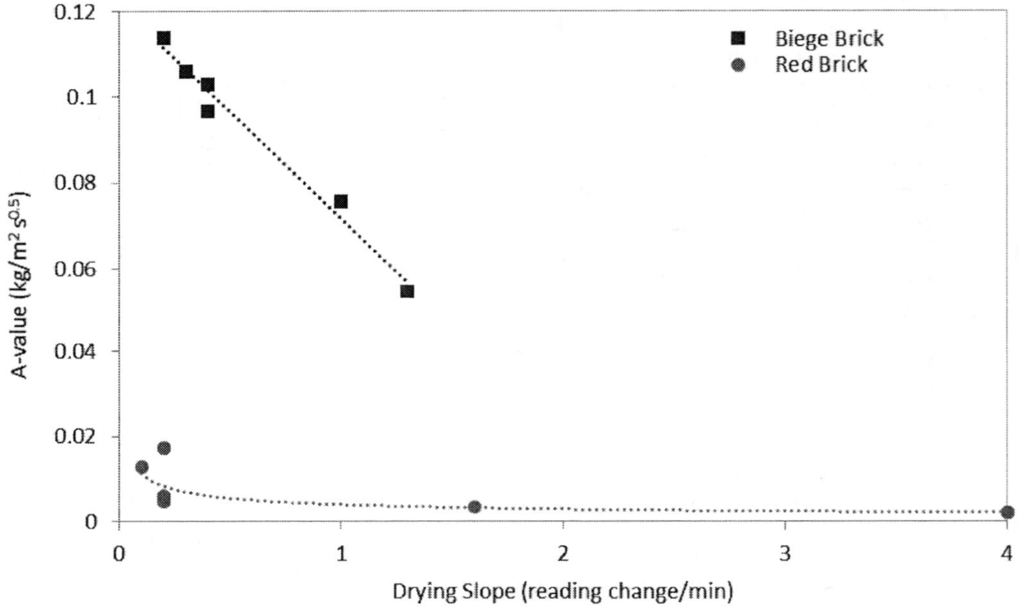

Figure 5 *Comparison of water uptake and drying measurement results for selected bricks.*

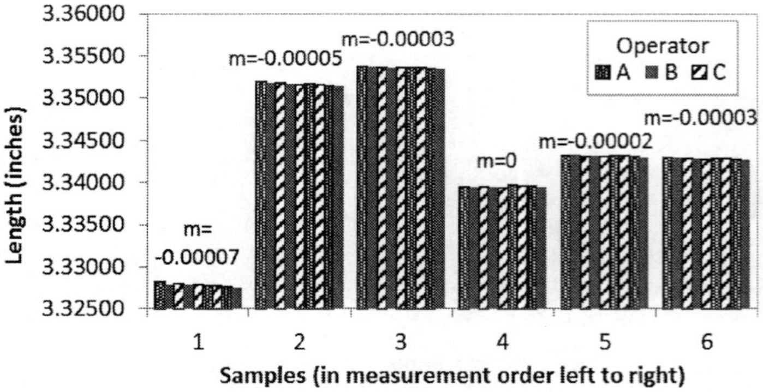

Figure 6 *Mean repeated length measurements of six sample limestone slices.*

To estimate the repeatability of length measurements, it was assumed that mechanical wear could be resolved in the future by repeating sample measurements until a consistent measurement was found. Equivalent strain values were adjusted for this rate of wear and calculated as follows:

$$\Delta'_{eq} = \frac{l_n - m(n-1) - l_1}{l_1} \quad (3)$$

The results of this calculation are shown in Figure 7 and suggest a repeatability of ±100 equivalent μstrain.

To account for potential mechanical wear effects, the testing methodology was modified as follows:

1. initial micrometer length measurements are repeated until consistent within ± 0.00005 in. (0.001 mm) to remove or set loose and/or soft surface materials, and
2. post freeze/thaw cycling slice length is only measured once to limit mechanical wear on the material surfaces by the micrometer.

The precision of critical saturation measurements including these modification were evaluated through a further within-lab critical saturation measurement reproducibility study presented in the next subsection.

Within-Lab Critical Saturation Measurement Reproducibility

Two operators measured critical saturation for slices from the same clay brick. For one operator, 12 slices were saturated to 80%, 90%, and 100%, for a total of 36 slices. The other operator used four slices at the same saturation levels. Testing of four slices at a number of saturation levels is more feasible for most projects, as collection of 12 slices would be expensive and may not be physically possible for some brick types. However, a larger number of samples was included in this study to explore variations in measurements.

Dilation measurements are given in Figure 8 for the slices after 12 cycles. Significant strain and variance were found for the 90% and 100% saturated samples. The 80% saturated samples show strain just above the length measurement repeatability and less variance. The graph includes an approximate plot of the Student T Test distribution (similar to Gaussian but adjusted for small sample lot sizes) based on the data set's mean strain, $\bar{\Delta}$, standard deviation, Δ, and number of samples. This variation in strain is thought to be due to lack of material homogeneity. The 95th percentile values, Δ, were calculated in an attempt to capture the materials within the brick that were sensitive to freeze/thaw degradation. The samples were subjected to a further 12 freeze/thaw cycles (24 in total). The results shown in Figure 9 show increased strain and variance.

The mean and 95th percentile strains are plotted relative to water saturation in Figure 10. Applying a method of least-squares fit resulted in a similar zero-strain intercept as for linear curve fits. The resulting critical saturation values are summarized in Table 1 within the "All Data" row. The same critical saturation value of 77.4% was found using the mean data while slightly lower critical values of 76.3% and 77.2% were found when the 95th percentile was used.

The results were further split between the two operators with results given in Table 1. Using the 77.4% critical saturation mean value measured from the whole dataset as a reference, the largest variation was 0.9% saturation. For the 95th percentile values and same reference, the largest variation was 2.6%.

CONCLUSIONS

This study reports developments in field sampling and critical saturation testing of in-service masonry. Two general approaches to sampling are explored. A method for prescreening samples from bulk sampling by water uptake testing is demonstrated. Within a case study of this approach, a number of issues are raised revolving around the low likelihood that actual random sampling will be applied in such sampling. A new alternative field testing approach for sample selection is presented. This nondestructive field drying rate measurement technique shows good correlation between measurements and water uptake. The technique used has an opportunity for further optimization, which should be explored in future stud-

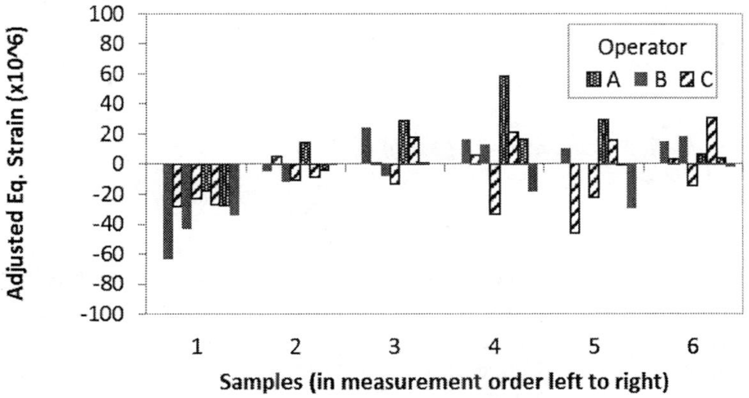

Figure 7 *Adjusted equivalent strain relative to first Operator A measurement.*

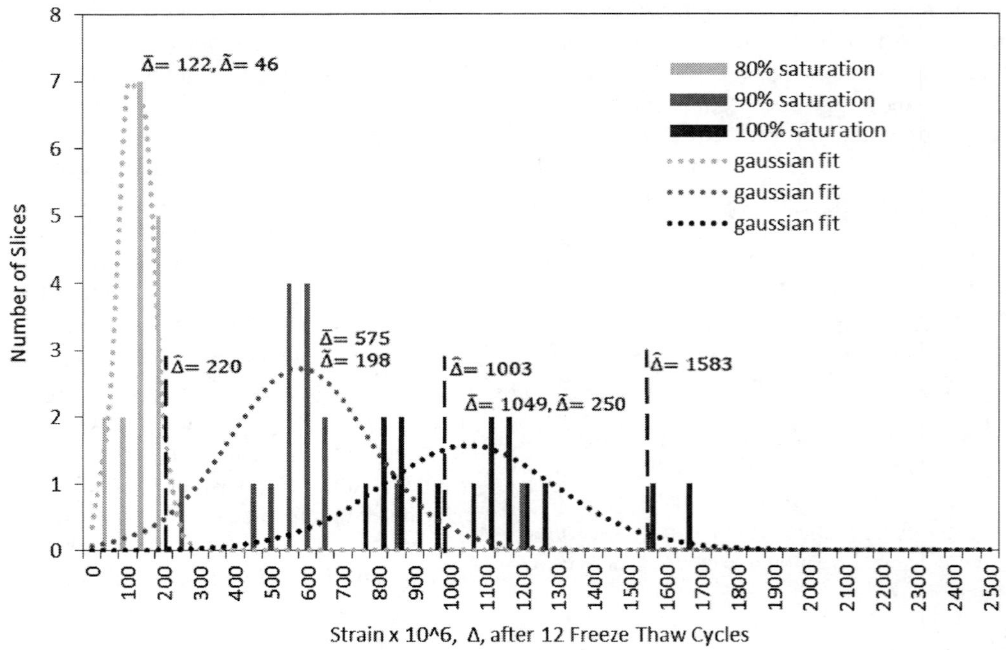

Figure 8 *Histogram of slice dilation measurements after 12 cycles.*

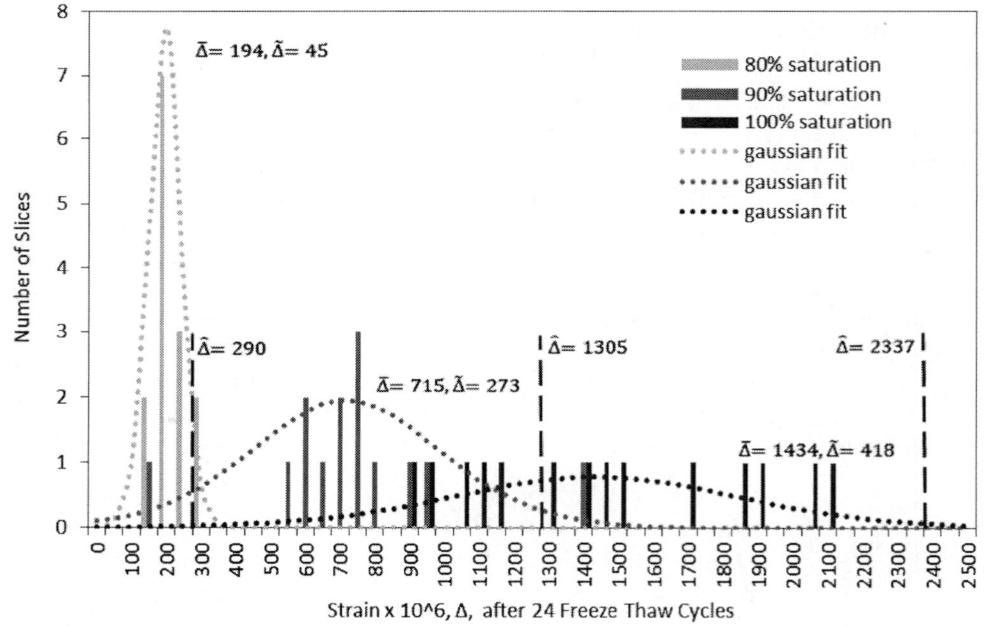

Figure 9 *Histogram of slice dilation measurements after 24 cycles.*

ies. It also presents an opportunity to provide initial moisture measurements, which may prove useful in sample selection and hygrothermal model calibration.

Significant variation in strain was found for freeze/thaw cycle slices at similar moisture saturation levels above critical saturation. It is thought that this variation is due to lack of material homogeneity within the brick resulting in variations between sample slices. The results suggest that use of a 95th percentile strain value for each saturation level will result in a slightly lower (more conservative) estimate of critical saturation.

It appears from the study that use of the mean strain for as little as four samples at each saturation level will allow repeatable determination of critical saturation. Overall, the study showed a high level of critical saturation measurement repro-

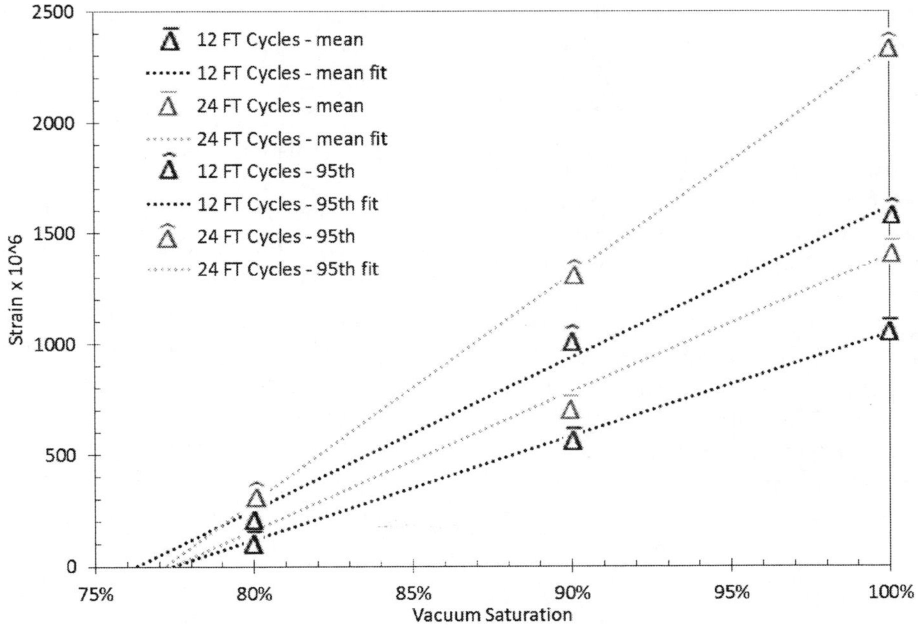

Figure 10 *Mean and 95th percentile strain after 12 and 24 freeze/thaw cycles.*

Table 1. Critical Freeze/Thaw Saturation Measurements

Operator	12 Freeze/Thaw Cycles			24 Freeze Thaw Cycles		
	# of Saturation Samples	Mean	95th Percentile	# of Saturation Samples	Mean	95th Percentile
All Data	16	77.4%	76.3%	14	77.4%	77.2%
A	4	76.5%	74.8%	2	77.0%	76.3%
B	12	77.4%	76.5%	12	77.6%	77.3%

ducibility using the new methodology described in the paper. This study should be repeated for a number of different masonry materials to allow greater confidence in the methodology.

ACKNOWLEDGMENTS

We would like to thank Kohta Ueno from Building Science Corporation for his involvement with some of the case studies. His continued interest in providing greater insights and service to clients by bringing scientific research into projects has greatly supported our efforts over the years.

NOMENCLATURE

m = rate of mechanical wear, in./measurement

M = moisture meter measurement value, dimensionless

$\dot{M}$ = drying rate, dimensionless/min.

n = total number of consecutive measurements

Δ = strain, in./in.

Δ' = adjusted strain, in./in.

$\bar{\Delta}$ = mean of strain measurements, in./in.

$\tilde{\Delta}$ = standard deviation of strain measurements, in./in.

$\hat{\Delta}$ = 95th percentile of strain measurements using Student T distribution, in./in.

Subscripts

eq = equivalent

ft = post freeze/thaw cycling

REFERENCES

Driscoll, M.E., and R.E.Gates. 1993. A comparative review of various test methods for evaluating the water penetration resistance of concrete masonry wall units. *Masonry: Design and Construction, Problems and Repair, ASTM STP 1180*, John M, Melander and Lynn R. Lauersdorf, Eds., ASTM, Philadelphia. 228–39.

Fagerlund, G. 1977. The critical degree of saturation method of assessing the freeze/thaw resistance of concrete. *Journal of Structures and Materials* 10(58):217–29.

Mensinga, P., J. Straube, and C. Schumacher. 2010. Assessing the freeze-thaw resistance of clay brick for interior insulation retrofit projects. *Proceedings of Performance of the Exterior Envelopes of Whole Buildings XI*, Atlanta, GA.

RILEM. 1980. Recommended tests to measure the deterioration of stone and to assess the effectiveness of treatment methods/Tentative Recommendations, RILEM Commission 25-PEM. *Materiaux et Constructions* 13(75):175–253.

Straube, J., and C. Schumacher. 2006. *Assessing the durability impact of energy efficient enclosure upgrades using hygrothermal modeling*. Research report 0605. Somerville, MA: Building Science Corporation.

Ueno, K., P. Kerrigan, H. Wytrykowska, and R. Van Straaten. 2013a. Retrofit of a mulitfamily mass masonry building in New England. Washington, DC: Building America Building Technologies Program, Office of Energy Efficiency and Renewable Energy U.S. Department of Energy.

Ueno, K., R. Van Straaten, and C. Schumacher. 2013b. Interior insulation of mass masonry walls: Joist monitoring, material test optimization, salt effects. Washington, DC: Building America Building Technologies Program, Office of Energy Efficiency and Renewable Energy U.S. Department of Energy.

Van Straaten, R. 2014. Improving access to the frost dilatometry methodology for assessing brick masonry freeze thaw degradation risk. *Proceedings, Canadian Conference on Building Science and Technology*, Toronto, ON.

Wilson, M., M. Carter, W. Hoff. 1999. British Standard and RILEM water absorption tests: A critical evaluation. *Materials and Structures* 32:571–78.

Thermal and Moisture Properties of Extruded Polystyrene from Inverted Roofs: A Pilot Study

Leslie B.B. Peer, PhD, PEng, RRC, FEC
Member ASHRAE

Fitsum Tariku, PhD
Member ASHRAE

Wendy Ying Simpson, MASc

ABSTRACT

Closed-cell polyethylene insulation has been installed in inverted roof systems for more than 40 years. In some of these applications the insulation is exposed to humidity and gradually absorbs moisture while in service. This paper presents data about wet and dry densities and aged U-factors of insulation samples retrieved from roofs in service in Western Canada for between 11 and 35 years. The data indicate that closed-cell polyethylene insulation exposed to wet service conditions absorbs sufficient amounts of moisture to significantly degrade the expected insulation U-factor and also increase roof loading. A wider study is required to isolate field variables such as assembly type, exposure history, insulation type and frost action to generalize findings.

INTRODUCTION

Inverted roof systems, also called protected membrane roof (PMR) systems, protect roofing and waterproofing membranes from exposure to weather stress such as thermal stress, ultraviolet radiation, and mechanical damage. (RCABC 2016). PMR systems are commonly used for low-slope roofs, plaza roofs, and green roof assemblies, placing the membrane on the structural deck, then covering it with insulation and ballast consisting of gravel, concrete, pavers, or soil. The materials above the membrane help to maintain a temperature biased toward the interior temperature to increase the membrane's life expectancy (Dechow and Epstein 1978; Basham et. al. 1999). Basham stated that closed-cell insulation is suitable in a wet environment above the roof membrane and that such roofs can provide energy conservation by providing a thermal barrier and insulating the thermal mass of the structure. Extruded closed-cell polyurethane insulation (XPS) has been installed in inverted roof systems for around 45 years.

In some applications, XPS insulation is exposed to humidity and gradually absorbs moisture while in service. Previous research includes laboratory and field performance measurements of insulation behavior in exposure to exterior environments (Dechow and Epstein1978; Sandberg 1986; Tobiasson et al. 1991; Kunzel and Kiesl 1997; Crandell. 2010). Through calculation results with measurement support, Kunzel and Kiesl (1997) found that moisture performance of XPS depends on mean relative humidity (RH) directly above the insulation layer. Exposure of XPS should be below 90% in summer to prevent cumulative moisture absorption year on year. At higher RH, moisture accumulation will be faster. Tobiasson et al. wetted insulation in the laboratory to study a thermal resistance ratio relating wet material to dry conditions, and found that as moisture accumulates in insulation its thermal resistance ratio decreases. Dechow and Epstein (1978) reviewed laboratory experiments and field service data up to 10 years on many insulation types and concluded that XPS had the best thermal performance due to its resistance to water absorption. They noted that XPS insulation would reduce in insulation value as it absorbed water, and that XPS should maintain at least 80% of its thermal resistance for a 10-year period. Kehrer and Christian (2012) retrieved field samples from foundation walls after 15 years to study energy-saving performance and concluded that moisture accumulation in XPS had an influence for long-term energy efficiency. They also concluded that XPS and expanded polystyrene (EPS) recovered their properties after drying, meaning no

Leslie B.B. Peer is a principal of RJC BC Restoration and Building Science Group, Vancouver, BC, Canada. **Fitsum Tariku** is the director of and **Wendy Ying Simpson** is a research analyst at the Building Science Centre of Excellence (BSCE), British Columbia Institute of Technology (BCIT), Burnaby, BC, Canada.

significant aging process occurred except moisture change. Overall the researchers concluded that elevated moisture content in XPS should be considered as a durability factor in design for performance.

Field observations of PMR assemblies have established that components of the roof below the ballast are often damp or wet in service, and that the density of insulation is raised because of water absorption. Assessment of the time of wetness, degradation of thermal resistance value, and increase in roof loading due to retained water warrant quantification, particularly because engineering designs rely upon values for thermal conductance and density of insulation. This preliminary study presents data gathered about the physical properties of insulation samples retrieved from inverted roofs in service for between 11 to 30 years. Measured properties include wet and dry densities, aged U-factors in wet states, and moisture distributions through the samples. The main objectives of the study were to assess the level of moisture accumulation in XPS insulation boards in service in inverted roof systems and to investigate the thermal resistance changes of the insulation boards after relatively long field service periods. To achieve these objectives, the following three sets of experimental tests were conducted on samples collected from the field:

1. Thermal conductivity tests of field samples: to measure the actual thermal conductivity of XPS insulation board retrieved from field service.
2. Moisture content of field samples: to determine how much water was absorbed by the XPS samples in service. The moisture content data are used to develop a relationship with the thermal conductivity measurement results.
3. Moisture content of sliced samples: to assess moisture distribution of absorbed water throughout the thickness of the samples, and if possible identify the direction of wetting, and if possible predict the moisture content dependent diffusion coefficient.

TEST SAMPLES

Test samples were retrieved from a sampling of roof renovation projects in western Canada during Summer 2015 (Figure 1). Scrap insulation was cut from test sites during demolition work at a variety of inverted roof assemblies. Table 1 describes the service conditions for each sample set.

The samples had similar construction conditions:

- They were installed over membranes over heated enclosures under concrete toppings.
- The assemblies did not have drainage mats above or beneath the insulation to provide a “diffusion open” environment.

Figure 1 *Samples retrieved from a rooftop parking area.*

Table 1. Descriptions of Field Insulation Samples Retrieved for Testing

Location, Date Constructed	Years in Service	Description of Location	Assembly Details	Roof Area Sampled, m^2
Calgary, 2003	11	Exterior pedestrian plaza	50 mm pavers 25 mm sand 38 mm insulation (extruded polystyrene [XPS]) Single-ply asphalt membrane Concrete 0% slope	300
Nanaimo, 1993	22	Apartment decks, 4th to 7th floor	70 mm concrete brick paver Filter fabric 38 mm insulation (pink) Liquid applied urethane membrane Concrete 2% slope	20
New Westminster, 1988	27	Rooftop parking at mall	100 mm cast in place (CIP) concrete 38 mm insulation (extruded polystyrene [XPS] HI) 1.6 mm bituthene membrane Concrete 1.5% slope	50
Burnaby, 1985	30	5th floor covered parking garage	90mm CIP concrete 38 mm insulation Ethylene propylene diene monomer (EPDM) membrane Concrete 2% slope	25

- The climate conditions likely kept the assemblies at high RH year round.

The Roof Area Sampled column indicates the representative area from which the samples were cut. Samples taken from larger areas were further apart when sampled, and were exposed to potentially more varied local conditions in service. Local conditions might include the amount of sun or shade per day, height of sample above drain elevation, cracks in concrete topping (Figure 1), and other similar variables.

EXPERIMENTAL SETUP

Sample Preparation

The XPS board samples were separately sealed in plastic bags and delivered to the Property Measurement Laboratory immediately after being removed from the roof systems. Test samples for thermal conductivity and moisture content measurements were prepared from each board as shown in Figure 3. A total of 52 samples for from 13 XPS insulation boards were obtained for this study. The 300 by 300 mm thermal conductivity test samples, shown in Figure 4, were wrapped with 0.15 mm (0.006 in.) polyethylene film and sealed with tape to avoid moisture loss from the samples before and during measurements. The three moisture content measurement samples, shown in Figure 5, had dimensions of 25 mm width and 200 mm length. The fourth piece was sliced across its thickness into three layers with approximate thicknesses of 12 mm for insight into moisture distribution across the insulation thickness (Figure 6). Once the samples were cut, their initial weights and their length, width, and thickness were measured at four points in each dimension.

While the samples were being cut into the required sizes for the different tests, water appeared on the fresh-cut lines, surfaces, and knife blade in some samples. This water was likely stored in pores that were cut, and beaded on the surface. In some samples water appeared on the surface for some time after cutting, suggesting that an air/water mixture was stored in the matrix at some pressure slightly above ambient atmospheric pressure, and was moved to the surface when the matric pressure was relieved at the cut surface, as shown in

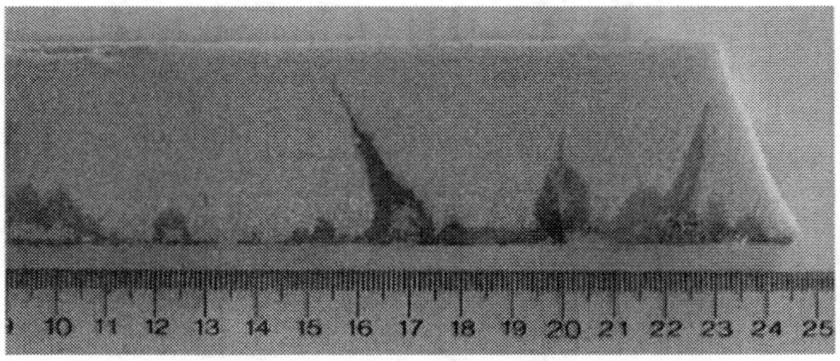

Figure 2 *Freshly cut cross-section of insulation retrieved during another study, showing strong directional absorption from surface and cracks and flaws in the foam material (Peer 2013).*

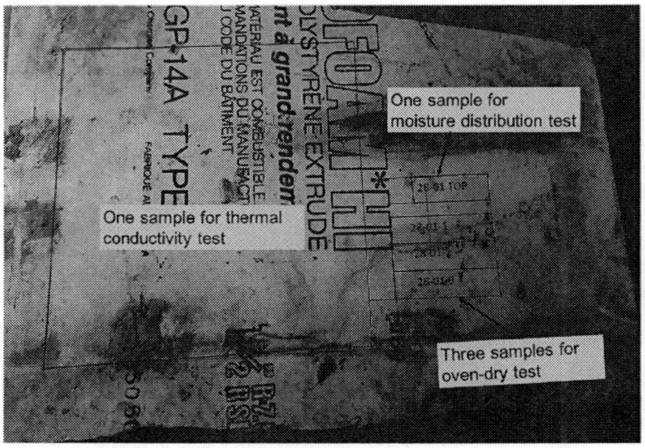

Figure 3 *Thermal conductivity and moisture content measurement sample preparation.*

Figure 4 *Thermal conductivity test samples wrapped with 6 in. thick polyethylene film and sealed with tape to avoid moisture loss.*

Figure 5 *Three moisture content measurement samples.*

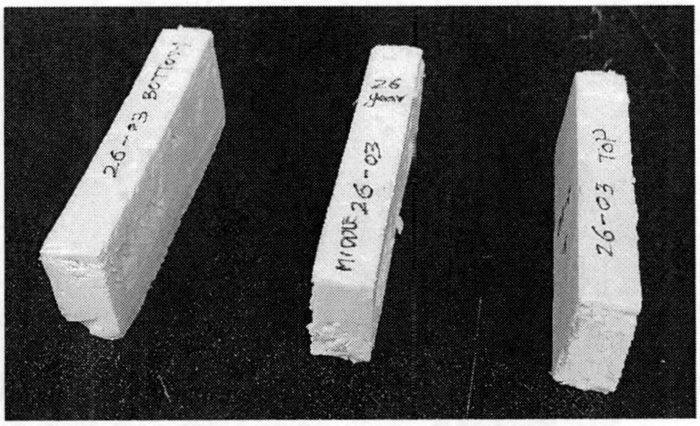

Figure 6 *Top, middle and bottom slices of an XPS sample to assess moisture distribution across the insulation thickness.*

Figure 7. The samples were not stored beneath the water table, so the pressure is thought to be due to compressed occluded air.

Moisture Content Measurement

The moisture content of the samples was determined by gravimetric means using an analytical balance, accurate to 0.0001 g, and a forced-air convection oven, shown in Figure 8.

After taking the initial weights of the samples, they were placed in the oven and left to dry at 70°C. Every 24 hours the samples were taken out of the oven, weighed, and put back into the oven for further drying until the samples were declared dry. The samples were assumed to be at dry state when three successive weight measurements were within 0.1% in conformance with the ASTM Standard 1498 (ASTM 2010). The dry weights of three samples cut out from each board were averaged to obtain the average dry mass of each piece of insulation board.

Moisture content is reported in this paper mostly as percent by volume of the sample porosity. Even though the moisture content was assessed by mass, this convention is found in other papers on this topic, and the authors converted the measurements to percent by volume during discussions for several reasons. Firstly, the dry densities of the insulations were different, so the mass percent gain over initial mass would have hindered comparisons. Secondly, the high measured moisture content relative to the low dry densities result in percent mass gains in the thousands of percentages, which are difficult to relate to a physical reality. The percent by volume is a useful indicator of the degree of partial saturation of the sample, in effect the amount of porosity in the samples that is occupied by water.

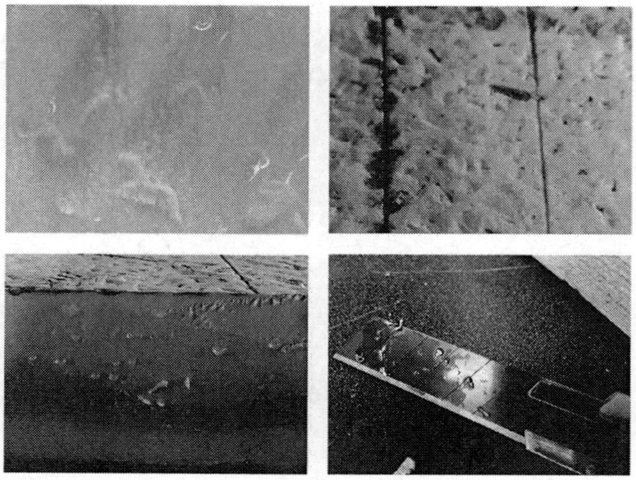

Figure 7 *Water appeared on the surface of cutting line and the cutter, water drops on the cut edges of samples.*

Thermal Conductivity

The thermal conductivity of the 300 mm long × 300 mm wide test samples was measured using a bench top heat flowmeter. According to the manufacturer's specification, the heat flowmeter has an accuracy of ±1% and repeatability ±0.3%. The measurements were carried out in accordance with ASTM C518 (ASTM 2012), at a mean temperature of 24°C and temperature difference of 20°C. The test setup temperatures are listed in Table 2.

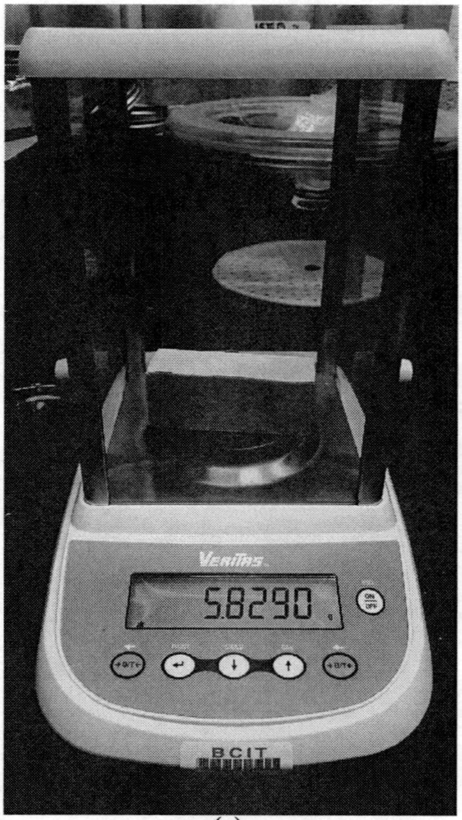

(a)

(b)

Figure 8 *(a) Scale for measuring masses of samples and (b) convection oven with drying samples.*

Prior to each test, the heat flowmeter was calibrated at the test temperature conditions using a standard reference material (SRM 1450C432). The thermal conductivity of the samples were measured while they were wrapped with polyethylene sheets to restrict moisture from leaving the samples. Weight measurements before and after the tests confirmed that there were no loss of moisture. A specialized software was used to set up the test parameters, measure the thickness and the thermal conductivity of the samples and record the intermediate and final test results in a computer. The thermal conductivity test ran until the software detected steady-state heat flow through the sample in successive measurements. The test results of at least three repeated measurements were used to determine the average thermal conductivity value of each sample.

RESULTS

Moisture Content

In the same roof of a building, moisture content in the insulation boards appear to differ significantly from location to location. This is apparently caused by varying microclimate conditions to which they are exposed, because the time of exposure is the same. For example, one of the samples from the 11-year-old roof weighed 141.36 g whereas another from the same roof measured 2496.27 g, as shown in Figure 9. This represents a weight ratio approximately 17 times higher than the lighter sample.

The moisture content of the, 11-, 22-, 27-, and 30-year-old roof samples, along with their corresponding oven dry weights, are presented in Table 3. The moisture content values in Table 3 represent an average of three or four small samples (see Figure 3). The samples' densities are determined from their oven dry mass and volume measurements. Samples from the same roof have similar dry densities as would be expected from manufacturing consistency. The dry densities of the XPS board in the 11-year-old and 22-year-old assemblies are between 26 to 33 kg/m^3, whereas the XPS boards on the 27- and 30-year-old roofs have density between 41 to 43 kg/m^3. Of the 15 XPS samples, which were taken from roofs of different years of service, geographic location, and roof assembly type, the samples from the 11-year-old roof represent both the maximum and the minimum moisture content by volume, 0.03% and 74%, respectively. About half of the samples have moisture content between 30% to 36%. The roof with the relatively long service life, 30 years, had roof areas with both dry and wet XPS board, 0.4% and 32%–36% by volume.

Because of this data from the 11- and 30-year-old roofs, where both the wettest and driest XPS samples were obtained in the same roof, it is suggested that local exposure conditions are more important that service years when it comes to moisture accumulation in the XPS board.

The increase in mass of insulation containing stored water is of significant concern to the structural design of roofs. The increase in weight measured in this study indicates the potential for loading increases of 0.5 to 1 kPa (10 to 20 psf).

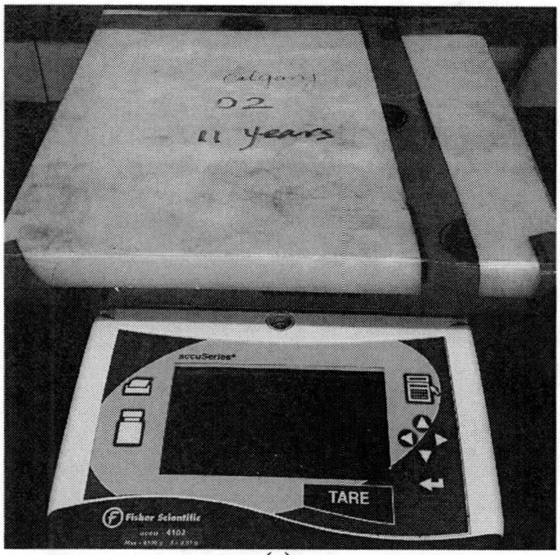

(a)

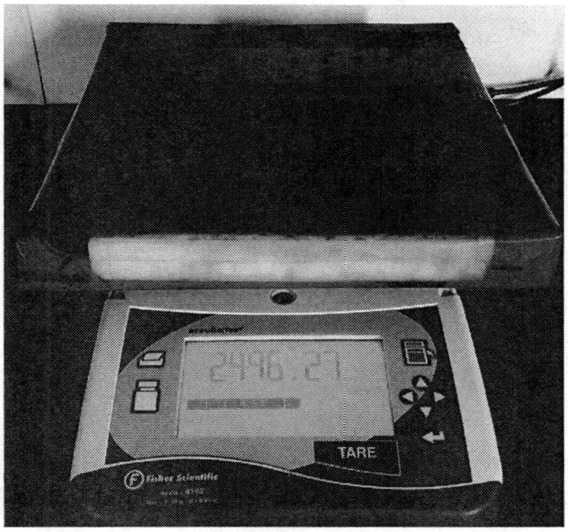

(b)

Figure 9 *Weights of two samples from the same 11-year-old roof plaza.*

Table 2. Testing Setup of Temperature and Accuracy

Mean Temperature	24.0°C	75.2°F
Temperature Difference	20.0°C	68.0°F
Upper Temperature	34.0°C	93.2°F
Lower Temperature	14.0°C	57.2°F

Overall, the moisture content in the XPS samples are significantly higher than considered normal, and their thermal values are significantly degraded due to the absorbed moisture. The results are also significantly higher than reported in previous papers. These results indicate that exposure condi-

Table 3. Moisture Contents of XPS Samples from 11-, 22-, 27-, and 30-Year-Old Roofs

Sample Label	Dimensions, mm			Volume, mm^3	Weights, g		Dry Density, kg/m^3	Moisture Content by Volume, % of Vol.
	Length	Width	Thickness		Wet	Oven Dry		
Samples from 11 year old roof								
11-01	100	32	38	122 770	94.44	3.13	26	74
11-02	99	29	36	103 852	3.32	3.29	32	0.03
11-03	102	30	35	105 839	35.03	3.08	29	30
11-04	100	29	37	108 005	41.18	2.90	27	35
Samples from 22 year old roof								
22-01	103	38	38	148 112	57.91	4.73	32	36
22-02	101	38	37	142 660	54.75	4.65	33	35
22-03	102	38	37	143 808	53.70	4.76	33	34
Samples from 27 year old roof								
27-01	102	38	36	140 805	39.81	5.81	41	24
27-02	102	37	37	139 810	40.81	5.73	41	25
27-03	101	38	37	140 392	16.38	5.80	41	8
Samples from 30 year old roof								
30-01	102	101	36	369 051	136.96	15.49	42	33
30-02	103	104	37	394 432	156.89	16.49	42	36
30-03	101	100	37	377 024	136.86	16.08	43	32
30-04	101	101	36	366 323	162.06	15.54	42	0.4
30-05	101	101	37	383 625	177.18	16.61	43	0.4

tions and absorption properties need to be more carefully studied to predict water absorption rates into closed-cell polystyrene insulation, so that assemblies can have reliable thermal insulation values over their expected service life. A wider study that includes samples taken from assemblies with measured *in situ* moisture exposure conditions over the years of service, and in particular Dow "diffusion open" conditions (Dow Building Solutions n.d.) is required to generalize findings about moisture absorption rates.

Moisture Distribution in the Samples

Observation of some wet sample edges revealed light to dark color water marks across the thickness, similar those shown in Figure 2. To better understand the moisture distributions across the thickness of the samples, from the pavement/ballast side to the roof membrane contact side, each sample was sliced into 12 mm thick along its length into three layers—a top, middle, and bottom layer. The weight of the sliced samples and their corresponding dry weight after oven drying were measured to calculate the moisture content in the three layers and characterize the moisture distribution across the sample XPS board. Figure 10 shows the moisture content distributions of 10 wet samples from the 11-, 22-, 27-, and 30-year-old roofs. Moisture distribution in the samples is not uniform. The steepest moisture content gradient was observed in the 11-year-old roof (Sample-01, Figure 10 a), with 82% moisture content in the middle layer and 24% moisture content by volume at the bottom layer. The second steepest gradient was in Sample-04 from the same roof, but the gradient existed between the top and middle slices. In the 22- and 27-year-old roof samples (Figure 10 b and c), the moisture content difference between the layers with the highest and the lowest moisture contents are 16% and 22%, respectively. In the 30-year-old samples (Figure 10d), the moisture content seems to decrease gradually with depth, with the top layer at about 50% and the bottom layer at about 30% moisture content by volume.

In general, the moisture content of the top layers is high, and in some cases slightly higher moisture content was measured in the middle layers. The relatively low moisture content in the bottom layers when compared to the middle and top layers in all except the 27-year-old roof (Figure 10c) suggest that the XPS boards are wetted from the top. The insulations were all situated directly on the roof membranes and may have been effectively sealed to the membrane by the ballast.

The data do not allow for calculations of moisture diffusion rates as a function of moisture content because of the uneven moisture content gradients in the samples. The moisture content distributions seem to indicate that the samples went through wetting and drying cycles as discussed by Kunzel and Kiesl (1997), and as might be expected in field service with weather-related wetting and drying cycles. The samples collected between May and August were subjected to some drying weather before sampling.

Thermal Conductivity

The physical and thermal characteristics of the different types of XPS and EPS insulation materials are defined in ASTM C578, *Standard Specification for Rigid, Cellular Polystyrene Thermal Insulation* (ASTM 2015). According to the standard, and on the basis of the measured samples dry densities, the samples from the 11- and 22-year-old roofs belong to Type VI Class (requiring a minimum density of 25 kg/m^3), whereas the other two roofs' samples are in Type VII Class (requiring a minimum density of 35 kg/m^3). Although these two XPS types have different densities, their nominal dry thermal conductivity values are the same, 0.0288 W/(m·K). In this section, this value is used as a reference to which the measured thermal conductivity values of the samples are compared to get insight into the degree of deviation from the dry thermal

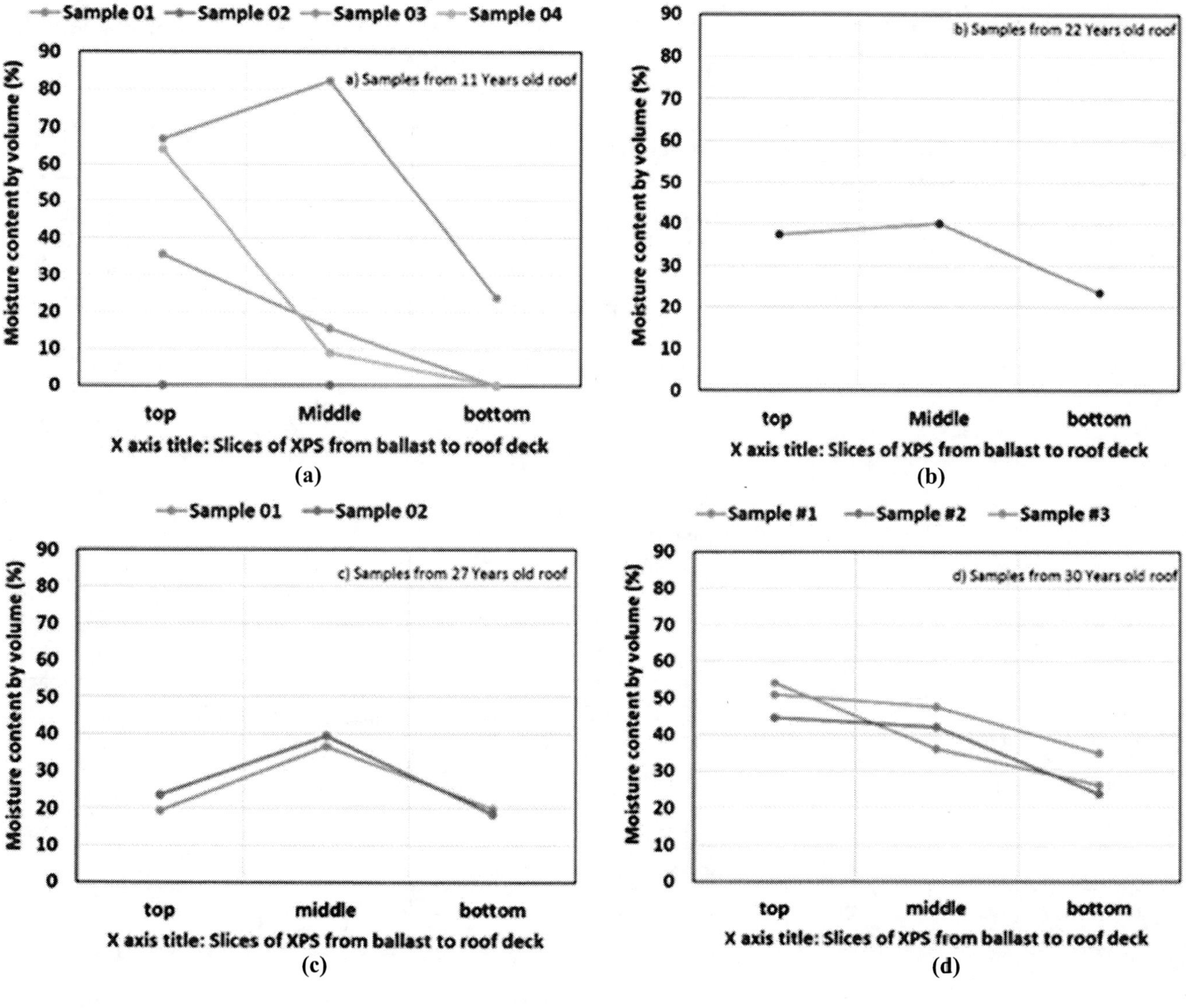

Figure 10 *Moisture distribution within the XPS samples: samples from (a) 11-, (b) 22-, (c) 27-, and (d) 30-year-old roofs shown.*

conductivity value. Of course it would be preferable to have original samples for control comparison, but these are not available for this long-term field study. The thermal conductivity and thermal resistance measurement results of the 13 XPS samples collected from the four buildings are presented in Table 4.

Overall in the study, the thermal conductivity values of field samples vary from 0.0295 to 0.1752 W/(m·K), which corresponds to between 1.02 and 6.07 times the ASTM reference thermal conductivity value. While two samples retained over 80% of their expected thermal resistance value, about two-third of the samples' thermal resistance values are less than half the value specified in the ASTM standard. Similar to the moisture content measurement results, the highest and the lowest thermal resistance retention, 94% and 16%, respectively, are obtained in the 11 year old roof.

Figure 11 graphs the thermal resistance ratio and the corresponding volumetric moisture content of each sample. The thermal resistance ratio (TRR in Equations 1 and 2) indicates the percentage decrease of thermal resistance of a sample from the reference (dry) thermal resistance value (as described by Tobiasson and Ricard [1979]). In general, and as has been shown in the other work referenced, as the moisture content increases, the thermal resistance decreases. The 27- and 30-year-old sample data lie close to a curve based on Equations 1 and 2, which were developed by Tobiasson et al. (1991) through series of combined vapor diffusion based wetting and thermal resistance measurements taken from XPS samples in a controlled laboratory setting. In the equations *MC* is moisture content by mass in percent.

$$TRR = 100 - 0.12 \times MC \text{ for } MC \leq 84\% \quad (1)$$

$$TRR = 137.37e^{-0.0008} \times MC - 39.47 \text{ for } MC > 84\% \quad (2)$$

The trend curve in Figure 11 shows that the Type VI XPS boards in the 11- and 22-year roofs may have relatively better thermal performance at high moisture content. According to Dechow and Epstein (1978) and Tobiasson et al. (1991), only two samples can be deemed to have retained their insulating characteristics. The other samples' thermal resistance reduction values are below the acceptable threshold, which they suggested to be 80%. Overall thermal resistance may also be related to moisture distribution in the samples.

Table 4. Thermal Conductivity Measurement Results of Samples from 11-, 22-, 27-, and 30-Year-Old Roofs

Sample Label	Thickness, mm	Dry Density, kg/m^3	Thermal Conductivity, W/(m·K)	Thermal Resistance, m^2·K/W	Thermal Conductivity Ratio, K_sample/K_ref	Thermal Resistance Ratio, R_ref/R_sample
			Samples from 11-year-old roof			
11-01	39	26	0.1752	0.219	6.07	0.16
11-02	37	32	0.0295	1.220	1.02	0.94
11-03	37	29	0.0560	0.741	1.94	0.57
11-04	38	27	0.0530	0.732	1.84	0.55
			Samples from 22-year-old roof			
22-01	37	32	0.0702	0.528	2.43	0.41
22-02	37	33	0.0669	0.447	2.32	0.35
22-03	34	33	0.0633	0.536	2.19	0.45
			Samples from 27-year-old roof			
27-01	38	41	0.0556	0.680	1.93	0.51
27-02	38	41	0.0629	0.605	2.18	0.45
27-03	38	41	0.0323	1.177	1.12	0.89
			Samples from 30-year-old roof			
30-01	39	42	0.0864	0.446	2.99	0.33
30-02	38	42	0.0948	0.402	3.28	0.30
30-03	39	43	0.0782	0.492	2.71	0.36

Figure 12 shows the thermal conductivity and the corresponding moisture content measurements of the samples from the four roofs. The thermal conductivity of the XPS samples was found to increase with moisture content and show a similar linear trend as other researchers found in general. Based on a laboratory experiment, Dechow and Epstein (1978) provided the following empirical equation for the relationship between thermal conductivity and moisture contents of XPS board:

$$K_{wet} = 0.0288 + 0.00115 * V \tag{3}$$

where

K_{wet} = thermal conductivity of wet XPS board, W/m·K

V = moisture content,% by volume

Based on Equation 3, sets of thermal conductivity values are calculated for different moisture contents and superimposed on the measured data in Figure 12 for comparison

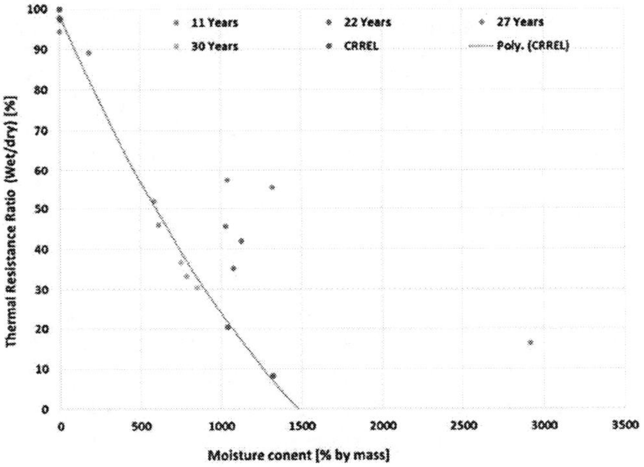

Figure 11 *Thermal resistance ratio and moisture content (by mass) relationship for the XPS samples.*

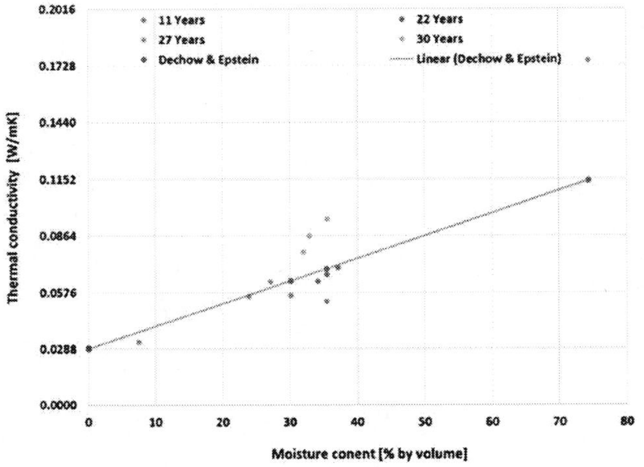

Figure 12 *Thermal conductivity of XPS samples as a function of moisture contents.*

purposes. As shown in Figure 12, the Dechow and Epstein trend curve passes through the measured data at low and medium moisture contents, but underestimates the conductivity in the high moisture content sample by about 35%. The equation thus appears to be applicable beyond experiments with controlled laboratory test conditions, but deserves review at high moisture contents. The sample size in this experimental series is not sufficiently large to provide conclusive statements about the behaviors observed, but does show some conformance with trends established in previous work, and does warn about potential service issues with permeable insulation materials.

CONCLUSIONS

This work relates field observations made with respect to absorption of water by closed cell insulation, and through some testing establishes the need for further study of PMR roof assemblies to allow reliable design in the future. Installation of vapor relief layers recommended by Kunzel and Kiesl (1997) and Dow (Dow Building Solutions n.d.) appear to be important and their effectiveness needs to be verified. Some specific conclusions from this study are:

- Insulation samples retrieved from all four roofs in the study showed measurable increase in density caused by moisture absorption.
- Density of wet roof insulation increased to levels where it is a concern for structural design of roofs.
- Thermal values of the wet insulation are significantly degraded due to levels of moisture absorbed in service.
- The relatively low moisture content in the bottom layers when compared to the middle and top layers of the insulation samples suggest that the XPS boards are wetted from the top.
- The moisture contents differed from location to location on the same roofs, indicating that the local exposure conditions are important as well as time in service when it comes to moisture accumulation.
- The sample with highest moisture content and shortest exposure time was from a Calgary roof, possibly implicating freeze-thaw as an mechanism in active wetting of insulation
- The moisture distributions within the samples seem to indicate that the more active surface of the insulation is the top. Further sampling and analysis is required to confirm this indication. Similarly, the effect of capillary action and drainage under the insulation should be explored.
- The type of ballast, with different moisture storage and access and drying characteristics, likely has an effect on RH in the assemblies, but was not quantified in this study.

The limited scope of this study informs that the effect of exposure conditions need to be more thoroughly studied in order to predict water absorption rates into closed cell polystyrene insulation, so that assemblies can have reliable thermal insulation values over their expected service life. A wider study is required to isolate field variables such as exposure history, insulation type, frost action, and assembly variations such as drainage layers that provide RH relief, in order to generalize findings. Hygrothermal design of PMR assemblies seems to be required in order to define assemblies in which XPS insulation can survive in wet climates. At a minimum and for the time being, it is recommended that designers follow Dow (Dow Building Solutions n.d.) recommendations with respect to inclusion of a vapor relief medium in the assembly to improve service conditions as much as possible. In the extreme, plastic closed-cell insulation should not be used where long-term moisture absorption would defeat engineering designs.

REFERENCES

ASTM. 2010. Standard C1498. *Test method for hygroscopic sorption isotherms of building materials.* West Conshohocken, PA: ASTM International.

ASTM. 2012. Standard C518. *Standard test method for steady state thermal transmission properties by means of the heat flow meter apparatus.* West Conshohocken, PA: ASTM International.

ASTM. 2015. Standard C578-15b. *Standard specification for rigid, cellular polystyrene thermal insulation.* West Conshohocken, PA: ASTM International.

Basham, D.L. et. al. 1999. *Commentary on roofing systems.* Technical instructions. Washington, D.C.: U.S. Army Corps of Engineers, Engineering Division, Directorat of Military Programs.

Crandell, J. 2010. Below-ground performance of rigid polystyrene foam insulation: Review of effective thermal resistivity values used in ASCE Standard 32-01—design and construction of frost-protected shallow foundations. *Journal of Cold Regions Engineering* 10.1061/(ASCE)CR.1943-5495.0000012, 35-53.

Dechow, F.J. and K.A. Epstein. 1978. Laboratory and field investigations of moisture absorption and its effect on thermal performance of various insulations. *Thermal Transmission Measurements of Insulations,* Special Technical Publication 600, American Society for Testing and Materials 234–260.

Dow Building Solutions. n.d. *Protected membrane roof installation guidelines (Canada and US).* Midland, MI: The Dow Chemical Company. http://msdssearch.dow.com/PublishedLiteratureDOWCOM/dh_093f/0901b8038093fbd5.pdf?filepath=styrofoam/pdfs/noreg/178-00554.pdf&fr omPage=GetDoc.

Kehrer. M., and J. Christian. 2012. *Measurement of exterior foundation insulation to assess durability in energy-saving performance.* ORNL/TM-2012/159. Oak Ridge, TN: Oak Ridge National Laboratory.

Kunzel H.M., and K. Kiesl. 1997. Moisture behaviour of protected membrane roofs with greenery. CIB W40 Proceedings Tokyo.

RCABC. 2016. *Roofing Practices Manual (RPM Online).* Lanley, BC: Roofing Contractors Association of British Columbia. http://rpm.rcabc.org/index.php?title=3.13_PROTECTED_MEMBRANE_AND_MODIFIED_PROTECTED_MEMBRANE_ROOFING_SYSTEMS.

Sandberg, P.I. 1986. Moisture content and thermal conductivity in soil insulation. *Journal of Thermal Insulation* 10:124–41.

Tobiasson, W. and J. Ricard. 1979. Moisture gain and its thermal consequence for common roof insulations. *Proceedings of 5th Conference on Roofing Technology.* Oak Park, IL: National Roofing Contractors Association.

Tobiasson, W., A. Greatorex, and D. Van Pelt. 1991. New wetting curves for common roof insulations. *International Symposium on Rooting Technology* 383–90.

Commercial Building Airtightness Testing—Lessons Learned from the Red River College Airtightness Testing Program

Gary Proskiw, PEng
Member ASHRAE

Rob Spewak, EIT

Kevin Knight

Cory Carson

ABSTRACT

In 2012, Red River College in Winnipeg, Canada, embarked on a program to expand the knowledge base on the airtightness characteristics of commercial buildings by conducting tests on a cross section of Manitoba's commercial building stock. This provided quantitative data on the performance of these buildings and offered critical, practical experience with airtightness testing procedures for commercial buildings.

Over three years, airtightness tests were carried out on a diverse sample of 26 commercial buildings ranging in age from 1 to over 100 years in age, floor areas of 150 to 19,788 m^2 (1615 to 212,918 ft^2), and building heights from 1 to 16 stories.

Using the normalized leakage rate (NLR_{75}) as the primary metric, the measured airtightness was found to vary by a factor of 18:1 between the tightest and loosest buildings in the sample (0.19 to 3.44 L/s·m^2 [0.037 to 0.678 ft^3/min]), with a mean of 1.70 L/s·m^2 (0.335 ft^3/min). The tightest building in the survey, a one-year-old middle school, achieved an airtightness of 0.19 L/s·m^2 (0.037 ft^3/min), better than almost all commercial building results reported in the literature. Overall, these results are particularly interesting given that current standards for new construction recommend (maximum) leakage rates of 1.27 to 2.03 L/s·m^2 (0.250 to 0.400 ft^3/min·ft^2.).

The study also revealed that then-current testing standards had serious weaknesses in their treatment of the building's intentional openings (such as HVAC inlets and outlets). Rather than employing a single sealing schedule, separate envelope and energy schedules are required depending on whether envelope integrity and durability or energy performance is of primary interest. This insight was subsequently incorporated into ASTM WK35913, New Test Method for Whole Building Enclosure Air Tightness Compliance.

PROJECT BACKGROUND, DESCRIPTION, AND OBJECTIVES

The overall goal of the Red River College program on the airtightness characteristics of Manitoba's commercial building stock was to significantly expand the knowledge base on airtightness characteristics and testing techniques for commercial buildings. Specific objectives were as follows:

- Develop practical airtightness testing protocols within the context of current standards.
- Determine baseline airtightness rates in a sample of commercial buildings in Manitoba representing a diverse range of building ages and types.
- Compare building pre- and post-retrofit airtightness rates for a subset of these buildings to determine the feasibility of cost-effective air-sealing strategies.
- Compare the baseline airtightness rates to the recommendations of the *National Building Code of Canada 2010* (NBC) (NRC 2010) and *National Energy Code of Canada for Buildings 2011* (NRC 2011) to provide rationale for potential inclusion of mandatory airtightness rates in subsequent codes.

Gary Proskiw is president of Proskiw Engineering Ltd., Winnipeg, MB, Canada. **Rob Spewak** is technology centre manager and **Kevin Knight** and **Cory Carson** are research professionals at the Building Envelope Technology Access Centre, Red River College, Winnipeg, MB, Canada.

To put the project's goals into perspective, in Manitoba alone more than 30,000 houses (approximately 10% of the total population) have had their airtightness measured over the last 20 to 30 years. In contrast, prior to this project fewer than 10 commercial Manitoba buildings had been tested and all were relatively modest in size. Yet over this same time period hundreds of millions of dollars were spent in the province repairing building envelopes of commercial buildings that had experienced premature failure due to excessive air leakage.

A key challenge in commercial building airtightness testing is the lack of experienced and technically proficient personnel capable of undertaking this complex task. Current standards for commercial building testing are relatively new and were largely derived from comparable residential testing standards. While the basic protocols used for airtightness testing of large and small buildings are similar, in practical terms, commercial buildings pose significantly greater challenges.

A literature search was conducted in 2001 wherein Proskiw Engineering Ltd. identified airtightness data for 192 large buildings worldwide (PEL 2001). Of these, 62 were in Canada, with the bulk situated in Central Canada. A wide range of airtightness rates were recorded, with some measured rates being 10 to 50 times greater than recommended values.

In the United States, Wiss, Janney, Elstner Associates, Inc., recently completed a study of large-building airtightness (WJE 2014). While this project underscored the importance of quantitative testing, the data was of limited value to the Canadian market because the focus was on new construction (less than five years old) and included only one building in a climate typical of that found in Manitoba.

AIRTIGHTNESS TEST PROTOCOLS, METRICS, AND STANDARDS

Airtightness Test Protocols

Numerous testing standards have been developed over the last 30 years to define the methodology for measuring a building's airtightness. While these are targeted at slightly different applications or use slightly different approaches, they all basically follow a similar protocol, as discussed in the following paragraphs.

Step One—Building Preparation. The building is first prepared for testing: all mechanical systems that move air into or out of the building are shut off, windows and exterior doors are closed, interior doors are opened, and intentional openings (such as air inlets and outlets as well as other mechanical penetrations) are sealed.

Step Two—Building Pressurization/Depressurization. Once the building preparation is complete, a high-capacity blower, or blowers, is installed in a suitable exterior doorway(s) and used to pressurize or depressurize the structure to a series of predefined indoor-to-outdoor pressure differentials. These are sufficiently large that they will normally overpower the naturally induced pressure differentials that the building may be experiencing due to stack effect and wind action (there should not be any mechanically induced pressure differentials since the mechanical system is disabled).

Step Three—Airflow Measurement. Once the indoor-to-outdoor pressure differential has stabilized at each test condition, the airflow rate through the blowers is measured—typically using orifice plates mounted on the blowers. This process is then repeated over a range of indoor-to-outdoor pressure differentials. Once all the flow rate and pressure differential data pairs have been recorded, along with various environmental information such as indoor and outdoor air temperatures and wind conditions, a mathematical analysis is used to develop a regression equation as shown in Equation 1:

$$Q = C\Delta P^n \tag{1}$$

where

Q = flow rate, L/s (ft^3/min)

C = flow coefficient, L/s·Pa^n (ft^3/min·Pa^n)

ΔP = indoor-to-outdoor pressure differential, Pa

n = flow exponent, dimensionless

Notice that Equation 1 is an empirical relationship that has to be measured on the actual building; it cannot be calculated or otherwise predicted using construction drawings, on-site inspections, computer modeling, or other means. While this statement applies to all buildings, it is particularly relevant when discussing larger commercial structures of the type investigated in this study.

Step Four—Calculation of Airtightness Results. The test results are combined with information on the building's size (envelope area and building volume) to generate a quantified airtightness result using one or more metrics. Some of the more significant standards commonly used for performing building airtightness tests are the following:

- CAN/CGSB-149.10-M86, *Determination of the Airtightness of Building Envelopes by the Fan Depressurization Method* (CGSB 1986)
- ASTM E779-10, *Standard Test Method for Determining Air Leakage Rate by Fan Pressurization* (ASTM 2010)
- *U.S. Army Corps of Engineers Air Leakage Test Protocol for Building Envelopes* (USACE 2012)

METRICS FOR REPORTING AIRTIGHTNESS TEST RESULTS

One of the problems encountered with discussions of air leakage is that several different metrics are commonly used for reporting building airtightness data. The most basic approach is to report the overall air leakage at some specified pressure differential (e.g., *X* L/s at a building pressure differential of 75 Pa). The primary limitation of this approach is that building size is not factored into the equation. A leakage rate of 1000 L/s in a large commercial building could indicate a very tight structure, whereas the same leakage rate in a small house would indicate a very loose building. Building size is normally introduced into airtightness metrics using either the building volume or the envelope surface area. For example, residential airtightness tests are commonly reported using "air change rate per hour at an indoor-to-outdoor pressure differential of 50 Pa," as shown in Equation 2. This is a volume-based metric and has been used extensively since the earliest days of airtightness testing.

Air change rate at 50 Pa:

$$ACH_{50} = \frac{\text{Total leakage rate at 50 Pa (expressed in building volumes)}}{\text{Building volume}} \quad (2)$$

For larger, commercial building tests (such as those used in this study), the most common method of expressing results is "normalized leakage rate at 75 Pa" (NLR_{75}), which is a building envelope area-based metric as defined by Equation 3:

Normalized leakage rate at 75 Pa:

$$NLR_{75} = \frac{\text{Total leakage rate at 75 Pa}}{\text{Envelope area}} \quad (3)$$

In North America, the *envelope area* in Equation 3 is defined to include both the above-grade and below-grade portions of the building envelope. In Europe (or at least some of its constituent countries), only the above-grade portions of the envelope are included in the term. In all cases, however, the dimensions used to calculate the envelope area are based on the building's interior, not exterior, dimensions. This is noted because it is contrary to common practice in the real estate industry, which often uses exterior dimensions to describe a building's size.

One point seldom recognized about airtightness testing metrics is as follows:

> Volume-based metrics (such as ac/hr_{50}) provide a commentary on the energy load which air leakage creates in the building since it describes the volumetric flow rate that the building experiences and the volumetric flow rate defines how much air has to be heated or cooled. In contrast, area-based metrics (such as the NLR_{75}) provide a commentary on the threat posed to the building envelope by air leakage-induced moisture deposition, since it is the moisture loading per unit area of building envelope which has to be controlled if building durability is to be safeguarded. (PEL 2004)

Of course, because both metrics are basically describing the same thing (how leaky is the building?), it should come as no surprise that both the NLR_{75} and ACH_{50} tend to generally behave in a similar—although not exact—fashion.

Obviously, both energy performance and envelope durability are of concern for buildings. However, observing the respective metrics used to describe residential versus commercial building leakage, it is apparent that residential airtightness results are designed (intentionally or otherwise) to emphasize energy conservation whereas commercial airtightness results focus on envelope protection and durability.

In this report, airtightness data is primarily reported using the normalized leakage rate at an indoor-to-outdoor pressure differential of 75 Pa (i.e., Equation 3) since this is standard practice in the literature. Because NLR_{75} is commonly reported using both Systéme International (SI) and Inch-Pound (I-P) units, both are shown in this report, with SI results reported first followed by the equivalent I-P units. In addition, the results are presented using the air change rate at 50 Pa (ACH_{50}). While this metric is most commonly used for residential buildings, it is also reported herein because of the commentary it provides for energy purposes.

Airtightness Standards for Commercial Buildings

At present, there are no mandatory or voluntary airtightness requirements in Canada for commercial buildings of the type studied in this project.

In the United States, the most widely used energy code is the *International Energy Conservation Code* (IECC) (ICC 2012). The 2012 edition introduced requirements for commercial building airtightness that can be met either by using materials and assemblies that have demonstrated low leakage characteristics or by testing the completed building. If the test option was selected, the maximum measured NLR_{75} could not exceed 2.03 $L/s{\cdot}m^2$ (0.40 cfm/ft^2).

Perhaps the most aggressive promotion of commercial building airtightness requirements has been provided by the U.S. Army Corps of Engineers (USACE), who are responsible for establishing building standards for U.S. military construction. In 2012 they introduced an NLR_{75} airtightness requirement of 1.27 $L/s{\cdot}m^2$ (0.25 cfm/ft^2) at 75 Pa for all new construction. Further, it is believed that they are considering lowering this target, possibly to 0.76 $L/s{\cdot}m^2$ (0.15 cfm/ft^2). See Table 1.

In Canada, the NBC requires that materials used as part of the air barrier assembly must have a measured leakage rate that does not exceed 0.02 $L/s{\cdot}m^2$ (0.004 $ft^3/min{\cdot}ft^2$) at 75 Pa. However, it should be noted that (at the time of this writing) consideration is being given to the possible inclusion of quantitative airtightness requirements for commercial buildings in a future edition of the NBC.

While the 2010 NBC does not contain any whole-building airtightness requirements for commercial structures, it does contain recommendations for commercial building air barrier systems (NRC 2010). The NBC describes an air barrier system as those portions of the building envelope that are opaque and insulated (i.e., walls, ceilings, foundations, etc., but excluding windows and doors and other openings). The recommendations are shown in Table 2 and illustrate how the system requirements vary depending on the anticipated indoor relative humidity—as the relative humidity increases, the maximum recommended system air leakage decreases, thereby providing additional protection against moisture transport through the building envelope.

Comparing the recommendations for air barrier systems to those shown in Table 1 for whole buildings, it is apparent

Table 1. Current Airtightness Performance Standards for Commercial Buildings

Airtightness Requirement	Maximum Permitted NLR_{75}	
	L/s·m²	cfm/ft²
2012 *International Energy Conservation Code*	2.03	0.40
2012*U.S. Army Corps of Engineers Air Leakage Test Protocol for Building Envelopes*	1.27	0.25
Proposed *U.S. Army Corps of Engineers Air Leakage Test Protocol for Building Envelopes*	0.76	0.15

Table 2. NBC Recommended Maximum Air Leakage Rates for Air Barrier Systems

Warm Side Relative Humidity at 21°C	Recommended Maximum System Air Leakage Rate at 75 Pa
< 27%	0.15 L/s·m² (0.030 ft³/min·ft²)
27% to 55%	0.10 L/s·m² (0.020 ft³/min·ft²)
> 55%	0.05 L/s·m² (0.010 ft³/min·ft²)

that the system recommendations are far more rigorous than those for the entire building. This is to be expected, since most of the air leakage does not occur through the large expanses of the walls, ceilings, or foundations but rather through the joints, intersections, and penetrations that interrupt these surfaces. Most Canadian buildings operate with indoor relative humidity levels in the middle range of Table 2 (i.e., 27% to 55% rh) during the heating season, and for that reason a system leakage value of 0.10 L/s·m² (0.020 ft³/min·ft²) is often used when discussing typical air barrier system requirements. This value will be referenced later when we discuss the measured airtightness results from this project.

Comparative Airtightness Testing Data—Results from the Literature

As mentioned, prior to this project, only a handful of commercial buildings in Manitoba had ever had their airtightness measured, meaning that the knowledge base was very limited. Fortunately, the literature contained some useful information. To provide some comparative data to interpret the project's results, three primary references were identified and used as benchmarks for interpreting the results of this study:

- **ASHRAE RP-1478, *Measuring Airtightness of Mid- and High-Rise Non-Residential Buildings* (WJE 2014).** This project measured the airtightness of 16 new buildings, with heights ranging from 4 to 14 stories and constructed since 2000. The tested buildings were located in the United States, specifically in climate zones 2 through 7 of the IECC climate zone map. (*Note:* Manitoba's climate zones consist of 7a, 7b, and 8.)
- ***Air Leakage Control in Multi Unit Residential Buildings: Development of Testing and Measurement Strategies to Quantify Air Leakage in MURBS* (RDH 2013).** This literature survey, commissioned by Canada Mortgage and Housing Corporation (CMHC), was primarily interested in documented airtightness results for multi-unit residential buildings (MURBs) since that is CMHC's primary focus. The report contains results from 296 buildings in Canada and the United States, although only 43 of these were MURBs. Most of the results (245 out of 296, or 83%) came from tests performed for the USACE on American military buildings.
- **Air Leakage Characteristics, Test Methods and Specifications for Large Buildings (PEL 2001).** This literature survey, also carried out for CMHC, was a precursor to the RDH study described above. It identified airtightness results from 192 buildings in Canada (including five in Manitoba), the United States, Great Britain, and Sweden. It should be noted that some of the results in the PEL report (approximately 5% to 15%) were also referenced in the RDH report.

It is also worth noting that the first two studies (ASHRAE RP-1478 and RDH) were both released within the last few years, while the third report (PEL) is 15 years old.

TEST PROTOCOL

Building Selection

For the purposes of this project, *commercial* buildings were deemed to include office buildings, schools, churches, industrial structures, and multi-unit residential buildings (MURBs) that also included institutional buildings such as schools and churches as well as light industrial buildings, especially those that included other uses such as an office component.

Test Equipment

Red River College's Sustainable Infrastructure Technology Research Group (RRC-SITRG) acquired a high-performance multifan blower door system capable of testing a wide range of commercial buildings (Figure 1). When weather and timing permitted, this multifan blower door system was supplemented by other diagnostic tools (i.e., thermal imaging

Figure 1 *Multifan blower door system used for airtightness testing.*

camera and a video borescope) for assessing building envelope performance.

Selection of an Airtightness Test Protocol

The previous discussion briefly reviewed the evolution of airtightness testing standards, particularly those intended for larger, commercial buildings. Based on this information, it was decided that the most appropriate methodology for the project was the USACE standard (USACE 2012). However, as mentioned, this document was under active review by its technical committee when the selection was made. Further, several of the project team members were also actively participating in the review process.

In some respects, the RRC-SITRG project was used as a test bed for possible changes to the USACE standard. One of these issues, the treatment of intentional openings, had great significance to the current project (see the Conclusions and Recommendations section) because it impacted the testing protocol eventually used in this project.

It is also standard practice to perform commercial building airtightness tests by conducting separate pressurization and depressurization tests. The final building airtightness is then calculated as the average of these two sets of results. This practice was used on 23 of the 26 buildings tested in this project. The remaining three were tested under depressurization only.

AIRTIGHTNESS TEST RESULTS AND DISCUSSION

Description of Buildings Tested

Table 3 summarizes the various types of commercial buildings encountered in the project. Note that some results appear in more than one category. The 26 buildings were located across Manitoba with 18 (69%) situated in the greater Winnipeg area. They ranged in age from 1 to over 100 years. In terms of size, floor areas ranged from 150 to 19,788 m^2 (1615 to 212,918 ft^2), a variation of 132:1 between the largest and the smallest. Volumes varied from 587 to 66,304 m^3 (20,710 to 2,340,079 ft^3), a variation of 113:1. Building height ranged from 1 to 16 stories (approximately 58 m or 189 ft). Manitoba Hydro owned five of the buildings, while the remainder were occupied by a variety of private and public owners.

Table 3. Categorization of Buildings Tested

Category	Number of Buildings
Complete sample	26
New construction	5
Warehouses/light industrial	11
Office buildings	7
Schools	4
Churches	2
Greenhouses	2
Retrofits	3

Measured Airtightness Results

Table 4 and Figure 2 summarize the results, while Table 5 provides comparative results from the literature using the previously mentioned primary references.

Also shown in Figure 2 are the 2012 IECC, 2012 USACE, and 2010 NBC recommendations. These are not suggested as definitive, reasonable, or even desirable goals, but rather as indications of what others have suggested or proposed as being appropriate. Further, it should be appreciated that the two building targets (2.03 and 1.27 L/s·m^2 [0.400 and 0.250 ft^3/min·ft^2]) are American standards developed largely on American experiences and their current air barrier technology. Canadian experiences and knowledge are generally more advanced in this field than those in the United States, so the appropriateness of the American targets has yet to be vetted against Canadian capabilities. Finally, it is important to recall that the final target shown, 0.10 L/s·m^2 (0.02 ft^3/min·ft^2), is a recommended (not mandatory) system requirement and is not a whole-building target. As described previously, it is used here to represent a theoretical, even ultimate, airtightness target.

Finally, the six buildings tested using the "energy" sealing protocol are identified by the downward-facing blue arrows positioned on the graph's x-axis.

As described previously, when a building is tested using the energy sealing protocol, the leakage will always be higher than if it had been sealed using the envelope protocol. So, for the six buildings tested using the energy protocol, the leakage rates shown on the graphical summaries represent the maximum possible leakage. Had these tests been conducted using

Table 4. Summary of Results

	Number of Buildings	NLR_{75} Mean (NLR_{75} Range)		ACH_{50} Mean (ACH_{50} Range)
		L/s·m²	cfm/ft²	
Complete sample	26	1.70 (0.19–3.44)	0.34 (0.04–0.68)	2.33 (0.19–5.92)

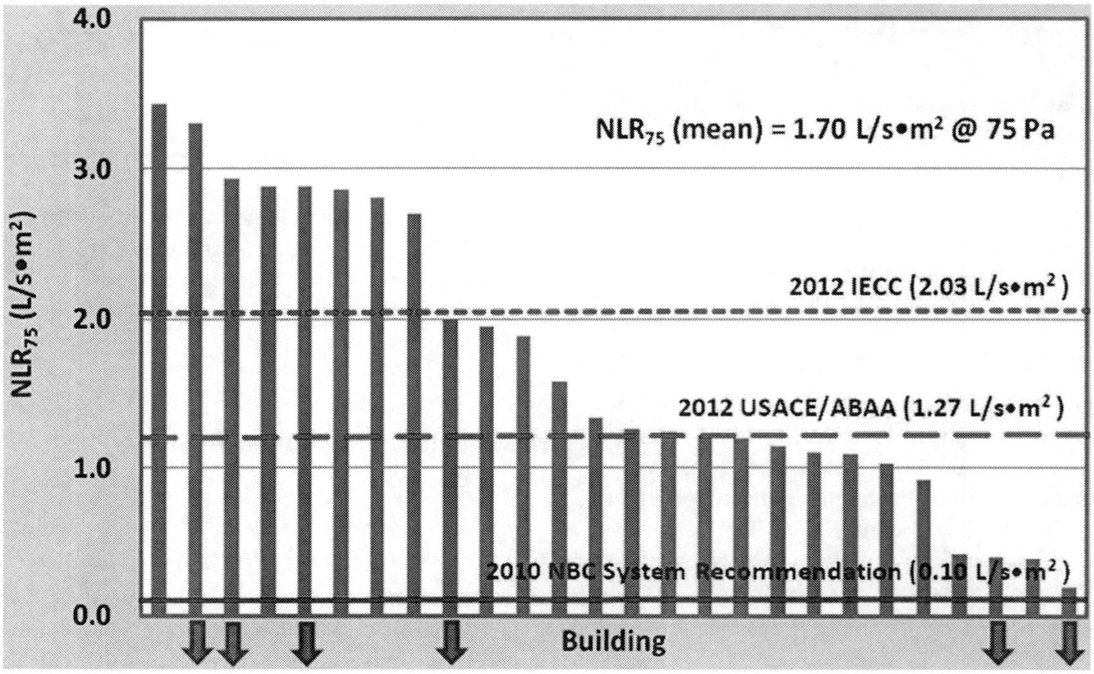

Figure 2 *NLR_{75} results for sample.*

Table 5. Comparative Results from the Literature

	Number of Buildings	NLR_{75} Mean (NLR_{75} Range)		Notes
		L/s·m²	cfm/ft²	
ASHRAE RP-1478 (Field tests)	16	1.47 (0.30–3.81)	0.29 (0.06–0.74)	Buildings less than 5 years old
RDH (Literature survey)	40	3.96 (0.81–10.00)	0.78 (0.16–1.97)	MURBs only
PEL (Literature survey)	41	2.12 (0.23–6.37)	0.42 (0.05–1.26)	Canadian buildings only

the envelope sealing protocol, the measured leakage would have been lower than that shown.

The sample of 26 buildings displayed a wide variation in all relevant factors that can influence airtightness (e.g., age, year of construction, building height, wall construction, etc.). The measured NLR_{75} values ranged from 0.19 to 3.44 L/s·m² (0.037 to 0.678 ft³/min·ft²), a variation of almost 18:1 between the leakiest and tightest buildings tested. The mean NLR_{75} for the entire sample was 1.70 L/s·m² (0.335 ft³/min·ft²), which is surprisingly similar to the mean value reported in the ASHRAE RP-1478 survey of new buildings (1.47 L/s·m² [0.290 ft³/min·ft²)]), even though most of the buildings in the

ASHRAE RP-1478 study were designed and built to comply with some voluntary standard such as that described by the Leadership in Energy and Environmental Design® (LEED®) Green Building Rating System. Only a handful of the buildings in this present study were designed and constructed with any explicit concern for airtightness. Further, the Manitoba buildings tested for this project performed significantly better than those identified in the previous literature reviews by RDH or PEL.

OBSERVATIONS AND LESSONS LEARNED ABOUT AIRTIGHTNESS TESTING OF COMMERCIAL BUILDINGS

This project afforded the RRC-SITRG team a rare opportunity to learn about both the airtightness of commercial buildings and the nuts and bolts of conducting these types of tests. The following sections discuss some observations compiled during the airtightness testing program that address some of the practical aspects of commercial building testing.

Occupied versus Non-Occupied Buildings

Whether the building's owners have or have not (yet) occupied the structure has no direct impact on the test protocol, but it does have considerable effect on the logistics and scheduling of the test. Because airtightness testing requires those performing the work to have absolute control over operation of the building for several hours (including mechanical systems, windows, doors, etc.), this can pose scheduling issues for the owner. If the building has a fairly conventional occupancy schedule (say 10 to 12 hours per day, 5 days a week), the testing agency can usually schedule the work during evenings, nights, or weekends. Evening and nighttime tests also have the advantage that winds are usually lighter than during the day. However, if the building is occupied on a continuous basis, then scheduling becomes very problematic. Hospitals, personal care homes, hotels, and prisons are all examples of structures that never really close. Depending on the type of business, testing of commercial buildings can usually be conducted after normal business hours. This requires the testing agency to be able to work on a 24/7 schedule. If the test includes sealing the intentional openings using the full building envelope schedule, days of planning and preparation may be required to efficiently execute this work in the field (and possibly in the dark).

Security is also of utmost importance, and an agreed-upon plan with the owner must be in place prior to the test. Environmental conditions may negatively affect the occupied space due to changes in temperature when introducing cold/hot air during the test. With new construction, it is critical that all building components that could influence airtightness be completed. This includes all doors and windows and all of the mechanical, electrical, and plumbing penetrations through the building envelope.

Testing Limitations Created by Different Forms of Occupancy

Testing agencies may have to respect restricted areas within commercial, office, workshop, warehouse, government, and utility buildings due to facility management policies and security issues. For example, special conditioned spaces with restricted access, such as data, communication, or archive areas, may have to be excluded from the test and be accounted for in the test calculations and reporting.

Occupied schools operate with scheduled in-service days and holidays that can often be used to perform the test. Churches are usually easier to schedule since they tend to be lightly utilized during weekdays and evenings. Hospitals and medical centers are extremely problematic and, in most cases, cannot be tested once the building is occupied.

One of the outcomes of RRC's research has been the recognition of the challenges involved with airtightness testing of MURBs, which are unique from those encountered testing other types of large buildings. The airtightness test methods and standards are based upon the building being vacant or the testing team having control over the movement and actions of the occupants during the test (e.g., restricting entry and exit, keeping windows closed, not operating mechanical systems, etc.). This introduces a major problem with testing buildings such as MURBs once they are occupied. As a result, airtightness testing of the over 3 million dwelling units in Canadian MURBs is rare.

Testing Limitations Created by Environmental Conditions

All airtightness testing standards set limits on the maximum wind speed that can be experienced during the test. For example, ASTM WK35913, *New Test Method for Whole Building Enclosure Air Tightness Compliance* (ASTM n.d.), stipulates a maximum site wind speed of 20 km/h (12 mph). Exposure and building height will also influence the maximum wind speed that can be endured during the test. Generally, taller structures and those unshielded by neighboring buildings or terrain are more vulnerable to high winds.

Rain, frost, and condensation also create problems when sealing the HVAC openings. Wet, damp, or icy surfaces can create major problems when sealing intentional openings since moisture and frost can cause de-bonding of the tape under sustained positive pressure differentials. The lowest temperature encountered during the 26-building test program was about –10°C (14°F). Although tests were successfully completed at this temperature, considerable care had to be exercised to ensure that all the seals remained in place.

Testing Limitations Created by Safety Concerns

Safety is of the utmost importance, and all workplace and safety regulations must be followed. Some rooftop HVAC equipment (which has to be masked) may be situated in roof locations that are outside the unrestricted work area of the roof and therefore require fall protection. During cold weather, areas of the roof and sidewalks may be ice and snow covered, which presents slip and fall hazards. Ladders and lifts are often needed and must only be used by workers who have taken appropriate safety training. An activity hazard analysis may find that the building does not have adequate fall arrest equipment mounted on the building. As a result, safe preparation can be difficult to achieve. In any event, testing should not be initiated if environmental conditions make the building dangerous to work on.

Limitations of Airtightness Testing for Code Compliance and Commissioning Purposes

While whole-building airtightness testing is often promoted as part of a building's commissioning plan, additional measures are also required during construction since sealing leakage sites is more difficult after the building is finished and may—in some cases—not even be possible. At all stages of the construction, testing of the materials and individual assemblies is strongly recommended. Whole-building airtightness testing does not eliminate these requirements, it complements them.

CONCLUSIONS AND RECOMMENDATIONS

Conclusions

Overall, the findings from this project indicate that designing and constructing commercial buildings with a high level of airtightness is achievable using knowledge and technology that currently exists within the Manitoba design and construction industries. Further, these results demonstrate that very low air leakage rates can be attained in commercial construction—even if the building is architecturally complex.

From the buildings tested in Manitoba, it is also clear that there is a broad variation in airtightness. One of the most encouraging observations from this project was the newly constructed middle school that demonstrated what is achievable when due diligence is taken during the design and construction periods (NLR_{75} = 0.19 L/s·m^2 [0.037 ft^3/min·ft^2]). The school was constructed with a full building envelope commissioning program that incorporated a detailed design review and testing of building assemblies throughout the project. The end result was that the whole-building airtightness was nearly as tight as the NBC recommendations for a building system. However, results indicate there is still need for improvement in the design and construction of air barriers and building envelopes.

During testing of the 26 buildings it became evident that two distinct sealing methods were required for intentional openings in the building. One method is employed for an energy test (mechanical system unsealed) and one for a building envelope test (mechanical system sealed). These changes have been incorporated into ASTM WK35913 (ASTM n.d.).

It was also observed that while mechanical system damper air leakage poses a potential energy liability for commercial buildings, it also represents a possible opportunity for improving building energy performance in both new and existing structures.

Recommendations

Recommendation #1—Review Potential Opportunities for Low-Leakage HVAC Dampers. Given the project's findings about the large impact that HVAC dampers have on the airtightness characteristics of commercial buildings, it is recommended that additional attention be paid to the selection and installation of mechanical dampers in commercial-building HVAC systems.

Recommendation #2—Consider Mandatory Airtightness Testing in the NBC of Buildings with High-Threat Indoor Environments or Those that Cannot be Tested after They are Occupied. A final recommendation from this project deals with the possible adoption of mandatory airtightness testing requirements for commercial buildings in the NBC or other codes. At present, this issue is being discussed by the Task Group on Building Envelopes for consideration in the 2020 NBC. While this project has helped to identify some of the issues associated with such a code change, it should also be clear that this is not a simple matter and the task group will need to consider a variety of factors, including cost, availability of testing agencies and trained personnel, scheduling implications for the normal construction process, weather and climate-related issues, etc.

However, one recommendation that can be offered deals with the types of buildings which should given first priority for possible mandatory airtightness testing. Based on our experiences with this project, we believe there are two types of buildings that should be considered as prime candidates.

The first type would be those which, from a moisture perspective, are exposed to severe indoor environments. This includes structures such as indoor swimming pools in which the indoor relative humidity is considerably and consistently higher than normally encountered in most commercial buildings. Referring to Table 2, these can be defined as buildings that operate with relative humidity levels greater than 55%. This would apply to both new construction and to buildings undergoing major building envelope retrofits.

The second type of building that should be considered as a prime candidate for mandatory airtightness testing is that which, once occupied, cannot realistically be considered for future testing due to its occupancy or use. Hospitals, personal care homes, hotels, and prisons are prime examples. Once they are occupied, they are never unoccupied, they never close, their mechanical systems can never be shut down, and in most cases they can never be surrendered to an airtightness testing organization for the 6 to 12 hours necessary to complete a test. If they are not tested at the completion of construction, it is unlikely they can ever be tested.

ACKNOWLEDGMENTS

The authors wish to thank Manitoba Hydro for their generous support for this project. It would not have been possible without them.

REFERENCES

ASTM. n.d. ASTM WK35913, *New Test Method for Whole Building Enclosure Air Tightness Compliance*. West Conshohocken, PA: ASTM International. https://www.astm.org/DATABASE.CART/WORKITEMS/WK35913.htm.

ASTM. 2010. ASTM E779-10, *Standard Test Method for Determining Air Leakage Rate by Fan Pressurization*. West Conshohocken, PA: ASTM International.

CGSB. 1986. CAN/CGSB-149.10-M86, *Determination of the airtightness of building envelopes by the fan depressurization method*. Ottawa, ON: Canadian General Standards Board.

ICC. 2012. *2012 International Energy Conservation Code*. Washington, DC: International Code Council.

NRC. 2010. *National Building Code of Canada 2010*. Ottawa, ON: National Research Council of Canada, Canadian Commission on Building and Fire Codes.

NRC. 2011. *National Energy Code of Canada for Buildings 2011*. Ottawa, ON: National Research Council of Canada, Canadian Commission on Building and Fire Codes.

PEL. 2001. *Air leakage characteristics, test methods and specifications for large buildings*. Report prepared for Canada Mortgage and Housing Corporation. Winnipeg, MB: Proskiw Engineering Ltd.

PEL. 2004. *A review of airtightness requirements in the Super E® standard for the UK and Ireland*. Report prepared for Natural Resources Canada. Winnipeg, MB: Proskiw Engineering Ltd.

RDH. 2013. *Air leakage control in multi-unit residential buildings: Development of testing and measurement strategies to quantity air leakage in MURBS*. Report prepared for Canada Mortgage and Housing Corporation. Vancouver, BC: RDH Building Engineering Ltd. http://rdh.com/wp-content/uploads/2014/04/Air-Leakage-Control-in-Multi-Unit-Residential-Buildings.pdf.

USACE. 2012. *U.S. Army Corps of Engineers air leakage test protocol for building envelopes*, ver. 3. Champaign, IL: U.S. Army Research and Development Center, Construction Engineering Research Laboratory. http://www.wbdg.org/pdfs/usace_airleakagetestprotocol.pdf.

WJE. 2014. *ASHRAE RP-1478, Measuring airtightness of mid- and high-rise non-residential buildings*. Report prepared for ASHRAE. Northbrook, IL: Wiss, Janney, Elstner Associates, Inc.

Impact of Large Building Airtightness Requirements

Lorne Ricketts, MASc, PEng
Associate Member ASHRAE

ABSTRACT

Air leakage into and out of buildings affects building durability, occupant comfort, indoor air quality, and energy consumption. Recently, in response to increasing societal concern regarding the environment and rising energy costs, building enclosure airtightness is garnering more attention. Various jurisdictions in North America and worldwide have implemented or are considering implementing airtightness testing and/or quantitative airtightness performance requirements into building regulations. This paper summarizes select results from a study into the state of the industry with respect to the airtightness of large buildings in North America.

As part of this study, airtightness test results from more than 500 buildings located primarily in North America were compiled into a one-of-a-kind database. The results are presented to demonstrate historical and current airtightness levels as well as to quantify the impact or whole building airtightness regulations on the measured airtightness of building enclosures. Additionally, the results of a survey that asked industry members working in a jurisdiction with mandatory whole-building airtightness testing to respond to questions regarding the impact of the regulations as well as their perception of the regulations are presented. Finally, the paper provides commentary on the difference between air leakage and airtightness and identifies where further research is required to allow for more accurate determination of large building air leakage rates.

Overall, this paper provides the foundational information necessary to assess current airtightness performance and the impact of existing requirements so as to allow for evidence-based widespread adoption of whole-building airtightness regulations for large buildings.

INTRODUCTION AND BACKGROUND

Air leakage into and out of buildings affects building durability, occupant thermal and acoustical comfort, indoor air quality, and energy consumption. In response to increasing societal concern regarding these building performance characteristics, and in particular energy consumption, improving building enclosure airtightness to reduce air leakage is receiving increased attention. In various North American jurisdictions and worldwide, this has led to a shift in the way the industry designs, specifies, builds, and measures airtightness. One important change is a shift away from relying solely on material and component airtightness performance towards a whole building airtightness approach. This move by the industry recognizes that airtightness performance relies heavily on detailing and quality control through the entire construction process and that a method which accounts for these aspects is valuable. While whole-building airtightness testing of single-family homes is common, airtightness testing of large buildings is still relatively uncommon. In some jurisdictions mandatory whole-building airtightness testing and performance standards have already been implemented and many others are investigating and planning for implementation in the near future.

In support of potential widespread adoption of airtightness performance and testing requirements for large buildings, a study was conducted to better understand the current state of

Lorne Ricketts is a building science engineer at RDH Building Science, Vancouver, BC.

the industry with respect to airtightness of large buildings and to evaluate the impact of implementing testing and performance standards. This study compiled the results of more than 500 airtightness tests and surveyed industry members where whole-building airtightness regulations have already been implemented to assess the quantitative performance impact of these types of regulations as well as other factors such as the cost of construction and perceived value. This paper is a summary of select findings from a larger study by RDH Building Science (2015).

MEASURED PERFORMANCE

To quantify the airtightness performance of large buildings, a database of test results was compiled. The database is populated with data from various sources available in literature, from industry members that contributed data, and from unpublished test results available directly through members of the project team. Data collection efforts focused on results from the United States and Canada; however, available results from other jurisdictions were also collected, in particular from the United Kingdom. A complete set of references for the airtightness data is provided in the RDH Building Science report (2015).

In total, 721 airtightness tests were collected, within which there were 584 unique buildings, as many buildings were tested multiple times. Test results that had inconsistent reporting, unclear test boundaries, or used unclear methods were removed from the dataset, resulting in 566 test results of unique buildings that were deemed suitable for analysis and included in the database. Figure 1 presents the distribution of (a) building type and of (b) geographical location for the buildings included in the database.

Figure 1 illustrates that the majority of buildings in the database are "military" and located in the United States. This finding is because one of the major jurisdictions currently requiring airtightness testing is the United States Army Corps of Engineers (USACE & ABAA 2012). Hundreds of buildings have been tested for compliance with USACE regulations and these results form approximately 46% of the total database population. The impact of this large sampling of USACE buildings can also be seen in Figure 2, which shows the distribution of airtightness performance for all of the buildings in the database. A significant decrease in frequency is noticeable at 1.25 L/(s·m^2) (0.25 cfm/ft^2), which is the USACE performance requirement. This significant decrease suggests that buildings are often retested so that they achieve the requirement.

As part of the data collection process, information on various building characteristics was collected, including age, height, floor area, enclosure area, general construction type, and air barrier type. An analysis of this data is possible; however, this paper focuses on changes in airtightness performance over time because of gradual evolution in construction materials and practices as well as because of the implementation of testing requirements. The database contains buildings constructed between 1956 and 2014. To assess the change in the airtightness of large buildings over this time period, the test results were plotted versus the year of construction in Figure 3. This figure indicates a general trend of improving airtightness in modern construction and, in particular, this figure illustrates a narrowing of the expected range in performance values.

Buildings in the database may have been tested for a number of reasons, whether as part of investigations, as part of voluntary high-performance building programs, for general research, or for compliance with mandatory testing requirements. Different reasons for testing would likely also relate to expected airtightness of the building. For example, if a building is being designed and constructed with the knowledge that it will be tested upon completion, it is reasonable to expect that it would be more airtight than the general population of build-

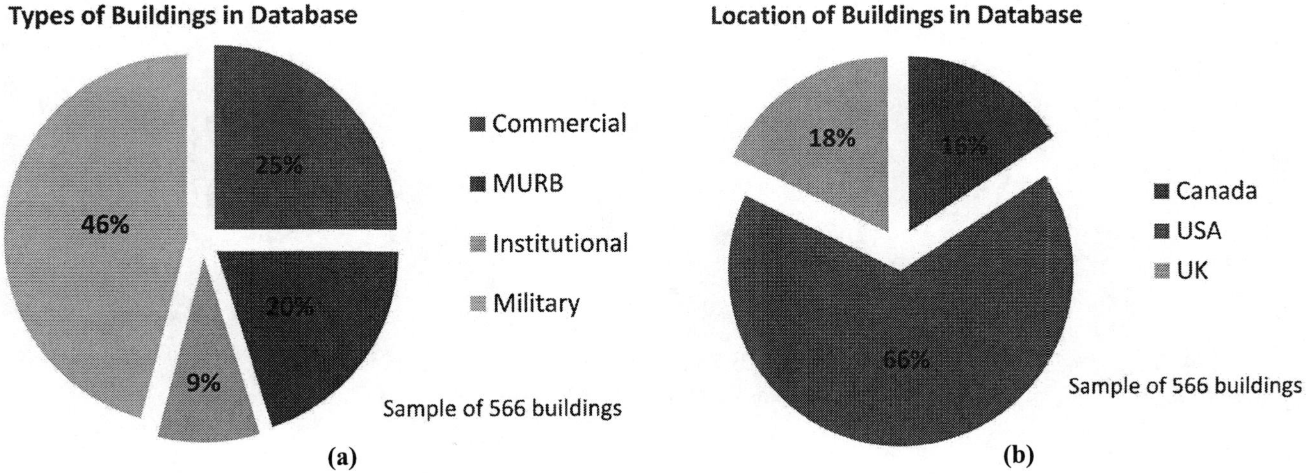

Figure 1 *Charts of (a) distribution of building types in the airtightness database and (b) geographical location of buildings in the airtightness database.*

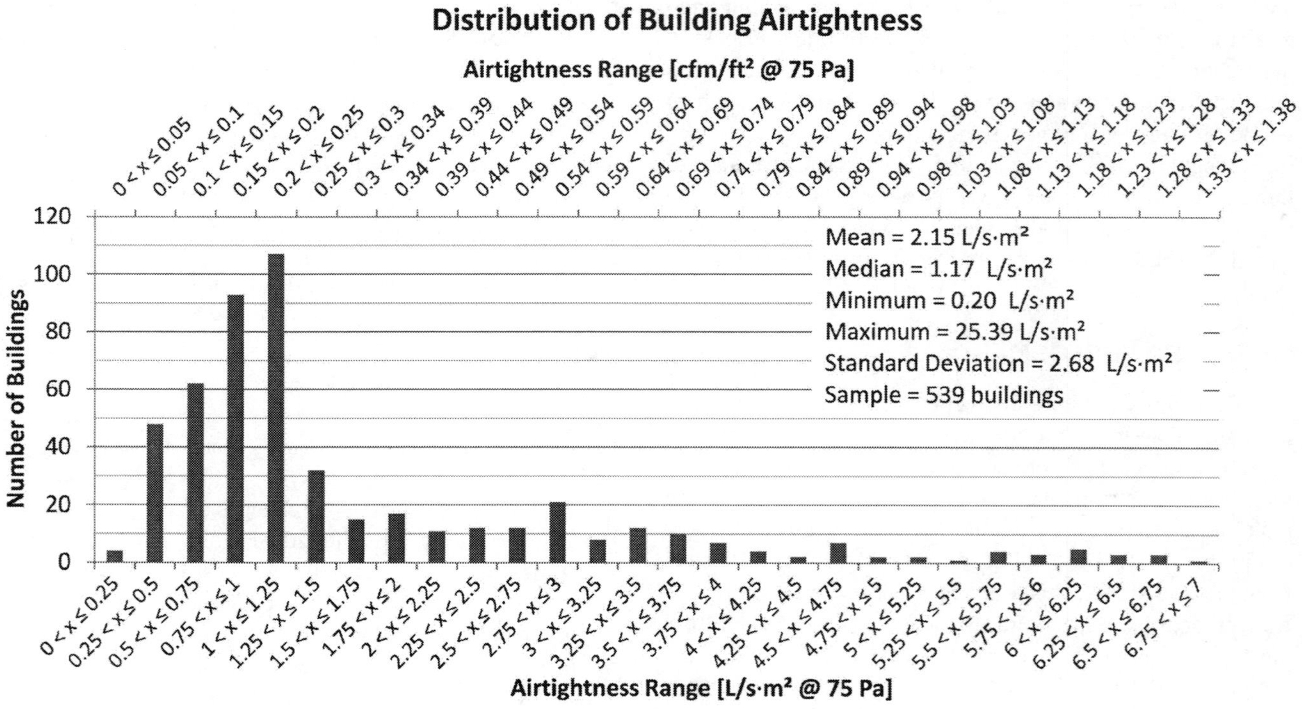

Figure 2 *Distribution of airtightness of all of the buildings in the database. Note the significant reduction in frequency at approximately 1.25 L/(s·m²) (0.25 cfm/ft²), which corresponds to the USACE performance requirement.*

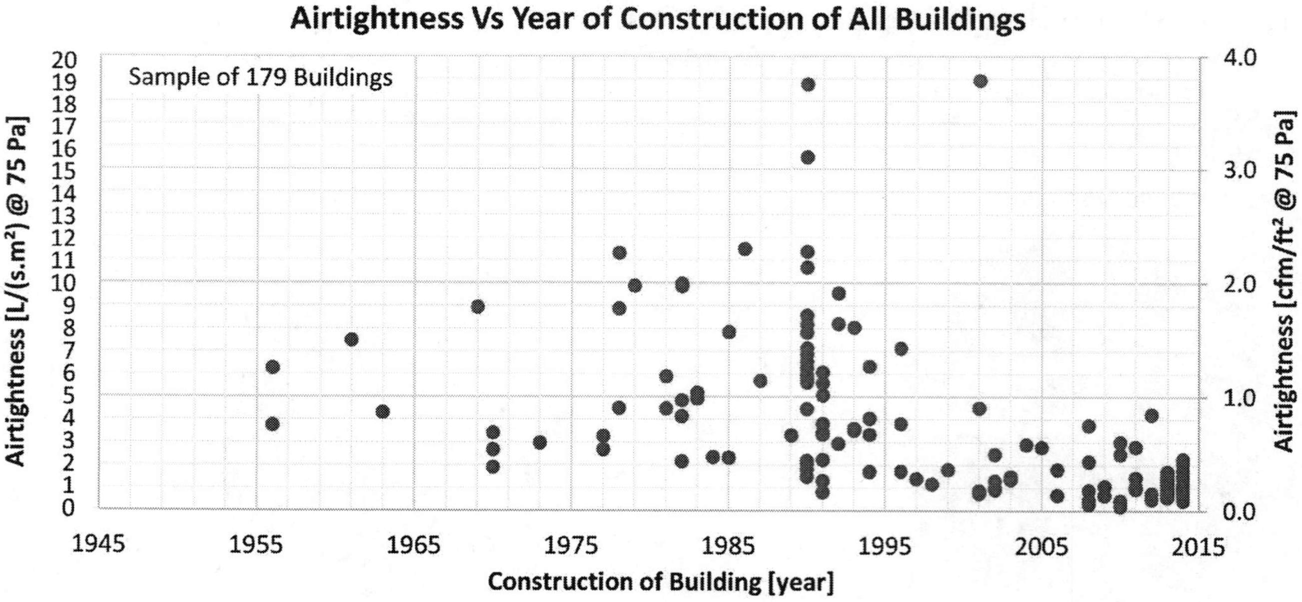

Figure 3 *Chart of airtightness of all buildings in the database versus the year of construction. In this case year of construction refers to either the year the building was originally built or the most recent year at which modifications to the air barrier were performed.*

ings. To specifically assess the airtightness of buildings that were designed to meet mandatory performance and/or testing requirements, Figure 4 plots the distribution of buildings tested for research purposes (Research), tested for compliance with USACE (USACE), or tested for compliance with State of Washington (Washington State 2012) or Seattle Energy Code (City of Seattle 2014) requirements (Washington). Buildings categorized as Research include all buildings for which testing was voluntary (i.e., investigation, green building program, etc.).

Figure 4 clearly indicates a significant reduction in the mean average airtightness of buildings that were built to meet a performance and/or testing requirement. The mean average airtightness for Research, USACE, and Washington buildings are 3.52, 1.00, and 1.30 L/(s·m^2) (0.69, 0.20, and 0.26 cfm/ft^2) respectively. Interestingly, the most airtight of the Research buildings has similar or even better performance to that of the best performing USACE or Washington buildings; however, the range of performance for the Research buildings is much larger with the least airtight building two orders of magnitude leakier than the most airtight (approximately 0.2 L/(s·m^2) [0.04 cfm/ft^2] versus 25 L/(s·m^2) [5.0 cfm/ft^2]). Overall, this chart suggests that the implementation of airtightness requirements by the USACE and Washington/Seattle has had a substantial impact on the airtightness of buildings and has reduced the mean average airtightness by 72% and 63% respectively.

The data for the Research buildings presented in Figure 4 includes data from buildings which are much older than the population of buildings in the USACE and Washington datasets. As a result, it is possible that the perceived impact of the airtightness regulations is the result of gradual evolution in building materials, design, and construction practices rather than the result of more stringent airtightness regulations. To provide a fairer comparison, buildings that were constructed prior to the year 2000 were removed from the datasets and the distribution of these results is provided in Figure 5. Removing older buildings from the analysis ensures the comparison is for buildings using similar techniques and materials; consequently, any improvement in airtightness can be attributed to the regulation of whole-building airtightness.

Figure 5 illustrates that when buildings of similarly modern construction are compared, buildings in which whole-building airtightness is regulated are noticeably more airtight. The mean average airtightness of USACE and Washington buildings are 0.94 L/(s·m^2) (0.19 cfm/ft^2) and 1.12 L/(s·m^2) (0.22 cfm/ft^2) respectively. These are 55% and 43% improvements in performance as compared to the average airtightness of 2.12 L/(s·m^2) (0.42 cfm/ft^2) for buildings constructed in the year 2000 or later for which there was no testing requirement. It is also apparent that in both jurisdictions with regulations, buildings are consistently able to meet and exceed the current performance targets.

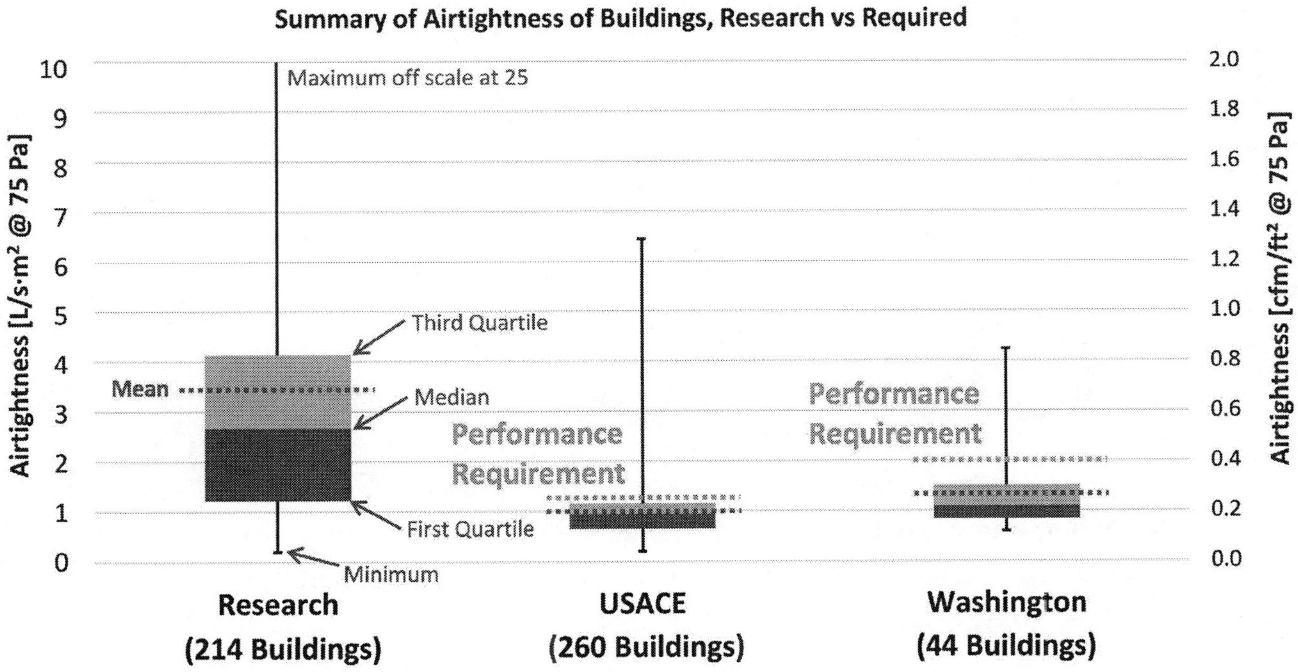

Figure 4 *Chart of the distribution of airtightness performance of buildings tested for research purposes, for compliance with USACE requirements and for compliance with Washington State or Seattle Energy Code requirements. Note that all of the testing performed for the latter was at a time when testing and reporting of the result was required, but compliance with a performance target was not yet required.*

Even with the older buildings removed from the dataset, it is apparent that some of the modern buildings in jurisdictions without airtightness requirements continue to exhibit relatively poor airtightness performance, with the maximum two orders of magnitude higher than the minimum. In addition to reducing the average airtightness, regulations have also decreased the expected range of performance which substantially improves the ability to predict airtightness performance. This improved predictive capability is important for efforts such as energy modeling for code compliance and design of mechanical systems where it is necessary to make reasonable assumptions about building performance during the design phase.

Overall, the quantitative airtightness testing results collected in the database provide strong support for the conclusion that airtightness regulations have improved the average airtightness of buildings in those jurisdictions as compared to similar buildings in other nonregulated jurisdictions, and that they have substantially reduced the typical range of performance.

QUALITATIVE PERCEPTION

To supplement the quantitative comparison of airtightness performance, a survey was conducted of industry members working in the State of Washington where airtightness performance requirements have been implemented. For reference, the first versions of the Washington State Energy Code (WSEC) and Seattle Energy Codes (SEC) to include a whole-building airtightness testing requirement were dated 2009, and came into effect July 1, 2010 and January 1, 2011 respectively. The 2009 SEC required testing of all new commercial buildings, while the 2009 WSEC required testing of all new commercial buildings over 5 stories in height. (Commercial buildings are defined in the same manner as for the 2012 International Building Code.) In the 2009 version of these codes, only testing of the buildings was required, and the result had to be reported, but a performance target did not have to be achieved.

The new 2012 WSEC and SEC also require whole-building airtightness testing, and both are applicable to all new commercial buildings. Both codes specify a minimum airtightness of 2.0 L/(s·m^2) (0.40 cfm/ft^2) at 0.3 inches of water gauge (75 Pa). Under both codes, if the building fails to meet the airtightness performance target, measures must be taken to inspect and seal the air barrier "to the extent practicable." Once a report documenting this sealing work has been submitted to the code official and owner, there is no need to retest the building to ensure compliance with the performance target. These codes came into effect in December 2013; however, only recently have buildings to which the new codes apply started to reach completion and, consequently, relatively few buildings have been tested that require compliance with the performance target.

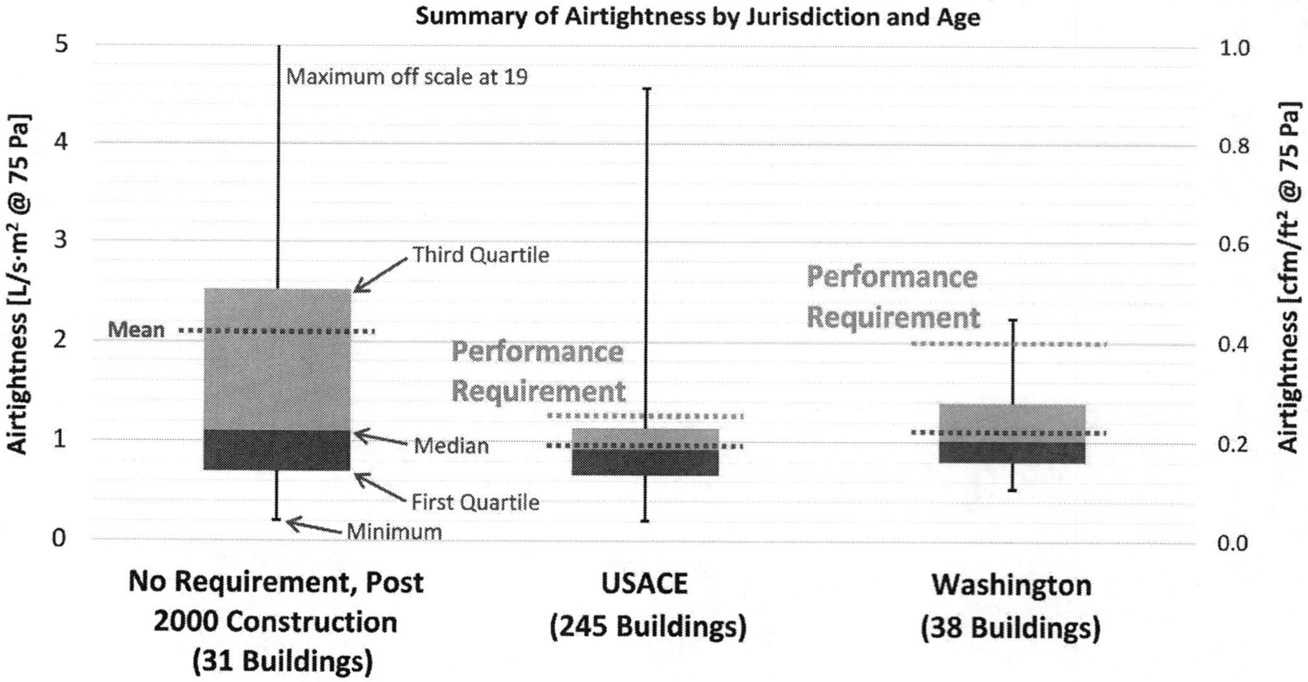

Figure 5 *Distribution of airtightness performance of buildings tested for research purposes (i.e., no requirement), for compliance with USACE requirements, and for compliance with Washington State or Seattle Energy Code requirements but limited to buildings tested more recently than the year 2000. By filtering out test results of older buildings, a more direct comparison is possible between typical modern construction practices and the jurisdictions with mandatory testing. Note the change in y-axis scale for this figure as compared to Figure 4.*

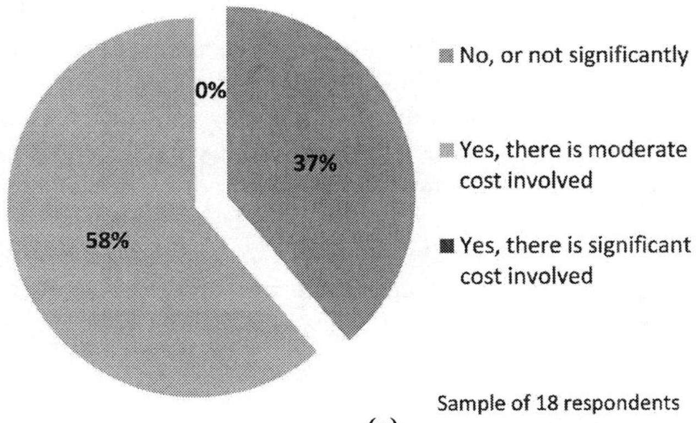

(a)

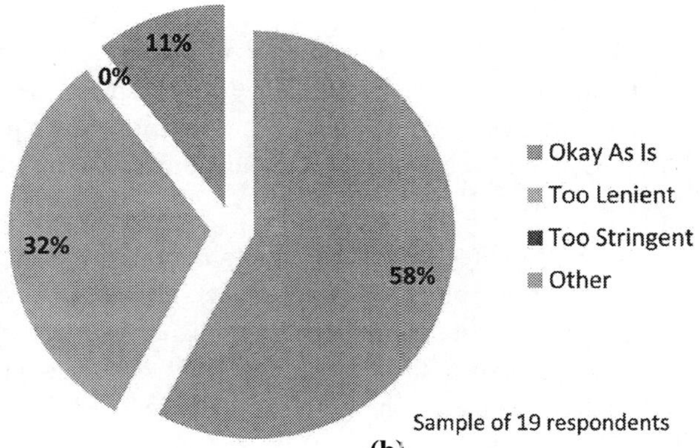

(b)

Figure 6 *(a) Chart of responses to the question, "Aside from the cost of the test itself, do you feel that whole-building air leakage requirements increase the total cost of construction?" and (b) chart of responses to the question, "Choose one of the following statements that best represents your opinion of the current whole-building air leakage target in your jurisdiction." This survey was distributed in the State of Washington where the guideline performance testing requirement is set at 2.0 L/(s·m^2) (0.40 cfm/ft^2).*

To gain an understanding of how these regulations have impacted the building industry in Washington State, and specifically Seattle, survey participants were asked a series of question pertaining to impact and perception of the requirements. The responses to three of these questions are summarized in Figure 6 and Figure 7.

Figure 6a clearly indicates that respondents generally felt that the cost implications of airtightness testing on the project were either minor or not significant, and Figure 6b indicates that respondents generally felt that the current airtightness performance target of 2.0 L/(s·m^2) (0.40 cfm/ft^2) is achievable. The latter finding is consistent with the testing results, which indicate that buildings are consistently able to achieve the stated airtightness performance target in Washington/Seattle despite there being no consequence for failure until recently.

Figure 7 illustrates the responses to the question "In general, do you feel that whole building air leakage requirements are beneficial and worthwhile in terms of increased building performance and quality of design/workmanship?" This question is of particular importance as it provides the respondents with an opportunity to holistically evaluate whether the regulated airtightness requirements are worth the additional effort and expense. The responses to this question indicate that the vast majority of respondents (84%) feel that the airtightness regulations are both beneficial and worthwhile, while only 5% of respondents felt they were neither beneficial nor worthwhile.

Overall, the survey results indicated a strong positive perception of the current airtightness regulations in the State of Washington and Seattle, both with respect to limited cost, achievability, and benefit to the project. These findings provide qualitative indication that perceived barriers to implementation can be readily overcome to the net benefit of building projects.

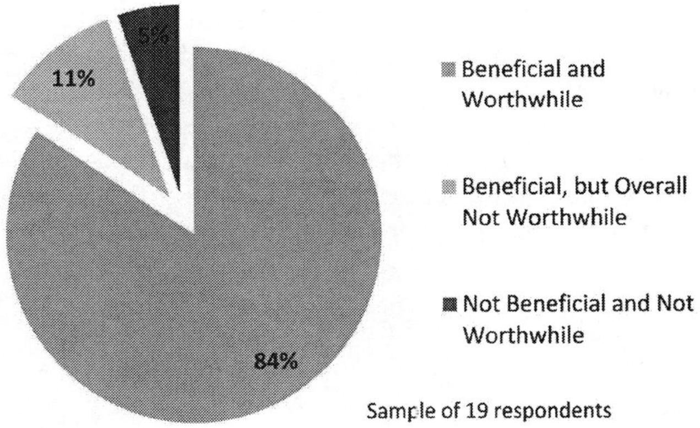

Figure 7 *Chart of responses to the question "In general, do you feel that whole-building air leakage requirements are beneficial and worthwhile in terms of increased building performance and quality of design/workmanship?"*

AIRTIGHTNESS AND ENERGY

One of the key drivers for improving the airtightness of buildings is to reduce the energy consumption associated with conditioning air that infiltrates through the building enclosure. That said, it is important to recognize the difference between air leakage and airtightness. For the purpose of this paper, air leakage is defined as *the in-service exfiltration or infiltration of air through the building enclosure* and airtightness is defined as *a physical property of the building enclosure that represents the resistance to airflow that it provides*. The amount of air leakage that occurs is dependent on a number of factors, including climate (wind and temperature), local exposure, building height, mechanical system operation, occupant behavior, and the airtightness of the building enclosure. It is air leakage, rather than airtightness, which is of significance with respect to energy consumption as well as various other performance considerations including air quality, moisture durability, and occupant comfort.

As an example, it is necessary to accurately predict air leakage rates for predictive energy modeling exercises such as those used for code compliance and sizing of mechanical systems. Analysis of the airtightness database has shown that airtightness regulations have substantially reduced the range in expected performance, which consequently allows designers to more accurately predict the airtightness of a building prior to construction; however, airtightness of the building enclosure alone is insufficient to predict air leakage. In large part this is because of variable pressure differences across the building enclosure as the result of wind, stack effect, and mechanical ventilation systems. These pressure differences can commonly cause building enclosures to experience a range of –25 to 25 Pa (–0.1 to 0.1 in. w.c.) from the interior to the exterior. (Ricketts 2014). Figure 8 illustrates the range of typical in-service pressure differences that buildings experience as well as the pressure flow relationships for building enclosures with a range of airtightness performance levels. This figure clearly illustrates that despite knowledge of the airtightness of a building enclosure, it is still difficult to accurately predict the in-service air leakage of buildings.

Because of both the complexity and quantity of contributing factors, calculation of building infiltration rates is difficult and often inaccurate with limited measured results available for large buildings to allow for validation of existing infiltration models. As other aspects of building energy consumption are improved, and as energy codes move towards performance-based targets such as energy use intensity (EUI) to define compliance, the importance of accurate in-service air leakage rates will increase. Further research is required in this area to validate and, as necessary, create, infiltration models that can account for the various factors that affect large building air leakage.

CONCLUSION

In recognition of the broad impact of air leakage on building performance, the building industry in North America is moving towards whole-building airtightness regulations in place of existing material, assembly, and component requirements. Based on compilation of more than 500 airtightness test results, it is clear that adoption of airtightness regulations has substantially improved the airtightness of large buildings. Furthermore, the expected range of performance has been dramatically reduced, allowing for more reliable prediction of performance. Surveying industry members in the State of Washington found that implementation of these regulations had minor cost implications for building projects, that the current target of 2.0 L/(s·m^2) (0.40 cfm/ft^2) is readily achievable, and that overall requirements are perceived as both beneficial and worthwhile.

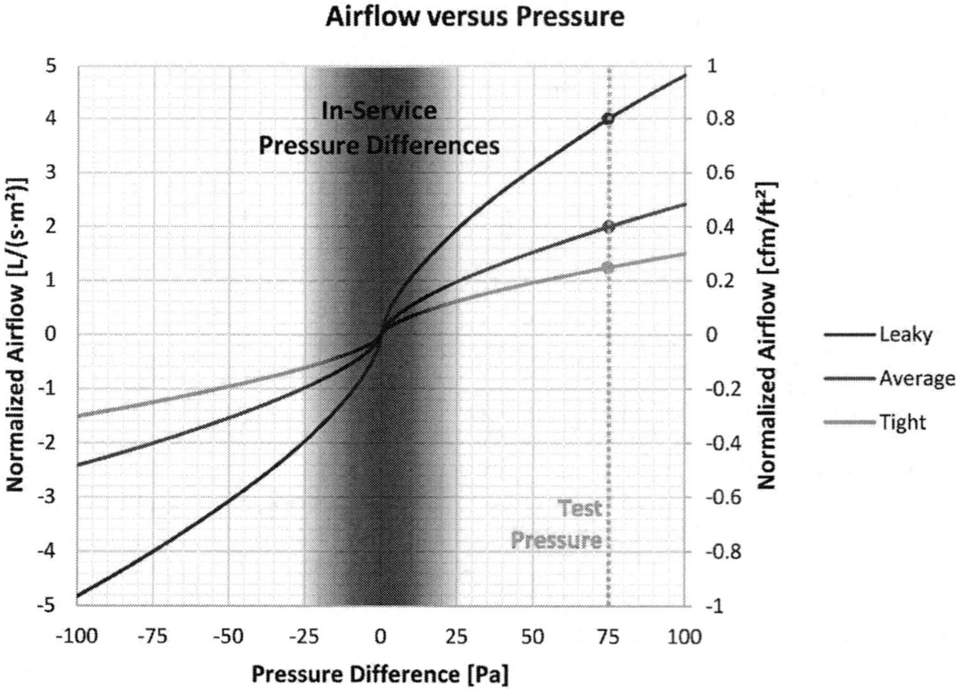

Figure 8 *Distribution in-service pressure difference that can be expected at buildings with pressure flow curves for building enclosures with different levels of airtightness. This chart shows how knowledge of the airtightness of a building is insufficient to accurately predict in-service air leakage rates.*

These documented improvements in airtightness will undoubtedly provide performance benefits; however, it is difficult to quantify these benefits as air leakage occurs based on a combination of factors, only one of which is building enclosure airtightness. While the narrowed expected range of performance allows for more reliable predictions of building airtightness during the design phase, further work is required to more accurately correlate airtightness with air leakage such that accurate predictions of performance can be made.

ACKNOWLEDGMENTS

The larger study on which this paper is based was funded and reviewed by the National Research Council of Canada as part of efforts to implement whole building airtightness testing requirements in the National Energy Code of Canada for Buildings.

REFERENCES

City of Seattle. 2014. *Seattle adoption and amendment of the 2012 edition of the International Energy Conservation Code, commercial*. Seattle: City of Seattle.

RDH Building Science. 2015. Study of part 3: Building airtightness. Vancouver, BC: RDH Building Science, Inc.

Ricketts, L. 2014. *A field study of airflow in a high-rise multi-unit residential building*. Master's thesis. Civil Engineering Department. Waterloo, ON: University of Waterloo.

USACE & ABAA. 2012. *U.S. Army Corps of Engineers air leakage test protocol for building envelopes*. Champaign, IL: U.S. Army Corps of Engineers & Air Barrier Association of America.

Washington State. 2012. *Washington State energy code, commercial provisions*. Olympia, WA: Washington State.

Air Sealing Tight Commercial and Institutional Buildings

David Bohac, PE
Member ASHRAE

James Fitzgerald

Martha Hewett

Joshua Novacheck

Andrew Lutz

ABSTRACT

Air leakage investigations were conducted on 26 commercial and institutional buildings located in Minnesota. The buildings were constructed from 1936 to 2009 and ranged in size from 23,920 to 484,290 ft² (2220 to 44,990 m²). The investigations found an average of 13.3 ft² (1.24 m²) of leakage to seal with the cost per area to be sealed ranging from \$937 to \$8490/ft² (\$10,100 to \$91,400/m²). There was no significant relationship between the relative amount of sealing identified and the year of construction or the energy utilization index. Approximately 59% of the area to be sealed occurred at wall/roof joints, 17% at exterior doors, and the remaining 23% distributed fairly evenly between soffits, windows, mechanical penetrations, and other locations. The specified gap width for the leaks was typically 1/16 to 1/8 in. (1.6 to 3.2 mm) for roof/wall joints, 1/16 in. (1.6 mm) for doors, and an average of 1/6 in. (4.2 mm) for soffits. The presealing envelope leakage tests conducted on eight buildings determined that they were relatively tight with normalized leakages that ranged from 0.09 to 0.23 cfm/ft² (1.64 to 4.18 m³/h·m²). The sealing areas estimated from the investigations were optimistic. The ratio of estimated sealing area to measured equivalent leakage area ranged from 0.76 to 1.76 with a median of 0.90. The measured leakage reduction ranged from 2.6% to 17.2% with an average of 9.0%. Post-inspections indicated that the lower measured reductions were not due to incomplete air sealing, but to an inability to properly estimate leakage areas. The lower than expected leakage reductions resulted in an average cost per area sealed of \$10,060/ft² (\$108,000/m²) and long paybacks.

INTRODUCTION

Most building architects, engineers, contractors, and facility managers used to think of large buildings as fairly airtight and assumed that enclosure air leakage did not have a significant impact on energy use in commercial and institutional (C&I) buildings (Persily 1988). Emmerich's (2005) modeling results contradicted that conventional wisdom, estimating that air infiltration can account for 33% of total heating energy use in U.S. office buildings. Air infiltration is caused by wind and thermal stack effect pressures, and HVAC system flow imbalances can increase or decrease infiltration. Most systems are designed to pressurize buildings, but often they either do not operate properly or they create less pressurization than needed to significantly reduce infiltration (Bohac et al. 2016).

There have been numerous efforts to improve the tightness of larger new buildings in the United States. Thirteen states, including Minnesota, have recently incorporated air barrier requirements into their energy codes for new C&I buildings (ABAA 2013). The U.S. Army Corps of Engineers has established a performance airtightness requirement of 0.25 cfm/ft² (4.5 m³/h·m²) of total enclosure surface area at a pressure of 75 Pascals (Pa) for both new buildings and major retrofits (USACE 2010), and other codes and organizations have adopted a standard of 0.40 cfm/ft² (7.3 m³/h·m²) (Anis 2013). It is becoming standard practice to require testing of new buildings. For example, the City of Seattle requires a whole-building leakage test for all new buildings and the State of Washington requires a whole-building leakage test for buildings over five stories. In addition, the EPA ENERGY

David Bohac is the director of research and **James Fitzgerald** is a senior building analyst at the Center for Energy and Environment (CEE), Minneapolis, MN. **Martha Hewett** was previously the director of research and **Joshua Novacheck** and **Andrew Lutz** were research analysts at CEE.

STAR® Multifamily High Rise Program for new construction requires a measured leakage rate no greater than 0.30cfm/ft^2 at 50 Pa. In addition to prescriptive and performance requirements, the air barrier needs to be properly specified, implemented, and commissioned. A recent ASHRAE study of newer C&I buildings found that buildings that had an air barrier consultant involved in the design and/or construction process were generally tighter than those that did not (Anis et al. 2013).

While these efforts will improve the performance of new construction, the building stock that existed prior to this increased emphasis on envelope tightness contributes much more to overall energy use than buildings built since these changes, and will continue to do so for many years as the C&I building stock turns over slowly (between 2% and 2.5% a year) (CBECS 2012). Reducing air leakage energy use in existing buildings requires effective screening tools to identify buildings with a higher probability of having treatable leaks, investigation methods to identify the precise location of key envelope air leakage deficiencies and/or mechanical system pressure issues, and reliable procedures to estimate cost and savings.

METHODOLOGY

Air Leakage Investigations

An air sealing consultant and contractor led the investigations with assistance from project staff. The building air leakage investigations used an exterior infrared (IR) survey and interior leak investigation with smoke visualization. The exterior IR surveys were conducted using the building's HVAC system or temporary fans to pressurize the buildings (ASTM 2009) at a minimum inside to outside temperature difference of 18°F (10°C) and with no direct solar radiation on the surfaces for four to six hours prior to the survey. The results of the IR survey and experience of typical leakage sites helped guide a visual envelope leakage investigation. From the building interior, a smoke puffer was used to release chemical smoke near suspected leakage sites. The velocity and volume of smoke movement indicated the relative magnitude of air leakage.

The investigations typically focused on wall/roof intersections, roof elevation changes, exterior doors, mechanical system penetrations, and windows. Air sealing recommendations targeted larger gaps that were accessible with little or no potential for damage to surrounding materials. The project consultant and contractor stated that the simple payback on energy savings is typically too long for owners for leaks that require removal and refinishing of building materials in order to create access to conduct sealing. The exceptions are very large air leaks or those that can create additional "high value" issues, such as moisture damage or frozen pipes. Investigators typically check the leakage or gap width at a sample of representative locations for a type of leak and assume the findings are typical for the entire leak. The leakage findings from all the locations would then be used to determine whether the entire joint should be sealed and to estimate the typical gap width for the joint. It is not common to specify sealing of only portions of joints, because a complete barrier is usually necessary for effective sealing. Doors are typically evaluated and sealing measures specified on an individual basis.

Envelope Airtightness Tests

Multipoint, pressurization, and depressurization whole-building envelope airtightness tests were conducted before and after air sealing work was performed. The tests generally followed the requirements of ASTM E779-10 (2010), with additions to address the complexities of testing larger buildings. The average of four ground level in/outdoor pressure measurements placed on different sides of the building was used to indicate the building indoor-to-outdoor pressure difference. Building baseline pressures were measured for at least five minutes before and after both the pressurization and depressurization tests. Multiple calibrated fans were used to vary the baseline adjusted building in/outdoor pressure at evenly spaced increments from approximately 0.06 to 0.30 in. wc (15 to 75 Pa) for 60 seconds at each level. All mechanical dampers were closed, and the dampers or terminations of the outside air ducts, exhaust air ducts, and exhaust fans were temporarily sealed. Project staff used commercially available software to record building pressure differences, record fan flow rates, control test fan speeds, graphically display the measurements, and compute air leakage values. Data quality was optimized by using distributed gages to minimize tube lengths and real-time regression analysis to identify erroneous measurements. Project staff conducted all of the envelope leakage tests.

Air Infiltration and Energy Modeling

CONTAMW airflow models (Walton and Dols 2013) were developed to evaluate the impact of air infiltration on building space heating and cooling energy consumption. CONTAMW is a multizone indoor air quality and ventilation analysis computer program developed by the National Institute of Standards and Technology to calculate time varying infiltration, exfiltration, and area-to-area building airflow rates. The model includes driving forces due to HVAC flow imbalances, wind pressures on the building exterior, and thermal buoyancy effects induced by the indoor and outdoor temperature differences. The driving forces are applied to a network of flow path elements defined by the user to compute airflow rates between the outside and building interior and within the building interior. The flow path elements (e.g., envelope air leaks and mechanical system leakage) are distributed vertically and on different building faces to account for thermal buoyancy and wind direction effects on building pressures.

Building envelope air leakage measurements were used to determine the sum of the leakage for all of the flow paths assigned to the building exterior. Building pressure measurements and physical dimensions were used to distribute the leakage by building height and orientation. The building indoor to outdoor pressure measurements were filtered by HVAC operating mode and binned by outside temperature to establish algorithms for the HVAC flow imbalance. Hourly simulations were performed using annual hourly typical meteorological year outside air and wind data from the National Oceanic and Atmospheric Administration weather station closest to the building. A fixed heating and cooling balance point method was used to compute the hourly space heating and cooling loads and energy use. Additional energy computation information is available from a companion conference paper (Bohac et al. 2016).

RESULTS AND DISCUSSION

AIR LEAKAGE INVESTIGATIONS

Air leakage investigations were conducted on 26 C&I Minnesota buildings with 16 of them institutional, five office, three medical, and two public. The buildings were constructed from 1936 to 2009 and ranged in size from 23,920 to 484,290 ft^2 (2220 to 44,990 m^2) with an average floor area of 149,070 ft^2 (13,850 m^2). An air sealing consultant/contractor with experience in specifying and sealing hundreds of large buildings led the investigation, work scope preparation, sealing cost estimate, and energy savings estimate at 22 of the buildings. A second air sealing contractor completed the same tasks for four other buildings. Project staff conducted follow-up investigations at those four building and the contractor expanded the initial work scope based on staff feedback. Air sealing and pre/post-testing was completed for three of those buildings (Elementary School 1, Elementary School 2, and Middle School). Because the same project staff was involved with the investigations for all 26 buildings, the approach for identifying air sealing opportunities was consistent across all project buildings. The investigation reports specified the type of sealing to be performed (e.g., caulk, foam, weatherstripping/sweep), quantity (length of gap or quantity of leakage sites), leak or gap width, and leakage area to be sealed for each leakage site. The reports also included the total cost and estimated energy savings for all sealing work on the building. The savings estimates used a heating and cooling degree day method with a proprietary conversion of leakage area to average infiltration rate based on building type and other factors.

The leakage area to be sealed ranged from 1.6 to 46.5 ft^2 (0.15 to 4.32 m^2) with a median value of 10.0 ft^2 (0.93 m^2) and interquartile range (IQR) of 5.5 to 16.5 ft^2 (0.51 to 1.53 m^2). The sealing area was divided by the floor area to evaluate the relative amount of sealing that could be identified in buildings of various sizes. This normalized sealing area was fairly evenly distributed from 0.01 to 0.28 $\times 10^{-3}$ with a median of 0.11 $\times 10^{-3}$ and IQR of 0.06 to 0.17 $\times 10^{-3}$. Least-square linear regressions were performed separately for four different variables (construction year, number of stories, site energy use intensity[1] [EUI], and site EUI divided by a benchmark for the building type) to determine whether any of the characteristics could help identify buildings that are more likely to require air sealing work. The number of stories was the only variable that provided a significant relationship (R^2= 0.24 and slope p-value = 0.01), and the regression indicated a 0.007 $\times$ 1010^{-3} decrease in sealing per story (95% confidence interval: –0.0118 to –0.0017 $\times 10^{-3}$). This might be expected, because roof/wall sealing was the most frequent measure (see results later in this section) and shorter buildings have a higher fraction of roof/wall length per floor area than taller buildings. It is somewhat surprising that construction year, EUI, and EUI/benchmark were not useful predictors of the relative amount of sealing work. However, Emmerich and Persily (2014) reported that a national database of air leakage test results shows little or no correlation of normalized building leakage with construction year. Some of the possible explanations for the poor correlation of sealing area with total building energy use include the following: (1) there is significant variety in the fraction of envelope leakage that can be sealed compared to the total envelope leakage, (2) air infiltration does not typically have a large impact on energy use, and (3) the estimates of area to be sealed were not highly accurate.

The air sealing cost for the 26 buildings ranged from \$2240 to \$200,508 with a median of \$18,805. There was little difference in the cost per area sealed for the project consultant and contractor. The median cost per area sealed provided by the contractor for the four schools was only 10% less than the median for the other 22 buildings. The cost normalized by the floor area ranged from \$8 to \$541 per 1,000 ft^2 (\$8.6 to \$582 per 1000 m^2) with a median of \$208 (\$224/1000 m^2). There was no apparent relationship between the relative sealing cost and construction year. Regression analysis confirmed that there was no significant relationship between relative sealing cost with the construction year and relative EUI. There was a relatively narrow span in the cost per area sealed with 80% of the values ranging from about \$1250 to \$2500 per ft^2 (\$1,345 to \$2,690/m^2). The sealing cost per area sealed can be used along with an estimate of energy cost savings per area sealed to determine whether the work will be cost effective. The average sealing cost per area sealed was \$2195/ft^2 (\$23,600/m^2). Similar to the other sealing areas and cost, there was no statistically significant relationship between the cost of sealing per area sealed with construction year or EUI/benchmark. The median simple payback of air sealing based on energy savings alone was 5.9 years, and 80% of the values ranged from 4.0 to 8.7 years. This is towards the upper range of what most C&I building managers will consider for an energy improvement when other benefits are not considered. The relatively narrow range of cost per area to be sealed and payback should be expected because the investigators have an understanding of the cost effectiveness required by owners and are not likely to

1. The EUI and relative EUI values were only available for 17 of the buildings, and the regressions for the EUI variables were conducted for only those 17 buildings.

include work that exceeds those limits. There was no significant relationship between payback and construction year. There was a statistically significant relationship between the relative EUI and payback ($R^2 = 0.30$ and slope p-value = 0.02) with an unexpected result of shorter paybacks for lower EUI buildings. However, the relationship is largely driven by three buildings with atypical sealing or energy use situations and the relationship is not statistically significant ($R^2 = 0.14$ and slope p-value= 0.16) when those three buildings are removed from the relative EUI versus payback analysis.

The investigator specified the type and area of leakage for each leakage site. Almost all of the leaks were linear gaps with the area specified by a length and average gap width. Over the 26 buildings 59% of the area to be sealed occurred at wall/roof joints, 17% at exterior doors, and the remaining 23% distributed fairly evenly between soffits, windows, mechanical penetrations, and other locations. Only two buildings had no roof/wall sealing, while roof/wall sealing was responsible for over half of the sealed area for 20 (77%) of the buildings (see Figure 1). The results are consistent with the findings from an ASHRAE study of 16 recently constructed U.S. buildings (Brennan et al. 2013) and confirms that air leakage investigations should prioritize the top of the building, in particular where the exterior walls meet the roof. Exterior door weatherstripping, sweeps, and/or stops were specified for every building. However, that typically accounted for only 18% of area to be sealed, and the most it accounted for in a building was 45%. Applying weatherstripping and sweeps to single doors accounted for 80% of the door sealing. Overhead doors are usually leaky, but sealing overhead doors was only identified for four buildings. Mechanical penetration, window, and soffit sealing was specified for 46%, 38%, and 31% of the buildings respectively.

Sealing cost effectiveness is governed by the sealing cost per length of leakage and the average width of the gap being sealed. Data from the leakage reports on the specified gap width by type of area to be sealed was compiled to evaluate the frequency of the widths by leak type. Soffits were expected to have the largest gaps with an average value of about 1/6 in. (4.2 mm). In contrast, window gaps were specified to be 1/64 in. (0.4 mm) 64% of the time, and, on average, the gap was about 1/30 in. (0.8 mm). Exterior doors to be sealed were almost

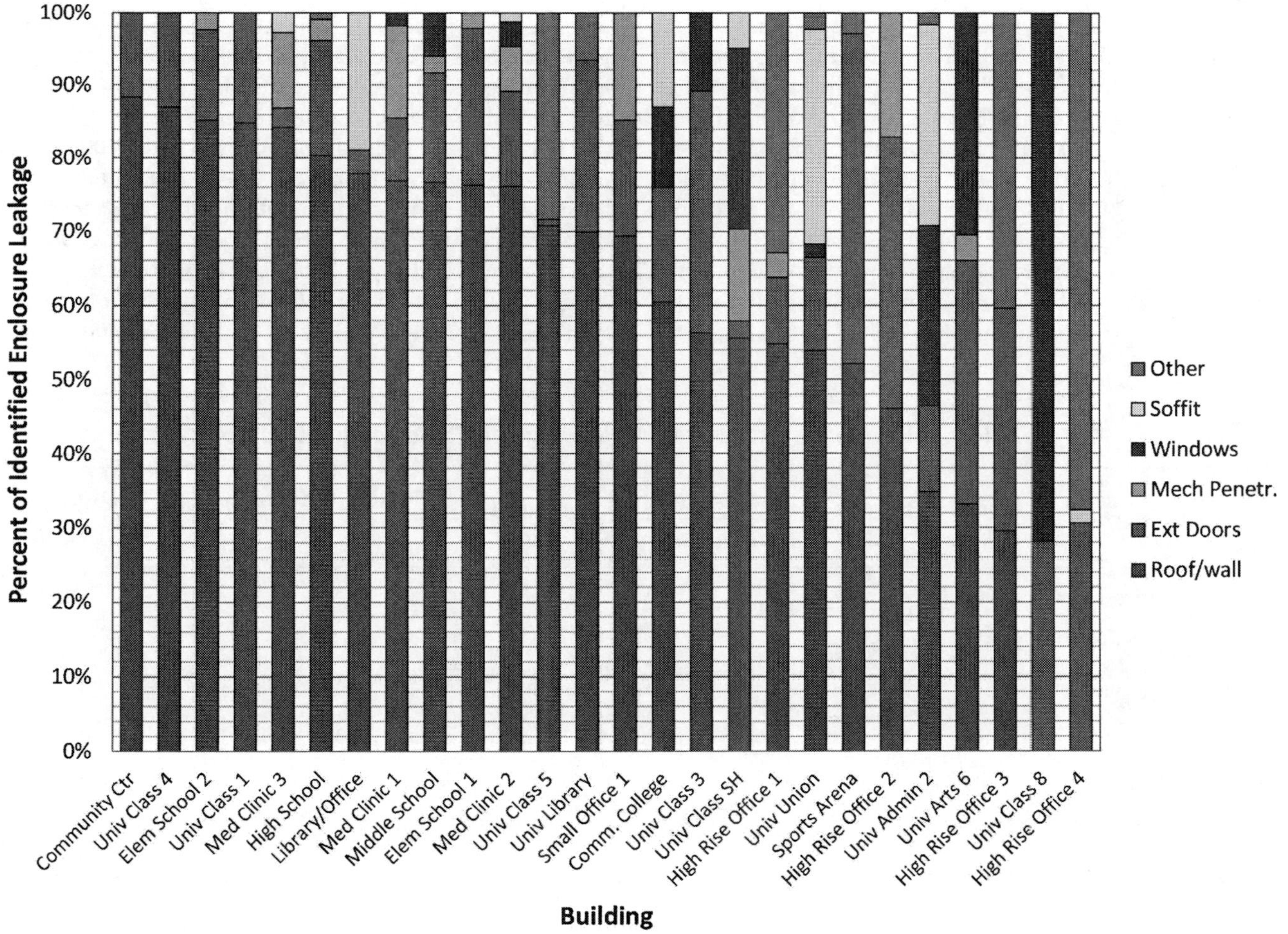

Figure 1 *Distribution of leakage type for each building.*

always specified to have a gap of 1/16 in. (1.6 mm). On average, the gap for mechanical penetrations and roof/wall sealing was about 1/10 in. (2.5 mm). There was more variety in the roof/wall gap, with a minimum of 1/32 in. (0.8 mm) and maximum of 1/8 in. (3.2 mm). It is important to note that these are estimated values based on the consultant or contractor's experience and are not measured values.

Evaluation of the total costs along with the quantity and type of sealing performed was used to estimate the following range of typical costs per length of sealing: roof/wall, \$8 to \$16/ft (\$26 to \$52/m); single doors weatherstrip and sweep, \$160 to \$240/door; window glazing, \$2 to \$6/ft (\$6.6 to \$20/m); window weatherstrip, \$3 to \$5/ft (\$9.8 to \$16.4/m); soffit, \$40 to \$100/ft (\$131 to \$330/m); and mechanical penetrations, \$40 to \$60/ft (\$130 to \$200/m). Information for the range of sealing costs was used with the results for the frequency of gap width to compute the cost per area sealed. Roof/wall sealing tended to have the lowest cost per area sealed. Roof/wall gaps were specified to have a gap width from 1/16 to 1/8 in. (1.6 to 3.2 mm) with a typical cost of \$8 to \$10 for "one-line" sealing and up to \$16/ft (\$53/m) for three-line or more difficult sealing. Consequently, wider (1/8 in. [3.2 mm]) gaps that only required one-line sealing only cost \$960/ft^2 (\$10,300/m^2). Exterior door sealing typically cost \$10/ft (\$32/m) to seal and had a specified gap of 1/16 in. (1.6 mm), resulting in a typical cost of \$1920/ft^2 (\$20,700/m^2). Windows tended to be more expensive to seal. They were most often specified as having a 1/64 in. (0.4 mm) gap, resulting in a reasonable cost of \$1536/ft^2 (\$16,500/m^2) for less expensive (interior) sealing and a high cost of \$4,608/ft^2 (\$49,600/m^2) for more expensive (exterior) sealing. However, window sealing is typically a required maintenance item that also improves building durability. Soffit and mechanical system penetration sealing is four to ten times more expensive per length than roof/wall sealing. However, this is due to the fact that the gap widths are wider, resulting in typical sealing costs from \$2000/ft^2 to \$4000/ft^2 (\$21,500 to \$43,100/m^2). Although none of the soffit gaps for these buildings was specified to be wider than ¼ in. (6.4 mm), those gaps can be ½ in. (12.7 mm) or greater in leaky buildings, resulting in more cost-effective sealing.

Existing Envelope Air Leakage and Infiltration

Whole-building envelope air leakage tests were conducted on eight buildings before air sealing work was performed. The buildings were one to three story tall and built from 1936 to 2007 with average floor and six-sided envelope areas of 86,877 ft^2 (8071 m^2) and 133,151 ft^2 (12,370 m^2), respectively (see Table 1). The buildings were much tighter than expected, even though none were required to meet an envelope tightness test standard and half were built before there were any air barrier code requirements. The average normalized tightness of 0.16 cfm/ft^2 (2.84 m^3/h·m^2) is 83% less than the average value of 0.92 cfm/ft^2 (16.7 m^3/h·m^2) reported by Emmerich and Persily (2014) for a convenience sample of 228 U.S. C&I buildings that were not built to a tightness standard. All of the buildings were tighter than the 0.25 cfm/ft^2 (4.54 m^3/h·m^2) USACE tightness standard. It is possible that buildings in cold climates are tighter due to greater concern for cold drafts, frozen pipes, and higher space heating costs. For the buildings in the January 2014 version of the NIST database, the median leakage was 0.43 cfm/ft^2 (7.82 m^3/h·m^2) for buildings in a climate with 3600 or more heating degree days (Emmerich 2014). Linear regressions were used to determine that for the eight buildings there was no significant correlation between envelope leakage and construction year, floor area, and number of stories. Emmerich and Persily (2014) found a relationship between the number of stories and floor area with leakage for the U.S. building data-

Table 1. Pre-Sealing Measured Air Leakage and Modeled Infiltration

Building ID	Floor Area, ft^2	Envelope Area, ft^2	Leakage at 75 Pa		Sealing/ EqLA	Model Infilt		Rqd Ventil, ach	Infilt/ Ventil
			cfm	cfm/ft^2		cfm	ach		
Small Office 2	11,904	21,945	5,056	0.23	NA	NA	NA	NA	NA
Elementary School 1	59,558	146,977	27,425	0.19	0.76	1296	0.09	1.41	0.06
Community College	95,000	164,844	28,881	0.18	0.78	1730	0.08	1.65	0.05
Middle School	138,887	208,733	32,818	0.16	0.92	4330	0.14	1.71	0.08
Small Office 1	26,927	65,267	9,177	0.14	0.90	964	0.13	0.64	0.21
University Library	246,365	171,712	23,356	0.14	0.82	249	0.01	1.26	0.004
Elementary School 2	60,968	145,766	17,602	0.12	1.76	1453	0.12	2.22	0.06
Library/Office	55,407	139,965	12,321	0.09	0.95	1477	0.10	0.85	0.12
Average	86,877	133,151	21,654	0.16	0.99	1643	0.10	1.39	0.08

base and no significant relationship between construction year and building leakage. A larger sample for a greater variety of Minnesota buildings may produce significant correlations.

The ratio of the specified air sealing area to measured EqLA[2] was used to evaluate the fraction of existing building leakage that was expected to be sealed (see column Sealing/EqLA of Table 1). The ratio varied from 0.76 to 1.76 with a median of 0.90. This indicates that for all of the buildings at least 76% of the leakage would be sealed and typically 90% would be sealed. For one of the buildings (University Library) the amount of sealing was 75% greater than the total building leakage. This high fraction of sealing seems overly optimistic and suggests that the specified sealing areas may be greater than can actually be achieved. Since the roof/wall sealing was responsible for 75% of the specified sealing area, the gap designations for that leak may be greater than actual. If investigators use the gap width of the observed opening at the interior portion of the leak, they may tend to overestimate the effective gap width for the leak. The flow restriction across an entire leakage path may be only partially due to the restriction at the interior surface. There may be other restrictions along the leakage path that are not visible from the building interior. Finally, leakage investigations do not often include a leakage test before or before and after sealing. Most air leakage consultants and contractors do not get feedback on the comparison between their specified leakage areas and measured values that would help them better estimate areas for future projects.

Air sealing was performed at seven of the eight buildings. CONTAMW airflow models were generated for those seven buildings to evaluate the impact of air infiltration on building space heating and cooling energy consumption. The average infiltration rates include the time periods when the HVAC system was off and on. Table 1 lists the annual average infiltration rates along with the measured envelope leakage and computed code required ventilation rate for fully occupied conditions. The model infiltration rates vary from 249 to 4330 cfm (7050 to 123,000 Lpm) and average 1643 cfm (46,500 Lpm) or 0.10 air changes per hour (ach). The ratio of air infiltration to code-required ventilation was used as a reference for excess outside air due to infiltration. The ratios of model infiltration to code required ventilation varied from 0.004 to 0.21 and averaged 0.08 (see Table 1). This indicates that, on average, the infiltration was responsible for adding about 8% to the outside air flow beyond code requirements. That value is a low estimate of the relative amount of added outside air, because the ventilation requirement is only necessary when the building is occupied. The buildings are typically occupied and require ventilation about one third of the time. Consequently, the average impact on annual average outside airflows is about a 25% increase over ventilation requirements (0.08 divided by a third).

2. The equivalent leakage area (EqLA) is calculated at a 10 Pa pressure difference assuming the leakage area is a single, sharp-edged orifice with a discharge coefficient of 0.61. It is approximately equal to the physical area of the building leakage, but the relationship depends on the type of leakage.

Envelope Air Sealing

The entire air sealing work scope was completed at the three schools and more targeted work was performed at the other four buildings. Figure 1 displays a breakdown of the type of sealing that was completed at the schools. There were minor adjustments to the school work scopes due to access issues and a reassessment of the magnitude of leakage at some locations. The final work scopes included estimated sealing areas of 8.5, 15.6, and 16.9 ft^2 (0.79, 1.45, and 1.57 m^2) for the three buildings (see Table 2). Lower than expected leakage reductions at the three schools, pre-sealing tests that showed that the envelopes were tighter than expected, and limits on project funds led to a reduction of the scopes at the other four buildings. That work targeted sealing of larger gaps that were expected to produce more cost-effective leakage reductions. For example, sealing of the Community College focused on 140 ft (43 m) of roof/wall sealing at nine locations where the investigation found greater air movement through the leaks. An additional 1050 ft (320 m) of roof/wall sealing was excluded due to indications that there was less leakage at those locations. The scope for the Small Office 1 building was reduced to only include sealing in the unoccupied portion of the building: ten truss pockets, the ductwork perimeter of five RTUs, and one powered roof ventilator. This totaled 45 ft (18 m) of sealing. The roof/wall joint and two RTUs in the occupied space were not sealed. The sealing of the University Library focused on about 140 ft (43 m) of roof/wall sealing at the stairwell bump outs. The investigations indicated higher leakage in those locations, and there had also recently been extensive moisture damage to masonry work in those areas that had required expensive repairs. Project staff completed the limited sealing of a 26 ft (7.9 m) long soffit or canopy leak at the Library/Office building. Staff monitored sealing work at all of the buildings, and post-sealing IR and smoke puffer visual inspections confirmed that the sealing either eliminated or greatly reduced air movement at all of the sealing locations except a portion of the wall/roof at the University Library.

The post-sealing envelope leakage tests determined that the sealing resulted in significantly less leakage reduction than expected. The leakage reduction ranged from 2.6% to 17.2% with an average of 9.0% and average reduction in normalized leakage of 0.013 cfm/ft^2 (0.24 $m^3/h{\cdot}m^2$). The sealing results are evaluated separately for two groups of buildings based on the sealing approach for each group. The three school buildings had comprehensive work scopes that were specified and completed by an experienced air sealing contractor who also generated the estimate of leakage area reduction and energy cost savings. It is encouraging that a moderate amount of reduction (average = 13.1%) was possible even though the buildings were all at least 25% tighter than the USACE standard. However, the ratio of the measured change in the equivalent leakage area (ΔEqLA) to the contractor-estimated sealing area was only 0.04 to 0.32. The sealing area estimates were based on their visual inspections and experience from previous projects. The estimates

Table 2. Measured Air Leakage and Sealing Costs

Building ID	Leakage at 75 Pa, cfm/ft²		Red, %	Measured EqLA, ft²		Sealing Cost	Cost/ Reduction $/ft²	Est. Seal, ft²	ΔEqLA/ Est. Seal
	Pre	Reduction		Pre	Reduction				
Elementary School 1	0.19	0.032 (0.003)	17.2%	15.18	2.72	$18,550	$6,822	8.51	0.32
Middle School	0.16	0.019 (0.002)	12.0%	16.64	2.81	$23,700	$8,434	15.61	0.18
Elementary School 2	0.12	0.012 (0.002)	10.0%	9.61	0.70	$26,700	$38,132	16.94	0.04
Community College	0.18	0.005 (0.002)	2.6%	17.24	1.03	$14,445	$13,982	2.50	0.41
Small Office 1	0.14	0.011 (0.002)	7.7%	4.59	0.47	$4,768	$10,058	NA	NA
University Library	0.14	0.008 (0.002)	6.0%	13.06	0.24	$15,918	$65,159	0.73	0.34
Library/Office	0.09	0.007 (0.002)	7.7%	6.85	0.89	$1,152	$1,297	3.25	0.27
Average	0.14	0.013 (0.002)	9.0%	11.88	1.27	$15,033	$20,555	7.92	0.26
Median	0.14	0.011 (0.002)	7.7%	13.06	0.89	$15,918	$10,058	5.88	0.30

were generated before the pre-sealing leakage tests were conducted. Those tests showed that the estimated sealing area was 56%, 94%, and 176% of the EqLA for the three schools. In order for the sealing work to achieve the estimated sealing area at the Middle School, the sealing would have had to eliminate almost all of the leakage. For the Elementary School 2 building, it was not possible to achieve the estimated sealing, because the estimate was 76% greater than the measured envelope leakage. The high fraction of leakage that was estimated to be sealed and post-sealing inspections that indicated that the sealing work significantly reduced leakage at the treated areas indicate that the low ratios of measured to estimated leakage reductions were due to overly high estimates of sealing area.

More targeted sealing work was performed at the other four buildings. The project consultant led the leakage investigation, work scope generation, and savings estimates. Experienced local contractors bid and completed sealing work at three buildings and project staff completed the canopy sealing at the Library/Office building. The average leakage reduction of 6.0% was about half that for the school buildings, but the average ratio of measured ΔEqLA to estimated sealing area of 0.34 was almost twice that of the average for the school buildings. This indicates that limiting the sealing to areas where qualitative methods indicate greater leakage can improve sealing area estimates. However, sealing areas were still overestimated by a factor of three.

The challenge with estimating leakage areas is that unless you can see through the leak to the outside, there may be other portions of the leakage path that have equal or greater restriction than the interior gap. For example, a large gap width of 1.5 in. (38 mm) was observed for the open flutes at the corrugated roof of the canopy for the Library/Office building. When the area was sealed, the ΔEqLA indicated that the gap width was about 0.4 ± 0.1 in. (10 ± 2.5 mm), or about 3.7 times less than the physical opening at the point of sealing. The velocity of puffer smoke movement through a gap provides an indication of the leakage size. However, the smoke velocity is also dependent on the pressure across the leak, and that is typically not controlled and measured during the inspection process. Since these buildings were much tighter than typical U.S. buildings, it is possible that the project contractor's leakage area estimating experience from leakier buildings caused them to overestimate leakage areas for these tighter buildings. This highlights the concerns with using the current qualitative area estimating techniques and the need to develop cost-effective quantification methods.

There was a large variation in the sealing cost per ΔEqLA ($1152 to $65,159/ft² [$12,400 to $701,000/m²]). However, for the three school buildings with comprehensive work scopes, the sealing cost per estimated area to be sealed was fairly consistent with a range from $1518 to $2180 and average of $1758. The largest values of sealing cost per ΔEqLA were due to high estimates of sealing area or gap widths of the linear leaks. Better methods to estimate gap widths could have helped prioritize sealing of larger leaks, which would have resulted in more cost-effective sealing. Also, envelope leakage tests would have shown that the estimated sealing areas were not realistic.The most cost-effective sealing of $1297/ft² was for the tightest (0.09 cfm/ft² [1.6 m³/h·m²]) Library/Office building with an isolated but large gap leak.The roof/wall sealing of the University Library building was the least cost-effective sealing. The expensive repairs to exterior masonry at the site of the leakage suggested that there was significant leakage. While it appears the leakage was not as large as anticipated, it is hoped that the sealing will reduce the likelihood of future masonry damage.

Post-sealing envelope air leakage measurements were used to modify the pre-sealing CONTAMW airflow models to evaluate the impact of envelope sealing on air infiltration flow rates and space conditioning energy consumption. The annual average air infiltration reductions varied from 15 to 504 cfm (425 to 14,300 Lpm) and averaged 154 cfm (4360 Lpm) (Table 3). There was almost a five to one range in the infiltration reduction per ΔEqLA and an average value of 99 cfm/ft^2. This indicates that there is no simple relationship between area sealed and infiltration savings, even for buildings in similar climates. In general, infiltration will be greater for taller buildings with less wind shielding and HVAC systems that do not impact building pressure or use economizing strategies. The space heating annual energy use savings ranged from a low of $9 for the University Library to a high of $758 for the Middle School, with an average for all seven buildings of $240. Including the space cooling savings improves the annual energy cost savings to an average of $309. On average, the space cooling savings accounted for 14% of the total savings. The measured reduction in the envelope leakage was significantly less than expected, which caused higher costs per area sealed and longer than expected paybacks. The simple paybacks ranged from 19 to 800+ years with a median of 84 years. Given that the average annual savings was about $190 per ft^2 ($2045/m^2) of leakage reduction, the sealing cost would have needed to be from $950 to $1800 per ft^2 ($10,200 to $19,400/m^2) for five to ten year paybacks. The only building that had less than a $5000 per ft^2 ($54,000/m^2) sealing cost was the Library/Office with a large, accessible leak of a corrugated roof at the canopy.

The contractor and consultant estimates of energy cost savings per area of leakage sealed were relatively consistent with an average of $441 per ft^2 and standard deviation of $183 per ft^2. The average savings per area was 2.3 times that for the CONTAMW modeling, but two of the values were within 12% of those for the modeling. The CONTAMW model included the impact of HVAC pressurization to reduce infiltration and the energy calculation accounted for lower space heat load during occupancy. Those were the two most significant factors in the difference between the contractor and modeled energy savings per area of leakage sealed.

CONCLUSION

The leakage investigations of 26 C&I buildings resulted in identified sealing areas divided by floor area that were fairly evenly distributed from 0.01 to 0.28 × 10^{-3} with a median of 0.11 × 10^{-3} and an IQR of 0.06 to 0.17 ×10^{-3}. A total of 59% of the area to be sealed occurred at wall/roof joints, 17% at exterior doors, and the remaining 23% distributed fairly evenly between soffits, windows, mechanical penetrations, and other locations. Least-square linear regressions were performed separately of construction year, number of stories, site EUI, and site EUI divided by a benchmark for the building type with the relative amount of sealing. The number of stories was the only variable that provided a significant relationship (R^2 = 0.24). The median sealing cost per 1000 ft^2 was $208 ($224 per 100 m^2), and the median payback was 5.9 years with 80% of the values ranging from 4.0 to 8.7 years. The narrow range in cost per floor area and payback most likely occurred because the investigators excluded work that they knew exceeded owner's cost effectiveness limits.

The buildings were much tighter than expected, even though none were required to meet an envelope tightness test standard and most were built before there were any air barrier code requirements. The average normalized tightness was 0.16 cfm/ft^2 (2.84 m^3/h·m^2), and all of the buildings were tighter than the USACE tightness standard. It is possible that buildings in cold climates are tighter due to greater concern for

Table 3. Modeled Infiltration Energy Savings

	Infiltration Reduction		Heating Savings		Cooling Savings		Total Savings		Payback, yrs
Building ID	cfm	cfm/ft^2	therm/yr	$	kWh/yr	$	$	$/ft^2	
Elementary School 1	285	105	549	$460	1,034	$121	$581	$214	32
Middle School	504	179	905	$758	2,523	$295	$1,053	$375	23
Elementary School 2	120	171	223	$187	487	$57	$244	$349	110
Community College	84	82	174	$146	232	$27	$173	$168	84
Small Office 1	17	37	39	$33	18	$2	$35	$74	137
University Library	15	60	11	$9	79	$9	$18	$75	861
Library/Office	55	62	107	$89	-232	-$27	$62	$70	19
Average	154	99	287	$240	592	$69	$309	$189	181
Median	84	82	174	$146	232	$27	$173	$168	84

cold drafts, frozen pipes, and higher space heating costs. The ratio of the specified air sealing area to measured EqLA varied from 0.76 to 1.76 with a median of 0.90, which seemed overly optimistic. CONTAMW model infiltration rates varied from 249 to 4,330 cfm (7050 to 123,000 Lpm) and averaged 1643 cfm (46,500 Lpm) or 0.10 air changes per hour (ach). The ratio of air infiltration to code required ventilation varied from 0.004 to 0.21 and averaged 0.08. The average impact on annual average outside airflows was about a 25% increase over ventilation requirements.

Even though the buildings were already quite tight, the leakage was reduced a moderate amount in all of the buildings. The measured change in building leakage from air sealing was much less than estimated by the investigators, and the cost per area sealed was higher than expected. The leakage reduction ranged from 2.6% to 17.2% with an average of 9.0% and average reduction in normalized leakage of 0.013 cfm/ft^2 (0.24 m^3/h·m^2). The ratios of ΔEqLA with revised estimates of sealing area ranged from 0.04 to 0.32 for the three schools with comprehensive work and was slightly better for the three buildings with more focused work (0.30, 0.34, and 0.41). It is possible that the project contractor's leakage area estimating experience from leakier buildings caused them to overestimate leakage areas for these tighter buildings. The space conditioning annual energy use savings ranged from a low of $18 for the University Library to a high of $1052 for the Middle School, with the average for all seven buildings being $309. The simple paybacks ranged from 19 to 800+ years with a median of 84 years. Given that the average annual savings was about $190 per ft^2 ($2045/m^2) of leakage reduction, the sealing cost would have needed to be from $950 to $1800 per ft^2 ($10,200 to $19,400/m^2) for five to ten year paybacks. The only building that had less than a $5000 per ft^2 ($54,000/m^2) sealing cost was the Library/Office with a large, accessible leak of a corrugated roof at the canopy.

ACKNOWLEDGMENTS

This project was supported in part (or in whole) by a grant from the Minnesota Department of Commerce, Division of Energy Resources, through the Conservation Applied Research and Development (CARD) program, which is funded by Minnesota ratepayers.

REFERENCES

ABAA. 2013. Air Barrier Association of America Codes, Standards & Test Methods. www.airbarrier.org/codes/index_e.php.

Anis, W. 2013. The changing requirements of airtightness in the U.S. Presented at AIVC Airtightness Workshop, Washington DC, April 18–19.

Anis, W., T. Brennan, D.L. Bohac, G. Nelson, C. Olson, and H. Durschlag. 2013. Measuring the air tightness of mid and high rise non-residential buildings. Presented at AIVC Airtightness Workshop, Washington DC, April 18–19.

ASTM. 2010. ASTM E779-10, *Standard test method for determining air leakage rate by fan pressurization*. West Conshohocken, PA: ASTM International.

ASTM. 2009. ASTM E1186-03, *Standard practices for air leakage site detection in building envelopes and air barrier systems*. West Conshohocken, PA: ASTM International.

Bohac, D., J. Quinnell, M. Hewett, J. Fitzgerald, J. Novacheck, and A. Lutz. 2016. Measured commercial and institutional building pressures. Presented at Thermal Performance of the Exterior Envelopes of Whole Buildings XIII International Conference.

Brennan, T., W. Anis, G. Nelson, and C. Olson. 2013. ASHRAE 1478: Measuring airtightness of mid- and high-rise non-residential buildings. Presented at Thermal Performance of the Exterior Envelopes of Whole Buildings XII International Conference.

CBECS. 2012 Commercial Buildings Energy Consumption Survey. Accessed on July 30, 2016. https://www.eia.gov/consumption/commercial/reports/2012/buildstock/.

Emmerich, S.J. 2014. Telephone communications, January 28, 2014.

Emmerich, S.J., and A.K. Persily. 2014. Analysis of U.S. commercial building envelope air leakage database to support sustainable building design. *International Journal of Ventilation* 12(4).

Emmerich, S.J., T.P. McDowell, and W. Anis. 2005. Investigation of the impact of commercial building envelope airtightness on HVAC energy use. Report NISTIR 7238. Gaithersburg, MD: National Institute of Standards and Technology.

Persily, A.K. 1998. Airtightness of commercial and institutional buildings—Blowing holes in the myth of tight buildings. Presented at Thermal Envelopes VII Conference, Clearwater, FL, December 6–10.

USACE. 2010. *Air leakage test protocol for building envelopes*. Air Barrier Association of America. U.S. Army Corps of Engineers ERDC.

Walton, G.N., and W.S. Dols. 2013. CONTAMW User Guide and Program Documentation, National Institute of Standards and Technology, NISTIR 7251. National Institute of Standards and Technology. Downloaded at http://www.bfrl.nist.gov/IAQanalysis/docs/CONTAM_31.pdf on February 28, 2013.

Protocol to Evaluate the Moisture Durability of Energy-Efficient Walls

Philip Boudreaux, MS **Simon Pallin, PhD, CEM** **Diana Hun, PhD, PE**

Manfred Kehrer, Dipl-Ing
Member ASHRAE

Roderick Jackson, PhD
Associate Member ASHRAE

Andre Desjarlais

ABSTRACT

Walls account for about 8% of the energy used in residential buildings. This energy penalty can be reduced with higher insulation levels and increased airtightness. However, these measures can compromise the moisture durability and long-term performance of wall assemblies because they can lead to lower moisture tolerance due to reduced drying potential. To avert these problems, a moisture durability protocol was developed to evaluate the probability that an energy-efficient wall design will experience mold growth. This protocol examines the effects of moisture sources in walls through a combination of simulations and laboratory experiments, uses the mold growth index as the moisture durability indicator, and is based on a probabilistic approach that utilizes stochastically varying input parameters. The simulation tools used include a new validated method for taking into account the effects of air leakage in wall assemblies. This paper provides an overview of the developed protocol, discusses the probabilistic simulation approach, and describes results from the evaluation of two wall assemblies in Climate Zones 2, 4, and 6. The protocol will be used to supply builders with wall designs that are energy efficient, moisture durable, and cost-effective.

INTRODUCTION

Energy transfer through walls in residential buildings accounts for about 1.8 quads per year of primary energy consumption in the United States (Sawyer 2014). To meet the U.S. Department of Energy (DOE) goal of doubling energy productivity by 2030, it is critical that walls are addressed (Keyser et al. 2015). Building codes and standards have started to address this by requiring energy-efficient walls with high levels of insulation and better air sealing than walls of the past (ICC 2012). While these walls have a high potential to save energy, their moisture durability compared to traditional walls needs to be investigated and communicated to the building industry to aid in the market adoption of energy-efficient, code-level walls. Some of the potential concerns about these walls include a lower drying potential due to high levels of insulation and airtightness, materials that have lower water vapor permeance, and colder outside or inside wall surfaces (depending on climate zone and season) that may increase chances for condensation.

To better evaluate and predict the risk of moisture durability issues in energy-efficient wall assemblies, Oak Ridge National Laboratory (ORNL) developed a protocol to assess different wall designs. This protocol includes probabilistic hygrothermal simulations, laboratory testing, and field testing of a given wall assembly. Once the simulation tools are validated against the laboratory and field testing, the probabilistic simulations alone can be used to assess the moisture durability of wall assemblies, decreasing the time and cost of testing new designs. In this paper, we describe the protocol while focusing on discussing the probabilistic hygrothermal simulations. Results from two wall assemblies that were evaluated using the protocol are presented.

METHODS

A brief description of the protocol for assessing the risk of moisture issues for a given wall is shown in Figure 1 and is discussed briefly in this section. More detailed discussion of Step 3 of the protocol follows this introduction. This protocol

Philip Boudreaux, **Simon Pallin**, and **Diana Hun** are members of the Research and Development Staff, **Roderick Jackson** is Building Envelope Systems group leader, and **Andre Desjarlais** is Building Envelopes Sub program manager with Oak Ridge National Laboratory, Oak Ridge, TN. **Manfred Kehrer** is president of SmartSolutions, Oak Ridge, TN.

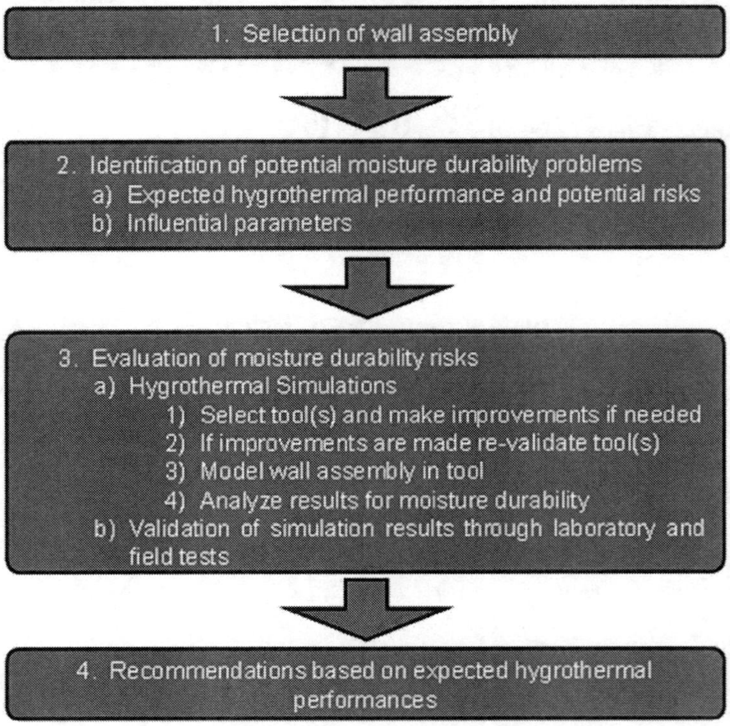

Figure 1 *Protocol for assessing potential risks from moisture issues.*

(Pallin et al. 2014) is meant to be used by building science researchers to evaluate the moisture durability of new energy-efficient wall designs in different climate zones. Based on the results, researchers will recommend energy-efficient, moisture durable wall designs to the building community.

After identification of the wall assembly with appropriate materials defined in Step 1, the expected risks and influential parameters are identified in Step 2 for the specific wall assembly. The potential risks identified for walls in U.S. homes include decay and corrosion that can lead to structural failure and mold growth that can lead to respiratory-related health problems for the occupants. These risks lead to two available metrics for assessing hygrothermal performance: moisture content of building components and the mold growth index inside an assembly. For the purpose of this paper, the mold growth index is used; it is discussed in the following subsection. The influential parameters for moisture durability of a wall include any variables that might effect moist air movement through or beside the wall or the temperature and moisture content of that air. The parameters that were selected for this study are outdoor climate, indoor climate, and wall construction. Of these parameters, many of the variables affecting the indoor climate are handled in a probabilistic manner and are discussed in the following subsection. The workmanship variability of the wall is not directly included in this protocol, although some of the effects of workmanship deficiencies are captured, for example, by stochastically varying the airtightness of the home.

The outdoor climate is dependent on the climate zone in which the home is located. For each climate zone, five consecutive years of actual meteorological weather data for the representative city were used. So for each wall in a specific climate zone the simulation was run for five years, capturing the wall's moisture performance due to outdoor weather variability within a single climate zone. The indoor climate (temperature and relative humidity) depends on many parameters, including but not limited to the following:

- Outdoor climate
- Indoor heat generation
- Indoor moisture generation
- Moisture buffering of materials
- Airtightness of the home
- Thermostat set point
- Duct leakage
- House characteristics (insulation levels in wall and attic, window thermal characteristics and foundation type, etc.)

Step 3 includes identifying tools available for accomplishing the hygrothermal simulations. For our purposes, we selected EnergyPlus (DOE 2015) and WUFI® Pro (Fraunhofer IBP 2015). EnergyPlus is used to simulate the indoor environment of the home, and WUFI® is used, in combination with the indoor climate from EnergyPlus and actual weather files, to assess the hygrothermal performance of the wall assembly. Both of these simulation tools needed improvements. The effective moisture penetration depth (EMPD) model of EnergyPlus was modified by the National Renewable Energy Laboratory (NREL) to use two different moisture buffering time scales for materials, long-term and short-term time constants. This custom version of EnergyPlus 7.1 is used to improve the hygrothermal modeling of the home to yield more accurate indoor-air moisture conditions. This version of the EMPD model has been validated by NREL against the more complicated and accurate combined heat and moisture transfer (HAMT) model (Woods et al. 2013).

WUFI® was also modified to include a more accurate method for accounting for the hygrothermal effects of airflow in an assembly in one dimension. The results from the WUFI® analysis will yield a mold growth index (MGI) as the performance indicator for the wall assembly in question (Ojanen et al. 2011). This indicator ranges from 0 to 6, with each value corresponding to a characteristic level of mold growth, as shown in Table 1. The definition in the second column of Table 1 is a qualified assessment based on the MGI at interfaces and surfaces that are not in direct contact with the interior air (Viitanen et al. 2015). Although MGI is as indicator of mold, MGI does not specifically define the risk of moisture durability problems or indoor air quality.

The risk of mold was evaluated at a single position inside the simulated wall assembly. Since a one-dimensional simulation tool was used, no other areas inside the wall, other than this single position, were evaluated using the MGI. Therefore, no other analyses were done to estimate how widespread the simulated mold growth was.

Another tool that was used is an ORNL-developed Generation of Indoor Heat and Moisture (GIHM) tool. This tool yields hourly indoor moisture and heat generation for a home in a probabilistic way. Each individual profile of hourly heat and moisture is based on a stochastically chosen set of influencing parameters from real probability distributions of occupant behavior and appliance and miscellaneous electric load (MEL) characteristics available in the literature. If the GIHM tool is run 1000 times, then the resulting set of hourly heat generation profiles represent the U.S. residential building stock (as far as inputs of the GIHM tool reflect the building stock).

Steps 3.a.3 and 3.a.4 shown in Figure 1 involve modeling the wall assembly and analyzing the results; these steps are described in the Results and Discussion section for two specific wall assemblies. Step 3.b involves comparing the modeling results to laboratory tests and field tests of the wall assemblies. This comparison is used to validate the modeling efforts and, if needed, guide modifications in the simulation protocol. After sufficient validation, the protocol is used without Step 3.b. Finally, Step 4 involves taking the modeling results and making recommendations for the most moisture-durable wall assemblies for each climate zone.

Hygrothermal Probabilistic Simulations of Indoor Climate and MGI of Walls

To properly model the indoor climate of a home and the hygrothermal performance of walls, ORNL created, used, or modified three tools. These tools can be separated into the tasks they accomplish and are as follows:

1. Indoor climate
   a. GIHM tool
   b. Custom EnergyPlus
2. Hygrothermal performance of wall assembly
   a. Custom WUFI®

Indoor Climate. To estimate varying hourly indoor temperatures and relative humidities requires the consideration of any variable that affects these parameters. These variables include interior moisture (latent heat) and sensible heat generation, ventilation rate, climate, and dehumidification rate during cooling or dedicated dehumidifier or humidifier operation (TenWolde 1994). The ventilation rate includes both intentional (mechanical ventilation) and unintentional air (through leakage paths in walls) moving from the interior to the outside or vise versa. The interior latent and sensible heat generation include many sources, such as occupants, pets, plants, cooking, showering, bathing, clothes washing and

Table 1. Mold Growth Index (Hukka and Viitanen 1999; Ojanen et al. 2011; Viitanen et al. 2015)

Index	Description of Growth Rate	Risk Level
0	No growth	Low
1	Small amounts of mold on surface (microscopic), initial stages of local growth	Low
2	Several local mold growth colonies on surface (microscopic)	Low
3	Visual findings of mold on surface, <10% coverage or <50% coverage of mold (microscopic)	Moderate
4	Visual findings of mold on surface, 10% to 50% coverage or >50% coverage of mold (microscopic)	High
5	Plenty of growth on surface, >50% coverage (visual)	High
6	Heavy and tight growth, coverage approximately 100%	High

drying, MELs, etc. Another important variable is the interzonal leakage between the interior and other semi- or unconditioned zones, such as the attic, which can carry moisture to and from the interior (TenWolde and Walker 2001).

GIHM Tool. The GIHM tool was developed to simulate residential generation of moisture and heat by using statistical data for residential user behaviors together with moisture and heat production rates from occupant activities and appliances inside homes. Because the amount of moisture generated in homes is building- and climate-dependent, the tool also uses the type of building (multifamily or single-family) and location as inputs. The tool is a probabilistic instrument that simulates hourly variations of moisture and heat generation in homes using a stochastic approach. So for each climate zone, an output set of hourly profiles captures the range and distribution of moisture and heat generation in real homes. Figure 2 shows the probability distribution of average daily moisture production simulated by the tool as a function of occupants compared with the deterministic ANSI/ASHRAE Standard 160 (ASHRAE 2011).

At the time of this research, the tool gave only moisture generation profiles. For the simulations presented in this study, the indoor sensible heat generation was based directly on the indoor latent heat (moisture) generation. To estimate the sensible heat generation, a multiplier of 2.7 was used with the simulated latent load. This is based on an estimation of the latent/sensible total load split from the appliances, MELs, and occupants from the Building America research benchmark (Hendron and Engebrecht 2010). Currently the GIHM tool is being updated to include heat generation as well.

Custom EnergyPlus. EnergyPlus was used with an improved version of the EMPD model (Woods et al. 2013). EnergyPlus models of an average-size single-family detached home (2367 ft^2) for select U.S. climate zones were completed. The house characteristics depend on the climate zone in which the house is located and are shown in Table 2. The wall, ceiling, and fenestration R-values and U-factors are based on the *2015 International Residential Code* (ICC 2015). The foundation type is based on data from ORNL and Residential Energy Consumption Survey (Labs et al. 1988; EIA 2015). For each climate zone the model home included the most probable foundation type found in these data.

The other stochastically varying EnergyPlus inputs, besides the indoor latent and sensible heat generation that are generated by the GIHM tool, include the thermostat set point and the airtightness of the house. For this study, NREL provided data on indoor temperatures measured in 132 houses across the United States (Booten and Norton 2014). The data included the average and standard deviations in indoor temperature for cooling and heating seasons and in different climate zones. Based on these data and a survey of more than 50 homes in West Point, New York (Sullivan 2015), a normal probability distribution was built and sampled to create matched set point pairs (where the heating and cooling set

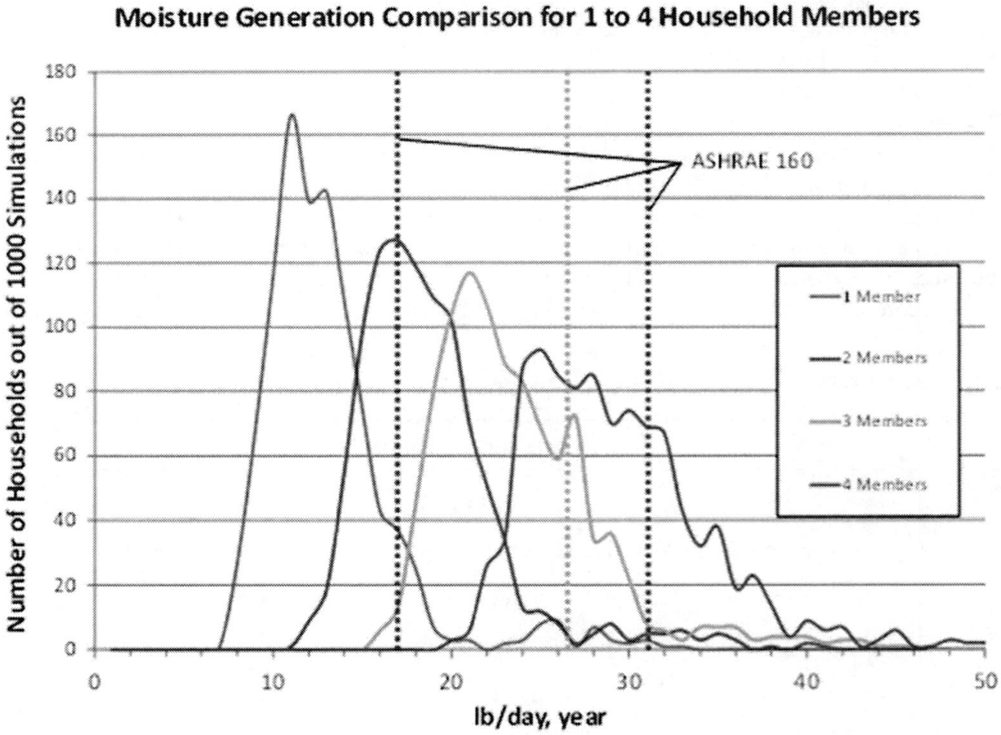

Figure 2 *Probabilistic distribution of indoor moisture generation in homes as a function of household members. The probabilistic curves are depicted together with recommended generation rates (ASHRAE 2011).*

points fell at the same location on each respective probability distribution) for each climate zone. The pairs were selected in this way to avoid having cooling set points that were lower than the heating set points, which would have confused the "auto" setting of the modeled thermostat. Table 3 shows the average thermostat set point and the standard deviation for location and season. Figure 3 shows an example distribution of set points for climate zones 5A–6A.

To determine the building air leakage rate, the Lawrence Berkeley National Laboratory (LBNL) Residential Diagnostics Database (LBNL 2015) was used, which contains whole-house air leakage data from about 147,000 homes. These data can be filtered by climate zone and house characteristics to yield a probability and cumulative distribution function for a particular home type in a particular climate zone. For the Residential Diagnostics Database, the following inputs were used:

- Floor area: 2376 ft^2
- Ceiling height: 16 ft (two floors)
- Year built: after 2000

Table 2. Input Values for EnergyPlus Model to Simulate Indoor Climate

Climate Zone	Ceiling R-Value	Fenestration U-Factor	Glazed Fenestration Solar Heat Gain Coefficient	Foundation
2A	38	0.4	0.25	Slab
4A	49	0.35	0.4	Crawlspace
5A	49	0.32	No requirement	Basement
6A and 7	49	0.32	No requirement	Basement

Table 3. Thermostat Set Points and Standard Deviations for Selected Climate Zones

Climate Zone	Sample Size	Heating Season Average T, °F	Heating Season σ, °F	Cooling Season Average T, °F	Cooling Season σ, °F
2A	20	71.3	3.5	78.3	2.8
4A	50	68.4	2.7	72.4	2.9
5A and 6A	8	69.8	2.8	73.5	3.0
7	20	65.0	2.5	76.2	2.2

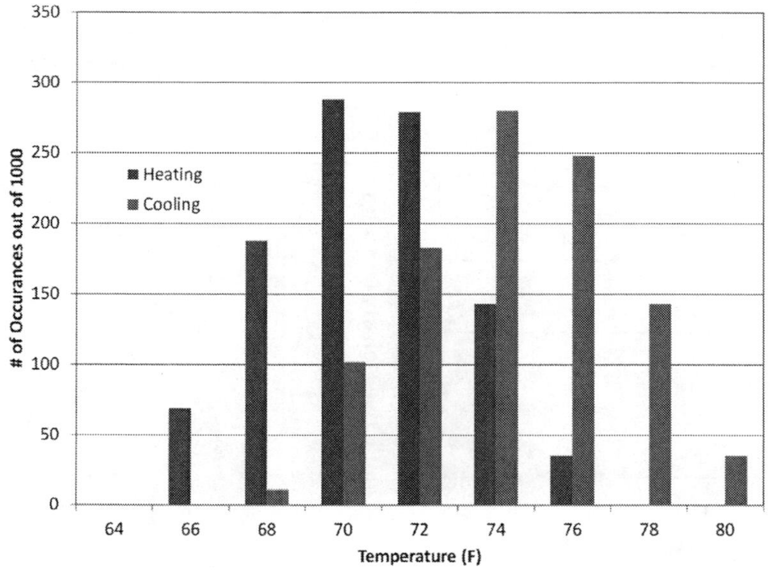

Figure 3 *Distribution of heating and cooling set points for climate zones 5A and 6A.*

- Weatherization Assistance Program? No
- ENERGY STAR certified? Yes
- Region: climate zone
- Foundation type: depends on climate zone (see Table 2)
- Duct system location: inside the conditioned space

After these parameters where input into the online calculator, a probability density function of normalized leakage and a cumulative distribution function of air changes per hour at 50 Pa (ACH_{50}) were given as shown in Figure 4 for climate zone 4A. Data from the ACH_{50} cumulative distribution function was extracted from the html code and then used to generate an effective leakage area for 1000 homes for each climate zone. To accomplish this, a random number generator was used to probabilistically sample the discrete cumulative distribution function with linear interpolation used between the discrete points. The ACH_{50} was then converted to an effective air leakage area (ELA) at 4 Pa for use in EnergyPlus's Zone Infiltration ELA inputs. For this conversion the ACH_{50} was converted to cubic feet per minute at 50 Pa (CFM50) using Equation 1:

$$\text{CFM50} = \text{ACH50}\frac{V}{60} \qquad (1)$$

where V is the house volume in ft^3.

Then the CFM50 was converted to leakage area using Equation 2 (ASHRAE 2001):

$$A_L = 0.186 Q_r \frac{\sqrt{\frac{\rho}{2\Delta p_r}}}{C_D} \qquad (2)$$

where A_L is the effective air leakage area in $in.^2$, ρ is the air density (0.0749 lb/ft^3 at 70°F or 21°C), C_D is the discharge coefficient and is equal to 1, Δp_r is the reference pressure of 4 Pa (0.01606 in. H_2O), and Q_r in ft^3/min is the predicted airflow rate at 4 Pa. To get Q_r Equation 3 is used:

$$Q = C\Delta p^n \qquad (3)$$

where C is the flow coefficient and the exponent n is assumed to be 0.65. To find C the initial CFM50 computed from Equation 1 is used with a Δp of 50 Pa. Once C is found, Q at 4 Pa can be computed and input into Equation 2 as Q_r.

This stochastic approach yields a data set that reflects the U.S. building stock described in the LBNL database. An example of the resulting leakage areas for climate zone 4 is shown in Figure 5.

Hygrothermal Performance of Wall Assembly. The indoor climate variables discussed previously serve as inputs to EnergyPlus, which simulates the indoor boundary conditions (temperature and humidity) of the wall assembly. The wall assembly is built in WUFI®, with these interior boundary conditions and the local weather that was used in Energy Plus as the exterior boundary condition, to estimate the moisture durability of the wall assembly. To do this in a probabilistic way, two considerations were made when using the hygrothermal modeling software WUFI®. The way the software models the airflow in the cavity needed to be investigated and improved and the critical orientation of the wall in question needed to be chosen based on weather data. From the WUFI® outputs the MGI is computed per the literature (Ojanen et al. 2011).

Custom WUFI®. WUFI® handles air leakage as a point source or sink of indoor or outdoor air that can be placed at any position in the wall assembly. This point-source approach has been questioned because it does not address the interactions of the air with the materials as it travels through the assembly (Künzel et al. 2011; Lstiburek 2015). To increase the accuracy of the point-source approach, a correction factor that is dependent on location in the airflow path and the airflow rate has been multiplied with the air-source equation (Pallin et al. 2014, 2015). This approach is called the *η-method*. The method takes into account the interaction of the air with the materials on its path to the point-source location. Figure 6 shows the measured temperature in an airflow path through a wall with predictions using the traditional WUFI® method and then with WUFI® using the η-method. The η-method

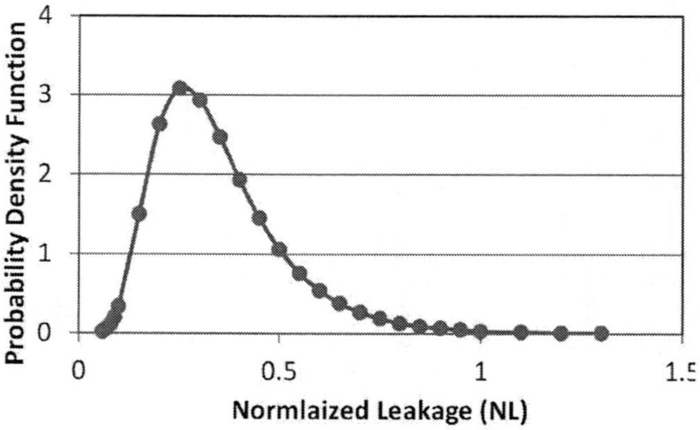

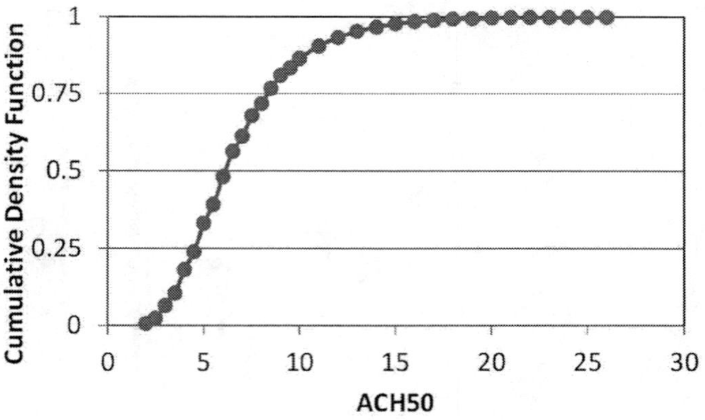

Figure 4 *Normalized leakage and ACH_{50} for climate zone 4A.*

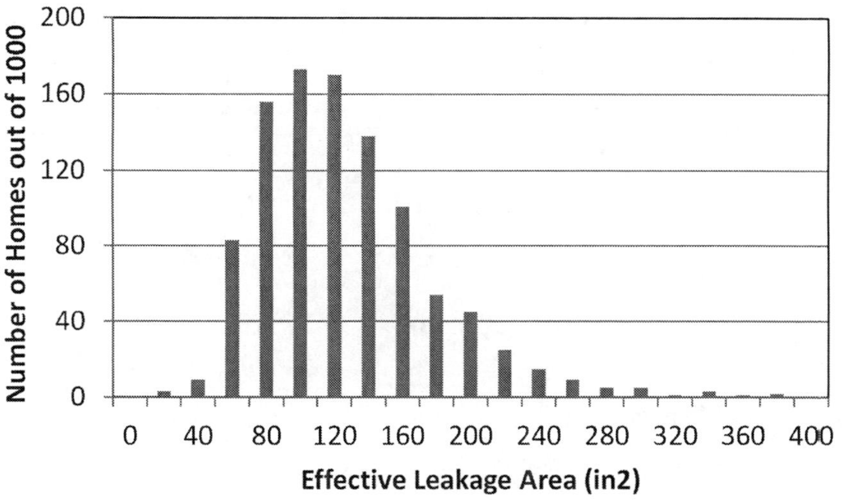

Figure 5 *Probabilistic distribution of air leakage area found by converting ACH_{50} from the LBNL Residential Diagnostics Database (LBNL 2015).*

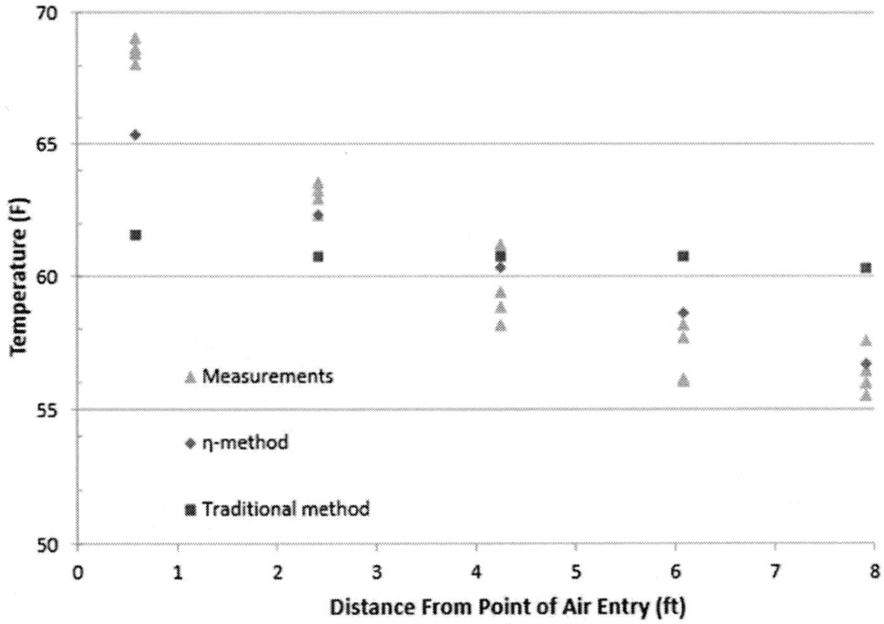

Figure 6 *Comparison of measured temperature along an airflow path in a wall with simulated temperature from WUFI® with the traditional method and from WUFI® with the η-method. The η-method improves the predicted temperatures by three times.*

improves the temperature prediction error by about three times, from ±3.7 to ±1.3°F.

The orientation of a wall impacts the exterior heating and wetting loads, which is why the most critical orientation needs to be determined to best predict the most decisive outdoor conditions. In this study, ORNL created a preprocessing tool to determine the critical orientation in each specific climate zone based on hourly rain and wind loads from reigning climate conditions during the simulation time frame.

RESULTS AND DISCUSSION

Two wall assemblies were evaluated using the probabilistic risk assessment presented previously, where key inputs (indoor heat and moisture generation, outdoor climate, whole-house leakage area (wall air permeability), thermostat setting, and air leakage paths in the wall) were varied stochastically (Pallin et al. 2015). The results were evaluated using a performance indicator for mold (MGI) as described in the Methods section. Table 4 shows the details of the two wall assemblies.

Table 4. Cavity and Exterior Insulation for Each Simulated Wall and Climate Zone

Wall Assembly 1	Houston, TX Climate Zone 2A	New York, NY Climate Zone 4A	Minneapolis, MN Climate Zone 6A
Siding	Vinyl	Vinyl	Vinyl
Air/water barrier	Spun-bonded polyolefin	R-3.75 extruded polystyrene (XPS)	R-11.25 XPS
External continuous insulation	—	R-3.75 XPS (same as air/water barrier)	R-11.25 XPS (same as air/water barrier)
External sheathing	7/16 in. oriented strand board (OSB)	7/16 in. OSB	7/16 in. OSB
Framing	2×6 studs	2×6 studs	2×6 studs
Cavity insulation	R-20 fiberglass batts	R-20 fiberglass batts	R-20 fiberglass batts
Interior sheathing	1/2 in. drywall	1/2 in. drywall	1/2 in. drywall
Vapor retarder	Latex paint	Latex paint	Latex paint
Wall Assembly 2	**Chicago, IL Climate Zone 5A**	**Minneapolis, MN Climate Zone 6A**	**Anchorage, AL Climate Zone 7**
Siding	Vinyl	Vinyl	Vinyl
Air/water barrier	Spun-bonded polyolefin	R-5 XPS	R-5 XPS
External continuous insulation	—	R-5 XPS (same as air/water barrier)	R-5 XPS (same as air/water barrier)
External sheathing	7/16 in. OSB	7/16 in. OSB	7/16 in. OSB
Framing	2×6 studs	2×6 studs	2×6 studs
Cavity insulation	R-20 fiberglass batts	R-20 fiberglass batts	R-20 fiberglass batts
Vapor retarder	Kraft paper	Kraft paper	Kraft paper
Interior sheathing	1/2 in. drywall	1/2 in. drywall	1/2 in. drywall

Wall Assembly 1

For the first wall assembly, 3000 simulations were run, 1000 for each of the three climate zones—2A (Houston), 4A (New York City), and 6A (Minneapolis). For each climate zone EnergyPlus was used to probabilistically model 1000 interior climate conditions, each with hourly temperature and relative humidity for five years. These interior condition profiles were used as inputs to WUFI® simulations of the wall assembly. The WUFI® results were used to complete a hygrothermal assessment of the wall assembly for each of the 3000 different indoor climates from EnergyPlus.

The risks of mold growth for all three climate zones are presented in Figure 7. For climate zone 2A, the wall was evaluated with an orientation that resulted in the highest exterior air infiltration rates and for climate zone 6A, the orientation that resulted in the highest for indoor air exfiltration. In climate zone 4A, both infiltration and exfiltration were evaluated, thus resulting in two different wall orientations. As seen in Figure 7, there could be a large mold risk associated with wall assembly 1 in climate zone 2A. According to the results, about 10% of the walls can be expected to have visual evidence of mold at the accessed location. For climate zones 4A and 6A, the risks of mold growth are substantially lower. The results from simulating the risk of mold growth in climate zone 4A reveal that 1.6% of the walls have a high risk of visual mold; this is the result of including the moisture leaks from exfiltration of indoor air. For the case of air infiltration, the risks were negligible. In climate zone 6A, high-risk MGI is lower than 1% for all simulated scenarios.

To understand why there is a higher MGI for climate zone 2, it is beneficial to investigate how much each varying parameter affects the MGI using a sensitivity study. Figure 8 shows the sensitivity study for wall assembly 1 in climate zone 2. The correlations between the varying parameters and the MGI are described by a number from • 1 to 1, where 0 means no correlation, • 1 means a full negative correlation, and 1 means a full positive correlation.

Figure 8 shows that MGI is well correlated with the cooling set point—when the cooling set point goes down, the risk of mold growth goes up. This is due to the increase in condensation potential or relative humidity because the lower indoor temperature during the summer drives the temperature inside

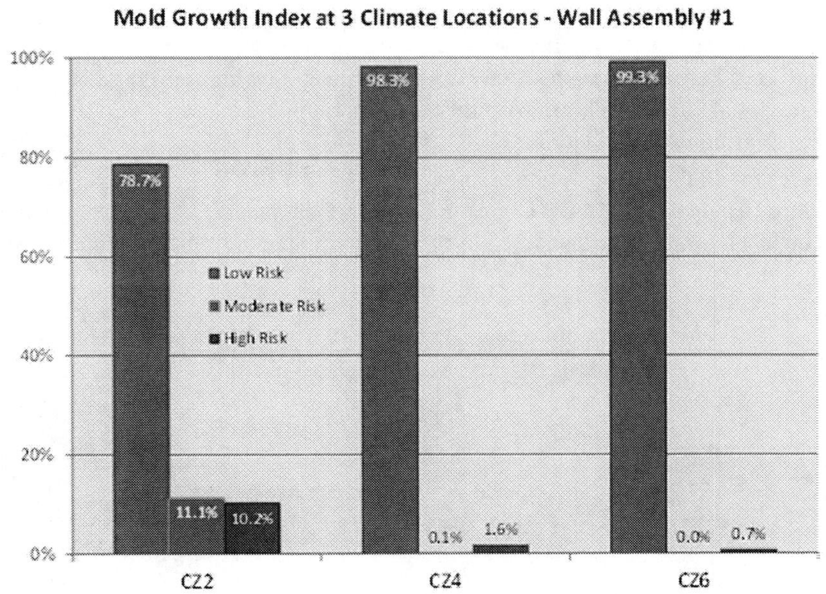

Figure 7 *Hygrothermal performances for wall assembly 1, based on the MGI: climate zone 2 (Houston), climate zone 4 (New York), and climate zone 6 (Minneapolis).*

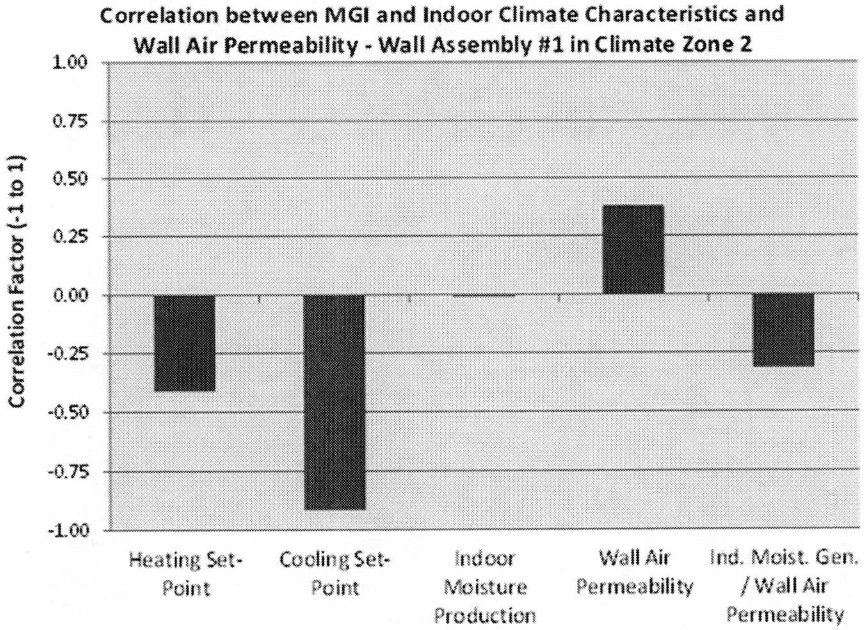

Figure 8 *Sensitivity analyses for wall assembly 1 in climate zone 2 (Houston).*

the wall lower. As expected, indoor moisture generation does not impact the risk of mold growth, since infiltration only is being evaluated in this climate zone. The wall permeability also correlates with the MGI; the correlation is not very strong but is still recognizable. The rightmost column in Figure 8 is a parameter created by considering the ratio between indoor moisture generation and wall permeability. The reason air permeability has a positive correlation while the ratio has a negative correlation is that wall air permeability is applied as the denominator. This ratio isn't strongly correlated to MGI in this case, mainly due to this wall being evaluated under infiltration only.

Wall Assembly 2

For the second wall, the risks of mold growth were evaluated in climate zones 5A, 6A, and 7A, since the assembly is appropriate for cold climates. As seen in Figure 9, the risk of visual mold coverage is 8.1% and 10.4% for the walls in climate zones 5A (Chicago) and 7 (Anchorage), respectively. In climate zone 6A (Minneapolis), the risk is lower. All three walls were evaluated due to exfiltration of indoor air.

The sensitivity analyses for all three climate zones show very good agreement. Figure 10 indicates that indoor moisture production will have an impact on wall performance, slightly more than air permeability. The strongest correlation is seen for high indoor moisture production together with low wall air permeability (high airtightness). On average, the correlation is as strong as 0.9.

CONCLUSION

This paper presents a protocol (Pallin et al. 2014) for determining the moisture durability risk of walls in residential

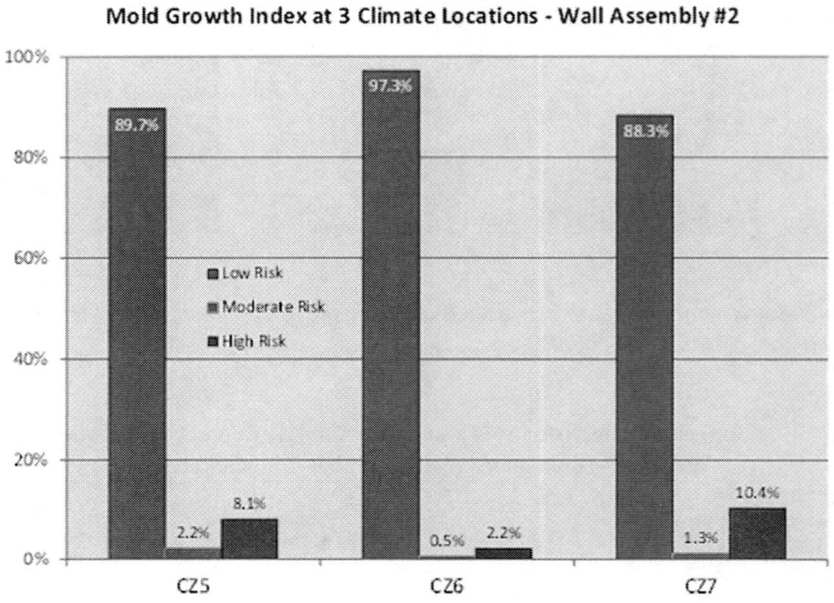

Figure 9 *Hygrothermal performances for wall assembly 2, based on the MGI: climate zone 5 (Chicago), climate zone 6 (Minneapolis), and climate zone 7 (Anchorage).*

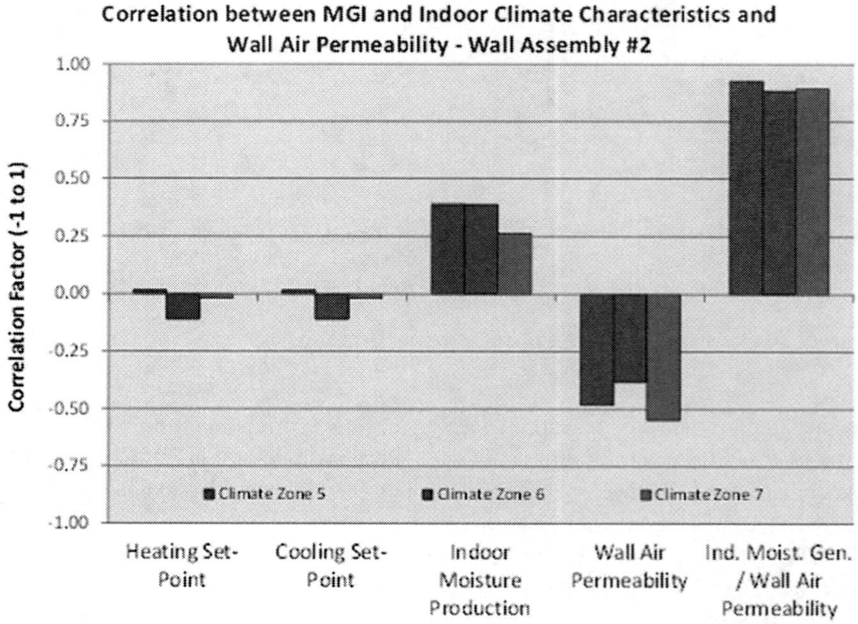

Figure 10 *Sensitivity analyses for wall assembly 2 in the climate of Chicago, Minneapolis, and Anchorage.*

buildings. Two wall assemblies were brought through steps 1–3 of this protocol. Using this protocol, two wall assemblies were evaluated. Accomplishing this required a new tool for probabilistically determining hourly indoor moisture generation to be developed, called the Generation of Indoor Heat and Moisture (GIHM) tool. Currently the indoor (sensible) heat aspect of this tool is being completed. This protocol also required two modeling packages to be customized. EnergyPlus was customized to enable a more accurate effective moisture penetration depth model, which was validated against the more advanced combined heat and moisture transfer model. This enables a more accurate simulated indoor environment since the transfer of moisture between the building materials and the air is modeled more realistically. After this customization, EnergyPlus was used to model the indoor climate in a probabilistic manner, yielding a set of hourly indoor temperature and relative humidity profiles that represent the U.S. residential building stock. Stochastic inputs into EnergyPlus, for modeling the indoor climate, included the already mentioned moisture (latent) generation profile, heat (sensible) generation profile, whole-house air tightness/wall air permeability, and thermostat set point for each climate zone.

These indoor climate profiles were input into a custom version of WUFI® that was modified to more correctly model airflow in a wall assembly. This improvement increases the accuracy of the temperature simulations of the wall materials along the airflow path in WUFI®. The batch WUFI simulations were used to compute a set of mold growth indices (MGIs) for a given wall assembly in a given climate zone. In our case this was 1000 simulations per wall assembly per climate zone. Each of these simulations considers one home environment, but taken together as a set the output represents how the wall will perform in a sample of 1000 homes in the climate zone in question.

The WUFI® analysis showed that for each wall there was at least one climate zone where the MGI showed the assembly had a high risk for mold growth in about 10% of the homes in that climate zone. This is the power of this protocol and approach: being able to compare the moisture durability performance of different wall assemblies in order to make more informed decisions during the selection of a wall assembly. The next step in the protocol for assessing the risk of moisture issues in walls is to test these wall assemblies in the laboratory and at field sites to validate the results from the simulations.

ACKNOWLEDGMENTS

The authors would like to thank the U.S. DOE Building America program for support and funding of this research.

REFERENCES

ASHRAE. 2001. Chapter 26, Ventilation and infiltration. *ASHRAE Handbook—Fundamentals*. Atlanta: ASHRAE.

ASHRAE. 2011. ANSI/ASHRAE Addendum a to ANSI/ASHRAE Standard 160-2009, *Criteria for moisture-control design analysis in buildings*. Atlanta: ASHRAE.

Booten, C., and P. Norton. 2014. *The indoor temperature and humidity study: Status and preliminary results*. Golden, CO: National Renewable Energy Laboratory.

DOE. 2015. *EnergyPlus*, ver. 7.1. Washington, DC: U.S. Department of Energy, Building Technologies Office.

EIA. 2015. Residential Energy Consumption Survey (RECS). Washington, DC: U.S. Energy Information Administration. http://www.eia.gov/consumption/residential/.

Fraunhofer IBP. 2015. *WUFI® Pro*, ver. 5.0. Stuttgart, Germany: Fraunhofer-Institut für Bauphysik.

Hendron, R., and C. Engebrecht. 2010. *Building America research benchmark definition*. Golden, CO: National Renewable Energy Laboratory.

Hukka, A., and H.A. Viitanen. 1999. A mathematical model of mould growth on wooden material. *Wood Science and Technology* 33.

ICC. 2012. Residential energy efficiency. In *2012 International energy conservation code*. Washington, DC: International Code Council.

IRC. 2015. *2015 International residential code for one-and two-family dwellings*. Washington, DC: International Code Council.

Keyser, D., J.M. Mayernik, C. McMillan, J. Agan, N. Kempkey, and J. Zweig. 2015. *Accelerate energy productivity 2030: A strategic roadmap for American energy innovation, economic growth, and competitiveness*. Washington, DC: U.S. Department of Energy.

Künzel, H.M., D. Zirkelbach, and B. Schafaczek. 2011. Vapour control design in wooden structures including moisture sources due air exfiltration. Presented at the 9th Nordic Symposium on Building Physics, Tampere, Finland.

Labs, K., J. Carmody, R. Sterling, L. Shen, Y.J. Huang, and D. Parker. 1988. *Building foundation design handbook.* ORNL/Sub/86-72143/1. Oak Ridge, TN: Oak Ridge National Laboratory.

LBNL. 2015. Residential Diagnostics Database. Berkeley, CA: Lawrence Berkeley National Laboratory. http://resdb.lbl.gov/.

Lstiburek, J. 2015. WUFI: Barking up the wrong tree? *ASHRAE Journal* 57(10).

Ojanen, T., R. Peuhkuri, H. Viitanen, K. Lähdesmäki, J. Vinha, and K. Salminen. 2011. Classification of material sensitivity—New approach for mould growth modeling. Presented at the 9th Nordic Symposium on Building Physics, Tampere, Finland.

Pallin, S.B., D.E. Hun, R.K. Jackson, and M. Kehrer. 2014. *Risk assessment of energy-efficient walls*. Report ORNL/TM-2014/676. Oak Ridge, TN: Oak Ridge National Laboratory, Building Technologies Research and Integration Center—Energy and Transportation Science Division.

Pallin, S., P. Boudreaux, M. Kehrer, D.E. Hun, R.K. Jackson, and A.O. Desjarlais. 2015. Moisture durability assessment of common well-insulated wall assemblies. Report ORNL/TM-2015/645. Oak Ridge, TN: Oak Ridge National Laboratory, Building Technologies Research and Integration Center—Energy and Transportation Science Division.

Sawyer, K. 2014. *Windows and building envelope research and development: Roadmap for emerging technologies*. Washington, DC: U.S. DOE EERE Building Technologies Office.

Sullivan, P. 2015. Survey results of thermostat set point for homes in West Point, NY. Personal correspondence with S. Pallin.

TenWolde, A. 1994. The moisture balance: Fundamentals. *Bugs, mold and rot II*, pp. 67–69 Washington, DC: National Institute of Building Sciences.

TenWolde, A., and I.S. Walker. 2001. Interior moisture design loads for residences. *Performance of Exterior Envelopes of Whole Buildings VIII: Integration of Building Envelopes* [CD].

Viitanen, H.A., M. Krus, T. Ojanen, V. Eitner, and D. Zirkelbach. 2015. Mold risk classification based on comparative evaluation of two established growth models. Presented at the 6th International Building Physics Conference (IBPC).

Woods, J., J. Winkler, and D. Christensen. 2013. *Evaluation of the effective moisture penetration depth model for estimating moisture buffering in buildings.* Report NREL/TP-5500-57441. Golden, CO: National Renewable Energy Laboratory.

A Systematic Method of Assessing the Durability of Wood-Frame Wall Assemblies: Towards the Limit-States Design Approach

Michael A. Lacasse, PhD, PEng **Martin Morelli, PhD**

ABSTRACT

The long-term performance in respect to moisture management within any wall assembly depends on the hygrothermal response of the wall. Critical factors in estimating the longevity of wood-frame structures include limiting the temperature range, wood moisture content, and time of exposure to conditions suitable for the onset, growth, and propagation of mold and rot to occur. The intent in constructing highly insulted wood-frame walls is evidently to reduce energy usage in buildings, but the energy savings as might accrue necessarily cannot be achieved if these walls fail prematurely due to the effects of moisture accumulation in wall cavities. Several approaches to assessing the vulnerability of wood-frame structures to deterioration have been developed in recent years, some of which suggest applying a limit-states design approach to the performance assessment of the assembly. In this paper, a limit-states design approach is described that forms the basis of a performance assessment method for wood-frame wall assemblies. The approach is based on the requirements set out in ISO 13823. The approach, developed for the Moisture Management of Exterior Wall Systems (MEWS) project, is described; in it, the relative humidity temperature index (RHT index) is used as a basis for evaluating the long-term performance of wood-frame assemblies. This index captures the duration of the coexistence of moisture and thermal conditions above a set of threshold levels for which the risk to the formation of mold growth and wood rot is unacceptably high. An example is given to illustrate the application of the approach using the RHT index when assessing the moisture management performance of a North American stucco-clad wood-frame wall assembly in relation to wood rot.

INTRODUCTION

Given the importance of moisture to bring about deterioration in building materials over time, very generally, the long-term performance of a building component depends on the hygrothermal response of the component when subjected to interior environmental and exterior climatic loads. In respect to the durability of components having a wood-frame structure, this depends on whether the wood components remain dry and, if not, the time over which they are exposed to conditions that generate temperatures and elevated levels of moisture content suitable for the onset of the formation of wood-damaging fungi. Critical factors in estimating the longevity of wood-frame structures include assessing conditions suitable for the onset, growth, and propagation of damage to occur, more specifically, the temperature range, wood moisture content, and time of exposure.

In ISO 13823 (ISO 2008), recommendations are made on the use of different methods for the design and verification of structures for durability. For example, verifying a component of a structure for durability can be done by applying service-life formats such as the factor method or the limit-states format.

When employing the factor method, the design life is specified and the predicted service life is thereafter determined for the component or structure. The predicted service life is based on sets of factors that relate service life as determined from field studies to the expected life of a component for a similar use and subjected to a similar environmental load. However, according to Marteinsson (2003a, 2003b) one of the

Michael A. Lacasse is a senior research officer, NRC–Construction, National Research Council Canada, Ottawa, ON, Canada. **Martin Morelli** is a researcher at the Danish Building Research Institute, Aalborg University, Copenhagen, Denmark.

key issues regarding the use of the factor method is whether it provides reasonable estimates for service life, taking into account the probabilistic nature of the variables affecting service life in the field. Another issue is the difficulty in determining the value of factors and the consequences of changing these factors on the service life estimate (Listerud et al. 2011). In contrast to the service-life format, the limit-states format consists of checking the performance of a component or structure against a performance, or various limit states.

Different approaches to assessing the vulnerability of wood-frame structures to deterioration have been developed in recent years. Some of these approaches suggest applying a limit-states design approach to the performance assessment of the assembly. For example, a generalized limit-states method was presented by Bomberg and Allen (2006) as a systematic framework for limit-states design in respect to the durability of building envelopes. In addition, the limit-states format has been applied to structures containing wooden components regarding moisture durability. Mao et al. (2011) describe an approach based on the "in-cavity evaporation allowance," which is an experimental method to evaluate the moisture load from rain penetrating a wall. However, moisture management in wood-frame structures is often investigated by hygrothermal simulations. In these simulations, the threshold values for mold growth and wood decay are expressed in terms of relative humidity and moisture content, respectively. However, the influence of temperature and exposure time is often not included. The results from hygrothermal simulations can be transformed into an evaluation of limit states using, for example, mold isopleths (Sedlbauer 2002) or mold growth and wood decay models (Hukka and Viitanen 1999; Viitanen 1997a, 1997b). A performance-based model presented by Isaksen et al. (2010) uses a dose-response function to predict the onset of mold growth, which was used as a limit state.

The intent in this paper is to demonstrate a systematic method for the performance assessment of wood-frame wall assemblies that follows a limit-states design (LSD) approach and additionally respects the requirements set forth in ISO 13823, *General Principles on the Design of Structures for Durability* (ISO 2008). The use of the method is highlighted by applying it to the Moisture Management of Exterior Wall Systems (MEWS) project (Beaulieu et al. 2002; Kumaran et al. 2002, 2003), in which the concept referred to as the *relative humidity temperature index* (*RHT index*) is used as a basis for evaluating the response of wood-frame assemblies to climate loads from which, in turn, the long-term performance could be estimated. This index captures the duration of the coexistence of moisture and thermal conditions above a set of threshold levels for which the risk to the formation of mold growth or wood rot is unacceptably high. An example is given to illustrate the application of the RHT index in assessing the performance in respect to moisture management of a stucco-clad wood-frame wall assembly based on the results of simulations. The values for the RHT index given in the results are related to the risk of occurrence of wood rot. The notional concept of the RHT index and its application to informing on the hygrothermal response of wall assemblies has been previously presented in several other publications (Kumaran et al. 2002, 2003; Mukhopadhyaya et al. 2003a, 2003b, 2003c, 2004, 2006). This paper is intended to demonstrate that the approach used in the MEWS project, in which the RHT index is applied as a measure of the response of the wall to hygrothermal effects, is consistent with the broad precepts described in ISO 13823 (ISO 2008). As such, the methodology described in this paper offers a standard and systematic method for assessing the long-term performance (i.e., durability) of wood-frame wall assemblies. It is nonetheless relevant to note that in the approach described in this paper, wall configurations are evaluated against a reference wall of known and acceptable performance, as compared to inferring results from a damage function for the wall.

LIMIT-STATES DESIGN—ISO 13823

The principle of the limit-states method for any component or structure is to understand the structure environment, the transfer mechanisms, and the environmental action leading to action effects that over time result in the failure of the component or structure, as shown in Figure 1.

The durability requirements for structures or components of structures are maintenance of their required minimum level

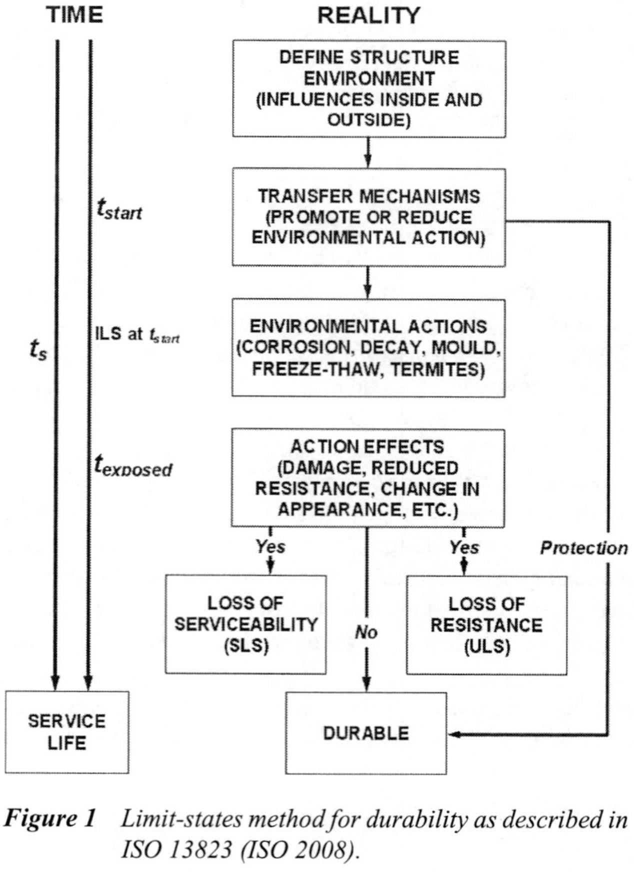

Figure 1 *Limit-states method for durability as described in ISO 13823 (ISO 2008).*

of performance during their design life with sufficient reliability for their intended use. The concept of the limit-states method is to verify the performance against a threshold value (limiting performance state). The effect of exceeding a limit state may be irreversible or reversible. An irreversible case relates, for example in respect to structural design, to collapse or similar structural failure, whereas a reversible case relates to requirements to maintain the component, assembly, or structure serviceable over time. Hence, serviceability limit states are not as severe as those that define an irreversible limit state or what might be referred to as *the ultimate limit state*. In ISO 13823 (ISO 2008), ultimate limit states (ULSs) and serviceability limit states (SLSs) are interpreted as follows:

- Ultimate limit state (ULS) is an irreversible process and is typically a condition state associated with collapse; in respect to material deterioration resulting in failure due to loss of resistance, the ULS is defined as when the resistance of the component or structure becomes equal to or less than the internal mechanical force; in the context of wood-frame structures, material deterioration relates to the onset of wood rot where after initiation of the rot process, loss of wood mass ensues (Viitanen et al. 2010).
- Serviceability limit state (SLS) can be either an irreversible or a reversible process; it is a condition state that corresponds to no longer providing the specified serviceability requirements for a structure or its components; for material degradation, the serviceability limit states are defined by either (i) the relative displacements that affect the function or appearance of either structural or nonstructural components or (ii) local damage or change in appearance that likewise affect the function or appearance of either structural or nonstructural components; in respect to wood-frame structures, changes in material appearance or local damage due to mold growth would constitute a limiting condition state.

Thus, mold growth is understood as a reversible process where mold growth can be initiated under certain conditions and later cease to grow; thus the condition state may temporarily exceed the SLS until over time the conditions are altered whereby the condition state no longer exceeds the limiting value.

The basic requirement for ULS is defined in Equation 1 and expresses that the load effect $S(t)$ must be smaller than the resistance. SLS is defined in Equation 2 and expresses that the load effect must be smaller than the limit indicating onset of serviceability failure (ISO 2008). These two requirements must be satisfied at any time t during the design life t_s of the component.

$$R(t) \geq S(t) \tag{1}$$

$$SLS > S(t) \tag{2}$$

where

$R(t)$ = resistance capacity of the structural component at time t

$S(t)$ = action effect, e.g., an internal force, stress, deformation, at any time t

SLS = serviceability limit state

In Figure 1, t_{start} is the time to reach the initiation of deterioration of a component, i.e., the initiation limit state (ILS). For a component protected against agents, e.g., preservative treatment of wood, the service life, t_s, can be determined as in Equation 3:

$$t_s = t_{start} + t_{exposed} \tag{3}$$

where $t_{exposed}$ is the service life after initiation of deterioration.

The initiation time for the degradation of wood-based products has been empirically defined by Viitanen et al. (2010) whereby the initiation time for the activation of the wood rot process for pine sapwood is given in terms of the temperature and relative humidity conditions sustained during the initiation period.

MEWS PROJECT—AN EXAMPLE OF THE LSD APPROACH FOR PERFORMANCE ASSESSMENT OF WOOD-FRAME WALL ASSEMBLIES

Project Overview

A limit state design approach has been used for performance assessment of wood-frame wall assemblies in the MEWS project. The intent of the MEWS project was to develop guidelines for moisture management strategies for wall assemblies to meet user requirements of long-term performance (durability) for the wide range of climate zones to which the walls could be subjected across North America (Beaulieu et al. 2002; Kumaran et al. 2002, 2003). The MEWS project was about defining the ability of specific wall systems to manage moisture sources, including construction moisture, humid indoor and outdoor air, precipitation, and moisture derived from indoor human activities. Given that moisture may enter wall assemblies in many ways, including vapor diffusion, air movement, rain penetration, and seepage, the envelope design strategy must therefore address all such processes and must control moisture accumulation throughout the annual climatic cycle, over many years of service life. The focus of this present study is on wood-frame buildings of four stories or less exposed to a range of outdoor climates found in North America. However, the main emphasis of the MEWS project was to estimate the hygrothermal responses of several different wall assemblies that were exposed to various climate loads as occur in North America and that engendered a range of rain penetration loads within the wall; such estimates of response were derived from the outputs of a hygrothermal simulation model. The simulation model results were obtained as values of temperature (T) and relative humidity (RH) for each of the discrete elements as configured for the model. In the MEWS project, a moisture index (MI) and a relative humidity temperature index (RHT index) were both developed as a means to assess the performance of wall assemblies.

Moisture Index (MI) and Climate Loads on Wall Structures

The MI was representative of the climate loads to which a given wall would be subjected in simulation and was based on the potential offered for wetting and drying of wall assemblies for a given climate. As such, and given the broad range of climates as might be found in North America, the climate was categorized into five severity classes, from low (MI < 0.7) to severe (MI ≥ 1.0) (Cornick et al. 2003; Cornick and Dalgliesh 2003a, 2003b). The MI thus was representative of the climate load to which wall assemblies were subjected and refers to the risk to the onset of moisture problems within a wall assembly; a map of North America showing these different MI classes is given in Figure 2. The RHT index was used for evaluating the responses of wall assemblies to climate loads and from which could be inferred the long-term performance of wood-frame structures. The MI together with the RHT index revealed the hygrothermal wall response under different climate conditions and differing wall configurations in relation to the degree of water ingress to the wall assembly due to inadvertent rain penetration of the cladding.

Wall Moisture Response Indicator—RHT Index

The RHT index is a moisture response indicator derived from simulation of a selected period of time (e.g., the last two years of a three-year simulation period) for any specified area of a wall cross section (Beaulieu et al. 2002; Kumaran et al. 2002). The index captures the duration of the coexistence of moisture and thermal conditions above set threshold levels for which the risk to the formation of, for example, mold growth or wood rot is considered too high, and is given as

$$\text{RHT (Cumulative value, second and third years of simulation)} = \Sigma(\text{RH} - \text{RH}_x) \times (T - T_x) \quad (4)$$

where $\text{RH} > \text{RH}_x$ (%) and $T > T_x$ (°C) are values averaged over 10-day intervals of simulation.

Hence, a single-numbered indicator RHT is formed by summing only non-zero values over a two-year simulation period for a target region in the wall. During any time step when either or both $\text{RH} < \text{RH}_x$ and $T \leq T_x$, the RHT value for that time step is zero. The user-defined values used in the MEWS project and as nominal conditions for the initiation of damage to wood were $\text{RH}_x = 95\%$ and $T_x = 5$°C; the limiting values for the initiation of mold growth were $\text{RH}_x = 80\%$ and $T_x = 5$°C. The higher RHT index values indicate an increased

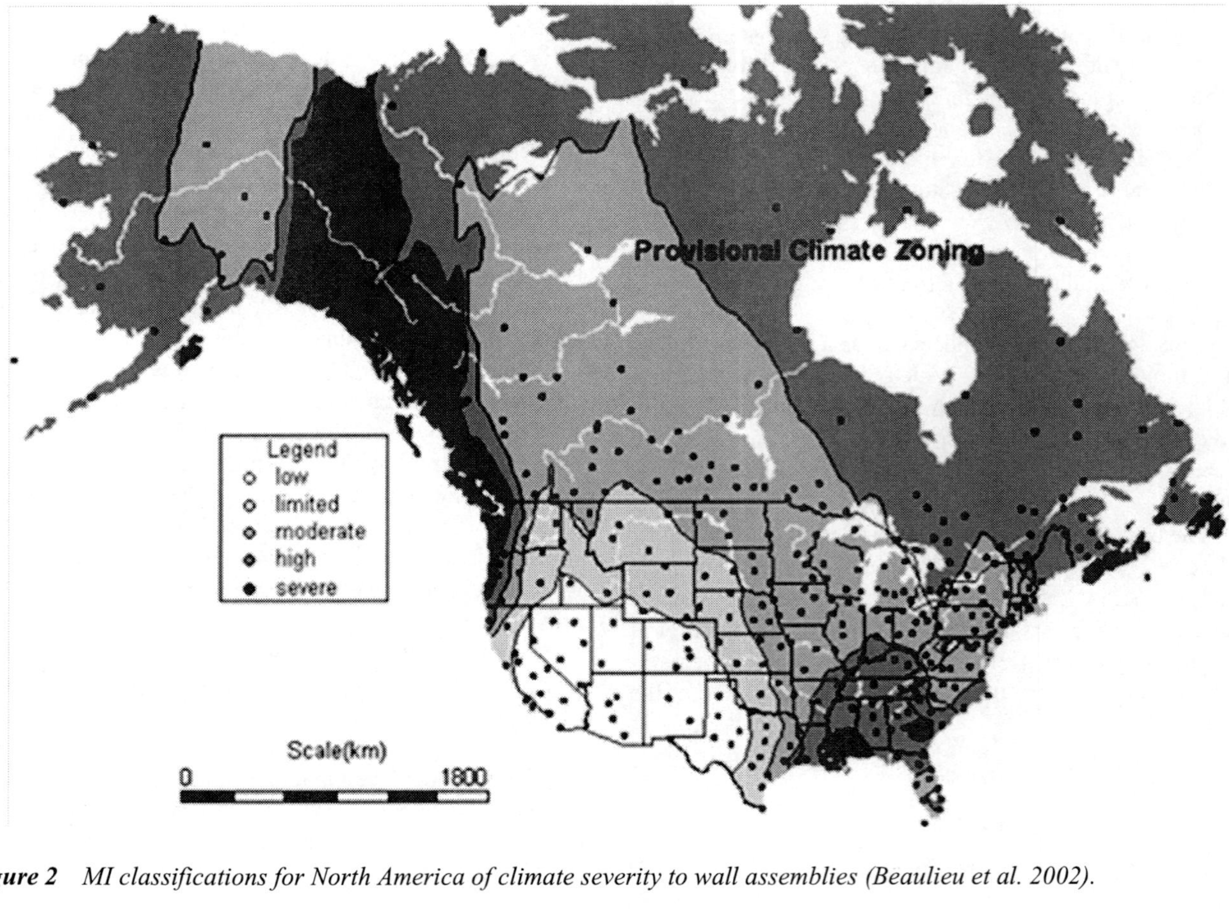

Figure 2 MI classifications for North America of climate severity to wall assemblies (Beaulieu et al. 2002).

severity of the hygrothermal response—a greater risk to the formation of damage. As such, the RHT index can be used for a comparison of different wall configurations in specific climates or the same configuration in various different climates; these elements are discussed in the subsequent section.

LSD Approach as Applied to a Stucco-Clad Wood-Frame Wall Assembly

In our example, consideration is given to the attack mechanism by decay fungi in a stucco-clad wood-frame wall assembly. Two different scenarios were simulated: one without water ingress from rain and one with a well-defined rain ingress through inadvertent leakage in the wall. The sources of data for this example were obtained from the MEWS project based on laboratory studies (Beaulieu et al. 2002; Mukhopadhyaya et al. 2003a, 2003b, 2003c).

Defining "Structure" and "Environment". Referring to Figure 1, the two primary and initial elements of the LSD method for durability are defining the structure and the environment. In this instance the structure is defined as the stucco-clad wood-frame wall assembly, as configured for simulation, whereas the environment relates to the conditions and loads to which the structure is to be subjected, through which environmental actions (e.g., wood decay, mold growth) arise and from which the effects of these actions manifest themselves (heightened risk to formation of damage, health issues); both of these elements are further described in the subsequent sections.

Stucco-Clad Wood-Frame Wall Assembly. The wall configuration is shown in Figure 3 together with the region of focus that in the simulations was a thin slice (5 mm) of the top surface of the bottom plate, extending 53 mm from the sheathing board. It was understood that the onset of wood damage or mold formation in this particular location would be equal to an insufficient wall configuration.

Boundary Conditions. The exterior environment influencing the wall structure was described by hourly values of exterior temperature and relative humidity, solar radiation, cloud index, and wind-driven rain (WDR) (i.e., for WDR: hourly rainfall, wind speed, and wind direction) for specified climatic locations of interest in the study; these included, amongst a set of seven locations, Wilmington, NC; Ottawa, ON; Winnipeg, MB; and Seattle, WA. The respective MI values for each of these locations were, respectively, 1.13, 0.99, 0.93, and 0.86. From multi-year weather data (>30 years), two years were selected for each location defining a "wet" and an "average" year. A wet year represented the wettest year that also exceeded the average year by at least one standard error of the mean (i.e., 1σ). These two years were combined to define a three-year weather cycle (wet-wet-average) to which the wall assemblies were subjected in the simulation. Derived values for the RHT index were, however, based on the results of the last two years of simulation.

The interior climate was represented by two parameters: temperature (T) and relative humidity (RH). Interior room T and RH were switched from winter values of 22°C, 25% rh (when mean monthly outdoor temperature was less than 11°C) to summer values of 25°C, 55% rh for the warmer months, following the criterion given in *ASHRAE Handbook—HVAC Applications* (ASHRAE 1999).

Defining Transfer Mechanisms. Transfer mechanisms are mechanisms by which loads (e.g., structural, environmental) are transferred to the structure. In respect to a wood-frame wall assembly, the hygrothermal simulation model, a descrip-

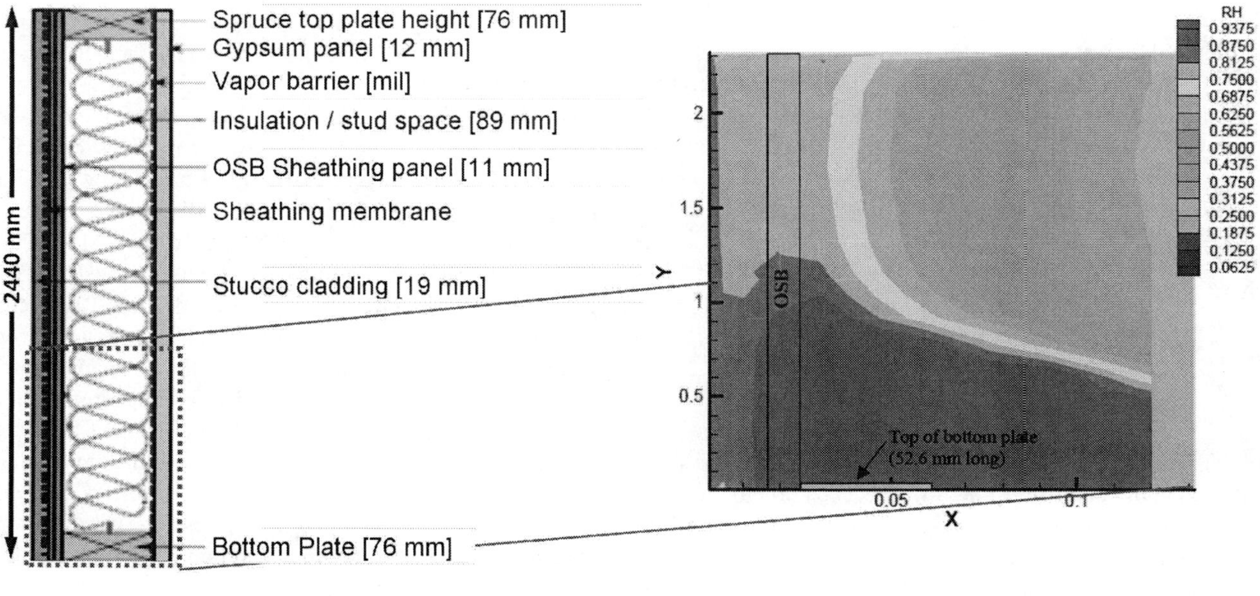

Figure 3 *Stucco-clad wall cross section subjected to Ottawa climate; two-year simulation. The dark (red) areas are regions for which the model predicted RH > 87% (Beaulieu et al. 2002).*

tion of which is given by Nofal et al. (2001), is capable of transferring moisture loads to the structure provided these are compatible with the basic variables that affect heat, air, and moisture transfer across the wall. As such, and given that the hygrothermal model used in these studies could manage the transfer of heat air and moisture through the usual mechanisms, there were nonetheless a number of elements that required specific mechanisms to be developed: (i) WDR to be converted to a rain load on the vertical surface of the wall, (ii) proportion of water at the surface of the wall that might enter the wall assembly, and (iii) location of water deposition in the wall.

In regard to the a rain load on the wall due to WDR (i), Straube's method (Cornick et al. 2002) for calculating the amount of WDR impinging on a wall was selected for use; it was chosen because it generally provided the most conservative values of the several different methods considered in the MEWS study; it took the following form:

$$WDR = RAF \times DRF(r_h) \times \cos(\theta) \times V(h) \times r_h \qquad (5)$$

where

WDR	=	wind-driven load, L/m^2·h
RAF	=	rain admittance factor
DRF	=	driving rain factor
r_h	=	horizontal rainfall intensity, mm/m^2·h
$V(h)$	=	wind speed at the height of interest, m/s
θ	=	angle of the wind to the wall normal

Derived values for the RAF and DRF can be found in the work by Cornick et al. (2002).

The proportion of water at the surface of the wall that might enter the wall assembly (ii) was based on the results of water penetration tests of a series of stucco-clad wall assemblies (Lacasse et al. 2002) from which an empirical function was derived for water entry to the wall that considered both the quantity of water being deposited on the wall and the pressure difference across the assembly. This function was given as

$$Q\ (\text{L/h}) = R_w \times f(\Delta P)$$
$$= R_w \times \{0.0314 + 7.74 \times 10^{-5}\,\Delta P - 8.14 \times 10^{-8}\,\Delta P^2\} \qquad (6)$$

where

Q	=	rate of water entry to the wall
R_w	=	rate of water striking the wall
ΔP	=	pressure difference across the wall assembly

The third derived mechanism for the transfer of moisture to the assembly is the location of water deposition in the wall proper (iii). For the study being used as an example of the LSD approach, the water was placed in two different locations on the interior surface of the wood-based wall assembly. Based on observations of experiments of water entry conducted during this study, water was observed to accumulate along the lower portion (ca. 5 mm × 0.8 m height) of the sheathing panel (oriented strand board, OSB) in proximity to the bottom plate and on the horizontal surface of the bottom plate over a length of 5 × 56.4 mm). Both these surfaces were wood-based and are depicted in the RH contour plot provided in Figure 3. In this figure, the different levels of RH in the respective elements of the wall configuration are shown following a simulation. As is perhaps evident, those locations having the highest RH formed regions of focus for which the results of T and RH were compiled to provide the hygrothermal responses of the most vulnerable points in the wall assembly given the assumptions made in respect to the deposition of water on wall components. Evidently, these locations were also those where water tended to accumulate.

Defining Environmental Actions. Environmental actions arise as a result of the response of a wall to the effects of climate loads and interior environmental conditions based on, as was previously described, the different transfer mechanisms that affect moisture transfer to the structure. Actions such as damage to wood-based components or the formation of mold result from specific conditions of T and RH in the wood-based component for a period sufficiently long to effect an increase to the risk in formation of rot or mold. The RHT index is a moisture response indicator derived from the result of simulation that captures the degree of risk over time; in this study, threshold values for the initiation of mold growth were RH_x = 80% and T_x = 5°C and for damage of wood were RH_x = 95% and T_x = 5°C. Using the values of RHT index derived from simulation results, one can readily compare the performance of different wall assemblies for a given climate or the effect of different climates on a specific assembly.

Hygrothermal Simulation of Wall Response

The results of hygrothermal simulation of the stucco wall assembly are provided in terms of values for T and RH for each of the discrete elements of which the wall is configured (or for the moisture content if desired). Examples of plots of T and RH from which were subsequently derived values for the RHT index are provided in Figures 4 and5. Figure 4 shows the T and RH profiles predicted for the stucco wall assembly (Figure 2) for Winnipeg, MB (cumulative RHT(95) = 1337), and Figure 5 shows them for Ottawa, ON (cumulative RHT(95) = 1536). The plots show average values of T and RH over the two-year simulation period (730 days) in the region of focus as depicted in Figure 3, in this instance the 5 × 56 mm portion of the bottom plate. The threshold RH values in both these plots represent RHT(95); a solid line is evident at that the 95% rh threshold value. Likewise for the T threshold value, a line at 5°C has been traced. The amount of water (Q) dosed to the bottom plate is provided in decilitres·m^{-2}; values range from 0 to 20 decilitres·m^{-2} for the Winnipeg climate and to 13 decilitres·m^{-2} for Ottawa. Results for RHT(95) for four locations in North America having different values for MI are given in Table 1.

As is perhaps evident from the values provided in Table 1, there is little or no hygrothermal response when no water is admitted to the stud cavity, with the exception of the Wilmington, NC, location, where there is nonetheless a small response (RHT(95) = 9) at the base of the wall. When the full dose of water is admitted to the stud cavity (i.e., 1 Q), the increases in values of RHT(95) are all evident; increases in values of RHT(95) correspond to increases in values of MI—e.g., for MI = 0.86, RHT(95) = 1337 and for MI = 1.13, RHT(95) = 3213. Thus, the use of this approach offers a measure of the relative hygrothermal response of the stucco-clad wall for different climate locations in respect to the risk to the formation of wood damage at the bottom plate.

A comparison of values for RHT(95) for walls incorporating different types of vapor barrier membranes can be found across the lower three lines of Table 1 (i.e., 1 Q (VB1), 1 Q

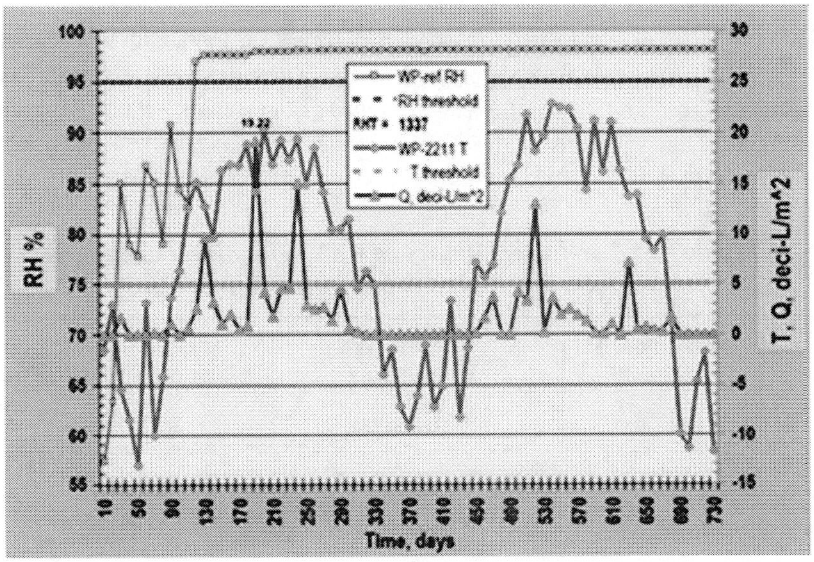

Figure 4 T *and* RH *profiles predicted for the wall described in Figure 3 for Winnipeg. Cumulative RHT(95) = 1337.* WP *refers to Winnipeg and* WP-ref RH *is the average RH at 10-day intervals in the region of focus for the reference wall.* WP-2211 T *refers to the average* T *at 10-day intervals in the region of focus for the reference wall (i.e., wall 2211).*

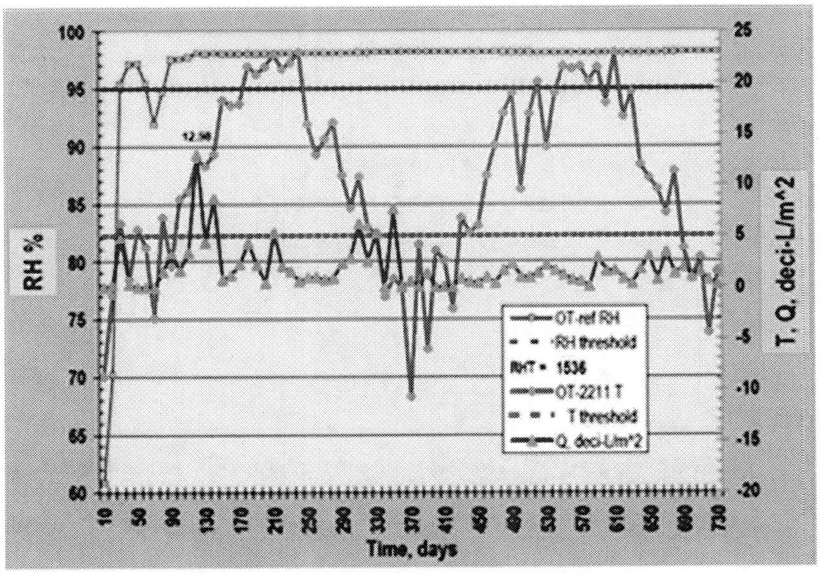

Figure 5 *Predicted* T *and RH profile for the wall described in Figure 3 for Ottawa climate. Cumulative RHT(95)=1536.* OT *refers to Ottawa;* OT-ref RH *and* OT-2211 T*, respectively, are the average RH and average* T *at 10-day intervals in the region of focus for the reference wall (i.e., wall 2211).*

(VB2), and 1 *Q* (VB3)). The relationships between the vapor permeability and relative humidity of these three types of membrane are provided in Figure 6. Type 1 and type 2 vapor barrier membranes are polymer-based products whose permeability to vapor is invariable to changes in RH, whereas the type 3 membrane is a paper-based product that shows a marked change in vapor permeability with an increase in RH. These results simply serve to illustrate that one can differentiate between the relative performance of walls having different components in a given climate.

CONCLUDING REMARKS

There are several other examples of how the RHT index of wall response can be used to assess the relative performance of wall assemblies; these can be found in reports by Beaulieu et al. (2002) and Mukhopadhyaya et al. (2003) and the several additional papers previously cited that have been published on these results. The intent of this paper was not to delve into an in-depth explanation of the results but to demonstrate that the proposed approach is consistent with the broad precepts described in ISO 13823 (ISO 2008) and, as such, offers a systematic method of assessing the long-term performance (i.e., durability) of wall assemblies. The limit-states method is probabilistic, i.e., it is based on the estimation of the probability of failure, which is a function of the loads imposed on the components and the probability of occurrence of these loads. However, the probability of occurrence of loads still needs to be considered; this is discussed below. The use of the RHT index is a means of relating how specific components of the structure respond to these loads; the variability of the material characteristics are not explicitly taken into consideration.

Table 1. Cumulative Values of RHT(95) for Four Locations Having Varying Moisture Loads to the Stud Cavity

	Wilmington, NC MI = 1.13	Seattle, WA MI = 0.99	Ottawa, ON MI =0.93	Winnipeg, MB MI = 0.86
Q Dose Fraction	RHT(95)	RHT(95)	RHT(95)	RHT(95)
0	9	0	0	0
1/4 *Q*	2841	1979	864	697
1/2 *Q*	3008	2177	1434	1190
1 *Q* (VB1)	3213	2290	1536	1337
1 *Q* (VB2)	3261	2245	1517	1321
1 *Q* (VB3)	3080	2148	1482	1295

*MI = moisture index; VB1–3 = difference vapor barrier as shown in Figure 6.

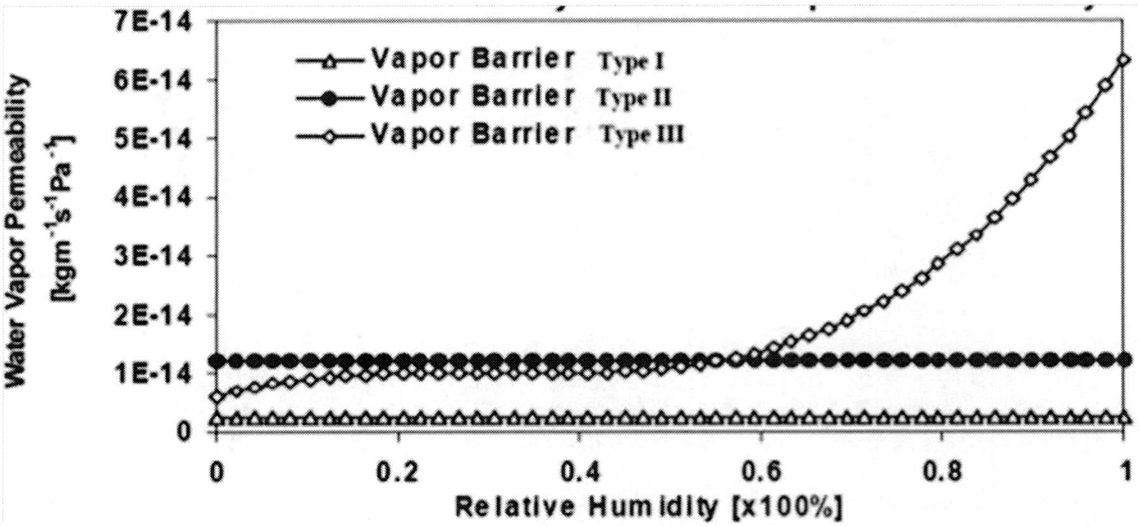

Figure 6 *Relationships between vapor permeability and relative humidity of three types of vapor barrier membranes.*

Does the proposed approach provide a basis for assessing the SLS or indeed the ULS? Given that the ULS is related to the structural integrity of the wall assembly, the proposed approach, as described in this paper, does not currently provide a means to determine whether the ULS is attained. It was earlier proposed by Nofal (1998) that damage functions for wood could readily be adapted to the results of hygrothermal simulation, and the wood decay models developed by Viitanen (1997a, 1997b) were indeed implemented within a version of National Research Council Canada's (NRC's) simulation model but not in fact used in the MEWS study previously described. However, recent developments by Bastidas-Arteaga et al. (2015), Isaksson et al. (2013), Saito et al. (2012), and Viitanen et al. (2010) as relate to wood decay, mass loss, and loss in structural resistance suggest that such models could be implemented in the proposed approach.

The approach as currently developed more appropriately relates to assessing the SLS of components in the wall assembly. How then is the long-term performance determined on the basis of the proposed approach? Two methods are proposed: (i) by demonstrating performance of a wall assembly equivalent to that of a wall having known or accepted long-term performance as obtained from observations of in-service performance in the field or (ii) by subjecting the wall assemblies to climate loads of a specified return period.

The first of the two proposed methods is what is currently being used at NRC as the basis for assessing the performance of wall assemblies when technical opinions are rendered on new construction products to demonstrate compliance to the *National Building Code of Canada* (NRC 2015) in respect to hygrothermal performance of the assembly as a whole. A comparison is made between the hygrothermal performance of a code-complaint reference wall of known or accepted minimum level of performance to that of the wall whose performance is to be evaluated. Provided the performance of the wall being evaluated equals or is better than the reference wall, the "evaluated" wall is deemed to have met the minimum level of conformance to the building code.

The second method has not yet been fully explored, but it nonetheless has merit, as structural loads that arise from the effects of climate (e.g., wind loads, snow loads) are typically based on a given return period. Clearly for a given climate location, the greater the return period the more severe the climate loads. The severity of climate loads in respect to their return period and as regards the water-tightness performance of wall assemblies has been explored by Cornick and Lacasse (2009) and more recently by Van Den Bossche et al. (2013) and Pérez-Bella et al. (2015). This indicates that the SLS can be adapted to the proposed approach in the context of an adequate in-service hygrothermal performance.

ACKNOWLEDGMENTS

Thanks are extended to the Construction program of the National Research Council Canada for support of this work and to the following personnel of the Heat and Moisture Performance of the Building Envelope Group who contributed to the MEWS project: Mr. S.M. Cornick, Mr. W.A. Dalgliesh, Dr. M. K. Kumaran, Dr. W. Maref, Dr. P. Mukhopadhyaya, Dr. M. Nofal, Dr. D. Quirt, Ms. M.Z. Rousseau, and Mr. M.C. Swinton.

REFERENCES

ASHRAE. 1999. Chapter 3, Commercial and public buildings. In *ASHRAE Handbook—HVAC Applications*. Atlanta: ASHRAE.

Bastidas-Arteaga, E., Y. Aoues, and A. Chateauneuf. 2015. Optimal design of deteriorating timber components under climate variations. Presented at the 12th International Conference on Applications of Statistics and Probability in Civil Engineering, ICASP12, Vancouver, Canada, July 12–15.

Beaulieu, P., M.T. Bomberg, S.M. Cornick, W.A. Dalgliesh, G. Desmarais, R. Djebbar, M.K. Kumaran, M.A. Lacasse, J.C. Lackey, W. Maref, P. Mukhopadhyaya, M. Nofal, N. Normandin, M. Nicholls, T. O'Connor, J.D. Quirt, M.Z. Rousseau, M.N. Said, M.C. Swinton, F. Tariku, and D. van Reenen. 2002. Final report from Task 8 of MEWS Project (T8-03)—Hygrothermal response of exterior wall systems to climate loading: Methodology and interpretation of results for stucco, EIFS, masonry and siding clad wood-frame wall. IRC Research Report IRC-RR-118. Ottawa, Ontario: National Research Council Canada, Institute for Research in Construction.

Bomberg, M., and D. Allen. 1996. Use of generalized limit states method for design of building envelopes for durability. *Journal of Thermal Insulation and Building Envelopes* 20:18–39.

Cornick, S.M., W.A. Dalgliesh, M.N. Said, R. Djebbar, F. Tariku, and M.K. Kumaran. 2002. Report from Task 4 of MEWS Project: Task 4—Environmental Conditions final report. Research Report RR-113. Ottawa, Ontario: NRC Institute for Research in Construction, National Research Council Canada.

Cornick, S., R. Djebbar, and W.A. Dalgliesh. 2003. Selecting moisture reference years using a moisture index approach. *Building and Environment* 38(12):1367–79.

Cornick, S.M., and W.A. Dalgliesh. 2003a. A moisture index approach to characterizing climates for moisture management of building envelopes. *Proceedings of the 9th Canadian Conference on Building Science and Technology*, pp. 383–98.

Cornick, S.M., and W.A. Dalgliesh. 2003b. A moisture index to characterize climates for building envelope design. *Journal of Thermal Envelope & Building Science* 27(2):151–78.

Cornick, S.M., and M.A. Lacasse. 2009. An investigation of climate loads on building façades for selected locations. *Journal of ASTM International* 6(2)1–17.

Hukka, A., and H. Viitanen. 1999. A mathematical model of mould growth on wooden material. *Wood Science and Technology* 33(6):475–85.

Isaksson, T., S. Thelandersson, A. Ekstrand-Tobin, and P. Johansson. 2010. Critical conditions for onset of mould growth under varying climate conditions. *Building & Environment* 45(7):1712–21.

Isaksson, T., C. Brischke, and S. Thelandersson. 2013. Development of decay performance models for outdoor timber structures. *Materials and Structures* 46:1209–25.

ISO. 2008. ISO 13823:2008, *General principles on the design of structures for durability*. Geneva: International Organization for Standardization.

Kumaran, M.K., P. Mukhopadhyaya, S.M. Cornick, M.A. Lacasse, W. Maref, M.Z. Rousseau, M. Nofal, J.D. Quirt, and W.A. Dalgliesh. 2002. A methodology to develop moisture management strategies for wood-frame walls in North America: Application to stucco-clad walls. *Proceedings of the 6th Nordic Building Physics Symposium*, pp. 651–58.

Kumaran, M.K., P. Mukhopadhyaya, S.M. Cornick, M.A. Lacasse, M.Z. Rousseau, W. Maref, M. Nofal, J.D. Quirt, and W.A. Dalgliesh. 2003. An integrated methodology to develop moisture management strategies for exterior wall systems. *Proceedings of 9th Canadian Conference on Building Science and Technology*, pp. 45–62.

Lacasse, M.A., T. O'Connor, S.C. Nunes, and P. Beaulieu. 2003. Report from Task 6 of MEWS Project: Experimental assessment of water penetration and entry into wood-frame wall specimens. Final report, research report RR-133. Ottawa, Ontario: NRC Institute for Research in Construction, National Research Council Canada.

Listerud, C.A., S. Bjørberg, and P.J. Hovde. 2011. Service life estimation of facades—Use of the factor method in practice. *Proceedings of XII DBMC International Conference on Durability of Building Materials and Components*, Porto, Portugal, April 12–15.

Mao, Q., P. Fazio, and J. Rao. 2011. A limit state design (LSD) approach for comparing relative drying performance of wood-frame envelope systems with full-scale lab testing. *Building and Environment* 46(3):797–806.

Marteinsson, B. 2003a. Assessment of service lives in the design of buildings—Development of the factor method. Licentiate thesis, KTH Research School—HIG Centre for Built Environment, Gavle, Sweden.

Marteinsson, B. 2003b. Durability and the factor method of ISO 15686-1. *Building Research & Information* 31(6):416–26.

Mukhopadhyaya, P., K. Kumaran, F. Tariku, and D. van Reenen. 2003a. Long-term performance: Predict the moisture management performance of wall systems as a function of climate, material properties, through mathematical modelling. Final Report from Task 7 of MEWS Project, research report IRC-RR-132. Ottawa, Ontario: Institute for Research in Construction, National Research Council Canada.

Mukhopadhyaya, P., R. Djebbar, M.K. Kumaran, and D. Van Reenen. 2003b. Moisture management strategy in wood-frame stucco wall—Observations from hygrothermal simulation. Presented at XXX IAHS 31, World Congress on Housing, Housing Process & Product, June 23, Montreal, Quebec.

Mukhopadhyaya, P., K. Kumaran, M. Rousseau, F. Tariku, D. van Reenen, and W.A. Dalgliesh. 2003c. Application of hygrothermal analyses to optimize exterior wall design. *Proceedings of Research in Building Physics*, pp. 417–26.

Mukhopadhyaya, P., K. Kumaran, and D. van Reenen. 2004. Vapour barrier and moisture response of wood-frame stucco wall—Results from hygrothermal simulation. *Proceedings of CIB World Building Congress 2004*, pp. 1–10.

Mukhopadhyaya, P., K. Kumaran, F. Tariku, and D. van Reenen. 2006. Application of hygrothermal modeling tool to assess moisture response of exterior walls. *Journal of Architectural Engineering* 12(4):178–86.

Nofal, M. 1998. Hygrothermal damage of building materials and components: State-of-the-art report on studies of hygrothermal damage and proposed approach for damage assessments. Internal report No. 757. Ottawa, Ontario: Institute for Research in Construction, National Research Council Canada.

Nofal, M., M. Straver, and M.K. Kumaran. 2001. Comparison of four hygrothermal models in terms of long-term performance assessment of wood-frame constructions. Conference publication NRCC 44690. *8th Conference on Building Science & Technology, Solutions to Moisture Problems in Building Enclosures*. Ottawa, Ontario: National Research Council Canada, NRC-Construction; p. 1–19.

NRC. 2015. *National Building Code of Canada*. Ottawa, Ontario: National Research Council Canada.

Pérez-Bellaa, J.M., J. Domínguez-Hernándeza, E. Cano-Suñéna, J.J. del Coz-Díazb, and F.P. Álvarez Rabanalb. 2015. Improvement alternatives for determining the watertightness performance of building facades. *Building Research & Information* 43(6):723–36.

Saito, H., K. Fukuda, and T. Sawachi. 2012. Integration model of hygrothermal analysis with decay process for durability assessment of building envelopes. *Building Simulation* 5:315–24.

Sedlbauer, K. 2002. Prediction of mould growth by hygrothermal calculation. *Journal of Building Physics* 25(4):321–36.

Van Den Bossche, N., M.A. Lacasse, and A. Janssens. 2013. A uniform methodology to establish test parameters for watertightness testing: Part I: A critical review. *Building and Environment* 63:145–56.

Viitanen, H.A. 1997a. Modelling the time factor in the development of brown rot decay in pine and spruce sapwood—The effect of critical humidity and temperature conditions. *Holzforschung-International Journal of the Biology, Chemistry, Physics and Technology of Wood* 51(2):99–106.

Viitanen, H.A. 1997b. Modelling the time factor in the development of mould fungi—The effect of critical humidity and temperature conditions on pine and spruce sapwood. *Holzforschung—International Journal of the Biology, Chemistry, Physics and Technology of Wood* 51(1):6–14.

Viitanen, H., T. Toratti, L. Makkonen, R. Peuhkuri, T. Ojanen, L. Ruokolainen, and J. Räisänen. 2010. Towards modelling of decay risk of wooden materials. *Eur. J. Wood Prod.* 68:303–13.

Risk of Condensation and Mold Growth in Highly Insulated Wood Frame Walls

Hamed H. Saber, PhD **Michael A. Lacasse, PhD, PEng** **G. Ganapathy, BEng (ME)**
Silvio Plescia, PEng **Anil Parekh, MASc**

ABSTRACT

A research study was conducted to investigate the risk of condensation and mold growth in 2 × 6 wood framing wall assemblies associated with increasing the thermal resistance (R-value) of cavity insulation for various scenarios of exterior insulation products. Based on the current construction practices, a set of three wall assemblies with different types of exterior insulation systems were chosen for field study with different R-values. In the first phase of this study, the hygrothermal model was benchmarked against the test data of full-scale wood framing wall systems. The predications of the model were in good agreement with the test data. Thereafter, the model was used to conduct parametric study to assess the risk of condensation of these wall assemblies when they were subjected to different air leakage rates for various climate zones. Both the numerical results and the field monitoring data showed different behaviors of exterior insulation strategies. The results of the hygrothermal performance were expressed using the mold index criteria, which allowed sufficient resolution to assess the risk of moisture condensation and related risk of mold growth in the wall assemblies. The results showed, as well, that adding exterior insulation of different water vapor permeance has resulted in lower risk of condensation and mold growth than the reference wall system (i.e., without exterior insulation).

INTRODUCTION—PROJECT OVERVIEW

Given the heightened interest of homebuilders to provide homes that met or exceeded ENERGY STAR® requirements, and their voiced concerns regarding "super-insulated" homes, the intent of this work was to demonstrate compliance of a set of highly insulated wall assemblies as compared to a code-compliant reference wall in respect to their anticipated thermal and hygrothermal performance when subjected to Canadian climate extremes. On the basis of providing useful information to building practitioners, the National Research Council of Canada (NRC) undertook field monitoring and numerical modeling to investigate the risk of condensation in wall assemblies having different combinations of increased thermal resistance (R-value) of insulation for selected insulation products.

The field monitoring of different wall assemblies was undertaken in the NRC's Field Exposure of Walls Facility (FEWF) in a two-phase, two-year project. Three wall assemblies were tested in the initial year of Phase 1 of the project, and another three wall assemblies were tested in Phase 2 of the second project year. A detailed description of the wall assemblies of Phase 2 are available in Lacasse et al. (2016); this paper focuses on describing the results derived from the initial three wall assemblies evaluated in Phase 1.

In Phase 1, three 1219 × 1829 mm (4 × 6 ft) test specimens having conventional 38 × 140 mm (2 × 6 in.) wood frame walls were constructed using different types of exterior insulation products that included 25 mm (1 in.) EPS, 51 mm (2 in.) XPS, and 76 mm (3 in.) mineral fiber (MF) insulation. The three specimens were installed side-by-side in the FEWF and exposed to local climate conditions of Ottawa over a one-year period; the test period started in August 2013 and ended in October 2014.

On the basis of results obtained from monitoring the response of the respective wall assemblies to local climate conditions, the numerical hygrothermal model hygIRC-C was benchmarked against selected experimental data. Thereafter,

Hamed H. Saber is an associate research officer, **Michael A. Lacasse** is a senior research officer, and **G. Ganapathy** is a technical officer at NRC-Construction, National Research Council Canada Ottawa, ON, Canada. **Silvio Plescia** is a senior researcher in the Policy and Research Division of Canada Mortgage and Housing Corporation Ottawa, ON, Canada. **Anil Parekh** is a senior researcher at the Buildings and Renewables Group, Natural Resources Canada Ottawa, ON, Canada.

the model was used to conduct a parametric analysis to investigate the risk of condensation and mold growth in the respective wall assemblies subjected to different climatic conditions for a select set of locations in Canada. A description of the hygIRC numerical hygrothermal simulation model and record of benchmarking exercises are available in Saber and Ganapathy (2016).

In this paper, results are provided for benchmarking the hygIRC-C numerical simulation model against experimental test data. Thereafter, information derived from simulations is provided on the risk of formation of condensation and subsequent mold growth within the three configurations of wood frame wall assembly as were tested in the FEWF. The simulations considered subjecting each of the three walls to the climatic conditions of five locations in Canada and for which the climate varied as being mild-wet to cold-wet and cold- wet to cold-dry.

DESCRIPTION OF WALL SPECIMENS

The three wall test specimens of Phase 1 (1219 × 1829 mm [4 × 6 ft]) consisted of 38 × 140 mm [2 × 6 in.]) wood frame walls installed side-by-side in the FEWF (Figure 1). The different material layers and the dimensions of the wall specimens are given in Figure 2 to Figure 4. The backup wall for all three design strategies consisted of interior drywall (12.7 mm [0.5 in.] thick), polyethylene air and vapor barrier (6 mil thick), 38 × 140 mm (2 × 6 in.) wood frame wall with friction-fit glass fiber batt insulation of R-24, and oriented strand board (OSB) (11 mm [7/16 in.] thick). The test specimens were constructed by adding different types of exterior insulation products of different thicknesses to the backup wall. The first wall (Wall 1; Figure 2,) was constructed by adding an expanded polystyrene (EPS) layer of 25 mm (1 in.) thick onto the OSB sheathing. The second wall (Wall 2; Figure 3) was constructed with a 51 mm (2 in.) thick extruded polystyrene (XPS) panel. The final wall (Wall 3; Figure 4) was constructed with 76 mm (3 in.) thick mineral fiber insulation.

As a part of the test protocol, more fully described in Lacasse et al. (2016), all heat flux transducers (HFTs) used in the three test specimens were calibrated according to ASTM C-1130, *Standard Practice for Calibrating Thin Heat Flux Transducers* (ASTM 2009). The locations of the HFTs in the respective test specimens are shown in Figures 2 to 4. The uncertainty of the heat flux measurements was ±5%.

MODEL BENCHMARKING

Having previously benchmarked the present model to several tests undertaken in controlled laboratory conditions as described in Saber and Ganapathy (2016) and Saber et al.

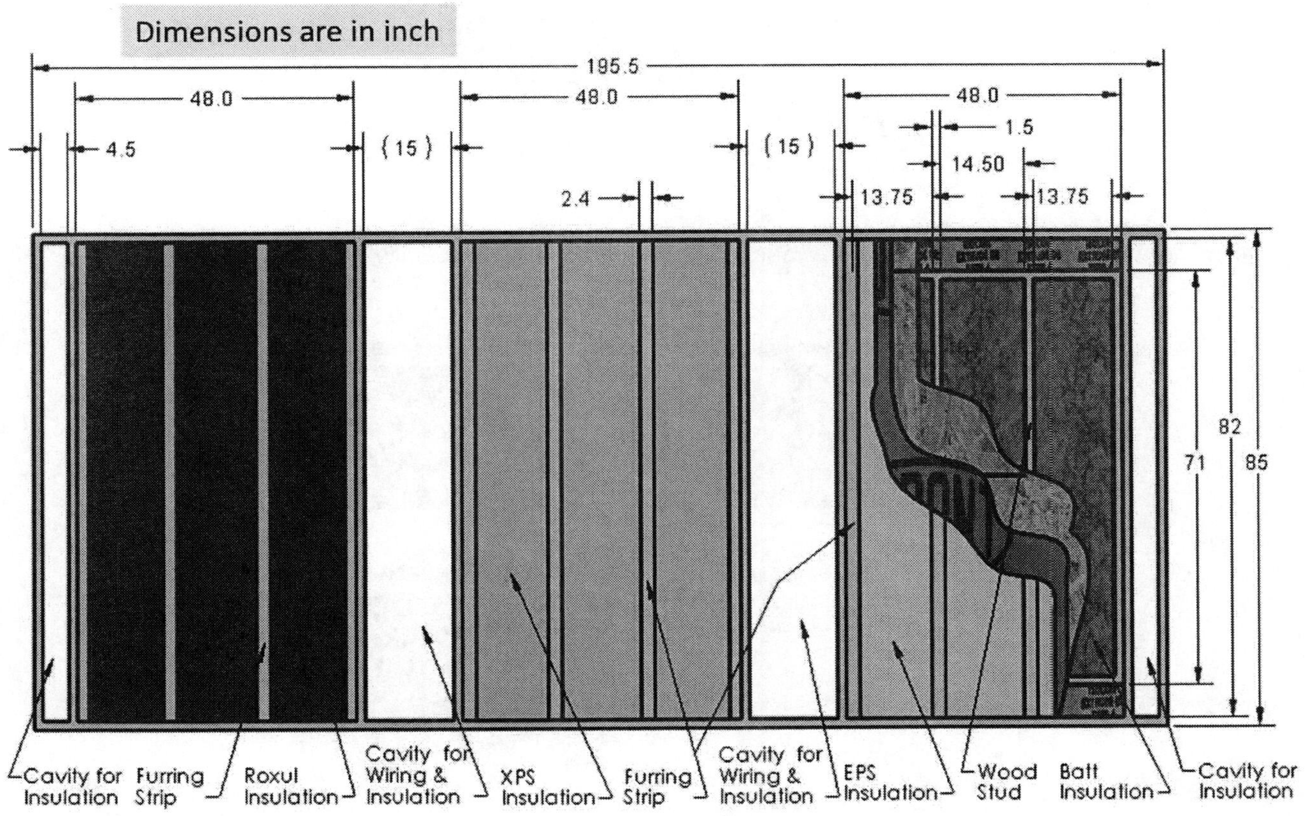

Figure 1 *Schematic of three residential 38 × 140 mm (2 × 6 in.) wood frame wall test specimens installed side by side in the FEWF.*

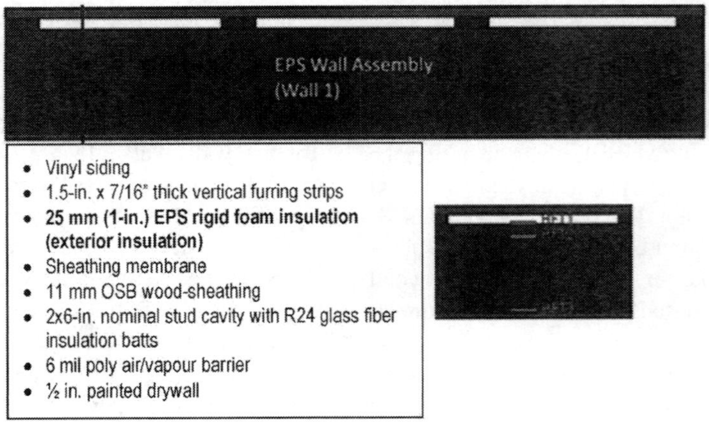

Figure 2 *Horizontal cross section through EPS wall assembly showing locations of heat flux transducers (HFTs) (Wall 1).*

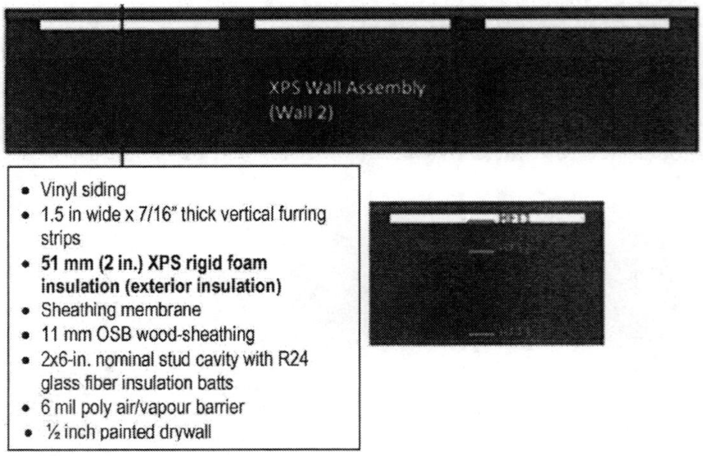

Figure 3 *Horizontal cross-section through XPS wall assembly showing locations of HFTs (Wall 2).*

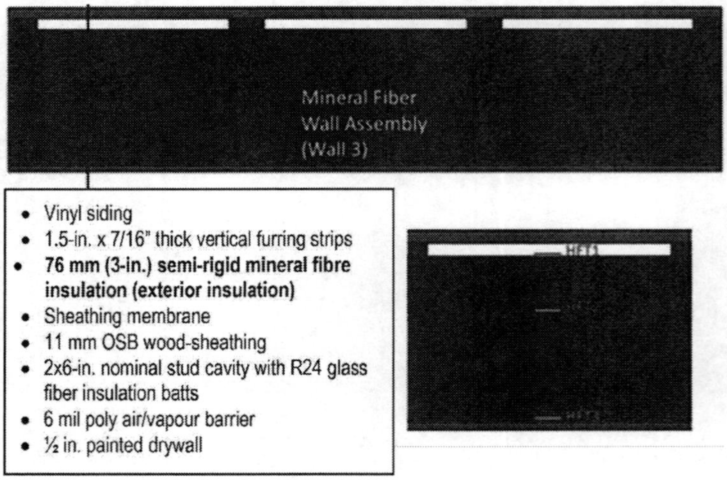

Figure 4 *Horizontal cross-section through mineral fiber wall assembly showing locations of HFTs (Wall 3).*

2014, a subsequent and important step was to benchmark the present model against the field measurements for the three wall systems previously described. Information is provided regarding assumptions in completing the transient numerical simulations and initial and boundary conditions that were used in conducting model benchmarking.

Transient Numerical Simulations

This section presents the assumptions and initial and boundary conditions that were used in conducting the numerical simulations for respective wall test specimens. The hygrothermal properties of insulation and other wall components were either measured or taken from the in-house database of hygrothermal material properties Mukhopadhyaya (2007).

Assumptions. It was assumed that all material layers were in direct contact with one another (i.e., the interfacial thermal resistances between all material layers were neglected). The emissivity of all surfaces that bounded the airspaces (i.e., airspaces between exterior insulations, furring strips, and vinyl siding) was taken equal to 0.9 (ASHRAE 2009b). The effects of heat transfer by conduction, convection, and radiation within these airspaces on the thermal performance of wall assemblies were also determined.

Initial and Boundary Conditions. The initial temperature in all material layers of the respective wall specimens (Wall 1, 2, and 3) was assumed uniform and equal to 10.0°C. Because this initial temperature was not the same as that of the test, it was anticipated that the predicted dynamic response of the different wall specimens in the initial test period (e.g., the first 24–48 h) would be different from that obtained in the test itself. The boundary conditions on the top and bottom surfaces of the wall systems were assumed to be adiabatic (i.e., no edge heat losses). The outdoor surface of the vinyl siding and the indoor surface of the gypsum board were respectively subjected to a temperature boundary condition for all wall systems. The temperatures on the outdoor and indoor surfaces of the respective wall specimens, and that changed over time, were taken equal to that measured on these surfaces.

Comparison between Model Predictions and Measurements

To benchmark the hygIRC-C model, transient numerical simulations were conducted for the three wall test specimens (Figure 1). A full description of all instrumentation (i.e., thermocouples, heat flux transducers (HFTs), pressure (P) sensors, and relative humidity (RH) sensors) and experimental data are available in Lacasse et al. (2016). In each wall system, three HFTs were used to measure the heat flux at the middle (mid-height and mid-width, as given in Figure 2, Figure 3, and Figure 4) of each of the respective wall test specimens at three interfaces, namely

- HFT1 at the interface between the airspace and exterior insulation,
- HFT2 at the interface between the exterior insulation and OSB panel, and
- HFT3 at the interface between the polyethylene air barrier membrane and gypsum panel.

For the EPS wall specimen (Wall 1), Figure 5(a), (b) and (c) show comparisons between the measured and predicted values of heat flux during the test period. In these figures, time = 0 at which experimental data was collected corresponded to August 11, 2013 at 12:37 a.m. Unlike the other wall specimens, the HFT1 located at the interface between the airspace and exterior insulation in Wall 1 was not functioning properly during the entire test period (Figure 5[a]). However, as shown in Figure 5(b), the predicted heat flux at the EPS-OSB interface is in good

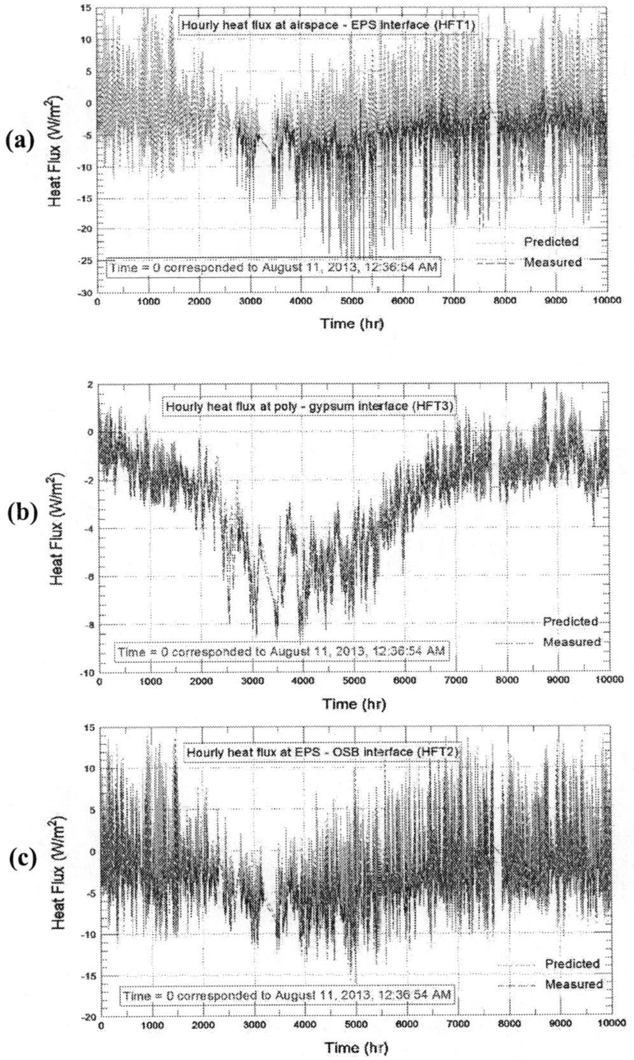

Figure 5 *Wall 1 (EPS)—Comparison between predicted and measured heat fluxes at interface: (a) airspace-EPS, (b) EPS-OSB, and (c) polygypsum.*

agreement with the measurements from HFT2. Similarly, Figure 5(c) shows that the predicted heat flux at the polygypsum interface is in good agreement with measurements from HFT3.

Figures 6(a), 6(b), and 6(c) show comparisons between the measured and predicted values of heat flux during the test period at different locations for the XPS wall specimen (Wall 2). As shown in these figures, the predicted values of heat flux were in good agreement with the measured values at each of the respective interfaces, specifically at the: airspace/XPS interface (Figure 6[a]), XPS/OSB interface (Figure 6[b]), and the polygypsum interface (Figure 6[c]).

Finally, for the wall specimen with mineral fiber insulation (Wall 3), comparisons are provided in Figures 7(a), 7(b) and 7(c) and in which are shown the predicted values of heat flux that were in good agreement with the measured values at each of the respective interfaces, specifically at the airspace/mineral fiber interface (Figure 7[a]), the mineral fiber/OSB interface (Figure 7[b]), and the polygypsum interface (Figure 7[c]).

In summary, the results presented in this section show that the predicted values of heat flux at different locations in the wall assembly are in good agreement with the measured values of heat flux for the three walls, namely, the EPS wall specimen (Figure 5), XPS wall specimen (Figure 6), and mineral fiber wall specimen (Figure 7).

It is not possible to complete the benchmarking of the model in respect to moisture transport on the basis of the field trials undertaken in the FEWF. This requires conducting experiments in a controlled environment after first conditioning the test specimens to known levels of moisture content. In fact, the model has previously been benchmarked in controlled conditions as is described in more detail in Saber and Ganapathy (2016) and Saber et al. 2014.

Thus, after benchmarking the present model in respect to heat flux, and having previously benchmarked the model to several tests undertaken in field and controlled laboratory conditions as indicated in Saber and Ganapathy (2016) and Saber et al. 2014, this model was used with confidence to investigate the risk of condensation and mold growth in different wall assemblies with exterior insulations when these walls were subjected different Canadian climatic conditions.

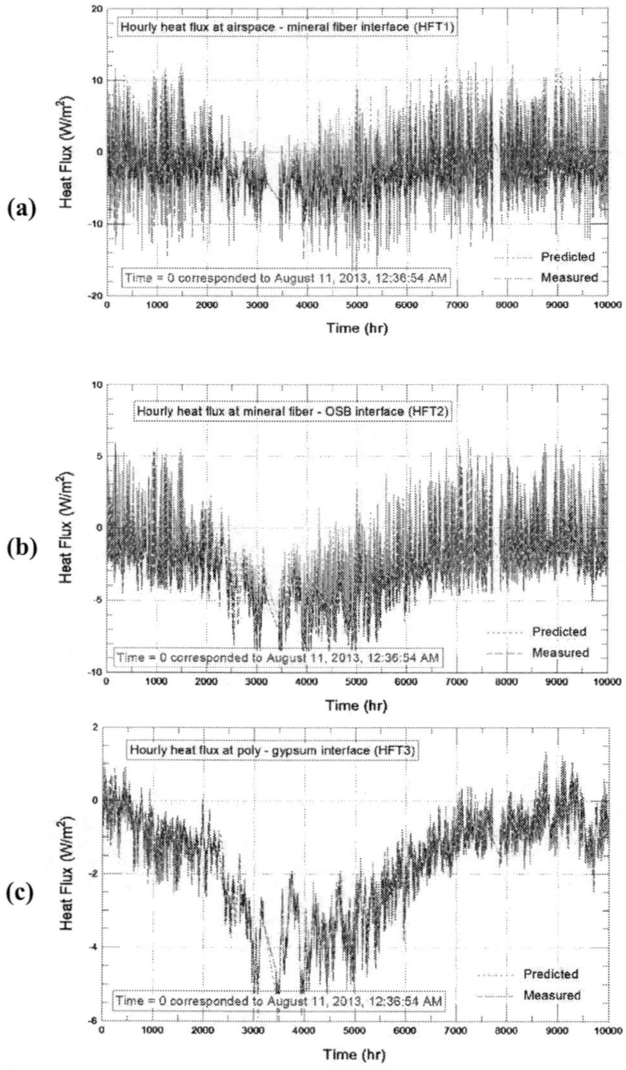

Figure 6 *Wall 2 (XPS)—Comparison between predicted and measured heat fluxes of interface: (a) airspace/XPS, (b) XPS/OSB, and (c) polygypsum.*

SIMULATION OF WALL ASSEMBLIES

Wall Assembly Configurations for Simulation and Simulation Parameters

Hygrothermal simulations of all wall assemblies were conducted using the hygIRC-C model and using the construction details common to all wall assemblies to be modeled as listed in Table 1. For each of the materials or components specified, the rationale for the selection of specific materials is also given in this table.

Figure 8 shows a schematic of the code-compliant reference wall assembly to which the performance of test specimens Wall 1, 2, and 3 was compared. Whereas the simulated results for moisture content and temperature are produced for every location within the wall system and at every 1 hour, an analysis of results was performed to establish those locations in the wall that had the greatest susceptibility to risk of condensation; this permitted rationalizing the presentation of results. Post processing of simulation results and reporting thus focused on the locations shown in Figure 8.

Simulation Conditions for Parametric Study

In this section, the different simulation conditions are discussed, and that were used to conduct the numerical simulations for all wall assemblies shown in Figure 2 to Figure 4 and in Figure 8 (see Table 2).

Vapor Barrier Conditions. As provided in Subsection 9.25.4 of the NBCC (2010), the current maximum allowable WVP value for vapor barriers is 60 ng/(Pa·s·m^2). The parametric study was conducted using a vapor barrier with a WVP of 60 ng/(Pa·s·m^2) as provided in Table 1.

Air Leakage Conditions. All cases were modeled with some airflow introduced through openings into the assembly, as this is a likely scenario given the imperfections of the air barrier system of wall assemblies. Additionally, completing the investigation without considering the effects of air leakage would not create useful results in terms of assessing the risk to the formation of condensation in wall assemblies,

(a) (b) (c)

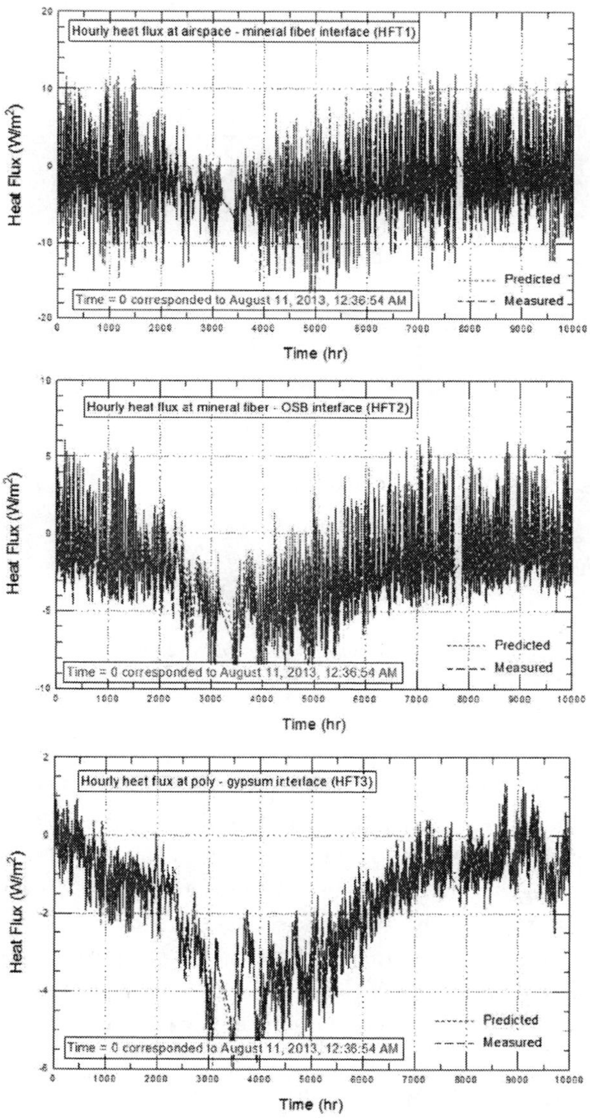

Figure 7 *Wall 3 (Mineral fiber)—Comparison between predicted and measured heat fluxes of interface: (a) airspace/mineral fiber, (b) mineral fiber/OSB, and (c) polygypsum.*

given that exfiltration of indoor air to the wall assembly is the primary cause for the formation of condensation in the assembly itself.

It was assumed that the path for air movement (shown in Figure 8) was initiated at the interior of the assembly at the bottom of the wall, and, thereafter, moisture was deposited along the interior face of the sheathing panel and exited through the top of the wall. This air leakage path was one of the scenarios used in the study by Ojanen and Kumaran (1995) in which it was assumed that air would move through imperfections that existed at the wall top plate and specifically at the interface between the interior face of the exterior sheathing and the exterior of the top plate. For the code-compliant reference wall (i.e. no exterior insulation), this air leakage path is shown Figure 8; a similar path was assumed for Walls 1, 2, and 3.

The air leakage rate for all cases in all locations was set to 0.1 L/(s·m^2) at 75 Pa, which was an assumption used in at least one previous study (see Saber et al. 2014, Ojanen and Kumaran 1995, and Saber and Maref 2015). These air leakage conditions are those assumed for the wall components of a home and would be roughly an order of magnitude smaller than the whole house air leakage of an R-2000 or ENERGY STAR qualified home, as provided by Natural Resources Canada (NRCan).

The impact of this assumption on the hygrothermal performance was investigated in a sensitivity analysis by modeling a wall assembly with different air leakage rates ranging between 0 and 0.1 L/(s·m^2) at 75 Pa. The results from such an analysis permitted deriving the least performing and most vulnerable wall assembly with respect to the formation of condensation and the risk to the formation of mold within the assembly. In this paper, rates of air leakage referred to as 0%, 50%, and 100% of air leakage, respectively, corresponded to 0.0, 0.05, and 0.1 L/(s·m^2) at 75 Pa. As indicated later, the results of the sensitivity study supported the selection of 0.1 L/(s·m^2) at 75 Pa as a means of challenging the wall system with moisture ingress from air leakage.

Approach to Simulation of Air Leakage. In the present study, the air leakage rate (Q) as a function of the total pressure differential across the wall assemblies (ΔP_{tot}) is given as follows:

$$Q = a\Delta P_{tot}^{n} \quad (1)$$

In a previous NRC project "Wall Energy Rating, WER" (Elmahdy et al. 2010, Elmahdy et al. 2009; Saber et al. 2010; Saber et al. 2012), the air leakage rates were measured for a number of 2 × 6 in. wood frame walls having different types of thermal insulation in the wall cavities, including open-cell spray foam, closed-cell spray foam, and glass fiber. For the full-scale wall systems with and without penetrations and having glass fiber insulation, the average value of the exponent n in Equation 1 was 0.7; this value was used in this study. The value of coefficient a in Equation 1 was determined to satisfy the condition at which the air leakage rate is 0.1 L/(s·m^2) at ΔP_{tot} = 75 Pa when the exponent n = 0.7. As such, the value of coefficient a is equal 0.00487 L/(s·m^2·Pa$^{0.7}$) where Q in is L/(s·m^2) and ΔP_{tot} in Pa.

Table 1. Construction Details Common to All Wall Assemblies to be Modeled

Material Selection	Rationale
Exterior finish consisting of vinyl cladding installed on 19 mm strapping	To minimize impact of water ingress from the exterior
Weather-resistive barrier (WRB) with WVP of 1400 ng/(Pa·s·m^2) (25 U.S. perm) such as spun-bonded polyolefin membrane	Common construction and highly permeable so as not to limit application of insulation materials for which selection of a more vapor-tight material might otherwise affect intent of project
Wood frame construction (2 × 6 in.); framing at 16 in. on center	Currently, most common construction framing used in housing
Vapor barrier with WVP of 60 ng/(Pa·s·m^2)	NBCC 2010 minimum requirement 9.25.4.2 (see NBCC 2010)
Interior finish of 12.5 mm gypsum board	Currently most common construction method for interior finish

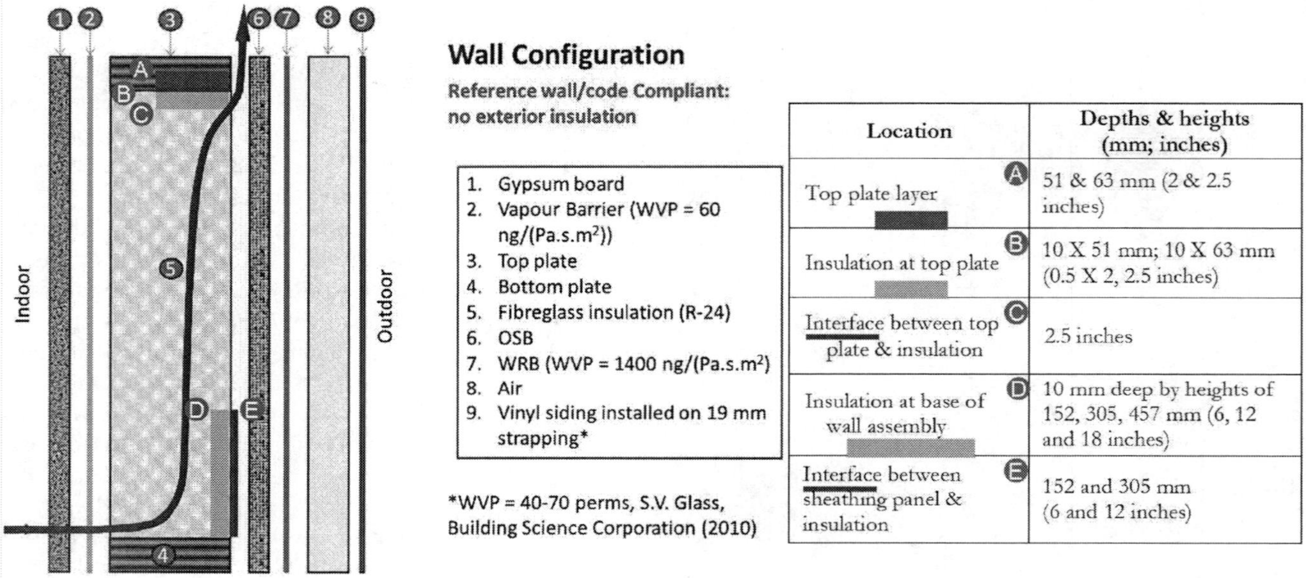

Figure 8 *Schematic of reference wall assembly/code-compliant configuration showing different component layers and assumed path of airflow through assembly (no exterior insulation).*

In Equation 1, the total pressure across the building envelope can be calculated as:

$$\Delta P_{tot} = \Delta P_{wind} + \Delta P_{st} + \Delta P_{ven} \quad (2)$$

where

ΔP_{wind} = pressure differential due to wind

ΔP_{st} = pressure differential due to the stack effect

ΔP_{ven} = pressure differential due to mechanical ventilation system (i.e., pressurization or depressurization due to heating and cooling conditions); ΔP_{ven} was neglected and thus $\Delta P_{ven} = 0$

The full details for calculating both the pressure differential due to the stack effect and wind pressure differential given different climatic conditions are available in Saber et al. (2014). As indicated in Saber et al. (2014), the greater the exfiltration rate, the higher the risk of formation of condensation and mold growth within the wall assembly; the wall is located at the third story and facing the direction of the highest exfiltration rate. This would thus represent the worst-case scenario for the risk of formation of condensation and mold growth within the wall cavity. As such, all wall assemblies that were investigated in this study represent wall assemblies of the third story of low-rise buildings.

Exterior Climatic Conditions

Locations. All wall assemblies were subjected to different climatic conditions of five different locations within Canada and having differing values of heating degree days (HDD) and moisture index (MI), namely,

Table 2. Wood-Framed (2 × 6 in.) Wall Systems with Exterior Insulations*

	Wall 1	Wall 2	Wall 3
		Exterior Insulation Type	
Parameter	EPS	XPS	Mineral Fiber
Thickness, in. (mm)	1 (25)	2 (51)	3 (76)
Dry density, kg/m^3	18	26	122
Dry thermal conductivity, W/m·K	0.0369	0.0290	0.0347
Vapor permeability, $kg·m/(s·m^2·Pa)$	2.91E-12	1.39E-12	1.62E-10
Total vapor permeance, $ng/(s·m^2·Pa)$	114.4	27.4	2129
Vapor permeance, $ng/s·m^2·Pa)/25$ mm	114.4	54.9	6390
Total RSI value, $m^2·K/W$	0.69	1.75	2.20
Total R-value, $ft^2·h·°F/Btu$	3.91	9.95	12.47
R-value, $m^2·K/W/25.4$ mm	0.69	0.88	0.73
R-value, $ft^2·h·°F/Btu/in.$	3.9	5.0	4.2

*2 × 6 in. wood frame cavity insulation: batt insulation of R-24 (RSI-4.2).

- Vancouver, BC (mild, wet, HDD18 = from 2600 to 3100, MI = 1.44)
- St. John's, NL (cold, wet, HDD18 = 4800, MI = 1.41)
- Ottawa, ON (cold, dry, HDD18 = 4440 – 4500, MI = 0.84),
- Edmonton, AB (cold, dry, HDD18 = 5120, MI = 0.48)
- Yellowknife, NT (cold, dry, HDD18 = 8170, MI = 0.58)

Orientation of Wall Assemblies. The wall assemblies of the third story of low-rise buildings were modeled in the orientation showing the highest average annual air exfiltration rate. The walls were assumed to be shaded to minimize the impact of solar-driven moisture ingress into the assembly and likewise to minimize the solar drying effect on the wall. However, diffuse radiation was taken into consideration.

Weather Data. Hygrothermal simulations were conducted for a period of two years. The first year corresponded to an average year (conditioning year, where equal drying and wetting potential exists), and the second year corresponded to a wet year. Note that the wet year in each location was the worst-case situation given that the drying potential for a wall assembly is limited during a wet as compared to a drier year. The weather data of the different locations were obtained from the NRC's weather database.

Indoor Conditions

Regarding the indoor moisture load, the water vapor pressure differential across the wall assembly (from indoor to outdoor) corresponded to a moisture load of 5.2 g/m^3. This indoor moisture load is consistent with that used in previous studies, in which a moisture load of 7.1 L/day was chosen for a one story, 80 m^2 house, with an indoor temperature of 21°C, water vapor pressure differential close to 700 Pa ($\Delta P_v = P_{v,\,indoor} - P_{v,\,outdoor} = 700$ Pa), and 0.3 ach by mechanical ventilation; this is referred to as Option A in Figure 9 in which the indoor relative humidity (RH_{ind}) for the climatic conditions of Ottawa, as an example, are provided. As shown in this figure, Option A resulted in a high value for RH_{ind}, that at times exceeded 100%. Three additional options for the indoor RH profile were also considered within a period of one year, as shown in Figure 9; these were:

Option A = $\Delta P_v = 700$ Pa

Option B = this option was based on the method given in ASHRAE Standard 160 (2009a)

Option C = similar to Option A (i.e., $\Delta P_v = 700$ Pa), but the value of RH_{ind} was capped at 70%

Option D = based on ASHRAE Standard 160; interior RH during winter reduced with cold temperature

Option C was the recommended option to evaluate the indoor relative humidity (Saber and Ganapathy 2016; Saber et al.

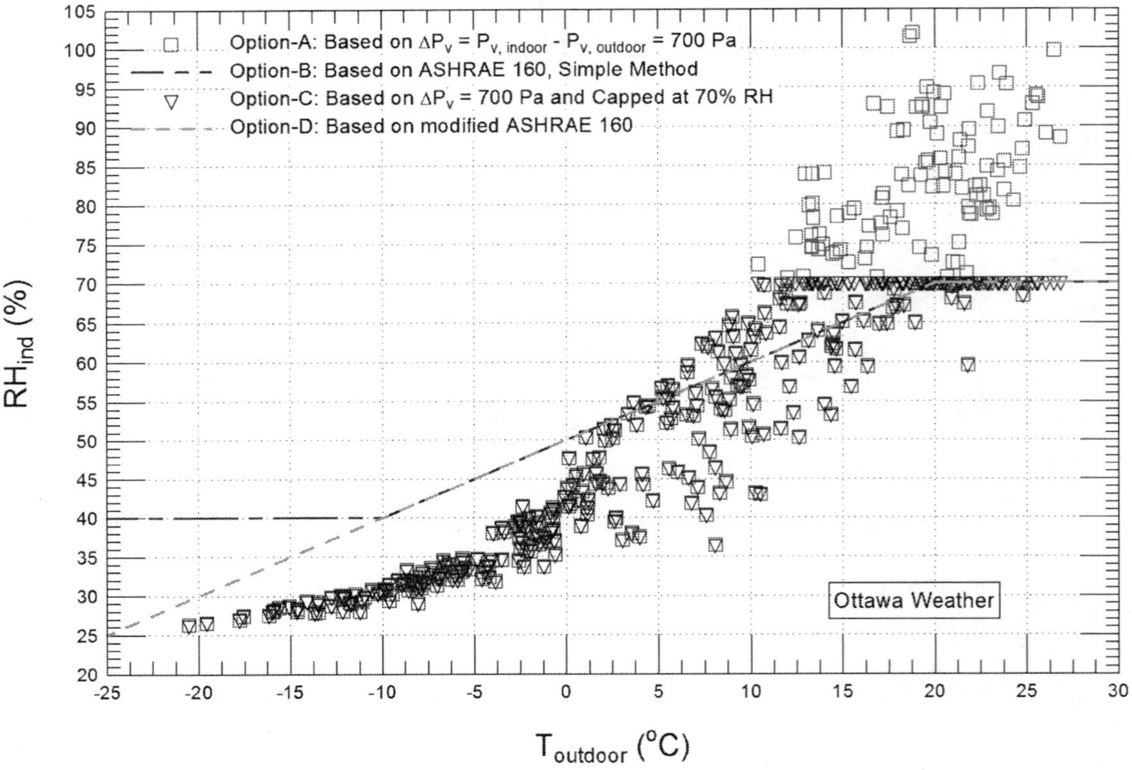

Figure 9 *Different options for indoor relative humidity (Ottawa weather).*

2014) and was used in this study to conduct all numerical simulations of different wall assemblies that are listed in Table 2. Other indoor conditions (i.e., indoor temperature) were set according to that provided in the ASHRAE Standard 160 (ASHRAE 2009a) with respect to recommendations for conditioned space.

Initial Conditions

The initial temperature in all layers of the wall assemblies were taken equal to 21°C, and the initial moisture content of all material layers corresponded to a relative humidity of 50%.

Material Properties

The hygrothermal properties for the exterior insulations in Wall 1 (EPS), Wall 2 (XPS), and Wall 3 (mineral fiber) were measured in NRC's laboratory. The different types of exterior insulation products used in this study are listed in Table 2. All wall assemblies incorporated a sheathing panel made of 11 mm (7/16 in.) thick OSB; of importance was the choice of WVP for the panel. Glass (2013) has compiled the available data for the WVP of OSB (11 mm thick). The recommended values by Glass (2013) for the WVP of OSB as a function of RH were used in the numerical simulations. The hygrothermal properties of the other material layers shown in Figure 8 were obtained from NRC's material database (Mukhopadhyaya 2007).

Summary of Simulated Conditions

A summary of the simulated conditions for all wall assemblies is provided in Table 3.

ACCEPTABLE PERFORMANCE

The modeling results for each case were expressed using the mold index (M) criteria developed by Hukka and Viitanen (1999), Viitanen and Ojanen (2007), and Ojanen et al. (2007). The descriptions of mold index levels are provided in Table 4. The most recent mold model by Ojanen et al. (2010) was used in this study to determine the mold index of different materials of the wall assemblies. In that model (Ojanen et al. 2010), the sensitivity of different construction materials for mold growth was classified in four sensitivity classes (Table 5): very sensitive, sensitive, medium resistant, and resistant.

Table 6 provides the assumed correspondence of sensitivity class for materials located within the wall assembly modeled in this study. More specifically, the sensitivity class for the top and bottom plates, OSB and foam, was considered "Sensitive," whereas the sensitivity class of the materials for cavity insulation (fiber based), drywall, and membranes was considered "Medium Resistant." The selected mold index criteria allowed sufficient resolution to assess the risk of moisture condensation.

Table 3. Summary of Simulated Conditions

Criteria	Assumptions/Conditions
Pressure exponent, n	0.7 (see Equation 1)
Predominant wall orientation	Facing the highest exfiltration rate
ΔP for stack effect	Top story of a three-story building to maximize effect of exfiltration
ΔP for ventilation	Assume depressurization/pressurization from ventilation source is negligible
Air leakage rate, Q	Corresponds to 0.0, 0.05, and 0.1 L/(s·m^2) at 75 Pa
Interior moisture load	Constant water vapor pressure differential, ΔP_v = 700 Pa and capped at 70% rh
Water vapor permeance of OSB	Function of RH ranging from 0%–100% as recommended by Glass (2013)
Modeling period	Two years, January to December: one "average" year followed by one "wet" year
Geographical locations	Ottawa (ON), Edmonton (AB), Vancouver (BC) Yellowknife (NT), and St. John's (NL)

Table 4. Description of Mold Index (M) Levels (Hukka and Viitanen 1999; Viitanen and Ojanen 2007; Ojanen et al. 2010)

M	Mold Index (M) Description of Growth Rate
0	No growth
1	Small amounts of mold on surface (microscope), initial stages of local growth
2	Several local mold growth colonies on surface (microscope)
3	Visual findings of mold on surface, < 10% coverage, or <50% coverage of mold (microscope)
4	Visual findings of mold on surface, 10%–50% coverage, or >50% coverage of mold (microscope)
5	Plenty of growth on surface, >50% coverage (visual)
6	Heavy and tight growth, coverage about 100%

Table 5. Mold Growth Sensitivity Classes and Some Corresponding Materials (Hukka and Viitanen 1999; Viitanen and Ojanen 2007; Ojanen et al. 2010)

Sensitivity Class	Materials	*RH_{min}, %
Very sensitive	Pine sapwood	80
Sensitive	Glued wooden boards, PUR with paper surface, spruce	80
Medium resistant	Concrete, aerated and cellular concrete, glass wool, polyester wool	85
Resistant	PUR with polished surface	85

* Minimum relative humidity needed for mold growth.

Approach for Assessing the Overall Performance

The overall hygrothermal performance of the different wall assemblies (i.e., Wall 1 [EPS]; Wall 2 [XPS], and Wall 3 [mineral fiber]) evaluated in this study was assessed against the performance of the code compliant reference wall. The values of the mold index as determined from simulation results of the three walls were compared against the values of the mold index of the reference wall; adequate performance was achieved provided values for mold index were equal to or lower than that of the reference wall.

Table 6. Mold Growth Sensitivity Classes for Materials of Wall Assemblies

Sensitivity Class	Material Layers of Wall Assemblies	*RH_{min}, %
Very sensitive		80
Sensitive	Top plate, bottom plate, OSB, foam	80
Medium resistant	Fiberglass insulation, gypsum, membranes	85
Resistant		85

*Minimum relative humidity needed for mold to grow.

The primary locations within the wall assemblies at risk for condensation of moisture and possible mold growth were identified as those located at either the top or bottom portion of a wall assembly. All of these "at risk" locations are shown in Figure 8. More details of these locations are available in Saber and Ganapathy (2016). For the purpose of comparing the performance of different wall assemblies, these results are presented on basis of a simplified form using the following two parameters (Saber and Ganapathy 2016; Saber 2014):

- M_{AVG}: Overall average value of mold index at different locations in the wall; locations are shown in Figure 8
- M_{MAX}: Overall maximum value of mold index at different locations in the wall.

The two parameters above were determined based on a simulation period of two years (i.e., simulation of the average year followed by a wet year for the location of interest) (see Figure 8). Both these values are provided for each of the different wall configurations having nominal insulation in the stud cavity (referred to as *inboard insulation*) of R-24, as well as for each of the exterior insulation conditions (referred to as *outboard insulation*); exterior insulation conditions varied from R-0 (i.e., reference wall) to values of R-3.91, R-9.95, and R-12.47 (see Table 2). However, for the climatic conditions of Edmonton, stud cavity insulation of R-19 was also considered in this study to allow comparison of the risk of condensation and mold growth in these walls to the case of a wall having stud cavity insulation of R-24.

RESULTS AND DISCUSSION

In this section, the effects of different parameters that affect the hygrothermal performance of wall assemblies are discussed. The list of wall assemblies is provided in Table 2. In this paper, in instances where the units for water vapor permeance (WVP) and R-value are not reported, the units for each of these parameters are, respectively, ng/(Pa·s·m^2) and ft^2·h·°F/Btu. Note that all results derived for the reference wall having stud cavity insulation of R-19 and R-24 were obtained from a previously published project report (see Saber 2014 for more details). The different parameters affecting the hygrothermal performance of walls and discussed in this section include air leakage rate, R-value of stud-cavity insulation, and geographical locations.

Effect of Air Leakage Rate

The effect of the air leakage rate on the hygrothermal performance of wall assemblies listed (Table 2), as determined by the risk of mold growth in a wall assembly, permitted identifying the locations within the assembly of likely mold growth given the different air leakage rates to which it was subjected. In these analyses, the full amount of the air leakage given by Equation 1 (i.e., = 100%) and different percentages of that value (= 0% and 50%) were considered.

The mold index (M) was calculated for different wall assemblies on the basis of the mold sensitivity classes of the different material layers within the wall assembly, as provided in Table 6. It is important to point out that the locations within the wall assemblies at risk of condensation and mold growth are based on the respective air leakage paths that were considered in this study and shown in Figure 8. It is noted that a different air leakage path, however, would result in different locations within the wall assemblies at risk of condensation and mold growth.

Figures 10(a) and 10(b), respectively, show the overall average mold index (M_{AVG}) and overall maximum mold index (M_{MAX}) for different percentages of the air leakage rate of $\xi = 0\%$ (no air leakage), 50%, and 100% for the wall system with exterior insulation of EPS (RSI-0.69 and WVP = 114 [R-3.9 and WVP = 2.0 US perm]) subjected to different climatic conditions. As shown in these figures, decreasing the air leakage rate resulted in decreasing the risk of condensation and mold growth, whereas no risk of condensation occurred at $\xi = 0\%$. For example, the values of M_{AVG} for the case of $\xi = 50\%$ are 29%, 44%, 75%, 56%, and 24% of the values of M_{AVG} for $\xi = 100\%$, respectively, for Ottawa, Edmonton St. John's, Vancouver, and Yellowknife (Figure 10[a]). Note that at a pressure difference (ΔP) of 75 Pa, $\xi = 50\%$ corresponds to an air leakage of 0.05 L/(s·m^2), whereas $\xi = 100\%$ corresponds an air leakage rate of 0.1 L/(s·m^2). Furthermore, the values of M_{MAX} for the case of $\xi = 50\%$ are 42% and 56%, 83%, 70%, and 37% of the values of M_{MAX} for the case of $\xi = 100\%$ for Ottawa, Edmonton St. John's, Vancouver, and Yellowknife, respectively (Figure 10[b]).

For the wall system with XPS exterior insulation of R-10 and WVP = 27, Figure 11(a) and 11(b) show the effect of air leakage rate on the overall maximum and average mold index. Similar to the results obtained for the EPS wall, decreasing the

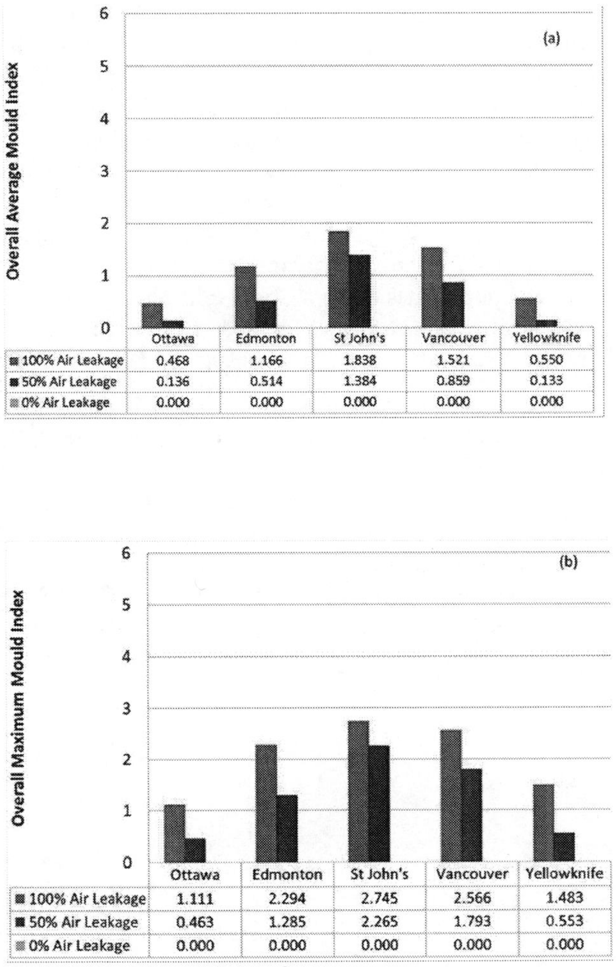

Figure 10 *Wall 1 (EPS)—Effect of air leakage rate on overall (a) average and (b) maximum mold index.*

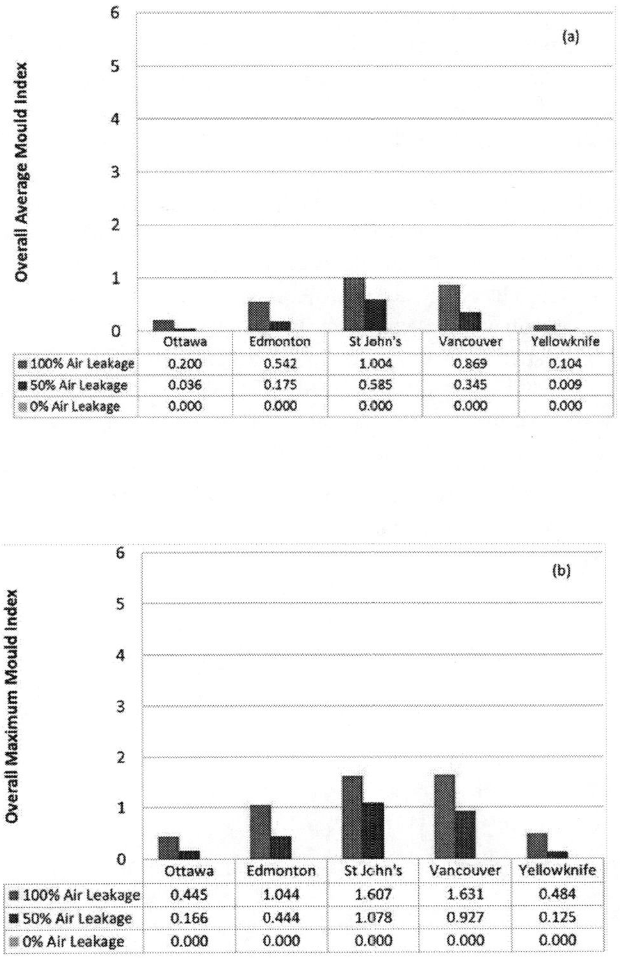

Figure 11 *Wall 2 (XPS)—Effect of air leakage rate on overall (a) average and (b) maximum mold index*

air leakage rate resulted in decreasing the risk of condensation and mold growth. As shown in Figure 11(a), the values of M_{AVG} for the case of $\xi = 50\%$ are 18% and 32%, 58%, 40%, and 9% of the values of M_{AVG} for $\xi = 100\%$, respectively, for Ottawa, Edmonton St. John's, Vancouver, and Yellowknife. The corresponding values of M_{MAX} for the case of $\xi = 50\%$ are 56% and 52%, 90%, 79%, and 26%, respectively, of the values of M_{MAX} for $\xi = 100\%$ for Ottawa, Edmonton St. John's, Vancouver, and Yellowknife (Figure 11[b]).

Figures 12(a) and 12(b) show M_{AVG} and M_{MAX} for different air leakage rates for the wall system with mineral fiber exterior insulation of R-12.5 and WVP = 2130, when this wall was subjected to different climatic conditions. As shown in these figures, the values of M_{AVG} for $\xi = 50\%$ are 31% and 40%, 79%, 66%, and 19% of the values of M_{AVG} for $\xi = 100\%$, respectively, for Ottawa, Edmonton St. John's, Vancouver, and Yellowknife (Figure 12[a]). Also, the values of M_{MAX} for $\xi = 50\%$ are 56% and 52%, 90% 79%, and 43% of the values of M_{MAX} for $\xi = 100\%$ for Ottawa, Edmonton St. John's, Vancouver, and Yellowknife, respectively (Figure 12[b]).

The air leakage in buildings has consistently improved since the 1990s when the NBCC began to mandate a "designated" air barrier system. The air barrier system is to provide a designated element to be the "principal plane of airtightness" and accessories and components to maintain continuity of the airtightness across joints and penetrations. The value of the air leakage rate for an air barrier system was recommended as 0.1 $L/(s{\cdot}m^2)$ at a pressure difference of 75 Pa for typical indoor winter conditions of 27% to 55% rh at 21°C. This recommended air leakage rate was introduced in the 1995 NBCC and continues to be referenced in the 2010 NBCC (A-5.4.1.2. [1] and [2]).

Effect of Inboard and Outboard Insulation

In a previous NRC project (Saber et al. 2014) and for the case of a 100% air leakage rate, numerical simulations were conducted to investigate the effect of the inboard (stud cavity) insulation on the risk of condensation and mold growth for different wall assemblies. In that project (Saber et al. 2014), two types of stud cavity fiber insulation products having nominal R-values of RSI-3.35 (R-19) and RSI-4.23 (R-24) were investigated for wall systems with and without outboard insulation and having of wide range of R-values (R-4, R-5 and R-6) and values of WVP (2 to 1800 ng/(Pa·s·m^2)) when these walls were subjected to the climates of Ottawa, Edmonton, St. John's and Vancouver. The results of that study (Saber et al. 2014) showed that the wall systems having stud cavity insulation of R-24 have a higher risk of condensation and mold growth than that for wall systems with stud cavity insulation of R-19. The other NTRC project (Saber and Ganapathy 2016) focuses on wall systems with stud cavity insulation of R-24. In this paper, however, as an example, simulations were conducted when the EPS wall, XPS wall, and mineral fiber wall have stud cavity insulation of R-19 and R-24 when these walls were subjected to the climate of Edmonton.

Figures 13(a) and 13(b), respectively, show the overall average mold index (M_{AVG}) and overall maximum mold index (M_{MAX}) in different wall assemblies. For the purposes of comparison, the performance of the code-compliant reference wall that was obtained in a previous NRC project (Saber et al. 2014) is included in these figures. As shown in these figures, increasing the R-value of the stud-cavity insulation resulted in a slightly higher risk of condensation and mold growth. For example, the values of M_{AVG} for walls having R-24 stud cavity insulation were 5% (reference wall), 4% (EPS wall), 14% (XPS wall), and 15% (mineral fiber wall) higher than that for a wall having R-19 stud cavity insulation (Figure 13[a]). The corresponding values of M_{MAX} for walls having R-24 stud-cavity insulation were 5%, 4%, 14%, and 15% higher than that for walls having R-19 stud-cavity insulation (Figure 13[b]).

The higher the R-value of the outboard, exterior insulation (R-0, R-3.9, R-10.0, and R-12.5 for reference wall, EPS wall, XPS wall, and mineral fiber wall, respectively), the warmer the wall cavity temperature, the less likely the formation of condensation occurring during the cold periods and consequently, the lower the value for mold index. As shown in Figure 13(a), the values of M_{AVG} for the EPS wall, XPS wall, and mineral fiber wall, respectively, were 59%, 27%, and 23%

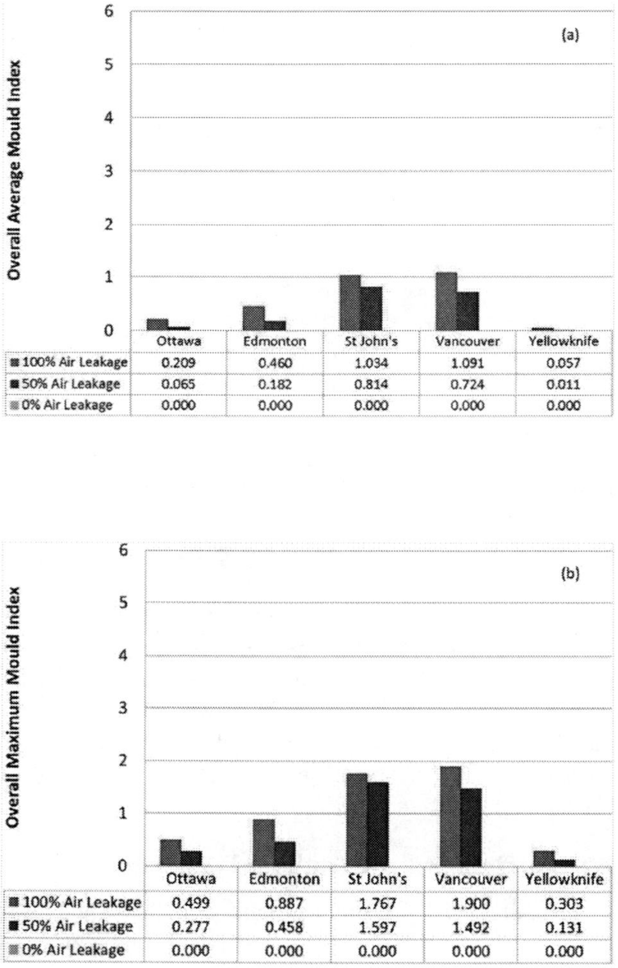

Figure 12 *Wall 3 (Mineral fiber)—Effect of air leakage rate on overall (a) average and (b) maximum mold index.*

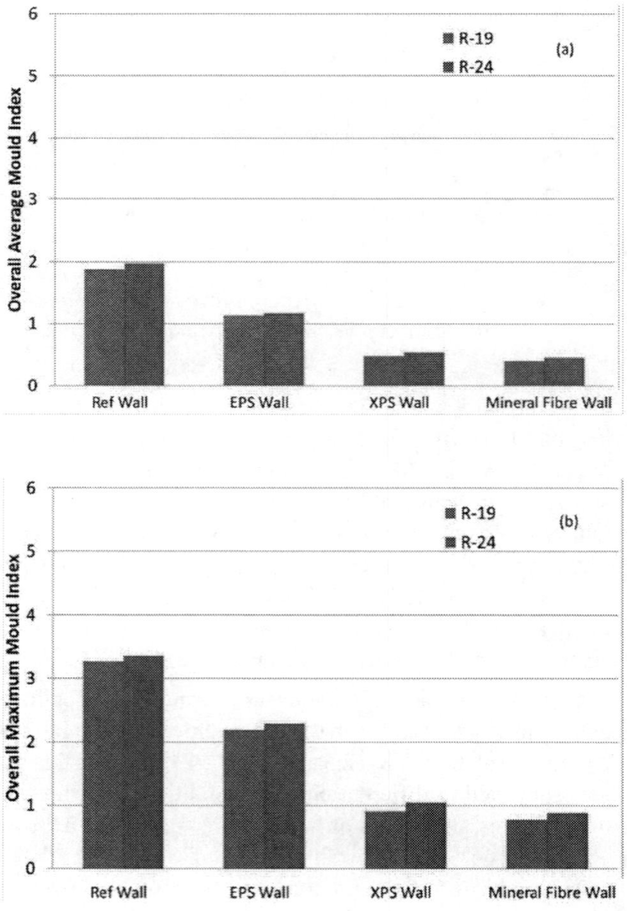

Figure 13 *Effect of stud cavity insulation on the overall maximum and average mold index at 100% air leakage rate (Edmonton weather).*

of the value of M_{AVG} for the reference wall. Also, the corresponding M_{MAX} values for these walls were 68%, 31%, and 26% of that for the reference wall (Figure 13).

Comparisons for the yearly heat losses in winter season and yearly heat gains in the summer season for the reference wall and walls with exterior insulations with stud cavity insulations of R-19 and R-24 are provided in Figure 14 and Figure 15, respectively, for both the average year and wet year. It is important to point out that the yearly heat losses and yearly heat gains shown in these figures are the heat losses in the wall systems and are not necessarily the heat losses one can expect in a house. Note that the summer season in this paper was defined as when the outdoor temperature was greater than the indoor temperature and vice versa for the winter season.

As shown in Figure 14 and Figure 15, increasing the stud-cavity thermal resistance resulted in a decrease in both yearly heat loss and yearly heat gain. For example, the yearly heat losses of walls with R-24 stud-cavity were 22% (reference wall), 19% (EPS wall), 15% (XPS wall), and 14% (mineral fiber wall) lower than that for walls with R-19 stud-cavity. Conversely, as indicated above, the risk of condensation and mold growth for walls with R-24 stud-cavity was 4%–15% higher than the walls with R-19 stud-cavity. Note that the percentage decrease in the heat loss for the with R-24 as compared to wall with R-19 was for the case of 100% air leakage rate. Considering a lower air leakage rate, however, would result in a greater percentage decrease in the heat loss for wall with R-24 as compared to wall with R-19.

Adding outboard, exterior insulation not only decreases the yearly heat losses and yearly heat gains (see Figure 14 and Figure 15) but also decreases the risk of condensation and mold growth as explained earlier (see Figure 13). For example, the yearly heat losses of walls with R-19 stud-cavity were 19% (EPS wall with R-3.9), 47% (XPS wall with R-10.0), and 58% (mineral fiber wall with R-12.5) lower than that for the reference wall (no exterior insulation). The corresponding yearly heat losses of walls with R-24 stud-cavity were 16%, 39%, and 48%, respectively, lower than that for the reference wall.

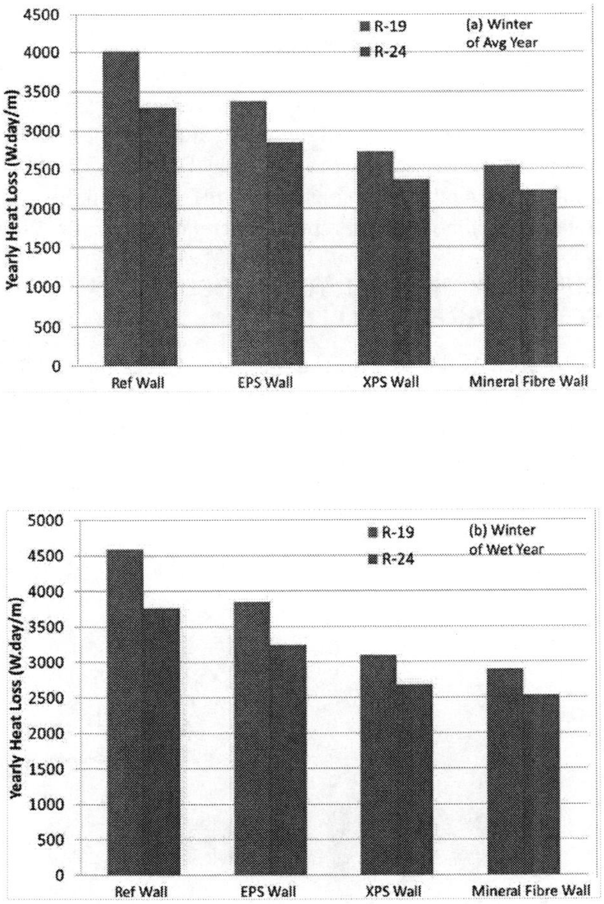

Figure 14 *Effect of stud cavity insulation on the yearly heat loss at 100% air leakage rate (Edmonton weather).*

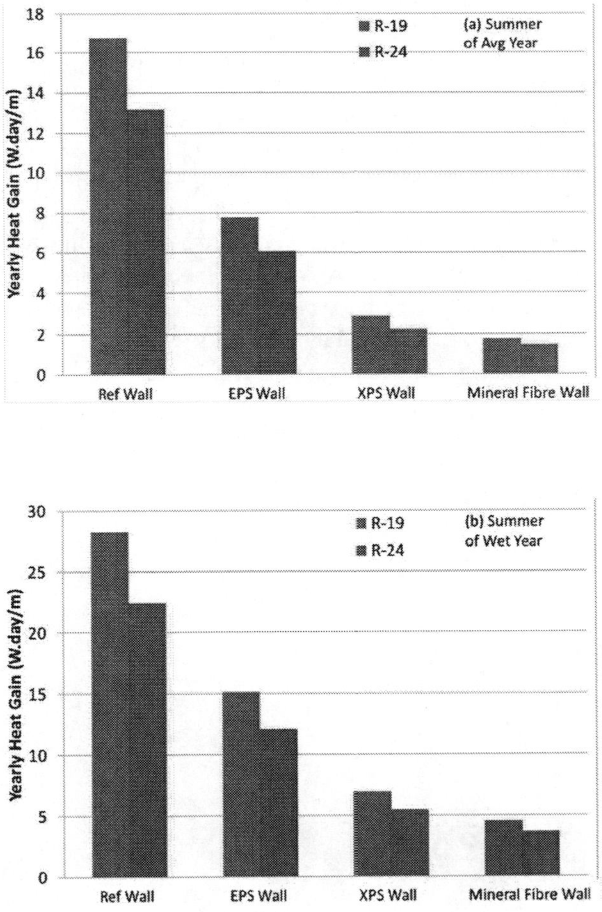

Figure 15 *Effect of stud cavity insulation on the yearly heat gain at 100% air leakage rate (Edmonton weather).*

For the case of stud cavity insulation of R-24, similar comparisons for the yearly heat loss and yearly heat gain for the reference wall and wall systems with exterior insulations are provided in Saber and Ganapathy (2016).

In summary, the yearly heat loss and heat gain are lower for walls with R-24 stud cavity insulation than that for walls with R-19 stud cavity insulation, but the risk of condensation in the former is higher than that in latter. For a given stud cavity insulation, wall systems having additional exterior insulation resulted not only in higher energy performance but also lower risk of condensation and mold growth than walls without exterior insulation.

Effect of Geographical Locations

The hygrothermal performance for different wall assemblies (see Table 2) were obtained when these walls were subjected to the climates of five Canadian cities each differing in geographical location, including Ottawa (ON), Edmonton (AB), Vancouver (BC), St. John's (NL), and Yellowknife (NT). The primary environmental parameters that greatly affected the hygrothermal performance related to the risk of condensation and mold growth were as follow:

Outdoor Temperature. This can be represented by the heating degree days (HDD). The greater the number of HDD, the higher the risk for mold growth in a wall assembly. Amongst the geographical locations considered in this study, Yellowknife had the highest HDD (HDD = 8170), followed by Edmonton (HDD = 5120).

Outdoor Relative Humidity. This can be represented by the moisture index (MI). The higher the MI value, the smaller the drying potential of a wall assembly and hence, the higher the risk of mold growth. Amongst the geographical locations considered in this study, Vancouver had the highest value of MI (MI = 1.44), followed by St. John's (MI = 1.41).

Wind Speed. The higher the wind speed, the greater the air leakage rate across the wall assembly, and, hence, the higher the risk for mold growth within the wall assembly, as indicated earlier.

Details for determining the air leakage rates of the different geographical locations are provided in Saber et al. (2014). Also, the air leakage rates of the different locations are provided in Saber and Ganapathy (2016) and Saber et al. (2014). Amongst the geographical locations considered in this study, St. John's provided the highest air leakage rate across wall assemblies (see Saber and Ganapathy 2016; Saber et al. 2014).

Figures 16(a) and 16(b) show a comparison of the overall average value for mold index (M_{AVG}) and overall maximum value of mold index (M_{MAX}) for different wall assemblies having R-24 as stud cavity insulation at a 100% air leakage rate when subjected to different climatic conditions. As shown in these figures, the combined effects of the three environmental parameters, listed above, have brought about, in the case of walls subjected to the climatic conditions of Ottawa and Yellowknife, the lowest value of mold index.

For example, the overall average values for mold index for the reference wall were 1.18, 1.98, 2.41, 3.05, and 1.06, respectively, for the climatic conditions of Ottawa, Edmonton, Vancouver, St. John's, and Yellowknife (Figure 16[a]). The highest value of overall average mold index can be found for the reference wall, EPS wall, and XPS wall subjected to the climatic conditions of St. John's. For the mineral fiber wall subjected to the climatic conditions of St. John's and Vancouver, the values of overall average mold (1.03 and 1.09, respectively) were approximately the same. Details about the risk of condensation and mold growth in wall assemblies with exterior insulation over a broad range of values for thermal resistance (e.g., R-values of 4, 5, and 6 $ft^2 \cdot h \cdot °F/Btu$) and WVP (2 to 1800 $ng/Pa \cdot s \cdot m^2$) and subjected to different climatic conditions are available in reference (Saber and Ganapathy 2016).

SUMMARY OF SIMULATION RESULTS FOR DIFFERENT WALLS

For all wall assemblies having R-24 stud cavity insulation and with 100% air leakage rate, the results of the risk of mold growth are presented in the following order: Edmonton, Ottawa, Vancouver, Yellowknife, and St. John's.

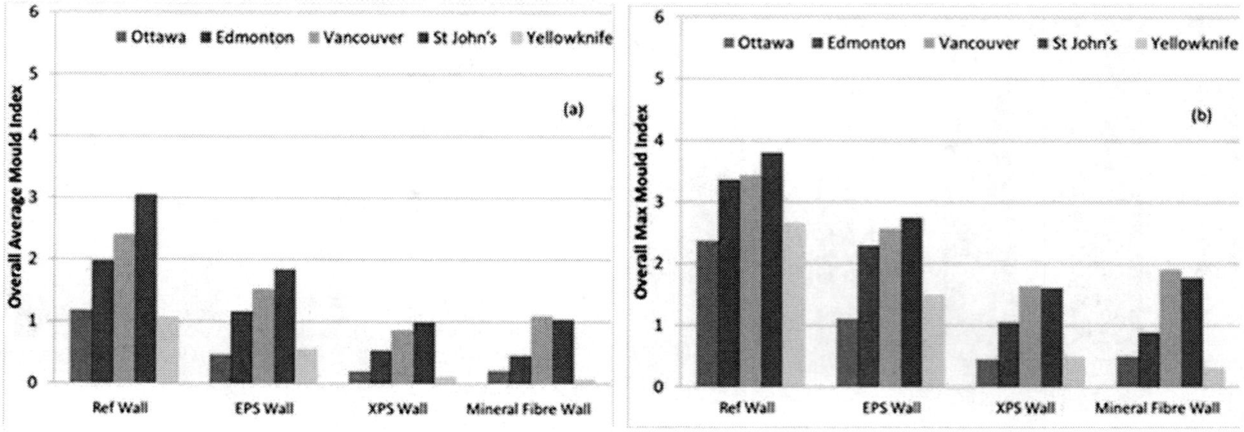

Figure 16 Effect of geographical locations on overall (a) average and (b) maximum mold index at 100% air leakage rate.

- Figure 17 for Edmonton, AB (cold, dry climate with HDD18 = 5120, MI = 0.48)
- Figure 18 for Ottawa, ON (cold, dry climate with HDD18 = 4440 to 4500, MI = 0.84)
- Figure 19 for Vancouver, BC (mild, wet climate with HDD18=2600 to 3100, MI = 1.44)
- Figure 20 for Yellowknife, NT (cold, dry climate with HDD18 = 8170, MI = 0.58)
- Figure 21 for St. John's, NL (mild, wet climate with HDD18 = 4800, MI = 1.41)

The results of thermal performance of these wall systems, expressed in terms of the yearly heat losses and gains, are available in Saber et al. (2014).

Edmonton, AB

As might be expected, increasing the R-value of the outboard, exterior insulation (Table 2) decreases the overall average value of M_{AVG} and overall maximum value, M_{MAX}. Figure 17 shows that in all instances, the values derived for both M_{AVG} and M_{MAX} for wall assemblies with exterior insulation are less than those for the code-compliant reference wall (Figure 17[a]). The values for M_{AVG} for these walls range from 1.98 to 0.46; corresponding values for M_{MAX} of these walls range from 0.89 to 3.36 (Figure 17[b]).

Ottawa, ON

As was the case for Edmonton, in all instances, the values derived for the overall average mold index and overall maximum mold index for walls with exterior insulations are less than those for the reference wall. The values derived for M_{AVG} for all wall assemblies range from 1.18 to 0.20 (Figure 18[a]). The corresponding values for M_{MAX} for these walls range from 2.37 to 0.44 (Figure 18[b]). As compared to the results obtained for Edmonton, however, the values for MAVG and MMAX for these sets of walls are lower. For example, the values of M_{AVG} of reference wall and EPS wall are 1.18 and 0.47, respectively, for Ottawa (Figure 18[a]); whereas, for

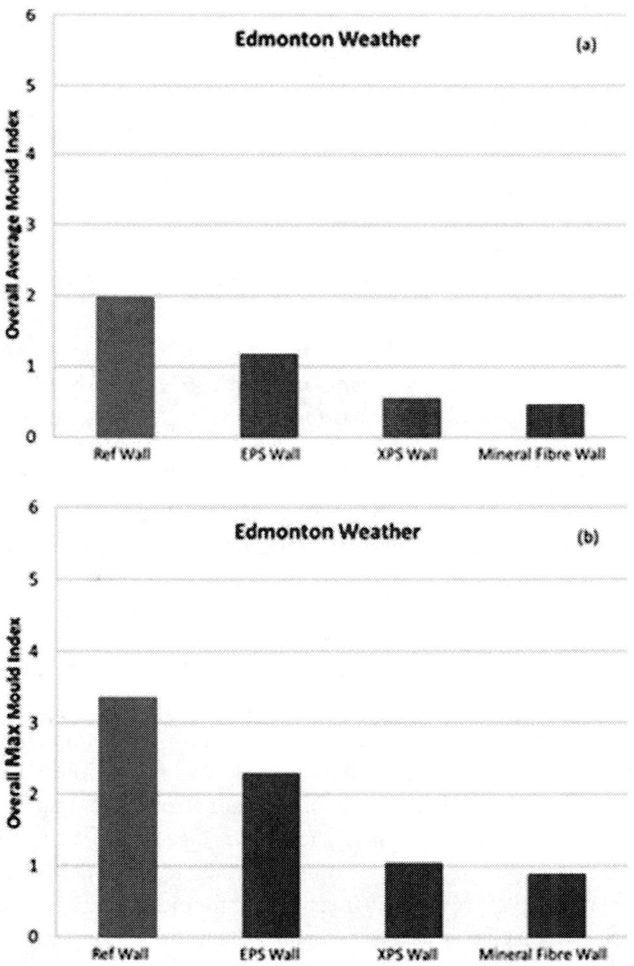

Figure 17 *Effect of R-value and WVP on overall maximum and average mold index at 100% air leakage rate (Edmonton weather).*

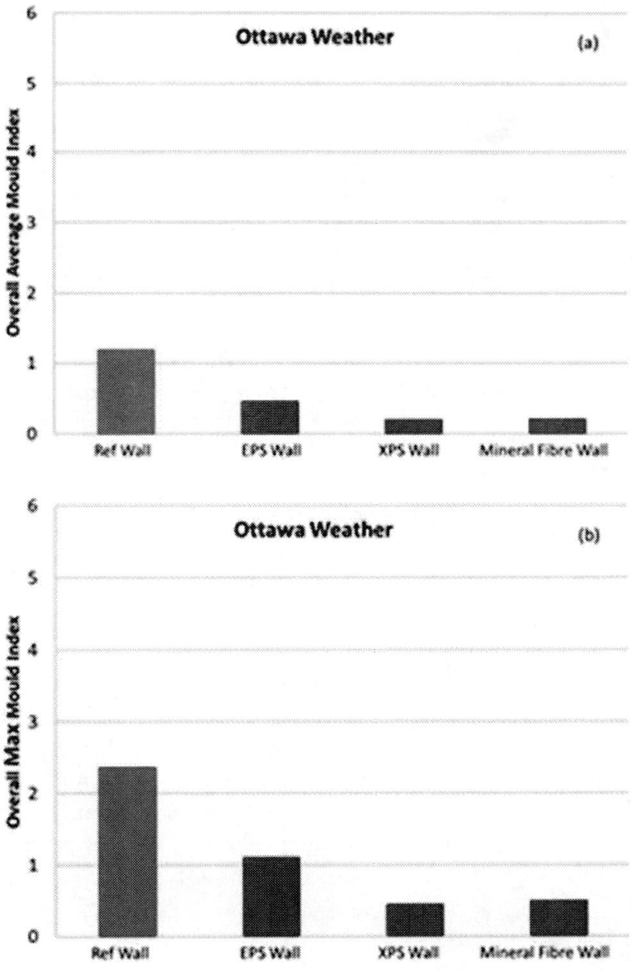

Figure 18 *Effect of R-value and WVP on overall maximum and average mold index at 100% air leakage rate (Ottawa weather).*

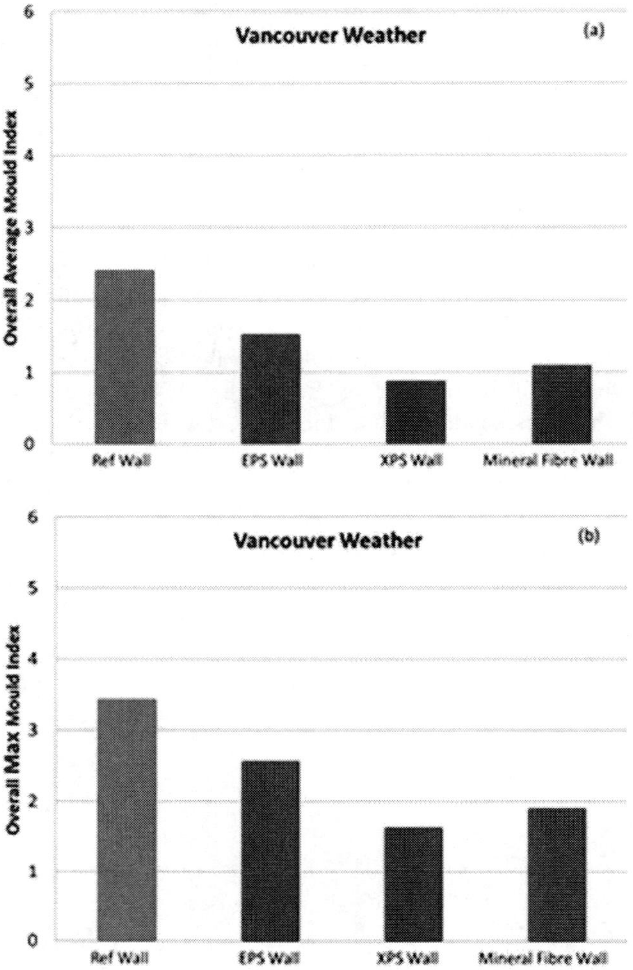

Figure 19 *Effect of R-value and WVP on overall maximum and average mold index at 100% air leakage rate (Vancouver weather).*

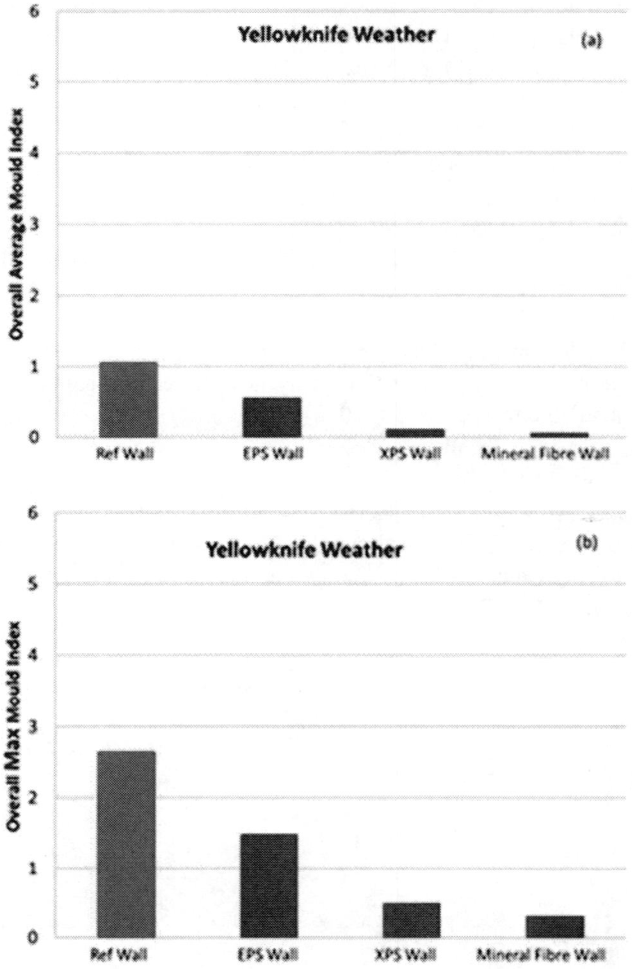

Figure 20 *Effect of R-value and WVP on overall maximum and average mold index at 100% air leakage rate (Yellowknife weather).*

Edmonton these values are 1.98 and 1.17 (Figure 17a). The corresponding values of M_{MAX} are 2.37 and 1.11 for Ottawa (Figure 18[b]); whereas for Edmonton (Figure 17b) these values are 3.36 and 2.29.

Vancouver, BC

Values of M_{AVG} and M_{MAX} for walls subjected to a Vancouver climate are comparatively greater than that of Ottawa and slightly greater than those of Edmonton. As shown in Figure 19(a), the values for M_{AVG} for these walls range from 2.41 to 0.87. The corresponding values for M_{MAX} for these walls range from 3.43 to 1.63 (Figure 19b).

Yellowknife, NT

Values of M_{AVG} and M_{MAX} for walls subjected to a Yellowknife climate are approximately the same as that of Ottawa. As shown in Figure 20(a) for Yellowknife, the values for M_{AVG} for different wall systems range from 1.06 to 0.06 compared to the range of 1.18 to 0.21 for Ottawa (Figure 18[a]). The corresponding values for M_{MAX} for these walls range from 2.65 to 0.30 for Yellowknife (Figure 20[b]) and from 2.37 to 0.44 for Ottawa (Figure 18[b]).

St. John's, NL

The greatest values for the overall M_{AVG} of the reference, EPS, and XPS wall configurations occur in St. John's, as compared to the other cities investigated. Note that the St. John's climate has the highest air leakage rate compared to the other geographical locations investigated (see Saber and Ganapathy 2016 and Saber et al. 2014 for more details). Also, St. John's climate is a wet climate with moisture index (MI = 1.41), slightly lower than that of Vancouver climate (MI = 1.44). Figure 21(a) shows that the values for M_{AVG} for all walls range from 3.05 to 1.0, whereas the corresponding values for M_{MAX} for these walls range from 3.8 to 1.61 (Figure 21b).

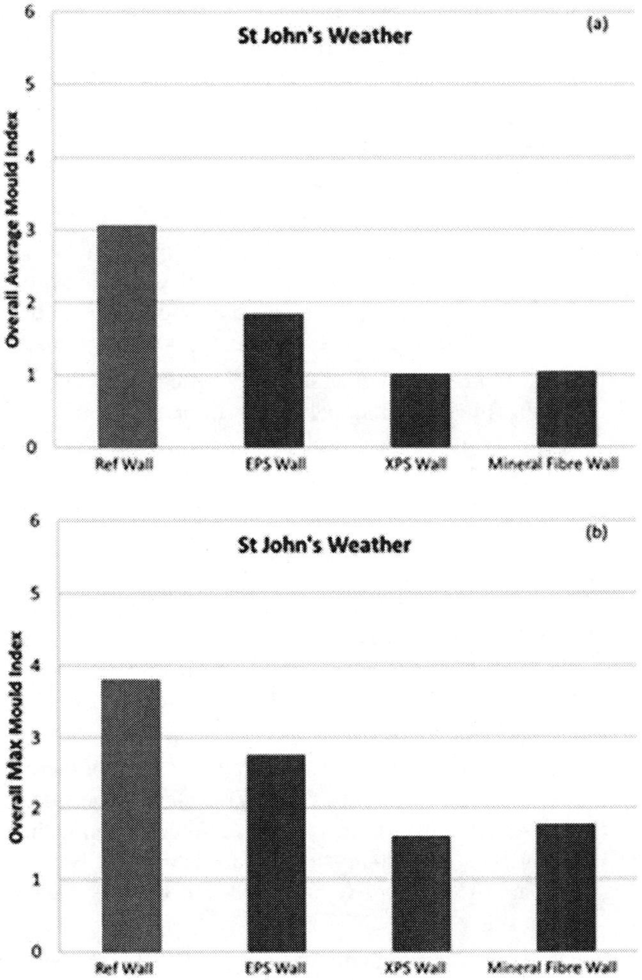

Figure 21 *Effect of R-value and WVP on overall maximum and average mold index at 100% air leakage rate (St. John's weather).*

CONCLUDING REMARKS

NRC's hygrothermal numerical model, hygIRC-C, was first benchmarked against experimental work carried out in the FWEF; the benchmarking exercise consisted of completing transient numerical simulations, as was done in a number of other studies, to benchmark the model against experimental work. The numerical results derived for values of heat flux attained at different locations within the respective wall assemblies were compared with the measured values for assemblies incorporating different exterior insulation products that included the following: (a) EPS of 1in. thick (R-3.9 and WVP = 114), (b) XPS of 2 in. thick (R-10 and WVP = 27), and (c) mineral fiber insulation products of 3 in. thick (R-12.5 and WVP = 2130). The results showed that the comparison between the present model predictions and experimental data were in good agreement.

Following the benchmarking of the present model, it was then used to conduct a parametric study to investigate the risk of condensation and mold growth in different wall assemblies when these assemblies were subjected to climatic conditions in Canada, specifically, that of Vancouver (BC), Edmonton (AB), Ottawa (ON), St. John's (NL), and Yellowknife (NT). The modeling results for the different walls were expressed using values of M based on the most recent mold growth model developed by Ojanen et al. (2010) and from which were obtained values of M of different materials within the respective wall assemblies.

Based on the air leakage path that was considered in this study, the simulation results showed that the critical locations inside the wall assembly at risk of mold growth are the top and bottom portions of the wall assembly. However, had a different air leakage path been considered in this study, it would have resulted in different locations within the wall assemblies being at risk of condensation and mold growth.

Similar to a previous NRC study (Saber et al. 2014), the simulation results were presented on the basis of a simple form using the following two parameters:

- Average value of mold index (MAVG): average value of M at all locations within the assembly
- Maximum value of mold index (MMAX): average value of maximum value of M at all locations within wall

A sensitivity analysis was conducted to investigate the effect of different air leakage rates of 0% (no air leakage), 50%, and 100% on the hygrothermal performance of wall assemblies. No risk of condensation occurred in the wall assemblies for the case of no air leakage. The case of 100% air leakage rate (i.e., 0.1 L/(s·m^2) at 75 Pa) resulted in higher risk of condensation and mold growth than that for the case of 50% air leakage rate (i.e., 0.05 L/(s·m^2) at 75 Pa).

The values for the overall average mold index and overall maximum mold index in different wall assemblies with the inboard stud cavity insulation of R-24 were higher than that of walls having stud cavity insulation of R-19. The values for the overall average mold index and overall maximum mold index in walls with different types of outboard, exterior insulations were lower than that of the code-compliant reference wall.

For the code-compliant reference wall, the EPS wall, and XPS wall, St. John's (NL) appeared to have the most severe climate in comparison to the other four locations investigated (Vancouver, BC, Ottawa, ON, Edmonton, AB, and Yellowknife, NT); the greatest values of the overall average mold index in the wall configurations amongst the five locations occurred in this location. For the wall having mineral fiber as exterior insulation, the values of the overall average mold index were approximately the same for St. John's and Vancouver.

ACKNOWLEDGMENTS

The authors wish to thank Canada Mortgage and Housing Corporation (CMHC) and Natural Resources Canada (NRCan) for contributing funding for this project. The authors wish to thank the Project Advisory Committee that included: Constance Thivierge (formerly with FP Innovations), Doug Tarry (Doug Tarry Homes), Rick Gratton (CHBA), Christopher McLellan (CHBA), Chris Mattock (Habitat Design and Consulting), Salvatore Ciarlo (Owens Corning Canada), and Robert Jonkman (Canadian Wood Council). Also, the authors wish to thank NRC-Construction for providing the funding to enable researchers to build, operate and maintain a state-of-the-art field exposure of walls facility (FEWF) that was used in this project.

REFERENCES

ASHRAE 2009a. ANSI/ASHRAE Standard 160-2009, *Criteria for moisture-control design analysis in buildings*. Atlanta: ASHRAE.

ASHRAE. 2009b. Chapter 26, *ASHRAE Handbook—Fundamentals* (SI). Atlanta: ASHRAE.

ASTM. 2009. ASTM Designation C-1130-07, *Standard practice for calibrating thin heat flux transducers*. Annual Book of ASTM Standards, sec 4, Construction, vol. 04.06, Thermal Insulation; Building & Environment Acoustic, 577–84.

Elmahdy, A.H., W. Maref, M.C. Swinton, H.H. Saber, and R. Glazer. 2009. Development of energy ratings for insulated wall assemblies. Building Envelope Symposium, San Diego, California, October 26, pp. 21–30.

Elmahdy, A.H., W. Maref, H.H. Saber, M.C. Swinton, and R. Glazer. 2010. Assessment of the energy rating of insulated wall assemblies a step towards building energy labeling. 10th Int. Conference for Enhanced Building Operations (ICEBO2010), Kuwait, October.

Glass, S.V. 2013. Hygrothermal analysis of wood-frame wall assemblies in a mixed-humid climate. United States Department of Agriculture, Forest Service, Forest Products Laboratory, Research Paper FPL–RP–675, pp. 1–25, April 2013.

Hukka, A., and H.A. Viitanen. 1999. A mathematical model of mold growth on wooden material. *Wood Science and Technology* 33(6):475–85.

Lacasse, M.A., H.H. Saber, G. Ganapathy, and M. Nicholls. 2016. *Evaluation of thermal and moisture response of highly insulated wood-frame wall assemblies? Phase 1; Part I: Experimental trials in field exposure of walls test facility*. Report No. A1-000444.5. Ottawa: NRC-Construction, National Research Council of Canada. 12 January.

Mukhopadhyaya, P., M.K. Kumaran, J. Lackey, N. Normandin, D. van Reenen, and F. Tariku. 2007. Hygrothermal properties of exterior claddings, sheathing boards, membranes and insulation materials for building envelope. *Proceedings of thermal performance of the exterior envelopes of whole buildings X*; 1–16. Clearwater, Florida, December 2–7.

NBCC. 1995. National Building Code of Canada, Canadian Commission on Building and Fire Codes, National Research Council of Canada, Ottawa, Canada.

NBCC. 2010. National Building Code of Canada. Section 9.25, Canadian Commission on Building and Fire Codes National Research Council of Canada, Ottawa, Canada.

Ojanen, T. and M.K. Kumaran. 1995. Effect of exfiltration on the hygrothermal behaviour of a residential wall assembly: results from calculations and computer simulations. International Symposium on Moisture Problems in Building Walls, Porto-Portugal. 11–13 September. 157.

Ojanen, T., H.A. Viitanen, R. Peuhkuri, K. Lähdesmäki, J. Vinha, and K. Salminen. 2010. Mold growth modeling of building structures using sensitivity classes of materials. 11th International Conference on Thermal Performance of the Exterior Envelopes of Whole Buildings XI. Clearwater, FL. December 05(10).

Saber, H.H., and G. Ganapathy. 2016. Evaluation of thermal and moisture response of highly insulated wood-frame wall assemblies? Phase 1; Part II: Numerical modeling. Report No. A1-000444.4; NRC-Construction, National Research Council of Canada, Ottawa, Canada; 8 January.

Saber, H.H., W. Maref, and K. Abdulghani. 2014. *Properties and position of materials in the building envelope for housing and small buildings*. Report No. A1-004615.1, NRC-Construction, National Research Council of Canada, Ottawa, December 31.

Saber, H.H., and W. Maref. 2015. Risk of condensation and mold growth in wood-frame wall systems with different exterior insulations. BEST Building Enclosure Science & Technology Conference (BEST4), held in April 12–15, 2015. Kansas City, Missouri. 19.

Saber, H.H., W. Maref, A.H. Elmahdy, M.C. Swinton, and R. Glazer. 2010. 3D Thermal model for predicting the thermal resistances of spray polyurethane foam wall assemblies. Building XI Conference, Clearwater, Florida.

Saber, H.H., W. Maref, A.H. Elmahdy, M.C. Swinton, and R. Glazer. 2012. 3D Heat and air transport model for predicting the thermal resistances of insulated wall assemblies. *International Journal of Building Performance Simulation* 5(2):75–91.

Viitanen, H.A., and T. Ojanen. 2007. Improved model to predict mold growth in building materials. *Proceedings of Thermal Performance of the Exterior Envelopes of Whole Buildings X*, 8.

Practical Applications and Case Study of Temperature-Based Smart Ventilation Controls

Michael Lubliner
Member ASHRAE

Paul Francisco
Member ASHRAE

Eric Martin

Iain Walker, PhD
Fellow ASHRAE

Brennan Less

Robin Vieira
Member ASHRAE

Rick Kunkle, PE
Member ASHRAE

Zachary Merrin
Associate Member ASHRAE

ABSTRACT

Smart ventilation controls (SVCs) provide energy and indoor air quality (IAQ) advantages through dynamic real-time control of mechanical ventilation rates. Section 4.6 of ASHRAE Standard 62.2-2013 allows for alternative SVC strategies if equivalent annual pollutant exposure is provided. This paper presents results from U.S. Department of Energy (DOE) Building America program stakeholder collaborations on temperature-based smart ventilation control (TSVC) as one potential option. Unlike the "continuous" ventilation typically implemented to comply with ASHRAE Standard 62.2-2013, temperature-based ventilation controls take advantage of the dynamics of stack infiltration, which varies with weather, house height, and envelope leakage. The TSVC strategy turns off mechanical ventilation at times of large temperature difference and moves ventilation to periods with smaller temperature differences. To ensure equivalent annual pollutant exposure, mechanical airflow rates must be increased for temperature-based systems. In this paper, we compare the energy and IAQ performance of a TSVC that shuts the fan off at a specified single cutoff temperature to a continuously operated ventilation system fan that is sized to comply with ASHRAE Standard 62.2-2013. This paper presents simulation analysis results using two simulation tools: 1) REGCAP and its advanced ventilation model for detraining ventilation TSVC equivalency, 2) a new beta version of EnergyGauge® USA for estimating energy savings. Relative humidity (RH) and carbon dioxide (CO_2) concentrations were monitored in two occupied case study homes in cold and marine climates involving weekly "flip flop" tests between continuous and single temperature cutoff control scenarios. Preliminary investigations of market costs and modeled energy savings suggest simple paybacks of 4–11 years from TSVCs for these case study homes.

BACKGROUND

Residential buildings account for 22% of total United States energy consumption, with most attributed to HVAC. Residential energy-efficient construction has focused on tighter envelopes to save energy; however, this also creates the potential for underventilation and compromised indoor air quality (IAQ) (Offerman 2009). The air quality inside a space is a complex result of occupant activities, human responses, source emissions, and contaminant removal processes. The purpose of ventilation is to remove contaminants generated inside a space, preferably to levels constituting "good" IAQ. Because people typically spend 80% to 90% of their time indoors, residential building science principles and ASHRAE Standard 62.2 (ASHRAE 2013a) recognize that energy conservation activities should be carried out in ways that do not negatively impact occupant exposure to air pollutants. ASHRAE Standard 62.2 assumes that there will be a constant ventilation rate from a purpose-provided mechanical ventilation system, for every hour of the day, with guidance for intermittent system operation. Ventilation fans are usually operated for 24 hours per day at the same rate regardless of the outdoor environmental conditions. These continuous systems ignore weather-driven natural infiltration, which can lead to excess ventilation during cold periods and unnecessary energy expense. Weather-driven natural infiltration through openings in the building envelope is caused by pressure differences due

Michael Lubliner is a senior building science specialist and **Rick Kunkle** is a research engineer with the Washington State University Energy Program, Olympia, WA. **Paul Francisco** and **Zachary Merrin** are research engineers with Indoor Climate Research & Training at the University of Illinois at Urbana–Champaign, Champaign, IL. **Eric Martin** is a senior research engineer and **Robin Vieira** is a researcher at Florida Solar Energy Center, University of Central Florida, Cocoa, FL. **Iain Walker** is a scientist and **Brennan Less** is with Lawrence Berkeley National Laboratory, Berkeley, CA.

to indoor-outdoor temperature differences ("stack") and wind. The magnitude of infiltration depends on many factors, including environmental conditions, building airtightness (as determined by a blower door test), and stack height. In this study, the effect of wind was neglected because on-site temperature measurements were used only for control. In the future, it is possible that controls will use Internet connections to determine local weather that includes wind speed, allowing inclusion of both wind and stack effect in the infiltration estimates.

MODELING

Two different modeling efforts were conducted in this project. REGCAP (Walker and Sherman 2006) was used for ventilation equivalency and EnergyGauge® USA (Parker et al. 1999) was used for energy modeling. These were both used for the case study homes described in the Field Monitoring section.

REGCAP Ventilation Equivalency Modeling

Previous work by Lawrence Berkeley National Laboratory (LBNL) has built the capacity for performing detailed simulation studies of combined energy and IAQ performance in residences. This work has primarily relied on the REGCAP simulation tool, which combines models for mass-balance ventilation (including envelope, duct, and mechanical flows), heat transfer, HVAC equipment, and moisture. Simulations are performed with a 1 min time resolution to allow for dynamic HVAC system operation and use of real-time ventilation controllers. The details of this model are presented elsewhere (Walker et al. 2005; Walker and Sherman 2006; Walker 1993), and it has been validated in other contexts (Walker et al. 2005; Walker 1993). Recent work has developed the concept of ventilation equivalence (Sherman et al. 2012), which can be used to ensure that variable ventilation rates give the same pollutant exposures as constant ventilation rates. Algorithms for calculating pollutant exposure and ventilation equivalence have been incorporated in the REGCAP model. These have been used to develop smart ventilation controllers that provide annual equivalence while varying ventilation rates with occupancy (OSVC), time-of-day demand (DSVC), and operation of other intermittent exhaust (IESVC) devices such as bath and kitchen fans and clothes dryers in the home (Turner and Walker 2012; Walker et al. 2012).

In this research, LBNL used REGCAP to simulate the performance of temperature-based smart ventilation controllers (TSVCs) in two prototypical test homes across United States climate zones, with varying envelope airtightness and numbers of stories (Less et al. 2014). The simulations compared energy and IAQ performance of a TSVC to a base case of a continuously operated ventilation fan sized to comply with ASHRAE Standard 62.2-2013 whole-house (WH) ventilation requirements, including the ASHRAE Standard 62.2 infiltration credit. The LBNL simulations of the simple TSVCs satisfied ASHRAE Standard 62.2 by ensuring that annual pollutant exposure was equivalent between the dynamic temperature-controlled and continuous fan cases. This required that WH mechanical ventilation systems are oversized relative to ASHRAE Standard 62.2 requirements for continuous systems. Determining the appropriate temperature cutoff and required fan oversizing is critical if TSVC systems are to provide equivalence with ASHRAE Standard 62.2. Four steps were used in the LBNL analysis to ensure the TSVCs provided "equivalent ventilation" in the two test homes:

1. Outdoor temperature cutoffs were calculated using the stack part of the enhanced ventilation model (AIM-2) in *ASHRAE Handbook—Fundamentals* (ASHRAE 2013b) so that predicted stack infiltration was equal to the required total ventilation rate in ASHRAE Standard 62.2. These cutoffs varied by climate zone, house air leakage, and number of stories. Other cutoffs were also developed, including a simple fixed 41°F value and 25th percentile values based on outdoor temperature distributions in TMY3 data files.
2. The REGCAP simulation tool was used to estimate the energy savings and IAQ impacts of using these cutoffs to control a mechanical exhaust fan. The simulated fans were not oversized and as a result were not compliant with ASHRAE Standard 62.2.
3. To ensure equivalence, exhaust fans were resized to maintain equivalence using an iterative optimization procedure that incorporated both the stack and wind parts of the AIM-2 ventilation model.
4. The REGCAP simulation tool was used to calculate energy and IAQ performance using these oversized exhaust fans.

These simulations considered only annual ventilation equivalence. Short-term fluctuations were not controlled for, such that acute exposures may present some health concerns. While annual values represent "chronic" or long-term exposures, short-term peaks in pollutant levels can also present "acute" risks to occupant health. These were not considered. For example, during extended periods of low outdoor temperature, the ventilation system may have been continuously off for long periods of time. Ventilation was operated at a higher airflow during warmer periods of the year to ensure annual equivalence, but pollutant levels could have reached unacceptable short-term levels during the off periods. In our temperature-based work, these concerns should be mitigated by the fact that periods of low temperature are associated with higher natural infiltration due to stack pressures. Other smart approaches will not have such a benefit because they are not necessarily associated with times of higher infiltration. It is also notable that acute pollutant exposures in homes are typically caused by occupant activities that release large quantities of pollutants over a short period of time, such as cooking or cleaning. Marginally lower or higher continuous ventilation rates will have little effect on these localized events.

In light of these acute exposure concerns, ASHRAE Standard 62.2-2013 has adopted a change for ventilation equivalence that includes a maximum pollutant exposure value (C_{max}) that smart controllers can use to provide annual equivalence while also protecting against acute hazards. Annually, smart controls will target a C-value of ≤1. TSVC approaches that guard against unacceptable acute exposures may result in some minimum amount of hourly or daily ventilation system runtime, or other provisions, all of which might reduce predicted energy savings for TSVCs. As noted above, the TSVCs in this project did not explicitly include a C_{max} limit. Modeling shows that the highest peak exposures were found in the most airtight homes and those in mild climates.

EnergyGauge® USA Energy Modeling and Simple Payback of TSVC

Using the TSVC strategies assessed by LBNL, the EnergyGauge USA software was used with new beta features to assess potential energy savings for the advanced controls. Simulation test cases were developed that reflect the field-tested homes described in the Field Monitoring section. EnergyGauge USA analysis was employed to estimate the energy use of TSVCs as compared to continuous operating systems in these example homes. Baseline cases with continuous WH exhaust fans were sized per ASHRAE Standard 62.2-2013 (with infiltration credit). The two test homes using TSVCs used oversized WH exhaust fans, and the installed airflow rates were used in conjunction with the iterative fan-sizing tool (see item 3 in the list in the previous subsection) to identify custom cutoff temperatures with the highest anticipated energy savings. These cutoffs are different from those used in the LBNL modeling described in the previous subsection. House parameters used in EnergyGauge USA modeling are summarized in Table 1.

FIELD MONITORING

Field monitoring was used to verify the indoor environmental performance of the TSVCs developed through LBNL simulations described above. Two homes were outfitted with TSVCs and monitored over a period of several months. These were in two distinct climates: the marine climate of Olympia, Washington, and the cold climate of central Illinois. One primary question addressed by the field testing was whether the fan control operated as intended. Two issues are considered: 1) fan cycling and 2) whether the temperature control is working properly. The case studies suggest that the TSVCs seem to operate as intended in both the Washington and Illinois homes. Another primary aim is to determine how the operation of the TSVCs impacts each home's carbon dioxide (CO_2) concentration and relative humidity (RH).

Occupants in the Washington and Illinois homes were not able to activate the fan on demand when the fan was in the TSVC mode and the temperature was below the cutoff temperature due to wiring retrofit constraints. Occupants in both homes were free to operate their windows as they normally would.

Case Study Homes

Washington Case Study. The Howard residence test house (hereafter referenced as the *WA home*" is a two-story, 1640 ft^2 historical house located in Olympia, Washington. It underwent a deep energy retrofit in 2010 that significantly upgraded the building envelope (R-38/R-49 ceiling, R-15/R-26 wall, R-30 floor insulation) and reduced the air change rate from 20 to 5 ACH_{50}. Space conditioning is provided by a 1 ton single-head ENERGY STAR (HSPF = 12, SEER 25) ductless heat pump with the indoor unit located on the first floor. The TSVC outdoor temperature sensor was located outside the bathroom behind the ductless heat pump outdoor unit. The WH exhaust fan with TSVC was located in the first floor bathroom. During the weekly flip flop testing in the TSVC mode, the WH fan operated at 90 cfm whenever the outdoor temperature reached 57°F and did not operate at all below that temperature. During the ASHRAE Standard 62.2 mode, the WH exhaust fan ran continuously at 40 cfm regardless of outdoor temperature. EnergyGauge USA analysis suggested that for the WA home, the energy savings from enacting the cutoff strategy is estimated at $23.00/yr of the HVAC energy use, as shown in Table 1. Table 1 also provides a four-year simple payback estimate, assuming a cost to the consumer of $80 (the installed TSVC cost is $20 for the temperature relay control and $60 in labor for an electrician to wire the controller to the WH fan). This simple payback is well within the expected useful life of the TSVC, which is at least 15 years.

Table 1. ASHRAE Standard 62.2 Ventilation/Infiltration Parameters, Requirements, Energy Savings, and Simple Payback

Location	ACH_{50}	Standard 62.2		TSVC		TSVC	TSVC
		CFM_{50}	cfm	cfm	Off > °F	$/year saved*	Payback
Olympia, WA 2 stories, 1640 ft^2	5	1152	40	90	57	$23.00	4 years
Champaign, IL 1 story, 900 ft^2	9	1073	30	80	55	$7.30	11 years

* Assumes 0.10/kWh and $1.00/therm. TSVC is assumed to have a retail installed cost of $80.

Illinois Case Study. The test home in Champaign, Illinois, (hereafter referred to as the *IL home*) is a single-family, one-story home on a full unoccupied and unfinished basement. The home floor area is 900 ft^2. Blower door tests measured the home at about 9 ACH_{50}. The WH exhaust fan was installed in the bathroom. The TSVC outdoor temperature sensor was located in the gable vent of the attic. During the weekly flip flop testing in the TSVC mode, the WH fan operated at 80 cfm whenever the outdoor temperature reached 55°F and did not operate at all below that temperature. During the ASHRAE Standard 62.2 mode, the WH exhaust fan ran continuously at 30 cfm regardless of outdoor temperature. EnergyGauge USA analysis suggested that the energy savings from enacting the cutoff strategy is estimated at $7.30/yr of the HVAC energy use, as shown in Table 1. Table 1 also provides a simple payback estimate using the same assumptions as for the WA home. This simple payback is well within the expected useful life of the TSVC.

Table 2 provides the breakdown on the modeled energy savings associated with the space-conditioning HVAC blower and WH fan. For comparison, the LBNL simulations (Less et al. 2014) showed very similar results for the WA home and for the IL home. Given that the LBNL study did not use the specific characteristics of these two homes, the two modeling exercises show reasonable agreement.

Methodology

Homes were instrumented with fans connected to a controller that could either run the fan continuously or use the TSVC control algorithm using the selected cutoff temperature. The ventilation strategy was alternated between these two modes in one-week periods. Indoor and outdoor temperatures were monitored, along with indoor RH and CO_2 throughout the study period.

Fan Cycling and Temperature Control Operation Results

There is a 1 min delay in the fan controller and a 0.5°F dead band between when the fan turns on and off to avoid rapid cycling of the fan. A review of 1 min fan on-time data for the WA house for the first three months of operation (May–July) indicates there were a little over 40 instances of on-off cycles during the analysis period (for 1466 h of temperature-based fan control). This means the fan had an on-off-on or off-on-off cycle in consecutive 1 min periods less than 0.05% of the time. While these instances were isolated, they indicate that some undesirable fan cycling did occur. We expect cycling to occur more frequently during spring and fall periods, when temperatures are more often around 57°F, but because of a data filtering problem, 1 min fan operation data were not available after July. This suggests further research and discussions with fan manufacturers are needed to evaluate a longer delay period on

Table 2. Modeled Energy Savings

Washington Case Study

Case	AC, kWh	AC Blower, kWh	AC WH, kWh	Heat, kWh	Heat Blower, kWh	Heat WH, kWh	Total, kWh	Total, Therm	Total, $
Standard 62.2-2013, 24/7/365	229	82	23	1700	273	26	2333	—	—
TSVC mode	216	77	22	1566	217	5	2103	—	—
Energy savings	13	5	1	134	56	21	230	—	—
Utility cost saved, $	1.3	0.5	0.1	13.4	5.6	2.1	—	—	$23
% Savings, kWh	5.7	6.1	4.3	7.9	20.5	80.8	9.9	—	—

Illinois Case Study

Case	AC, kWh	AC Blower, kWh	AC WH, kWh	Heat, Therm	Heat Blower, kWh	Heat WH, kWh	Total, kWh	Total, Therm	Total, $
No override	905	201	32	866	660	38	1836	866	—
Override	989	218	39	852	646	11	1903	852	—
Energy savings	–84	–17	–7	14	14	27	–67	14	—
Cost savings, $	–8.4	–1.7	–0.7	14	1.4	2.7	—	—	$7.30
Percent savings	–9.3	–8.5	–21.9	1.6	2.1	71.1	—	—	—

the controller. A 15 min delay has been proposed by LBNL for Residential Integrated VEntilation Controller (RIVEC) (Turner and Walker 2013).

In the WA home, the thermostat controlling the fan is located on the west side of the house behind the outdoor unit of the ductless heat pump. This provides shading from direct sun and some air movement when the heat pump is operating. Comparing the average hourly outdoor temperature with fan runtime during the hour indicates if the temperature control was functioning properly. Review of the TSVC operation indicates the fan temperature control seems to be working properly. The fan was always off when the average hourly outdoor temperature was less than 57°F. The fan was always on when the average hourly outdoor temperature on the porch was greater than 60°F, with a handful of exceptions. These exceptions mostly occurred on summer mornings when the temperature was rising rapidly, so that temperatures during a portion of the hour were below the fan set point. For hourly average temperatures between 55°F and 60°F, there is a mix of operation, as expected, because the outdoor temperature during an hour could be above or below the fan control set point but the fan is increasingly on as the outdoor temperature rises. A preliminary analysis of energy use data indicates that the energy savings are zero above the cutoff temperature of 55°F, about 25% at 45°F, and about 15% at 40°F. A more detailed analysis of energy use will be performed in the future.

Another key issue for temperature control is the location of the temperature sensor controlling the fan. For the IL home, the sensor was located in the gable vent to protect it from rain and direct solar gain. However, a comparison of gable temperature with local airport temperature showed that the gable vent temperature was elevated because of solar gain. The analysis suggests the fan operated about 7% more than it would have using the airport temperature. At the WA home, three different sensors were used to collect outdoor temperature. The measured temperatures are very similar and there is no evidence of fan control problems due to the location of the temperature sensor.

Carbon Dioxide Results

WA Home. The residents of the WA home indicated that windows were closed throughout the heating season but were often open at night in the summer months. Therefore, the following analysis screens out June–September. The highest CO_2 level in the WA home when all windows were closed was 2155 ppm (some isolated higher levels occurred in the second bedroom). Levels of CO_2 tended to be higher with TSVC, particularly at lower temperatures. A two-sample t-test shows that the difference is statistically significant, with the TSVC being statistically higher (average 748 ppm) than the continuous ventilation (average 666 ppm) at the 95% confidence level ($p<0.0001$). Above 57°F, the CO_2 levels in TSVC mode should be lower because of higher ventilation from fan operation at 90 cfm instead of 40 cfm. However, this is not the case. A two-sample t-test shows that CO_2 in TSVC mode in the main living area when the outdoor temperature is above 57°F averaged 567 ppm compared to 527 ppm with continuous ventilation, and that this difference is statistically significant at the 95% confidence level ($p=0.0024$). This is likely due to differences in occupancy, air mixing, and possibly some window opening. The small differences in CO_2 levels during all conditions for the two modes of operation suggest that other influences on CO_2 were greater than the control mode. Further study regarding other contaminants, particularly when the ventilation system is off in TSVC mode, seems warranted.

IL Home. The highest CO_2 level measured during periods when all windows were closed in the IL home was just above 1600 ppm. CO_2 levels tended to be higher with TSVC at lower temperatures. A two-sample t-test shows that the difference is statistically significant, with the temperature-based control being statistically higher (average 966 ppm) than the continuous ventilation (average 868 ppm) at the 95% confidence level ($p=0.0038$). For data above 55°F, the two-sample t-test shows that there is a reduction in CO_2 with the temperature-based control (average 643 ppm versus 692 ppm for continuous fan operation), which is not quite statistically significant at the 95% level but is at the 90% level ($p=0.0592$).

As is the case for the WA home, the differences in CO_2 levels in the IL home between the two operation modes are small relative to other influences. We might expect greater variation in CO_2 with TSVC operation, and this tends to be the case. Further study regarding other contaminants in the two modes of operation is needed. The combination of results is consistent with the concept of targeting the annual dose to contaminants with different control strategies.

Humidity Results

Analysis of humidity is not straightforward for a variety of reasons, including that RH is temperature dependent, outdoor air can be a significant source, and there can be significant moisture storage in building materials and furnishings. The first two of these issues can be addressed by using the moisture balance technique (Francisco and Rose 2010). This technique, also used by IEA Annex 41 in the development of ISO 13788 (IEA 2008), compares the indoor vapor pressure (which is temperature independent) to the vapor pressure outdoors over a range of outdoor temperatures. The method involves performing a linear regression on the data between 68°F and 32°F and forcing the regression through zero at 68°F. The value at the 32°F intercept is the vapor pressure excess value used to characterize the home.

Figure 1 shows the data from the WA home (Figure 1a) and the IL home (Figure 1b) for temperatures of 68°F and colder. Vertical lines mark 55°F, where the temperature-based control changed fan operation modes, and the 32°F intercept. A number of points have negative values, indicating that there was more moisture in the outdoor air than in the indoor air (such as during precipitation events).

The WA analysis found that with the temperature-based control, the vapor pressure excess was 616 Pa; for continuous

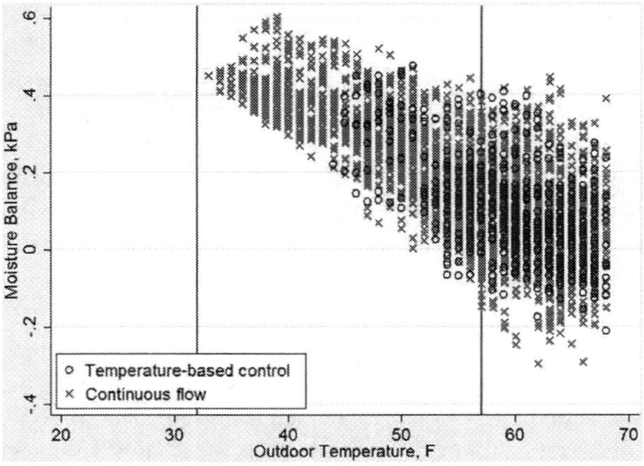

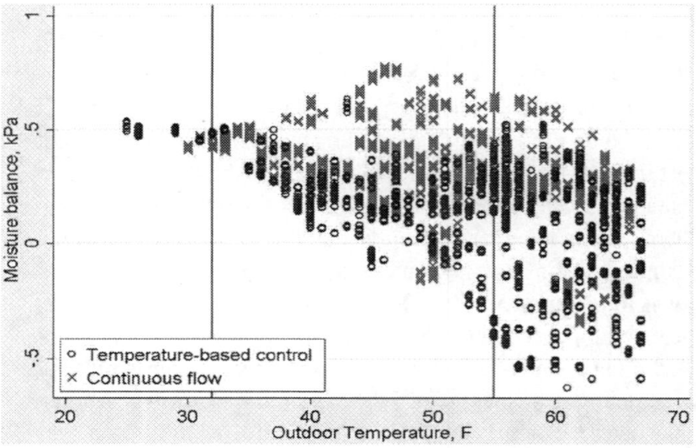

Figure 1 *Moisture balance versus outdoor temperature for the WA and IL homes.*

ventilation, the vapor pressure excess was 472 Pa. Qualitatively, the direction of this change is as expected.

The IL analysis found that with the temperature-based control, the vapor pressure excess was 234 Pa, compared to 586 Pa for continuous ventilation. These differences are statistically significant at the 95% confidence level. This result is counter to expectations. The difference is largely attributable to the data in the 44°F to 54°F range, when the temperature-based control turned the ventilation fan off and therefore did not bring in additional cool (and likely dry) outdoor air or exhaust indoor air.

CONCLUSIONS

The field testing of the TSVC successfully demonstrated proof of concept and the viability of simple controls in two homes where WH mechanical ventilation systems were installed. One home was located in the marine climate of Olympia, Washington, and the other was in the cold climate of Champaign, Illinois. Data were collected in 2015–2016. During this monitoring period, the research plan employed flip flop tests, which varied the WH ventilation system between continuous operation per ASHRAE Standard 62.2-2013 and outdoor-temperature-based operation. Findings and recommendations are listed below. Detailed simulations of ventilation performance were used to develop smart ventilation controls based on outdoor temperature that provided equivalence with a continuous fan sized according to ASHRAE Standard 62.2. The energy performance of these strategies was then assessed using the EnergyGauge® USA simulation tool. IAQ measurements were used to assess potential moisture and CO_2 issues in two occupied homes.

1. The TSVC prototype control is estimated to cost $80 (installed), and it functioned as intended to control the ventilation system lockout at the targeted temperature, based on actual field conditions in the Washington and Illinois case studies. Two case study findings suggest the following:
   a. Some limited cycling of the ventilation fan may indicate the need for increasing the fan control delay time period. The LBNL control strategy employs a 15 min dead band control algorithm to limit cycling, which may also be appropriate for TSVC. Fan cycling can cause reduced service life and create occupant perceptions of noise.
   b. Guidance on the appropriate location and long-term reliability of the temperature sensor is an area that could benefit from continuing research and product development. Utilization of Internet-based temperature data can eliminate this issue.
2. The TSVC device for the marine and cold climate test homes is estimated to save as much as $7.30 to $23.00 per year, based on EnergyGauge USA modeling. Given the low first cost of the system and its modeled energy savings, a large potential market may exist in retrofit and new homes.
3. The impact of the TSVC on CO_2 and humidity was not significant compared to other factors, such as occupancy, window opening, and wind effects.

RECOMMENDATIONS

1. A simple tool is needed for users to determine appropriate fan sizing and cutoff temperatures that will ensure IAQ equivalence. This tool must also consider IAQ implications of extended periods of ventilation system lockout, based on home characteristics and climate.
2. New ventilation equivalency procedures introduced into ASHRAE Standard 62.2-2013 provide the guidance required to develop Standard 62.2-compliant SVC ventilation systems, including those that employ temperature-based controls. Continued support and dialogue with ASHRAE Standard 62.2 will help refine the emerging definition of ventilation equivalency. These developments will be critical in creating a regulatory environ-

ment in which TSVCs and SVCs can be used for compliance purposes in a growing market.

3. Ongoing Building America research will also evaluate other SVC strategies, such as occupancy (OSVC), intermittent exhaust operation (IESVC), time-of-day demand (DSVC), and combinations of these.
4. TSVC-measured energy savings research will be addressed in the future. It is suspected that the flip flop monitoring of a larger sample of homes will be required to validate the relatively small energy savings signal from the "noise" that is associated with larger uncertainty inherent in space-heating regression analysis.

ACKNOWLEDGMENTS

The authors would like to acknowledge the following individuals who contributed to this effort: members of the ASHRAE Standard 62.2 committee, Gary Nelson of The Energy Conservatory, Don Stevens of Panasonic USA, Luke Howard and Melinda Spencer of Washington State University Energy Program, David Beal of the Florida Solar Energy Center at the University of Central Florida, and Eric Werling of U.S. DOE Building America.

REFERENCES

ASHRAE. 2013a. ANSI/ASHRAE Standard 62.2-2013, *Ventilation and acceptable indoor air quality in low-rise residential buildings*. Atlanta: ASHRAE.

ASHRAE. 2013b. Chapter 16, Ventilation and infiltration. In *ASHRAE Handbook—Fundamentals* (SI). Atlanta: ASHRAE.

Francisco, P., and W. Rose. 2010. Temperature and humidity measurements in 71 homes participating in an IAQ improvement program. *Proceedings of Thermal Performance of Exterior Envelopes of Whole Buildings XI.*

IEA. 2008. IEA/ECBCS Annex 41—Whole building heat, air and moisture response (MOIST-ENG)—Proceedings of the closing seminar. Presented at IEA/ECBCS Annex 41, June 19, Copenhagen, Denmark.

Less, B., I.S. Walker, and Y. Tang. 2014. *Development of an outdoor temperature-based control algorithm for residential mechanical ventilation control*. No. LBNL-6936E. Berkeley, CA: Lawrence Berkeley National Laboratory. http://eetd.lbl.gov/publications/development-of-an-outdoor-temperature.

Offerman, F.J. 2009. *Ventilation and indoor air quality in new homes*. PIER collaborative report, California Energy Commission & California Environmental Protection Agency Air Resources Board.

Parker, D., P. Broman, J. Grant, L. Gu, M. Anello, R. Vieira and H. Henderson. 1999. EnergyGauge USA: A residential building energy design tool. *Proceedings of Building Simulation '99*.

Sherman, M.H., I.S. Walker. and J.M. Logue. 2012. Equivalence in ventilation and indoor air quality. *HVAC&R Research* 18(4):760–73.

Turner, W.J.N., and I.S. Walker. 2012. *Advanced controls and sustainable systems for residential ventilation*. No. LBNL-5968E. Berkeley, CA: Lawrence Berkeley National Laboratory. http://eetd.lbl.gov/sites/all/files/publications/lbnl-5968e.pdf.

Turner, W.J.N., and I.S. Walker. 2013. Using a ventilation controller to optimise residential passive ventilation for energy and indoor air quality. *Building and Environment* 70:20–30.

Walker, I.S. 1993. Attic ventilation, heat and moisture transfer. Doctoral thesis, University of Alberta, Edmonton, Alberta.

Walker, I.S., and M.H. Sherman. 2006. *Evaluation of existing technologies for meeting residential ventilation requirements*. No. LBNL-59998. Berkeley, CA: Lawrence Berkeley National Laboratory. https://buildings.lbl.gov/sites/all/files/lbnl-59998.pdf.

Walker, I.S., T.W. Forest, and D.J. Wilson. 2005. An attic-interior infiltration and interzone transport model of a house. *Building and Environment* 40(5):701–18. http://doi.org/10.1016/j.buildenv.2004.08.002.

Walker, I.S., M. Sherman, and D. Dickerhoff. 2012. *Development of a residential integrated ventilation controller*. No. LBNL-5554E. Berkeley, CA: Lawrence Berkeley National Laboratory. http://homes.lbl.gov/sites/all/files/lbnl-5554e.pdf.

Smart Ventilation Control of Indoor Humidity in High-Performance Homes in Humid U.S. Climates

Brennan Less

Iain Walker
Fellow ASHRAE

ABSTRACT

Elevated indoor humidity levels in homes represent a risk to occupant thermal comfort and health, as well as building durability. The improved thermal properties of high-performance homes lead to less cooling system runtime and associated moisture removal. High-performance homes often have elevated indoor humidity as a result. Current technologies for addressing this high humidity include dehumidifiers, energy recovery ventilators, and enhanced cooling strategies. A strategy that has not been assessed to date is the smart control of ventilation systems to better manage indoor moisture and reduce humidity loads. Such smart controls time shift ventilation to reduce the duration and number of hours of high indoor humidity, while providing annual pollutant exposure equivalent to a continuously operated fan sized to ASHRAE Standard 62.2-2013. The REGCAP simulation tool was used to assess 13 smart ventilation control strategies. Elevated indoor humidity was mostly problematic only in smaller homes with higher moisture gains. The best controls were able to achieve significant reductions in indoor humidity without excessive energy penalties (e.g., 16% of annual hours reduced below 60% rh in a small Miami home, using 277 kWh annually). They also maintained equivalent air quality to a continuous ASHRAE Standard 62.2-2013 fan. In the cases with highest indoor humidity, smart ventilation controls did not eliminate the need for supplemental dehumidification, with 2%0 to 25% of annual hours remaining >60% rh.

INTRODUCTION

High relative humidity (RH) levels have been documented in mechanically vented, high-performance homes in monitoring studies (e.g., Rudd and Henderson, Jr. 2007), as well as predicted in building simulations (Lstiburek et al. 2007; Martin 2014; Walker and Sherman 2007; Rudd et al. 2013; Fang, Winkler, and Christensen 2010). In standard (non-high-performance) homes, indoor humidity is kept at least partially in check by operation of the central cooling system and its associated moisture removal. The improved thermal properties of high-performance homes lead to less cooling load and less associated moisture removal. While ventilation is often cited as a contributor to higher indoor RH in high-performance homes in hot-humid climates (where outdoor air can at times be more humid than inside), it is a secondary factor along with home moisture capacitance. Primary factors include varying internal moisture generation rates, sensible gains (as they impact cooling system runtime), thermostat set points, and duct location (Henderson and Rudd 2010). The impact of ventilation on humidity levels in high-performance homes is unclear, because the effects depend on many other factors, though some have suggested that mechanical ventilation clearly contributes to high humidity during shoulder seasons.

Simulation efforts and field studies have described and assessed the costs and effectiveness of strategies to reduce indoor humidity levels in high-performance, humid climate homes (Kerrigan and Norton 2014; Rudd and Henderson, Jr. 2007; Withers and Sonne 2014). The main goal of these efforts was to reduce the number of hours above 60% rh to an unspecified, "acceptable" level. Strategies have included: dehumidifiers, including stand-alone and integrated with ventilation

Brennan Less is a residential buildings engineer in the Residential Building Systems (RBS) group and **Iain Walker** is a staff scientist and RBS group leader at Lawrence Berkeley National Laboratory (LBNL), Berkeley, CA.

and central HVAC systems; energy recovery ventilators; and enhanced cooling strategies (e.g., reduced airflow per ton, subcooling of space, and subcooling plus reheat).

Key findings from this past work include the following: high indoor humidity generally does not occur during cooling system operation, and most problems occur during winter and shoulder season transitions or during late evening and early morning hours; internal moisture generation has a strong impact on indoor humidity; sensible cooling load drives cooling system moisture removal, in particular duct location (house versus attic) and thermostat setting; mechanical ventilation has non-negligible but secondary impacts on indoor humidity levels; supplemental dehumidification is required in high-performance homes in humid climates, irrespective of mechanical ventilation rates; and homes using supplemental dehumidification strategies are able to reduce but not eliminate hours of indoor relative humidity above 60% (on average from around 30% of annual hours to 15% of hours >60%; dehumidifier capacity and set points interact such that all high humidity hours are not eliminated). Supplemental humidity control strategies have mixed effectiveness and first costs from $150 to $2000 (Rudd 2013a; Kerrigan and Norton 2014). Rudd (2013a) estimated that supplemental dehumidification in high-performance homes requires approximately 170 kWh per year with a 60% rh set point (five times that with a 50% rh set point), and Kerrigan and Norton (2014) estimated that dehumidifiers operate 10% of the year in high-performance homes with annual energy use of 976 kWh/year. Field research in conventional homes suggests that dehumidifiers use between 300 and 2000 kWh annually, averaging 1000 to 1200 kWh per year (Mattison and Korn 2012; Whitehead et al. 2013).

The current study used simulations to examine several approaches to smart ventilation controls aimed at improving humidity control. The ventilation strategies were evaluated in several climates, house sizes, and moisture generation rates to ensure that recommended strategies have robust performance across a wide range of homes. The performance of each control strategy was determined by comparing to baseline simulations with a constant fan.

SMART VENTILATION CONTROLS

Smart ventilation strategies have been previously applied to energy conservation and peak demand reduction using time-based controls and sensors for other ventilation fans (Sherman and Walker 2011) and based on indoor/outdoor temperatures (Less et al. 2014). The same principles will be applied in this study for humidity control.

The objectives of the smart ventilation control strategies were to reduce the number of hours of high indoor humidity, while avoiding excessive energy use and maintaining acceptable IAQ. *Excessive energy use* is defined relative to the energy use of other means of providing moisture control in high-performance homes, namely supplemental dehumidification (roughly 1000 kWh/year). *Acceptable IAQ* is defined as providing annual pollutant exposure equivalent to a continuously operated ventilation fan sized to ASHRAE Standard 62.2-2013. Recent work has developed the concept of ventilation equivalence (Sherman et al. 2012; Sherman et al. 2011), which can be used in smart controls to ensure that variable ventilation rates give the same pollutant exposures as constant ventilation rates. Equivalence is assessed relative to a generic pollutant generated indoors at a constant rate—indoor moisture is not the pollutant considered in the equivalence calculations presented in this work. Equivalence allows a controller to change the time that ventilation occurs to better control indoor humidity. Annual equivalence is determined using real-time calculations of turnover, relative exposure, and relative dose (see Equations 1, 2, and 3)—relative dose is roughly the relative exposure over a 24-hour integration time. Dose and exposure equal one when ventilation is equivalent to a continuous fan sized to ASHRAE Standard 62.2-2013. Values greater than one mean dose and exposure are higher (worse IAQ), and values below one mean dose and exposure are reduced relative to a constant ASHRAE Standard 62.2 fan (better IAQ). In order to avoid acute exposures, the smart controllers developed in this study used real time relative exposure and dose calculations to limit relative exposure to a maximum of 2.5 (i.e., pollutant exposure roughly two-and-a-half times the reference case). This is considered a conservative, protective value for 24-hour calculation periods, based on an assessment of the ratios of acute-to-chronic health-based exposure limits (Sherman et al. 2011).

$$\tau_i = \frac{1 - e^{-A_i \Delta t}}{A_i} + \tau_{i-1} e^{-A_i \Delta t} \tag{1}$$

where

τ_i = turnover at time-step, i

τ_{i-1} = turnover at the previous time-step, $i-1$

A_i = air exchange rate at time-step i, h^{-1}

Δt = time-step, 1/60 hours

$$\varepsilon_i = A_{eq} \tau_i \tag{2}$$

where

ε_i = relative exposure at time-step i

A_{eq} = target steady-state ventilation rate (Q_{tot} from ASHRAE 62.2-2013 Equation 4.1b), h^{-1}

$$d_i = A_{eq} \tau_i (1 - e^{-\Delta t / 24\text{ h}}) + d_{i-1} e^{-\Delta t / 24\text{ h}} \tag{3}$$

where

d_i = relative dose at time-step i

In addition to time varying ventilation, this study also investigated the influence of ventilation system design by including systems that tied operation of the ventilation system to operation of the cooling system (that includes incidental dehumidification). Six climates (Miami, Houston, Orlando, Memphis, Charleston, and Baltimore) were selected to represent a range of climates that include locations where outdoor humidity is high year-round (Miami) and ones where outdoor humidity is only high in summer (Baltimore). TMY3 data were used in this humidity analysis. Daily and hourly variations in outdoor humidity were found to be progressively smaller than the monthly seasonal variation by factors of about 3 and 4–9 respectively, indicating that there is more to be gained by ventilation timing strategies that are seasonal.

Control Algorithm Descriptions

A total of 13 controls were developed and tested for humidity, energy, and IAQ performance. In all cases, ventilation rates were controlled using a real-time calculation of relative exposure and relative dose. Targets or limits were set depending on the control condition (i.e., month of year, indoor/outdoor humidity difference, etc.), and the ventilation fan was operated to meet those targets. The key characteristics summarized in Table 1 are schedule, sensors, relative dose target, and cooling tie-in. Full details for all control strategies are documented in Less et al. (n.d.).

Schedule. Control strategies are considered scheduled if the controls function based solely on the month of the year, or on the hour of the day, with no sensor inputs. Scheduled controls included Controls 6, 8, and 12. These approaches generally relied on the consistent and predictable changes in outdoor humidity that occur over the course of the year. We used the monthly average differences in indoor and outdoor humidity from baseline simulations to determine the months targeted for over- or underventilation, and ventilation rates during these predetermined months controlled using either a high or low relative dose target.

Sensors. Sensor-based strategies used either an outdoor humidity sensor or two sensors—one indoors and one outdoors. All homes had a single HVAC system. Control 10 used a simple outdoor humidity cutoff. Control 13 used the annual outdoor median humidity ratio for each climate zone. The controller overvented when the real-time value was below the annual outdoor median, and it undervented when above the real-time annual median. Control 14 provided a similar but more complicated example, where the 25th and 75th percentile outdoor humidity ratios were calculated for each month of the year, and these were used in real time to either over- or underventilate. Two sensor strategies (Controls 2, 3, 4, 5, 7, and 9) varied in terms of their relative dose targets and cooling tie-in features. Two types of two-sensor approaches were tried, one with a simple on/off, indoor/outdoor humidity balance (called "fixed sensor" in the Control Name column of Table 1) and the other with a proportional control approach ("proportional sensor" in Table 1). In "proportional sensor" control cases, the amount of over- or underventilation was proportional to the real-time difference between indoor and outdoor humidity ratios. This was done so that when humidity differences were large, large changes in ventilation rate were allowed, and when humidity differences were small, ventilation rates were adjusted only minimally. This was an attempt to avoid excessive over- or underventilation when there was little anticipated value from a moisture-control perspective. "Fixed sensor" controls over- or underventilated to their maximum allowed levels whenever the humidity balance shifted between indoor and outdoor.

Table 1. Comparison of Key Elements of 13 Smart Ventilation Control Strategies (Bold Entries Were Identified as Best Performers and Were Tested in Other Home Configurations)

ID	Control Name	Schedule	Sensors	Rel Dose Target	Cooling Tie-In
1	Cooling system tie-in	N	0	Fixed	Y
6	Monthly seasonal	Y	0	Variable	N
8	Monthly seasonal + hourly	Y	0	Variable	N
12	Monthly seasonal + cooling system tie-in	Y	0	Variable	Y
10	Fixed outdoor HR cutoff	N	1	Fixed	N
13	Annual medians	N	1	Variable	N
14	Monthly quartiles	N	1	Variable	N
2	Fixed sensor	N	2	Fixed	N
3	Fixed sensor + cooling system tie-in	N	2	Fixed	Y
4	Proportional sensor	N	2	Fixed	N
5	Proportional sensor + cooling system tie-in	N	2	Fixed	Y
9	Fixed sensor + variable-dose target	N	2	Variable	N
7	Fixed sensor + cooling system tie-in + variable-dose target	N	2	Variable	Y

Relative Dose Targets. While all controllers calculated relative exposure and relative dose in real time, the targets used varied substantially and were either fixed or variable. When using fixed relative dose targets, a threshold value of 1 was always used. Variable-dose target controls set the dose target to either above or below 1, depending on the control conditions. This leads to sometimes extended, continuous periods of either over- or underventilation (still with the exposure limit of 2.5). For example, in Control 6, the relative dose target took one of two values, depending on whether the indoor/outdoor moisture difference was expected to be positive or negative for that month (this expectation was determined from baseline simulations). So, for months on end, the ventilation rate was either below or above the ASHRAE Standard 62.2-2013 target rate, such that annual relative dose was less than one. A similar high or low dose target approach was used with the one-sensor Control 13 based on the annual median outdoor humidity ratio. The dose target was 0.5 during dry periods and 1.5 during humid outdoor periods (roughly equal to doubling ventilation rate and reducing by 33%). A more dynamic approach was used in the two-sensor, variable-dose controls (Controls 2, 3, 4, 5, 7, and 9). Again, two relative dose targets were determined, and the controller targeted one or the other depending on the direction of the indoor/outdoor moisture difference.

Cooling System Tie-in. This approach was used as an isolated control strategy, as well as in combination with most of the other controls. The ventilation system was controlled to operate at full capacity during space cooling operation, sometimes subject to other constraints (e.g., only vent during cooling operation if indoor RH >55%). This approach was useful for two reasons. First, introducing high humidity outside air into this airstream before mixing with house air results in higher humidity ratio air passing over the cooling coil. This leads to more moisture removal for a given sensible load. Second, the sensible cooling load is increased through introduction of more outside air, and this increases cooling system runtime. The relative dose and exposure are reduced during this period of overventilation (at ~300% of ASHRAE Standard 62.2 rate), such that the system does not ventilate for substantial remaining portions of the day.

SIMULATIONS

The REGCAP simulation tool was used to provide estimates of indoor humidity, energy use, air exchange rates, and relative dose and exposure for each control strategy. The model has recently been used in development of other smart ventilation controls that provide annual equivalence to ASHRAE Standard 62.2, while varying ventilation rates with occupancy, time-of-day, outdoor temperature, and operation of other exhaust devices in the home (Turner and Walker 2012; Walker et al. 2012; Less et al. 2014). The REGCAP simulation combines detailed models for mass balance ventilation (including envelope, duct, and mechanical flows), heat transfer, HVAC equipment, and moisture. Two zones are simulated: the main house and the attic (the separate attic is important if the HVAC system is located in the attic). REGCAP was implemented using a one-minute time step to capture sub-hourly fan operation and the dynamics of cycling HVAC system performance. TMY3 weather data were linearly interpolated from one-hour to one-minute time steps for use in the REGCAP. The decision to turn the whole-house fan on or off using the smart ventilation controls was made once every ten minutes.

All simulations were of a high-performance, single-family home that meets the U.S. DOE zero net-energy ready home requirements (DOE 2013). The whole-house ventilation systems were central fan integrated supply (CFIS) ventilations systems sized to 300% of the airflow requirement of ASHRAE Standard 62.2-2013, including infiltration credit (required airflow rates differed by house sizes and occupancy rates). The CFIS operated 20 minutes of every hour irrespective of heating or cooling demand. Other local exhausts were scheduled on a semi-random basis. House sizes of 100, 200, and 300 m^2 (referred to as small, medium, and large) were studied together with moisture generation rates of 3, 6.5, and 11.8 kg/day (referred to as low, medium, and high), respectively. These moisture generation rates were associated with two, four, and six occupants, respectively. Occupancy was assumed to be continuous. More details of house characteristics can be found in Less et al. (n.d.). Fixed cooling and heating set points of 24.4°C and 21.7°C were used, which match assumptions of the Building America reference home (Engebrecht and Hendron 2010). Simulations were first performed for baseline cases. These baseline simulations provide the comparison cases for all of the smart control cases. All 13 control strategies were tested for small homes with high moisture gains, medium sized homes with medium gains, and large homes with low gains.

RESULTS

Baseline Simulations

For all baseline cases with non-smart ventilation, we evaluated the distribution of indoor relative humidity, the fractions of the year exceeding 60% and 70% rh, and the longest continuous periods exceeding these thresholds. The results are summarized for each climate zone in Table 2. The variation in high humidity by house size and moisture gains is pictured in Figure 1. A summary of energy, IAQ, and humidity performance is provided for all baseline cases in Less et al. (n.d.). Key conclusions from this baseline simulation analysis include the following:

- Annual hours of elevated indoor humidity above 60% rh were highly variable and were strong functions of the house size and moisture generation rates (see Figure 1), with smaller homes and higher moisture generation rates leading to higher indoor humidity. Fully 69% of all baseline simulations had less than 5% of annual hours above 60% rh, and 95% of cases had less than 5% of

Table 2. Annual Humidity Summary for Cases with Fan Sized to 100% of ASHRAE Standard 62.2-2013, Averaged across House Size and Moisture Generation Rate

Climate Zone	Indoor Relative Humidity (%)						Annual Fraction		Maximum Duration (days)	
	Min	25th	Median	Mean	75th	Max	>60% rh	>70% rh	>60% rh	>70% rh
Miami	35	48	51	51	55	87	10%	1%	1.2	0.5
Orlando	31	45	49	49	53	77	8%	1%	2.5	0.6
Houston	26	44	49	49	54	77	9%	1%	2.0	0.3
Charleston	24	44	49	48	54	71	8%	0%	1.7	0.2
Memphis	20	36	44	42	49	63	2%	0%	0.2	0.0
Baltimore	17	28	40	38	46	62	1%	0%	0.2	0.0

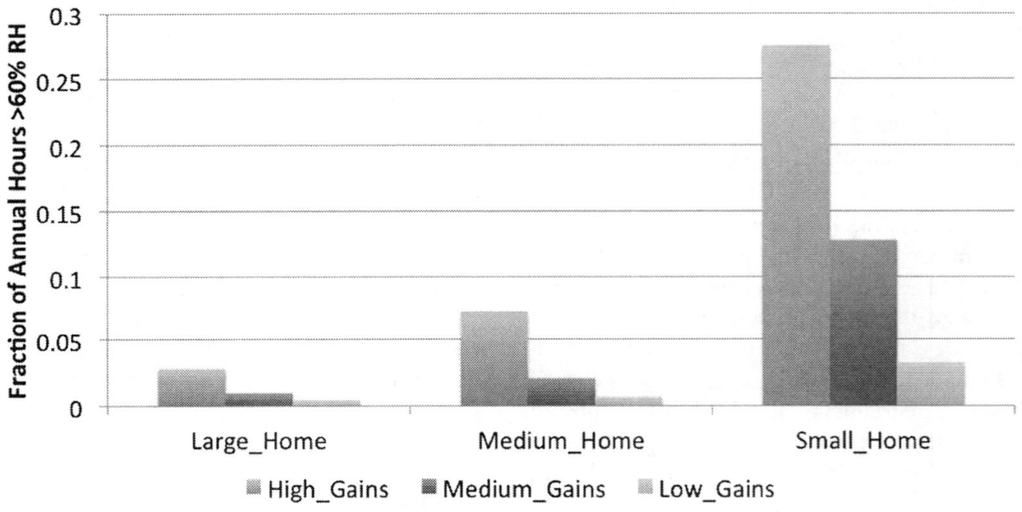

Figure 1 *Summary of annual hours >60% rh as they vary with home size and indoor moisture gains, averaged across climate zones.*

annual hours exceeding 70% rh. We consider indoor humidity to be of little concern in these cases, and subsequent analysis focuses on the high humidity cases.

- Maximum duration periods above 60% rh varied between one and eight days, by climate zone.
- Some locations had high indoor humidity all year (e.g., Miami and Orlando), whereas others experienced it only during summer months (e.g., Memphis and Baltimore).
- Shoulder seasons had the highest humidity, due to low sensible cooling loads and similar indoor and outdoor absolute humidity.
- Few high humidity hours occurred during either heating or cooling system operation (<10%).

Smart Ventilation Control

Figure 2 shows reductions in annual hours >60% rh averaged across the six climate zones for each of the 13 control strategies. To focus on cases with the highest humidity, these values include only the small homes with high moisture gains. Key conclusions from analysis of these smart control simulations include the following:

- All smart controls increased HVAC energy use, but they also decreased hours of high humidity and shifted overall indoor humidity distributions downward (see Figure 3). Energy use was strongly dependent on the heating demand of the climate zone and the ventilation strategy during heating periods (overventilation during winter led to large energy use increases in some cases).

In general, smart controls used the least energy and were the most effective in the hottest climates with the highest indoor humidity.

- When averaged across climate zones, the best performing strategy (Control 7, two sensor, variable-dose target with cooling tie-in) was able to shift roughly 10% of annual hours from above to below the humidity thresholds, with energy consumption that varied significantly by climate zone (see black triangles in Figure 4). This energy use was largely driven by the heating demand in a given location. In the most humid location (Miami), Control 7 reduced 16% of annual hours from above to below 60% while increasing energy use by only 277 kWh in the small home with high moisture gains.
- In cases of most concern (i.e., small homes with high moisture generation rates), there were still substantial numbers of hours of elevated indoor RH. As pictured in Figure 4, between 20% and 25% of annual hours remained >60% in the most humid locations. Mechanical dehumidification may still be required in these homes. Future work should investigate the interactions

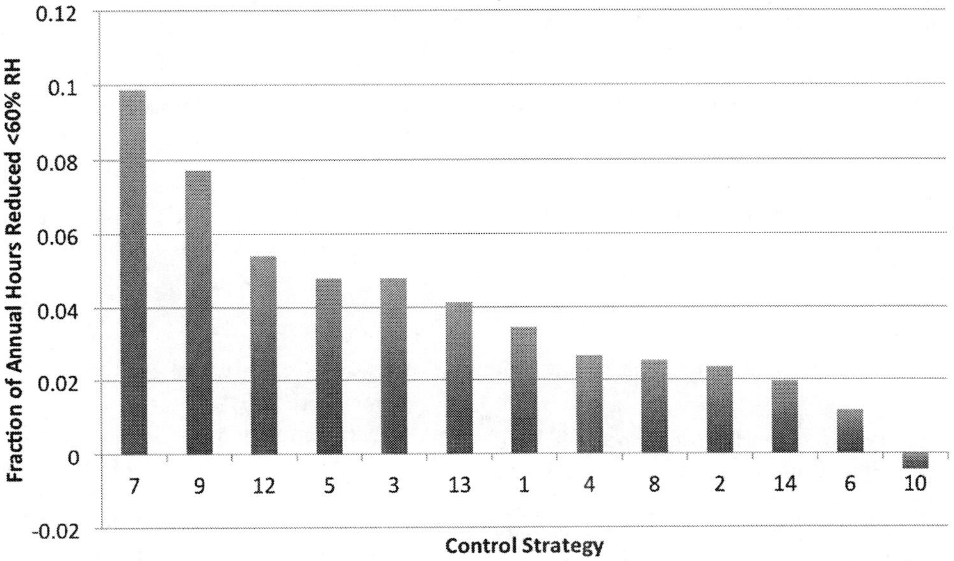

Figure 2 *Reductions in high humidity hours for the 13 strategies in small homes with high moisture gains, averaged across all climate zones. Reductions are expressed as fractions of the year, such that a value of 0.10 means that 0.10 × 8760 = 876 hours were reduced from above to below 60% rh.*

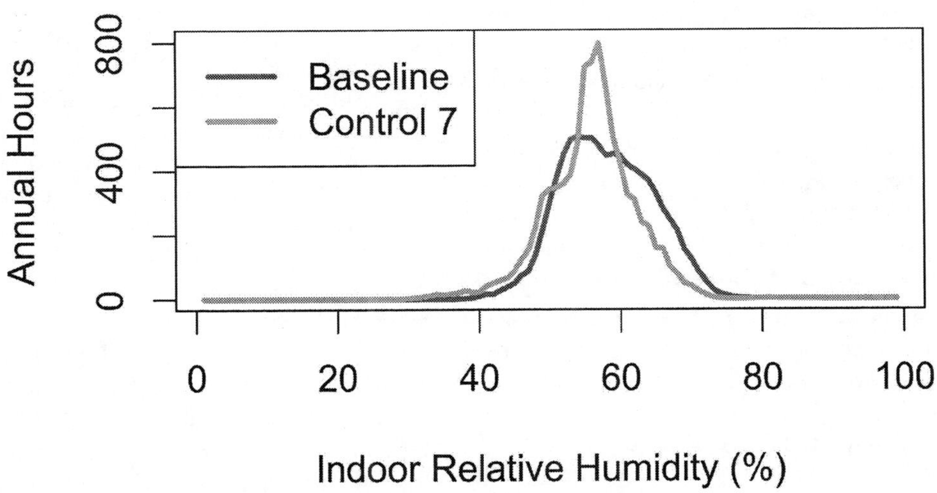

Figure 3 *Histograms showing annual distributions of indoor RH in baseline and Control 7 cases, for a small home with high moisture gains in Miami, FL.*

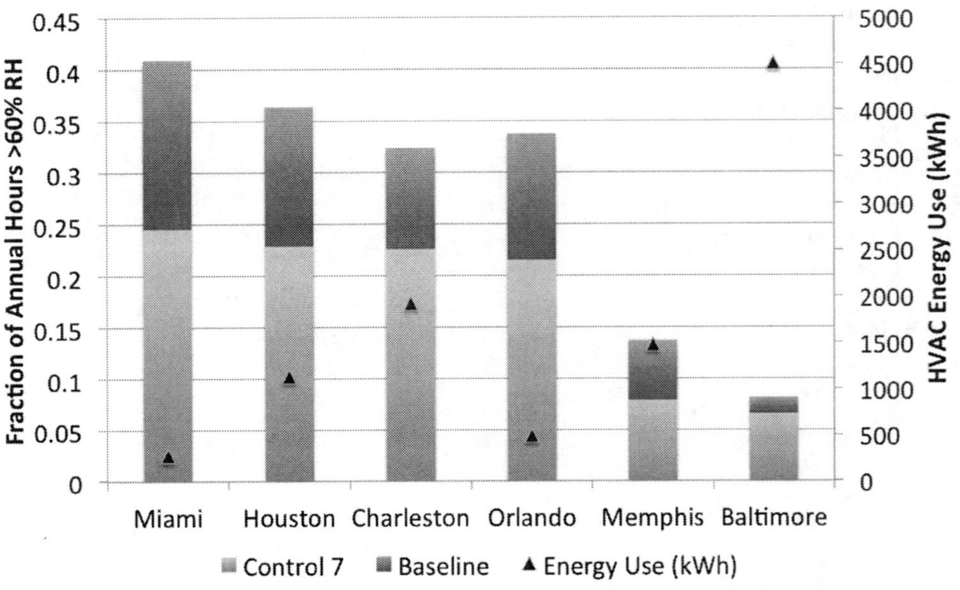

Figure 4 *Fraction of the year >60% rh in Baseline (red) versus Control 7 (blue) cases and changes in annual HVAC energy consumption from baseline when using Control 7 (black triangles, 2nd* y*-axis). Includes only small homes with high moisture gains.*

between smart controls and supplemental dehumidification systems.

- Sensor-based strategies outperformed schedule-based approaches, although Control 12 (which combined monthly relative dose targets with a cooling tie-in feature) was the third best in reducing high humidity hours in small homes with high moisture gains (see Figure 2).
- Two sensors were generally better than one, as they were able to respond to real-time changes in indoor and outdoor humidity. The one-sensor Control 13 based on annual median outdoor humidity ratios was reasonably effective, though less so than the no-sensor Control 12. Given the desirability of using web-based weather data for smart ventilation control, future work should investigate ways to improve these outdoor-only sensor approaches.
- The cooling tie-in feature had varied results, but it generally led to better performance with a small energy penalty (roughly 450 kWh in small homes and 580 kWh in medium homes). We recommend this approach in combination with schedule- or sensor-based controls.
- Controls using variable-dose targets were more effective, but fixed-dose approaches worked well in locations with substantial heating demand.

SUMMARY AND CONCLUSIONS

- High indoor humidity was not an issue in many combinations of location, house size, and moisture gains. The most problematic cases were small homes with high moisture gains, where between 5% and 40% of annual hours were >60% rh.
- Smart ventilation controls were effective at reducing indoor humidity levels, and they maintained air quality equivalent to or better than a continuous fan sized to ASHRAE Standard 62.2-2013. The best performing strategy was Control 7 that used both indoor and outdoor sensors and a cooling system tie-in. It was able to reduce 16% of annual hours <60% rh in a small Miami home using under 300 kWh.
- Estimated energy use for smart controls was in the same range as that used by mechanical supplemental dehumidification strategies.
- In the most challenging cases, indoor humidity remained >60% for 20% to 25% of annual hours despite use of smart controls, and use of supplemental dehumidification in humid climates may be necessary to achieve acceptable levels in these high-performance homes. Our next steps are to evaluate how smart ventilation controls interact with and compare to a supplemental mechanical dehumidification strategy.

REFERENCES

ASHRAE. 2013. ANSI/ASHRAE Standard 62.2-2013, *Ventilation and acceptable indoor air quality in low-rise residential buildings*. Atlanta: ASHRAE.

BSC. 2005. Analysis of indoor environmental data. Building America Report 0601. CDH Energy Group, and Building Science Corporation. http://buildingscience.com/documents/bareports/ba-0601-analysis-of-indoor-environmental-data/view.

Engebrecht, Cheryn, and Bob Hendron. 2010. Building America Simulation Protocols. DOE/GO-102010-3141. U.S. Department of Energy. www.nrel.gov/docs/fy11osti/49246.pdf.

Fang, Xia, Jon Winkler, and Dane Christensen. 2010. Advanced dehumidification analysis on Building America Homes using EnergyPlus. International High Performance Buildings Conference. Paper 7. http://docs.lib.purdue.edu/ihpbc/7

Fisk, W.J., Q. Lei-Gomez, and M.J. Mendell. 2007. Meta-analyses of the associations of respiratory health effects with dampness and mold in homes. *Indoor Air* 17(4): 284–96. doi:10.1111/j.1600-0668.2007.00475.x.

Henderson, Hugh I., and Armin Rudd. 2010. Energy efficiency and cost assessment of humidity control options for residential buildings. In Thermal Performance of the Exterior Envelopes of Whole Buildings XI Conference. Clearwater Beach, FL: ASHRAE.

Kerrigan, P., and P. Norton. 2014. Evaluation of the performance of houses with and without supplemental dehumidification in a hot-humid climate. Golden, CO: National Renewable Energy Laboratory. http://apps1.eere.energy.gov/buildings/publications/pdfs/building_america/houses_supplemental_dehumidification.pdf.

Less, Brennan, Iain S. Walker, and Yihuan Tang. 2014. Development of an outdoor temperature-based control algorithm for residential mechanical ventilation control. LBNL-6936E. Berkeley, CA: Lawrence Berkeley National Laboratory. http://eetd.lbl.gov/publications/development-of-an-outdoor-temperature.

Less, Brennan, Sara Ticci, and Iain Walker. In Process. Development of smart ventilation control algorithms for humidity control in high-performance homes in humid U.S. climates. Berkeley, CA: Lawrence Berkeley National Lab.

Lstiburek, Joseph, Betsy Pettit, Armin Rudd, Max Sherman, and Iain Walker. 2007. Whole house ventilation system options—Phase 1 simulation study. ARTI Report 30090-01. Arlington, VA: Air-Conditioning and Refrigeration Technology Institute. http://www.energy.ca.gov/title24/2008standards/prerulemaking/documents/2007-06-13-15_workshop/comments/AIR-CONDITIONING_REFRIGERATION_TECHNOLOGY_INST_2007-03-05.PDF.

Martin, Eric. 2014. Impact of residential mechanical ventilation on energy cost and humidity control. NREL-60675. Golden, CO: National Renewable Energy Laboratory. http://www.fsec.ucf.edu/en/publications/pdf/NREL-60675.pdf.

Mattison, Lauren, and Dave Korn. 2012. Dehumidifiers: A major consumer of residential electricity. In Summer Study for Energy Efficiency in Buildings, 9–204 to 9–215. Pacific Grove, CA: American Council for an Energy-Efficient Economy. http://www.aceee.org/files/proceedings/2012/data/papers/0193-000291.pdf.

Moyer, Neil, Dave Chasar, Dave Hoak, and Subrato Chandra. 2004. Assessing six residential ventilation techniques in hot and humid climates. In Pacific Grove, CA: American Council for an Energy-Efficient Economy. http://www.fsec.ucf.edu/en/publications/pdf/FSEC-PF-378-04.pdf.

Rudd, Armin. 2013a. Expert meeting: Recommended approaches to humidity control in high performance homes. NREL-57483. Golden, CO: National Renewable Energy Laboratory. http://www.nrel.gov/docs/fy13osti/57483.pdf.

Rudd, Armin. 2013b. Supplemental dehumidification in warm-humid climates. Building America Report 1310. Somerville, MA: Building Science Corporation. http://buildingscience.com/documents/bareports/ba-1310-supplemental-dehumidification-warm-humid-climates/view.

Rudd, Armin, Hugh I. Henderson, Daniel Bergey, and Don B. Shirey. 2013. ASHRAE 1449- RP, *Energy efficiency and cost assessment of humidity control options for residential buildings*. Research Report RP-1449. Atlanta, GA: ASHRAE. http://www.techstreet.com/products/1856921.

Rudd, Armin, and Hugh I. Henderson, Jr. 2007. Monitored indoor moisture and temperature conditions in humid-climate US residences. *ASHRAE Transactions* 113(1):435–49.

Rudd, Armin, J.W. Listiburek, and Kohta Ueno. 2005. Residential dehumidification systems research for hot-humid climates. September 1, 2001–December 30th, 2003. NREL/SR-550-36643. Golden, CO: National Renewable Energy Laboratory. http://www.nrel.gov/docs/fy05osti/36643.pdf.

Sherman, Max H., D.K. Mortensen, and Iain S. Walker. 2011. Derivation of equivalent continuous dilution for cyclic, unsteady driving forces. *International Journal of Heat and Mass Transfer* 54 (11–12):2696–702.

Sherman, Max H., and Iain S. Walker. 2011. Meeting residential ventilation standards through dynamic control of ventilation systems. *Energy and Buildings* 43(8):1904–12. doi:10.1016/j.enbuild.2011.03.037.

Sherman, Max H., Iain S. Walker, and Jennifer M. Logue. 2012. Equivalence in ventilation and indoor air quality. *HVAC&R Research* 18(4):760–73. doi:10.1080/10789669.2012.667038.

Sherman, M.H., J.M. Logue, and B.C. Singer. 2011. Infiltration effects on residential pollutant concentrations for continuous and intermittent mechanical ventilation approaches. *HVAC&R Research* 17(2):159–73. doi:10.1080/10789669.2011.543258.

Turner, William J.N., and Iain S. Walker. 2012. Advanced controls and sustainable systems for residential ventilation. LBNL-5968E. Berkeley, CA: Lawrence Berkeley National Laboratory. http://eetd.lbl.gov/sites/all/files/publications/lbnl-5968e.pdf.

DOE. 2013. DOE Challenge Home National Program Requirements (Rev. 03). U.S. Department of Energy. http://www1.eere.energy.gov/buildings/residential/pdfs/doe_challenge_home_requirementsv3.pdf.

Walker, Iain S., and Max Sherman. 2007. Humidity implications for meeting residential ventilation requirements. In Thermal Performance of the Exterior Envelopes of Whole Buildings. Clearwater Beach, FL: ASHRAE. https://www.researchgate.net/publication/242726385_Humidity_Implications_for_Meeting_Residential_Ventilation_Requirements.

Walker, Iain S., Max Sherman, and Darryl Dickerhoff. 2012. Development of a residential integrated ventilation controller. LBNL-5554E. Berkeley, CA: Lawrence Berkeley National Laboratory. http://homes.lbl.gov/sites/all/files/lbnl-5554e.pdf.

Whitehead, Camilla Dunham, Henry Willem, Bereket Beraki, Tom Burke, Moya Melody, Mythri Nagaraju, Chun Chun Ni, Stacy Pratt, and Hung-Chia (Dominique) Yang. 2013. Field-metered data from portable unit dehumidifiers in the U.S. residential sector: Initial results of a pilot study. LBNL-6093E. Berkeley, CA: Lawrence Berkeley National Laboratory. https://escholarship.org/uc/item/13q224sc#page-7.

Withers, Charles R. Jr., and Jeff Sonne. 2014. Assessment of energy efficient methods of indoor humidity control for Florida Building Commission research. FSEC-CR-1976-14. Cocoa, FL: Florida Solar Energy Center. http://www.floridabuilding.org/fbc/commission/FBC_0614B/Energy/Energy_Efficient_RH_control_Draft_Final_06_15_14.pdf.

Measured Commercial and Institutional Building Pressures

David Bohac, PE
Member ASHRAE

Joshua Quinnell, PhD
Member ASHRAE

Martha Hewett

James Fitzgerald

Joshua Novacheck

Andrew Lutz

ABSTRACT

Continuous building pressure monitoring was conducted for two weeks to a year at 16 commercial and institutional (C&I) buildings in Minnesota (Climate Zone 6). Building automation system trend data and occupancy schedules were used to identify HVAC operation modes. The measurements were used to evaluate the impact of HVAC system operation on building pressures. The building pressures and pressure changes from HVAC operation were compared to the typical design level of 0.05 in. wc (12.5 Pa). Only one building consistently had an average ground level pressure above the design level, while the median across buildings ranged from 0.016 in. wc (4 Pa) for warm weather and –0.04 in. wc (–10 Pa) for the coldest conditions. Only 10% of the data periods had an average ground level pressure during HVAC operation that was greater than the design level and just 24% had an average greater than half the design level. Excluding two buildings, HVAC operation had a positive pressurization effect. However, for all of the data periods the median pressure increase was 0.02 in. wc (5.0 Pa), the pressurization effect was above design for only 15% of the periods, and above half of design for just 40% of the periods. This suggests that most buildings would have significant ground level air infiltration most of the year and greater HVAC pressurization would reduce air infiltration in those buildings. CONTAMW models were used for six of the buildings to determine the effect of HVAC operation on building air infiltration rates. For four of the buildings, HVAC operation decreased modeled infiltration rates by 433 to 2210 cfm (12,300 to 62,600 L/m) and for two it increased infiltration by 261 and 484 cfm (7390 to 13,700 L/m).

INTRODUCTION

Air movement through the building envelope requires a leak in the envelope and a pressure driving force across the leak. The driving forces are caused by wind, thermal gradients (stack effect), and flow imbalances of air brought into and exhausted out of the building by the HVAC system. The system air flows are either caused by active movement from supply, return, and/or exhaust fans or they can be from passive vents or relief openings. Some systems rely on passive relief or a specified flow imbalance (based on design assumptions) to provide pressure control. Other systems actively monitor the building pressure and adjust flow rates to achieve the desired set-point. Proper building pressure control has the potential to reduce energy costs associated with air infiltration at relatively low cost with the added benefits of improved occupant comfort and reduced liability for cold air infiltration freezing pipes in building cavities.

Several ASHRAE Handbooks provide designers with the background physics and design considerations on building pressurization, including variations because of wind and thermal buoyancy effects as well as the impact of mechanical systems and potential strategies for building pressure control (ASHRAE 2013, 2015). Most of the empirical literature is focused on single-family residential buildings and, to a lesser extent, high-rise multifamily buildings. While the basic concepts are the same there is an absence of pressure data in commercial and institutional (C&I) building like those included in this study, which, with only two exceptions, were large well-connected spaces under 60 ft (18 m) tall.

David Bohac is the director of research, **Joshua Quinnell** is a senior research engineer, **Martha Hewett** was previously the director of research, **James Fitzgerald** is a senior building analyst, **Joshua Novacheck** was a research analyst, and **Andrew Lutz** was a research analyst at the Center for Energy and Environment (CEE), Minneapolis, MN.

ASHRAE recommends slight positive building pressurization of between 0.005 and 0.08 in. wc to reduce infiltration (ASHRAE 2015). More specifically, a 2002 Trane Engineers Newsletter (Stanke and Bradley 2002) provides useful information on building pressure control benefits, strategies, and set-point options. They suggest summer pressurization of 0.05 in. wc or up to 0.10 in. wc for hot and humid climates. It is noted that in cold climates a d pressurization strategy can be used in the winter to reduce exfiltration and envelope moisture concerns. They suggest wintertime pressurization of 0.0 to –0.02 in. wc or "perhaps –0.10 in. wc on a windy day for a building of only four stories" (Stanke and Bradley 2002) may be necessary to control the winter stack effect. However, our experience is that Minnesota building HVAC systems are typically designed to achieve a building pressurization level of ΔP_{des} = 0.05 in. wc (12.5 Pa) and the primary goal is to reduce cold drafts, space heating energy use, and concerns for frozen pipes during the heating season. This pressurization strategy is also used for the cooling season. Building automation systems (BAS) are sometimes configured to monitor the building indoor/outdoor pressure difference and actively adjust flows to achieve a desired set point pressure difference. Despite design advice acknowledging variations of building pressure, all but two buildings in this study do not have active building pressure control. Instead the BAS or individual air handler controls were designed to maintain a constant difference between the outdoor air and exhaust/relief flow rates with a goal to produce a similar increase in building pressure caused by HVAC system operation. While this strategy is easy to implement, it does not assure a positive building pressure for all weather conditions. For example, low outdoor temperatures or windier conditions may require greater HVAC pressurization to produce a positive building pressure throughout the building. The continuous building pressure measurements presented in this paper provide data with which to test this simple building pressurization strategy under variable weather conditions.

METHODOLOGY

Building Pressure Monitoring

Pressure monitors were installed to measure the difference between the building indoor and outdoor pressure. The pressure differences were measured at least 5 Hz and one minute averages were recorded. The pressure monitors typically consisted of a digital manometer[1] connected to a microcontroller board for control and storage. The manometer had two independent, bidirectional pressure channels, each with a 0.0004 in. wc (0.1 Pa) resolution and accuracy of ±1% of the reading. A few of the sites were monitored using a multi-channel digital manometer[2]. The multichannel digital manometers had two to eight independent, bidirectional pressure channels with a resolution of 0.0004 in. wc (0.1 Pa) and sensor accuracy of the greater of 0.0008 in. wc (0.2 Pa) or ±1% of reading. These manometers were either used with the above microcontroller board or a PC laptop running manufacturer provided software for recording measurements. Both gages have an auto-zero function that was activated at 2 to 5 min intervals.

The tubes for the outdoor pressure reference exited the buildings at either the roof or ground level depending on accessibility. An outdoor static pressure sensor mounted on a vertical pipe approximately 2 ft (0.6 m) above the roof surface was used for roof locations. The sensor was typically located near the center of the roof away from rooftop units, exhaust fan terminations, and other obstructions. During warmer months the outdoor pressure termination was sometimes located at ground level. The indoor pressure reference tube was run through ceiling and other building cavities as necessary to place the termination in a common area. Almost all of the buildings had open hallways and/or stairwells between each section of the building so a pressure measurement at a single location was representative of the indoor pressure for that level of the building.

BAS trend data and occupancy schedules were used to identify HVAC operation modes. Because BASs often do not include the status and control of all HVAC equipment, it is possible some equipment operated during periods specified as *HVAC OFF*. This is more likely for smaller exhaust fans controlled either manually or with time clocks. In addition, during preparation for air leakage tests, equipment that the building manager had specified to be off was found operating. Equipment operation was noted for those cases, but this suggests that it was likely that there were situations in other buildings where equipment was operating during HVAC OFF periods without the knowledge of the building operators.

The 1 min average pressure data were converted to 15 min averages and merged with HVAC status and local weather station outdoor temperature, wind speed, and wind direction data. The data were analyzed to evaluate typical pressures acting on envelope leaks at the ground and roof levels of the building, as well as the effect of HVAC operation on building pressures and their variation with outdoor temperature. The thermal buoyancy caused by the difference between the indoor and outdoor air dry-bulb temperatures was used to estimate the pressure difference for other vertical locations. For example, when the pressure difference was measured at the roof, the pressure difference at ground level was estimated from the vertical height difference and the indoor and outdoor air dry-bulb temperature. Some data were removed from the analysis because of monitoring equipment malfunctions, unavailability of BAS trend data, transient HVAC operation (i.e., start-up or shut-down periods), and envelope air sealing that changed the building pressure.

1. Energy Conservatory model DG-700 digital micromanometer
2. Energy Conservatory model Automated Performance Testing

Air Infiltration and Energy Modeling

CONTAMW airflow models (Walton and Dols 2013) were developed to estimate building space heating and cooling energy consumption caused by air infiltration. CONTAMW is a multi-zone indoor air quality and ventilation computer model developed by the National Institute of Standards and Technology (NIST) to calculate time varying infiltration, exfiltration, and area-to-area building airflow rates. The model includes driving forces caused by HVAC flow imbalances, wind pressures on the building exterior, and thermal buoyancy effects induced by the indoor and outdoor temperature differences. The driving forces are applied to a network of user-defined flow path elements to time varying air flows. The flow path elements (e.g. air leaks and other flow paths) are distributed vertically and on different building faces to account for thermal buoyancy and wind direction effects on building pressures.

Building envelope air leakage measurements were used to determine the sum of the leakage for all building exterior flow paths. Building pressure measurements and physical dimensions were used to distribute the leakage by building height and orientation. Envelope leaks were first distributed according to the location of exterior doors and mechanical systems. The remaining leaks were characterized as 1) roof/wall interface leaks located at the building height and 2) wall/window leaks distributed by wall area. The proportion of these two leakage paths—the vertical leakage distribution—was adjusted so that the modeled neutral level matched the level measured for the building when the HVAC system was off. The building indoor to outdoor pressure measurements were filtered by HVAC operating mode and binned by outdoor temperature to establish equations for the HVAC flow imbalance. Weather data was used to validate the model leakage paths by comparing modeled and measured building pressures.

Hourly simulations were performed using annual, hourly typical meteorological year (TMY) outdoor air and wind data from the National Oceanic and Atmospheric Administration (NOAA) weather station closest to the building. The flow path results of the CONTAMW simulations were summed to generate hourly air infiltration rates. The simulations were performed for the measured HVAC flow imbalances and repeated using increased flow imbalances to evaluate the impact of imbalances on building air infiltration and space conditioning loads. A fixed heating and cooling balance point method was used to compute the hourly space heating and cooling loads. For each hour that the outdoor temperature was below the heating balance point, the heating load caused by infiltration was computed from the sum of the building air infiltration rate and the difference between the balance point and outdoor temperatures. In some instances economizer operation caused space heating requirements when the outdoor temperature was below the balance point. In these cases it was assumed that a fraction of the infiltration load resulted in increased space heating energy use. The fraction varied from zero at the balance point to one at the outdoor temperature when the building no longer economized. The annual space heating energy use was computed from a sum of the infiltration space heating loads divided by the heating system efficiency. A similar process was used for space cooling except no multiplier was applied for economizer operation because the cooling balance point should be set at the outdoor temperature when mechanical cooling is required and the economizer operation is not used.

RESULTS AND DISCUSSION

Building Pressure Monitoring

General characteristics of the buildings for which building pressure data were monitored are given in Table 1. The data duration for the 16 buildings varied from 8.5 to 129 days with a median duration of 46 days. The buildings generally had pressure measurements for outdoor temperatures from 0°F to 70°F (–17.8°C to 21.1°C), but few had measurements for more extreme cold or warm temperatures. About half of all data fell between outdoor temperatures 20°F to 45°F (–6.7°C to 7.2°C). All except two buildings had at least 12 h of data in the three temperature ranges between 0°F and 70°F (–17.8°C to 21.1°C). Five of the buildings had 12+ h of data at temperatures below 0°F (–17.8°C) and only seven had 12+ h of data at temperatures above 70°F (21.1°C). Most of the buildings had open floor plans that allowed a simple thermal stack effect calculation to be used to compute the building pressures at the top and bottom of the building. The five buildings with potential restriction to vertical air movement were: High Rise Office 3, University Class SH, and the three medical clinics.

The average pressures at ground and roof levels were computed for five temperature ranges and two HVAC operating modes (OFF/ON). As expected, the ground level HVAC OFF pressures were generally slightly negative with a trend of lower (i.e., more negative) values for colder temperatures and taller buildings. There were two low-rise buildings with lower than expected pressure for warm outdoor conditions, indicating operation of unidentified exhaust fans in those buildings. Figure 1 displays the ground-level pressure averages for HVAC ON periods for each outdoor air temperature bin. Only one building (University Library) consistently had an average ground level pressure above the typical ΔP_{des} and the median across buildings ranged from 0.016 in. wc (4 Pa) for warm weather to –0.04 in. wc (–10 Pa) for the coldest conditions. Only 10% of the data periods had an average HVAC pressure greater than ΔP_{des} and just 24% had an average greater than $\Delta P_{des}/2$. Because of the stack effect, the roof level pressures were slightly higher. Two of the 16 buildings achieved the typical ΔP_{des}s for most of the year and the median of the buildings was 0.030 to 0.038 in. wc (7.5 to 9.5 Pa) for most of the year, or about 70% of ΔP_{des}. Only 24% of the data periods had an average pressure greater than ΔP_{des} and 60% had an average greater than $\Delta P_{des}/2$. Thus, throughout the year most of the buildings had significant ground-level air infiltration, which could be reduced by greater HVAC pressurization.

Table 1. Characteristics of Monitored Buildings

Site ID	Type	No. of Stories	Construction Year	Floor Area, ft^2	Outside Location	Building Height, ft	Days Monitored
University Class 5	Inst	3	2008	66,783	Ground	52	8.5
Sports Arena	Inst	2	2000	142,951	Ground	25	47.7
University Union	Inst	2	1972	210,388	Ground	32.5	51.6
University Library	Inst	3	1967	246,365	Roof	40	96.9
Elementary School 1	Inst	1	1951	59,558	Roof	21.5	88.8
Middle School	Inst	3	1936	138,887	Roof	36	40.2
High School	Inst	1	1976	289,909	Roof	28	12.3
Elementary School 2	Inst	1	1965	60,000	Roof	18.25	48.6
Library/Office	Public	1	2007	55,407	Ground	21	129.0
Community College	Public	2	1996	108,102	Roof	26	12.6
University Class SH	Inst	3	1948	177,951	Roof	36	41.1
High Rise Office 3	Office	23	1985	484,290	4th Floor	312	53.9
Medical Clinic 1	Medical	2	1960	47,805	Roof	26	32.2
Medical Clinic 3	Medical	4	1994	56,803	Roof	52	41.5
Medical Clinic 2	Medical	7	1968	268,408	Roof	78	44.0
Small Office 1	Office	1	1998	26,732	Ground	16.25	88.5

The difference between the HVAC ON and HVAC OFF average pressure for the same outdoor temperature period for each building is shown in Figure 2. HVAC operation pressurized all of the buildings except Medical Clinic 3 and Small Office 1. However, for all of the data periods the median pressure increase was 0.02 in. wc (5.0 Pa), and the building pressurized above ΔP_{des} for only 15% of the periods and above $\Delta P_{des}/2$ for only 40% of the periods. Thus, it is rare for HVAC systems to lower the building pressure and also rare for HVAC systems to achieve ΔP_{des} (1 in 16 buildings). For the entire population of buildings there was no strong trend of HVAC pressurization with outdoor temperature.

For each building, the pressure measurements were filtered by HVAC operating mode and sorted into 5°F (2.8°C) outdoor air temperature bins to explore the variation in building pressure with outdoor temperature. The red area in Figure 3(a) shows the interquartile range (IQR) of the roof level pressures and Figure 3(b) shows the ground level pressures for Elementary School 1. The red area designates HVAC ON periods, while the blue area designates the HVAC OFF periods. The variation in the pressures within a temperature bin is indicated by the vertical height of the filled area. This building has three air handling units, six rooftop units, and six fan coil units for the original building and the four additions. Some units have outdoor air and some use an economizer operation. None of the units or exhaust fans operate at night and there is no active building pressure control. The system relies on a predetermined balance of outdoor and exhaust air to maintain the building design pressure.

As expected, with the HVAC system off (blue area) the IQR for the roof pressure is inversely proportional to the outdoor air temperature. The opposite trend occurs for the ground level pressure. HVAC operation increased the building pressure from about 0.004 in. wc (1 Pa) for warm weather and up to 0.02 in. wc (5 Pa) for cold weather. For both the roof and ground level, the 25th percentile pressures (lowest portion of red area in Figure 3) were neutral or positive for the entire monitoring period. The pressure variation during HVAC ON is greatest for outdoor temperatures from about 25°F to 65°F (–3.9°C to 18.3°C), possibly because of variations in economizer or other HVAC operation. It is also possible that the outdoor air and exhaust air flow imbalance varies with the degree of economizer cooling. This would occur if the outdoor air and exhaust dampers do not consistently track each other over the entire range of damper settings. For constant outdoor air temperature, economizer cooling may vary if it is dependent on internal occupancy heating loads and solar gains.

Figure 3 shows the interquartile range (IQR) of measured building pressures as a function of outdoor air temperature for all wind conditions at Elementary School 1. We note that when the data are filtered for wind speeds less than 5 mph (2.2 m/s), the IQR changes by less than 0.001 in. wc (0.25 Pa), except for temperatures between 65°F and 70°F (18.3°C and 21.1°C), where there is a large (1.75 Pa) IQR for the case of wind speeds less than 5 mph (2.2 m/s). Thus while wind effects exaggerate minimum and maximum pressures, the variations shown here are not predominately caused by wind effects. These data show the uncertainty of using a short-term building pressure

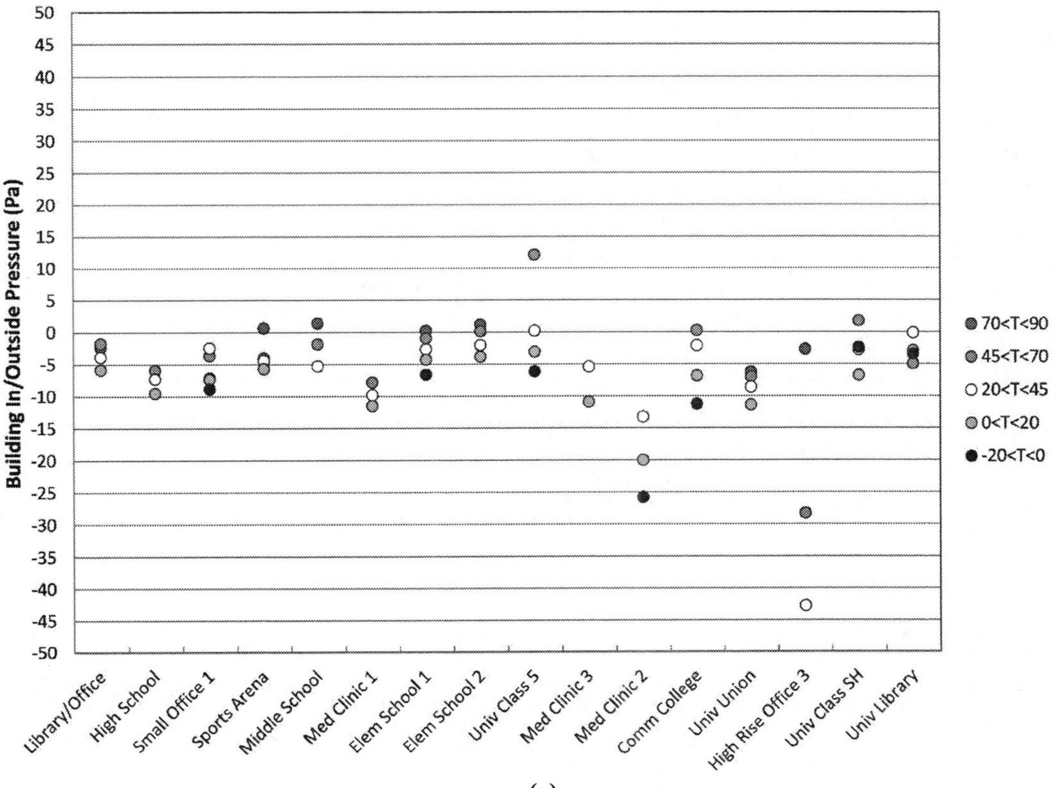

(a)

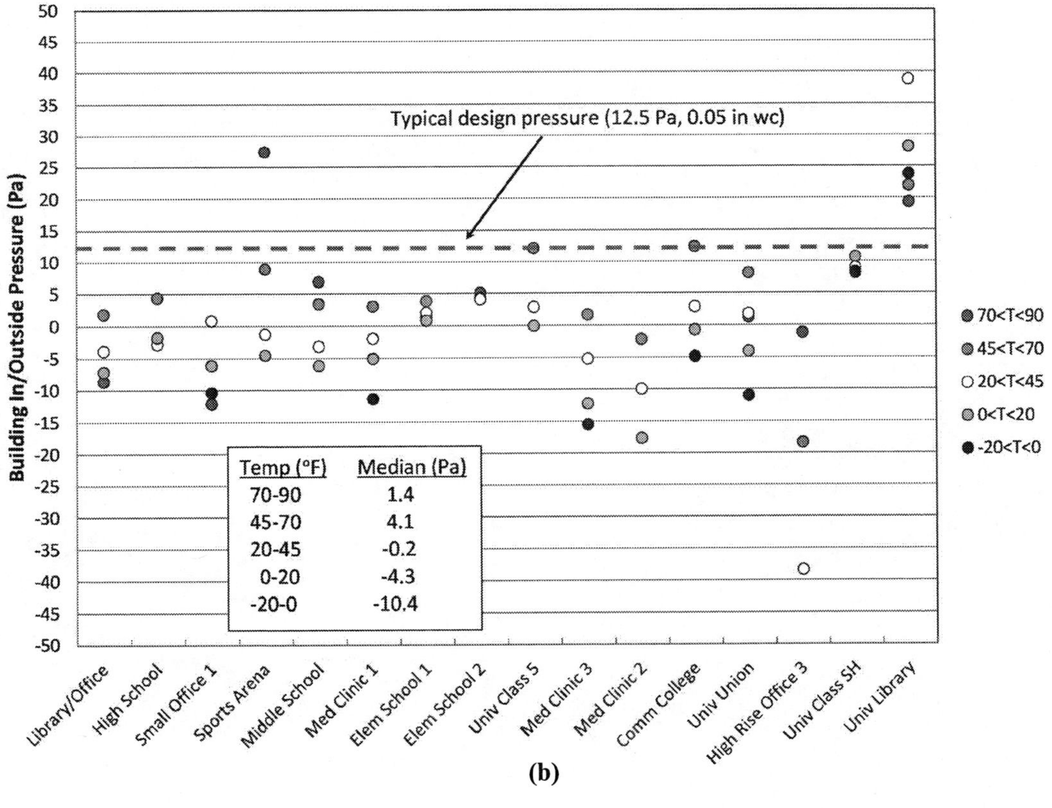

(b)

Figure 1 *Average building pressure at ground level by outdoor temperature bin for (a) HVAC OFF and (b) HVAC ON. The red circles indicate warmer outdoor temperatures and the blue circles indicate colder temperatures. The buildings are sorted by the HVAC ON pressure for the 20°F to 45°F (–6.7°C to 7.2°C) temperature range.*

measurement taken at a single visit to indicate a long-term average. While the HVAC OFF IQR was typically less than 0.004 in. wc (1 Pa), it was as high as 0.016 in. wc (4 Pa) for HVAC ON periods. There was about a 0.008 in. wc (2 Pa) trend in ground level pressure with outdoor temperature and a greater trend in roof pressure with outdoor temperature. While some of the increase in roof level pressure with colder outdoor conditions is expected, there are also nonlinear variations that appear to be caused by changes in HVAC flow imbalances.

The pressure variations for Elementary School 1 are generally representative of those for low-rise buildings without pressure control and moderate HVAC flow imbalance. The pressure variations for taller buildings or those with larger HVAC flow imbalances are often more extreme than this example. The results suggest that short-term building pressure measurements at a single visit can help identify pressurization problems, but long-term measurements over a range of outdoor air temperature are necessary to accurately determine HVAC operation effects on building pres-

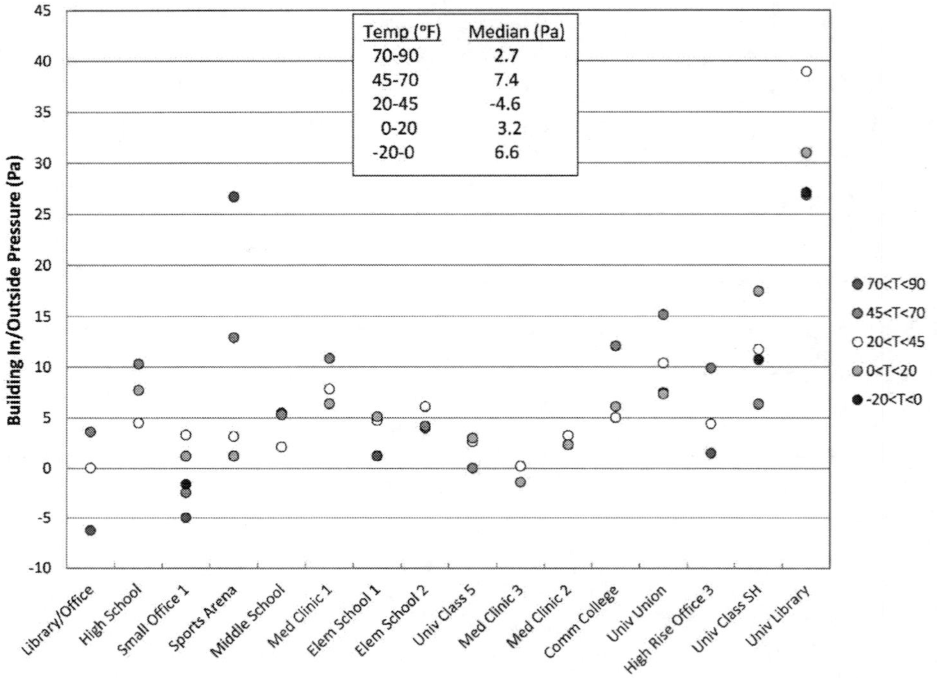

Figure 2 *Difference in building pressure for HVAC ON/HVAC OFF.*

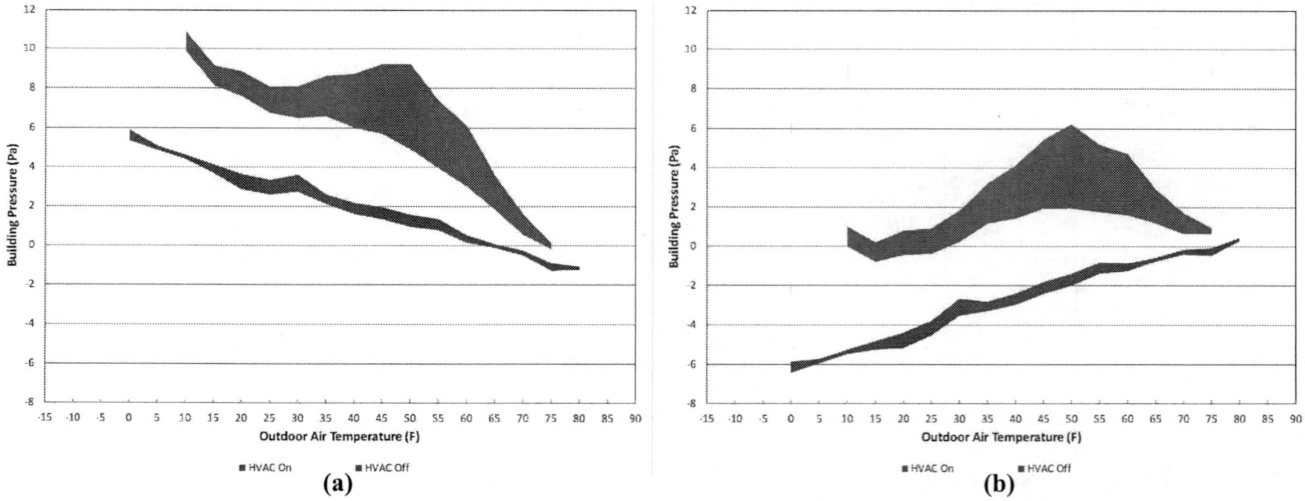

Figure 3 *Interquartile range (a) roof and (b) ground level pressures for Elementary School 1 HVAC ON (red) and off (blue).*

sure. For most buildings it is particularly important to monitor during periods of economizer operation because this tends to generate the greatest deviations from expected pressure trends.

Two buildings demonstrate how neglecting the HVAC operation pressure effects can greatly overestimate or underestimate calculated savings from envelope air sealing. Figure 4 displays the building pressure IQR charts for the University Library and Library/Office buildings. For University Library in milder weather, the pressure at the roof typically varied from 0.06 to 0.141 in. wc (15 - 35 Pa) during daytime operation, but when the outdoor temperature was lower than 40°F (4.4°C) the building pressure was consistently above 0.1 in. wc (25 Pa) and sometimes as high as 0.241 in. wc (60 Pa). Even the ground level pressure was consistently above 0.02 in. wc (5 Pa) and sometimes reached 0.181 in. wc (45 Pa). None of the building's four air handlers had active building pressure control. There was insufficient or malfunctioning exhaust capacity that produced high positive building pressures, often causing ground floor entry doors to blow open. Air infiltration into this building during HVAC operation would be extremely low, possibly limited to isolated regions facing high wind speeds.

The Library/Office is a one-story building that had independent active building pressure control on each of the five air handlers. Each handler had its own building pressure sensor with an outdoor reference tube that terminated to the roof or side of the building and a relief damper adjusted to achieve the desired building pressure. The pressure sensor for one of the air handlers was malfunctioning and there was no delay or smoothing of the pressure response. Consequently, the relief damper position was not adjusted properly for the system with the malfunctioning pressure sensor and on moderate to high wind days the dampers for the other four systems fluctuated drastically between fully closed and fully open. The building did not often achieve its pressure set point of +0.04 in. wc (10 Pa) and typically only maintained a positive building pressure at outdoor temperatures between 40°F and 70°F (4.4°C to 21.1°C). The remainder of the time the building was depressurized, with significant depressurization in warmer

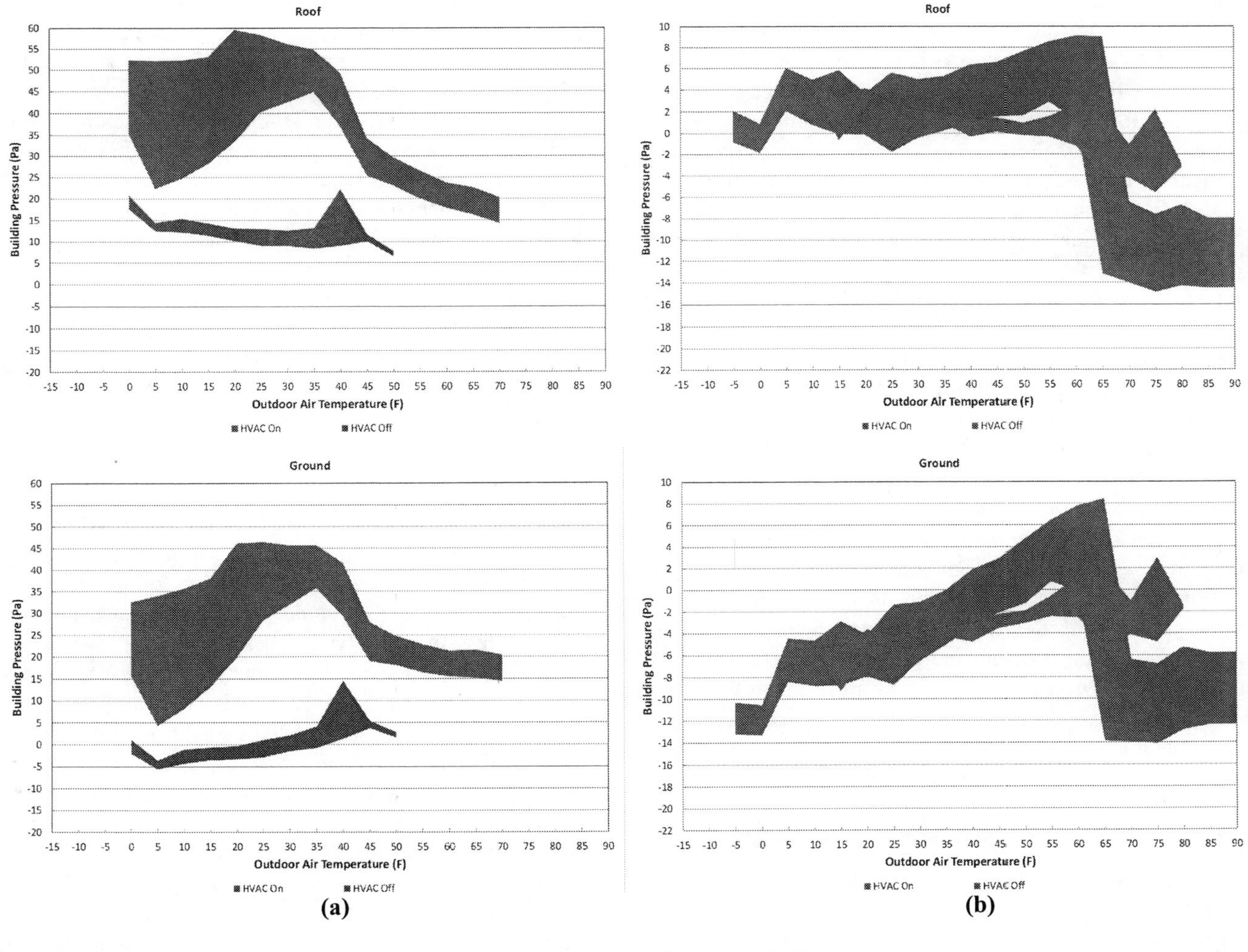

Figure 4 *Building pressure charts: (a) University Library and (b) Library/Office.*

weather. Neglecting the HVAC pressure effect in this building would cause underestimation of air infiltration and air sealing savings.

Pressure control improvements were identified for monitored buildings. The inadequate relief air for the University Library building was because of either improperly actuated relief dampers or restricted relief air capacity. Further investigation was required to determine a specific retrofit and that was not possible before the end of the project. The malfunctioning pressure sensor at the Library/Office needed a repair and an activation delay or pressure smoothing function was suggested for the pressure control process to improve the active pressure control response. Instead the building managers abandoned active control and implemented a fixed lag between the outdoor air and relief dampers to provide a fixed flow imbalance. For the other low rise buildings pressurization could be increased to reduce infiltration during HVAC operation. That would have typically required reduction in the lag between the outdoor air and relief dampers. For buildings that were sealed and modeled with CONTAMW, the impact of greater flow imbalance on air infiltration was evaluated by varying the flow imbalance in the models. Building pressure control changes for the high rise buildings were not considered as part of this project.

An investigation on the use of HVAC operation information to predict building pressures was completed because long-term building pressure monitoring is not cost effective for envelope air sealing energy savings estimates on typical projects. Project staff compiled HVAC system operation information from building plans, BAS control sequences, and BAS trend data. Engineering calculations from these data sources produced estimates of the net flow imbalance. Building envelope leakage measurements were performed over a range of building pressures to produce a characteristic curve of envelope leakage as a function of pressure using a power law relationship (Walker et al. 1998) for each building. The net flow imbalance was input into the characteristic power law relationship to compute the expected change in building pressure caused by HVAC operation. These results were compared with the measured building pressure during HVAC operation and the modeled baseline pressure for HVAC OFF conditions based on outdoor air temperature. For numerous buildings there was a poor correlation between building pressure changes because of estimated HVAC flow imbalance and the measured building pressure above the baseline. In general, the uncertainty in the estimated flow imbalance was much greater than the actual value. Thus, while HVAC system information can help understand the relationship between operating modes and pressurization, it alone cannot be used to estimate building pressures.

Impact of Additional Building Pressurization

There was an opportunity to reduce air infiltration through increased building pressurization in all seven buildings where air sealing was conducted except the University Library building. This building had an extremely high level of pressurization when the HVAC system was operating. For the remaining buildings the pressurization because of HVAC operation typically ranged from 0.004 to 0.02 in. wc (1 to 5 Pa), well below the ΔP_{des} recommended to reduce air infiltration. An additional series of CONTAMW infiltration modeling was conducted for five of the buildings to evaluate the effect of additional pressurization. In post-sealing simulations, the HVAC imbalance was varied from 1000 to 4000 cfm above the current level of flow imbalance by reducing the relief airflow, therefore achieving building pressurization without introducing more supply air. The reductions in annual average air infiltration for the increased flow imbalances are displayed in Figure 5 along with the annual space conditioning energy savings for a flow change of 4000 cfm (113,300 L/m) (right side of the chart). For four buildings with HVAC pressurization from 0.008 to 0.024 in. wc (2 to 6 Pa) infiltration was reduced between 100 and 400 cfm (2830 to 11,330 L/m), resulting in annual space conditioning savings between $159 and $667. However, it is likely those changes in the HVAC flow imbalance could be achieved with straightforward changes to the outdoor air/relief damper sequences of operation and the paybacks would be five years or less.

The annual savings were much larger ($1683) for the Library/Office building that had active building pressure control. Because of problems with one of the pressure sensors and an overly quick response to wind induced pressure differences, the actual building pressures did not meet the design levels and were negative in warmer weather. Addressing negative building pressures would have required pressure sensor repairs and changes to the building pressure control algorithm. The simple payback for that modification would likely have been two to three years. These examples demonstrate possible energy cost savings from improved building pressure control for relatively tight buildings that have a moderate level of HVAC pressurization. The savings would be even greater for buildings with leakier envelopes or HVAC systems that cause building depressurization.

CONCLUSION

The pressure monitoring results show that for Minnesota C&I buildings HVAC system operation almost always produces moderate building pressurization for both the heating and cooling seasons. For almost all buildings the pressurization is less than the typical design level of 0.05 in. wc (12.5 Pa). The median average ground level pressure across buildings ranged from 0.016 in. wc (4 Pa) for warm weather to –0.04 in. wc (–10 Pa) for the coldest conditions. Only one of the 16 buildings consistently had a ground level pressure above recommended design. These data demonstrate the building pressure variations that can be expected when relying on simple supply/exhaust flow imbalances for building pressure control. Pressure control improvements were identified for monitored buildings. One building had excessive pressurization because of inadequate relief airflow. That minimized air infiltration, but often blew open exterior doors. Another building had a malfunctioning pressure sensor for one system and overly responsive control algorithms that sometimes resulted in depressuriztion. For all buildings with a constant offset between outdoor and relief air flow, the relief air flow could have been reduced to improve pressurization. Modeling showed that this could save $160 to $1,600 per year for these relatively tight buildings with moderate pressurization. In these cases winter time exfiltration concerns are limited because of low indoor humidity. In buildings or environments wither higher humidity pressurization, energy savings must be balanced against the consequences of moisture transfer into the envelope via increased exfiltration. Relatively large variations in building pressure demonstrated that short-term building pressure measurements at a single visit can help identify pressurization problems, but long-term measurements over a range of outdoor air temperature are necessary to accurately determine HVAC operation effects on building pressure.

ACKNOWLEDGMENTS

This project was supported in part (or in whole) by a grant from the Minnesota Department of Commerce, Division of Energy Resources, through the Conservation Applied Research and Development (CARD) program, which is funded by Minnesota ratepayers.

REFERENCES

ASHRAE. 2013. ASHRAE Handbook—Fundamentals. Atlanta: ASHRAE.

ASHRAE. 2015. ASHRAE Handbook—HVAC Applications. Atlanta: ASHRAE.

Stanke, D., and B. Bradley. 2002. Managing the ins and outs of … commercial building pressurization. *Trane Engineering Newsletter* 31(2).

Walker, I.S., D.J. Wilson, and M.H. Sherman. 1998. A comparison of the power law to quadratic formulations for air infiltration calculations. *Energy and Buildings* 27(3):293–99.

Walton, G.N., and W.S. Dols. 2013. *CONTAMW user guide and program documentation*. NISTIR 7251. Gaithersburg, MD: National Institute of Standards and Technology. Downloaded at www.bfrl.nist.gov/IAQanalysis/docs/CONTAM_31.pdf on February 28, 2013.

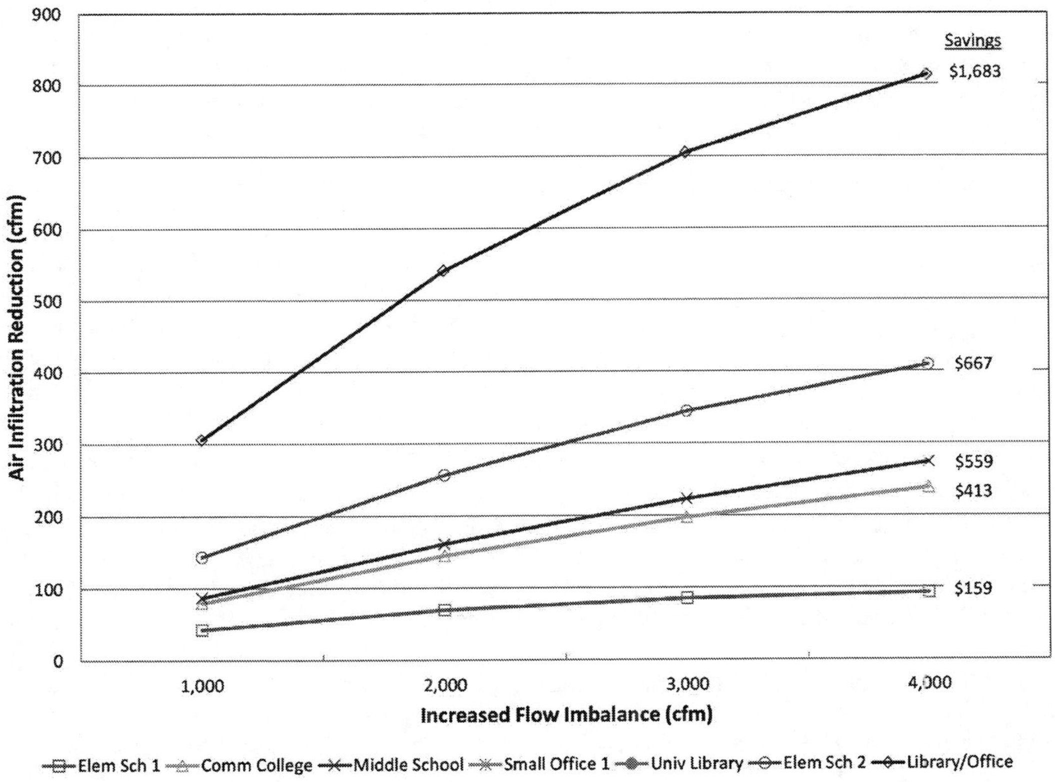

Figure 5 *Infiltration reduction from additional building pressurization.*

Sealed Attics Exposed to Two Years of Weathering in a Hot and Humid Climate

William Miller, PhD
Member ASHRAE

Sudhir Railkar, PhD
Member ASHRAE

Ming Shiao, PhD
Member ASHRAE

Andre Desjarlais, PE

ABSTRACT

Field studies in a hot, humid climate were conducted to investigate the thermal and hygrothermal performance of ventilated attics and nonventilated semi-conditioned attics sealed with open-cell and with closed-cell spray polyurethane foam insulation. In the ventilated attics the relative humidity drops as the attic air warms; however, the opposite was observed in the sealed attics. Peaks in measured relative humidity in excess of 80%–90% and occasionally near saturation (i.e., 100%) were observed from roughly solar noon till about 8 p.m. on hot, humid days. The conditioned space of the test facility is heated and cooled by an air-to-air heat pump. The space was not occupied and had no interior moisture load due to occupancy. Moisture pin measurements made in the sheathing and absolute humidity sensor data from inside the foam and from the attic air show that moisture is being stored in the foam and the roof sheathing. The moisture in the foam diffuses to and from the sheathing dependent on the pressure gradient at the foam-sheathing interface, which is driven by the irradiance and night-sky radiation. Ventilated attics in the same hot, humid climate showed less moisture movement in the sheathing than those sealed with either open- or closed-cell spray foam. The temperature measured on the underside of the sheathing was 10°C cooler (north-facing roof deck) and 20°C cooler (south-facing roof deck) for the ventilated attic as compared to the sheathing temperatures in the sealed attics. Foam was physically removed and moisture was observed around the rafter and sheathing interface. Observations, sheathing temperature, and partial pressure measurements suggest two-dimensional heat and moisture flow that is cyclic moving in and out of the depth of the roof and that also moves along the plane of the sheathing towards the rafters. The use of permeable spray foam in a hot, humid climate inadvertently allows moisture to be held against the roof deck. The moisture transfers back to the attic air as solar irradiance bears down on the roof.

INTRODUCTION

Roughly half of the 112 million U.S. homes were built from about 1960 until 1979 (U.S. Census Bureau 2013), and the prevailing building code was ASHRAE 1980. Also single-family homes built with slab-on-grade foundations rose from 34% of all homes constructed in 1971 to 45% of all homes built by 2013 (U.S. Census Bureau 2013). In other words, by 2013, 45% of all new and existing homes in the southern and western U.S. had slab-on-grade foundations. In the colder northern climates, basements dominate about 80% of the residential market and a smaller percentage of HVAC systems are placed in the attics (U.S. Census Bureau 2013). Although as the demand and price for land rose, the percentage of two-story homes built in the northeast jumped from 44% in 1971 up to 70% by 2013. These homes typically have an HVAC system per floor, and the unit conditioning the upstairs has its duct system in the attic, which raises the statistic for HVAC and ducts in attics to about 52% of all single-family homes.

The convenience of the attic space appeals to builders, who all too often install the HVAC unit and the ducts in the attic to conserve living space while completing the rough-in at a low first cost. However, installing the HVAC and ducts inside an unconditioned and ventilated attic is not the most energy efficient option because of the extreme summer and winter

William Miller is a building scientist and **Andre Desjarlais** conducts program development on behalf of the Energy and Transportation Division at the Oak Ridge National Laboratory (ORNL), Oak Ridge, TN. **Sudhir Railkar** is the research and development manager for Steep Slope Roofing and **Ming Shiao** is a materials scientist at GAF in Parsippany, NJ.

operating temperatures occurring in the attic. Parker, Fairey, and Gu (1993) simulated the effects of ducts on space conditioning Florida homes and observed that air leakage and heat transfer to the duct were major contributors to the peak electrical burden on the Florida utility. Walker (1998), in his bibliography of duct air leakage, reports average measured leakage rates of roughly 7% to a maximum of 20% of the supply airflow. To improve envelope performance, researchers opted to literally encapsulate the HVAC and ducts by moving the boundary of the insulating planes to the roofline, gables and eaves of the attic. The concept was first introduced by Building Science Corp (Rudd and Lstiburek 1998). They built and monitored test homes in a hot, dry climate and demonstrated that the prototype homes with unvented attics yielded significant cooling and heating energy savings over a conventional home with ventilated attic. Transforming residential attics into a nonventilated, semiconditioned attic space has therefore gained approval among builders (Chasar et al. 2010). Boudreaux, Pallin, and Jackson (2013), however, recommend that sealed attics be conditioned to mitigate the moisture buildup issue as do Roppel, Norris, and Lawton (2013). Building Science has successfully demonstrated a unique ridge vent that is vapor permeable but air impermeable, and can be used to vent moisture in sealed attics (Lstiburek 2015). However, field data demonstrating successful implementation in all climates are sparse and there still remains an educational gap in the best practices for builders. The dearth of data in hot, humid climates has caused confusion among builders and code officials because of the confounding variables affecting the hygrothermal performance of sealed attics.

NATURAL EXPOSURE TEST FACILITY

General Aniline and Film (GAF) and the Oak Ridge National Laboratory (ORNL) conducted field experiments at a natural exposure test facility (NET) located in the hot and humid climate of Charleston, SC, Figure 1. The study was conducted to better understand the effectiveness of various ventilation strategies and its impact on temperature and moisture management (Railkar et al. 2010). Two full years of field-testing were completed on different roof and attic constructions including two attics sealed with open-cell spray polyurethane foam (ocSPF) insulation and closed-cell spray polyurethane foam (ccSPF) insulation. All other attics were equipped with soffit and ridge ventilation set at an area ratio of 1:300 (i.e., 1 unit area of vent opening to 300 units of attic footprint).

The attic of the NET was subdivided into seven attic modules. Barrier walls insulated and air sealed adjacent attics from each other. The barriers were made of R-15 expanded polystyrene insulation sandwiched between two pieces of oriented strand board (OSB). Different wall systems with brick and stucco claddings were similarly tested in previous field tests on the NET, Figure 1. During the course of the experiments an air-to-air heat pump maintained the interior ambient conditions at 23.9°C and 45 percent relative humidity. Salient features of the two sealed attics and a conventionally ventilated attic follow and are described in the sequence that they appear in Figures 1 and 2 (starting from the left).

Figure 1 *The southern exposure of the NET Facility in Charleston, SC.*

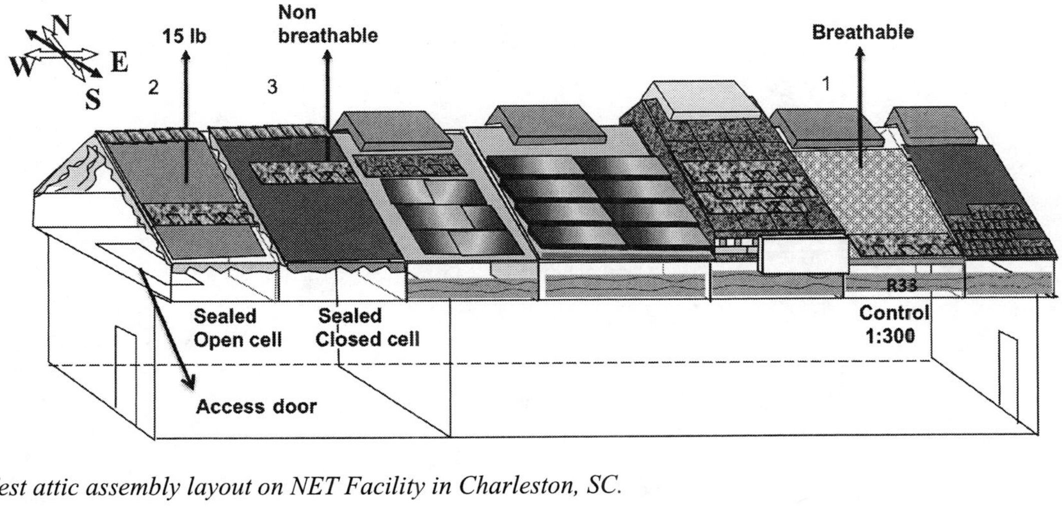

Figure 2 *Test attic assembly layout on NET Facility in Charleston, SC.*

Attic 2 [Sealed Attic (ocSPF)]. This attic had 6 inches of open-cell spray polyurethane foam (ocSPF) adhered to the underside of plywood roof sheathing. The plywood was protected by 15 pound felt paper (8 Perm) and traditional dark-colored shingles (impermeable) with solar reflectance of about 3%. The attic was not ventilated; it is sealed with R-22 ocSPF. The foam extended about ½ in. past the 2 by 6 roof rafters to completely encapsulate the roofline. Foam was also applied to the west-facing gable end of the attic bay. There is no insulation on the attic floor.

Tracer gas testing was conducted on the attic sealed with ocSPF to determine its airtightness to the conditioned space and to the outdoor environment under natural atmospheric pressures. The ASTM E 741 (2000) procedure was used to measure the airflow leakage by monitoring the decay rate of the tracer gas R-134a over time. The testing took six hours and yielded an air exchange rate of 0.25 per hour. Parker in his literature review reported that the Conservation Service Group (EDU 2005) evaluated airtightness of some 466 homes; 18 of the homes had attics sealed with spray polyurethane foam. The ACH of sealed homes was about 0.24, which is consistent with the present study.

Attic 3 [Sealed Attic (ccSPF)]: This attic is similar to Attic 2 but is sealed with 4 inches of closed-cell spray polyurethane foam (ccSPF) installed to the underside of plywood roof sheathing. The ccSPF covered the 2 by 4 rafters of this attic bay. The plywood was protected by a nonbreathable underlayment having a vapor permeance of 0.04. Traditional dark-colored shingles with solar reflectance of about 3% were used as roof cover. The attic was not ventilated; it is sealed with R-22 ccSPF. There is no insulation on the attic floor. Barrier walls with R-15 insulation seal Attic 3 from the adjacent attic assemblies #2 and #4. Airtightness of Attic 3 has not been verified as of this report.

Attic 1 [Control Attic]: This attic had R-38 fibrous batt insulation installed on the attic floor. The sheathing was covered with a permeable membrane underlayment (16 perms) and traditional dark-colored shingles with solar reflectance of about 3%. Appropriate length of soffit and ridge venting are added to obtain 1:300 ventilation. Tracer gas testing conducted on the ventilated attic yielded an ACH of 1.63.

INSTRUMENTATION

Micro-loggers were setup for remote acquisition and recording of field data. The loggers are equipped with 4 MB of memory, 32bit relay multiplexers, rechargeable battery, a 115 Vac-to-24 Vdc transformer, modem, modem surge protector, and associated cables. The micro-loggers scan all instrument data every 15 seconds and reduce analog signals to engineering units. Averages of the reduced data were written electronically to an open file every 15 minutes. Averages are calculated over the 15-minute interval and are not running averages; they are reset after each 15-minute interval. The electronic format is comma delimited for direct access by spreadsheet programs.

During the construction of the NET facility, thermistors, humidity probes, heat flux transducers, and static pressure sensors were installed at locations consistent from attic to attic to compare the performance of each attic bay. Figure 3 shows the placement of sensors in the two sealed attics. Table 1 describes the attic measurements, instrument range, and uncertainty of measurement.

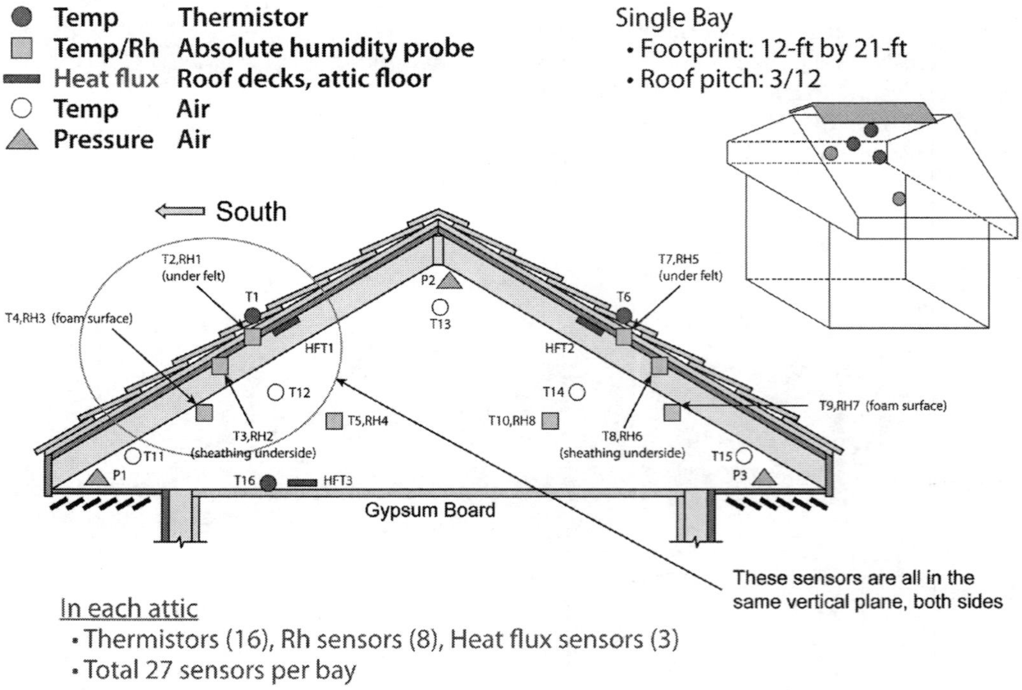

***Figure 3** NET facility setup of instruments in the sealed Attics 2 and 3 in Charleston, SC.*

Absolute humidity sensors were fabricated using the technique described by Straube, Onysko, and Schumacher (2002). A thermistor and relative humidity probe were packaged together in a vapor-permeable and liquid-water-repellent cover fabricated from commercially available weather-resistive barriers designed to allow passage of water vapor but not liquid water. Thermistor and humidity sensors were calibrated by ORNL Metrology in the above-described vapor permeable protective sack.

The response of the thermistors was measured at the following nominal conditions: 15.8°C, 18.5°C, 21.2°C, 23.9°C, and 26.6 °C. All probes met the manufacturer's specification for the temperature response of ±0.2°C. Humidity sensors were checked at 25%, 50%, 75%, and 90% rh. The error in RH ranged from 2% of reading at 25% RH and 15°C to 6.5% of reading at 90% rh and 26°C.

Table 1. Descriptive and Location of Instruments in an Attic Bay of the NET Facility in Charleston, SC

Sensor ID	Sensor Type	Location	Measurement Uncertainty	Measurement Range
South-Facing Roof deck				
T1	Thermistor	Between asphalt shingle overlap	±0.2°C	–60 to 150°C
T2, RH1	Absolute Humidity Thermistor Relative Humidity	Between underlayment and sheathing	±0.2°C ±5%	–60 to 150°C 25 to 100% RH
T3, RH2	Absolute Humidity Thermistor Relative Humidity	Between underside of sheathing and spray foam	±0.2°C ±5%	–60 to 150°C 25 to 100% RH
T4, RH3	Absolute Humidity Thermistor Relative Humidity	At foam surface adjacent the attic air	±0.2°C ±5%	–60 to 150°C 25 to 100% RH
T5, RH4	Absolute Humidity Thermistor Relative Humidity	Attic Air	±0.2°C ±5%	–60 to 150°C 25 to 100% RH
T11	Thermistor	Attic air near soffit	±0.2°C	–60 to 150°C
T12	Thermistor	Attic air midway between soffit and ridge	±0.2°C	–60 to 150°C
North-Facing Roof Deck				
T6	Thermistor	Between asphalt shingle overlap	±0.2°C	–60 to 150°C
T7, RH5	Absolute Humidity Thermistor Relative Humidity	Between underlayment and sheathing	±0.2°C ±5%	–60 to 150°C 25 to 100% RH
T8, RH6	Absolute Humidity Thermistor Relative Humidity	Between underside of sheathing and spray foam	±0.2°C ±5%	–60 to 150°C 25 to 100% RH
T9, RH7	Absolute Humidity Thermistor Relative Humidity	At foam surface adjacent the attic air	±0.2°C ±5%	–60 to 150°C 25 to 100% RH
T10, RH8	Absolute Humidity Thermistor Relative Humidity	Attic Air	±0.2°C ±5%	–60 to 150°C 25 to 100% RH
T13	Thermistor	Attic air near ridge	±0.2°C	–60 to 150°C
T14	Thermistor	Attic air midway between soffit and ridge	±0.2°C	–60 to 150°C
T15	Thermistor	Attic air near soffit	±0.2°C	–60 to 150°C
T16	Thermistor	On Gypsum board facing attic	±0.2°C	–60 to 150°C

The instrument setup for the ventilated attic differed slightly from that of the two sealed attics. The absolute humidity probe (T5, RH4) was placed on top of the R-38 fibrous batt insulation installed on the attic floor. The probe (T4, RH3) measured the temperature and relative humidity on a truss adjacent a roof rafter. All other instrument locations were the same as those placed in the two sealed attics.

Heat flux transducers were placed on the underside of the south- and north- facing roof decks and on the attic floor, sandwiched between the gypsum ceiling and the R-38 batt insulation. A heat paste is used to reduce contact resistance between the HFT and the gypsum board. All transducers were calibrated in accordance with ASTM C 518-10 (ASTM 2010). R&D Services from Cookeville, TN used a FOX heat flow (FHF) meter having a 0.305 by 0.305 m footprint to measure the thermal resistance of batt insulation insulating the attic floor of Attic 1 (see Table 2). Each HFT was placed in a 0.305 by 0.305 m (30 by 30 in.) guard made of gypsum board and from the same lot of batt insulation calibrated in the FHF meter. The HFT and checked batt insulation were later placed in the Attic 1 bay. Use of the guard corrects for the shunting of heat flow around the transducer. The manufacturer states accuracy of their sensor as ±1% of full-scale reading with a sensitivity of about 4.1 W/m^2 per mV of signal (1.3 Btu/h·ft^2 per mV). The stated time constant is ≤5s. R&D calibrations showed them to be accurate within ±5% of the instrument's reading.

Measuring the moisture content of wood is a key fundamental measure of hygrothermal performance, because the measure is directly relatable to the potential for mold growth leading to the decay of wood components. Viitanen (1997) reported fungus requires moisture content (MC) of 25%–28% for growth. Details for the formulation and benchmark of an algorithm to compute moisture content is presented in Appendix A.

FIELD EXPERIENCE

Field testing and data collection started September 2013 and continued through December 2015. During the course of study two anomalies were observed in reduced data for the sealed attics: (1) during the warmer months the relative humidity of the attic air increased as the temperature of the attic air increased, and (2) the heat flow crossing the ceiling was more than double the ceiling heat flow measurements for ventilated attics having R-38 (IECC 2012 code) levels of insulation. Traditional dark-colored asphalt shingles having a solar reflectance of 3% cover both sealed attics and the ventilated attic.

Sealed Attic—Relative Humidity Trends

Trends in attic moisture for the sealed attics differed from trends for the conventionally ventilated attics (see Figure 4). The relative humidity in the attic sealed with ocSPF showed values in excess of 80% to 90%, and occasionally near-saturated air was measured during late afternoons on warm-to-hot days. In other words, the moisture content in the attic sealed with ocSPF was consistently high from solar noon to around 8 p.m.; example trends are shown for seven contiguous June/July days in Figure 4. The attic sealed with ccSPF shows less of an increase in attic air relative humidity than that observed for the attic with ocSPF. Cloud cover dropped the solar irradiance during the week, and the attic relative humidity also dropped below 70% for both sealed attics because the sun did not drive stored moisture from the plywood and ocSPF into the attic air. The trend in peak relative humidity was not observed in either sealed attic during the winter months. In comparison, the ventilated attic (Figure 4) shows drops in relative humidity as the attic air warms.

As temperature increases, the moisture-bearing capacity of the air increases, and, by definition of relative humidity (partial pressure of water vapor in air to the partial pressure of water vapor at saturation for the same temperature and pressure of the actual state), it should drop not rise as observed in the sealed attics. Instrument measurements and data acquisition channels were checked for consistency. A portable relative humidity sensor yielded measures within 1% point relative humidity of the field sensor. Next the portable sensor was held close to the ocSPF adhered to the south-facing roof deck. The test was made in October 2014 around 2 p.m. The humidity readings began to steadily climb from 55% to 65% over a brief ten-minute period. The result indicated that moisture was being driven out of the foam as the sun heated the south-facing roof structure.

Table 2. Calibration of Fiberglass Batt Insulation Used as Guard for the Ceiling Heat Flow of Attic 1

Specimen	Thickness, in.	Density, lbf/ft^3	Apparent Thermal Conductivity, Btu in./ft^2·h·°F	Thermal Resistance, h·ft^2·°F/Btu
13-3411	12.000	0.53	0.3056	39.3
13-3422	12.001	0.53	0.3417	35.1
13-3423	12.000	0.54	0.3448	34.8
13-3423	12.000	0.54	0.3448	34.8

1 in. = 0.0254 m; 1 lbf/ft^3 = 16.02 kg/m^3; 1 (Btu·in.)/(ft^2·h·°F) = 1.442 W/(m·K).
1 (h·ft^2·°F)/Btu = 6.7 (m^2·K)/W.

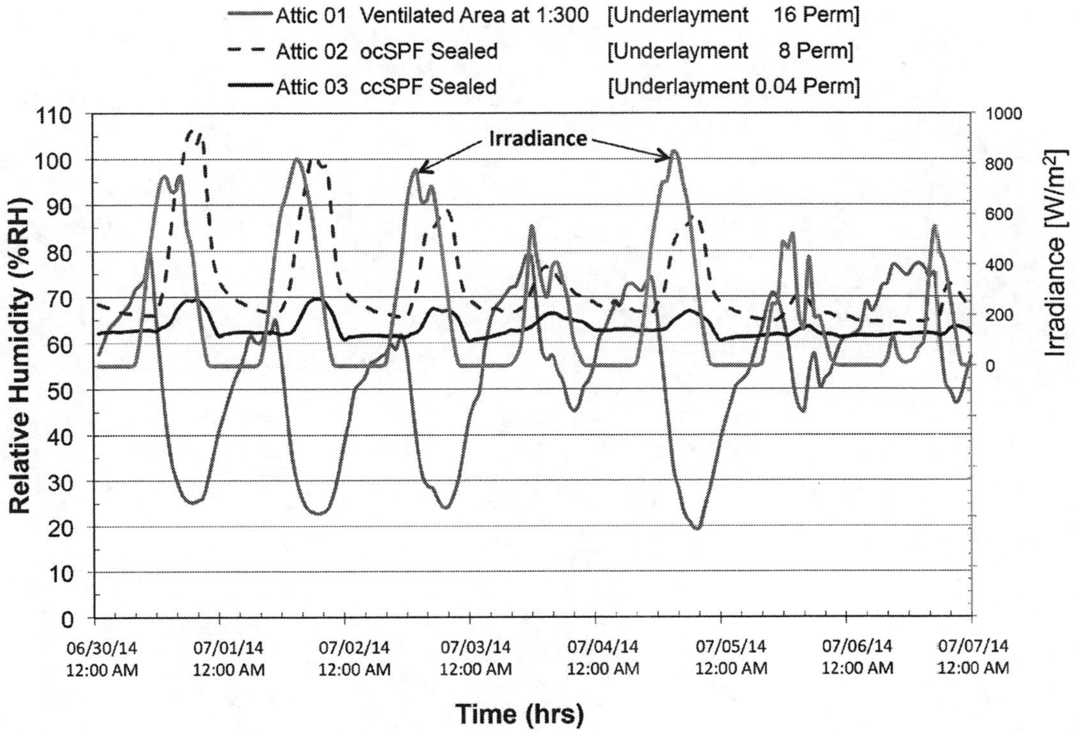

Figure 4 *Relative humidity of the attic air measured in the attics sealed with ocSPF and with ccSPF and also in the conventionally ventilated attic.*

A section of the open-cell spray foam was removed in August 2013 from between 2 by 6 rafters on the north-facing roof deck (see Figure 5). The sheathing's underside was slightly wet to the touch. It is keenly interesting that water marks are observable around the exposed rafters. The thermal resistance through the depth of the rafters is much less than that of the foam insulation. Therefore, the sheathing temperature at the rafter would be less than that measured between the sheathing and foam during daylight hours. Measurements show the underside sheathing in contact with foam to reach 60°C to 70°C during the summer. Hence, there is a strong potential for two-dimensional heat and moisture transfer with movement into the depth of the roof deck and along the plane of the sheathing toward the rafters.

Spray foam has a moisture capacity of only 0.33 kg per cubic meter of foam, while plywood at about 60% relative humidity has a hygroscopic storage capacity of 48 kg per cubic meter of wood. The wood is far more absorbent than the foam and as the sun drives moisture from the wood into the foam there forms a moisture layer at the wood-foam interface. Therefore moisture pins were nailed into the underside of the 0.5 in. plywood prior to refilling the cavity with ocSPF (see Figure 5). Absolute humidity probes were also installed to the underside of the plywood, midway through the depth of a new application of ocSPF foam, and at the foam's surface adjacent the attic air.

The living space of the NET facility is comfort conditioned by a split-system air-to-air heat pump, which maintains moisture level in the comfort zone. Figure 6 provides measures of the partial pressure of water vapor computed from the absolute humidity probes placed in the living space, the outdoor air, the attic air, the surface of the ocSPF, mid-depth of the foam, the sheathing's underside, and between the underlayment and topside of the plywood sheathing. Moisture content values of the sheathing are also illustrated on the secondary ordinate axis of Figure 6. Moisture content is adjusted for temperature (see Appendix A).

The absolute humidity probe yielding partial pressure of water vapor in the living space shows very little potential for moisture to diffuse from the conditioned space into the attic across the ceiling plane. For this study occupancy, habits are not a factor because there were no internal moisture loads. The building owner had closed the water main during the field study because of construction and renovation of the high school adjacent the NET. It is also very interesting that around 4 to 8 p.m. the partial pressure of water vapor in the outdoor air is less than that of the attic air and also for that of the measures on or in the foam. At all other times, the partial pressure of the outdoor air exceeds that of the attic air and that of the foam surface (see Figure 6). Therefore, during hours from about solar noon till about 8 p.m., there is a potential for moisture to diffuse from the plywood and foam into the attic air as

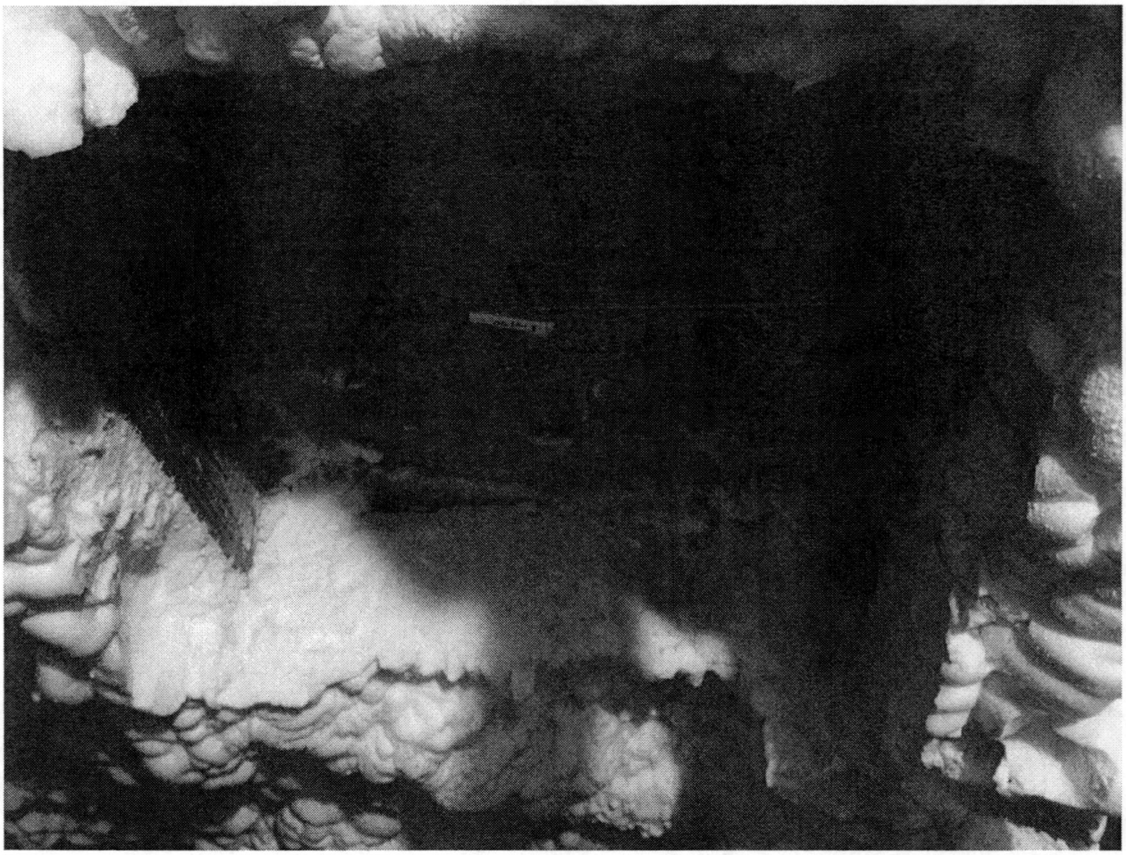

Figure 5 *The sheathing on the north-facing roof deck. Open-cell spray polyurethane foam insulation removed to inspect sheathing and add instruments after two years of climatic exposure.*

actually measured earlier with a portable humidity sensor. During the evening, night-sky radiation cools the roof below the outdoor ambient and the lowest partial pressure is seen at night below the underlayment at the topside of the sheathing. At night the attic moisture is transported by air back into the foam and diffuses through the foam and into the plywood deck (see Figure 6). The following day the sun heats the roof, and the moisture is driven back into the attic air, thus explaining why the attic air moisture exceeds the moisture content of the outdoor air during the late afternoon immediately following peak irradiance.

The moisture content of the plywood (as measured with moisture pins) also verifies these moisture movements. During the heat of the day, the moisture content of the sheathing drops, which implies that moisture moves from the sheathing into the foam. As the sun sets, the roof cools and moisture is reabsorbed from the attic air and the foam back into the roof deck. It is also keenly interesting to compare these moisture content measurements to the visual observation made when a portion of ocSPF was removed from the deck (Figure 4). The moisture pins are measuring the average moisture content through the thickness of the 0.5 in. plywood. Removing the foam revealed the sheathing to be wet, especially near the rafters. After completion of the field study, the roof shingles and underlayment were removed to verify workmanship of the roof (see Figure 7).

A Delmhorst meter was used to measure the moisture content on the top side of the bare plywood deck. All meter readings showed both the south- and north-facing decks dry. Wood resistance was too high for the Delmhorst to yield a measurement. However, the Delmhorst showed 10% to 14% moisture content at the abutting of two 4 by 8 plywood sheets where the plywood is nailed to the roof rafters. Water damage was observed below and to the right of instrumentation placed in the north-facing roof deck (see Figure 6). The damage was near the same rafter shown in Figure 4 but was about 1.22 m below the instrumentation and about 1.6 m up from the soffit.

Parsons and Drzyzga (2013) reviewed the challenges to sealing existing residential attics and concluded that air sealing the joint between the exterior sheathing and top plate when the exterior sheathing was not flush with the top of the top plate was problematic. They concluded it best to seal the gap from the exterior rather than from the confined interior space of the attic. Their observations are consistent with independent work conducted by ORNL on wall assemblies. The work of Hun, Spafford, and Desjarlais (2014) concluded that the inter-

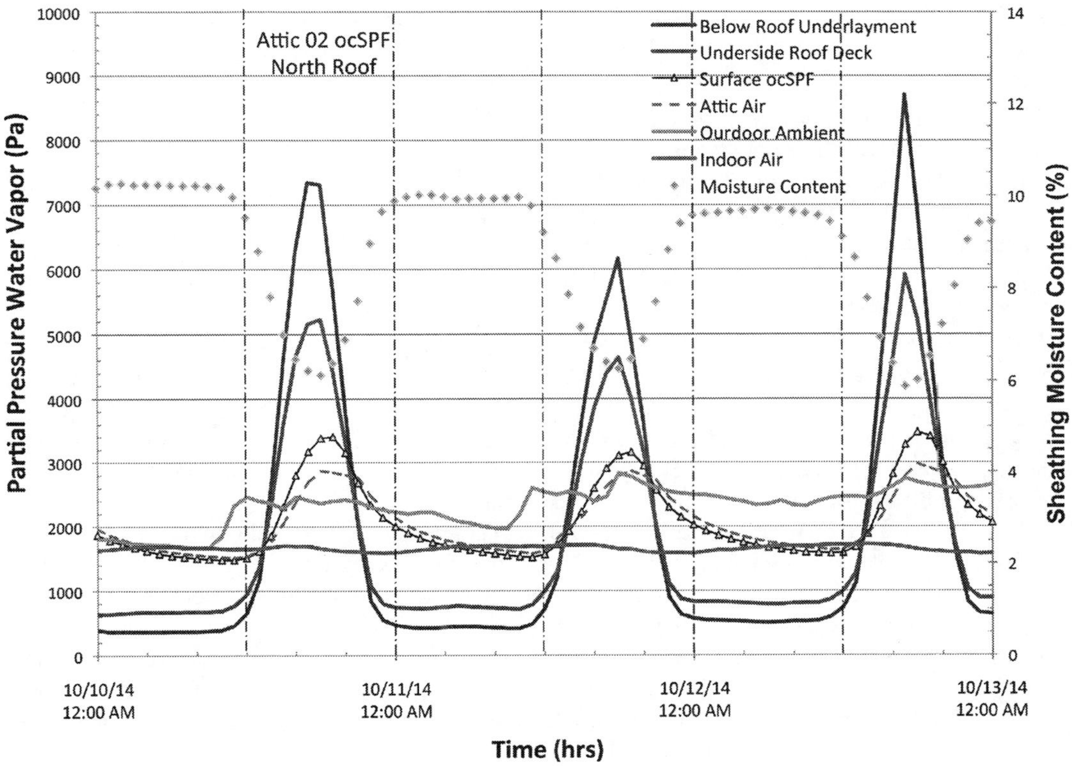

Figure 6 *The partial pressure of water vapor measured at various locations from the roof underlayment to the surface of the foam adjacent the attic air. Moisture content of the sheathing is shown on the secondary* y*-axis.*

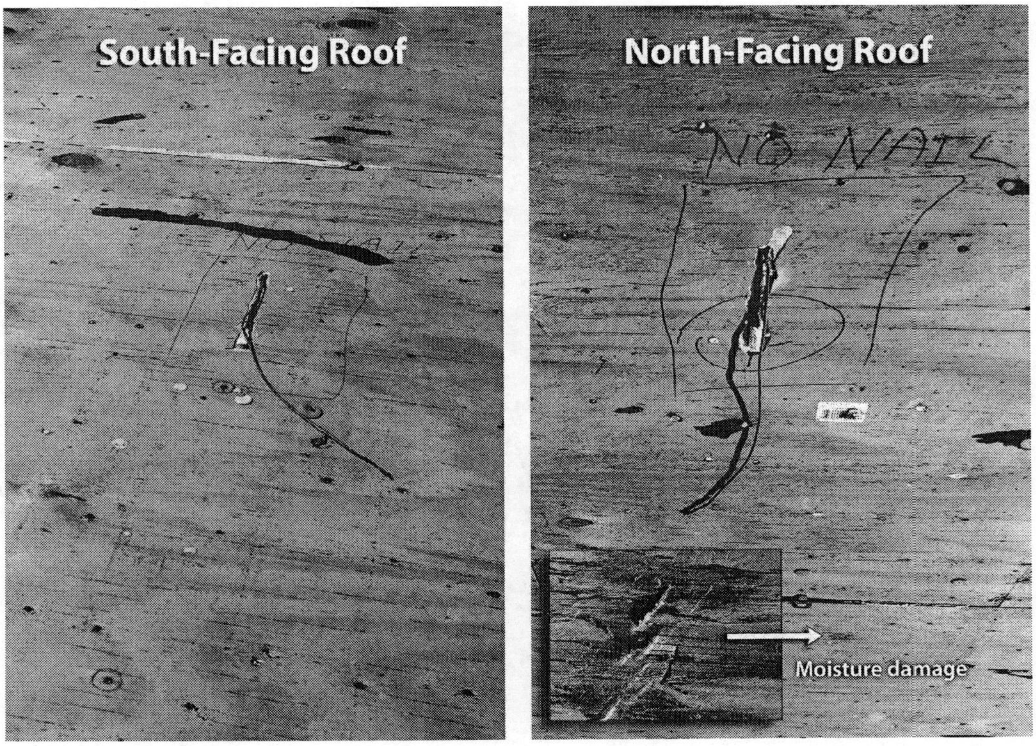

Figure 7 *Condition of the 0.5 in. plywood deck after two years of exposure. Replacement shingles and underlayment replaced February 2016.*

face between the OSB and the top plates was responsible for most of the air leakage at the wall-to-roof joint and allowed the entry of unwanted moist air into the sealed attic.

Open-cell SPF is very permeable, being about 19 perms at a 3 in. thickness; however ccSPF is only 0.5 perms for the same 3 in. thickness of closed-cell SPF, Kumaran (2008). Therefore, it is very plausible that the ccSPF, being semi-impermeable, does not release as much moisture during times of peak irradiance as observed for the ocSPF, which, in turn, helps explains the differences in humidity levels between the two sealed attics, shown back in Figure 4. Data similar to that in Figure 6 is highlighted in Figure 8 for the attic sealed with ccSPF.

The two graphs of the partial pressure of water vapor in the attic air and at the interior surface of the ccSPF show there is very little pressure gradient during the three contiguous days (see Figure 8). Hence, the flux of water vapor leaving the ccSPF is much less than that observed for the case with ocSPF. Also, the moisture content of the plywood sheathing (sealed with ccSPF) is seen to vary similar to that observed in Figure 6 for the plywood sheathing sealed by ocSPF. The variation in moisture content is about the same, changing from about 10% at night to 6% in the afternoon. There appears to be some moisture movement and storage in the ccSPF, but it is not as large as that observed in the ocSPF case. The attic sealed with ccSPF had a nonbreathable underlayment whose vapor permeance was only 0.04. Therefore, moisture did not diffuse from above the asphalt shingles. Boudreaux, Pallin, and Jackson (2016) conducted an experimental study on a sealed attic home and compared the moisture content below the roof sheathing before and after installing a vapor impermeable underlayment. They observed no statistically significant difference in absolute humidity before and after the vapor barrier installation.

Building Science recommended a vapor barrier be installed between the asphalt shingles and the roof deck (Building Science Press 2003). They also stated that moisture condensing on the roof can be pulled by capillary forces through the gaps made by overlapping shingles and is then driven inwards as the sun heats the roof. Boudreaux, Pallin, and Jackson (2016) examined the necessary physics for capillary movement of water within gaps made by shingle overlaps. They concluded the conditions are not favorable for the occurrence of solar-driven moisture. The water stored in the plywood and foam does not come from the roof top. Plywood sheathing contains moisture; however, moisture is also coming from stagnant humid air within the attic that originates from the indoor conditioned space and/or the outdoor ambient. In the NET case, the moisture originates from the outdoor ambient and leaks into the wall cavity of the exterior wall, around the top plate, and into the attic.

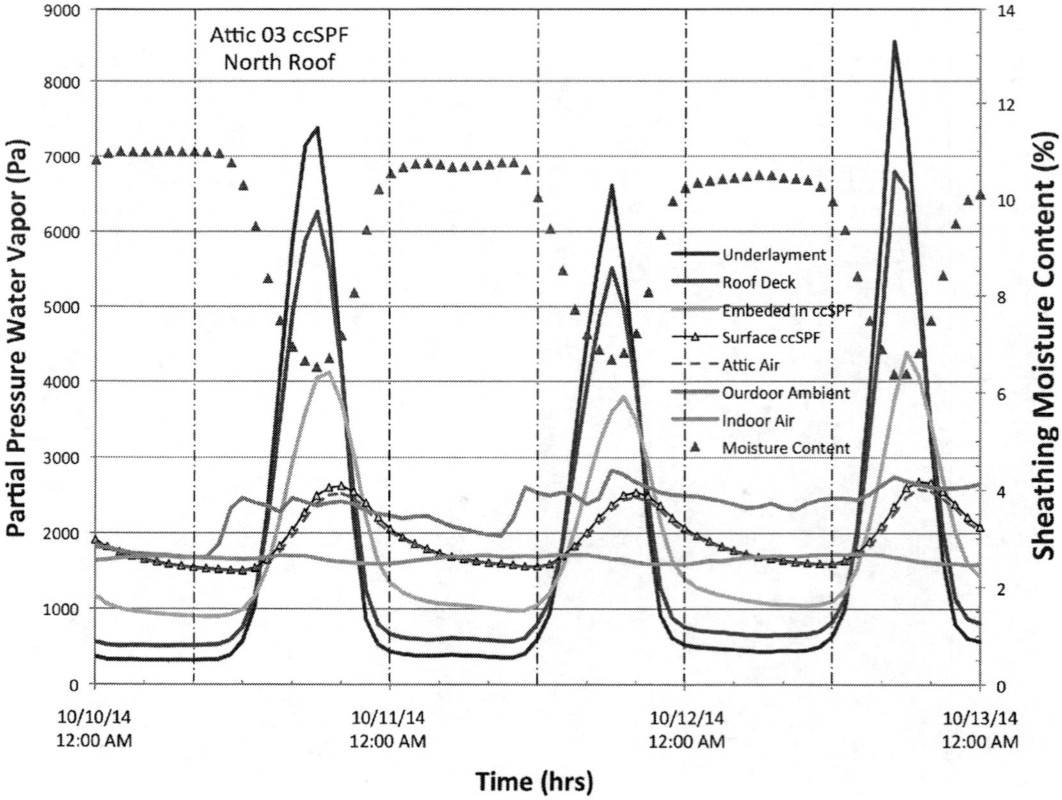

Figure 8 *The partial pressure of water vapor measured at various locations from the roof underlayment to the surface of the foam adjacent the attic air. Moisture content of the sheathing is shown on the secondary* y*-axis.*

Moisture pin measurements are shown in Figure 9 for three contiguous days for the three attic systems to better understand the differences in water movement in the sheathing of the two sealed attics and the attic equipped with ventilation at 1:300. The data is corrected for temperature effects (see Appendix A). The moisture pin data for the two sealed attics is the same data as shown in Figure 6 for the attic sealed with ocSPF and in Figure 8 for the attic sealed with ccSPF. As stated previously, the movements of moisture in the roof decks of the two sealed attics are very similar in value and trends, dropping as the decks are heated by the sun and rising during the late evening/early morning hours as the night-sky cools the decks.

The sheathing of the ventilated attic also shows about the same moisture content during the late evening/early morning hours as seen in the sealed attics. However during periods of peak irradiance the ventilated attic does not drop as much as that measured for the sheathings of the two sealed attics. The underside of the sheathing in the ventilated attic is exposed to air while the sealed attics are covered by foam having a higher thermal resistance than air. Therefore the ventilation in Attic 1 causes the temperature of the sheathing's underside to be about 10°C cooler than that of the sheathing covered by foam insulation. For the south-facing deck, the sheathing is 20°C cooler for the vented attic as compared to the sealed attics. There is less drive for diffusion from the plywood in the ventilated attic. Also, the underside of the sheathing in the ventilated attic is exposed to air and its vapor diffusion resistance is of the order 0.006 (Pa × s × m) per ng as compared to a diffusion resistance for foam of about 0.4 (Pa × s × m) per ng (ASHRAE 2009). Therefore, water vapor more readily moves from the sheathing into the air of the ventilated attic as compared to the two sealed attics. Temperature is therefore the driver. What is also very interesting is that all three sheathings have about the same moisture content during the late evening/early morning hours and the trends are repetitive over the three contiguous days of data. The moisture pins provide an average resistance measure of the wood that is made over the 0.5 in. depth of the plywood sheathing. Moisture pin data for the south-facing roof decks are very similar to that measured on the north-facing roof deck. However, the sheathing's underside temperature in the sealed attics was 20°C warmer on the south-facing deck than that of the sheathing in the ventilated attic.

Sealed Attic—Ceiling Heat Flows

Contractors sealing and insulating attics in ASHRAE Climate Zones 1 through 4 often apply spray polyurethane foam (SPF) insulation to a thickness of R-20 (Holladay 2010; Dillon 2013) and argue that there are "diminishing returns" for adding more insulation, which increases material costs and limitations due to fire code compliance. This level of insulation does not meet the IECC 2015 prescriptive code levels for attics. Building America recommends that sealed attic R-value should follow table N1102 of the 2012 International Residential Code (IRC).

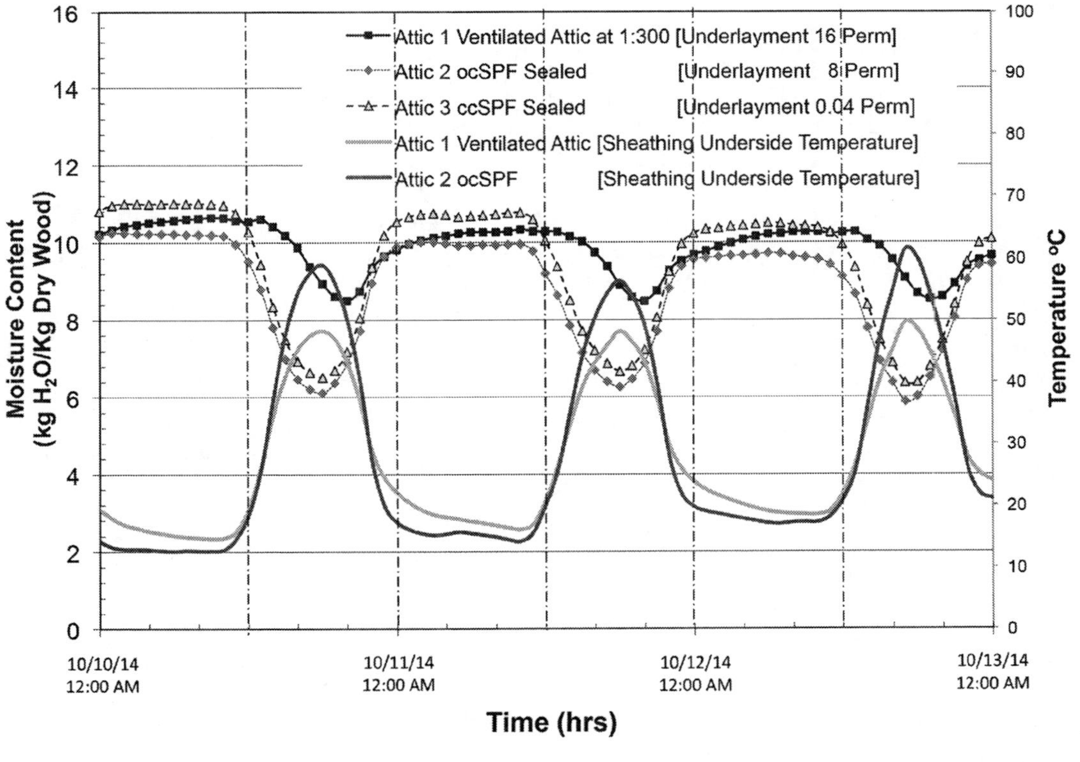

Figure 9 *The moisture content of the roof deck sheathing measured on the attics sealed with spray foam insulation (R-22 insulation adhered to sheathing) and on the attic with 1:300 ventilation (R-38 on attic floor).*

Field results from the NET facility reveal an interesting trend regarding the true behavior of attics with both open-cell and closed-cell SPF, as well as code levels of attic floor insulation (see Figure 10). The heat fluxes crossing the attic floor of the two different sealed attics insulated with R-22 of spray foam were measured using heat flux transducers (see section on Instrumentation). The flux in both sealed attics is literally double the flux of a ventilated attic having IECC 2015 code levels of insulation on the attic floor. Twice as much heat would be entering the home through the attic due to radiation heat transfer from the decks to the bare gypsum board floor.

The attic air temperatures in the sealed attics were less than that of the ventilated attic; they were only about 2°C to 3°C warmer that the outdoor air. However, ceiling drywall has a low thermal resistance, about (≅R-0.45) and is susceptible to the higher heat flows despite the lower attic air temperatures in the sealed attics as compared to the ventilated attic with R-38 insulation on the attic floor. Parker (2005) conducted field tests of ventilated and sealed attics on the FSEC Flexible Roof Facility (FRF). He tested a ventilated attic having R-19 on the attic floor against a sealed attic with 6 inches of ocSPF (R-22) in the roof rafters. His results were similar to the results by ORNL with exception of the magnitude in difference. The peak day flux of the sealed attic at the FRF was 60% higher than that of the R-19 ventilated attic. In comparison, the NET showed fluxes 150% higher than the ventilated attic with R-38 insulation. The results should be investigated analytically for the trade-off between cost effectiveness and energy savings to show what the optimal level of insulation should be for sealed attic systems.

CONCLUSIONS

Field experiments were conducted on a building located in the hot, humid climate of Charleston SC. Several attic systems were tested for two full years under a well-controlled indoor environment to observe the thermal and hygrothermal effects of a hot and humid climate. Three major findings were observed for attics sealed with open-cell spray polyurethane foam and closed-cell spray polyurethane foam as compared to a conventionally ventilated attic with R-38 insulation on the attic floor.

The attic sealed with ocSPF showed higher moisture movement during the warmer months of the year than observed in the colder months. The attic with ccSPF showed similar trends, but the moisture transfer crossing the ccSPF was much less than that observed for the ocSPF because ccSPF is less permeable than ocSPF.

Moisture was physically observed along the 2 by 6 rafters of the attic sealed with ocSPF. Water damage was observed about 1.6 m up from the soffit. Checks made at the completion of the field study did not show evidence of moisture near the ridge; rather, water marks were observed along the rafters about mid span of the roof.

Sealed attic insulated with spray polyurethane foam that is less than code levels of insulation will have significantly higher heat flows crossing the attic floor, because the differ-

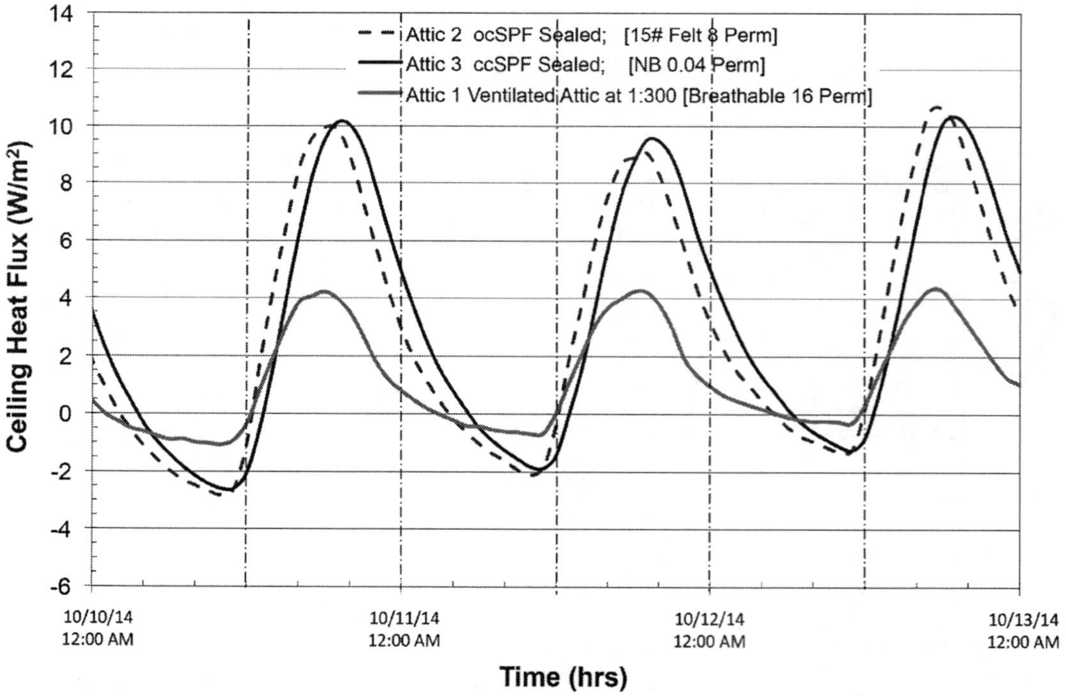

***Figure 10** The heat flux crossing the attic floor of the sealed attic (no insulation on the floor) and the conventionally ventilated attic having R-38 insulation lying on the floor.*

ence in thermal resistances of bare drywall as compared to an insulated ceiling predominates over the benefit of the reduced attic air temperature.

RECOMMENDATIONS

There is limited information on the flexural properties (bending stress and modulus of elasticity) of plywood and OSB. Green et al. (1999) and Ayrilmis, Buyuksari, and As (2010) demonstrated that the mechanical properties of plywood and OSB decrease as the temperature of the wood increases. The repetitive drying and wetting of plywood and OSB roof sheathing should be studied to document the wood's fatigue life when subjected to the stress and strain of roof loads while exposed to the diurnal temperature and humidity levels occurring in hot and humid climates. It is clear that sealed attics have many advantages for high-performance, low-load homes. It is not clear how to achieve this end with a high-R, moisture-managed solution that is acceptable to production builders. Rose (1998) recommended an air gap be provided between the sheathing and the top of the insulation in unvented attics to allow ventilation to carry the moisture out of the roof system. Lstiburek (2015) has successfully demonstrated a vapor diffusion ridge vent that is watertight but vapor open to expel moisture from the sealed attic. His results show all the moisture migrates toward the ridge. It is the author's belief that both an inclined cavity and the vapor diffusion ridge vent are needed to solve this problem, which is technologically complex, requiring consideration of both thermal and hygrothermal behavior in natural conditions, as well as requiring complex integration into builders' and trade subcontractors operations.

NOMENCLATURE

R-38 = thermal resistance to heat transfer [38×(h·ft^2·°F)/Btu], I-P units

= thermal resistance to heat transfer [6.7 (m^2·K)/W], SI units

APPENDIX—MISCELLANEOUS INSTRUMENTATION

Weather Station

Weather data collected on the roof of the NET facility include wind speed and direction, solar irradiance, night-sky radiation, rainfall, outdoor ambient temperature, relative humidity, and barometric pressure. These data are logged by the DAS and written to the same attic data files on a weekly basis. The outdoor air temperature was measured using a shielded thermistor accurate to ± 0.2°C from 0°C to 70°C. Total global solar irradiance was also measured on the pitched roofs (one pyranometer for each roof slope). Night-sky radiation was measured using a pyrgeometer. Spectral range of the pyrgeometer is 4500 to 42,000 nm with a nonlinearity less than 1% of measure over the range –250 to 250 W/m^2.

Moisture Pins

The moisture content of wood is directly relatable to the electrical resistance of the wood, which in turn, can be related to the moisture content of the wood. The resistance of wood is typically so high that it usually cannot be measured directly without specialized equipment. Moisture pins, circuits, and conversions of resistance to moisture content were developed by ORNL. A circuit for measuring the very high resistance typical of wood with a standard DAS or hand-held multimeter is shown schematically in Figure A1, as reported by Straube, Onysko and Schumacher (2002).

R_w is the resistance of the wood; R_p is the protection resistor (in the event that pins are short-circuited by moisture on the wood surface); the triangle is an optional diode that protects the data acquisition equipment by "clamping" the voltage to a specific maximum if the pins are shorted; R_s is the sensing

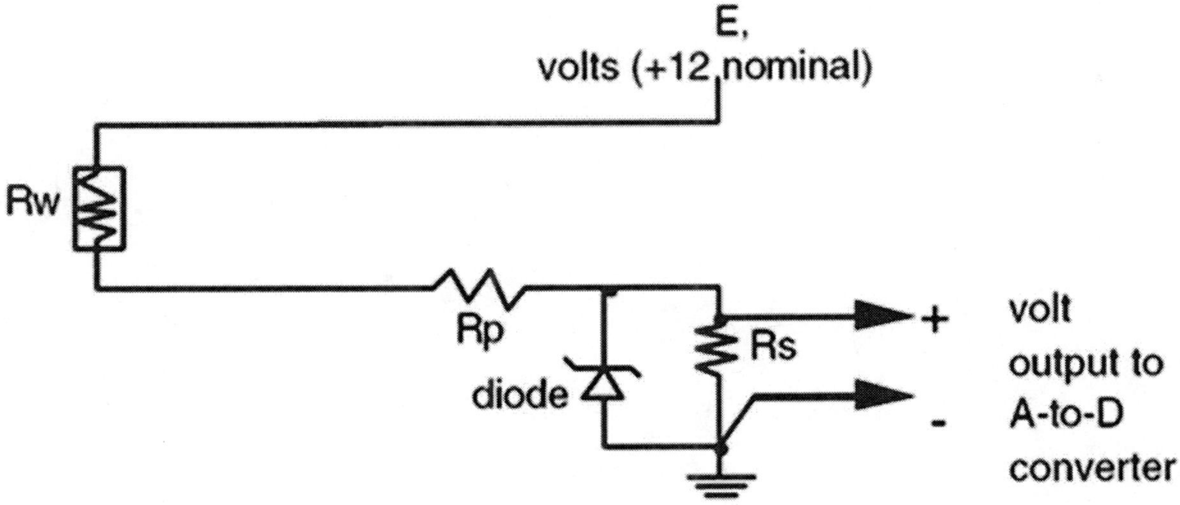

***Figure A1** Circuit to measure resistance of wood with DAS.*

resistor; and E is the supply voltage. Typically $R_p = R_s = 100\ k\Omega$ and $E = 12V$ combined with a 16-bit analog-to-digital converter (Straube, Onysko, and Schumacher 2002).

Using Ohm's Law, we can say that:

$$I = E/(R_W + R_P + R_S) \quad (1)$$

and

$$V = I \cdot R_S \text{ and } I = V/R_S \quad (2)$$

For the same current (I), equating Equations 1 and 2 we have:

$$\frac{V}{E} = \frac{R_S}{(R_W + R_P + R_S)} \quad (3)$$

Therefore, in terms of the resistance of the wood in ohms:

$$R_W = R_S\left(\frac{E}{V}\right) - (R_P - R_S) \quad (4)$$

An empirical fit of moisture content to electrical resistance with and without temperature effects was formulated from literature data gleaned from ASTM D4444-13 (2013) and James (1975; 1988) and compared to data from Garrahan (1989), Carli et al. (2007), and Huber Engineering (2013). The formulation of the ORNL equation takes the following form:

$$MC_u = 10^{[a + b \cdot Log_{10}\{Log_{10}\{R_W\}\}]} \quad (5)$$

where the a and b coefficients of Equation 5 are derived from linear fits to data from James (1988) and ASTM D4444-13 for 20 different species of wood, including oriented strand board data gleaned from Huber Engineering (2013). Figure A2 shows the consistency of the correlation against literature data.

The results in Figure A2 are for a wood temperature of about 20°C. The ORNL correlation was made more robust by using data from James (1988) and Garrahan (1989) to account for temperature effects in the range of roof deck temperatures experienced on the NET facility. Note that information by James omits derivation of his correction for temperature compensation. The correction to temperature takes the following form:

$$MC_{cor} = \frac{a + bx + cx^2 + dx^3 + ey}{1 + fx + gx^2 + hx^3 + iy}$$

where

x = wood temperature (°F) as measured by a thermistor adjacent each pair of pins

y = MC_u, the uncorrected moisture content from Equation 5

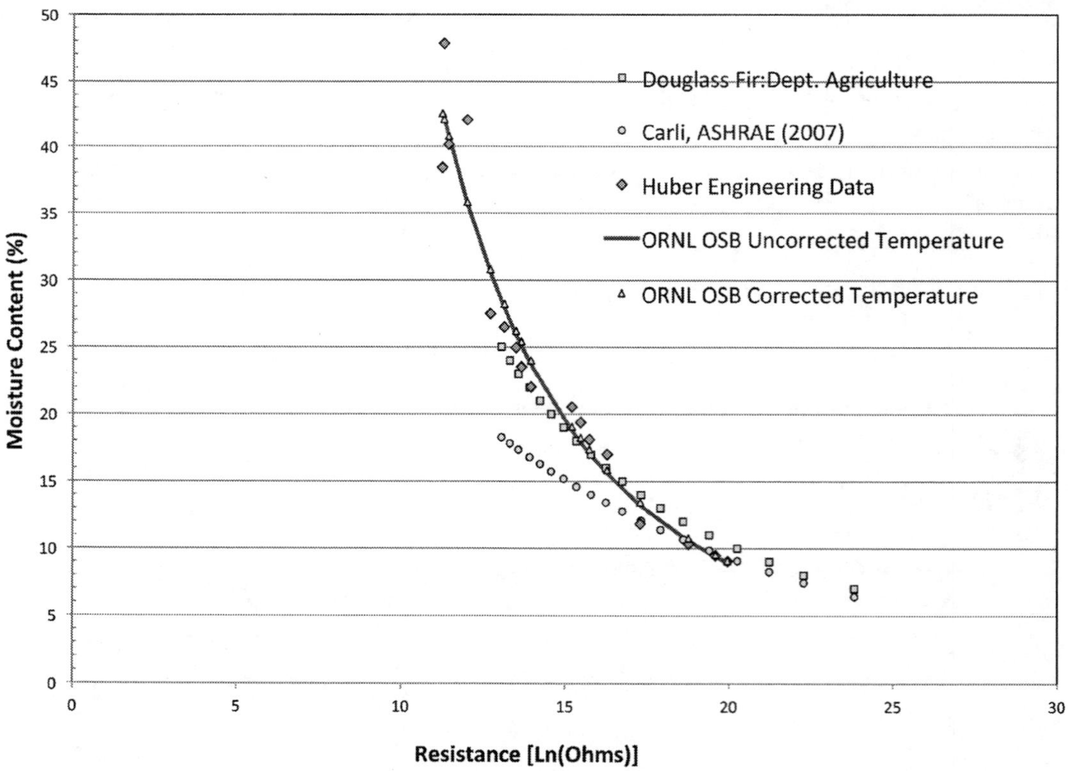

***Figure A2** Empirical fits of moisture content of Douglas fir and oriented strand board (OSB) plotted from data gleaned from the open literature. Wood measurements made at about 20°C.*

Garrahan (1989) conducted an extensive series of moisture measures for Douglas fir and provided correction tables usable in field and laboratory measurements. Garrahan's results as reported by Straube, Onysko, and Schumacher (2002) yielded an average error in the order of 0.5% MC for an identified species, although individual readings may be as much as 2% MC different from the oven-dry values as reported by Straube, Onysko, and Schumacher. Results of the ORNL correlation based on James (1988) and ASTM D4444-13 (2013) is shown against Garrahan (1989) for wood temperatures of –5°C, 22.8°C, and 80°C (see Figure A3).

The average absolute error at the three temperature levels are +1.8% difference at 80°C wood temperature, 0.2% difference at 22.8°C wood temperature, and –1.6% difference at –5°C.

REFERENCES

ASHRAE. 2009. *ASHRAE Handbook—Fundamentals*. Atlanta: ASHRAE.

ASHRAE. 1980A. ASHRAE 90-1980, *Energy efficient design of low-rise residential buildings*. Atlanta: ASHRAE.

ASTM. 2000. ASTM E741-00, *Standard test method for determining air change in a single zone by means of a tracer gas dilution*. West Conshohocken, PA: ASTM International.

ASTM. 2004. ASTM Standard D 4444-13, *Standard test method for laboratory standardization and calibration of hand-held moisture meters*. West Conshohocken, PA: ASTM International.

ASTM. 2010. ASTM C518-15, *Standard test method for steady-state thermal transmission properties by means of the heat flow meter apparatus*. West Conshohocken, PA: ASTM International.

ASTM. 2013. ASTM D4444-13, *Standard test method for laboratory standardization and calibration of hand-held moisture meters*. West Conshohocken, PA: ASTM International.

Ayrilmis, N., U. Buyuksari, and N. As,. 2010. Bending strength and modulus of elasticity of wood-based panels at cold and moderate temperatures. *Cold Region Science and Technology* 63:40–43.

Boudreaux, P., S. Pallin, and R. Jackson. 2016. Investigation of the proposed solar driven moisture phenomenon in asphalt shingle roofs, in Thermal Performance of the

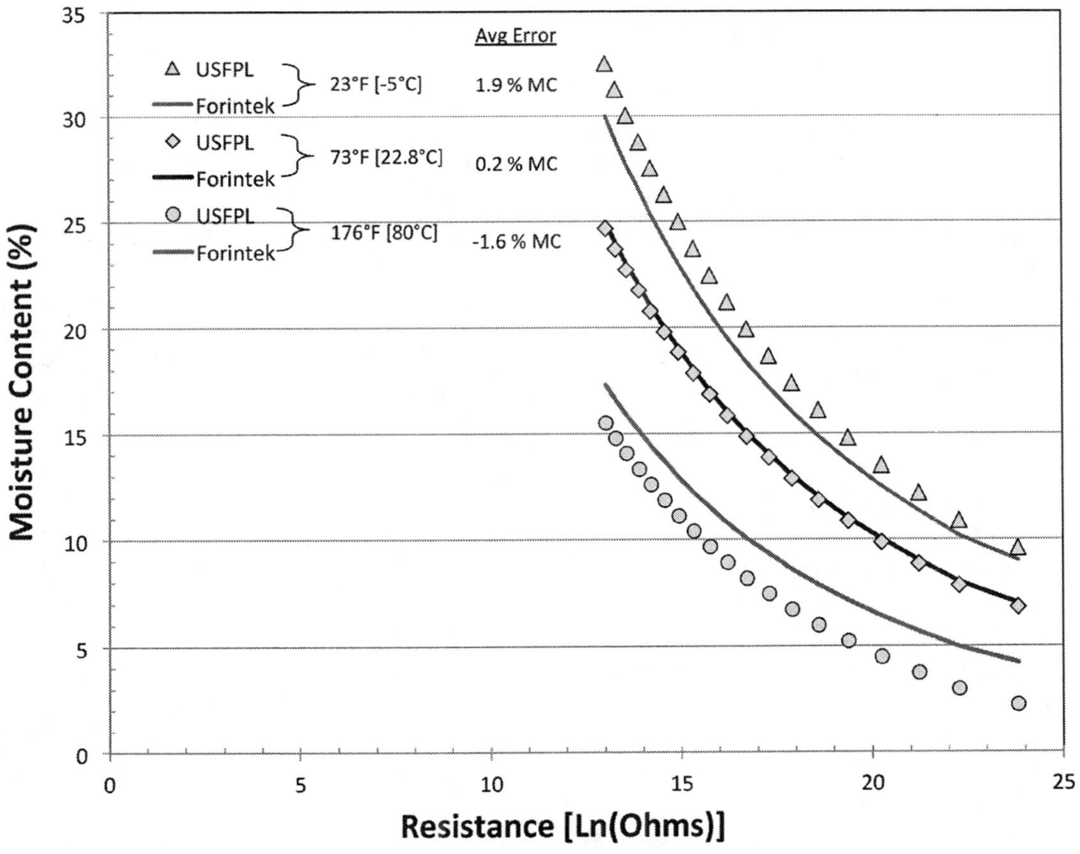

Figure A3 *ORNL fit of moisture content for Douglas fir plotted from data gleaned from the open literature (symbols are for fit to USFPL data) compared Garrahan's results (solid lines are Forintek data).*

Exterior Envelopes of Buildings, XIII, Proceedings of ASHRAE THERM XII, Clearwater, FL., Dec. 2016.

Building Science Press. 2008. *Unvented roofs, hot-humid climates, and asphalt roofing shingles*. Building Science Corporation, Research Report 0306, Feb. 2003.

Carli, C., A. TenWolde, and R. Munson. 2007. Moisture performance of a contemporary wood-frame house operated at design indoor humidity levels, in Thermal Performance of the Exterior Envelopes of Buildings, X, proceedings of ASHRAE THERM XII, Clearwater, FL., Dec. 2007.

Chasar, D., J. Sherwin, V. vonSchramm, and S. Chandra. 2010. Measured performance of side-by-side South Texas homes. *ASHRAE Thermal Performance of the Exterior Envelopes of Whole Building XI International Conference.*

EDU. 2005. Fiberglass-Insulated Homes are the Leakiest. *Energy Design Update*, Aspen, CO.

Dillon, B. 2013. Unvented attics: Encapsulating with spray polyurethane foam. *Home Energy Magazine*, Nov/Dec 2013, 30(6):16–18.

Garrahan, P. 1989. Moisture meter correction factors, In: Green, D.W., Shelley, B.E. and Vokey, H.P. (eds), *In-Grade Testing of Structural Lumber, Proceedings* 47363:39–43. Madison, WI: Forest Products Society.

Green, D., J. Evans, J. Logan, and W. Nelson. 1999. Adjusting modulus of elasticity of lumber for changes in temperature. *Forest Products Journal* 49(10):82–94.

Holladay, M. 2010. It's OK to skimp on insulation, Icynene says. http://www.greenbuildingadvisor.com/blogs/dept/musings/it-s-ok-skimp-insulation-icynene-says.

Hun, E., P. Spafford, and A. Desjarlais, 2014. *Evaluation of air barriers for residential buildings*. Oak Ridge National Laboratory, ORNL/TM-2013/604.

Huber Engineering. 2013. Personal communication.

IECC. 2012. International Energy Conservation Code. International Code Council.

IRC. 2012. International Residential Code. Table N1102. International Code Council.

James, W.L. 1975. *Electrical moisture meters for wood*. U.S. Department of Agriculture Forest Service, General Technical Report, FPL-6, Madison, WI.

James, W.L. 1988. *Electric moisture meters for wood*. Gen. Tech. Rep. FPL-GTR-6. Madison, WI: U.S. Department ofAgriculture, Forest Service, Forest Products Laboratory.

Kumaran, M. K. 2008. *A thermal and moisture property database for common building and insulating materials*. ASHRAE RP-1018. Atlanta: ASHRAE.

Lstiburek, J.W. 2015. Venting vapor. *ASHRAE Journal* Aug, 2015, 46–51.

Parker, D., P. Fairey, and L. Gu. 1993. Literature review of the impact and need for attic ventilation in Florida homes. Florida Solar Energy Center, FSEC-CR-1496-05.

Parker, D. 2005. Literature review of the impact and need of attic ventilation in Florida homes. Florida Solar Energy Center publication FSCE-CR-1496-05.

Parsons, G. and M. Drzyzga, M. 2013. Insulating and air sealing low-pitch residential attic spaces: Cost-effectiveness evaluation, in Thermal Performance of the Exterior Envelopes of Buildings, XII, proceedings of ASHRAE THERM XII, Clearwater, FL., Dec. 2013.

Railkar, S., A. Desjarlais, A. Chich, and W.A. Miller. 2010. Thermal and hygrothermal performance of sealed and ventilated attics with and without breathable membranes in a hot and humid climate, in Thermal Performance of the Exterior Envelopes of Buildings, XII, proceedings of ASHRAE THERM XII, Clearwater, FL., Dec. 2013.

Roppel, P., N. Norris, and M. Lawton, M. 2013. Highly insulated, ventilated, wood-framed attics in cool marine climates. ASHRAE Thermal Performance of the Exterior Envelopes of Whole Building XI International Conference.

Rose, J. D., and American Plywood. 1998. Preliminary testing of wood structural panel shear walls under cyclic (reversed) loading. American Plywood Association: Tacoma, WA.

Rudd, A. and Lstiburek, J. 1998. Vented and sealed attics in hot climates. *ASHRAE Transactions* 104(2).

Straube, J. and E. Burnett. 2005. Building Science for Building Enclosures. Building Science Press, Westford, MA 01886. ISBN: 0-9755127-4-9.

Straube, J., D. Onysko, and C. Schumacher. 2002. Methodology and design of field experiments for monitoring the hygrothermal performance of wood frame enclosures. *Journal of Building Physics* 26:123. http://jen.sagepub.com/content/26/2/123.

Viitanen, H.A. 1997. Modeling the time factor in the development of mold fungi—The effect of critical humidity and temperature conditions on pine and spruce sapwood. *Holzforschung—International Journal of the Biology, Chemistry, Physics and Technology of Wood* 51(1):6–14.

U.S. Census Bureau. 2013. Characteristics of New Housing. http://www.census.gov/construction/chars/completed.html.

Walker, I.S. 1998. Technical background for default values used for forced air systems in proposed ASHRAE Standard 152P. *ASHRAE Transactions* 104(1). Presented at ASHRAE TC 6.3 Symposium, January 1998. LBNL 40588.

Field Study on the Thermal and Hygrothermal Performance of Insulated Ventilated Nail Base System

Ming L. Shiao
Member ASHRAE

Sudhir Railkar
Member ASHRAE

William A. Miller
Member ASHRAE

Andre Desjarlais

ABSTRACT

The thermal and hygrothermal performance of a ventilated attic with and without an insulated ventilated nail base (IVNB) system over the roof deck was studied in a hot, humid climate. Seven test attics were constructed and instrumented for monitoring their field performance. Attics with the same configuration but having shingles with increased solar reflectance are also included for comparison. These test attics are instrumented to provide field data of temperature, heat flux, relative humidity, moisture content, and pressure differential across the roofing assemblies. Field data collected over a year suggests that the IVNB system over the roof deck can drastically reduce heat influx during summer and prevent heat loss in winter, which may provide a cost-effective solution for roof retrofitting to achieve high energy savings. During the winter, the IVNB system was found to prevent moisture buildup and, hence, may lower the condensation potential. Hygrothermal analysis of moisture content in roof deck and HVAC energy consumption calculations in various climate zones will be discussed to shed light on the total benefits of IVNB in the scenario of an above-sheathing roof retrofit solution to achieve higher energy savings.

INTRODUCTION

Attic construction in a steep sloped residential dwelling is a common practice to provide a buffer between the roofing assembly and the living space beneath, where the effect of heat influx/outflux and moisture buildup from various sources can be effectively managed (Rowley et al. 1939; Lstiburek 2006). On the other hand, unvented attics or attics sealed via spray foam applied directly to the underside of the roof sheathing have been gaining significant tractions as energy saving alternatives over the more traditional vented attics (Chasar et al. 2010). However, a recent study (Desjarlais et al. 2012) has shown that the energy required to condition the living space underneath a sealed attic was found to be significantly higher than that of a typical vented attic having standard 1/300 ventilation ratio. This study concluded that the heat gain during summer season and heat loss in winter season from attic into the living space via the attic floor were significantly higher in the sealed attic, because the sealed attic only has spray foam in the underside of roof deck and no insulation on attic floor, as compared to a vented attic that has insulation on attic floor to prevent heat flux through ceiling. The sealed attic was also found to have significantly high attic humidity and saturated moisture level in plywood decks, which is in agreement with other studies revealing significant moisture buildup in a sealed attic (Straube et al. 2009; Lstiburek 2015) due to the fact that air leakage can easily be present even in a well-sealed attic (Derome 2005; Prahl and Shaffer 2014). As a result, moisture from various sources can enter the unvented space either via the air leakage pass way or by ceiling penetrations, which can then be trapped within the sealed attic roof top where the temperatures tend to be highest. This can lead to significant moisture accumulation and deterioration of roof decks over time (Lstiburek 2015). Hence, it has been recommended that sealed attics should be conditioned to mitigate the moisture buildup issue (Boudreaux and Pallin 2013; Roppel, et al. 2013) or a so-called "moisture diffusion port" (Lstiburek 2015) can be used to vent moisture in sealed attics around the roof top area. In the first scenario, the total energy benefit of

Ming L. Shiao is the principal scientist at GAF and **Sudhir Railkar** is the director of Product Development Steep Slope at GAF, Parsippany, NJ. **William A. Miller** is an active member of ASME and ASHRAE. He formerly chaired ASHRAE's T.C. 1.2 Instruments and Controls. He leads the ASHRAE student body at the University of Tennessee, TN. **Andre Desjarlais** is the group leader at Oak Ridge National Laboratory in Oak Ridge, TN.

a sealed attic may be reduced, since additional HVAC energy is needed to condition the attic space that may not be occupied. For the second remedy, although the moisture diffusion port has been used in wall cavities (Plescia 2003), it is unclear that the moisture diffusion port can be an effective long-term solution for sealed attic without creating other unforeseen roofing issues (Ueno and Lstiburek 2015), such as wind uplift resistance or fire resistance of the roofing assembly due to the partial removal of the roof deck.

On the other hand, reduction of energy consumption in the built environment is a key area in the designing of new residential housing and many new technologies, such as ground-source heat pump, SIP walls, building integrated PV, or highly insulated windows, have been proposed (McNabb 2013). For roofing assemblies, the use of highly reflective roof coverings, radiant barriers, or above-sheathing ventilation (Miller et al. 2007) have also been adopted as ways to help reducing building energy consumption towards a net zero energy goal. In a recent study, Miller and Herman (2012) have shown that the combination of above-sheathing ventilation and insulation above roof deck can result in highly energy-efficient attic performance. Such an approach is especially suited for roof retrofit construction, where adding above-sheathing ventilation can be a very attractive solution due to the fact that the roof surface is readily accessible and the system can be installed without the need to access the interior of building. This is particularly true in retrofitting complex roof structures or cathedral ceilings where the attic spaces can be hard to reach for applying spray foam near the underside of sheathing. In this study, we have conducted a field study to examine the thermal and hygrothermal performance of a standard ventilated attic, attic with increased solar reflectance, and attic with an above-sheathing insulated vented nail base (IVNB) system in a hot, humid climate to compare and contrast their differences on overall attic performance.

ATTIC CONFIGURATION AND CONSTRUCTION

This field study was conducted in South Carolina (Climate Zone 3) using the existing test facility known as the Natural Exposure Test Facility (see Figure 1). The facility consists of a building approximately 25 × 80 ft (7.6 × 24.4 m) having attic space and pitched roof at 3/12 slope in a north/south orientation. The attic and the roof deck were subsequently subdivided into seven bays for various attic/roof configuration studies. Each bay is approximately 11 ft (3.4 m) in width and is separated by using barrier walls thermally insulated to about R-15. The barrier walls were air sealed using caulk and spray foam sealants.

These bays were then configured with combinations of attic ventilation, permeable and impermeable underlayment, above-sheathing ventilation, and solar reflectance of the shingle roof covering to study their effect on attic performance. The space below the attic floors was maintained at constant temperature of 70°F (21°C) to prevent any heat exchange between the bays. A schematic of the attic assemblies is shown in Figures 2 and 3. Configurations for each attic/roof system are listed in Table 1.

The Attic 1 is the control case and has standard attic construction with 1:300 soffit/ridge ventilation. The roof deck has a permeable underlayment (16 perms) and is covered in dark-colored asphalt shingles having solar reflectance of 0.03. The soffit and ridge openings are of equal area. For this study, we are focused on Attic 6 and Attic 4 to understand their field performance for the effect of above-sheathing ventilation and the increased solar reflectance. Attic 4 is identical to Attic 1 except it has cool shingles (solar reflectance of 0.28) and a nonpermeable underlayment (0.04 perms). It also has a nonactive building-integrated photovoltaic (BIPV) installed on the south-facing roof deck. The Attic 6 also has the same attic construction and soffit/ridge ventilated at 1:300 ratio as the Attic 1 and an insulated ventilated nail base (IVNB) system is added to the roof deck to provide the above-sheathing ventilation and insulation. A 1 in. ISO board was installed directly on top of a peel and stick roof underlayment and sheathing (OSB). Nominal 2 by 4 lumber was placed on top of the ISO board and

Figure 1 *The southern exposure of the NET Facility in Charleston SC.*

nailed through the insulation to the roof joists to form an insulated and ventilated nail base (IVNB). A second sheathing (OSB) was fastened and covered with 15 pound felt paper and dark, heat-absorbing shingles with solar reflectance of 0.03. R-38 fiberglass batt insulation was placed on the attic floor.

THE INSTRUMENTATION

Temperature sensors, relative humidity (RH) sensors, and heat flux transducers (HFT) were attached at pertinent locations to analyze the *in situ* performance of the individual attic bays. The sensors were placed in an identical pattern for each attic cavity as shown in Figure 3 to standardize the comparison of results. For the Attic 6 with the IVNB system, the temperature sensor locations are the same except additional temperature and percent RH sensors were added in the air gap within the IVNB system.

Thermistor and humidity sensors were calibrated by ORNL Metrology. The response of the thermistors was measured at the following nominal conditions: 15.8°C, 18.5°C, 21.2°C, 23.9°C, and 26.6°C. All probes met the manufacturer's specification for the temperature response of

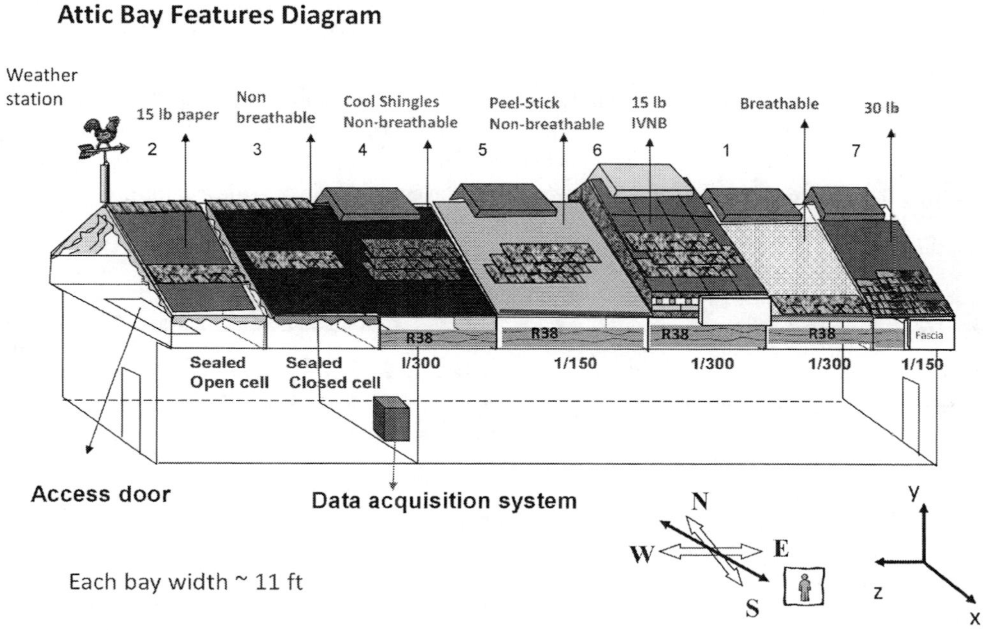

Figure 2 *Test attic assembly layout on NET.*

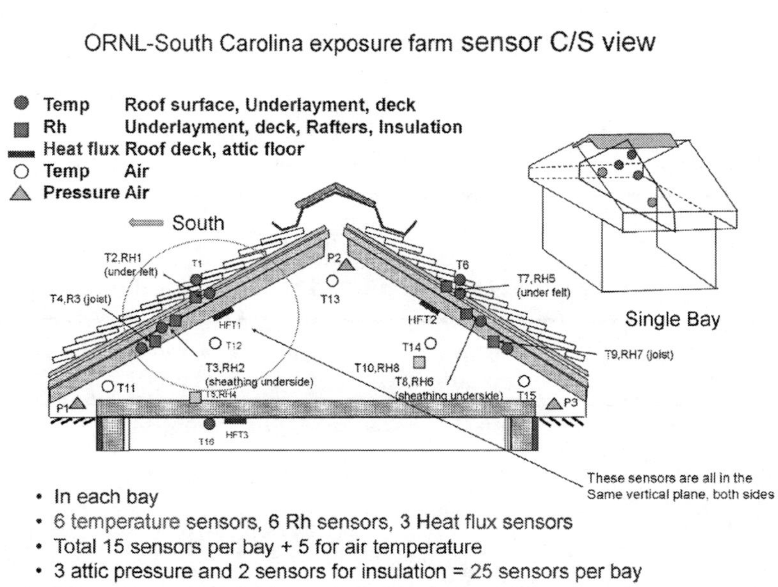

Figure 3 *Instrumentation layout for each attic facility in Charleston, SC, assembly.*

±0.2°C. Humidity sensors were checked at 25%, 50%, 75%, and 90% rh. The error in RH ranged from 2% of reading at 25% rh and 15°C to 6.5% of reading at 90% rh and 26°C.

All heat flux transducers were calibrated in accordance with ASTM C 518-10 (ASTM 2010). R&D Services from Cookeville, TN used a fox heat flow (FHF) meter having a 0.305 by 0.305 m footprint to measure the thermal resistance of batt insulation insulating the floor of Attic 1 (see Table 2). Each HFT was placed in a 0.305 by 0.305 m (30 by 30 in.) guard made of gypsum board and from the same lot of batt insulation calibrated in the FHF meter.

In the field instruments are scanned every 15 seconds and the raw data reduced to engineering units. Averages of the reduced data are written electronically to an open file every 15 min. Averages are calculated over the 15-min interval and are not running averages; they are reset after each 15-min interval. The electronic format is comma-delimited for direct access by spreadsheet programs. *In situ* checks were made of the instruments during the course of the field study.

RESULTS AND DISCUSSION

The temperature profiles across the south-facing roof decks for the control Attic 1 in a typical week during summer season are shown in Figure 4. The temperature data for Attic 4 having shingles with higher solar reflectance and Attic 6 with the IVNB system are plotted in Figure 5 and Figure 6 for direct comparison during the same time frame. As can be seen, the shingle temperatures in Attic 4 showed a slight drop of about 5°C at peak temperatures as comparing to the control attic, since the solar reflectance was increased from 0.03 in Attic 1 to 0.28 for Attic 4. Consequently, the peak attic air temperatures also showed a moderate reduction at about 3°C between the control and Attic 4. On the other hand, we observed a very significant temperature drop of ~10°C in attic air between control attic and Attic 6, where the IVNB system is used (see Figure 6). Also, variations in the temperature profile across the roof deck of Attic 6 were found to be much less than those of the control attic or Attic 4. This is clearly due to the combined

Table 1. Attic Configuration Summary

Attic No.	Attic Insulation	Attic Ventilation	Underlayment	Roof Deck Configuration	Shingle Solar Reflectance
1	R38 on attic floor	1/300 soffit-ridge	Synthetic breathable film (16 perm)	Asphalt shingle	TSR = 0.03
2	5.5 in. open-cell spray foam on roof sheathing	None	15 lb felt (8 perm)	Asphalt shingle	TSR = 0.03
3	5.5 in. closed-cell spray foam on roof sheathing	None	Synthetic nonbreathable (0.04 perm)	Asphalt shingle	TSR = 0.03
4	R38 on attic floor	1/300 soffit-ridge	Synthetic nonbreathable (0.04 perm)	Asphalt shingle; BIPV on south deck	TSR = 0.28
5	R38 on attic floor	1/300 soffit-ridge	Peel and stick nonbreathable	Asphalt shingle; BIPV on south deck	TSR = 0.03
6	R38 on attic floor	1/300 soffit-ridge	15 lb felt (8 perm)	IVNB (1/100) and shingle	TSR = 0.03
7	R38 on attic floor	1/150 intake-ridge ventilation	30 lb felt (5 perm)	Metal shingles	TSR ~ 0.10

Table 2. Calibration of Fiberglass Batt Insulation Used as Guard for the Ceiling Heat Flow of Attic 1

Specimen	Thickness, in.	Density, lbf/ft^3	Apparent Thermal Conductivity, $Btu \cdot in/(ft^2 \cdot h \cdot °F)$	Thermal Resistance $(h \cdot ft^2 \cdot °F)/Btu$
13-3411	12.000	0.53	0.3056	39.3
13-3422	12.001	0.53	0.3417	35.1
13-3423	12.000	0.54	0.3448	34.8
13-3423	12.000	0.54	0.3448	34.8

1 in. = 0.0254 m; 1 lbf/ft^3 = 16.02 kg/m^3; 1 $(Btu \cdot in)/(ft^2 \cdot h \cdot °F)$ = 1.442 $W/(m \cdot K)$
1 $(h \cdot ft^2 \cdot °F)/Btu$ = 6.7 $(m^2 \cdot K)/W$

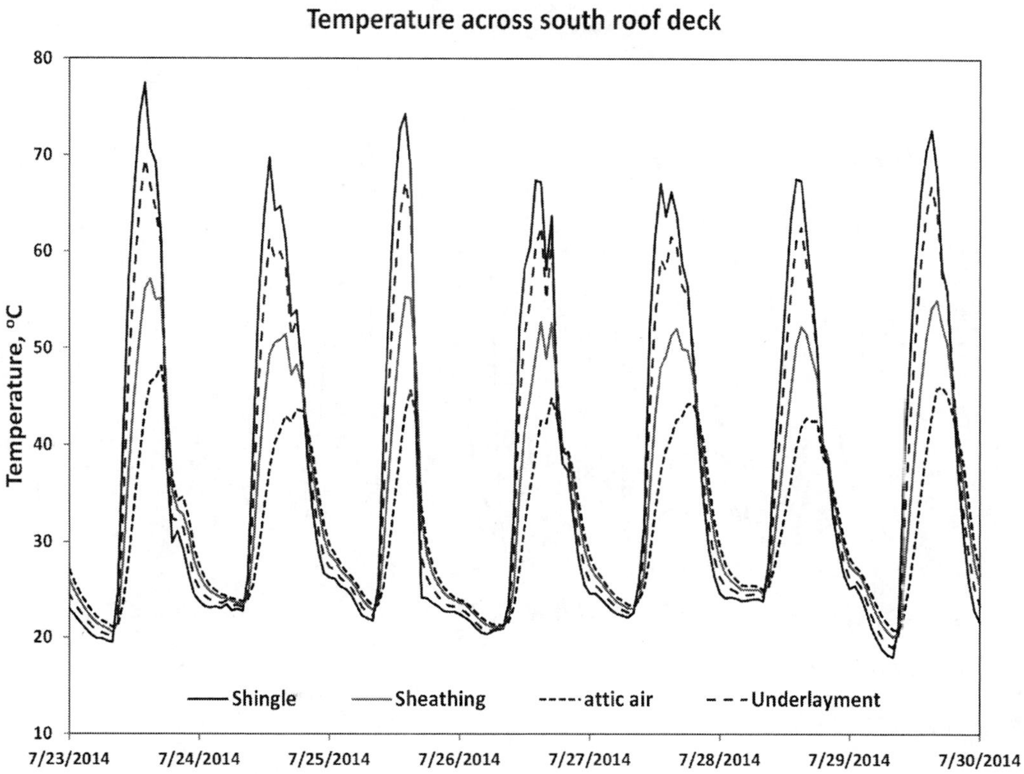

Figure 4 *Temperature in roof deck for Attic 1.*

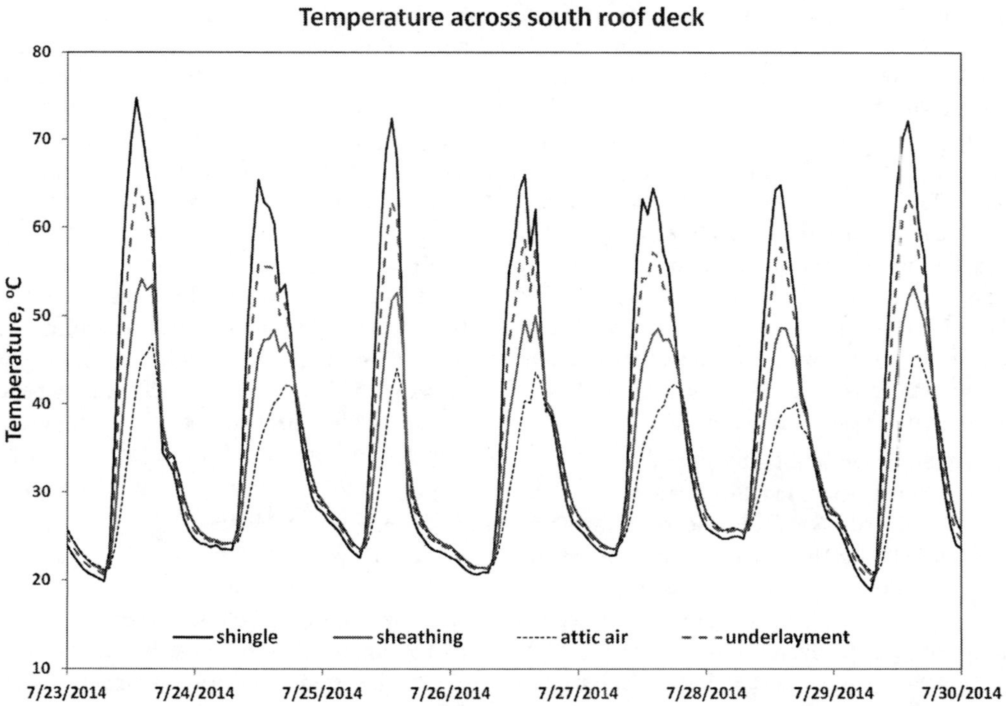

Figure 5 *Temperature in roof deck for Attic 4.*

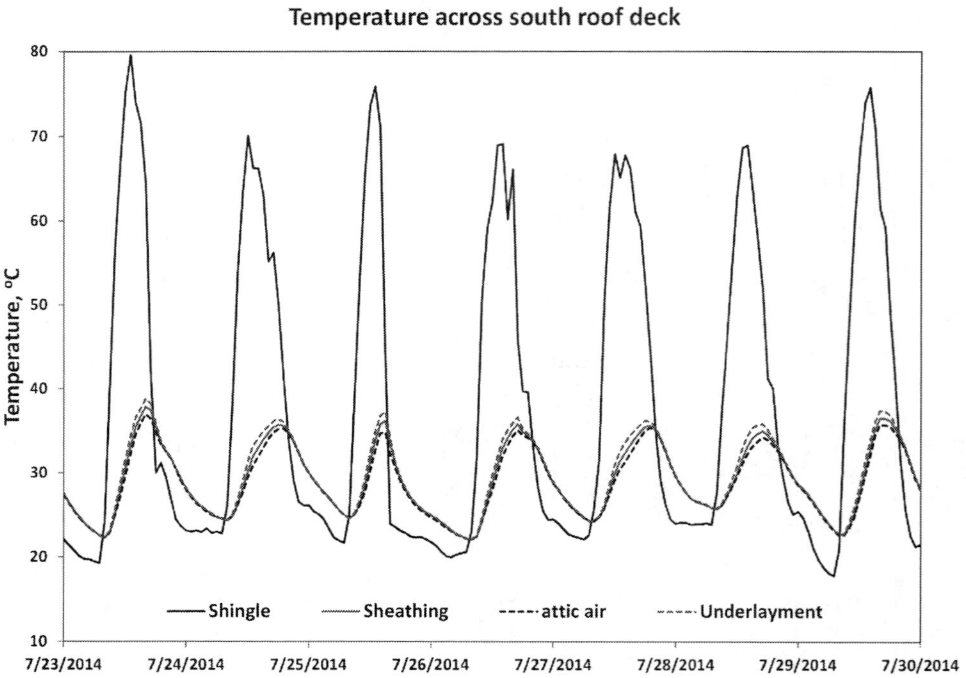

Figure 6 *Temperatures across roof deck in Attic 6.*

ventilation effect under the shingle and the insulation in the IVNB system to prevent heat flux into attic. Such an above-sheathing system in Attic 6 appeared to provide an effective means to mitigate the solar heat influx that is only marginally reduced by the cool shingles as in Attic 4. As a result, there is a drastic reduction in heat flux through roof deck by the IVNB system (see Figure 7), which showed that over 90% of solar heat was effectively blocked from entering the roof deck, as compared to 25%~30% drop in peak heat flux by increasing shingle solar reflectance from 0.03 to 0.28 (see Figure 7). The data revealed that the IVNB can be highly effective in reducing heat flux into attic space. Since the IVNB system is designed as an above-sheathing installation to avoid change in sub-attic structure, it can also provide a practical solution in situations where deep energy retrofit is needed by simple re-roofing or requiring minimum impact on interior change.

During the winter season, similar benefit of the IVNB on effectively managing energy loss in attics was also observed. This can be seen in Figure 8, where the roof sheathing temperatures and the heat flux through roof deck in the north-facing side were plotted during the weeks in January. As can be seen, the Attic 6 with the IVNB was found to have the modest sheathing temperature change during each day cycle (i.e., the lowest sheathing temperature swing from high point at noon to lowest at night was observed as compared to the control and Attic 4). As a result, the heat flux data in Figure 8 showed that Attic 6 resulted in very small heat flux through roof deck during this period of time, where nighttime radiation cooling can lead to significant heat loss. On the other hand, the attic with cool shingles only showed a loss in solar heat influx during the daytime, which has been noted as the heating energy penalty when higher solar reflective materials are used (Saber et al. 2011). Although the IVNB was also found to reduce the benefit of solar heat during daytime in winter, similar to the heating energy penalty of a cool roof, the prevention of the heat loss during the nighttime black body radiation was found to outweigh the daytime solar heat loss in this case. Since this study was carried out in Climate Zone 3 where the solar heating penalty is less severe, one may suspect that the IVNB system may not be as energy efficient in colder regions. However, the total energy benefit from preventing summer solar heat influx and the nighttime heat loss during winter can be very noticeable in the IVNB system, which may outweigh the heating penalty even in cold regions. This aspect will be further discussed in the Annual Energy Consumption Simulation section by modeling the energy use in various cities across U.S. to shed light on the benefits of the IVNB system.

For moisture performance, the percent relative humidity in attic air across the north roofing assembly for the three attics are shown in Figure 9 for the winter season and in Figure 10 for the summer season. As can be seen in Figure 9, Attic 6 with the IVNB system showed a consistently lower value in percent RH across the roof deck when compared to the other two attics during the winter months. We noticed that the attic with cool shingles was found to have higher value in percent RH, particularly at the underside of sheathing during winter months. Since the sheathing temperature is also lowered slightly due to

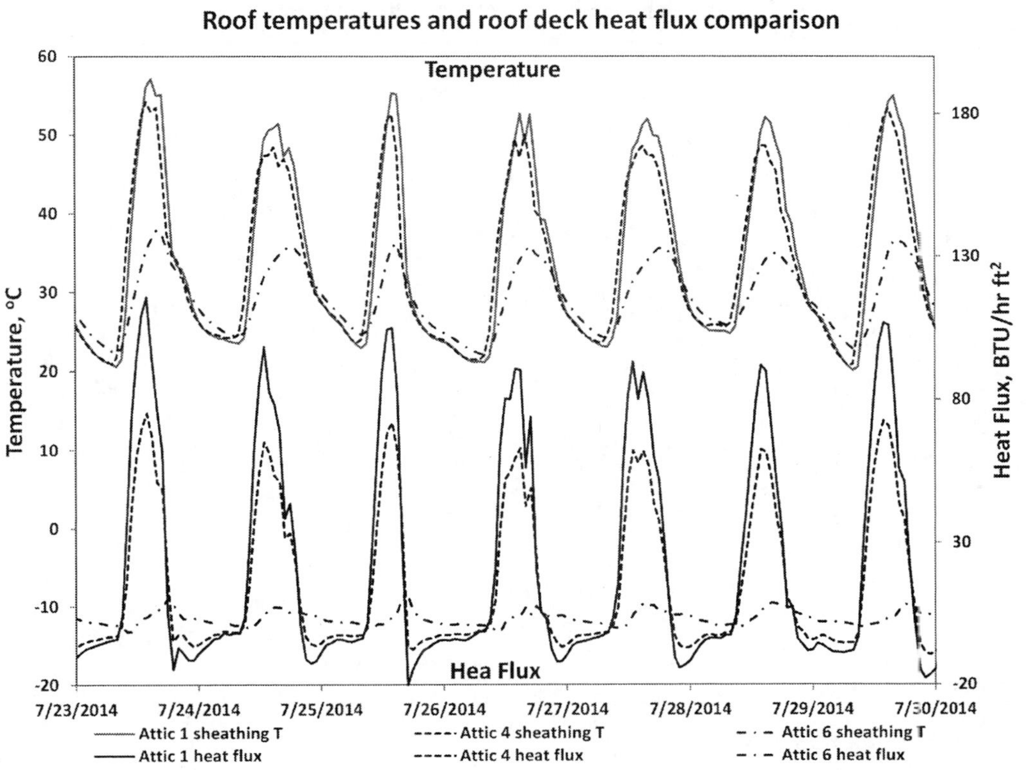

Figure 7 *Summer heat flux in south-facing roof deck.*

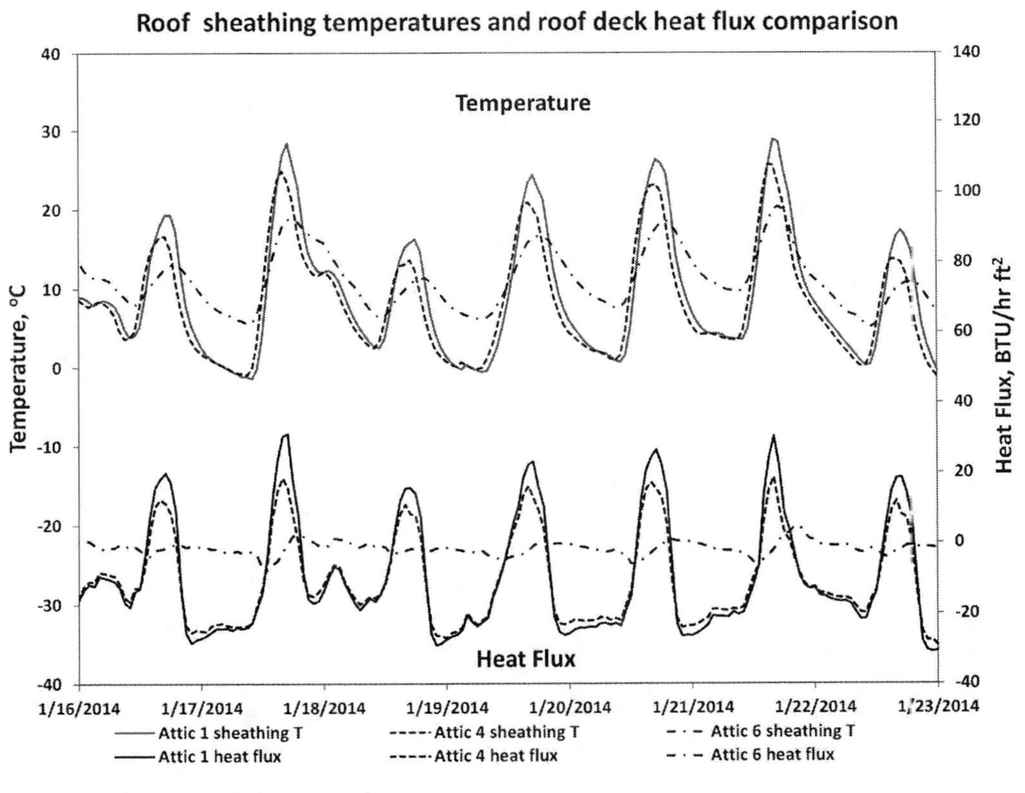

Figure 8 *Winter heat flux in north-facing roofs.*

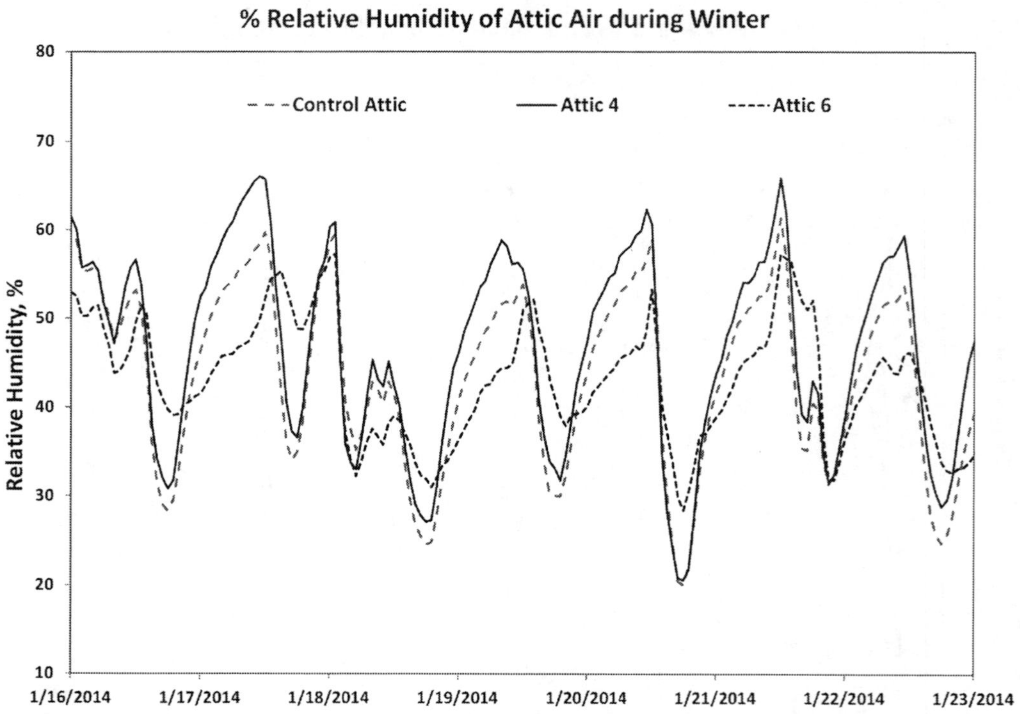

Figure 9 *Percent RH in attic air during winter.*

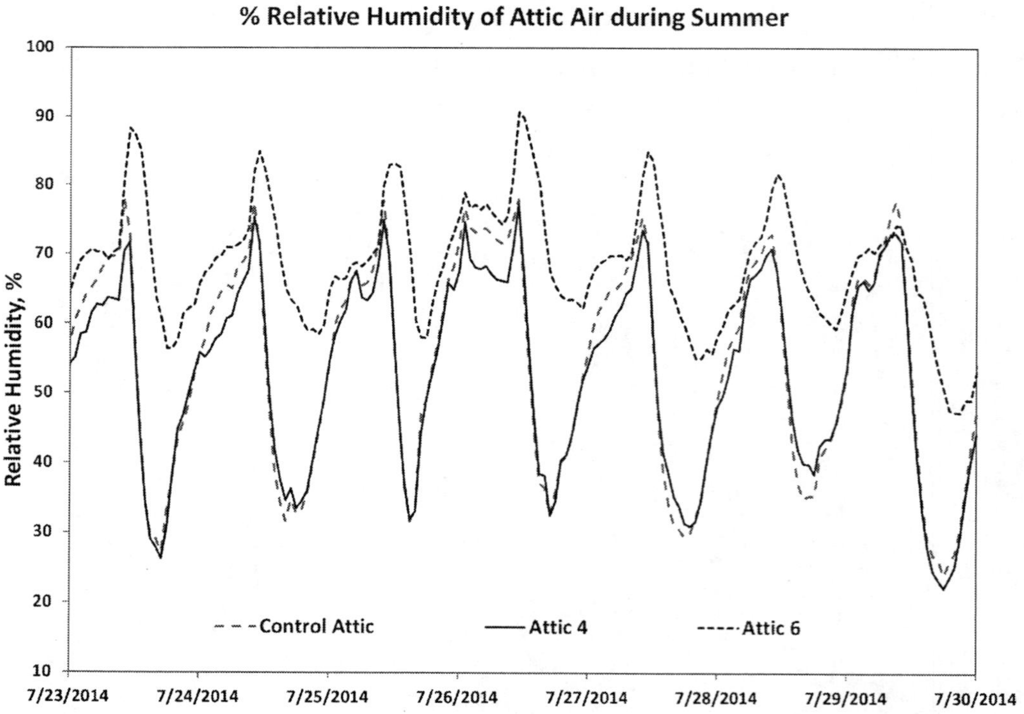

Figure 10 *Percent RH in attic air during summer.*

the presence of cool shingles, the actual moisture level may not be as high. This is plotted in Figure 11 by calculating the specific humidity for the three test attics at the same period. The specific humidity indicated that the attics with cool shingles have slightly higher moisture level during midday as compared to the control and the attic having the IVNB system. However, the differences are not very significant. It is suspected that this increase may due to the lowered solar heat drive in daytime as the shingle solar reflectance increases. As a result, one may postulate that attic with increased solar reflectance may have a higher potential for condensation to occur, particularly in cold regions, due to the loss of solar heating. For the IVNB system, the specific humidity at sheathing and roof joists was found to remain low, and this could result from the reduced heat loss at nighttime where the roof deck temperature was not significantly affected by the night time radiation cooling. Hence, the IVNB system may also appear to help reduce the condensation potential in the attic during winter times.

For the summer season, the percent RH shown in Figure 10 revealed that the attic air in the IVNB system appeared to have higher percent RH values compared to the control attic, whereas the Attic 4, roofed with cool shingle, had similar or slightly lower percent RH level compared to the control attic. This is somewhat unexpected, because the attic air temperatures in Attic 6 showed a significant drop as compared to the other two attics. One may postulate that this may result from the significant reduction of solar heat influx that can drive the moisture out via ventilation at ridge, because Attic 6 has the added insulation and ventilation above roof deck. A closer look at the specific humidity in Figure 12 also confirms that higher moisture level in the attic with IVNB system was found during the summer season. One plausible explanation for this observed higher moisture level may be related to the fact that the attic temperatures are significantly lowered due to the IVNB system, and the temperature striation along the roof line could also be reduced. Hence, the effect of natural ventilation via air buoyancy to mitigate moisture accumulation may be more limited. However, it is unclear if that is the case and more study is needed to reveal the reason to this observed phenomenon.

To further evaluate the impact of long-term moisture, the moisture content of the roof deck was measured over a year period for the three attics. The data is shown in Figure 13 for the north-facing roof deck. One can see that the moisture content showed a seasonal trend and a daily variation about 5%~10%, but the attic with IVNB system was found to have slightly higher moisture content, although the differences were not drastic and the moisture level is still well below the 28% threshold (Canadian Wood Council 2000) for deterioration or mold growth concern. Because the IVNB attic has both attic and above-sheathing ventilations, it is believed that long-term moisture buildups are unlikely to occur. On the other hand, the moisture vapor transport between the roof deck and attics (see Figure 14) revealed that pressure differential for all

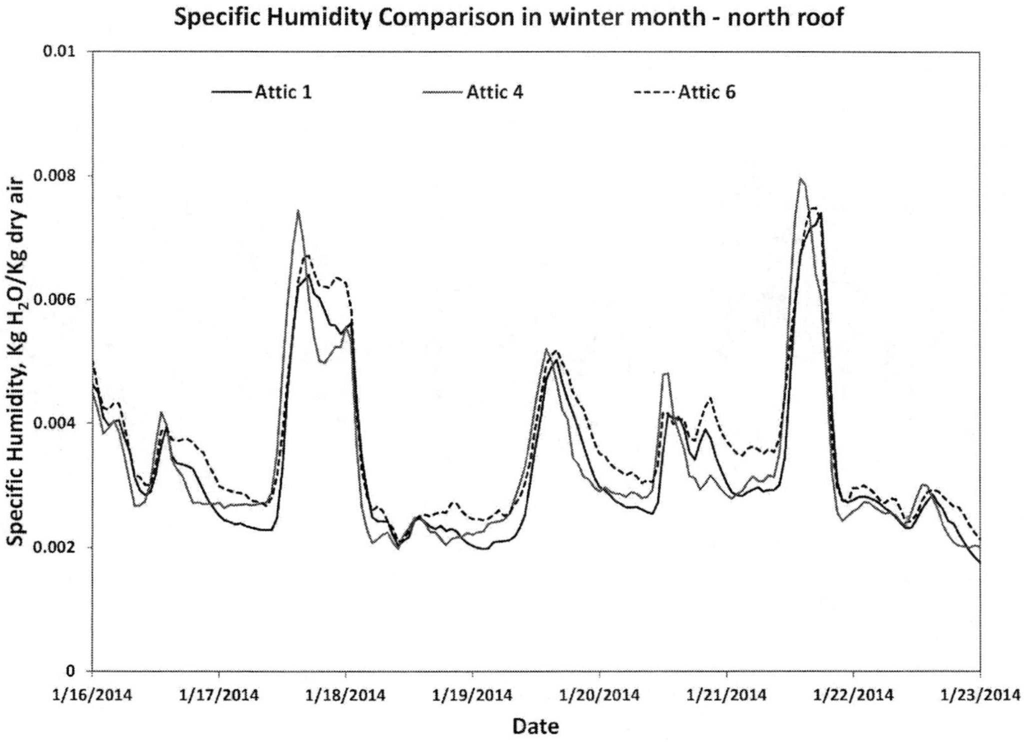

Figure 11 *Specific humidity of attic air in winter.*

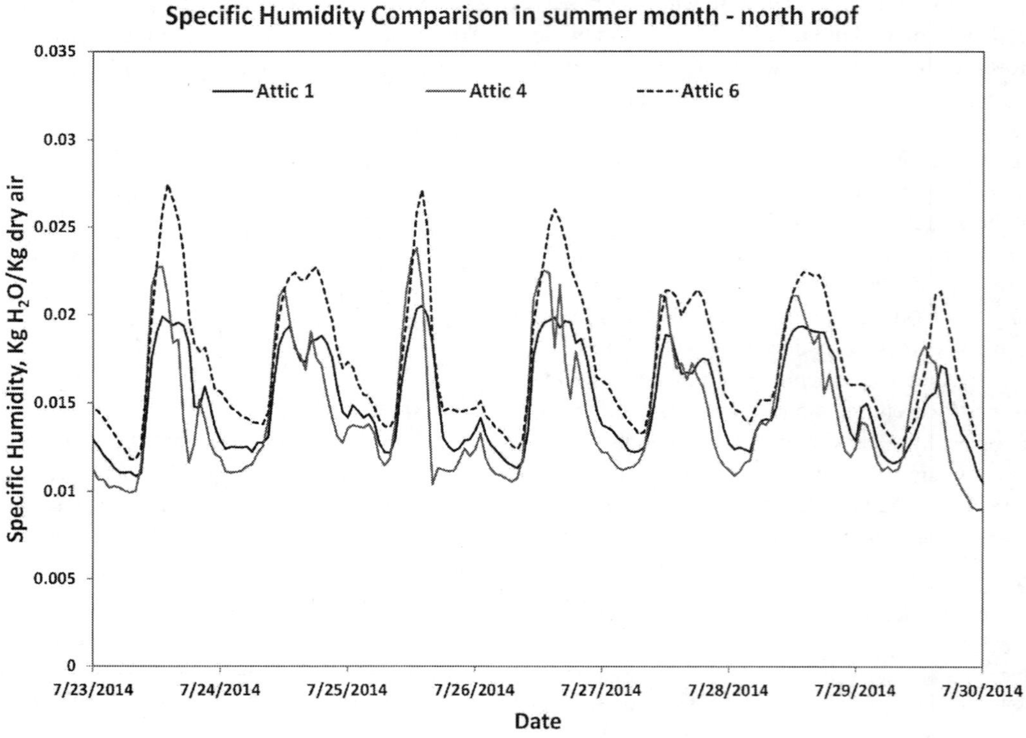

Figure 12 *Specific humidity of attic air in summer.*

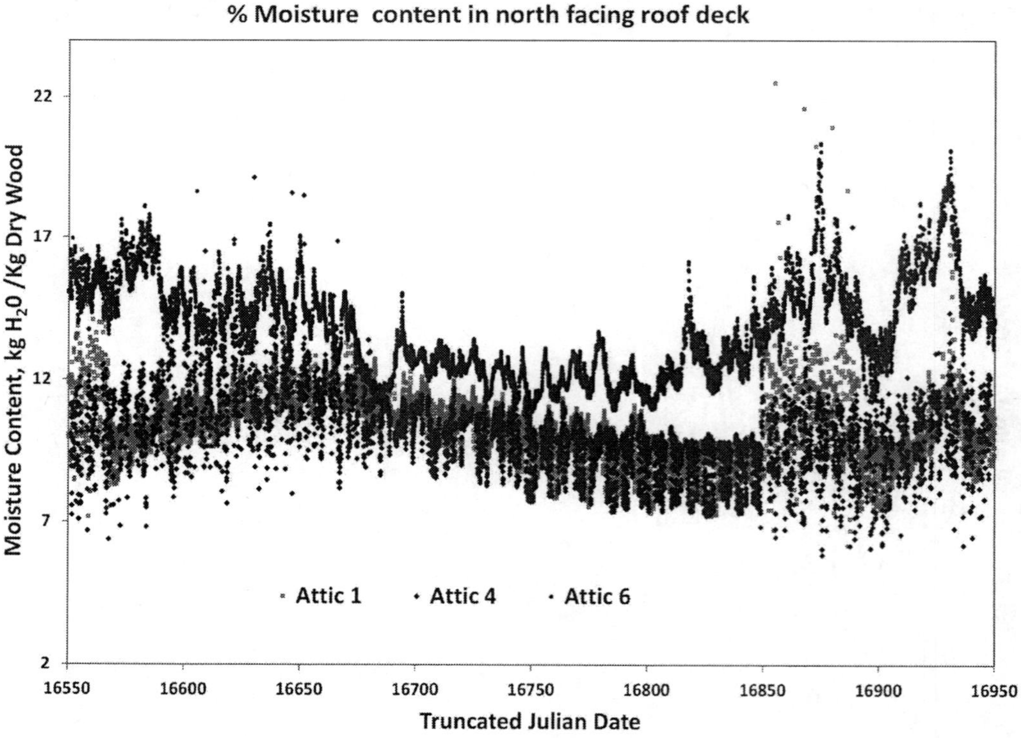

Figure 13 *Moisture content of roof decks for the three attics.*

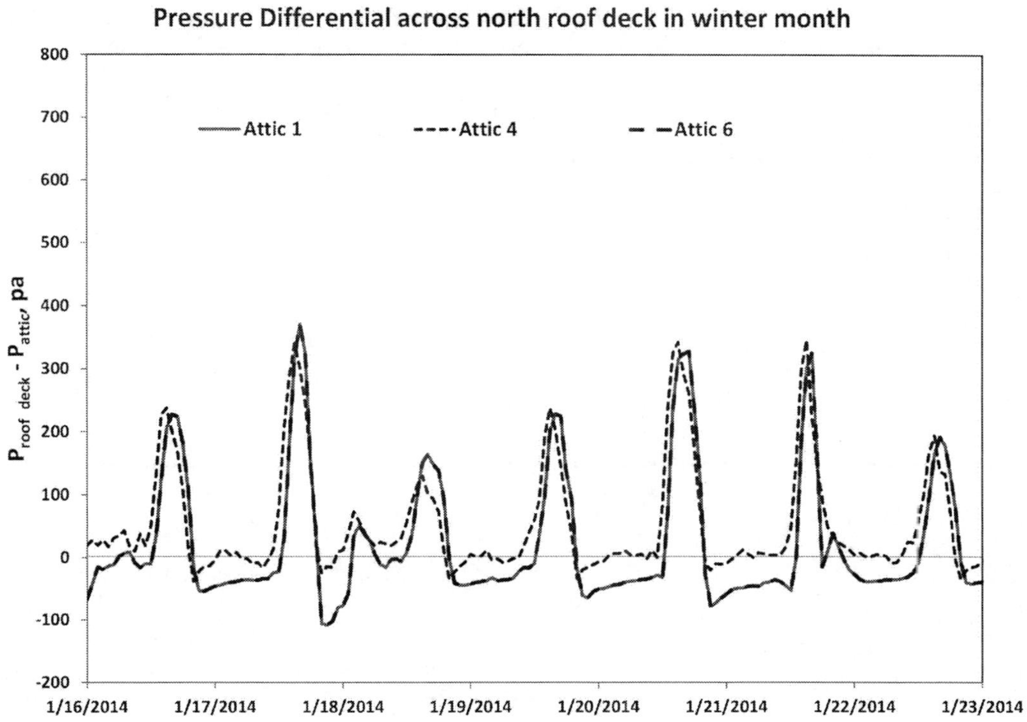

Figure 14 *Pressure differential across roof deck for the three attics.*

three attics during winter daytime tends to drive the moisture from roof decks into the attic space and vice versus. For the Attic 4 with cool shingles, however, little pressure differentials were found during the nighttime hours. Hence, the ability for moisture exchange was lowered, and, thus, moisture accumulation in Attic 4 during cold nights may represent a higher condensation potential.

ANNUAL ENERGY CONSUMPTION SIMULATION

Energy calculations were also done to estimate the potential annual energy savings for the IVNB attic as compared to the conventional attic design (Attic 1). The simulation was carried out using the ORNL AtticSim model for two ASHRAE cities, one in Zone 2 for mostly cooling and one in Zone 6 for mostly heating energy consumption. All attics assumed a roof SR of 0.10 and attic ventilation of 1:300. The IVNB roof system assumed 1 in. of ISO board above the deck. The attic was ventilated to 1:300 and R-40 insulation level was assumed on the attic floor. The simulation results are tabulated in Table 3 and show that the IVNB system can drop energy use attributed to the attic by 16.5% in a hot climate and by 11% in the predominantly heating load climate of Minneapolis, MN. As a result, adding an IVNB system during a roof retrofit as an above-sheathing solution can be a cost-effective solution and should be considered as part of the deep energy retrofit strategy for building envelopes.

CONCLUSIONS

The field performance of attics with increased solar reflectance and with an above-sheathing insulated ventilated nail base system was monitored and studied in the ORNL Natural Exposure Test Facility. Data on their annual thermal and hygrothermal performance were recorded and were compared to a standard 1/300 vented attic having dark shingles as control case. The results showed that the attics with the above-sheathing IVNB system significantly outperform other attics in managing the reduction of solar heat flux into attic during summer and the heat loss during the winter season. Also, the attic with the IVNB system was found to reduce the moisture level in the attic space during winter time and can reduce the potential of condensation. The moisture level for the attic with IVNB system was found to be slightly higher during summer season, and further study is needed to fully understand this observed behavior. The attic with cool shingles was found to have slightly lowered attic air temperatures and similar moisture performance as the control case; however, the reduction in heat flux during summer was found to be only marginal.

Energy consumption simulations showed that the IVNB system can significantly reduce cooling and heating energy compared to the control attic which has R40 attic insulation and standard 1/300 ventilation. Hence, the IVBN system should be considered as an effective above-sheathing solution for a deep energy retrofit of the building envelope.

Table 3. Simulations ASHRAE Climate Zones 2 and 6 for the IVNB Attic Containing a 10 in. Diameter Duct Warped in R-8 Insulation and Leaking 8%, Attic Ventilation Set at 1:300

Attic	Attic Floor/Roof R-Value	Ceiling Heat Flow (kBtu/yr)		Duct Heat Losses (kBtu/yr)		Annual (kBtu/yr)
		Cooling	Heating	Cooling	Heating	
Climate Zone 2, Austin, TX						
Control	R40/R0	2123	–1324	3525	–4137	20,089
IVNB	R40/R06	404	–858	2991	–6551	16,736
Climate Zone 6, Minneapolis, MN						
Control	R40/R0	453	–5021	1390	–22,143	36,755
IVNB	R40/R06	36	–4519	774	–21,431	32,704

REFERENCES

Boudreaux, P., S. Pallin, and R. Jackson. 2013. Moisture performance of sealed attics in the mixed-humid climate. Oak Ridge, TN: Oak Ridge National Lab. ORNL/TM2013-525.

Canadian Wood Council. 2000. Moisture and wood-frame buildings. *Building performance* Series No. 1. Ottawa: Canadian Wood Council.

Chasar, D., V.V. VonSchramn, J. Sherwin, and S. Chandra. 2010. Measured performance of side-by-side South Texas homes. *ASHRAE Thermal Performance of the Exterior Envelopes of Whole Building XI International Conference*. FL: Clearwater.

Derome, D. January 2005. Moisture accumulation in cellulose insulation caused by air leakage in flat wood frame roofs. *Journal of Building Physics* 28 (3):269–87.

Desjarlais, A., W. Miller, S. Railkar, and A. Chich. 2012. Energy and moisture performance of attic assemblies. Symposium on Building Envelope Technology. AZ: Phoenix.

Lstiburek, J. 2006. Understanding attic ventilation. *Building science digest*. MA: Westford.

Lstiburek, J. 2015. Venting Vapor. *ASHRAE Journal* (August):46–51.

Lstiburek, Joseph, Achilles Karagiozis, and Paul Gassman. 2011. Vapor permeability provides no performance benefit for roofing underlayments in ventilated attics. OH: Toledo: Owens Corning.

McNabb, N. 2013. Strategies to achieve net-zero energy homes: A framework for future guidelines. Washington, DC: National Institute of Standards and Technology Special Publication 1140.

Miller, W., M. Keyhani, T. Stovall, and A. Youngquist. 2007. Natural Convection Heat Transfer in Roofs with Above-Sheathing Ventilation. *ASHRAE Building X*.

Miller, William A., and Ethan Herman. 2012. The tradeoff between solar reflectance and above-sheathing ventilation for metal roofs on residential and commercial buildings. Oak Ridge, TN: ORNL.

Plescia, S. 2003. Evaluation of vapour diffusion ports on wood-frames walls under controlled conditions. Ottawa, Ontario: CMHC Technical Series 02-130.

Prahl, D., and M. Shaffer. 2014. Moisture risk in unvented attics due to air leakage paths. Oak Ridge, TN: Department of Energy.

Railkar, Sudhir, Adem Chih, Ming Shiao, Andre Desjarlais, and William Miller. 2015. Thermal and hygorthermal performance of ventilated attics with and without breathable underlayments. KS: Kansas City. Building Envelope Science and Technology Conference.

Roppel, P., N. Norris, and M. Lawton. 2013. Highly insulated, ventilated, wood-framed attics in cool marine climates. Thermal Performance of The Exterior Envelopes of Whole Buildings XII International Conference. Clearwater, FL.

Rowley, F.B., C.E. Algren, and C.E. Lund. 1939. Condensation of moisture and its relation to building construction and operation. *ASHVE Transactions* 44:1115.

Saber, H.H., M.C. Swinton, P. Kalinger, and R. Paroli. 2011. Hygrothermal simulations of cool reflective and conventional roofs. 2011 International Roofing Symposium. Washington D.C.

Straube, J., R. Smith, and G. Finch. 2009. Spray polyurethane foam: The need for vapor retarders in above-grade walls. Canadian Urethane Foam Contractors Association.

Ueno, Kohta, and Joseph Lstiburek. 2015. Field testing unvented roofs with asphalt shingles in cold and hot-humid climates. Building America Report - 1409. Westford, MA: Building Science Corporation.

Annual Energy and Heat Flows in Vented and Sealed Attics—Parametric Study: Climate Zone 2A

Mikael Salonvaara
Member ASHRAE

Achilles Karagiozis, PhD
Member ASHRAE

William Miller, PhD
Member ASHRAE

ABSTRACT

Placing ducts in conditioned or semiconditioned space can be one of the most energy-saving as well as potentially cost-effective measures for high-performance buildings. The moisture management in a building with a sealed attic is very different when compared to conventional vented attics, with respect to the mold and mildew exposure of the occupants as well as maintaining the structural integrity of the home (Salonvaara et al. 2013). It is also important to fully quantify the thermal benefits of different advanced attic systems and then to analyze the applicable moisture strategies for high performance. This paper provides a summary of an exhaustive analysis comparing the energy performance of three attic systems: unvented attic, vented attic, *and the* hybrid attic. *The computer tool AtticSim (ASTM 2004) was used to simulate vented, sealed, and semiconditioned attics to evaluate the energy and economic trade-off between sealing the attic as compared to sealing and better insulating an HVAC duct system contained in the attic (vented). The duct diameter, R-value of duct and roof insulation, leakage rate of ducts and attics, roof slope and orientation, attic size, and climate were simulated parametrically to judge the potential for making ducts in attics operate as if they were operating in the conditioned space. More than 250,000 simulations were performed in 33 locations spanning Climate Zones 1 to 6. This paper is focused on the performance of attics in International Energy Conservation Code (IECC) Climate Zone 2A. Unvented attics outperformed the other configurations when simulated with current duct-air sealing technologies and duct insulation as per the ICC (2015). At the same time, it is also demonstrated that when a better designed, insulated, and air-tight ducts were considered, vented attics can perform as well as the sealed, semiconditioned attic.*

INTRODUCTION

Contractors continue to use the attic space for installing ductwork at a low price because of its convenience. Therefore, computer simulations were developed for documenting the potential energy savings for better designed duct systems operating in the extreme temperatures of attics.

The AtticSim computer tool (ASTM 2004) was used to compare the thermal performance of semiconditioned attics with ventilated attics. Simulations were conducted using Typical Meteorological Year 2 (TMY2) weather data for 33 cities in California, Texas, Nevada, Georgia, New Mexico, and Florida. Two different sizes of attics were simulated: One with a footprint of 1567 ft^2 and a roof slope at 22.6° (5/12 pitch) and another with a footprint of 3200 ft^2 and a roof slope at 30.2° (7/12 pitch). The features of these two attics were varied in the simulations to analyze the parametric thermal performance based on orientation, roof slope, solar reflectance of the roof shingles, approach to insulating the roof or ceiling (R-value and location of insulation), ventilation rate of the attic, duct leakage rates, duct R-values, and duct size.

The vented attics were insulated at various thickness with blown-in fiberglass on the ceiling level. The sealed attics had different designs and insulation materials: open-cell spray foam (ocSPF), draped cellulose (CFI), and continuous blown fiberglass (FG). Additionally a hybrid system (vented) with R-30 on ceiling and R-10 on the roof deck was simulated. Results were generated for 33 locations. More than 250,000 simulations were performed and all results

Mikael Salonvaara is a building science leader and **Achilles Karagiozis** is global director of building science at Owens Corning LLC, Granville, OH. **William Miller** is a building scientist at Oak Ridge National Laboratory, Oak Ridge, TN.

have been collected in spreadsheet files, which allowed for comparing the performance in more ways than can be presented in a printed report.

AtticSim is a computer tool formulated and validated for conducting attic simulations that was initially developed at Owens Corning (Wilkes 1991). AtticSim carries out hourly simulations, unlike some home energy rating tools that use seasonal weather data. The model can account for different insulation R-values placed on the attic floor and/or roof deck and has an algorithm for predicting the effect of air-conditioning ducts placed in the attic as reported and benchmarked by Petrie et al. (2004). Wilkes (1991) conducted validations for attics having direct-nailed roof products. Miller et al. (2014) also validated AtticSim against a steep-slope attic assembly having direct-nailed asphalt shingles. Miller et al. (2007) validated the model for roofs exploiting above-sheathing ventilation. AtticSim is capable of estimating the potential energy savings for semiconditioned attics and ventilated attics.

PROJECT GOALS

The project conducted 241,920 simulations (and additional simulations for more focused parametric analyses) exercising AtticSim for various leakage rates from the supply and return duct, for various diameters of circular metal ducts having various levels of duct insulation, and other parameters as listed in Table 1.

The heating and cooling system type and efficiencies were assumed the same in all the locations. All houses were equipped with a heat pump. The size varied depending on the locations but the efficiencies were set to heat seasonal performance factor (HSPF) = 8.5 for heating and seasonal energy-efficiency ratio (SEER) = 13 for cooling.

PARAMETRIC INPUTS AND THEIR EFFECT ON PERFORMANCE

Locations

A total of 33 locations covering Southern states in the United States were included in the project. All results have

Table 1. Simulations for Semiconditioned and Ventilated Attics

Parameters Analyzed	Vented Attic	Sealed Attic	Naming in Charts
Attic leakage to outside, leakage area*	1:300 1:150 1:75	1:1567 1:3000 0	As in "Vented" and "Sealed"
Solar absorptivity of shingles	0.97 0.90 0.75	0.97 0.90 0.75	0.97 0.90 0.75
Attic insulation	R-30 R-38 R-49 R-30 + R-10 (deck)	R-30 boxed FG R-30 draped CFI R-22 ocSPF R-30 ocSPF (Owens Corning n.d.)	As in "Vented" and "Sealed"
Duct diameter, in.	6 8 10	6 8 10	d06 d08 d10
Duct leakage, % of flow†	0 2 4 6 10	0 2 4 6 10	L0 L2 L4 L6 L10
Duct R-value,‡ (hr·ft^2·°F)/Btu	R-6 R-8 R-12 R-16	R-4 R-6 R-8	R04 R06 R08 R12 R16

* Attic leakage to outside represents the airtightness of the attic through the roof deck. The vented attics have a designed vent area per roof area. The air exchange characteristics of a sealed attic is represented with leakage area that can replicate the blower door test performance in the simulation model, in this case AtticSim. The values are shown as vent area ratios both for the vented and the sealed attics.

† Duct leakage per conditioned floor area (CFA) varies depending on location and attic size. The airflow rates in the different locations vary depending on the system sizes; every 2% duct leakage based on flow corresponds to 1% to 2% duct leakage based on duct leakage per 100 ft^2 of floor area. The latter method is commonly used by raters and in building codes.

‡ The duct thermal resistances (R-values) listed in these parametric analyses follow the commonly used naming for the duct R-values, i.e., the R-values are listed for the flat insulation and not for the effective R-value for the cylindrical duct. The attic insulation values are shown above in terms of their respective R-values.

been collected per city in individual spreadsheet files that can be used to plot and analyze the results in multiple ways with the help of pivot tables and charts. The annual purchased energy of the heat pump is shown in kilowatt-hours for heating and cooling. Locations simulated were selected such that they would cover most Southern states and include some cold-climate locations (Table 2).

Indoor Conditions

AtticSim simulates only the attic and not the whole house. The indoor temperature set points were 68°F and 78°F for heating and cooling, respectively. AtticSim does not calculate the overall energy balance of the house, but instead the runtime for heating and cooling is either calculated based on design loads and indoor to outdoor temperature difference or the runtime information for heat pump can be read from an input file. A base case house (vented attic R-30, R-6, 4% leaky ducts) was simulated with EnergyPlus (Crawley et al. 2000) to get the hourly indoor temperatures and the hourly runtime (on time) values for the heat pump for the reference house. The resulting indoor temperatures are based on the energy supplied by the heating and cooling system and the heat loss or gain in that reference house. If the heat loss (or gain) in the heating and cooling system itself is different in the designed house than in the reference house (e.g., leakier or tighter ducts), the supplied energy to the house would be different because of different airflows that end up in the living space or because of a different temperature at the air inlet than in the reference house. To have the same energy balance satisfy the same indoor temperatures, the runtime of the heat pump has to change in the design house. AtticSim by default does not change the runtime but instead only calculates the heat loss and gain in the duct system and through the roof using the fixed runtime of the heat pump. The results from AtticSim were post-processed in such a way that all homes in the location would have the same heat gain or loss at a given hour of the year, thus providing the exact same conditions in the house regardless of the attic system. The heat gains and losses include the heat flow via conduction through the ceiling (between attic and the interior of the home) and the heat provided by the airflow into the room(s). Different attic and duct layouts end up with different runtime for the air conditioner, which in turn results in different energy use; however, all the attics provide the exact same conditions in the living space for every hour of the year. Figure 1 displays the energy flows and the airflows to correct the heat pump runtime. The indoor temperature for any given hour is given as input as is fixed which means that all the other building components such as walls, floor, doors and windows will have the same heat flows in all the different simulations (AtticSim simulates only the attic). The heat flow through the ceiling to the controlled indoor environment can change, however, depending on the

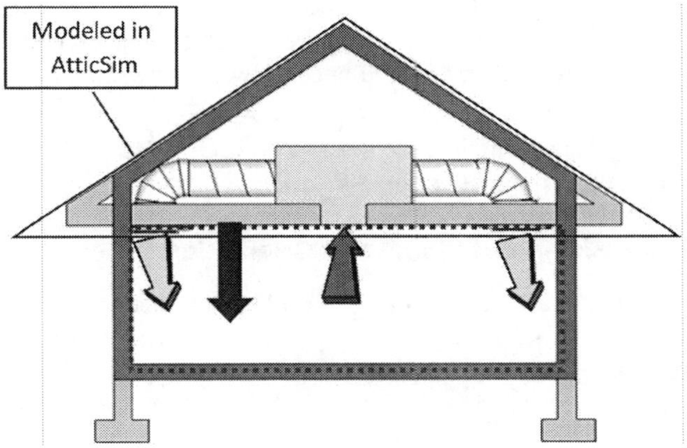

Figure 1 *Energy flows to correct the AtticSim runtime. Dashed line shows the control boundary for indoor conditions and the triangle the area that is modeled by AtticSim.*

Table 2. Locations Where The Houses Were Simulated

Southern States Other Than California	California Climate Zones	
TX—Austin, Houston, Dallas, El Paso, San Antonio, Lubbock	CA1 Arcata	CA9 Pasadena
AZ—Flagstaff, Phoenix	CA2 Santa Rosa	CA10 Riverside
NM—Albuquerque	CA3 Oakland	CA11 Red Bluff
LA—New Orleans	CA4 Sunnyvale	CA12 Sacramento
GA—Atlanta	CA5 Santa Maria	CA13 Fresno
SC—Charleston	CA6 Los Angeles	CA14 China Lake
FL—Orlando, Gainesville	CA7 San Diego	CA15 El Centro
NV—Las Vegas	CA8 El Toro	CA16 Mount Shasta
MN—Minneapolis		
IL—Chicago		

attic conditions and insulation levels either at the roof deck or on the ceiling. Likewise, the duct losses caused by duct leakage and duct insulation thermal resistance will affect the airflow rate as well as the supply air temperature. If the airflow rate is $q_{m,\,air}$ when the air handler is running and the heat flow through the ceiling is $Q_{ceiling}$, then in both the design house and in the reference house this equation—(Run time$_i$) * $q_{m,\,air}$ * c_p * ($T_{supply\ at\ inlet} - T_{indoor}$) + $Q_{ceiling}$—must be equal.

In a simple vented attic case, assume that due to thermal inefficiencies in the duct system the supply air temperature in the design house is 100°F and in the reference house it is 105°F. Everything else is the same (attic temperature, ceiling insulation, etc.) and the indoor temperature is 68°F. The duct leakage is the same, which means that the supplied energy to the design house is (100 – 68)/(105 –68) = 86% of that in the reference house. Therefore, in the design house the heating system must run 16% longer to provide the same energy to the design house as in the reference house. The longer runtime is needed to satisfy the energy balance that provides the given indoor temperature. The energy use of the heat pump reported by AtticSim is then adjusted linearly by the new runtime.

It would be possible to modify AtticSim to correct the energy balance in the code itself, however, this was not available at the time of this paper's writing.

Attic Size, Roof Slope, and Orientation

The main simulations with all parametric analyses—except roof slope and orientation—were carried out for two attic sizes. The roofs had one ridge and two gable ends. The smaller attic was 57 × 27.5 ft (1567.5 ft^2) with a slope of 5/12. The larger attic was 80 × 40ft (3200 ft^2) with a slope of 7/12. Figure 2 displays a comparison of the vented attic with R-38 insulation and sealed attics with R-30 or R-22 insulation in the small home in Houston, TX. *Kilowatt-hours* (kWh) refer to purchased electricity to run the heat pump in cooling or heating mode.

The impact of the roof slope (5/12, 7/12, 9/12) and the orientation (ridge in north-south or east-west orientation) were analyzed in a smaller set that did not include Houston, TX and therefore the results are shown for San Antonio, TX. The climate in San Antonio, TX, is thermally very similar to that in Houston, TX. The roof slope affects the insulated surface area in the sealed attic by increasing the sloped area but also by increasing the surface area of the gable ends. The area multiplier for the sloped roof area is 1.08, 1.16, and 1.25 for 5/12, 7/12, and 9/12 pitch, respectively. In addition to the increase in the roof deck are, the gable ends (two of them) add 10%, 15%, or 19% (5/12, 7/12, and 9/12, respectively) of the attic floor area to the area that needs to be insulated in the attic when the attic aspect ratio is 2:1 (length:width) and when the ridge runs along the length of the roof.

The vented attic (R-38, on the right in Figure 3) is almost unaffected by the roof slope with slightly increasing heat loss and gain the steeper the roof slope. The east-west oriented roofs have more heat loss but less heat gain than the north-south oriented roofs. The sealed attic (R-30 boxed fiberglass) has more heat loss and gain at a higher roof pitch than at a lower roof pitch. The east-west orientation gives slightly lower heat gain (cooling loss) than the north-south orientation.

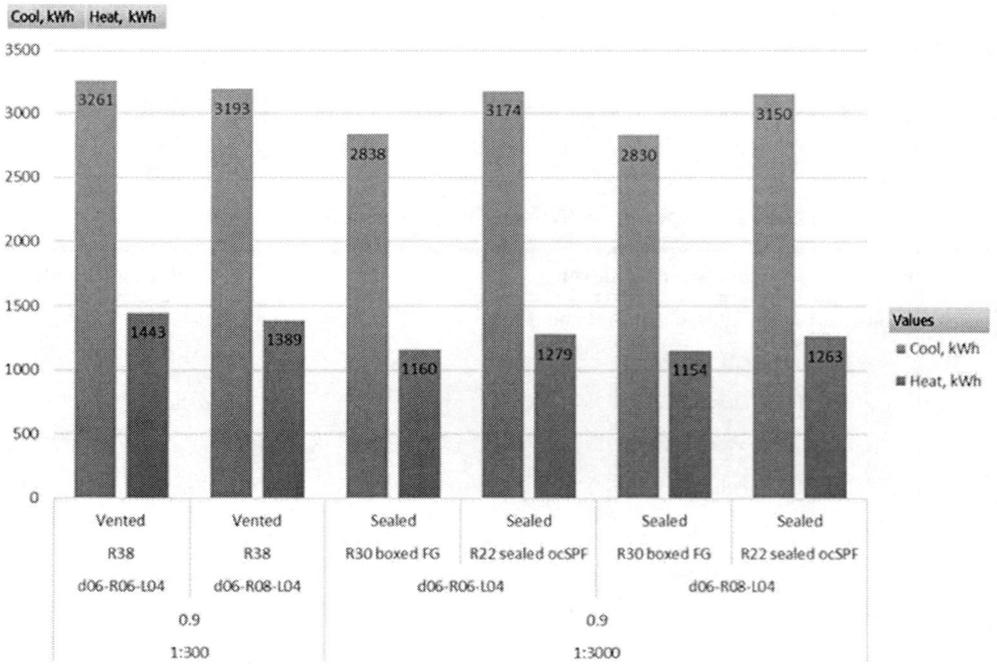

Figure 2 *Comparison of vented R-38 attic and sealed R-30 and R-22 attic in a small home in Houston, TX.*

Attic Insulation Thermal Resistance—Structure

Both vented and sealed attics were simulated. The roof sheathing in vented attics had a radiant barrier surface with emissivity = 0.05. The ducts in the vented attics also had the reflective surface (emissivity = 0.05). The ducts in sealed attics had surface emissivity = 0.90. Based on a separate analysis, the duct surface emissivity had almost no effect on the results in sealed attics. The low emissivity is beneficial in vented attics because of the low and high temperatures in the roof deck, whereas in sealed attics the interior surface of the insulated roof deck is close to the attic temperature, which is close to the indoor temperature. The insulation on the ceiling had either R-30, R-38, or R-49 thermal resistance in the vented attic. In Houston, TX the increase of the ceiling insulation from R-30 to R-49 resulted in 5% reduction in cooling and 2% reduction in heating energy use (Figure 4).

Increasing the thermal resistance of the roof deck insulation in the small sealed attic from R-22 to R-30 (open cell) reduced the cooling energy use by 10% and the heating energy use by 8% in Houston TX (Figure 4).

Attic Venting

Attic vent areas (total area of vents/attic floor area) for the vented attics were the standard 1:300 and 1:150 or, for an advanced vent area, 1:75. The sealed attic is not vented but instead has air leakage between the attic and outdoors. The vent areas were adjusted to account for the air leakage that is reasonable and that has been estimated in many sealed attics based on blower door tests isolating the attic to outdoor air leakage. The vent areas were based on 0, 475, and 910 cubic feet per minute at 50 Pa pressure (cfm_{50}) (three airtightness levels) for the small sealed attic and 0, 950, and 1820 cfm_{50} for the large sealed attic. These values represent 0, 0.3, and 0.6 cfm/ft^2 air leakage per attic floor area. The natural air exchange between the attic and the outdoors is calculated based on stack and wind pressures. The air leakage expressed in terms of air changes per hour will depend on the volume, which in turn is affected by the roof slope.

Figure 5 shows the heating and cooling energy use of the house with the small attic in Houston, TX when the attic ventilation is provided with 1:75, 1:150, or 1:300 vents for the vented attic. The sealed attic similarly has three different air leakage characteristics. Attic venting increases heat losses in the winter but provides some relief in the summer by reducing the attic temperatures during sunny days. Air leakage in the sealed attic has a much more pronounced effect on heating energy use than what the different vent sizes have on the vented attic.

Solar Absorptivity of Shingles

Three different solar absorptivity values were used for the shingles: 0.97, 0.9, and 0.75. These represent the typical ranges of solar absorptivity for shingles from dark shingles to cool roof shingles. The roof with shingles with low solar absorptivity (0.75) has higher heat loss but lower cooling demand than the roof with high solar absorptivity (0.9 or 0.97) (Figure 5).

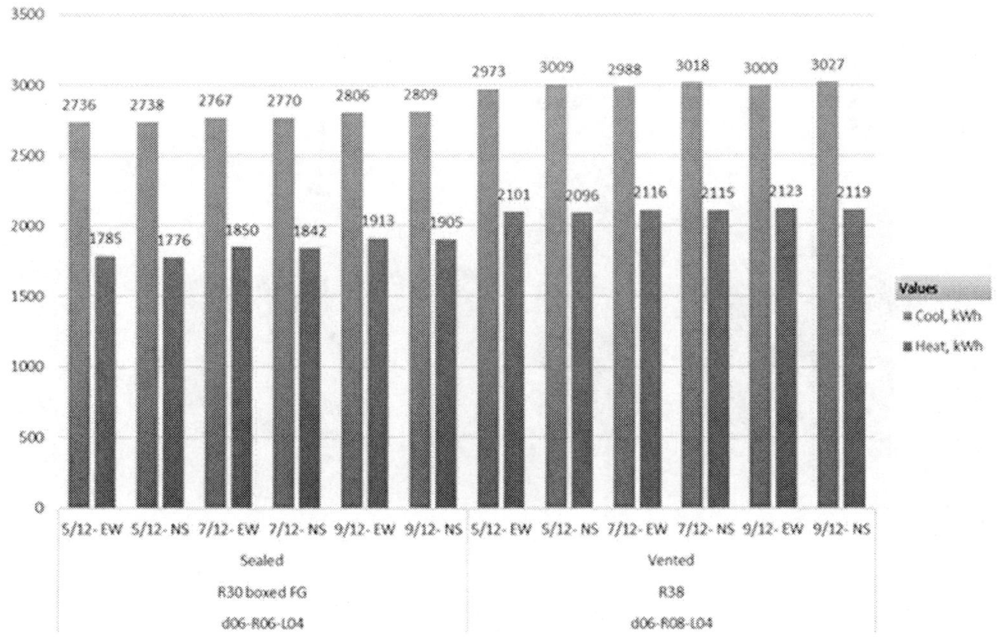

Figure 3 *Effect of roof pitch and orientation (slope orientations: NS = north-south, EW = east-west) on cooling and heating demand in San Antonio, TX, for the 1567 ft^2 attic.*

Duct Size and Surface Area

The design for duct sizes is based on needed airflow rates and pressure loss calculations so that the airflow to the spaces is delivered with the capacity of the air handler and its fans. There are several possible designs for duct systems for a single home, and in energy-efficient homes today the duct systems can be designed to be more compact than before. One reason for this is that the air inlets do not necessarily need to be next to the windows anymore, which allows for more flexible designs. The air inlets have been traditionally placed under or above the windows when windows were cold or hot because of their inefficient performance factors. The better windows

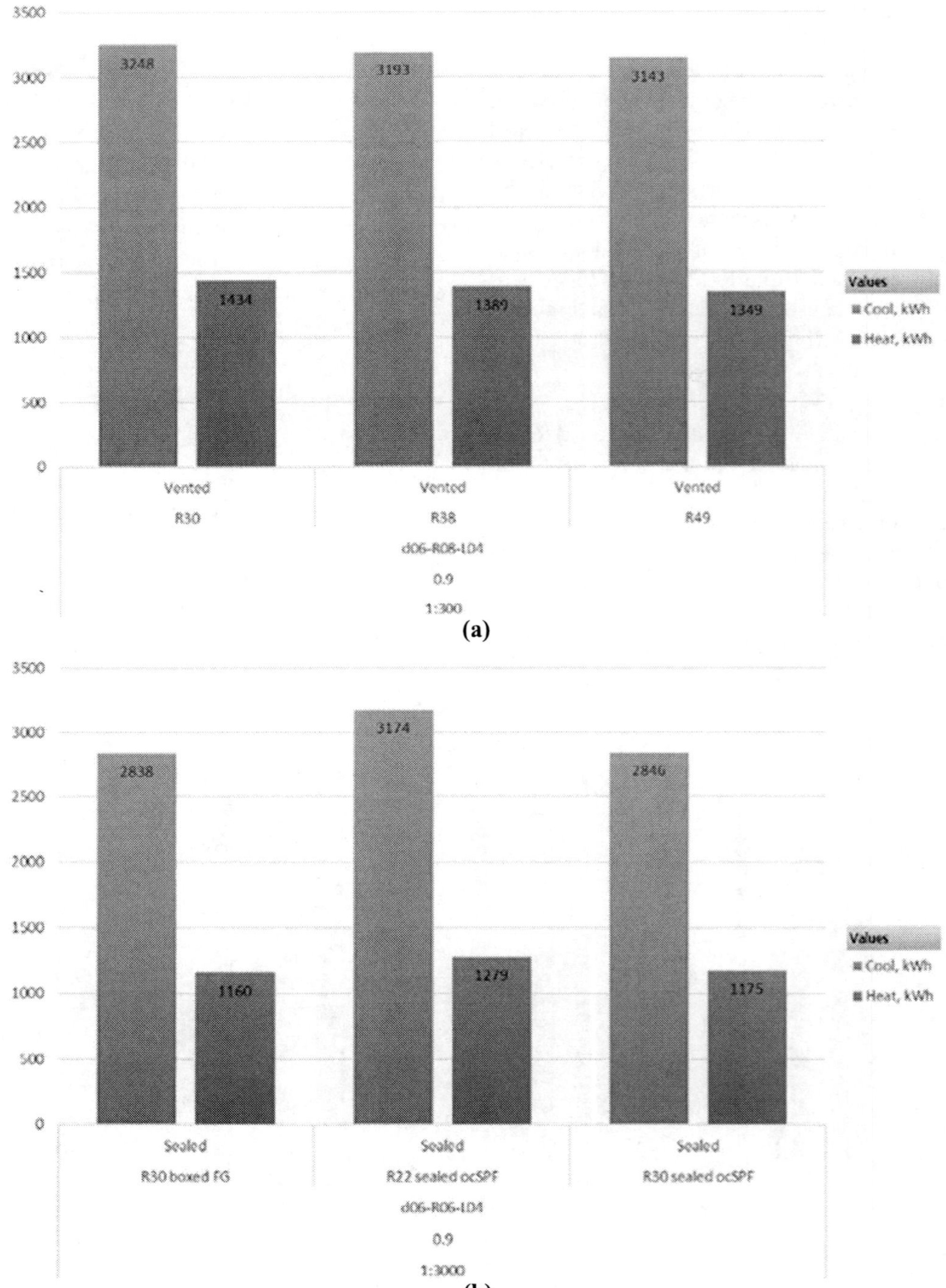

Figure 4 *(a) Effect of ceiling insulation on the energy losses of a small vented attic. (b) The effect of R-value of the insulation on the roof deck in the small sealed attic. Location is Houston, TX.*

(higher U-factors, less solar heat gains) available today do not necessarily provide such thermal imbalance that would require conditioning near the windows. Even in existing homes the duct designs can often be compacted by designing larger ducts to take air to rooms that are adjacent to each other and splitting the flow only near the end of the main trunk. The compact design of the duct system can result in less surface area and thus in less heat loss by conduction.

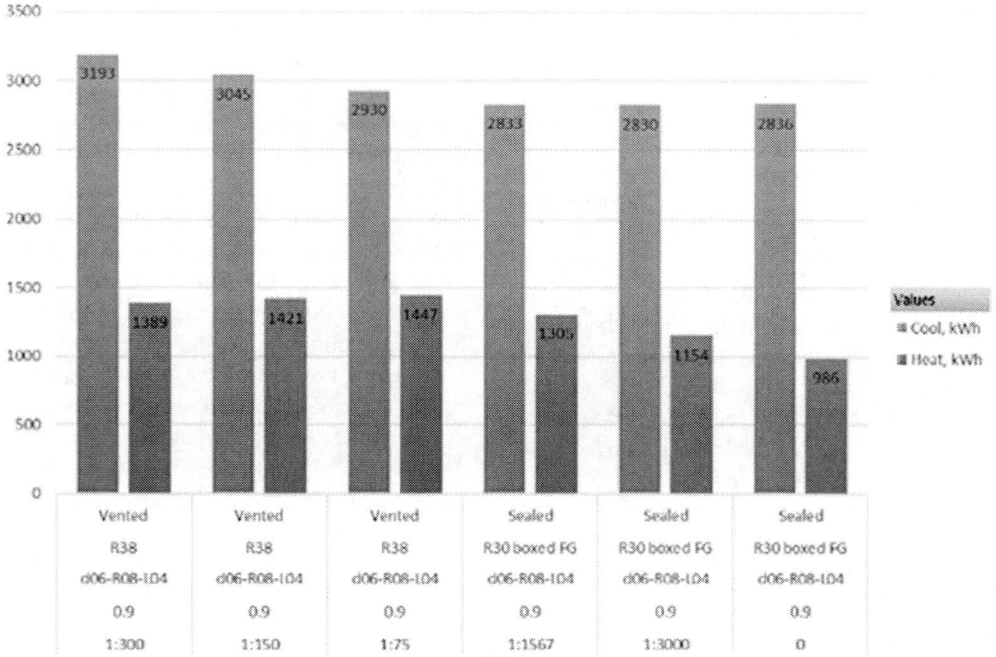

(a)

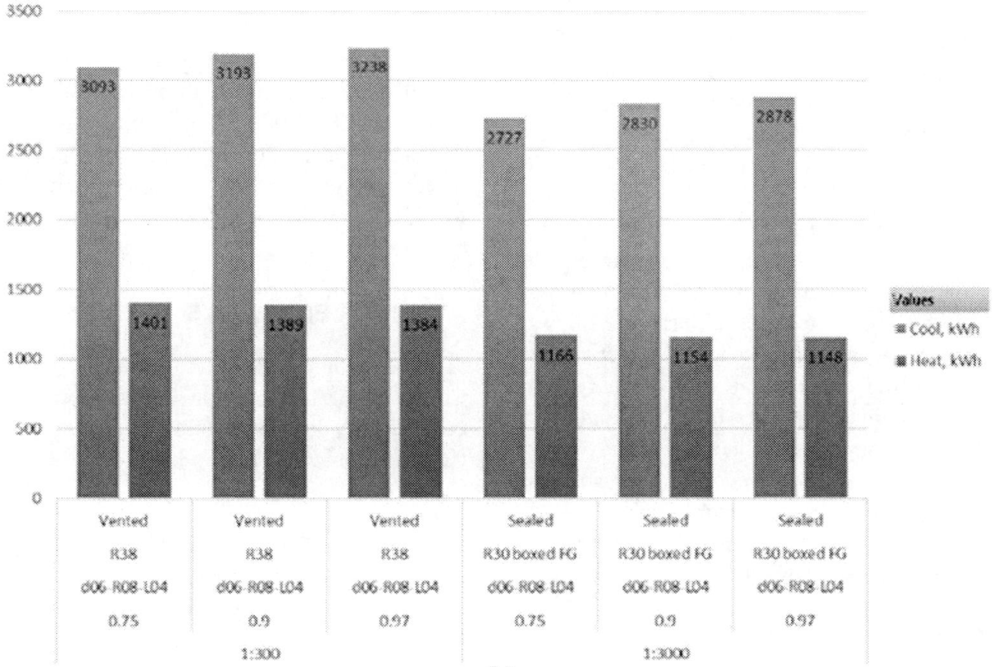

(b)

Figure 5 *(a) The effect of the attic venting on the heat loss and gain in the house with the small attic. For sealed attic, the venting is actually air leakage between the attic and the outdoors because venting is not provided on purpose. (b) The effect of solar absorptivity on the heat loss and the heat gain of a large vented and a small sealed attic. Location is Houston, TX.*

The basic duct system in these simulations was modeled using the assumptions used by the rating software. The rating software use the exterior surface area of the ducts as an input. In California the supply duct surface area is 27% of conditioned floor area and for return ducts 5% (CEC 2014 suggested value). In all other locations the surface area for the supply ducts is equivalent to 20% of conditioned floor area and the surface area for return ducts is 5% (RESNET default). Both homes—one with the small attic and one with the large attic—were one-story buildings and all ducts were assumed to be in the attic.

Designing for a more compact duct system in the attic results in shorter overall duct length and much less duct surface area than the common "spider" style duct layout. The duct system can be compacted by making use of larger trunk lines and having only short branches out of the main trunk. The surface area can often be reduced to half of the worst design with many long ducts. In the set of simulations carried out here, the base surface area is for a 6 in. duct diameter (inner dimension of flow duct) with R-8 insulation. If the airflow can be designed to use fewer, larger ducts, the exterior surface area of the ducts could be reduced by 19% and 29% with 8 and 10 in. flow duct diameters, respectively. A compact duct system (10 versus 6 in. duct diameter in the charts) would save about 6% in cooling and 10% in heating with the vented attic in Houston, TX (Figure 6). The savings in a sealed attic are 3% and 6% for cooling and heating, respectively.

Duct Thermal Resistance

The duct thermal resistance (R-values) listed in these parametric analyses follow the commonly used naming for the duct R-values, i.e., the R-values are listed for the flat insulation and not for the effective R-value for the cylindrical duct. The real performance of a cylindrical duct is calculated internally in AtticSim.

The duct thermal resistance affects the total duct surface area; the higher thermal resistance ducts will have larger exterior surface area due to the cylindrical form of the ducts. The AtticSim model does not use the exterior surface area as an input parameter, instead it calculates the actual effective thermal transmittance coefficients for each foot of duct length based on the thermal conductivity of the insulation and cylindrical shape of the duct. Therefore, the increase in the exterior surface area is mentioned only in reference to rating software inputs. Note: home energy raters commonly select just the 'default' button when doing the ratings and the inputs are typically only R-value, air leakage, and duct surface area.

Duct Air Leakage

Duct air leakage is balanced and the return side leaks are equal to the amount of air leaks from the supply side.

The air leakage rates are shown as percentage of flow in Figure 7. For example, if the flow through the ducts is 800 cfm and duct leakage is 10%, then the 80 cfm of air is leaking into the return duct and 80 cfm of air is leaking out of the supply ducts to the attic. Note: a common assumption in rating software is that 60% of leaks occur in the return side.

CONCLUSIONS

The computer tool AtticSim was used to simulate ventilated attics and sealed semiconditioned attics to compare the energy use when improving the systems or the attics. The improvements in the attics included air sealing the attic, sealing and better insulating the duct system contained in the attic, changing the duct size, changing the attic type, and changing the insulation level at the roof deck or at the ceiling. The results show that unvented attics mostly perform better than standard vented attics in the analyzed climates. However, vented attics with exceptionally well-insulated airtight ducts in the attic and ducts with higher thermal resistance can perform equally as sealed attics. Sometimes the results are mixed: one performs better in cooling season and one performs better during heating season. The common assumption that leaky ducts in the sealed attics don't affect the performance is not true: The performance of the sealed attic is affected by the leaky ducts or the air exchange between the attic and the indoors. The attic conditions differ from the indoor conditions and using the attic space as a buffer zone adds resistance between the living space and the outdoors. However, sealed attics may need ventilation to provide acceptable moisture performance. Moisture performance was not part of this study, but unvented attic designs require moisture engineering analysis so that they perform well in all climates zones (Karagiozis et al. 2002).

The parametric study showed that better designed, highly insulated, and more airtight ducts in vented attics can perform as well as sealed semiconditioned attics. The choice of the type of attics used in a building goes beyond just the energy performance such as additional conditioned storage space, wildfire protection, wind-driven rain protection (high wind cases), increased comfort, builder market differentiation, or homeowner preferences.

ACKNOWLEDGMENTS

The authors would like to thank the High Performance Conditioned Attic team at Owens Corning for support and constructive discussions, in particular Abe Cubano, Elaina Carpino, Don Kosanka, and Neil Freidberg.

REFERENCES

ASTM. 2004. Standard C 1340-04. *Standard practice for estimation of heat gain or loss through ceilings under attics containing radiant barriers by use of a computer program.* West Conshohocken, PA: American Society for Testing and Materials.

CEC. 2014. Title 24. *Residential compliance manual.* California: California Energy Commission. http://www.energy.ca.gov/2013publications/CEC-400-2013-001/chapters/04_Building_HVAC_Requirements.pdf.

Crawley, D.B., C.O. Pedersen, L.K. Lawrie, and F.C. Winkelmann. 2000. EnergyPlus: Energy simulation program. *ASHRAE Journal* 42:49–56.

ICC. 2015. *International energy conservation code.* Washington, D.C.: International Code Council. http://codes.iccsafe.org/app/book/toc/2015/I-Codes/2015%20IECC%20HTML/index.html.

Karagiozis, A., A. Desjarlais, and M. Salonvaara. 2002. Sensitivity of low sloped roofs designs to initial water and air leakage. Hot and Humid Symposium, ESL-HH-02/05-47, Houston, TX, pp. 547 – 554, May 2002.

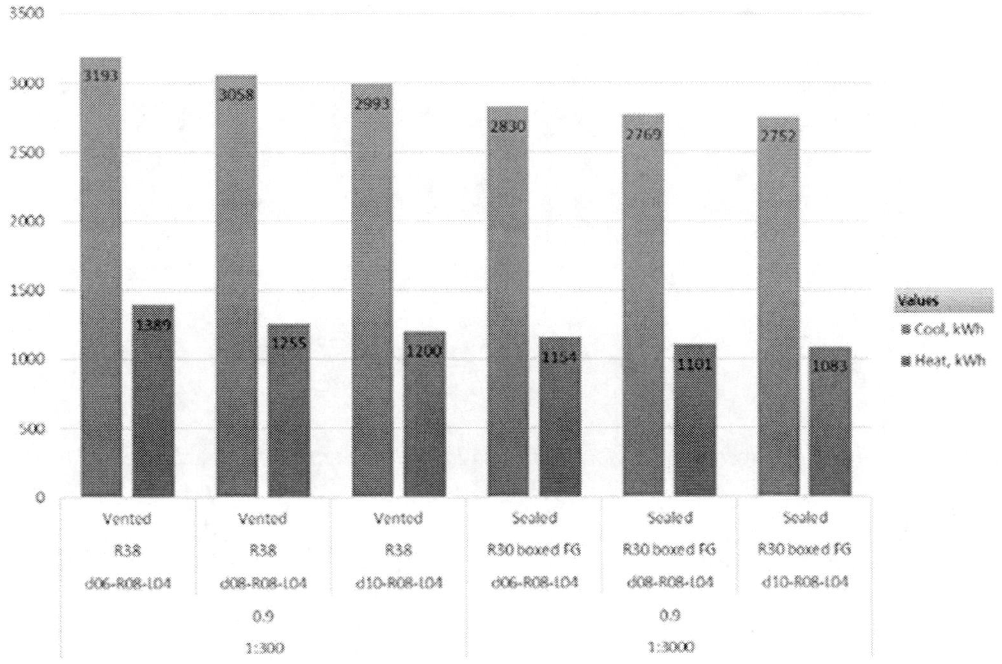

(a)

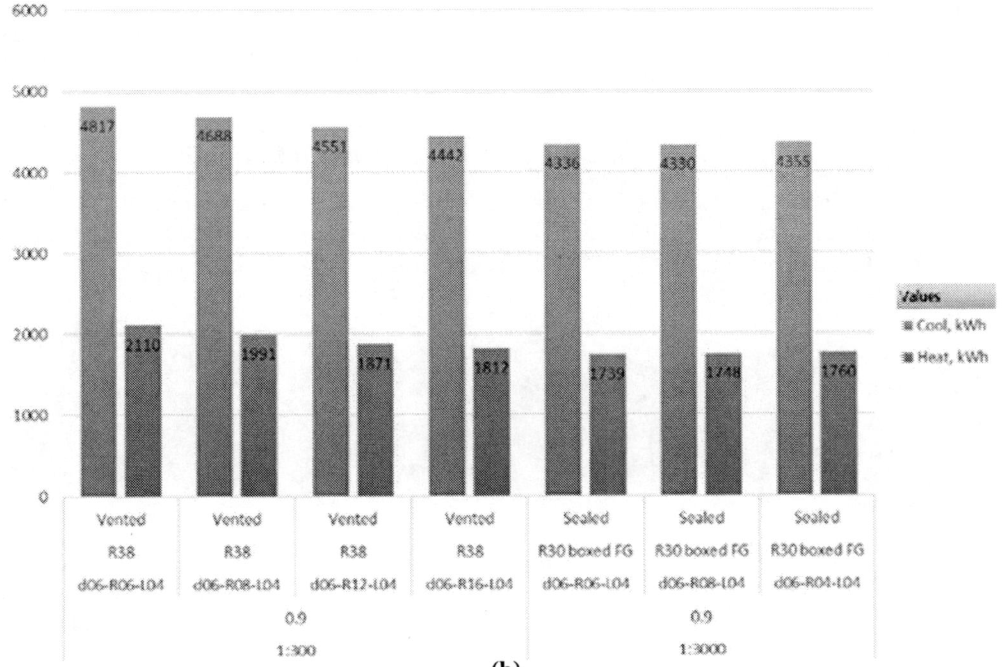

(b)

Figure 6 *(a) The effect of duct size on the performance of a large vented and a small sealed attic. (b) The effect of the thermal resistance of the ducts on the performance of a large vented and a large sealed attic. Location is Houston, TX.*

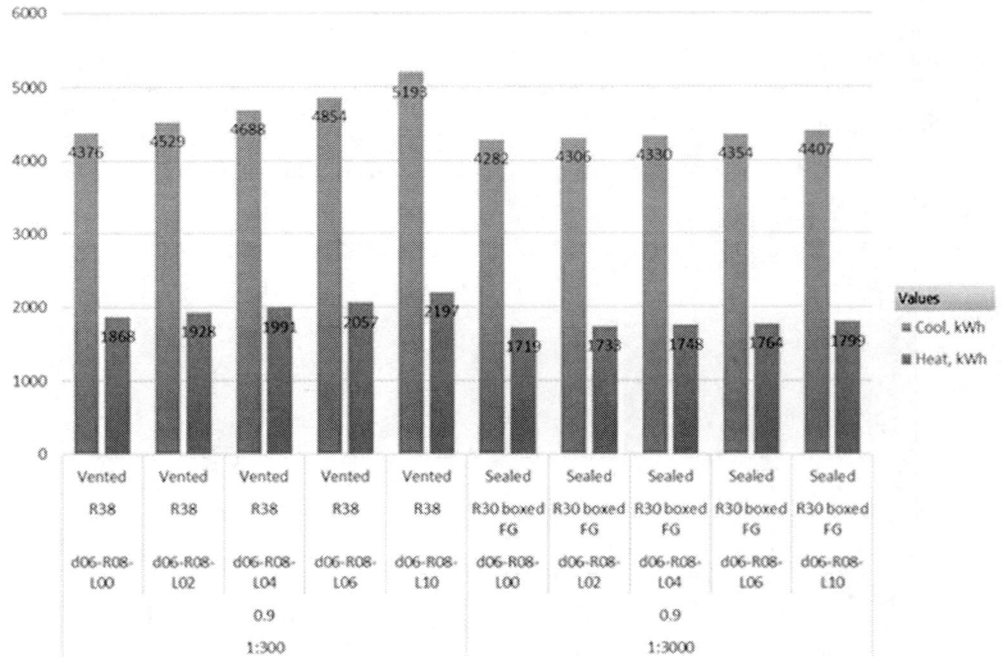

***Figure** 7 The effect of duct air leakage on the performance of a large vented and a large sealed attic in Houston, TX.*

Miller, W.A., M. Keyhani, T. Stovall and A. Youngquist. 2007. Natural convection heat transfer in roofs with above-sheathing ventilation. Thermal Performance of the Exterior Envelopes of Buildings X. Proceedings of ASHRAE THERM X, Clearwater, FL., Dec. 2007.

Miller, W.A., A. Desjarlais, and M. LaFrance. 2014. Roof and attic design guidelines for new and retrofit construction of homes in hot and cold climates. Thermal Performance of the Exterior Envelopes of Buildings XII. Proceedings of ASHRAE THERM XII, Clearwater, FL., Dec. 2013.

Owens Corning. PROPINK® high performance conditioned attic system installation guidelines. Toledo, OH: Owens Corning. www2.owenscorning.com/literature/pdfs/HPCA%20Installation%20Instructions.pdf.

Petrie, T.W., T.K. Stovall, and A.O. Desjarlais. 2004. Comparison of cathedralized attics to conventional attics: Where and when do cathedralized attics save energy and operating costs? Thermal Performance of the Exterior Envelopes of Buildings IX. Proceedings of ASHRAE THERM VIII, Clearwater, FL, December 2004.

Salonvaara, H.M., A.N. Karagiozis, and A. Desjarlais. 2013. Moisture performance of sealed attics in climate zones 1 to 4. Thermal Performance of the Exterior Envelopes of Buildings XII. Proceedings of ASHRAE THERM XII, Clearwater, FL., Dec. 2013.

Wilkes, K.E. 1991. *Thermal model of attic systems with radiant barriers.* ORNL/CON-262. Oak Ridge, TN: Oak Ridge National Laboratory.

Innovative Roof Design for Passive Low-Income Housing in Western India

Madeline Gradillas, RA **Qin Zhang, MS** **Emma Nelson, BSc**

Bradley Tran **Leon Glicksman, PhD**
Fellow ASHRAE **Tejas Kotak, BE Civil**

ABSTRACT

India is undertaking a large-scale program to provide housing for low-income residents throughout the country. Many parts of the country are in tropical high-temperature climates, but air conditioning is not economically feasible in these parts. In cooperation with a local research foundation in Western India, the authors have been participating in the analysis, design, and testing of several prototype housing units. The research has stressed the integration of thermal mass, insulation, roof design, and natural ventilation strategies to achieve thermal comfort without the use of mechanical cooling. Full-size test structures have been constructed and detailed measurements have been made of thermal conditions. The designs have included several concepts for thermal insulation using indigenous low-cost materials. A focus of the research has been on the development of innovative roof systems. These include multilayer structures that allow for air circulation, radiation heat transfer control, and providing mechanical strength. Experimental results have shown that even in severe arid climates the interior conditions can be substantially moderated through most of the operating hours. The results of this research will be an important contribution to the designs of the initial phase of large-scale housing construction over the next few years in Western India.

INTRODUCTION

India is undertaking a large-scale program to provide housing for low-income residents throughout the country: the Housing for All program (MHUPA 2007). The program envisions the construction of 20 million new homes for low-income citizens, many of them current slum dwellers, by 2022. The housing needs to be economical and designed to meet the occupants' needs and conform to acceptable local design traditions. The latter requires engaging the would-be homeowners through neighborhood planning. Many parts of the country are in tropical high-temperature climates, but air conditioning is not economically feasible in these parts. With current temporary housing (see Figure 1 for examples), the structures are not only very uncomfortable but they also put occupants in danger due to the excessive interior temperature conditions. Considering future climate change, these issues become even more serious. An important characteristic of future housing is the need to provide comfortable and safe interior environments without the use of substantial energy for conditioning. This requires the development of passive housing designs. These designs must be achieved at very low cost. In addition, some regions of India have the potential for severe earthquakes, so designs for these regions must also take this into account. Designs that utilize waste material in the structure are also appealing not only is a cost-reduction feature but also as a vehicle for capturing and removing potential pollutants from the landscape.

If the occupants of these passive houses are satisfied with their performance, then as their incomes and lifestyles improve over the years, they may opt for the same building technologies. The result would be the minimization of future energy consumption growth for air conditioning in India.

In cooperation with a local research foundation, Hunnarshala Foundation in Bhuj-Kutch Gujarat, India, the authors have been participating in the analysis, design, and testing of several prototype housing units. The research has stressed the

Madeline Gradillas is a former fellow of the Massachusetts Institute of Technology (MIT) Tata Center for Technology and Design and a SMArchS Graduate from MIT's Building Technology program, **Qin Zhang** is a doctoral student, **Emma Nelson** is a master's candidate in the Department of Mechanical Engineering and a fellow of the Tata Center for Technology and Design, **Bradley Tran** is a graduate student in the Building Technology program, and **Leon Glicksman** is a professor at MIT, Cambridge, MA, USA. **Tejas Kotak** is executive director of the Hunnarshala Foundation, Bhuj-Kutch, Gujarat, India.

Figure 1 *Examples of temporary housing.*

Figure 2 *Building 1: single-room dwelling with 15 cm thick solid CMU walls and an uninsulated asbestos cement corrugated sheet roof.*

integration of thermal mass, insulation, roof design, and natural ventilation strategies to achieve thermal comfort without the use of mechanical cooling. Full-size test structures have been constructed and detailed measurements have been made of the thermal conditions. The designs have included several concepts for thermal insulation using indigenous low-cost materials. A focus of the research has been on the development of innovative roof systems. These include multilayer structures that allow for air circulation, radiation heat transfer control, and providing mechanical strength. Experimental results have shown that even in severe arid climates the interior conditions can be substantially moderated through most of the operating hours (Gradillas 2015). The results of this research will be an important contribution to the designs of the initial phase of large-scale housing construction over the next few years in Western India.

EXISTING HOUSING IN GUJARAT

The authors monitored seven different existing vernacular houses that are occupied during various periods of the day. These include houses that would be typical of current low-income housing as well as housing for families of higher income levels. For the purposes of this paper, our focus is on dwelling construction types that are commonly seen in low-income settlements in the area. Two of the monitored houses fall under this category: building 1 and building 7. In both of these we monitored the ceiling and wall temperatures along with the interior air temperature in the approximate middle of the space over a number of weeks. We also recorded concurrent half-hourly outdoor ambient air temperatures with on-site sensors for the entire test period.

Building 1 is a single room dwelling with 15 cm thick solid concrete masonry unit (CMU) walls and an uninsulated asbestos cement corrugated sheet roof, as shown in Figure 2. It is largely unprotected from direct solar gain. Though of typical construction for informal housing communities, this particular dwelling is unoccupied for all but a few hours during midday. The house is equipped with an electric ceiling fan, used when occupied on warm days. Two small unglazed windows on the east and west facades are kept open on a similar schedule.

Building 7 is part of a series of clustered buildings on the Gujarat Industrial Development Corporation (GIDC) 2001 earthquake redevelopment site outside of Bhuj City. It is a one-room dwelling with 23 cm thick uninsulated compressed stabilized earth block (CSEB) walls and 12 cm reinforced cement concrete (RCC) slab roof, as shown in Figure 3. The dwelling shares its north wall with its mirrored neighbor; otherwise, wall surfaces are generally exposed to solar radiation. The CSEB walls are plastered and are light in color. The room where the thermal sensors are installed is equipped with an electric ceiling fan, used when occupied during warm days and nights. All windows are unglazed and are covered by sheets of paper and interior curtains. The house is irregularly occupied during the day by three family members.

Figure 3 *Building 7: one-room dwelling with 23 cm thick uninsulated CSEB walls and 12 cm RCC slab roof.*

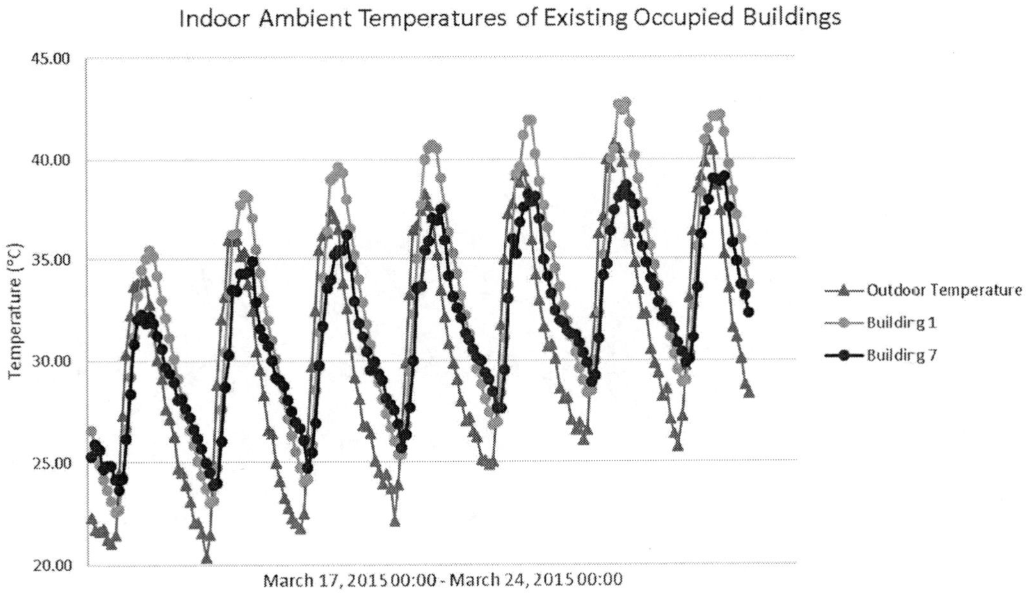

Figure 4 *Indoor air temperatures of buildings 1 and 7 along with the ambient air temperature.*

Figure 4 shows the measured indoor air temperatures of buildings 1 and 7, located near Bhuj, Gujarat, along with the outdoor ambient air temperature for seven days in March. During two days of this period the outdoor temperature exceeded 40°C. For all days during this test period, the peak inside temperature of building 1 exceeds the outdoor temperature, while building 7 is not far behind; both, when combined with wall and ceiling temperatures, lead to operative temperatures that are well in excess of normal comfort standards (Gradillas 2015). Figure 5 shows the ceiling temperatures of the two buildings, with building 1 exceeding 50°C and building 7 at 47°C. These high levels are the result of solar gains on the uninsulated roofs and also explain the elevated interior air temperatures. The hot ceiling will also result in an elevated operative temperature in the building interior.

ROOF DESIGNS

Our initial assessment of existing vernacular dwellings confirmed known key design strategies in an effective passive building design. By using night ventilation combined with interior thermal mass, the cool nighttime temperatures can be utilized to minimize peak temperature increase during the daytime. To be done properly, the ventilation rate during the daytime must be minimized. The third key element to achieving effective performance is the roof design. It must function to minimize internal heat gains due to solar radiation and elevated ambient temperatures. A well-insulated roof deck with reflective coating is the usual solution in the developed world.

In our work we were looking for innovative solutions that not only incorporate the aforementioned passive cooling tech-

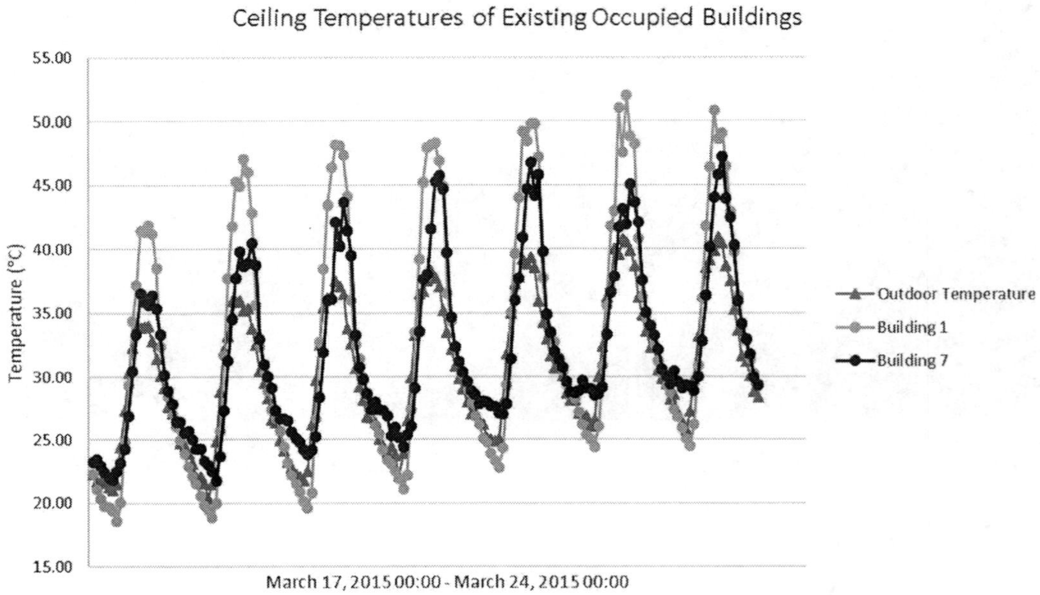

Figure 5 *Indoor ceiling temperatures of buildings 1 and 7 along with the ambient air temperature.*

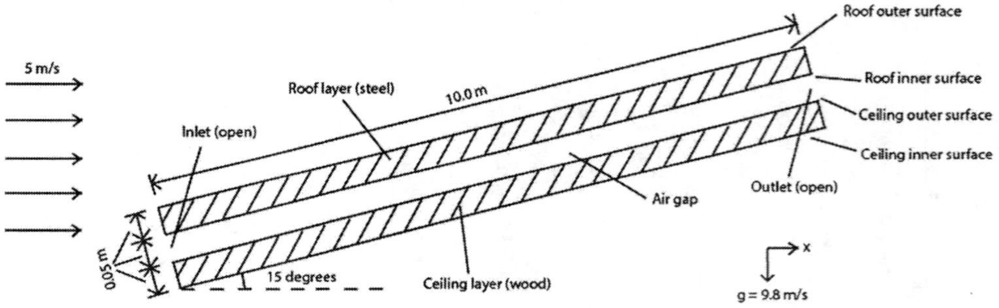

Figure 6 *Domain of the CFD model for the case with 15 degrees inclination and wind velocity of 5 m/s.*

niques but also are cost-effective for low-income housing. The design we are investigating is a two-layer ventilated roof. In this case, both layers are made of inexpensive corrugated metal sheet. The inside surface of the upper sheet is covered with very thin aluminum foil to minimize infrared radiation heat transfer between the two sheets. The air gap allows for ventilation and may provide insulation without additional material costs. Our primary work recorded in this paper deals with the thermal performance of the roof and of the test buildings using this concept. In ongoing research, we are investigating the use of the double layer not only for good thermal performance but also as a low-cost structural system. The dual use of the double-layer system makes it attractive for the current application in India.

We carried out several computational fluid dynamics (CFD) studies to investigate the influence of the roof slope on its thermal performance. We also investigated whether the roof cavity should be closed or left open at both ends. Based on our analysis, we arrived at a roof design that allowed a seasonal switch between open and closed cavities. This way, during warmer seasons the air gap n the summer is ventilated to flush out hot air, while in cooler seasons the air gap can act more as an insulation layer.

We simulated the roof performance for a typical summer day and a typical winter day. Three different roof angles were investigated for operation during the middle of the day and the middle of the night. A two-dimensional layered roof-ceiling model was built, shown in Figure 6, based on the typical building geometries in India. There is an air gap between the ceiling and the roof to create additional insulation. Both the ceiling and the roof are 10 m long and 0.05 m thick. Three inclination angles were considered: 0, 15, and 30 degrees. The model has an inlet and an outlet at the edges, which can be either fully open or fully closed. The inlet and outlet were defined as zero gage pressure openings to an environment, if fully open. The ambient domain was substantially larger than the ceiling-roof system. An ambient velocity of 5 m/s was set for cases with a windy environment. A reflective material with an emissivity of 0.1 was applied to the interior surface to reduce radiation between the ceiling and roof. The emissivity for the roof was different for different wavelengths: 0.3 for the visible part and 1.0 for the infrared part. Therefore, for the day cases, a solar heat flux of 400 W/m^2 was applied to the exterior roof surface and 30% (120 W/m^2) was absorbed. The interior ceiling surface, the ambient, and the sky were set at different fixed temperatures, as shown in Table 1. The sky temperature was set to be the same as the ambient temperature. The other surfaces were modeled using "coupled" thermal conditions.

All the cases were simulated under steady-state mode using Fluent (part of the ANSYS 14.5 suite) (ANSYS 2013). A pressure-based solver was adopted. The viscous model used is the re-normalization group (RNG) k–ε, and the radiation model used is surface-to-surface. The near-wall regions were meshed using very fine elements ($y^+ \leq 5$), as required by the "enhanced wall treatment" used in this simulation.

Figure 7 gives the total heat flux though the inner ceiling surface for the 4 cases at different inclination angles. The inclination angle of 30 degrees has the best performance in the summer given that there was less heat gain in the day and more heat loss in the night. However, this inclination angle performs badly on winter nights with an open inlet/outlet. The presence of the wind plays a major role in the performance, suggesting that if possible the open inlet should be aligned with the prevailing wind direction. Therefore, to achieve a balanced thermal performance for all four cases, it is recommended to adopt a large inclination angle (30 degrees) but be flexible about the operation of the inlet/outlet—that is, open the inlet/outlet during the summer yet close them in the winter.

FIELD TESTS

We constructed a thermal field research laboratory southwest of the Bhuj city center to test the performance of various roof types (see Gradillas 2015 for more details). For this paper

Table 1. Boundary Conditions for CFD Simulations of Ventilated Roof Concept, Summer and Winter Conditions

Season	Time	Boundary Conditions		
		Solar Heat Gain at Roof Outer Surface, W/m^2	Ceiling Inner Surface Temperature, K	Ambient Temperature, K
Summer	Day	120	302.15	302.15
	Night	0	298.15	293.15
Winter	Day	120	293.15	288.15
	Night	0	293.15	283.15

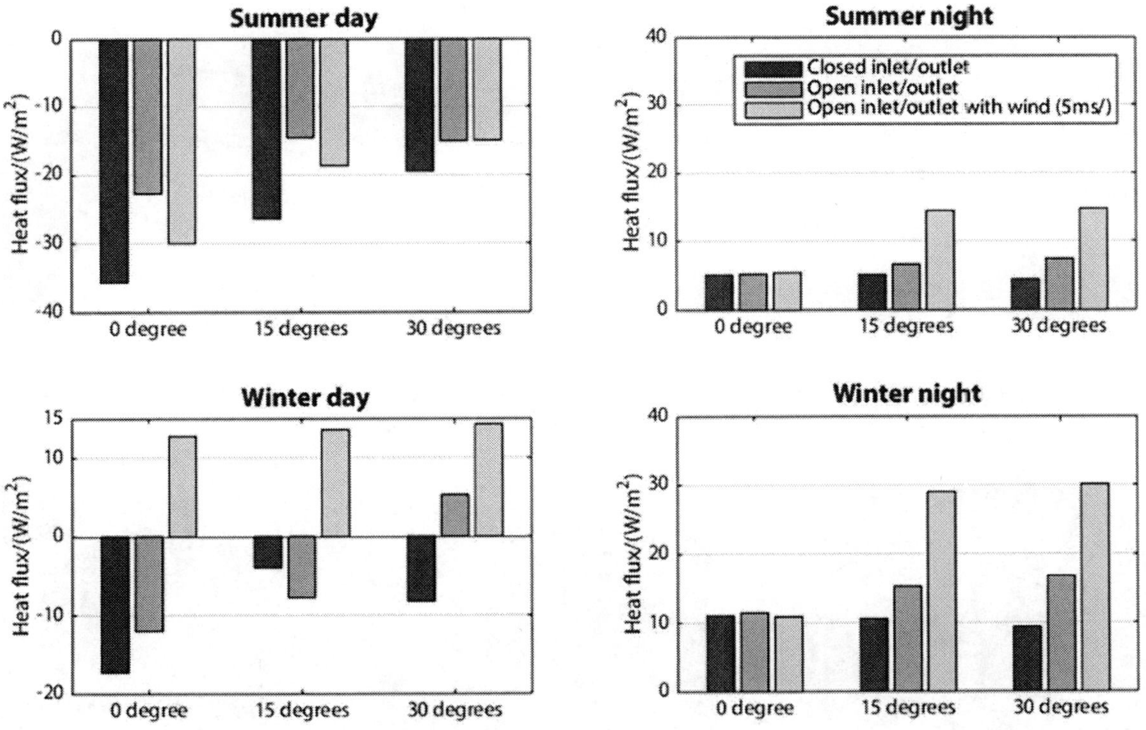

Figure 7 *Comparison of the CFD results for total heat flux through interior surface (negative = heat flow into room, positive = heat flow out of the ceiling for summer/winter, day/night cases at three different inclination angles [0, 15, and 30 degrees] with inlet and outlet closed, with outlets open, with and without wind).*

we concentrate on the results of the double-layer ventilated roof, shown in Figure 8, and compare it to a single-layer corrugated metal uninsulated roof. The foundation for the laboratory consists of five 2.4 × 2.4 × approximately 2.4 m chambers. Construction of all chambers is identical save the roof assemblies. Chamber walls are constructed from 23 cm thick local sandstone block and mud mortar and are externally insulated with proximately 15 cm of straw thatch. The insulated block walls are further protected from direct solar radiation—and cows' mouths—with an outer layer of burlap fabric coated with lime plaster (see Figure 9). Each chamber is oriented to the four cardinal directions, having a 0.75 m wide by 0.5 m high screened opening on the north- and south-facing walls. Each opening has an outer hinged solid panel that can be manually operated according to the desired ventilation schedules. The openings are oriented north/south for two reasons. The first is so that the hinged panels can provide full shade to the southern opening, to minimize direct solar gain to the interior. The second is to take advantage of Bhuj's summer evening prevailing winds from the southwest. These openings are offset vertically by 1.5 m so that at times of little or no wind, buoyancy-driven ventilation can be maximized for the volume.

In addition to indoor ambient temperatures, temperatures were also measured on the interiors of the walls and on the underside of the ceiling, and for the ventilated roof the temperatures in the air gap were measured. The wall, ceiling, and air temperatures were used to determine operative temperatures for calculating hours within thermal comfort standards. In this series of experiments, during the summer-like seasons the ventilated roof was left open. In addition, the temperature of the air in roughly the middle of the room was also measured. All temperatures were measured using a two-channel waterproof HOBO U23 Pro External Temperature Data Logger (U23-003). The wall and ceiling temperatures were measured by making a small hole and inserting the temperature probe tightly in place.

Experiments were carried out over three different time periods, with temperatures recorded hourly. The three periods corresponded to three different ventilation strategies. In the first the door and windows were closed. In the second the doors and windows were kept open 24 hours a day. In the third the windows were kept open at night and closed during the day. The windows were not tight fitting, so even when closed they admitted a modest amount of airflow.

TEST RESULTS

Figure 10 shows the interior air temperatures for the ventilated and the single-layer uninsulated roof when the test chamber windows are kept open day and night. Not surprisingly, the indoor air temperature closely followed the outdoor ambient temperature. When night ventilation is used there is a substantial improvement in the indoor air temperature during

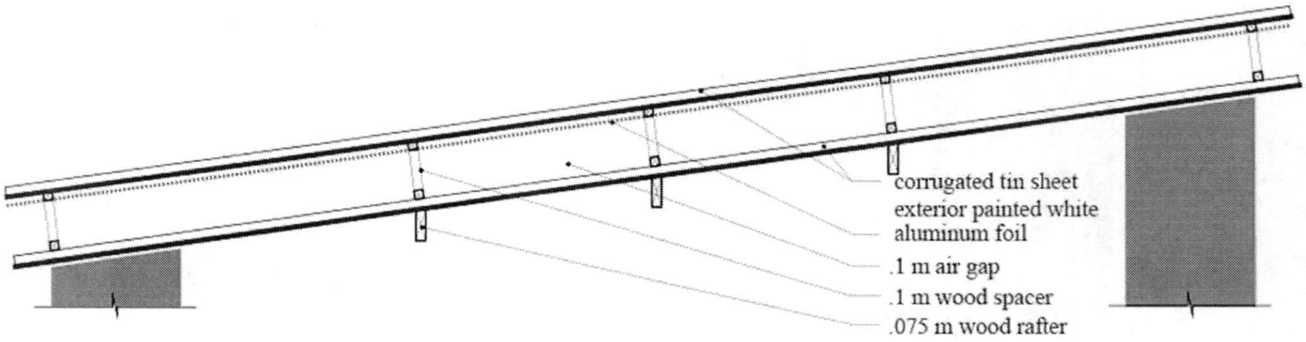

Figure 8 *Double-layer ventilated roof with internal radiation shield.*

Figure 9 *Construction of 2.4 m square test chamber.*

the peak daytime hours. The night ventilation does a good job of cooling the thick walls, whose interior surfaces are in good contact with the interior airflow. During the daytime when the ventilation is suppressed, the thick walls maintain the interior temperature well below the outdoor ambient air temperature for the test module with the ventilated roof, as shown in Figure 11. The module with a single uninsulated roof element fares somewhat worse because of the heat gains through the roof during the daytime. Because of the limited floor dimensions of the test chambers, the interior air temperature is dominated by the wall conditions. For residences with larger floor plans, the ceiling temperatures would be more dominant and the difference between the air temperatures for the two roof designs would be much larger.

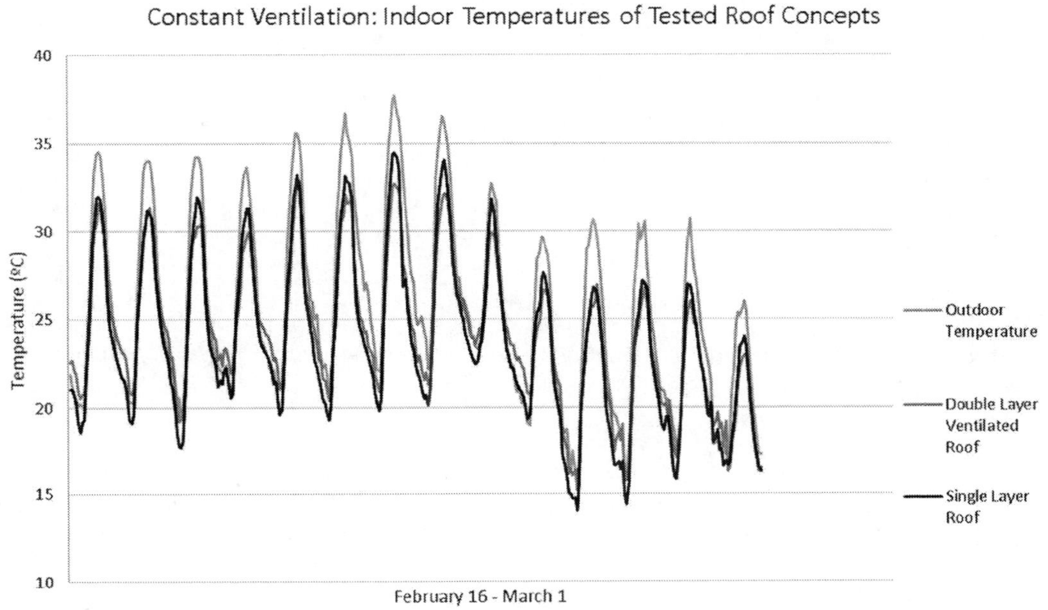

Figure 10 *Indoor air temperatures and ambient air temperature for ventilated roof and for single roof with windows open 24 hours per day.*

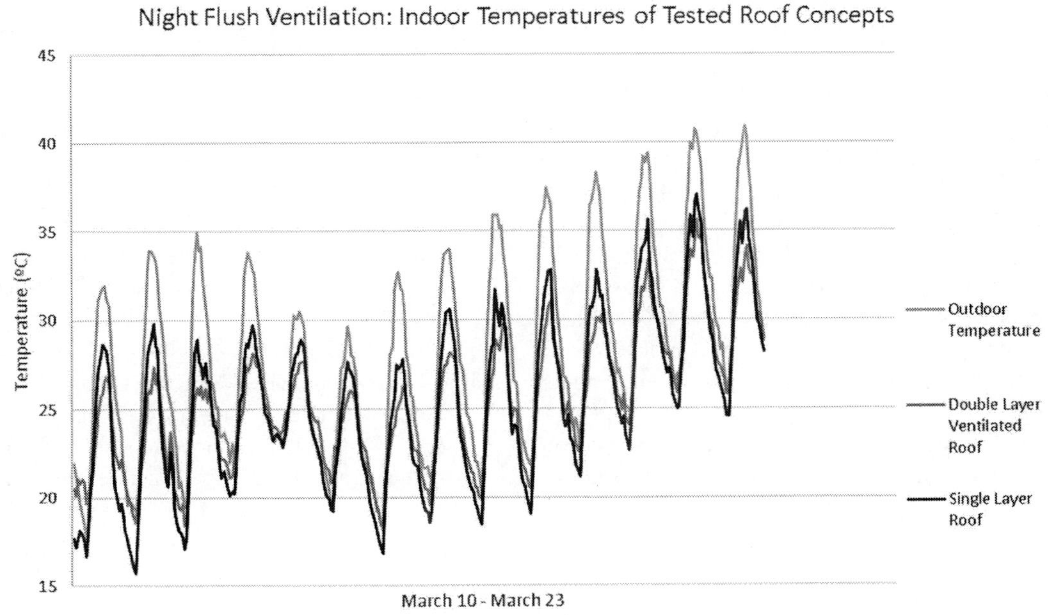

Figure 11 *Indoor air temperatures and ambient air temperature for ventilated roof and for single roof with windows open only at night and night ventilation.*

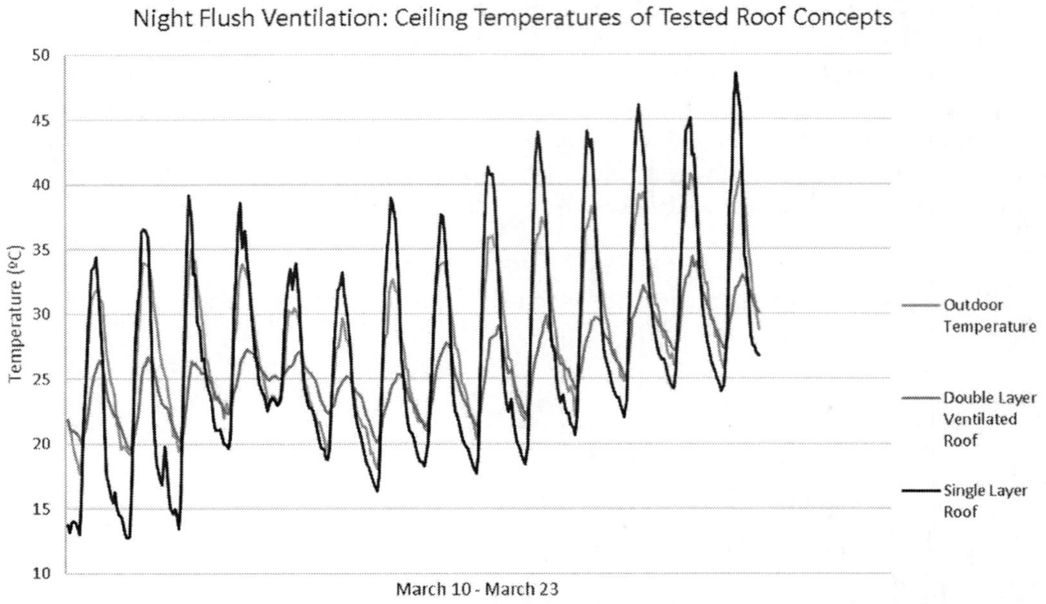

Figure 12 *Ceiling temperatures and ambient air temperature for ventilated roof and for single roof with windows open only at night and night ventilation.*

The ceiling conditions are evident in Figure 12, which shows the ceiling temperatures for the two roof systems. Although the outdoor temperature peaks well above comfort or even thermal safety levels, the interior temperature with the ventilated roof remains much more livable, while the single uninsulated roof exhibits excessively high temperature levels. Moreover, if the occupants in the uninsulated roof building used a ceiling fan during the day it would increase the indoor air temperatures as a result of recirculating the hot air surrounding the ceiling.

CONCLUSIONS

Existing low-income residential buildings in India subject occupants to excessive temperature levels. One primary cause is the uninsulated single-layer roofs. The roof temperatures reach unacceptable levels due to solar gains and high ambient temperatures during midday.

Double-layer ventilated roof prototypes were tested in the field to develop low-cost, better-performing roofs. Field research results as well as simulations indicate that these double-layer roofs, combined with night ventilation and exposed thermal mass in walls, significantly lowered indoor air temperatures.

The innovative roof design uses a double layer of corrugated metal sheets to create an air circulation space. The double-layer design also offers the possibility of a stand-alone structural system. Several new housing developments in Gujarat will utilize the innovative features to judge performance and user satisfaction.

ACKNOWLEDGMENTS

This research was sponsored by the Massachusetts Institute of Technology (MIT) Tata Center for Technology and Design. The field studies were made possible by the invaluable contributions of many researchers at the Hunnarshala Foundation in Bhuj-Kutch Gujarat, India

REFERENCES

ANSYS. 2013. ANSYS Fluent. Canonsburg, PA: ANSYS, Inc. http://www.ansys.com/Products/Fluids/ANSYS-Fluent

Gradillas, M.S. 2015. Analysis and design for thermally autonomous housing in resource constrained communities: A case study in Bhuj, India. Master's Thesis, Massachusetts Institute of Technology, Cambridge, MA.

MHUPA. 2007. Policies: National Urban Housing & Habitat Policy (NUHHP). Ministry of Housing and Urban Poverty Alleviation, Government of India. http://mhupa.gov.in/User_Panel/Userview.aspx?TypeID=1414.

Vegetative Roof Performance during Summer—Critical Analysis of the Irrigation Effect

S. de Freitas
FEUP-CONSTRUCT—LFC

N.N.M. Ramos
FEUP-CONSTRUCT—LFC

V.P. de Freitas
FEUP-CONSTRUCT—LFC

H.M. Künzel
Member ASHRAE, IBP

D. Zirkelbach
IBP

ABSTRACT

Vegetative roofs are nowadays a widespread technology due to many benefits provided both at building and city scale. Namely, they reduce buildings' energy consumption and improve cities' urban environment by heat island mitigation. They may also increase rooftop comfort conditions for the users during summer days. Therefore, the assessment of the hygrothermal vegetative roof behavior is very important for a correct quantification of its impacts.

Irrigated vegetative roofs are necessary for certain climates and plant species during summer; however, data concerning the ideal water requirements are still scarce. This study assesses the vegetative roof performance for a temperate climate concerning the effect on the outdoor environment, based on surface temperatures and heat fluxes, with and without the contribution of irrigation.

The experimental vegetative roof setup located in Porto provides data for the hygrothermal analysis. The monitored parameters—temperature profiles, heat fluxes and water content profiles, together with outdoor conditions—allow a detailed analysis of the real behavior in summer with varying irrigation. The air temperature, solar radiation, and irrigation amounts, together with irrigation schedules and antecedent irrigation periods, were found to be determinant on the heat fluxes variation and on surface temperature reduction.

The available hygrothermal simulation tools for vegetative roof, key for studying the roof response, were briefly assessed. The major limitations are related to the water balance and the properties of the vegetation and substrate.

INTRODUCTION

During the last decades there has been unprecedented urban growth, with 50% of the world population living in urban spaces in 2008 and with a predicted 70% rate in 2050 (UNPD 2012). This growth changes the urban landscape and the radiation exchanges, hence modifying the local urban environment, with consequences on the users' comfort and on the urban climate. Vegetative roofs can change the urban microclimate thanks to evapotranspirative cooling and roof shading, thus reducing the roof surface temperature and, hence, the surrounding air temperature (EPA 2008).

Vegetative roofs, also called *green roofs*, are being progressively applied in new and renovated buildings due to the multiple benefits provided for the building and also, at a larger scale, for the city itself (Zinzi and Agnoli 2012). The knowledge available so far on vegetative roofs comprises a wide range of research areas. This technology is considered to have a positive effect in improving buildings' energy efficiency (Olivieri et al. 2013; Spala et al. 2008; Virk et al. 2015), mitigating urban heat island effect by reduction of the roof surface temperature and the surrounding air (Rosenzweig et al. 2006; Gill et al. 2007; Qin et al. 2012) and consequently increasing the comfort of the roof users (Peng and Jim 2013). The urban water balance is also improved due to the attenuation of runoff flows and minimization of flooding risks (Fioretti et al. 2010). Nonetheless, the studies available are reporting contradictory results, and the comparison between them is sometimes difficult due to different vegetative roof

S. de Freitas is a doctoral candidate, **N.N.M. Ramos** is an assistant professor, and **V.P. de Freitas** is a full professor in the Department of Civil Engineering at the Faculty of Engineering of University of Porto, Porto, Portugal. **H.M. Künzel** is an honorary professor at the University of Stuttgart and head of the Department Hygrothermics at the Fraunhofer-Institute for Building Physics, Germany. **D. Zirkelbach** is group manager of Moisture Control and Climate Specific Design at the Fraunhofer-Institute for Building Physics, Holzkirchen, Germany.

design, study conditions, and length of the studies. For instance, in terms of energy improvements, some uncertainty still exists concerning the energy performance of vegetative roofs in winter with both positive and negative energy consumption values being reported (Berardi et al. 2014; Sailor et al. 2012).

Plants, despite having maintenance requirements, have an important role in reducing the heat flux by regulating latent heat flux through better water management and additional water storage in the plant leaves/roots and sensible heat flux through additional shading provided by the plant leaves (Tabares-Velasco and Srebric 2009). The assessment of plants' contribution to the heat flux reduction is therefore influenced by many factors: extra shading, reduced radiation absorption, wind shielding, additional water storage, and a water-controlled mechanism based in photosynthesis and evaporation (Liu 2004).

Traditionally, vegetative roofs are designed to keep their performance without additional water besides precipitation (Mechelen et al. 2015); however, the environment conditions constraints, especially in summer conditions, may impose additional water supply for the vegetative roof's survival. Many studies have focused on the water content in vegetative roofs and found that it plays a crucial role through its influence on the heat fluxes and surface temperature at the external surface (Lazarrin et al. 2005; Sun et al. 2014l; Jaffal et al. 2012). The extra evapotranspiration induced by irrigated vegetative roofs contributes to the urban heat island mitigation (Berardi et al. 2014; Liu and Bass 2005). Simulation studies concluded that the impact of green surfaces on the urban heat balance depends on water availability; hence, if a vegetative roof is not watered, it quickly experiences water stress and the contribution to the heat balance is unchanged compared to the reference situation (Salagnac et al. 2013).

The water demand for plant survival and summer performance enhancement increases the maintenance costs and is an added difficulty for cities or areas with water limitation regulations. Hence, a careful study on irrigation specifications for vegetative roofs is mandatory to maximize roof performances, whilst maintaining comfort conditions for users on top of the roofs with minimum irrigation costs.

This study focuses on the analysis of irrigation effects on vegetative roofs during summer. A case study of a vegetative roof for a temperate climate, Porto, is presented concerning heat fluxes and surface/near surface temperatures. A selection of numerical models to simulate vegetative roofs with inherent limitations is also presented and, finally, a discussion section summarizes the results and states the future developments.

EXPERIMENTAL VEGETATIVE ROOF MONITORING

The test building setup is located in an urban area in Porto, Portugal. The two roofs monitored (Figure 1) present a similar composition up to the thermal insulation system, as schematized in Figure 2, and are located over a detached dwelling. The vegetative roof (20 m^2) comprises a lightweight, expanded clay drainage layer, a sand layer, a substrate, and a grass lawn (grass seed mix). On the other hand, the traditional roof (30 m^2) has white concrete slabs (pavers).

The vegetative roof has installed a permanent sprinkler irrigation system that ensures survival in hot and dry periods. Several parameters are being monitored in 10-minute intervals on the vegetative and traditional roof, namely, temperature profiles on the vegetative roof, surface temperature on the vegetative and traditional roof, water content profiles on the vegetative roof, and exterior heat fluxes (Figure 3). Additionally, relevant weather data such as air temperature above the roof, precipitation, and solar global radiation are being measured on site. The sensors characteristics are described in Table 1.

(a) (b)

Figure 1 *Upper levels of the (a) vegetative and (b) traditional roofs.*

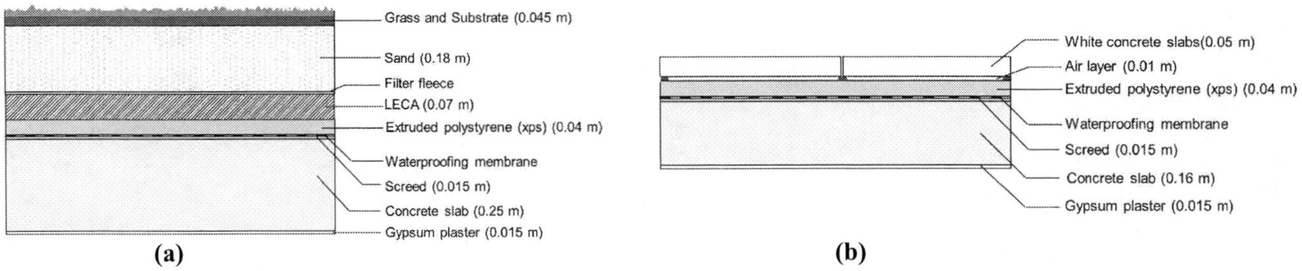

Figure 2 *Roof compositions: (a) vegetative and (b) traditional.*

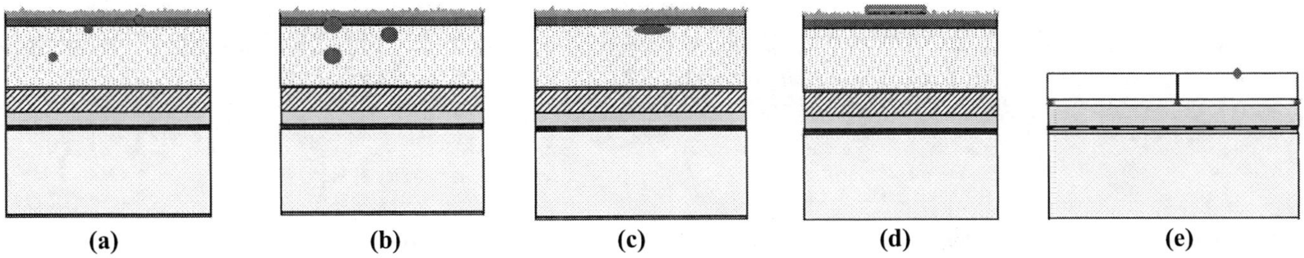

Figure 3 *Monitoring positions: (a) vegetative roof temperature profile at 0.02, 0.045, and 0.135 m below the surface; (b) vegetative roof water content profile at 0.045 m (WC0), 0.09 m (WC1), and 0.135 m (WC2) below the surface; (c) vegetative roof heat flux sensor at the beginning of the sand layer (0.045 m below the surface); (d) vegetative roof infrared surface temperature sensor, placed at 0.6 m above the surface; and (e) traditional roof surface temperature.*

Table 1. Accuracies and Ranges of the Installed Sensors

Reference	Type	Accuracy	Range
Temperature probes	Pt100 - AC1904 (Rotronic)	±0.15°C at 0°C	–50°C to 110°C
Water content probes	ThetaProbe Soil Moisture Sensor ML2x and ML3 (Delta-T)	Vol WC: ± 1%	0 to 100%
Infrared temperature sensor	PyroEpsilon – PE21LT (Calex)	±1% or ±1°C	–20°C to 100°C
Heat flux sensor	Heat Flux HFP01 (Hukseflux)	+5% to –15%	–2000 to +2000 W/m^2
Air temperature sensor	HOBO U12 (Onset)	±0.35°C	–20°C to 70°C

VEGETATIVE ROOF PERFORMANCE IN SUMMER CONDITIONS

The experimental analysis for a temperate climate focuses on the vegetative roof summer performance with and without irrigation in terms of heat fluxes, surface/near surface temperatures, water content distribution, and, additionally, comparison with a traditional roof.

Irrigation and Heat Fluxes

The heat flux affects directly the comfort of the vegetative roof user as well as the urban environment. Figure 4 shows the exterior heat flux near the surface (at 0.045 m below the roof surface) measured from July 9 to 19, 2015. The corresponding air temperature (measured in a shaded device at 1.5 m), solar global radiation, and irrigation amount are provided in Figure 5. The irrigation was not running during the day to minimize the strong evaporation during those hours.

It is possible to observe a change in the heat flux with irrigation that corresponds to 6.3 L/m^2 applied at 8 a.m. and 11 p.m., whereas it does not exist on the first day without irrigation (Figure 4). This means that the irrigation works as a passive cooler, due to evapotranspiration. Figure 6 shows two consecutive periods, one without irrigation (and with the previous 15 days without irrigation) followed by an irrigated period corresponding to 2.5 L/m^2 (at 8 a.m. and 11 p.m.). The weather data for this period is represented on

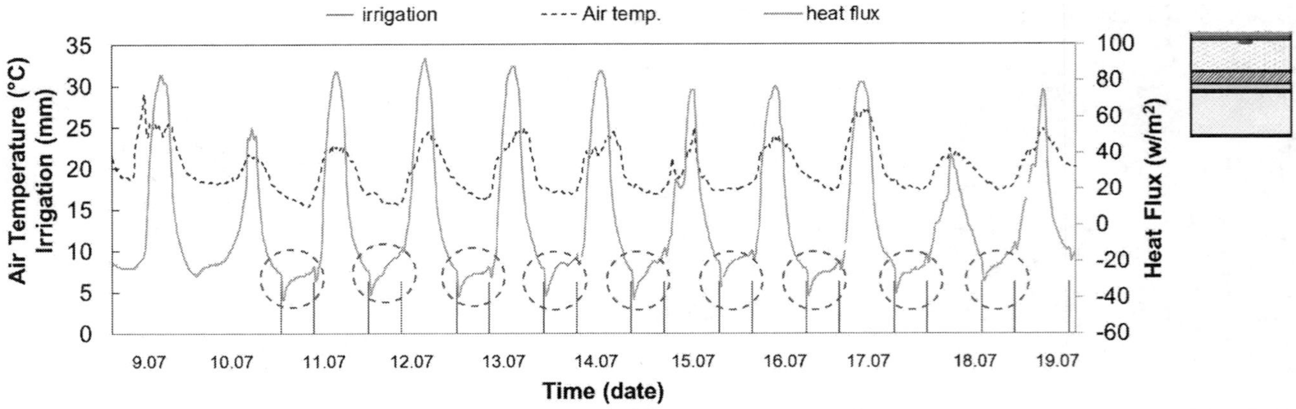

Figure 4 *Impact of the irrigation (6.3 L/m²) on the heat flux measured at 4.5 cm below the vegetative roof surface from July 9 to 19, 2015 (10-min data), marked with the red circles.*

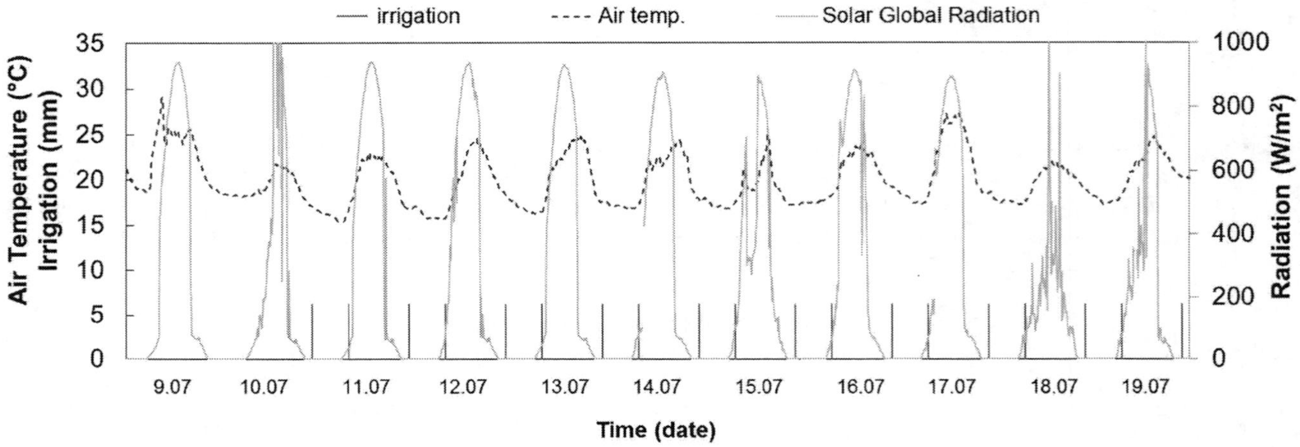

Figure 5 *Measured weather parameters on the vegetative roof: air temperature at 1.5 m and solar global radiation from July 9 to 19, 2015 (10-min data).*

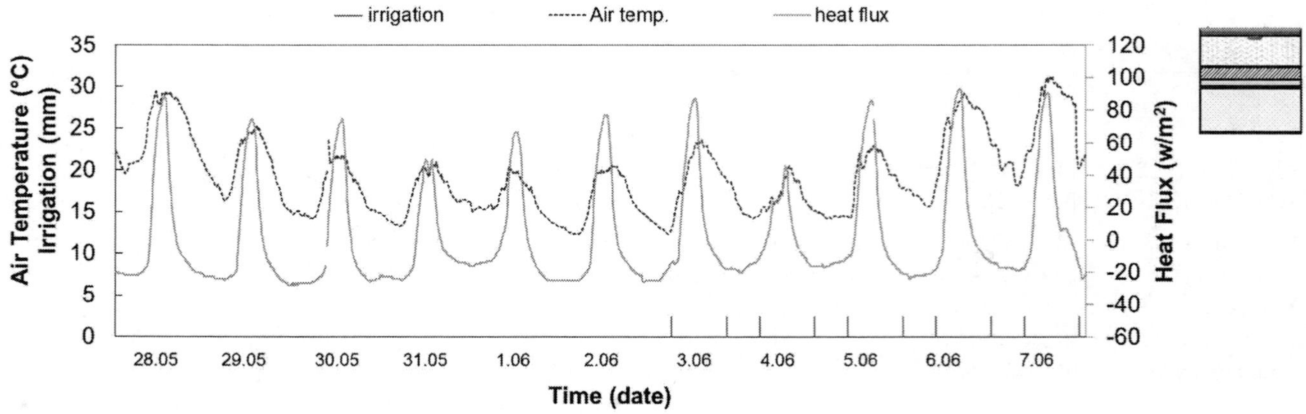

Figure 6 *Impact of the irrigation (2.5 L/m²) on the heat flux measured at 4.5 cm below the vegetative roof surface from May 28 to June 7, 2015 (period from May 13 to June 2 without irrigation) (10-min data).*

Figure 7. This small irrigation level and the previous 21 days without irrigation, which lead to very low water content values (0.02 to 0.04 kg/kg), did not result in a visible heat flux variation as it was observed with a 2.5 times higher irrigation regime.

A closer analysis of the behavior monitored for Porto on four different days with two irrigation quantities—2.5 L/m^2 and 6.3 L/m^2—presented in Figure 8 implies that the smaller the water amount, the lower the impact. However, other aspects apart from the irrigation amount also may have a significant influence in the heat fluxes as can be observed in Figure 8[a]) for the same irrigation quantity. Therefore, the daily schedule for the irrigation, the air temperature, and solar radiation evolution prior to the irrigation event are important in terms of the heat flux impact.

Complementary studies that focus on accounting all the determinant parameters for the heat fluxes will be performed in the future.

Irrigation and Surface/Near Surface Temperatures

The expected irrigation effect during summer periods is temperature reduction. Figure 9 shows the influence of the irrigation during the same 11 days presented before on the roof surface temperature (measured with an infrared sensor with an emissivity of 0.95), temperature at 0.02 m below the roof surface (measured with a Pt100 probe), temperature at 0.045 m between the substrate and the sand layer (measured with a Pt100 probe), and the temperature at 0.135 m, in the middle of the sand layer (measured with a Pt100 probe).

The measurements show a daily amplitude decrease and a peak delay as the distance from the surface increases. The roof surface peak temperature values are on average 6°C higher than 2 cm below in the substrate layer during the day, while during the night the values are approximately 6°C lower. A temperature reduction is clear after the irrigation at 11 p.m., with a value lower than the air temperature, but this effect at 8 a.m. is not as visible. An instant vegetative roof surface

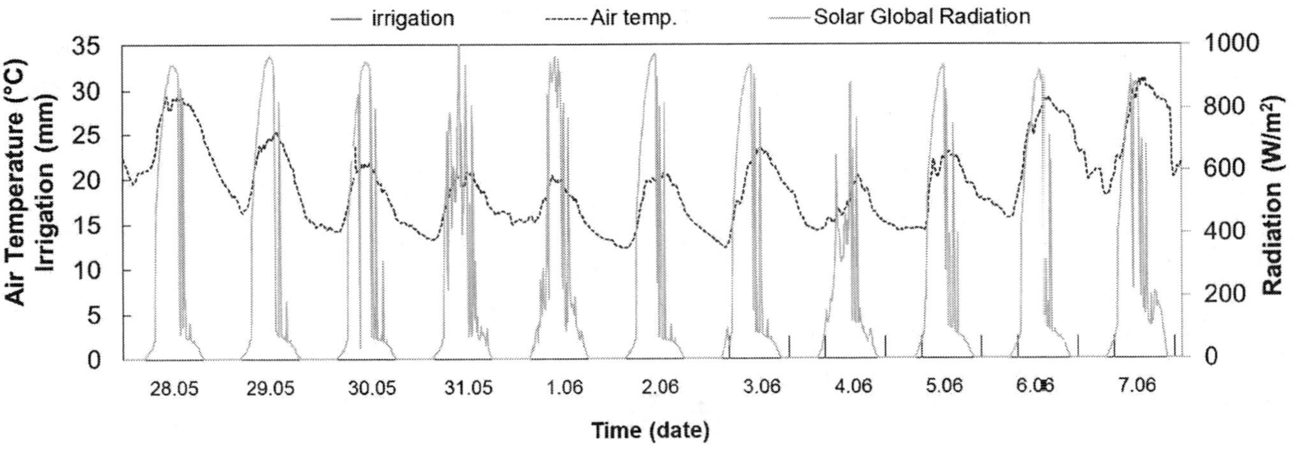

Figure 7 Measured weather parameters on the vegetative roof: air temperature at 1.5 m and global radiation from May 28 to June 7, 2015 (10-min data).

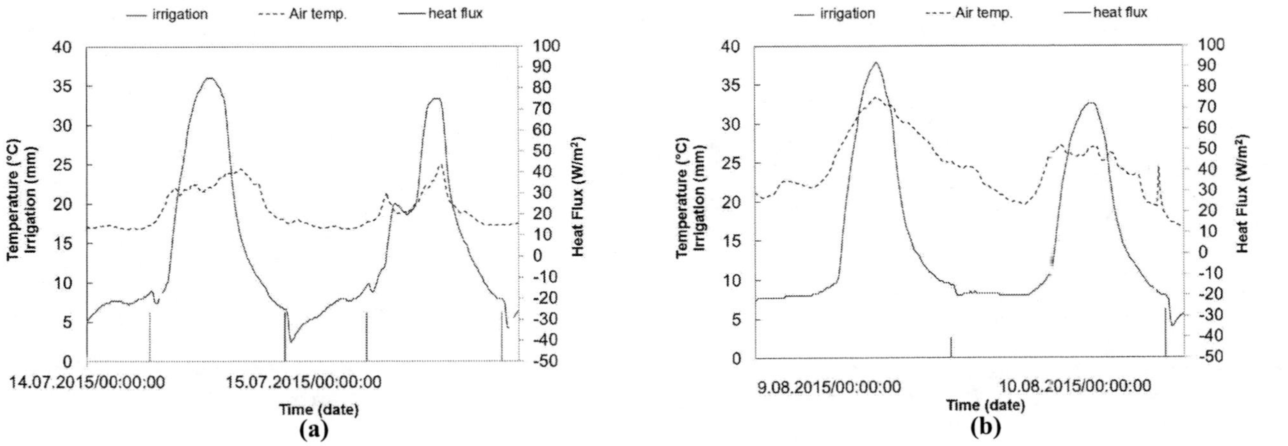

Figure 8 Irrigation impact in terms of heat fluxes: (a) July 14 and 15, 2015, and (b) August 9 and 10, 2015 (10-min data).

temperature reduction ranging from 1°C to 3°C is observed, while at 2 cm from the surface the decrease is less than 1°C (Figure 10).

In this case, the air temperature for the nighttime irrigation seems to be an important factor in the roof response with irrigation [Figure 10(b)]. Supplementary analysis will be required to relate all the involved parameters (radiation, wind speed, relative humidity, air temperature, and water content) and evaluate which are the most important for the enhancement of the roof temperature performance.

Irrigation and Water Content Distribution

The water balance of a vegetative roof that includes evapotranspiration, runoff, and drainage is connected to the energy balance and is particularly important for the summer performance (Palla et al. 2009; Berretta et al. 2009). Soil moisture, together with the following three factors, plant type, stage of plant development (dormant or actively growing and size), and weather (amount of energy available for evaporation dependent on solar radiation, wind speed, humidity, and temperature), affect the rate of evapotranspiration—soil evaporation and plant transpiration (FAO 1998; Bureau of Meteorology 2012).

Figure 11 presents the water content evolution at three levels (WC0, WC1, and WC2), taking in account the occurrence of intermittent irrigation during the morning (8 a.m.) and late in the evening (11 p.m.). The water supply to the roof results in a punctual increase of the water content values. The water is then transported mainly by gravity to the bottom as a function of the retaining capacity. Therefore, the sand water content is higher as the depth increases and a water content stratification is observed.

The absence of water supply in the summer, as can be observed for the period between July 7 and 10 [Figure 11(b)], results in a progressive decrease of the water content for all the three monitoring positions on the sand layer, with consequences on the vegetation survival.

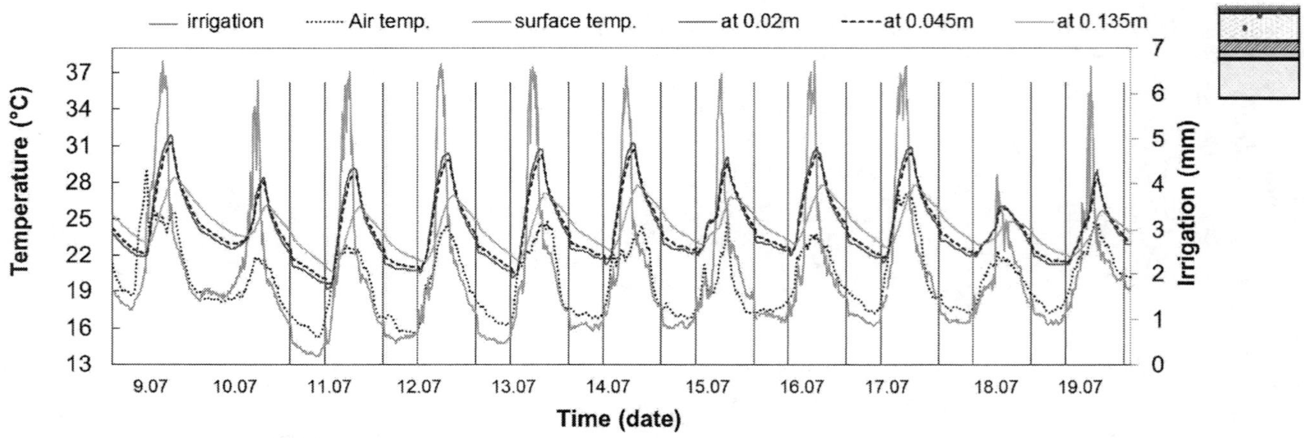

Figure 9 *Impact of the irrigation (6.3 L/m² at 8 a.m. and 11 p.m.) on the vegetative roof temperature (surface temperature and at 0.02, 0.045, and 0.135 m below the vegetative roof surface) from July 9 to 19, 2015.*

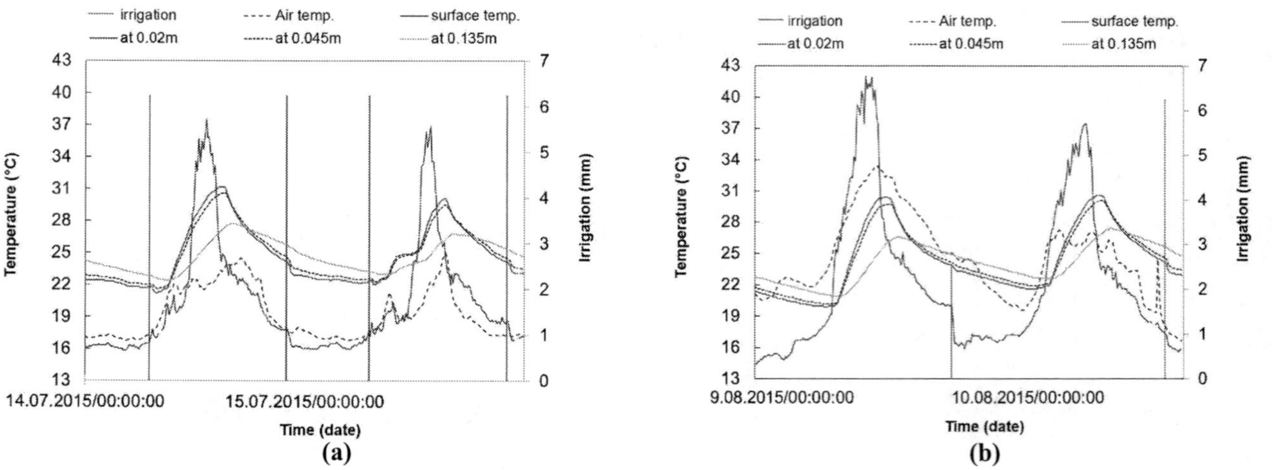

Figure 10 *Irrigation impact in terms of temperature: (a) July 14 and 15, 2015, and (b) August 9 and 10, 2015 (10-min data).*

A coupled analysis of the water supplied with the drained water is a topic that should be studied for irrigation events and would allow a better understanding of the vegetative roof water balance.

Traditional Versus Vegetative Roof Performance

An additional comparison of the vegetative roof with the concrete slab roof was performed during summertime with and without irrigation applied to the vegetative roof. In summer, vegetation absorbs most of the incoming solar radiation, which is used for biological functions (photosynthesis and evapotranspiration), thus decreasing the leaf temperature (Takakura et al. 2000). Figure 12 shows the surface temperature on the vegetative roof obtained with an infrared thermometer, while, for the traditional roof, the temperature is measured with a Pt100. The graph shows higher peaks during the day for the traditional roof (around 10°C above the vegetative roof) and also equal or lower temperatures during the night compared to the vegetative roof, if irrigation is applied.

During a summer night, a steep surface temperature decrease (approximately 1.5°C) on the vegetative roof is obtained (Figure 13) for a 6.3 L/m^2 irrigation at 11 p.m. This represents for the studied day a maximum 3°C difference after 50 minutes, compared to the concrete slab roof.

The potentialities of the vegetative roof compared to the traditional slab roof have been demonstrated in what concerns the summer roof temperature. A long-term analysis for the summer period in different climates with different irrigation amounts would emphasize the different vegetative roof impacts.

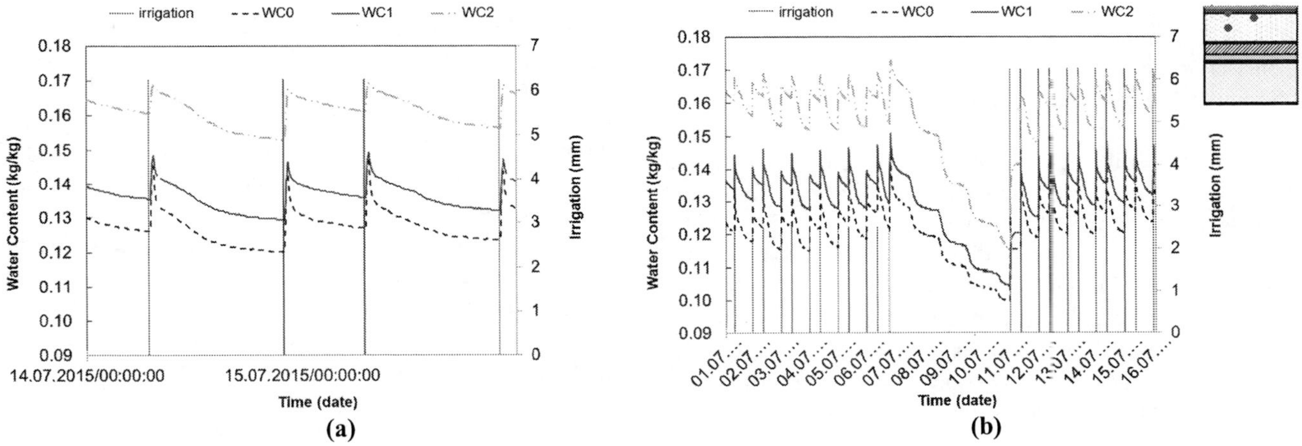

Figure 11 *Water content (WC0, WC1, WC2) evolution in kg/kg: (a) July 14 and 15, 2015, and (b) July 1 to 15, 2015, including periods with and without irrigation (10-min values).*

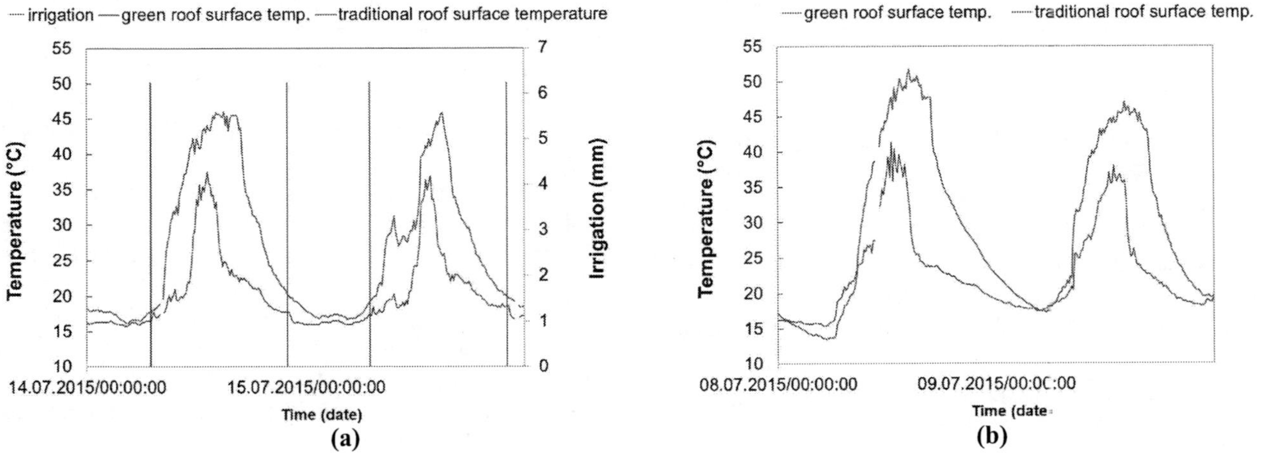

Figure 12 *Surface temperature—vegetative roof and traditional roof: (a) two summer days with irrigation (July 14 and 15), and (b) two summer days without irrigation (July 8 and 9).*

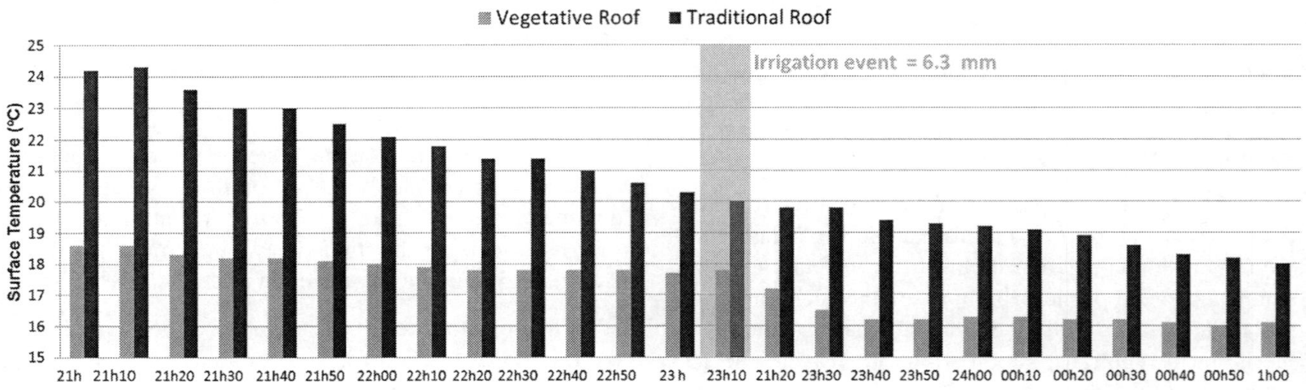

Figure 13 *Surface temperature: vegetative roof with an irrigation event at 11 p.m. versus traditional roof for July 14, 2015.*

VEGETATIVE ROOF MODELING

In a simplified way, a vegetative roof energy balance consists of the following aspects: short-wave radiation from the sun and long-wave radiation emitted by the environment are balanced by sensible and latent heat flux from the vegetation and the soil, together with heat conduction into the substrate and long-wave radiation to and from the soil and vegetation. The energy balance is particularly influenced by the vegetation as a living shading device and the vegetation-substrate-induced evaporative cooling.

Modeling the hygrothermal behavior of vegetative roofs is a challenge but is also decisive for a correct roof hygrothermal performance evaluation. The simulation of vegetative roofs requires the study of several interacting phenomena, including the contribution of the vegetation to the heat and moisture balance. Since the first simplified theoretical model was developed (Nayak et al. 1982), many models have emerged, some simpler (Niachou et al. 2001) and some more detailed, where more complex parameters and their variability are considered, including plant cover effect (Liu and Baskaran 2004; Jaffal et al. 2012) and energy and mass balance across the roof along with calculation of evapotranspiration rates (Del Barrio 1998; Onmura et al. 2001). Some studies evaluated heat flux reduction and resulting R-values and evapotranspiration processes (Tabares-Velasco and Srebric 2012). Sailor (2008) presented a vegetative roof model that is implemented in Energy Plus[1] that applies the vegetative roof as an outer layer of the roof structure and takes into account the convective, conductive, radiative, and moisture effects. Zirkelbach et al. (2010) implemented a green modeling approach in the WUFI[2] software. There is work done by some authors considering the model integration into a building thermal program (Ouldboukhitine et al. 2011; Del Barrio 1998; Tabares-Velasco and Srebric 2012). Laboratory and field experiments are also abundant, although some simplifications are still being made in the models for evapotranspiration effects and time-varying soil properties (Sailor and Hagos 2011). Some optimization models have been developed like Chang et al. (2011) that focus on costs, considering water control, and energy savings.

The modeling approaches of the numerical simulation tools are very different, due to the large number of parameters that are required to describe as accurately as possible the physical phenomena that take place within and underneath the vegetation layer and the consequent assumptions and simplifications. For example, some studies consider a constant thermal conduction, while others use constant properties for the vegetation throughout the year. Some do not account for the drainage layer, whereas others use a constant level of water in the drainage layer. Table 2 summarizes the major models, depending on the dimensional analysis (1D or 2D), scale (at building or city level), coupling with other simulation models (yes or no), existence of a simplified hydrological model (simple, S, or complex, C), and heat and mass transfer (simple, S, or complex, C) for the support, substrate, vegetation, drainage layer, and waterproofing membrane. It also considers the existence of an air stratification model for the climate above the vegetative roof.

The assessment of irrigation scenarios through numerical simulation is essential in vegetative roofs. Previous studies showed that hygrothermal vegetative roof models were able to reproduce the real roof behavior, with measured and simulated data showing analogous reproduction of the diurnal variation for periods without irrigation. However, only a limited number of studies that account for the complete roof composition focused on scenarios where irrigation is present. Preliminary simulation studies performed for a vegetative roof located in Italy show a good agreement for non-irrigated periods but a slight difference is obtained for days with irrigation (Figure 14) (de Freitas et al. 2014).

1. EnergyPlus is a whole building energy simulation program developed by the U.S. Department of Energy Building Technologies Office.

2. WUFI is a software family for dynamic simulations of coupled heat and moisture transfer developed by the Fraunhofer-Institute for Building Physics, Germany.

Table 2. Vegetative Roof Models

Authors	Dimension		Scale		Model		Hydrological Model		Support				Substrate				Vegetation				Drainage Layer				Waterproofing Membrane		Air Stratification
	1D	2D	Building	City	Coupled				Heat Transf.		Mass Transf.		Heat Transf.		Mass Transf.		Heat Transf.		Mass Transf.		Heat Transf.		Mass Transf.		Heat Transf.		
					Yes	No	S	C	S	C	S	C	S	C	S	C	S	C	S	C	S	C	S	C	S	C	
Künzel (1995)	x		x			x	x			x		x		x	x		x		x			x	x			x	
Del Barrio (1988)	x		x			x	x			x				x		x		x		x							
Lazzarin et al. (2005)	x		x		x		x		x				x		x		x		x		x		x		x		
Kumar and Kaushik (2005)	x		x		x		x			x				x		x		x		x							
Alexandri and Jones (2007)	x			x		x	x			x		x		x		x		x		x							x
Sailor (2008)	x		x		x		x						x		x			x		x							
Palla et al. (2008)	x		x			x		x						x		x											
Feng et al. (2010)	x		x			x							x				x		x								
He and Jim (2010)	x			x		x												x									x
Sailor and Hagos (2011)						x							x		x												
Ouldboukhitine et al. (2011)	x		x			x		x						x		x		x		x							
Jaffal et al (2012)	x		x		x								x		x		x		x								
Tabares-Velasco and Srebric (2012)	x		x		x		x						x		x			x		x							
Djedjig et al. (2012)	x		x		x			x						x		x		x		x							
Munck et al. (2013)		x		x	x			x	x				x		x		x		x		x		x				
Sun et al. (2013)	x		x		x			x		x		x		x		x		x		x							
Olivieri et al. (2013)	x		x			x							x				x										

The major limitations of the models are connected with the hydrological balance and the simplified or more complex form in which the properties are considered. According to Table 2, only 5 out of 17 models consider a complex hydrological model. This type of modeling strategy usually requires the knowledge of very complex parameters that are difficult to access. On the other hand, the simplified hydrological models that translate the water balance on the roof in a simpler way may introduce some approximation errors, for instance, on the evaporation effect due to the water content roof distribution. The second aspect mentioned reflects the simplified or more complex way in which the vegetation and substrate layer with

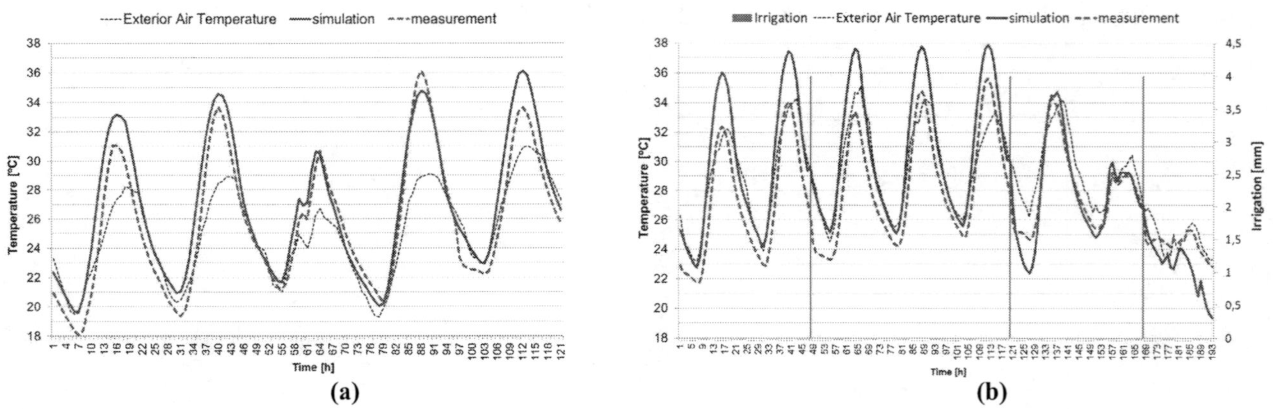

Figure 14 *Simulated vegetative roof near surface temperatures at 0.02 m below the vegetation: (a) without night irrigation and (b) with night irrigation.*

varying properties are accounted. Often, if a model is too detailed and considers a wide range of parameters, it can be difficult to obtain the correct values for the specific vegetative roof, though a more simplified approach can also lead to less accurate results.

Future developments for the validation of a numerical simulation tool that responds well on summer conditions with irrigation will allow the test of different watering scenarios and the exploration of the most viable solutions for improving the roof performance. A comparison of the models' response to irrigation should also be very interesting.

CRITICAL ANALYSIS AND CONCLUSIONS

Although vegetative roofs are designed to depend primarily on precipitation for water supply, some vegetation species and some climatic conditions, such as warm, drought, or extreme hot climates require irrigation for the plant survival after the crucial establishing period. The maximization of the water efficiency is imperative to reduce the water waste, minimize the costs for water and energy, and enhance the overall roof performance during summer, due to the ability to produce evaporative cooling. Nonetheless, the ideal or minimum amount of water requirement and the respective schedules of irrigation during summer are not yet clearly defined. For instance, the daily irrigation amounts found in literature for a Csb climate (warm-summer Mediterranean climate, according to Köppen and Geiger classification), which is the same as in Porto, suggest that for keeping plants alive, the irrigation regime should be 1.1 mm/day for sedum and moss vegetation and for grasses and herbs 2.9 mm/day. Other studies refer values up to 9 mm/day (Mechelen et al. 2015).

Another important aspect to take into account, besides the climate and plant characteristics, is the soil type. Usually, sandy soils absorb water faster, while clay soils have a lower absorption rate, which impacts the plant survival and the evapotranspiration effect. The depth of the substrate layers is found to influence the redistribution of water, thus limiting surface evaporation (Sun et al. 2014). The experimental analysis here presented concerned a typical grass lawn with a small substrate thickness above a sand layer.

The experimental data obtained from the Porto prototype showed that a night irrigation of 6.3 L/m^2 changed the heat fluxes as well as reduced the surface and near surface temperatures. On the other hand, a smaller irrigation amount of 2.5 L/m^2, with an antecedent low-water content period, had little impact. Moreover, watering the vegetative roof is necessary for the preservation of a water content value, compatible with the survival of the grass on the vegetative roof. A sharp decrease of the sand moisture content was observed after four days without irrigation in Porto.

The main performance determinants for the two parameters analyzed—surface temperature and heat flux—are solar radiation, air temperature, irrigation amounts, irrigation schedules, antecedent irrigation periods, and subsequent water content values. The effect of wind and relative humidity should be also a point of study in the future.

Beyond the imperative analysis of experimental data, it is important to simulate irrigation in controlled scenarios and on different vegetative roofs. The literature provides a considerable amount of vegetative roof models capable of simulating the complex hygrothermal behavior of vegetative roofs with good results. An accurate irrigation approach with numerical models is critical once the vegetative roofs are applied in even harsher climate conditions where irrigation is fundamental and should be possible to quantify. The soil and, specifically, the sand substrates used on grass lawns are extremely affected by water penetration due to gravity. This transport phenomenon is difficult to incorporate in the models and is totally dependent on the input data, as it cannot be simulated directly. Moreover, the material properties are dynamic and change over time. Preliminary simulation results showed the estimated irrigation impact and the difficulty in obtaining an agreement with experimental data.

Further investigations with complementary studies to define the best modeling approach for irrigation will provide more accurate hygrothermal simulations approaches for the long-term analysis of vegetative roofs applied in different climates and its effect on the roof microclimate. Consequently, the test of irrigation schedules and the assessment of comfort conditions on top of the vegetative roof should be possible in the near future and will allow the design of the most appropriate irrigation schemes.

ACKNOWLEDGMENTS

Sara Freitas would like to thank Fundação para a Ciência e a Tecnologia (FCT) for financial support through the grant SFRH/BD/85838/2012. A special gratitude to Riccardo Paolini from Politecnico di Milano for his help with the numerical simulation programs and for supplying his experimental vegetative roof data.

REFERENCES

Alexandri, Eleftheria, and Phil Jones. 2007. Developing a one-dimensional heat and mass transfer algorithm for describing the effect of green roofs on the built environment: Comparison with experimental results. *Building and Environment* 42 (8): 2835–49. doi:10.1016/j.buildenv.2006.07.004.

Berardi, Umberto, AmirHosein GhaffarianHoseini, and Ali GhaffarianHoseini. 2014. State-of-the-art analysis of the environmental benefits of green roofs. *Applied Energy* 115: 411–28. doi:10.1016/j.apenergy.2013.10.047.

Berretta, Christian, Simon Poë, and Virginia Stovin. 2014. Moisture content behaviour in extensive green roofs during dry periods: The influence of vegetation and substrate characteristics. *Journal of Hydrology* 511: 374–86. doi:10.1016/j.jhydrol.2014.01.036.

Bureau of Meteorology. 2012. Average annual & monthly evapotranspiration. Australian Government. Last modified 6 May 2015. http://www.bom.gov.au/jsp/ncc/climate_averages/evapotranspiration/index.jsp?maptype=1&period=may

Chang, N., B. Rivera, and M. Wanielista. 2011. Optimal design for water conservation and energy savings using green roofs in a green building under mixed uncertainties. *Journal of Cleaner Production* 19 (11):1180–88. doi:10.1016/j.jclepro.2011.02.008.

de Freitas, S.S., N. Ramos, D. Zirkelbach, H. Künzel, B. Stöckl, M. Fiori, R. Paolini, and C. Corbari. 2014. Assessment of green roofs performance with numerical simulation tools and experimental measurements, Third International Conference on Countermeasures to Urban Heat Island, Venice, October 13–15, 2014.

Del Barrio, E. 1998. Analysis of the green roofs cooling potential in buildings. *Energy and Buildings* 27(2): 179–93. doi:10.1016/S0378-7788(97)00029-7.

Djedjig, R., S. Ouldboukhitine, R. Belarbi, and E. Bozonnet. 2012. Development and validation of a coupled heat and mass transfer model for green roofs. *International Communications in Heat and Mass Transfer* 39(6):752–61. doi:10.1016/j.icheatmasstransfer.2012.03.024.

EPA. 2008. Reducing urban heat islands: Compendium of strategies—Green roofs, climate protection partnership division—U.S. Environmental Protection Agency's Office of Atmospheric Programs. Environmental Protection Agency. http://www.epa.gov/heatisland/resources/pdf/GreenRoofsCompendium.pdf.

FAO. 1998. Weather parameters, crop characteristics, management and environmental aspects are factors affecting evaporation and transpiration. http://www.fao.org/docrep/x0490e/x0490e04.htm

Feng, C., Q. Meng, and Y. Zhang. 2010. Theoretical and experimental analysis of the energy balance of extensive green roofs. *Energy and Buildings* 42(6):959–65. doi:10.1016/j.enbuild.2009.12.014.

Fioretti, R., A. Palla, L.G. Lanza, and P. Principi. 2010. Green roof energy and water related performance in the Mediterranean climate. *Building and Environment* 45(8):1890–1904. doi:10.1016/j.buildenv.2010.03.001.

Gill, S.E., J.F. Handley, A.R. Ennos, and S. Pauleit. 2007. Adapting cities for climate change: The role of the green infrastructure. *Built Environment* 33(1):115–133. http://www.coolrooftoolkit.org/wp-content/uploads/2012/04/Gill_Adapting_Cities.pdf.

He, Hongming, and C.Y. Jim. 2010. Simulation of thermodynamic transmission in green roof ecosystem. *Ecological Modeling* 221(24):2949–58. doi:10.1016/j.ecolmodel.2010.09.002.

Jaffal, I., S. Ouldboukhitine, and R. Belarbi. 2012. A comprehensive study of the impact of green roofs on building energy performance. *Renewable Energy* 43:157–64. doi:10.1016/j.renene.2011.12.004.

Kumar, R., and S.C. Kaushik. 2005. Performance evaluation of green roof and shading for thermal protection of buildings. *Building and Environment* 40(11):1505–11. doi:10.1016/j.buildenv.2004.11.015.

Künzel, H. 1995. Simultaneous heat and moisture transport in building components—One and two dimensional calculation using simple parameters. PhD Thesis. IBP Verlag, Stuttgart, Germany.

Lazzarin, R., F. Castellotti, and F. Busato. 2005. Experimental measurements and numerical modeling of a green roof. *Energy and Buildings* 37(12):1260–67. doi:10.1016/j.enbuild.2005.02.001.

Liu, K. 2004. Sustainable building envelope—Garden roof system performance. (NRCC-47354) Paper presented at RCI Building Envelope Symposium, 4–5 November, pp. 1–14, New Orleans, Louisiana.

Liu, K. and B. Baskaran. 2004. Green roof infrastructure—Technology demonstration, monitoring and market expansion project. (B-1054.1) NRC Institute for Research in Construction.

Liu, K. and B. Bass. 2005. Performance of green roof systems. Paper presented at the Cool Roofing Symposium. Atlanta, GA. http://nparc.cisti-icist.nrc-cnrc.gc.ca/npsi/ctrl?lang=en.

Mechelen, C., T. Dutoit, and M. Hermy. 2015. Adapting green roof irrigation practices for a sustainable future: A review. *Sustainable Cities and Society* 19: 74–90.

Munck, C.S., A. Lemonsu, R. Bouzouidja, V. Masson, and R. Claverie. 2013. The GREENROOF Module (v7.3) for modeling green roof hydrological and energetic performances within TEB. *Geoscientific Model Development* 6(6):1941–60. doi:10.5194/gmd-6-1941-2013.

Nayak, J.K., A. Srivastava, U. Singh, and M. S. Sodha. 1982. The relative performance of different approaches to the passive cooling of roofs. *Building and Environment* 17(2):145–61.

Niachou, A., K. Papakonstantinou, M. Santamouris, A. Tsangrassoulis, and G. Mihalakakou. 2001. Analysis of the green roof thermal properties and investigation of its energy performance. *Energy and Buildings* 33(7):719–29. doi:10.1016/S0378-7788(01)00062-7.

Olivieri, F., C. Di Perna, M. D'Orazio, L. Olivieri, and J. Neila. 2013. Experimental measurements and numerical model for the summer performance assessment of extensive green roofs in a Mediterranean coastal climate. *Energy and Buildings* 63:1–14. doi:10.1016/j.enbuild.2013.03.054.

Onmura, S., M. Matsumoto, and S. Hokoi. 2001. Study on evaporative cooling effect of roof lawn gardens. *Energy and Buildings* 33(7):653–66. doi:10.1016/S0378-7788(00)00134-1.

Ouldboukhitine, S., R. Belarbi, I. Jaffal, and A. Trabelsi. 2011. Assessment of green roof thermal behavior: A coupled heat and mass transfer model. *Building and Environment* 46(12):2624–31. doi:10.1016/j.buildenv.2011.06.021.

Palla, A., I. Gnecco, and L.G. Lanza. 2009. Unsaturated 2D modeling of subsurface water flow in the coarse-grained porous matrix of a green roof. *Journal of Hydrology* 379(1–2):193–204. doi:10.1016/j.jhydrol.2009.10.008.

Palla, Anna, Luca G. Lanza, and P. La Barbera. 2008. A green roof experimental site in the Mediterranean climate. Paper presented at the 11th International Conference on Urban Drainage, 31/8–5/9, Edinburg, Scotland, UK. doi:10.2166/wst.2013.196.

Peng, Lilliana L.H., and C.Y. Jim. 2013. Green-roof effects on neighborhood microclimate and human thermal sensation. *Energies* 6(2):598–618. doi:10.3390/en6020598.

Qin, X., X. Wu, Y. Chiew, and Y. Li. 2012. A green roof test bed for stormwater management and reduction of urban heat island effect in Singapore. *British Journal of Environment and Climate Change* 2(4):410–20. doi:10.9734/BJECC/2012/2704.

Rosenzweig, C., S. Gaffin, and L. Parshall. 2006. *Green roofs in the New York Metropolitan Region: Research report*. Columbia University Center for Climate Systems Research and NASA Goddard Institute for Space Studies. New York. 59 pages. Available online: http://pubs.giss.nasa.gov/abs/ro05800e.html.

Sailor, D.J. 2008. A green roof model for building energy simulation programs. *Energy and Buildings* 40(8):1466–78. doi:10.1016/j.enbuild.2008.02.001.

Sailor, D.J., and M. Hagos. 2011. An updated and expanded set of thermal property data for green roof growing media. *Energy and Buildings* 43(9):2298–303. doi:10.1016/j.enbuild.2011.05.014.

Sailor, D.J., T.B. Elley, and M. Gibson. 2012. Exploring the building energy impacts of green roof design decisions—A modeling study of buildings in four distinct climates. *Journal of Building Physics* 35(4):372–91. doi:10.1177/1744259111420076.

Salagnac, J., J. Desplat, R. Kounkou-Arnaud, A. Lemonsu, and V. Masson. 2013. Assessment of adaptation strategies to climate change impacts in a big city: the case of Paris. Paper presented at the World Building Congress, May 5–9, Brisbane, Australia.

Spala, A., H.S. Bagiorgas, M.N. Assimakopoulos, J. Kalavrouziotis, D. Matthopoulos, and G. Mihalakakou. 2008. On the green roof system. selection, state of the art and energy potential investigation of a system installed in an office building in Athens, Greece. *Renewable Energy* 33(1):173–77. doi:10.1016/j.renene.2007.03.022.

Sun, T., E. Bou-Zeid, Z. Wang, E. Zerba, and G. Ni. 2013. Hydrometeorological determinants of green roof performance via a vertically-resolved model for heat and water transport. *Building and Environment* 60:211–24. doi:10.1016/j.buildenv.2012.10.018.

Sun, Ting, Elie Bou-Zeid, and Guang Heng Ni. 2014. To irrigate or not to irrigate: Analysis of green roof performance via a vertically-resolved hygrothermal model. *Building and Environment* 73:127–37. doi:10.1016/j.buildenv.2013.12.004.

Tabares-Velasco, P., and J. Srebric. 2009. The role of plants in the reduction of heat flux through green roofs: Laboratory experiments. *ASHRAE Transactions* 115(2):793–802.

Tabares-Velasco, P., and J. Srebric. 2012. A heat transfer model for assessment of plant based roofing systems in summer conditions. *Building and Environment* 49(1):310–23. doi:10.1016/j.buildenv.2011.07.019.

Takakura, T., S. Kitade, and E. Goto. 2000. Cooling effect of greenery cover over a building. *Energy and Buildings* 31(1):1–6. doi:10.1016/S0378-7788(98)00063-2.

UNPD. 2012. World urbanization prospects: The 2011 revision, 2012. Department of economic and social affairs of the united nations population division (UNPD). http://esa.un.org/unpd/wpp/ppt/CSIS/WUP_2011_CSIS_4.pdf.

Virk, G., A. Jansz, A. Mavrogianni, A. Mylona, J. Stocker, and M. Davies. 2015. Microclimatic effects of green and cool roofs in London and their impacts on energy use for a typical office building. *Energy and Buildings* 88:214–28. doi:10.1016/j.enbuild.2014.11.039.

Zinzi, M., and S. Agnoli. 2012. Cool and green roofs. An energy and comfort comparison between passive cooling and mitigation urban heat island techniques for residential buildings in the Mediterranean region. *Energy Build* 55:66–76.

Zirkelbach, D., B. Schafaczek, and H. Künzel. 2010. Long-term hygrothermal performance of green roofs."Paper presented at the Thermal Performance of the Exterior Envelopes of Whole Buildings XI, International Conference, December 5–9, Clearwater Beach, Florida, International Conference.

Effective Floor Cavity and Knee Wall Construction Techniques in Two-Story Homes in Hot Climates

Charles R. Withers, Jr.

ABSTRACT

It is understood that the floor cavity between the first and second floors of residential construction should be isolated from unconditioned space by an effective air and thermal barrier. Although the construction materials utilized likely have reasonable air and thermal resistance, the application may not be applied correctly or some applications may not hold up well over time. A research project of existing homes (Withers and Kono 2015) measured impacts from this problem and identified durable construction techniques and cost-effective repairs. Floor cavity pathways to unsealed attics combined with natural or mechanical driving forces result in air transport between the attic and the floor cavity. Unconditioned air leakage into residential floor cavities has been found to cause elevated space-conditioning costs, increase peak electric utility demand, and cause moisture-related problems.

This paper provides new research on measured cooling and heating impacts from floor cavity and knee wall repair. Overall, 56 multistory homes were inspected and had performance tests completed, with 12 monitored for energy savings from floor cavity and some knee wall repairs. This paper describes practical inspection, testing, and dependable construction methods. Construction methods discussed consider challenges such as contractor experience, limited working access, and complex structural geometry. The measured energy savings from floor cavity repairs combined with cost-effective repair methods in 12 homes was found to be able to pay for repairs in about 3 to 4 years. Evidence from one home has shown that significant building moisture damage was eliminated as a result of floor cavity repair.

INTRODUCTION

Floor cavities are assumed to be an interstitial space within the primary air and thermal barrier of the home; however, even relatively minor defects around this space can have negative consequences. Portions of floor cavities are often exposed to attic spaces. Attic air can be transported into floor cavities through natural or mechanical forces. Natural forces such as wind can drive air through attic venting and create pressure differentials able to drive unconditioned air into floor cavities (Figure 1). The term *wind washing* is sometimes used to describe this. Wind washing is a general term referring to diminished thermal control caused by air movement over or through an intended thermal barrier. Given the extreme thermal conditions of vented attics, convective air transport can still transport air into floor cavity openings, even without wind forces. Mechanical forces induced by duct leakage can also create pressure differentials that will move unconditioned air into and out of floor cavities. During hot humid weather, condensation may occur on cold supply duct surfaces within the floor cavity, resulting in ceiling moisture damage. In cold climates, cold air from wind washing can chill surfaces within the interior floor space and result in frozen water pipes in severe cases. Besides potential building damage, diminished comfort and increased space-conditioning energy use may also result from poorly insulated and sealed floor cavities. Enlightened contractors already understand the importance of creating a barrier between floor cavities and attics, but failures continue to occur, and older construction has a legacy of fail-

Charles R. Withers, Jr. is a senior research analyst at Florida Solar Energy Center, a research institute of the University of Central Florida, Cocoa, FL.

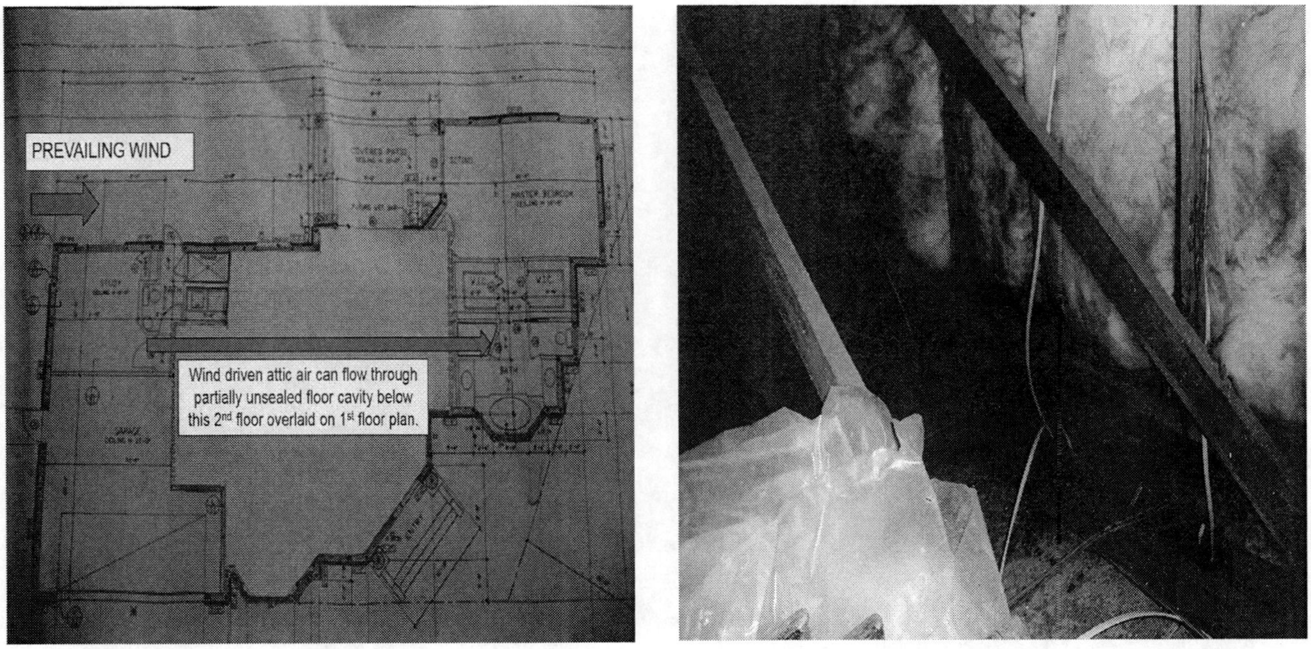

(a) (b)

Figure 1 *(a) Conceptual illustration of wind washing through floor cavity shown on a floor plan. (b) View from within attic of poorly insulated knee wall and floor cavity with areas of missing and compressed batts. Batts were dirty from years of air movement across them. Attic soffit vent is nearby in dark left region of photo.*

ures. An adequate air and thermal barrier may have been installed but another trade penetrated or removed portions without repairing it. In the worst cases there are no air or thermal barriers installed at all.

BACKGROUND

The past emphasis on wind washing has concentrated on U-factor impacts and specific types of construction or scenarios (Bankvall 1978; Harrje et al. 1985; Lecompte 1990; Silberstein et al. 1991; Uvslokk 1996; Janssens and Hens 2007). Knee wall, floor cavity, and wind washing problems have been recognized in other published literature, such as that by Sidall (2009), Lstiburek (2005), and EERE (2012), which provided some information on best practices in specific applications. While wind washing has been known about for years, there had been no obtainable published results associated with measured energy impacts or cost-benefit analysis of retrofit opportunities. This paper is based upon results from more recent wind washing research in Florida carried out in two phases. The first phase resulted in published results covering details on the study method and results but was limited to measured cooling energy impacts in six homes (Withers and Cummings 2010; Cummings and Withers 2012). A second phase was added to increase sample size and measure heating and cooling impacts in six more homes (Withers and Kono 2015). The primary goal of the Florida study was to evaluate inspection methods and determine the cost-effectiveness of retrofit solutions for wind washing in two-story homes in hot humid climates. The retrofit solutions covered here may also be applied to new construction primarily in hot- or warm-dominant climate regions.

INSPECTION METHODS

House performance-related tests were completed in 56 Florida single-family multistory homes of various ages. The purpose of the house tests was to gather data relevant to wind washing and to find reasonable candidates for monitoring and retrofit. The field assessments characterized wind washing failures of the house air and thermal boundary. Assessments included a house tightness test, primary air boundary evaluation, air differential pressure mapping, air leakage assessment, duct leakage assessment, infrared scans of house surfaces, and visual inspections.

A comprehensive visual inspection completed within the attic focused around floor cavities or dropped ceiling areas was found to be the most important and effective method of identifying homes needing floor cavity repair. Figure 1b shows a commonly found issue of poorly fit knee wall insulation and areas of floor cavity open to attic spaces commonly over garage areas. Inspections should focus efforts on locating an effective air and thermal barrier over the floor cavity. Some homes have very small attic areas that were either inaccessible or had limited accessibility for inspection. Equipment such as a bore scope may be needed in such cases, or material disassembly may be warranted in some instances, such as moisture damage issues. Figure 2 shows an example where soffit venting was removed for inspection in areas where floor cavities terminated nearby in a home with severe moisture damage found to be related to floor cavity wind washing. Spray foam applications can also have failures if not applied carefully. Figure 3a shows an area of leakage within an attic that was found and painted so the contractor could come back to seal it.

Infrared (IR) thermography and pressure differential measurements were useful diagnostics under specific circumstances. IR thermography worked best with large temperature

(a) (b)

Figure 2 *(a) Sections of vented soffit were removed to inspect floor cavities not accessible from attic space. (b) View within cavity sections found pathways about 0.05 m (2 in.) high under floor joist supports that connect to small attic space beyond.*

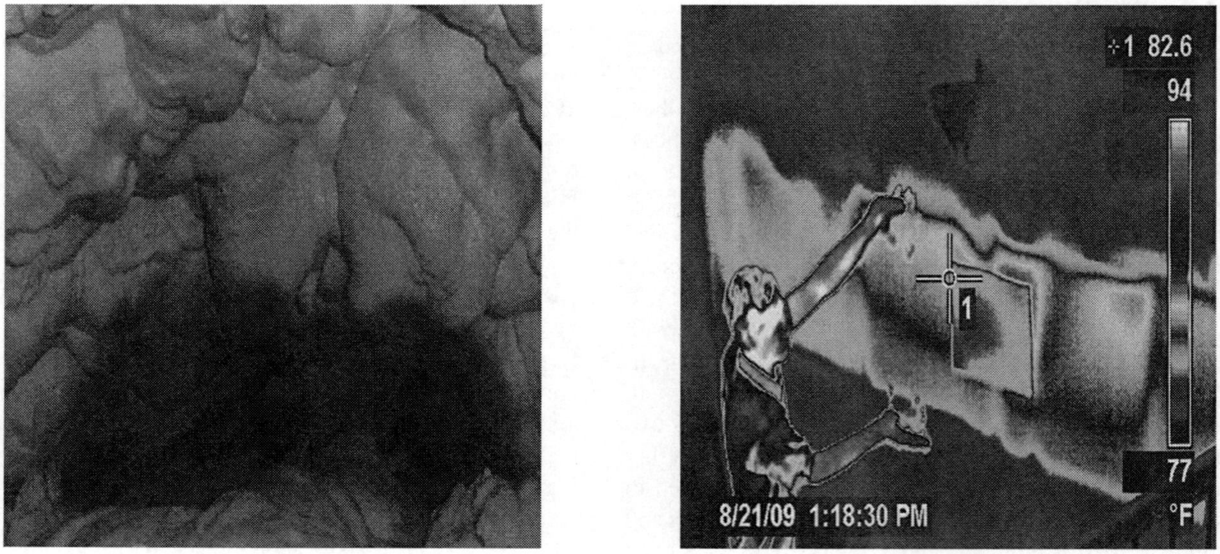

(a) (b)

Figure 3 *(a) Air leak in spray foam painted for contractor to come back and seal it. (b) Inspection using IR camera shows elevated wall temperature at floor cavity area.*

differentials occurring between attic and indoors for at least a few hours. Figure 3b shows warm summer indoor wall temperatures due to wind washing into the floor cavity. This is the same home of the floor plan shown in Figure 1a. At times this author found it more useful to use the IR camera from within the attic looking at knee wall and floor cavity areas instead of from indoors.

An indication of whether the floor cavity is more indoors than outdoors regarding the primary air barrier was evaluated using differential pressure (dp) measurements while the house was depressurized to 0.2 in. w.c. (50 Pa) with reference to outside. This diagnostic can be performed quickly if a house tightness test is already planned. The project evaluation results indicated that this diagnostic was clearly not adequate on its own to be a replacement of visual inspection to determine if a floor cavity has a significant potential for wind washing (Withers and Kono 2015). Measuring the house pressure with reference to the floor cavity could be considered a simple and quick diagnostic test that can be done to determine if more detailed visual investigation of the floor cavity is warranted, particularly if the measurement is between –0.136 in. w.c. (–34 Pa) to 0 in. w.c. (0 Pa) (house depressurized). The dp diagnostic could be useful to prompt a more careful visual inspection and to consider possible leakage to areas not easily accessible that may require some disassembly.

WHY TIGHT INSULATED FLOOR CAVITIES MATTER

Inadequately insulated and sealed floor cavities can result in occupant health and comfort problems, building damage, and elevated heating and cooling costs. Sealed floor cavities also help minimize duct leakage impacts occurring within these spaces. Return duct leakage decreased 37% and supply leakage decreased by 38% from sealing floor cavities and making no direct repairs to ducts in 10 homes. This is not surprising since ducts were located within the floor cavity of all these homes. Sealing the floor cavity is a more cost-effective way to limit duct leakage impacts within the floor cavities of existing constructions. House airtightness, ACH_{50}, became 11% tighter and homes with vented recessed can lights in the floor cavity had house tightness reductions double (23% tighter). ACH_{50} is an industry measure of building airtightness measured at a pressure of 0.2 in. w.c. (50 Pa) and normalized by the house conditioned volume. Implementing good practice in new construction would not necessarily add cost if current building codes are followed and enforced. Failure to implement good practice or damage created by trades may have an average cost of repair around $350. The range in homeowner repair cost can vary widely depending upon particular circumstances and materials used—from as little as $170 up to $800 or more in very large homes.

Moisture Issues

In heating-dominant climates, water damage within the floor cavity is most likely to occur due to water pipe freeze damage resulting in water leakage or condensation on cold building surfaces. Weatherization programs have worked at improving this in older homes over the years, but a breach or degradation over time can increase the potential of this problem during prolonged severe winter conditions. Cooling-dominated climates are more likely to experience moisture issues in the summer due to condensation on cool duct or building surfaces and material damage associated with moisture absorption under specific circumstances. One of the homes had severe moisture damage exhibited by mold and mildew stains on first-floor ceilings occurring under cold supply air ducts, severely warped wood flooring on the second floor, and cracked wood and paint on sections of stair risers. Photos and additional details on this can be found in the paper by Withers and Cummings (2010). Prior to repair, a section of faced batt that had been wrapped around a cold air supply duct where it entered the floor cavity was found to be drenched in condensation. Wood moisture content was measured before and after floor cavity repairs in this home during hot and humid weather. Moisture content in wood stair sections closest to the floor cavity were at 13.1% and gradually decreased to 9.5% on lower sections farther away from the floor cavity. Interior air conditions were cool and dry and wood furniture had a moisture content around 7% to 8% by comparison, which is expected for an air-conditioned space. The average stair moisture content decreased to 9.0% after repair and cracks in paint and wood stair risers decreased within a few weeks of repair. Evidence of duct condensation also stopped after repair. Other homes in the Florida study did not exhibit such severe moisture problems. The severe moisture issues in this home were attributed to severe floor cavity leakage on opposing sides of the home (complimentary pathways) oriented in line with daily occurring wind combined with the occupant preference for low interior summer temperatures (73°F [22.8°C] overnight and 76°F [24.4°C] daytime).

Energy

Overall cooling energy savings resulting from floor cavity repair varied from 1% up to 33%, and heating savings varied from 0% to 39%. It was not surprising that some homes did not demonstrate high energy savings, since the study homes had varying degrees of thermal bypass issues. The case with no heating savings is most attributed to mild winter monitoring conditions and the preference for low heating set point, which did not provide much heating data for analysis. The previously discussed house with severe moisture issues had 8% cooling savings, but it was a large upscale home that used a lot of cooling energy, so the annual cooling energy savings of 2770 kWh/yr and peak cooling reduction of 1.8 kW were quite large. Table 1 shows the predicted annual and peak energy savings based on 12 homes; heating results are based on 5 homes. Regression least-squares best-fit analysis was performed from monitored heating and cooling data along with local monitored indoor and outdoor conditions for each home to create equations that could be used with typical meteorological year (TMY) data of four Florida cities to predict the annual cooling and heating savings for typical weather conditions. The average predicted annual cooling and heating energy savings was 1034 kWh per home from wind washing repairs. This was potentially $119 saved each year by each homeowner based on an average rate of $0.115/kWh. The savings can be expected to last as long as the home, because the repair lifetime has no limit under reasonable circumstances. These results are limited here to represent hot and humid climate regions; monitoring in colder climate regions should be done to more accurately reflect regional construction practice and measurement under much more extreme winter conditions.

Table 1. Predicted Average Annual and Peak Energy Cooling and Heating Impacts from Floor Cavity Repairs

	Annual Cooling kWh	Summer Peak Hour Cooling kW	Annual Heating kWh	Winter Peak Hour Heating kW
Pre-repair	9392	3.66	502	1.86
Post-repair	8483	3.27	378	1.53
Savings	909	0.39	125	0.33
Savings %	9.7%	10.7%	24.9%	18.7%

SEALING FLOOR CAVITY AREAS EXPOSED TO ATTIC SPACES

New Construction

Good air and thermal barriers begin with a good design. First, designs should not place structural members within about 18 in. (0.45 m) of the floor cavity or knee wall to allow room to correctly install continuous sections of air and thermal barrier materials. This distance is less critical if spray foam is used to insulate and seal the floor cavity and knee walls. Figure 4a shows structural members directly against the floor cavity, which makes it difficult to install a continuous air and thermal barrier with faced batts. Figure 4b shows the roof truss further away from the plywood placed over the knee wall and floor cavity. The continuity of air barrier should be maintained from the roof deck at top of knee wall, covering the knee wall, and continuing down over the floor cavity opening and onto drywall of first floor below. If rigid panels or fabric-based air barrier are used, all panel seams should be sealed with materials appropriate for attic conditions. Insulation should be applied with approved mechanical fastening materials and methods. This is especially important for thick batts. If applied meticulously, kraft-faced batts have the potential to perform well and to be the lowest-cost option. However, the long-term durability is less certain than rigid or spray-applied materials. New homes with batt insulation are more likely to use R-30 batts. Batts are prone to sag in vertical installations, do not hold up well to physical impacts, and are not typically repaired after intentional displacement or penetration. Since ducts,

(a) (b)

Figure 4 *(a) Existing construction showing multiple truss sections tightly against each other and the knee wall and floor cavity structure that does not allow sheet stock to simply be applied continuously as cover. (b) New construction shown uses plywood sheet over knee wall and floor cavity sections with room to seal seams and add insulation to improve wall R-value.*

wires, and pipes are typically run inside floor cavities, each trade should be held responsible for maintaining the integrity of the air and thermal barrier around their work. Protect cold supply ducts from condensation in vented attics whenever envelope insulation is placed in contact with the exterior duct surface (see Figure 5), particularly in cooling-dominated climates. Condensation potential can be decreased by wrapping ducts at contact points with a thin insulative wrap having an exterior vapor barrier before insulation is placed in contact with it. This cannot guarantee that condensation will never occur, but it has been demonstrated to work in homes with cooling set points maintained above 74°F (23.3°C). In regard to limiting general water vapor issues, materials with low vapor permeability, less than 1, should not be placed on the dominantly coldest side of a wall/cavity barrier assembly.

Existing Construction

Implementing repairs in existing construction can be more challenging. Two basic methods were implemented within the Florida study. One method used low-density spray foam and the other used well-placed kraft-faced insulation batts. Floor cavity and knee wall wind washing repairs typically use much less insulation material than is common in conventional insulation jobs. The average repair area was 169 ft^2 (15.7 m^2) with a range of 48–510 ft^2 (4.5–47.4 m^2). The relatively small repair area and high mobilization cost of spray foam rigs can make repair by spray foam twice as costly as using batts or insulated board stock. Low-density spray foam had a starting cost of at least \$370 but is more likely to be around \$650, compared to batts, which could average more around \$300. Homes with garage attic space next to second-story floor space (the most common location) generally have easy access and often require less than 40 ft^2 (3.7 m^2) of material to seal and insulate the floor cavity from the attic. If the knee wall is also insulated, the total area typically increases to 70–100 ft^2 (6.5–9.3 m^2). Repair cost of an easy-to-access floor cavity and knee wall area of about 100 ft^2 (9.3 m^2) and 2.8 person-hours could cost as little as \$173 and even less, at about \$85, if only the floor cavity is sealed with faced batt and sealant. Of all homes having floor cavity pathways, about 25% of those inspected had very small leakage pathways (longitudinal seams) that could be sealed inexpensively using canned expansive foams for perhaps as little as \$20–\$40 in retail material costs.

An example of a repair using faced batts is shown in Figure 6. The kraft face is intended to perform as the air barrier so it must be carefully sealed to the rigid materials around the perimeter of the floor cavity. Figure 5b shows the kraft facing the attic. To comply with fire-related building codes, the facing should not be left exposed to the attic. The facing was placed in this way to show how it must be able to seal around the perimeter and framing members. The correct way to install is to generously apply an appropriate construction adhesive, caulk, or foam on the entire perimeter of floor cavity and apply the kraft face against the adhesive towards the inside of the cavity with the unfaced insulation side visible from the attic. The batt should also be mechanically fastened to the perimeter using staples. If preferred, an air-barrier wrap or rigid board stock could also be used as the air barrier in place of faced batt as long as it is also covered by an appropriate code-approved ignition and thermal barrier.

(a)

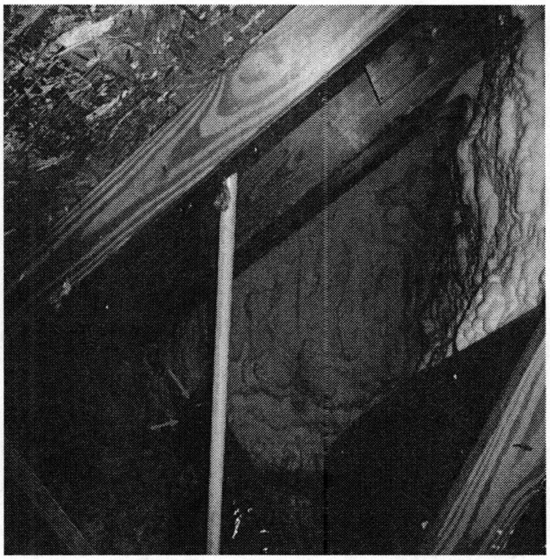

(b)

Figure 5 *(a) The cold air supply duct is first wrapped using a thin insulation sheet with a vapor barrier on the exterior face. (b) Then insulation can be placed around and in contact with the barrier wrap, where a low-density spray foam was used.*

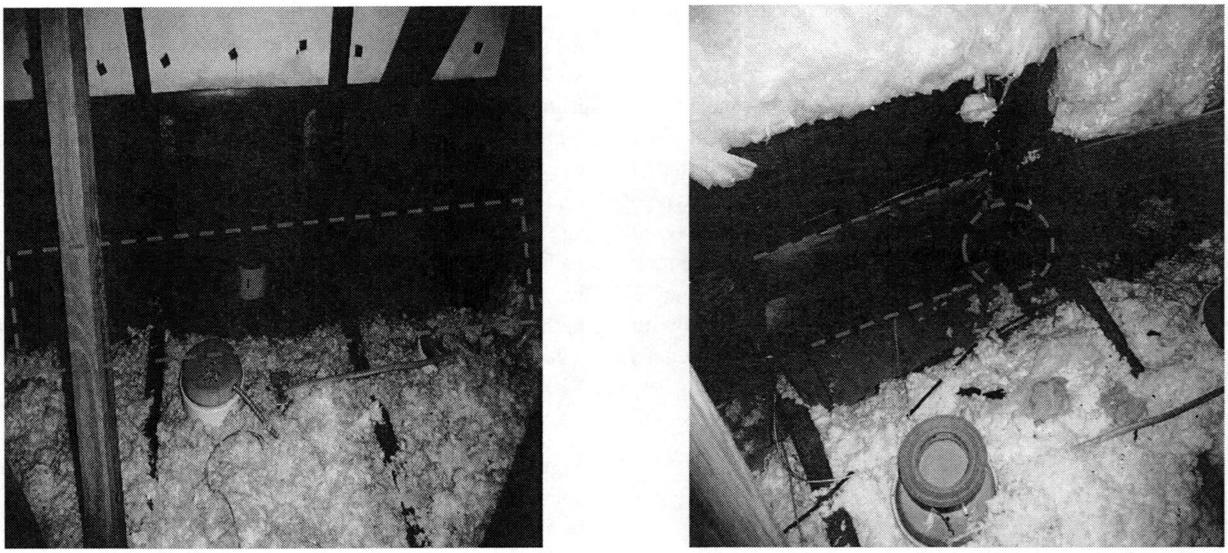

(a) (b)

Figure 6 *(a) An open dropped ceiling connected to a large open floor cavity was sealed and insulated using faced batts. (b) The batt paper facing should not be left exposed to the attic. Consult local codes to address approved fire safety practices for each specific type of repair.*

While using relatively inexpensive materials (such as kraft-faced insulation batts) can result in lower installation costs, specific installation details must be carefully followed to ensure effective air and thermal barrier performance. Materials have to be carefully cut to fit within framing, be fastened mechanically, and have a functional air seal. Tight attic spaces or complex framing require much more labor and skill to apply the inexpensive materials and the final results are less likely to be successful compared to spray foam. If duct leakage is known to exist within the floor cavity, an airtight seal is even more important. Therefore, spray-applied low-density foam may be the best choice in certain circumstances. *2012 International Residential Code*, Section R316.5.3 (ICC 2012), requires a fire ignition barrier on foam insulation exposed in attic spaces. Some manufacturers may accomplish this by spraying the attic exposed foam with a code-approved fire ignition barrier coating, also referred to as an *intumescent material*. Mineral fiber or cellulose insulation 1.5 in. (0.038 m) thick or 1/4 in. (0.0064 m) thick hardboard or wood structural panel are some other examples of approved ignition barrier materials noted in this code. One should be aware of local code requirements and practices that may vary from this.

CONCLUSION

Floor cavities can be insulated and sealed with a variety of materials relatively inexpensively in well-designed new construction. In retrofits, the primary consideration is cost and limitation of the work space. Using materials readily available at home improvement stores will likely be less expensive than using spray-applied foams. However, that alone does not imply the lowest-cost materials are the best choice. Complex structural geometry and tight restrictive working environments may make it difficult and even impossible to make good airtight installations that maintain continuity of the air and thermal barriers. Low-cost material such as faced batt insulation is more prone to damage in areas occasionally accessed by trades or homeowners. Therefore, spray-applied foams may be the best or only choice in certain circumstances. Methods discussed in this paper managed moisture, decreased air leakage, and reduced energy in a hot-and-humid-dominant climate. However, local codes and best local practices should be followed along with application by knowledgeable contractors to help ensure a safe, healthy, and durable result in other climate regions.

Based on monitoring results in Florida, the total annual estimated energy savings from repair was 1034 kWh/year. At a cost of $0.115/kWh, this equates to $119/year saved in consumer energy costs. It is believed that the average cost of repair could be $350, assuming about 75% of homes can be sealed by professional contractors using off-the-shelf air and thermal barrier products and the other 25% are sealed using professionally applied low-density spray foam. The average simple payback would be about three years. Average peak cooling and heating reduction were about 0.3 kW and 0.4 kW, respectively, providing additional benefit to electric utilities and consumers with peak use charges.

ACKNOWLEDGMENTS

This paper would not have been possible without research support from the U.S. Department of Energy Building America Program and Florida Power & Light. Jim Cummings, retired PI from Florida Solar Energy Center, is acknowledged for initiating the interest in this research and project management during the Florida research study.

REFERENCES

Bankvall, C.G. 1978. Forced convection: Practical thermal conductivity in an insulated structure under the influence of workmanship and wind. In *Thermal transmission measurements of insulation.* ASTM STP660, pp. 409–25. West Conshohocken, PA: ASTM International.

Cummings, J., and C.R. Withers, Jr. 2012. Envelope thermal failures due to wind washing in two-story homes. *Proceedings of the 3rd Conference on Building Enclosure Science and Technology, Atlanta, GA.*

EERE. 2012. *Energy renovations: Volume 17: Insulation—A guide for contractors to share with homeowners.* Washington, DC: U.S. Department of Energy, Energy Efficiency and Renewable Energy. http://apps1.eere.energy.gov/buildings/publications/pdfs/building_america/insulation_guide.pdf.

Harrje, D.T., G.S. Dutt, and K.J. Gadsby. 1985. Convective loop heat losses in buildings. *ASHRAE Transactions* 91(2):751–60.

ICC. 2012. *2012 International Residential Code.* Geneva: International Code Council.

Janssens, A., and H. Hens. 2007. Effects of wind on the transmission heat loss in duo-pitched insulated roofs: A field study. *Energy and Buildings* 39(9):1047–54.

Lecompte, J. 1990. The influence of natural convection on the thermal quality of insulated cavity construction. *Building Research and Practice* 6:349–54.

Lstiburek, J. 2005. Chapter 8. In *Builder's guide to hot-humid climates*, pp. 219–23. Westford, MA: Building Science Press.

Sidall, M. 2009. The impact of thermal bypass. *Green Building Magazine* 19(1):16–23.

Silberstein, A., E. Arquis, and D.J. McCaa. 1991. Forced convection effects in fibrous insulation. In *Insulation materials: Testing and applications*, 2nd vol. ASTM STP1116, pp. 292–309. West Conshohocken, PA: ASTM International.

Uvsløkk, S. 1996. The importance of wind barriers for insulated timber frame constructions. *Journal of Thermal Insulation and Building Envelopes* 20:40–62.

Withers, C., and J. Cummings. 2010. Opportunities for energy conservation and improved comfort from wind washing retrofits in two-story homes—Part II. *Proceedings of the 17th Symposium on Improving Building Systems in Hot and Humid Climates, Austin, TX.*

Withers, C., and J. Kono. 2015. Investigating solutions to wind washing issues in two-story Florida homes—Phase 2. Report DOE/GO-102015-4595. Golden, CO: National Renewable Energy Laboratory.

Performance Factor for Floor Heating Systems Using New Analytical Formula

Henrik Karlsson, PhD **Carl-Eric Hagentoft**

ABSTRACT

A recently developed analytical formula for the self-regulation ability of low temperature water-based floor heating systems is experimentally validated. By controlling the feed-forward supply temperature, keeping it rather constant during the year, any thermal perturbance is counteracted by a corresponding shift in the opposite direction of the supplied heat flux. This will give a more stable indoor temperature and an enhanced use of heat gains.

Buildings with small heat losses, such as well insulated single-family houses, in combination with a high equivalent thermal conductance from the supply of the pipe circuit towards the interior, yield a high self-regulation utilization factor. Hence, self-regulation is an integrated phenomenon that depends on both the design of the floor heating element and the building. The theory behind self-regulation floor heating is outlined and an experimental setup using climate chambers is presented. Both light and heavy floor heating systems with embedded hydronic pipes are studied. Steady-state and dynamical tests are used in the validation.

INTRODUCTION

Background

During the last two decades, water-based floor heating systems have frequently been installed in residential and single-family houses in Sweden and Denmark. Floor heating is commonly used in the Nordic countries and central Europe and extensively used in Korea as the primary system for heat distribution (Weitzmann 2004; Olesen 2001). The main characteristic of floor heating is that space heating is provided at a low supply fluid temperature (i.e., 25°C to 35°C). Hence, a flexible choice of the heat source is possible. The energy efficiency benefits from such low supply temperatures because the losses in generation and distribution phases decrease with the supply temperature. As an example, when the floor heating system is coupled to a heat pump (ground, outdoor air, or exhaust air), the low supply temperature of the floor heating circuit yields a beneficial temperature difference between source and destination—the coefficient of performance will ultimately be high.

The control of a floor heating system is generally divided into a central feed-forward control and a local feedback room control (Olesen 2001). The common supply water temperature in the pipe circuits is controlled according to the outdoor conditions (feed-forward control) by means of the heating curve. The slope of the heating curve will be very small in the case of floor heating in energy efficient residential buildings. The feedback control at room level consists of temperature sensors connected to the manifold assembly. Generally, each thermal zone is fitted with a sensor where the user can select a set point temperature. The room sensor will control the water flow rate in the floor heating pipes. As an example, the room control will slow down or stop (depending on the applied control method) the heat supply to the pipe circuit if a heat source is introduced in the room. This active room control is dependent on sensors, communication systems, control algorithms, actuators, etc.

However, there is also a second (intrinsic) feedback system present that will also counteract any thermal perturbance in the zone. The indoor temperature will slowly rise as heat is emitted in the room. Over time, the return temperature

Henrik Karlsson is a researcher at the Technical Research Institute of Sweden, Borås, Sweden. **Carl-Eric Hagentoft** is a professor of Building Physics in the Department of Civil and Environmental Engineering, Chalmers University of Technology, Gothenburg, Sweden.

in the in-built floor heating circuit will correspondingly rise due to the influence of the thermal perturbance. Increased return water temperatures will ultimately result in a decreased heat flow from the heat source to the floor heating circuit. Hence, the outcome is a negative feedback loop that always counteracts any thermal perturbance. If the perturbance is negative (e.g., after opening the window in the heating season), the negative feedback yields a decreased return temperature and an amplified heat supply. This phenomenon is the subject of this paper; we call this negative feedback *self regulation*.

Floor heating systems with a high degree of self regulation are very robust, even without the active room control (sensors for active feedback). The intrinsic self-regulation ability may, to a high degree, counteract any thermal perturbance. Hence, there is a possibility for a simplified space heating system at lower cost (Karlsson and Hagentoft 2015).

The self-regulation process is briefly described by Fort (1989). Based on an experiment, Fort (1989) concludes that the closer the supply water temperature is to the indoor temperature, the more significant is the self-regulation ability. The same conclusion is also given by Olesen (2001, 2002) and Weitzmann (2004). The magnitude of the self regulation is given in a few examples by Olesen (2001, 2002). Furthermore, Olesen (2001, 2002) states that self regulation is only possible for water-based floor heating system and not for electrically heated floors.

Self regulation is a main advantage of floor heating systems used in houses with a small energy demand, as the low over-temperature in the floor construction ensures that when there are solar gains, the heating system stops to supply heat to the room (Weitzmann 2004). The given references quantify the self-regulation process based on an induced indoor temperature shift. Hence, the interaction between the floor heating system and the building is not taken into consideration; such a utilization factor does not directly link the intensity of the thermal perturbance with the subsequent system response.

In more recent research, Karlsson (2008, 2010) studied the dynamic response of the self-regulation process by means of numerical simulations. Karlsson (2010) also defined a utilization factor that takes into account the basic properties of the space heating system as well as the basic properties of the building. With such an utilization factor, there is a certain relation between the heat content of the thermal perturbance and the subsequent shift in heat supply (heat content).

Scope

The validity of Karlsson's utilization factor, γ, will be tested. For this purpose, a test room is prepared where the self-regulation process will be measured. The input parameters for the analytical formula will also be measured in separate measurements. Three different floor heating systems will be considered in the experiments: one lightweight system and two embedded heavy concrete systems.

ANALYTICAL FORMULA

Equation 1 displays a formula that states an utilization factor, γ, which quantifies the ratio between the net shift in supplied heat ΔE_s and the heat content of an arbitrary thermal perturbance ΔE_p. The derivation of γ is given in detail by Karlsson (2010). The utilization factor is based on three fundamental parameters; γ is closer to –1 if K_e is low, K_{up} is high, and η is close to 1.

$$\gamma = -\frac{K_{up}}{K_e + \eta \cdot K_{up}} \quad , \quad -1 < \gamma < 0 \qquad (1)$$

η quantifies the ratio between the upward heat flux and the totally supplied heat flux from the floor heating circuit (see Roots [1998] and Roots and Hagentoft [2005]). Because there is a thermal resistance above the floor heating pipe, there is also a heat loss downwards, which must be compensated for. The useful heat flux is, of course, the upward heat flux, which is transferred to the intended space. In the case of a well-designed floor heating system, only a few percent of the supplied heat will be conducted downwards (approximately 2%–5%). The ability to exchange heat from the entire floor heating circuit upwards to the indoor air temperature is given by the equivalent thermal conductance K_{up}, taking into account the water flow rate, center-to-center distance of the pipe circuit, the thermal resistance of floor-covering material, and the surface heat transfer. The total heat loss of the considered zone is quantified by the thermal conductance K_e. The heat loss comprises both conductive and ventilation heat losses.

As a consequence, a significant effect of self-regulation arises if

- The room has a low total heat loss
- Large area is covered by floor heating circuits
- The thermal resistance between the pipe circuit and the indoor space is low
- A small center-to-center distance is applied
- High fluid flow rate is applied

EXPERIMENTAL VERIFICATION

Introduction

The utilization factor, γ, has been determined in three different lab tests:

1. **Analytical solution based on measured system and building properties.** The utilization factor can be calculated according to Equation 1. Input data are the measured properties of the floor heating system and the test room.
2. **Step-response test.** This test involves measuring the stationary shift in heat supply due to the introduction of a stationary thermal perturbance. The shift in heat supply is a direct measurement of the utilization factor, γ.

3. **Pulse-response test.** This test involves measuring the heat content of the shift in heat supply due to a thermal perturbance in the shape of a pulse. The heat content of the pulse response swing of the heat supply is a direct measurement of the utilization factor, γ.

Design of Test Room

A test room (11.13 m^2) equipped with a water-based floor heating system has been constructed inside a two-sided climate chamber. Next to the external wall is the secondary chamber with constant outdoor condition (T_e = –12.5°C). The external wall has a 1 m^2 window (U = 0.8W/m^2K). The walls and ceiling of the test room have 13 mm internal gypsum boards on top of sandwich elements (steel sheets and thermal insulation). Supply air is provided to the test room by means of a fan. Hence, the test room has a higher air pressure than the adjoining lab hall and the secondary chamber. This prevents unwanted infiltration of air. The supply air is carefully supplied along the surface of the ceiling with a low air velocity. The supply airflow (10 L/s) and supply air temperature was measured continuously. We tried to simulate a mechanical air-to-air heat exchanger with temperature efficiency around 80%. Therefore, exhaust air from the test room was mixed with cold air from the secondary chamber before the supply air was distributed to the test room. The design heat demand was measured to be in the range of 12–13 W/m^2 as the outdoor temperature was –12.5°C in the secondary chamber.

The complete test room was located in a large lab hall (>100 m^2). The air temperature in the lab hall was around +20°C. Hence, all sides of the test room are facing the lab hall—except the exterior wall, which faces the secondary chamber. At steady-state initial condition, the heat flux from test room to lab hall is small, because the temperature difference is minimal.

Three different floor heating systems have been applied in the test room: one lightweight system and two heavy embedded systems. The lightweight system consists of PEX pipes fitted in plates of aluminum for lateral heat distribution (see Figure 1). The center-to-center distance between the loops in the lightweight system is 300 mm. The heavy systems are embedded in a glass fiber reinforced cement based floating screed with a thickness of 104 mm. The two embedded systems are fitted in the same slab: a 300 mm center-to-center circuit in the bottom of the slab and an upper 200 mm circuit located as close as possible to the surface of the slab (see Figure 1). In order to attain homogeneous properties, there is no steel reinforcement in the slab. The thermal conductivity of the screed is 0.7 W/mK; thus, it is considerably lower then a steel-reinforced concrete slab. The two heavy systems are used one at a time.

The same 12 mm chipboard on top of a thin rubber impact sound insulation membrane has been applied on top of all three floor heating systems. This represents a typical situation with a wooden floor. Furthermore, the 150 mm of expanded polystyrene (EPS) insulation between the existing chamber floor and the floor heating construction has been the same for all three systems.

The test room has been equipped with 27 thermocouples in the air volume. The measurement shows that the air temperature has been very homogeneous within the test room. The average of these thermocouples has been used as the average air temperature in the evaluation. Furthermore, numerous thermocouples have been applied on top of the floor surface and other internal surfaces within the test room in order to monitor the conditions during the experiments.

An electrical heater, a circulation pump, and a water flow meter have been fitted outside the test room in sequence with the floor heating circuit. The water temperature has been measured at the point were the pipe circuit enters/exits the test

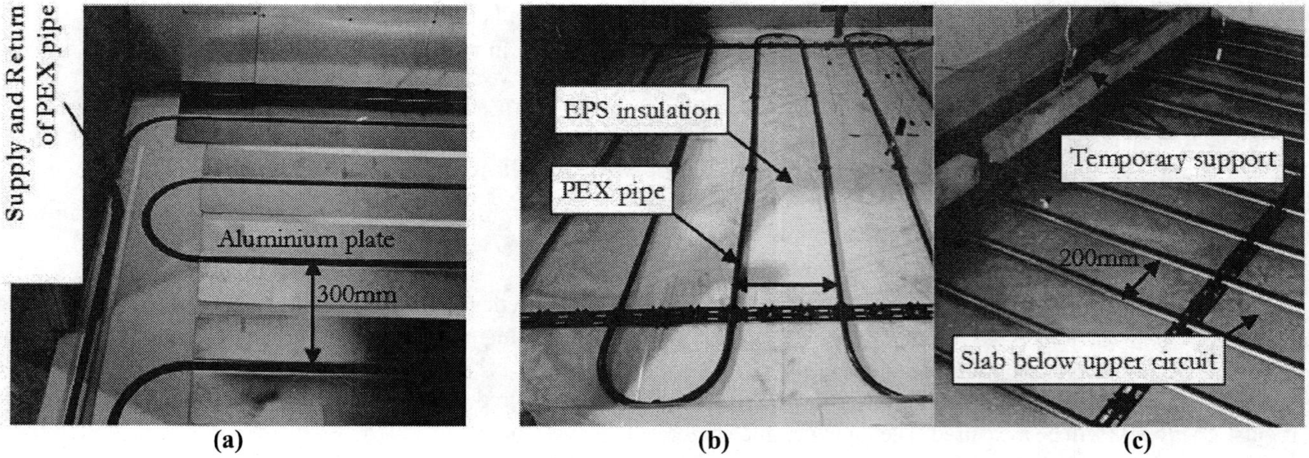

Figure 1 *(a) Lightweight system with aluminum plates directly on top of EPS insulation. (b) Bottom embedded system (CC300) with pipe circuit directly on top of EPS before casting of slab. (c) Upper embedded system (CC200) before the final layer of screed is casted.*

room. PT100 temperature sensors have been installed for accurate measurements of the heat that is supplied to the test room by means of the floor heating pipe circuit. There is also a PI regulator installed that controls the electrical heater in order to attain a constant supply water temperature. The supply temperature has been very stable during the experiments (see, for example, Figure 2).

A thin electrical resistance cable has been fitted within the test room at heights between 0.25 m and 0.75 m. The heat cable is evenly distributed above the whole floor surface, as far as possible from the air thermocouples. The electrical power emitted in the test room can directly be measured. The heat cable represents a thermal perturbance.

Analytical Solution Based on Measured System and Building Properties

We have measured the input parameters of the analytical formula in order to calculate the utilization factor, γ, according to Equation 1. The total conductive heat loss (all building elements) has been measured in a separate stationary measurement. Heat was released inside the room and compared to the measured temperature difference between the indoor air and the surrounding temperature in order to quantify the conductive heat loss. The ventilation of the test room was sealed during this measurement. The ventilation heat loss is instead measured directly by measuring the supply airflow and the supply air temperature, as well as the room air temperature. The sum of conductive and ventilation losses defines the heat loss conductance, K_e.

The total heat exchange, K_{up}, from the supply water temperature, T_s, towards the average indoor air temperature, T_i, is measured for each of the three floor heating systems. The product $\eta\ Q_S$ defines the upward heat flux from the floor heating circuit that actually heats up the floor surface of the test room. η is measured separately (see below). Q_S is measured directly in the test setup by means of the water flow rate and the temperature difference between supply and return ends of the pipe circuit. K_{up} is defined according to the following:

$$K_{up} = \frac{\eta \cdot Q_S}{T_s - T_i}\ , \text{W/K}$$

The lightweight system and the embedded CC300 have similar heat exchange properties (see Table 1). The CC200 embedded system has much higher heat exchange potential. This is caused by the fact that the pipe circuit is longer and tighter; the pipe is located close to the floor surface and the water flow rate is greatly increased.

The insulation efficiency, η, was determined in two separate tests. In the first test is the heat demand delivered through the floor heating system; in the second test is the heat demand delivered entirely by a heater entirely located inside the test room (η = 1). The boundary conditions are the same in both cases. These measurements show that 89% of the supplied heat to the floor heating circuit, Q_S, actually is conducted upwards towards the test room. 11% is conducted downwards to the lab hall. Hence, the insulation efficiency,

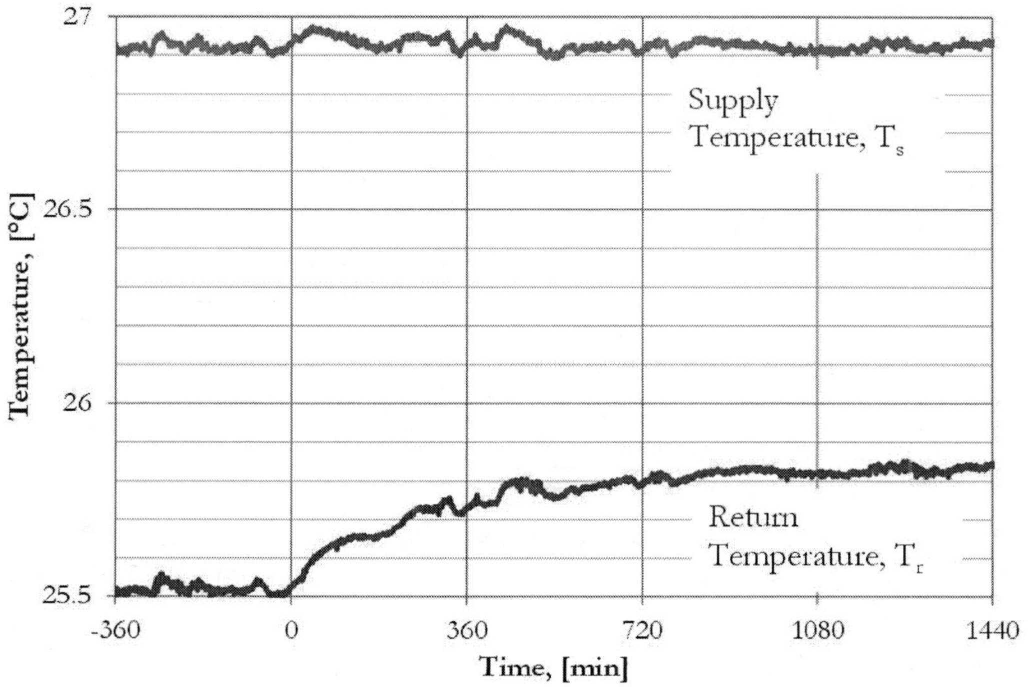

Figure 2 *Measurement of the supply and return water temperatures in the step response test for the lightweight floor heating system. The perturbance is activated at zero time.*

η, is 0.89. These tests were only conducted with the lightweight floor heating system. The embedded system with CC200 (pipe close to surface) was assumed to have the same η as the lightweight system because the same floor covering is applied and the same insulation is applied underneath. The CC300 embedded system is calculated to have η equal to 0.85 based on the thermal conductivity of the slab and the position of the pipe circuit.

In the end, we found similar utilization factors, γ, for the lightweight and embedded CC300 systems. Due to the higher heat exchange capability (K_{up}), the embedded CC200 show significant higher ability for self-regulation.

Step Response Test

In this test we will trigger the self-regulation response in the floor heating circuit by activating a step thermal perturbance. At steady-state conditions, heat is emitted from the resistance wire inside the test room. The self-regulation response can directly be measured as a shift in the supplied heat flux, Q_S. All boundary conditions are constant during this test, including the supply water temperature in the floor heating circuit.

The dynamic response of the lightweight floor heating system will be presented in Figure 2 and Figure 3. Before activation of the thermal perturbance, the indoor temperature is steady at 20.2°C, the outdoor temperature is –12.5°C, and the

Table 1. Measured Input Data and Resulting Self-Regulation Utilization Factor

	Lightweight CC300	Embedded CC300	Embedded CC200	Expanded Measurement Uncertainty (k = 2)
η, —	0.89	0.85*	0.89*	±0.024
K_c, W/K	16.2	16.2	16.2	
K_v, W/K	2.5	2.5	2.6	
K_e, W/K	18.7	18.7	18.8	±0.60 W/K
K_{up}, W/K	19.2	19.6	27.9	±4.3%/±4.2%/±5.2% **
γ, —	–0.54	–0.55	–0.64	±0.016/±0.017/±0.019 **

* Assumed/calculated based on the measured value for lightweight system and the thermal properties of the embedded slabs.
** Uncertainty for each system: lightweight 300/embedded 300/embedded 200.

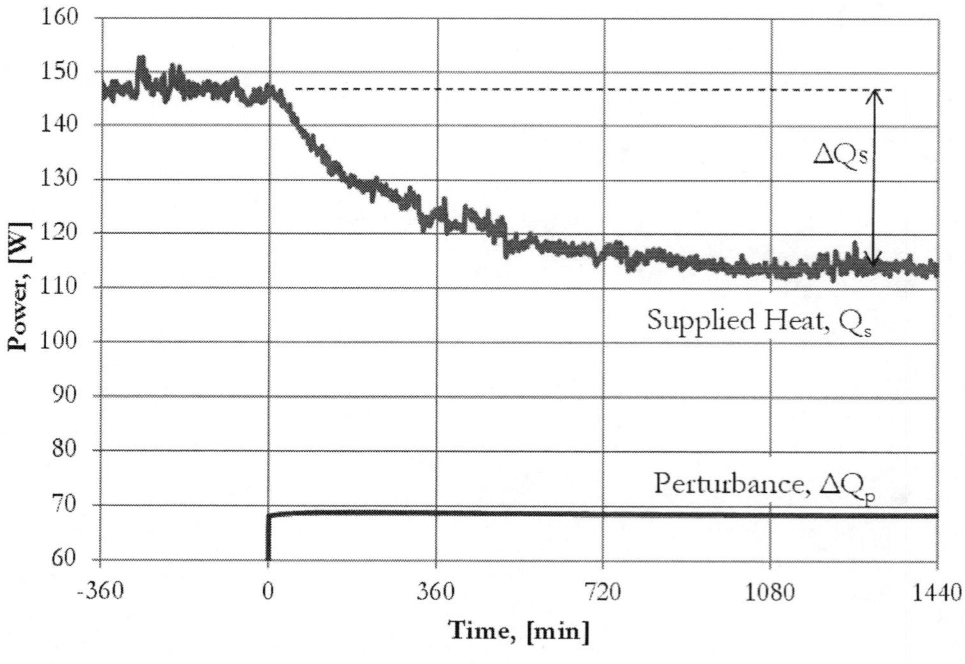

Figure 3 *Measurement of the supplied heat to the floor heating pipe in the step response test for the lightweight floor heating system.*

lab temperature is 20°C. In order to have balance, the supply water temperature T_s is 26.92°C and the return water temperature is 25.52°C (see Figure 2). This yields a heat supply, Q_s, of 147.1 W as the initial condition (see Figure 3).

The return water temperature and indoor temperature are seemingly affected as the perturbance, ΔQ_p, (68.5 W) is activated; see Figure 2 and Figure 3. In the case of the lightweight system, the time delay between activation of the perturbance and an observable shift in return water temperature is very short. For the heavy concrete systems, this delay time is much greater.

After roughly 24 h, the test room has reached a new steady-state condition, see Figure 2 and Figure 3. At the final steady-state condition we found that the return water temperature has increased to 25.85°C; and subsequently, the supplied heat to the circuit has fallen to 111.2 W. Thus, the shift ΔQ_s is –35.9 W.

The utilization factor, γ, is found as the ratio between the stationary shift in supplied heat, ΔQ_s, and the amplitude of the perturbance, ΔQ_p. Because the self-regulation process is affecting the supplied heat, we will utilize 52% of the heat released from the thermal perturbance.

The step response test is repeated for the two heavy embedded systems as well. The main difference is the delay time between activation of the perturbance and the settling of a new steady-state condition. All results are given by Table 2. The embedded CC200 system has the water fluid flow rate increased in order to further boost the self-regulation effect.

Pulse Response Test

The pulse response test is performed with a similar procedure as the step response test. The heat content of any perturbance pulse is a bounded figure that we can directly measure. The heat content of both the perturbance and the response in the supplied heat flux are measured simultaneously. The ratio between these figures quantifies the self-regulation ability. The pulse test only considers the embedded CC200 system.

The perturbance is activated at zero time and is active for 4 days. As seen in Figure 3, there is a negative shift in the supplied heat, Q_s. The floor heating circuit is affected even longer than 4 days due the thermal inertia of the test room and the concrete slab. The self-regulation response is reducing the heat supply for about 1.5 days longer than the perturbance pulse is active (see Figure 4). When the supplied heat is back at the reference level (Q_s = 163 W), the test is completed.

The utilization factor, γ, is found as the ratio between the heat content of the shift in supplied heat ΔE_s and the heat content of the perturbance pulse, ΔE_p (see Equation 2). Hence, γ quantifies how much heat of the perturbance is utilized by means of the reduction of the supplied heat flux.

$$\gamma = \frac{\Delta E_s}{\Delta E_p} \tag{2}$$

The total heat content of the response in the floor heating circuit, ΔE_s, is –4.53 kWh, while the perturbance pulse, ΔE_p, is measured to be 6.62 kWh. Therefore, the utilization factor, γ, is measured to be –0.68 ± 0.09 for the embedded CC200 system.

DISCUSSION

The measured utilization factor has been measured in the range of –0.5 for the lightweight and the embedded 300 mm system and around –0.7 for the embedded 200 mm system. The self-regulation ability is not so significant in this test case. The form factor of the test room is not beneficial for an energy-efficient building in general; the heated floor area is small in comparison to the total envelope of the test room (including walls towards the lab hall). Additionally, the existing chamber probably has some major thermal bridges; at least, the separate measurements of the conductive heat losses, K_c, indicate this. For the two embedded systems, the choice of a floating screed is not beneficial for the upward heat exchange capability, K_{up}, as the thermal conductivity is considerably lower than a regular concrete slab.

Table 2. Results of the Step Response Test

	Lightweight CC300, Initial/Final	Embedded CC300, Initial/Final	Embedded CC200, Initial/Final	Expanded Measurement Uncertainty, $k = 2$
T_s,°C	26.92/26.92	27.17/27.17	25.26/25.26	±0.02 K
T_r, °C	25.52/25.85	25.70/26.01	24.34/24.62	±0.02 K
V_f, L/h	89.9/89.9	92.4/92.4	155.9/155.9	±0.25%
Q_s, W**	147.1/111.2	156.7/123.3	163.4/115.5	±2.1 W/±2.2 W/±3.6 W *
ΔQ_p, W	+68.5	+68.3	+68.6	±0.22%
ΔQ_s, W	–35.9	–33.3	–47.9	
γ, —	–0.52	–0.49	–0.70	±0.05/±0.07/ ±0.09 *

* Uncertainty for each system: lightweight 300/embedded 300/embedded 200
** The measurements are performed with slightly different conditions in the lab hall (different time of the year), which may affect Q_s.

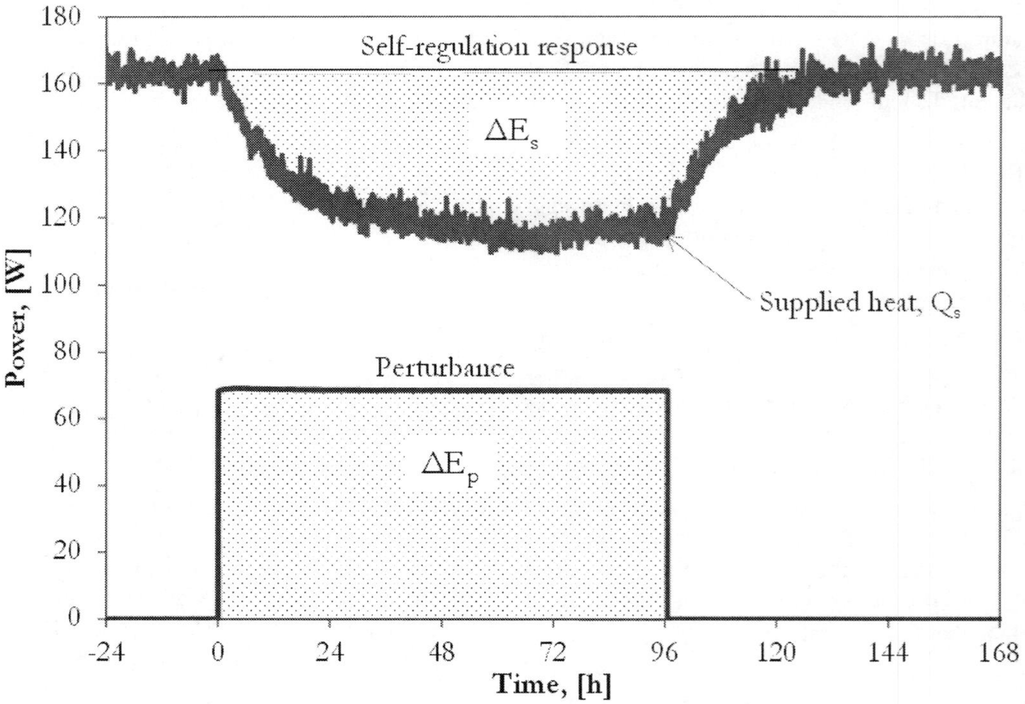

Figure 4 *Results from pulse response test of the embedded CC200 system.*

Similar systems as the tested CC300 screed embedded system have shown γ in the range of –0.85 for a well-insulated, single-family house (see Karlsson and Hagentoft 2015) and in the range of –0.90 in the case of a multi-family house.

A temperature potential between T_s and T_i is selected when quantifying K_{up}. However, the thermal radiation at the floor surface dominates over the convective heat transfer. Nevertheless, when measuring/calculating the view factor weighted radiation temperature, T_{rad}, of walls, windows, and ceiling, T_i and T_{rad} are very close to each other (the difference is much less than the measurement uncertainty of ±0.15 K). Consequently, in this test situation K_{up} is approximately valid for a temperature potential between T_s and T_{rad} or T_s and the equivalent indoor temperature (≈ average between T_i and T_{rad}).

CONCLUSION

In a specific test room, we have found a good agreement between calculated self-regulation utilization factor, γ, and direct measurement of the same property. When calculating γ by means of the analytical formula earlier published by Karlsson (2010), the input parameters of the tested systems are measured separately.

ACKNOWLEDGMENTS

This project has been financed by the Development Fund of the Swedish Construction Industry (FoU-Väst). The experimental tests have been conducted at SP Technical Research Institute of Sweden in Borås, Sweden.

NOMENCLATURE

γ	=	utilization factor for self-regulation, dimensionless
η	=	insulation efficiency, dimensionless
K_c	=	conductive heat loss of test room, W/K
K_v	=	ventilation heat loss of test room, W/K
K_e	=	total heat loss of test room, W/K
K_{up}	=	ability to exchange heat from supply water temperature to indoor air temperature, W/K
Q_s	=	heat flux supplied to floor heating circuit, W
ΔQ_p	=	thermal perturbance, W
ΔQ_s	=	shift in heat flux supplied to floor heating circuit, W
T_s	=	supply water temperature, °C
T_r	=	return water temperature, °C
T_i	=	average indoor air temperature, °C
T_{rad}	=	average radiation temperature, °C
ΔE_p	=	heat content of thermal perturbance, kWh
ΔE_s	=	heat content of the shift in supplied heat, kWh
V_f	=	water flow rate, L/s
k	=	coverage factor, number of standard deviations

REFERENCES

Fort K. 1989. Dynamisches Verhalten von Fussbodenheizungen. Juris Druck + Verlag, Zürich, Switzerland.

Karlsson H. 2008. Self-Regulating Floor Heating Systems in Low Energy Buildings. Proceedings of the 8th Symposium on Buildings Physics in the Nordic Countries, 519–26. Copenhagen, Denmark.

Karlsson, H. 2010. Thermal Modeling of Water-Based Floor Heating Systems - supply temperature optimization and self-regulating effects. PhD thesis, Chalmers University of Technology, Gothenburg, Sweden.

Karlsson, H. and Hagentoft, C.-E. 2015. Experimentell verifiering av förenklade och självreglerande golvvärmesystem för välisolerade bostäder (in Swedish). Report 2015:10, Department of Civil and Environmental Engineering, Chalmers University of Technology, Gothenburg, Sweden.

Olesen B. W. 2001. Control of floor heating and cooling systems. 7th World Congress Clima 2000. Napoli, Italy.

Olesen B. W. 2002. Radiant Floor Heating In Theory and Practice. ASHRAE Journal.

Roots P. 1998. Värmeförlust från en grund som utföres med golvvärme (in Swedish). Working paper 52. University of Gävle, Department of Technology. Gävle, Sweden.

Roots P. and Hagentoft C.-E. 2005. Released heat from a Line Source in a Foundation Construction. Proceedings of the 7th Symposium on Buildings Physics in the Nordic Countries, 839–44. Reykjavík, Iceland.

Weitzmann P. 2004. Modeling Building Integrated heating and cooling systems. PhD thesis, Department of Civil Engineering, Technical University of Denmark, Lyngby, Denmark.

Measured Differences of Ground and Space Temperatures for Side-by-Side Slab-on-Grade Residences with and without Carpet

Robin K. Vieira **Danny S. Parker** **Jamie Kono**
Eric Martin **John Sherwin**

ABSTRACT

A particularly suspect aspect for building simulations has been the ability to predict ground heat transfer. In Florida, slab-on-grade construction dominates. To better understand ground heat transfer, as well as the differences between uncovered slab and applying an insulation layer (carpeting) over a slab in a mild climate, the Florida Solar Energy Center built two identical residential laboratory buildings with 164 embedded slab and ground thermocouples. In July 2014, an experiment began comparing the thermal performance of carpet to uncovered slab flooring. The buildings were cooled to 77°F (25°C) in summer and heated to 73°F (22.8°C) in winter. The thermostats were set to either cooling or heating, as during Florida's winters interior temperatures sometimes drift above the cooling set point and occasionally when set to cooling they drift below the winter set point. Each laboratory home is unoccupied with automated internal sensible and moisture loads provided hourly to represent human, appliance, and lighting loads. The hypothesis is that in Central Florida, where year-round ground temperatures are between winter and summer set points, the non-carpeted slab should have an advantage.

The paper presents findings for a year's worth of data collection, differences in heating and cooling loads on each home, and images of temperature differences through the matrix of slab measurements. Net heat transfer in Central Florida was small during the cooling season. There was some benefit available during early spring time. Results are sensitive to geographic location and interior set points.

INTRODUCTION

In warm regions bordering the Mediterranean or the Gulf of Mexico, tiled or terrazzo slab-on-grade floors are a popular residential option. With tile on a slab floor there is more heat transfer than with carpet. Is this beneficial? Empirical data are somewhat limited in warm climates.

Past research has focused on the impact of slab-on-grade foundations and insulation schemes in predominantly heating-dominated climates. Bareither et al. (1948) at the University of Illinois compared the performance of seven types of slab floor insulation. Perhaps the most rigorous work has been that of Adjali et al. (2000), at Cardiff School of Engineering, comparing numerical simulations with measured slab performance of detailed measurements in Wales.

Simulation research is often used by researchers or practitioners to compare options. However, models using EnergyPlus and DOE2 engines indicate cooling savings of 4% to 7% for single-family homes with tile versus carpet. Are those results accurate?

EXPERIMENTAL DESIGN

Experiments in this research were performed over an entire year from 2014–2015 in Cocoa, Florida, at the Florida Solar Energy Center (FSEC) Flexible Residential Test Facility (FRTF) and intended to assess slab-on-grade influence in a cooling-dominated climate. This facility consists of two 1536 ft^2 (142.7 m^2) identical buildings that are extensively metered (Figure 1). The FRTF was completed in 2010 and is more thoroughly described elsewhere (Vieira and Sherwin 2012).

Robin K. Vieira, **Danny S. Parker**, **Eric Martin**, and **John Sherwin** are researchers at the Florida Solar Energy Center, a research institute of the University of Central Florida, Cocoa, FL. **Jamie Kono** is a graduate student at Georgia Institute of Technology, Atlanta, GA.

Figure 1 *Side-by-side FRTF buildings at FSEC in Cocoa, Florida.*

Table 1. Locations of Sensors

Configuration	Quantity	0 ft (0 m)	1 ft (0.30 m)	2 ft (0.61 m)	5 ft (1.5 m)	10 ft (3.0 m)	20 ft (6.1 m)	Moisture at 1 and 5 ft (0.30 and 1.5 m)	Location
A	3	x	x	x	x	x	x	x	Center of homes and midway between homes
B	6	x	x	x	x	x	x		Footer midway on east and west sides and 2 ft (0.61 m) out from home
C	12	x	x	x	x	x			Corners of home and midway on north- and south-side footers
D	12	x	x	x					8 ft (2.44 m) in from each corner in both directions and 8 ft (2.44 m) in from midway edge points on north and south sides

Each building has a standard 4 in. (0.10 m) concrete monolithic slab with perimeter footers poured over sand fill at the time of construction. The east test building has a standard 3/8 in. (0.0095 m) rubber pad and 1/2 in. (0.15 m) synthetic carpet (which has compressed somewhat after installation) installed in late 2013 with an estimated combined R-value of approximately 2.4 h·ft$^{2}\cdot$°F/Btu (0.423 m$^{2}\cdot$°C/W). The west test building has an unmodified uncovered concrete slab floor. The envelope for each is meant to represent older building stock in Florida with R-19 h·ft$^{2}\cdot$°F/Btu (3.35 m$^{2}\cdot$°C/W) ceiling, uninsulated concrete block walls, and single-pane glass windows with blinds.

Identical datalogger systems and associated instrumentation were installed in each building. Data were processed and archived using a pair of networked Campbell Scientific dataloggers (CR3000 and CR1000) along with associated peripherals. The data set included pertinent building materials' surface temperatures, interior space conditions (temperature and relative humidity), attic space conditions (*T*/RH), HVAC energy use, and infiltration rate via carbon dioxide injection and monitoring. An on-site meteorological weather station monitored irradiance, wind speed, ambient temperature, humidity, and carbon dioxide concentrations. Measurements were taken every 10 s and stored at 15 min intervals (Vieira and Sherwin 2012). Heating and cooling set points in each building were set to 73°F and 77°F (22.8°C and 25°C), respectively, by a pair of identical digital thermostats. A schedule of internal sensible and latent heat gains was introduced to both buildings to simulate occupied loads.

A grid of 156 temperature measurements (see Table 1) were taken using copper-constantan (type T) thermocouples with a read resolution of 0.1°F (0.56°C) and an uncertainty of +0.9°F (0.5°C). Temperatures were affixed to the slab surface using thermocouple epoxy and buried at depths of 0 (i.e., ground surface under concrete slab polyethylene vapor barrier), 1, 2, 5, 10 and 20 ft (0, 0.3, 0.61, 1.5, 3 and 6.1 m). A schematic diagram of the measurement locations at the two FRTF buildings, starting at the surface and descending down

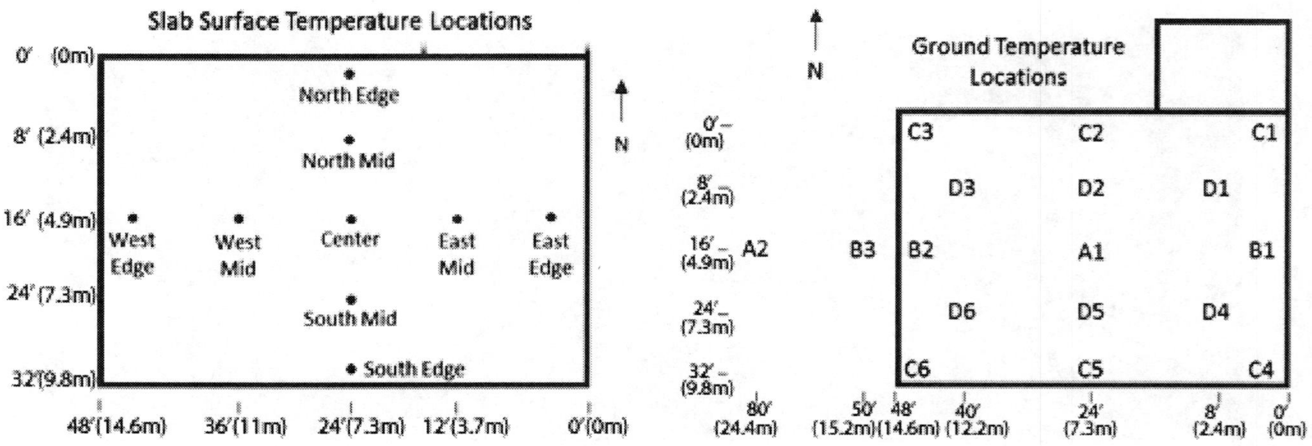

Figure 2 *Schematic diagram of slab temperature measurement points at FRTF.*

through the ground, is shown in Figure 2. Point A2 is taken 32.5 ft (9.9 m) from the west and east walls, respectively, of the two buildings. Point B3 is taken 2 ft (0.61 m) from the outside edge of the slab.

EXPERIMENTAL RESULTS

Floor Slab Thermal Performance

Various center-of-building slab temperatures were examined at both surface and depth, located at A1 in Figure 2. Examples for the center slab profiles are shown in Figure 3 and show strong thermal heat flow from the interior building to the ground during the uncooled spring period, which is quite apparent to a depth of 5 ft (1.52 m). Figure 3 also shows widely varying shallow temperatures between October 31, 2014, and April 2015, during Florida's heating season when the cooling system is not available and the temperature inside the building floats upward above 73°F (22.8°C) but not below. However, this period of time with higher floating interior temperatures creates prominent heat flows, evident down through the slab and into the ground layers as shown by the data. In particular, the data show that the uncovered slab floor in the west building is creating surface temperatures approximately 1°F (0.56°C) lower than those shown for the east carpeted building. The trend remains the same down to 5 ft (1.5 m) below the surface.

Estimated Surrogate Heat Fluxes

To address the lack of physical measurement of heat fluxes, surrogate heat fluxes were estimated from the available temperature data. This involved subtracting the measured slab surface temperature from the measured interior air temperature. While they are not true heat fluxes, these temperature differences give an indication of heat flow direction and order of magnitude.[1] The measured values on the east carpeted section reflect the measured temperature under the carpet and pad and have been divided by the R-value of that assembly, 2.4 h·ft^2·°F/Btu (0.423 m^2·°C/W), to yield a comparative value shown for the west house, which is exposed slab.

Table 2 shows the surrogate heat fluxes for the carpeted East building and the uncarpeted West building with the exposed slab. Positive numbers indicated heat gain to the space; negative values indicate heat losses from the slab. Note some missing data for January for the north slab data due to sensor/communication issues.

Evaluating the surrogate heat flux data reveals the following:

- *Heating:* Winter heat losses are seen at all measurement points; greatest at edges. Heat gains and losses are lowest slab floor center at the midpoints between the center and edges.
- *Heating:* Carpet attenuates winter heat losses seen relative to the unsurfaced slab.
- *Cooling:* Heat gains, adding to cooling load, are seen across the slab in summer and are also greatest at the edges—particularly the north edge by the unconditioned garage (which becomes very hot in summer).
- *Cooling:* Heat gains to the interior in summer are reduced by the carpet.
- *Floating Conditions:* Heat fluxes change (including flow direction) over the seasons, particularly at the slab center and midpoints.

Heating Results

For heating, we examined 14 days (Table 3) over the year where the outdoor temperature was less than 55°F (12.8°C), which have been shown to clearly be heating periods in Florida

1. As an approximation, it can be considered that the horizontal heat transfer surface conductance for still air from the 2009 *ASHRAE Handbook—Fundamentals* (Table 1, p. 26.1) (ASHRAE 2009) are 1.63 Btu/h·ft^2·°F (9.25 W/m^2·°C) for heat transfer upwards and 1.08 Btu/h·ft^2·°F (6.13 W/m^2·°C) for heat transfer downwards.

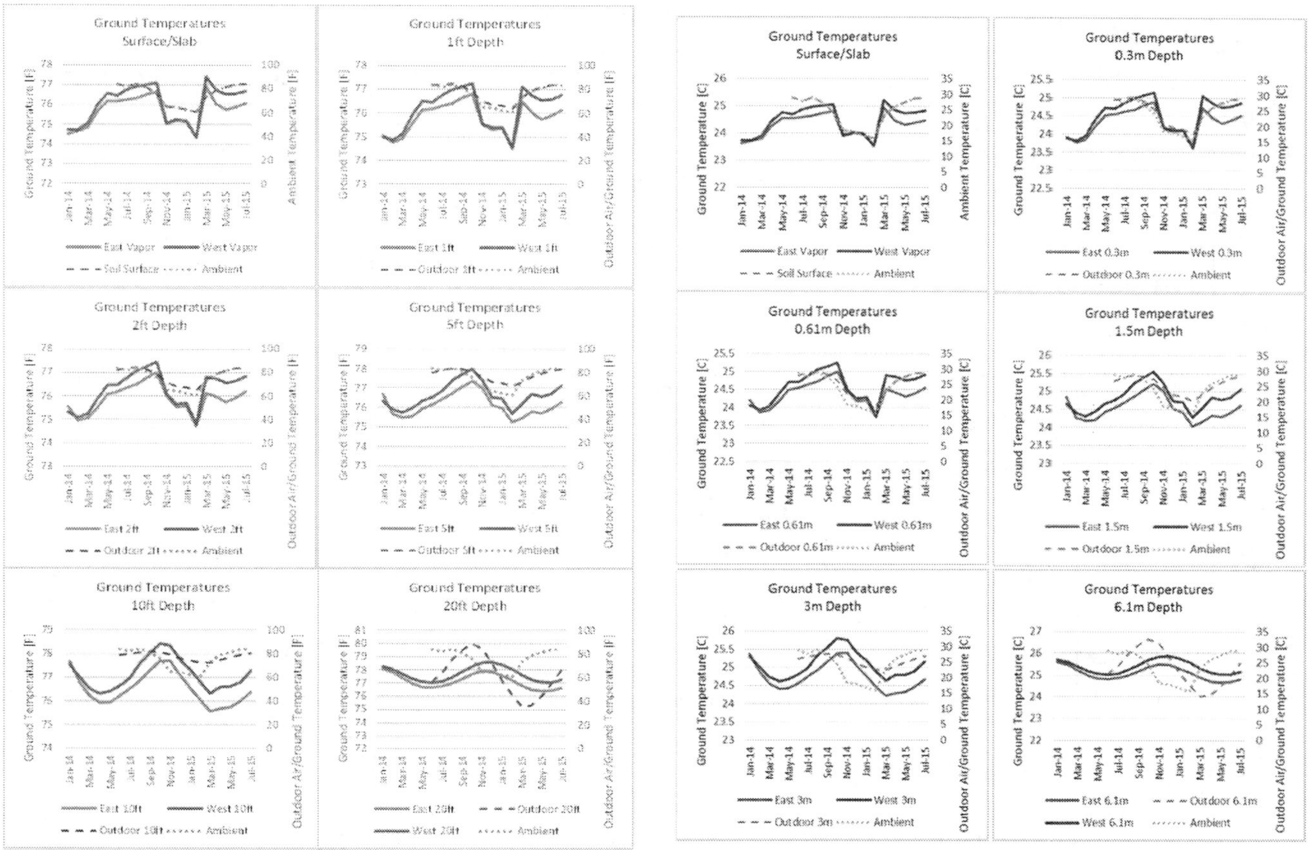

(a) (b)

***Figure 3** Ground temperatures for the carpeted east building and uncarpeted west building: (a) °F and (b) °C.*

homes (Parker et al. 2015). The carpeted east building showed lower energy consumption by approximately 0.65±0.15 kWh/day (4.3%). Resistance heat consumption was 15.16 kWh/day in the carpeted east building and 15.81 kWh/day in the west exposed slab building. While this difference was measurable and statistically significant, the energy quantity is very small. Relative humidity in the two buildings was virtually identical. Similar differences in energy use (4.1%) were found limiting the data to the four days of temperature less than 50°F (10°C).

Contour plots presented in this and following sections were generated using the average measured data for the period indicated with the points shown (small black dots). Thermal contours were generated using the Plotly.js software package (https://plot.ly/feed/), which interpolates contours using a rectangular mesh. The contours of heating period data (Figure 4) show heat loss to the ground, with particularly striking changes seen from the slab edge. Note that these changes at slab edge are not an artifact of sensor space, as the temperature measured at the slab edge are 1 ft (0.3 m) inside the interior of the wall and the estimate outside the slab edge is only 2 ft (0.61 m) from the wall.[2] This finding is in agreement with experimental measurements of slab floors stretching back to work done post-World War II by the University of Illinois (Bareither et al. 1948). The plots show solar heating of the soil surface outside the building profile as well as the profile across the building interior the ground temperatures down to 10 ft (3 m). Note the very different thermal domains under and outside the building in winter.

Performance During Floating Conditions

The performance of the two buildings was evaluated during the fall and spring for a period of 55 days when the HVAC system was inactive. During this 55-day spring and fall period, the carpeted east building ran 0.72°F ± 0.5°F (0.40°C ± 0.28°C) warmer than the uncarpeted west building (77.89°F versus 77.17°F [25.5°C versus 25.1°C]). Interior relative humidity was slightly higher in the carpeted east building

2. The contour plots shown should be considered approximate. While spacing of 2 ft (0.61 m) or less would provide much more precise indication of thermal anomalies and intervals, it was decided that such a representation as that made in the report must be used to help the reader visually understand the relationship of the eight odd measurement points being represented for each slab floor.

Table 2. Monthly Surrogate Heat Flux Data ΔT for Measured Locations

Month	NE	NE	SE	SE	WE	WE	EE	EE	CNT	CNT	NM	NM	SM	SM	WM	WM	EM	EM
	°F	°C	°F	°C	°F	°C	°F	°C	°F	°C	°F	°C	°F	°C	°F	°C	°F	°C
West—Exposed Slab (Uncarpeted)																		
Jan	—	—	–2.38	–1.32	–3.83	–2.13	–4.06	–2.26	–1.04	–0.58	—	—	–0.94	–0.52	–0.79	–0.44	–0.62	–0.34
Feb	–2.84	–1.58	–3.64	–2.02	–4.63	–2.57	–4.21	–2.34	–1.37	–0.76	–1.69	–0.94	–1.22	–0.68	–1.10	–0.61	–0.79	–0.44
Mar	–1.56	–0.87	–2.82	–1.57	–3.18	–1.77	–2.44	–1.36	–0.97	–0.54	–1.11	–0.62	–0.95	–0.53	–0.79	–0.44	–0.50	–0.28
Apr	–1.14	–0.63	–2.31	–1.28	–1.61	–0.89	–0.65	–0.36	–0.71	–0.39	–0.44	–0.24	–0.85	–0.47	–0.76	–0.42	–0.44	–0.24
May	2.99	1.66	0.39	0.22	1.41	0.78	1.61	0.89	–0.35	–0.19	0.21	0.12	–0.45	–0.25	–0.39	–0.22	–0.43	–0.24
June	4.46	2.48	1.62	0.90	3.05	1.69	3.34	1.86	0.33	0.18	1.03	0.57	0.29	0.16	0.30	0.17	0.13	0.07
July	5.22	2.90	2.33	1.29	3.77	2.09	4.10	2.28	0.26	0.14	1.06	0.59	0.26	0.14	0.25	0.14	0.00	0.00
Aug	5.52	3.07	3.19	1.77	3.73	2.07	6.09	3.38	0.17	0.09	1.47	0.82	0.88	0.49	0.75	0.42	0.47	0.26
Sept	3.06	1.70	1.46	0.81	0.87	0.48	1.12	0.62	0.71	0.39	1.02	0.57	0.92	0.51	0.61	0.34	0.51	0.28
Oct	–0.51	–0.28	–0.89	–0.49	–2.42	–1.34	–1.65	–0.92	–0.45	–0.25	–0.60	–0.33	–0.35	–0.19	–0.44	–0.24	–0.25	–0.14
Nov	–2.61	–1.45	–1.72	–0.96	–4.62	–2.57	–2.94	–1.63	–0.83	–0.46	–1.23	–0.68	–0.66	–0.37	–0.82	–0.46	–0.45	–0.25
Dec	–3.12	–1.73	–2.28	–1.27	–4.46	–2.48	–4.06	–2.26	–1.01	–0.56	–1.50	–0.83	–0.82	–0.46	–0.92	–0.51	–0.61	–0.34
East—Carpeted																		
Jan	—	—	–1.05	–0.58	–1.78	–0.99	–2.41	–1.34	–0.23	–0.13	—	—	–0.22	–0.12	–0.16	–0.09	–0.26	–0.14
Feb	–1.06	–0.59	–1.63	–0.91	–2.30	–1.28	–2.40	–1.33	0.25	0.14	–0.13	–0.07	–0.51	–0.28	–0.46	–0.26	–0.25	–0.14
Mar	–0.61	–0.34	–1.37	–0.76	–1.68	–0.93	–1.47	–0.82	0.08	0.04	–0.14	–0.08	–0.34	–0.19	–0.33	–0.18	–0.24	–0.13
Apr	0.72	0.40	–1.36	–0.76	–0.87	–0.48	–0.38	–0.21	0.64	0.36	–0.14	–0.08	–0.44	–0.24	–0.45	–0.25	–0.29	–0.16
May	1.68	0.93	–0.13	–0.07	0.54	0.30	0.68	0.38	0.93	0.52	–0.10	–0.06	–0.53	–0.29	–0.51	–0.28	–0.44	–0.24
June	3.05	1.69	0.99	0.55	2.03	1.13	2.00	1.11	1.22	0.68	0.96	0.53	0.38	0.21	0.45	0.25	0.40	0.22
July	3.28	1.82	1.19	0.66	2.27	1.26	2.24	1.24	0.52	0.29	0.77	0.43	0.17	0.09	0.29	0.16	0.17	0.09
Aug	3.73	2.07	1.78	0.99	2.54	1.41	2.55	1.42	1.17	0.65	1.29	0.72	0.66	0.37	0.76	0.42	0.61	0.34
Sept	2.80	1.56	1.27	0.71	1.54	0.86	1.50	0.83	1.23	0.68	1.77	0.98	1.23	0.68	1.33	0.74	1.17	0.65
Oct	0.23	0.13	–0.48	–0.27	–0.95	–0.53	–0.81	–0.45	0.25	0.14	0.39	0.22	0.01	0.01	0.15	0.08	0.08	0.04
Nov	–0.66	–0.37	–0.71	–0.39	–2.03	–1.13	–1.54	–0.86	–0.76	–0.42	0.24	0.13	–0.04	–0.02	0.03	0.02	0.02	0.01
Dec	–1.17	–0.65	–1.03	–0.57	–2.31	–1.28	–2.26	–1.26	–0.28	–0.16	–0.01	–0.01	–0.20	–0.11	–0.16	–0.09	–0.14	–0.08

Codes are CNT/Center, NE/North Edge, NM North Mid, and so on

(0.51% ± 0.22%), and the dew point in the carpeted east building was 0.48°F ± 0.07°F (0.27°C ± 0.039°C) higher.[3]

The interior temperature profile for a springtime floating period, March 13–24, 2015, shows a clear rise in interior temperature in the carpeted building, as shown in Figure 5. The temperature of the carpeted home averaged over 86°F (30.0°C) in the afternoon, whereas the exposed-slab home kept the peak average around 84°F (28.9°C). During such periods in spring while the ground is near its coolest temperatures the exposed slab would provide an advantage.

3. Uncertainties estimated considering stated sensor accuracy. For multiple measurement points, such as dew point, the estimate assumes stated accuracies of temperature and relative humidity measurement with errors propagated considering the root mean square error (RMSE) of the resulting error stream.

Thermal contours shown in Figure 6 for the March 13–24, 2015, period reveal that most of the free cooling during spring is actually coming from the slab edge at a depth down to approximately 2 ft (0.61 m). Of particular note, the deep ground temperature is now actually higher than that at shallow depths (as seen in the undisturbed ground temperature profiles). Also, the magnitude of the temperature differences seen from the slab surface down to the ground at a 10 ft (3.05 m) depth are modest—just over 6°F (3.3°C). This is quite different from the larger differences seen during the heating periods. Homes with carpeted slab floors in central Florida will tend to run slightly warmer during the floating unconditioned season than those with strongly earth-coupled floors.

Light Cooling Period Results

Periods of light cooling were defined as the 132 days where the average daily dew-point temperature was less than 70°F (21.1°C) but the building still required the air conditioning system

Table 3. Interior Conditions and Space Heating Energy Use for the 14 Days of Heating Below 55°F (12.8°C)

	East (Carpeted)	West (Exposed Slab)	Difference	% Difference	Standard Deviation	Standard Error	T-Stat (df = 313)	P-Value
Interior temperature, °F (°C)	74.99 (23.9)	74.49 (23.6)	0.50 (0.3)		1.27	0.07	6.95	<0.001
Interior relative humidity, %	39.16	39.20	–0.04		7.64	0.43	–0.09	0.927
Interior dew point, °F (°C)	47.86 (8.81)	47.45 (8.58)	0.42 (0.23)		2.55	0.14	2.89	0.004
Air-handling unit energy use, kWh/day	15.16	15.81	–0.65	–4.3%	2.65	0.15	–4.35	<0.001

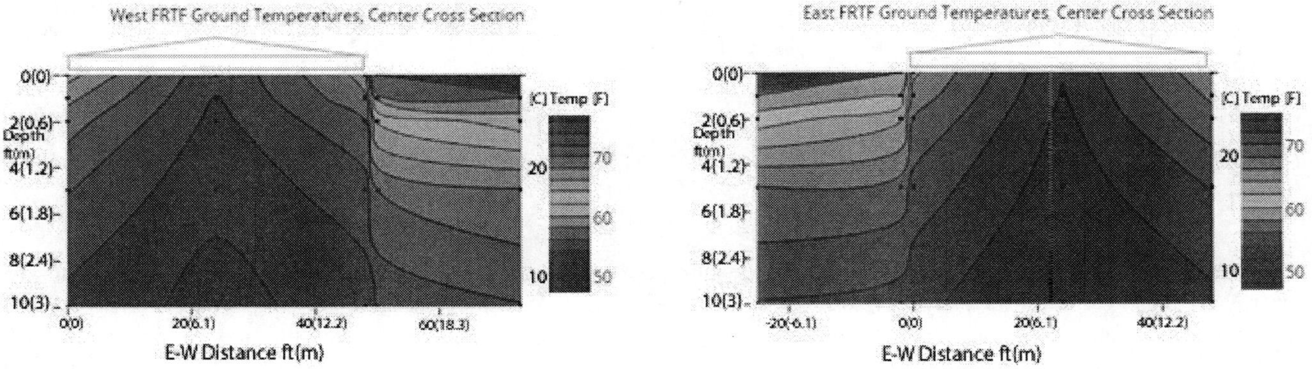

Figure 4 Thermal performance contours of slab and ground thermal performance during peak winter day, February 20, 2015. Note that the distance between houses has been shortened from 65 to 50 ft (19.8 to 15.2 m) to simplify display.

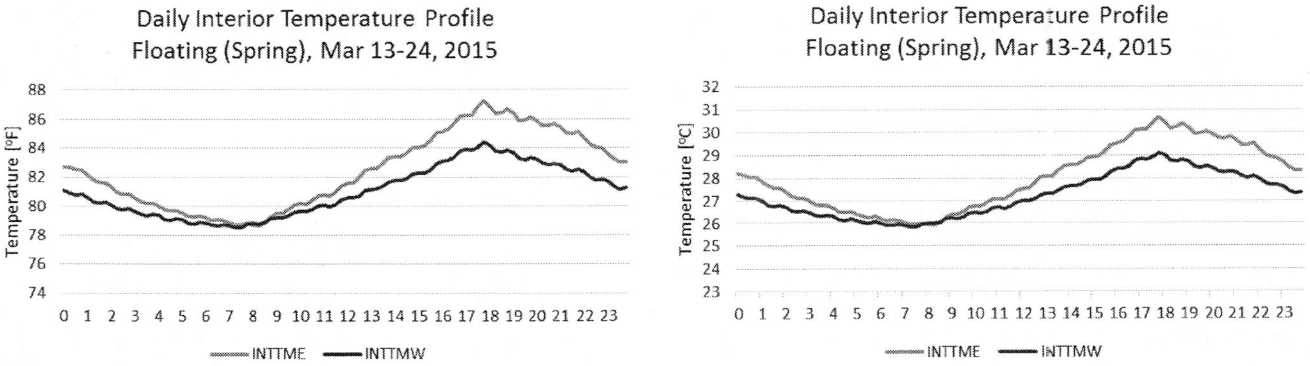

Figure 5 Interior temperatures while floating during a period of warm spring weather, March 13–24, 2015.

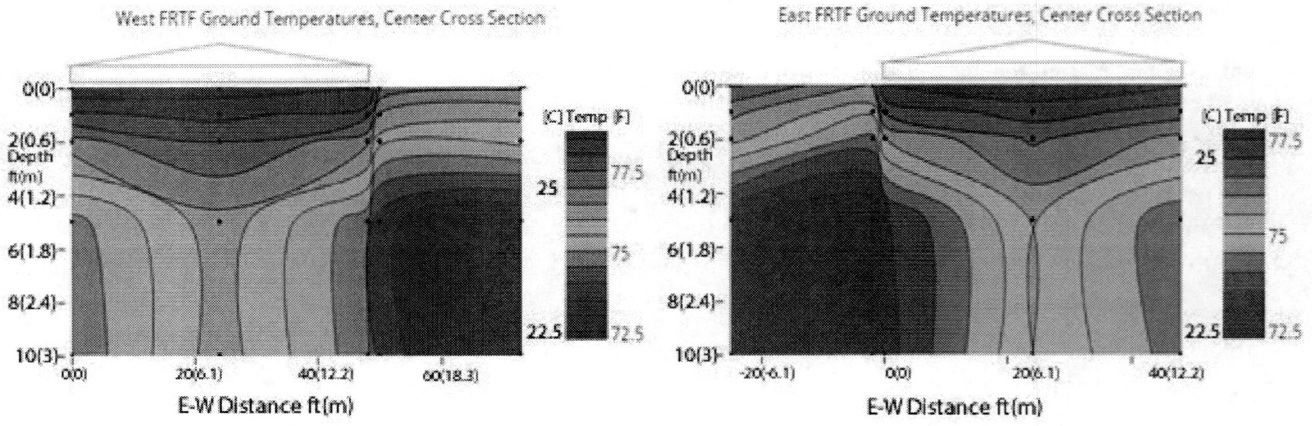

Figure 6 *Thermal performance contours during spring floating condition: March 13–24, 2015. Note the distance between houses has been shortened from 65 to 50 ft (19.8 to 15.2 m) to simplify display.*

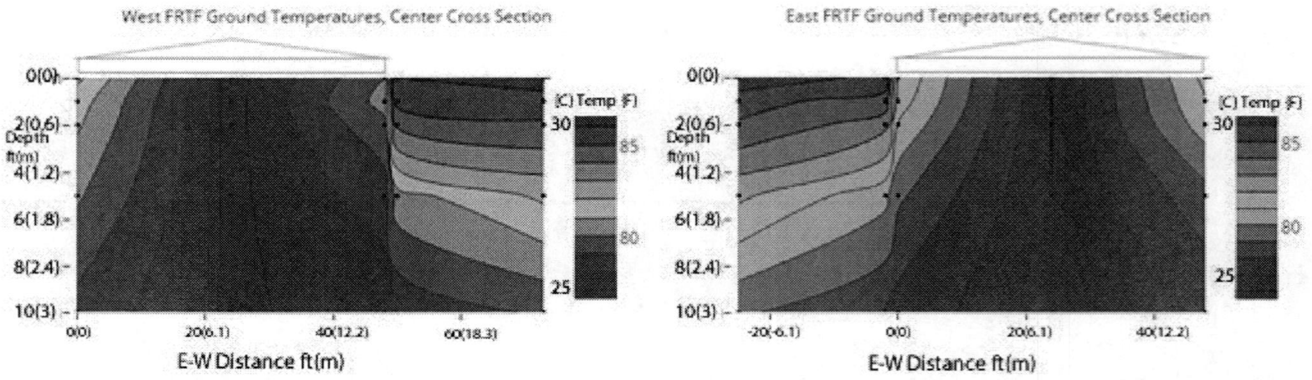

Figure 7 *Thermal performance contours during an extended period of heavy cooling: June 13–28, 2015. Note that the distance between houses has been shortened from 65 to 50 ft (19.8 to 15.2 m) to simplify display.*

to maintain the interior set point of 76°F (24.4°C). These cooling energy needs were typically low; however, the uncarpeted west building showed an average measured daily space cooling of 16.94 kWh against 16.64 kWh in the carpeted east building. This results in a small but statistically significant increase in energy use for the carpeted home (0.29±0.03 kWh, 1.7%).

Heavy Cooling Period Results

During days where the average dew-point temperature was greater than 70°F (21.1°C) (115 heavy cooling days), we found that the uncarpeted west building tended to run about 0.35°F (0.194°C) cooler than the carpeted east building, with a 0.43°F (0.239°C) lower dew-point temperature as well. Cooling energy use in the two homes was nearly identical (28.18 and 28.27 kWh for east and west, respectively) and showed no statistically significant difference. This tends to reinforce the idea that for summer conditions in Central Florida, the slab floor has only very minor influence on cooling and the center of the slab is largely adiabatic with the deep ground temperature condition and across the rest of the slab floor the temperature differences remain very modest.

During an extended maximum cooling period, June 13–28, 2015, where the ambient temperature averaged 81.7°F (27.6°C) compared to the overall heavy cooling period ambient temperature of 80.2°F (26.8°C), the carpeted building showed a significant decrease in daily cooling energy, 0.73±0.15 kWh (2.1%). This supports the idea that carpet helps to control slab edge heat gains under very hot weather conditions and was reinforced by time-lapse thermography taken of the interior of the uncarpeted west building over a series of cooling days that showed heat gain from the slab edge.

Thermal performance contours during the June 13–28, 2015, period (Figure 7) showed greatest thermal gains from the slab edge.

Comparison to Simulation Models

The FRTF measurements contradict the National Renewable Energy Laboratory (NREL) BEopt foundation model (https://beopt.nrel.gov/) using the Winkelmann (2002) approximations within EnergyPlus (DOE 2015). The measured heating savings of carpeted floors on heating does not show up in the simulation even though we would expect envelope differences that have the savings shown in the experiment to be represented in a building simulation result. Furthermore, the simulation predicts cooling savings for uncarpeted floors, and that was not observed. This same limitation existed with other software often used for residential analysis. For instance, EGUSA (http://www.energygauge.com/) shows the same behavior as BEopt, which not surprising since it also uses the Winkelmann foundation model as linked with DOE-2.1E (http://doe2.com/doe2).

Sensitivity to Soil Temperature Assumptions in BEopt

To explore whether the soil temperatures assumed were to blame for the poor correspondence between measurements and the BEopt results, we took the EnergyPlus files for the FRTF and edited the monthly ground temperature input file of the software. The conclusion of this exercise is that altering the ground temperatures in BEopt will not necessarily solve the heat transfer phenomenon observed. A more sophisticated analytical simulation method appears to be needed.

CONCLUSIONS

In experiments from 2014 to 2015, the impact of slab floors on space heating and cooling in the Central Florida climate of two extensively instrumented residential buildings was examined in detail. One building (east) had standard carpet and pad over the monolithic slab, while the other building (west) had unsurfaced concrete exposed. Both buildings were carefully configured to be otherwise nominally identical. A cooling set point of 77°F (25°C) and a heating set point of 73°F (22.8°C) were used in both buildings.

A fundamental finding is that uncarpeted floors have relatively minor thermal influence on building heating and cooling loads in Central Florida's climate under the temperatures examined. In particular, they show low influence on space cooling, with the center slab floor being nearly adiabatic with the deep ground temperature over the summer season. Carpeted floors were shown to have small (4.3%) energy-savings advantages over uncarpeted slab floors for space heating.

The most important caveat is that we would expect results to differ considerably for different interior temperature conditions, as the ground temperature and interior set points in Central Florida are very close. For instance, a household maintaining 80°F (26.7°C) inside during the cooling season would likely see an air-conditioning advantage from the uncovered slab floor. Similarly, occupants maintaining a cooling set point of 73°F (22.8°C) would likely find the carpeted floor to yield significant performance advantages during the cooling season. Because the floor is largely adiabatic in summer in Central Florida at a set point of 77°F (25°C), uncarpeted slabs may prove beneficial in summer for houses located in Georgia or other locations with lower ground temperatures.

Another finding is that the results are not readily replicable in typical software used for modeling residential building energy use. More empirical data in other climates should be collected and models should be refined to match the data.

ACKNOWLEDGMENTS

The authors thank the Florida Energy Systems Consortium and the State of Florida for the initial funding of the facilities. This experiment was funded by the Department of Energy (DOE) Building America program led by Eric Werling. The full report (Parker, et. al. 2015) on which this paper was based is available at the Building America website.

REFERENCES

Adjali, M.H., M. Davies, S.W. Rees, and J. Litter. 2000. Temperature in and under a slab-on-grade ground floor: Two and three dimensional simulations and comparison with experiment data. *Building and Environment* 35:655–52.

ASHRAE. 2009. Chapter 26, Table 1. In *ASHRAE handbook—Fundamentals*, p. 26.1. Atlanta: ASHRAE.

Bareither, H.D., A.D. Fleming, and B.E. Alberty. 1948. *Temperature and heat loss characteristics of concrete floors laid on the ground.* Report 48-1. Urbana, IL: University of Illinois, Small Homes Council.

DOE. 2015. *EnergyPlus*, ver. 8.1. Washington, DC: U.S. Department of Energy. https://energyplus.net/.

Parker, D., J. Kono, R. Vieira, and L. Gu. *Evaluation of the impact of slab foundation heat transfer on heating and cooling in Florida.* Final report for NREL Contract KNDJ-0-40339-05. Golden, CO: National Renewable Energy Laboratory.

Vieira, R., and J. Sherwin. 2012. *Flexible residential test facility instrumentation plan.* Report to the Building America Building Technologies Program. Washington, DC: U.S. Department of Energy, Office of Energy Efficiency and Renewable Energy.

Winkelmann, F. 2002. Underground surfaces: How to get a better underground surface heat transfer calculation in DOE-2.1E. *Building Energy Simulation User News* 23(6):19–26.

Characterizing Indoor Humidity for Comparison Studies: The Moisture Balance Approach

William B. Rose
Fellow ASHRAE

Stacy L. Gloss

ABSTRACT

If an intervention aims to reduce indoor humidity, how can the pre-intervention humidity be compared to the post-intervention humidity if these measurements are taken at different times under different outdoor conditions? For a set of houses, how can their interior humidities be compared if they are measured with different outdoor weather conditions? The approach discussed in this paper is the moisture balance approach, described and presented previously (Rose and Francisco 2004, 2010).

This approach was applied to homes in a weatherization and ventilation study in the U.S. Midwest. The results seem to show that weatherization provided a statistically significant lowering of indoor humidity and that there was a difference between the control and treatment groups. However, on close review, a bias was noted when measurements were taken during mild weather, with outdoor temperatures between 10°C and 15°C. Data with outdoor temperatures >15°C were excluded from this data review.

This paper explores the basis for the findings. It identifies alternative regression methods besides the method previously applied. It investigates the possible error with measurement periods of 1-week, 2-week, and 1-month measurement intervals. It closes with cautionary recommendations regarding methods for characterizing wetness with outdoor temperature corrections.

INTRODUCTION

In building science we often measure and monitor buildings for temperature and humidity. Temperature measurements serve purposes that are usually obvious. However, interpretation of humidity measurements is not at all straightforward. The primary purpose of this paper is to review the methods that allow someone analyzing temperature (T) and relative humidity (RH) data to draw reasonable conclusions regarding the wetness or dryness of the space as a function primarily of infiltration, ventilation, source strength, and occupancy.

Damp Indoor Spaces and Health (NAP 2004) finds associations between the dampness of the indoor environment and respiratory health. Table 3-1 of that report lists the dampness definitions in selected cross-sectional and analytic epidemiologic studies in which "dampness" was a key risk factor in health outcomes. There are 25 studies in this list; in 22 of these studies, dampness was assessed by self-reporting or by interview, and the remaining three were by surveyor assessment. Measured values of temperature and humidity were not part of any of the studies. We need to inquire—why are we bothering to measure T and RH in indoor spaces if we use subjective judgments to determine if spaces are humid or not? Can we not apply measured values and psychrometrics to address and compare space wetness?

Relative humidity is a temperature-dependent characterization of humidity in a space. A higher or lower RH measurement may indicate higher or lower humidity in the space, but it may also indicate a lower or higher temperature. For this reason, humidity analysis typically begins with a conversion of T and RH to some measure of absolute humidity. In this paper, vapor pressure (VP, in Pa) is the measure of absolute humidity adopted. Formulas to convert T and RH to VP are found in *ASHRAE Handbook—Fundamentals*, Chapter 1 (ASHRAE 2013), and elsewhere.

William B. Rose is a senior research architect and **Stacy L. Gloss** is a research specialist with Indoor Climate Research and Training, University of Illinois at Urbana-Champaign, Champaign, IL.

Outdoor air has vapor pressure. Indoor air VP would equal outdoor air VP except for the influences exerted by the building, indoor processes, foundation processes, and occupancy. For this paper we use dVP to refer to *vapor pressure excess* or *vapor pressure difference*, that is, the difference between indoor and outdoor vapor pressures. During winter, dVP in a house or occupied building is usually positive because of humidity generation in the space. During summer with air conditioning in operation, the dVP is likely to be negative.

For side-by-side houses we may easily answer the question of which house is drier and which is wetter by comparing the dVP for each house. But suppose instead we wish to determine if an intervention in a given house has led to drier or wetter conditions when the two measurement periods—pre-treatment and post-treatment—occur at different times with differing outdoor VPs. How does one normalize for outdoor conditions in making a judgment regarding indoor dampness?

The intent of this paper is to address the following questions:

- How do we interpret T and RH data taken at different measurement periods to draw a conclusion regarding wetness?
- Measurements of dVP are useful during cold weather to characterize the humidity effects of infiltration, ventilation, occupants, and other sources. They are not useful during warm weather. How and where do we draw the line?
- How well can sampling intervals of one week, two weeks, and one month characterize long-term humidity in a space?
- Are we ready to propose a method for characterizing humidity in a space?

ISO 13788

What is a typical dVP profile in a space as a function of outdoor temperature? This question is addressed by ISO 13788, but for a different purpose than the purpose of this paper. ISO 13788 (ISO 2012) sought to determine what should be the appropriate indoor humidity to use when conducting hygrothermal analysis of buildings. Their results are shown in Figure 1. We may imagine that at 20°C, homeowners keep their windows open, so the difference between indoor and outdoor vapor pressure is zero. Their window-opening diminishes linearly at lower temperatures, and at 0°C, the windows are closed. Below freezing, the windows stay closed and the dVP remains constant. Figure 1 shows four different profiles representing four climate classes—2, 3, 4, and 5. Higher classes represent higher indoor humidity than lower classes. What is characteristic of these profiles is that the defined profile of dVP as program input is undefined at outdoor temperatures above 20°C, is 0 at 20°C, rises to a value at outdoor 0°C and is horizontal below 0°C. With this profile, a single value—the value at 0°C—can define the entire profile. That single value is assigned the name *moisture balance*, or MB, in this paper. A higher MB represents a wetter space.

This description of dVP also provides no assistance for assigning indoor dVP at temperatures >20°C. For much of Northern Europe, in buildings without air conditioning, dVP may be presumed to be 0.

If dVP works as shown in ISO 13788, then the dogleg profile shown in the standard might be a satisfactory regression for actual data, delivering a single value to characterize the wetness of a space, in a way that uses all of the measured data. But, do actual dVP data follow the line shown in ISO 13788 (ISO 2012)? If they do, then the MB process would be considered a success, and measurements taken during a cold period and during a mild period (warm-weather data remain excluded) would provide similar values for MB.

Figure 2 shows a dogleg regression of data from one house in the study, prior to treatment. This regression line is constructed as follows:

1. For outdoor temperatures <0°C, the hourly dVP values are averaged (MB1) and the count of those values is taken.
2. For outdoor temperatures between 0°C and an upper allowable temperature (Figure 2), a sloped regression line that is fixed through 20°C is constructed. The intersection of that sloped line with the y-axis at 0°C is recorded (MB2) and the count of those values is taken.
3. A single MB value is derived from the values MB1 and MB2 weighted by the counts of each component. The intersection of the regression line with 0°C can then be used to characterize the house wetness with a single number.

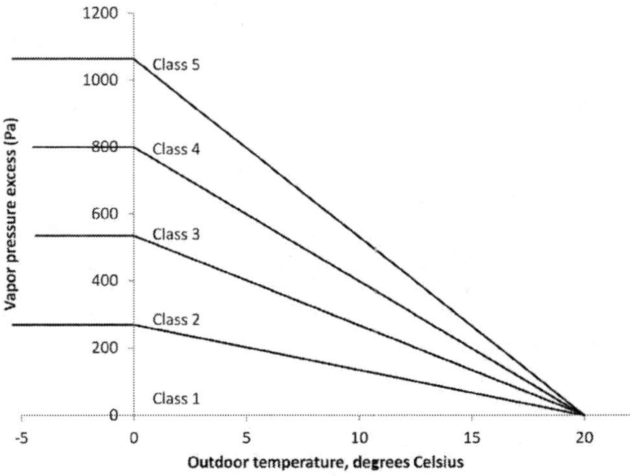

Figure 1 *Description of climate classes in ISO 13788, using vapor pressure excess plotted against outdoor temperature. "dVP" is an alternative expression for vapor pressure excess.*

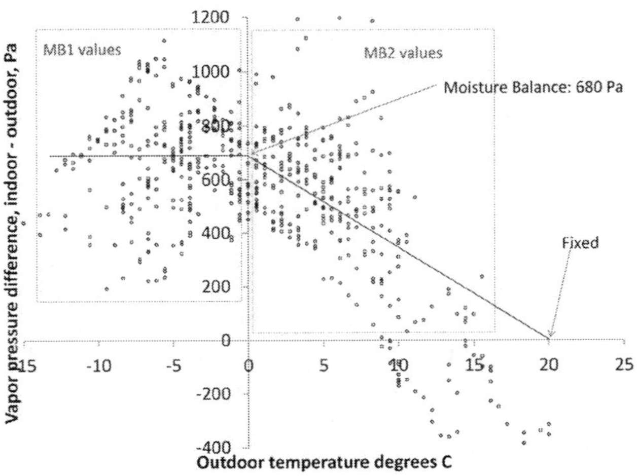

Figure 2 *Data from house IL-049 pre-treatment. A dogleg regression line is estimated, and Moisture Balance is the value of the regression at 0°C.*

WARM-WEATHER DATA EXCLUSION CRITERION

During seasons with air conditioning in operation, the dVP is strongly a function of air conditioning operation factors, including run time and coil temperature. This paper focuses on cold-weather performance. It excludes warm-weather data from the analysis. However, the question of identifying the criterion that distinguishes acceptable from non-acceptable data must be addressed.

In previous work by Rose and Francisco (2004, 2010), measured data with outdoor temperature >20°C were excluded. In more recent work (Rose et al. 2015), the data exclusion criteria applied were as follows: 1) all data where outdoor temperature was >20°C, and 2) all data where 16 or more hours of the most recent 24 hours show outdoor temperature >20°C. The reason for the latter exclusion is predicated on the assumption that if air conditioning is in operation, there will be an artificial lowering of moisture balance, even if some of the hours (nighttime, for example) have outdoor temperatures < 20°C.

In this paper, data are excluded from MB calculation when the hourly outdoor temperature meets the "too warm" criterion of >15°C. In addition, MB calculations are not made when more than half the hourly outdoor temperatures are >15°C.

MEASURED DATA

Initial Study

The study by Rose et al. (2015) was done to characterize indoor conditions, including temperature and humidity, before and after weatherization. The weatherization included applying two different ventilation standards—ASHRAE Standard 62-1989 (ASHRAE 1989) and ASHRAE Standard 62.2-2010 (ASHRAE 2010). The study was conducted in Chicago, Illinois, and in various cities and towns in Indiana. Moisture results were published for 51 temperature (*T*) and relative humidity (RH) sensors located in 32 of the homes. The hourly measured *T* and RH data were used to compute VPi (indoor vapor pressure). The outdoor temperature and humidity conditions were taken from local airports; Chicago Midway International Airport data were used for all of the Chicago homes. The results appeared to show drying post-weatherization, with a stronger and more significant drying effect with ASHRAE Standard 62.2 ventilation compared to ASHRAE Standard 62-1989 ventilation.

Given that the moisture balance analysis is not a standard method of assessing wetness of a space from measured data (there is no standard method), the Chicago portion of the data from the Rose et al. (2015) study was subjected to a review for purposes of this paper. This paper aims to assess the reliability of assigning a single value from the regression to characterize the wetness of the space. All data used in this analysis are from the data set developed for Rose et al. (2015).

House Studies

Data from two houses were used to investigate the reliability of MB analysis. Results from the first house, identified as IL-043F, are presented in Figures 3 and 4. The pre-treatment monitoring period was from 11/14/2013 to 12/12/2013. The post-treatment monitoring period extended from 12/12/2013 to 6/23/2014. The hourly samples of dVP are shown in Figure 3, plotted against outdoor temperature.

What is evident in Figure 3 is that the data show a general shape similar to that shown in Figure 2, with the higher values of dVP occurring at cold temperatures. Also evident is the fact that the dVP values have a far wider distribution at warm temperatures. Can a regression of these two data sets be used to determine if the treatment resulted in dryness improvements following treatment? And can data taken during warm or mild weather predict the cold-weather performance of the building? How should one normalize data where different slopes and shapes of the data set are seen in the two pre- and post-treatment cases?

Moisture balance was calculated using data sets of one-week, two-week, and one-month durations, to represent likely measurement periods. The MB calculation was conducted as described previously (dogleg regression), with all dVP values excluded when outdoor temperature was >15°C. The MB was calculated for each hour of the study, with the hour value taken at the middle of the averaging interval. The results are shown in a timeline in Figure 4.

Figure 4 shows measurement over winter 2013–2014 into spring 2014. During the coldest periods, the MB values correspond closely to the average dVP values. This is to be expected with all dVP values at temperatures below 0°C being averaged. During the mild weather of April and May, the MB values climb and no longer represent the average MB values taken during colder temperatures. It is the aim of the method to provide somewhat the same value for MB regardless of

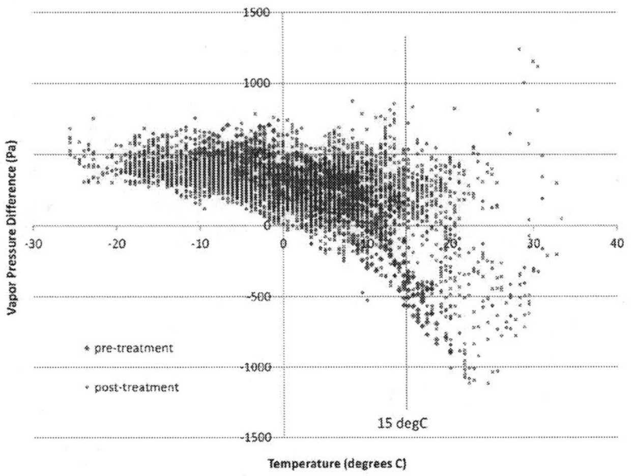

Figure 3 *Values of dVP for house IL-043F. Large blue points represent pre-treatment dVP values, and red points represent post-treatment values. Values of dVP are plotted against outdoor temperature.*

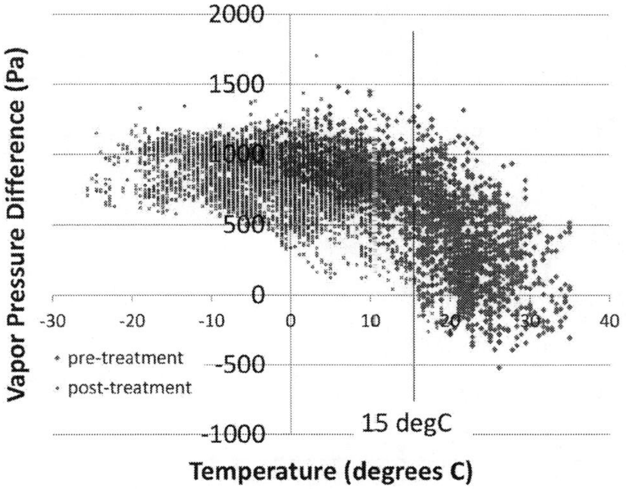

Figure 5 *Hourly sample dVP values plotted against outdoor temperature for both pre- and post- treatment for house IL-029.*

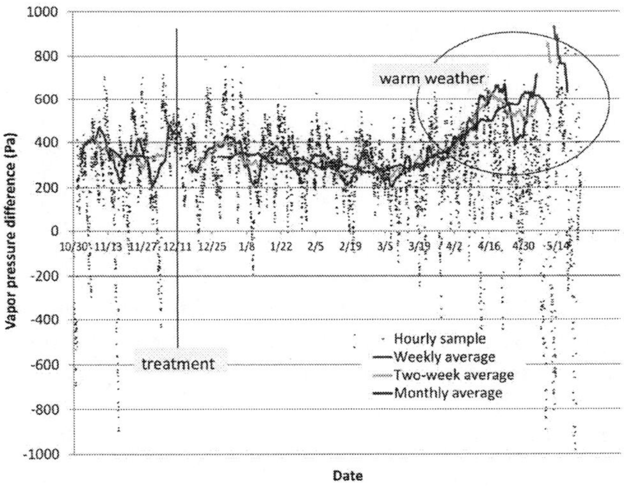

Figure 4 *Timeline of house IL-043F. One-week, two-week, and one-month averages of the sample data are shown.*

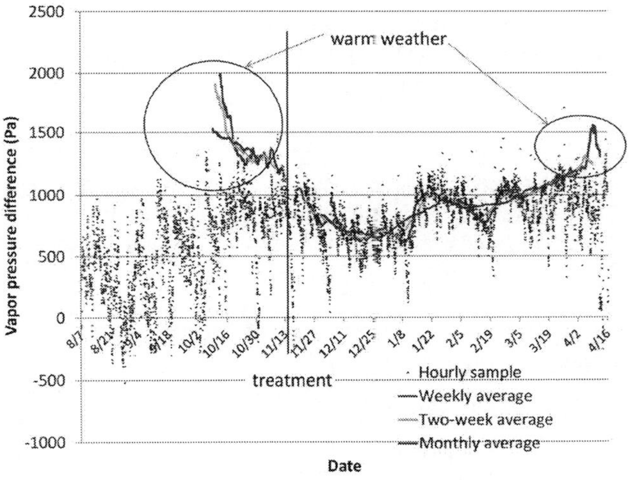

Figure 6 *Timeline with dVP and MB values plotted, as well as running averages of MB, for house IL-029.*

whether the measurements are taken during cold weather or mild weather. This rise in MB values during mild weather undermines the utility of the method; it suggests that even mild-weather monitoring periods may do a poor job of characterizing the cold-weather MB.

Data from a second house (IL-029), with different measurement periods, show generally the same outcome (see Figures 5 and 6). For this house, the pre-treatment measurement period was during the mild weather of fall. Once again, the MB calculated values are high and are not consistent with the cold-weather MB values.

It is clear that mild-weather MB values should not be used in comparison with cold-weather values. The reason for the disparity seems to be this: the MB dogleg regression has two parts. The cold-weather part uses the average of all dVP values, and the intersection of this average value with the y-axis is a one-to-one correspondence. The mild-weather part creates a sloped regression line that is fixed through 20°C. A cluster of MB data points far from the 0°C axis (warmer than freezing) will lead to possible varying slopes of the regression line. The effect of the varying slopes will have a magnified effect on the intersection of that sloped line with the y-axis.

SHORT-TERM MONITORING TO REPRESENT LONG-TERM CONDITIONS

Short-term monitoring campaigns are valuable if they satisfactorily represent long-term conditions. How well do

Table 1. Moisture Balance Statistics for Seven Houses

	All	Pre	Post
MB total	494	568	451
One-week average	497	583	475
One-week SD	167	135	145
Two-week average	483	556	469
Two-week SD	132	82	121
One-month average	481	543	469
One-month SD	125	47	113

measurement terms of one week, two weeks, and one month represent long-term values?

Moisture balance was calculated for each hour of the study. The ranges for the MB calculation were one week, two weeks, and one month, centered on the reported hour. The results for seven houses are shown in Table 1. The means and standard deviation (SD) for the three measurement terms, and for all data, pre-treatment data and post-treatment data, were compiled.

What appears in Table 1 is that the means of the different sampling intervals do not deviate strongly from the overall MB. There is a wide variation as represented by the high values for standard deviation. As expected, standard deviation declines with longer averaging interval.

ALTERNATIVE MOISTURE BALANCE CALCULATIONS

If the ISO 13788-based method of calculating MB gives results that are biased high at mid-temperatures, how should the project of characterizing wetness in a space proceed? Note the assumptions in the ISO 13788-based approach: 1) at temperatures <0°C, the dVP values are averaged, and 2) at temperatures >0°C, the linear regression of the values is sloped, passing through 0MB at 20°C. Both of these assumptions pertain to the slope of the data regression. Judging roughly by the data patterns in Figures 3 and 5, the two assumptions do not appear out of line—the data itself takes a rough dogleg shape.

Further study was undertaken to determine the slopes of the two regression segments. This was done for 45 houses in the study. The results are given in the appendix. The four lines in each chart (one chart per house) represent the linear regression lines for four sample data sets:

- Pre-treatment, outdoor temperatures <0°C (blue)
- Pre-treatment, outdoor temperatures ≥0°C and ≤15°C (red)
- Post-treatment, outdoor temperatures <0°C (green)
- Post-treatment, outdoor temperatures ≥0°C and ≤15°C (purple)

Missing lines indicate no data points meeting those conditions. It is apparent that the slopes vary widely by house. They also represent differing numbers of hourly data points.

The "average" of all the slopes was calculated, using the averages for temperature and dVP. The results are

- for outdoor temperatures <0°C, average slope is –5.4 Pa/K, and
- for outdoor temperatures ≥0°C and ≤15°C, average slope is –11 Pa/K.

We may suppose that a regression using the slopes given above for the two data segments may lead to improved predictive capability—that is, that similar values for wetness of a space may be attained whether the monitoring period was in the coldest weather or in mild weather. It is impossible to claim, from this limited data set, that a regression of all data to the slopes given above would lead to improved predictive capability of the model.

A further approach would be to further limit the data to be used in calculating wetness of the space. It may be that only data taken at outdoor temperatures below 10°C can be used to determine wetness with a greater degree of replicability between the coldest periods and the mildest periods. However, this speculation was not tested in this analysis.

CONCLUSIONS

This paper seeks to address how to characterize the wetness (or dampness or dryness) of a space by analyzing data of indoor temperature and humidity. The starting point for the analysis is the application of a vapor pressure excess profile presented in ISO 13788 (ISO 2012) to calculate moisture balance. Warm-weather data must be excluded from the analysis, because vapor pressure excess at high temperatures in the United States is driven primarily by use of air conditioning, which gives information about equipment operation, not about cold-weather wetness of the space.

The MB approach has been used in previous research. In the previous studies, the data cutoff was 20°C. It became apparent in reviewing the previous data that that cutoff was too high—it included much data that skewed the results. In this paper a cutoff of 15°C was used.

Close analysis of the resulting data shows that mild-weather estimates of MB taken with a 15°C cutoff do not correspond to the MB of the same house for measurements during cold weather.

Estimates of the standard deviation for MB measurements taken at one-week, two-week, and one-month intervals are presented. Of course, longer terms of measurement show less variance in the calculations.

This paper introduces a caution in the use of the MB method for data where the outdoor temperature is >10°C. Improved methods that achieve greater correspondence between MB results for the same house at cold temperatures and at mild temperatures must await further investigation.

REFERENCES

ASHRAE. 1989. ANSI/ASHRAE Standard 62-1989, *Ventilation for acceptable indoor air quality*. Atlanta: ASHRAE.

ASHRAE. 2010. ANSI/ASHRAE Standard 62.2-2010, *Ventilation and acceptable indoor air quality in low-rise residential buildings*. Atlanta: ASHRAE.

ASHRAE. 2013. Chapter 1. In *ASHRAE Handbook—Fundamentals*. Atlanta: ASHRAE.

ISO. 2012. ISO 13788:2012, *Hygrothermal performance of building components and building elements—Internal surface temperature to avoid critical surface humidity and interstitial condensation—Calculation methods.* Geneva: International Organization for Standardization.

NAP. 2004. *Damp indoor spaces and health.* Washington, DC: National Academies Press, Institute of Medicine of the National Academies, Committee on Damp Indoor Spaces and Health, Board on Health Promotion and Disease Prevention.

Rose, W., and P. Francisco. 2004. Characterization of wetness in Providence RI homes. *Proceedings of the Performance of the Exterior Envelopes of Whole Buildings IX Conference* [CD].

Rose, W.B., and P.W. Francisco. 2010. Temperature and humidity measurements in 71 homes participating in an IAQ improvement program. *Proceedings of the Performance of the Exterior Envelopes of Whole Buildings XI Conference* [CD].

Rose, W., S. Gloss, P. Francisco, Z. Merrin, S. Cali, and D. Jacobs. 2015. Weatherization ventilation strategy as a factor in indoor humidity. Presented at Healthy Buildings 2015 America, Boulder, CO, July 19–22.

APPENDIX

house	slope	interce	count	
IL-001	-6.41	492	80	low pre- (blue)
	-12.20	446	1110	hi pre- (red)
	-5.72	369	1607	low post- (green)
	-10.19	378	2642	hi post- (purple)
IL-029	-3.23	992	64	
	-21.84	1040	882	
	-5.71	832	2283	
	-11.40	802	1292	
IL-043F	-10.39	391	385	
	-34.89	401	601	
	-9.12	281	1951	
	0.39	241	1513	
IL-043	-11.35	518	385	low pre- (blue)
	-36.05	544	601	hi pre- (red)
	-8.09	400	1951	low post- (green)
	-1.12	367	1513	hi post- (purple)
IL-049	3.09	555	879	
	-29.92	563	463	
	-6.74	325	1404	
	-0.22	267	1439	
IL-052F	3.84	436	914	
	-14.84	374	439	
	-0.59	342	1360	
	3.55	343	1362	
IL-053	-6.92	151	793	low pre- (blue)
	-19.60	91	359	hi pre- (red)
	-4.97	134	1361	low post- (green)
	-0.08	101	1387	hi post- (purple)
IL-41L	0.00	0	0	
	4.45	579	229	
	-3.58	515	2347	
	-11.52	533	1934	
IL-41F	0.00	0	0	
	-20.26	644	229	
	-6.36	305	2347	
	-11.59	314	1934	

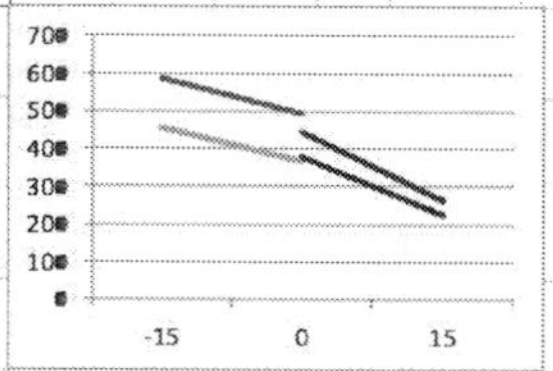

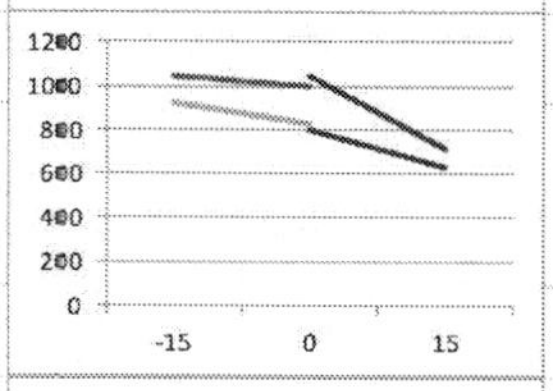

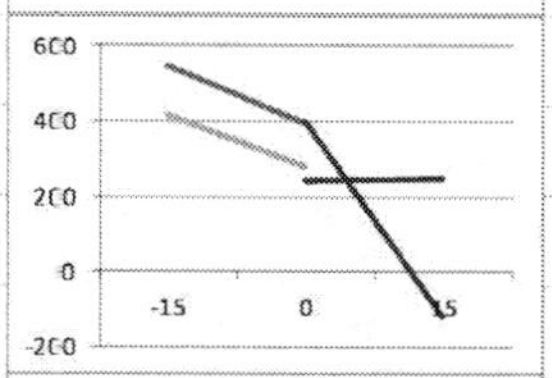

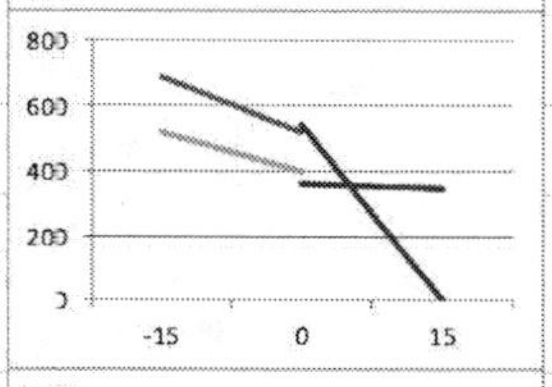

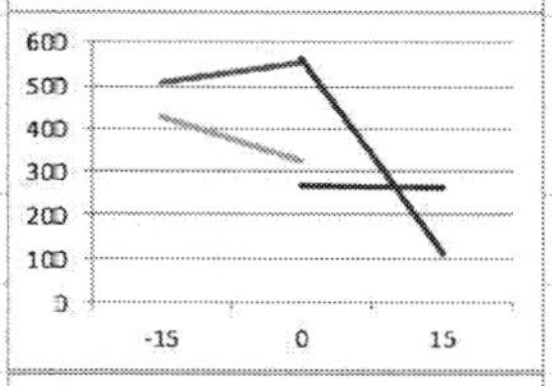

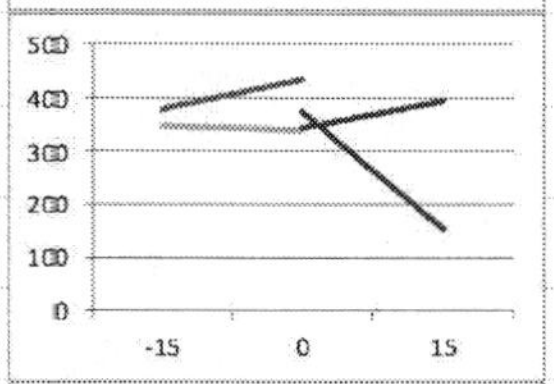

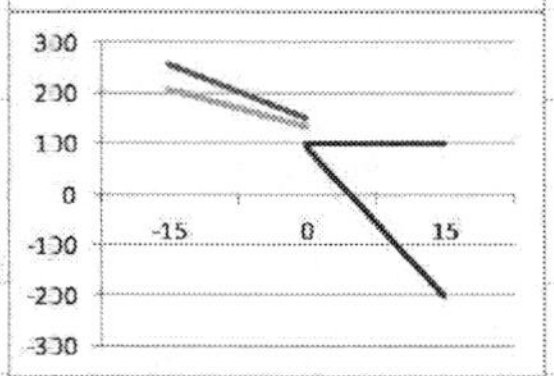

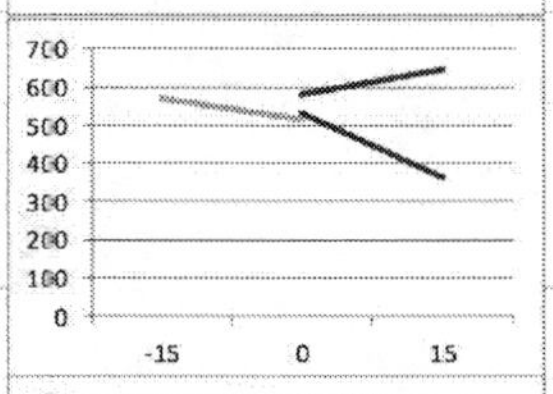

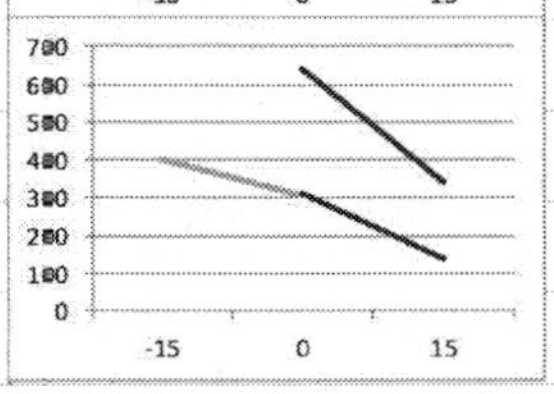

house	slope	interce	count	
IL-040	-30.54	819	73	low pre- (blue)
	-28.05	891	966	hi pre- (red)
	-10.79	869	2274	low post- (green)
	-15.95	769	1719	hi post- (purple)
IL-040F	-14.64	680	73	
	-25.08	735	966	
	-9.24	606	2274	
	-12.11	532	1719	
IL-038	-18.84	824	75	
	-41.25	928	1015	
	-8.72	615	2272	
	-26.02	555	1672	
IL-38F	-2.65	409	75	low pre- (blue)
	-17.52	406	1016	hi pre- (red)
	-6.50	185	2272	low post- (green)
	-5.96	148	1672	hi post- (purple)
IL-037	0.00	0	0	
	-29.77	500	53	
	-10.22	254	2326	
	-10.81	222	1868	
IL-032	9.84	1163	73	
	-50.75	1291	993	
	-15.80	891	2253	
	-22.73	847	944	
IL-004L	20.78	753	19	low pre- (blue)
	-34.23	893	1186	hi pre- (red)
	-3.58	518	826	low post- (green)
	-17.26	555	2002	hi post- (purple)
IL-004F	28.30	509	19	
	-17.75	572	1193	
	-3.88	502	826	
	-13.23	554	1995	
IL-005F	14.55	621	19	
	-25.13	640	1027	
	5.83	428	1284	
	-11.47	434	2336	

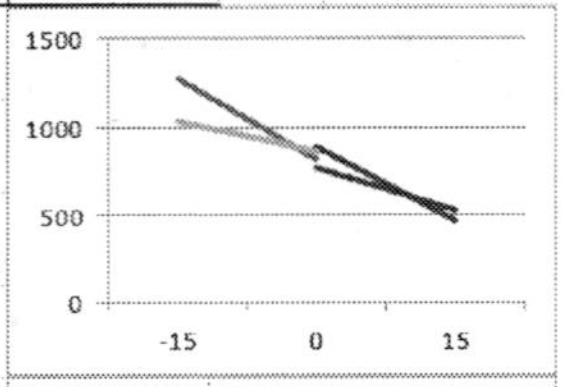

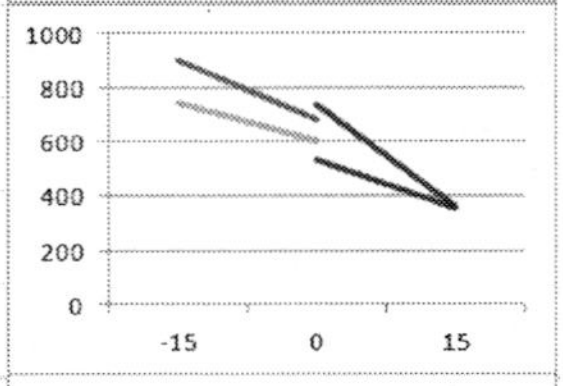

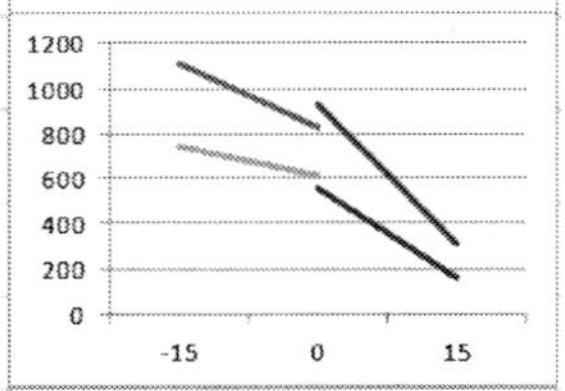

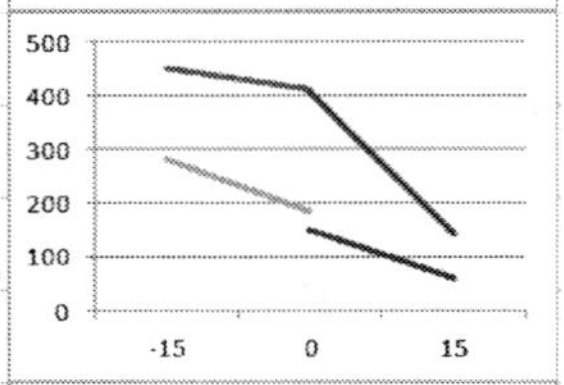

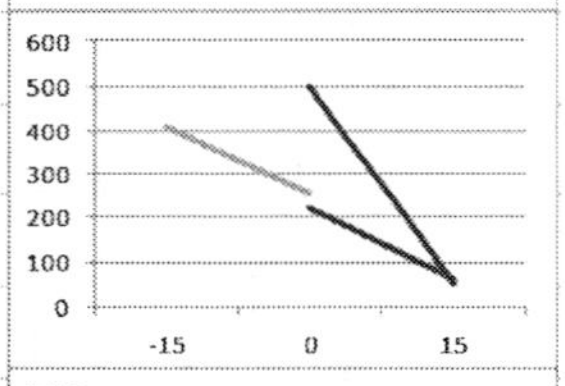

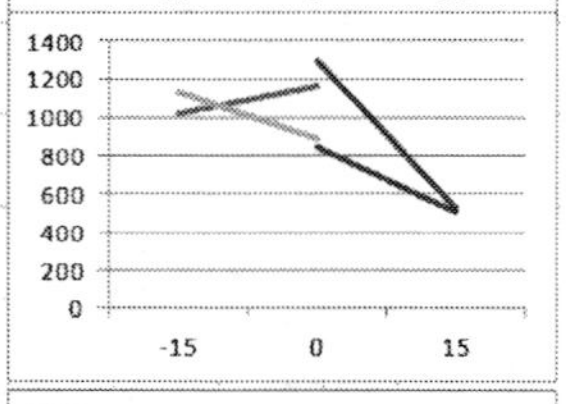

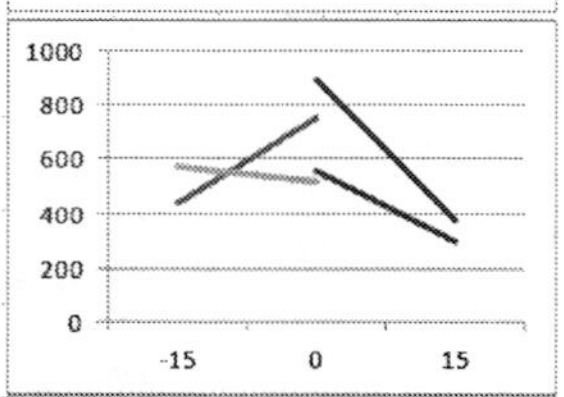

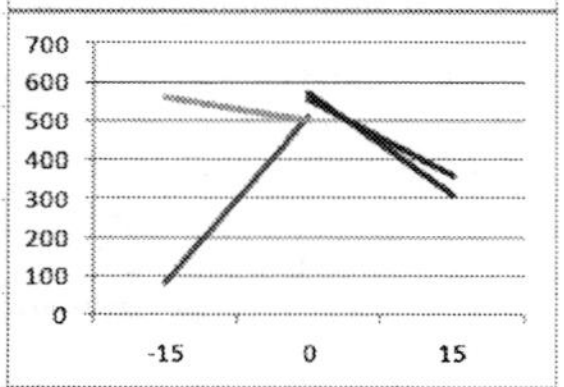

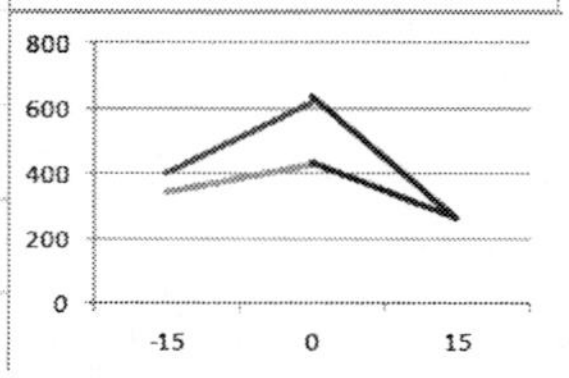

house	slope	interce	count	
IL-005L	16.07	642	19	low pre- (blue)
	-27.01	725	1027	hi pre- (red)
	4.23	388	1284	low post- (green)
	-9.19	399	2336	hi post- (purple)
IL-008L	0.00	0	0	
	-14.63	523	1090	
	-1.40	473	1669	
	-6.51	480	3610	
IL-009L	0.00	0	0	
	-32.61	726	811	
	-15.63	518	1607	
	-9.46	574	2719	
IL-010L	0.00	0	0	low pre- (blue)
	-17.82	606	733	hi pre- (red)
	3.92	301	1284	low post- (green)
	-1.90	336	2538	hi post- (purple)
IL-010F	0.00	0	0	
	-23.09	501	733	
	4.50	304	1284	
	-1.01	340	2538	
IL-013L	0.00	0	0	
	-15.44	610	998	
	-13.01	620	1607	
	-9.92	683	2440	
IL-014L	0.00	0	0	low pre- (blue)
	-24.97	744	756	hi pre- (red)
	-14.49	381	1607	low post- (green)
	-4.66	380	2680	hi post- (purple)
IL-014F	0.00	0	0	
	-28.92	878	756	
	-18.24	526	1607	
	-0.72	532	2680	
IL_015_L	0.00	0	0	
	-22.06	602	894	
	7.80	395	1284	
	-7.68	446	2277	

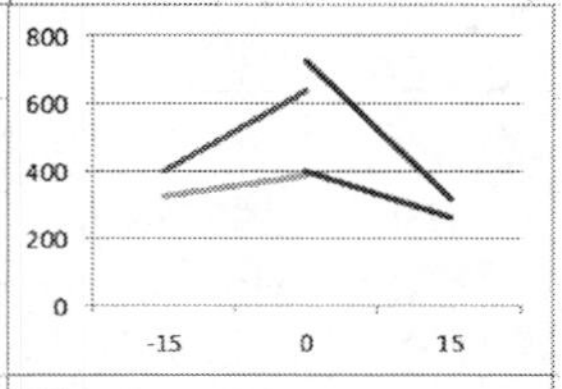

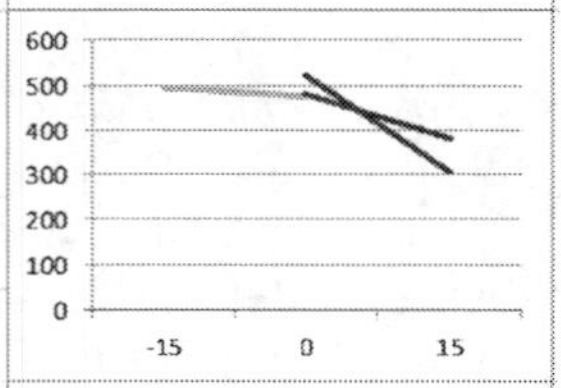

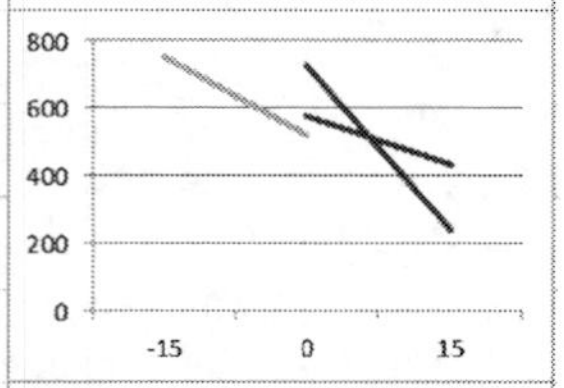

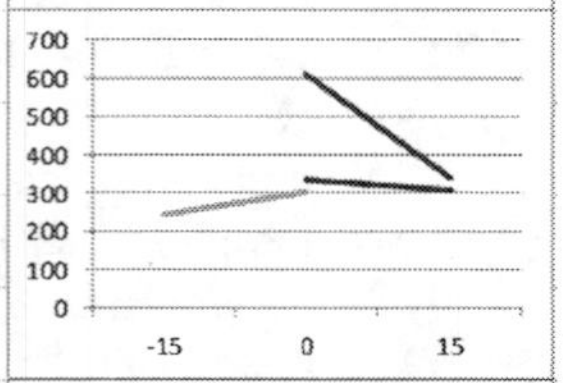

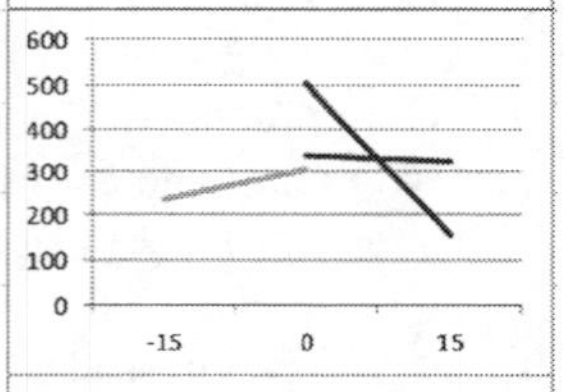

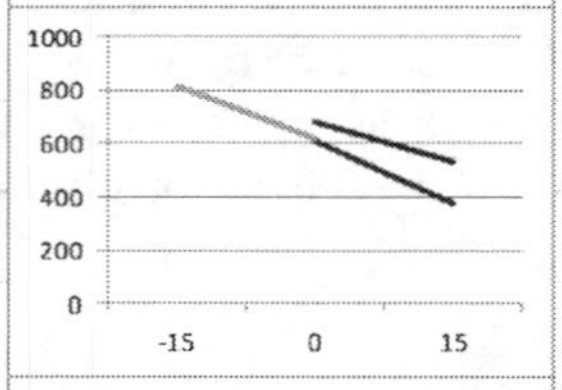

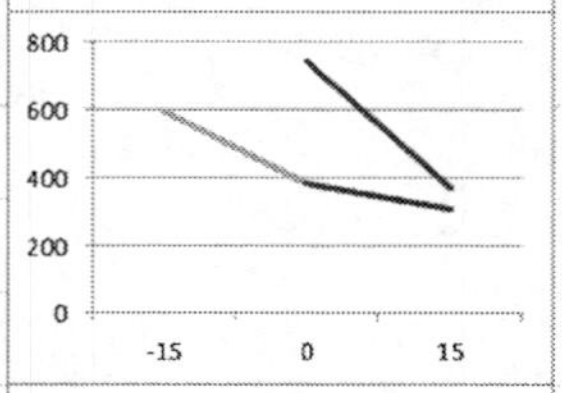

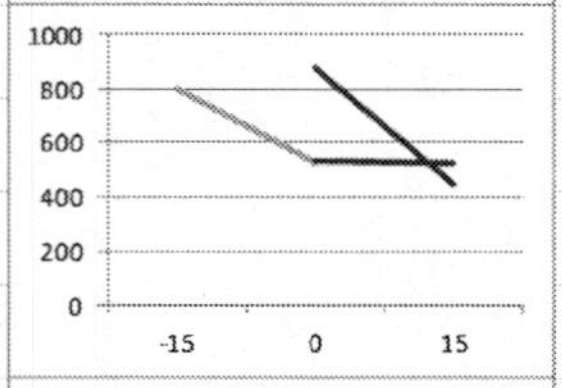

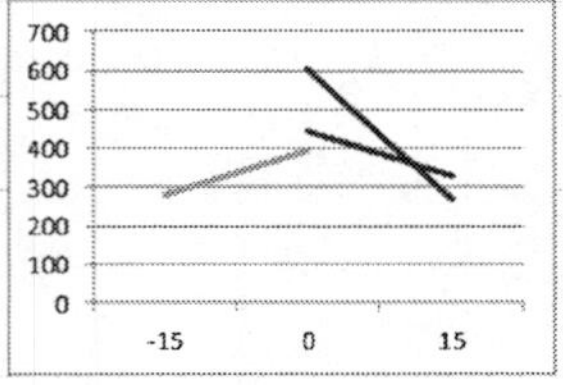

house	slope	interce	count		
IL_015_F	0.00	0	0	low pre- (blue)	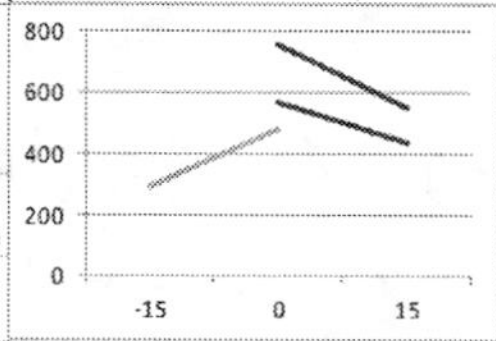
	-13.76	755	894	hi pre- (red)	
	12.59	482	1284	low post- (green)	
	-8.62	567	2277	hi post- (purple)	
IL_016_L	0.00	0	0		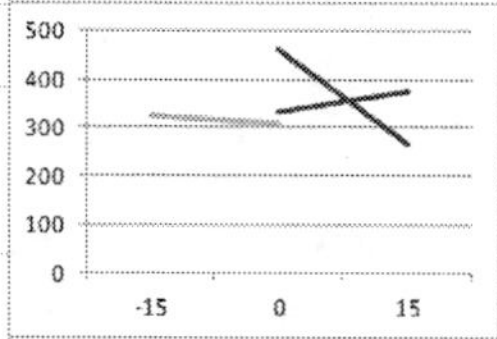
	-12.81	460	979		
	-0.94	311	1669		
	2.92	333	3620		
IL_018_L	0.00	0	0		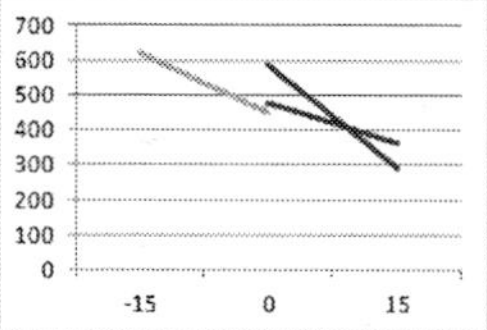
	-19.83	587	872		
	-11.29	452	1607		
	-7.75	476	2536		
IL_018_F	0.00	0	0	low pre- (blue)	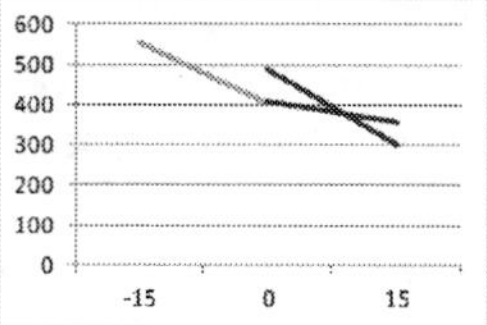
	-12.62	490	872	hi pre- (red)	
	-10.09	405	1607	low post- (green)	
	-3.41	410	2536	hi post- (purple)	
IL_019_L	0.00	0	0		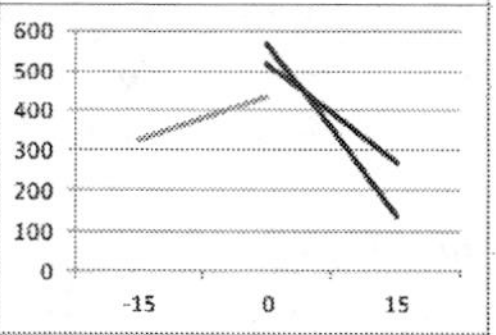
	-28.57	567	851		
	7.20	432	1284		
	-16.47	516	2192		
IL_019_F	0.00	0	0		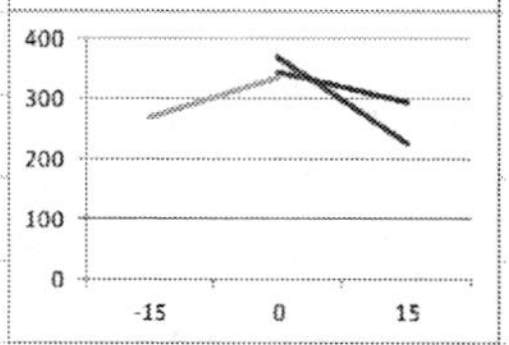
	-9.70	370	851		
	4.38	335	1284		
	-3.37	344	2192		
IL-020_L	0.00	0	0		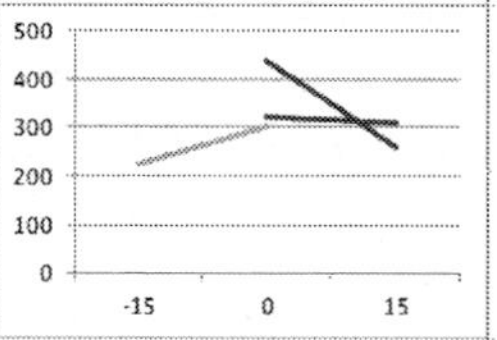
	-12.21	440	702		
	5.12	302	1284		
	-0.89	323	2322		
IL_021_L	0.00	0	0	low pre- (blue)	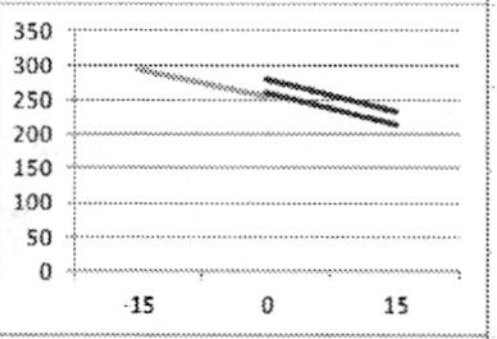
	-2.94	278	722	hi pre- (red)	
	-2.64	254	1607	low post- (green)	
	-3.15	260	2536	hi post- (purple)	
IL_022_F	0.00	0	0		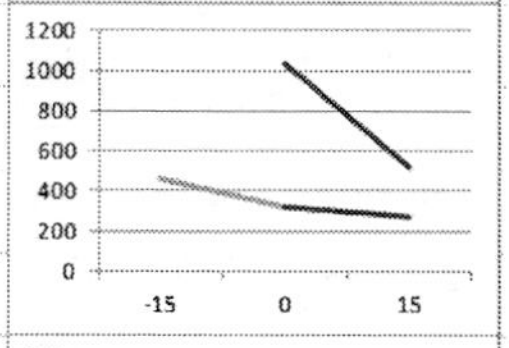
	-34.80	1037	361		
	-9.29	317	1607		
	-2.62	317	2535		
IL_023_F	0.00	0	0		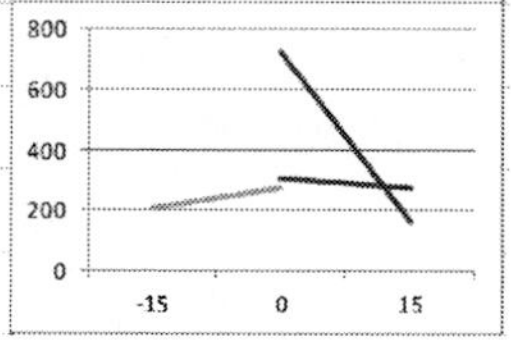
	-37.63	728	210		
	4.78	278	1284		
	-1.80	306	2277		

house	slope	interce	count		
IL_024_L	0.00	0	0	low pre- (blue)	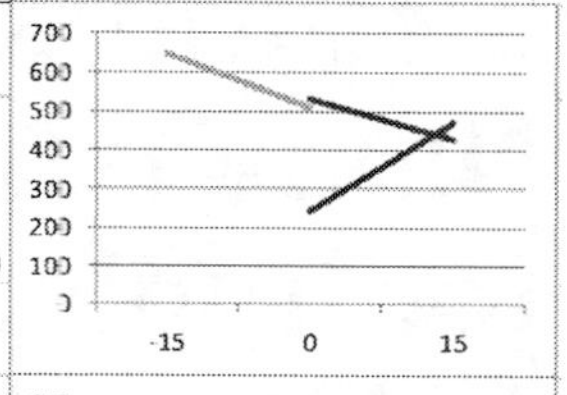
	15.29	242	166	hi pre- (red)	
	-9.16	511	1607	low post- (green)	
	-6.60	529	2469	hi post- (purple)	
IL_024_F	0.00	0	0		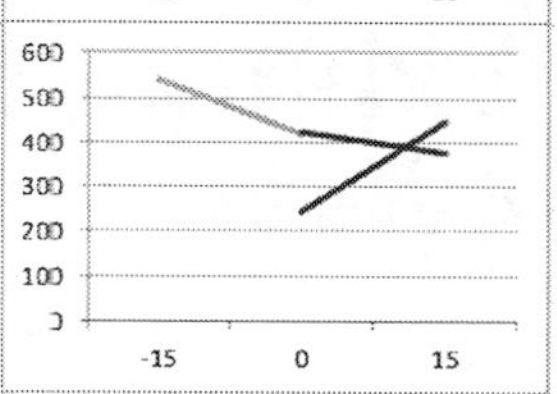
	13.36	245	166		
	-8.26	416	1607		
	-2.92	422	2469		
IL_025_L	0.00	0	0		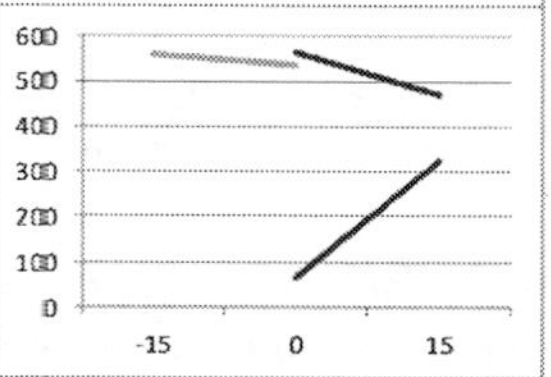
	17.27	65	136		
	-1.50	538	1607		
	-6.26	562	2451		
IL_025_F	0.00	0	0	low pre- (blue)	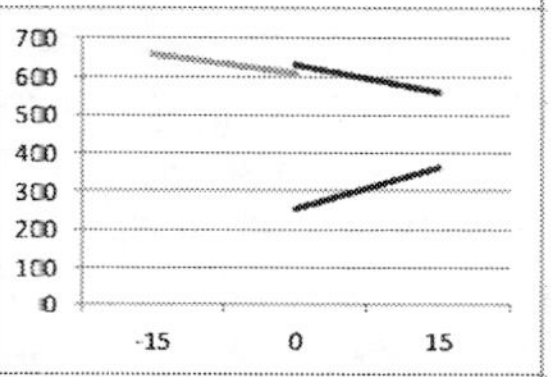
	7.41	250	136	hi pre- (red)	
	-3.37	608	1607	low post- (green)	
	-4.74	631	2451	hi post- (purple)	
IL_026_L	0.00	0	0		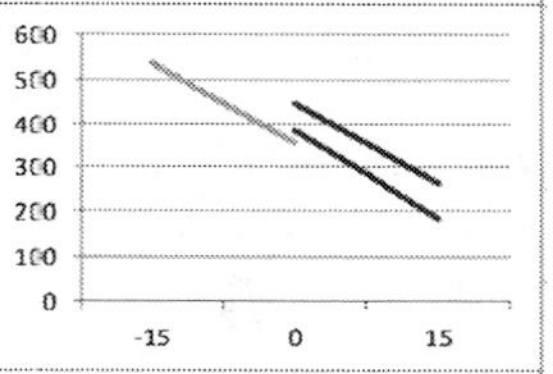
	-12.17	448	192		
	-11.96	357	1607		
	-13.37	385	2258		
IL_026_F	0.00	0	0		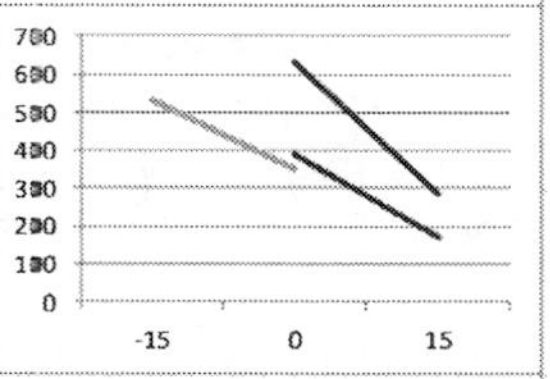
	-22.83	630	192		
	-11.78	354	1607		
	-14.82	392	2259		
IL_027_L	0.00	0	0	low pre- (blue)	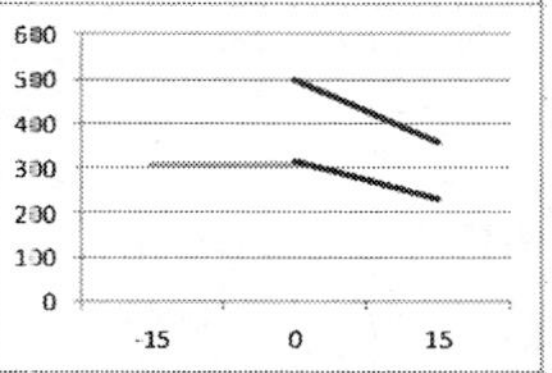
	-9.46	498	566	hi pre- (red)	
	-0.28	304	1669	low post- (green)	
	-5.70	317	3054	hi post- (purple)	
IL-028_L	0.00	0	0		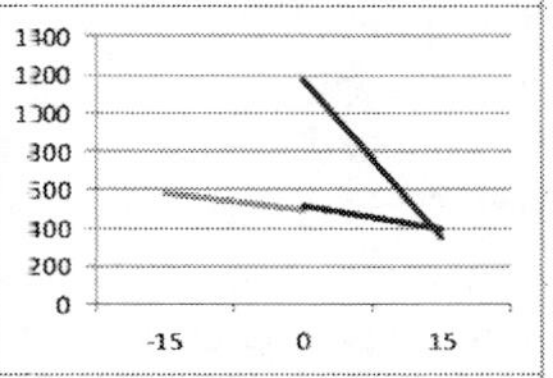
	-54.74	1178	356		
	-5.95	490	1669		
	-8.26	515	3264		
IL_028_F	0.00	0	0		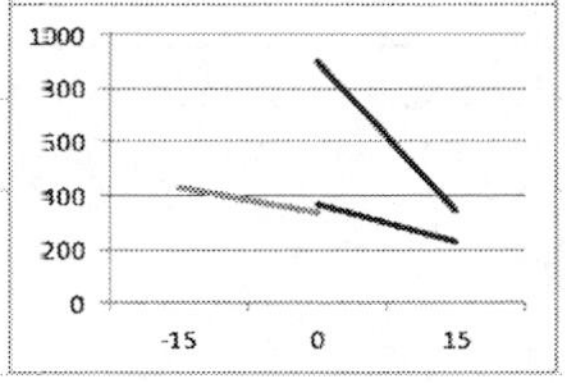
	-37.11	901	356		
	-6.42	337	1669		
	-9.27	370	3264		

Method to Evaluate and Develop Next-Generation Vacuum Insulation Panels for Implementation in the Retrofit of Existing Building Envelopes

Kyle Vansice

Ulrike Passe, Dipl.-Ing.
Associate Member ASHRAE

ABSTRACT

Vacuum insulation panels (VIPs) at the exterior envelope have the potential to significantly reduce whole-building thermal energy consumption in retrofit applications. Case studies of VIP implementation exist; however, the current state of the art fails to properly optimize their use and integration into the exterior wall assembly and understand their impact on whole-building energy consumption over the lifetime of a building or retrofit solution.

This research project addresses current vacuum insulation limitations and utilizes whole-building energy analysis, optimization, and the development of construction techniques as a means for providing best practices for the implementation of low-cost VIPs. Using an innovative new core material composition developed within our research team and in tandem with a manufacturing partner, the research is developed in a way that is congruent with current construction methods but that is also disruptive to current practices for the design and construction of building envelopes.

To capture retrofit potential, an existing U.S. Department of Energy reference building was analyzed as an energy consumption baseline utilizing EnergyPlus within a parametric software platform. Models for whole-building energy and hygrothermal transfer will be developed to include thermal conductivity, thermal bridging, and moisture/airflow within the envelope assembly. Coupling energy models with a parametric platform will facilitate inclusion of manufacturing and construction processes within design as well as allow for the optimization of cost, panel size, assembly, and energy-efficiency variables. Appropriate envelope assemblies will emerge from this research, which incorporate VIPs as a novel, highly energy-efficient construction method for existing building envelopes.

The findings of the paper suggest that significant thermal energy savings can be achieved through VIP retrofitting. Yet limitation such as barrier film properties and pressurization affecting the fidelity of the vacuum itself as well as thermal bridging undermining the overall thermal resistance of the panel as a unit could potentially reduce the cost viability of the panel versus more conventional retrofit methods. Analysis and development are required in these areas in order to ensure that VIPs are cost-effective from a life-cycle perspective.

INTRODUCTION

The application of vacuum insulation panels (VIPs) in building retrofits shows great promise and has the potential to reduce whole-building energy consumption significantly due to reduced thermal heat transfer through the building envelope, but panel manufacturing size, panel cost, and issues regarding unsolved integration into the building assembly are current obstacles. Due to their minimal thickness, VIPs present the opportunity to rethink the exterior wall assembly, decreasing its overall depth and thus retaining usable and rentable area during retrofit. This has the potential to open new markets for retrofit and will more broadly encourage the densification of existing sites by making possible the reuse of existing structures, thus dramatically reducing construction costs, emissions, and waste by minimizing the mobilization of equipment required for new construction.

Kyle Vansice is an architect with Skidmore, Owings & Merrill, Chicago, IL. **Ulrike Passe** is an associate professor in the Department of Architecture, Iowa State University, Ames, IA.

This research project addresses current vacuum insulation limitations during construction and utilizes whole-building energy analysis, optimization, and the development of construction techniques as a means for providing best practices for the implementation of low-cost VIPs. Using an innovative new core material composition developed within our research team and in tandem with a manufacturing partner (Chang et al. 2016), the research is developed in a way that is congruent with current construction methods but that is also disruptive to current practices for the design and construction of building envelopes.

To capture retrofit potential, an existing U.S. Department of Energy (DOE) pre-1980 reference building is analyzed as an energy consumption baseline utilizing EnergyPlus (DOE 2015) within a parametric software platform. Models for whole-building energy and hygrothermal transfer will be developed to include thermal conductivity, thermal bridging, and moisture/airflow within the envelope assembly. Coupling energy models with a parametric platform will facilitate inclusion of manufacturing and construction processes within the design phase as well as allow for the optimization of cost, panel size, assembly, and energy-efficiency variables. Appropriate envelope assemblies will emerge from this research project, which incorporates VIPs as a novel, highly energy-efficient construction method for existing building envelopes.

WHOLE-BUILDING ANALYSIS

Existing Reference Building

To create a baseline condition, DOE's existing residential apartment reference building case was modeled, employing EnergyPlus pre-1980 construction assemblies and HVAC systems (Deru et al. 2011). Table 1 shows the assemblies organized by their respective ASHRAE climate zone (ASHRAE 2007). It is important that the retrofit assemblies be developed according to these regions, as thermal and moisture transport differ widely and implementation varies accordingly. The construction type is a standard, cold-formed steel stud system composed of exterior wood siding, interior gypsum board finish, and varying levels of batt insulation in the stud cavity.

Two models were run for each climate zone, the first being the pre-1980 baseline and the second being the applied VIP retrofit construction (see Figure 1). The difference in performance between these two conditions is the projected reduction in thermal energy consumption.

The analytical model was constructed in Rhino Grasshopper (version 0.9.0076) using the Ladybug and Honeybee plugins (Sadeghipour Roudsari and Pak 2013) to link directly to the EnergyPlus analysis engine (version 8.1). EnergyPlus utilizes a linear thermal transfer coefficient model and does not account for nonlinear or moisture transport effects. The existing reference building as modeled recreates the EnergyPlus files given in a geometric capacity utilizing the Rhino 5 platform. The initial files provided are in .idf format and can only be run within the EnergyPlus engine. Ladybug and Honeybee allow for the extraction of this information into a geometric format; however, many of the assumptions and system presets must be manually emulated. These assumptions include glazing ratios, HVAC systems and distribution types, internal load schedules, and other site energy supply information (natural gas/electric). The analysis was also completed through Design Builder and the EnergyPlus standalone engine using the .idf file generated to ensure the agreement of the input geometry and output parameters within the Rhino Grasshopper environment.

Table 1. DOE Pre-1980 Residential Exterior Wall Assemblies Listed by ASHRAE Climate Zone (ASHRAE 2007)

ASHRAE Climate Zone	Reference City	U-Value, $Btu/ft^2·h·°F$	R-Value, $ft^2·h·°F/Btu$
2A	Houston	0.172	5.822
2B	Phoenix	0.240	4.169
3A	Atlanta	0.146	6.846
3B	Los Angeles	0.220	4.549
3C	San Francisco	0.130	7.699
4A	Baltimore	0.088	11.251
4B	Albuquerque	0.099	10.003
4C	Seattle	0.092	10.885
5A	Chicago	0.082	12.193
5B	Boulder	0.082	12.193
6A	Minneapolis	0.065	15.398
6B	Helena	0.072	13.892
7A	Duluth	0.058	17.270
8A	Fairbanks	0.045	22.195

Generalized Retrofit Case

The retrofit cases simulate the addition of VIPs into the existing wall assembly of the DOE reference building summarized in Table 1 to improve whole-building thermal performance (see Figure 1). The baseline condition varies by climate zone, but the basic composition is consistent. Figure 1 shows the existing baseline condition with the insertion of the VIPs between the stud layer and the gypsum finish board, creating the retrofit condition. The standard EnergyPlus pre-1980 construction (baseline) includes wood siding on steel framing with varying levels of interior insulation and gypsum wall board finish. The roof material is of a similar construction with an added layer of gypsum board and was held constant across both the existing and retrofit cases—it is assumed that the VIPs would only be installed at the exterior wall.

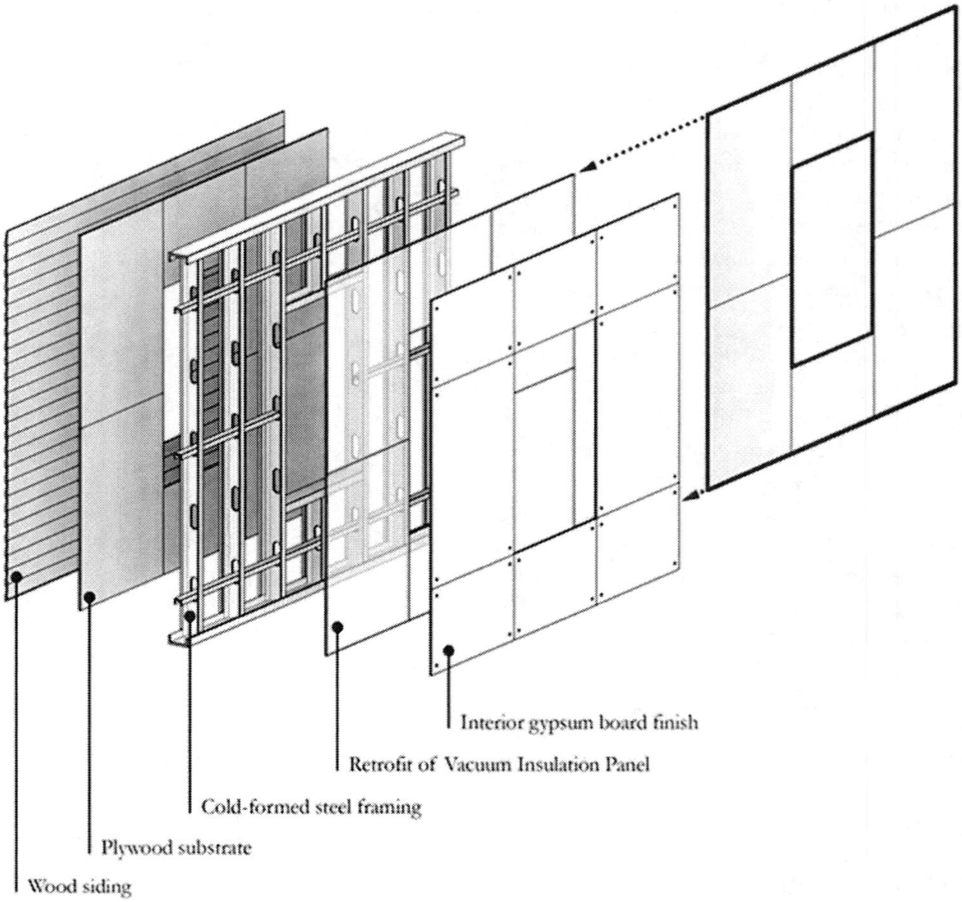

Figure 1 *VIP assembly showing elements of the construction, standard across all climate zones with varying levels of insulation in the steel stud cavity.*

Table 2. Material Properties of the Various Core Materials Used (Chang 2016)

Material Type	Diatomaceous	Fumed silica	Glass bubbles	Glass fiber
Thermal Conductivity, W/m·K @ 1 atm	0.0269	0.0175	0.0145	0.0106
Thermal Conductivity, W/m·K @ vacuum	0.0029	0.0042	0.0027	0.0041
Emissivity	0.930	0.930	0.930	0.930
Density, g/cm^3	0.400	0.037	0.150	0.255
Specific Heat, J/m^3·K	0.5×10^6	0.1×10^6	0.06×10^6	0.1×10^6
Moisture Diffusivity, cm^2/s	0.160	0.213	0.095	0.184

For the additional insulation layer, four VIP core materials were applied in the retrofit condition based upon laboratory testing completed by our team (Chang et al. 2016). Their relevant material properties are provided in Table 2 and their associated thermal resistance values across pressures ranging from 26 Pa (vacuum) to one atmosphere (1 atm) are shown in Figure 2 (Chang et al. 2016). Diatomaceous Earth was developed and tested by Chang et al. (2016); however, the other core materials (fumed silica, glass bubbles, and glass fibre) are already available as core materials for testing. The whole-building analysis considers only the 26 Pa vacuum case, however, implying an idealized vacuum. As the pressure increases to 1 atm, thermal resistance decreases by a significant magnitude—a problem that is discussed in Figure 6. The conductivity values shown in Table 2 represent properties at 1 atm per 1 in. of wall section. Thus, for a standard 1 in. thick VIP, one would expect an R-value of between 34 and 53 $ft^2·h·°F/Btu·in$. The average of these values represents a significant increase in resistance over the average of pre-1980 assemblies.

The existing pre-1980 models were modified to take into account the potential "retrofit" of their exterior wall assemblies to include VIPs. This is an idealized condition that modifies the thermal conductivity of the exterior assembly with the addition of one VIP layer but ignores issues of thermal bridging and moisture transport. Because this study does not examine the material composition or properties of the barrier film, hygrothermal modeling was not undertaken. While the exterior wall assemblies were modified, all other compositions including roof and floor assemblies were held constant. The HVAC system and internal load schedules were modeled to resemble a pre-1980 condition in order to be held constant with the existing reference building case.

RESULTS

Thermal Energy Consumption

Figure 3 shows the results of the analysis by ASHRAE climate zone. The values when compared with the existing baseline yield an average thermal energy reduction of 18.3% across all climate zones. Table 3 shows the energy reduction by climate zone. Furthermore, the average energy reduction across climate zones that are heating dominated is 36.3%. The change in thermal energy consumption in the baseline DOE exterior wall assemblies from the pre-1980 condition to the most current is a net reduction of 14.1%. This means that the implementation of VIPs creates the potential for a reduction of 4.2% over an update to the most current exterior wall assembly across all climate zones and a reduction of 22.2% when compared with heating-dominated climate zones.

VIP Construction Assembly

The retrofit exterior wall assembly utilizes a VIP thickness of 1 in. in the initial simulation. The insulating efficacy of the panel can be compromised by its condition within the assembly, however. There are many limitations to implementation, a few of which are noted in the Future Directions and Current Limitations section of this paper; however, analysis is provided here to address the concerns of thermal bridging and panel thickness. Figure 1 shows the development of the existing assembly with the inclusion of the vacuum panel.

This simulation does not take into account the properties of the barrier film or its interaction with other elements within the assembly. It is assumed that direct radiation does not affect the overall thermal resistance of the panel and that the core material provides all available resistance. This is a steady-state simulation, and therefore material and fluid degradation do not occur over time; thus, the VIP assembly is considered idealized. Air infiltration and thus increased pressure within the core dramatically affect the resistive potential of the assembly. The limitations of this approach to VIP implementation are discussed in the Future Directions and Current Limitations section.

Thermal bridging occurs at the panel edge where the barrier film is folded over and welded to complete a full pressure seal of the unit. The barrier film is most commonly composed of a metallic or aluminum material with a high thermal conductance value. This leads to a condition of thermal transfer through conduction at the joints between panels. To maximize the VIP's overall thermal resistance efficacy, the ratio between the panel joints and overall face area needs to be minimized. The panel width and height are limited by the core

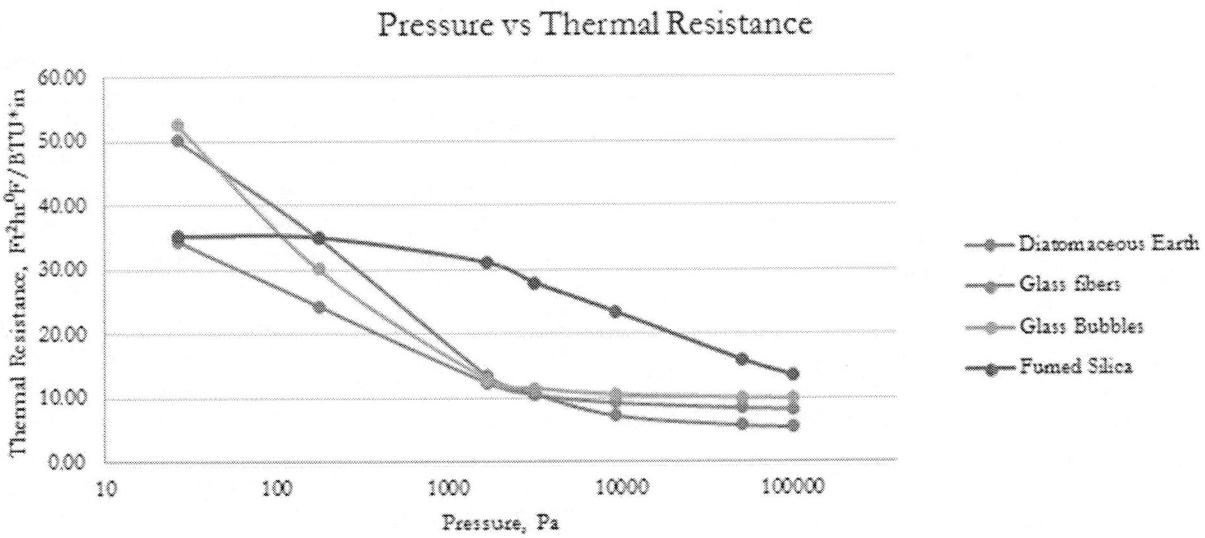

Figure 2 *Thermal resistances of various core material types as tested across pressures ranging from 26 Pa to 1 atm.*

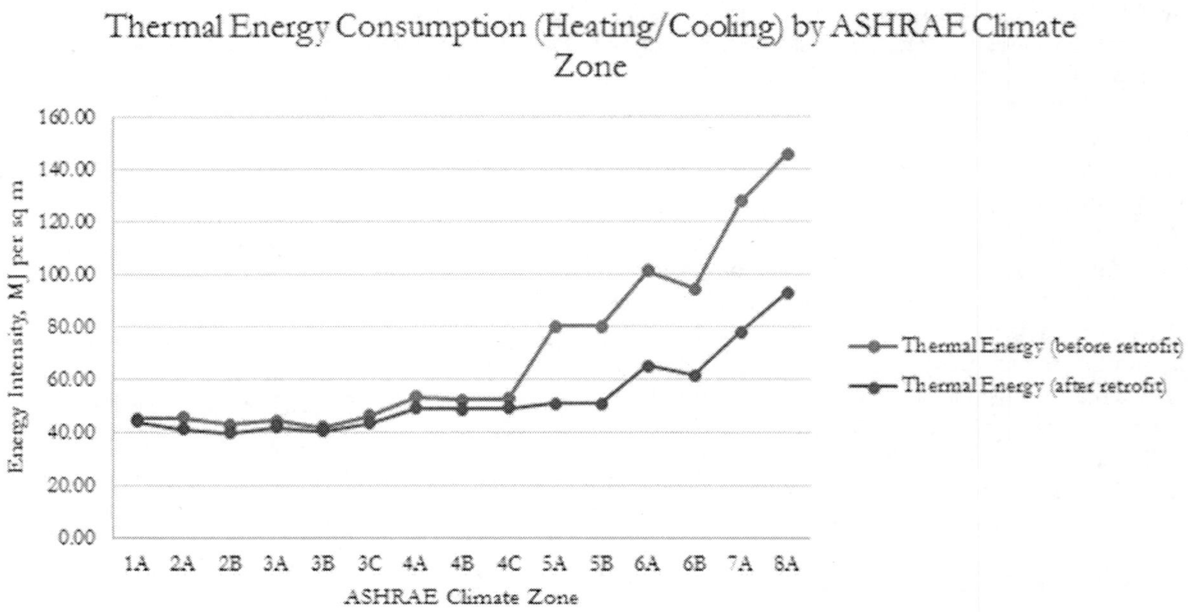

Figure 3 *Results of EnergyPlus simulation showing thermal energy consumption by ASHRAE climate zone.*

Table 3. Results of EnergyPlus Simulation Showing Thermal Energy Consumption by ASHRAE Climate Zone—Comparison between Pre-1980 and Retrofit Cases

Climate Zone	Thermal Energy before Retrofit, mJ/m^2	Thermal Energy after Retrofit, mJ/m^2	Net Reduction
1A	45.4	44.2	2.6%
2A	45.7	41.4	9.4%
2B	43.2	40.2	6.9%
3A	44.6	42.3	5.2%
3B	42.4	40.7	4.0%
3C	46.6	43.6	6.4%
4A	53.7	49.3	8.2%
4B	52.4	49.1	6.3%
4C	53.1	49.2	7.3%
5A	80.4	51.2	36.3%
5B	80.4	51.2	36.3%
6A	101.4	65.3	35.6%
6B	94.6	61.9	34.6%
7A	127.8	78.4	38.7%
8A	145.9	93.1	36.2%

material's hardness and by the panel edge bending resistance. At a certain dimension, the panel's thermal integrity and mechanical stability can be compromised by either bending or buckling at the panel center. This leads to either complete failure of the panel or an increased rate of pressure infiltration. To modify the initial whole-building analysis to include individual panel joints at a number of size iterations would have been unnecessarily computationally heavy; therefore, a lesser model was developed that parametrically varied the panel length and width to determine an overall thermal resistance average. To further simplify the analysis, the panel dimensions were assumed to be equal (square), and only conductance was considered.

Figure 4 shows the results of varied panel sizes on whole-building energy consumption. This captures the effects of thermal bridging at the building scale and suggests that the magnitude of these effects is significant. The implementation of VIPs yields an energy consumption reduction of 25% to 35%, with the 10% difference between bounds accounting for the implications of various panel sizes. The effects of panel size are much more pronounced at dimensions between 1 and 4 ft, after which thermal energy consumption is unaffected. This suggests that the ratio between surface area and perimeter is in fact important and that a minimum panel size of 4 ft square is most advantageous if manufacturing to that scale is feasible. This reduction accounts for ASHRAE climate zones that are heating dominated. These results are expected, as they somewhat diverge from the first simulation, which does not consider panel size. However, the results for the upper bound of panel sizes is within 3% of the original average for these climate zones, which supports the finding that once the panel

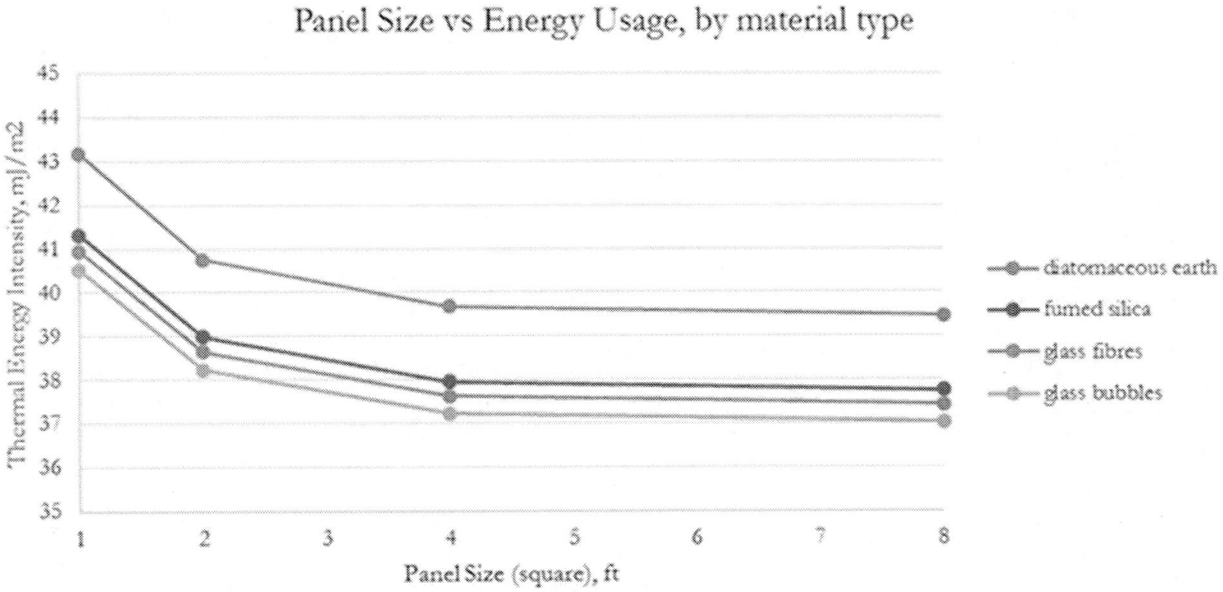

Figure 4 *Change in performance across core materials and panel size.*

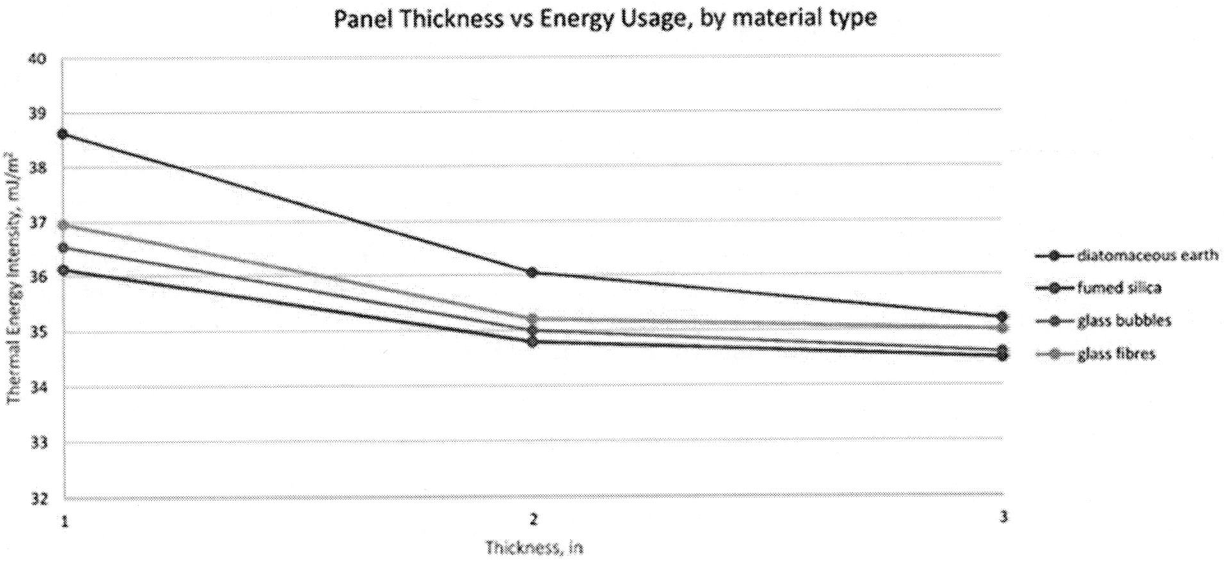

Figure 5 *Effects of panel thickness on whole-building thermal energy consumption when compared with the baseline condition.*

size increases beyond 4 ft square, panel size has a negligible impact on overall thermal energy consumption.

Figure 5 presents the effects of increased panel thickness on whole-building energy consumption. This result ignores the effects of thermal bridging and simply assumes the exterior envelope is continuous. The energy consumption reduction is between 34% and 38%. The range between these bounds is much smaller than that of the effects of panel size and suggests that little is gained from increased VIP thickness beyond 1 in. This result is somewhat expected, as ASHRAE recommendations for maximum R-value are much lower than the R-value of VIPs at 1 atm.

ASHRAE recommendations are based on climate-zone-specific analysis, which establishes maximum resistance values; thereafter, limited gain is achieved.

A clear convergence appears between the results shown in Figures 4 and 5. The maximum reduction due to panel size is similar to the average reduction realized across all panel thicknesses, where thermal bridging is not considered. This suggests that the effects of thermal bridging can be nearly eliminated at larger panel sizes. Existing research has discussed the potential for staggered, multilayered paneling and other means for thermal bridging mitigation (Wakili et al. 2011); however, the finding presented in this paper suggests that a panel thickness of 1 in. may be completely sufficient. This has significant impacts on the development of construction assemblies and allows for greater flexibility in design. Overall, our research suggests that the potential for whole-building thermal energy reduction exists between the approximate bounds of 25% and 38%.

FUTURE DIRECTIONS AND CURRENT LIMITATIONS

Thermal Conductivity and Pressure Increase Over Time

An important property of VIPs is the level of pressure within the cavity, as it is directly proportional to the effective thermal resistance of the panel. At a value of zero, the thermal conductivity of the panel is minimized, but as pressure increases to 1 atm (normalized to sea level), thermal conductivity maximizes. This no-vacuum condition relegates the insulating properties of the panel to the resistance of the material within the cavity. When this occurs, there is little advantage to the implementation of VIPs over conventional means of envelope insulation.

It thus becomes necessary to develop a change-in-pressure model within the time domain. This provides a means for analyzing the feasibility of VIP implementation based on the useful life of the panel. Almost immediately after manufacturing, the panel will begin to pressurize—albeit very slowly. The point of pressure decay at which the panel resistance equalizes with conventional insulation must occur after a specified payback period. That is to say, the panel must retain an effective resistance for a period long enough for the energy savings gained from implementation to eclipse all costs associated with the initial VIP design, manufacture, and construction costs. It is likely that the effective life span will need to be many times greater than the initial payback for implementation to be truly advantageous.

VIPs are a relatively new technology and thus there is little research regarding vacuum decay over time. It is also difficult to develop a predictive model with any accuracy. Future research must be completed to understand this relationship to enable satisfying assembly design and optimization.

Preliminary results of this research have attempted to analyze thermal resistance across the three available and one novel core material as a function of time. Two relationships must be developed in further depth: conductivity versus pressure, and pressure versus time. Conductivity versus pressure is shown in Figure 6 and is based on laboratory tests completed by our research team. Pressure versus time is shown in Figure 7 and is based directly on the empirical research presented by both Fricke et al. (2007) and Simmler and Brunner (2005). The results are based upon a relatively short period of time—two years and less—and develop a linear relationship. A more parabolic (asymptotic) result is expected, and thus these models are likely problematic for longer durations of 25–50 years.

The resulting thermal resistance versus time graph is shown in Figure 8. It is important to note, however, that the accuracy of this result is in question due to the limited pressure-time information. Here it is shown that across all core material types, with the exception of fumed silica, there is a reduction in thermal resistance to conventional insulation levels at approximately 10–15 years—a result that would surely prohibit the use of VIPs in most applications. It is also important to note that the conductivity-versus-pressure tests in the referenced material testing performed by our team were

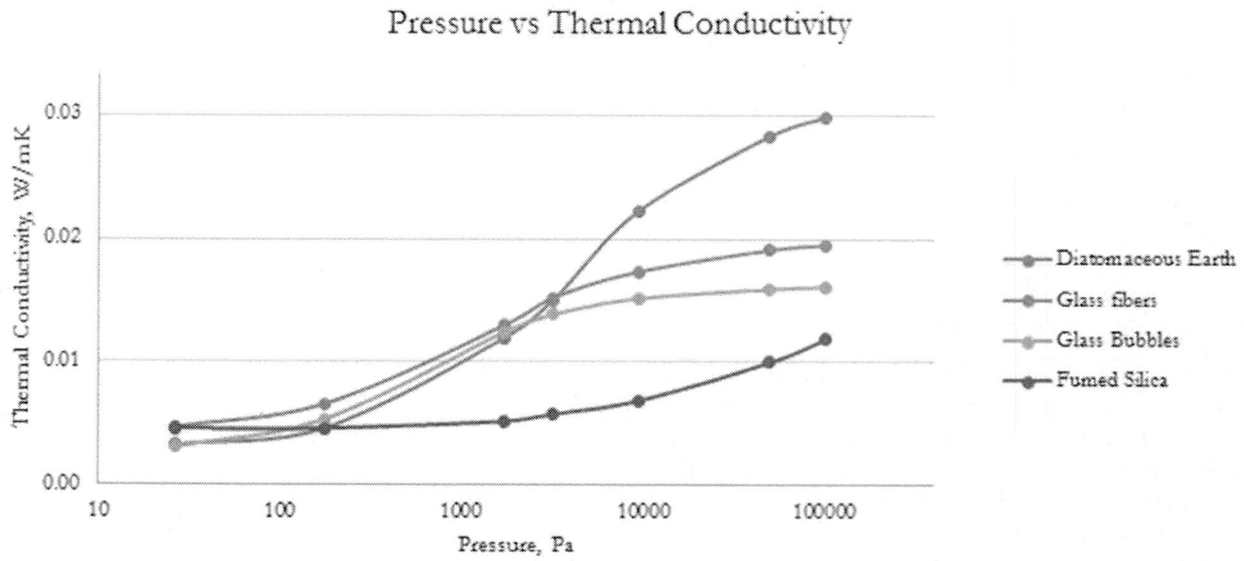

Figure 6 Increase in thermal conductivity within a pressure range corresponding with expected values over 50 years.

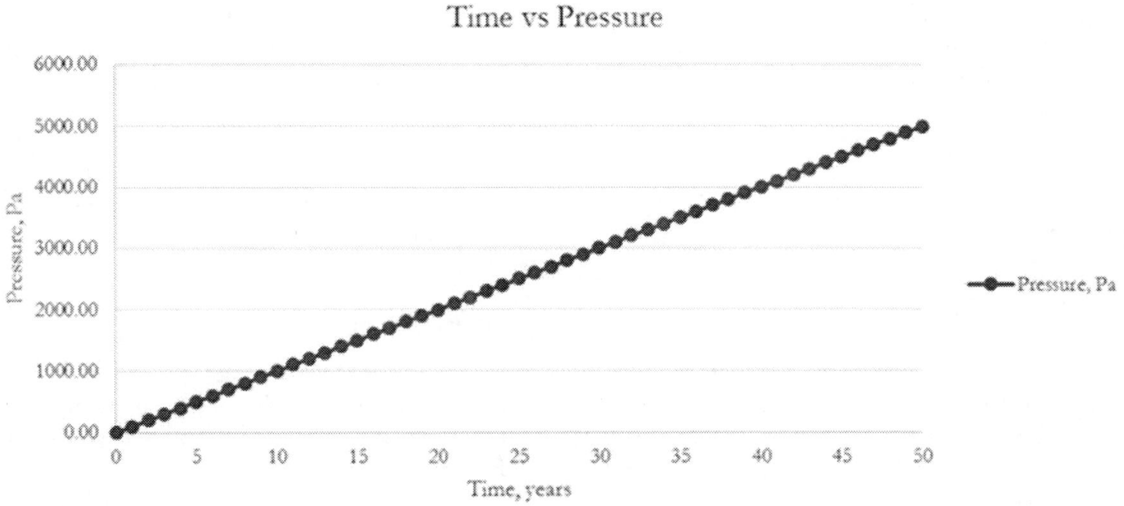

Figure 7 *Increase in pressure over time based on the model proposed by Fricke et al. (2007).*

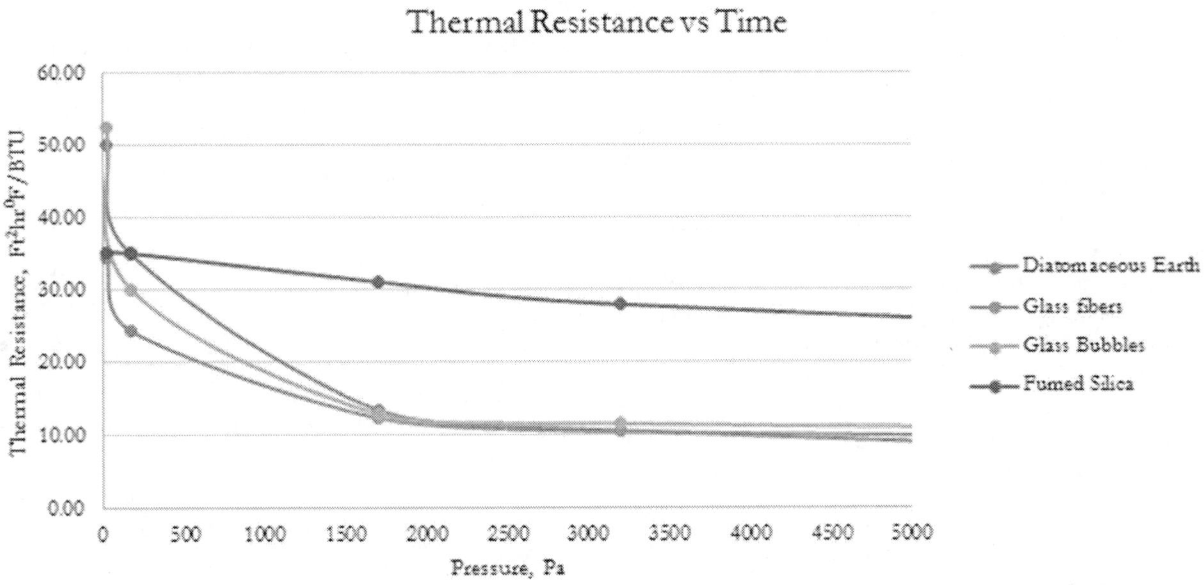

Figure 8 *Resulting decrease in thermal resistance over a period of 50 years due to pressure increase.*

done in a controlled setting where the specimen was composed only of the material itself and not the broader panel assembly, which would include the barrier film. This would likely exacerbate the issue of increased thermal conductivity, as barrier films are typically made of highly conductive metallic materials. Barrier films are important to the pressure-time relationship because their associated diffusivity—both vapor and water—govern the broader infiltration into the panel and thus the rate at which it pressurizes.

Thermal Transfer Modeling

It is clear that the development of the barrier film is as important as—if not more important than—the core material. If the core vacuum is pressurized at a rate too quickly, then the core material type makes little difference. Further research into barrier film materials and compositions must be completed, and accurate models for vapor and moisture diffusivity and emissivity must be developed. This is important in furthering understanding of the rate of panel pressurization over time.

Thermal Bridging Effects

It has been established that it is likely possible to mitigate the effects of thermal bridging at the panel edge by simply increasing the panel size beyond a minimum threshold. This is only true at the whole-building energy consumption scale, where bridging is only a small part of the broader thermal equation. However, when looking at the panel individually, it is clear that thermal bridging is related to the perimeter-surface area ratio. Moving forward, it will be useful to develop a model for associating an average VIP thermal resistance value to this ratio.

CONCLUSION

This paper presents the initial results of a broader research project seeking to provide best practices for the implementation of VIPs in the exterior building envelope. Whole-building thermal energy analysis using the EnergyPlus engine was undertaken to determine the potential thermal energy reductions possible when implementing a VIP retrofit to a pre-1980 residential building across all ASHRAE climate zones. The results of this analysis predict that VIP retrofits in existing buildings built before 1980 have the potential for reducing overall thermal energy consumption by up to 37%.

Stemming from this initial analysis, limitations are identified, two of which are analyzed utilizing a similar EnergyPlus analysis method. Thermal bridging occurs at the panel edges where the panel is sealed, creating a condition whereby direct conduction compromises the resistive capacity of the assembly as a whole. As such, it is inferred here that the ratio of the panel perimeter to its surface area is important, and the analysis provided suggests that this parameter has a measurable impact up to a panel size of approximately 4 ft square. The variance in thermal energy reduction at sizes between 1 and 4 ft is approximately 13%. It is thus suggested that the effects of thermal bridging can be mitigated through the use of larger panels—at least 4ft square. The thickness of the panel considered in this study is 1 in.; however, other studies have presented the potential for an increased panel thickness or for multiple panels layered to reduce the effects of thermal bridging. Analysis was completed to determine the opportunity costs of increased panel thickness or increasing the number of panels in depth, and it was determined that due to the high thermal resistance provided at 1 in., further increasing the thickness provided little benefit in the way of reduced thermal energy consumption.

The principal limitation of this study is the omission of the barrier film properties within the thermal energy analysis. As was shown in the discussion of the limitations, pressurization of the cavity is inevitable. Once the panel pressurizes to the level of 1 atm, the cavity material analyzed here behaves only nominally better than any other standard insulation material. If VIPs are to be an economically viable retrofit approach, the pressurization must occur at a rate slow enough that the initial construction costs are overcome by the thermal energy cost savings over time. The linear time-pressure model referenced previously creates a condition whereby the panel efficacy for all materials except fumed silica approaches more conventional insulation values within 15 years. A detailed life-cycle cost analysis would be necessary to determine if this is an acceptable case.

To fulfill the initial goals of the research with regards to optimization and assembly best practices, complete panel composition, an understanding of the rate of pressurization over time, and further thermal energy modeling at the material scale are necessary. Currently, the parametric model apparatus has been constructed and is tied directly to the EnergyPlus analytical model; however, these models must be populated with the correct cost data and thermal properties before further analysis and optimization can take place.

ACKNOWLEDGMENTS

The authors are very grateful to the Iowa Energy Center for funding this work as part of an opportunity grant with Dr. Mufit Akinc, Professor of Materials Science and Engineering at Iowa State University as principal investigator. The authors also thank Boyce Chang as well as Ganesh Balasubramanian and Mufit Akinc for sharing their knowledge about the core materials and for assistance in obtaining the data.

REFERENCES

ASHRAE. 2007. Normative Appendix B, Building envelope climate criteria. In ANSI/ASHRAE/IESNA Standard 90.1-2007, *Energy Standard for Buildings except Low-Rise residential buildings*. Atlanta: ASHRAE.

Chang, B., L. Zhong, and M. Akinc. 2016. Low cost VIP core materials for building applications. *Vacuum* 131:120–26.

Deru, M., K. Field, D. Studer, K. Benne, B. Griffith, P. Torcellini, B. Liu, M. Halverson, D. Winiarski, M. Rosenberg, M. Yazdanian, J. Huang, and D. Crawley. 2011. *U.S. Department of Energy commercial reference building models of the national building stock*. Tech. no. NREL/TP-5500-46861. Golden, CO: National Renewable Energy Laboratory.

DOE. 2015. *EnergyPlus*, ver. 8.1. Washington, DC: U.S. Department of Energy. https://energyplus.net/.

Fricke, J., U. Heinemann, and H.P. Ebert. 2007. Vacuum insulation panels—From research to market. *Vacuum* 82:680–90.

Sadeghipour Roudsari, M., and M. Pak. 2013. Ladybug: A parametric environmental plugin for Grasshopper to help designers create an environmentally-conscious design. In *Proceedings of BS2013: 13th Conference of International Building Performance Simulation Association*. http://www.ibpsa.org/proceedings/BS2013/p_2499.pdf.

Simmler, H., and S. Brunner. 2005. Vacuum insulation panels for building application basic properties, aging mechanisms and service life. *Energy and Buildings* 37:1122–31.

Wakili, K.G., T. Stahl, and S. Brunner. 2011. Effective thermal conductivity of a staggered double layer of vacuum insulation panels. *Energy and Buildings* 43:1241–46.

Development of High-Performance Composite Foam Insulation with Vacuum Insulation Cores

Kaushik Biswas, PhD **John Letts, PhD** **Andre Desjarlais** **Jennifer Yao, PhD** **Douglas Smith, PhD**

ABSTRACT

Development of a high-performance thermal insulation (thermal resistance or R-value per inch of R-12 h·ft²·°F/Btu·in. or greater), with twice the thermal resistance of state-of-the-art commercial insulation materials (≤R6/inch for foam insulation), promises a transformational impact in the area of building insulation. In 2010 in the United States, the building envelope-related primary energy consumption was 15.6 quads, of which 5.75 quads were because of opaque wall and roof sections; the total U.S. consumption (building, industrial and transportation) was 98 quadrillion Btus (98 quads). In other words, the wall and roof contribution was almost 6% of the entire United States primary energy consumption.

Building energy modeling analyses have shown that adding insulation to increase the R-value of the external walls of residential buildings by R10-20 (h·ft2·°F/Btu) can yield savings of 38% to 50% in wall-generated heating and cooling loads.

Adding R20 will require substantial thicknesses of current commercial insulation materials, often requiring significant (and sometimes cost prohibitive) alterations to existing buildings. This article describes the development of a next-generation composite insulation with a target thermal resistance of R25 for a 2 in. thick board (R12/in. or higher). The composite insulation will contain vacuum insulation cores, which are nominally R35-40/in., encapsulated in polyisocyanurate (PIR) foam. A recently developed variant of vacuum insulation, called modified atmosphere insulation *(MAI), was used in this research. Some background information on the thermal performance and distinguishing features of MAI has been provided. Technical details of the composite insulation development and manufacturing as well as laboratory evaluation of prototype insulation boards are presented.*

INTRODUCTION

Reroofing represents about 75% of the roofing market, with about 3 billion square feet of roofing being installed yearly. Existing roofs are difficult to bring up to code from an energy-efficiency perspective because equipment stands, scuppers, parapet walls, and countless other features can prevent the addition of large amounts of insulation. Similarly, wall retrofits typically do not result in a substantial increase in the amount of continuous insulation because of the additional work required to fit existing architectural features to the new wall thickness. A high R-value per inch product that can be used in these applications would alleviate these problems.

Polyisocyanurate (PIR) insulation has been successfully used in a variety of commercial and residential building applications, ranging from roofs to walls, for more than 40 years. It has the highest R-value per inch (~6 h·ft²·°F/Btu·in.) of any commercial insulation that is widely used and represents almost 80% of the insulation used in low-slope commercial roofing in the United States. Besides having a high R-value, PIR has a long history of use in the commercial environment. The board has demonstrated its toughness and durability and can successfully withstand the nicks and dents it is commonly subjected to in a construction environment with minimal loss of performance.

Retrofitting walls and roofs using thick layers of conventional insulation materials (≤ R5.5/inch) and traditional technologies has proven to be ineffective, especially in terms of cost and payback. Thinner insulation systems and newer tech-

Kaushik Biswas is a member of the research and development staff and **Andre Desjarlais** is a program manager at Oak Ridge National Laboratory, Oak Ridge, TN. **Douglas Smith** is president of NanoPore, Inc., Albuquerque, NM. **John Letts** is a technical director and **Jennifer Yao** is a scientist III at Firestone Building Products Company, Indianapolis, IN.

nologies are needed to make wall retrofits economically and technically feasible. According to the road map created by Department of Energy's (DOE) Building Technologies Office (BTO), the technical primary energy saving potential of 2 in. of R12/in. insulation added to existing residential walls is 1.1 quadrillion Btus (quads); additional savings can be expected from retrofitting commercial buildings (DOE 2014).

A vacuum insulation panel (VIP) offers very low conductivity (or high R/in.) compared to conventional insulation materials (Alam et al. 2011). It is the only technology that can enable the development of insulation materials with R-values of 12/in. or greater. This article describes the development and evaluation of a composite insulation material consisting of vacuum insulation cores completely encapsulated within PIR foam. The board is 2 in. thick with a target R-value of 25 h·ft^2·°F/Btu.

VIPs have existed for more than two decades, but their adoption in buildings has been limited, their high cost being one of the hindrances. A newly developed, lower-cost variant of vacuum insulation, called *modified atmosphere insulation* (MAI), was used in this research. MAI is created using a different processing method compared to VIPs, which leads to significant cost reductions. However, to achieve the stated R/in. of the composite insulation board, it was critical that MAI possess the same thermal performance as regular VIPs. In this article, some background information on the MAI technology, its distinguishing features, and its thermal performance are also provided.

VACUUM INSULATION FOR BUILDINGS

VIPs contain an evacuated nano/mircoporous core that is sealed with multilayered barrier films (Figure 1). Two major issues that have hindered the widespread use of VIPs in buildings are cost and integration into the building envelope. Protecting the VIPs during transpiration and handling as well as when it is installed in the envelope are also issues that need to be mitigated to successfully deploy VIPs in building envelopes. The newly developed insulation board contains VIPs that are fully encapsulated with foam. The composite board can be used as a surrogate for regular foam boards, with little or no change in handling and installation methods.

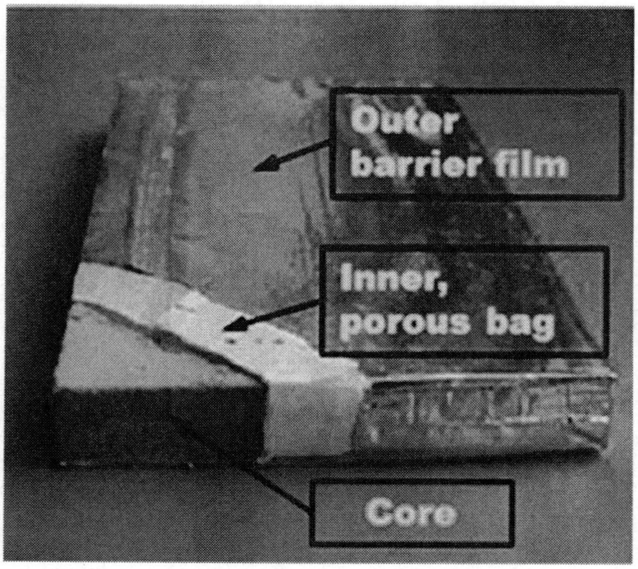

Figure 1 *Vacuum insulation panel.*

The cost of VIPs are primarily driven by the manufacturing process (up to 75% of the total cost). The composite boards contain vacuum insulation cores that are manufactured in a different manner compared to traditional VIPs. MAI is processed at ambient pressure using standard vertical form-fill-seal machines of the type used to package potato chips. Regular VIP processing, which is expensive and energy intensive, includes pressing fine powders into boards, which must then be cut, dried, evacuated, and sealed in a vacuum. MAI avoids these process steps by replacing the air in the fluidized powder core with a low conductivity vapor (steam, for example). Once the core is sealed within the barrier films and cool-formed to final shapes, the vapor condenses and creates the vacuum. On a commercial scale, MAI cost is projected to be \$0.10–0.13/ft^2 per R versus about \$0.25/ft^2 per R for regular VIPs.[1] In comparison, cost of 1 in. foam insulation boards is about \$0.06–0.10/ft^2 per R (RS Means 2014). Figure 2 illustrates the differences in processing of VIP and MAI.

R25 COMPOSITE BOARD DEVELOPMENT

The development team started by evaluating different multilayered barrier films. Three were metalized polymeric films and one was all polymeric film. The advantage of a completely polymeric barrier film compared to metalized films is the reduction in heat transfer along the edges of the MAI panels (thermal bridging) at the interfaces between adja-

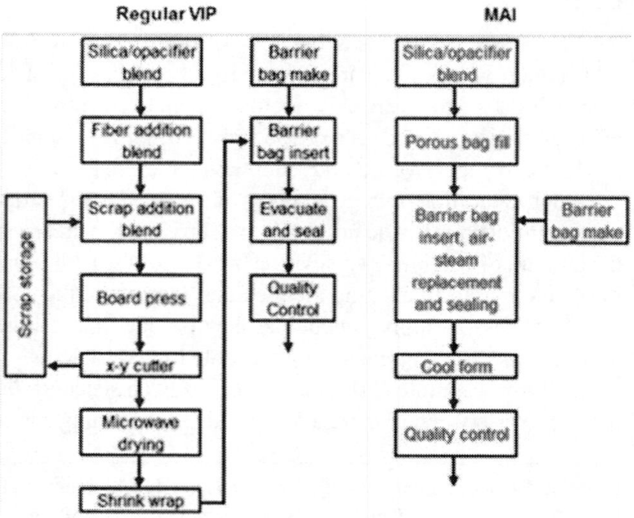

Figure 2 *Differences in MAI and VIP manufacturing.*

1. Cost based on R value in h·ft^2·°F/Btu.

cent MAI panels. One critical criterion for the barrier material was adhesion with the spray-applied PIR foam. Further, the foam needs to flow around the MAI panels and completely encapsulate the panels without any air pockets. Finally, the ability of MAI panels to withstand the high temperatures and pressures caused by the exothermic expansion of the spray-applied foam without getting damaged was another important success criterion.

Laboratory-Scale Trials

Laboratory-scale tests were performed to test the aforementioned success criteria. MAI panels were fixed within molds in different orientations and the foaming compounds were poured on top and around them. Some of the MAI panels were wrapped in thin plastic sheets so they could be extracted from the foam blocks for examination. Few other MAI panels were instrumented with temperature sensors to detect the rise in temperature during the foaming process.

Figure 3 shows a sample laboratory experiment in which one MAI panel was fixed into a mold, and a liquid foaming mixture was poured around it and allowed to expand. Different orientations and configurations of the MAI panels fit into molds were examined; foaming liquid was applied perpendicular and parallel to the larger MAI surfaces.

Examination of several foamed MAI panels revealed good foam coverage all around the panels. The examination of the MAI panels covered in plastic revealed no discernible change in dimensions and internal pressure (vacuum) caused by the pressure and high temperature of the foam expansion. The maximum observed temperatures stayed below 194°F, except one location in one test, which was deemed acceptable. It should be noted that the subject barrier films are exposed to accelerated aging tests at 176°F over many months.

Adhesion of the liquid-applied foam with the MAI barrier films were also tested. This was done by securing the MAI panels on a substrate and pouring liquid foaming compounds on the MAI. Small metal plates were glued on the liquid-applied foam surfaces and the samples were cured for 3 days. The foamed MAI specimens were secured in a holder and the foam was cut around the plates. Finally the metal plates were pulled away using a load cell till failure. All four barrier films passed the minimum required adhesion strength threshold; the samples with two metalized barrier films showed cohesive failure (foam break) rather than adhesive failure between the foam and the barrier film.

The main findings of the lab-scale tests were as follows:

- Satisfactory encapsulation of MAI panels by the liquid-applied foam.
- No discernible changes observed in the shape and size of MAI panels because of the exothermic foam expansion.
- Temperatures at the foam-MAI interface stayed below the tolerance level of all barrier materials.
- All barrier materials passed the threshold adhesive strength tests.

Manufacturing-Scale Trials

The all-polymeric barrier film was chosen for use in the full-scale composite boards. For the first set of manufacturing-scale trials, composite boards containing a 4 × 3 array of MAI panels were designed. The composite boards were of dimensions 8 × 4 ft and 2 in. thick. The MAI panels were of dimensions 22.75 × 14.67 in. This resulted in a design with 1 in. gaps around all MAI panels to be filled with spray-applied foam, which are useful as spacers for mechanical fasteners. The MAI panels were glued to a 0.5 in. high-density (HD) foam substrate (Figure 4), which were then fed into the manufacturing line for application of sprayed PIR foam.

The MAI containing HD boards were fed into the one of the manufacturing lines for regular PIR foam boards. Figure 5 shows the application of the spray-applied foaming compounds to the MAI containing HD foam substrate and the finished composite foam boards. Visual inspections revealed no defects in the spray-applied foam. Three composite boards were created during the first trial.

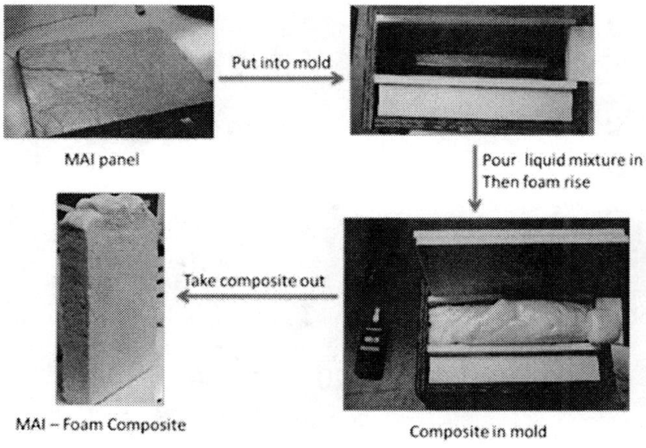

Figure 3 *Laboratory-scale foam encapsulation experiment.*

Figure 4 *MAI panels attached to a high-density foam substrate.*

EVALUATION OF R25 COMPOSITE BOARD INSULATION

Small-Scale Laboratory Tests

Several small-scale MAI panels and MAI-foam composites were tested in heat flow meter apparatus (HFMA) following ASTM C518 (ASTM 2015). An HFMA typically consists to two independently temperature-controlled plates that sandwich the specimen. The plates also contain transducers that can measure the heat flowing into or out of the plates, through the specimen. By measuring the plate temperatures, heat flow, and specimen thickness, the thermal conductivity (or R-value) of the specimen can be determined. These tests were done to verify the performance of the MAI panels to be the same as regular VIPs and also to help calibrate numerical models to predict the overall R-value of full-scale composites. Figure 6 shows the measured R/in. of MAI panels as a function of temperature; the horizontal error bars represent the HFMA plate temperatures. The measurements were done at the central 3 × 3 in. (7.6 × 7.6 cm) area of the MAI panels (i.e., center-of-panel [COP] R-value), which had typical dimensions of 12 × 12 in. and 1 in. thickness.

(a)

(b)

Figure 5 *(a) MAI-HD foam substrate passing under the foaming heads in the manufacturing line, and (b) finished composite insulation boards.*

In addition, composite sampled with MAI panels separated by foam strips were tested in a custom-built multi-transducer HFMA. The custom HFMA contains 15 transducers in each plate that can map the heat flux through nonhomogeneous samples. Figure 7 shows sample test data from one MAI-foam composite using the multi-transducer HFMA. The differences between the heat flow through the MAI panel versus the MAI-foam interface can be clearly observed. The data from the multi-transducer HFMA tests were used to calibrate numerical models.

Numerical Modeling

Finite element modeling was utilized to design the composite boards to achieve the desired R-value. First, small-scale models were created to match the specimens tested in the HFMA. Thermal conductivities of the MAI panel and foam were taken from the HFMA measurements to be used as model inputs. An effective thermal conductivity of the multilayer barrier film was assumed to be a variable that was tuned for model calibration. Figure 7 shows the calculated heat flows through a MAI-foam composite. The effective barrier conductivity was deduced by minimizing the root mean squared error between the calculated and measured heat flows.

Next, full-scale 4 × 8 ft models of the composite boards were created. A parametric simulation study was done to estimate the overall R-value of the composite board by varying the areal coverage of the MAI panels (i.e., number and size of the MAI panels). Figure 8 shows the simulated heat flux distribution through a composite foam board with MAI cores. The darker areas indicate higher heat flux at the interfaces between the MAI panels that were assumed to be filled with foam. Based on the simulations of the 4 × 8 ft composite boards, it

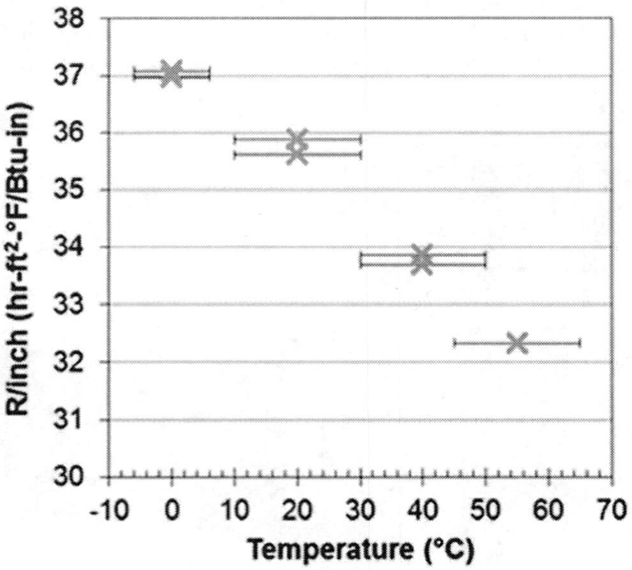

Figure 6 *Measured center-of-panel R-value of MAI panels.*

was estimated that an areal coverage of 86.9% of 1 in. thick MAI panels (sandwiched by 0.5 in. foam on either side) would provide the targeted R12/in. The 86.9% MAI coverage translated to a 4 × 3 matrix of 22.75 × 14.67 in. MAI panels with 1 in. gaps around all MAI panels (intended as space for mechanical fasteners).

Full-Scale Laboratory Tests

Three 4 × 8 ft composite boards were created during the first manufacturing-scale trial for testing. The overall R-value of the boards were measured by a guarded hot box test, according to ASTM C1363 (ASTM 2011). Figure 9 shows the guarded hot box and the specimen holder containing the MAI-foam composite boards. The hot box consists of hot (meter) and cold (climate) chambers that sandwich the test specimen. The test area is 8 × 8 ft. The overall wall R-value was calculated based on the heat input to the meter chamber (less heat loss to the surroundings) and the measured surface temperatures on the hot (meter) and cold (climate) sides of the test wall. The air temperatures in the climate and meter chambers were maintained at 50°F and 100°F, respectively. The perim-

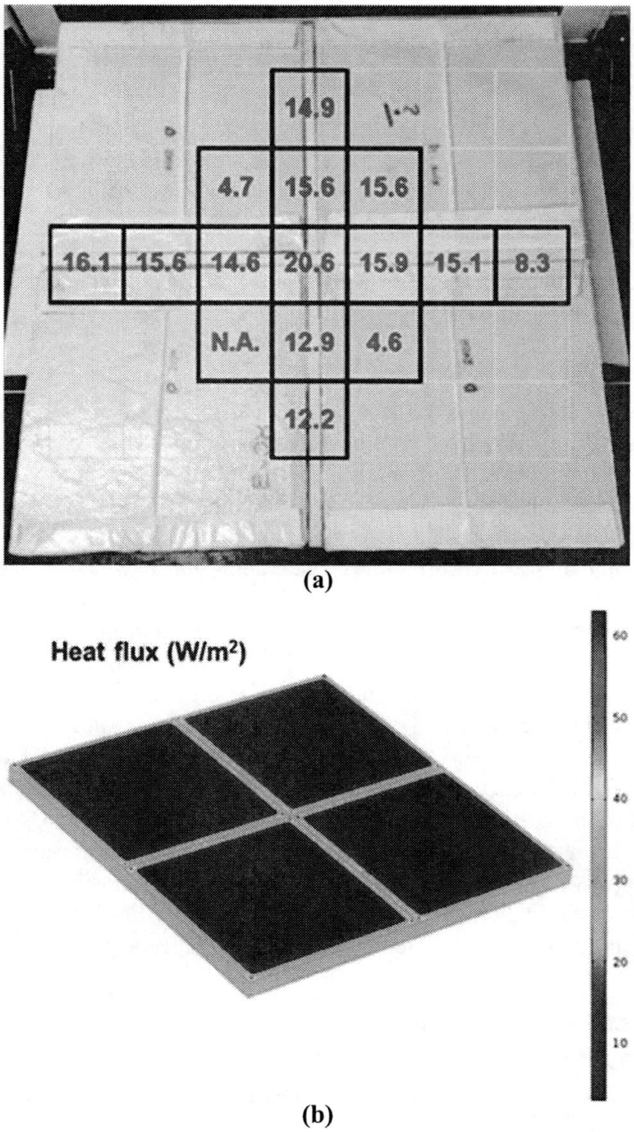

Figure 7 *(a) Measured and (b) simulated heat flux distribution through an MAI-foam composite.*

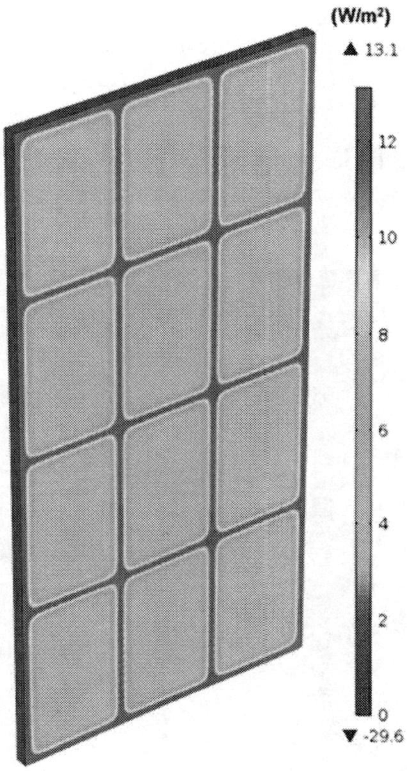

Figure 8 *Simulated heat flows through a full-scale MAI-foam composite board.*

Figure 9 *Guarded hot box with two MAI-foam composite boards installed in the specimen holder (center).*

eter of the aperture of the test specimen holder was insulated as shown in Figure 10. The insulated perimeter serves to reduce any flanking losses, which are defined by ASTM C1363 (2011) as "heat exchange from the metering chamber to the climatic chamber and or guard chamber that is due to the two-dimensional heat transfer at the interface of the test specimen and the surround panel or metering box wall." The guarded hot box used in this study is calibrated periodically using 5 in. thick, aged expanded polystyrene (EPS) foam boards. Further, this guarded hot box was part of an interlaboratory study (ILS) conducted by ASTM International in 2012–13.

The experimental and calculation details are briefly described below and results are tabulated in Table 1.

Figure 10 Insulated perimeter of the aperture holding the composite foam insulation test specimens.

1. The data (temperatures, power supply, etc.) were collected every 30 s.
2. The test wall was instrumented with 25 pairs of thermocouples distributed over the 8 × 8 ft test area, over foam borders and MAI panel sections. Each thermocouple pair was located across from each other on the opposite hot and cold test wall faces.
3. The overall R-value R was calculated as

$$R = \frac{\Delta t}{q_{meter} \cdot A} \quad (1)$$

where q = heat input.

4. In Equation 1, Δt is the area-weighted average temperature difference between the hot and cold surfaces, q_{meter} is the heat input to the meter chamber corrected for heat losses to the surroundings, and A is the test wall area.
5. Δt was calculated as:

$$\Delta t = \Delta t_{MAI} \cdot \left(\frac{A_{MAI}}{A}\right) + \Delta t_{Foam} \cdot \left(1 - \frac{A_{MAI}}{A}\right) \quad (2)$$

6. In Equation 2, Δt_{MAI} and Δt_{Foam} are the average temperature differences between the hot and cold surface temperatures over the MAI and foam sections. As mentioned earlier, each 4 × 8 ft composite board contained 12 22.75 × 14.67 in. MAI panels, yielding an A_{MAI}/A of 0.869.

CONCLUSION

In summary, the development and evaluation of a high R/in. composite insulation board are described. The composite board consisted of modified atmosphere insulation cores encapsulated in PIR foam. The development process consisted of laboratory-scale experimentation to test foam encapsulation of the MAI cores, MAI panels being able to withstand the high pressure and high temperature foaming process, and adherence between the spray-applied foam and MAI barrier films.

Laboratory tests and numerical simulations of MAI panels and small-scale MAI-foam composites were performed. HFMA tests of MAI panels (following ASTM C518 [2015]) yielded R-values in the range R32-37 per inch of thickness, similar to regular VIPs. Next, small-scale MAI-foam composite samples were tested in a multitransducer HFMA. Numerical models of MAI-foam composites were created, calibrated by comparison against the HFMA data, and used to estimate the MAI areal coverage needed in the

Table 1. Guarded Hot Box Test Details

Meter Side Surface Temperature, °F	Climate Side Surface Temperature, °F	Area-Weighted Δt	Heat Input, q_{meter}, Btu/h	Test Wall Area, ft^2	R-Value, h·ft^2·°F/Btu
98.1	50.9	47.6	141.4	64	21.6

composite insulation boards to achieve the target R/in. Finally, full-scale 4 × 8 ft composites were created in a manufacturing plant and tested for overall R-value. The tested R-value of the 2 in. thick composite was 21.6 h·ft^2·°F/Btu or R10.8/in. Research is ongoing to the improve the overall R-value to the desired R12/in.

ACKNOWLEDGMENTS

This work was supported by the Building Technologies Office of the U.S. Department of Energy under Contract No. DE-AC05-00OR22725 with UT-Battelle, LLC. The authors would like to acknowledge the support from Dr. Karma Sawyer and Dr. Bahman Habibzadeh of U.S. Department of Energy. The contributions of the following individuals are gratefully acknowledged: Mr. Ron Esparza of NanoPore in manufacturing the MAI panels and making the MAI-HD board panels; Mr. Randy Strauser, the plant manager of the Firestone plant where the manufacturing-scale trial took place; and Mr. Jerald Atchley of ORNL for performing the hot box tests.

REFERENCES

Alam, M., H. Singh, and M.C. Limbachiya, M.C. 2011. Vacuum insulation panels (VIPs) for building construction industry—A review of the contemporary developments and future directions. *Applied Energy* 88:3592–602.

ASTM. 2011. ASTM C1363-11 *Standard test method for thermal performance of building materials and envelope assemblies by means of a hot box apparatus.* Philadelphia: ASTM International.

ASTM. 2015. ASTM C518-15. *Standard test method for steady-state thermal transmission properties by means of the heat flow meter apparatus.* Philadelphia: ASTM International.

DOE. 2014. Windows and building envelope research and development: Roadmap for emerging technologies. http://energy.gov/BTO_windows_and_envelope_report_3.pdf. Washington, D.C.: U.S. Department of Energy, Buildings Technologies Office.

RSMeans. 2014. Table 07 21 13.10 Rigid Insulation.

Field Demonstration of Superior Performance of Advanced Building Envelope Technologies

Nitin Shukla, PhD **Alliston Watts** **Jan Kosny, PhD**
Member ASHRAE

ABSTRACT

Aerogels, phase-change materials (PCMs), and vacuum insulation panel (VIPs) are the three emerging technologies that have potential to significantly reduce building-envelope-generated (wall, roof, etc.) energy loads. Although performance characteristics of these technologies have been well established in numerical simulations and laboratory-scale experiments, there is a general lack of real application data demonstrating their energy-saving potential. In this work, we use a combination of outdoor test structures located in Albuquerque, New Mexico, and a façade panel platform located in Boston, Massachusetts, to investigate the field performance of these advanced technologies. The following configurations were examined: 10 mm (3/8 in.) thick aerogel blanket (R-0.7 $m^2 \cdot K/W$) with reflective facings in the roof and gable envelopes of a test structure in Albuquerque, PCM (5 mm [0.2 in.] thick sheets of a shape-stabilized PCM) in the ceiling of a test structure in Albuquerque, and a VIP assembly (25 mm [1in.] extruded polystyrene [XPS] + 25 mm [1in.] VIP + 25 mm [1in.] XPS; R-5.28 $m^2 \cdot K/W$) in the walls of a façade panel platform in Boston. The field data were collected during the summer season of 2014. Reflective aerogel and PCM configurations were found to save approximately 38% and 2.5%, respectively, in the cooling energy demand through the ceiling compared to the reference configuration built to standard insulation levels. In addition, PCM configuration showed as much as 4.5 hours of time shift in daily peak ceiling heat flux, with an average of 1.68 hours for the duration of the project—an attractive feature in regions with dynamic electricity tariffs. Façade test data showed a dramatic reduction in wall heat flux through the VIP wall assembly compared to a similar-thickness expanded polystyrene (EPS) insulated wall assembly. Wall-generated cooling energy loads were determined to be approximately 82% lower through the VIP wall assembly compared to the EPS wall assembly. This field study validates the superior thermal and energy performances of all the three novel technologies as established in numerical simulation and laboratory testing data reported in scientific literature.

INTRODUCTION

Space heating and cooling in buildings account for approximately 13 quads of primary energy consumption and $142 billion in energy expenditure in the United States each year (DOE 2011). This is equivalent to approximately 13% of both the total primary energy consumption and carbon dioxide emissions in the United States. Considering heat exchange between a building interior and the outdoor environment through the building envelope is the primary source of the space heating/cooling requirements, inserting thermal insulation in the building envelope has been a simple and straightforward way to retard heat flow and save on the space heating and cooling demands. However, conventional plastic foam and fiber-based insulations are limited in their performance—the lowest thermal conductivity achievable is only around 0.021 W/m·K (equivalent to R-7 $h \cdot ft^2 \cdot °F/Btu$ per in.). Moreover, conventional insulations perform poorly in fire conditions and are potential health hazards. Fire retardants added to improve the insulation's fire properties have been reported as potential carcinogens and produce toxic fumes that may cause developmental and neurological disorders (Messer 2010; Ether 2014). Recent years have witnessed an increasing

Nitin Shukla is a research scientist in the Building Energy Technology Group, **Alliston Watts** is on the technical staff, and **Jan Kosny** is the lead of the Building Enclosure Program at Fraunhofer Center for Sustainable Energy Systems (CSE), Boston, MA.

emphasis from the government and building code bodies on improving energy efficiency and the environmental, health, and fire safety of buildings, fueling the development of advanced envelope technologies that are significantly more energy efficient and fire resistant and pose lower health risk than conventional thermal insulations. High-R envelope designs and deep-energy retrofits are two major untapped markets that cannot be realized without application of advanced envelope technologies. It is estimated that a successful implementation of high-R building enclosures in new and existing homes nationwide can decrease energy use by about 2.75 quads per year (Werling 2009). The use of conventional insulations is often impractical in high-R projects because of the high thickness solutions required to meet the target R-values.

Aerogels, vacuum insulation panels (VIPs), and phase-change materials (PCMs) are three emerging building envelope technologies that have great potential to improve the energy efficiency of buildings. These three technologies have already found major applications in other industries: aerogels in subsea pipelines, VIPs as refrigerators, and PCMs as thermal storage materials in solar thermal and electronics. However, in the North American building sector these advanced technologies have remained largely a topic of academic and research interest so far. Several material-scale testing studies have established the superior thermal performance of these advanced technologies in laboratory conditions. Using these experimental data as inputs, numerous computer simulation studies have been conducted to show their energy-saving potential at the building scale in different climates. Surprisingly, despite the availability of a large body of material-scale data and building-scale energy simulations, there is a general lack of field (outdoor) demonstrations on the building-scale deployment and energy performance of these advanced technologies in North America, excepting a few field studies such as those by Parekh et al. (2013), Mukhopadhyaya et al. (2014), Fallahi et al. (2015), Muruganantham et al. (2010), Chen and Guo (2008), Kosny et al. (2012), and Urban et al. (2015). Historically, the high cost of these advanced technologies has been the major hurdle towards their market adoption in the buildings sector. Technology advancements and economy of scale have led to considerable drops in the cost of these technologies during the last decade, making them cost-competitive to conventional thermal insulations in several new and retrofit applications (Kosny et al. 2013a, 2013b). These facts stress the need to generate field data demonstrating the superior energy performance of these technologies. Such data not only is crucial in advancing the understanding of the functioning of these novel technologies in real-world conditions but also helps accelerate the market adoption of these technologies by filling the data gap between laboratory-scale research and market expectations.

The objective of this research was to deploy aerogel, PCM, and VIP technologies in the envelopes of building-scale structures exposed to outdoor conditions and evaluate and compare their thermal and energy performance with that of conventional insulation technologies. In this paper, we first provide a brief introduction of these three technologies. Next, we describe the outline of the outdoor field tests, present the field data, and analyze the performance of the three novel technologies. Finally, we discuss the conclusions gained from this research work.

Aerogel

Aerogel is a high-performance thermal insulation. Aerogel is a product of the sol-gel process where the liquid component of the gel is completely removed and replaced with air using supercritical drying. Samuel Kistler first synthesized aerogels from waterglass in 1931. Silica aerogel is the most common aerogel because of its easier processability, superior thermal performance, and lower cost than other types of aerogels such as carbon and metal oxides. Aerogel is a highly porous material consisting of > 90% air by volume and a mean pore size of around 10–100 nm. The highly nanoporous structure leads to minimized solid and gas conductivities and elimination of pore air convection, resulting in thermal conductivity as low as 0.014 W/m·K—significantly lower than the air thermal conductivity of 0.025 W/m·K and 1.5–3 times lower than conventional insulations. Silica aerogels are lightweight, nontoxic, and water repellent and exhibit excellent fire resistance: highly desirable properties for a building thermal insulating material. Aerogel has been proposed as a cost-effective solution in interior wall retrofit projects (Kosny et al. 2013a). In addition, other building applications such as in curtain walls, window panels, daylighting, and local insulation around window and steel frames to mitigate thermal bridging have been actively pursued (Duer and Svendsen 1998; Kim and Hyun 2003; Kosny et al. 2007; Baetens et al. 2011). Apart from building applications, aerogels have found applications in a variety of other areas, such as thermal insulation for subsea pipelines, space suits, dust collectors on NASA missions, and sound absorbers (Tang et al. 2006; Pierre and Pajonk 2002; Aegerter et al. 2011; Williams 2011).

Phase-Change Material (PCM)

A phase-change material (PCM) is defined as a material with a large latent heat of fusion. Based on material composition, PCMs are classified into two categories: organic and inorganic. Organic PCMs are composed of organic materials such as paraffins, fatty acids, and sugar alcohols. Paraffinic hydrocarbon PCMs are more commonly used in building applications because their melting temperature falls between approximately • 5°C and 60°C (23°F and 140°F) (Mehling and Cabeza 2008), a temperature range covering most of the global climates. In addition, paraffinic PCMs are chemically inert, nontoxic, reliable, and biocompatible and show negligible subcooling effect. Among inorganic PCMs, salt hydrates are the most popular choice for building applications as they have melting temperatures in the range of 5°C to 130°C (41°F to 266°F), a suitable range for building applications. Salt hydrates are a type of inorganic salts containing one or multiple water molecules such that the resulting crystalline solid has a chemical formula of AB.nH_2O. Salt hydrates are nontoxic and nonflammable and possess higher enthalpy, higher thermal conductivity, and lower corrosivity than organic PCMs. Because paraffins have one big disadvantage, that they increase the flammability of the building envelope (Kissock et al. 1998; Salyer and Sircar 1989), salt hydrates have been drawing significant interest both from the scientific community and the building industry in recent years.

PCMs have found applications in a wide array of areas, such as in thermal energy storage, cooling of food products, spacecraft thermal systems, solar power plants, microelectronics thermal protection, waste heat recovery, etc. (Khudhair and Farid 2004; Sharma et al. 2009). In buildings, a PCM may be used to mitigate the heat exchange through the envelope by storing the excess heat during the daytime and releasing it during the cooler nighttime (Kosny et al. 2013b). In addition, numerical studies have shown that PCM may affect shifting of the peak demand time, resulting in reduced cooling energy expenditures in locations with dynamic electricity tariffs (Tomlinson et al. 1992; Stovall and Tomlinson 1995; Kosny 2013b). PCMs have been integrated with different components of the building envelope such as drywall, thermal insulation, ceiling tiles, roof deck elements, etc., to show significant energy savings (Kissock and Limas 2006; Kosny et al. 2010, 2012; Tyagi et al. 2011; Shukla et al. 2012).

Vacuum Insulation Panel (VIP)

A vacuum insulation panel (VIP) consists of a nanoporous core that is evacuated and enclosed in a gas barrier film. Pressed fumed, precipitated, and powder silica are materials of choice as the core, while a thin metal or metallized polymer film is used as the barrier (IEA 2005; Simmler 2005). The nanoporous structure of the core causes minimization of the solid conductivity, while evacuation helps in eliminating the gas conductivity and convection effects within the pore, resulting in a typical VIP thermal conductivity of about 0.004 W/m·K (0.0023 Btu/h·ft·°F), which is almost 5–10 times lower than that of conventional insulation such as extruded polystyrene (XPS, 0.029 W/m·K [0.0168 Btu/h·ft·°F]), expanded polystyrene (EPS, 0.036 W/m·K [0.0208 Btu/h·ft·°F]), polyisocyanurate (PIC, 0.022 W/m·K [0.0127 Btu/h·ft·°F]), and polyurethane (PU, 0.024 W/m·K [0.0139 Btu/h·ft·°F]). VIP technology becomes attractive in building retrofit applications because its high thermal resistivity allows for thinner retrofit solutions, saving on usable space. Future success of VIP technology depends on its cost-effectiveness in addition to its thermal performance and durability as compared to existing conventional technologies (Kosny et al. 2013a; Shukla et al. 2013).

FIELD INVESTIGATION PLAN

For the purpose of this demonstration study, we made use of the Fraunhofer Center for Sustainable Energy Systems (CSE) testing facilities in Albuquerque, New Mexico (ABQ) and Boston, Massachusetts (BOS). The ABQ outdoor test structure facility serves to evaluate the thermal and energy performance of a building envelope technology in real-world conditions. A typical test structure experiment involves constructing two identical test hut structures except that the technology of interest is applied in the envelope of one of the test huts. Both test huts are oriented along the same direction and positioned to minimize solar shading to ensure exposure to similar outdoor conditions. The temperature of the conditioned interior space of each hut is maintained to an identical value using a portable air-conditioner unit. Key thermal and energy information of the test structures and exterior environments are recorded in each hut using thermocouples, heat flow meters, relative humidity (RH) sensors, HOBO temperature and humidity data loggers, weather stations, and pyranometers. After the completion of the experiment, thermal and energy data of the test huts are analyzed and compared to conclude about the thermal and energy performance of the tested envelope technology.

The BOS façade panel test platform facilitates demonstration of assembly-scale energy performance of a wall envelope technology. The test platform consists of a series of 2.44 × 1.22 m (8 × 4 ft) south-facing wall panel placeholders (plugs) located on the sixth floor of Fraunhofer CSE's Living Lab in Boston. In a typical experiment, a wall panel plug is retrofitted with the technology under investigation, exposing it to the cold, humid climate conditions of Boston. The interior space is conditioned to the desired set-point temperature. The test data are analyzed or compared with another technology to validate the performance in a manner similar to that described for the ABQ test facility.

Aerogel and PCM technologies were simultaneously tested in the ABQ test structure facility during the summer months of 2014, while VIP was tested in the BOS façade panel platform during August 2014. The following configurations were investigated:

- *Aerogel:* 10 mm (3/8 in.) thick aerogel blankets with reflective aluminum facings on both sides; total thermal resistance of R-0.7 m^2·K/W (R-4 h·ft^2·°F/Btu); applied to cover the roof deck and gable cavities of a test hut structure located at the ABQ test facility.
- *VIP:* 75 mm (3 in.) thick VIP assembly consisting of a 25 mm (1 in.) thick VIP sandwiched between two layers of 25 mm (1 in.) XPS (25 mm [1 in.] XPS + 25 mm [1 in.] VIP + 25 mm [1 in.] XPS); total thermal resistance of R-5.28 m^2·K/W (R-30 h·ft^2·°F/Btu); applied in the test panel walls of the BOS test platform.
- *PCM:* 5 mm (0.2 in.) thick sheets of a shape-stabilized PCM in a polymer matrix; total heat storage of 170 kJ/kg in the phase active range of 14°C to 30°C; applied in the ceiling of a test hut structure located at the ABQ test facility.

RESULTS AND DISCUSSIONS

Aerogel and PCM Demonstrations in the ABQ Outdoor Test Structure Facility

The thermal and energy performance of the aerogel and PCM technologies were evaluated at the ABQ test structure facility. Three identical test structures were built to standard construction practices and insulation levels. Internal dimensions were approximately 1.89 m (6.2 ft) high (floor to ceiling), 2.13 m (7 ft) long (north to south), and 3.35 m (11 ft) wide (east to west). Thus, the approximate interior floor and ceiling area were each 7.18 m^2 (77 ft^2), the north and south walls were each 6.5 m^2 (70 ft^2), and the east and west walls were each 4 m^2 (43 ft^2). The exterior walls and gable roof were composed of oriented strand board (OSB) attached to wood stud framing. The wall framing was 2–4 construction (actual size of 38 × 89 mm), spaced 0.41 m (16 in.) on center. The roof deck was composed of OSB attached to 2×6 wood framings (actual size of 38 × 140 mm), also spaced 0.41 m (16 in.) on center. Above this OSB roof deck was a three-tab charcoal shingle roof. Figure 1 shows the as-built test huts.

The attic design of the test huts was varied to simulate the following three configurations:

- *Hut 1:* reference hut. A ceiling structure with a 50 mm (2.5 in.) thick R-1.76 $m^2 \cdot K/W$ (R-10 $h \cdot ft^2 \cdot °F/Btu$) XPS board to separate the attic space from the conditioned zone below.
- *Hut 2:* PCM in the ceiling. A simple one-dimensional heat transfer model suggested that sandwiching the PCM layer across two layers of board insulation in the ceiling would maximize the effectiveness of the PCM in the test enclosure. A shape-stabilized PCM embedded in a high-density polyethylene (HDPE) matrix was chosen because this type of PCM does not leak from matrix pores even when melted. The PCM core is enclosed in an impermeable aluminized foil to further impede the chances of PCM seepage to the exterior environment.
- *Hut 3:* Reflective aerogel blanket below the roof deck cavity and behind the interior of the gable cavity (i.e., covering the surface of the roof deck and gable cavities). The reflective surfaces are intended to act as a radiation barrier in the attic space to further reduce heat flow.

Table 1 and Figures 2 and 3 provide more details of the insulation levels and locations of each technology in the different envelope components of each hut. Thermocouple and heat flux sensors were instrumented on key hut envelope locations—ceiling, roof, gables, walls, and floor—to obtain thermal information of conditioned and attic spaces and to ensure the thermal integrity of different components and spaces of the test hut. A data acquisition (DAQ) scheme was implemented to collect and record data in real time. The temperature of the conditioned space of each test hut was set and maintained at 70°F using portable air-conditioner units. The data were collected during summer between 06/01/2014 and 09/08/2014.

The analysis of the heat flux through the ceiling shows that the highest daily peak heat flux occurred in hut 1 (reference hut), followed by hut 2 containing PCM in the ceiling. Hut 3 with reflective aerogel showed the lowest daily heat flux. A weekly plot of ceiling heat flux in the three huts is shown in Figure 4 to confirm this observation. Monthly averages of the daily peak ceiling heat flux (MACHF) were calculated and plotted in Figure 5 to show a common trend that hut 1 has the highest MACHF, followed by hut 2 and hut 3. The MACHF of hut 2 and hut 3 were found to be 1% to 5% and

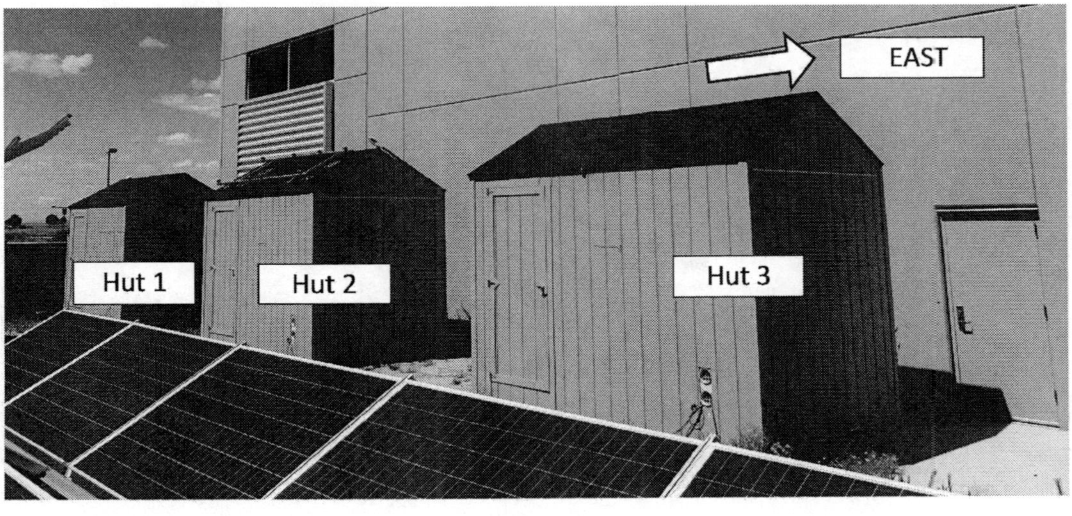

Figure 1 *As-constructed test hut structures.*

Table 1. Thermal Insulation Used in the Three Test Hut Enclosures

Test Structure	Walls	Roof	Gables	Ceiling (Top to Bottom Layers)
Hut 1 (Reference)	R-13 fiberglass batt in cavity + 50 mm (2 in.) XPS	None	None	50 mm (2 in.) XPS + 11 mm (7/16 in.) OSB
Hut 2 (PCM)	R-13 fiberglass batt in cavity + 50 mm (2 in.) XPS	None	None	25 mm (1 in.) XPS + 5 mm (0.2 in.) PCM + 25 mm (1 in.) XPS + 11 mm (7/16 in.) OSB
Hut 3 (Aerogel)	R-13 fiberglass batt in cavity + 50 mm (2 in.) XPS	10 mm (3/8 in.) reflective aerogel	10 mm (3/8 in.) reflective aerogel	50 mm (2 in.) XPS + 11 mm (7/16 in.) OSB

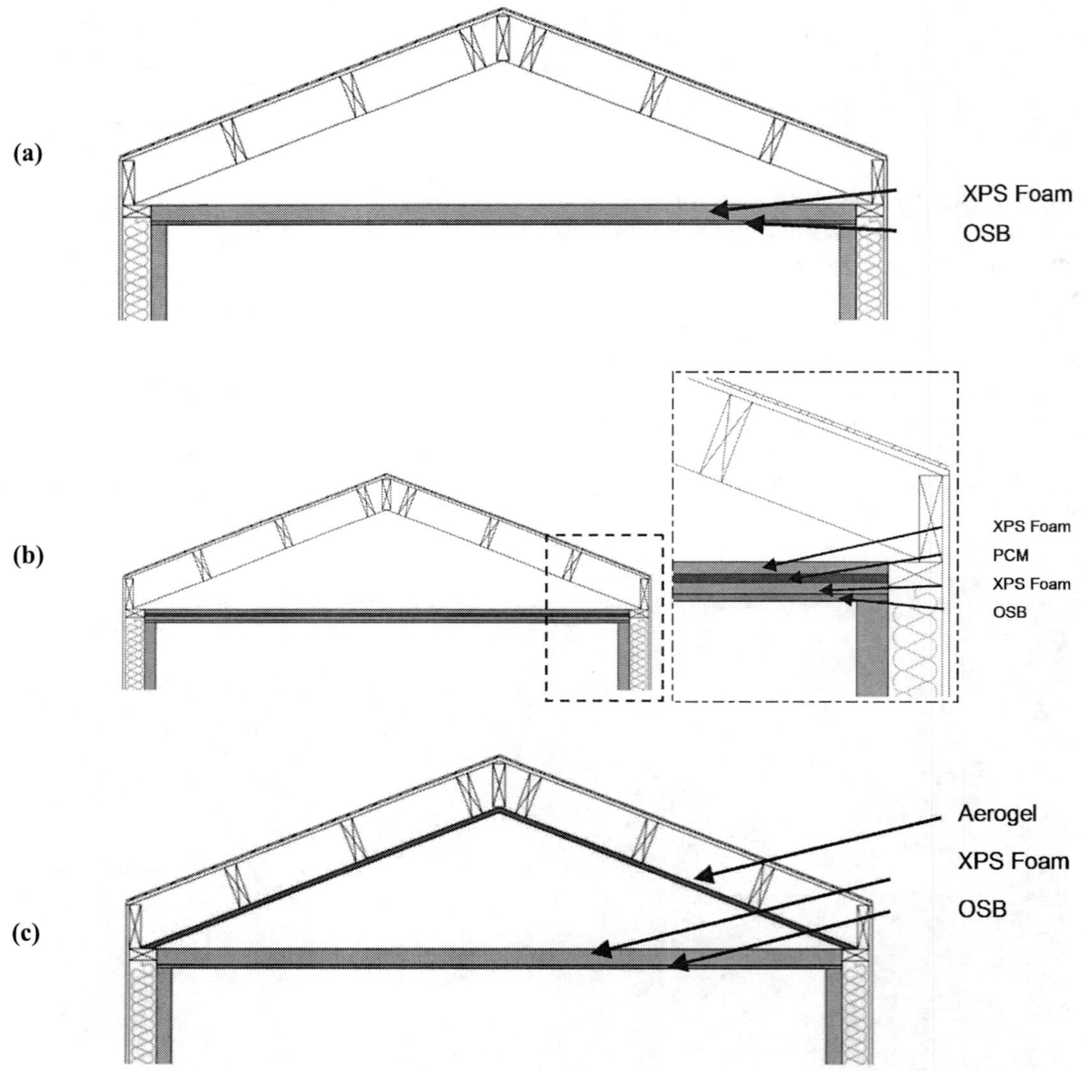

Figure 2 *Test hut configurations: (a) hut 1 (reference hut) containing XPS insulation in ceiling, (b) hut 2 containing PCM in the ceiling structure, and (c) hut 3 containing reflective aerogel in roof and gables.*

Figure 3 *Interior of test hut structures during construction: (a) hut 1, 50 mm fiberglass installed inside wall cavity, (b) hut 2, PCM ceiling installed, and (c) hut 3, reflective aerogel installed under roof rafters and gable cavity.*

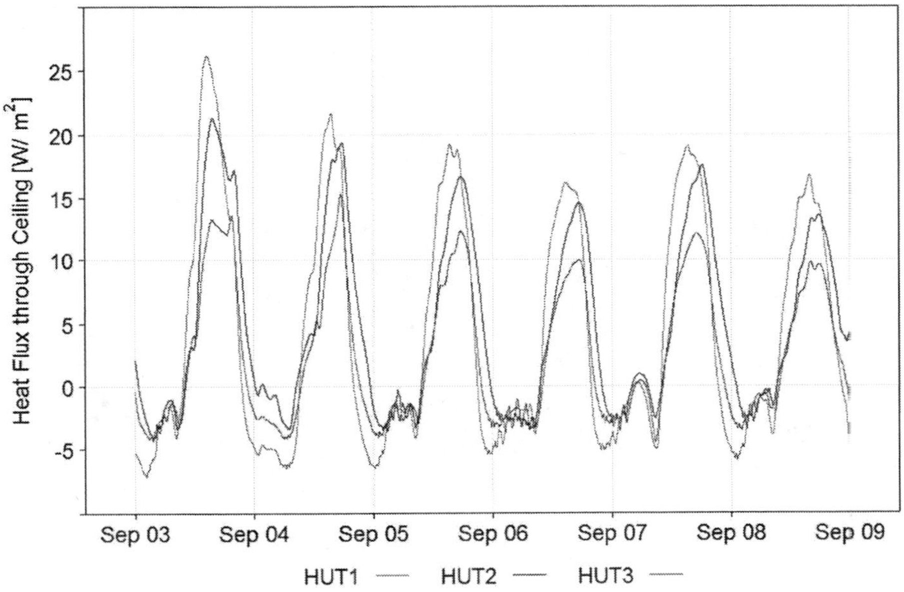

Figure 4 *Ceiling heat flux for hut 1, hut 2, and hut 3.*

33% to 35% lower, respectively, than that of hut 1 during the investigation period, as shown in Figure 6. These results indicate that the PCM in the ceiling of hut 2 caused a reduction, although small, in the ceiling heat flow compared to hut 1. On the other hand, the reflective aerogel insulation in hut 3 was found to affect significant reduction in MACHF compared to hut 1. To estimate the attic-generated cooling energy demand (i.e., the cooling load) in each test hut, positive heat flux (heat gain) values through the ceiling were integrated over the entire time period of the testing. Figure 7 shows the daily cooling demand through the ceiling for a week in June 2014. Hut 3 was found to have the lowest daily ceiling cooling load, while hut 2 showed a slightly lower attic-generated cooling energy consumption than hut 1. When considering the entire duration of the project, the energy analysis showed that hut 1, hut 2 and hut 3 generated 56.6, 55.2, and 34.9 MJ/m^2 of heat gains, respectively, through their ceiling envelopes. In other words, this demonstration study shows that PCM and reflective aerogel technologies as deployed in the test huts save 2.5% and 38%, respectively, of attic-generated cooling energy demand compared to the reference ceiling consisting of conventional insulation.

Interestingly, the presence of PCM in the ceiling has been found to cause a time shift in the occurrence of peak heat flux on a daily basis, which can be seen clearly in Figure 4. This time shift in daily ceiling peak heat flux was calculated on a daily basis by subtracting the time occurrence of the peak ceiling heat flux in hut 2 from the time occurrence of the same in hut 1 (see Figure 8). Time shifts as large as 4.5 h were achieved during the experiment. On an average, a daily time shift of 1.68 h (101 min) was observed by applying the PCM technology in the ceiling for the summer period investigated.

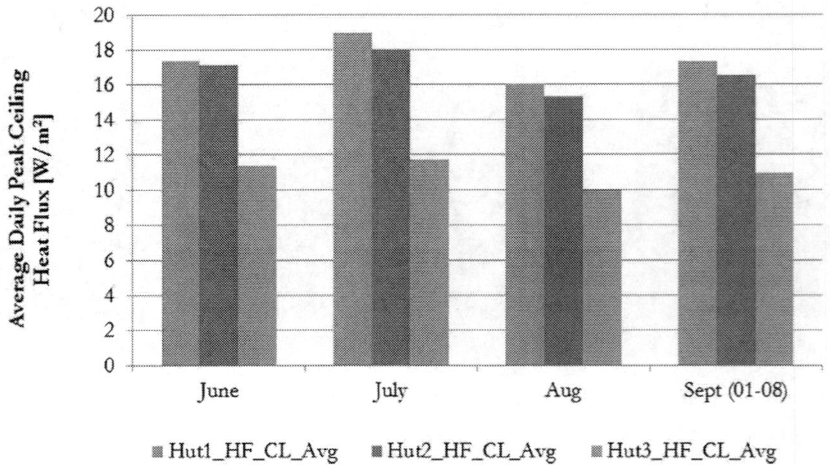

Figure 5 *Monthly average of daily peak heat flux through the ceiling of each hut.*

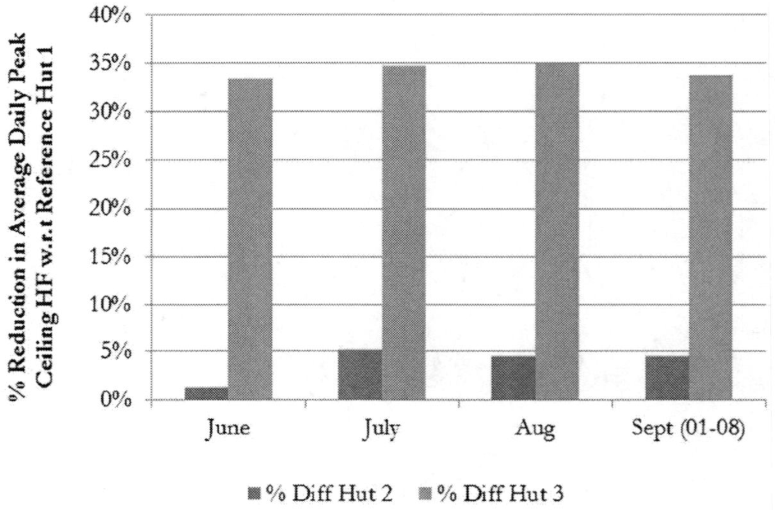

Figure 6 *Percentage reduction in monthly average of daily peak ceiling heat flux of hut 2 and hut 3 compared to hut 1 (reference hut) during phase I experiments.*

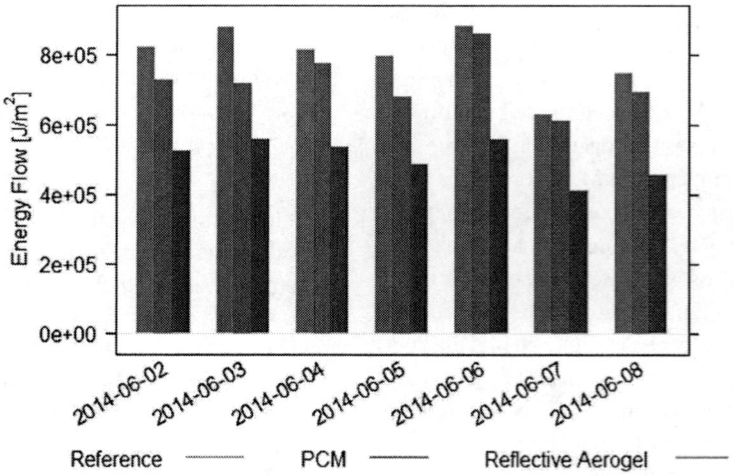

Figure 7 *Thermal energy (heat) gains through the ceiling of the three huts during a week of June 2014.*

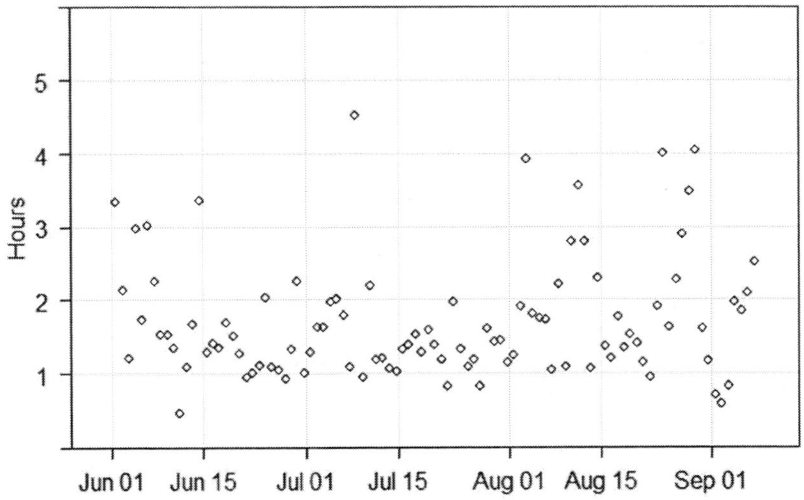

Figure 8 *Time shift in daily peak ceiling heat flux of hut 2 compared to hut 1 (reference hut) for the duration of the experiment.*

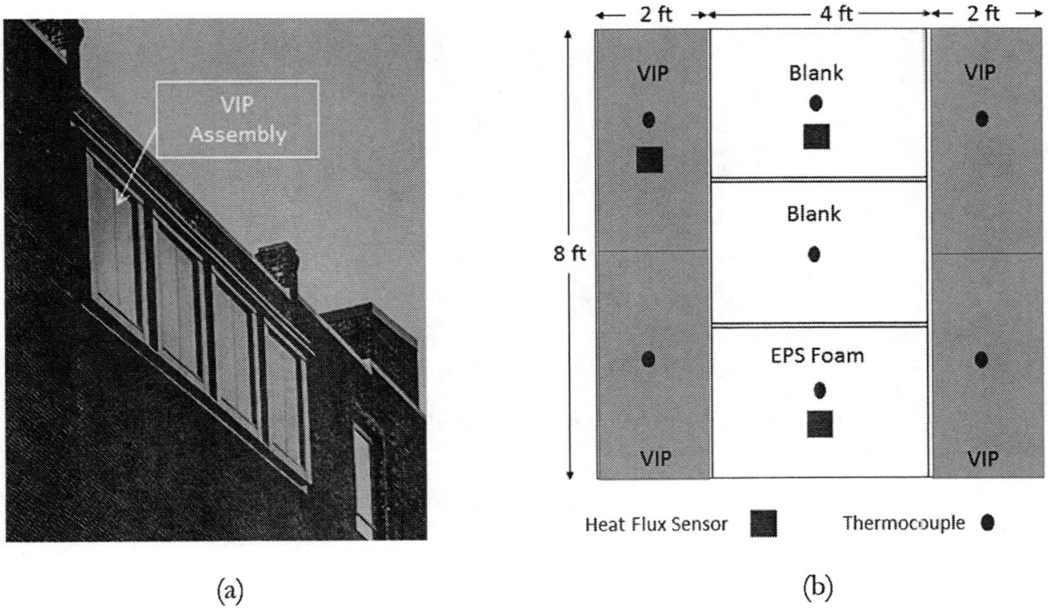

Figure 9 *(a) VIP assembly located in BOS façade panel test platform and (b) schematic of the design and sensor placement in the tested VIP wall panel.*

VIP Demonstration in BOS Façade Panel Testing Facility

To perform this study, one of the wall panel plugs was retrofitted with the 75 mm (3 in.) thick VIP assembly described in the previous section. The plug consisted of a 25.4 mm (1 in.) thick insulated metal wall panel that was used to support the interior facing VIP assembly and was exposed to the exterior environment as shown in Figure 9a. To compare the performance of the VIP assembly with uninsulated and conventional foam configurations, part of the wall panel was left uncovered (blank containing insulated metal wall panel only) and part was insulated with 75 mm (3 in.) EPS (RSI-0.7 $m^2 \cdot K/W$ [R-4 $h \cdot ft^2 \cdot F/Btu$ per in.]) sheets, as shown in Figure 9b. The arrangement of the heat flow sensors and thermocouple is shown in Figure 9b. The interior space was conditioned to 21.1°C (70°F) for the purpose of the experiments. The data were collected for the summertime period of 08/01/2014 to 08/21/2014.

Figure 10 shows the heat flux through the VIP assembly, EPS, and blank (uninsulated) sections of the tested panel. Typical daily peak heat fluxes of 10–18 W/m^2 (3.17–5.71 Btu/$h \cdot ft^2$) were observed for the blank section. The EPS-insulated section showed significant reduction of 30% to 40% in the daily peak heat flux compared to the blank. On the other hand, the daily peak heat flux through the VIP assembly was found to be almost completely eliminated—a clear demonstration of the energy-saving potential of the VIP technology in buildings. Next, daily cooling energy loads were calculated by integrating the positive heat flow values (i.e., heat gain) on a daily basis as depicted in Figure 11. It was noticed that on most days the daily cooling loads were decreased by 90% to 95% for the VIP section and 50% to 60% for the EPS section, compared to the blank section. During three weeks of the experiment, the VIP, EPS, and blank sections generated cooling loads of 0.27, 1.48, and 3.03 MJ/m^2, respectively, which is equivalent to approximately 91% and 51% reductions in the cooling demand for the VIP assembly and EPS sections, respectively, compared to the blank section. In other words, this field study demonstrates a superior energy efficiency of the tested VIP technology—approximately 82% in cooling energy requirement compared to a similar-thickness conventional EPS foam insulation, when applied in the wall envelope.

CONCLUSIONS

Aerogels, PCMs, and VIPs are three important emerging building technologies with their promise to significantly

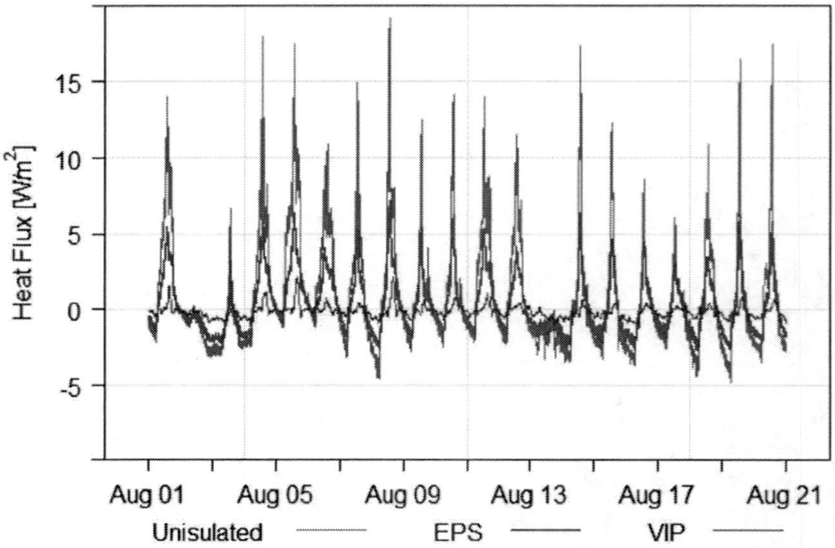

Figure 10 *Heat flow through the VIP assembly, EPS, and uninsulated sections of the test wall panel.*

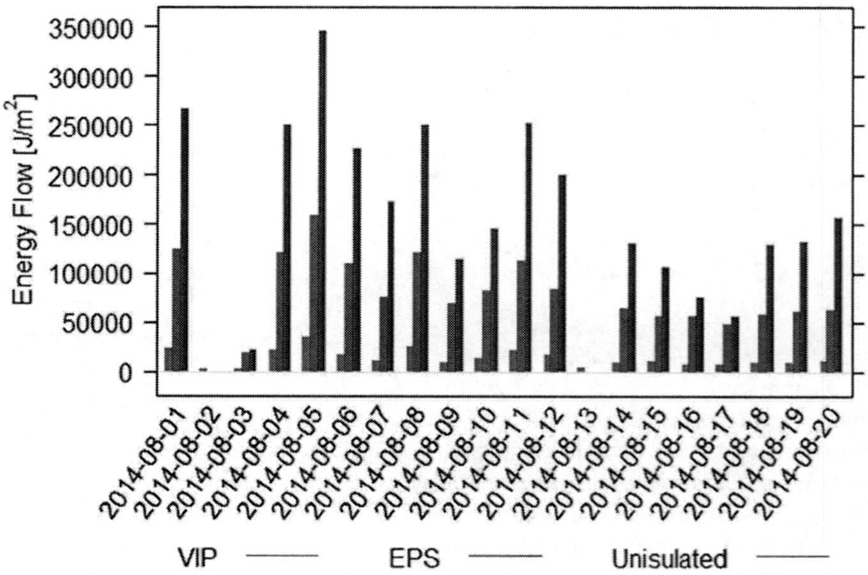

Figure 11 *Daily cooling energy consumption through the VIP assembly, EPS, and uninsulated sections of the test wall panel.*

improve the energy efficiency of buildings. The high installed cost of these advanced technologies has been a major barrier towards their widespread acceptance in the building sector; however, recent technology and production breakthroughs have brought down the product cost, making these technologies attractive in several new and retrofit building projects. Several testing studies have established the superior material-scale thermal performance of these technologies compared to conventional insulations, and numerical simulations have shown significant energy savings of these technologies at the building scale. However, there is a general lack of field data demonstrating their superior performance when implemented in a real-world building application.

This research was intended to deploy these technologies in the envelope of building-scale structures and demonstrate their superior performance. For this purpose, we employed Fraunhofer CSE's test structure facility located in Albuquerque, New Mexico (ABQ), and façade panel test platform located in Boston, Massachusetts (BOS). Reflective aerogel (consisting of an aerogel blanket laminated with aluminum foil on both sides) and PCM technologies were tested using three identical test hut structures at the ABQ facility. The first test hut served as the reference. A thin 5 mm (0.2 in.) PCM sheet was located in the ceiling of one of the huts, and a 10 mm (3/8 in.) thick reflective aerogel blanket was applied in the roof and gables of another hut. Experiments conducted during three summer months of 2014 showed that the addition of the PCM layer in the ceiling structure caused 1% to 5% reduction in the monthly daily ceiling heat flux. Daily peak heat flux was also found to be time-shifted by as much as 4.5 h, with an average of 1.68 h during the course of the experiment—a very beneficial feature in locations with dynamic electricity tariffs. For the reflective aerogel case, the decrease in the monthly daily peak ceiling heat flux was even higher—as much as 30% to 40% lower compared to the reference test hut. The cooling loads through the ceiling decreased by approximately 2.5% with the addition of PCM in the ceiling and by 38% with the application of the reflective aerogel blanket in the roof and gable envelopes.

VIP technology was evaluated at the BOS facility during three weeks of summer in 2014. A 75 mm (3 in.) thick VIP assembly consisting of 25 mm (1 in.) thick VIP sandwiched between two layers of 25 mm (1 in.) thick EPS boards was inserted into the south-facing wall plugs. For comparison purposes, part of the wall plug was left blank and part was insulated with 75 mm (3 in.) thick EPS boards. Heat flux through the VIP assembly was found to be negligible compared to the blank section. Further analysis of field data indicated approximately 82% reduction in wall-generated cooling energy demand through the VIP assembly when compared to the EPS board.

In conclusion, the two outdoor field studies performed in this work demonstrate significantly superior thermal and energy performance of all the three novel technologies investigated in this work. Considering a current lack of the performance data on the real-building application of these technologies, the data generated in this work serve to gain technical confidence in these technologies.

REFERENCES

Aegerter, M.A., N. Leventis, and M.M. Koebel. 2011. *Aerogels handbook*. New York: Springer-Verlag.

Baetens, R., P.J. Bjørn, and A. Gustavsend. 2011. Aerogel insulation for building applications: A state-of-the-art review. *Energy and Buildings* 43:761–69.

Chen, C., and H. Guo. 2008. A new kind of phase change material (PCM) for energy-storing wallboard. *Energy and Buildings* 40:882–90.

DOE. 2011. *DOE energy data book*. Washington, DC: U.S> Department of Energy. http://buildingsdatabook.eren.doe.gov.

Duer, K., and S. Svendsen. 1998. Monolithic silica aerogel in superinsulating glazings. *Solar Energy* 63(4):259–67.

Ether, R.D. 2014. *An alternatives assessment for the flame retardant decabromodiphenyl ether*. Final report. Washington, DC: U.S. Environmental Protection Agency.

Fallahi, A., J. Kosny, N. Shukla, et al. 2015. Field thermal performance of EIFS using vacuum insulation panels. Presented at the Buildings Enclosure Science & Technology (BEST) Conference 4, Kansas City, MO, April 13–15.

IEA. 2005. ECBCS/IEA Annex 39. Vacuum insulation panels, study on VIP-components and panels for service life prediction of VIP in building applications (Subtask A). Paris: International Energy Agency. http://www.ecbcs.org/annexes/annex39.htm.

Khudhair, A.M., and M.M. Farid. 2004. A review on energy conservation in building applications with thermal storage by latent heat using phase change materials. *Energy Conversion and Management* 45(2):263–75.

Kim, G.S., and S.H. Hyun. 2003. Synthesis of window glazing coated with silica aerogel films via ambient drying. *Journal of Non-Crystalline Solids* 320(1-3):125–32.

Kissock, K., and S. Limas. 2006. Diurnal load reduction through phase-change building components. *ASHRAE Transactions* 112(1):509–17.

Kissock, J., J. Kelly, M. Hannig, and I. Thomas. 1998. Testing and simulation of phase change wallboard for thermal storage in buildings. *Proceedings of 1998 International Solar Energy Conference*.

Kosny, J., T. Petrie, D. Yarbrough, P. Childs, A. Syed, and C. Blair. 2007. Nano-scale insulation at work: Thermal performance of thermally bridged wood and steel structures insulated with local aerogel insulation. *Proceedings of Thermal Performance of the Exterior Envelopes of Whole Buildings X.*

Kosny, J., D. Yarbrough, W. Miller, S. Shrestha, E. Kossecka, and E. Lee. 2010. Numerical and experimental analysis of building envelopes containing blown fiberglass insulation thermally enhanced with phase change material (PCM). *Proceedings of the 1st Central European Symposium on Building Physics.*

Kosny, J., K. Biswas, W. Miller, and S. Kriner. 2012. Field thermal performance of naturally ventilated solar roof with PCM heat sink. *Solar Energy* 86:2504–14.

Kosny, J., A. Fallahi, and N. Shukla. 2013a. *Cold climate building enclosure solutions.* Report prepared for the U.S. Department of Energy Building Technology Program. Washington, DC: U.S. Department of Energy.

Kosny, J., A. Fallahi, and N. Shukla. 2013b. *Cost analysis of simple phase change material-enhanced building envelopes in southern U.S. climates.* Report prepared for the U.S. Department of Energy Building Technology Program. Washington, DC: U.S. Department of Energy.

Mehling, H., and L.F. Cabeza. 2008. *Heat and cold storage with PCM: An up to date introduction into basics and applications.* Berlin: Springer.

Messer, A. 2010. Mini-review: Polybrominated diphenyl ether (PBDE) flame retardants as potential autism risk factors. *Physiology & Behavior* 100:245–49.

Mukhopadhyaya, P., D. MacLean, J. Korn, et al. 2014. Building application and thermal performance of vacuum insulation panels (VIPs) in Canadian subarctic climate. *Energy and Buildings* 85:672–80.

Murugananthama, K., P. Phelan, P. Horwath, D. Ludlam, and T. McDonald. 2010. Experimental investigation of a bio-based phase-change material to improve building energy performance. *Proceedings of ASME 2010 4th International Conference on Energy Sustainability, ES2010.*

Parekh, A., C. Mattock, and C.J. McLellan. 2013. Construction and field monitoring of exterior walls using vacuum insulation panels (VIPs) in a net zero home. *Proceedings of the Thermal Performance of the Exterior Envelopes of Whole Buildings XII International Conference.*

Pierre, A.C., and G.M. Pajonk. 2002. chemistry of aerogels and their applications. *Chemical Reviews* 102:4243–65.

Salyer, I., and A. Sircar. 1989. Development of PCM wallboard for heating and cooling of residential buildings. *Thermal Energy Storage Research Activities Review*, pp. 97–123.

Sharma, A., V.V. Tyagi, C.R. Chen, and D. Buddhi. 2009. Review on thermal energy storage with phase change materials and applications. *Renewable Sustainable Energy Rev.* 13:318.

Shukla, N., A. Fallahi, and J. Kosny. 2012. Performance characterization of PCM impregnated gypsum board for building applications. *Energy Procedia* 30:370–79.

Shukla, N., A. Fallahi, and J. Kosny, J. 2013. VIP—Technology review and cost study for building retrofits in northern US locations. *Proceedings of Thermal Performance of the Exterior Envelopes of Whole Buildings XII International Conference.*

Simmler, H., S. Brunner, U. Heinemenn, et al. 2005. *Vacuum insulation panels: Study of VIP-components and panels for service life prediction of VIP in building applications.* IEA/ECBCS Annex 39• Implementing agreement, Report on Subtask A. Paris: International Energy Agency.

Stovall, T.K., and J.J. Tomlinson. 1995. What are the potential benefits of including latent storage in common wallboard? *Journal of Solar Energy Engineering* 117:318–25.

Tang, H.H., E.S. Orndoff, and L.A. Trevino. 2006. Thermal performance of space suit elements with aerogel insulation for moon and mars exploration. Presented in the EVA–PLSS & Support Equipment I session at the International Conference on Environmental Systems, Norfolk, VA, July.

Tomlinson, J., C. Jotshi, and D. Goswami. 1992. Solar thermal energy storage in phase change materials. *Proceedings of Solar '92: The American Solar Energy Society Annual Conference.*

Tyagi, V.V., S.C. Kaushik, S.K. Tyagi, and T. Akiyamac. 2011. Development of phase change materials based microencapsulated technology for buildings: A review. *Renewable and Sustainable Energy Reviews* 15(2):1373–91.

Urban, B., N. Shukla, A. Fontanini, and J. Kosny. 2015. Field performance of two aerogel blankets in a hot, dry climate. Presented at the Buildings Enclosure Science & Technology (BEST) Conference 4, Kansas City, MO, April 13–15.

Werling, E. 2009. Building America research-to-market plan. Technical report DOE--EE 1285 7251. Washington, DC: U.S. Department of Energy, Office of Energy Efficiency and Renewable Energy (EERE), Building Technologies Office.

Williams, J. 2011. Aerogel-based insulation for industrial steam distribution systems. Technical Report submitted to US-DOE Office of Industrial Technology Program (EE-2F) Report Number DOE/GO16056. Washington, DC: U.S. Department of Energy.

Analysis of Wind-Driven Rain Exposure Based on Long-Term Monitoring

Fitsum Tariku, PhD
Member ASHRAE

Wendy Ying Simpson, MASc

Elsa Ngudjiharto, MASc

ABSTRACT

Building envelopes are subjected to highly variable climatic loadings on the exterior and relatively stable conditions in the interior. These loadings result in transportation of heat, air, and moisture across the building envelope, and, therefore, its components may experience variances in wetting or drying depending on the building design and construction, material use, exterior climate and interior conditions. The durability of these components will be compromised if there is excessive moisture accumulation for extended periods. Wind-driven rain (WDR) is a major moisture load that affects the durability of building envelope components. To control and manage rain load, building designers incorporate overhangs in their roof design and introduce an air cavity between cladding and sheathing membrane with the objective of creating a capillary break. To properly design for WDR load, however, one has to have information on the relative magnitude and frequency of the wind-driven load on facades. In this paper, the WDR measurements that the authors carried out for four years and nine months at a two-story level test building in a mild coastal climate are presented. The measured WDR exposures on different orientations of the building's facades are compared and the rain admittance factors (RAFs) for different locations on the building are calculated using datasets with low and high rainfall intensities and are compared.

INTRODUCTION

Building envelopes are exposed to varying climatic loadings on the exterior, while the interior conditions are typically stable. Depending on the building design, construction, materials, and these conditions, building envelope assemblies may experience wetting and drying from heat, air, and moisture transfer across the assemblies. When the wetting of the assembly occurs more frequently, for longer periods of time, or at a higher quantity than the drying capability, the accumulation of moisture may cause deterioration of envelope components, such as corrosion of steel studs or rotting of wood frames. In a West Coast climate, such as in British Columbia, Canada, the source of exterior moisture loading causing the most durability issues to wall assemblies is wind-driven rain (WDR) (Morrison Hershfield 1996). Improvements to building envelope design have since included considerations to increase the drying capability of the assemblies against moisture loading, in particular from wind-driven rain. Determining the amount of wind-driven rain that a building envelope assembly is exposed to is therefore a critical step in this improved design effort.

With regards to building envelope design, *wind-driven rain* is defined as the amount of rain drops that are impinging on the vertical building surfaces due to wind forces blowing towards the building. Many factors affect the amount of wind-driven rain, including rain intensity, wind speed and direction, topography, building geometry, orientation of the building assembly, and the location of area of interest on the building elevation. However, the primary components to wind-driven rain intensity are rain intensity, wind speed, and wind direction.

Estimation formulas on wind-driven rain have approached the subject using semi-empirical and numerical analysis methods. The numerical analysis method most commonly referred to today was mainly developed in the

Fitsum Tariku is the director and **Wendy Ying Simpson** is a research analyst at the Building Science Centre of Excellence (BSCE), British Columbia Institute of Technology (BCIT), Burnaby, BC. **Elsa Ngudjiharto** is a project coordinator at IRC Building Sciences Group, Richmond, BC.

1990s (Choi 1994a, 1994b, 1994c, 1999). On the other hand, the semi-empirical analysis method was developed in the 1950s by Lacy (Hens 2015; Lacy 1965) using experimental-based relations of raindrop size, raindrop speed, and horizontal rain intensity by Best (1950a, 1950b). A more recently developed semi-empirical estimation formula is by Straube and Burnett (2000), based on Lacy's work. Much research on the validation of these wind-driven rain estimation methods has been published and reviewed (Blocken and Carmeliet 2010; Hens, 2015; van Mook 2002).

To estimate the amount of wind-driven rain without direct field measurement at the location of interest, historical weather data can be used. Thirty-year meteorological data for various cities in Canada have been studied (Zhu et al. 1995). To account for other factors, such as the building geometry, orientation of the building assembly, and the area of interest on the wall facade, semi-empirical studies have provided coefficients to be used in calculations. One such factor is the wall factor, *W* (BSI EN 1997), or, in other literature, *rain admittance factor* (RAF) (Straube and Burnett 2000), which considers the building aspect ratio, the presence of a roof overhang, and the area of interest within the wall facade.

In this paper, the dependency of one these coefficients, RAF values, on rainfall intensity and two facade orientations exposed to different levels of WDR loads are examined.

WDR CALCULATION

In addition to obtaining the horizontal rain intensity and wind data, several parameters are first calculated. First is the median raindrop diameter, based on the relations by Best (1950a), which is a function of the horizontal rain intensity:

$$d_{50} = 1.30(R_h^{0.232} \times 0.69^{1/n}) \quad (1)$$

where

n = coefficient, 2.25

d_{50} = median raindrop diameter, mm

R_h = horizontal rainfall intensity, mm/h·m^2

Second, the terminal velocity of raindrops, $V_t(d)$ is based on a regression fit developed by Dingle and Lee (1972) and is a function of the median raindrop diameter, limited to diameters ranging from 0.1 to 5.8 mm, as follows:

$$V_t(d) = -0.166033 + 4.918441d - 0.888016d^2 + 0.054888d^3 \quad (2)$$

where

$V_t(d)$ = terminal velocity of raindrops

d = median raindrop diameter, m

Using the calculated terminal velocity, the wind-driven intensity on a vertical surface according to Straube and Burnett's method (Straube and Burnett 2000) is

$$R_{wdr} = \text{RAF} \times \text{DRF}(V_t) \times \cos(\theta) \times U(h) \times R_h \quad (3)$$

where

R_{wdr} = wind-driven rain intensity, mm

RAF = rain admittance factor

DRF(V_t) = driving rain factor = $1/V_t$

θ = angle of the wind to the wall's normal

$U(h)$ = wind speed at the height of interest, m/s

R_h = horizontal rainfall intensity, mm/h·m^2

Factors [$\cos\theta \times U(h) \times R_h$] represent free-field wind-driven rain, the rain intensity through a 1 m^2 area of unobstructed vertical field. Driving rain factor, or DRF (V_t), is the inverse of raindrops' terminal velocities. RAF accounts for the obstructions to the airflow around the wall facade due to the building's aspect ratio and other wall geometry features, as well as the location of interest within the facade. According to Van Den Bossche et al. (2013), using the median drop diameter introduces an error ranging from 15% up to 20%.

METHODOLOGY

Weather data, such as horizontal rain intensity, vertical rain intensity at four wall elevations, wind speed, and wind direction, were collected from January 2009 to September 2013 using a horizontal rain gauge, 15 wall rain gages, and an anemometer at the Building Envelope Test Facility (BETF) in Burnaby, British Columbia, Canada. The horizontal rain intensity, the wind speed and direction data were used to calculate the free-field wind-driven rain and the driving rain factor. Based on Equation 3, the rain admittance factor (RAF) at each wall rain gauge location was determined by the slope of the plot of wind-driven rain intensity against the calculated free-field wind-driven rain (Ngudjiharto et al. 2014). The RAF values at the different locations on the walls were then compared to the RAF values in literature, as shown in Figure 1.

Building Envelope Test Facility

The BETF is located on the Burnaby campus of the British Columbia Institute of Technology (BCIT). As shown in Figure 2, the topography around the test facility is an open field area, surrounded by parking lots on the west and south side and a grassy area on the north side. The closest trees, which may become obstacles to wind-driven rain, are at approximately 24.4 meters (80 ft) from the building's south east corner. The interior environment of the test facility was maintained at approximately 21°C and 50% rh on the interior throughout the measurement period. A weather station located on the roof of the building includes horizontal rain gages and an anemometer. At the test location, the south easterly wind is the most dominant wind direction (Tariku et al. 2015).

Horizontal Rain Gages and Anemometer

The type of horizontal rain gauge installed on the roof of the building is a tipping-bucket rain gauge. For every 2.0 grams of rain collected by the gauge, the horizontal rain gauge tips. The data acquisition system counts the total number of times the bucket tips each minute. The minute totals

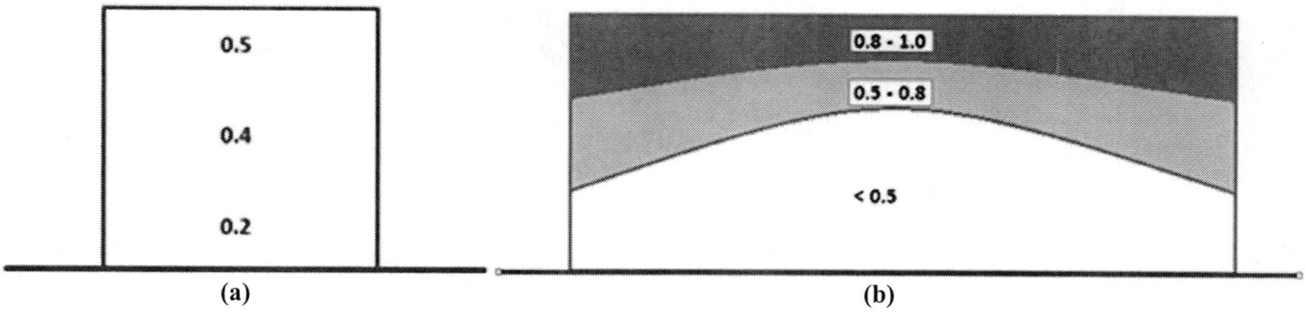

Figure 1 *Wall factor (W) or rain admittance factor (RAF) on a wall facade (a) for a two-story flat roof with a slope less than <20° (BSI EN 1997) and (b) for a low-rise building with an aspect ratio much less than 1 (Straube and Burnett 2000).*

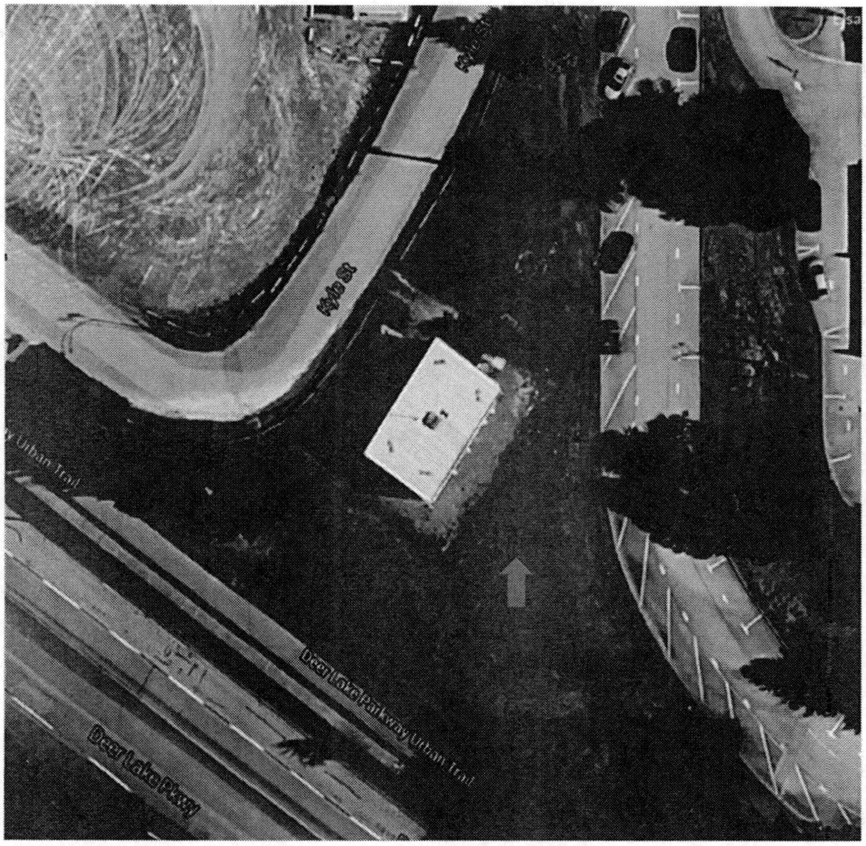

Figure 2 *Satellite image of the BETF from Google Maps (Google Maps 2015).*

are then averaged for hourly calculations. Based on the catchment area, each 2.0 grams or mL of rainwater collected is equal to 0.1 mm of rainwater. The anemometer was also installed on the roof of the building to measure wind speed and wind direction each minute.

Wall Rain Gages. A wall rain gauge is made up of a diamond-shaped catchment metal plate connected to a tipping bucket rain gauge apparatus via a plastic tubing (Figure 3). The diamond shape of the metal plate was designed to ensure that all rainwater on the plate is directed to a single point where the plastic tube is. Each tip of WDR drip collected is equal to 0.038 mm of rainwater on the wall.

RESULTS

The annual horizontal rain and wind speed data for the five years (2009 to 2013) are summarized in Table 1. The data for 2013 is based on nine months' measured values (January to September). The total rainfall and number of rain hours in the first four years are within 8% and 15% difference, respectively. The distributions of the horizontal rain and the wind speed data can be represented by gamma function as the probability plot in Figure 4 shows (linear trend). The plot helps to examine if the selected distribution function is appropriate to use for the data. Discussion on probability plot can be found in Montogomery and Runger (2011). The mean and the median rain intensities for the five years varies from 0.88 to 0.96 mm/h and 0.5 to 0.6 mm/h, respectively, whereas the mean and median wind speed during rainfall varies from 0.79 to 0.89 m/s and 0.56 to 0.72 m/s, respectively. This statistical summary shows that the average rain intensities and wind speeds do not differ significantly from year to year.

The directional wind-driven rain for the five years in the direction of the test building's facade orientations; northeast, northwest, southwest, and southeast; are presented in the bar chart in Figure 5. The predominant wind-driven rain direction for the location is the southeast orientation, which is consistent for all years, followed by the northeast orientation. The northwest orientation has consistently the least potential for wind-driven rain exposure compared to the other orientations.

Figure 3 *A typical wall rain gauge installed on the walls of the BETF.*

RAF Values

Figure 6 shows a typical plot of the hourly measured wind-driven rain against the hourly free-field wind-driven rain calculated using the hourly total amount of horizontal rain and the hourly average value of normalized wind speed, based on Equation 3. As previously discussed, the slopes of these plots are the RAFs for the particular location on the particular wall orientation. In this work, the RAF values of ten wind-driven rain locations on the northeast, northwest, and southeast facades are determined for each year of the measurement period. The results from the 48 plots for years 2009 to 2013 are summarized in Table 2. Because of the presence of obstructions around the Southwest orientation, which would have an impact on the measured wind-driven rain load on the wall and subsequently lead to incorrect RAF values, RAF values for the southwest orientation are excluded from the table. In general, the RAF values of the different gages (orientations and locations) are fairly close across the five years. The five years' average RAF values for the ten locations vary from 0.19 to 0.55. The typical RAF value is approximately 0.3. In general, the RAF values appear to be more comparable to the wall factor, *W*, as in Figure 1a.

To investigate the sensitivity of the RAF values to rain fall intensity, the measured and the calculated free-field wind-driven rain data of each year are grouped into high and low horizontal rain categories. The 90 percentile horizontal rainfall value, which is 2 mm/h, is used as a cut-off value to sort and group the data into two categories: rain events with an hourly rainfall of 2 mm/h and above are grouped into the high rainfall category and rain events below 2 mm/h are in the low rainfall category. Figure 7 shows plots of hourly measured WDR at the

Table 1. Summary of Annual Horizontal Rain and Wind Speed Data

	2009	2010	2011	2012	2013*
Total horizontal rain, mm	1426	1475	1517	1539	992
Total number of rain hours	1478	1638	1593	1736	885
Average rain fall intensity, mm/h	0.96	0.9	0.95	0.88	0.96
Median rain intensity, mm/h	0.6	0.5	0.6	0.5	0.6
Average wind speed for the year, m/s	0.82	0.85	0.81	0.81	0.84
Average wind speed during rain, m/s	0.86	0.89	0.85	0.79	0.82
Median wind speed during rain, m/s	0.65	0.72	0.68	0.56	0.57

*The data for 2013 are based on nine months' measured values (January to September).

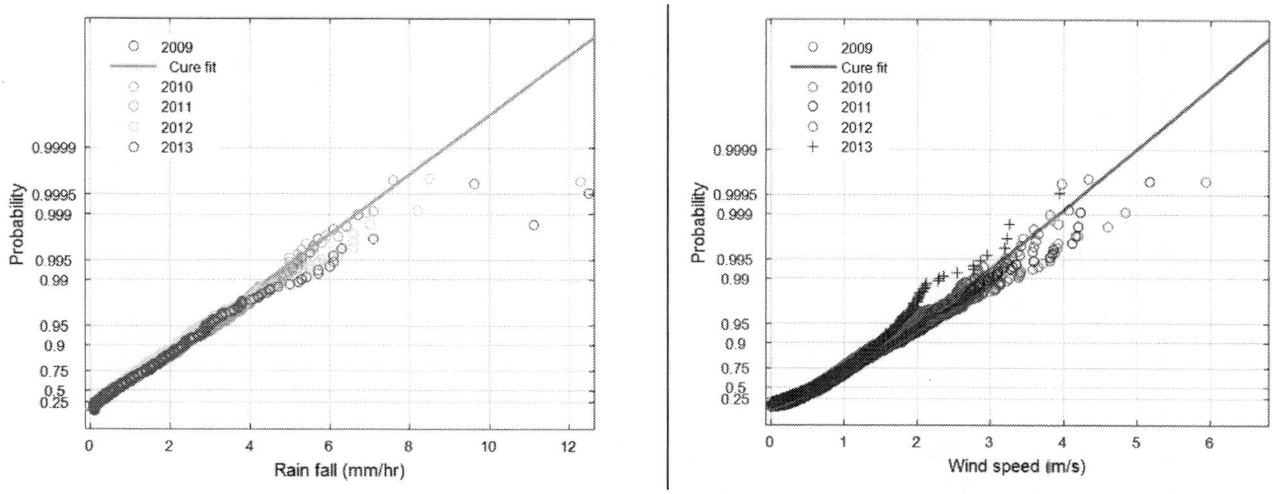

(a) (b)

Figure 4 *Probability plot of rainfall and wind speed data for years 2009 to 2013.*

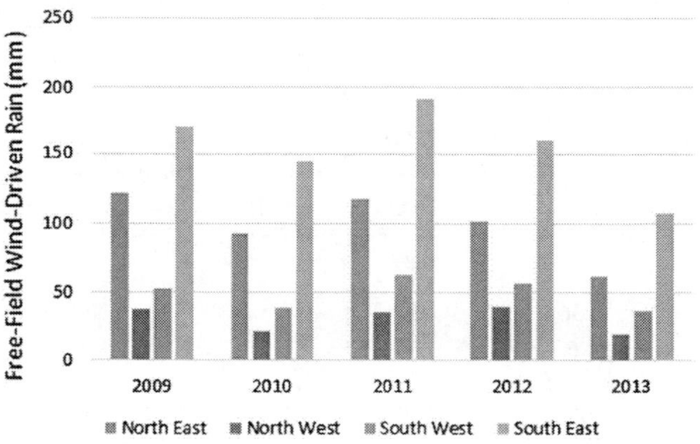

Figure 5 *Directional free-field wind-driven rain data collected from four facade orientations.*

top-center and mid-center of the southeast facade against the calculated hourly free-field WDR using the two data sets: low and high horizontal rain events are used as examples. As expected in both figures, the RAF values at the top are slightly higher than at the mid-height of the facade. The figures also show that the RAF value of a location can vary according to the rainfall intensity category (low or high). In fact, during the low horizontal rain events (less than 2 mm/h of rain), the wind-driven rain that is measured by the rain gages is less than 0.1 mm/h for 99% of the time. The figures suggest that at high horizontal rain intensity, the free field and measured wind-driven rain values are also higher. According to Blocken (2004) CFD simulation results, the catch ratios (WDR on the wall divided by horizontal rain) remain the same for rain intensities higher than 1 mm/h, which leads to a similar curve as presented in Figure 7, given that the wind speed at the monitoring location is generally low.

The RAF values of the ten rain gages' locations obtained by using the discrete data sets (low and the high horizontal rain data sets) and the combined data sets are presented in Figure 8 and Table 3. In Figure 8, the values in the yellow boxes are for the low horizontal rain data set, green boxes for the high horizontal rain data set, and red boxes for the combined data set. The RAFs in the table and the figure are the average values for the location over the study period of 2009 to 2013. As can be seen in the table and figure, in general, the RAF values based on the entire data set are in between the low and high horizontal rain data sets' RAF values. Moreover, the differences in the RAF values of the low and high horizontal rain data sets are higher at the upper locations compared to that of the mid section of the facade, which suggests that the RAF values of the upper locations seem to be more sensitive to the horizontal rain intensity than that of the mid height of the facade.

CONCLUSION

The wind-driven rain measurements and free-field wind-driven rain have been presented here and the rain admittance factors have been calculated. The average rainfall, wind speed, and the relative wind-driven rain measurements in the four orientations were consistent from year to year. Slight variations in the RAFs' pattern on the different wall orientations were observed. Overall, the RAFs for higher horizontal rainfall category are higher than the lower horizontal rainfall category. The authors are aiming to conduct sensitivity analysis to further investigate the effect of using rainfall intensity dependent RAF factors in a hygrothermal performance assessment of building envelope components.

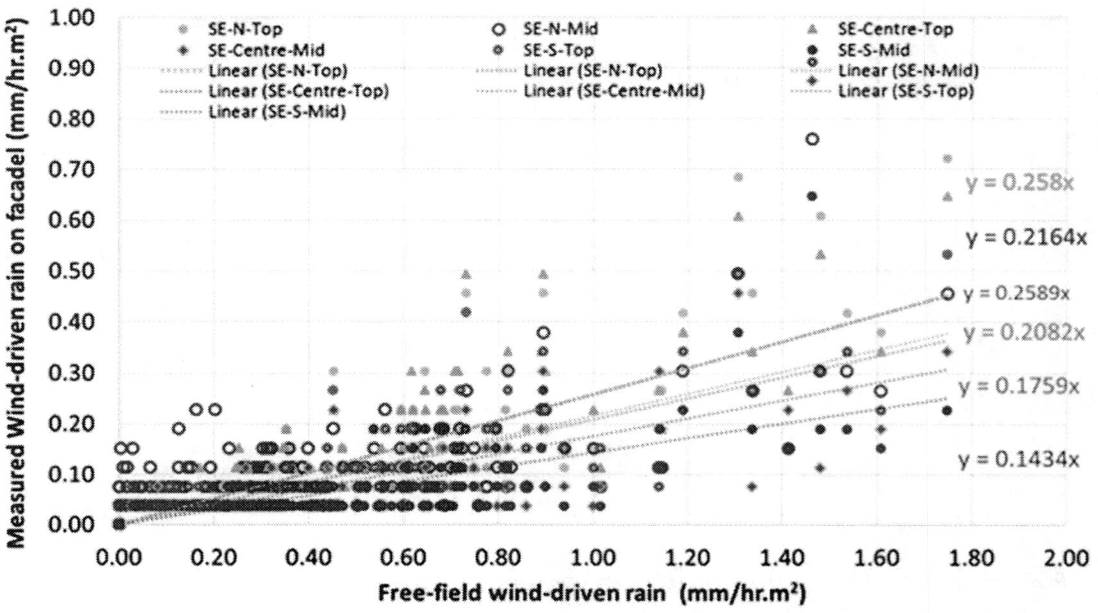

Figure 6 *Typical plot of hourly measured WDR by wall rain gages against hourly free-field WDR for the southeast facade in 2009.*

Table 2. Rain Admittance Factors at Wall Rain Gages

Wall Rain Gage Name	2009	2010	2011	2012	2013	Average
NE center-top	0.27	0.35	0.32	0.29	0.34	0.31
NW N-top	0.38	0.40	0.36	0.36	0.36	0.37
NW center-top	0.30	0.35	0.31	0.34	0.30	0.32
NW S top	0.53	0.62	0.46	0.53	0.62	0.55
SE N top	0.26	0.32	0.32	0.24	0.24	0.28
SE center-top	0.26	0.34	0.23	0.17	0.19	0.24
SE S top	0.21	0.32	0.30	0.26	0.33	0.28
SE N mid	0.22	0.37	**	0.31	0.31	0.30
SE center-mid	0.18	0.36	**	0.22	0.23	0.25
SE S mid	0.14	0.21	0.18	0.14	0.27	0.19

** The north and the center rain gages at the mid-height of the southeast facade were malfunctioning in 2011 and therefore the RAF values for those locations are not reported in the table.

REFERENCES

Best, A. 1950a. The size distribution of raindrops. *Quarterly Journal of the Royal Meteorological Society* 76(327):16–36.

Best, A.C. 1950b. Empirical formulae for the terminal velocity of water drops falling through the atmosphere. *Quarterly Journal of the Royal Meteorological Society* 76(329):302–11.

Blocken, B. 2004. Wind-driven rain on buildings: Measurements, numerical modeling, and application. Doctoral Thesis, Katholieke Universiteit Leuven, Belgium.

Blocken, B., and J. Carmeliet. 2010. Overview of three state-of-the-art wind-driven rain assessment models and comparison based on model theory. *Building and Environment* 45(3):691–703.

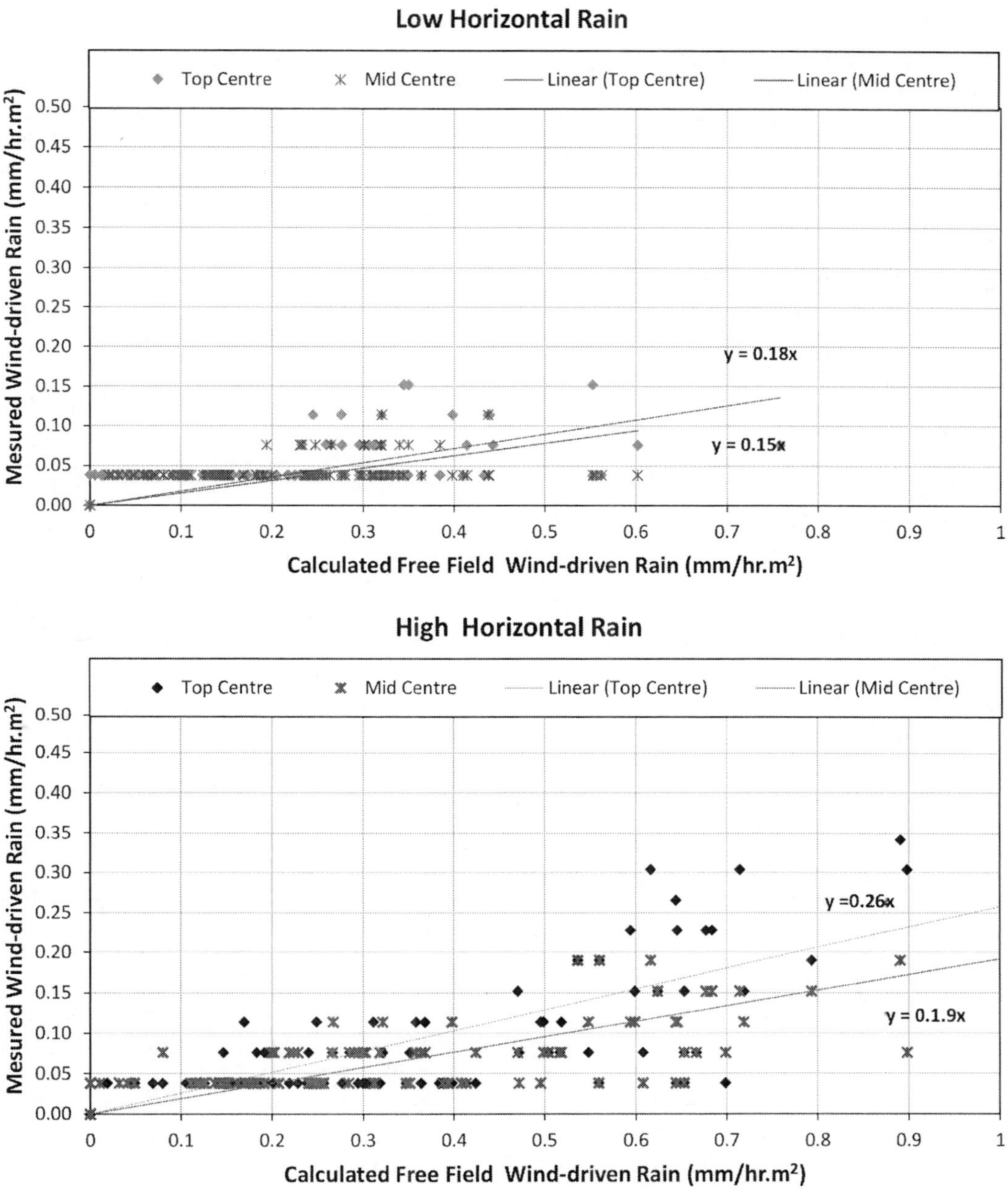

Figure 7 *Plots of hourly measured WDR at the top center and mid center of the southeast facade against the calculated hourly free-field WDR in 2009. Figures a and b represent cases with low and high horizontal rain events, respectively.*

Choi, E. 1994a. Determination of wind-driven-rain intensity on building faces. *Journal of Wind Engineering and Industrial Aerodynamics* 51(1):55–69.

Choi, E.C. 1994b. Characteristics of the co-occurrence of wind and rain and the driving-rain index. *Journal of Wind Engineering and Industrial Aerodynamics* 53(1):49–62.

Choi, E.C. 1994c. Parameters affecting the intensity of wind-driven rain on the front face of a building. *Journal of Wind Engineering and Industrial Aerodynamics* 53(1):1–17.

Choi, E.C. 1999. Wind-driven rain on building faces and the driving-rain index. *Journal of Wind Engineering and Industrial Aerodynamics* 79(1):105–22.

BSI EN 13013-3. 1997. Hygrothermal performance of buildings—Climatic data—Part 3: Calculation of a driving rain index for vertical surfaces from hourly wind and rain data. British Standards Institution, Committee B/540/1.

Dingle, N., and Y. Lee. 1972. Terminal fallspeeds of raindrops. *Journal of Applied Meteorology* 11(5):877–79.

Google Maps. 2016. British Columbia Institute of Technology, Burnaby Campus [Satellite Map]. Retrieved from

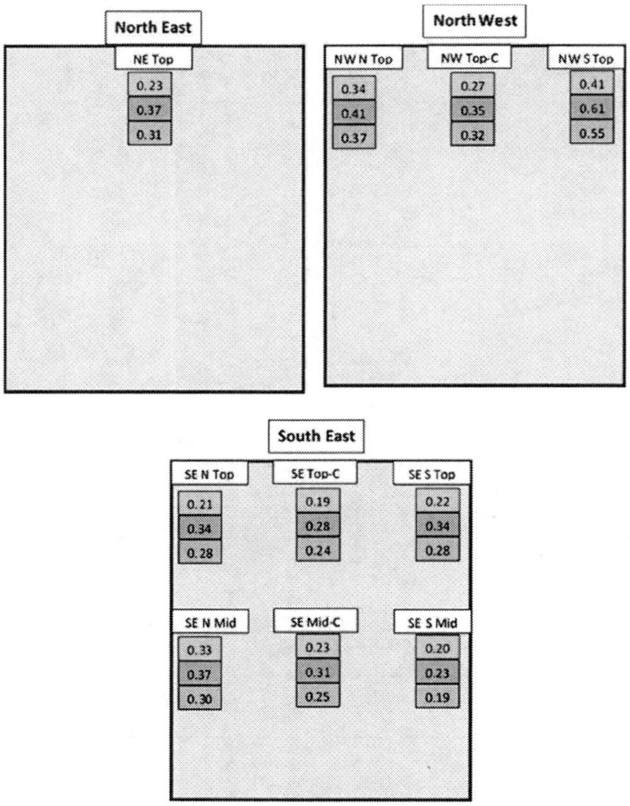

Figure 8 *The average RAF of rain gages on the northeast (NE), northwest (NW), and southeast (SE) facades over the study period of January 2009 to September 2013. RAF values in yellow boxes are for the low horizontal rain data set, green boxes for the high horizontal rain data set, and red boxes for the combined data set.*

Table 3. Average RAFs at Wall Rain Gages with Low and High Rainfall Data Sets

Wall Rain Gage Name	Low Horizontal Rain (< 2 mm/h)	High Horizontal Rain (< 2 mm/h)	All Data
NE center-top	0.23	0.37	0.31
NW N-top	0.34	0.41	0.37
NW center-top	0.27	0.35	0.32
NW S top	0.41	0.61	0.55
SE N top	0.21	0.34	0.28
SE center-top	0.19	0.28	0.24
SE S top	0.22	0.34	0.28
SE N mid	0.33	0.37	0.30
SE centre-mid	0.23	0.31	0.25
SE S mid	0.20	0.23	0.19

https://www.google.ca/maps/place/British+Columbia +Institute+of+Technology+(BCIT+Burnaby)/ @49.2423218,-123.0000113,111m/data=!3m1!1e 3!4m2!3m1!1s0x548676e1e49ab735:0x3acab7711e1f1 0d!6m1!1e1.

Hens, L.S.C.H. 2015. Combined heat, air, moisture modelling: A look back, how, of help? *Building and Environment* 91:138–51.

Lacy, R.E. 1965. Driving-rain maps and the onslaught of rain on buildings. Presented to the RILEM/CIB Symposium on Moisture in Buildings, August 16–19, Helsinki, Finland. Building Research Station.

Montogomery, D.C. and G.C. Runger. 2011. *Applied statistics and probability for engineers*, 5th ed. New York: John Wiley & Sons, Inc.

Morrison Hershfield. 1996. Survey of building envelope failures in the coastal climate of British Columbia. Report for Canada Mortgage and Housing Corporation, Ottawa, Ontario.

Ngudjiharto, E., F. Tariku, and P. Fazio. 2014. Preliminary results from field experimental study of rain load and penetration into wood-frame wall systems at window sill defects. 14th Canadian Conference on Building Science and Technology, Oct 28–30, Toronto, Ontario.

Straube, J.F., and E.F.P. Burnett. 2000. Simplified prediction of driving rain on buildings. *Proceedings of International Building Physics Conference*, Eindhoven, The Netherlands: 375–82.

Tariku, F., Y. Simpson, and E. Iffa. 2015. Experimental investigation of the wetting and drying potentials of wood frame walls subjected to vapor diffusion and wind-driven rain loads. *Building and Environment* 92:368–79.

Van Den Bossche, N., M. Lacasse, and A. Janssens. 2013. A uniform methodology to establish test parameters for watertightness testing. *Building and Environment* 63:145–56.

Van Mook, F.J.R. 2002. Driving rain on building envelopes. PhD Thesis. Faculty of Architecture. Planning and Building of the Eindhoven University of Technology.

Zhu, D., Mallidi, S.R., Fazio, P. 1995. Quantitative driving rain exposure on a vertical wall at various Canadian cities. *Building and Environment* 30(4):533–44.

Laboratory Study of Rates of Inward Leakage in Seven Different Gaps in a Facade Exposed to Driving Rain or Water Splash

Lars Olsson, Lic. Eng.

ABSTRACT

Water or rainwater can leak into outer walls and facades to a greater or lesser extent, but there is a general lack of specific information how much water usually penetrates. If we simply rely on theoretical analyses with only limited knowledge of the amounts involved in such inward leakage, we run serious risks of damp-related damage, indoor health problems, and failure to meet requirements in terms of energy efficiencies. The purpose of this laboratory study was to examine the significance of water splash, different wind pressures, and different rain loads, as well as combinations of all three, on inward leakage through various gaps in a facade. The study was restricted to seven small or invisible gaps created in a facade element. The results show that the proportion of inward leakage for several gaps was around 2% (0.03 L/min per gap) with an applied rain load equivalent to heavy driving rain on multistory buildings. These figures include conditions where no wind load was present. In cases where there is a risk exist of inward leaking water collecting locally or becoming concentrated inside the wall, the results may be applied as a point load in two-dimensional, or mainly three-dimensional, moisture calculations. The results also indicate significant volumes of inward leakage from water splash alone, meaning that inward leakage in facades may occur every time it rains, regardless of whether the rain is driving rain.

INTRODUCTION

Water can leak into outer walls and facades (Samuelson et al. 2008; Jansson 2014; Woodbury 2009) to a greater or lesser degree. Window connections represent the most common item of facade detail, and a laboratory study has shown that the proportion of leaking joints around windows is approximately 70% (Olsson 2014). So, it is relatively common for joints around windows to leak, but it has also been found that the window structure itself can leak (Gustavsson 2009; CMHC 2003). There is however a lack of specific information as to the exact amounts of water that penetrates through window joints and other penetrations in building facades.

There is currently a lack of reliable theoretical analytic tools capable of fully dimensioning and assessing the risk to an outer wall's moisture safety when facade details are present and specific data is not available (Bednar and Hagentoft 2015). Often ballpark type assessments are used, based on the percentage of driving rain that penetrates (TenWolde 2011) distributed per square meter. This is an important factor indicating that more detailed information is needed in order to assess moisture safety. It would appear the volume of inward leakage might well represent a very significant unknown and risk. More knowledge is therefore needed in order to more accurately account for it (Ngudjiharto et al. 2014; Van Den Bossche et al. 2014; TenWolde 2011; De Meersman et al. 2014). To be able to use the correct input data in theoretical simulations, more data and findings based on both field and laboratory experiments are probably required, as well as tolerance assessments or specific tests for the solution in question.

There is therefore a need for more knowledge if we are to have the requisite theoretical background needed to dimension, plan, and assess new solutions and designs, or those commonly used, in a reliable manner. Given the need for more knowledge about how facades and solutions work, both gener-

Lars Olsson is an industrial doctoral student in the Department of Civil and Environmental Engineering, Chalmers University of Technology, Göteborg, Sweden, and a senior engineer in the Department of Sustainable Built Environment, SP Technical Research, Institute of Sweden, Stockholm, Sweden.

ally and specifically during driving rain and water splash conditions, an extensive laboratory experiment has been conducted. The purpose of this study was to examine the significance of water splash, different wind pressures, and different rain loads, as well as combinations of all three, on inward leakage through various gaps in a facade.

MATERIALS AND METHODS

Experimental Designs

The study was restricted to laboratory experiments involving seven gaps in connections to different facade details, such as exterior window sill, a circular and a square duct, and integrated flashing, created in a facade element, as shown in Figure 1. The rear of the facade or wall element was not provided either with a wind barrier and insulation or with an internal surface, allowing the rear of the facade to be accessible for visual inspection. Therefore, the facade element and all details and gaps were exposed to presented wind pressure load. The size and shape of the gaps involved in the details are indicated in Figure 2 and Table 1. The facade was constructed from a timber stud frame with facade material of water-repellent plaster-based boards. The board was impregnated and the surface coating is similar to acrylic paint. The surface of the material is considered to be superficially similar to a plaster or concrete facade painted or treated to render it hydrophobic, painted cement elements or composite panels, or other smooth and water-repellent surface material. All joints between the boards, as well as other connecting joints and penetrations, were carefully sealed. The gaps were selected based on findings from investigations in both the laboratory and the field and were restricted to small or extremely small gaps (Olsson 2014, 2015; Jansson 2014). The sizes of theses gaps were measured by feeler gauge and calipers, apart from Detail 6 where the gap is not possible to measure. Different types of sealing solutions or their function were not part of the study.

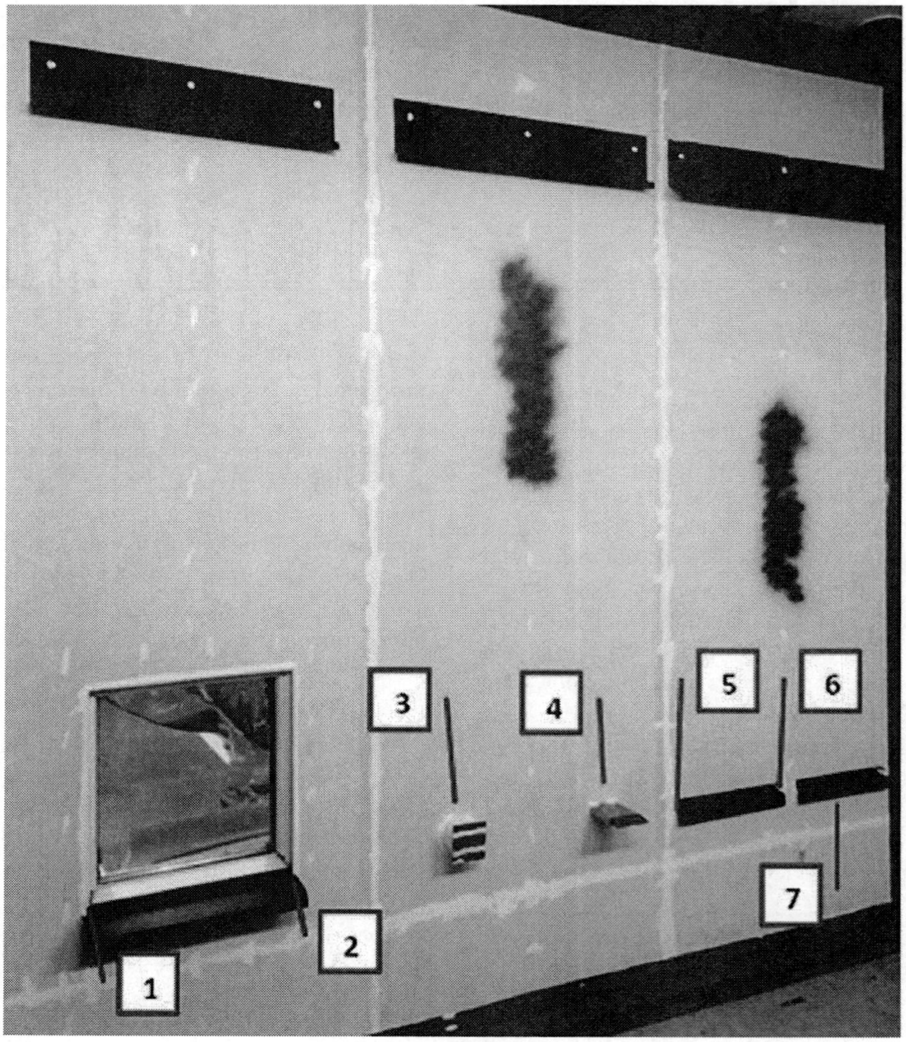

Figure 1 *Photo of the front of the wall element with the gaps in the details numbered.*

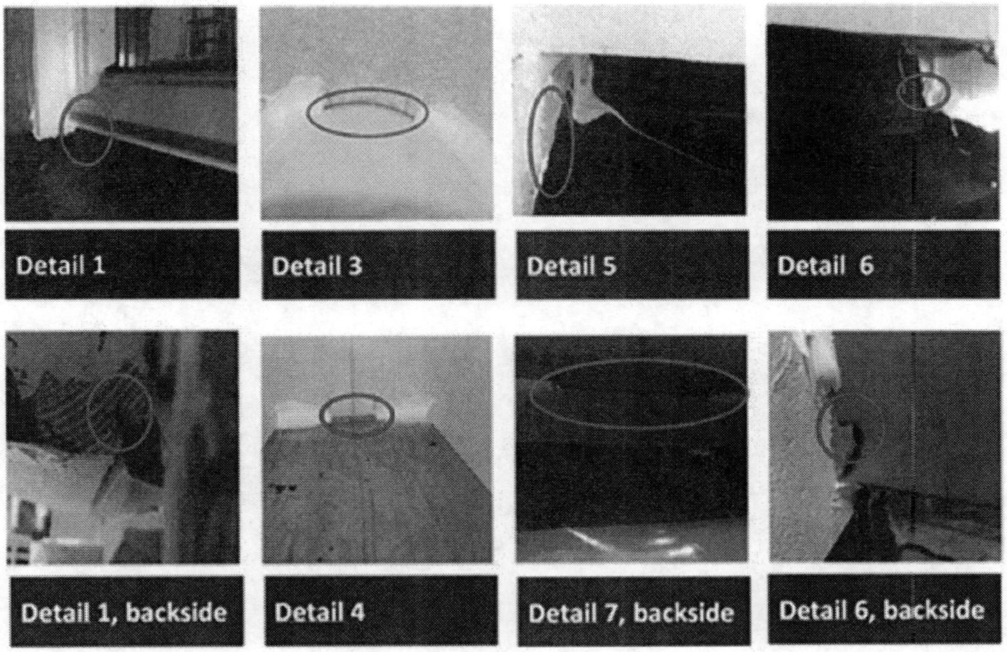

Figure 2 *Photos of the gaps, except for Detail 2, which is similar to Detail 1.*

Table 1. Descriptions of the Gaps in the Details in Terms of Hole or Aperture Dimensions, plus Remarks as to Whether the Gaps are Concealed, Invisible, or Visible from the Outside

Detail	Type of Detail	Gap Dimensions, mm	Comments
1	Connection with exterior window sill, corner and rear edge	(1.5 × 1.5) + (9 × 0.2) + (50 × 0.1), 9 mm^2	Concealed position
2	Connection with exterior window sill, corner	2 × 2, 4 mm^2	Concealed position
3	Circular duct, connection	35 × 0.9, 32 mm^2	Visible
4	Rectangular duct, connection	30 × 2, 60 mm^2	Visible
5	Connection with flashing, end	35 × 0.1, 3.5 mm^2	Not visible
6	Corner fold, flashing	Small gap, not measurable	Not visible
7	Connection with flashing, underneath	120 × 0.3, 36 mm^2	Concealed position, not exposed to driving rain, only exposed to water splash from below.

Special external flashings were fitted high up on the facade, as shown in Figure 1, to counteract water drip. These created water splash on Details 1 through 4 (not 5 through 7) and on the pavement underneath Details 5 through 7. To prevent reciprocal effects of water splash at Details 6 and 7, a separator was positioned between them, projecting from the facade by 120 mm, as shown in Figure 3. Under Details 5, 6, and 7, a pavement of concrete was created at a distance of 30 mm respectively, 160 mm from the details, as shown in Figure 3. The purpose of the pavement is to represent a ground surface that butts up against the facade.

Test Procedures

The experiment was carried out partly using the standardized test method of EN 12865, *Determination of the Resistance of External Wall Systems to Driving Rain Under Pulsating Air Pressure*, but was extended to include additional load combinations and repetitions (BSI 2001). Simulation of driving rain was obtained by using specified water spray nozzles and dynamic pressure loading with compressed air at successive pressure steps, such as 0 Pa, 0–75 Pa, 0–150 Pa, 0–300 Pa, 0–450 Pa, and 0–600 Pa. For certain rain loads, the experiment was restricted to a few commonly occurring pressure differences (0–75, 0–150, and 0–300 Pa).

This method requires that Procedure A is applied if the facade consists of non-moisture-absorbing material, with a test period of 10 minutes per pressure step. In these experiments, a test period of 20 minutes was used. This was chosen in order to reduce the significance of the start and stop proce-

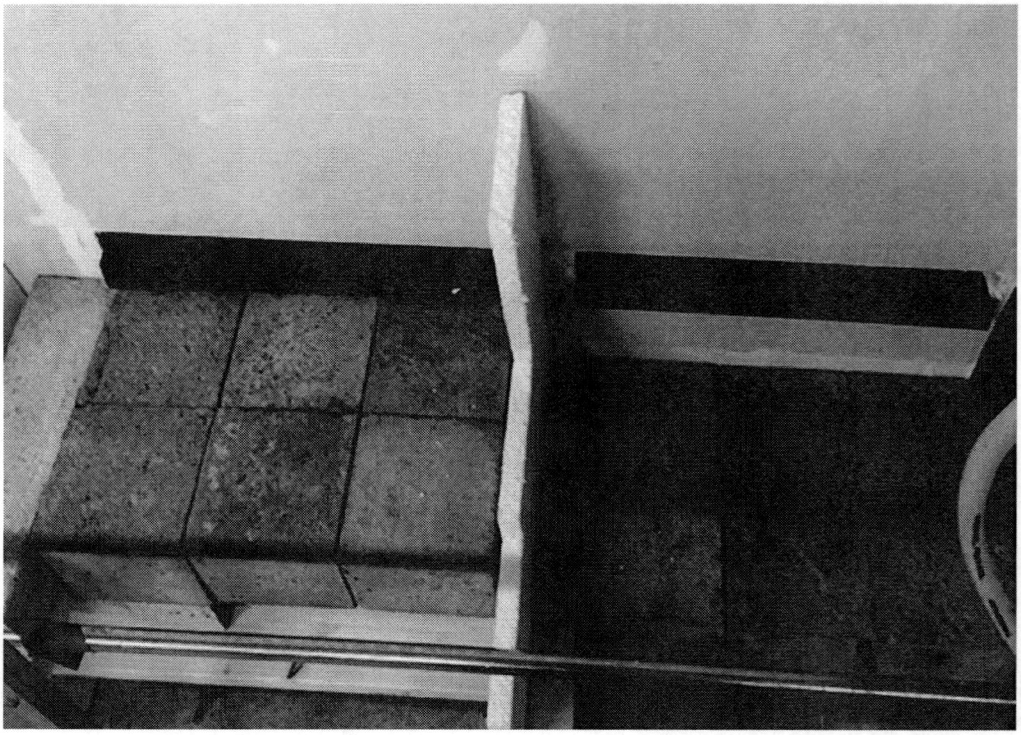

Figure 3 *Photos of Details 5 through 7.*

dure and continued dripping that occurred on each experiment. Before the first experiment, conditioning was performed with full rain load for 20 minutes. This was repeated before each experiment in cases where there had been a break of more than 15 minutes before the start of the experiment, the idea being to fully saturate the facade surface.

The test method requires that a given volume of water should be applied to the facade element in a given distribution pattern. In a number of these experiments, the specified rain intensity was applied, but the rain load was also reduced by disconnecting water nozzles a row at a time. Where water splash was being studied, water was applied above the outer flashing. In several cases, the outer flashing was removed when all nozzles were used simultaneously. Where experiments involved water splash, the top row of nozzles, intended to simulate water runoff from above, were angled so that all the water was applied above the flashing. A steady stream of water droplets was created as the water ran off the flashing. This water has therefore not been included in the presentation of applied rain load, as it was conducted off, away from the facade layer.

Under each detail, collection funnels made from plastic foil were fitted. These were taped up and sealed against the rear of the facade. Each funnel emptied into a glass bowl to collect the water, as shown in Figure 4, and weighed.

In the description of the method, the option is provided of weighing the wall element before and after each test procedure so as to be able to determine the total amount of water absorbed. This was not applied, as any water leaking into the wall was caught and weighed separately. Moreover, it was felt that it would not have resulted in an exact answer as to where the water had been absorbed or collected, nor as to its distribution in the wall element or the pressure step at which this might have happened.

In total, around 100 tests were performed that include three to seven repeated cases per rain load and wind pressure (more repetitions in cases with comparisons of different rain load at the same wind pressure).

The estimated uncertainty of measurement of the pressure difference (Pa) was a maximum of ±5%, for the weighing of bowls with water a maximum of ±1g, and for evaporation from bowls a maximum of ±1g. The time taken for starting and stopping the rain and wind load could vary by almost 30 seconds, which would give a maximum variation in leakage flow (L/min) of ±1%. The measurement uncertainty of the applied rainfall intensity and its distribution is unknown according to EN 12865. However, the flow of water from the nozzles was sample tested, and these tests showed that the specified flow had a measurement uncertainty of a maximum of ±3%. The uncertainty of the sizes of gaps are ±0,1 mm.

RESULTS AND COMMENTS

The specified rain load is the water applied to the facade above the gaps and the water to which the gaps were, in principle, exposed. Additionally, the runoff pattern differed (due to surface tension) between the tests, which created unstable results. The water applied above the external flashing has not been included in the specified rain load, as the water was conducted off, away from the facade surface, and made to drop down to the ground. This water has not been included and should not be confused with the concept of driving rain (wind induced).

The gap in Detail 1 resulted in inward leakage (0.027 and 0.04 L/min) at a wind pressure of 75 to 600 Pa, with an applied rain load of 2.93 L/min-m plus water splash, as shown in Figure 5. At a wind pressure of 0 Pa, inward leakage was 0.011 L/min, which may also be considered a leakage flow of significant proportions. Comparing the bars with and without wind pressure, it is evident that inward leakage was almost three to four times greater when wind pressure was present. It is also clear that the significance of the extent of the difference in pressure at pressures above 75 Pa was comparatively limited.

Figure 4 *Photo of the rear of the wall element with collection funnels and bowls.*

The gap in Detail 2 resulted in inward leakage (0.003 and 0.013 L/min) at a wind pressure of 75 to 600 Pa, as shown in Figure 5. Without wind pressure load (0 Pa), inward leakage of 0.0002 L/min was obtained, which may be considered a difference of marginal proportions when compared with the figures for leakage with wind pressure.

The gap in Detail 3 resulted in inward leakage of approximately 0.036 L/min, regardless of the amount of wind pressure, as shown in Figure 5. This implies that large volumes of water could penetrate without wind pressure if the facade is exposed to heavy rain loads. The driving force behind inward leakage was not, therefore, a difference in wind pressure.

The gap in Detail 4 resulted in inward leakage (0.06 and 0.11 L/min) at a wind pressure of 300 to 600 Pa, as shown in Figure 5. At lower wind pressure, inward leakage was comparatively small, indicating that there needs to be a sufficient amount of wind pressure if significant inward leakage is to occur for this gap.

The gap in Detail 5 resulted in inward leakage (0.001 and 0.008 L/min) with the leakage increasing more or less in proportion to increases in wind pressure (see Figure 5). Without wind pressure (0 Pa), inward leakage of 0.0002 L/min occurred, which is small in comparison to the figures with wind pressure.

The gap in Detail 6 resulted in marginal inward leakage.

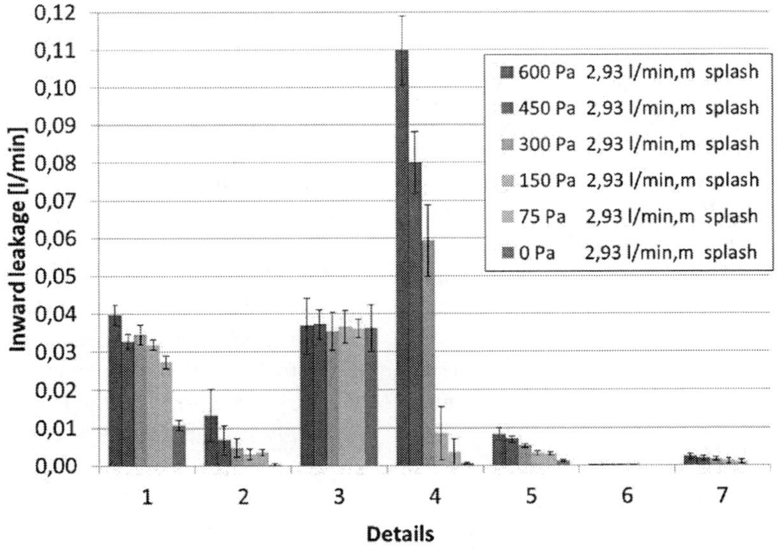

Figure 5 *The bars show inward leakage (mean value of three to seven repetitions) for the gap in the relevant detail for six pressure steps/wind pressure with pulsation. The applied water load was rain at 2.93 L/min-m and water splash. The standard deviation is also shown.*

The gap in Detail 7 resulted in inward leakage (0.001 and 0.002 L/min), with the leakage increasing more or less in proportion to increases in wind pressure, as shown in Figure 5. Without wind pressure (0 Pa) no leakage occurred.

At a lower rain load of 1.22 L/min-m, inward leakage in Detail 4 was greater without water splash than with, as shown in Figure 6a. For this gap, water splash appears to reduce the amount of inward leakage by one third, compared to the result without water splash. When drops from water splash hit the inlet, there is a disturbance of inward leakage. Overall, the measurements indicate somewhat higher inward leakage with water splash in Detail 1, but the opposite in Details 2 and 3. One reason is added water from water splash. In Detail 7 a small amount of inward leakage (0.002 L/min) occurred when there was water splash, but no leakage worthy of mention when there was no water splash. Despite the fact that the rain load had halved (1.22 L/min-m) in these load cases compared to the load case in Figure 5 with a rain load of 2.93 L/min-m, almost the same amount of inward leakage was obtained at several gaps. This indicates that gaps may have a limited or optimum degree of inward leakage with respect to percentage of total rain exposure. Furthermore, about 2% inward leakage of the applied rain load was obtained when the rain load was halved.

Further measurements show in principle that the amount of inward leakage with water splash only is the same as when a rain load of 0.055 L/min-m is applied to the facade, as shown in Figure 6b. Additionally, water from the upper part of the window created water splash for Details 1 and 2, which was unavoidable. Inward leakage occurred in four out of seven details, and the greatest amount of inward leakage was approximately 0.007 L/min for Detail 1. Water splash on its own can therefore cause inward leakage to exactly the same degree as driving rain.

DISCUSSION

Previous studies have shown that water tends to follow the types of joints and angles often found in facade details, thus resulting in a build-up of water, which means the water flow can be several times greater at these details (Garden 1963). Water that hits the facade runs down and the greatest flow may be expected at the very bottom of the facade. This is provided that it is not diverted away at obstacles or details, absorbed by capillary action, or does not leak through the facade at some point above (Van Den Bossche 2013). Furthermore, it implies that all of the water that hits the facade and which runs down to the gap becomes the actual load (L/min-m) to which gaps are exposed. This means that the water volumes that run along the lower part of nonabsorptive facades and fully absorbed facades may be ten times greater than the volume of driving rain per square meter of facade surface.

Sandin (1987) demonstrated driving rain intensity per hour for Göteborg and measured it at around 2 kg/m^{2}h, which was judged to be normal driving rain, and 4 kg/m^{2}h, which was judged to be extreme driving rain. Driving rain is at its most intense high up on facades but also further down near the corners. As far as climate data with 15-minute measurements is concerned, rain intensity is often twice that measured in one-hour data (Olsson 2015). This has also been shown by (Högberg 2002). If extreme driving rain based on the volume above is first doubled based on 15-minute data and then multiplied by a 10 meter vertical facade surface for instance, a water flow of approximately 1.3 L/min-m is obtained and similar to present experiment (i.e., 1.22 L/min-m).

How representative the gaps examined in this laboratory study are in comparison with other gaps and those occurring in actual buildings remains up for discussion. The inward leakage obtained in these experiments is however pretty consistent with other studies (Lacasse 2003; Lacasse et al. 2012) especially with 0 Pa pressure difference over the exposed facade. Comparisons have, for instance, been made between vertical and horizontal cracks and gaps, and these showed significant differences in inward leakage (Lacasse et al. 2009). Field studies of moisture levels as well as damage surveys have moreover produced a lot of evidence pointing to considerable variation in inward leakage, as well as extensive inward leakage (Jansson and Hansén 2015). Notice that inward leakages are concentrated to gaps (small) in the facades and the most damages are usually seen in specific points rather than evenly distributed over the wall. Consequently, it is not appropriate to carry out one 1-D calculations of inward leakages in specific points.

CONCLUSIONS

The results show that, under heavy rain loads, there may be continuous inward facade leakage of significant volumes of water (0.03 L/min) per gap, including where gaps are small. This is also the case where there is no pressure load, such as on well-ventilated facades, for instance. Measurements show that, for several gaps, inward leakage amounts to around 2% of the total applied rain load of 1.22 L/min-m. The applied rain load is estimated to correspond to heavy driving rain in Sweden on the facade of a multistory building. Where there is a risk that water collects locally or becomes concentrated inside the wall, the results may be applied as a point load in 2-D or mainly 3-D moisture calculations.

In many cases, the results show that increases in wind pressure lead to increased inward leakage. There are, however, several exceptions where increased wind pressure results in relatively little increase in inward leakage or no increase at all. Nevertheless a certain amount of wind load is still needed for the rain to be able to hit the façade.In certain cases there was relatively little difference in inward leakage within the 75 to 600 Pa pressure difference range.

Measurements indicate that lower rain load results in a significantly higher proportion of inward leakage in percentage terms than does greater rain load. This might be explained by the fact that not all the water that arrives at the gap will penetrate it—some will run past it. Gaps therefore have a limited or optimum degree of inward leakage.

The shape of the gaps is very significant when combined with specific rain, wind, and water splash loads, and combinations of these. Small gaps are also capable of resulting in relatively large amounts of inward leakage.

Since water splash on its own can bring about inward leakage, it is evident that driving rain is not needed for inward leakage to occur. This means that inward leakage may occur every time it rains or water drips from melting snow or condensation, provided there are details in the facade that both create and contribute to splashing.

ACKNOWLEDGMENTS

The support from the Swedish Energy Agency and the SBUF (the Swedish construction industry's organization for research and development) is gratefully acknowledged.

REFERENCES

Bednar, T., and C.E. Hagentoft. 2015. Annex 55, *Reliability of energy efficient building retrofitting—Probability assessment of performance and cost (RAP-RETRO)*. Report 2015:7. In International Energy Agency (IEA)

(a)

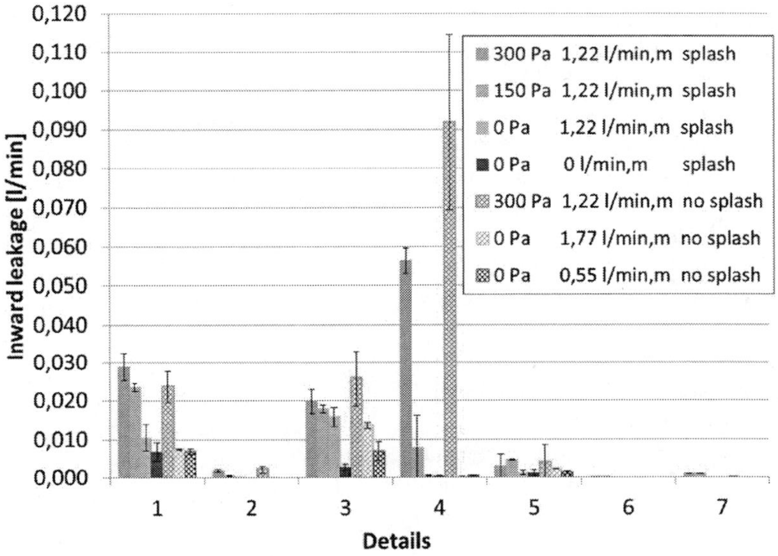

(b)

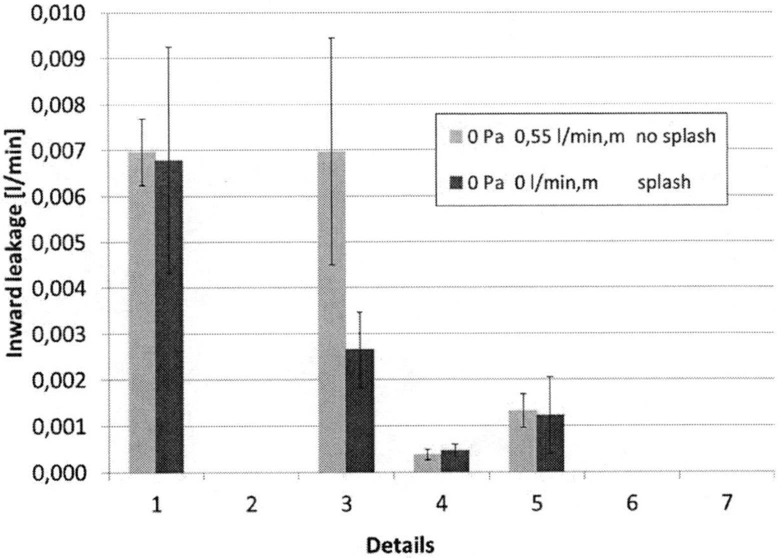

Figure 6 *(a) The bars show inward leakage (mean value of three to seven repetitions) for the relevant detail and pressure differences with pulsation. The solid bars show inward leakage with an applied rain load of 1.22 L/min-m in combination with water splash (the red bar shows inward leakage from water splash only). The patterned bars show inward leakage on application of three different volumes of water without water splash. The standard deviation is also shown. (b) The red bars indicate inward leakage due to water splash on its own at 0 Pa. The orange/yellow bars show inward leakage with an applied rain load of 0.55 L/min-m at 0 Pa.*

och Energy in Buildings and Communities Programme (EBC). Gothenburg: Chalmers University of Technology.

BSI. 2001. EN 12865, *Determination of the resistance of external wall systems to driving rain under pulsating air pressure*. London: British Standards Institution.

CMHC. 2003. *Water penetration resistance of windows-study of manufacturing, building design, installation and maintenance factors*. Technical Series 03-124. Ottawa: Canada Mortgage and Housing Corporation.

De Meersman, G., N. Van Den Bossche, and A. Janssens. 2014. A parametric HAM study of the risk assessment of water ingress in wooden constructions. In 10th Nordic Symposium on Building Physics, Lund, Sweden.

Garden, G.K. 1963. Rain penetration and its control. CBD-publication 40. http://web.mit.edu/parmstr/Public/NRCan /CanBldgDigests/cbd040_e.html.

Gustavsson, Börje. 2009. Fönstermontage (SP rapport 2009:35). Borås: SP Sveriges Tekniska Forskningsinstitut.

Högberg, A. 2002. Microclimate load: Transformed weather observations for use in design of durable buildings, building physics. Gothenburg: Chalmers University of Technology.

Jansson, A. 2014. Actions against moisture damage in ETIC walls. Paper read at 10th Nordic Symposium on Building Physics, Lund, Sweden, June 15–19 2014.

Jansson, A., and M. Hansén. 2015. Putsade enstegstätade regelväggar- Erfarenheter från undersökningar som SP har utfört. Borås: SP Sveriges Tekniska Forskningsinstitut.

Lacasse, MA. 2003. *Recent studies on the control of rain penetration in exterior wood-frame walls*. NRCC-46889. Ottawa: National Research Council.

Lacasse, Michael A., Hiroyuki Miyauchi, and J. Hiemstra. 2009. Water penetration of cladding components-results from laboratory tests on simulated sealed vertical and horizontal joints of wall cladding. *Journal of ASTM International* 6(6).

Lacasse, Michael A., Travis Moore, and Nathan Van Den Bossche. 2012. Water infiltration through openings in a vertical plane under static and dynamic pressure conditions. Paper read at Building Enclosure Science & Technology. BEST3-2012.

Ngudjiharto, E., F. Tariku, and P. Fazio. 2014. Window sill defects. Paper read at 14th Canadian Conference on Building Science and Technology, Toronto, Canada.

Olsson, Lars. 2014. Results from laboratory measurements of wind driven tightness in different types of facades. In 10th Nordic Symposium on Building Physics. Lund, Sweden.

Olsson, Lars. 2015. Long-time field moisture measurements in wooden walls with different types of façades: Focus on driving rain tightness. In 6th International Building Physics Conference, IBPC 2015. Torino, Italy.

Samuelson, I., K. Mjörnell, and A. Jansson. 2008. Moisture damage in rendered, undrained, well insulated stud walls. *Proceedings of 8th Symposium of Building Physics in the Nordic Countries*. Copenhagen: Denmark.

Sandin, Kenneth. 1987. The moisture condition in aerated lightweight concrete walls—In situ measurements of the effect of the driving rain and the surface coating. Lund, Sweden: Division of Building Materials Lund Institute of Technology.

TenWolde, A. 2011. A review of ASHRAE Standard 160, Criteria for Moisture Control Design Analysis in Buildings. *Journal of Testing and Evaluation* 39(1). doi: 10.1520/JTE102896.

Van Den Bossche, Nathan. 2013. Watertightness of building components: Principles, testing and design guidelines. Faculty of engineering and architecture. Ghent, Belgium: University Ghent, Belgium.

Van Den Bossche, N., G. De Meersman, L. Devos, S. Maertens, and A. Janssens. 2014. Watertightness of window-wall interfaces in wood-frame construction. In 10th Nordic Symposium on Building Physics. Lund, Sweden.

Woodbury, City. 2009. Stucco in new residential construction—A position paper. Includes updates to original information. Edited by Building Inspection Division. Woodbury, Minnesota: City of Woodbury.

Thermal and Moisture Performance of Buried Ducts

Mikael Salonvaara
Member ASHRAE

Katrina Keeley

Achilles Karagiozis, PhD
Member ASHRAE

ABSTRACT

Simulation, laboratory, and field studies in a hot, humid climate were conducted to investigate the thermal and hygrothermal performance of ducts buried in blown-in insulation in vented attics. Ducts that leak in vented attic can carry a large energy penalty. The U.S. Department of Energy's Zero Energy Ready Home National Program Requirements ask for ducts to be located within the homes' thermal and air barrier boundary. An alternative to placing the ducts in conditioned space is to bury the ducts inside the insulation. The surface temperature of the duct insulation that includes a vapor barrier may fall below the dew point of the attic air and cause condensation. This paper includes details of a study performed to evaluate the condensation potential with different duct R-values, duct size, ceiling R-values, and air leakage characteristics. Both flexible ducts and square ducts with duct board were investigated. The measured relative humidity around the ducts suggests that current encapsulation recommendations may be overly conservative and that ducts buried in the insulation with proper duct R-value for each climate do not experience condensation that could cause moisture damage. The research for this paper focused on Climate Zones 5A (Cleveland, OH) and 3A (Charleston, SC).

INTRODUCTION

Conventional vented attics can be constructed to be very well insulated and airtight at the ceiling level. Separation of the roof deck from the insulation and access to the attic space allows for easy inspection and repair of the roof. The attic space can be used for routing ducts to the interior spaces; however, the extreme temperatures in the attic cause a penalty to the ducts unless the ducts are highly insulated and very well air sealed. A seemingly obvious solution to insulate the ducts better is to place them inside (buried in) the loose fill insulation above the ceiling (i.e., by bringing the ducts closer to the interior space and thermally isolating the ducts from the attic). This method has been investigated in several projects (Griffiths and Zuluaga 2004; Griffiths et al. 2004; Vineyard et al. 2004; Griffiths et al. 2002). The results have shown significant reduction in duct losses. The moisture performance of buried ducts has not been studied as much as the thermal performance. Recently Malley (2016) investigated buried ducts in hot/humid climate in a field study and noted that the currently recommended duct insulation levels to prevent condensation are overly conservative.

This paper focused on analyzing moisture performance of buried ducts in weather conditions requiring cooling. Burying the ducts in the attic insulation can cause the temperature of the surface of the vapor barrier on the duct insulation to drop, which can result in condensation. Previous studies have proposed a solution to add 1.5 in. of spray foam on top of the duct to increase the surface temperature and to reduce the chance of condensation. The duct systems studied in this paper used no spray foam. The paper includes a combination of advanced hygrothermal numerical simulations, laboratory tests and field monitoring.

THERMAL AND MOISTURE ANALYSIS FOR CONDENSATION

Simulating Condensation on Buried Ducts

The ducts in the attic, whether hanging or buried, are exposed to the conditions in the attic air and indoor conditions below the ceiling drywall. The conditions in a vented attic can be very different from those in the outdoor air: The attic

Mikael Salonvaara is building science leader, **Katrina Keeley** is an advanced scientist, and **Achilles Karagiozis** is global director of Building Science, Owens Corning Science & Technology, Granville, OH.

temperature can get hot during the sunny days, and, while attic ventilation will allow for some removal of moisture from the attic, there can be times when moisture movement between the roof structures and the attic air can create humid conditions. Previous studies have looked at buried ducts assuming that the design condition for preventing condensation is the hottest time of day with a dew-point temperature slightly higher than that outside the house. These studies have only looked at calculated temperature fields and compared the duct surface temperature to the dew-point temperature inside the blown-in insulation without the duct. This past approach gives a first look at the risk but does not address the true dynamics involved in moisture transport around ducts.

Figure 1(a) shows the relative humidity field around a buried round duct as simulated with a two-dimensional hygrothermal model when the attic is at 130°F and 31% rh (= 90°F dew-point temperature) while the duct temperature is 55°F. This condition has been used as the design condition in previous studies; however, the time when the highest dew-point temperature occurs may not be the most severe condition that results in condensation. Higher temperature may allow for higher duct surface temperature, which reduces the potential for moisture accumulation. In the following simulations, the attic, duct, and indoor conditions were kept constant for a full week. Many different simulations were carried out and the location where a buried duct might condense is typically at about the mid-depth of the blown insulation or above the duct where the vapor drive is coming from. Figure 1(b) shows how the vapor pressure is slightly higher on top of the round duct, while the bottom side of the duct has a lower vapor pressure than in the free field in the blown insulation. The duct insulation on a round duct results in a different effective R-value as for a square duct. Figures 1(c) and 1(d) compare square duct to the round duct in Figure 1(a). Figure 1(e) shows how the larger 8 in. square duct experiences more potential for condensation on the top of the duct than the smaller 4 in. duct.

Temperature controls most of the condensation behavior. In an attic when the dry-bulb temperature drops after sunset, the dew-point temperature typically drops similarly. Figure 2 shows an example of an hourly cycle of dry-bulb and dew-point temperature in an attic in Charleston (SC). It is important to notice that the cooling system may run not only during the daytime but also in the evening or night when the attic becomes cooler. On/off cycling could be enough to prevent (net) condensation even if steady-state analysis claims otherwise.

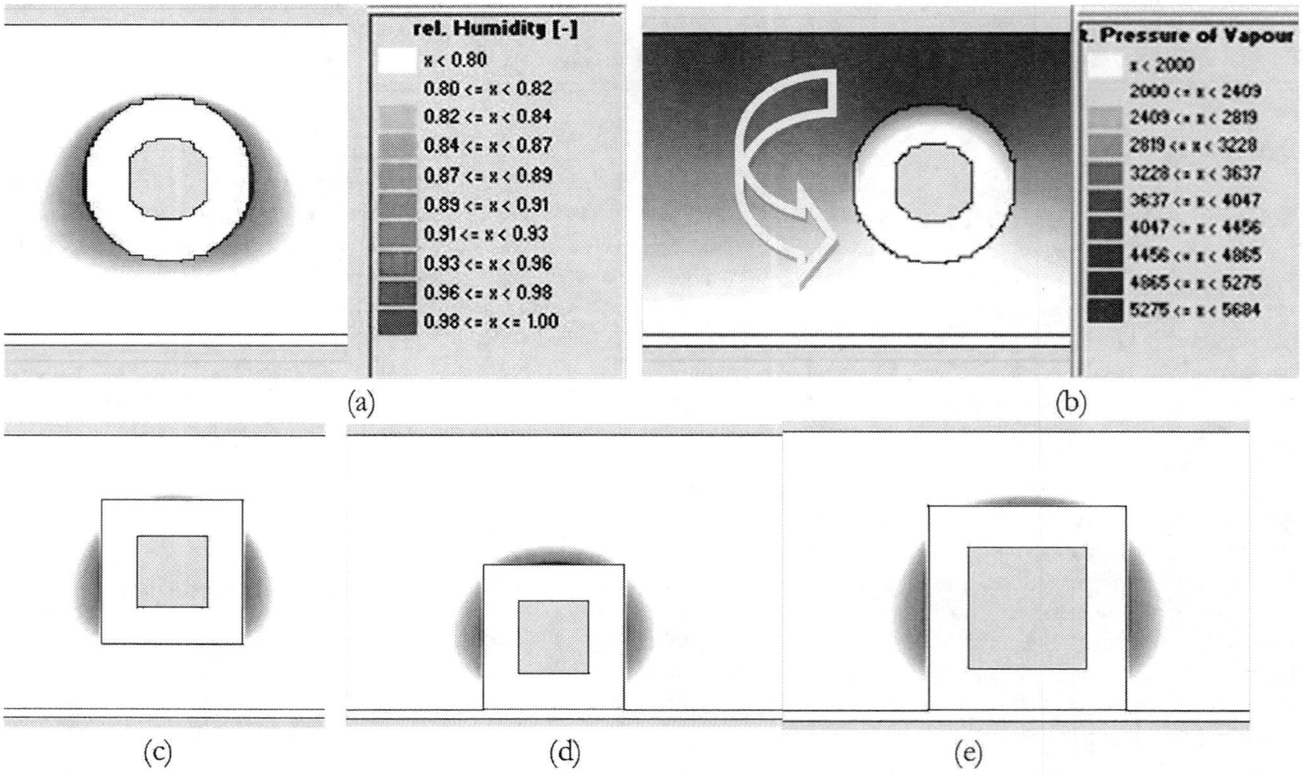

Figure 1 *(a) Relative humidity field (dimensionless) and (b) vapor pressure (Pa) field around a round, 4 in. duct with R-8 (exact R-8, not R-8.7 as commonly available) insulation when attic is at 130°F dry bulb (T_{db}) and 90°F dew-point temperature (T_{dp}), duct temperature is 55°F, and the indoors is at 75°F and 55°F dew point. The 4 inch square duct placed in the (c) mid-height or (d) deeply buried shows how the likely condensation occurs at the mid-depth of the insulation. (e) Deeply buried 8 inch square duct. All configurations have at least 3.5 in. of blown-in insulation above the duct.*

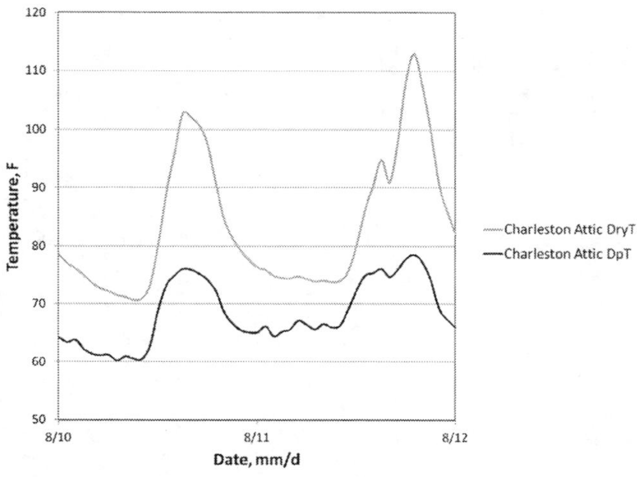

Figure 2 *Hourly dry-bulb and dew-point temperatures in an attic in Charleston, SC.*

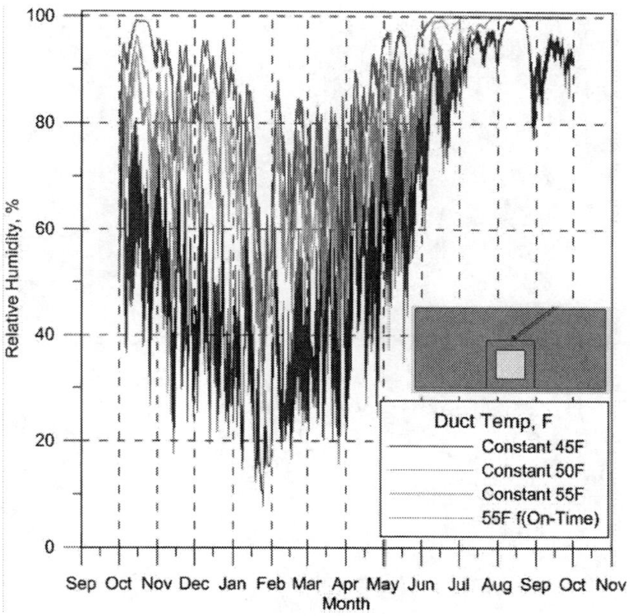

Figure 3 *Simulated relative humidity on top of a square duct (R-8) in Charleston, SC. Attic conditions were taken from field data in a test hut in Charleston, SC. The colder the duct temperature, the higher the surface relative humidity.*

A round duct and a square duct have small differences not only in the consistency of the R-value (compression of insulation affects the R-value locally) but also due to the shape: Square ducts provide more consistent R-value (less potential for condensation), while the flat top surface of a square duct makes it harder for the vapor drive to bypass the duct.

The binderless blown fiberglass insulation and the vapor retarder on the duct are mold resistant and the relative humidity does not indicate damage yet. There has to be enough time to accumulate moisture that could drip or not dry out when cooling is not on or when the changed conditions do not allow condensation before the next cycle. Full year simulations were carried out with constant duct temperature (assuming cooling is on all the time every hour of the year) and finally with an on-off schedule for cooling received from a whole-building energy model to investigate the dynamics of thermal and moisture movements. Figure 3 shows the relative humidity on top of a square duct in Charleston, SC, when the duct temperature is 45°F, 50°F, 55°F, or varying as a function of on/off cycling of the cooling system.

The simulated results indicate that midsummer is not necessarily as severe a time of year for condensation as early fall (August).

LABORATORY TESTING FOR CONDENSATION

Buried Duct Test Fixture

A buried duct test fixture was constructed in the lab for experimental testing of fiberglass-insulated ducts buried beneath fiberglass loose fill. The fixture enables control of air temperature through the ducts as well as temperature and humidity conditioning of the "attic" space (4 × 4 ft space containing buried ducts and airspace above the loose fill). Additionally, the bottom of the test fixture is enclosed in an attempt to mimic "inside the home" temperature and humidity conditions but is not controlled to such.

Temperature and relative humidity sensors are used for monitoring and controlling temperature and humidity within the test fixture. The "attic" space is heated via a light source installed in the lid of the test fixture and humidified by the flow of air through a heated water bucket. The ducts traverse in a U-fashion through the fixture, and air through the ducts is temperature controlled via a water chiller. Fiberglass loose fill is blown into the test fixture to a defined R-value. Images of the test fixture are shown below in Figure 4(a) through Figure 4(f).

Buried Duct Experimental Lab-Scale Testing

The buried duct test fixture enables the flexibility for a range of experiments studying buried ducts. For example, the following variables can be adjusted within an experiment: insulated duct type, insulated duct R-value, fiberglass loose fill R-value, "attic" temperature and humidity, inside duct temperature, as well as hold time at specific conditions. The test fixture, however, is not currently capable of cycling the ducts (i.e., delivery of air-conditioned air for specified times) as may likely be the case in a residential setting. Nor was it capable of being utilized for daily cycling of attic conditions that would be expected in a real-life scenario.

An example of an experiment completed with the test fixture is one in which extreme South Carolina conditions were studied for buried ducts. Based on modeling data, worst-case conditions for condensation potential occur during the fall in South Carolina, and, thus, the attic conditions were chosen to represent that time frame (77°F,

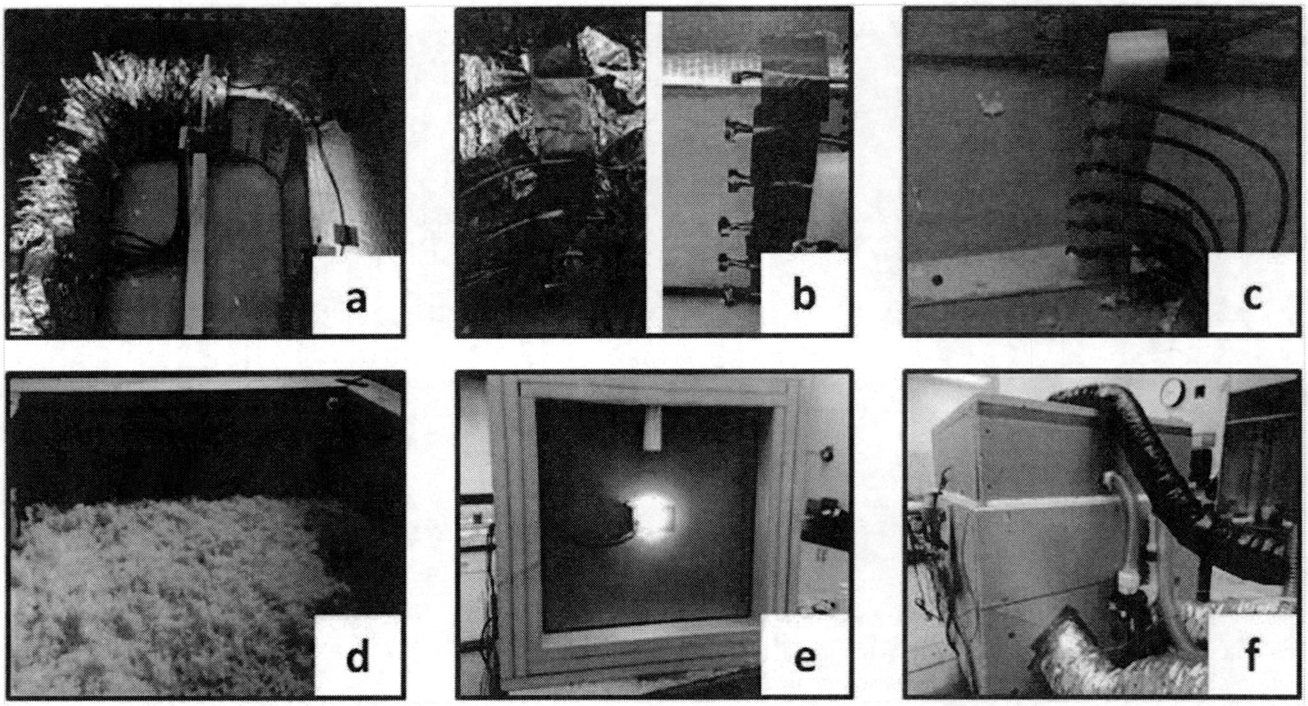

Figure 4 *Experimental lab-based buried duct test fixture. The individual images represent (a) fiberglass insulated ducts run along the drywall floor through the "attic" space, (b) sensors placed every 2 in. vertically along the duct surfaces, (c) sensors mounted 2 in. vertically through the loose fill pack, (d) "attic" filled with fiberglass loose fill insulation, (e) heat source for "attic" conditioning, and (f) overall setup of the test fixture.*

80% rh). Cooling air was supplied through the ducts at 55°F for 100% of the time, and testing was maintained for 22 days total runtime. The bottom of the test fixture was closed (as described above) to achieve higher humidity below the ceiling drywall to better represent South Carolina home conditions. R-8 fiberglass insulated flexible duct and R-8.7 fiberglass duct board were studied under these conditions with R-38 fiberglass loose fill level in the test fixture (no mounding of insulation over the ducts).

During the 22 days of runtime, and once the cold air duct was turned on, attic temperature and humidity averaged 78.4°F and 80.7% rh, duct temperature averaged 54.9°F, and "indoor" conditions averaged 68.5°F and 50.7% rh with minimum RH of 38% and maximum RH of 73.2%. Figure 5 below shows plots of the relative humidity around the (a) R-8 flex duct and (b) R-8.7 duct board as measured by the sensors placed around the ducts. The colored lines represent the placement of the sensor around the duct (i.e., 0 in. above ceiling bottom of duct to 10 in. above ceiling top of duct). The black line represents the tracking of humidity within the enclosed "indoor" space.

Both the 8 and 10 in. sensor for the R-8 flex duct were observed to maintain maximum humidity throughout the duration of the test. For the R-8.7 duct board, the 10 in. sensor maintained maximum humidity throughout the duration of the test, while the 8 in. sensor reached maximum humidity during periods of high "indoor" humidity. In addition to the sensors, moisture indicator tape pieces were placed near each sensor and throughout other areas of the flex duct and duct board. These tape pieces are white when dry, turn red upon contact with moisture, and remain red even after the moisture is no longer present (irreversible color change). Figure 6 below depicts the process of color change with the presence of moisture for illustration purposes.

After completion of the extreme 22-day test period, visible moisture was observed on the surface of the R-8 flex duct and the loose fill felt moist in the vicinity, but this was not observed in all areas. Some of the moisture indicator tape pieces remained unchanged, while others turned red in color, indicating moisture. It was also noted that no dripping onto the drywall was observed prior to moving the ducts during observation.

Observation of the R-8.7 duct board after the extreme 22-day test period showed no visible moisture in the area of the sensors and the duct board did not feel wet to the touch in this area. This is also noted by observing no color change in the moisture indicator tape (Figure 7[a]). The indicator tape pieces did show a color change to red in the top middle of the duct board (Figure 7[b]) and moisture was observed along the

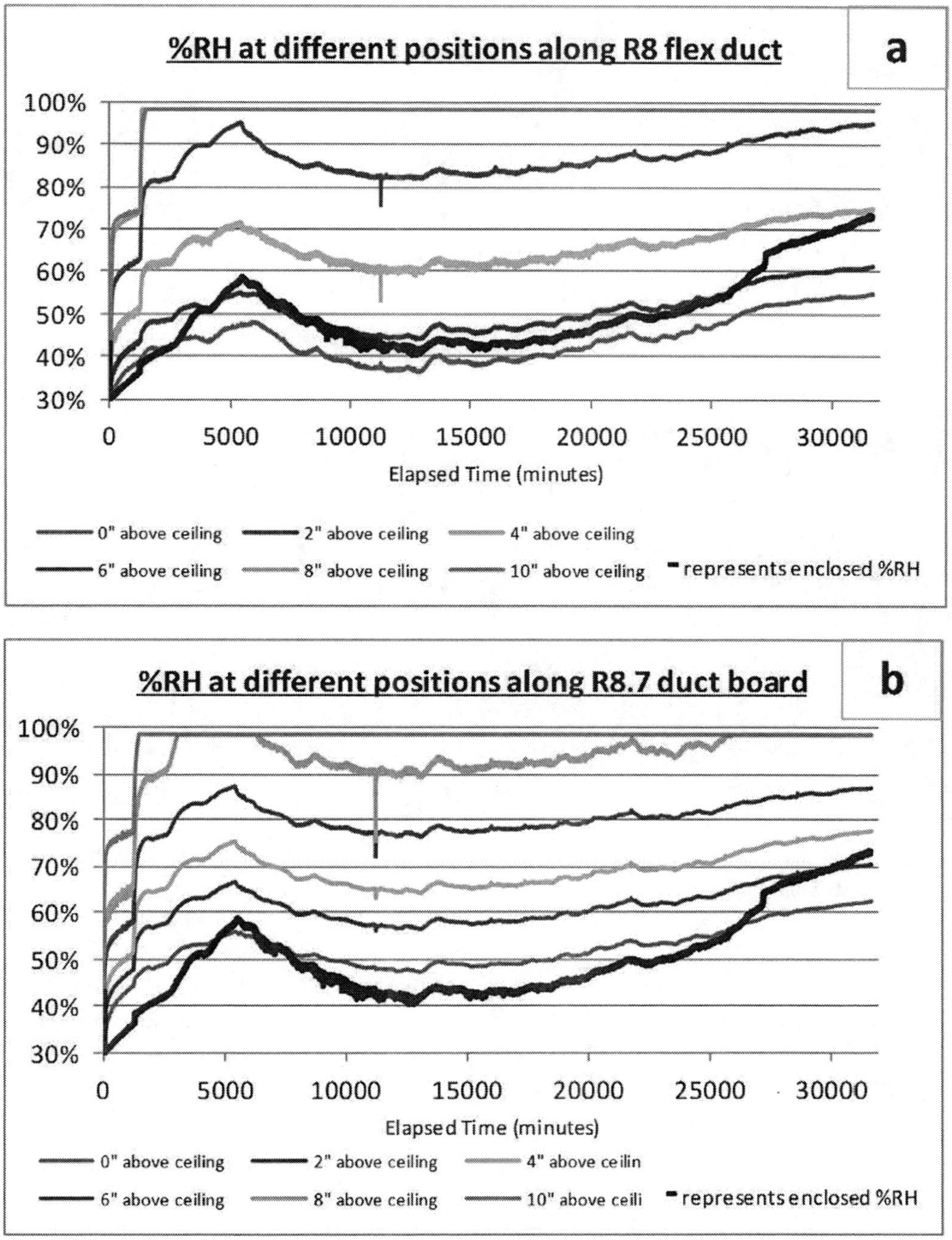

Figure 5 Relative humidity around the (a) R-8 flex duct and (b) R-8.7 duct board during the described conditions tested.

top and edge at the elbow. As was the case with the flex duct, no dripping of condensation was observed onto the drywall beneath.

Buried Duct Field Testing—Cleveland, OH

Field monitoring of buried ducts is being completed in Cleveland, OH, with collection and analysis of temperature and humidity data over the course of a year. R-8 fiberglass insulated ducts were installed on the attic floor with R-49 fiberglass loose fill installed throughout the attic. Prior to ceiling drywall being installed, temperature and humidity probes were placed around a portion of a R-8 flexible duct as well as through the depth of the loose fill. Figure 8 shows a picture of the sensors installed around the duct (Figure 8[a]) and mounted to a post for loose fill monitoring (Figure 8[b]).

The sensors were installed around the flex duct and through the loose fill at 0, 3.5, 7, and 9.5 in. from the bottom of the duct for flexible duct and at the same distances from the

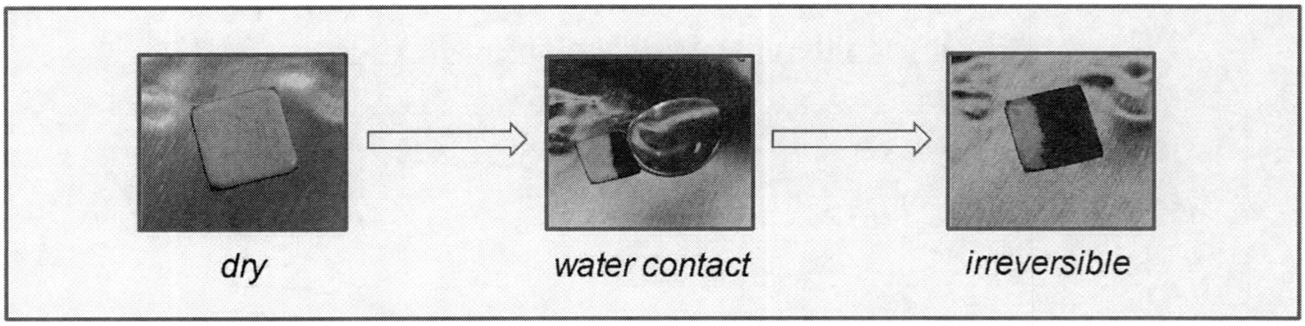

Figure 6 *Illustrative process showing irreversible color change with a moisture indicator tape piece.*

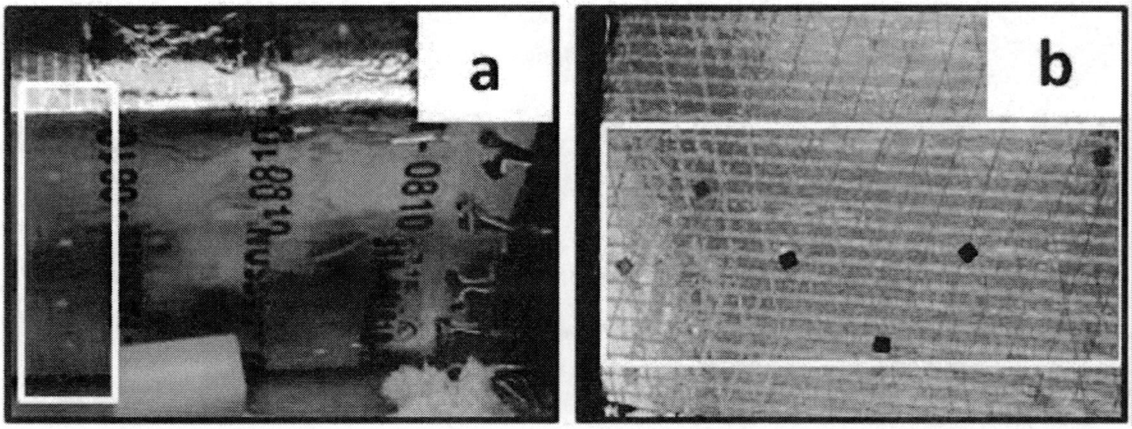

Figure 7 *Moisture indicator tape observed on R-8.7 duct board after 22-days testing under South Carolina attic conditions with constant cooling air supplied through the ducts: (a) side and (b) top.*

top surface of the ceiling drywall for the loose fill. The supply air temperature was about 50°C during cooling at the air inlet close to the air-handling unit and about 1°C higher at the far away inlet grill, indicating fairly constant duct temperature along the duct length.

In addition, temperature and humidity sensors were installed next to the flex duct core for monitoring approximate air temperature traveling through the duct, in the attic air space to represent attic ambient conditions, within an inside bedroom to represent indoor conditions, and under an exterior deck to represent outdoor conditions. These sensors are used to capture boundary condition data.

Data collection for buried ducts in the Cleveland climate began mid-February 2015 with temperature and relative humidity data recorded every 30 minutes. Figure 9 shows percent relative humidity around the flex duct beginning February 19, 2015 and running through November 9, 2015 (data collection will continue through February 2016).

During periods of heating for the climate, the relative humidity remains low around the flex duct. During the cooling seasons, relative humidity increases with elevated humidity (≥90% rh) observed during limited days May through September. Measured periods of high humidity are then followed by lower humidity. It is noted that at no point was the humidity on the bottom of the flex duct ≥90%. The maximum humidity observed for the 0 in. (bottom) sensor was 88%; for the 3.5 in. sensor was 96%, for the 7 in. sensor was 96%, and for the 9.5 in. sensor (top) was 97%.

Additional analysis of the data can be done to represent the hours above a specified humidity for each of the sensors. Note the data records every 30 minutes so each data point above the specified humidity represents 30 minutes and the overall hours is thus an approximation. The total hours during the data collection period was 6812.5. The hours of time that each sensor was above the humidity are shown in Table 1.

Moisture indicator tape pieces were applied around the buried duct during an intermediate observation of the duct on July 15, 2015. Observations were made of the ducts, including the moisture indicator tape pieces, on November 11, 2015 after completion of the cooling season. No signs of condensation were observed around the duct (no change in moisture indicator tape color) or on the paper drywall (no visible staining of the paper). Figure 10(a) through Figure 10(c) show

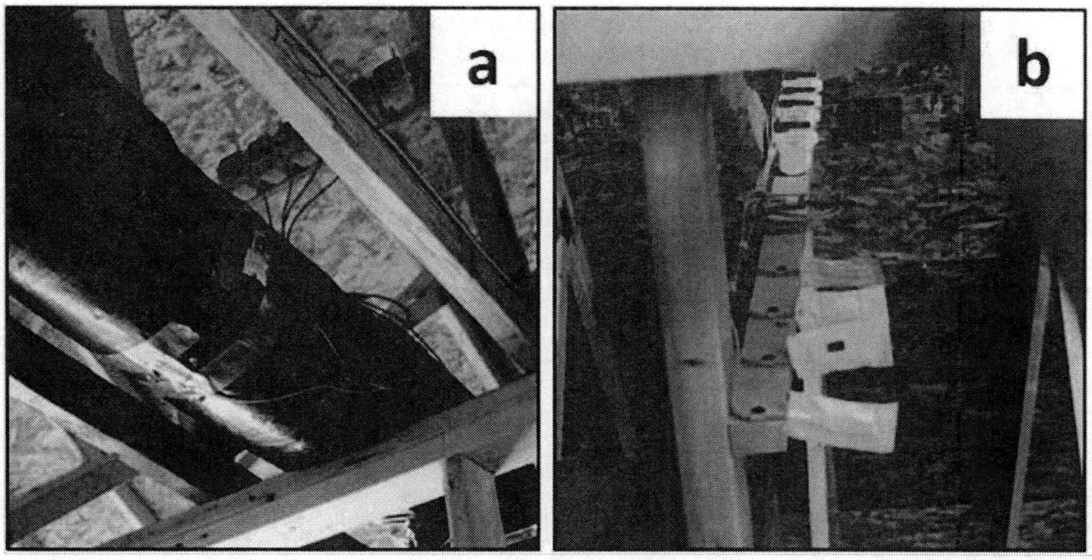

Figure 8 *Temperature and humidity sensors installed around (a) R-8 flex duct and (b) mounted to post for monitoring of loose fill.*

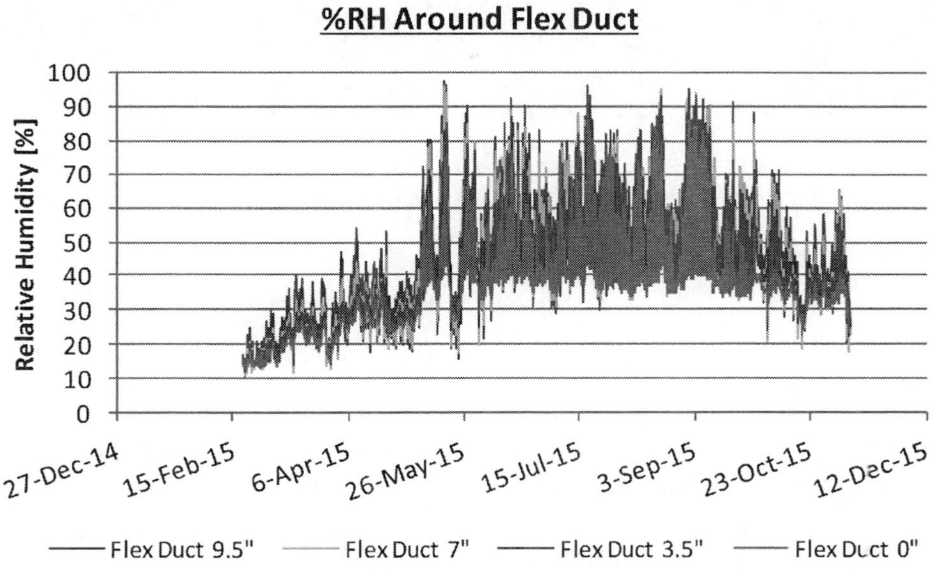

Figure 9 *Percent RH around R-8 flexible duct buried under R-49 loose fill in Cleveland, OH, climate from February 19, 2015, to November 9, 2015.*

pictures taken of the flex duct and drywall on November 11 after pulling back the fiberglass loose fill revealing the white indicator tape pieces with no signs of color change, thereby indicating no visible signs of condensation.

CONCLUSIONS

Attics experience very dynamic conditions, and existing steady-state analyses used to design duct systems to avoid condensation can be misleading and result in overly conservative designs.

The laboratory and field tests as well as numerical simulations indicate that R-8 duct insulation could be sufficient to avoid condensation problems in buried ducts. Simulations and field tests did not show moisture accumulation that would have resulted in dripping water or wet insulation. However an acceptance criteria needs to be defined. Currently there is no consensus whether we can allow for any condensation or high humidity conditions adjacent to the ducts and what should those limit state values be? This paper provides new insight by integrating hygrothermal modeling

Table 1. Hours above Specified Humidity for Each of the Sensors Installed around an R-8 Buried Flexible Duct with 6812.5 Total Hours of Data Collection

	Hours ≥80% rh	Hours ≥85% rh	Hours ≥90% rh	Hours ≥95%rh
Flex 0 in. (bottom)	37	9.5	0	0
Flex 3.5 in. (middle)	176	74.5	22	2
Flex 7 in. (middle)	122.5	39.5	12	1
Flex 9.5 in. (top)	126.5	49.5	13.5	1.5

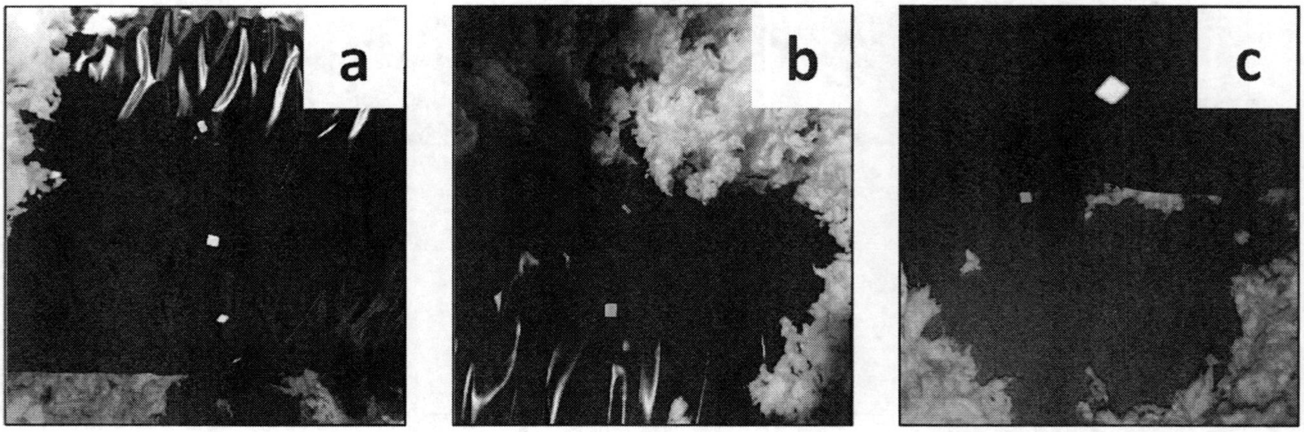

Figure 10 *Observations of buried R-8 flex duct on November 11, 2015, showing no color change observed for moisture indicator tape pieces installed on July 15, 2015, around flex duct and on paper drywall.*

analysis, substantiated by laboratory testing coupled with real life field testing indicating that buried ducts can perform well even at R-8 duct insulation levels.

Currently the majority of residential homes that have ducts in vented attics have buried ducts: These are the hanging ducts that connect to the duct boots through the insulation layer. These buried duct sections are performing well even at substantially lower duct R-values than previous proposed. The results from the work suggest that R-8 buried duct system performs well in Climate Zone 5A. More research should be focused on an R-8 buried duct system in humid southern climate zones (4A and warmer) to investigate if higher R-value is warranted.

ACKNOWLEDGMENTS

The authors would like to acknowledge Dean Potter VP Quality and Home Production process at K. Hovnanian for the field test home used in this study. Furthermore, the sharing of both data and scientific insights from Home Innovation Research Labs (HIRL) by Dave Malley is very much appreciated and is valued enormously.

REFERENCES

Griffiths, D., and M. Zuluaga. 2004. An analysis of the effective r-value for insulation buried attic ducts. *ASHRAE Transactions* 110(2):721–26.

Griffiths, D., M. Zuluaga, R. Aldrich, and D. Springer. 2004. Insulation buried attic ducts: Analysis and field evaluation findings. *Proceedings of the 2004 ACEEE Summer Study on Energy Efficiency in Buildings*. Washington, DC: ACEEE:1.81–1.90.

Griffiths, D., R. Aldrich, W. Zoeller, and M. Zuluaga. 2002. An innovative approach to reducing duct heat gains for a production builder in a hot and humid climate—How we got there. *Proceedings of the 2002 ACEEE Summer Study on Energy Efficiency in Buildings*. Washington, DC: American Council for an Energy-Efficient Economy (ACEEE):1.81–1.90.

Malley, D. 2016. Compact buried ducts in a hot-humid climate house. Home Innovation Research Labs, National Renewable Energy Laboratory & Building America Program (January 2016).

Shapiro, C., A. Magee, and W. Zoeller. 2013. Reducing thermal losses and gains with buried and encapsulated ducts in hot-humid climates. Consortium for Advanced Resi-

dential Buildings. National Renewable Energy Laboratory & Building America Program.
Vineyard, E.A., T.W. Petrie,. J.A. Atchley, and R.L. Linkous. 2004. Experimental measurements of the effective r-value of a duct buried under various levels of insulation. To be published.

Field Test on Two Interior Insulation Systems with Large Thickness—Influence of Orientation and Airtightness

M. Stelzmann, Dipl.Ing. **B. Nusser, PhD** **U. Möller, PhD**

ABSTRACT

From the hygrothermal point of view, exterior insulation finishing systems (EIFS) are the preferred way to retrofit old buildings. However, in some cases EIFS are not possible to apply (e.g., for historic buildings, individual flats, or technical reasons). An interior insulation can be applied in such cases. Due to the dew-point shift and changed drying capacity of the wall, this might result in moisture and durability problems. There are different kinds of interior insulation systems and each one has its own way of moisture management. In this paper, the hygrothermal behavior of two different interior insulation systems with moderate vapor permeability and large insulation thickness will be discussed. A 200 mm EPS (expanded polystyrene) system and a 160 mm mineral wool system were monitored during a 18-month field test. Each system had been wall mounted on hollow block concrete masonry with orientation to South, West, and North. Temperature and relative humidity were monitored between interior insulation and the outer wall. The moisture content of the old plaster as well as the heat flux was measured. To test the impact of air leakage in the air barrier of the mineral wool system two elements with different numbers and sizes of leaks were investigated. Furthermore, hygrothermal simulations were verified with the collected data.

INTRODUCTION

From the building physics point of view, external insulation is preferred. However, in some cases external insulation is not possible either because of design, historic preservation, or technical or economic aspects. Interior insulation may be chosen in these cases. In contrast to a retrofitted wall with an exterior insulation, a subsequent interior insulation changes the building physics of an outer wall construction significantly. The dew-point zone moves inside the construction and increases the risk of condensation between the interior insulation and the original exterior wall (WTA 2009). Another cause for moisture entering in the construction is wind-driven rain. Because of reduced heat flux during winter months, the temperature in the outer wall decreases. Thus, the drying potential of the wall is reduced and the risk of freeze increased. For moderate vapor-permeable interior insulation systems, correctly dimensioned vapor barriers were used to guarantee airtightness and control the water vapor transport into the construction. The risk of condensation depends on the level of the quality of the airtightness of the inside insulation. In case of damaged airtightness, the humid warm air can flow into the interior insulation and condensate on the cold wall. The moisture performance of different inside insulation systems can be affected by the wind-driven rain and solar radiation in the different orientations. With the help of computer simulation of coupled heat and moisture transport (DIN 2007), building physic computations should be provided in any case while using interior insulation with large thickness. If the relative humidity turns under 95% in steady state, the evidence is provided (WTA 2014). This article describes the monitoring of a 200 mm EPS system (EPS) and a 160 mm mineral wool system (MW) under field conditions over a 18-month period. Further, the experimental set up, measurement results, and a comparison with hygrothermal simulations are presented. The research questions were: Did the moisture performance of the selected inside insulation systems work in large thickness, and is it possible to confirm the results in a simple and pragmatic simulation model?

M. Stelzmann is a doctoral student and **U. Möller** is a professor in the Institute of Building Construction, Structural Design and Building Physics, University of Applied Sciences, Leipzig, Germany. **B. Nusser** is the head of the building physic group at Holzforschung Austria, Vienna Austria (during project conduction he was employed at Saint-Gobain Isover, Ladenburg Germany).

EXPERIMENTAL SETUP

Construction of Test Walls

The test house is located in Ladenburg, Germany. The single-story building covered a ground floor area of 5.60 × 5.90 m. The exterior walls were built of hollow block concrete masonry, plastered on both sides with a thickness of 27.5 cm. The south, west, and north exterior test walls consist of six separate 65.5 cm (South and North) or 60.5 cm (West) wide wall sections. The individual wall sections were separated by a 6 cm gap. The sides and top of the wall sections were covered with a continuous air barrier. A bituminous sheeting was placed at the bottom side. The remaining gaps between the wall sections were filled with mineral wool. The wall sections were insulated with two different interior insulation systems. A picture of the test house and the ground plan with details are shown in Figure 1. Both interior insulation systems were installed on the south, west and north walls. Two additional wall sections were equipped with the mineral wool system on the north wall. Each of these two additional systems had one, respectively two, systematic holes in the smart vapor retarder and the drywall installation. The holes had a diameter of 5 mm. A system without air leak was used as reference. A historic wall construction with an heat transfer coefficient (U-factor) of 1.7 W/m^2·K was simulated with hollow block concrete masonry. The EPS system upgraded the U-factor of the construction to 0.15 W/m^2·K and the mineral wool system to 0.17 W/m^2·K (DIN 2008).

EPS System. The EPS system is based on expanded polystyrene bonded to a gypsum board. The insulation material has a thermal conductivity of 0.032 W/m·K. The diffusion equivalent air layer thickness (s_{di} value) of the EPS system was 11.1 m. The installation of the system had been carried out in a point bead method. Thereby the insulation board is fixed to the wall by a continuous mortar beat at the edge and mortar clump at the center of the boards back. For this field test, the interior insulation system had been installed with a insulation thickness of 200 mm. A sketch of the EPS system is shown in Figure 2.

Mineral Wool System. The mineral wool system is based on the insulating material mineral wool with a thermal conductivity of 0.032 W/m·K. A polyamide-based smart vapor retarder with variable vapor diffusion resistance reduces the entry of moisture from the room air. The variable s_{di} value of the mineral wool system was in the range between 0.5 and 25 m. An additional layer of a metal stud framing with drywall installation level protects the smart vapor retarder from damages. For this field test, the interior insulation had been installed with a mineral wool thickness of 160 mm. A sketch of the mineral wool system is shown in Figure 3.

(a)

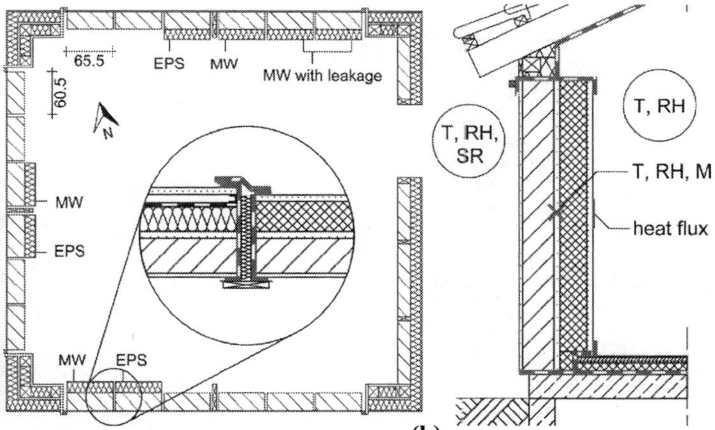

(b)

Figure 1 *(a) A picture of the test house in Ladenburg and (b) ground plan and sectional view with wall sections, marked positions of sensors (see Monitoring System section), and sealing layer.*

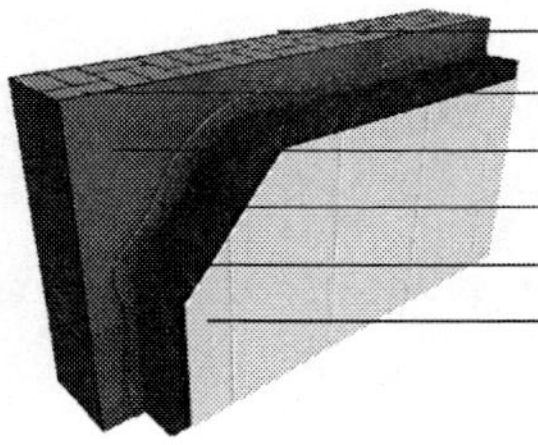

Figure 2 *Sketch of the EPS system with material legend (Bäumler 2013).*

Wind-Driven Rain Resistance

Referring to DIN (2014), the building was located in Stress Zone 2 with medium loads of wind-driven rain (WDR). Possible influences of moisture from WDR were therefore excluded by measuring the water absorption coefficient (W_w value) of the external plaster system. According to the German standard (DIN 2014) and technical bulletin (WTA 2009), a W_w value of the external plaster with 0.5 kg/(m$^2 \cdot \sqrt{h}$) should not be exceeded. The water absorption coefficient was measured by the help of a nondestructive *in situ* technology presented in Möller and Stelzmann (2013) and Stelzmann et al. (2015). The measured W_w value of the external plaster system was 0.3 kg/(m$^2 \cdot \sqrt{h}$). Thus, with a low W_w value of the external plaster in consideration with a single-story test building and an existing overhang of the roof, effects of WDR could be neglected in this study.

Monitoring System

To study the heat transmission resistances, heat flux plates (Q; sensor type Almemo FQA019C) were attached to the inner surface of the interior insulation at walls on the south, west and north side (see Figure 4). For the investigation of the insulation systems, a combined sensor for temperature (T) and relative humidity (RH; sensor type Almemo FHAD46-2), as well as a sensor for determining the moisture content (M; sensor type Brookhuis FME moisture meter) of the old plaster, were installed at the top, center, and bottom of the wall sections. In this paper, only the center position was discussed. The measurement of the relative humidity was carried out through a capacitive measuring principle. To make limited statements of moisture content of the old plaster, the electrical-conductivity based FME sensor was used. However, in laboratory tests, the FME with the calibration curve plaster was able to differentiate between the capillary and sorption moisture water region of plaster material. Thereby, the FME was able to assess the moisture content in case of relative humidity of over 95% rh. As seen in Figure 4, the sensor wiring was guided out laterally inside the old interior plaster from the wall sections. Thus, the insulation layer was not penetrated. Diffuse and direct solar radiation (SR) was determined using a sensor type Almemo FLA613GS-SDEK. With the exception of moisture content, the Almemo 5690-1 was used for automatically recording data. The moisture content was recorded by manual reading once a week. By an air-conditioning system (heating and cooling) combined with a relative-humidity-controlled humidifier (type Seliger Fogstar 300), the indoor climate was adjusted to standard climate in accordance to (DIN 2007).

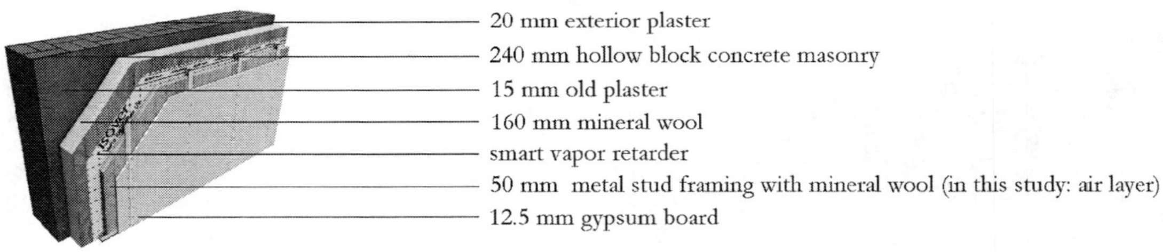

Figure 3 *Sketch of the mineral wool system with material legend (Bäumler 2013). The cavity between the metal stud framing was not insulated in these investigations.*

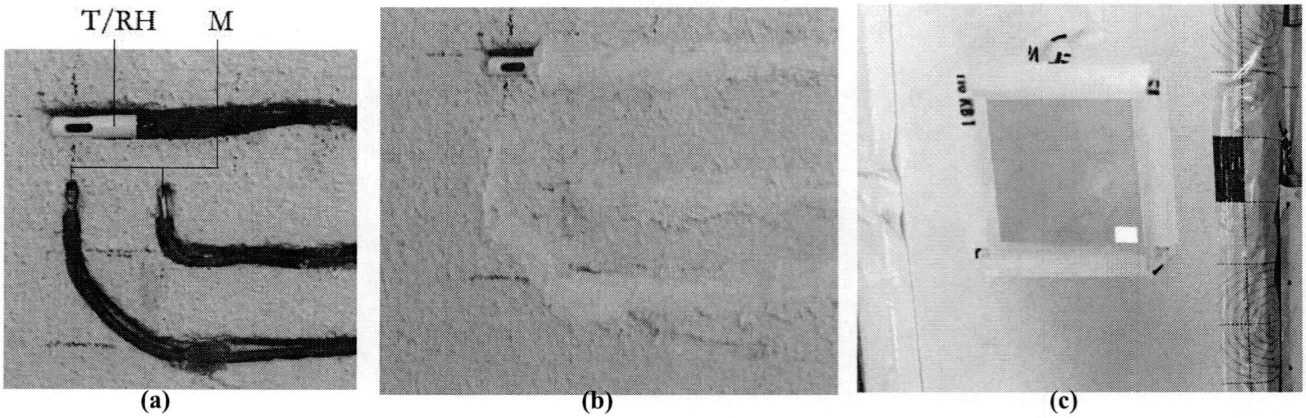

Figure 4 *(a) Embedded sensors in plaster, (b) cable slots were filled with plaster, and (c) selected heat flux plate.*

EXPERIMENTAL RESULTS

In the following section, a few results from the field test are presented (Stelzmann and Möller 2015). The climatic conditions inside and outside as well as the moisture behavior between the old plaster and insulation are shown. The results consider a 18-month period from October 1, 2014 to April 1, 2016. The installation of the surveyed interior insulation systems took place on October 17, 2014.

Conditions

The measurement data of the indoor and outdoor conditions are shown in Figure 5. The outdoor climate in 2014/2015 showed a typical trend for the location Ladenburg, Germany. The average outdoor temperature was 12°C with an average relative humidity of 78 %. The room temperature varied between 20°C and 25°C. Inside the test house, the relative humidity followed a sinusoidal course of a normal occupation in accordance to DIN (2007). The air-conditioning system was deactivated from October 1, 2014 to November 27, 2014. The results of relative humidity tests in test wall sections did not show a significant drying or wetting in this time. However, the inside climate nearly followed the standard climate in accordance to DIN (2007).

Thermal Performance

The heat transfer coefficients (U-factor) of the studied wall sections were determined by measuring the heat flux density from January 20, 2015 to January 28, 2015. Only data between 18:00 and 5:50 with a temperature difference of ≥20 K between inside and outside were considered. For calculating the U-factor in $W/(m^2{\cdot}K)$, the mean of measured heat flux in W/m^2 was divided by the mean of temperature difference between inside and outside in K. These measured results are compared to the theoretical calculated U-factor in Figure 6. The U-factor of the original wall was not tested in field. The theoretical transmission heat loss has been reduced by more than 90 % using interior insulation. The calculated U-factors were higher due to production-related material differences.

Moisture Performance

The moisture performance of interior insulation was assessed by the presence of liquid water on the old plaster. Usually this is the case at a relative humidity higher than 95%. The direction that moisture vapor flowed in inside the insulation is mainly from room humidity into the condensation zone at the old plaster in winter and from there to the inner and outer wall surfaces in summer. The results of the relative humidity and moisture content of the plaster at EPS system and mineral wool system are shown in Figure 7.

The EPS system shows an increased relative humidity of over 95% on the western and northern building sides from January to June in 2015. The moisture content of the plaster remains at a low level of under 1 M% over the entire study period. The installation moisture from the glue of the EPS system was responsible for the higher relative humidity in the first year. In March 2016 the 95% rh level was exceeded. It remains to be seen whether the relative humidity stays permanently below 95% for the remainder of 2016. The relative humidity of the mineral wool system remained below 95% over the entire period of investigation (see Figure 8). The lowest values were detected at the southern wall. The moisture content of the plaster also remained at a low level of under 1 M %. In contrast to the EPS system, there was no additional installation moisture brought into the construction.

Effects of Orientation

For testing effects of different orientations to the hygrothermal behavior, the two studied interior insulation systems were installed on the south, west and north wall (see Figures 7 and 8). As a result, the relative humidity tended to show lower values in the South and higher values in the North. The values of the western side tended to vary in between. Differences in relative humidity were attributable to temperature influence. By higher temperatures in the wall, moisture was driven inwards. Differences of measurement results in moisture content were rather small. The effect was largely solar radiation based.

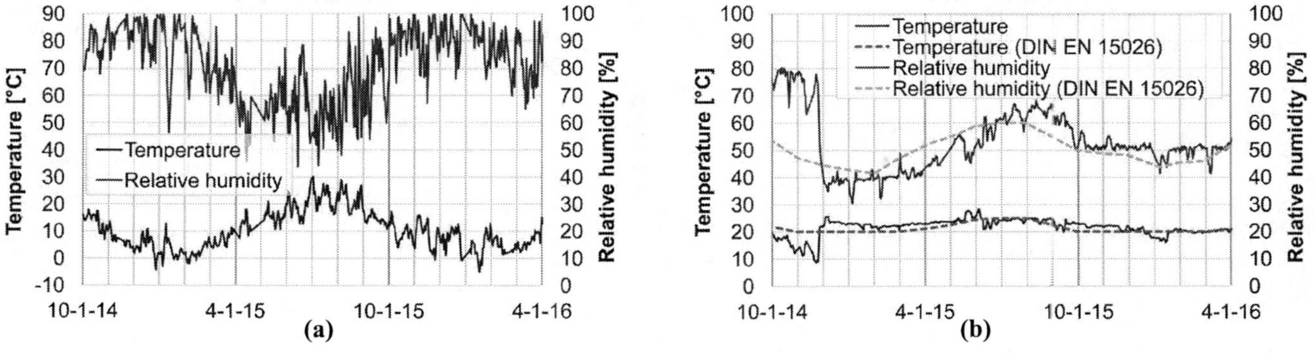

Figure 5 *(a) Outdoor and (b) room climate.*

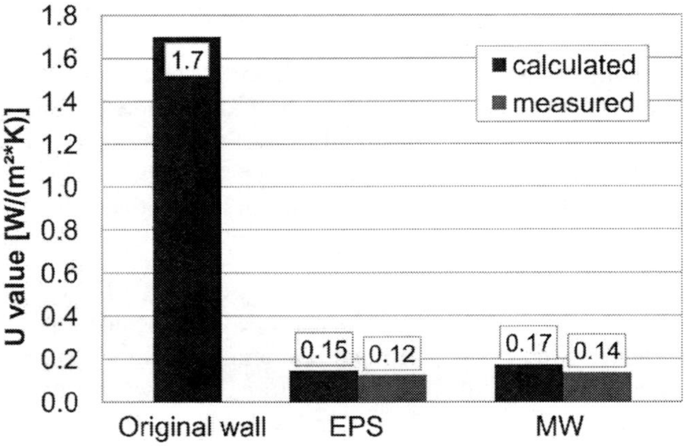

Figure 6 *Comparison of calculated and measured heat transfer coefficients.*

Effects of Leakage in Airtightness of Mineral Wool System

Two wall sections of the mineral wool system were modified by 5 mm holes in the smart vapor retarder and inner lining. Figure 9(a) shows a sketch for testing leakage in the airtightness layer of the insulation system. The airtightness of the wall is already guaranteed by interior and exterior plaster. A fan moved the air in the room. Differences in relative humidity might justified by an airflow from the room to condensation zone. In contrast to Fox et al. (2014), in this study no air pump was used to inject an airflow inside the insulation. An airflow could drive temperature-based convection between top and bottom hole (two leaks). Figure 9(b) gives the conditions in the mineral wool system with holes and thermal-stack-based air pressure difference (P) between the holes in two leaks test. The examined variants behaved similarly. The system with no leakage showed mostly higher relative humidities than the other two systems with leaks. In the used test setup, an effect

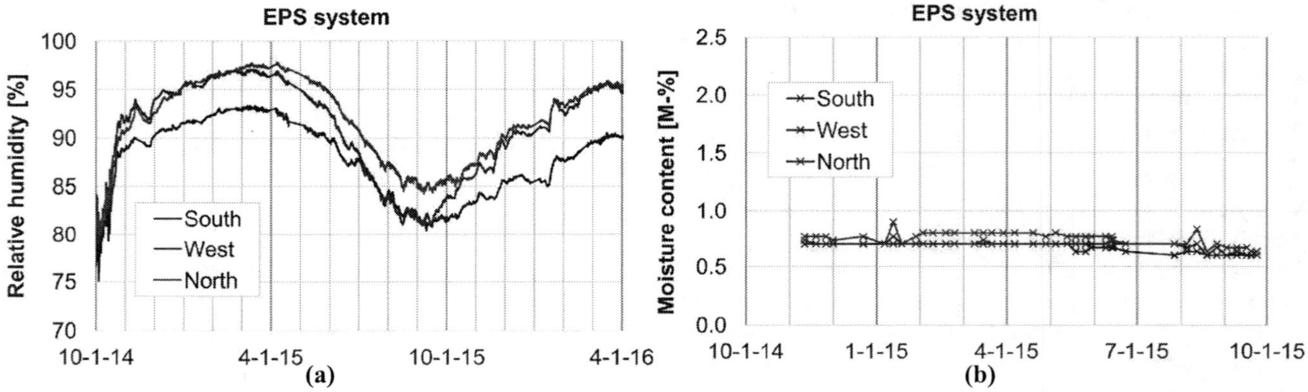

Figure 7 *Comparison of (a) relative humidity and (b) moisture behavior of the EPS system for the orientations South, West, and North.*

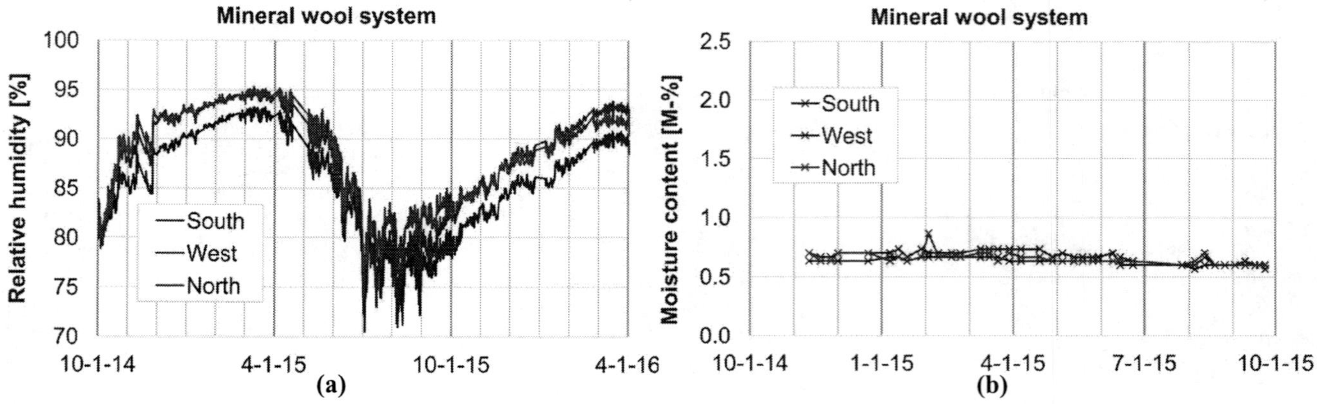

Figure 8 *Comparison of (a) relative humidity and (b) moisture behavior of mineral wool system for the orientations South, West, and North.*

of the holes in the smart vapor retarder and inner lining could not be found. The measured temperature difference between the top and bottom hole was about 1 K in winter. From these data the pressure difference (P) was approximated by Equation 2.14 in Hens (2012). In this study, air pressure differences by thermal stack were small (two holes). The airtight exterior plaster prevented an airflow between inside and outside and the diameter and numbers of holes were rather small. The airflow resistance of the mineral wool could explain the results too. As a result, the collected data showed an existing hygrothermal robustness of the studied mineral wool system.

LONG-TERM TRENDS WITH HYGROTHERMAL SIMULATION

If the relative humidity at the condensation zone turns under 95% in the long term, the inside insulation system is proved (WTA 2014). Long-term simulation can indicate wetting or drying trends. To conduct the hygrothermal simulation (DIN 2007) the program Delphin 5.8 (Nicolai et al. 2010) was used in one-dimensional mode. Measured data of indoor and outdoor climate (temperature and relative humidity) as well as material data of the tested inside insulation systems was available. For material data of the hollow block concrete masonry the Delphin database (Scheffler and Plagge 2009) was used. For the simulation model of mineral wool system, a vapor retarder with fixed vapor permeability was used (s_{di} = 2.5 m). The influence of wind-driven rain (WDR) was not included in the simulations. For the consideration of initial conditions or the installation moisture, experience-based assumptions were made (DIN 2010). Due to lack of exact input data (e.g., initial conditions, material data) the simulations could only provide limited statements. However, with the conducted simulation, trends could be stated. The simulation results of the two studied interior insulation systems for five years are shown in Figure 10. Additionally, the measurement data was overlaid onto the graphs. Simulation and measurement of the mineral wool system demonstrated a similar trend. The EPS system revealed a poor correlation of simulation and measurement results. There were doubts as to whether the simulation results of EPS

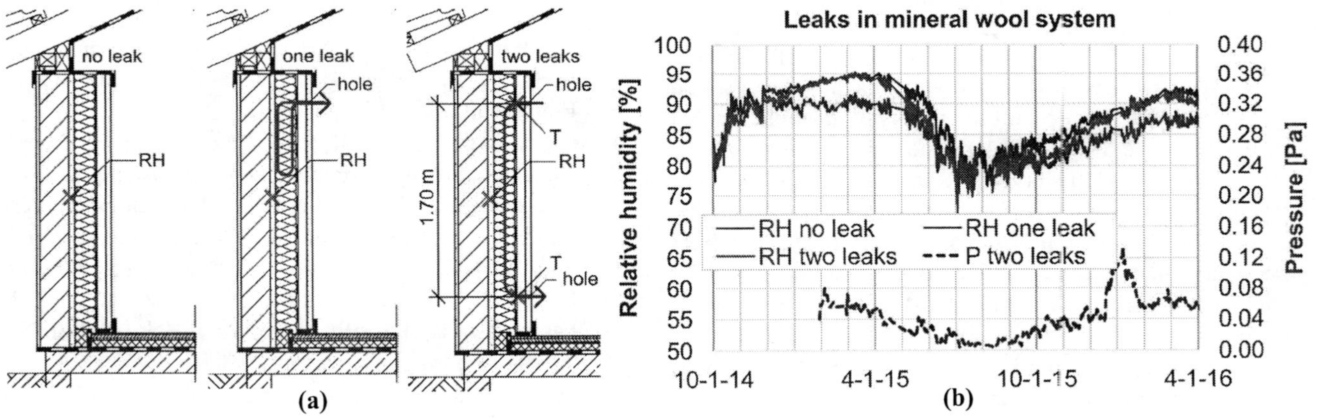

Figure 9 *(a) A sketch of test setup with leaks and (b) behavior of relative humidity for inner plaster of mineral wool system with and without leaks and vertical air pressure difference between top and bottom hole.*

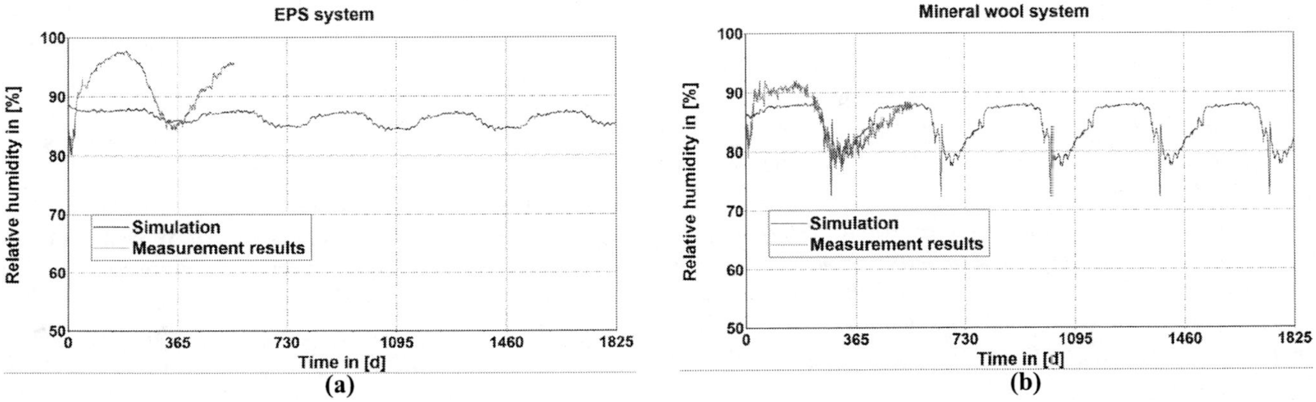

Figure 10 *Comparison of simulation and measurement results for relative humidity of (a) EPS and (b) MW.*

system could predict the moisture behavior. Deviations at the beginning were caused by the installation duration. Assumptions of material data can be responsible for differences in moisture behavior. The simulation results indicated a trend towards lower relative humidities. Already, starting from the second simulated year, both interior insulation systems reached a steady state. As a result, the created simple hygrothermal simulation did not match with the recorded data. The explanation for this could be the fact that material data was used from database, simulation was designed in one-dimensional mode, and no WDR data was included.

CONCLUSIONS

An EPS and a mineral-wool-based interior insulation system with large insulation thicknesses were monitored in a test house for an 18-month period. In the study, exterior walls without thermal bridges were tested. The test setup was realized with a high wind-driven rain resistance; thus, entering moisture by wind-driven rain was significantly limited. Still, outdoor relative humidity, surface condensation, and inward vapor flow could have some wetting impact on the moisture of the masonry. The results demonstrated a theoretical reduction of transmission heat loss by the interior insulation systems of more than 90%. The EPS system showed relative humidity above 95 % in the condensation zone in both years. The relative humidity of the mineral wool system remained below 95% over the entire period of investigation. Moisture effects by different orientations in both systems were detected. The south wall showed the lowest and the north walls the highest relative humidities. This effect was solar radiation based. Furthermore, effects of leaks in the inner airtight layers of the mineral wool system were investigated. In the specific experimental design with an airtight exterior plaster, no significant hygrothermal differences between the airtight and leaky systems were found. Results of a simple hygrothermal simulation study did not confirm with the measured data. To improve the simulation, some more climate and material data should be implemented. The collected and analyzed data showed that the highly insulated interior systems retrofit strategies are possible.

REFERENCES

Bäumler, S. 2013. Aktuelle Untersuchungen der SAINT-GOBAIN Innendämm-Systeme. wksb – Zeitschrift für Wärmeschutz Kälteschutz Schallschutz Brandschutz 69:43–48.

DIN. 2007. DIN EN 15026, *Hygrothermal performance of building components and building elements—Assessment of moisture transfer by numerical simulation*. Berlin: Deutsches Institut für Normung e.V.

DIN. 2008. DIN EN ISO 6946, *Building components and building elements—Thermal resistance and thermal transmittance—Calculation method*. Berlin: Deutsches Institut für Normung e.V.

DIN. 2010. DIN EN ISO 10456, *Building materials and products—Hygrothermal properties—Tabulated design values and procedures for determining declared and design thermal values*. Berlin: Deutsches Institut für Normung e.V.

DIN. 2013. DIN 4108-2, *Thermal protection and energy economy in buildings—Part 2: Minimum requirements to thermal insulation*. Berlin: Deutsches Institut für Normung e.V.

DIN. 2014. DIN 4108-3, *Thermal protection and energy economy in buildings—Part 3: Protection against moisture subject to climate conditions—Requirements and directions for design and construction*. Berlin: Deutsches Institut für Normung e.V.

Fox, M., J. Straube, H. Ge, and T. Trainor. 2014. Field test of hygrothermal performance of highly insulated wall assemblies. In 14th Canadian Conference on Building Science and Technology, Toronto. October 28–30.

Hens, H. 2012. *Building physics—Heat, air and moisture: Fundamentals and engineering methods with examples and exercises*, 2nd edition. Berlin: Ernst & Sohn. ISBN 978-3-433-03027-1, 134–35.

Möller, U., and M. Stelzmann. 2013. In-Situ-Messgerät für die zerstörungsfreie Messung der Wasseraufnahme. in: 2. Internationaler Innendämmkongress vom 12. bis 13. April 2013 in Dresden, Tagungsunterlagen, Grunewald, J., Plagge, R. (Hrsg.), Dresden: TU-Dresden, Institut für Bauklimatik, 188–97.

Nicolai, A., J. Grunewald, R. Plagge, and G. Scheffler. 2010. Development of a combined heat, moisture, and salt transport model for unsaturated porous building materials. In Simulation of Time Dependent Degradation of Porous Materials, Research Report on Priority Program DFG SPP 1122, eds. L. Franke, G Deckelmann, and R. Espinosa-Marzal, Cuilliver Verlag. ISBN 978-3-86727-902-4. 67–84.

Scheffler, G. and R. Plagge. 2009. A whole range hygric material model: Modeling liquid and vapour transport properties in porous media. *Int. J. Heat Mass Transfer*, doi:10.1016/j.ijheatmasstransfer. 2009.09.030.

Stelzmann, M., U. Möller, and R. Plagge. 2015. Water-absorption-measurement instrument for masonry façades, in: ETNDT6, *Emerging Technologies in Non-Destructive Testing* 6, 27–29 May 2015, Brussels, Belgium.

Stelzmann, M., and U. Möller. 2015. Saint-Gobain Innendämmsysteme im Freilandversuch–Abschlussbericht.

WTA. 2009. WTA Guideline 6-4, Inside insulation according to WTA I: Plenary guide. Stuttgart: Wissenschaftlich-Technische Arbeitsgemeinschaft für Bauwerkserhaltung und Demkmalpflege e.V.

WTA. 2014. WTA Guideline 6-5, *Interior insulation according to WTA II: Evaluation of internal insulation systems with numerical design methods*. Stuttgart: Wissenschaftlich-Technische Arbeitsgemeinschaft für Bauwerkserhaltung und Demkmalpflege e.V.

Approach to Assessing Continuous Insulation Products

Jonathan Smegal, MASc

John Straube, PhD, PEng
Associate Member ASHRAE

Theresa Weston, PhD
Member ASHRAE

Aaron Grin, MASc, PEng

ABSTRACT

More rigorous energy codes and requirements for energy-efficient construction are spurring interest in new insulation products, specifically those that provide continuous exterior insulation. New exterior insulation products, especially those that have nontraditional forms, need to be evaluated not only for basic material properties such as R-value, but also for their installed assembly performance and compatibility with other system components. This paper discusses several methods that can be used for system-level product evaluation and demonstrates the value of this approach.

Full-scale laboratory testing was conducted on wall assemblies constructed using two different types of exterior insulation. The exterior insulation products compared were a low vapor permeance XPS insulation and a high vapor permeance fibrous exterior insulation relatively new to the market. Evaluation methods included thermal imaging analysis using a climate simulator and structural performance testing of cladding attached to strapping that was fastened to the structure through the continuous exterior insulation. The thermal imaging analysis included side-by-side wall construction in the climate simulator and infrared thermal images taken of both the interior drywall and exterior insulation or cladding while operating the simulator in a winter climate condition. The structural assembly testing consisted of deflection measurement of both short- and long-term loading of various simulated cladding weights to the strapping, which was fastened to the structure through the exterior insulation.

Conducting these laboratory tests and analysis produced a more realistic in-service performance prediction of exterior insulated wall assemblies that accounts for the interaction of the various assembly components.

INTRODUCTION

Construction practices have evolved alongside the development of construction materials to meet the demands of the industry today. Many historical buildings were built with load-bearing masonry or solid wood, with little or no intentional insulation. Most modern enclosure wall systems employ a layered approach, with a variety of engineered materials that are designed to fulfill specific functions (such as structure, finish, water control, and thermal control) that have the advantage of lighter weight and a wider range of exterior finish and enhanced thermal and moisture performance. As part of this evolutionary change in enclosure walls, new products are being developed and deployed to provide continuous insulation, airflow control, and liquid water control.

Some amount of insulation is required in all new building projects. The amount of thermal resistance is based on the climate region for the project and the specific building or energy code in force. The thermal performance of an insulation product is defined by its resistance to heat flow, or R-value (RSI in SI). The R-value of the insulation is the most widely used metric to differentiate between individual insulation materials, although there are several other functional characteristics that should be considered. In many cases, insulation products are chosen without adequate knowledge of how the insulation will perform as part of a system with other components of the enclosure. It is important to consider system effects in order to meet specific project criteria such as price, wall thickness, R-value, and control of air, vapor, fire, noise

Jonathan Smegal and **Aaron Grin** are associates at RDH Building Science Laboratories, Waterloo, ON. **John Straube** is a principal at RDH and an associate professor at the University of Waterloo, Waterloo, ON. **Theresa Weston** is a research fellow at DuPont Building Innovations, Wilmington, DE.

etc. Standard tests are often required for compliance to building codes and standards, but in-depth research and nonstandard tests are required to predict how materials and systems interact. Systems testing is not a new concept, as manufacturers have long conducted tests to compare their products to other products, but because of the ever-increasing number of products, types of materials, and myriad combinations, systems testing is now even more important.

This paper discusses several methods that have been used to evaluate how a new continuous insulation product performed in terms of its functional requirements within a building system. The product consists of polyester-blend fibrous matt with a vapor permeable air and water barrier membrane integrated on the exterior face of the matt. The product differs from the majority of exterior continuous insulation products in that it is more flexible and so is provided in rolls that can be installed and detailed in a similar fashion to building wrap. Vinyl siding can be attached directly through the fibrous insulation; however, when heavier claddings are installed, such as wood or fiber cement claddings, the fibrous exterior insulation alone cannot structurally support cladding loads. To provide additional support, insulated battens made from plywood and XPS are first installed against the sheathing and attached to the structure. The fibrous insulation is then rolled out over top of the insulated battens, and heavier claddings are installed into the nailing strip on the battens, providing a thermal break and preventing a loss in R-value where the fibrous insulation is compressed.

Because this product differed in form from standard exterior continuous insulation products, its evaluation serves as a case history to understand the many system attributes that need to be considered as part of understanding installed assembly performance.

FUNCTIONAL REQUIREMENTS

Depending on the individual project, there are many functional requirements an insulation product may be required to meet in addition to thermal resistance. The broad categories of functions of all products in the building enclosure can be broken into three main categories: finish, support, and control (Straube and Burnett 2005). Generally speaking, the insulation is not the intended finish, although in some cases it might be visible through an open-jointed cladding system. Some products used on the interior must perform an interior finish function (e.g., roll batts and rigid board insulation for some industrial buildings). In terms of the support function, the insulation may be required to transmit gravity loads from cladding, wind loads from the cladding, and air barrier, seismic loads, etc., from the cladding to the structure. There are four primary control-related functions: water, air, thermal, and vapor. The thermal control function is the primary function for insulation materials. Some materials can perform multiple functional requirements. For example, closed-cell spray foam could be the air, thermal, and vapor control if it is installed in the stud cavity and could be a part of the water control layer if it is installed on the exterior of the sheathing. Another example is that in some locations, polyethylene sheeting is installed on the interior as both the air and vapor control layer. There are two additional control attributes that are sometimes considerations when selecting a product for a particular application: fire spread and sound attenuation. This paper focuses on evaluation of the support and thermal control functions.

Thermal Transport Control

The most critical function of the insulation is to resist the movement of heat energy through the enclosure. The United States Federal Trade Commission requires that R-value testing be conducted at a mean temperature of 75°F (24°C), although it does not specify a temperature difference or an orientation for the sample (FTC 2005). In the case of closed-cell foam insulations blown with special gases, long-term thermal resistance testing (as per ASTM C1303 [2015c]) is also required to determine the long-term R-value, because the R-value of these products is known to change over time. Insulation products are usually tested according to ASTM C518/ASTM C177 (2015a) material test standards at one temperature difference across the insulation product. This common procedure for measuring R-value provides the ability to compare different products but also has some disadvantages. The R-value of insulation materials changes depending on the mean temperature and, to a lesser extent, the temperature difference across the insulation. Many insulation products have a higher R-value at lower mean temperatures (because of the reduced role played by radiation transfer), but there are some materials where the R-value degrades significantly at lower temperatures. These differences are not considered in the standard reporting of material R-value. It is also important to realize that when insulation is part of an enclosure system, the enclosure does not have the same R-value as the insulation. The most common cause of this change in R-value is thermal bridging (caused by building structure [framing and other elements], fasteners, windows, etc. penetrating the insulation) and convection loops (caused by air moving around insulation). Provided the installation of the insulation layer does not allow significant air loops to bypass the product, the advertised R-value can only be realized when the insulation material is continuous over the enclosure without interruptions. It is important to realize that all insulation materials are affected in this way. The total R-value of the fibrous exterior insulation material is R-5.4 for the entire thickness (approximately 1.5 in. [38 mm]) (DuPont 2013).

Vapor Diffusion Control

A specific level of vapor permeance may also be a functional requirement for insulation. Insulation may be specified as vapor permeable or vapor impermeable (to either increase drying or limit vapor diffusion), depending on where the insulation is located in the enclosure (closer to the interior or exterior) as well as the exterior climate, the intended indoor environment, and other layers in the enclosure system. The vapor permeance of a material is usually determined using the ASTM E96 test method. The most common tests determine how much water vapor moves across a material from 50% relative humidity to 0% relative humidity (desiccant or dry cup method) and from 100% relative humidity to 50% relative humidity (wet cup method). In some cases, the vapor permeance is the same for both conditions, but some materials exhibit different vapor permeances at the two different test conditions. For many materials, the vapor permeance will decrease with increased thickness of the material, such as foam-based insulation products for example. For the fibrous exterior insulation material being reviewed, the ASTM E-96 wet cup and dry cup results are 269 perms and 228 perms, respectively (DuPont 2013). Knowing the vapor permeance of a material will help determine climate regions and locations within the enclosure where the material may be used correctly without increasing moisture durability risk.

Air Leakage Control

The insulation material may serve the functional requirement of an air barrier material and be part of the air barrier system in the enclosure. The airtightness of a material can be determined based on ASTM E2178 (2011b). The air permeance of insulation may be important if the insulation product is to be part of the air barrier system, but also important is the quality of the installation and the transitions between all components of the air barrier system. System airtightness can be evaluated using mock-up assemblies and testing by ASTM E283 (2012b). Mock-up assemblies can be designed to evaluate installation details such as window integration, pipe or duct penetrations, and cladding attachments. ASTM E2357 (2011c) provides a standardized assembly for air leakage evaluation. Because the product used as the example in this study had an air barrier membrane as its exterior face, it met industry and code requirements as an air barrier material. Installation practices for standard mechanically attached air barrier membranes were modified to account for the increase in product thickness. The creation of flaps allowed for shingling and sealing. Several mock-up assemblies were tested, and installation practices were developed that allowed for the system to meet industry and code assembly air leakage requirements.

Water Control

Some insulating products used on the exterior of buildings can also function as the water control layer. This water control layer is typically installed behind a cladding that is presumed to leak rain water (i.e., a rainscreen) but which acts as the primary water-shedding layer. It is key that the water control layer be properly detailed to ensure rain water is directed back out of the enclosure at regular intervals and at key junctions with penetrations. To evaluate the ability of a product to act as a water control layer, numerous standard (such as ASTM E331 [2016e], ASTM E547 [2016f], or AAMA 508 [2014]), and nonstandard tests of the product installed in a mock-up with typical details are commonly undertaken.

Sound Control

In some cases, insulation products are selected based on their sound attenuation or dampening characteristics. The effect on sound transmission of an individual material may be important, but airtightness levels and the combined effects of different materials will have a more significant impact on noise control.

Fire Control

Some fire-related testing is conducted on individual materials, but many systems of materials such as roof and wall assemblies must be fire tested as complete systems before they can be allowed in construction of different residential and commercial buildings. Many common fire control standards (e.g., ASTM E84 [2016c]) recognize that assemblies will perform differently than materials considered separately, by protecting other materials in the assembly or providing paths for fire and smoke within the assembly.

Summary

To ensure that a specified product will meet all of the necessary functional requirements for a specific project, material property and performance data can be very useful, especially when used in combination with thermal or hygrothermal modeling. However, system-level testing can confirm and demonstrate that the performance will meet the functional requirements of a project. If unique assemblies and nonstandard applications are considered, system-level testing becomes more important.

PREDICTION OF IN-SERVICE PERFORMANCE USING MATERIAL PROPERTY DATA

Typically, evaluation of insulation performance is based on standardized material testing. This testing would include, as a minimum, determining the R-value or thermal conductivity in the ASTM C518/C177 (2015a) test and determining the vapor permeance in the ASTM E96 (2016d) test. Other tests that manufacturers commonly conduct include compressive strength (e.g., ASTM D1621 [2016b]), water absorption (e.g., ASTM D2842 [2012a]), and coefficient of linear expansion (e.g., ASTM D696 [2016a]). Individual material property testing is routinely done and is therefore relatively quick, straightforward, and inexpensive. Once the individual material properties have been identified, thermal and hygrothermal simulations of assemblies may be conducted to predict performance based on both the material properties of the several materials used in a modern enclosure and the geometry of the intended applications. Computing power has increased in the last several years, and simulations have improved in their predictions. Using material data to build a model may provide a reasonable basis for making recommendations and explaining system performance.

However, there are limitations to this approach. Individual material property testing does not take into account any of the system effects that occur when different materials are combined together, such as accurately determining the effective R-value of an assembly with different materials or accounting for air leakage. Also, performance may be affected by quality of installation or incompatibility of different products. For example, some foam plastics such as EPS foam insulation can be dissolved by some solvents. These solvents are used in solvent-extended asphalt emulsions for waterproofing as well as some other products. Simulations can address some of these issues, but these too have their own limitations. Models are easily affected by inexperienced users or incorrect material data (for example, when a generic data set is substituted for a material that does not have data available or the temperature dependent R-value may be assumed to be linear when it is not). In some situations, the factor of safety and/or risk can be inflated or deflated sufficiently that applications of that system in the real world fails. Most software packages also have inherent limitations, including the inability to accurately predict air leakage effects on R-value, condensation, and moisture accumulation. To ensure that models and simulations reflect reality and can be relied upon, measurement in the field or laboratory testing must be completed.

MEASURING PERFORMANCE AT THE SYSTEM LEVEL

A systems approach to testing considers several aspects of performance simultaneously, instead of a single performance variable. Several test protocols are described below that illustrate the systems approach. Results from most of the tests compare a low vapor permeance XPS insulation and the high vapor permeance fibrous exterior insulation that is used as the example. These tests generally do not meet any current standardized test method. They were developed to meet a need for manufacturers to answer industry questions and perceptions or compare their products to other products in a way that is more representative of how the product will actually be used and installed. Some of the tests are more qualitative and some are more quantitative in nature, but each demonstrates some aspect of insulation system performance.

Deflection Testing

There are many questions, and even more perceptions, in the industry regarding the attachment of cladding through thick layers of exterior insulation. These questions began with the trend towards effective R-values of wall assemblies and the application of continuous exterior insulation to minimize thermal bridging. It is common to install vertical strapping or furring on the exterior of the insulation attached back to the wood frame structure and then attach the cladding to the vertical strapping. Strapping or furring strips are usually 1 × 3 nominal lumber or, in some cases, plywood sheathing cut into strips of the desired width and required thickness. In the case of the fibrous insulation considered here, insulated battens were provided as furring to compensate for R-value loss due to compression of the low-density fibrous insulation material. Potential deflection of the cladding is a system effect of the cladding, strapping, insulation type, fastener type, and attachment method, which is why it is critical to test the assembled wall. There are currently no standard test methods for determining the deflection of cladding using the entire wall assembly.

A test method was developed to answer questions regarding potential deflection or movement of cladding attached through the exterior insulation to the structure to minimize the thermal bridging through the insulation. To interpret the deflection data, some typical weights of cladding were calculated based on product manufacturer documentation to estimate the approximate square foot weight of each cladding type. In some cases, it was necessary to use a range of weights, as it was found that similarly advertised products varied in weight. The weights are shown in Table 1. Deflection testing was completed in the laboratory on a short-term basis and a long-term basis. Long-term testing was also completed in an outdoor environment under full exposure to the elements.

Laboratory Short-Term Deflection Testing.

To conduct the testing, a 4 × 8 ft (1.2 × 2.4 m) wall frame with 16 in. (406 mm) stud spacing was securely fastened to a concrete block wall in the laboratory. The 7/16 in. (11 mm) OSB sheathing was installed over the framing. The insulated battens (plywood nailer on an XPS base) for the exterior insulation system were installed through the framing into the structure with 3 in. (76 mm) framing nails (11 in. [280 mm] o.c. vertically). The exterior insulation system was installed over the battens, and 2 × ½ in. (50 × 12.5 mm) plywood strapping was installed through the insulation to the battens, compressing the insulation between the strapping and the batten. A 4 × 8 ft (1.2 × 2.4 m) cladding panel was installed to help distribute the point load at the bottom of the wall assembly to each of the three strapping pieces, so the force was equally applied. The cladding was attached with 2 in. (50 mm) nails through the insulation into the wood surface of the battens. The steel straps on the exterior were only attached to the face of the cladding and not to the strapping.

Table 1. Approximate Cladding Weights

Cladding Type	Typical Mass Density, psf	Equivalent Weight for 4 × 8 ft Test Panel, lb
Vinyl siding	0.6–1.0	20–32
Wood siding	1.0–1.5	32–48
Fiber cement siding	3–5	96–160
Cement stucco	10–12	320–384
Adhered stone veneer	17–25	544–800

This report only includes one comparison between similar wall assemblies with XPS exterior insulation and fibrous exterior insulation (Figure 1), although the initial research program also investigated different lengths of fastener without strapping, as well as a plastic rainscreen strapping product, all with fibrous exterior insulation.

Both of the comparison walls used 0.095 in. (2.4 mm) gauge galvanized siding nails to attach the cladding to the strapping 8 in. (203 mm) o.c. vertically on each row (three vertical rows) of strapping, for a total of 39 fasteners.

A 4000 lb (18 kN) capacity hydraulic ram was used to apply force to a metal angle in connection with the bottom edge of both metal strapping pieces attached to the surface of the cladding. To measure the applied force, a 1000 lb (4500 N) strain gauge load cell (with ±0.4 lbf rated accuracy) was placed between the angle and the ram (Figure 2).

Deflection gages (with a resolution of 1/1000 in. [0.025 mm]) were used to measure the movement of the wall sheathing and cladding on both the left and right side. Measurements of the framing were also taken to verify that the movement occurred between the sheathing and the vertical strapping.

Loads were applied in increments of 100 lb (45 kg) between 100 lb and 1000 lb (approximately 45–450 kg). The four deflection readings were recorded at each increment. Each load increment was applied over about 30 to 60 seconds, and the readings taken within 30 seconds. All of the tests were conducted three times on the same test specimen. The wall was loaded to 1000 lb (454 kg), unloaded of all weight, and reloaded two more times to 1000 lb (454 kg). The analysis of the effect of loading only compares the first test for each construction type, but the second and third test can be useful to show any permanent deflection in the wall assembly following a high loading event. For the test wall results in Figure 3, the total deflection at 1000 lb (453 kg) was approximately 0.03 in. (0.8 mm). After the initial load was removed in each of these tests, the wall returned to approximately 0.015 in. (0.4 mm), so there was a permanent deformation after the load

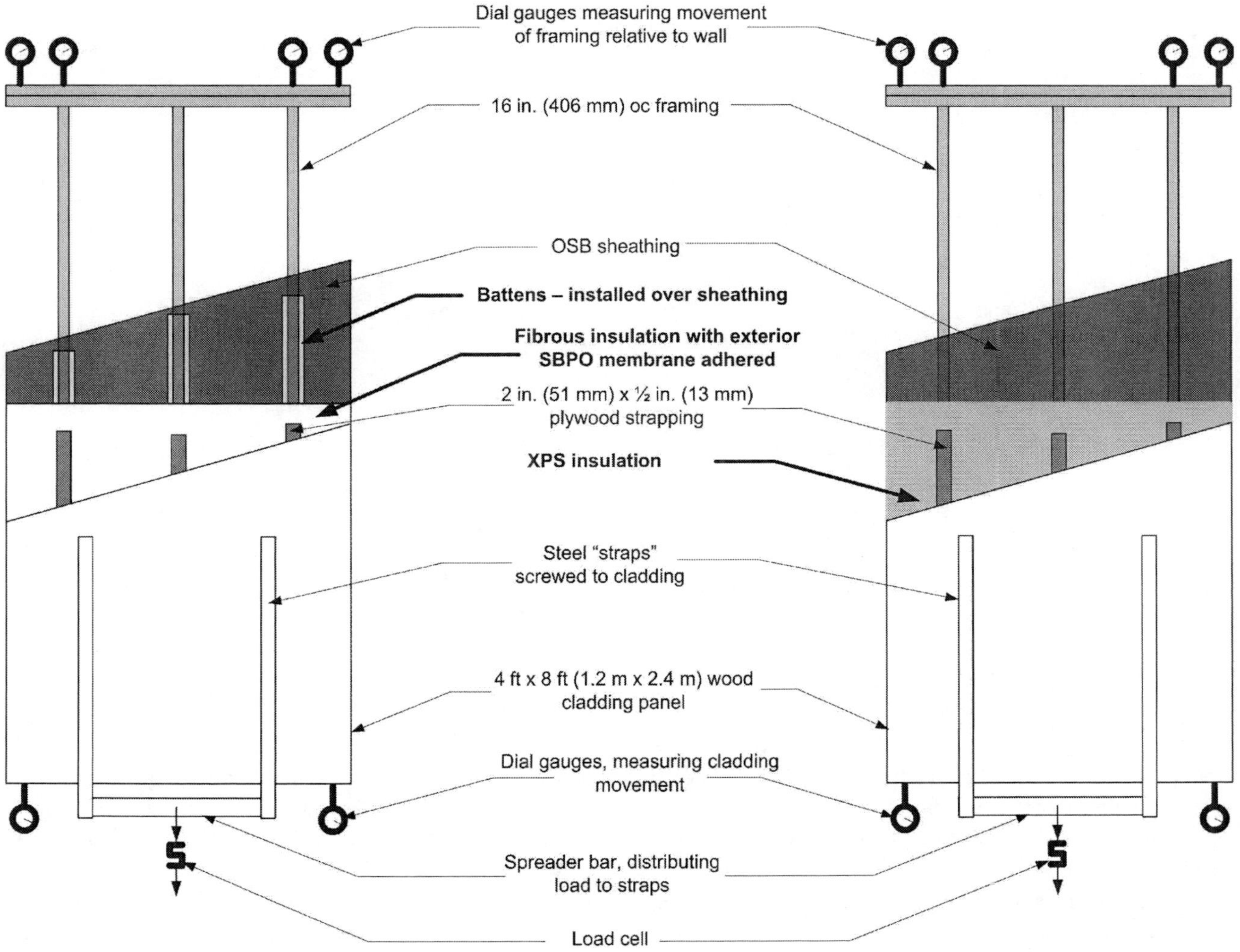

Figure 1 *Schematic construction drawing of the fibrous insulation and XPS wall assemblies for short-term deflection testing.*

Figure 2 *Short-term deflection test setup and measurement.*

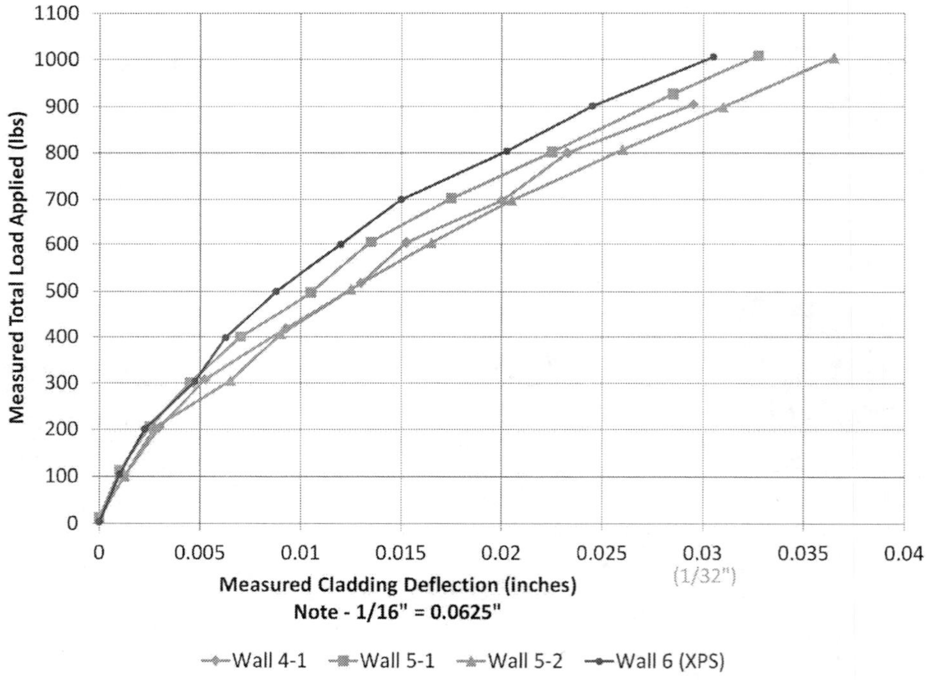

Figure 3 *Short-term deflection comparison data.*

of 0.015 in. (0.4 mm). Following the second loading, the wall returned to the same location (0.015 in. or 0.4 mm) for each of the tests, showing that there was no further permanent movement once the initial loading had settled the wall assembly.

Experimental variables in the short-term deflection testing included fastener type, length and spacing, insulation type and thickness, and strapping type. During construction of test wall assemblies, products from a local hardware store were used wherever possible, and specialized fasteners and materials were not used, to be able to determine the practical worst case scenario that may be constructed.

The comparison results of the wall assembly with XPS continuous insulation (blue line), and fibrous exterior insulation (green lines) in Figure 3 shows almost no difference in deflection results. At 1000 lb of total force (31 lb/ft^2), the deflection in all of the assemblies was approximately 0.03 or 1/32 in. (0.76 mm). The fibrous insulation wall assembly test was repeated three times to show the repeatability of the measurements. All other variables were the same for the comparison tests in Figure 3.

Based on the initial short-term deflection testing, questions arose regarding the effects of a long-term sustained loading of the cladding and the possible deflection results that would occur (Smegal and Straube 2011). Long-term deflection testing for at least 12 months was conducted in response to these industry questions.

Laboratory Long-Term Deflection Testing

Long-term deflection testing has been conducted on many insulation materials to address the questions about long-term loading of cladding weights attached to strapping or furring strips. The results of the fibrous exterior insulation long-term testing was conducted in a similar way so that comparisons with previous test results were possible. A wood-framed wall section measuring 16 in. (406 mm) wide and 8 ft (2.4 m) tall with a single vertical stud in the center was securely fastened to a concrete block wall in the lab. The 7/16 in. (11 mm) OSB was attached to the framing, and an insulating batten was installed into the framing with 3 in. framing nails spaced 11 in. (280 mm) o.c. vertically. The fibrous insulation was installed, and a 2 in. (51 mm) wide ½ in. (13 mm) thick plywood strapping was installed the full height of the test wall. Cladding was installed through the strapping to the wood layer of the batten with 2 in. (51 mm) 0.095 gauge galvanized ring shank nails spaced 8 in. (203 mm) vertically. A metal strap identical to the short-term testing was installed against the cladding and only attached to the cladding, so that the long term load was held by the cladding and transferred to the structure through the cladding attachment fasteners. The XPS long-term deflection test was part of a different research project, and the load was applied directly to the strapping, because no cladding was installed on the test assembly. However, this is not expected to change the results for the XPS wall assembly, because, due to the rigidity and compressive strength of the foam insulation, the load would be transferred directly from the cladding to the strapping and to the surface of the insulation and fasteners.

A weight of 210 lb (almost 100 kg) was attached to a single strapping member attached through different exterior insulations (shown in Figure 4) simulating a weight of 20 psf, greater than relatively heavy standard three-coat cement stucco (Table 1). The perception with long-term deflection is that it is only a concern with heavy claddings, which is why a weight of 20 psf was chosen. Past research conducted with a 4 psf weight indicated there was no long-term deflection concerns. This weight was suspended from the strapping for more than a year, and deflection gauge readings were taken periodically.

Figure 4 *Long-term deflection testing weights.*

Initial observations from the long-term laboratory deflection testing suggested there was a measured deflection response to changes in laboratory relative humidity and temperature conditions. The movement of the wood-based furring and framing of the wall system is assumed to explain these observations. The other observation was that with a 20 psf cladding weight, there was very little long-term deflection following the immediate initial deflection measured when applying the loading of the weight on the wall (see also Baker and Lepage 2013). Typically, the heavier cladding systems are also not applied in a single coat; they are applied in layers and the system has the opportunity to adjust between each layer.

Figure 5 shows long-term deflection data for two assemblies with exterior fibrous insulation and one assembly with exterior continuous XPS insulation. The XPS and fibrous insulation wall assemblies were not conducted simultaneously and were exposed to slightly different laboratory conditions during the testing. There was a correlation between the movement of the XPS wall assembly to the fluctuations in interior relative humidity, although the changes in deflection were very small. This occurs because wood materials shrink and expand in response to changes in the surrounding moisture conditions; so, as the relative humidity in the lab changes, there will be a response in all wood materials, and this shows up as a change in the deflection of a wood-framed test assembly. The entire y-axis of the plot spans a total of only 1/32 of an inch. For this comparison, the first hour of testing was not plotted; so, the immediate initial deflection was removed and the long-term effects of the 20 psf cladding emphasized. The long-term deflection results in Figure 5 show approximately the same deflection for the XPS exterior insulation wall assembly and the fibrous insulation exterior wall assembly.

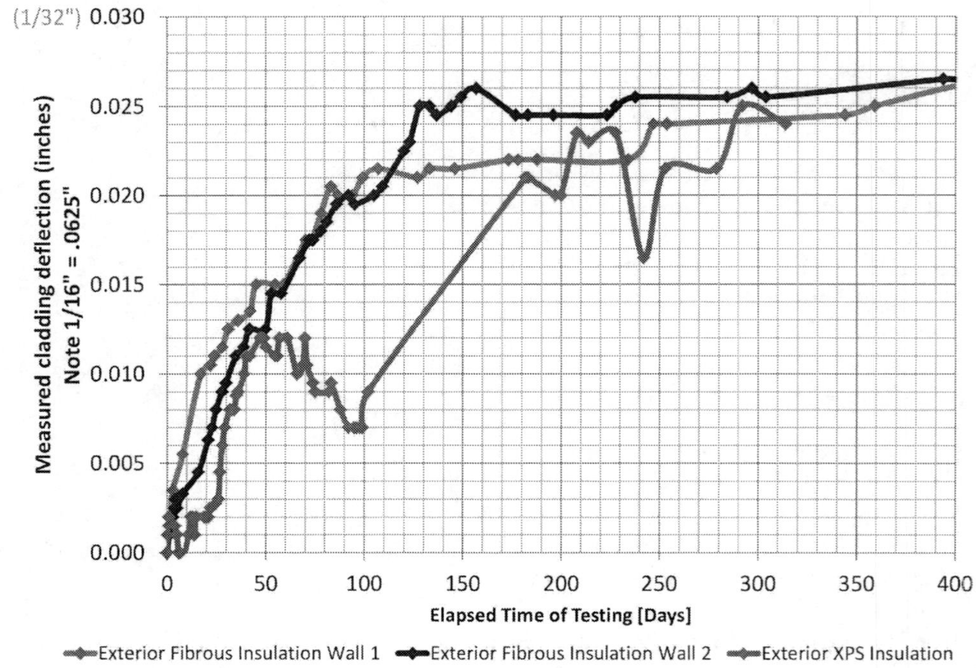

Figure 5 *Long-term deflection comparison data excluding initial deflection.*

Based on the small changes in deflection due to changes in laboratory RH, it was concluded that long-term deflection should be conducted fully exposed to the weather outdoors for the period of a year. The goal was to measure the deflection as a result of both moisture and temperature fluctuations of the assemblies and investigate if the cyclic nature of weather (wind, temperature, and humidity) would cause more deflection than testing in the protected lab environment. This testing was conducted, but is not included in this paper; more information can be found in previously published reports (Baker 2014).

Deflection testing demonstrates the measured deflection with a specific set of installation criteria, and, by modifying the experimental criteria, the effects on deflection can be determined. Typically, fewer parameters are tested during long-term testing because of the space required for each test over a long period of time. It is difficult to predict the deflection based only on the individual material testing because the interactions between the materials, such as friction effects between the layers and effects of moisture, are not included in material testing. Deflection testing may also be used for quantifying the impact of construction defects, such as missing fasteners and screw attachment to the sheathing rather than the framing, for example.

Limitations of Deflection Testing

Although deflection testing provides valuable information regarding the load capacity and stiffness of attaching cladding through layers of continuous insulation, there is no agreement on how much deflection is allowable. Although lap siding systems can tolerate relatively large movements without problems, the allowable settling and movement of rigid cladding systems, such as three-coat cement stucco, has not been determined. It is unclear how much movement hard-coat stucco can experience before surface cracking occurs. Although adhered stone veneers are heavy, and should generate the most deflection, even when cracks occur they are not as visible because the surface is very irregular with both mortar and stone (instead of smooth and continuous as with stucco). This will act to camouflage any small cracks that might occur. If cracks are large enough, they may increase the absorption rate of water into the cladding. There are decades of experience with millions of buildings that show the normal movement of wood structures (due to seasonal expansion and contraction due to temperature and humidity effects) can be accommodated successfully. The deflection at sawn timber rim joists are significantly more than the deflections measured in the tests reported. Nevertheless, the industry would be well served by better allowable design criteria for cladding movements.

Thermal Imaging Analysis

Infrared thermal imaging analysis is another method that can be used to determine the relative performance between two different wall assemblies. It has the advantage that it can visually demonstrate spatial performance differences of relatively high resolution (i.e., fractions of an inch). Because the thermal performance of a wall assembly is a function of the interrelated performance of all components, infrared imaging of enclosure thermal performance requires all of the components to be assembled in a realistic manner and exposed to air pressure differences to ensure air leakage is accounted for. As with many other methods, a comparison between two samples provides a much more reliable and accurate means of measuring performance.

As an example of the use of infrared imaging, two walls were constructed in a climate chamber capable of maintaining constant interior and exterior temperatures as well as imposing air pressure differences. The similarities between the two walls were: taped interior drywall, 2 × 4 wood framing at 16 in. (406 mm) o.c., and R-13 fiberglass cavity batt. One wall had R-5 of continuous XPS insulation with taped seams installed to the manufacturer's specifications. The other had R-5 continuous fibrous insulation that included a spun bonded polyolefin (SBPO) exterior air and water control layer. The walls were thermally isolated from each other to avoid any heat transfer. Each test wall consisted of a rim joist section with short floor joists, subfloor, and ceiling drywall. An electrical outlet was installed in each wall in an identical manner. All of the wall framing was sealed to the edge of the climate chamber to stop the air from bypassing the wall assembly during the pressurization and depressurization tests. Because the battens are not used for vinyl siding installations, they were not used in this side-by-side comparison. Another wall was constructed that did use the insulated battens for a different thermal comparison.

Thermal imaging of the test pairs was conducted following approximately 24 hours of conditioning at the interior and exterior setpoints of 72°F (22°C) and 19.5°F (–7°C), respectively. Thermal imaging was first conducted with no applied pressure difference across the wall, then with positive 10 Pa, and finally with negative 10 Pa. In real buildings there is often a pressure induced across a wall either from wind effects, natural stack effects, or internal HVAC systems. The pressures across low-rise residential enclosures are typically 3 to 10 Pascals. To provide an easy-to-control and meaningful signal, the pressure on the test walls was maintained using a blower connected to the cold side of the climate chamber to both pressurize and depressurize the climate side. A Magnahelic® differential pressure gauge was connected to a pressure tap on the climate box to show the pressure in the climate chamber.

On the 2 × 4 R-5 XPS wall (Figure 6), the average of the temperatures between the studs is 71.3°F (21.8°C) and over the studs is 70.4°F (21.3°C). On the 2 × 4 wall with R-5 fibrous exterior insulation (Figure 7), the average of the temperatures between the studs is 71.7°F (22.1°C) and over the studs is 70.9°F (21.6°C). During this test, the measurements showed very small but consistently warmer surface temperatures both on and between the studs of the continuous R-5 fibrous insulation wall assembly compared to the R-5 XPS wall assembly. These results imply that the two approaches to continuous insulation provide similar performance, with the fibrous roll insulation performing slightly better. The results may be caused by the slightly higher C518 testing results of the fibrous insulation (R-5.4) compared to XPS (R-5.0).

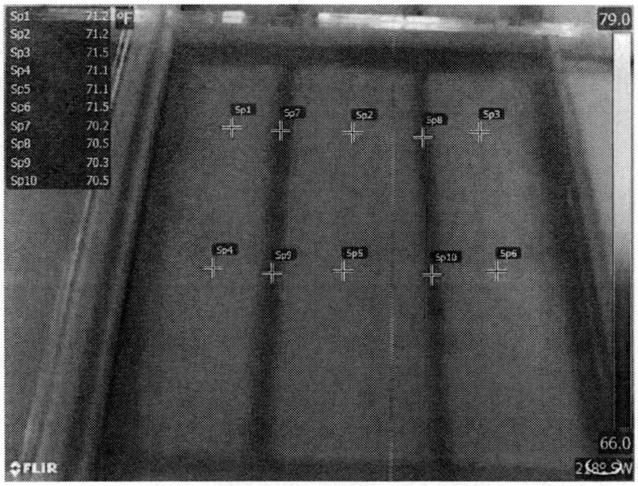

Figure 6 *Drywall surface temperatures of 2 × 4 framed wall with continuous exterior R-5 XPS.*

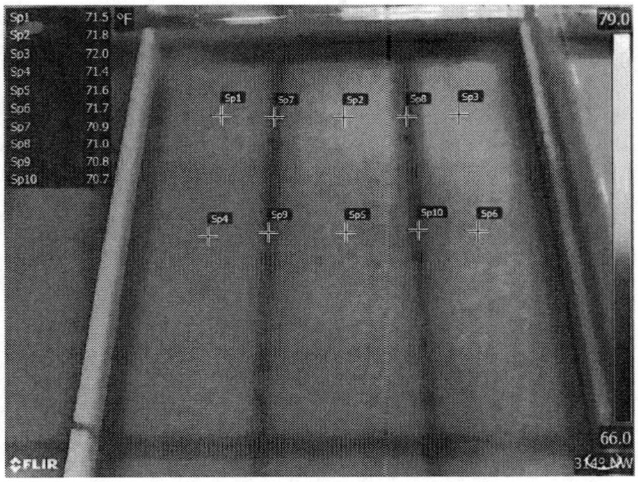

Figure 7 *Drywall surface temperatures of 2 × 4 framed wall with continuous exterior R-5 fibrous insulation.*

Guarded Hot Box Testing

RDH Building Science Laboratories in Waterloo, ON (formerly Building Science Consulting Inc.), in partnership with Building Science Corporation, has conducted guarded hot-box testing on many types of insulation and wall assemblies based on ASTM C1363 (2011a) with some modifications (BSC 2015). The modified test measures the overall R-value of a 12 ft wide by 8 ft high test sample. Framed walls have been tested that include all of the enclosure construction material from the interior drywall to the exterior vinyl siding. The R-value that is calculated is based on the heat flow through the wall assembly and all of its components, such as framing, exterior insulation, air leakage, and temperature-dependent conductivities of the materials. The standard RDH protocol employs numerous outdoor air temperature set points with a fixed realistic indoor temperature to allow for the temperature dependence of all components (including shrinkage and swelling of joints) and insulation materials to be assessed. Example results from two walls are plotted in Figure 8 (Schumacher et al, 2013). The assemblies consisted of all enclosure materials including drywall, wood framing, OSB sheathing, housewrap, and vinyl siding. One of the test walls had R-5 of exterior XPS insulation. The walls shown in Figure 8 were tested in a "sealed" condition, with a thin layer of adhered plastic film on each side air sealing both sides of the wall assembly, eliminating air leakage from either side. The airtightness was confirmed with tracer gas testing. Other testing has specifically looked at the effects of air leakage on the effective R-value of different wall assemblies.

These types of impacts on advertised R-value were expected based on thermal simulations and physics, but, until recently, there was almost no measurement results on full-scale wall assemblies that are able to measure the heat loss/gain through the air transfer system when applying controlled pressures to the full-scale wall assembly.

This testing has reinforced that critical factors for insulation effectiveness include the amount of thermal bridging, temperature-dependent effects, and, in other testing, the amount of air leakage across the assembly. These factors can significantly impact both the thermal and moisture performance of an enclosure assembly but cannot be determined by any individual material tests.

Full-Scale Field Testing and Moisture Monitoring

All of the previous assessment methods investigated the thermal impacts of the assembly and its components. It is often more important to understand the moisture behavior of the wall assembly to predict durability and common in-service problems, such as mold. Hygrothermal modeling, material testing, laboratory subassembly testing, climate chamber exposure, and field exposure are all methods available to assess moisture performance. Realistic full-scale heat and moisture testing of an assembly exposed to a real exterior climate with controlled interior conditions over at least one year provides a robust method. Test huts are often preferred over the use of whole buildings because test hut walls can be stressed to failure with moisture without the cost of replacing whole buildings (Straube et al. 2002).

Moisture performance is more difficult to predict than thermal performance, because it depends on a greater number of physical properties as well as the moisture conditions on the interior and exterior of the wall assembly in combination with all of the thermal considerations.

Figure 9 is a photo of the exterior of a series of residential and commercial test walls during construction at a test hut in Vancouver, BC. The test walls have exterior insulation (some are covered by a drainage plane or UV barrier). By instrumenting the walls, and monitoring them under normal operating conditions, the side-by-side performance of the insulation systems can be measured and compared on all four orientations. In some cases, water is injected into the wall assemblies in a controlled manner to simulate a rain leak through the

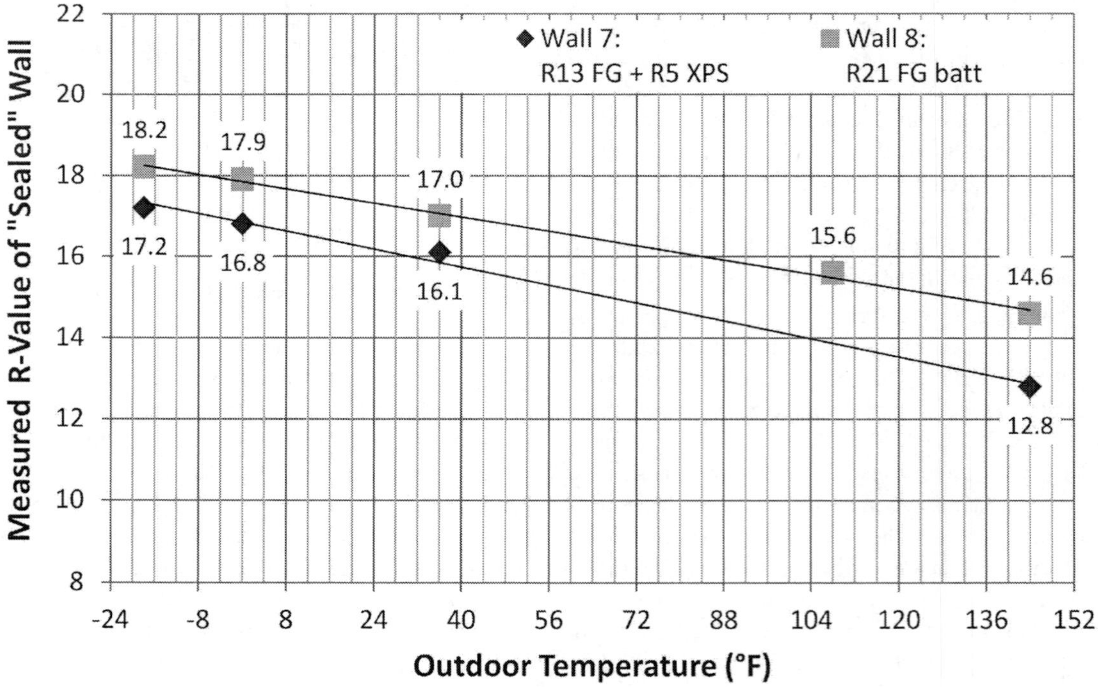

Figure 8 *Results of guarded hot-box testing of two wall assemblies (Schumacher et al. 2013).*

Figure 9 *Full-scale field test assemblies.*

enclosure or to add stress to the system. In other applications indoor or outdoor air can be injected into the wall system at a known rate to simulate an air leak into the enclosure and measure the system's response.

The two comparison walls included in the test hut research relevant to this study are two residential wood-framed wall assemblies, one with exterior R-5 exterior XPS and one with R-5 of exterior fibrous insulation with insulated battens. The common construction variables in the wall assembly include the following: painted drywall, 2 × 6 wood framing with R-21 Kraft-faced fiberglass batt, 7/16 in. (11 mm) OSB sheathing, 2 × ½ in. (51 × 12 mm) wood strapping, and fiber cement siding. Both wall assemblies consist of an SBPO membrane outboard of the exterior insulation as the air and water control layer. The only difference in the comparison walls was the type of exterior insulation.

Two wood-framed test walls, one with R-5 exterior XPS insulation and one with R-5 exterior fibrous insulation, were compared in the Vancouver test hut. Figure 10 shows the measured OSB sheathing moisture. The red dashed lines indicate two instances when a controlled volume of water was injected into a controlled location on the exterior of the sheathing between the sheathing and the exterior insulation. The moisture content measurements shown in Figure 10 are measured at two depths in the OSB sheathing immediately adjacent to the wetting system on the north orientation. It can be seen that the fibrous insulation with the highly vapor-permeable air-water barrier facing (green line) allows the sheathing to dry much more quickly than the comparison wall (red lines), both in the summer and the fall. The drying rates are a direct result of the material properties of the exterior insulation layers, as anticipated.

This means that if there were a leak beyond the exterior insulation at a window, or other penetration, a higher vapor permeance fibrous insulation would allow the water to dry out, decreasing the moisture in the OSB sheathing much more quickly than a lower vapor permeance XPS insulation.

With measurement data such as this from field test hut monitoring, it may be possible to correlate measured data, including the interior and exterior conditions, to results from hygrothermal simulations, although this has not been done yet.

CONCLUSIONS

Standardized material tests are a good starting point for a new product to establish comparison criteria with other accepted building materials. It is generally necessary to measure material characteristics such as vapor permeance and R-value for building code compliance. These properties are useful to help predict performance and to provide material property data for hygrothermal simulations as well as marketing. However, systems testing is required to understand performance under more realistic installed conditions and to compare predicted performance to measured performance. There are currently fewer standardized tests for assemblies (e.g., ASTM C177 [(2015a]; ASTM C1363 [2011a]; ASTM E283 [2012b]; ASTM E331 [2016e]), especially for insulation

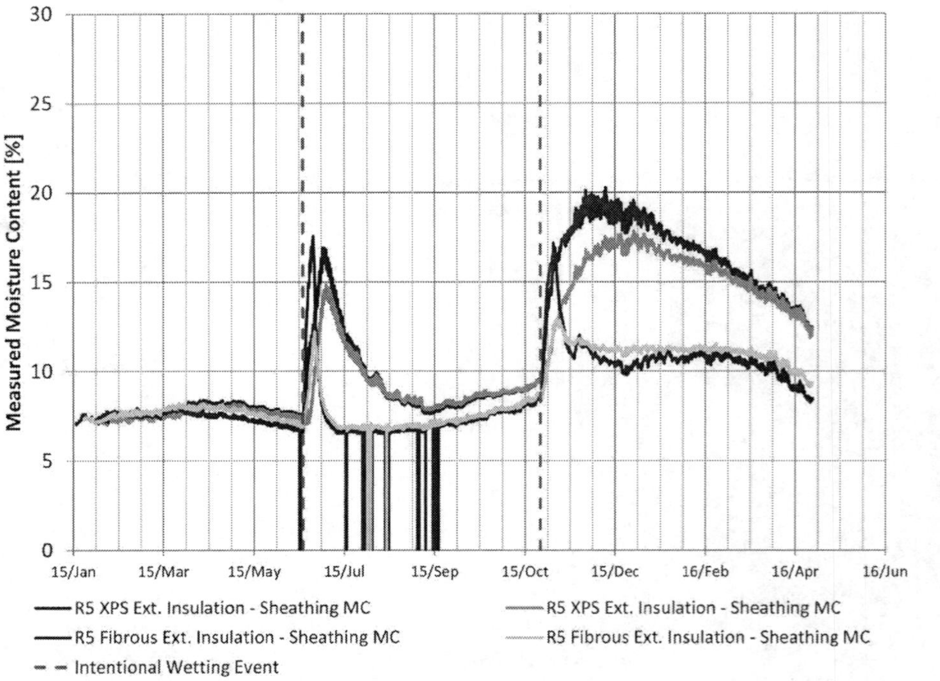

Figure 10 *Example full-scale wall performance comparison graph.*

products. However, this may change as building codes focus more on effective R-values and U-factors. In the meantime, nonstandard tests can also be developed and used.

The example system testing of the product used throughout this paper showed that the relatively new R-5 fibrous exterior insulation performed very similarly to the industry-accepted and widely used R-5 XPS. In the short- and long-term deflection testing, the flexible fibrous insulation system with insulated battens had essentially the same amount of deflection as the XPS wall assembly, because the fibrous insulation was compressed at the battens and behaved more like a rigid material. In the thermal imaging assessment, both test wall assemblies had very similar interior surface temperatures, indicating that the air and vapor permeable fibrous insulation performed as well at insulating the exterior of the wall assembly as the XPS insulation. During the field testing of the full-scale wall assemblies, it was demonstrated that the R-5 fibrous continuous insulation performed very similarly to the R-5 XPS under normal operating conditions in terms of measured sheathing moisture content and moisture-related durability of the wall assembly. However, during intentional wetting of the walls when water was trapped intentionally between the exterior surface of the sheathing and the exterior insulation, the fibrous continuous insulation wall assembly allowed faster drying, as demonstrated by the decreasing measured sheathing moisture content.

Based on the wall systems testing results from both the laboratory and field tests that were conducted, it was shown that the structural, thermal, and moisture performance of the exterior fibrous insulation is similar to the results of the XPS wall assembly even though their individual material properties are quite different from each other. This shows the benefits of testing systems or assemblies using tests that more closely simulate real world conditions.

REFERENCES

AAMA. 2014. AAMA 508-14, *Voluntary test method and specification for pressure-equalized rain screen wall cladding system*. Schaumberg, Illinois: American Architectural Manufacturers Association.

ASTM. 2011a. ASTM C1363, *Standard test method for thermal performance of building materials and envelope assemblies by means of a hot box apparatus*. West Conshohocken, PA: ASTM International.

ASTM. 2011b. ASTM E2178-13, *Standard test method for air permeance of building materials*. West Conshohocken, PA: ASTM International.

ASTM. 2011c. ASTM E2357-11, *Standard test method for determining air leakage of air barrier assemblies*. West Conshohocken, PA: ASTM International.

ASTM. 2012a. ASTM D2842-12, *Standard test method for water absorption of rigid cellular plastics*. West Conshohocken, PA: ASTM International.

ASTM. 2012b. ASTM E283-04 (2012), *Standard test method for determining rate of air leakage through exterior windows, curtain walls, and doors under specified pressure differences across the specimen*. West Conshohocken, PA: ASTM International.

ASTM. 2015a. ASTM C177, *Standard test method for steady-state heat flux measurements and thermal transmission properties by means of the guarded-hot-plate apparatus*. West Conshohocken, PA: ASTM International.

ASTM. 2015b. ASTM C518-15, *Standard test method for steady-state thermal transmission properties by means of the heat flow meter apparatus*. West Conshohocken, PA: ASTM International.

ASTM. 2015c. ASTM C1303/C1303M-15, *Standard test method for predicting long-term thermal resistance of closed-cell foam insulation*. West Conshohocken, PA: ASTM International.

ASTM. 2016a. ASTM D696-16, *Standard test method for coefficient of linear thermal expansion of plastics between –30°C and 30°C with a vitreous silica dilatometer*. West Conshohocken, PA: ASTM International.

ASTM. 2016b. ASTM D1621-16, *Standard test method for compressive properties of rigid cellular plastics*. West Conshohocken, PA: ASTM International.

ASTM. 2016c. ASTM E84-16, *Standard test method for surface burning characteristics of building materials*. West Conshohocken, PA: ASTM International.

ASTM. 2016d. ASTM E96/E96M-16, *Standard test methods for water vapor transmission of materials*. West Conshohocken, PA: ASTM International.

ASTM. 2016e. ASTM E331-00(2016), *Standard test method for water penetration of exterior windows, skylights, doors, and curtain walls by uniform static air pressure difference*. West Conshohocken, PA: ASTM International.

ASTM. 2016f. ASTM E547-00(2016), *Standard test method for water penetration of exterior windows, skylights, doors, and curtain walls by cyclic static air pressure difference*. West Conshohocken, PA: ASTM International.

Baker, Peter and R. Lepage. 2013. *BA-1314: Cladding attachment over thick exterior insulating sheathing*. Building America Report–1314. Building Science Corporation.

Baker, P. 2014. *Initial and long term movement of cladding installed over exterior rigid insulation*. Building America Report–1404. Westford, MA: Building Science Corporation.

BSC. 2015. *Thermal metric summary report*. (June 2015 Update). Westford, MA: Building Science Corporation. http://buildingscience.com/documents/special/thermal-metric-documents/thermal-metric-summary-report.

DuPont. 2013. DuPont™ Tyvek® ThermaWrap™ R5.0 Physical Properties Data Sheet. K-27286.

FTC. 2005. 16 CFR Part 460c, Labeling and advertising of home insulation: Trade regulation rule; Final rule. Federal Trade Commission. May 31, 2005.

Schumacher, C., D. Ober, J. Straube, and A. Grin. 2013. Development of a new hot box apparatus to measure building enclosure thermal performance. Thermal Performance of the Exterior Envelopes of Whole Buildings XII International Conference, Clearwater, FL.

Smegal, J., and J. Straube. 2011. Roxul ComfortBoard™ insulating sheathing (IS) deflection testing. Test report completed by Building Science Consulting Inc. Available from http://www.roxul.com/files/5p_animation/RX-NA-EN/pdf/tech%20data/110816%20Roxul%20ComfortBoard%20Final%20Report.pdf

Straube, J., and E. Burnett. 2005. *Building Science for Building Enclosures*. Westford, MA: Building Science Press.

Straube, J., D. Onysko, and C. Schumacher. 2002. Methodology and design of field experiments for monitoring the hygrothermal performance of wood frame enclosures. *Journal of Thermal Envelope and Building Science* 26(2).

Duct Leakage Repeatability Testing

Iain Walker, PhD
Fellow ASHRAE

Max Sherman, PhD
Fellow ASHRAE

Paul W. Francisco
Member ASHRAE

ABSTRACT

Duct leakage often needs to be measured to demonstrate compliance with codes and standards requirements or to determine energy or indoor air quality impacts. There are several test methods available and documented in standards. Repeatability errors need to be understood because standards and codes need consistent evaluation procedures and because repeatability is also an indication of the resolution of the test method. This study focused on those test methods that are commonly used and are required by various organizations such as BPI, ACCA, RESNET, DOE Weatherization, and many other energy efficiency programs and in various test standards such as ASHRAE Standard 62.2 and ASHRAE Standard 152, and California Title 24. The three duct leak measurement methods assessed were two duct pressurization methods that are commonly used by many practitioners and the DeltaQ technique. This study assessed each measurement technique using data from the literature and recently obtained field data. The field data were taken in 30 homes in three locations in the U.S. Each home was tested multiple times with each measurement technique. The results of this study can be used by organizations such as BPI, ACCA, RESNET, ASHRAE, and other building codes and standards in determining what tests are appropriate for their specific application.

BACKGROUND

Until now there has been little investigation of the repeatability of the test methods and previous studies have only done a few tests in each home. In order to better estimate repeatability and hopefully capture a wider range of weather conditions at each home, this study planned for a whole day of testing with a target of ten repeat tests in each home performed throughout the day. Ten homes were tested in each of three geographical locations to capture a range of home and duct system design and construction. The initial plan was to only test tight duct systems[1] to evaluate the tests under the conditions that they are most used for (i.e., to confirm low leakage). However, as the results showed, there was a large range in duct leakage among the tested homes. Another requirement was that the homes have at least part of their duct system outside the conditioned space of the home. The overall intent was to produce a data set that robustly evaluated the repeatability of the three test methods so that we could draw conclusions about test applicability and possible repeatability-associated limitations.

PREVIOUS REPEATABILITY STUDIES

For DeltaQ testing, several previous studies have examined repeatability. Because the DeltaQ test was substantially modified in 2006 by adopting a "ramping" technique, only repeatability studies evaluating that method are discussed here. Walker and Dickerhoff (2006, 2008) performed the DeltaQ test five times each in two homes and then, in another two homes with the air handler turned off,[2] found that the root mean square of the variability for the leaks scaled with envelope leakage. Because the major source of repeatability uncer-

1. A tight duct system is one that would meet the 6% of total system flow requirement that is used in several codes and standards, for example ASHRAE Standard 62.2-2013 and California Title 24-2013 (CEC 2012).

Iain Walker and **Max Sherman** are scientists at Lawrence Berkeley National Laboratory, Berkeley, CA. **Paul Francisco** is a research specialist at the University of Illinois at Urbana-Champaign, Champaign, IL.

tainty is wind-induced pressure and flow fluctuations for the building envelope, we expect the repeatability errors to scale with the envelope leakage (i.e., for a fixed pressure fluctuation due to a wind gust, the change in flow rate is proportional to the envelope leakage.) To examine this, a rough rule of thumb was proposed to estimate repeatability uncertainty as a fraction of the envelope air leakage at 50 Pa (Q50). The magnitude of the fraction was determined from measured data to be roughly 1%, such that a home with Q50 of 1000 cfm (500 L/s) would have a repeatability uncertainty of 10 cfm (5 L/s). An additional seven homes were tested for DeltaQ repeatability by Dickerhoff and Walker (2008). All these homes roughly conformed to the 1% Q50 rule of thumb. Francisco (2009) tested 15 homes in which a minimum of five repeats were performed for each home. These tests showed that 1% of Q50 is a reasonable typical value but that there is substantial variation around this value. The middle half of the data are within about 30% of the rule of thumb and the most extreme values are about 4% of Q50.

For fan pressurization studies, Walker et al. (1998) tested nine new homes in southern California with predominantly attic ducts using pressurization testing. The tests were repeated two to four times—with most homes tested three times. The repeatability uncertainties averaged 16 cfm (8 L/s) (3% of measured leakage flow—these were very leaky duct systems).

FIELD TESTING

New field data on repeatability was generated for the purposes of this study by three different Building America teams, each in a different location: In the Champaign, IL area by PARR; in the Minneapolis, MN area by CEE; and in the Atlanta, GA area by Southface. Ten homes were tested in each location. At each test house roughly a day was spent repeating each test as many times as possible. Typically, seven to ten test repeats were achieved. Tests were performed from early morning through the afternoon in an attempt to capture a range of weather conditions in terms of temperature differences and wind speeds that are often diurnal. Additional measurement protocol details are given in Walker and Sherman (2014). In addition to the duct leakage flows, house and duct system characteristics were recorded, as well as ambient wind and temperature conditions. The furnace blower flow was recorded in order to normalize the leakage flows because the typical metric used to evaluate duct leakage is the flow as a fraction of the system blower flow. The time to perform each test was also recorded.

RESULTS

The results are presented by location and for all tests combined and as an absolute flow in cubic feet per minute (cfm) and as a fraction of blower flow (%). For each home the average (avg.) and standard deviation (s.d.) of the duct leakage were determined. In the following tables the following abbreviations were used:

Ptot = pressurization total leakage at 25 Pa

Pout = pressurization leakage to outside at 25 Pa

DQs = DeltaQ supply leakage

DQr = DeltaQ return leakage

House and System Summary Characteristics

Table 1 shows that these homes had an average leakage of 1939 cfm at 50 Pa (corresponding to about 7.5 ach_{50} [air changes per hour at 50 Pa] or normalized leakage [NL] of 0.6). The results of Chan and Sherman (2013) gave a geometric mean NL = 0.61 for air leakage tests of 134,000 U.S. homes—indicating that the homes in this study were typical of existing homes in the U.S. There were large variations from home to home with a standard deviation in envelope leakage of about ±50% and a wide range from minimum to maximum. This wide range is good for the purposes of this study because it ensures that the tests are exercised over a wide range of envelope leakage. The average blower flow varied by about 20% from region to region and also had significant house-to-house variability. There was a large range of house size from 800 to 3450 ft^2 with an average of 1829 ft^2. For the PARR and Southface tests, the wind speeds were low—typically 1 to 2 mph, but the CEE tests reported much higher wind speeds of about 10 mph. Previous work by Dickerhoff and Walker (2008) showed that there was no trend in increased or decreased DeltaQ test results with increasing wind speed. However, they also showed that the DeltaQ sensitivity to wind is predominantly from short-term gusts that disrupt only part of the test, and this may not be reflected in single point or average wind speeds for a test.

Duct Leakage

For the pressurization tests, the result is the combination of supply and return leaks. Because the DeltaQ test separates supply and return leakage, the DeltaQ test results will be presented as both supply and return separately (DQs and DQr) and with the supply and return combined for a total that can be compared to the other tests. In some cases of pressurization testing the ducts were so leaky that the air flow required to get to 25 Pa exceeded the capacity of the test equipment. For these ducts that were not fully pressurized to 25 Pa, the results were extrapolated to 25 Pa assuming a pressure exponent of 0.5. The results in Tables 2 and 3 (and Figure 1) show that the PARR and Southface homes had quite leaky ducts, and the CEE ducts are much tighter. The results show that generally the test with the lowest repeatability uncertainty is the pressurization testing to outside with an overall uncertainty of about 12 cfm or 1% of furnace blower flow. The pressurization and DeltaQ repeatability results are very close to each other at about 40 cfm and 6% of blower flow.

[2] Although this is not strictly a DeltaQ test, it is included for completeness.

Low-Leakage Results

An initial goal of this study was to focus on low-leakage ducts. This was because key applications for these test methods are those that are testing low-leakage ducts—typically less than 6% of blower flow (as used in ASHRAE Standard 62.2 and California Title 24-2013 [CEC 2012]). For the pressurization test results, no duct systems met the 6% limit. For the pressurization to outside test results, five systems met the 6% limit. For four out of the five homes, the DeltaQ results were also less than 6%—showing some agreement between these two approaches as a screening for tight ducts. However, the DeltaQ results were less than 6% in 12 homes and in only four of these were the pressurization to outside test results less than 6%. For the 12 homes with DeltaQ test results less than 6% of blower flow, Tables 4 summarizes the test results.

Other Testing Issues

In addition to repeatability, there are other aspects of duct testing reliability that need to be considered. In some homes the ducts were so leaky that they could not be pressurized to 25 Pa or pressurization testing produced leakage flows greater than the total blower flow. Seven of the thirty homes had systems that could not be pressurized to 25 Pa, and there was one home where pressurizing to outside also could not reach 25 Pa. Eighteen of the 30 homes had pressurization leakage greater than the blower flow. These results significantly limit the quantitative applicability of these pressurization tests. When testing duct systems that exhibit such high leakage levels, it is best to avoid pressurization testing if metrics such as potential energy savings are sought. Instead, these results are primarily useful only to indicate that the ducts are leaky and may warrant repair.

For DeltaQ testing, there are also occasional unrealistic results. For DeltaQ the results were occasionally negative. This occurs when the duct leakage is low and a negative leakage result falls within the bounds of uncertainty. There were two occurrences of negative supply leakage and two of negative return. In one home these combined to give a negative total—albeit only 1 cfm. Although the DeltaQ test has fewer issues than pressurization testing in this study, it is important to note that other stud-

Table 1. House and System Characteristics

	Envelope Leakage, Q50 (cfm)				Blower Flow (cfm)				Avg. Floor Area, ft^2	Avg. Wind Speed, mph
	Average	s.d.	Min.	Max.	Average	s.d.	Min.	Max.		
PARR	2002	921	940	3445	1030	771	380	2942	1364	2
CEE	1936	587	945	2911	891	205	704	1424	2465	10
Southface	2879	1181	1219	4456	706	281	296	1128	1658	1
All	2272	1022	940	4456	876	490	296	2942	1829	4

	Envelope Leakage, ach_{50}			
	Average	s.d.	Min.	Max.
PARR	12.3	6.1	4.2	21
CEE	6.2	2.1	3.6	10.9
Southface	14.1	5.8	8.1	23.9
All	11	6	3.6	23.9

Table 2. Duct Leakage Results, cfm

	DeltaQ avg.	DeltaA Q s.d.	Ptot avg.	Ptot s.d.	Pout avg.	Pout s.d.	DQs avg.	DQs s.d.	DQr avg.	DQr s.d.
PARR	190	32	666	30	492	18	107	14	82	21
CEE	39	36	1001	8	87	2	16	16	24	21
Southface	220	46	1075	92	442	15	95	21	126	26
All	150	38	914	43	340	12	73	17	77	23

Table 3. Duct Leakage Results, Fraction of Blower Flow, %

	DeltaQ avg.	DeltaQ s.d.	Ptot avg.	Ptot s.d.	Pout avg.	Pout s.d.	DQs avg.	DQs s.d.	DQr avg.	DQr s.d.
PARR	21	3	72	3	56	2	12	2	9	2
CEE	4	4	110	1	9	0	2	2	3	2
Southface	45	9	181	15	80	2	19	4	26	5
All	23	5	121	6	48	1	11	3	13	3

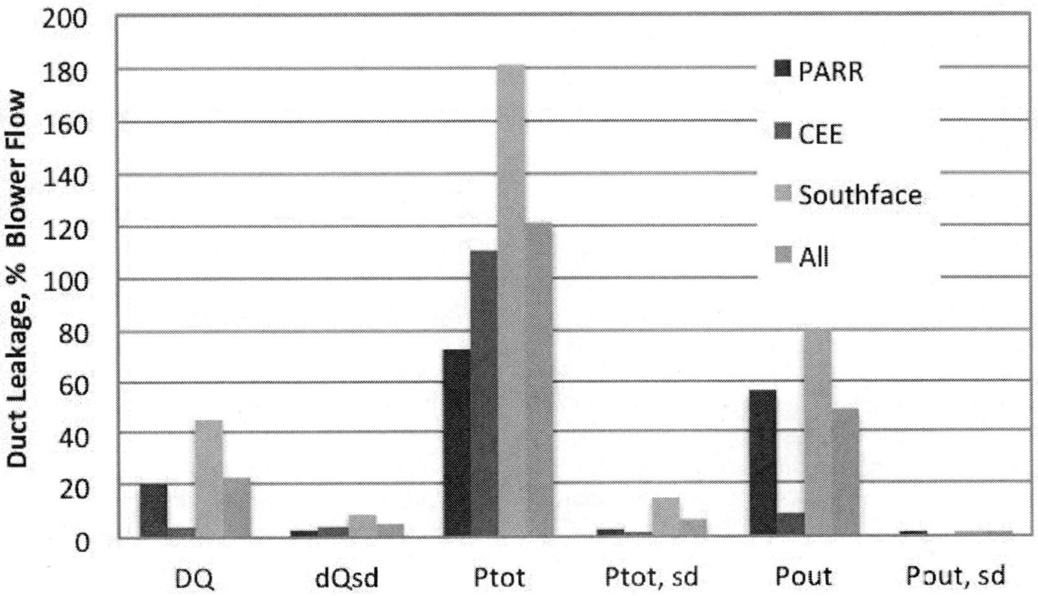

Figure 1 *Comparison of average and standard deviations of duct leakage flows for the three separate study groups and all groups together.*

Table 4. Low Leakage (<6%) Duct Leakage Results

	DeltaQ avg.	DeltaQ s.d.	Ptot avg.	Ptot s.d.	Pout avg.	Pout s.d.
cfm	29	28	832	10	115	3
% blower flow	3	3	91	1	14	0.3

ies have occasionally shown anomalous DeltaQ results. For example, Dickerhoff and Walker (2008) showed that large DeltaQ errors were usually the result of wind gusts during testing and that, if there were discrete large wind gusts (as opposed to generally blustery conditions), repeating the part of the test with the wind gust could reduce these problematic test results. The important issue is that, although DeltaQ testing showed far fewer anomalous test results in the current study, all three test methods are susceptible to producing nonsensical results.

The tests that include an envelope pressure measurement, DeltaQ and pressurization to outside, might be expected to show greater sensitivity to wind fluctuations that change the envelope pressures during the test. However, the results show that the much higher wind speeds reported by CEE did not result in higher test variability for either test procedure. This agrees with the discussion and analysis in Dickerhoff and Walker (2008), who also found that increased wind speed itself was not necessarily an indicator of increasing DeltaQ errors; rather it is the presence of sustained gusts over several seconds (i.e., changing wind speeds) during the measurements that can lead to large errors. Francisco (2009) showed that it is important to have the outdoor pressure reference be on the same side of the home as the blower door so as to reflect the same pressure as is seen by the blower door in order to reduce errors in the flow reported by the blower door.

In order to examine the plausibility of using 1% of Q50 as a rule of thumb for estimating the leakage uncertainty, Figure 2 combines the results from previous studies with the new tests performed for this study. The linear least square fit to all the data shows that 1% of Q50 is a reasonable rule of thumb to use for DeltaQ repeatability uncertainty but that in individual houses the error may be substantially larger or smaller depending on how gusty the wind is for a given test. The mean and RMS differences between the actual repeatability standard deviation and the 1% rule of thumb are 1 cfm and 11 cfm (0.5 and 5 L/s), respectively.

Time Required to Perform the Tests

When analyzing the time required to perform the tests, one result was immediately obvious: the first test in each home took much longer than subsequent tests. This is due to one-time setup tasks that were required for the first test but not subsequent tests. In any real testing environment, testers will only test the house a single time. Therefore, the first test time results are more indicative of what testers would experience in practice. Table 5 summarizes the time taken to perform the tests split into the first test and subsequent tests. The first test generally took double the time of subsequent tests—roughly thirty minutes longer. The DeltaQ test is the fastest on average for the first test by about 10 to 15 minutes, with the average for the two pressurization tests only three minutes different. The variability in time to perform tests for all three procedures was about the same for the first test—about 20 minutes. This variability indicates that some houses are easier to set up than others. This is reinforced by the variability in time required to do the tests also being significantly reduced after the first test, dropping from 20 minutes for the first test to only 3 minutes variability for subsequent tests.

DISCUSSION

Repeatability

In terms of repeatability of the test results, the data clearly shows that the pressurization to outdoors tests are the most repeatable with a standard deviation of only 1% of the blower flow. The other two methods are approximately 6%. If repeatability were the only criterion, this result would suggest that pressurization to outdoors should be the only method used. This method, however, is the costliest because it requires the most equipment (both blower door and duct tester) and the most time in the field. The other two methods provide roughly the same repeatability at 6% of blower flow. The DeltaQ technique separates out supply and return leakage at 3% error each, which is of interest if separate supply and return leakage estimates are required (e.g., for better targeting of particular parts of the duct system for sealing).

Looking at tighter duct systems only (DeltaQ <6% of blower flow), the repeatability improves for all the test methods. Pressurization to outside has the lowest average standard

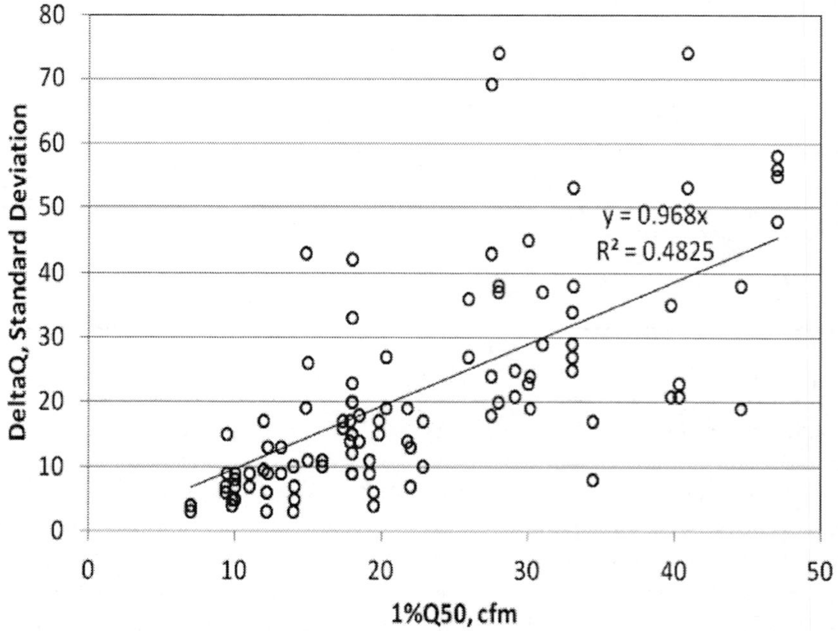

Figure 2 *Repeatability uncertainties for DeltaQ testing compared to 1% Q50 combining test results from the current study with previous studies, including least squares fit line.*

Table 5. Time Taken to Perform Each Test, Minutes

	DQ, first avg.	DQ, first s.d.	DQ avg.	DQ s.d.	Ptot, first avg.	Ptot, first s.d.	Ptot avg.	Ptot s.d.	Pout, first	Pout, first s.d.	Pout avg.	Pout s.d.
PARR	54	15	31	4	60	27	27	13	63	24	26	13
CEE	53	24	19	2	76	15	34	6	67	17	38	4
ALL	54	20	25	3	68	21	31	10	65	21	32	9

deviation of 0.3%, with pressurization improved to 1%. The DeltaQ test showed the least improvement for the tight duct subset with a standard deviation of 3%. It should be noted that even for these low-leak tests, the indicated leakage flows varied considerably. The DeltaQ test results averaged 30 cfm (15 L/s), the pressurization to outside tests averaged 115 cfm (54 L/s), and the pressurization test results were still very high at 800 cfm (376 L/s).

Based on repeatability alone, we cannot conclude that any method is too poor to use, but the selection of which test to use will depend in large part on the purpose for which it is intended. For example, for tight specifications, high repeatability can be critical. On the other hand, for applications with a primary focus on the magnitude of the leakage estimates, the repeatability issue may not be that important. Looking to the future and applying these test methods to high performance homes, these repeatability uncertainty estimates could be improved by testing more homes with tighter envelopes and duct systems, because for this data set the majority of homes had very poorly sealed systems.

Bias and Accuracy

The discussion above ignores the fact that each of the methods is measuring a different quantity. Because they are, it is important to address how well the quantity that is measured meets the needs of those who require the test and how much extra uncertainty is associated with turning what is measured into what is needed. There are typically two schools of thought. If one is going to use the value as part of some estimation of energy or indoor air quality, then it is normally clear what the parameter is. Energy and airflow models, for example, usually need to know the supply and return leakage at operating condition to estimate duct leakage impacts. In this case, test methods that measure leakage at a fixed pressure have an additional uncertainty because of the need to extrapolate to the desired quantity.

In order to provide more context for these uncertainties, we need to compare the total uncertainty of DeltaQ and pressurization testing. Because pressurization testing measures an airflow at a fixed pressure, the airflow at operating conditions may not be the same as the tested airflow. This issue of not having the same pressure across the leaks during the duct pressurization test as during normal operation is a bigger source of uncertainty than repeatability. Several studies have been performed where the true leakage was known: a laboratory study by Walker and Dickerhoff (2008) has shown duct pressurization errors (40 to 60 cfm [20 to 30 L/s]—or about 5% of furnace blower flow for these tests) that were double those of DeltaQ testing (20 to 30 cfm [15 – 25 L/s]). An earlier laboratory study by Andrews (2002) also showed that pressurization errors were double those for DeltaQ testing. Francisco et al. (2003) reported similar differences of a factor of two increase in uncertainty for pressurization compared to DeltaQ in a field study of 51 homes. Note that the latter two studies evaluated a version of the DeltaQ test that used fixed pressure stations rather than the ramping approach used in the current study.

The second school of thought is that a requirement is set at some stringent level, and the leakage test is more of a pass/fail than a quantitative measurement. In such a case, larger biases can be tolerated because the energy or IAQ impact of those biases would be small, but there may be cost or repeatability benefits of other methods. It is then reasonable to ask the question of whether a test method might be sufficiently poor in low-leakage cases as to be precluded. At the current state of the art, 6% of blower flow is what is generally considered the boundary of low leakage.

The results of this study showed that at the 6% limit the pressurization to outside and DeltaQ tests agreed on four out of five homes for homes where the pressurization results were less than 6%. However, the DeltaQ test identified an additional eight homes where there was less than 6% leakage. These results suggest that if a home were to pass the screening test using pressurization to outside, it is likely to also pass using the DeltaQ test. The DeltaQ test will show that more homes pass than using pressurization to outside. It is unclear whether these homes should have passed or not. Given that the DeltaQ test had the largest standard deviation among the repeated tests for those homes with DeltaQ <6% of blower flow (at 3% of blower flow the standard deviation is half of the target signal), it should be expected that some of these homes should not have passed based on a strict 6% passing limit.

These results are close to previous comparisons of DeltaQ to other duct leakage test methods examining pass/fail testing criteria by Walker et al. (2001) and Francisco et al. (2003). Note that these previous studies used slightly different DeltaQ techniques but achieved similar results.

An opposite concern, which appeared in the current field data, can happen in leakier duct systems with fan pressurization techniques. If the leakage is too high, it may not be possible to reach the reference pressure (25 Pa) and that requires extrapolation to the nominal rating point. This induces an additional source of bias both from the extrapolation and from calibration errors with equipment operating at their extreme limits. As shown in this study, in some cases the results are completely unrealistic for estimates of duct leakage under normal operating conditions.

CONCLUSION

Our objective in this study was to determine if the repeatability errors of any of the three common duct leakage measurement techniques were sufficiently poor that it should be *a priori* eliminated as an option for use. The answer to that question is "no." The field test results showed that the duct pressurization to outside test method had superior repeatability overall. The acceptability of the repeatability uncertainty for any of the methods will depend on the particular application. Each measurement technique has pros and cons, depending on the application, the result needed, and the specific environment. Therefore, the results of the current repeatability study combined with other field and laboratory testing indicate that there is no way to *a priori* eliminate any of these test methods from being used. Selection of the test method to be used should remain flexible, with selection based on home and duct characteristics, other testing to be undertaken, and the objectives of the testing.

ACKNOWLEDGMENTS

This work was supported by the assistant secretary for Energy Efficiency and Renewable Energy, Building Technologies Program, of the U.S. Department of Energy under Contract No. DE-AC02-05CH11231. The authors would like to thank the field testing teams for performing the field test and providing input and insight for interpreting the testing results, in particular, Kirk Kolehma and Dave Bohac from the Center for Energy and Environment in Minneapolis; MN, Zachary Merrin and Stacy Gloss from the University of Illinois Sustainable Technology Center in Champaign, IL; and Amelia Godfrey, Sydney Roberts and Eyu-Jin Kim of Southface in Atlanta, GA.

REFERENCES

Andrews, J.W. 2002. Laboratory evaluation of the DeltaQ test for duct leakage. ACEEE Summer Study. Pacific Grove. Ca. American Council for an Energy Efficient Economy, Washington, D.C.

ASHRAE. 2013. ASHRAE 62.2-2013, *Ventilation and Acceptable Indoor Air Quality in Low-Rise Residential Buildings*. Atlanta: ASHRAE.

ASTM. 2013. E1554, *Standard Test Methods for Determining Air Leakage of Air Distribution Systems by Fan Pressurization*. American Society for Testing and Materials.

BPI. 2010. BPI-101, *Home Energy Auditing Standard*. Building Performance Institute.

CEC. 2012. *Reference Appendixes—The 2013 Building Energy Efficiency Standards For Residential and Non-Residential Buildings*. CEC-400-2012-005-CMF-REV2. California Energy Commission.

CEC. 2013a. *Residential Alternative Calculation Method Reference Manual For The 2013 Building Energy Efficiency Standards*. CEC-400-2013-003-CMF-REV. California Energy Commission.

CEC. 2013b. *Residential Compliance Manual For The 2013 Building Energy Efficiency Standards*. CEC-400-2013-001-CMF. California Energy Commission.

Chan, W.R., and M.S. Sherman. 2013. Analysis of air leakage measurements of US houses. *Energy and Buildings* 66. 616–25. Elsevier.

Dickerhoff, D.J., and I.S. Walker, 2008. Evaluation of the repeatability of the DeltaQ duct leakage testing technique including investigation of robust analysis techniques and estimates of weather induced uncertainty. LBNL 1209E. Lawrence Berkeley National Laboratory, Berkeley, CA.

Francisco, P.W., L. Palmiter, E. Kruse, and B. Davis. 2003. *Field evaluation of improved methods for measuring the air leakage of duct systems under normal operating conditions in 51 homes.* National Energy Technology Laboratory Report. Contract # DE-FC26-01NT41257.

Francisco, P.W. 2009. Understanding and reducing uncertainties in the Delta-Q test during field testing. University of Illinois at Urbana-Champaign.

RESNET 2014. ANSI/RESNET 301-2014, *Standard for the Calculation and Labeling of the Energy Performance of Low-Rise Residential Buildings using the HERS Index.* Oceanside, CA: Residential Energy Savings Network.

Walker, I.S., M.H. Sherman, M.P. Modera, and J. Siegel. 1998. Leakage diagnostics, sealant longevity, sizing and technology transfer in residential thermal distribution systems. LBNL 41118. Berkeley, CA: Lawrence Berkeley National Laboratory.

Walker, I.S., M.H. Sherman, J. Wempen, D. Wang, J.A. McWilliams, and J.D. Dickerhoff. 2001. Development of a new duct leakage test: DeltaQ. LBNL 47308. Berkeley, CA: Lawrence Berkeley National Laboratory.

Walker, I.S. and D.J. Dickerhoff. 2006. Evaluation of a new ramping technique for duct leakage testing. LBNL 61743. Berkeley, CA: Lawrence Berkeley National Laboratory.

Walker, I.S. and D.J. Dickerhoff. 2008. Field and laboratory evaluation of a new ramping technique for duct leakage testing. *ASHRAE Transactions* 114(2):494–504. LBNL 62262. Berkeley, CA: Lawrence Berkeley National Laboratory.

Walker, I.S and M.H. Sherman. 2014. Duct leakage repeatability testing. LBNL-175563. Berkeley, CA: Lawrence Berkeley National Laboratory.

Airflow through Lightweight Wall Assemblies—Influence of Size and Location of Leakages

Philipp Kölsch

Andreas Zegowitz

Bernd Nusser, PhD

Daniel Zirkelbach

Hartwig Künzel, PhD
Member ASHRAE

Ralf Wagner

ABSTRACT

Compared to the overall building envelope airtightness concerning energetic aspects, the condensation caused by air exfiltration is still a challenge today. For a better understanding of the processes inside lightweight constructions, hot- and cold-box experiments and airflow measurements on 14 different full-scale wall assemblies under realistic climatic conditions were performed. The influences of different mineral wool airflow permeabilities and leak configurations, as well as different pressure differences were analyzed. Despite a certain level of uncertainty, the results show some interesting aspects about the influencing factors of the convective moisture entry. The distributions of the water content between sheathing and insulation samples show that more water remains inside the mineral wool with low air permeability, while in case of the more air permeable mineral wool, more water accumulates in the sheathing. Moreover, the findings indicate that the moisture distribution over the height of the sheathing is more influenced by the pressure difference level than by insulation air permeability or arrangements of the small openings. For the tested configurations, a lateral flow due to openings near the studs shows the highest total moisture contents inside the assemblies. In contrast, we could not observe significant differences concerning the moisture content for the different leak arrangements and diameters with the same total leakage area under humidification test conditions, although differences in the flow behavior are clearly detected. However, another important finding of these experiments is that, despite good workmanship, a small amount of air and moisture seams to penetrate the sealing tape joints.

INTRODUCTION

Under cold climatic conditions, airflow through lightweight building constructions also causes moisture entry into the assemblies. This airflow is driven by the stack effect and wind pressure differences over the building envelope. If warm and humid indoor air flows through the construction and reaches parts whose temperature lies below its dew-point temperature, condensation occurs. As a worst-case consequence, this can end up in mold growth and wood decay, causing structural damages. Unlike the well-known water vapor diffusion processes, the influencing factors concerning convective moisture entry are manifold and need to be further analyzed. One fundamental assumption can be summarized: "It is to be expected that when fairly airtight materials are assembled by sealing, taping, screws, etc., that the assembly will leak more air than the original material being used as the basic material. Also, when these assemblies are joined together into a whole building, the building enclosure will leak more air than the individual assemblies" (Wagdy 2015). By the assembly of materials, inevitable leakages arise, and here, generally, two different convective situations can be distinguished: On the one hand, airflow through larger visible singular leakages, leading to localized moisture accumulation. On the other hand is an "unavoidable" airflow through the standard cross-section by invisible imperfections of joints and sealings, such as tiny leakages at adhesive tape connections. In order to create moisture-tolerant assemblies, these "unavoidable" leakages should be accounted by hygrothermal simulation tools. The IBP infiltration model (Zirkelbach et al. 2009) or the model by Nusser (2012) derive the convective moisture

Philipp Kölsch is a research associate and **Daniel Zirkelbach** is a group manager at Moisture Control & Climate Specific Design, Department of Hygrothermics, Fraunhofer Institute for Building Physics IBP, Holzkirchen, Germany. **Bernd Nusser** is the head of the Building Physic Group at Holzforschung Austria, Vienna. **Ralf Wagner** is a research associate and **Andreas Zegowitz** is a group manager at Heat Coefficients and Climate Simulation in the Department of Hygrothermics, Fraunhofer Institute for Building Physics IBP, Stuttgart, Germany. **Hartwig Künzel** is head of department of Hygrothermics, Fraunhofer Institute for Building Physics IBP, Holzkirchen, Germany.

entry from the overall building airtightness. This kind of approach just considers the moisture-relevant fraction of the total exfiltration airflow, while size and location of the different leakages are neglected. Nevertheless, the detailed factors influencing the airflow through building components and the resulting condensation amount are still unknown and a matter of discussion. A more detailed understanding of the airflow and condensation processes inside lightweight building assemblies was the aim of several airflow and humidification experiments on full-scale components with mineral wool batt insulation under realistic climatic and pressure conditions.

DESCRIPTION OF THE FULL-SCALE TEST SETUP

Climatic Chamber. To perform the experiments, a climatic chamber with a hot and a cold box was built up in the lab of Fraunhofer-Institute for Building Physics (IBP) in Stuttgart, Germany. The whole chamber consists of two parts, which are separated by the test assembly. The climatic chamber is thermally insulated and airtight. Doors on both sides of the chamber allow access to the tested samples. The doors are insulated and well-sealed with adhesive tape during the experiments. Both sides of the chamber have dimensions of approximately 2.3 × 1.62 × 2.3 m (height × width × depth). The warm side of the chamber is conditioned with a heater and an air humidifier to +22°C and 50% rh. The cold side is conditioned with a cooling register to +2°C to avoid freezing. The relative humidity on the cold side cannot be controlled and reaches values around 90% rh during the test cycles. On the cold side of the chamber, a camera was installed to take photographs of the condensate forming on the transparent sheathing at the cold side of the assembly every 30 minutes. Figure 1(a) shows the inside of the climatic chamber without the test element, the view from the cold-box side into the hot box. Figure 1(b) shows the whole chamber from the outside.

Test Assembly. The test assembly between the climatic chambers is a full-scale timber frame wall element with two cavities (Test Fields A and B) for insulation batts. The framing consists of solid timber rafters (140 × 200 mm) wrapped in plastic film to avoid moisture absorption. The overall dimensions of the test construction are 2.52 × 1.62 × 0.2 m, with two cavities 2.24 × 0.6 × 0.2 m (A = 1.34 m^2). Construction details are displayed in Figure 2.

As interior lining of the timber frame construction, a 2 mm aluminum plate is used on the warm side to prevent any moisture entry due to diffusion. On the cold side of the construction, a 5 mm transparent Perspex (PMMA) plate is used. On the back side of the Perspex (inside the construction) we placed six paper towels in each field. As the towels become clearly darker in wet status, it is possible to make the condensation water visible. By using this transparent material, it is possible to take pictures of the paper behind and create a movie which shows the chronological sequence of the moisture accumulation. Weighing the paper sheets after the test provides information on total amount

(a) (b)

Figure 1 Climatic chamber (a) from the inside and (b) from the outside. Black is the cold side and white the warm side of the chamber.

and moisture distribution. Figure 3 shows the test construction without taped edges before the installation between the hot and cold box. In Figure 3(a), the paper sheets behind the Perspex plate are visible. Figure 3(b), shows the aluminum plate on the warm side.

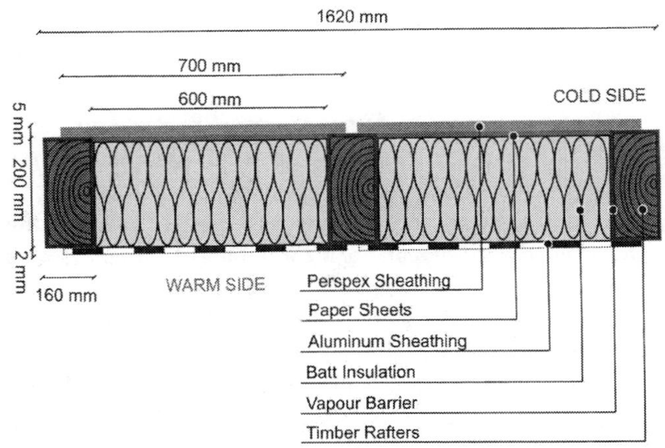

Figure 2 *Cross section of the lab test assembly.*

The aluminum and Perspex plates are mounted to the timber frame by screws. At Experiment 1 and 2, the screws and the edges of the sheathing plates are sealed with one layer of adhesive tape for vapor barriers and with one layer of aluminum adhesive tape on top. Starting from Experiment 3, an additional aluminum self-adhesive foil on top of the vapor barrier at the rafters inside the test construction has been installed to increase the airtightness of the taped seams. This sealing in combination with an aluminum plate on the warm side should be so vapor and airtight that the moisture accumulating inside the construction only can get in by the intended leakage holes. In the following, it becomes clear that this is hard to accomplish, but all attempts are made to get the assemblies as airtight as possible. The Perspex plate was mounted on the timber frame construction and remained there for the whole test cycle. From the warm side with the aluminum plate it is possible to change the setups.

Using two test cavities allows comparing different setups under the same climatic and pressure conditions simultaneously. For the different setups inside the cavities, mineral wool batt insulation with two different densities is used. The batt

(a) **(b)**

Figure 3 *Timber frame lab test construction from (a) the cold side and (b) the warm side before the mounting between the hot and cold box.*

insulation samples are cut 10 mm wider and 20 mm longer than the cavity dimensions, as under real construction site conditions, to achieve a close fit.

On the cold side of the assembly, an air outlet hole with a diameter of 5 mm is drilled in the Perspex sheathing. On top of this hole, an outlet nozzle is installed for the humidification test phase. Figure 4 shows the test assembly from the cold side with the Perspex and the paper towels during the humidification test phase with the mounted outlet nozzles at the bottom near the middle stud. The outlet nozzle is made of an insulated hose with an interior diameter of 25 mm to reduce the flow resistance. The whole nozzle is bent downward to prevent cold air entering the nozzle and the construction. Inside the nozzle we attached a temperature and relative humidity sensor close to the outlet hole to record the conditions of the air leaving the assembly. The arrangement of the intended leakage openings on the warm side varies.

Figure 4 *Test assembly from the cold side with the Perspex and the paper towels during the humidification test phase (outlet nozzles at the bottom near the middle stud).*

Sensors. The monitoring equipment consists of sensors for the relative humidity and temperature in the air of the hot and the cold box as well as sensors inside the test assemblies and in the outlet nozzle. The sensor positions inside the test construction are shown in Figure 5. Also, the absolute air pressure and the pressure difference between the hot and cold box were measured. With the Pressure Load Cell 1, the pressure difference between the hot and the cold box is monitored during the humidification test phase. The second load cell is used to reset the first load cell to zero, based on the absolute ambient pressure. Summarizing a measurement uncertainty of about ±1.0 Pa at the pressure stage of 5 Pa and ±1.5 Pa at a pressure stage of 15 Pa is reached during humidification. All data were recorded every ten minutes during the conditioning phase and the experiment. For the weighing of the paper sheets and the mineral wool samples an electronic scale is used. The air mass flow in the first test phase is recorded with a calibrated laminar flow element (LFE) with integrated pressure load cells, especially for the detection of very low air volume flows. The accuracy of the used calibrated LFE is, with a total uncertainty of ±0.25% M.V., significantly higher (factor of 10 to 20) than float-type flowmeters mostly used in similar experiments. The measuring devices and the uncertainty are listed in Table 1.

With ±0.25% M.V., a very good accuracy is reached by the air mass flow. The weighing of the different paper sheets has an uncertainty of absolute ±0.1 g; which is in a range between ±0.17% to ±0.35%, depending on the paper sheet weight. The weight of the high-density mineral wool samples is approximately 9 kg; the one of the low-density mineral wool approximately 4 kg. The measurement uncertainties are pursuant to ±0.6 g, respectively, approximately ±0.007% for the high density mineral wool and ±0.3 g, respectively, approximately ±0.0075% for the low-density mineral wool. The accuracy of the calculated absolute humidity during the humidification test phase, based on the temperature and humidity measurement as well as the ambient pressure, is about ±6.2% M.V. at 1.8°C to 2.5°C and 10% rh to 100% rh. On the warm side between 21°C to 24°C and 40% rh to 70% rh, the accuracy is about ±3.65% M.V.

The pure measurement device uncertainty concerning the calculated water content by fluxes and condensed water rates can be specified with approximately ±6.7%. This value is based on the combined uncertainties of all relevant measurement data. However, there are further uncertainties caused by hand disassembly of the test setups, aging of the mineral wool sample due to multiple uses and moisture influence, aging or partial delamination of adhesive tape under pressure impact, as well as the air mass flow measurements only before and after the humidification experiment. These influences are difficult to quantify but can clearly exceed the above-mentioned uncertainty of the measurement devices itself. The uncertainties may vary between the different test setups in a way that limits the comparability.

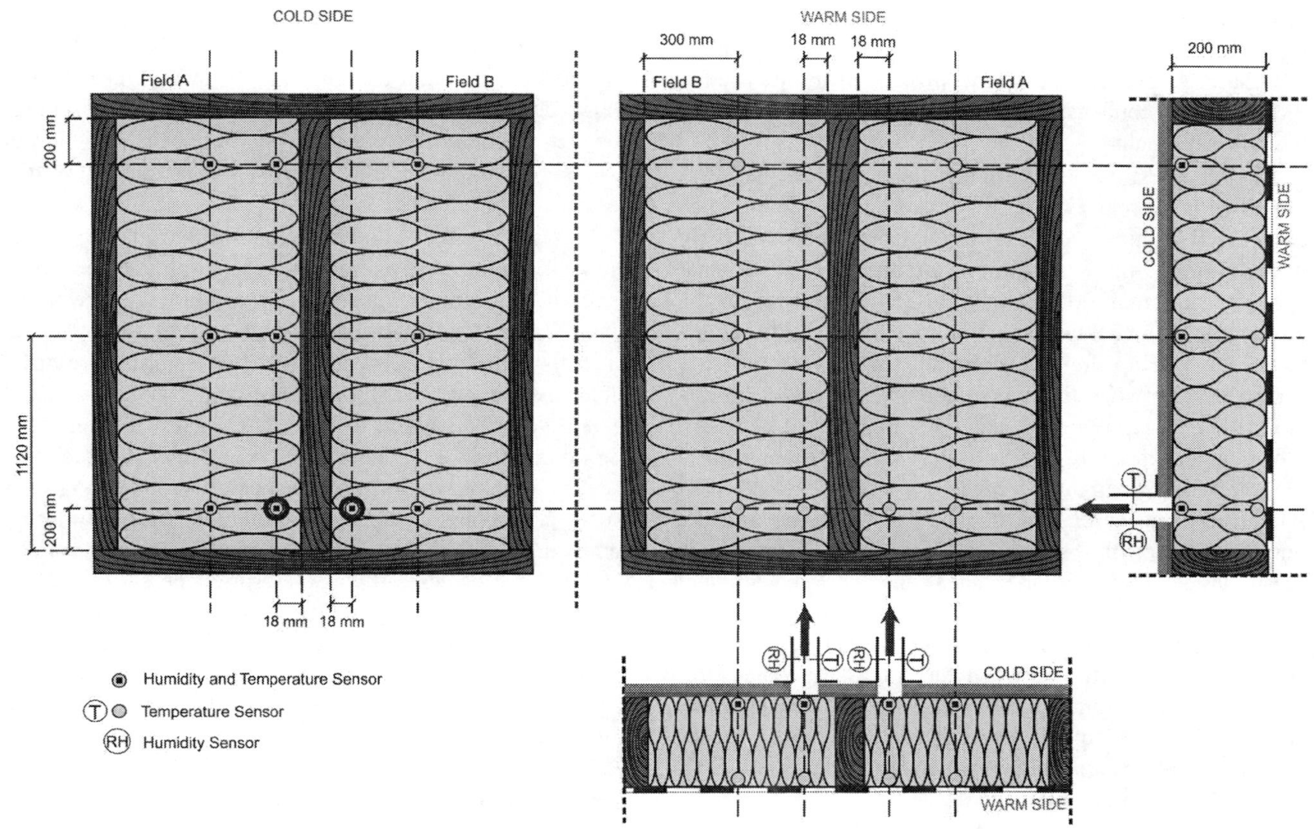

Figure 5 *Sensor positions inside the test assembly.*

Table 1. Measuring Devices and Accuracies

Measuring Device	Manufacturer	Type	Measuring Range	Measurement Uncertainty
Air mass flow (LFE)	Meriam Process Tech.	50MJ10-14	0 to 3.4 g/min	± 1% F.S. (data sheet)
Air mass flow (LFE)	Meriam Process Tech.	50MW20-01	3.1 to 260 g/min	± 1% M.V. (data sheet)
Both LFEs are calibrated to a total uncertainty of the measured air mass flow:				± 0.25% M.V.
Electronic scale	Mettler Toledo	CP34001P	0.3/3/8/16/34 kg	±0.1/0.1/0.3/0.6/0.8 g
Temperature	Newport	Cu-CuNi	–10°C to 120°C	± 0.12 K
Humidity	Driesen + Kern	DKRF-400-EXT	0% rh to 100% rh	± 2% at 20% rh to 90% rh ± 3.5% at >90% rh
Pressure difference 1	Driesen + Kern	DK 84		± 1.0 Pa
Pressure difference 2	TetraTec Instruments		0.1 to 200 Pa	± 0.5 Pa

Test Procedure. At first, the batt insulation samples are stored for a few weeks under the climatic conditions in the lab. This is necessary to ensure that the mineral wool can expand from the compressed "delivery condition" to its normal thickness and adapt to the lab climate conditions. Before installation, the batts are weighed by an electronic scale. Also, the paper towels and adhesive tapes are cut and weighed before the attachment on the Perspex plate. After placing the sensors and the mineral wool samples, the test cavities are closed with aluminum plates and sealed.

The whole experiment cycles are always separated in two different test phases. The first part is the airflow measurement on the different setups; the second part is the humidification test phase. Afterwards the airflow measurements are used to recalculate the condensation water amount, detected after the humidification phase.

For the first part, on both test fields airflow measurements with open and closed intended leakages (reference air mass flow measurements) are performed successively under room climate conditions on both sides. For that purpose a laminar flow element (LFE) is attached to the outlet hole in the Perspex sheathing on the cold side of each test field. During the measurement, the desired pressure differential between the warm side and the outlet hole is controlled by applying a negative pressure at the LFE outlet. Approximately 15 air mass flow measurements are recorded for pressure difference steps from 0.5 to 20.0 Pa. After these measurements, the LFE is removed and the outlet nozzles with the humidity and temperature sensors are mounted on the Perspex sheathing. Due to technical reasons, it was not possible to record the air mass flow during the humidification test phase. Therefore, the pressure fields partly differ between the air mass flow measurements and the humidification tests. This may lead to slightly different flow rates between the two test phases and, therefore, to some additional uncertainties in the evaluation of the experiments.

After installing the outlet nozzle, a preconditioning phase started. Here, the intended leakage openings in the aluminum sheathing on the cold side are closed again, the chamber is sealed, and the climatic conditions on the warm and cold side are applied. This tuning phase without any pressure difference is kept for a few days to reach nearly steady-state start conditions for the different experiments. The humidification test phase start criteria is a constant temperature and relative humidity detected at the cold side sensors, as well as a constant temperature on the warm side. Afterwards the intended leakage holes are opened again and the second test phase of humidification starts. During this phase of the experiment, a negative pressure of 5 or 15 Pa is applied to the whole cold part of the chamber. The experiments run between 15 to 21 days, except for Experiment 1, which runs 31 days.

At the end of each humidification test period the pressure difference is switched off, and the cold side of the chamber is warmed up to avoid additional condensation when opening the construction. Before dismantling the assemblies, a final airflow measurement is performed. Then both fields are opened from the cold side and the moist batt insulation is weighed. The paper towels are removed and packed in pre-weighed and numbered plastic bags. These bags with the wet towels are also weighed and the paper towels dried afterwards to confirm the reference value.

It was intended to take the same insulation samples for the subsequent experiments again to minimize the influence of inhomogeneous material properties as well as unavoidable edge inaccuracy from cutting the samples. Therefore, the mineral wool insulation samples are dried under the climatic conditions in the lab (22°C to 26°C and 30% rh to 60% rh) for 2 to 4 weeks to ensure similar starting conditions for the next test cycle. However, the low-density mineral wool samples especially showed signs of partial delamination after a few times of installation and removal, and this may have had some influence on the results. In some cases it was therefore necessary to pick a new sample from the same production batch, which also may affect the results, because of the inhomogeneity of the insulation material and the cutting edge inaccuracies. The different insulation samples used in the experiments are numbered and given in Table 4.

Materials and Experiment Conditions. During the test campaign, different types of mineral wool insulation, pressure differences, and arrangements of the inlet and outlet openings are investigated. Two types of mineral wool insulation, one with a density of 15.3 kg/m^3 and an air permeability of 2.03×10^{-9} m^2 and one with nearly double the density of 32.8 kg/m^3 and a nearly six times lower air permeability of $0{,}339 \times 10^{-9}$ m^2, are used. The material properties of the two different types of mineral wool are summarized in Table 2. Both investigated materials show a typical air permeability value compared with other mineral wool samples (e.g. those summarized in Riesner 2009) providing a range between approx. 0.3×10^{-9} and 5.0×10^{-9} m². The airflow resistance respective permeability is measured on ten samples, according to DIN EN 29053: 1993-05. The thermal conductivity is measured according to DIN EN 12667: 2001-05 with the use of two samples.

The humidification experiments are performed under two different constant-pressure differences between the warm and the cold side of the climatic chamber. The lower pressure

Table 2. Material Properties of the Tested Batt Insulation

Material	Bulk Density φ, kg/m^3	Thermal Conductivity λ, W/m·K	Flow Resistance r, 10^3 (Pa·s/m^2)	Air Permeability K 10^{-9} (m^2)
ZKF 040	15.3	0.038	9.0	2.03
ZKF 032	32.8	0.031	53.8	0.339

difference of 5 Pa is a value that represents the maximum stack effect in single family dwellings during winter time in Central Europe. It prevails, for example, in a building with three floors and an air column of 9 m at an indoor temperature of 22°C and an outdoor temperature of –5°C. The higher pressure difference of 15 Pa can only occur in tall buildings or very cold climates. Applying different pressure differentials helps to determine the condensation rate as a function of airflow rate. For clarity it must be mentioned that the determined moisture entry under the steady-state conditions during the humidification test phase are higher than the expected moisture entry under real conditions. In reality, pressure fluctuations, periods without condensation, and some dryout due to diffusion and convection limit or reduce the condensation amount.

At the warm side of the chamber, the temperature is kept constant at +22°C and the humidity at 50% rh; the cold side is conditioned to +2°C. During the experiment phase of humidification, the temperature and the relative humidity in the hot and the cold box, as well as the pressure difference across the test assembly, are continuously recorded. The boundary conditions, mean values, and the standard deviations of the recorded data are presented in Table 3.

Test Setups. Altogether, seven test cycles with two test setups at once (one in each test field) are performed. In two test cycles, Field A is completely closed. The other test setups can be divided in two groups: In the first one, the intended inlet and outlet consists of just one hole with a diameter of 5 mm. For the second group, a row of 25 smaller inlet holes with a diameter of 1 mm (amounting to the same leakage area) in the middle of the warm side is tested.

In the first group of test setups the location of the inlet and outlet hole is changed to figure out if this has an influence on the moisture entry. Furthermore, the installation of the batt insulation is mounted once edge-recess free and once with recess at the edges to the frame construction. Edge recess like this can occur when the builder presses the batt insulation into the cavity and pushes a bit too far at the edges where the insulation is kept back by the rough, unplaned, wooden studs. In the lab tests, with studs wrapped in plastic film, this is done by pushing the insulation approximately 20 mm inward, as shown in Figure 6. It is assumed that the air gaps lead to a different moisture distribution or a higher condensation amount due to the lower flow resistance, especially when the inlet holes open out directly into the resulting air gap. Due to strength and lower flexibility of the high-density mineral wool, it is not possible to form durable edge recess. For this reason, experiments with edge recess are only performed with low-density mineral wool. The different setups of the first group of experiments are shown in Figure 7 and Figure 8.

Another test cycle was performed with a direct flow path from the warm to the cold side. Therefore the inlet and outlet holes are arranged at opposing positions on both warm and cold side, in the middle of each field. In the first case, the batt insulation is punched, providing a direct air channel with a diameter of approximately 5 mm to the cold-side hole, while in the other case the airflow must permeate the insulation material fibers. This setup should check the theory that a direct flow, with a reduced flow resistance, leads to a higher air mass flow. A high airflow should heat up the flow path and thus reduce the condensation amount.

Geißler and Hauser had done some CFD simulations assuming different realistic flow paths in the year 2002. They found that airflow through larger leakages only has a small effect on the warming up of the flow path. The simulation results also show that larger leakages with a direct flow path show a higher risk of moisture accumulation, especially with indoor air conditions above 50% rh (Geißler and Hauser 2002). Nonetheless, the converse is often claimed.

In practice. this situation can occur at the joints of two different pieces of batt insulation or due to electrical wiring. In this setup, the outlet nozzle at the cold side is placed in the middle of each field (not at the bottom corner near the middle stud, like in the other cases). Figure 9 shows the experiment test setups with (Figure 9(a)) and without (Figure 9(b)) punched mineral wool forming a predefined flow channel.

Table 3. Experiment Boundary Conditions (Mean Value and Standard Deviation)

Experiment	Warm Side Temperature, °C	Warm Side Relative Humidity, %	Warm Side Absolute Humidity, g/kg	Cold Side Temperature, °C	Pressure Difference, Pa
1	21.69 ±0.10	50.63 ± 0.71	8.66 ± 0.13	2.04 ± 0.45	15.1 ± 1.01
2	22.04 ± 0.05	50.12 ± 0.35	8.67 ± 0.09	2.07 ± 0.045	5.13 ± 0.59
3	22.19 ± 0.072	49.65 ± 0.55	8.65 ± 0.13	1.93 ± 0.07	4.62 ± 0.285
4	22.12 ± 0.049	50.08 ± 0.61	8.70 ± 0.15	2.03 ± 0.07	4.92 ± 0.68
5	22.15 ± 0.074	50.11 ± 0.43	8.74 ± 0.10	2.03 ± 0.056	15.43 ± 1.74
6	21.82 ± 0.155	51.05 ± 1.87	8.75 ± 0.36	2.02 ± 0.21	4.83 ± 0.54
7	21.84 ± 0.056	50.34 ± 0.42	8.58 ± 0.10	2.27 ± 1.17	14.99 ± 0.80

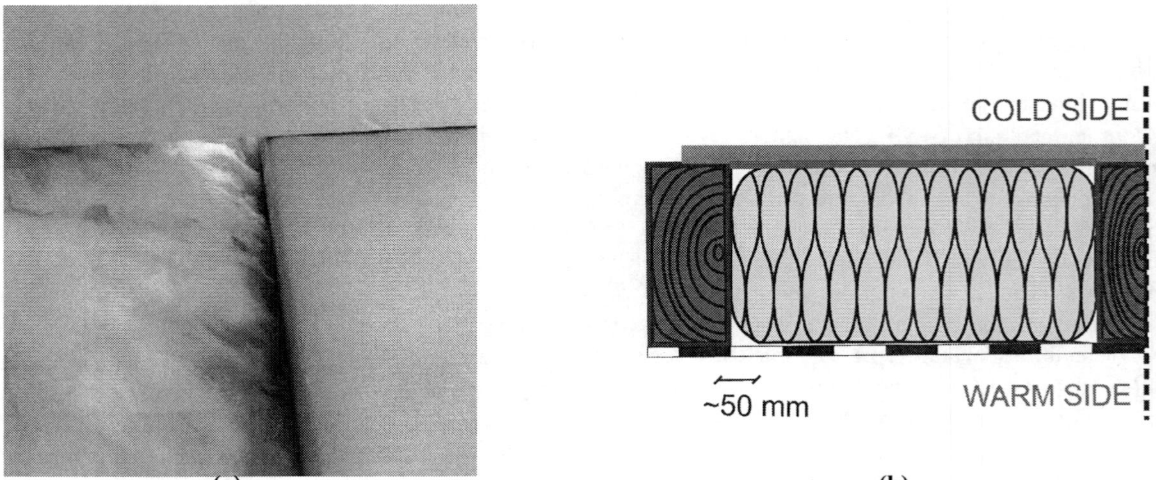

(a) **(b)**

Figure 6 *Edge recess of the mineral wool at the connection of insulation and stud: (a) at the test assembly and (b) principle at the cross section.*

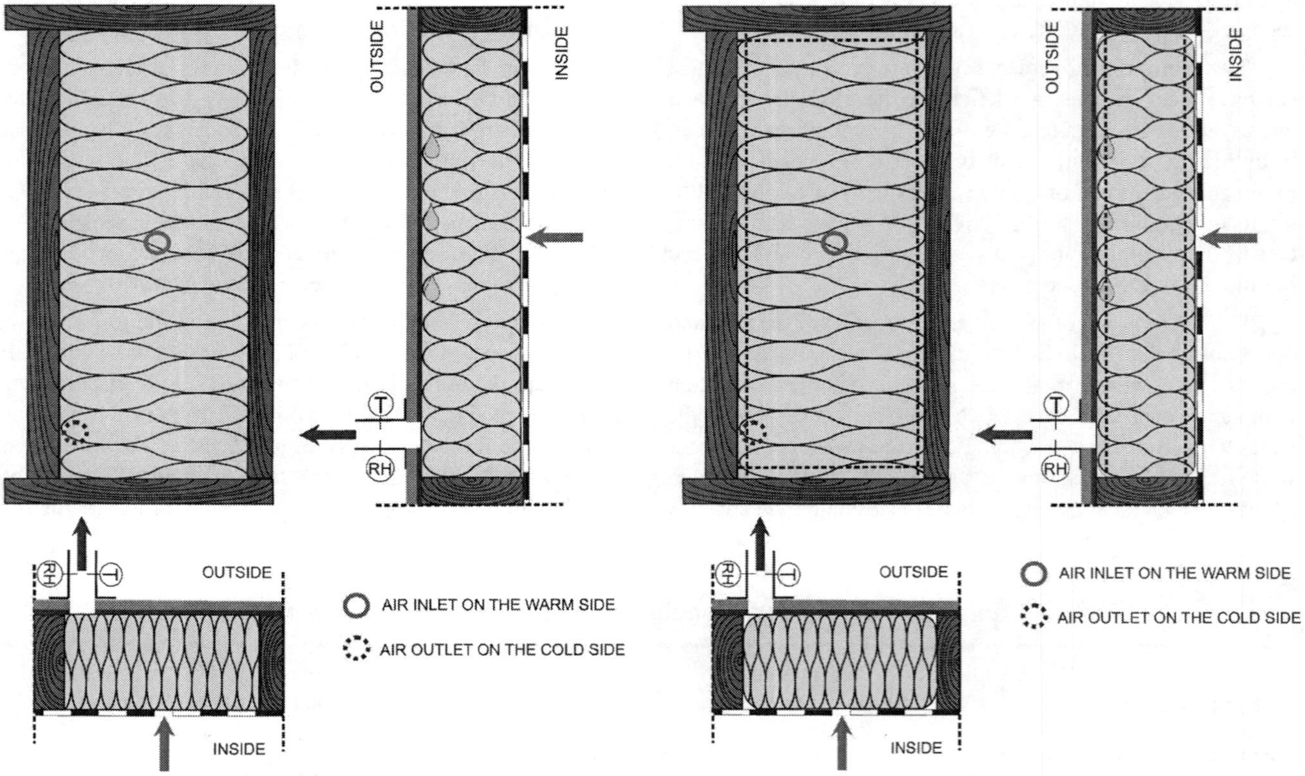

(a) **(b)**

Figure 7 *Flow from the center position of the warm side to the left bottom outlet hole near the middle stud on the cold side. Case (a) without and (b) with edge recess.*

The second group of test setups are the cases with a row of holes, consisting of 25 single holes with 1 mm diameter resulting in the same leakage area of approximately 20 mm^2 like the one single 5 mm hole, investigated before. The small holes were arranged in a vertical line in the middle of the aluminum sheathing. The distance between the single holes is 80 mm, which should avoid an interaction of the flows. The arrangement of the holes is displayed in Figure 10. For all

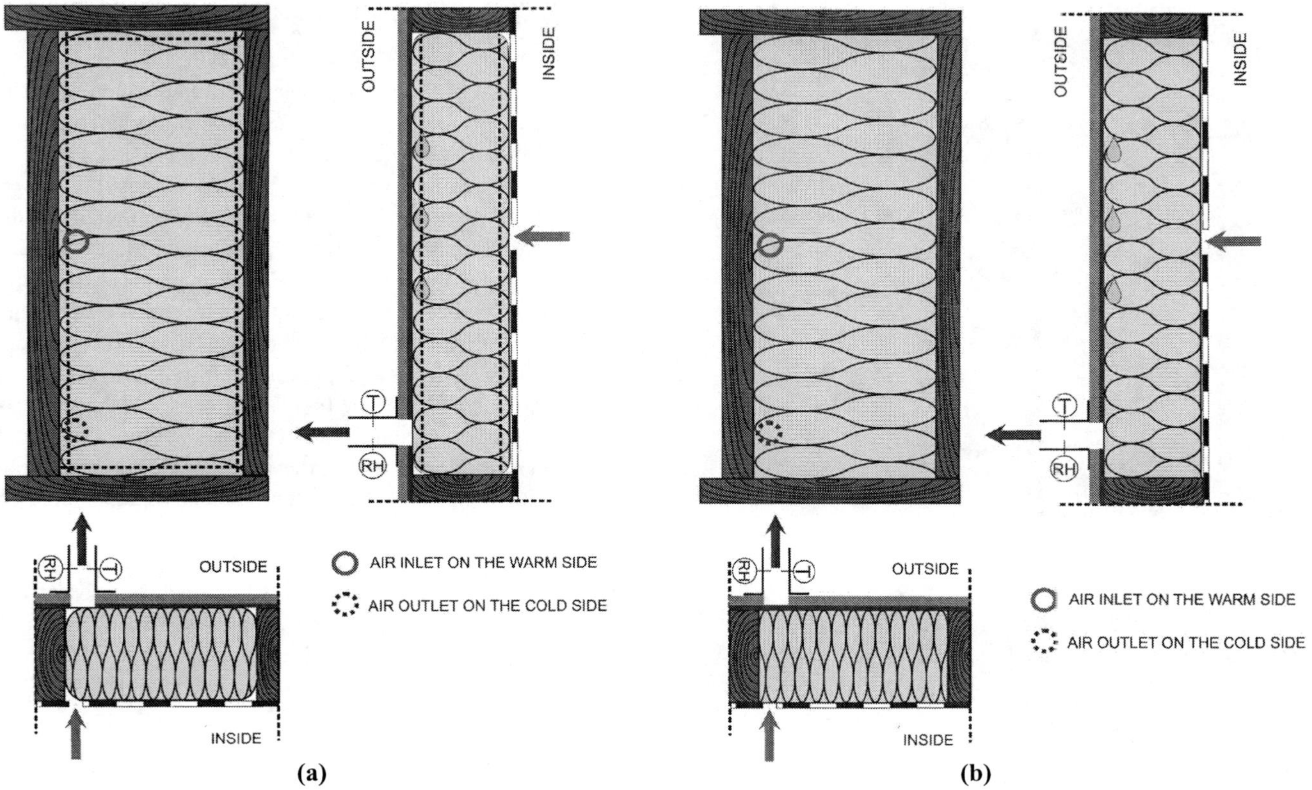

Figure 8 *Lateral flow near the center studs (a) with and (b) without edge recess.*

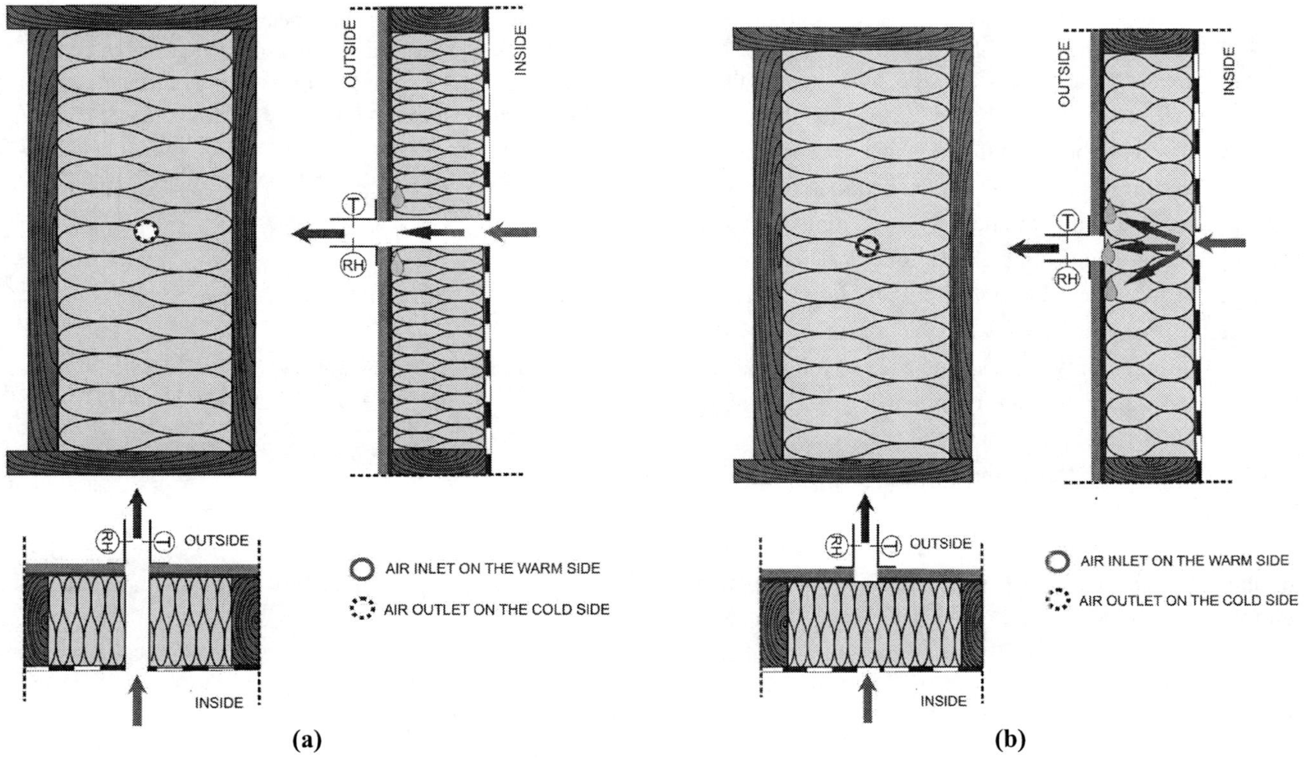

Figure 9 *Direct flow path through (a) the punched mineral wool and (b) without predefined flow channel.*

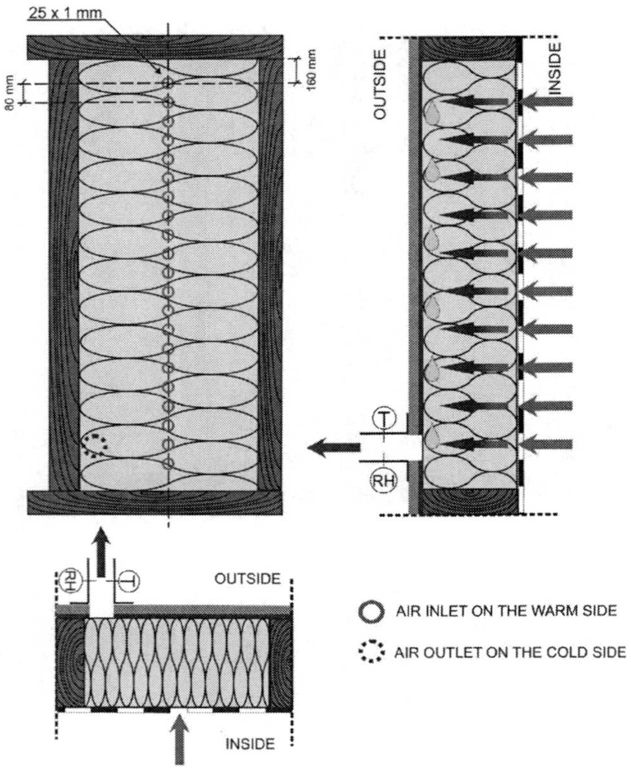

Figure 10 *Flow from the warm side through 25 holes of 1 mm diameter to the outlet hole near the middle stud on the cold side.*

cases with the small holes, the outlet hole in the Perspex sheathing is located at the bottom side, near the middle stud corner. These test setups should check the theory, that smaller holes with the same total leakage area lead—at the same pressure difference—to a more laminar flow regime and therefore to a higher air mass flow and moisture entry to the construction. Hints of this behavior can also be found in Schmidt (2014). According to the law of Hagen-Poiseuille, the volume flow increases with the fourth power of the radius. But this is only valid in the case of a laminar flow regime. The transition from a laminar to a turbulent flow regime itself is also dependent on the size of the leakage. At smaller leakages the transition occurs at higher air volume flows compared to larger leakages. At a laminar flow regime the pressure drop over a hole rises proportionally with the volume flow. The pressure drop of a turbulent flow increases with the square of the volume flow. This means, at a distinct pressure difference, the airflow through smaller holes becomes higher than the flow through a single large hole with the same total leakage area.

An overview of the seven performed experiments is given in Table 4.

RESULTS AND DISCUSSION

Air Mass Flow Measurements. The air mass flow measurements of the first test phase are performed with the LFE directly at the test assemblies between the hot and cold box. Neither the frame construction nor the insulation or sheathing plates are moved or changed in any other way between the air mass flow measurements and the air conditioned humidification experiment. Equation 1 describes as a power function the correlation between air mass flow and pressure difference. To determine the flow coefficient C and exponent n of Equation 1 by orthogonal distance regression, air mass flow points at 15 pressure differences from 0.5 to 20.0 Pa are recorded. At the first experiment only the air mass flow with the open intended leakage holes has been measured before the experiment. Using these values for the calculation of the water content leads to a significant deviation compared to the weighed water content. Hence, for the further experiments (starting from Experiment 2) the air mass flow after the experiment was also determined, which clearly differs from the flow before the experiment in most cases. However, different calculations using start, end, and average air mass flow for Experiment 2 still show unsatisfactory agreement with the weighed water contents. Therefore, starting with Experiment 3, also reference measurements with closed inlet leakages before the experiments are performed to determine the leakage rate of the assembly itself. Surprisingly such a high airflow through the sealing tape wasn't expected. Visible evidence for unintended openings could not be found. The edges of the sheathing were sealed and visibly tight but airflow in all reference cases could be measured. This was the reason for a verification of these measurements, which will be presented later. For Experiments 6 and 7, additional reference measurements were performed after the experiment. All resulting air mass flow coefficients and exponents are listed in Table 5. The air mass flow can be calculated according to Equation 1.

$$\dot{m}_a = C \cdot \Delta P^n \left[\frac{kg}{s}\right] \qquad (1)$$

In the following, the flow regime of the open leakage measurements, expressed by the flow exponent n, is regarded. As expected, all values are in the transition region between laminar ($n = 1.0$) and turbulent ($n = 0.5$). However, the values indicate the tendency of flow behavior. The measurements with mineral wool with low air permeability show a more laminar flow. This is also an expected behavior as the tighter fiber structure leads to finer flow channels in between the fibers.

Concerning the differences between the test assemblies with one intended single leakage and those with the row of holes, the flow regime also shows deviations. Comparing the constructions with the same air permeability, the row-of-holes assemblies show a more laminar flow regime. Experiment 3 Field A with a single hole and high air permeable mineral wool

Table 4. Experiment Setup Overview

Experiment	Test Field	Mineral Wool	Target Pressure, Pa	Description	Mineral Wool Sample
1	A	High air permeability	15	Closed assembly	Sample 1
	B	High air permeability	15	Flow from the middle of the warm side without edge recess (Figure 7a)	Sample 2
2	A	Low air permeability	5	Closed assembly	Sample 3
	B	Low air permeability	5	Flow from the middle of the warm side without edge recess (Figure 7a)	Sample 4
3	A	High air permeability	5	Flow from the middle of the warm side without edge recess (Figure 7a)	Sample 1
	B	High air permeability	5	Flow from the middle of the warm side with edge recess (Figure 7b)	Sample 2
4	A	Low air permeability	5	Direct flow with flow channel (Figure 9a)	Sample 5
	B	Low air permeability	5	Direct flow without flow channel (Figure 9b)	Sample 6
5	A	High air permeability	15	25 holes in the middle of the warm side (Figure 10)	Sample 7
	B	Low air permeability	15	Flow from the middle of the warm side without edge recess (Figure 7a)	Sample 4
6	A	High air permeability	5	Lateral flow near the middle stud with edge recess (Figure 8a)	Sample 1
	B	High air permeability	5	Lateral flow near the middle stud without edge recess (Figure 8b)	Sample 2
7	A	High air permeability	15	25 holes in the middle of the warm side (Figure 10)	Sample 7
	B	Low air permeability	15	25 holes in the middle of the warm side (Figure 10)	Sample 4

has a flow exponent of $n = 0.593$ (mean value of both measurements before and after the experiment). Experiment 5 Field A shows a value of $n = 0.654$ and Experiment 7 Field A shows a value of $n = 0.725$. In both cases, the exponent of the row-of-holes assemblies is more laminar than the one of the single hole leakage. To compare the cases with mineral wool of low air permeability, Experiment 5 Field B with a single hole and a mean value of $n = 0.725$ and Experiment 7 Field B with the 25 holes and $n = 0.832$ can be considered. Again, the flow regime is more laminar in the case of the smaller holes with the same leakage area.

A certain influence of the leakage area distribution on smaller or bigger hole diameters seems to be obvious. But considering the flow coefficient C of the different experiments, it becomes also visible that the coefficients decrease in all row-of-holes cases with a more laminar flow regime. Thus, it seems that, particularly at higher pressure differences, smaller holes lead to a higher air mass flow than one single hole with the same leakage area. The smaller the leakages, the more pronounced is this effect and the higher must be the pressure difference to get a turbulent flow regime.

Figure 11 shows the air mass flow pressure difference curves for the above-mentioned experiments with mineral wool of low and high air permeability. The curves are calculated by mean values of the flow coefficient and exponent before and after the experiment (compare Table 5). The row-of-holes experiments are displayed as solid lines and the measurements with single holes as dashed lines. The pressure

Table 5. Air Mass Flow Coefficients and Exponents

Exp.	Test Field	Measurement	Flow Coefficient C -10^{-6}, kg/(s Pa^n) before Experiment	Flow Exponent n, —, before Experiment	Flow Coefficient C -10^{-6}, kg/(s Pa^n) after Experiment	Flow Exponent n, — after Experiment
1	A	Reference	—	—	—	—
	A	Open leakage	Closed Leakages		—	—
1	B	Reference	—	—	—	—
	B	Open leakage	6.750	0.745	—	—
2	A	Reference	—	—	—	—
	A	Open leakage	Closed Leakages	—	—	
2	B	Reference	—	—	—	—
	B	Open leakage	2.483	0.777	0.842	0.854
3	A	Reference	3.270	0.827	—	—
	A	Open leakage	12.533	0.589	13.100	0.596
3	B	Reference	2.917	0.855	—	—
	B	Open leakage	12.250	0.565	15.850	0.565
4	A	Reference	5.533	0.758	—	—
	A	Open leakage	9.967	0.618	11.367	0.624
4	B	Reference	3.733	0.767	—	—
	B	Open leakage	6.800	0.662	5.367	0.687
5	A	Reference	6.467	0.788	—	—
	A	Open leakage	9.417	0.681	9.633	0.627
5	B	Reference	3.067	0.871	—	—
	B	Open leakage	7.317	0.729	6.283	0.721
6	A	Reference	7.883	0.752	10.067	0.743
	A	Open leakage	18.250	0.554	18.383	0.565
6	B	Reference	3.567	0.831	4.250	0.787
	B	Open leakage	16.600	0.561	15.533	0.592
7	A	Reference	6.933	0.749	6.133	0.856
	A	Open leakage	10.500	0.655	8.467	0.795
7	B	Reference	3.550	0.889	4.167	0.681
	B	Open leakage	6.350	0.8	5.433	0.864

difference range of the humidification test phase is presented as a gray area. At higher pressure differences it is clearly visible that the row of holes cause higher air mass flows than one single larger hole. This confirms the above-mentioned theory of turbulence occurring at larger leakages under lower pressures differences.

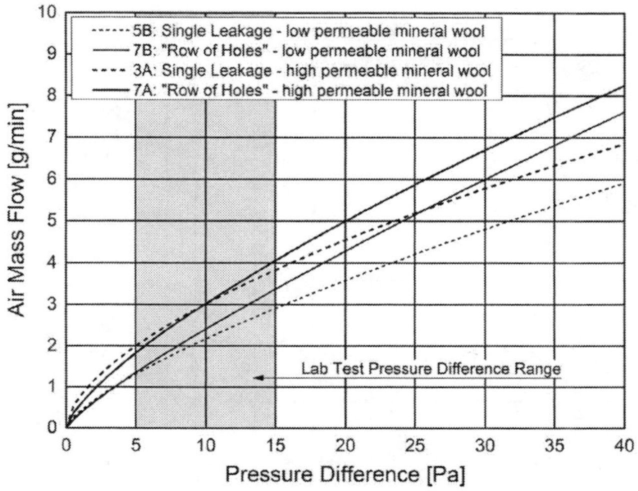

Figure 11 *Calculated air mass flow over pressure difference curves for single leakages and row-of-holes assemblies with different insulation air permeability based on the mean values presented in Table 5.*

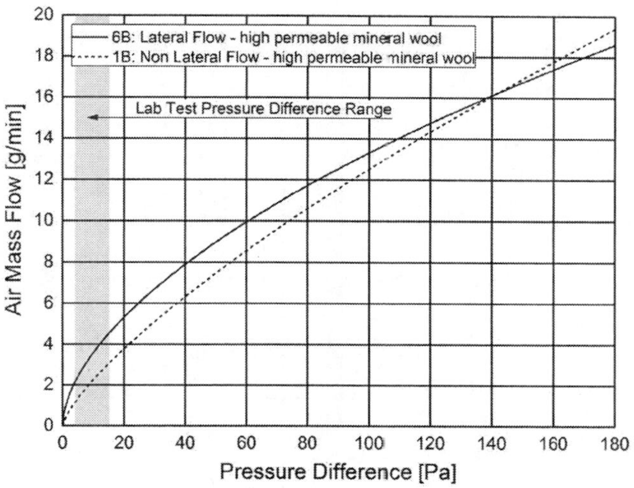

Figure 12 *Calculated air mass flow over pressure difference curves for a lateral and a non-lateral flow situation based on the values presented in Table 5.*

The lateral flow situations with the inlet openings near the studs are investigated in Experiment 6. All flow exponents show a value close to $n = 0.5$, meaning a flow situation more in the turbulent flow regime. This indicates a flow path with bigger diameters, probably at the interface between the insulation and the stud. In contrast, the flow coefficients of Experiment 6 are in the range around $C = 16$ to 18. This value is much higher than in all other experiments with low density insulation and a single hole. This leads at lower pressure differences to a higher air mass flow compared to the other experiments. However, due to the different shape of the air mass flow curve, depending on n and C, at higher pressure differences the air mass flow might also remain lower than in the other cases. While at $n = 1$, the air mass flow increases linearly with pressure difference; the increase remains much lower in case of $n = 0.5$.

Figure 12 shows, for example, the calculated air mass flow pressure difference curves for a lateral flow situation (Experiment 6 Field B) displayed as a solid line and a non-lateral flow situation (Experiment 1 Field B) with the dashed line. The curves are calculated based on the determined flow coefficient and flow exponent presented in Table 5. The pressure difference range of the humidification lab test is presented as a gray area. Both experiments are performed with high air permeable mineral wool. The average deviation between the curves in the lab test pressure range between 5 Pa and 15 Pa is about 60%. With a rising pressure difference, the deviation decrease until the point of intersection, here at about 140 Pa. Therefore, it can be assumed that a lateral flow situation leads to higher air mass flows and, thus, higher moisture entry rates under pressure conditions as they usually occur in single family dwellings.

Figure 13 shows the variation of the air flow rate after the experiments in percent compared to the initial value. The values are calculated on the basis of the air mass flow measurements before and after the humidification test phase with the respective target pressures of 5 or 15 Pa. The change of the air flow rate is related to the start value. The black columns are the values for the measurements with open leakages and the gray ones for the reference measurements. It was expected that, in all cases, the constructions become less permeable during the experiment because of the moisture accumulation inside the mineral wool. It is assumed that condensed water leads to an increase of the flow resistance because water droplets block the flow paths.

In the case of the low air permeable mineral wool, we actually can observe the expected behavior. Experiment 4 Field A, with its punched flow channel through the mineral wool, shows a different behavior, which is not surprising, as here no water droplets can block the flow channel. Furthermore, the airflow seems to increase during the humidification test phase.

The change of the airflow rate in case of the high air permeable mineral wool does not show clear tendencies. Here, in most cases, the test assemblies get more permeable. The analysis under the next heading shows that less moisture accumulation remains in the mineral wool with high air permeability; hence, the amount in this kind of mineral wool could be too small to influence the flow resistance. For Experiment 6 and 7, additional reference measurements after the experiments with closed intended leakages are performed; even these values vary in the same way as the measurements with open leakages. A plausible explanation for this behavior could not be found so far, but it shows that even the airflow fraction through the adhesive tape connections can vary in the range of approximately ±30%. This might be influenced by some changes inside assembly or at the adhesive tape connections itself.

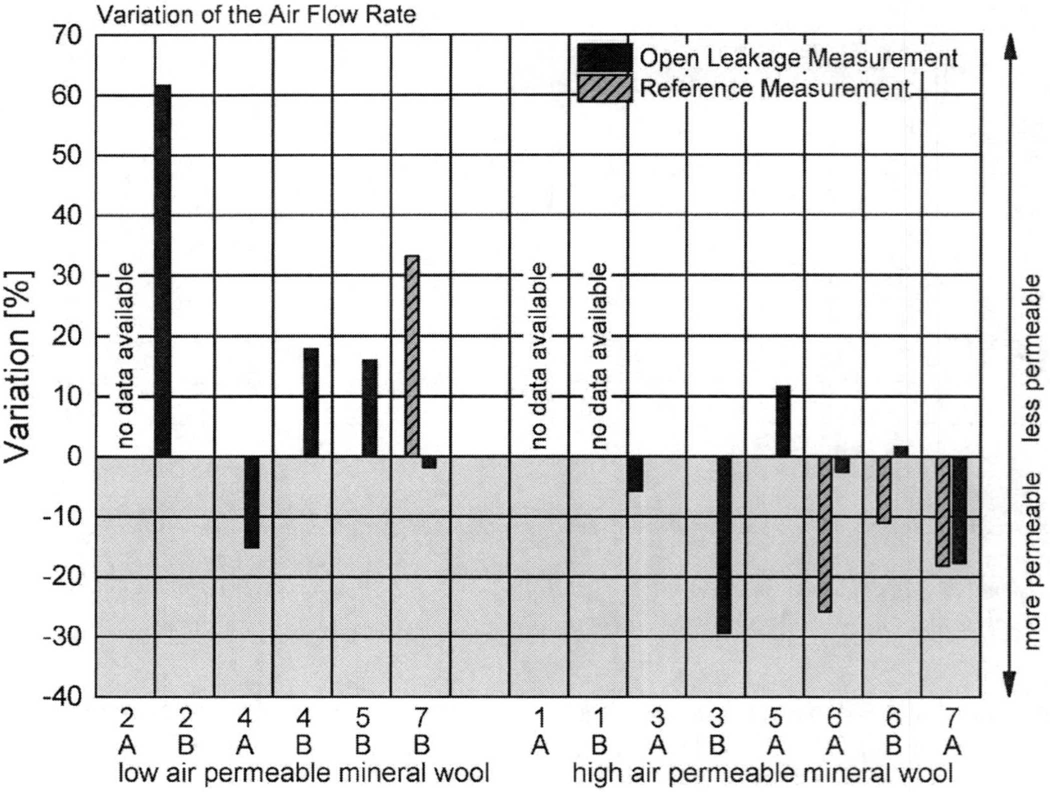

Figure 13 *Change of the airflow rate after the humidification experiments compared to the initial value. Each value is related to the target pressure conditions of 5 or 15 Pa of the respective humidification experiment (left: insulation with low air permeability, right insulation with high air permeability).*

Nevertheless, in the performed experiments it was observed that the low air permeable mineral wool is getting less permeable during the test period while the insulation with the high air permeability shows the opposite behavior. However, it is not possible to provide a consistent explanation for the observed behavior.

Verification of the Reference Measurements. Due to the unexpected detection of air mass flows during the reference measurements with closed intended test leakages ("airtight" case) a verification of these test results seems to be required. In fact, the aim of the investigations was to determine the convective moisture entry into a realistic full scale construction under lab conditions. Initially we assumed that the aluminum and Perspex sheathings with the taped edges are too airtight and the airflow and moisture entry would be too small to be measured. Therefore, additional holes were introduced to allow for a measurable airflow rate in a realistic magnitude. In this context it was also intended to analyze the impacts of different hole arrangements.

That the investigated assemblies are not completely airtight was an unexpected result and needed further verification. Actually, there are other publications reporting similar observations, for example Desmarais, who describes problems with not completely airtight adhesive tape connections. The author's theory was that dust on plastic film caused small leaks through the joints of the adhesive tape. However, this could not be confirmed because a second, dust-free joint assembly showed the same behavior (Desmarais 2000).

To investigate the reasons for the remaining leakages, a completely different experimental setup is devised. It consists of an absolutely airtight plastic chamber, with an open front for test specimens. At the test orifice an air barrier plastic film on a rigid board with slots is installed and tested up to a pressure difference of 50 Pa. No air mass flow through the plastic film can be measured, which means that the setup is completely airtight. Afterwards slots, having a total length of 5 m, are cut into the air-barrier film and then resealed with customary adhesive tape for air control layers. The experiments were performed several times with tapes from different manufacturers. Figure 14 shows this process exemplarily.

The tests are repeated with the different adhesive tapes up to a pressure difference of 20 Pa. The results show that for perfectly taped connections under lab conditions on a rigid substrate, the permeability per meter adhesive tape lies in a range between $1.9 \cdot 10^{-7}$ kg/(s m) up to $4.1 \cdot 10^{-6}$ kg/(s m) for 5 Pa pressure difference. But the measurements show also that only one small imperfection per meter adhesive tape can increase the permeability by a factor of 100.

Furthermore, it has been observed that some delamination between adhesive tape and plastic film occurred within only a few days. Therefore, an improvement of the airtightness is sought by pressing the adhesive tape to the substrate with a pressure roller. By this method complete airtightness can be achieved for the moment. However, also these connections become slightly leaky after a short period of time. The values obtained for the perfectly taped adhesive tape connection are under the limit of 0.1 $m^3/(m\ h\ (daPa^{2/3}))$, respectively ~ $2.08 \cdot 10^{-5}$ kg/(s m $(Pa^{2/3})$) at 5 Pa pressure difference mentioned in the German standard DIN 4108-2:2013-02 for connection joints. However, this limit is for connection joints concerning energy ratings disregarding convective moisture entry. Figure 15 shows the air mass flow per meter adhesive tape connection at the reference measurements as solid black columns and the air mass flow range of the verification measurements as a gray area. For the reference measurements the mean value is $1.4 \cdot 10^{-6}$ kg/(s m) at 5 Pa pressure difference. In summary, the detected air mass flows per meter adhesive tape are in a similar order of magnitude than those detected at the reference measurements but most likely these lab values are not achievable at the construction site.

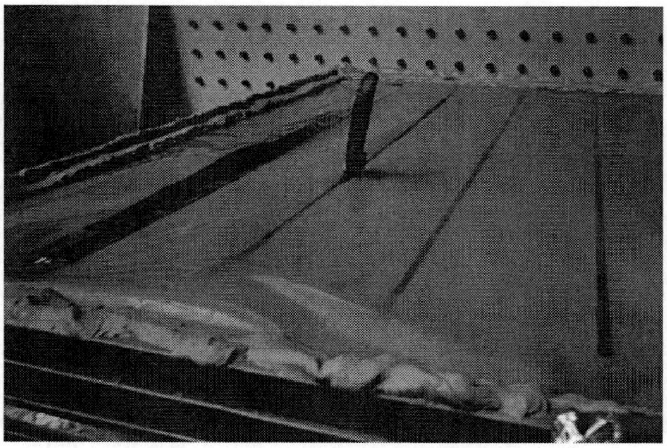

Figure 14 Test setup for verification of the reference measurements and determination of the air permeability per meter adhesive tape.

The International Energy Conservation Code also gives area-related limit values for the leakage through materials (0.02 $l/(s \cdot m^2)$ at 75 Pa), through assemblies (0.2 $l/(s \cdot m^2)$ at 75 Pa), and for the building envelope (2.0 $l/(s \cdot m^2)$ at 75 Pa) (IECC 2012). Taking the leakage rate through assemblies as a basis and assuming an air density of 1.2 kg/m^3 and a converting flow exponent of $n = 2/3$, the limit air mass flow results in $3.95 \cdot 10^{-5}$ kg/(s m^2 $(Pa^{2/3})$) at 5 Pa pressure difference. Compared with the mentioned limit for connection joints, this

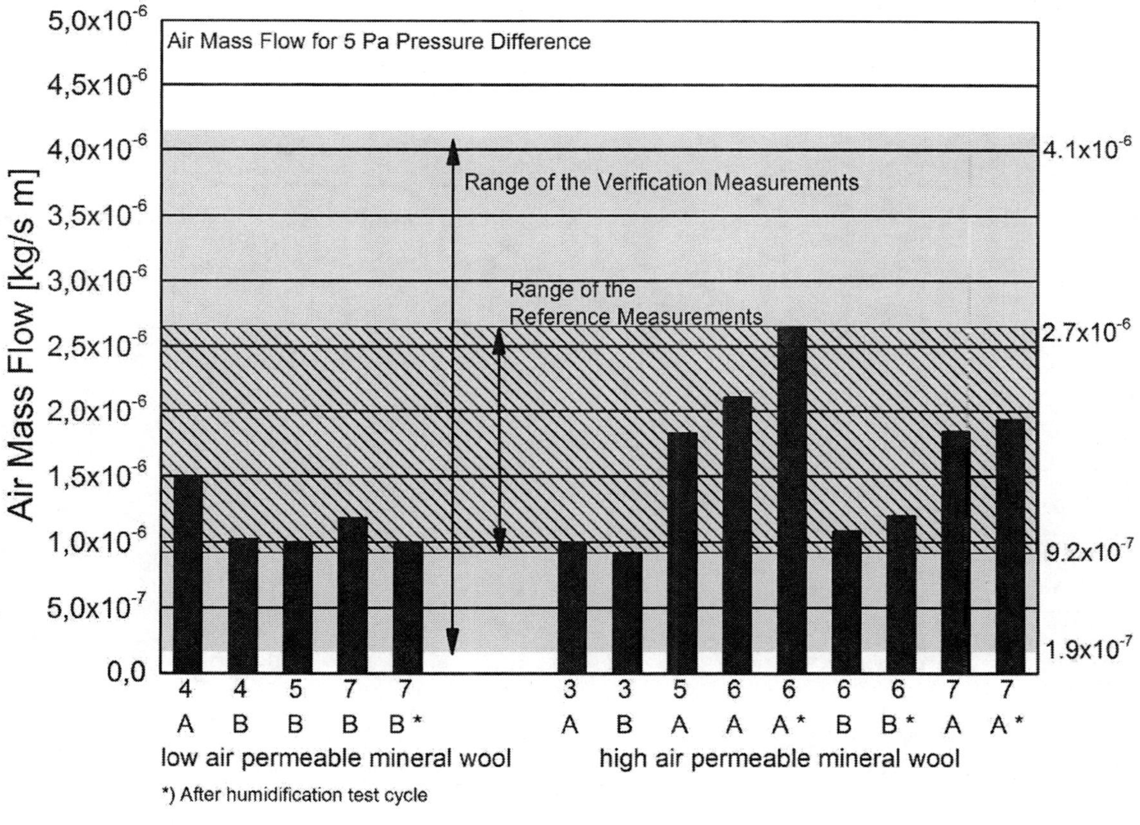

Figure 15 Air mass flow of the reference and the verification measurements per meter adhesive tape for 5 Pa pressure difference.

would result in approximately 2 m connection joints per square meter. This value seems to be in a realistic order of magnitude, but at the construction site this value varies as the case arises. Nevertheless the air mass flows through the sealing tape leakages are in a range, which is 5 to 22 times lower than the length related limits but not completely airtight. Regarding to energy losses due to exfiltration, the adhesive tape connections seem to be sufficiently airtight, but they may still contribute to convective moisture accumulation.

Weighed Moisture Contents. After the humidification test phase the distribution of condensation, water in the mineral wool batts and the paper sheets is determined by weighing. The results are summarized in Table 6. In the case of the mineral wool with high air permeability, only 30% of the water condenses in the insulation and about 70% in the paper. In case of the mineral wool with low air permeability, the distribution is about 50%/50%.

Moisture Mapping. Based on the weighed water content of the paper sheets, it is possible to visualize the vertical moisture distribution. The sheets at the top and at the bottom are always a little bit smaller than the four pieces in the middle. Figure 16 and Figure 17 show the moisture distribution in percent, based on the determined water content of the paper sheets in each test field. Therefore, each column sums up to 100%. Figure 16 displays the distribution for the experiments with mineral wool of low air permeability; Figure 17 displays the one for the experiments with high air permeable mineral wool. Darker colors refer to higher moisture content values. It should be mentioned that the displayed distribution of water may additionally be influenced by gravity.

As already mentioned, the closed assemblies without intended leakages, No. I and No. VII (Experiment 1A and 2A) are also not completely tight. In both cases approximately 30 g of water was detected after the experiments. The moisture

Table 6. Moisture Content by Weight

Experiment	Test Field	Mineral Wool	Target Pressure, Pa	Time, h	Water Content Paper, g	Water Content Mineral Wool, g	Water Content in Total, g
1	A	High air permeability	15	772	12.8	16.2	29.0
	B	High air permeability	15	772	193.9	150.1	344.0
2	A	Low air permeability	5	359	8.6	22.2	30.8
	B	Low air permeability	5	359	43.4	48.0	91.4
3	A	High air permeability	5	474	66.0	33.9	99.9
	B	High air permeability	5	474	106.9	42.3	149.2
4	A	Low air permeability	5	498	54.7	34.5	89.24
	B	Low air permeability	5	498	40.2	36.9	77.1
5	A	High air permeability	15	519	185.6	81.7	267.3
	B	Low air permeability	15	519	118.6	124.1	242.7
6	A	High air permeability	5	496	159.6	50.5	210.1
	B	High air permeability	5	496	127.3	46.3	173.6
7	A	High air permeability	15	521	160.6	60.8	221.4
	B	Low air permeability	15	521	121.9	123.8	245.7

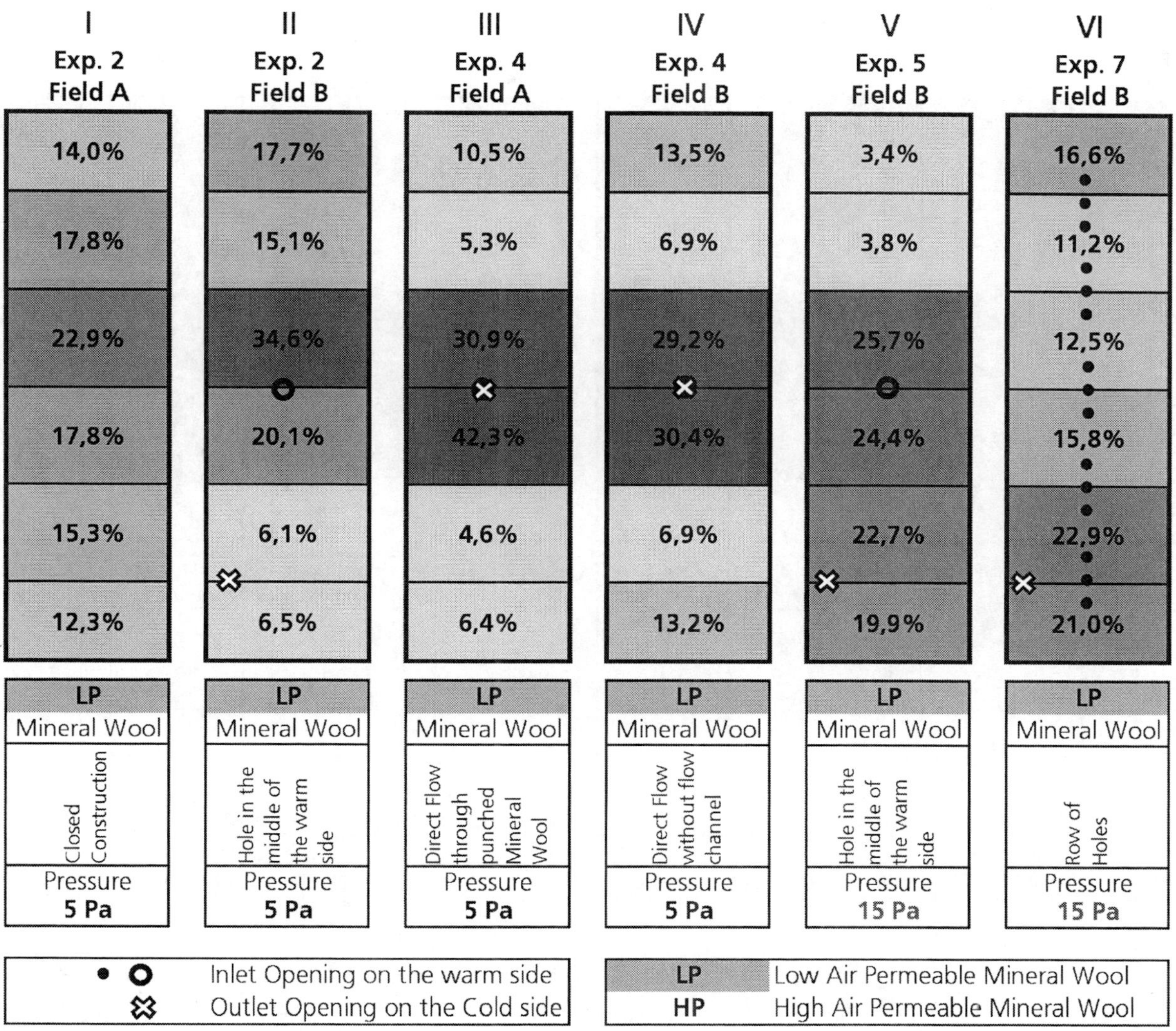

Figure 16 *Moisture distribution in the paper sheets for the experiments with low air permeable mineral wool (view from the cold side).*

distributions inside these fields are nearly homogeneous, as you can see in Figure 9 and Figure 10. Experiment 4 with the direct flow path, No. III and No. IV, shows a strong moisture concentration around the outlet holes in both cases. In practice, such leakages would lead to very high local moisture contents in the exterior sheathing.

Comparing the moisture distribution of No. II and No. V (Experiment 2 Field B and Experiment 5 Field B) with the same mineral wool with low air permeability and the same flow arrangement (intended inlet hole in the middle of the warm field) but triple pressure difference, it is visible that in both cases the highest moisture contents occur opposite of the intended inlet holes. In the case of 5 Pa pressure difference, the moisture seems to move slightly upward inside the construction, while the upper fields at 15 Pa remain dry. The same hole arrangement with the same pressure difference of 15 Pa but the high air permeable mineral wool is tested in No. 8 (Experiment 1, Field B). Considering this and No. V (Experiment 5, Field B), a similar effect of lower water content in the upper paper sheets can be observed. Compared to No. V, No. VIII does not show such a distinct moisture distribution, which may be caused by the longer time span (250 h) of No. VIII (Experiment 1).

The two different mineral wool air permeabilities can be compared at the same pressure difference of 5 Pa in No. II and No. IX (Experiment 2 Field B and Experiment 3 Field A). Here, in both cases, the moisture accumulates in the middle and top paper sheets. The bottom fields only show low water contents. At No. IX the highest moisture contents occur in the upper section while in the experiment with the low air permeable mineral wool No. II, the moisture concentrates more in the middle section. This slightly deviant behavior may be caused by the different flow resistance of the mineral wool samples. In summary, the moisture distribution in these four

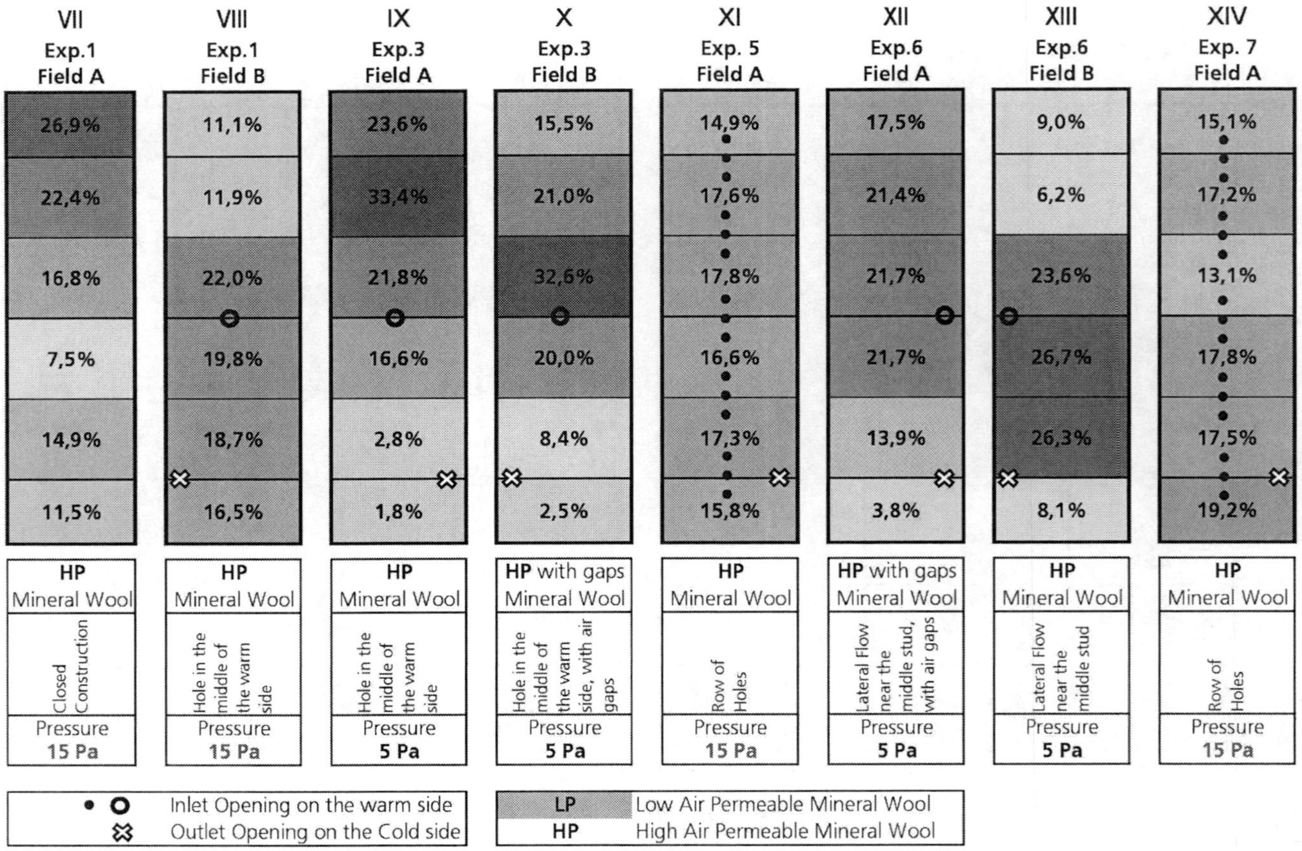

Figure 17 *Moisture distribution in the paper sheets for the experiments with high air permeable mineral wool (view from the cold side).*

experiments seems to be more influenced by the pressure difference than by the flow resistance of the mineral wool. The tendency is that at higher pressure differences, the moisture spreads more in the direction of the outlet hole (here towards the lower paper sheets) and a lower pressure difference leads more to an upward spread of the moisture. This finding could lead to the conclusion that buoyancy of moist air inside the insulation prevails in the case of very low pressure differences, leading to more condensation at the top of the constructions.

In Experiment 3 (No. IX and No. X), we investigate the influence of recesses at the edges of the mineral wool samples. Comparing Field A and B of Experiment 3, it becomes visible that with edge recesses (No. X), the moisture concentrates more in the middle sections. It is assumed that the compression of the mineral wool at the edges leads to a significantly higher flow resistance at these positions. As far as it is visible in the vertically arranged paper sheets, the increasing flow resistance towards the edges seems to also affect the moisture accumulation.

Experiment 6 (No. XII and No. XIII) compares the same arrangement with and without edge recess at a pressure difference of 5 Pa but with different positioning of the intended inlet leakages. Here the inlet leakages are close to the stud, so in No. XII (Experiment 6 Field A) the air can flow directly into the air gap formed by the edge recess. The air can also freely spread inside the circumferential gaps on the warm side. This seems to lead to a more homogeneous moisture distribution in comparison to Field B (No. XIII), where the moisture concentrates in the paper sheets between the inlet and outlet hole.

Experiment 5 Field A (No. XI) as well as Experiment 7 Field A (No. XIV) and B (No. VI) are performed with a row of 1 mm holes instead of one single hole with a 5 mm diameter. All experiments are performed with 15 Pa pressure difference and show a very homogeneous moisture distribution in the paper sheets. Only at No. VI (Experiment 7 Field B), with low air permeable mineral wool, is a slightly higher moisture content in the paper sheets near the bottom is recognized.

Hourly Condensed Water. All performed experiments have different durations. Therefore a comparison of the absolute weighed water content requires a time-dependent correction. Figure 18 shows the hourly condensed water rate in gram of water per hour based on the absolute weighed water content after the humidification tests and the experiment durations. The results are lumped together for the different mineral wool air permeabilities. The experiments with 15 Pa pressure difference are displayed as black solid columns and those with 5 Pa as grey columns.

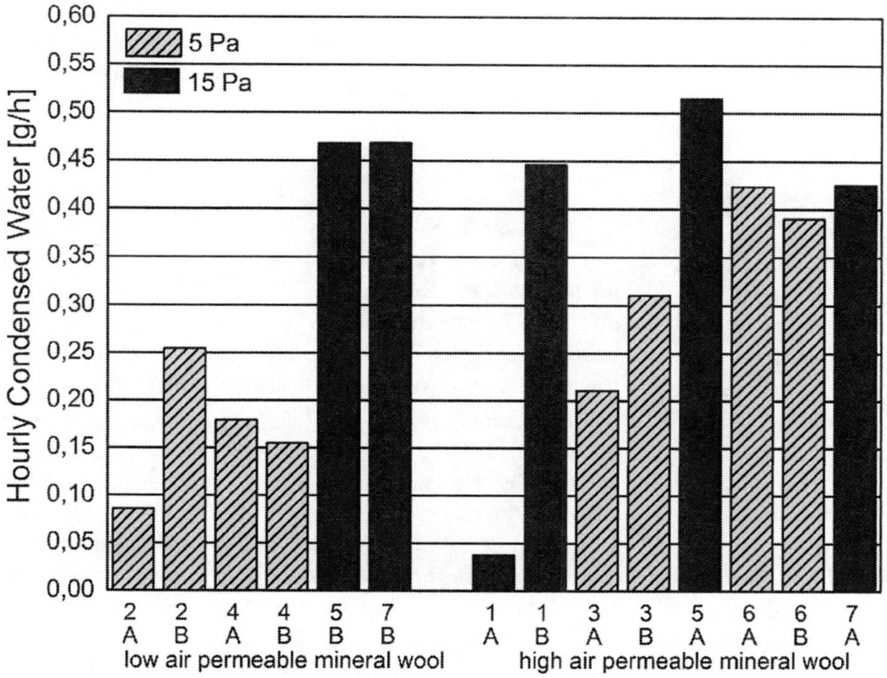

2A	Closed Assembly
2B	Flow from the middle of the warm side without edge recess
4A	Direct Flow with flow channel
4B	Direct Flow without flow channel
5B	Flow from the middle of the warm side without edge recess
7B	25 holes in the middle of the warm side
1A	Closed Assembly
1B	Flow from the middle of the warm side without edge recess
3A	Flow from the middle of the warm side without edge recess
3B	Flow from the middle of the warm side with edge recess
5A	25 holes in the middle of the warm side
6A	Lateral Flow near the middle stud with edge recess
6B	Lateral Flow near the middle stud without edge recess
7A	25 holes in the middle of the warm side

Figure 18 *Hourly condensed water based on the complete weighed water content inside the construction.*

With 0.515 g/h, the highest hourly condensed water accumulation can be observed at Experiment 5 Field A, with the row of holes. The repetition of this test setup in Experiment 7 Field A shows, with a moisture flow of 0.425 g/h, a slightly lower value, despite the higher permeability (compare Table 5). The difference is likely caused by the slightly higher pressure difference (~+0.44 Pa) and absolute humidity (+0.16 g/kg) at the warm side of Experiment 5 (compare Table 3).

Experiment 6 shows, in both fields, the highest hourly condensed water rate of the experiments performed with 5 Pa pressure difference. As described, the interaction of the high flow coefficient, the turbulent flow regime, and the low pressure difference leads to a high air mass flow and therefore to a high moisture entry in these both cases. The reason for this behavior could be found in a special airflow boundary effect reported for mineral wool insulation by Harreither et al (2015). Harreither et al. tested the airflow through mineral wool insulation with two different densities pressed into a pipe. From the results they concluded that the specific flow resistance at the edge region of the mineral wool insulation is about five to six times lower than in the core area. Weißinger had done similar experiments with different mineral wool densities before and received similar results (Weißinger 2012). In our Experiment 6, the intended leakages on the warm side are 18 mm beside the studs and, thus, are located close to the edge area where reduced flow resistance would be expected. Here, this effect leads to a hourly condensed water accumulation at a pressure difference of 5 Pa, which is nearly as high as in the cases of 15 Pa pressure difference without edge effect. Disregarding Experiment 6 and Experiment 2 Field A with the closed intended leakages, the hourly condensed water rate in case of 5 Pa pressure difference lies between 0.15 and 0.3 g/h. A slightly higher water accumulation can be observed in Experiment 3 with the high air permeable mineral wool. But here the differences are too small to draw a conclusion. At a pressure difference of 15 Pa all experiments with opened intended leakages are in a narrow range between 0.43 and 0.52 g/h. Here, at 15 Pa, it seems that neither the air permeability of the mineral wool nor the arrangements of the openings have a strong influence on the total condensation rate.

COMPARISON MASS BALANCE BY WEIGHT AND BY FLUXES

Calculation Method—Mass Balance by Fluxes. Theoretically, the moisture accumulation inside the construction can be calculated based on the difference of the absolute air humidity, the air mass flow, and the time span of the experiment (Equation 2):

$$m_w = \dot{m}_a \cdot \Delta x \cdot t \text{ [g]} \tag{2}$$

The air humidity is recorded on the warm side of the chamber and inside the outlet nozzle on the cold side. The difference is the amount of condensed water, which remained in the assembly. However, the airflow reference measurements show also that the setups with closed intended leakages are not completely airtight. Due to this fact, the calculation with the measured air mass flow through the intended leakages leads to deviating results. There occurs some side flow that escapes this calculation approach.

Therefore, an attempt is made to estimate the side flow by considering the airflow of the reference tests with sealed intended openings. During the airflow test phase, a negative pressure acted directly on the outlet nozzles; thus, the detected airflow through the unintended leaks in the tape sealing consists of air entering the construction from the cold and from the warm side of the chamber. During the humidification experiment, a negative pressure acted on the whole cold side of the chamber, drawing air from the warm side into the assembly. This air leaves the assembly by the intended outlet hole and the leaks in the tape sealing on the cold side. The principle behavior is displayed in Figure 19a for the case of the airflow test phase and Figure 19b for the humidification test phase.

For that reason, the measured reference air mass flow is split up evenly for the calculation. During the humidification test phase, only air from the warm side contributes to the moisture entry, while the airflow on the cold side is directed outwards (see Figure 19 [b]). Furthermore, the pressure field changes between the air mass flow measurement phase and the humidification test phase, which additionally affects the flow resistance. Because an accurate quantification of the leak air from both sides is difficult, a simple split of 50%/50% is assumed. However, as the total joint length is the same on both sides, this assumption seems rather reasonable but is, of course, afflicted with uncertainties. In the following considerations, the condensed water content is calculated according to Equation 3 based on the recorded data and the time interval of ten minutes.

$$m_w = \sum(\dot{m}_a - 0.5 \cdot \dot{m}_{a,ref}) \cdot \Delta x \cdot \Delta t \text{ [g]} \quad (3)$$

Results. Table 7 shows the weighed and calculated condensed water content as well as the deviation for the Experiments 3 to 7. For Experiments 1 and 2 we did not perform any reference measurements of the constructions with closed intended leakages. All calculated values are based on the air flow results prior to the humidification experiments.

Except for Experiment 3 Field A, Experiment 4 and Experiment 6 Field B, the deviation between water content balances obtained by weighing and that by integrating the measured moisture fluxes lies in a narrow range between –0.8 % and 12 %. Along with the assumed 50-50 leakage distribution, the main reason for the deviation between the weighed and the calculated water contents can be seen in the changing air mass flow during the experiment. As can be seen in Table 5, the flow coefficient and flow exponent vary during the experiments, and, therefore, the flow resistance also varies (compare to Figure 13). Both fields of Experiment 7 show a very low deviation within ±1.0%; in this case, the 50% approach seems to fit perfectly.

Comparison of the Moisture Shares. As shown in Table 7, the described approach leads in most cases to acceptable agreements of weight measurement and flux integration. Conse-

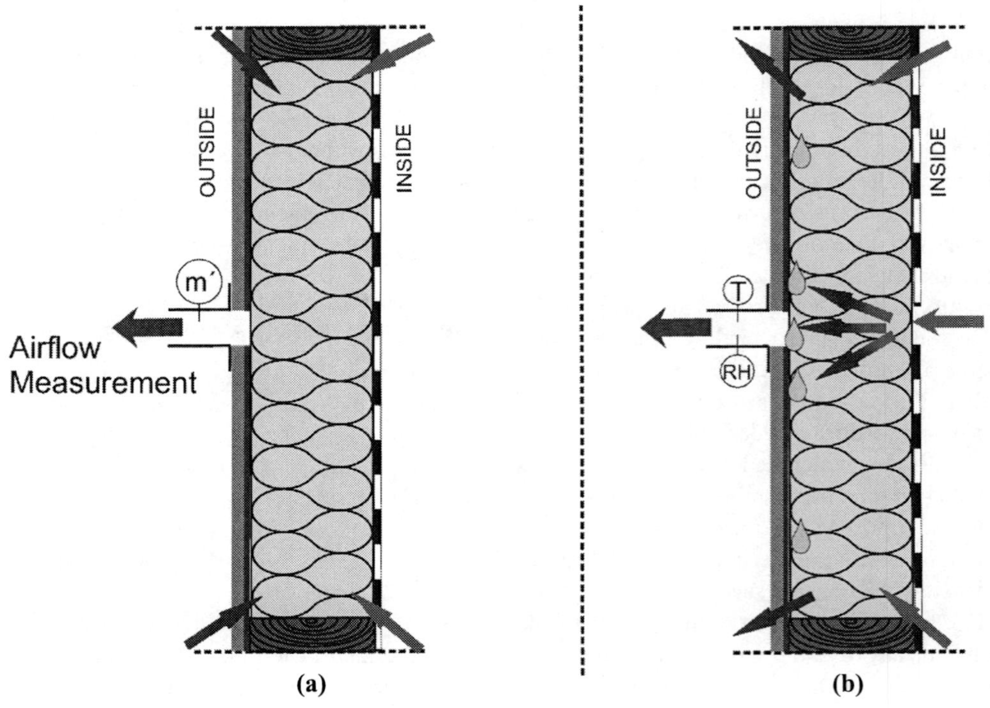

Figure 19 *Airflow through the assembly during (a) the airflow test phase and (b) the humidification test phase.*

Table 7. Mass Balance by Weight and by Fluxes

Experiment	Test Field	Mineral Wool	Target Pressure, Pa	Time, h	Water Content by Weight, g	Water Content by Fluxes, g	Deviation, %
1	A	High air permeability	15	772	29.0	No reference measurement	
	B	High air permeability	15	772	344.0	No reference measurement	
2	A	Low air permeability	5	359	30.8	No reference measurement	
	B	Low air permeability	5	359	91.4	No reference measurement	
3	A	High air permeability	5	474	99.9	185	85,19%
	B	High air permeability	5	474	149.2	163	9,25%
4	A	Low air permeability	5	498	89.24	103,4	15,87%
	B	Low air permeability	5	498	77.1	54,8	–28,92%
5	A	High air permeability	15	519	267.3	297,5	11,30%
	B	Low air permeability	15	519	242.7	254,4	4,82%
6	A	High air permeability	5	496	210.1	234,7	11,71%
	B	High air permeability	5	496	173.6	243,4	40,21%
7	A	High air permeability	15	521	221.4	220,12	–0,58%
	B	Low air permeability	15	521	245.7	243,85	–0,75%

quently, this means that the leak air, entering through the tape sealing of the assembly on the warm side, adds to the detected amount of condensed water. While certainly being of considerable importance in practice, the effects of unintended leak air through the tape sealing should be neutralized in order to compare the impacts of the intended leakages. Based on the weighed water content and the air mass flow calculated by the pressure difference and the determined coefficients and exponents, it is possible to divide the moisture part entering the construction through the intended openings and the part entering through the tape sealing at the warm side. The share of the water entering through the tape sealing is calculated by Equation 4.

$$m_{w,s} = 0.5 \cdot \sum C_{ref.} \cdot \Delta P_e^{n,ref.} \cdot \Delta x \cdot \Delta t \text{ [g]} \quad (4)$$

Figure 20 shows the shares of moisture entering the construction through intended leakage openings and through the tape sealing. The values are based on the weighed water content and the individual experiment conditions. The black columns show the share through the intended openings and the gray ones the share through the tape sealings. Experiment 7 shows within ±1.0% the best correlation between the weighed and calculated moisture content (compare Table 7). In both fields of Experiment 7 with the row-of-holes setup, the distribution is about 50% through the openings and 50% through the sealing tape. In all other cases, the shares of the moisture through the sealing tape vary between 25% and 70%. In all performed experiments, a large amount of condensed water can be assigned to the tape seams. On average 50% of the detected condensed water inside the test assemblies is caused by this unintended airflow.

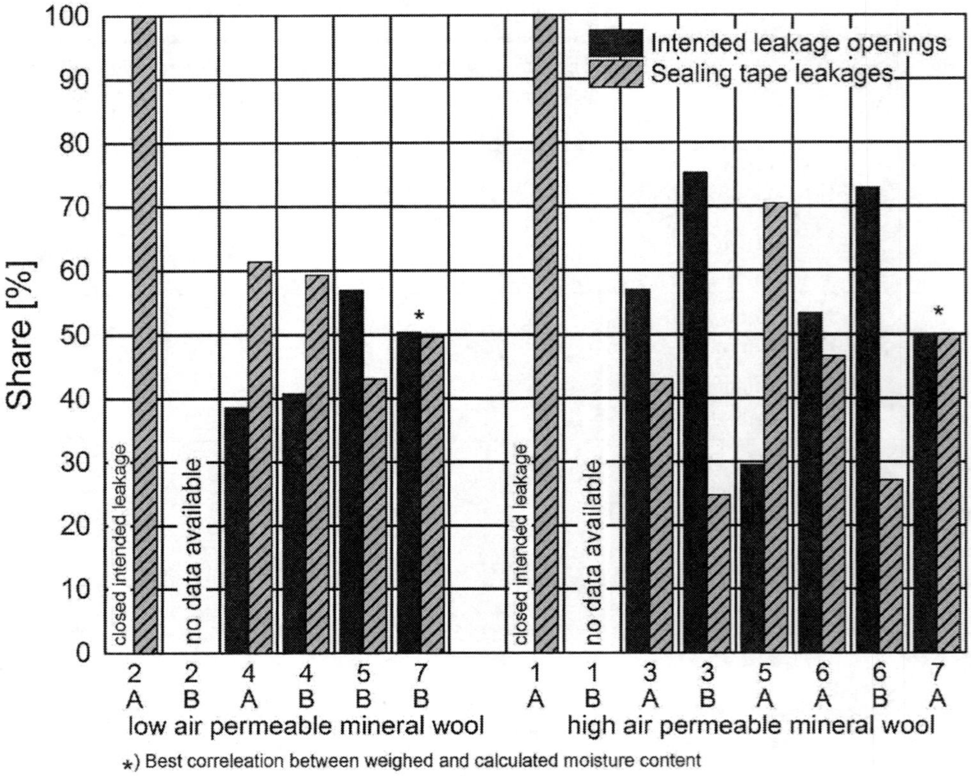

Figure 20 *Percentage share of the moisture entering the construction under lab conditions.*

Influence of the Setup Arrangement. In contrast to the different shares of the moisture entries, the evaluation of the impacts from the intended opening arrangements and the mineral wool permeabilities is more challenging. For a simple comparison of the values, neglecting the different pressure stages, the area-related specific hourly condensed water rate is calculated and presented for 5 Pa pressure difference. The calculation approach is based on the flow exponents of the air mass flow. For the flow through the sealing tape (Equation 6), the flow exponent of the reference measurement is used. For the calculation of the moisture flow through the intended openings (Equation 7) the resulting flow exponent, $n_{,res}$ is used. This flow exponent is based on the measurement with open leakages minus 50% of the reference measurement. All exponents are based on the airflow measurements performed before the humidification experiment starts (Table 5, Flow Exponent n before Experiment). The mass of water entering the construction through the intended holes is calculated by Equation 5 based on the total weighed mass of water and detected after the humidification test phase and the calculated share through the taped seams (Equation 4).

$$m_{w,h} = m_{weig.,tot} - m_{w,s} \text{ [g]} \quad (5)$$

$$\dot{m}_{w,s} = \frac{m_{w,s}}{A \cdot t_e \cdot \overline{\Delta P_e}^{n,ref}} \cdot 5^{n,ref}\left[\frac{\text{g}}{\text{h} \cdot \text{m}^2}\right] \quad (6)$$

$$\dot{m}_{w,h} = \frac{m_{w,h}}{A \cdot t_e \cdot \overline{\Delta P_e}^{n,res}} \cdot 5^{n,res}\left[\frac{\text{g}}{\text{h} \cdot \text{m}^2}\right] \quad (7)$$

The specific hourly condensed water rate of the closed constructions (Experiment 1 Field A and Experiment 2 Field A) is reduced by a flow exponent of $n = 0.8$. This is done because no reference measurements were performed for these experiments and all the other reference measurements show a flow exponent mean value of 0.8.

Figure 21 shows the specific hourly condensed water rate through the intended openings for 5 Pa as black columns. The gray area indicates the range between the minimum and maximum. Figure 22 shows the specific hourly condensed water rate for 5 Pa through the tape sealing as gray columns. The two experiments with closed intended leakages are displayed as white columns. The condensed water rate at Experiment 2 Field A (closed intended leakages) is in a similar order of magnitude as the rate through the sealing tape from most of the other experiments. This is evidence that the made approach is not that far from reality. In comparing the two figures it becomes visible that the hourly condensed water rate through the sealings is nearly in the same order of magnitude as the hourly condensed water rate caused by the intended openings. Due to the high deviation between the weighed and

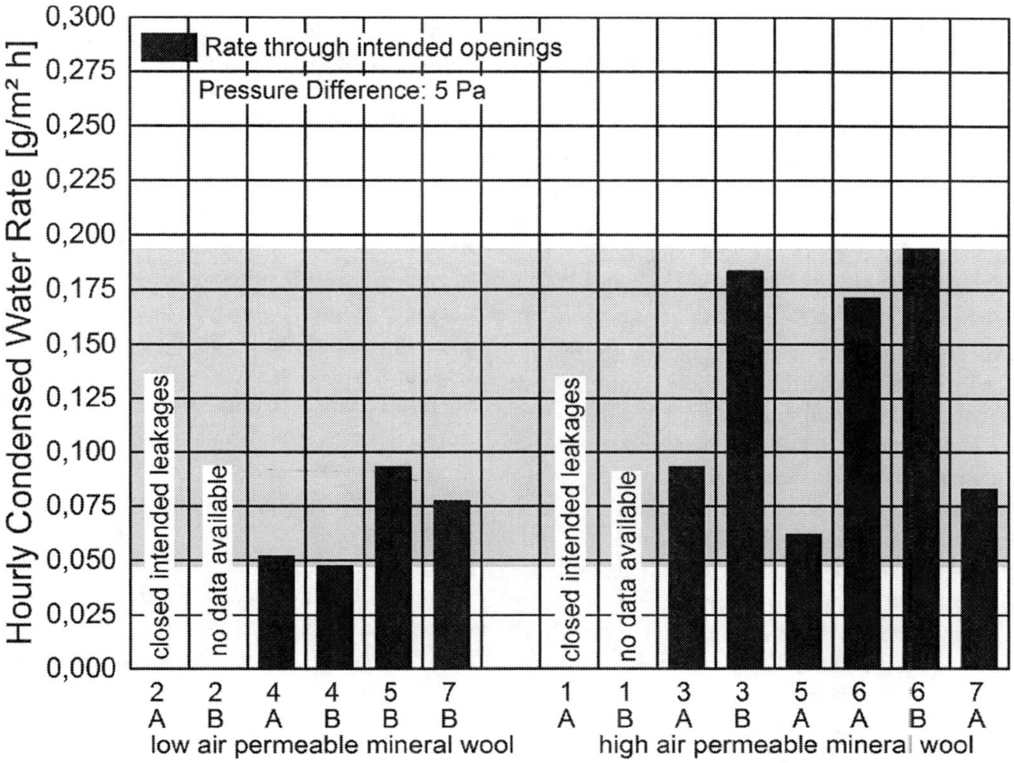

Figure 21 *Hourly condensed water through the intended openings for 5 Pa pressure difference.*

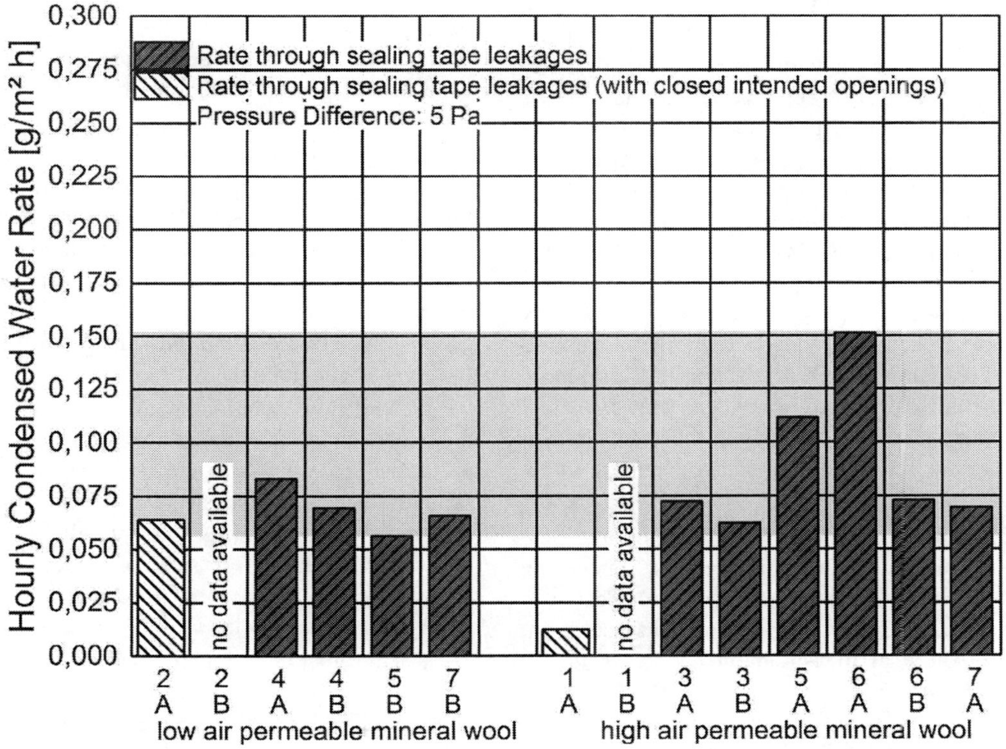

Figure 22 *Hourly condensed water rate through the sealings for 5 Pa pressure difference.*

calculated condensed water content (compare Table 7), especially in Experiment 3 Field A (but also in Experiment 4 Field A and Experiment 6 Field B), these results should be treated with caution.

Disregarding previously named experiments, the highest hourly condensed water rate through the intended openings can be found at Experiment 3 Field B and Experiment 6 Field A; both have mineral wool of high air permeability. Here the flow resistance of the mineral wool and maybe the described boundary effect influence the airflow and, therefore, the hourly condensed water rate.

The lowermost hourly condensed water rate caused by the openings can be observed at Experiment 4 Field B (direct flow without flow channel). One reason for that can be found in the strong moisture accumulation in the region around the openings. This concentrated water content on the flow path in the mineral wool leads to an 18% increase of the flow resistance over the experiment (compare Figure 13). At the constant pressure difference, the air mass flow decrease. This leads to a lower condensation rate below and, in consequence, a lower amount of water caused by the intended opening at this flow arrangement. An increase of the outlet air temperature at the outlet nozzle could not be observed either in Field B or in Field A. Both temperatures remain constant during the experiment so that in the investigated case a warming up of the flow channel can be excluded.

For the row-of-holes Experiment 5 Field A and Experiment 7 Field A and B, a higher hourly condensed water rate through the twenty-five 1 mm holes than through the intended holes with 5 mm diameter cannot be observed. Here, the diameter difference in correlation with the pressure difference seems too small, so that other influences and uncertainties overlap the estimated effect. But the fact that a high hourly condensed water rate occurs through the sealing tape is another hint of this behavior. At 5 Pa pressure difference, the hourly condensed water range through the intended openings is between 0.048 and 0.194 g/(m^{2}h), and the one through the sealing tape leakages between 0.056 and 0.152 g/(m^{2}h). These invisible leakages in the adhesive tape can only have a very small leakage, but it seems that they lead during the humidification tests to approximately 50% of the total moisture entry. Hence, the moisture entry rate through the approximately 6.3 m adhesive tape junctions is nearly in the same order of magnitude as the one through the intended openings with a leakage area of ~20 mm^2.

Conclusions. The present study provides some new insights on the possible influencing factors concerning airflows and, therefore, the convective moisture entry to lightweight constructions with mineral wool batt insulation. Despite a high degree of measurement device accuracy, there are factors that influence the results. Most of these influences are hard to quantify, starting with the used mineral wool and its hydrophobic treatment, which may have influence on the moisture distribution and water uptake. In addition, the samples are a bit strained after multiple installations and humidification, which can have influence on airflow rate, moisture distribution, and water content. Another thing to mention is the behavior of the adhesive tape connections during the humidification test phase. The verification measurements showed some delamination after a short test period, which surely occurs also during the humidification test phase. Concerning only this factor, the reference measurements give a possible range of uncertainty of approximately ±30%. Moreover, it would have been beneficial to repeat some of the tests in order to check the reproducibility, especially of the humidification experiments, but this was not possible within the frame of this project. However, with the experience from the performed lab test campaign, we do not see further accuracy improvement possibilities when using building assemblies close to reality.

Nevertheless, the study shows that in assemblies with low density batt insulation, most of the water condenses at the sheathing. In assemblies with mineral wool of higher density, a considerable amount of moisture is retained by the insulation material. Even in case of a direct flow channel through the mineral wool, the warming up of the flow path was not sufficient to reduce condensation. On the contrary, such a situation ends up in significant moisture concentration around the outlet opening in the sheathing, hence leading to an even more critical situation.

Furthermore, the experiments indicate that the vertical moisture distribution in the exterior sheathing is more influenced by the pressure difference than by the mineral wool density or the arrangement of the intended inlet leakage holes. With regard to the pressure fluctuations under climatic field conditions, a general prediction of critical positions is therefore rather difficult.

The experiments with edge recess of the mineral wool by the studs show no significant influence of this recess on the total condensed water rate. However, depending on the inlet opening position, the moisture distribution at the exterior sheathing can be influenced. An airflow directly into the air gap, formed by the recess of the edges, leads to a more homogeneous moisture distribution similar to the cases with the twenty-five smaller holes.

Lateral flow situations with the intended openings near the studs show the highest moisture entry rates of all performed lab tests. It is assumed that the flow resistance in the insulation at the interfaces between stud and insulation edges is somewhat reduced. This boundary effect seems to be more pronounced at lower pressure differences. Thus, imperfections in the air barrier close to the studs may be regarded as more critical than others.

For the other flow situations with 5 Pa pressure difference and the climatic conditions during the humidification test phase (+22°C and 50% rh on the warm and +2°C on the cold side), the weight-based, total hourly condensed water rate varies between 0.11 and 0.22 g/(m^{2}h). At a pressure difference of 15 Pa all experiments with open intended leakages remain in a narrow range between 0.32 and 0.39 g/(m^{2}h). Here, neither the mineral wool air permeability nor the arrangements of the openings seem to have a strong influence on the total condensed water rate.

Under the presented test conditions all moisture carried by the exfiltration air mass flow actually condensates inside the assembly. Size and arrangement of the intended inlet openings affect the airflow through the assembly and the vertical distribution of the condensed water, while the mineral wool permeability affects, in addition, the horizontal distribution over insulation layer and sheathing.

Nevertheless the most important factor is the pressure differential over the assembly. It mainly influences the flow regime prevailing at the leakages and, thus, in interaction with the flow regime the total air mass flow and the total condensation rate, as well as the moisture distribution at the cold side sheathing.

Another important finding of the lab tests is the fact that a significant fraction of the total moisture entry must be assigned to the taped joints of the test assembly. This confirms the assumption made in the beginning, that an "assembly will leak more air than the original material being used as the basic material" (Wagdy 2015). However, the factor of 10, given by the IECC (IECC 2012), cannot be confirmed. The airflow might be caused by very small, invisible leakages leading to a more laminar flow regime and, therefore, also to a higher area related air mass flow than at the intended leakage openings. Verification lab measurements on a totally different test setup confirmed that even well-executed adhesive tape connections are not totally airtight. Concerning energy aspects, the achieved airtightness level of the adhesive tape connections meets all possible requirements of different standards. Nevertheless, the air leakage rate through the tape connections (in the range between $1.9 \cdot 10^{-7}$ up to $4.1 \cdot 10^{-6}$ kg/(s m)) seems to be responsible for a considerable fraction of the total convective moisture entry. Furthermore, the tightness level from the lab will hardly be achievable on the construction site in practice, as already one tiny single imperfection can increase the permeability by a factor of 100.

With regard to the airtightness quality of the envelope, even better workmanship is required to limit the total convective moisture entry by eliminating at least all preventable leakages. It is also important to note that a higher pressure differential does not increase the amount of condensate proportionally. The experiments indicate that in the case of the sealing tape leakages, the flow exponent n is around 0.8. This value is between the approaches made by Nusser (2012) and Zirkelbach et al. (2009). Nusser assumed a flow exponent of $n = 2/3$ ($n \sim 0.66$), and Zirkelbach took a linear relationship as a basis, hence a flow exponent of $n = 1.0$.

Based on all these findings, the existing convection models for the use in hygrothermal simulation tools should be adapted accordingly. Additional experiments would be useful to clarify the presented relations more in detail. The airflow through the tape-sealed joints and other small imperfections especially should be further analyzed. In addition, not only the convective moisture entry but also the convective drying should be investigated to draw conclusions about the overall moisture balance in building elements caused by air convection.

NOMENCLATURE

A = mrea

C = flow coefficient

m = mass

n = flow exponent

x = absolute humidity

$\dot{m}$ = air mass flow

Subscripts

a = air

e = experiment

h = holes

ref = reference

res = resulting

s = sealing

tot = total

w = water

$weig$ = weighed

REFERENCES

Desmarais, G. 2000. Impact of added insulation on the hygrotherrnal performance of leaky exterior wall assemblies. Master's Thesis, Concordia University Montreal, Quebec, Canada.

DIN. 1991. DIN EN 29053: 1993-05: Materialien für akustische Anwendungen; Bestimmung des Strömungswiderstandes (ISO 9053:1991) (in German).

DIN. 2001. DIN EN 12667: 2001-05: Wärmetechnisches Verhalten von Baustoffen und Bauprodukten - Bestimmung des Wärmedurchlasswiderstandes nach dem Verfahren mit dem Plattengerät und dem Wärmestrommessplatten-Gerät (in German).

DIN. 2013. DIN 4108-2: 2013-02: Wärmeschutz und Energie-Einsparung in Gebäuden - Teil 2: Mindestanforderungen an den Wärmeschutz (in German).

Geißler, A., and G. Hauser. 2002. Abschätzung des Risikopotentials infolge konvektiven Feuchtetransports. (in german) Abschlußbericht; Juli 2002; AIF–Forschungsvorhaben Nr. 12764. Universität Kassel, Fachgebiet Bauphysik. Kassel. Germany. 2002.

Harreither, C., L. Krainz, J. Weissinger, and T. Bednar. 2015. Experimental determination of the influence of installing fibrous insulation materials in timber frames on the air permeability and convective air transport. *6th International Building Physics Conference*, IBPC 2015. June 14–16, Turin, Italy 2015.

IECC. 2012. 2012 International Energy Conservation Code. International Code Council. Country Club Hill, IL.

Nusser, B. 2012. Flachgeneigte hölzerne Dachkonstruktionen; Systemanalyse und neue Ansätze zur Planung hygrisch robuster flachgeneigter hölzerner Dachkonstruktionen unter Betrachtung konvektiver Feuchteeinträge und temporärer Beschattungssituationen. (in german) Dissertation. Vienna University of Technology. Vienna, Austria 2012. 33–40.

Riesner, K. 2009. Vermeidungsstrategien für Tauwasser- und Schimmelpilzrisiken in Außenwandgefachen, verursacht durch natürliche Konvektion in der Dämmung. (in german) Abschlussbericht Bau- und Wohnforschung Nr.: F2547 Universität Rostock, Germany 2009. Published by Fraunhofer IRB Verlag, Stuttgart, Germany 2010.

Schmidt, J. 2014. Konvektiver Feuchtetransport durch Leckagen in Holzleichtbaukonstruktionen mit permeablen Dämmstoffen. (in German), Dissertation. University of Kaiserslautern. Kaiserslautern, Germany 2014.

Wagdy, A.F. 2015. Airtight buildings: Past, present and future. In: *ASHRAE Journal* 57 (2015) 11:40–50.

Weißinger, J. 2012. Dauerhaftigkeit von Flachdächern im Holzbau - Experimentelle Bestimmung des Strömungswiderstands im Randbereich fasriger Dämmstoffe und Analyse kritischer Zustände mit hygrothermischer Bauteilsimulation. (in German) Master's Thesis. Vienna University of Technology. Vienna, Austria 2012.

Zirkelbach, D., H.M. Künzel, B. Schafaczek, and R. Borsch-Laaks. 2009. Dampfkonvektion wird berechenbar; Instationäres Modell zur berücksichtigung von konvektivem Feuchteeintrag bei der Simulation von Leichtbaukonstruktionen. (in German) 4th Intern. Symposium on Building and Ductwork Air Tightness – BUILDAIR 30th AIVC Conference. October 1–2, Berlin, Germany 2009.

Simulating Air Leakage in Walls and Roofs Using Indoor and Outdoor Boundary Conditions

Simon Pallin, PhD, CEM **Diana Hun, PhD, PE** **Philip Boudreaux, MS**

ABSTRACT

Air leakage through the building envelope can result in large energy penalties but also have detrimental effects on the moisture durability of materials. Researchers have utilized computer simulations to estimate the risk of moisture damages due to air leakage through wall and roof assemblies. However, the accuracy of these simulations has limitations, as they typically employ a one-dimensional approach even though airflow-related problems are two or three dimensional. Furthermore, the flow rate and direction of the air that are used in these assessments are usually subjective values. To improve on previous research, this paper describes how to calibrate one-dimensional simulation tools through theoretical equations so that they can be used to study two-dimensional effects.

Further, the present work provides guidance on how to use indoor and outdoor boundary conditions to estimate the air's flow rate and direction using existing hygrothermal simulation tools. The proposed method will allow researchers and designers to drastically improve their assessments of the effects of air leakage on the moisture durability of wall and roof assemblies.

INTRODUCTION

Improved building codes have increased the energy efficiency of buildings in the United States. Some of these energy savings come from decreasing the air leakage through building envelope components such as walls and roofs. From an energy savings perspective this has been very beneficial. From a moisture durability issue, airtight walls and roofs need more consideration and study. Care needs to be taken to make sure that, as walls become more airtight (keeping the conditioned air where it should stay), the risk of condensation, mold, and material damage are minimized by thoughtful envelope design.

One of the main reasons for high-R and airtight walls to be more at risk for moisture damage is the increased amount of thermal resistance. This causes a cold side of the exterior wall (either the inside or the outside, depending on outdoor climate conditions) to be colder, thus making condensation or critical relative humidity levels more likely.

Air leakage is a major cause of energy leaks in a building because it causes the space conditioning equipment to consume more energy. However, air leakage may also cause moisture problems, referred to as *moisture leaks* (Künzel et al. 2011). These moisture leaks are usually distinguished by having a diffuse air leakage path rather than effectively passing through the building envelope, as for an energy leak. The most significant characteristic of a moisture leak is that it changes the moisture content of the materials inside the airflow path but does not meaningfully change the temperature of the materials.

One of the main variables that determine whether a leak is a moisture leak or an energy leak is the flow rate of the air. Slow flow rates tend toward moisture leaks, and faster flow rates tend toward energy leaks. By these definitions it can be seen that a wall with a very small equivalent leakage area (Sherman and Chan 2006) or one where the air undergoes a tortuous path to get through the wall would have a slow leakage rate and therefore a high probability for moisture leaks.

Simon Pallin, **Diana Hun**, and **Philip Boudreaux** are members of the Research and Development Staff of Oak Ridge National Laboratory, Oak Ridge, TN.

The relation between the interior and the exterior climate conditions determines where condensation might occur inside a wall assembly from air leakage. In heating-dominated climates, the temperatures of materials closer to the exterior will be closer to those of outdoor temperatures, and with high-R wall assemblies these materials are relatively colder. As humid indoor air exfiltrates along colder surface materials, there is a greater risk of condensation. In cooling-dominated climates, this phenomenon is reversed, where moist outdoor air may infiltrate through the wall, along a colder inside surface of the wall, which in this case may result in condensation and moisture damage at interior parts of the wall.

Consequently, airflow, interior conditions, exterior climate, and wall material properties all affect the moisture durability of a high-R wall assembly. In this paper, we focus on the airflow in the wall and how to best model it to reflect reality in order to assess the moisture performance of different wall assemblies. We discuss a novel approach to modeling one-dimensional (1-D) airflow in a way that matches the results of a two-dimensional (2-D) model, as well as provide computational guidance and aid for considering air leakage in improved hygrothermal design of high-R walls. This will also significantly increase computation speed for assessing different wall assemblies.

METHOD

The Correction Factor

WUFI® is a frequently used calculation tool designed to estimate the hygrothermal performance of building materials and elements of the building enclosure (Kehrer and Schmidt 2008). It accounts for relevant phenomena of heat and moisture transfer through materials as well as air leakages. Unfortunately, the accuracy of the embedded feature to simulate air leakage is questionable (Künzel et al. 2011; Lstiburek 2015). WUFI® treats air leakage as a source/sink of indoor or outdoor air that can be placed by the user at any given position inside or between the materials of the simulation model.

In Figure 1, the hygrothermal conditions of an air leakage path in an arbitrary cold-climate wall assembly are illustrated when indoor air exfiltrates from the interior to the exterior. The two separate charts presented in the figure depict how temperature and vapor pressure conditions (steady state) at different locations inside an air leakage path change at different airflow rates. By comparing any of the colored curves in the two charts, at any given location inside the air leakage path, it becomes obvious that the temperature conditions of the material adapts less efficiently than vapor pressure to the incoming air conditions. This phenomenon is a result of how successfully changes in temperature and vapor pressure propagate through the materials adjacent to the air leakage path. Typically, at any given airflow rate and in a relative manner, heat escapes more efficiently, through the surface materials of the air leakage path, than vapor pressure. Obviously, this phenomenon has an impact on relative humidity levels, which is further explained in the next section and presented in Figure 2.

As Figure 1 depicts, even at low airflow rates and independently of where inside the air leakage path we are studying the moisture conditions, these conditions will reach levels close to that of the inflowing air. Temperature conditions are usually harder to predict because they are more dependent on the airflow rate, the studied location, and the materials surrounding the air leakage path. Therefore, following is an approach to better predict the impact on temperature conditions inside an air leakage path in WUFI®.

In WUFI®, a steady-state heat balance, including air leakage as a source or sink, can be defined as follows:

$$T_{12}(I) = \frac{\left(T_1 + \frac{1}{K_1}\right) \cdot K_1 + T_2 \cdot K_2}{K_1 + K_2} \tag{1}$$

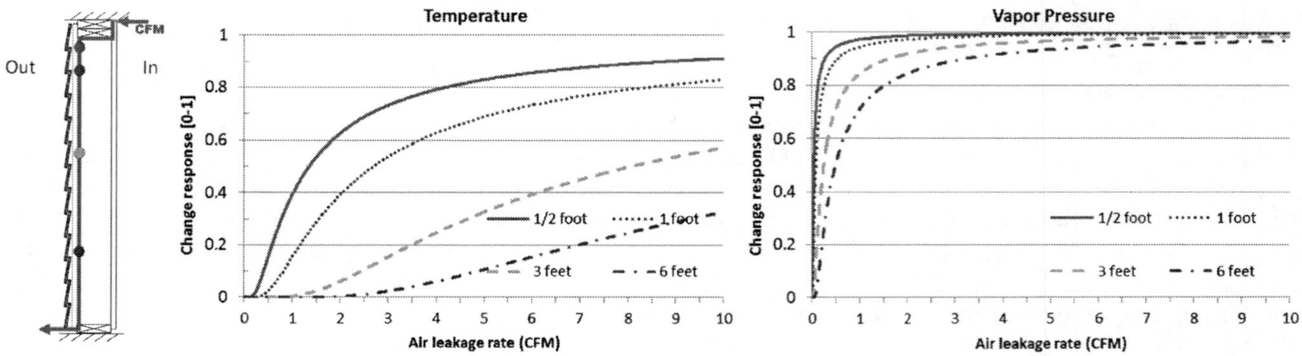

Figure 1 *Temperature and vapor pressure conditions of surface materials along an air leakage path with varying airflow rates from indoor air exfiltration. The adapting rate on the* y*-axis refers to how efficiently temperature and vapor pressure adapt to the conditions of the incoming air.*

$$I = \rho_{air} \cdot c_{air} \cdot (T_2 - T_{12}(I)) \cdot \dot{Q} \quad (2)$$

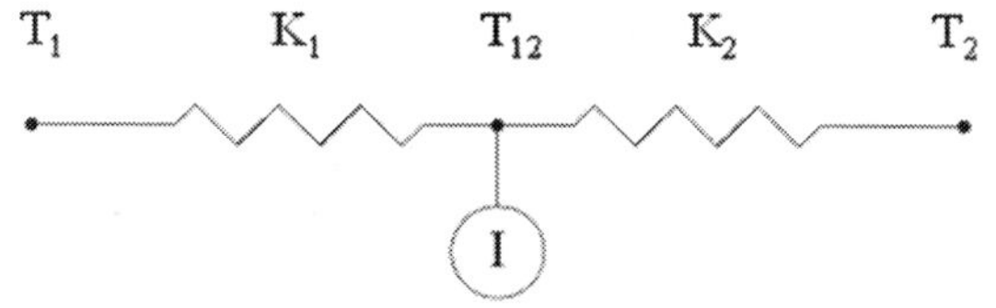

where

T	=	temperature, °F (K)
K	=	material thermal conductance, Btu/(h·°F) (W/K)
T_1, T_2, and T_{12}	=	the temperatures of the outdoor air, indoor air, and at an arbitrary point inside the building material, respectively
K_1 and K_2	=	thermal conductances of the materials between the temperature nodes

Air leakage is defined as a source/sink, I, and mainly depends on the temperature gradient between the reigning temperature conditions of where the source/sink is placed and the temperature conditions from which the air originates. It also depends on the airflow rate, $\dot{Q}$ (cfm [L/s]). Although with less varying dependency, I is also influenced by the volumetric heat capacity, $\rho_{air} \cdot c_{air}$ (Btu/(ft^3·°F) [J/(m^3·K)]).

In Equation 1, no consideration is taken to how and where air traveled to reach the location of where the source/sink is placed, thus representing a 1-D approach. In reality, though, air leakage is rarely a 1-D phenomenon, which makes this computational approach rather untrustworthy. However, an approach has been presented and validated (Pallin et al. 2015) that allows the embedded air leakage tool in WUFI® to be used with higher accuracy. In that work, the authors presented an approach to improve the accuracy of the air leakage model of WUFI® by allowing it to represent a 2-D airflow pattern. This approach is defined by allowing a 2-D solution of

$$T(x) = T_1 + (T_2 - T_1) \cdot e^{\left(\frac{-x}{l_c}\right)} \quad (3)$$

$$l_c = \frac{\rho_{air} \cdot c_{air} \cdot \dot{Q}}{L_{surf} \cdot K} \quad (4)$$

to equal that of the 1-D solution presented in Equation 1, under the following conditions:

$$T(x) = T_{12}(I_\eta) \quad (5)$$

$$I_\eta = \eta(x, \dot{Q}) \cdot I \quad (6)$$

In Equation 4, L_{surf} (ft [m]) is the surface length or the perimeter of the cross section through the air leakage path and K is the effective thermal conductance of the materials along the air leakage path (in this case $K_1 + K_2$). As seen in Equation 5, the condition is met by allowing the 1-D solution to be defined by a new source/sink, I_η, in which the correction factor, η, is a function of both the distance into the air leakage path and the airflow rate as seen in Equation 6.

The correction factor, η, allows Equation 5 to be valid and can be defined as

$$\eta(x, \dot{Q}) = \frac{\left(\frac{T(x) \cdot (K) - T_2 \cdot K_2}{K_1} - T_1\right) \cdot K_1}{\rho_{air} \cdot c_{air} \cdot (T_2 - T_{12}(I)) \cdot \dot{Q}} \quad (7)$$

Finally, the η can be used to determine which airflow rate to apply in the simulation model to obtain a result that better mimics realistic conditions. Because WUFI® uses air changes per hour (ACH) as an airflow rate, the corrected air leakage is defined as follows:

$$ACH_\eta = \eta(x, \dot{Q}) \cdot ACH \quad (8)$$

Time Optimization

The correction factor shown in Equation 8 allows simulation of 2-D air leakage in WUFI®. Depending on what the simulation model looks like and how and where the potential air leakage path is created, the hygrothermal performance of the materials surrounding an air leakage path can be evaluated. Unfortunately, analyzing the hygrothermal conditions of all the building materials along the air leakage path, and at every given location, is time consuming. Therefore, it becomes necessary to investigate whether the risk of moisture durability problems can be evaluated by analyzing a specific location of the air leakage path—a location along the path that is more prone to having higher levels of relative humidity.

By studying simulated relative humidity levels at different locations along an air leakage path, it is obvious that the relative humidity inside the leakage path depends on variables such as the airflow rate and where it is evaluated along the path, as seen in Figure 2. Naturally, it also depends on outdoor and indoor humidity conditions and the hygrothermal properties and conditions of the building material. The analysis presented in Figure 2 is made under reasonable indoor and outdoor conditions of a cold climate and reveals that no matter where inside the air leakage path the hygrothermal performance is evaluated, the relative humidity will peak at some given airflow rate under steady-state conditions. Interestingly, the peak is just as high for any location inside the air leakage path; only the air leakage rate determines when and where. For the benefit of optimizing simulation time, Figure 2 reveals that surface materials close to where air enters the air leakage path (solid and dotted) are of highest importance and that it is likely irrelevant to evaluate locations farther into the air leakage path since these peaks will occur at relatively higher airflow rates. As can be seen in Figure 4, air leakage rates above 1 cfm

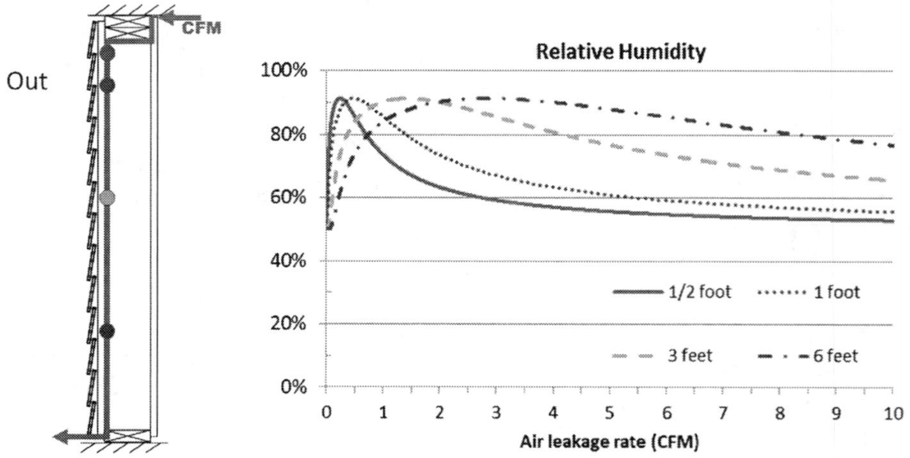

Figure 2 *Steady-state conditions for relative humidity at four locations along an arbitrary air leakage path with varying air leakage rates from indoor air exfiltration.*

(0.5 L/s) are rarely seen in moisture leaks, i.e., air leakages prone to cause moisture durability issues.

Because air leakage rates rarely exceed 1 cfm (0.5 L/s), locations close to where air enters an air leakage path are more likely to face moisture durability problems, if any. This means that analyzing the hygrothermal performance due to air leakage can sufficiently be executed within 1 ft (0.3 m) from the air's entry point and thus represent the design performance of the whole wall assembly. This conclusion is also beneficial for designers experiencing difficulties in predicting the physical air leakage path inside a wall. How and where the air travels beyond this critical point is therefore less relevant (Pallin et al. 2015). This paper relies on this analysis statement.

With a goal of optimizing computational effort, work has been done to evaluate whether a 1-D calculation tool actually can simulate 2-D air leakage using the presented method of this study. Because Figure 2 confirms that there is an area inside an air leakage path that is assumed most critical, it is of interest to investigate whether this area could be cut down to a 1-D strip that instead can be simulated in a 1-D tool. In order to confirm that we can utilize a 1-D tool to mimic 2-D conditions, a comparative analysis between WUFI 1D and WUFI 2-D was made. In the comparison, 10 equally large segments of the 2-D model were compared with ten separately simulated 1-D models, each representing a strip of a respective 2-D segment. For each element in the WUFI 2D simulation model and the separately simulated 1-D models, an air leakage rate was calculated using Equations 7 and 8. As seen in Figure 3, the comparison revealed good correlation, thus confirming the capability of using a 1-D tool to optimize simulation time.

Wind Loads

Wind loads are one of the major reasons air leakage occurs in wall assemblies. Together with buoyancy effects and ventilation systems, these mechanisms create air pressure differences between the interior and exterior surfaces of the building envelope. Wind loads can be converted to an air pressure difference by using code regulations for civil engineering design (ASCE 2010). Further, this air pressure gradient can be used to determine the air leakage rate if the permeability of the air leakage path is known. As explained in the Introduction, moisture leaks have much lower airflow rates in general than energy leaks. Since the airtightness of walls is rarely measured specifically, it is of value to find a correlation between the traditional airtightness value of buildings (air changes per hour at 50 Pa pressure difference, ACH_{50}) and the airtightness of moisture leaks. As shown in previous work (Lstiburek 2006; Künzel 2012; Pallin et al. 2015), a reasonable approach is to assume that moisture leaks consist of 10% of the total building envelope air leakage. The residual air leakage is considered pure energy leaks.

Even though ACH_{50} is a commonly used indicator for airtightness, it does not directly reflect the airtightness of the building envelope area, or actually the air permeability, and thus does not represent airtightness per area unit. In order to make use of ACH_{50} as a measure of areal airtightness, a plausible interpretation of what ACH_{50} represents in terms of building envelope air permeability is needed. Since ACH_{50} is a volumetric parameter rather than a surface area parameter, the geometry of a building is needed to correctly convert to air permeability per surface area. However, it is likely that this conversion will fall into a possible range. By converting ACH to cfm/ft^2 (L/s·m^2), a likely conversion value ranges somewhere between 1:10 to 1:15 (1:2 to 1:3)—i.e., expressing air leakage in cfm/ft^2 (L/s·m^2) is likely to represent a tenth to a fifteenth (or a half to a third in SI units) of the ACH value for typical buildings. Again, this is important if the actual air leakage rate needs to be known, and this cannot solely be converted from the ACH performance indicator. In this study, a conversion ratio of 1:12.5 is applied. Once the airtightness of the whole building is known, moisture leaks can be estimated using the previously presented 10% moisture leak ratio.

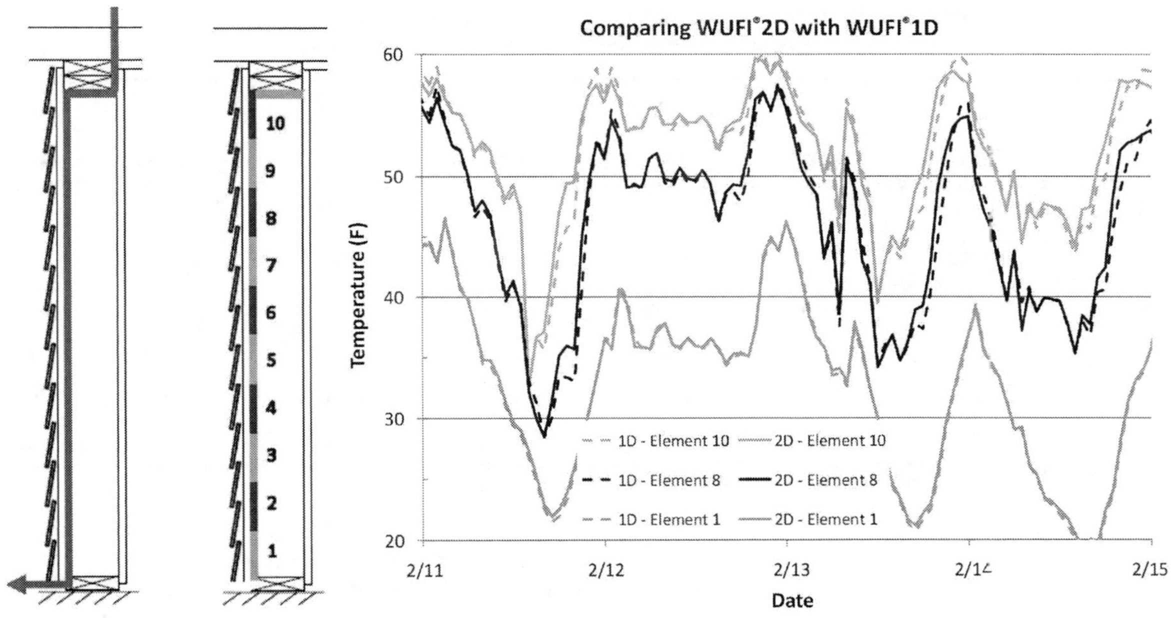

Figure 3 *A comparison between WUFI 2D and WUFI 1D reveals that the 1-D tool is capable of simulating air leakage at any location inside the air leakage path if using air leakage rates found by the correction factor,* η.

RESULTS

The purpose of this paper is to aid in the moisture durability design process when evaluating the risk of air leakage through wall assemblies. Therefore, empirical formulas are presented that help to convert wind loads to simulation input data for air leakage. These formulas have been found by using the correction factor, η, in Equation 7 and the qualified assumptions presented in the section Wind Loads. The first empirical formula looks as follows:

$$\dot{Q} = \\ = a + b \cdot W + c \cdot q_{50} + d \cdot W^2 + e \cdot W \cdot q_{50} + f \cdot q_{50}^2 \geq 0 \quad (9)$$

where

$\dot{Q}$ = airflow rate that should be implemented into the WUFI 1D simulation model to account for wind loads, cfm (L/s)

W = perpendicular wind speed at the wall surface, mph (m/s)

q_{50} = air permeability of the wall assembly at 50 Pa pressure difference, cfm/ft² (L/(s·m²))

a to f = factors that vary depending on the R-value of the wall and whether it is considered as a low or a high wind speed; these factors must be inserted into Equation 9 to calculate airflow rates for the simulation model at varying wind speed.

Tables 1 and 2 present a to f factors for wind speed below or above 5 mph. The values of a to f in Tables 1 and 2 depend on the R-value of the wall, and even more, the location of what is considered to be a critical location for moisture durability. For the walls given in these tables, a critical location is considered to be where the highest relative humidity in an organic material is most likely. In a cold climate, where indoor air exfiltration is the main hazard for high relative humidity levels, the critical location is likely somewhere on the interior side of the wooden sheathing, such as an oriented strand board (OSB). In a warm climate, though, outdoor air infiltration has the potential of increasing relative humidity levels on the exterior side of the drywall.

There are seven different R-value wall assemblies in Tables 1 and 2, and only a few of that could be reasonable to construct in a hot climate. This becomes relevant when reducing the number of factors presented in Equation 9. According to the current edition of *International Residential Code*, Section R402.4.1.2 (ICC 2015), a residential building or dwelling unit must be verified as having an ACH_{50} of 5 or less in climate zones 1 to 2. In climate zones 3 to 8, the required airtightness is 3 ACH_{50}. Using the conversion factor that was presented in the section Wind Loads, an ACH_{50} of 3 is representative of 0.4 cfm_{50}/ft^2 (2.0 L/(s·m²)). An ACH_{50} of 5 is representative of 0.24 cfm_{50}/ft^2 (1.2 L/(s·m²)). If the airtightness of a wall assembly is assumed to meet these code requirements, Equation 9 can be simplified, under the condition that q_{50} is known. Accordingly, the simplified formula is given as

$$\dot{Q} = A + B \cdot W + C \cdot W^2 \quad (10)$$

where A to C are the empirical factors under the conditions that q_{50} is constant, i.e., that the airtightness of the wall assembly

Table 1. Empirical Factors for Wind Speeds Below or Equal to 5 mph for Common Nominal R-Value Wall Assemblies

	Low Range (for $W \leq 5$ mph)						
	R-13	R-20	R-20 + 3.75	R-20 + 5	R-20 + 7.5	R-20 + 11.25	R-20 + 15
a	0.0721	0.0709	–0.0598	–0.0661	–0.0876	–0.0976	–0.1018
b	–0.1153	–0.1142	0.0112	0.0178	0.0416	0.0546	0.0612
c	–0.2478	–0.2439	0.1824	0.2029	0.2744	0.3094	0.3250
d	0.0218	0.0216	0.0033	0.0023	–0.0015	–0.0037	–0.0049
e	0.4424	0.4417	0.2812	0.2691	0.2196	0.1871	0.1671
f	0.0098	0.0075	–0.2210	–0.2306	–0.2608	–0.2722	–0.2751

Values followed by plus signs in the column header represent continuous exterior insulation.

Table 2. Empirical Factors for Wind Speeds Above 5 mph for Common Nominal R-Value Wall Assemblies

	High Range (for $W > 5$ mph)						
	R-13	R-20	R-20 + 3.75	R-20 + 5	R-20 + 7.5	R-20 + 11.25	R-20 + 15
a	–0.2534	–0.2599	–0.6423	–0.6214	–0.5036	–0.4064	–0.3419
b	0.0130	0.0147	0.1317	0.1292	0.1114	0.0950	0.0837
c	0.1563	0.1761	1.8678	1.8638	1.7311	1.5588	1.4249
d	0.0017	0.0016	–0.0046	–0.0045	–0.0039	–0.0032	–0.0028
e	0.5123	0.5094	0.1414	0.1277	0.0852	0.0670	0.0590
f	–0.6422	–0.6499	–1.2146	–1.2010	–1.0939	–0.9842	–0.9032

Values followed by plus signs in the column header represent continuous exterior insulation.

is known. As an example, two new tables, Tables 3 and 4, are provided where the airtightness of each wall assembly is presumed to meet code requirements (ACH_{50} = 5 and 3, respectively). In Tables 3 and 4, empirical factors to be implemented in Equation 10 are given for common R-value wall formations under presumed code compliance. For R-13 and R-20 walls, the factors are presented for both an ACH_{50} of 3 and of 5.

Figure 4 depicts a comparison between a traditional 1-D approach and the proposed empirical 2-D solution. As seen in the figure, there is a great disparity in the simulated air leakage rates. Consequently, this will affect the hygrothermal evaluation result due to air leakage, which is why the empirical formula is preferably used in favor of a traditional 1-D simulation approach.

In conclusion, Tables 1 and 2 provide empirical factors to aid in the design process when evaluating the risk of air leakage in common wall assemblies. By implementing the factors into Equation 9, the user will find appropriate air leakage rates to apply in WUFI® for different wind speeds for different wall assemblies. To further simplify, Equation 9 is shortened under the assumption of a known airtightness. Depending on the geometry of the simulation model, air leakage rates found by using Equation 10 can be converted to ACH values that are used by WUFI®.

DISCUSSION

The empirical 2-D air leakage modeling approach has great benefit over other more complicated methods, such as computational fluid dynamics (CFD), as well as the simplified 1-D methods traditionally used. Using the 1-D version of WUFI® with the proposed 2-D framework together with the empirical airflow equation can save both computation power and time needed to more accurately assess the impact of air leakage. Also, using a tool such as WUFI®, which is widely used in the building science community, can bring this type of assessment into the hands of more practitioners than if using complicated CFD modeling.

Furthermore, since the 1-D method was shown to greatly underestimate the airflow rate for a given wind speed, the

Table 3. Empirical Factors for Wind Speeds Below or Equal to 5 mph, Under the Conditions that the Airtightness of the Wall Assembly is Known (in this Table the Airtightness is Assumed to Meet Code—ACH_{50} of 3 or 5)

Low Range (for W ≤ 5 mph) for Known q_{50}									
	R-13		R-20		R-20 + 3.75	R-20 + 5	R-20 + 7.5	R-20 + 11.25	R-20 + 15
q_{50} (cfm_{50}/ft^2)	0.40	0.24	0.40	0.24	0.24	0.24	0.24	0.24	0.24
A	–0.025	0.013	–0.025	0.013	–0.029	–0.031	–0.037	–0.039	–0.040
B	0.062	–0.009	0.062	–0.008	0.079	0.082	0.094	0.099	0.101
C	0.022	0.022	0.022	0.022	0.003	0.002	–0.002	–0.004	–0.005

Values followed by plus signs in the column header represent continuous exterior insulation.

Table 4. Empirical Factors for Wind Speeds Above 5 mph, Under the Conditions that the Airtightness of the Wall Assembly is Known (in this Table the Airtightness is Assumed to Meet Code—ACH_{50} of 3 or 5)

High Range (for W > 5 mph) for known q_{50}									
	R-13		R-20		R-20 + 3.75	R-20 + 5	R-20 + 7.5	R-20 + 11.25	R-20 + 15
q_{50} (cfm_{50}/ft^2)	0.40	0.24	0.40	0.24	0.24	0.24	0.24	0.24	0.24
A	–0.294	–0.253	–0.293	–0.255	–0.264	–0.243	–0.151	–0.089	–0.052
B	0.218	0.136	0.218	0.137	0.166	0.160	0.132	0.111	0.098
C	0.002	0.002	0.002	0.002	–0.005	–0.005	–0.004	–0.003	–0.003

Values followed by plus signs in the column header represent continuous exterior insulation.

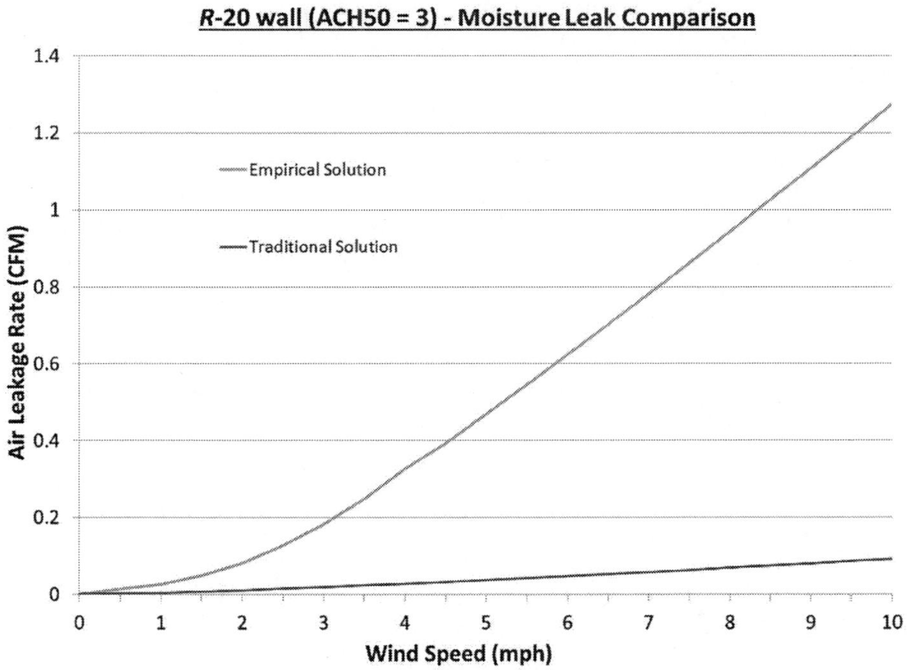

Figure 4 *A comparison between a traditional WUFI solution for air leakage and the proposed empirical solution reveals a great disparity. Apparently the traditional solution hugely underestimates the air leakage rate needed to represent realistic conditions.*

traditional method cannot be trusted to yield a correct assessment of the moisture durability when air leakage needs to be accounted for. In fact, since the traditional method typically underestimates airflow rates, the hygrothermal impact from air leakage, thus the risk of moisture problems, can be misjudged.

CONCLUSION

Increasing the airtightness and thermal resistance of walls is important to lower energy consumption of buildings. However, the hygrothermal performance of energy-efficient walls needs to be well understood so that they will not exhibit moisture durability issues when implemented in the field. To aid in this need, Oak Ridge National Laboratory (ORNL) developed a framework to determine the hygrothermal performance of energy-efficient walls, which involves using hygrothermal calculation tools to determine how the wall will perform.

There are a number of simulation tools that can be used to predict the hygrothermal performance, but these tools rarely handle air leakage well. Therefore, ORNL presented a novel approach for more accurately modeling a 2-D airflow in a wall assembly using WUFI 1D. This method takes into account wind, thermal wall characteristics, and the indoor and outdoor environments. When comparing the results with the typical method, the novel method was found to result in much larger air leakage rates than otherwise used in WUFI®, which obviously will affect the hygrothermal simulation results and thus the predicted performance.

However, calculating air leakage rates based on the proposed method can be quite complicated, which is why empirical formulas were developed to more accurately account for air leakage in the hygrothermal assessment of walls. These formulas were developed so that the user would need to provide the thermal resistance and airtightness of the walls, together with the wind speed perpendicular to the exterior wall surface. By using the empirical formulas, the user will obtain air leakage rates to be implemented in WUFI 1D or other tools where similar a simulation approach is used to account for air leakage. This approach will allow for more accurately accounting for air leakage and potential moisture durability risks in hygrothermal building envelope design.

REFERENCES

ASCE. 2010. ASCE/SEI 7-10, *Minimum design loads for buildings and other structures*. Reston, VA: American Society of Civil Engineers.

ICC. 2015. *2015 International residential code*. Geneva: International Code Council, Inc. http://codes.iccsafe.org/app/book/toc/2015/2015%20San%20Antonio/2015%20IRC%20HTML/.

Kehrer, M., and T. Schmidt. 2008. Radiation effects on exterior surfaces. *Proceedings of Nordic Symposium on Building Physics 2008, Copenhagen*.

Künzel, H.M. 2012. Modeling air leakage in hygrothermal envelope simulation. Presented at the BEST 3 Conference, Atlanta, GA.

Künzel, H.M., D. Zirkelbach, and B. Scfafaczek. 2011. Vapour control design in wooden structures including moisture sources due air exfiltration. Presented at the 9th Nordic Symposium on Building Physics—NSB 2011, Tampere, Finland.

Lstiburek, J. 2006. Understanding air barriers. *ASHRAE Journal* 47(7):24–30.

Lstiburek, J. 2015. WUFI: Barking up the wrong tree? *ASHRAE Journal* 57(10).

Pallin, S., P. Boudreaux, M. Kehrer, D. Hun, R. Jackson, and A. Desjarlais. 2015. Moisture durability assessment of selected well-insulated wall assemblies. Report ORNL/TM-2015/645. Oak Ridge, TN: Oak Ridge National Laboratory, Energy and Transportation Science Division. http://info.ornl.gov/sites/publications/files/Pub59790.pdf.

Sherman, M.H., and W.R. Chan. 2006. Building ventilation: The state of the art, building airtightness: Research and practice. Report LBNL-53356. Berkeley: Lawrence Berkeley National Laboratory. https://www.researchgate.net/publication/284044637_Building_ ventilation _The_state_of_the_art_building_airtightness_Research _and_practice.

Comparison of Measured Hygrothermal Performance of Wood-Frame Walls Built with Continuous Exterior Insulation versus Walls Built with Housewrap and Oriented Strand Board Sheathing in Single-Family Homes in a Cold Climate

Gary D. Parsons, LEED AP, CSI, RCI **Brian Lieburn**
Member ASHRAE
Greg Stewart **Peter Eastman, MS**

ABSTRACT

Many wall assembly options exist for wood frame construction and typically a choice is made between assemblies that include continuous exterior insulation and assemblies that include housewrap. Often this decision is made by considering a dew-point analysis of cavity condensation potential or by considering the prediction of wall drying mechanisms based on modeling. Three single-family occupied homes were constructed with continuous exterior insulation and six identical homes were constructed with housewrap in the same neighborhood in climate zone 5. Moisture content measurements of the condensation plane, continuously recorded for the past four years, in the rim joist and above-grade walls are compared. Additionally, similar test wall assemblies with moisture sensors were installed in a long-term exterior exposure test unit that allows comparison between long-term semi-controlled tests, i.e., those where the interior climate is carefully controlled, and homes where occupant behavior controls interior conditions and wall performance. Finally, the measured results are compared to predicted modeled performance. Forensic investigations and samplings of wall and rim joist areas of select homes in the study are presented for comparison of observed material conditions and expected conditions based on moisture exposure, both predicted and observed.

INTRODUCTION

Builders of single-family residential wood frame construction can achieve energy code compliance by combining many material and assembly choices. Two popular above-grade wall configurations include, from inside to outside, 1) drywall, 2×4 studs, 7/16 oriented strand board (OSB) structural sheathing, a layer of continuous insulation (c.i.), and an exterior cladding system and 2) drywall, 2×6 studs, 7/16 OSB, a layer of housewrap, and exterior cladding. Typically the cavity is filled with fiberglass, cellulose, or spray polyurethane foam insulation. The above-grade wall assembly choice can also impact the construction of the rim joist area of the building. Specifically, when c.i. is used on the opaque wall area then generally the c.i. layer also covers the rim (Lstiburek 2006). Often the assembly decision is based on the condensation potential within the wall cavity, normally paying most attention to the interior surface of the OSB layer, which is generally where condensation is most likely to occur (Lstiburek 2006). This condensation potential is estimated using simple dew-point analysis (ASHRAE 2013) or a more advanced hygrothermal analysis using WUFI® (Karagiozis et al. 2001). The results of hygrothermal analysis are highly dependent on input, such as starting moisture content assumptions, rendering this tool most useful for wall system comparisons, not as pass/fail criteria for wall selection.

Two condensation control strategies—moisture vapor permeance control and temperature control—can be used in wall design. First, walls using vapor permeance control, or breathable wall systems, can be constructed using combinations of cavity insulation and housewrap installed outboard of the structural sheathing (see Table 1). The second approach, temperature control, uses c.i. either with or without housewrap outboard of the structural sheathing combined with cavity insulation. Generally the c.i. is not considered vapor permeable and limits drying to the exterior. In this approach, c.i.

Gary D. Parsons is a research and development fellow, **Brian Lieburn** is a research scientist and residential applications development leader, and **Greg Stewart** is a research scientist with The Dow Chemical Company, Dow Building Solutions, Midland, MI. **Peter Eastman** is an associate scientist with The Dow Chemical Company, Collegeville, PA.

Table 1. Experimental Insulation Configurations

	Houses 1–3	Houses 4–6	Houses 7–9
Rim joist—Interior	R-19 fiberglass batt	R-19 fiberglass batt	R-16 closed-cell spray polyurethane foam
Rim joist—Exterior	None	None	R-5 extruded polystyrene
Above-Grade Wall Configurations			
Stud configuration	2×6 16 in. on center	2×6 16 in. on center	2×4 16 in. on center
Interior	R-19 fiberglass batt	R-19 fiberglass batt	R-16 closed-cell spray polyurethane foam
Exterior	Housewrap	Housewrap	R-5 c.i.
Continuous mechanical ventilation	No	Yes	Yes

causes increased cavity interior temperature during winter, which reduces both relative humidity within the cavity and condensation potential.

An objective of this research is to compare WUFI model predictions to measured wall moisture performance for the two suggested wall assemblies. A multiyear, multihome single-family home comparison project was constructed in Midland, Michigan, located near the boundary of climate zones 5 and 6 (6645 heating degree-days). The nine research houses are identically oriented on nine adjacent lots, with the front façades facing southwest. Instruments are installed in walls in common living spaces such as living or dining rooms on the northeast wall of each home. As occupant behavior is uncontrolled, but monitored, in these houses, the behavior impact on wall performance can be measured. Similar walls were also constructed in a long-term field exposure test facility (Test Hut), also located in Midland, Michigan, approximately 1 mile from the research homes, and are monitored in the same way (Smegal and Grin 2015). Unlike the occupied homes, the interior temperature and relative humidity (RH) conditions of the wall research laboratory are carefully controlled, which enables performance comparisons of similar wall sections in different situations.

Hygrothermal data, including data on moisture content, humidity, and temperature, are collected in the above-grade wall cavity and rim joist area. Table 2 details the instrumentation used for data collection.

Model predictions, using WUFI, are compared to measurements from both occupied homes and controlled experimental wall assemblies.

Table 2. Instrumentation

Measurement	Device	Description
Temperature	Type T thermocouple	
Moisture content	Pin resistance	Brass screws, 1 in. apart
Relative humidity	RH probe	Campbell Scientific CS210/10162
	Data logger	Campbell Scientific CR1000
	Multiplexer	Campbell Scientific AM 16/30

PREDICTED WALL PERFORMANCE

Comparison of Measured versus ASHRAE Standard 160 Predicted Interior Relative Humidity

WUFI was used to calculate the ASHRAE Standard 160 (ASHRAE 2009) predicted interior RH values, using the built-in weather data for Detroit, Michigan, which is approximately 130 miles southeast of the research houses' location. In addition to weather data, other inputs, such as occupant data and ventilation rates, are required to predict RH.

Data on the number of occupants for the year of October 1, 2014, through September 30, 2015, were available for four of the nine houses (see Table 3). Even though all of the houses have four bedrooms, and therefore five occupants assumed by ASHRAE Standard 160, the number of bedrooms used in the RH prediction was adjusted to account for the actual number of occupants. Mechanical and natural ventilation model inputs for each house are shown in Table 4.

Table 5 compares predicted and measured RH results for each of these four houses. In each case the model predictions are higher, substantially in some cases, compared to the actual.

Comparison of Measured versus WUFI-Predicted Sheathing Moisture Levels

WUFI 1D is often used as a moisture analysis tool, but many of the input variables can have a dramatic effect on the results. Guidelines for the setup of WUFI runs have been defined by Lstiburek et al. (2015); these guidelines were used in the modeling of the research houses in this work.

Table 3. Occupancy Data from October 2014 through September 2015

Wall Construction	House Number	Number of Bedrooms	ASHRAE Number of Occupants	Actual Number of Occupants	Number of Bedrooms Used in RH Predictions
2×6 Fiberglass	1	4	5	3	2
2×6 Fiberglass	3	4	5	5	4
2×6 Fiberglass	5	4	5	1	1
2×4 SPF w/c.i.	9	4	5	3	2

Table 4. Ventilation Rates, Natural and Mechanical

House Number	Natural Ventilation ACH	Mechanical Ventilation ACH	Total Ventilation ACH
1	0.14	0	0.14
3	0.15	0	0.15
5	0.15	0.21	0.36
9	0.12	0.20	0.32

Table 5. Actual versus Predicted Relative Humidity

House Number	Winter Average		Summer Average	
	Predicted % RH	Measured % RH	Predicted % RH	Measured % RH
1	42	29	60	39
3	43	29	60	59
5	22	17	60	51
9	25	23	60	49

For the WUFI analysis, ASHRAE Standard 160 was used to predict interior RH. Referring again to all nine study houses, Table 6 summarizes the measured and predicted sheathing moisture data. Some observations to make from the data are as follows: 1) the measured data in terms of days the sheathing is above 20% moisture is variable between three houses of the same construction type, and 2) the WUFI predictions for sheathing moisture are far from the measured data, with the predicted sheathing moisture being higher than measured for houses 1–3 and lower for houses 4–9.

To close the gap between measured and predicted sheathing moisture, an RH sine curve was developed to mimic the measured interior RH. The sheathing moisture predictions are shown in Table 7 and are now closer to actual sheathing moistures.

Table 6. Sheathing Moisture—Measured and Predicted using ASHRAE Standard 160 RH

House Numbers	Measured Sheathing Moisture		WUFI-Predicted Sheathing Moisture (Using ASHRAE Standard 160)	
	Maximum % Moisture	Number of Days >20%	Maximum % Moisture	Number of Days >20%
1–3	31%	71±35	50%	190
4–6	29%	42±35	19%	0
7–9	23%	5±5	12%	0

Table 7. Sheathing Moisture—Measured and Predicted Using Measured RH

House Numbers	Measured Sheathing Moisture		WUFI-Predicted Sheathing Moisture (Measured Interior RH)	
	Maximum % Moisture	Number of Days >20%	Maximum % Moisture	Number of Days >20%
1–3	31%	71±35	29%	46
4–6	29%	42±35	14%	0
7–9	23%	5±5	12%	0

MEASURED WALL PERFORMANCE

Semi-Controlled versus Uncontrolled Interior Conditions

In the long-term field exposure test facility (Test Hut), wall 3a (see Table 8) is of the same wall construction type as research houses 1–6.

Houses 1–6 had lower winter RH than was used in the Test Hut for the winter of 2014–2015. The sheathing moisture content and temperature for wall 3a of the Test Hut are shown in Figure 1. Test Hut wall 3a sheathing moisture content levels off close to 20% from mid-January through mid-February. The

Test Hut interior RH was intentionally increased to 40% in late January, leading to the increase in sheathing moisture in the red shaded area of the figure. This further emphasizes the need to maintain low interior RH.

Note that Test Hut wall 2 (see Table 8) has a maximum sheathing moisture content of 15%, even in these worst-case interior RH conditions (see Figure 2). Wall 2 has c.i., but unlike houses 7–9 it does not have spray foam insulation in the cavity.

Uncontrolled/Occupied Measured Hygrothermal Performance

Hygrothermal measurements are collected in both the above-grade frame walls within the cavity and in the rim joist area of each house, as the construction details differ in these areas of the home (as detailed in Table 1). This difference in the building component layers translates to potential differences in moisture performance of the surfaces being measured. Moisture content measurements are taken in the center of the cavity in the above-grade walls at the cavity side of the OSB layer and directly below in the center of the rim joist area of the basement.

Table 8. Test Hut Wall Configuration

Wall Component	Wall 3a	Wall 2
Framing	2×6	2×4
Cavity insulation	R-19 faced fiberglass	R-15 faced fiberglass
Sheathing	7/16 in. OSB	7/16 in. OSB
Exterior	Housewrap	R-5 c.i.

The moisture contents of individual houses and a group average for houses 1–6 are shown in Figure 3 from October 2012 through June 2015 and include three heating seasons. The red line indicates the threshold for elevated moisture content in wood. In each winter season the average moisture content of the OSB layer increases to over 20%. House-to-house variability in moisture performance can be seen, as house 5 moisture content varies seasonally between approximately 8% and 18% moisture content; however, others, such as house 2, rise to over 30% in winter. Differences in occupant behavior can be seen over the data collection period as well. In Figure 3, the moisture content lines for houses 2 and 5 are highlighted by bold dashed lines. Notice that house 5, indicated by the pale blue dashed line, did not change occupants over the experimental period, and a similar moisture performance repeats each season. In contrast, house 2 changed occupants twice and was vacant for several months. In this case, the

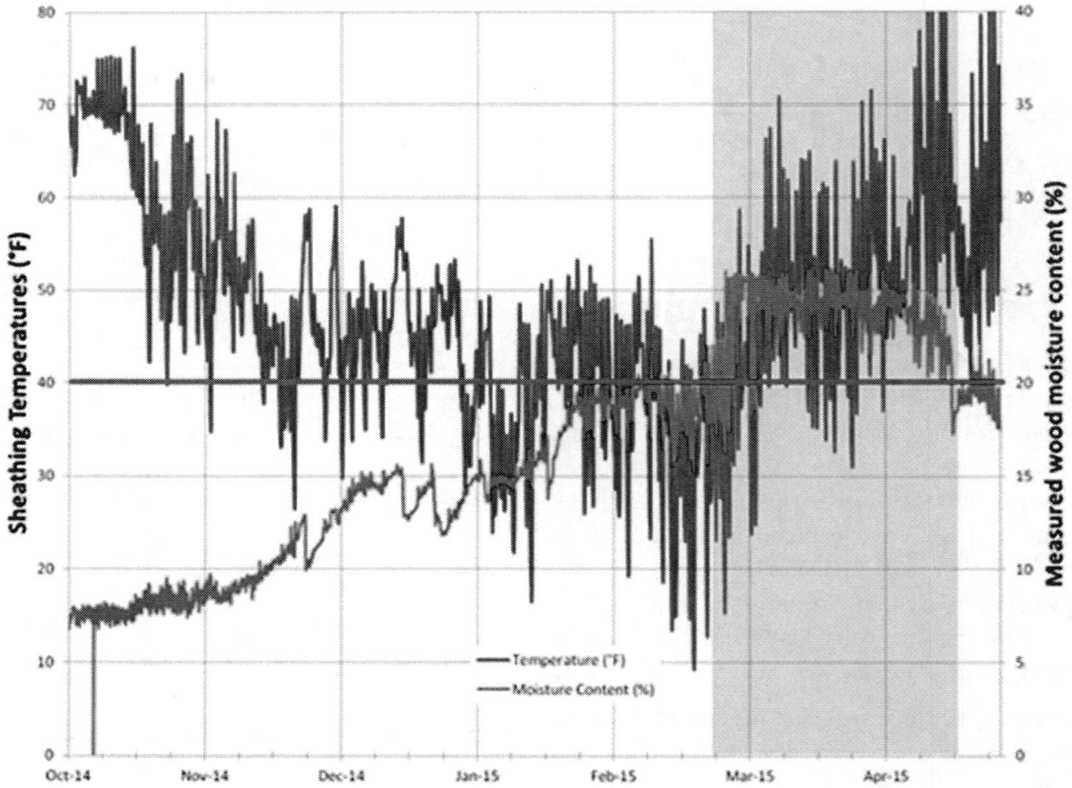

Figure 1 *Temperature and moisture content, Test Hut wall 3a (OSB/housewrap interior sheathing surface, above-grade cavity).*

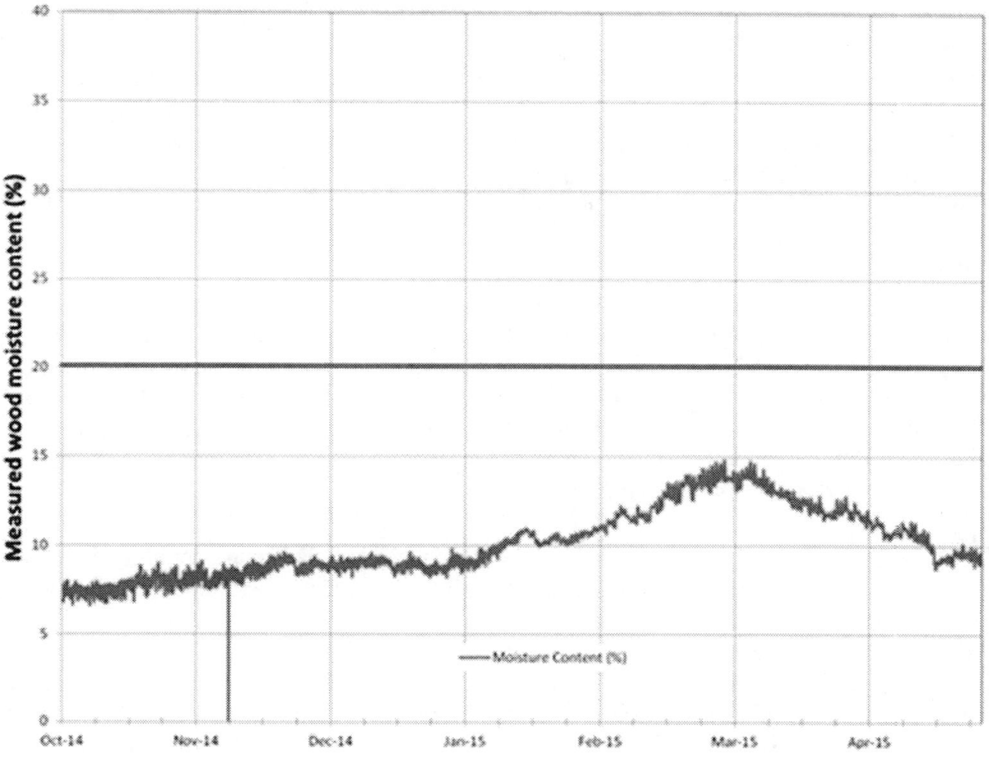

Figure 2 *Sheathing moisture content, Test Hut wall 2 (OSB/R-5 c.i. interior sheathing surface, above-grade cavity).*

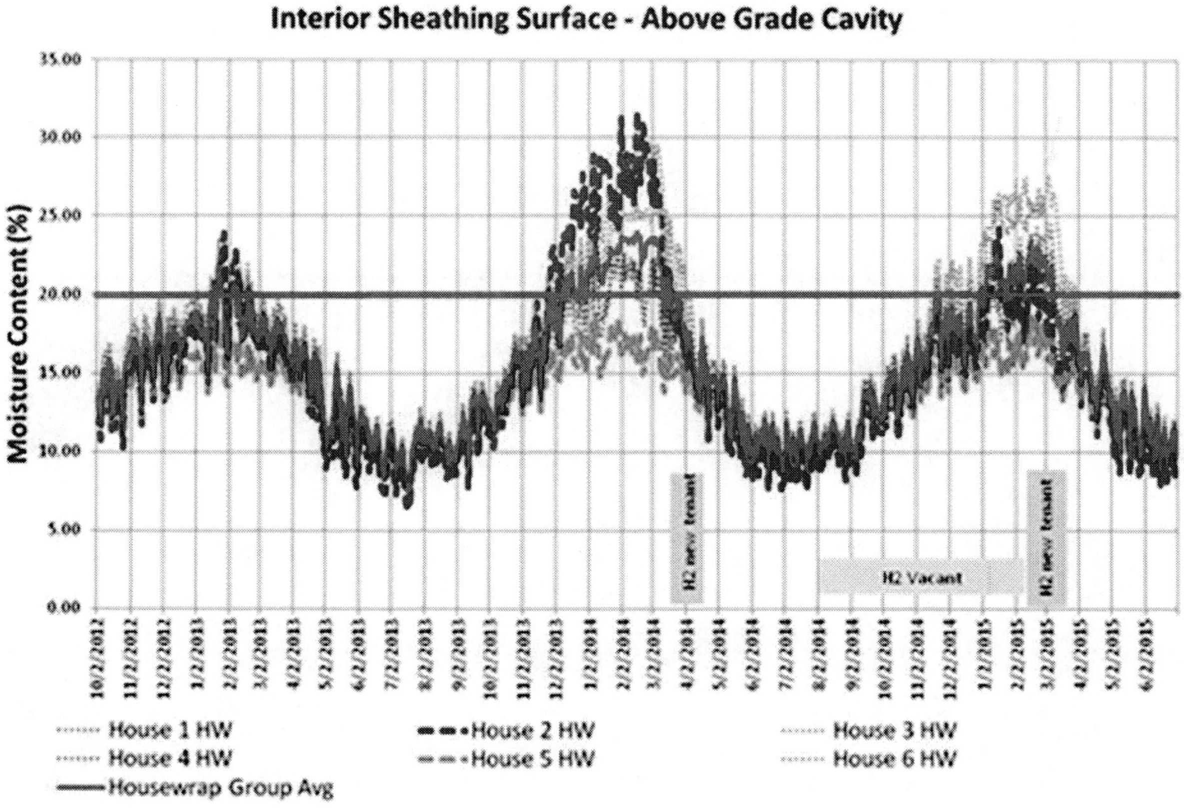

Figure 3 *Moisture content of above-grade sheathing in housewrap houses.*

maximum moisture content during winter 2014 was over 30%, but with a different occupant after a period of vacancy, the maximum moisture content remained less than 25% for winter 2015.

In contrast, Figure 4 shows that moisture content in the above-grade cavity walls of houses 7–9 peaked just over 20% for a few days in winter. Also, houses 7–9 performed similarly house to house, even though the occupancy is different between the houses.

Figures 5 and 6 show the measured moisture in the rim joists of houses 1–6 and 7–9, respectively. Similar to the above-grade wall measurement, the houses with fiberglass insulation in the rim and no exterior insulation also vary from house to house, with one house reaching 40% moisture content. The houses with spray foam in the rim and R-5 c.i. perform similarly and do not rise over 20% moisture content.

FORENSICS

After three winters of measured elevated moisture content on the interior surface of the OSB and rim joists on some of the homes, a forensic plan was developed to better understand what was happening in these wall cavities and rim joist spaces. The plan included removing vinyl siding and housewrap from the area on the wall where the data sensors are located. A 4 in. diameter hole saw was used to obtain a sample that could be microscopically examined. Microbial swabs were taken immediately from the various locations. A larger, approximately 16 in. wide by 50 in. tall sample of OSB was collected from the adjacent stud space for small-scale static bending tests. Engineered rim joists were inspected and microbial swabs were taken from the interior and exterior.

Observations

House 2 measured moisture content above 20% in the above-grade sheathing over all three winters. There was no evidence of moisture on the exterior of the OSB wall sheathing or engineered rim joist. When the sheathing section was removed, water staining was visible on the exterior of the wall plate, the stud, and the interior of the surface of the sheathing (Figure 7).

House 3 also measured moisture content above 20% in the above-grade sheathing over all three winters. Black staining was observed in an approximately 2 × 2 in. area on the interior of the OSB (Figure 8). The interior of the rim joist showed no signs of moisture.

House 5 measured consistent moisture content below 20% in both the rim joist and above-grade sheathing over all three winters. When the wall sheathing section was removed,

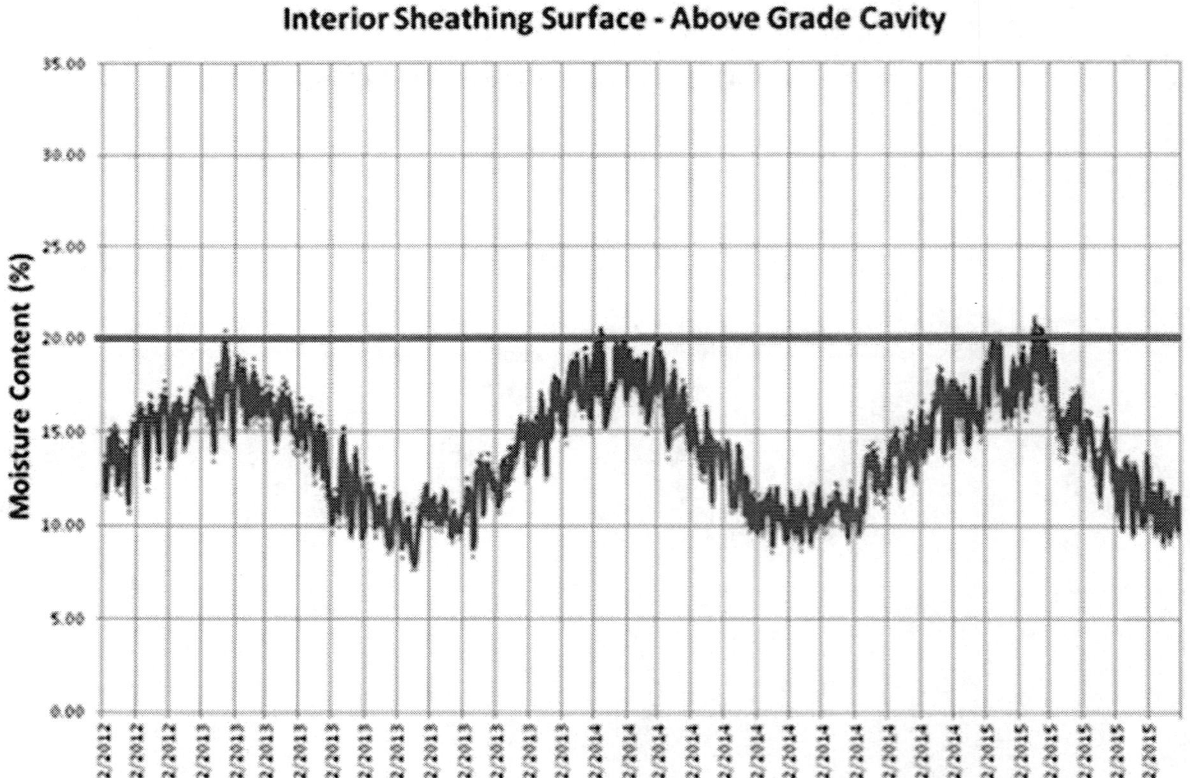

Figure 4 *Moisture content of above-grade sheathing in R-5 c.i. houses.*

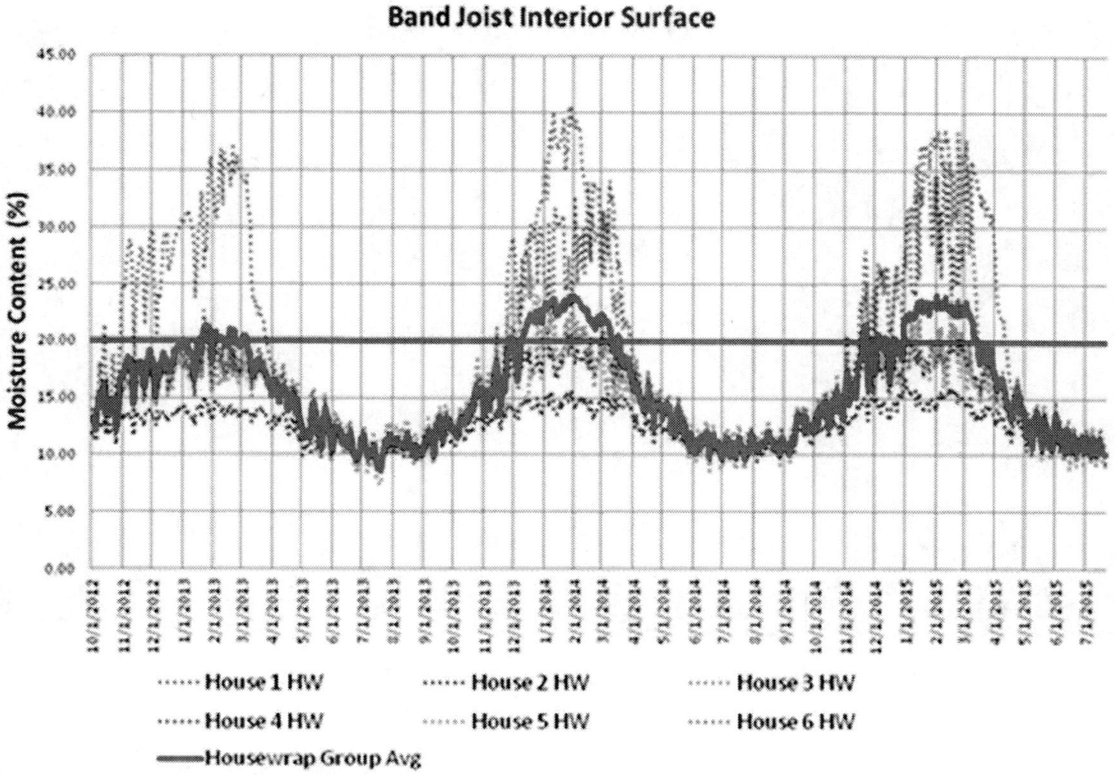

Figure 5 *Moisture content of band joists in housewrap houses.*

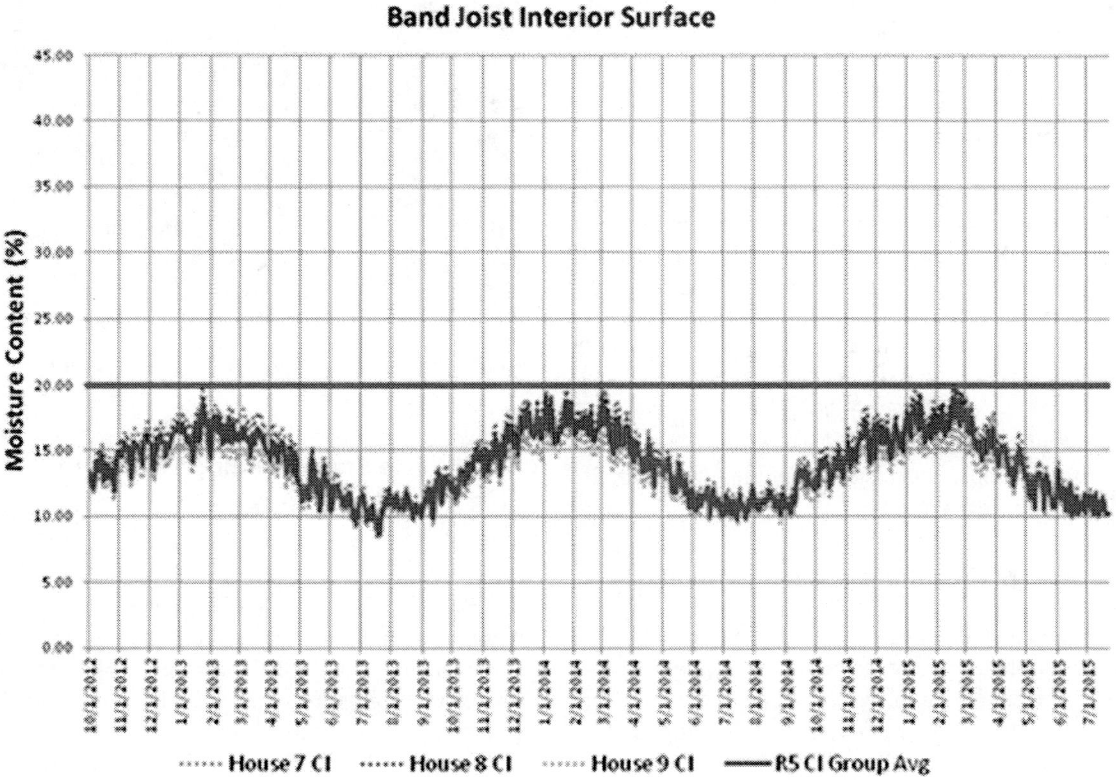

Figure 6 *Moisture content of band joists in R-5 c.i. houses.*

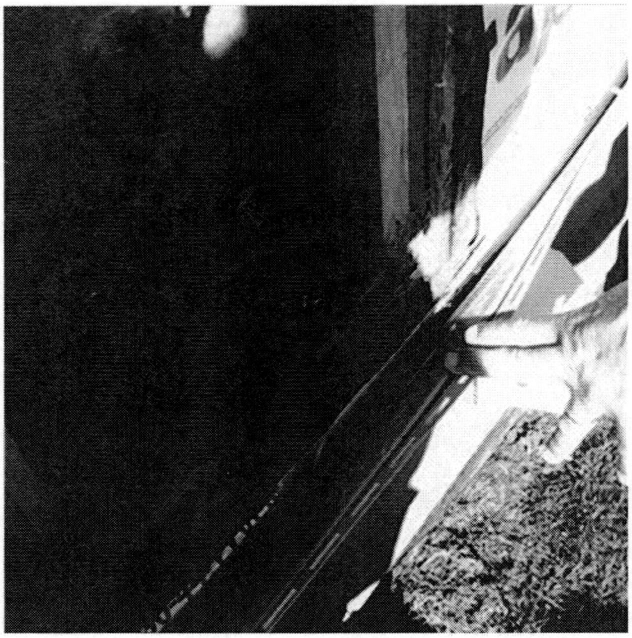

Figure 7 *House 2 water staining on plate and stud.*

Figure 8 *House 3 OSB staining.*

water staining was visible on the exterior of the wall plate (Figure 9).

House 6 measured consistent moisture content above 20% in both the rim joist and above-grade sheathing over all three winters. An inspection of the interior of the rim joist revealed staining on the bottom chord of the I-joist (Figure 10).

Microbial Observations

Twenty-four swab samples were collected at various locations on each of the homes investigated and were analyzed by sterility streak in the laboratory. The samples were analyzed primarily for fungal growth results. Moderate to heavy fungal contamination was observed on samples from house 2, where there was clear evidence of water damage (see Figure 7). Fungal contamination was also observed on swab samples taken from the interior of the rim joists on houses 3 and 6. Despite some evidence of water damage on house 5, no/trace fungal contamination was detected. All other swab samples showed little to no fungal growth.

Microscopy Experimental and Results Summary

Four nominal 4 in. plugs cut from OSB panels from sheathing of four houses were examined microscopically for the presence of fungal hyphae. No fungal hyphae were observed by light microscopy in any of the four plug samples by light or scanning electron microscopy (Figure 11). Samples were not examined for deterioration due to other factors such as freeze-thaw cycling or wet-dry cycling.

Figure 9 *House 5 water staining on exterior wall.*

Small-Scale Static Bending Testing

Small-scale static bending tests were performed on the larger OSB sheathing samples according to Section 7.6 of PS 2-10, *Performance Standard for Wood-Based Structural-use Panels* (APA 2010). All samples tested met minimum requirements in both strength axis parallel and strength axis perpendicular. While the measured OSB bending strength satisfied the minimum required by PS 2-10, OSB sheathing is commonly manufactured above the minimum target to ensure compliance. Thus, the measured strength does not necessarily indicate whether or not strength degradation may have occurred. However, if it did, it was not significant enough to

Figure 10 *House 6 I-joist staining.*

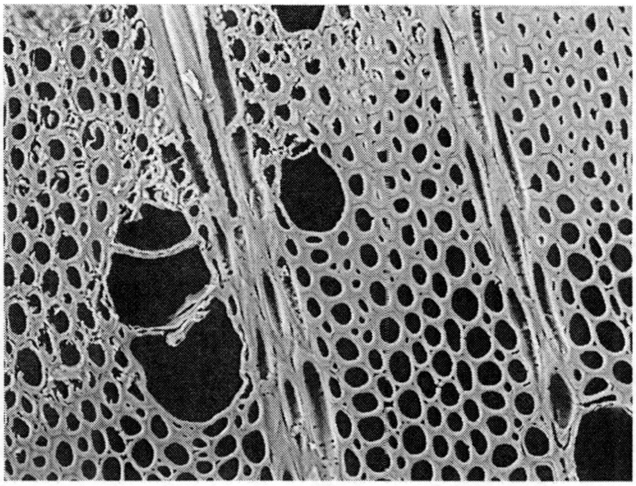

Figure 11 *Single scanning electron microscope (SEM) image showing no fungal evidence.*

reduce strength below minimum standard requirements within the four-year study period. In comparison, the OSB moisture contents measured in this study for houses 1–6 were within the range of moisture contents experienced in another study by HIRL (2013), which did show OSB bending strength degradation of 19% to have occurred in a two-year study period.

Forensic Results Summary

Measured moisture contents on the inside surfaces of the OSB wall sheathing and engineered rim joists measured well above 20% for significant periods of time on several of the homes each winter. Forensic analysis revealed signs of wetting and evidence of some fungal activity in certain locations. However, no significant decay was evident in the locations examined. Graphs overlaying surface temperature and moisture content were developed to understand the conditions in the areas where higher moisture contents were measured (Figures 12 and 13).

Highest moisture contents were measured at times when the surface temperatures were mostly below 40°F. This implies that while moisture contents were above 20% the surfaces were in a refrigerated state, minimizing the potential for fungal decay. Shaded areas on the graphs represent the periods of time when the moisture content was above 20% and the temperature was higher than 40°F. These conditions were limited to periods of time during spring and fall.

CONCLUSIONS

- Above-grade walls of exterior insulated walls (c.i.) with spray foam have lower and more consistent moisture content measured at the interior surface of the exterior sheathing compared to those without c.i.
- Band joists of exterior insulated walls (c.i.) with spray foam installed in the cavity have lower and more consistent moisture content measured at the interior surfaces of the band joists compared to those without c.i.
- In building enclosure assemblies that contain fiberglass cavity insulation along with housewrap, the impact of occupant behavior on sheathing moisture content is intensified. In comparison, homes built with c.i. and spray foam have moisture contents that are independent of occupant behavior.
- Seasonal moisture content changes of OSB have not caused short-term catastrophic deterioration of the wall wood components.
- In WUFI modeling, the interior RH input variable has a large impact on the output sheathing moisture content and increases the difficulty in accurately predicting the moisture content of exterior sheathing.
- Low-permeance foam insulation on both sides of the structural sheathing resulted in low moisture content, less than 20%, both in prediction and in measurement, during all seasons, suggesting that the double vapor barrier system is not problematic.
- Test Hut wall 2, which was built with R-5 c.i. and R-15 fiberglass, performed well with a maximum sheathing moisture content of 15%, even at elevated interior RH (40%).

REFERENCES

APA. 2010. PS 2-10, *Performance standard for wood-based structural-use panels*. Tacoma, WA: The Engineered Wood Association (APA).

ASHRAE. 2009. ANSI/ASHRAE Standard 160, *Criteria for moisture-control design analysis in buildings*. Atlanta: ASHRAE.

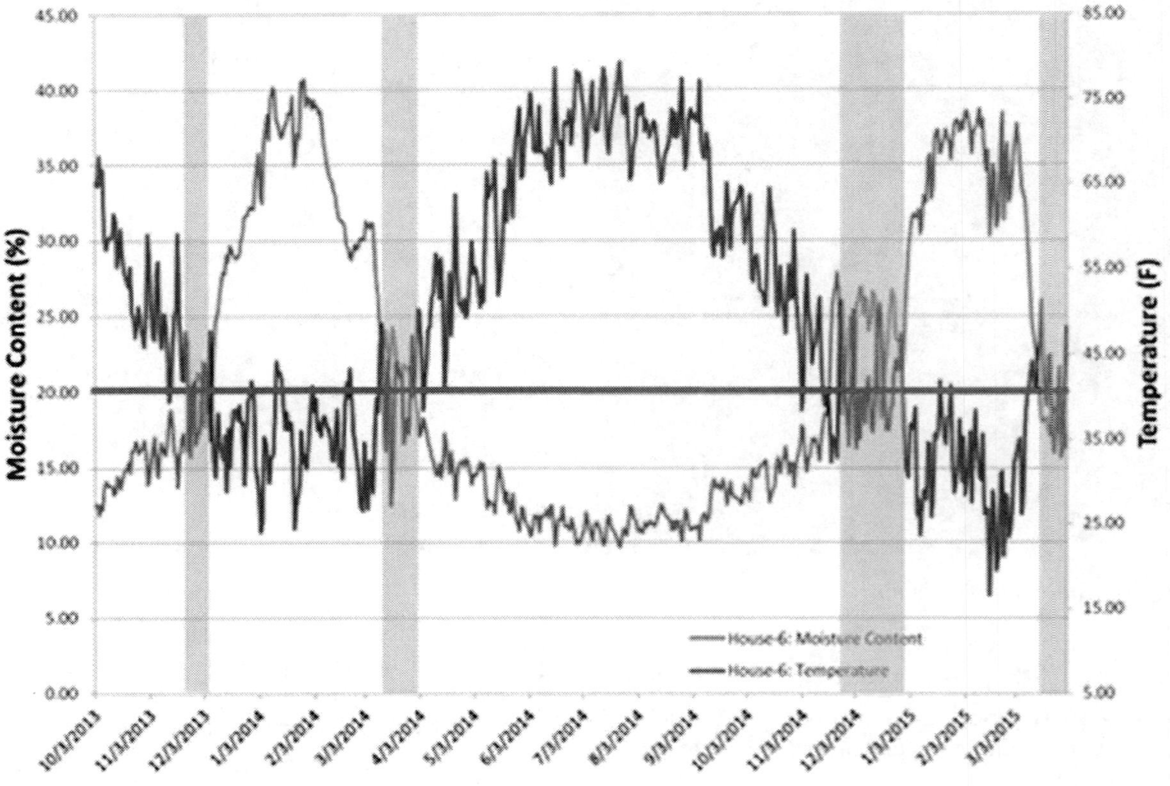

Figure 12 *Temperature and moisture content of band joist in house 6.*

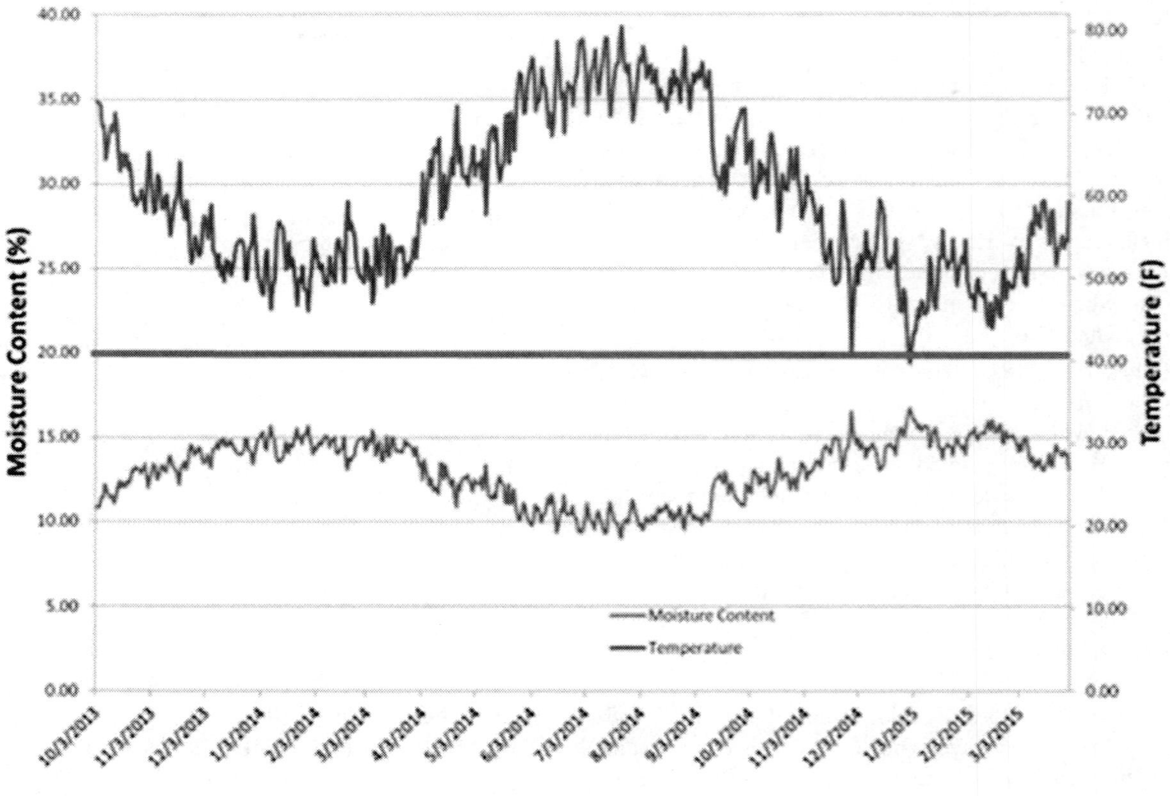

Figure 13 *Temperature and moisture content of band joist in house 2.*

ASHRAE. 2013. Chapter 25. In *ASHRAE Handbook—Fundamentals*. Atlanta: ASHRAE.

HIRL. 2013. Characterization of the moisture performance of energy-efficient and conventional light-frame wood wall systems. Joint Venture Agreement 12-JV-11111136-070, prepared for USDA Forest Products Laboratory by Home Innovation Research Labs, Inc.Karagiozis, A., H. Kunzel, and A. Holm. 2001. WUFI®— ORNL/IBPA North American Hygrothermal Model. *Proceedings of the Performance of Exterior Envelopes of Whole Buildings VIII: Integration of Building Envelopes.*

Lstiburek, J. 2006. *Builder's guide to cold climates*. Westford, MA: Building Science Press.

Lstiburek, J., K. Ueno, and S. Musunuru. 2015. *Modeling enclosure design in above grade walls*. Washington, DC: U.S. Department of Energy.

Smegal, J., and A. Grin. 2015 Inward vapor drives in adhered veneer wall assemblies with continuous exterior insulation. *Proceedings of Best4 Conference Building Enclosure Science and Technology.*

Measured and Predicted Moisture Performance of High-R Wall Assemblies in Cold Climates

Trevor M. Trainor

Jonathan Smegal

John Straube, PhD
Associate Member ASHRAE

Anil Parekh
Member ASHRAE

ABSTRACT

The wall system is a key component of an energy-efficient building. Incremental increases in building enclosure insulation levels have resulted in reduced heat loss and space heating loads. However, there are challenges that arise, including changes in the construction process and concerns about the long-term durability of these new assemblies. Of primary concern are the effects of higher insulation levels on the overall moisture performance and durability of the building enclosure. In the past, some practitioners and designers have resisted the use of low permeability exterior insulation as part of a high R-value wall system because of the concern that it will reduce drying and result in moisture accumulation within the wall system. An alternative approach to achieving high R-values is to use thick (20–50 cm [8–20 in.]) cavity spaces filled with insulation. A field study was conducted of four full-scale high-R (approximately R-30 effective) wall assemblies and one standard wall (for comparison), on both the north and south orientation, using both exterior insulation construction and thick insulation-filled wall systems. The study was performed in Waterloo, Ontario, which is on the border of IECC Climate Zones 5 and 6. The measured moisture performance data from the full-scale test walls along with interior and exterior boundary conditions was used to validate a hygrothermal model for the testing location. The model was then used to predict performance in Climate Zones 4 through 8. Both the field measurements and hygrothermal modeling demonstrated that exterior-insulated walls were much more tolerant to cold weather moisture accumulation as a result of interior air leakage into the assembly than thick cavity-insulated walls. There was also correlation between measurements and hygrothermal modeling on the slow drying of the wood sheathing following a rain leak with low permeance exterior insulation and interior vapor barrier. However, when the interior vapor barrier layer was removed, the assembly dried much more quickly. Both the field measurements and hygrothermal analysis from this study showed that the least risky wall with respect to moisture-related durability issues was an exterior-insulated wall with sufficient exterior insulation to allow a Class II or Class III vapor control layer to be used on the interior.

INTRODUCTION

Like all components of a building enclosure, above-grade wall systems must fulfill four important functions: to support, distribute (goods, people, and utilities), finish, and control. The most critical control layers, in order of importance, are as follows (Straube and Burnett 2005):

- Rain control layer
- Air control layer
- Vapor control layer
- Thermal control layer

Durability depends on the interaction between these layers. For example, increasing the thermal control layer (i.e., increasing insulation R-value) without considering the material properties and placement of other layers could lead to moisture accumulation and durability risks. In the development of high-R wall designs for cold climates, several approaches have been used, each with advantages and limitations.

Trevor M. Trainor is a building science researcher and **Jonathan Smegal** is an associate at RDH Building Science Laboratories, Waterloo, Ontario. **John Straube** is a principal at RDH and an associate professor at the University of Waterloo, Waterloo, Ontario. **Anil Parekh** is senior research manager at Natural Resources Canada, Ottawa, Ontario.

The *thick wall approach* relies on traditional methods of installing insulation between framing members and installing sheathing on the exterior. To allow for more insulation, the thickness of the framing is expanded. Double-stud walls, truss walls, and I-joist walls are all examples of the thick wall approach. The most critical performance concern with these thick wall systems is interstitial condensation. Because of the increased thermal resistance of the thick insulation layer, the outer wall sheathing will be colder than standard construction in winter and any construction moisture, air exfiltration, or vapor flow may result in sheathing condensation and higher moisture contents than typical 2 × 4 or 2 × 6 wood-framed assemblies (Arena et. al 2013; Ueno 2015). Because of the susceptibility of thick wall designs to condensation, an effective air barrier and good vapor control are especially critical in this type of wall system.

The *exterior insulation approach* involves the addition of a continuous insulation layer exterior to the building's structure. The benefits of the exterior insulation approach were first documented in North America in the early 1960s for masonry wall assemblies (Hutcheon 1964) and have more recently been analyzed by Kane and Titley (1987); Proskiw (1995); Maref et al. (2010); and Straube (2011). By locating insulation outboard of the sheathing, this strategy shelters an assembly's structural elements and control layers from temperature cycling and (assuming sufficient insulation is used) eliminates condensation on the inner surface of the wall sheathing. Both of these benefits can result in increased durability of the wall system. Designs that use exterior insulation also allow higher R-values to be achieved, especially when used in conjunction with a traditional insulation strategy with batt insulation between the studs. Some of the more common examples of the exterior insulation approach include insulated sheathing, Exterior Insulation Finish Systems (EIFS), and exterior spray foam.

Despite these advantages, some practitioners have resisted the use of exterior-insulated designs, out of concern that low permeability insulation outboard of the sheathing could trap moisture and result in moisture accumulation within the wall system.

This paper describes research that compared a commonly constructed 2 × 6 stud-bay insulated design to several different exterior-insulated R-30 (effective) wall assemblies and an R-34 (effective) double-stud wall design. Field measurements of full-scale wall assemblies were used to validate a hygrothermal model for a specific location in Southwestern Ontario (Climate Zone 5/6). The model was then used to predict performance for the field-tested assemblies (as well as an additional exterior-insulated assembly) in Climate Zones 4 to 8. The relative durability and thermal performance of high-R wall designs have implications for future research, practice, and code development.

Experimental Method

This experiment measured the performance of the test walls under ideal construction conditions (dry during construction, perfect airtightness and watertightness) and natural environmental exposure conditions. The experiment also measured the consequences of the two most important wetting mechanisms at different times: air leakage condensation and rain penetration.

Field measurements for this study were obtained from the University of Waterloo's natural exposure test facility, which is on the border of Climate Zones 5 and 6 (4200 heating degree days [HDD] °C / 7500 HDD °F [MMAH 2012]). Data were originally gathered as part of a larger collaboration between the Natural Sciences and Research Council of Canada (NSERC), the Network of Wood-Based Building Systems (NEWBuildS), the University of Waterloo Building Engineering Group, and Ryerson University's Department of Civil Engineering. The data from this study starts in October 2012, and ends August 2013.

Test Facility and Apparatus

The Building Engineering Group's outdoor full-scale permanent natural exposure test facility (the BEGHUT) is located on the University of Waterloo campus in Southwestern Ontario, Canada. It is approximately 34 × 34 ft. (10.5 × 10.5 m) in plan and 9.8 ft. (3.0 m) high floor to ceiling, with walls oriented in the four cardinal directions. The interior conditions were maintained at approximately 70°F (21°C) and 40% rh during this experiment. A ceiling-mounted air distribution system was used to distribute the conditioned air evenly. The roof overhang (8 in.[200 mm]) is sized to prevent shading from the sun under most conditions. The small overhang and the drip-edge in lieu of eavestrough, provide very little direct protection from rainfall. The test hut is sited on relatively flat land and is fully exposed to winds from most directions. The weather station on top of the BEGHUT measures exterior temperature, exterior relative humidity, horizontal incident solar radiation, and rainfall.

Two methods of introducing moisture into the wall systems were employed at different phases of the study: one to simulate air leakage from the interior and one to simulate rain leakage from the exterior.

To simulate an interior air leak from a point source such as an electrical outlet box, interior air (70°F [21°C], 40% rh) was pumped into the center test bay of each test wall. The injection of interior air acts as a potential wetting system during the coldest months of the year. When interior air flows past any surface within the wall assembly that is below its dew point, condensation will occur on that surface. The air injection system is composed of a high-pressure blower motor and fan, attached to a distribution system of 3/4 in. (19 mm) diameter polyethylene pipe. Manifolds are used to distribute air to each wall and airflow rotameters with integral needle valves are used to control and measure the flow entering each test wall panel. The air is introduced to the center of the test bay, at a height of 12 in. (305 mm) from the bottom of the wall through a 1 1/2 in. (38 mm) diameter plumbing cleanout. A flow rate of 0.24 L/s (30 cfh) was chosen based on air leakage rates expected in new residential construction during winter conditions (Trainor 2014). An air relief path was provided at each top plate to direct air back into the interior of the BEGHUT, as the exterior sheathing membrane acted as a reasonably effective air barrier in these test walls.

To simulate an exterior rain leak, wetting mats were applied to the exterior surface of the oriented strand board (OSB) sheathing board, interior to the sheathing membrane. Water was applied via plastic tubing with carefully punctured exit holes. The exit holes release water between a folded sheet of plastic-mesh-reinforced paper towel. This allows a controlled volume of water to be injected to a known area of the wall assembly and prevents liquid water from draining away from the site of application.

Assembly Design and Instrumentation

The datum wall for this study was a standard 2 × 6 assembly with R-24 fiberglass batt insulation, OSB sheathing, and an interior polyethylene sheet to provide air and vapor control. This type of wall has been used for almost 30 years in much of Canada and parts of the United States and is well accepted by the industry. Such a wall system provides an effective thermal resistance of approximately R-16 when accounting for the thermal bridging of the framing members (Smegal and Straube 2009).

The other four walls in the field study represent realistic construction techniques to reach higher wall R-values. A target R-value of approximately R-30 (effective) was chosen. For the high-R walls, exterior insulation was added to the 2 × 6 insulated (datum) wall design. Different insulation types with a range of vapor permeance (from very low to very high) were selected. The fourth wall used a thick wall approach with approximately 11 1/4 in. (286 mm) cellulose insulation cavity fill and double 2 × 4 framing.

Experimental variables included insulation strategy, framing thickness, vapor control strategy and air barrier strategy (Table 1). Elements common to all assemblies are listed in Table 2. The assemblies were tested on the north and south orientations of the BEGHUT in order to represent the most differentiated temperature conditions. Solar exposure on the south results in significantly higher peak and slightly higher average annual wall temperatures, while the north has the lowest winter and annual average temperature.

Each test wall was fitted with five resistance-based moisture content and temperature (MC/T) sensors: at the top and bottom plates and at three positions in the OSB sheathing. All MC/T sensors were installed along the center line of the test panels, within the test bay. As well, two relative humidity and temperature (RH/T) sensors were installed in each test wall, within the insulation at the center point of the test wall (midway through the depth of cavity), and in the air gap behind the siding. Two thermistor temperature (T) sensors were used to monitor temperature at two other positions in the assembly. All penetrations were air sealed with caulking or spray foam. Details of the instrumentation for this project and for similar

Table 1. Test Wall Variables

Test Wall	Insulation Strategy	Air/Vapor Control	Framing Thickness
1: Datum	Standard R-24 (RSI- 4.2) fiberglass batt insulation between studs	Air/vapor: 6 mil. polyethylene on the interior	Standard 2 × 6 wood studs, 24 in. (610 mm) on center
2: Mineral wool (MW)	Datum + 3 in. (76 mm) high-density MW (installed in two layers) exterior to the water control membrane	Air/Vapor: 6 mil. polyethylene on the interior	Standard 2 × 6 wood studs, 24 in. (610 mm) on center
3: Extruded polystyrene (XPS)	Datum + 2 1/2 in. (63.5 mm) of XPS (installed in two layers) exterior to the water control membrane	Vapor: XPS Air: airtight drywall	Standard 2 × 6 wood studs, 24 in. (610 mm) on center
4: Polyisocyanurate (PIC)	Datum + 2 in. (50 mm) foil-faced PIC (installed in two layers) exterior to the water control membrane	Vapor: PIC Air: airtight drywall	Standard 2 × 6 wood studs, 24 in. (610 mm) on center
5: Double stud	Cavity filled with dense-packed cellulose insulation at a density of approximately. 4 pounds per cubic foot (65 kg per cubic meter)	Air/vapor: 6 mil. polyethylene on the interior	Outer wall: 2 × 6 wood stud with a double top plate Inner wall: 2 × 4 wood stud with a single top plate Walls attached using 3/8 in. (9.5 mm) thick plywood gusset plates creating a 2 1/4 in. gap between inner and outer studs Total: 11 1/4 in. (286 mm) thick

monitoring projects can be found in the works by Trainor (2014) and Straube et al. (2002).

Field Measurement Results

Moisture Content Measurements. Moisture content (MC) sensors provide a direct measurement of moisture accumulation at the OSB sheathing and wood framing members. This moisture includes the built-in moisture, the absorption of condensation or bulk water, as well as the adsorption of moisture in the air and redistribution of moisture from surrounding materials. It is also important to consider that measured moisture content is the net effect of both wetting and drying mechanisms. Moisture content is reported as a percentage by mass. Wood moisture content measurements were used as the analysis criteria in this research to demonstrate the measured performance differences between the test walls and to correlate the measured performance to the hygrothermal analysis.

As shown in Figure 1, under the as-built conditions (Phase 1), the OSB sheathing moisture content remained primarily between 6% and 8% for the datum wall and the exterior-insulated walls, with a few brief peaks of 10% to 12% for the PIC wall during extended wet weather. The OSB moisture content for the double-stud wall was measurably higher than the other test walls, especially near the top of the walls on the south side. Moisture content readings at the upper sheathing position began to increase as the weather turned colder reaching as high as 17% without any intentional defects in construction. These results are likely because of the concentration of built-in moisture within the cellulose because of the buoyancy of the air within the solar heated (south side) double-stud wall. Many other sensor measurements are available in the work by Trainor (2014).

During Phase 2 (air injection phase), the air injection had little effect on the measured moisture content of any of the test walls at the top plate or upper sheathing positions. On the lower half of the walls near the air injection site, however, the air injection resulted in significant increases in moisture content in the double-stud and datum walls, as seen in Figure 2. The sheathing on the double-stud wall picked up moisture consistently over the air injection period and was not significantly affected by the exterior temperature variations over that time. This is likely due to the hygric buffering effects of the cellulose in the cavity. The sheathing of the datum wall, however, picked up moisture very quickly during the cold periods and dried down during periods of warmer weather. The sheathing moisture content of the exterior-insulated walls even with the injection of interior air did not exceed 10% MC.

Table 2. Test Wall Common Elements

Element	Description
Framing layout	Double top plates, single bottom plate; inner two studs of standard 2 × 6 framing lumber 16 in. on center
End studs	End studs of 3/8 in. (9.5 mm) thick plywood, attached to OSB sheathing and drywall with sheet metal L-brackets
Cladding	Prepainted fiber cement lap siding, applied continuously and sealed at the top edge
Air gap	To form a drainage and ventilation gap, the cladding was installed on vertical pine furring strips, 2 1/2 in. (64 mm) wide and 3/4 in. (19 mm) thick
Water-air control membrane	Spun bonded polyolefin, stapled to the OSB sheathing and sealed to the perimeter of the OSB and to the wall buck using 2 1/2 in. (60 mm) wide technical tape
Structural sheathing	7/16 in. (11 mm) thick OSB, attached to the wall framing using 1 1/2 in. (38 mm) wood screws at 16 in. (406 mm) on center
Interior finish	1/2 in. (12.7 mm) thick gypsum wallboard (drywall) installed with 1 1/4 in. (32 mm) drywall screws at 16 in. (406 mm) on center. and painted with two layers of latex paint

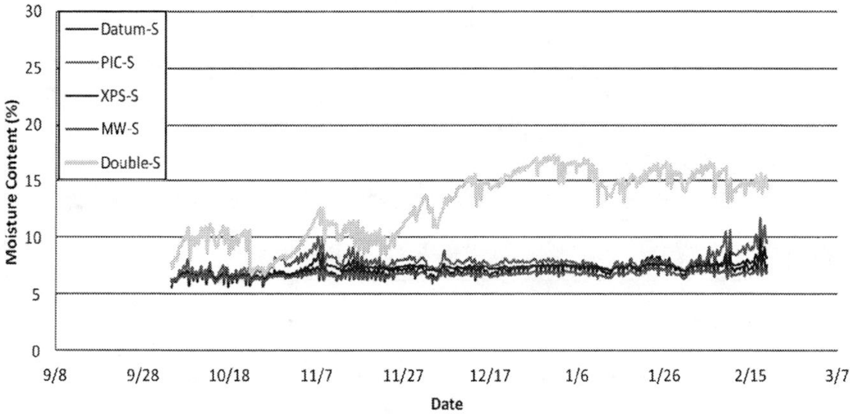

Figure 1 *Measured moisture content at upper sheathing during as-built Phase 1, south elevation.*

Rain Leaks and Drying Capability. The intentional wetting events were used to apply a controlled amount of water directly to the exterior surface of the OSB sheathing, simulating a rainwater leak at the corner of a window or other area of water concentration. It's important to note that the intentional wetting events simulated a construction failure in the water control layer of the enclosure. With a large enough failure of the water control layer, few walls are safe from durability-related issues such as mold and rot. However, drying potential can often manage small imperfections in moisture control. The primary method of quantifying drying capability in this study was to track the change in the measured sheathing MC during and following the wetting mat water application. The lower sheathing position was the focus of the assessment since it was within the field of the wetting mat.

Figure 3 shows the accumulation of moisture during and immediately following the wetting mat water injections and the subsequent drying phase (with the dashed line for each wall on the graph) at the lower sheathing position (immediately in line with the wetting mat). This graph shows that the datum and double-stud walls were the quickest to dry down to prewetting moisture content levels, requiring approximately a week and a half. The mineral wool (MW) exterior-insulated wall also dried down very quickly requiring only 2 weeks. The extruded polystyrene (XPS) and the polyisocyanurate exterior-insulated walls required 8 to 10 weeks to dry down to prewetting moisture content levels.

Clearly, the largest factor affecting time to dry is the vapor flow resistance of the assembly layers exterior to the sheathing. The XPS and PIC walls were only able to effectively dry to the interior and required approximately five times longer to dry down after wetting compared to the other test walls. These walls have not yet been dismantled to assess moisture damage visually.

HYGROTHERMAL MODELING

Combining measured data with hygrothermal modeling adds value to both research methods; measured data significantly improves the predictive validity of models, while well-designed and validated models can cost-effectively predict performance across a wide range of locations and other variables. In the current study, measured moisture performance

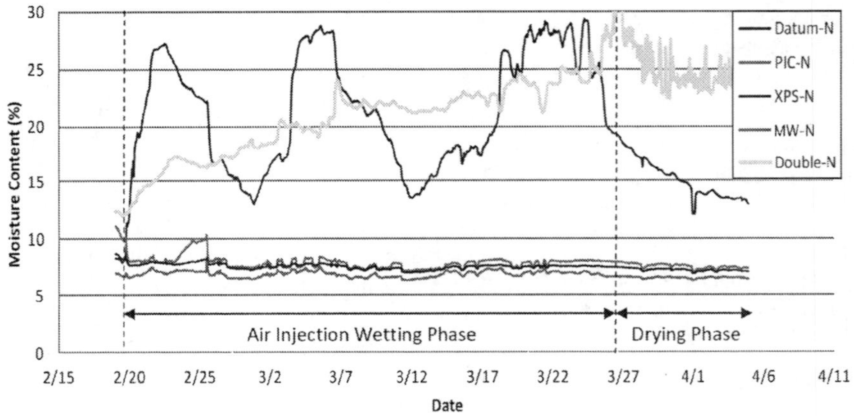

Figure 2 Measured moisture content at lower sheathing during air injection Phase 2, north elevation.

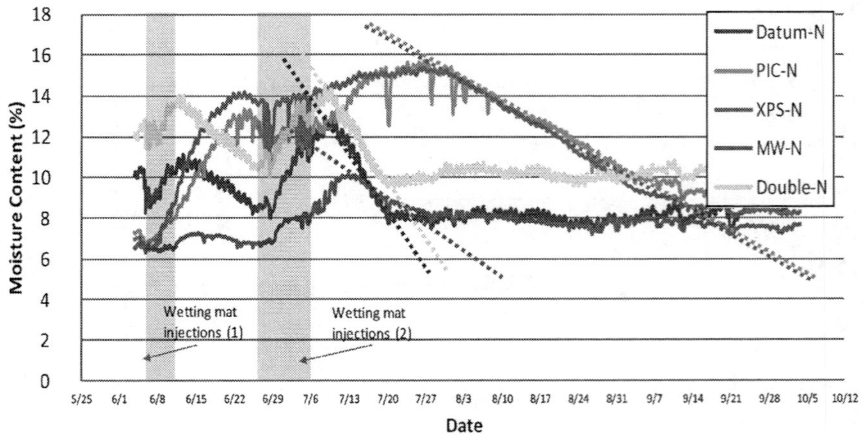

Figure 3 Measured moisture content at lower sheathing during wetting mat and subsequent drying phase, south side.

data from the full-scale test walls along with interior and exterior boundary conditions were used to validate a hygrothermal model for the testing location in Climate Zone 5/6. The model was then used to predict performance in Climate Zones 4 to 8.

Approach

Before hygrothermal simulations were conducted, simulations were set up in WUFI® Pro (Fraunhofer IBP 2011) for existing wall assemblies, using measured boundary conditions, to determine if the predicted moisture performance of the assemblies closely matched measured performance.

Model Validation. Measured data from the University of Waterloo NSERC-NewBuilds High-R Wall Study (Trainor 2014) was used to correlate the measured performance to the predicted performance. North-orientation wall data from November 1, 2012 through October 31, 2014 was used for this validation. The north orientation was selected because it experiences very little direct solar energy, which complicates the correlation. As well, elevated moisture content levels were required in the test walls to be able to correlate the predicted moisture content performance, and the north orientation provides the worst-case scenario for sheathing moisture content, due to cold weather moisture accumulation.

Each of the modeled assemblies was run with a custom two-year Waterloo, Ontario outdoor weather file and an interior conditions file which matched the time-frame of the monitoring data. The results from each simulation were then compared to monitoring data. Specifically, comparisons were made between measured and predicted cladding temperature, sheathing temperature, and sheathing moisture content. As an additional check of the models, a sensitivity analysis was performed for the following three variables most relevant to the predicted hygrothermal performance of these wall systems in cold climates:

- Exterior insulation level
- Climate zone
- Interior RH

The WUFI models were found to provide results that were consistent with field monitoring data and appropriately sensitive to critical modeling parameters. Figure 4 shows one example of the agreement of the WUFI predicted sheathing moisture content and the actual measured sheathing moisture content for the PIC wall over a two-year period.

Following the validation of the modeled wall assemblies, three rounds of simulations were conducted for the datum wall, the double-stud wall, and the three types of exterior-insulated wall systems (PIC, XPS, and MW exterior insulation). An additional exterior-insulated wall was also modeled using expanded polystyrene (EPS). Standard WUFI material data were used where available. Where standard data were unavailable, custom material files were created. The three rounds of hygrothermal simulations were conducted as follows:

1. Baseline simulations were first performed for each geographic location with the datum wall, the double-stud wall, and the exterior-insulated walls with approximately RSI 2.1 (R-12) of exterior insulation. These simulations assumed a perfect air barrier and no rain penetration and were used to determine the effect of climate on peak sheathing moisture content for these assemblies.
2. Simulations were then performed on the exterior-insulated wall systems with higher levels of exterior insulation in select cold climates and lower levels of exterior insulation in select warmer climates. These simulations also assumed a perfect air barrier and no rain penetration and were used to determine the interactive effect of climate and the exterior/interior insulation ratio on peak sheathing moisture content.
3. Finally, a third round of simulations was performed using the models from the baseline testing at select locations

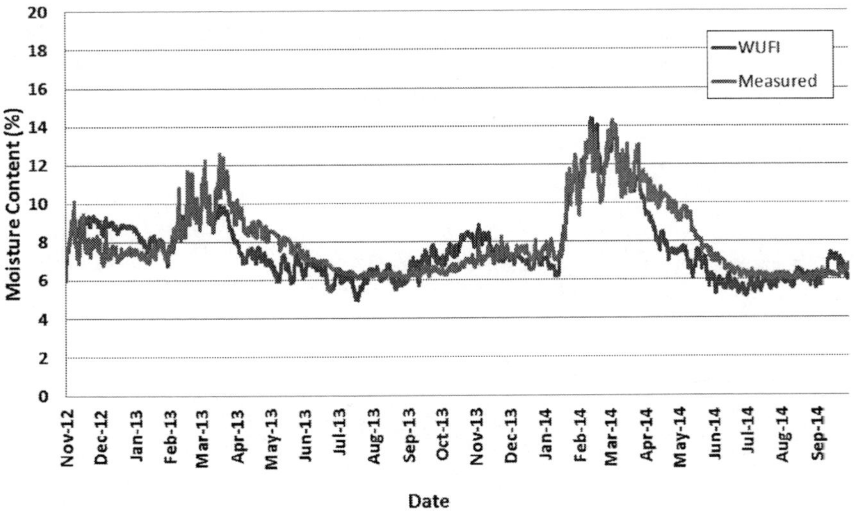

Figure 4 *Upper OSB moisture content: measured versus WUFI model for the PIC wall over a 2-year period.*

with the addition of air exfiltration or rain leakage to determine the effects of these conditions on peak sheathing moisture content.

Experimental Model Parameters. The parameters for the WUFI modeling runs were chosen to support the main objectives of the study: to evaluate the performance of wood-framed, high-R wall systems under a variety of realistic conditions. The eight locations in Canada that were chosen for WUFI modeling are shown in Table 3, along with some of the relevant climate parameters for each location.

The north orientation was used for all simulations that did not involve rain leakage. As mentioned above, the north orientation has the lowest drying capacity and is often the worst-case for cold weather moisture accumulation and elevated sheathing moisture content. For simulations that included rain leakage, the elevation with the highest driving rain exposure was used. There were two levels of interior RH used for simulations in all locations:

- Low: Sine curve function with 30% wintertime RH to represent an air-leaky house with low moisture generation and/or adequate mechanical ventilation.
- Medium: Sine curve function with 40% wintertime RH to represent a more airtight house with more moisture generation and/or less than adequate ventilation.

A third, lower RH level (20%) was used for simulations in Winnipeg and Yellowknife because of the extreme cold winter conditions. To compare the moisture durability of the wall assemblies, the MC of a thin slice (1/8 in. [3 mm]) of the structural sheathing on the interior and exterior faces was obtained from the simulations on an hourly basis. OSB sheathing moisture content was used as the performance criteria because this is generally where moisture will accumulate and wood sheathing is a moisture-susceptible material. The peak annual OSB sheathing MC was determined and the risk was assessed based on the following criteria:

- Low Risk: Peak OSB sheathing MC <20%, no mold growth.
- Moderate Risk: Peak OSB sheathing MC 20%–28%; potential for mold growth eventually, depending on frequency, length of wetting, and temperatures during wetting. This design can be successful, but conservative durability assessments usually require corrective action.
- High Risk: Peak OSB sheathing MC >28%; moisture-related problems are expected and this design is not recommended.

To test the effect of changing the ratio of exterior insulation to total R-value on sheathing moisture content, the simulations for the exterior-insulated walls were repeated with the same cavity insulation (R-24) and varying amounts of exterior insulation. In the very cold climates of Winnipeg and Yellowknife, exterior insulation values were increased to approx. RSI 3.1 (R-18) (43% insulation ratio) and RSI 4.2 (R-24) (50% insulation ratio). In the warmer climates of Vancouver, Toronto, and Ottawa, exterior insulation levels were reduced to approximately RSI 1.0 (R-6) (20% insulation ratio).

To test the effect of exterior insulation on the drying capacity of the wall systems, the simulations were repeated with two levels of simulated rain leakage. This was accomplished by creating a moisture source on the exterior surface of the sheathing layer. Simulations were performed with this moisture source at 1% and 2% of driving rain in Vancouver, Toronto, and St. John's. For these simulations, the elevation with highest driving rain exposure was used (east for Vancouver, south-east for Toronto, and south for St. John's).

To test the resistance of the wall systems to air leakage condensation, two levels of air leakage were tested: 0.02 L/s/m^2 (0.04 cfm/ft^2) and 0.08 L/s/m^2 (0.16 cfm/ft^2). These values represent the approximate natural air leakage rate for a very

Table 3. Locations and Relevant Climate Parameters in WUFI Analysis

Location	2.5% Design Temperature, °C	Heating Degree Days, HDD18	Annual Precipitation, mm	Annual Driving Rain, mm	IECC Climate Zone
Vancouver, BC	–7	2910	1169	950	4
St. John's, NL	–15	4881	914	900	6
Toronto, ON	–20	3800	606	225	5
Ottawa, ON	–25	4440	586	400	6
Quebec City, QC	–25	5080	807	375	6/7
Calgary, AL	–30	5000	304	250	6/7
Winnipeg, MB	–33	5670	309	200	7
Yellowknife, NWT	–41	8170	289	N/A	8

Note: 2.5% Design temperature and heating degree days data are from NRC (2012). Precipitation and driving rain are from WUFI Pro Standard Weather Files (Fraunhofer IBP 2011).

airtight building (approximately 0.6 ACH_{50} as required for Passive House certification [Passive House Institute 2014]) and a moderately airtight building (approx. 2.5 ACH_{50}, as required for ENERGY STAR® certification) respectively. These models were tested on all simulated wall assemblies under the climatic conditions of Vancouver, Toronto, and St. John's.

MODELING RESULTS

The results of the calibrated hygrothermal modeling study show the same general trends as the measured data. Figures 5–8 assume the following parameters:

- Walls are residential wood frame with light-colored thin cladding facing north. This is a worst-case scenario for cold-weather diffusion wetting.
- Results are for OSB sheathing. Plywood sheathing values will be equal or lower. OSB permeance is always over 60 ng/Pa·s·m^2 in exterior sheathing applications.
- Effective air barrier is assumed to be installed, as is proper rain control.
- MC values are for inner 0.12 in. (3 mm) OSB sheathing.

As shown in Figure 5, in the absence of air exfiltration and rain leakage, the double-stud wall consistently showed sheathing moisture content higher than the standard (datum) wall, but remained within the low-risk zone in all locations without air or water leakage. The results for the exterior-insulated walls demonstrate the importance of the exterior/total insulation ratio and its relationship with both climate and interior relative humidity. When the level of exterior insulation was not adequate for the climate and/or interior RH, sheathing moisture contents exceeded safe levels without sufficient interior vapor control. As shown in Figure 6, however, when a sufficient insulation ratio was used, the exterior-insulated walls remained within the low risk zone, even without an interior polyethylene vapor barrier. More specifically, results showed that:

(shading) = MC < 20%, no mold growth; (shading) = MC is 20 to 28%, potential for mold growth; (shading) = MC > 28%, moisture problems expected, this design is NOT recommended

Wall	Cavity Insulation	Cavity Depth	Exterior Insulation	Ext. Insul. Thickness	Vapour Control	Vancouver 2.5% Temp. = -7 HDD 2910 Low RH 30/55%	Vancouver Med. RH 40/60%	St. John's 2.5% Temp. = -15 HDD 4881 Low RH 30/55%	St. John's Med. RH 40/60%	Toronto 2.5% Temp. = -20 HDD 3800 Low RH 30/55%	Toronto Med. RH 40/60%	Ottawa 2.5% Temp. = -25 HDD 4440 Low RH 30/55%	Ottawa Med. RH 40/60%	Quebec City 2.5% Temp. = -25 5080 Low RH 30/55%	Quebec City Med. RH 40/60%	Calgary 2.5% Temp. = -30 HDD 5000 Low RH 30/55%	Calgary Med. RH 40/60%	Winnipeg 2.5% Temp. = -33 HDD 5670 Ex.Low RH 20/50%	Winnipeg Low RH 30/55%	Winnipeg Med. RH 40/60%	Yellowknife 2.5% Temp. = -41 HDD 8170 Ex.Low RH 20/50%	Yellowknife Low RH 30/55%	Yellowknife Med. RH 40/60%
Datum	Fiberglass	5.5"	none	0	Polyethylene sheet	10%	11%	11%	11%	11%	11%	10%	10%	11%	11%	8%	8%	10%	10%	10%	11%	11%	12%
PIC	Fiberglass	5.5"	polyiso-cyanurate	2"	Latex paint+primer	10%	11%	12%	15%	11%	13%	12%	18%	13%	19%	14%	21%	12%	20%	30%	15%	30%	35+%
					Polyethylene sheet												7%			8%		7%	
XPS	Fiberglass	5.5"	extruded polystyrene	2.5"	Latex paint+primer	10%	11%	12%	15%	11%	13%	12%	14%	13%	19%	14%	21%	12%	20%	30%	15%	30%	35+%
					Polyethylene sheet												7%			8%		7%	
EPS	Fiberglass	5.5"	expanded polystyrene	3.0"	Latex paint+primer	10%	11%	12%	17%	11%	14%	13%	20%	14%	20%	15%	24%	12%	21%	34%	17%	34%	35+%
					Polyethylene sheet								8%		8%		7%			8%		7%	
MW	Fiberglass	5.5"	mineral wool insulated sheathing	3.0"	Latex paint+primer		12%						12%		15%		16%	12%	17%	25%	13%	25%	35+%
					Polyethylene sheet	8%	9%	8%	8%	9%	9%	8%	8%	9%	9%	7%	7%	7%	6%	8%	7%	7%	8%
Double Stud	Cellulose	11.25"	none	0	Polyethylene sheet	12%	13%	13%	14%	13%	13%	14%	14%	14%	14%	13%	13%	15%	15%	14%	16%	16%	16%

Figure 5 *Modeling results: maximum sheathing moisture content for baseline conditions (no rain or air leakage).*

(shading) = MC < 20%, no mold growth; (shading) = MC is 20 to 28%, potential for mold growth; (shading) = MC > 28%, moisture problems expected, this design is NOT recommended

Wall	Cavity Insulation	Cavity Depth	Exterior Insulation	Ext. Insul. Thickness	Vapour Control	Winnipeg R-12 (RSI 2.1) HDD 5777 Low RH 30/55%	Winnipeg R-12 Med. RH 40/60%	Winnipeg R-18 (RSI 3.1) HDD 5777 Low RH 30/55%	Winnipeg R-18 Med. RH 40/60%	Winnipeg R-24 (RSI 4.2) HDD 5777 Low RH 30/55%	Winnipeg R-24 Med. RH 40/60%	Yellowknife R-12 (RSI 2.1) HDD 8166 Low RH 30/55%	Yellowknife R-12 Med. RH 40/60%	Yellowknife R-18 (RSI 3.1) HDD 8166 Low RH 30/55%	Yellowknife R-18 Med. RH 40/60%	Yellowknife R-24 (RSI 4.2) HDD 8166 Low RH 30/55%	Yellowknife R-24 Med. RH 40/60%	Vancouver R-6 (RSI 1.0) HDD 2910 Low RH 30/55%	Vancouver Med. RH 40/60%	Toronto R-6 (RSI 1.0) HDD 3800 Low RH 30/55%	Toronto Med. RH 40/60%	Ottawa R-6 (RSI 1.0) HDD 4440 Low RH 30/55%	Ottawa Med. RH 40/60%
PIC	Fiberglass	5.5"	polyiso-cyanurate	2.0" 3.0" 4.0"	Latex paint+primer	20%	30%	15%	23%		18%	30%	35+%	15%	30%	12%	15%	12%	14%	13%	19%	16%	26%
					Polyethylene sheet		8%		8%		8%	7%			7%		7%				9%		9%
XPS	Fiberglass	5.5"	extruded polystyrene	2.5" 3.5" 5.0"	Latex paint+primer	20%	30%	15%	23%		18%	30%	35+%	15%	30%	12%	15%	12%	12%	13%	20%	18%	28%
					Polyethylene sheet		8%		8%		8%	7%			7%		7%				9%		9%
EPS	Fiberglass	5.5"	expanded polystyrene	3.0" 4.5" 6.0"	Latex paint+primer	21%	34%	17%	24%		19%	34%	35+%	17%	34%	12%	17%	12%	12%	13%	20%	18%	28%
					Polyethylene sheet		8%		8%		8%	7%			7%		7%				9%		9%
MW	Fiberglass	5.5"	mineral wool insulated sheathing	3.0" 4.5" 6.0"	Latex paint+primer	17%	25%	12%	19%		14%	25%	35+%	13%	25%	10%	13%	9%	12%	12%	16%	15%	24%
					Polyethylene sheet	8%	8%		8%		8%	7%	8%		7%	7%	7%				9%	8%	9%

Figure 6 *Modeling results: maximum sheathing moisture content for higher and lower levels of exterior insulation.*

- When an interior vapor barrier was present, and there was no air exfiltration or rain leakage, all of the walls tested were in the low-risk zone for sheathing MC. The sheathing MC for the exterior-insulated walls remained very low at 8% or below, while the datum wall remained at 11% or lower and the double-stud wall ranged from 12% in warmer climates to 16% in the colder climates.
- The interior RH had very little effect on sheathing moisture content for walls with interior polyethylene. In the walls with low-permeance exterior insulation and no interior polyethylene (only latex paint) in colder climates (Ottawa and colder), the increase of interior RH from the low to the medium level resulted in a significant increase in sheathing moisture content, in some cases to risky levels. This shows that the durability of the wall system is closely related to the interior relative humidity, exterior climate, and vapor control performance.
- Higher levels of exterior insulation resulted in decreased sheathing moisture content. The addition of an inadequate level of exterior insulation without a polyethylene vapor barrier was found to result in increased sheathing moisture content risk levels compared to the standard (datum) wall in several cases.

The results of the simulations that included rain leakage can be seen in Figure 7. When rain leakage is included in the model, sheathing moisture content appears to be a function of the amount of rain exposure, the annual climatic temperature conditions, and the drying capacity of the wall system. The drying capacity of the wall system is significantly reduced with the addition of low-permeance exterior insulation; however, much of that reduction can be mitigated by removing the interior polyethylene vapor barrier to allow inward drying. Removal of the interior polyethylene vapor barrier should only be done when a sufficient amount of exterior insulation is used.

The results of the simulations that included air leakage can be seen in Figure 8. When air leakage was included in the simulation, the walls with adequate levels of exterior insulation were not significantly affected by the air leakage. Conversely, the datum wall and double-stud wall showed a significant increase in sheathing moisture content under the air leakage conditions. Air exfiltration at lower rate resulted in moisture contents at or near the moderate risk level, while air leakage at the higher rate resulted in moisture content levels at or near the high risk levels for these walls. A more thorough analysis of the modeling results including graphs can be found in the work by Trainor and Smegal (2015).

CONCLUSIONS

As building and energy codes continue to call for higher thermal performance, and move toward using effective R-values as the standard for construction, assembly designs with more insulation will be required. Abundant previous field research, dew point analysis calculations, moisture physics calculations, and hygrothermal simulations support the use of insulation to the exterior of the structure and control layers. Yet there are still moisture-related durability concerns about continuous exterior insulation from some practitioners and designers in the industry.

This study compared measured data from standard wood-framed walls without exterior insulation, high-R exterior-insulated walls, and a thick high-R wall design. Measured and modeled data showed that the thick high-R wall (with a high level of insulation between the sheathing and drywall) had significantly more risk and higher measured sheathing moisture content than the standard 2 × 6 wood-framed wall, both

= MC < 20%, no mold growth = MC is 20 to 28%, potential for mold growth = MC > 28%, moisture problems expected, this design is NOT recommended

	Wall Construction					Vancouver Ave. Annual Temp 9 °C HDD 2910, 304 mm Driving Rain			Toronto Ave. Annual Temp 7 °C HDD 3800, 77 mm Driving Rain				St. Johns Ave. Annual Temp 4 °C HDD 4881, 291 mm Driving Rain		
Wall	Cavity Insulation	Cavity Depth	Exterior Insulation	Ext. Insul. Thickness	Vapour Control	No Leakage	1% Rain leak	2% Rain leak	No Leakage	1% Rain leak	2% Rain leak	5% Rain leak	No Leakage	1% Rain leak	2% Rain leak
Datum	Fiberglass	5.5"	none	0	Polyethylene sheet	11%	15%	22%	11%	11%	12%	15%	11%	15%	22%
PIC	Fiberglass	5.5"	polyiso-cyanurate	2"	Latex paint+primer	11%	16%	24%	13%	13%	17%	23%	15%	25%	35+%
					Polyethylene sheet		35+%	35+%		11%	15%	35+%		35+%	35+%
XPS	Fiberglass	5.5"	extruded polystyrene	2.5"	Latex paint+primer	11%	16%	24%	13%	13%	17%	23%	15%	25%	35+%
					Polyethylene sheet		35+%	35+%			15%	35+%		35+%	35+%
EPS	Fiberglass	5.5"	expanded polystyrene	3.0"	Latex paint+primer	11%	17%	27%	14%	15%	18%	25%	17%	26%	35+%
					Polyethylene sheet		35+%	35+%			20%	35+%		35+%	35+%
MW	Fiberglass	5.5"	mineral wool insulated sheathing	3.0"	Latex paint+primer		13%	18%			12%	16%		13%	20%
					Polyethylene sheet	9%	12%	18%	9%	9%	9%	13%	8%		20%
Double Stud	Cellulose	11.25"	none	0	Polyethylene sheet	13%	15%	20%	13%	13%	16%	19%	14%	17%	20%

Figure 7 Modeling results: maximum sheathing moisture content including rain leakage.

= MC < 20%. no mold growth = MC is 20 to 28%, potential for mold growth = MC > 28%, moisture problems expected, this design is NOT recommended

Wall Construction						Vancouver			Toronto			St. Johns		
						2.5% Temp. = -7			2.5% Temp. = -20			2.5% Temp. = -15		
						HDD 2926, 304 mm Driving Rain			HDD 4065, 77 mm Driving Rain			HDD 5777, 291 mm Driving Rain		
Wall	Cavity Insulation	Cavity Depth	Exterior Insulation	Ext. Insul. Thickness	Vapour Control	No Exfiltration	Air Exfil. 0.5 ACH	Air Exfil. 2.0 ACH	No Exfiltration	Air Exfil. 0.5 ACH	Air Exfil. 2.0 ACH	No Exfiltration	Air Exfil. 0.5 ACH	Air Exfil. 2.0 ACH
Datum	Fiberglass	5.5"	none	0	Polyethylene sheet	11%	17%	29%	11%	22%	35+%	11%	24%	35+%
PIC	Fiberglass	5.5"	polyiso-cyanurate	2"	Latex paint+primer	11%	12%	13%	13%	15%	17%	15%	18%	20%
					Polyethylene sheet			13%			16%			
XPS	Fiberglass	5.5"	extruded polystyrene	2.5"	Latex paint+primer	11%	12%	13%	13%	14%	16%	15%	18%	20%
					Polyethylene sheet									
EPS	Fiberglass	5.5"	expanded polystyrene	3.0"	Latex paint+primer	11%	12%	13%	14%	18%	19%	17%	20%	22%
					Polyethylene sheet									
MW	Fiberglass	5.5"	mineral wool insulated sheathing	3.0"	Latex paint+primer									15%
					Polyethylene sheet	9%	10%	12%	9%	10%	13%	8%	11%	13%
Double Stud	Cellulose	11.25"	none	0	Polyethylene sheet	13%	25%	28%	13%	25%	29%	14%	31%	35+%

Figure 8 *Modeling results: maximum sheathing moisture content including air leakage.*

under normal operating conditions and when there was a source of interior air leakage into the assembly. The exterior-insulated wall assemblies had the lowest sheathing moisture content under normal operating conditions and the lowest sheathing moisture content levels when exposed to intentional air leakage from the interior. During simulated rain leakage, the standard 2 × 6 wall, the double-stud wall, and the wall with permeable MW exterior insulation had the highest drying rates and maintained the lowest sheathing moisture content, showing the importance of vapor movement to the exterior for rapid drying of sheathing moisture. The walls with low-permeance exterior insulation were significantly slower to dry and showed higher sheathing moisture levels as the moisture movement to the exterior was dramatically reduced. When an interior polyethylene vapor barrier was present, sheathing moisture content levels increased quickly to dangerous levels; however, when the interior polyethylene was removed, these walls were able to dry to the interior and could tolerate some level of rain leakage.

In both measured field testing and modeled hygrothermal simulations, using continuous exterior insulation was shown to be a more moisture-durable option than either standard 2 × 6 construction or thick high-R value wall assemblies, both under normal operating conditions and when subjected to an interior air leak of warm moist air at a failure in the interior air control layer. When sufficient exterior continuous insulation is installed, it is recommended to decrease the vapor resistance of the interior vapor control layer to allow the possibility of moisture redistribution to the interior if necessary.

REFERENCES

Arena, L., P. Owens, and P. Mantha. 2013. Measured performance of an R-40 double-stud wall in Climate Zone 5a. Presented at the Thermal Performance of the Exterior Envelopes of Whole Buildings XII International Conference, Clearwater, FL, December 1–5.

Fraunhofer IBP. 2011. *WUFI® Pro* v. 5.1. Holzkirchen, Germany: Fraunhofer Institute for Building Physics.

Hutcheon, N. 1964. Principles applied to an insulated masonry wall. *Canadian Building Digest* 50. Ottawa, ON: National research Council of Canada.

Kane, R., and G. Titley. 1987. Case histories of moisture monitoring in residential walls. *Thermal Insulation: Materials and Systems,* ASTM STP 922: 615–29.

Maref, W., M. Armstrong, M. Rousseau, and W. Lei. 2010. A field monitoring investigation of the effect of adding different exterior thermal insulation materials on the hygrothermal response of wood-frame walls in a cold climate. Presented at the Building Enclosure Science & Technology Conference, Portlant, OR, April 12–14.

MMAH. 2012. Ministry of Municipal Affairs and Housing. *Ontario building code 2012 compendium, supplementary standard SB-1.* Toronto, ON: ServiceOntario Publications.

NRC. 2015. National Building Code of Canada. Appendix C—Table C-2. Ottawa, ON: National Research Council.

Passive House Institute. 2014. *Passive house requirements.* Darmstadt, Germany: Passive House Institute. www.passiv.de/en/02_informations/02_passive-house-requirements/02_passive-house-requirements.htm.

Proskiw, G. 1995. The performance of energy-efficient residential building envelope systems. *Proceedings of Thermal Performance of the Exterior Envelopes of Buildings VI,* Clearwater, FL: 391–8.

Smegal, J., and J. Straube. 2009. *High-R walls case study analysis.* Building America Report 0903, Special

Research Project. Gaithersburg, MD: National Renewable Energy Laboratory.

Smegal, J., and J. Straube. 2014. *Building America special research project: High-R walls case study analysis* (BA-0903). Somerville, MA: Building Science Corporation.

Straube, J. 2011. Controlling cold-weather condensation using insulation. *Building Science Digest* 163.

Straube, J., and E. Burnett. 2005. *Building science for building enclosures*. Westford, MA: Building Science Press.

Straube, J., D. Onysko, and C. Schumacher. 2002. Methodology and design of field experiments for monitoring the hygrothermal performance of wood frame enclosures. *Journal of Building Physics* 26(2): 123-151.

Trainor, T. 2014. *The hygrothermal performance of exterior insulated wall systems*. Master's Thesis, University of Waterloo, Waterloo, Ontario.

Trainor, T., and J. Smegal. 2015. *Analysis of exterior insulated high-R wall systems*. Ottawa: Natural Resources Canada (publication pending).

Ueno, K. 2015. *Monitoring of double-stud wall moisture conditions in the northeast*. Building America Report. Gaithersburg, MD: National Renewable Energy Laboratory.

Field Evaluation of Thermal and Moisture Response of Highly Insulated Wood-Frame Walls

Michael A. Lacasse, PhD, PEng
Wahid Maref, PhD
Member ASHRAE
Silvio Plescia, PEng
Hamed H. Saber, PhD
G. Ganapathy, BEng (ME)
Anil Parekh, MASc
Member ASHRAE

ABSTRACT

Over the years, the energy efficiency of the North American housing stock has significantly improved, mainly because of higher insulation levels, more efficient windows, and, more importantly, adoption of various energy-efficiency measures by building codes. The increased insulation levels of building envelopes for homes leads to a multitude of opportunities as well as challenges. A major opportunity is to reduce heat losses and thereby significantly reduce the space heating loads. However, there are many challenges that necessitate changes in the current construction process, durability of products, and, more importantly, the effect of higher insulation levels on the overall moisture performance and expected long-term performance of the building envelope. A major barrier to the uptake of highly insulated homes is the lack of proven evidence of reliable thermal and moisture performance of these homes as might be achieved in various climates of Canada. Following a survey of current construction practices, a set of six wall assemblies with different types of exterior insulation systems were chosen for field monitoring study undertaken in two separate phases over a 2-year period in a test facility located in Ottawa, Canada. These wall assemblies ranged from total insulation value was R-27 to R-45 (RSI 4.8 to RSI 7.9). Full-scale testing included year-round performance monitoring. These wall specimens were installed in a side-by-side test bay and were subjected to local climate conditions of Ottawa, Canada; on the interior of the specimens conditions were nominally maintained at 20°C and 50% rh. This paper provides results of field trials of the six wall assemblies in terms of their hygrothermal performance and risk for condensation over a 2-year period of operation.

INTRODUCTION—PROJECT OVERVIEW

The Canadian market for new residential home construction and existing retrofits was about 79 million m^2, which represents an investment of about CDN $9.7 billion in 2012 (Canada Mortgage and Housing Corporation 2014). Conservatively assuming a 25% market share by 2025, highly insulated (enhanced) envelopes could save 11.6 PJ of energy and 1.2 megatonnes of greenhouse gas per year (Environment Canada 2013). This represents a significant economic opportunity for the housing industry within Canada and for value-added exports such as manufactured panelized wall systems. This project is needed to provide the scientific data and analysis necessary to characterize the thermal and moisture performance of advanced residential wall systems. Such information is needed to inform any considerations for further energy-efficiency improvements to building codes and voluntary residential energy-efficiency program requirements. Evidence was needed to guide the development of highly energy efficient, cost-effective, durable, and buildable solutions that are well integrated with Canadian housing system practices.

Given the heightened interest of homebuilders to provide homes that met or exceeded ENERGY STAR® requirements, and their voiced concerns regarding "super-insulted" homes, the intent of this work was to demonstrate compliance of a set of highly insulated wall assemblies as compared to a code-compliant reference wall in respect to their anticipated thermal and hygrothermal performance when subjected to Canadian climate extremes. On the basis of providing useful information to building practitioners, National Research Council Canada

Michael A. Lacasse and **Wahid Maref** are senior research officers, **Hamed H. Saber** is an associate research officer, and **G. Ganapathy** is a technical officer at NRC-Construction, National Research Council Canada, Ottawa, ON, Canada. **Silvio Plescia** is a senior researcher, policy and research division, at Canada Mortgage and Housing Corporation, Ottawa, ON, Canada. **Anil Parekh** is a senior researcher, buildings and renewables group, Natural Resources Canada, Ottawa, ON, Canada.

(NRC) undertook field monitoring and numerical modeling to investigate the risk of condensation in wall assemblies having different combinations of increased thermal resistance (R-value) of insulation for selected insulation products.

To achieve higher wall insulation levels in wood-frame wall assemblies, it has been commonplace to apply rigid or semi-rigid board insulation over the inside or outside of the framing. Alternatively, double wall or deep wall construction assemblies can be used to attain the desired level of insulation. However, concerns have been expressed in the building community regarding the use of progressively higher levels of insulation in wall assemblies as a means of attaining greater energy savings. It was felt that the use of higher insulation levels would increase the risk of incurring moisture problems given the perceived effect that highly insulated walls have on the wetting and drying characteristics of walls.

To help address the development of such unintended consequences, a collaborative research project with government and industry participation was initiated to explore, develop, test, and document the energy performance and moisture response of highly insulated residential wall systems. The information derived from the project will be used to guide the development of durable new wood-frame wall assemblies.

DESCRIPTION

The intent of the project was to develop information on the moisture and thermal performance of progressively higher insulated wood-frame wall assemblies. The information is to be used to: 1) support the evaluation of future code proposals regarding energy efficiency improvements to building envelope systems; 2) support the development of knowledge, details, and practices for advanced wall systems for voluntary residential energy-efficiency programs; 3) help the housing industry meet near and net zero energy targets; and 4) promote the deployment of highly energy-efficient wall details for new and retrofitted existing wood-frame construction. In essence, the primary outcome from this project was to facilitate the widespread adoption of high-performance residential wood-frame wall systems (EGH-83 and 86 [EnerGuide for Houses]) that are practical, buildable, durable, energy efficient, and affordable.

To achieve these objectives, the project focused on investigating the moisture and energy performance of thermally enhanced wood-frame walls with progressively higher levels of insulation from a baseline minimum (RSI 3.27–4.13) to new R-2000 performance levels (RSI 7.04–7.92) (National Building Code 2010). It permitted comparing the heat, air, and moisture response of the advanced wall systems to the baseline code walls and thereafter conduct the necessary analysis to develop meaningful information that could be used to address, and thereby alleviate, industry concerns regarding the unintended consequences of enhancing the thermal performance of walls with additional insulation on the moisture performance of these wall assemblies over time.

NRC, in collaboration with the Canada Mortgage and Housing Corporation and Natural Resources Canada, industry partners that included the Canadian Wood Council, Homeowners Protection Office of British Columbia, and the Cellulose Insulation Manufacturers Association of Canada, developed proposals for six wall assemblies that could potentially be used in progressively higher thermally performing houses between baseline code to R-2000 levels. The wall assemblies were exposed to outdoor conditions in NRC's Field Exposure of Walls test facility (FEWF) over a period of 9 months in a yearly cycle, with three walls tested every cycle.

In this paper, results are provided of the monitoring of the hygrothermal response of a set of six highly insulated wall assemblies to local climate conditions and when subjected to forced exfiltration conditions. Given that experimental tests to determine the hygrothermal performance of different wall systems in different Canadian locations is time consuming and expensive, an advanced hygrothermal modeling tool was used in this project. The modeling tool used in this study (hygIRC-C) was benchmarked against experimental data obtained from Phase 1 and Phase 2 of the project (Lacasse et al. 2016). Once the benchmarking of the modeling tool was completed, it was then used to investigate the effect of different outdoor and indoor conditions as may be found in the different regions of Canada on the moisture and thermal performance of the wall assemblies.

RESEARCH APPROACH

Scope of Testing

The field monitoring of the six different wall assemblies was undertaken in two separate phases over a 2-year period in the NRC's FEWF located at the NRC campus in Ottawa, Canada. Three wall assemblies were tested in the initial year of Phase 1 of the project (June 2013 to October 2014) and another three wall assemblies were tested in Phase 2 of the second project year (January 2015 to September 2015). A detailed description of the wall assemblies of Phases 1 and 2 are available in the work by Lacasse et al. (2016).

Field monitoring of the test specimens was conducted in each project phase nominally over a 9-month period; during this period test specimens were subjected to a broad range of weather conditions as was experienced in Ottawa over the respective time periods.

DESCRIPTION OF WALL SPECIMENS

Phase 1 Test Specimens

The three wall test specimens of Phase 1 (1219 × 1829 mm [4 × 6 ft]) consisted of 38 × 140 mm (2 × 6 in.) wood-frame walls installed side-by-side in the FEWF (Figure 1). The different material layers and the dimensions of the wall specimens are given in Figures 2 to 4. The backup wall for all three design strategies consisted of interior drywall (12.7 mm [0.5 in.] thick), polyethylene air and vapor barrier (6 mil thick), 38 × 140 mm (2 × 6 in.) wood-frame wall with friction-fit glass fiber batt

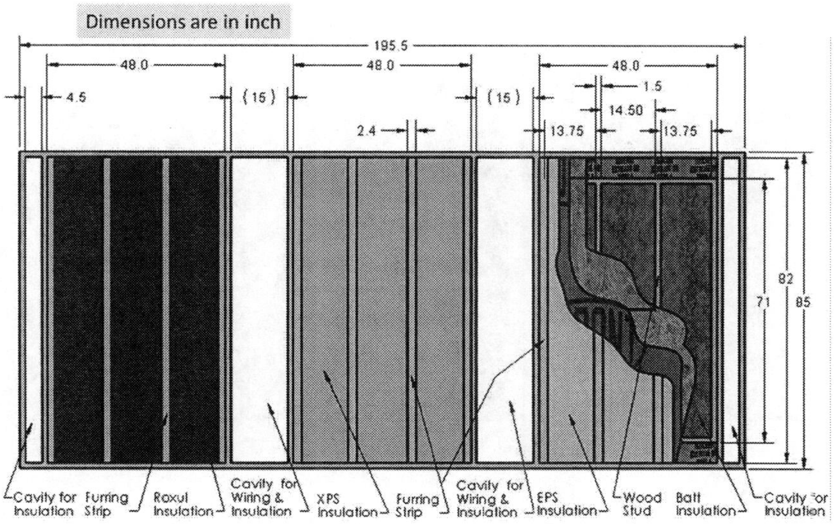

Figure 1 *Schematic of three 38 × 140 mm (2 × 6 in.) wood-frame wall (residential) test specimens installed side by side in FEWF.*

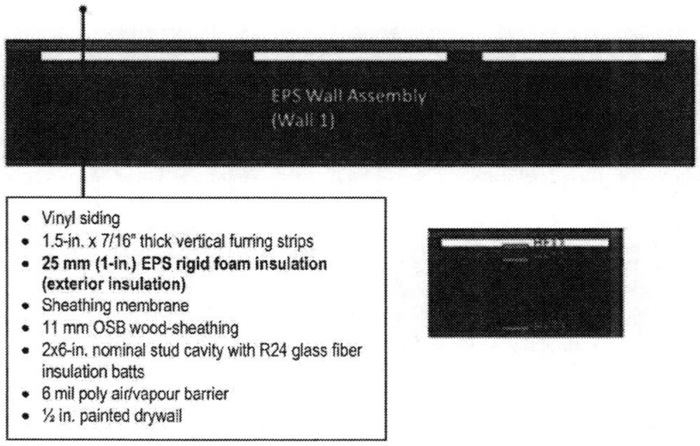

Figure 2 *Wall 1 (W1: EPS)—Horizontal cross section through expanded polystyrene (EPS) wall assembly showing locations of heat flux transducers (HFTs).*

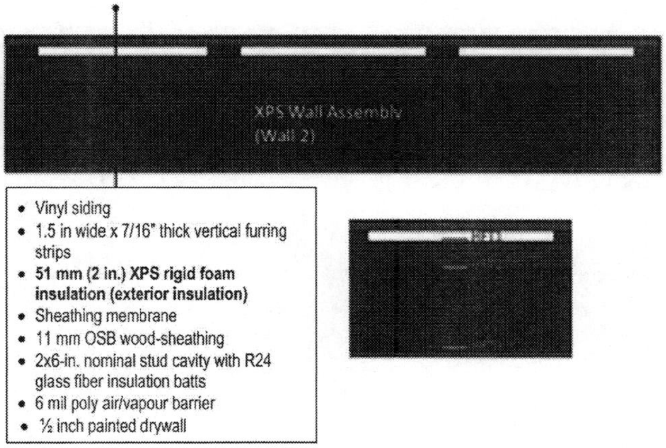

Figure 3 *Wall 2 (W2: XPS)—Horizontal cross section through extruded polystyrene (XPS) wall assembly showing locations of HFTs.*

insulation of R-24, and oriented strand board (OSB) (11 mm [7/16 in.] thick). The test specimens were constructed by adding different types of exterior insulation products of different thicknesses to the backup wall. The first wall (Wall 1 [Figure 2]) was constructed by adding an expanded polystyrene (EPS) layer of 25 mm (1 in.) thick onto the OSB sheathing, yielding a nominal R-27. The second wall (Wall 2 [Figure 3]) was constructed with a 51 mm (2 in.) thick extruded polystyrene (XPS) panel and nominally provided R-35. The final wall (Wall 3 [Figure 4]) was constructed with 76 mm (3 in.) thick mineral fiber (MF) insulation, providing an R-35.

Phase 2 Test Specimens

The three wall test specimens of Phase 1 (1219 × 1829 mm [4 × 6 ft]) are shown in Figures 5 to 8 and were, as before, installed side-by-side in the FEWF (Figure 5). The common elements of all three test specimens were, toward the

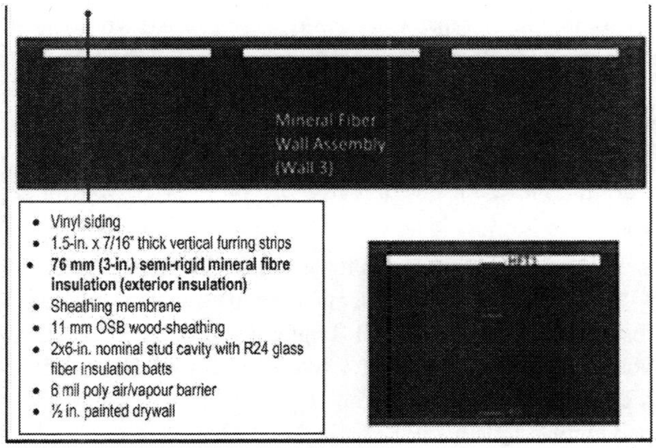

Figure 4 *Wall 3 (W3: MF)—Horizontal cross section through mineral fiber (MF) wall assembly showing locations of HFTs.*

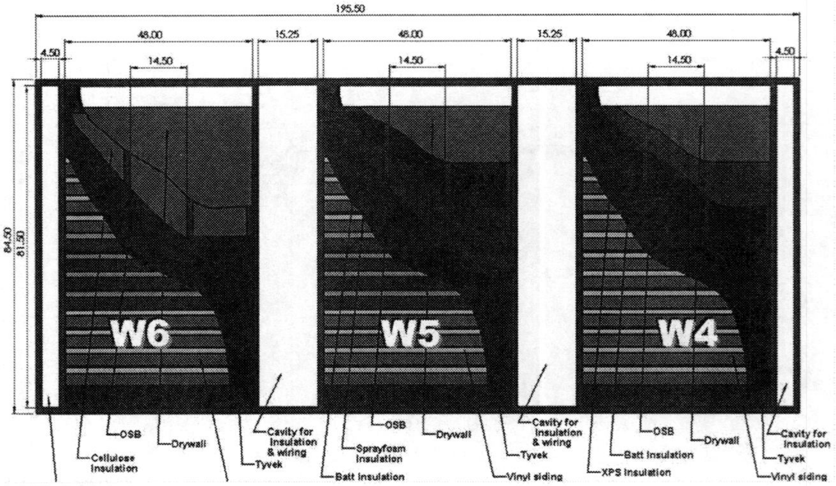

Figure 5 Schematic of three 38 × 140 mm (2 × 6 in.) wood-frame wall (residential) test specimens installed side by side in FEWF.

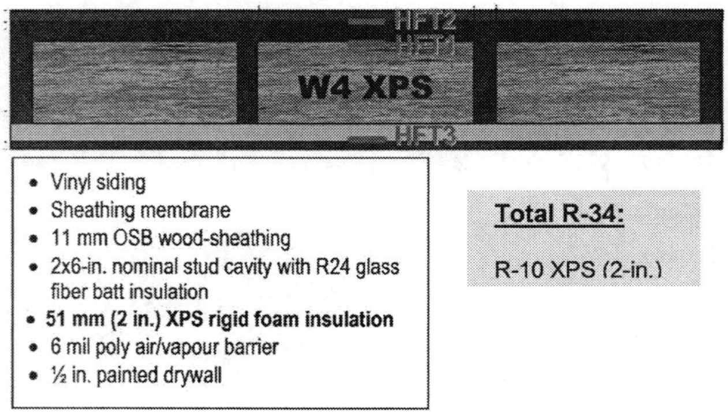

Figure 6 Wall 4 (W4: XPS)—Horizontal cross section through XPS wall assembly showing locations of HFTs.

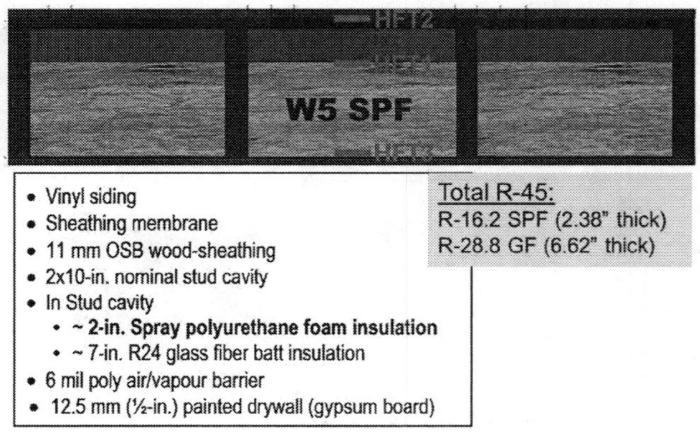

Figure 7 Wall 5 (W5:SPF)—Horizontal cross section through spray polyurethane foam (SPF) wall assembly showing locations of HFTs.

exterior of the assemblies, the vinyl siding, polymer-based sheathing membrane, and 11 mm (7/8 in.) OSB wood-sheathing panel and, for the interior finish, a 6 mil polyethylene air and vapor barrier and a 12.5 mm (1/2 in.) drywall panel. All walls were of wood-frame construction, the difference being the W4 frame nominally consisted of 38 × 140 mm (2 × 6 in.) framing, the W5 frame had 38 × 255 mm (2 ×10 in.) framing, and the W6 frame had 38 × 305 mm (2 × 12 in.) framing.

The primary insulation materials for W4: XPS were: R-24 glass fiber batt insulation over which and toward the interior was installed 51 mm (2 in.) of XPS rigid foam insulation; the overall nominal thermal resistance for W4: XPS was R-34. For W5: SPF the primary insulation materials were: ~60 mm (~2.4 in.) of spray polyurethane foam (SPF) applied on the interior surface of the OSB sheathing panel and 168 mm (6.6 in.) of glass fiber batt insulation; the overall nominal thermal resistance for W5: SPF was R-45. For W6: the insulation products consisted of cellulose fiber insulation (CFI) that provided an R-value of R-40.

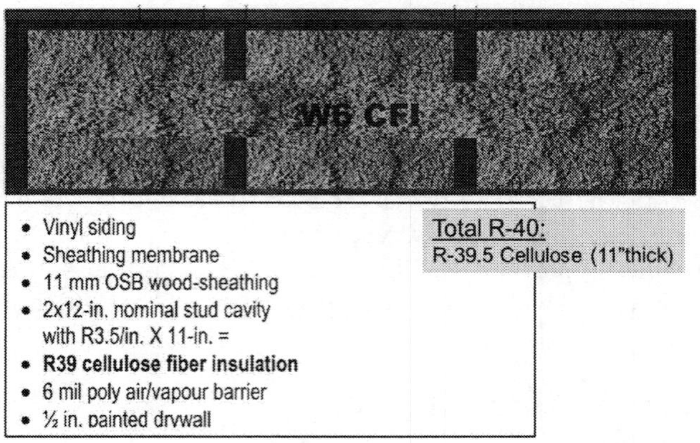

Figure 8 Wall 6 (W6: CFI)—Horizontal cross section through wall assembly having cellulose fiber insulation (CFI) and showing locations of HFTs.

Whereas for W4: XPS and W5: SPF the full depth of the vertical studs was used across the wall assembly, for W6: CFI, 38 × 100 mm (2 × 4 in.) framing was used in the central stud cavity on the exterior and interior of the wall. This arrangement is more clearly evident from the horizontal sectional view provided in Figure 8.

Instrumentation

For the characterization of the hygrothermal response of the walls, the test specimens were instrumented with different sensors to characterize heat transfer across the specimens and to establish values for surface temperature and local relative humidity (RH) as well as the presence of moisture accumulation. Heat flux transducers (HFTs), thermocouples, RH sensors, differential pressure transducers, and moisture detection tapes were deployed in a grid at each layer of the assembly. A local weather monitoring device that included a rain gauge and RH and temperature sensors was installed to provide information on hourly ambient exterior conditions in close proximity to the wall specimens. The combined package permitted determining the response of the respective wall assemblies to changing weather conditions over hourly, daily, and seasonal periods.

For each project phase, each of the three test specimens was isolated from each other by a vertical chase that provided effective control of heat, air and moisture flow and as well provided room for running wires for instrumentation (Figures 1 and 5). The test specimens were also isolated from the surrounding building envelope of the test house with effective heat, air, and moisture control materials (i.e., thermal insulation and self-adhered airtight membranes were used), as well as flashings where warranted.

As a part of the test protocol, more fully described in the work by Lacasse et al. (2016), all HFTs used in the test specimens were calibrated according to ASTM C-1130, *Standard Practice for Calibrating Thin Heat Flux Transducers* (ASTM 2009). The uncertainty of the heat flux measurements was ±5%. The locations of the HFTs in the respective test specimens are shown in Figures 2 to 4.

Test Protocol

The test protocol was conceived to ensure that the wall assemblies were monitored to assess their vapor diffusion characteristics over at least a 9-month period, the initial period (A) covering the coldest period of the year. The test protocol consisted of exposing the wall specimens to exterior weather conditions prevalent at the FEWF test site and a set of indoor controlled temperature and humidity conditions.

Indoor controlled conditions were based on that provided in *ASHRAE Handbook—Fundamentals* (ASHRAE 2009) as these relate to indoor winter and summer conditions (20°C [68°F]; 35% rh). To attain these conditions, an indoor climatic chamber was installed and sealed against the perimeter of the test opening, thereby allowing control of indoor conditions to which test specimens were subjected.

Test Periods. The monitoring program was set up in four periods (A, B, C, and D) as shown in Table 1 for Phase 1 and in Table 2 for Phase 2. The length of periods varied, but in essence the initial period, Period A, was intended as a period over which the fluctuations of heat flow, temperature, RH, and pressure differences within the assembly were monitored to gain an appreciation of the response of the wall to local weather conditions and when the interior conditions were nominally maintained at 21°C and 35% rh. Period B was meant to subject specimens to conditions where warm moist air (21°C, 50% rh) exfiltrated through intended deficiencies in the test specimens. Exfiltration was induced by applying a 10 Pa pressure across the test specimens from the interior enclosure. In this instance, there was interest in bringing about condensation on cold surfaces within the wall. The subsequent period, Period C, provided a short period over which the deficiencies remained in the assembly but no exfiltration pressure was applied to the enclosure and interior RH conditions were not controlled to a set RH level as they were for Periods A and B. As such, the wall might recover from moisture deposition within the wall and, in essence, "dry out". In this way, the propensity for the wall to dry out following occurrences of moisture deposition within the wall provided a means to assess the robustness of the moisture management ability of the wall assembly. The final period (Period D), was again intended as a period over which to gain an understanding of the response of the wall to increasingly cooler periods following the summer months; the deficiencies were not present in this instance and neither was the RH in the interior enclosure maintained to a set level.

Deficiencies in the Wall Assembly. Each test specimen included a deficiency through which air from the interior was meant to exfiltrate through the test specimens. The size and location of the respective deficiencies are shown in Figure 9. Pressure differences across the assemblies, occurring either naturally or through forced airflow to the interior enclosure, induced the flow of air through the deficiencies as suggested in the schematic on the right of the photos in Figure 9. The location of the deficiencies at the interior of the test specimens and at the exterior sheathing panel, in principal, permitted a path for air to flow. The air leakage rate at several different enclosure pressures for each of these deficiencies was determined the results of which are given in Table 3. These are given as the leakage rate (L/min) as a function of pressure difference (Pa) in the exfiltration mode. The coefficients for the best fit equation are given for each test wall as is the degree of fit (R^2). Of importance are the values of air leakage at the exfiltration mode when the deficiencies were present in the walls (see Figure 10, Period B—Deficiency State).

Monitoring

Data was gathered continuously over a 9-month period and data analysis consisted of reviewing relevant sensors within the assembly and determining whether there was risk to the formation of condensation within the respective assemblies over the evaluation period.

On the basis of results obtained from monitoring the response of the respective wall assemblies to local climate conditions, the numerical hygrothermal model hygIRC-C was benchmarked against selected experimental data. Thereafter, the model was used to conduct a parametric analysis to investigate the risk of condensation and mold growth in the respective wall assemblies subjected to different climatic conditions for a select set of locations in Canada. A description of the hygIRC-C numerical hygrothermal simulation model and record of benchmarking exercises is available in the work by Saber and Ganapathy (2016).

RESULTS

Results are provided for Phase 1 and thereafter for Phase 2 of the project. In each of these sections, the hygrothermal response of wall test specimens to local climate conditions is described and those instances when moisture from condensation within the wall assembly was observed are discussed.

Response of Wall Assemblies—Phase 1 (2013/14)

The hygrothermal response of those test specimens evaluated in Phase 1 over the monitoring period of ~11 months is shown in Figure 10 for the respective test specimens: (a) W1: EPS; (b) W2: XPS, and; (c) W3: MF. The location of the temperature and RH sensors for which values are provided in the figure are shown in the diagram adjacent to the plots. In the plot of Figure 10 the values recorded for the interior surface temperature and RH of the OSB sheathing panel at mid-height of the wall are shown over the four periods of the test protocol (i.e., A, B, C, and D). The daily fluctuations in temperature of both the interior surface of the OSB and well as the interior dew point temperature are superimposed on the seasonal variations; the OSB interior surface gradually rises from below

Table 1. Phase 1 (2013/14)—Test Protocol Over Monitoring Period

Period	Interior Conditions				Exterior Conditions	
	Temperature, °C	RH, %	Pressure, Pa	Deficiency, 3 mm slit	Deficiency, 3 mm slit	Temperature/RH
A (175 days)	21	35	0	Closed	Open	Ambient local
B (20 days)	21	50	10	Open	Open	Ambient local
C (40 days)	21	Variable/natural	0	Open	Open	Ambient local
D (100 days)	21	Variable/natural	0	Closed	Open	Ambient local

Table 2. Phase 2 (2014/15)—Test Protocol Over Monitoring Period

Period	Interior Conditions				Exterior Conditions	
	Temperature, °C	RH, %	Pressure, Pa	Deficiency, 3 mm slit	Deficiency, 3 mm slit	Temperature/RH
A (40 days)	21	35	0	Closed	Open	Ambient local
B (9 days)	21	Variable/natural	Variable	Open	Open	Ambient local
C (47 days)	21	Variable/natural	0	Open	Open	Ambient local
D (167 days)	21	Variable/natural	0	Closed	Open	Ambient local

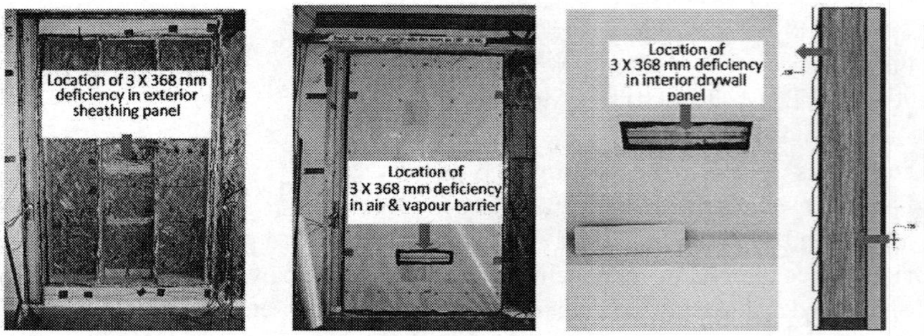

Figure 9 *Example of deficiencies in wall test specimens showing location and size of deficiency in W4:XPS.*

Table 3. Air Leakage (ξ) Characterization of Test Specimens

Phase	Wall	ξ (L/min) = $a\ \Delta P^{b}$*	
		a	b
Phase 1	W1	73.5	0.320
	W2	75.9	0.316
	W3	53.4	0.300
Phase 2	W4	0.685	0.989
	W5	0.654	0.766
	W6	0.593	0.953

*ΔP in Pa

0°C to upwards of 25°C (77°F) with maximum values attaining >30°C (86°F) in Period C (May–July) and the beginning of Period D (July–August). The RH at the interior surface of the OSB panel varies over the monitoring period between ~55% and 35%, the lowest values being recorded during Period B of the monitoring program.

Period B of the monitoring program was intended to bring about occurrences of condensation within the wall assembly. As noted in Figure 10, this was the "deficiency state" when warm moist air from the interior was to exfiltrate to the exterior of the test specimens under forced flow conditions. Period B was completed over a 3-week period at the beginning of May (2 May to 22 May 2014). This being the case, and given that

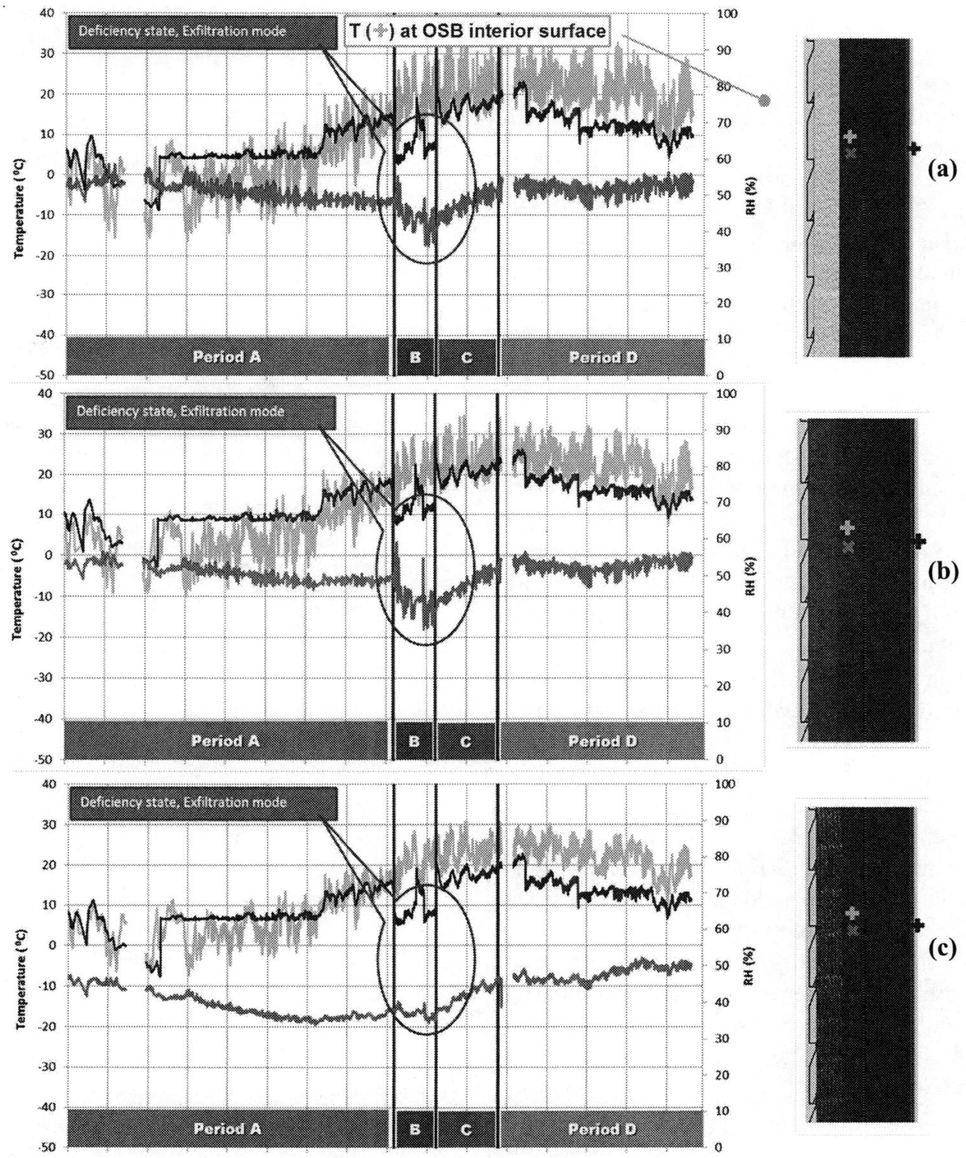

Figure 10 Response of respective wall assemblies (temperature [T], RH at interior OSB surface, and dew point temperature over the monitoring period (2013/14): (a) W1: EPS, (b) W2: XPS, and (c) W3: MF.

the surface temperature of the interior of the OSB sheathing panel had already attained a minimum temperature of ~7°C (45°F), it was unlikely that any surface condensation would occur given as well that this surface temperature was well above the interior dew point temperature. Consequently, no condensation was induced in this period for any of the three walls tested. Then again, neither was any condensation evident for any of the walls over all other periods monitored in Phase 1; this was also visually verified when the walls were taken down following the end of Phase 1.

Response of Wall Assemblies—Phase 2 (2014/15)

The information made available for Phase 2 is quite considerable and additional details of this work can be found in the work by Lacasse et al. (2016). The hygrothermal response of each of the test walls of Phase 2 (W4, W5, and W6) are dealt with in turn.

Response of W4: XPS. The response of W4: XPS is provided in plots of RH and temperature (RH & *T*) as a function of the monitoring period in days as provided in Figure 11. The different periods (A to D, inclusive) are highlighted at the base of the figure. The uppermost plot represents the RH (blue color) and *T* (red color) on the surface of the interior side of the OSB at approximately mid-height of the wall and in proximity to the 3 mm (1/8 in.) slit through which air could exfiltrate the wall. For the middle plot, the values for temperature given at the bottom of the wall were found to be much lower than that for the top of the wall (by perhaps 20°C), and given in the uppermost plot; the lower temperatures in the bottom plot were considered inaccurate and, consequently, in the middle plot it was deemed appropriate to use the values for temperature obtained from the top of the wall as given in the uppermost plot. The middle plot thus provides values of RH on the surface of the OSB toward the bottom of the wall; this location (marked with an X) is shown in the schematic of the test specimen to the right of each plot.

The propensity for the growth of mold at the top or bottom of the wall is shown in the lowermost plot of Figure 11, in which is given the mold index value as a function of the monitoring period, in days. The mold index value is based on a mathematical relationship developed by Viitanen et al. (Viitanen and Ojanen 2007; Hukka and Viitanen 1999; Ojanen et al. 2010), in which values for RH and *T* and information on the type of surface onto which mold may grow permits establishing the susceptibility to mold growth. Values for mold index may range between 0 and 6, where 0 represents no mold growth and 6 represents very heavy (100% coverage) and tight growth. The use of this equation in providing information on wall performance is more fully described in the work by Saber and Ganapathy (2016).

It is perhaps apparent that the RH (blue lines) at both the top and bottom locations in the wall experienced levels of RH ranging between 50% and 60% during Period A, whereas these levels during Period B increase significantly. Recalling that Period B of the test protocol (see Table 2) was one where purposely placed deficiencies in the wall were open to permit the passage of warm air (21°C [70°F]) from the interior and through the wall to the exterior. The RH in these instances attained ~85% rh at the top of the wall (uppermost plot) and ~100% rh (middle plot) at the bottom of the wall, suggesting that with a sufficiently low temperature, condensation could readily occur on the surfaces of the OSB. The values of *T* used to calculate the mold index are the same as those given in the uppermost plot (top off the wall) for reasons previously explained; these values range from –20°C to 0°C.

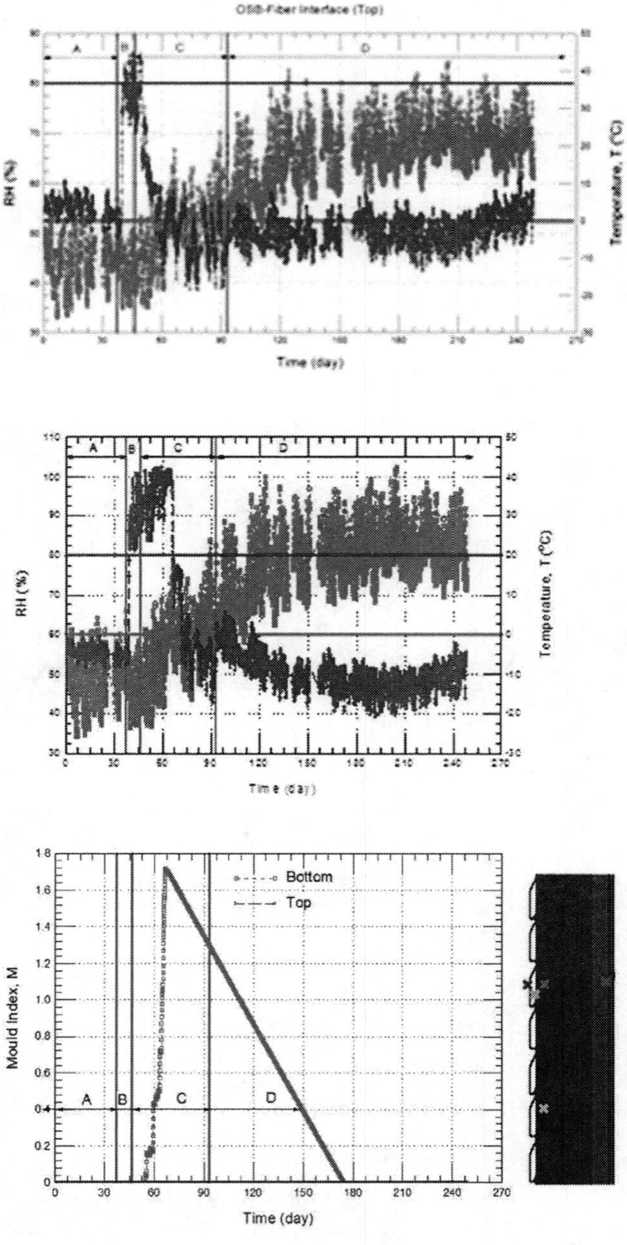

Figure 11 *Response of test wall W4: XPS to local weather load of Ottawa, ON, Canada, having a nominal R-value of R-33.*

Thus, condensation could occur at these locations, and this is evident from the information provided in Figure 12 which shows the response of electrical resistance to the presence of moisture at locations provided in the lowermost figure of Figure 12. The response of the moisture detection strips to the presence of moisture is marked with a decrease in electrical resistance which manifests itself by an increase in the value for conductivity. The increase in values for conductivity during Period B denote the presence of condensation whereas the dissipation of moisture from the same locations in Period C is evidence from decreases in values over time; no moisture is detected in Period D. This implies that all moisture that was deposited in the wall over this period could readily dissipate from the wall assembly.

In respect to the risk for mold growth, given the low temperatures on the wall surface, and moreover that mold does not grow at temperatures below 0°C, the propensity for mold growth during Period B was nil. However, as the temperature at these locations increased over Period C, the susceptibility to mold growth at these locations likewise increased, although, given the ever decreasing level of RH over the same period, the mold index values considerably abated in Period C. Overall, the susceptibility to mold growth in this wall was low.

Response of W5: SPF. The response of W5: SPF is provided in plots of RH and *T* as a function of the monitoring period in days as provided in Figure 13; as before, Periods A to D are highlighted at the base of the figure. The uppermost and middle plots of Figure 13 represent the RH (blue color) and *T* (red color) on the surface of the interior side of the OSB, respectively, at approximately mid-height and toward the bottom of the wall; both these locations (marked with an *X*) are shown in the schematic of the test specimen to the right of each plot. The propensity for the growth of mold at the top or bottom of the wall is shown in the lowermost plot of Figure 13, in which is given the mold index value as a function of the monitoring period, in days.

As was the case for W4, the RH at both the top and bottom locations in test specimen W5 experienced levels of RH ranging between 35% and 65% during Period A, whereas these levels during Period B, again, increased significantly. In Period B, values of RH attained upwards of 85% at the bottom location and 95% at the top location; the corresponding temperatures were again quite low, reaching –20°C. As such, the propensity of the formation for condensation was high whereas the risk of the formation of mold was low.

The information provided in Figure 14 shows the response of electrical resistance to the presence of moisture at locations shown in the figure. The increase in values for conductivity during Period B denote the presence of condensation whereas the dissipation of moisture from the same locations in Period C is evident from decreases in values over time; no moisture detected in Period D, as was the case for W4. This implies that all moisture that was deposited in the wall over this period could readily dissipate; this is the same response as found for W4.

The risk of the growth of mold diminished over Period C irrespective of the increase in temperature over this period and the subsequent period of monitoring Overall, the susceptibility to mold growth in this wall was low.

Response of W6: CFI. The response of W6 is provided in plots of RH and temperature (RH & *T*) as a function of the monitoring period in days as provided in Figure 15; as before, Periods A to D are highlighted at the base of the figure. The uppermost and middle plots of Figure 15 represent the RH

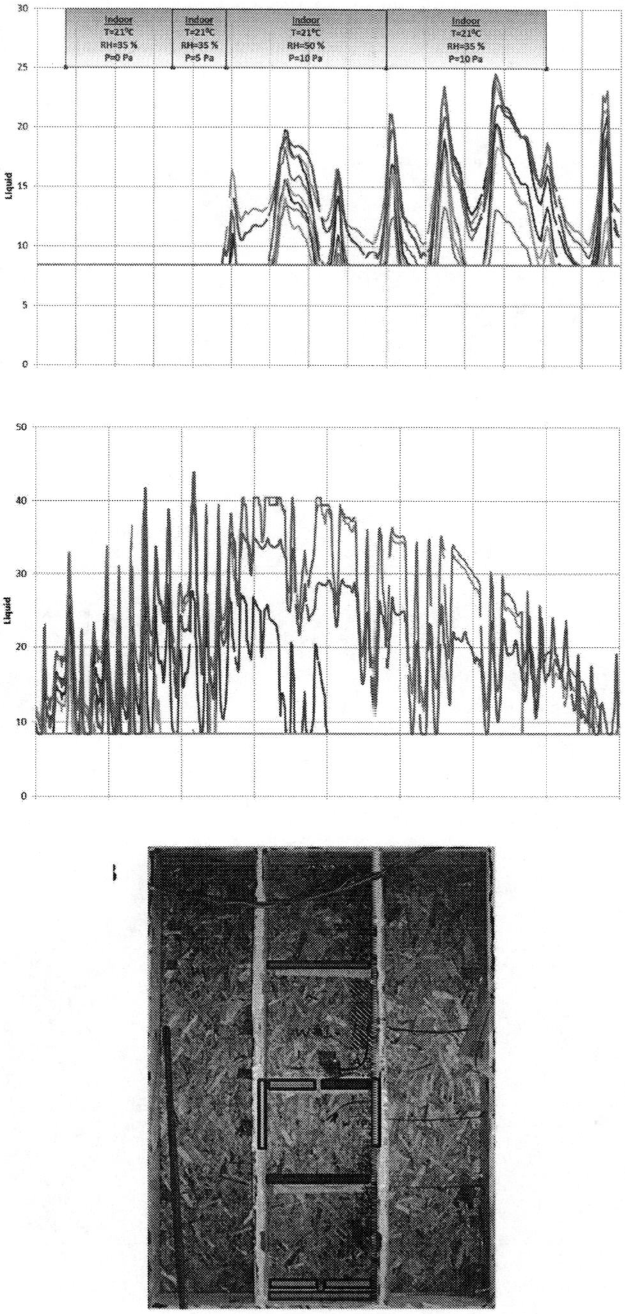

***Figure 12** W4 response of moisture detection strip to the presence of moisture in the wall.*

(blue color) and *T* (red color) on the surface of the interior side of the OSB, respectively, at approximately mid-height and toward the bottom of the wall; both these locations (marked with an X) are shown in the schematic of the test specimen to the right of each plot. The propensity for the growth of mold at the top or bottom of the wall is shown in the lowermost plot of Figure 15 and in which is given the value for the mold index as a function of the monitoring period in days.

As for W4 and W5, the RH at both the top and bottom locations in test specimen W6 experienced levels of RH ranging between 60% and 70% during Period A, whereas these levels during Period B increased, although not as significantly as in W4 and W5. In Period B, values of RH attained upwards of 85% at the bottom location and 75% at the top location; the corresponding temperatures in Period B were again quite low, reaching –20°C. As such, the propensity of the formation of condensation was high, although the risk of the formation of mold was low.

The information provided in Figure 16 shows the response of electrical resistance to the presence of moisture at locations provided in the figure. During Period B the increases in values for conductivity denote the presence of condensation; this is a

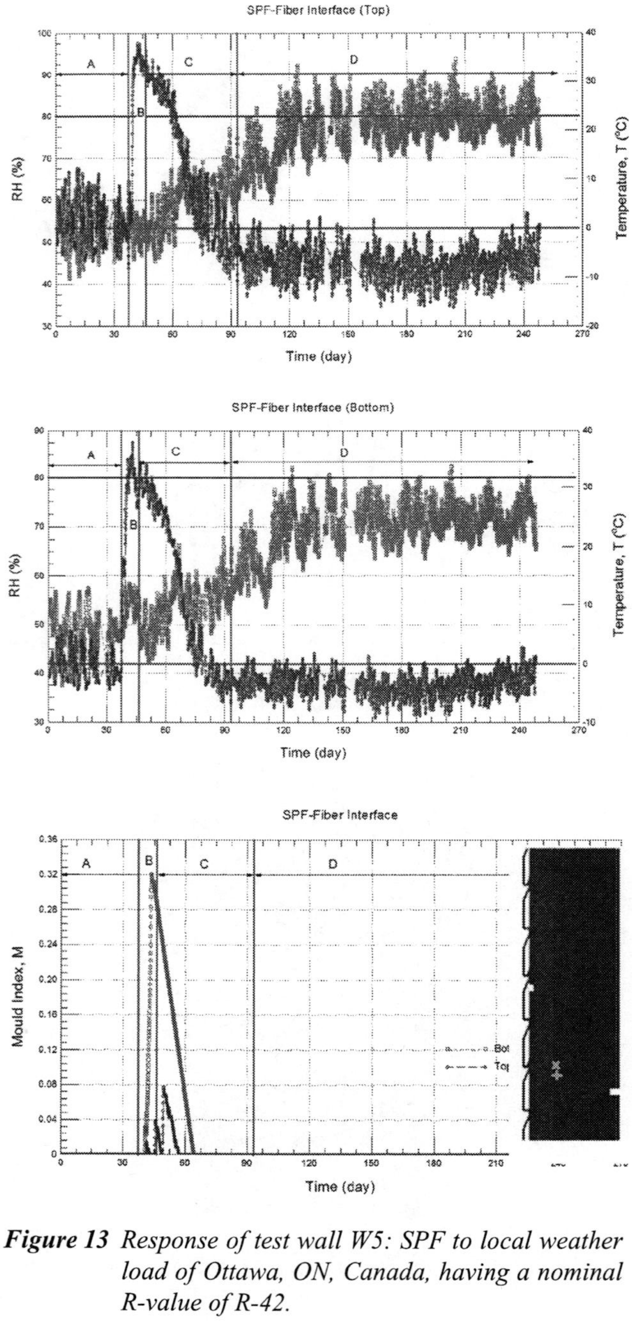

Figure 13 *Response of test wall W5: SPF to local weather load of Ottawa, ON, Canada, having a nominal R-value of R-42.*

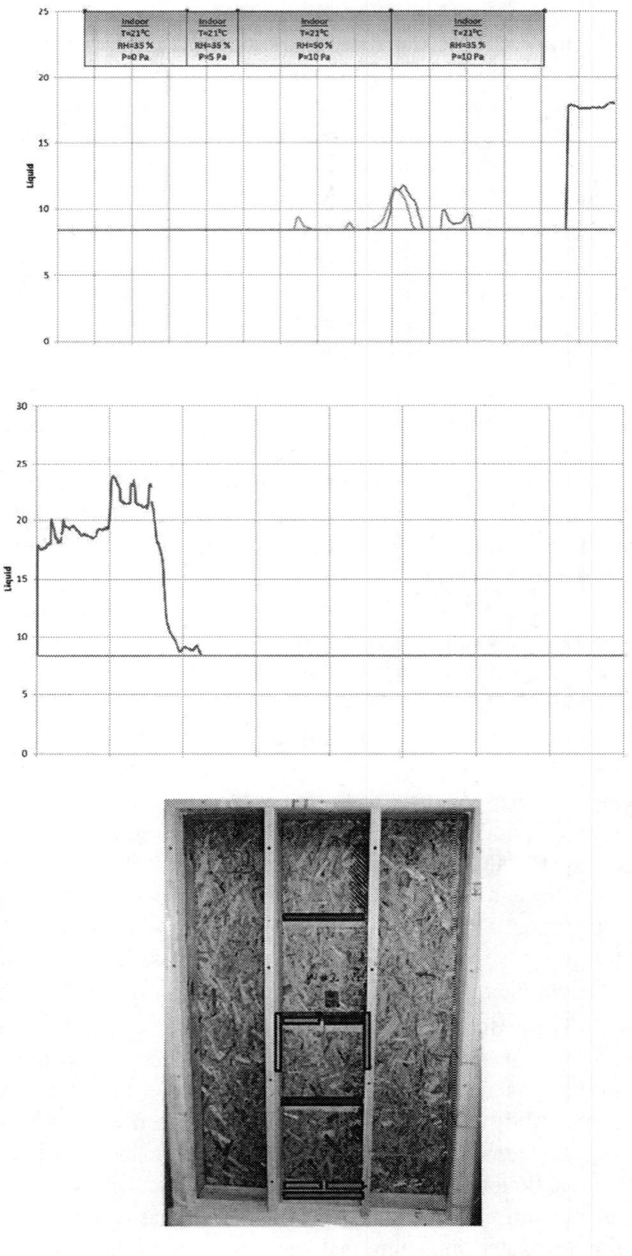

Figure 14 *W5 response of moisture detection strip to the presence of moisture in the wall.*

similar response as that determined from W4 and W5. The dissipation of moisture from the same locations in Period C is evident from decreases in values over time; no moisture detected in Period D as was the case for W4 and W5. Again, all moisture deposited in the wall over this period could readily dissipate, indicative of the same response as was found in W4 and W5.

The risk of mold growth diminished over Period C irrespective of the increase in temperature and RH over this period and the subsequent period of monitoring. Overall, the susceptibility to mold growth in this wall was low.

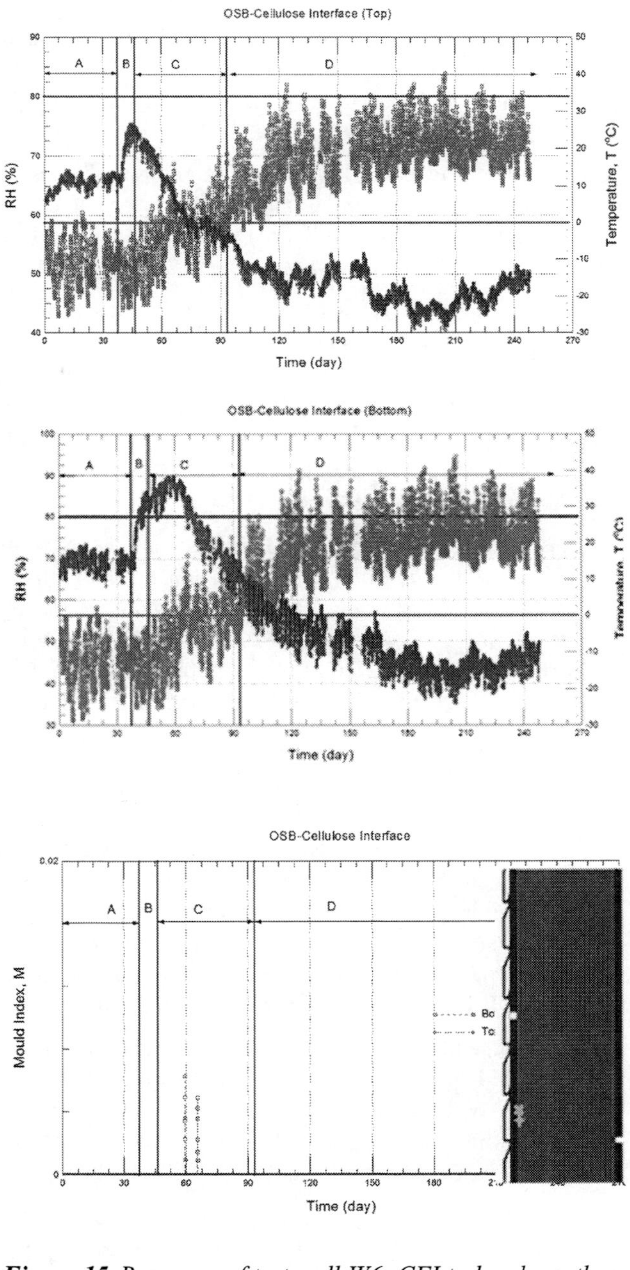

Figure 15 *Response of test wall W6: CFI to local weather of Ottawa, ON, Canada, having a nominal R-value of R-40.*

SUMMARY OF RESPONSE OF WALL ASSEMBLIES—PHASES 1 AND 2

The hygrothermal response of six highly insulated wood frame wall assemblies, a summary of which is provided in Figure 17, was monitored over a 2-year period (June 2013–September 2015) to determine whether any of these walls were susceptible to moisture problems as might accrue from inadvertent condensation in the wall assemblies. The walls were

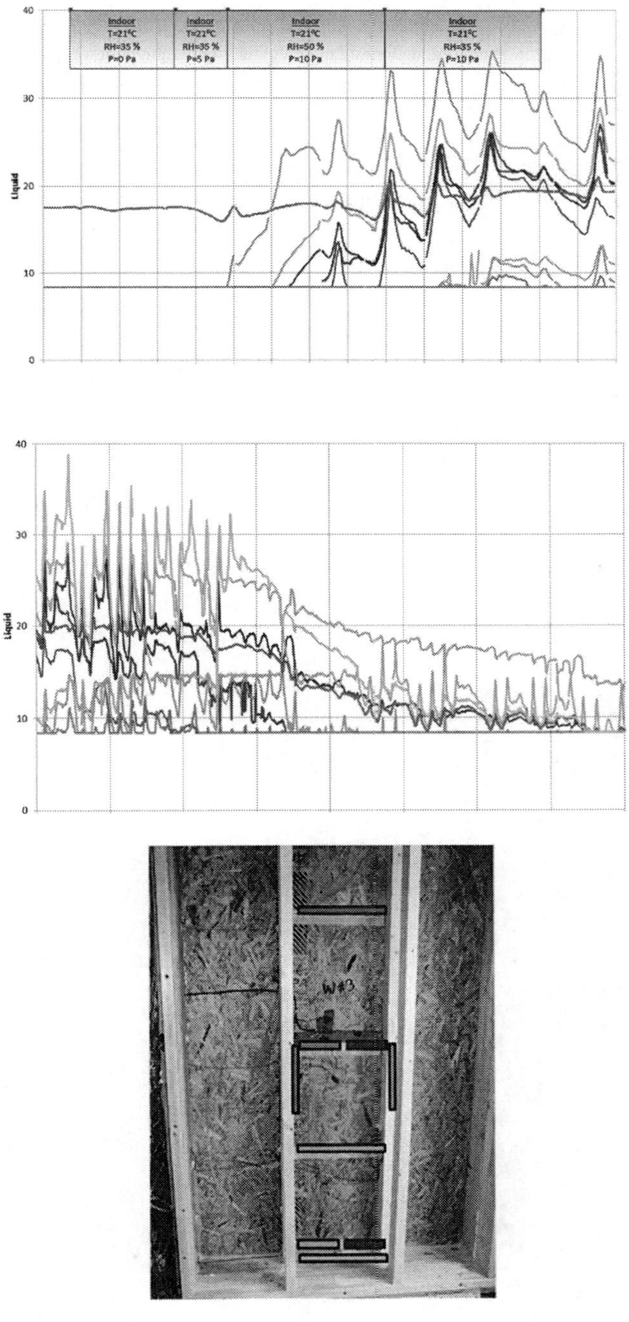

Figure 16 *W6 response of moisture detection strip to the presence of moisture in the wall.*

subjected to local climate conditions of Ottawa, Canada. To determine the vulnerability of the wall assembles and to assess their robustness to moisture ingress from air leakage, all assemblies were configured to include openings in the air barrier and exterior sheathing paneled to permit the passage of air exfiltration from or infiltration to an interior conditioned enclosure.

Results from this study show that in instances where openings for air leakage exist, there is a risk to condensation in wall assemblies. However, this risk is mitigated by 1) the low risk of the formation of mold (Viitanen mold index value less than 1.8 [Saber and Ganapathy 2016]), given the very low temperatures that exist on the interior surface of the exterior OSB sheathing panel and 2) the ability for moisture to dissipate from within the wall in a period of 2 to 3 months should moisture be present within the wall.

ACKNOWLEDGMENTS

The authors wish to thank Canada Mortgage and Housing Corporation and Natural Resources Canada for contributing funding for this project. The authors wish to thank the Project Advisory Committee that included: Constance Thivierge (formerly with FP Innovations), Doug Tarry (Doug Tarry Homes), Rick Gratton (CHBA), Christopher McLellan (CHBA), Chris Mattock (Habitat Design and Consulting), Salvatore Ciarlo (Owens Corning Canada), and Robert Jonkman (Canadian Wood Council). Also, the authors wish to thank NRC for providing the funding to enable researchers to build, operate and maintain a state-of-the-art field exposure of walls facility that was used in this project.

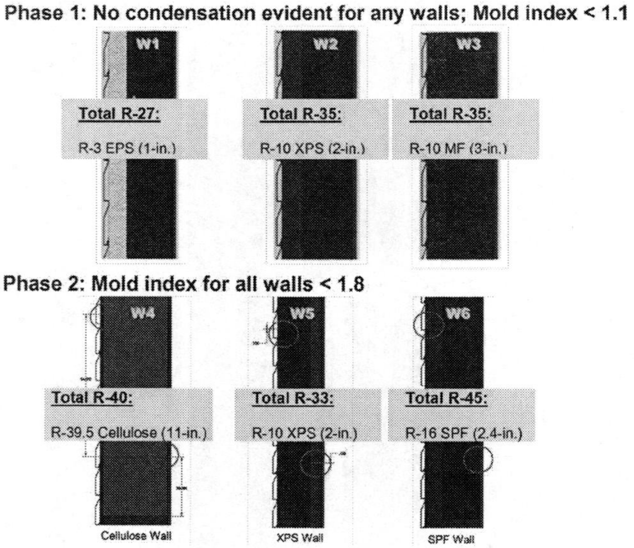

Figure 17 *Summary of response of wall assemblies—Phase 1 (W1–W3) and Phase 2 (W4–W6).*

REFERENCES

ASHRAE. 2009. 2009 ASHRAE Handbook –Fundamentals (SI), Chapter 26. Atlanta: ASHRAE.

ASTM. 2009. Designation: C 1130-07. Standard Practice for Calibrating Thin Heat Flux Transducers. Annual Book of ASTM Standards; Construction. Vol. 04.06, Thermal Insulation; Building & Environment Acoustic: 577–584.

Canada Mortgage and Housing Corporation. 2014. Canadian Housing Observe: 132. Ottawa: Canada Mortgage and Housing Corporation. www.cmhc-schl.gc.ca/odpub/pdf/68189.pdf.

Environment Canada. 2013. Canada's Emissions Trends: 71. Ottawa: Environment Canada. www.ec.gc.ca/ges-ghg/985F05FB-4744-4269-8C1A-D443F8A86814/1001-Canada's%20Emissions%20Trends%202013_e.pdf.

EnerGuide for Houses; Energy efficient new house: EGH 81 to 85; High-performance, energy efficient new house: EGH 86 to 99. www.nrcan.gc.ca/energy/efficiency/housing/home-improvements/5005.

Hukka, A., and H.A. Viitanen. 1999. A mathematical model of mold growth on wooden material. Wood Science and Technology 33(6): 475–85.

Lacasse, M.A., H.H. Saber, G. Ganapathy, and M. Nicholls. 2016. *Evaluation of thermal and moisture response of highly insulated wood-frame wall assemblies—Phase 1 and Phase 2: Experimental trials in field exposure of walls test facility*. Report No. A1-000444.5; NRC-Construction. Ottawa: National Research Council Canada.

National Building Code of Canada. 2010. NRCC 53301. Thirteenth edition. Ottawa: National Research Council of Canada.

Ojanen, T., H.A.Viitanen, R. Peuhkuri, K. Lähdesmäki, J. Vinha, and K. Salminen. 2010. Mold Growth Modeling of Building Structures Using Sensitivity Classes of Materials. 11th International Conference on Thermal Performance of the Exterior Envelopes of Whole Buildings XI: 10, Clearwater, FL., December 5–10.

Saber, H.H., and G. Ganapathy. 2016. Evaluation of Thermal and Moisture Response of Highly Insulated Wood-Frame Wall Assemblies—Phase 1, Part II: Numerical Modelling. Report No. A1-000444.4: 83. NRC-Construction. Ottawa: National Research Council Canada.

Viitanen, H.A., and T. Ojanen. 2007. *Improved model to predict mold growth in building materials*. Proceedings of Thermal Performance of the Exterior Envelopes of Whole Buildings X: 8.

Potential and Limitations of Infrared Thermography on Unventilated Walls

Katrien Maroy **Nathan Van Den Bossche, PhD** **Marijke Steeman, PhD**

ABSTRACT

Reducing the energy demand of existing buildings is one of the main ways to achieve a strong decrease in CO_2 emissions, as imposed by the EU. In Belgium, brick cavity walls are a widespread external wall type. Consequently, this offers a large potential in energy savings by cavity wall insulation. However, to avoid and track execution flaws, it is recommended to assess the thermal performance before and after the execution and to detect obstructions in the cavity (e.g., thermal bridges).

Infrared thermography offers a quick and nondestructive technique to fulfill the task. With an infrared camera, the surface temperature of a building component is instantly visualized, by which thermal bridges and execution flaws can be detected. To assess the potential of thermography in the evaluation of cavity wall insulation, dynamic numerical simulations were executed on a brick wall with cavity insulation. One model contained an air inclusion in the cavity insulation; in the second model the visibility of a thermal bridge was examined. Next to that, infrared inspections of over 30 in-situ cases were analyzed and compared with execution reports, flux measurements, and endoscopy inspections. The results show that thermal bridges and large insulation defects are detectable when direct sun radiance is avoided during and a day prior to the survey. Additionally, attention should be granted to the measurement distance and condition of the surface. The results have been converted into practical guidelines for the infrared inspection of filled cavity walls.

INTRODUCTION

By 2050, the CO_2-emissions in Europe need to be reduced by more than 80% compared to the situation in 1990. One of the key factors to achieve this goal is to reduce the energy demand of existing buildings. A popular measure in Belgium consists of providing insulation in existing brick cavity walls, by which unventilated walls are created (Figure 1). Foams (e.g., polyurethane [PUR]), fibers (e.g., mineral wool, glass fibers), and granulates (e.g., expanded polystyrene [EPS]) are common used materials to inject in the cavity wall (Beulque 2014).

In Belgium, brick cavity walls are a widespread external wall type because of the excellent moisture tolerance. Water that penetrates through the outer leaf of the cavity wall can drain back outside through the cavity and the wall dries faster because of the airflows in the empty cavity. In the mid 1970s, cavity wall insulation became a popular measure due to the oil crisis and corresponding high energy prices (Hens et al. 2007). However, the technique had to deal with several start-up issues (e.g., the use of inconvenient products and wrong installation methods). Frequent problems were moisture penetration, shrinkage, and degradation or sinking of the insulation material in the cavity. Consequently, quality control facilities for the insulation materials and execution were established (Beulque 2014).

Nowadays, wall cavity insulation still offers a large potential to reduce the energy demand of residential buildings in Belgium. In the northern part of the country, 59% of all residential buildings do not have sufficient insulated exterior walls nowadays, while 35% of all heat loss is transmitted through exterior walls. Due to the easy application, the limited incon-

Katrien Maroy is a doctoral student, **Nathan Van Den Bossche** is a doctor and professor, and **Marijke Steeman** is a doctor and professor at the Faculty of Engineering and Architecture, Ghent University, Ghent, Belgium.

venience, and the short runtime of wall cavity filling, this technique was used in 75% of all renovation projects to provide exterior wall insulation (European Commission 2011; Beulque 2014).

Despite of the current quality control framework, some execution failures arise frequently: (1) obstructions of the insulation material by mortar piles and rubble in the cavity, (2) accumulation of the glue that binds the EPS pearls in corners of windows and doors, and (3) wrong glue flow rate for the EPS pearls caused by the over- or underestimation of cavity widths locally. Often, these problems could have been prevented if the wall was inspected on these kinds of obstructions before injecting the insulation. A common method to track these obstructions locally is endoscopy (Figure 2a and 2b), but also infrared thermography is frequently used to evaluate the whole cavity wall (Figure 2c). Otherwise, execution flaws are only discovered once the consequences become visible, in the form of cracks or mold growth (Beulque 2014). Therefore, it is important to check the cavity wall on local thermal deficiencies before and after the execution of cavity wall insulation (e.g., concrete beams above windows, wall-foundation interface, intermediate concrete floors).

Infrared thermography could offer a quick, nondestructive method to track thermal bridges and other obstructions inside the cavity wall before the execution and to check the thermal performance after cavity wall insulation (see also Figure 2c). With an infrared camera, the surface temperature of a building component is instantly visualized. This might provide useful information to evaluate the thermal resistance on site and to detect thermal deficiencies (e.g., air leakages and thermal bridges) (Taylor et al. 2013). However, misinterpretation of infrared images is possible when no attention is paid to the surface characteristics of the cavity wall, camera use, and boundary conditions. For example, a warm spot on the infrared image of a wall surface can be caused by lacking insulation in the cavity, but this effect can also be generated by local sun radiation prior to the measurement, as was the case in Figure 3. Hence, it is necessary to generate practical guidelines to prevent these misinterpretations.

In this paper, the potential and limitations of infrared thermography on unventilated walls are discussed. First, an outline is given of the accuracy of infrared thermography. The influence of emissivity and reflected ambient temperature, camera characteristics, and changing weather conditions on the measured surface temperatures is illustrated. Subsequently, the use of infrared thermography in the inspection of unventilated walls is discussed. Through dynamic numerical simulations and the analysis of over 30 case studies, the potential and limitations of infrared thermography in the area of cavity filling inspection are examined.

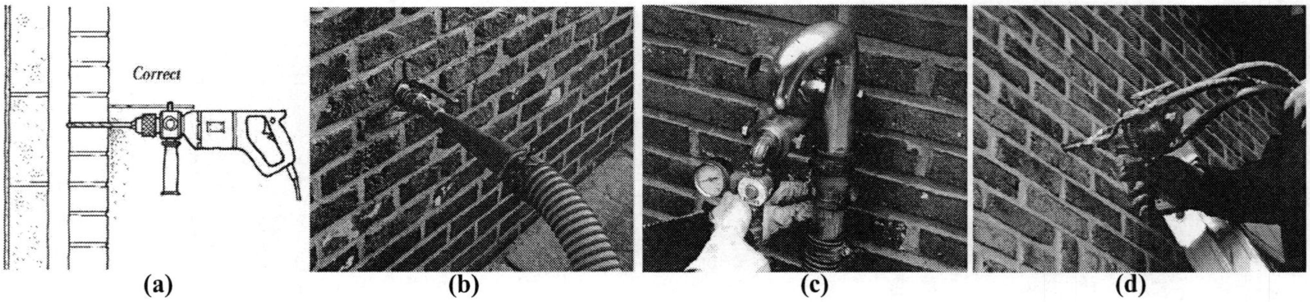

Figure 1 *(a) General principle of cavity wall insulation, (b) nozzles to inject mineral wool, (c) for EPS, and (d) PUR (Beulque 2014).*

Figure 2 *(a) Obstructions in the cavity by mortar piles or (b) wooden planks from the window junction. Both pictures were obtained with an endoscope (WTCB 2012). (c) Insufficient cavity filling in the cavity wall, probably due to an incorrect glue flow and local obstructions. Visualized with an infrared camera (Beulque 2014).*

(a) (b) (c)

Figure 3 *(a) IR picture of a cavity wall filled with EPS. The warm spot in the corner seems to signal a lack of insulation in that area. However, considering the geometry of the building (b and c) and the sunny weather on the day prior to the measurement, the warm spot is probably attributed to local heating by sun radiance (Beuque 2014).*

ACCURACY OF INFRARED THERMOGRAPHY

Main Principles: Emissivity and Reflected Ambient Temperature

All objects with a temperature above 0 Kelvin emit electromagnetic radiation, including infrared rays. In the electromagnetic spectrum, infrared radiation is situated between the visible light and microwaves, in the wavelengths from 0.75 to 1000 μm. The radiation emitted by an object (W_{obj}) is dependent of its temperature (θ, Kelvin), the Stefan-Boltzmann constant (σ), and the emissivity of that object (ε) (Jewett and Serway 2008; Eads et al. 2000):

$$W_{obj} = \varepsilon \cdot \sigma \cdot (\theta_{obj})^4 \text{ (W/m}^2\text{)}$$
$$\text{(with } \sigma = 5.67 \cdot 10^{-8} \text{ W/m}^2\cdot\text{K}^4\text{)} \quad (1)$$

Next to radiation originating from the objects temperature, the camera captures also radiation from the environment reflected on the object and radiation transmitted through the object in case of transparent materials. Fortunately, most building materials are opaque in the infrared field. With help of a thermographic camera, all infrared radiation emitted from an opaque object is recalculated to a matrix of surface temperatures, which composes the IR-image. The calculation is described by the main law of thermography (Dall'O' et al. 2013; Charlier 2007; FLIR Systems 2012):

$$W_{tot} = \varepsilon \cdot \tau_{atm} \cdot W_{obj} + (1-\varepsilon) \cdot \tau_{atm} \cdot W_{amb} + (1-\tau_{atm}) \cdot W_{atm} \text{ (W/m}^2\text{)} \quad (2)$$

W_{tot} is the total radiation captured by the thermographic camera (W/m^2); W_{obj} is the radiation of the object (Equation 1) (W/m^2). W_{amb} is the ambient radiation (with the reflected temperature of the environment) (W/m^2), τ_{atm} is the transmission of the atmosphere, and W_{atm} is the amount of radiation emitted by the atmosphere. Most infrared cameras operate in the range of 3–5 μm and 8–14 μm, which happens to be the range in which the transmission of the atmosphere τ_{atm} equals 1. Consequently, Equation 2 can be simplified to

$$W_{tot} = \varepsilon \cdot W_{obj} + (1-\varepsilon) \cdot W_{amb} \text{ (W/m}^2\text{)} \quad (3)$$

The surface temperatures calculated by the camera are thus dependent on the emissivity ε of the object and the reflected ambient temperature (via W_{amb}). For the surface of a brick wall, a standard value of 0.90 can be assumed (FLIR 2015). This means that 90% of the radiation from a brick wall surface originated from its surface temperature. Thermal deficiencies behind the brick wall facade (e.g. thermal bridges), will therefore be clearly visible. For that matter, most building materials show an emissivity above 0.90.

When the surface of the wall is covered with a reflecting material (e.g., a coating with an emissivity of 0.10), it is impossible to detect any thermal deficiency. The radiation coming from the wall surface then consists of only 10% of its own temperature. The remaining 90% is radiation from the environment, reflected onto the wall surface.

The emissivity of an object and the reflected temperature of the environment can be determined with ASTM Standard E1993-99a and ASTM Standard C1060-11a, respectively (ASTM 2005, 2002).

Camera Use

Next to the emissivity of the surface and the reflected ambient temperature, the camera characteristics determine the accuracy of the measured surface temperature. Each pixel represents a surface with a certain surface temperature. The real-life area represented by one pixel depends on three things (Charlier 2007):

- Measurement distance
- Field of view (FOV) of the lens (Figure 4)
- IR resolution

When the measurement distance and the field of view of the lens are large, the pixel will represent a larger area and the

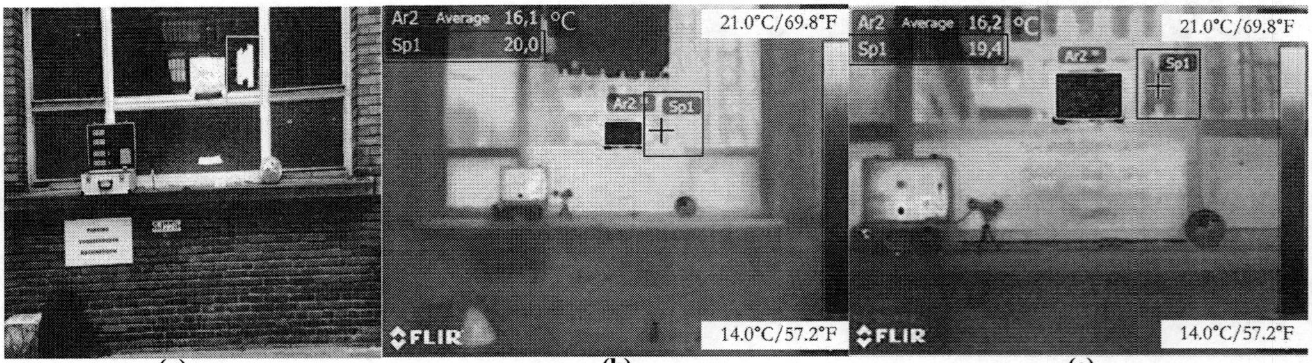

(a) **(b)** **(c)**

Figure 4 *(a) Window with aluminum foil and emissivity tape, measured with infrared thermography on Figures 4b and 4c. The measurement distance has an influence on the measured surface temperatures. On (b), the surface temperature at Sp1 is 20°C (68°F). This infrared picture was taken with a FOV of 45°×33.8° (c) The infrared picture, with a FOV of 25°×19°, shows a surface temperature of 19.4°C (67.28°F). Both pictures (b) and (c) were taken at 5 m distance from the object, with an IR-resolution of IR-resolution of 320×240 (Maroy et al. 2015).*

actual surface temperature will be more averaged (Figure 4b). Consequently, thermal deficiencies that cause small temperature differences on an IR image will be less visible when taking IR pictures from a large distance with a large field of view. On the other hand, when the IR resolution of the camera increases, the accuracy increases as well. After all, a large IR-resolution means that more pixels are available for the same area. Obviously, the actual surface temperatures will be less averaged.

Next to that, each camera is also characterized by a noise equivalent temperature difference (NETD). This number represents the smallest temperature difference necessary to generate a different signal from the sensors in the IR camera (Orlove 2012). For thermography on buildings, a NETD of 0.05°C–0.08°C (50–80 mK) is recommended (Dall'O' et al. 2013; Behaneck 2011). Most cameras offer a NETD of <0.045°C (45 mK) in an environmental temperature of 30°C (86°F).

The Influence of Weather Conditions

Lastly, the boundary conditions are determining factors in the measurement value of surface temperatures and, thus, in the application of IR for a thermal performance estimation of opaque elements (Lehmann and Ghazi Wakili 2013; Fokaides and Kalogirou 2011; Tanner et al. 2011; Albatici en Tonelli 2010).

The temperature difference between inside and outside should be a minimum 10°C (18°F), according to the literature (Charlier 2007; Eads et al. 2000; Fokaides and Kalogirou 2011; Pearson 2011) and the standard ASTM C1060-11A (ASTM 2011). Other sources state that a difference of 5°C (9°F) (Dall'O' 2013; NBN 1999; Taylor et al. 2013) is sufficient or that this should be at least 15°C (27°F) (VATh 2011). Nonetheless, the higher the temperature difference, the clearer a thermal bridge or insulation deficiency is visualized on an IR-image. In our case studies and dynamic simulation models, we respected the minimum temperature difference of 10°C (18°F) and this seemed to be sufficient for the visibility of thermal bridges and lacking cavity insulation. When more accurate measurements are demanded for IR images (e.g., representative IR images for a thermal performance estimation), a temperature difference of at least 15°C (27°F) is recommended (Maroy et al. 2015).

Next to temperature difference between inside and outside, the weather conditions have a large impact on the outside surface temperature and, thus, on the appearance of surface temperatures. For example, after sun or clear-sky radiation, the surface temperature of a building component is increased or decreased respectively, compared to the steady-state conditions. In these weather conditions, the surface temperatures of the cavity wall are not representative for the actual thermal performance. The duration of this effect depends on the thermal effusivity of the surface material. Materials with a high thermal effusivity act more as a heat storage medium (Hens et al. 2007). Consequently, the effect of sun or clear-sky radiation will last longer.

The period of time in which the sun or clear-sky radiation influences the outside surface temperatures was derived for six different types of walls frequently used in Belgium (Van De Vijver et al. 2014). Figure 5 shows the period of time in which heavy clouded windless weather should occur before it is possible to measure surface temperatures closer than ±0.5°C (±0.9°F) from the steady state. The limit of ± 0.5°C (±0.9°F) corresponds to the smallest accuracy of IR measurements, achievable in lab conditions (Griffith et al. 1995). The vertical axis shows the maximum temperature deviation between the surface temperature under a specific weather condition and under steady-state conditions. In general, 4 hours of clouded weather is sufficient for correct IR measurement on a timber frame wall. For a nonfilled cavity wall, 17 hours of heavy

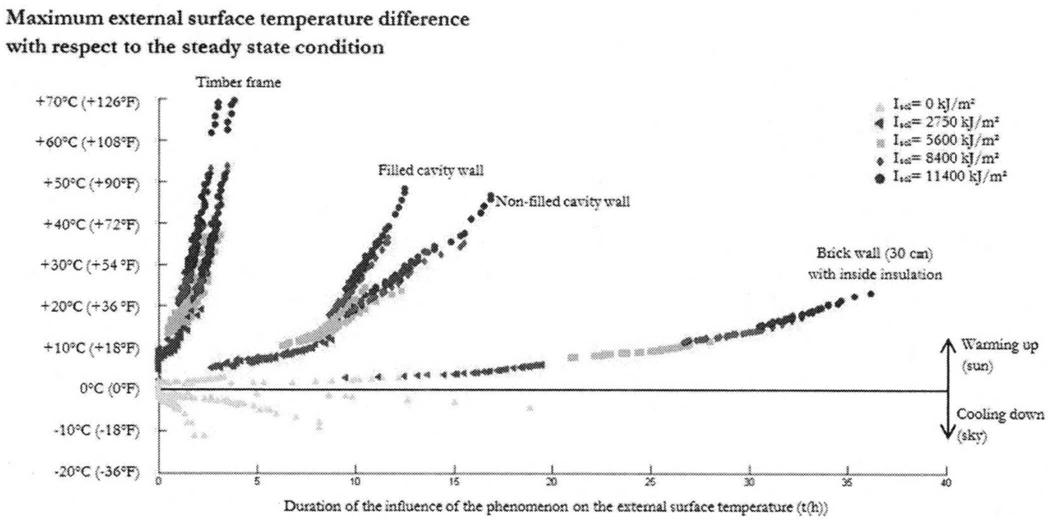

Figure 5 *Maximal external wall surface temperature deviation from the steady state and waiting time for six types of building components. The impact of sun radiation, clear-sky radiation, and varying wind velocity was taken in account. The horizontal axis shows the period of time in which heavily clouded weather should occur before it is possible to measure surface temperatures closer than 0.5°C (0.9°F) to the steady state (Van De Vijver et al. 2014).*

clouded weather should occur before IR measurements are possible that represent the thermal performance (Van De Vijver et al. 2014).

In practice, the steady-state condition is hard to reach. For example, for an empty cavity wall, it should be heavy clouded windless (<1 m/s) weather for 17 hours before the outside surface temperatures are representative for the thermal resistance of the entire wall. With these strict boundary restrictions in mind, only a few days in the winter season are suitable for IR measurements that should represent the steady-state conditions.

The impact of weather conditions could be somewhat less restricted when the infrared measurement only serves to detect thermal deficiencies (e.g. thermal bridges or air inclusions in an insufficient filled cavity wall). The sun or clear-sky radiation could possibly stimulate the visibility of thermal deficiencies. Because of the varying heat capacity of materials in the thermal deficiency and in intact parts, the surfaces at a thermal deficiency can heat up or cool down at a different rate under sun radiation or clear-sky radiation. For example, after sun radiation, an infrared picture of an insufficient filled cavity wall will show colder zones at air inclusions in the insulation layer. After all, the surface at the air inclusions will cool down faster than at areas with sufficient insulation. The thermographer should thus be aware of this effect. Next to that, the conscious use of sun radiation is limited because of the shadow of neighboring buildings, leaving random patterns on the IR image (see, for example, Figure 3).

In the next section, the potential and limitations of infrared thermography for the detection of thermal deficiencies in unventilated walls is researched. The influence of weather conditions on the visibility of thermal deficiencies is examined through a dynamic numerical simulation. Next to that, the conclusions of the simulations will be compared with the results from the case studies.

POTENTIAL AND LIMITATIONS OF INFRARED THERMOGRAPHY ON UNVENTILATED WALLS

Simulation Model: Boundary Conditions and the Visibility of Thermal Deficiencies

As stated above, certain weather conditions (e.g., sun and clear sky) could stimulate the visibility of thermal deficiencies in cavity walls. The difference in surface temperatures between an area with a thermal deficiency and a sufficiently insulated area can increase under certain weather conditions. However, the measured surface temperature is not representative for the steady-state condition and, thus, for the thermal performance.

To examine this potential, a brick cavity wall with an air inclusion in the insulation layer and with a concrete thermal bridge was modeled in VOLTRA (Figure 6). VOLTRA is a thermal analysis program for transient heat transfer in three-dimensional rectangular objects (Physibel 2008). The surface temperatures of building components were simulated under changing boundary conditions such as follow: (1) the inside and (2) outside temperature, (3) convective heat transfer coefficient (h_c) (dependent on the wind velocity), (4) global and (5) diffuse sun radiance, and (6) temperature from the sky. Figure 6 shows a section of the models used. In total, 2 × 200 simulations were executed with varying sky temperature, global and diffuse sun radiation and convective heat transfer coefficient h_c (dependent on the wind velocity). The difference between the inside and outside temperature was at least 10°C

(18°F). The simulation method and input data is included in Appendix A.

The air in the air inclusion is modeled as a transparent material in order to incorporate the radiation flux and the convection flux through the inclusion. The convective heat flux depended on the Nusselt number of the air in the inclusion, the temperature gradient across the inclusion, and the thermal conductivity (λ). The latter varied with the temperature. For example, the thermal conductivity of the air in the inclusion is 0.022 W/mK when the air has a temperature of 0°C (32°F) and 0.024 W/mK at an air temperature of 10°C (50°F) (Physibel 2008). These values are close to the thermal conductivity of the insulation layer in the model ($\lambda_{insulation}$ = 0.035 W/mK).

The outer surface was oriented to the South. The outside surface temperature at the thermal bridge ($\theta_{se,tb}$), at the air inclusion ($\theta_{se,air}$) and at a sufficient insulated area ($\theta_{se,wall}$) (dots on Figure 6) were the output data of the model and were simulated over the period of one week.

The effect of the sun radiation on the surface temperature difference can be divided in two parts (Figure 7). First, directly after the maximum amount of sun radiation (at noon, 13 h), the surface temperature difference diminished. In practice, this means that the thermal deficiency is no longer visible on an IR image. This first effect can be explained as follows: Before sunrise, the surface temperature at the thermal bridge is higher, because of the higher thermal conductivity. When the surface is exposed to the sun, the entire structure is heating up. In the case of the concrete thermal bridge, the surface at the insulated parts can "catch up" with the initial increased surface temperature at the thermal bridge. In the model with the air inclusion, the still air in the inclusion heats up at an equal rate compared to the insulated part.

When the model starts to cool down, the surface at the air inclusion seems to cool down faster, resulting in colder spots on a thermal image. The surface at the thermal bridge is also temporary lower in surface temperature than at a sufficient insulated part of the surface. In this simulation, the effect faded away 9 hours after the maximal sun radiation, according to Figure 7.

Theoretically, the thermographer could use the sun radiation in a conscious way: as a heat source to detect thermal deficiencies. However, one does not know beforehand what kind of thermal deficiencies are present. Hence, it is difficult to determine the optimal moment during cooling after the direct sun radiation has disappeared. Next to that, practical

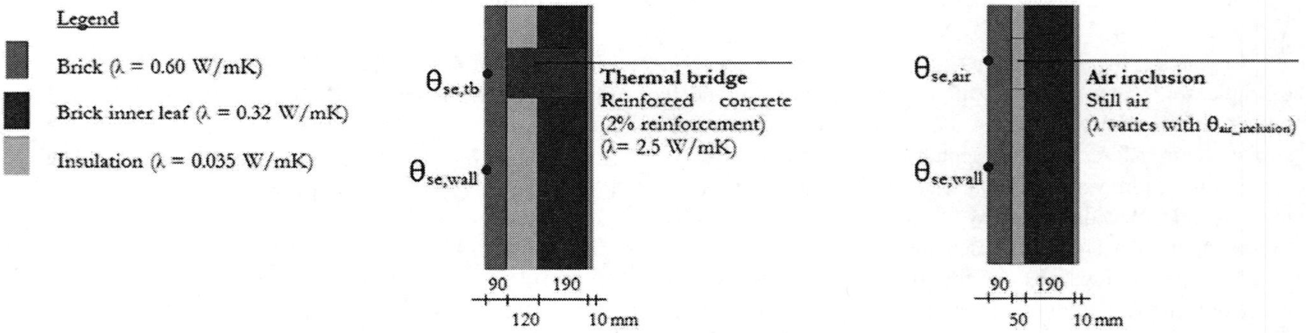

Figure 6 *Simulation models of an unventilated wall with a thermal bridge and an unventilated wall with an air inclusion.*

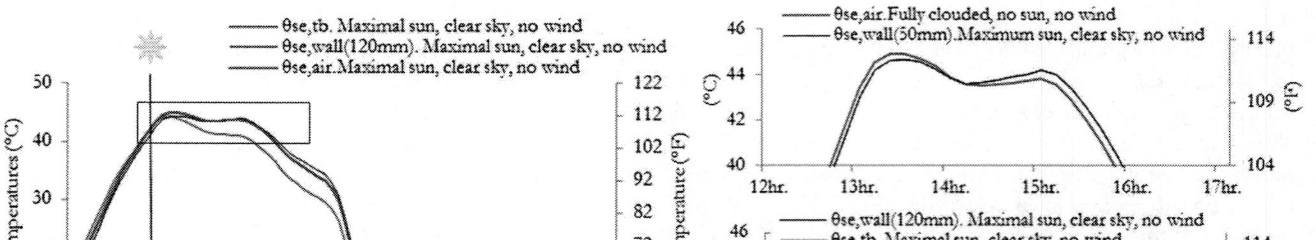

Figure 7 *Effect of direct sun radiance on the surface temperature difference between a thermal deficiency and an intact part of the insulated cavity wall. See Figure 6 for the locations of the output points.*

issues (e.g., shadows of neighboring buildings or the unknown composition of the wall) hinder a conscious use of sun radiance to visualize thermal deficiencies in unventilated walls.

Concerning clear-sky radiation (without sun radiation, at night), this boundary effect has a slightly positive effect on the visibility of thermal bridges (Figure 8a). The surface of the thermal bridge cools down slower; hence, the surface temperature difference increases during clear-sky conditions. Wind, on the other hand, decreases the surface temperature differences with increasing wind velocity. The same observations were made for the model with the air inclusion. (Figure 8b).

In general, the simulations showed that a thermal bridge causes higher surface temperature differences than air inclusions: 1.4°C (2.5°F) for the thermal bridge against 0.4°C (0.72°F) for the model with the air inclusion (Figure 8). This can be explained by the thermal conductivity of the air inclusion (between 0.022 and 0.024 W/mK), which is close to the thermal conductivity of the insulation (0.035 W/mK) in the model. In practice, this means that thermal bridges will be easier to detect than air inclusions.

The simulation results show that only a waiting period after sun radiation is necessary to visualize thermal deficiencies as warm areas on the outside surface. Table 1 suggests possible waiting times for IR detection of thermal deficiencies after sunset, based on the study of Van De Vijver et al. (2014). The waiting times correspond with the period of time after sunset before it is possible to measure surface temperatures closer than 2°C (3.6°F) to the steady-state. The value of 2°C (3.6°F) corresponds to the standard accuracy of most infrared cameras.

The simulations showed that thermal bridges are especially easy to detect when the influence of sun radiation is avoided and a temperature difference of 10°C (18°F) is present between inside and outside. These boundary conditions are fulfilled when the infrared survey takes place before sunrise, in the winter season. The *in-situ* measurements led to an identical observation. In both uninsulated and insulated cavity walls, the common thermal bridges are clearly visible when measuring before and during sunrise (Figure 9) (Beulque 2014; Maroy et al. 2015).

Case Studies: The Visibility of Insufficient Air Cavity Filling

From 2013 to 2015, over 30 cases were monitored to evaluate the potential and limitations of thermography for building inspection. Thermographic inspections were compared with heat flux measurements. Most of the cases regarded filled cavity walls. In the context of a master's thesis concerning the execution quality of wall cavity filling, the most common execution flaws in the case studies were listed as follow (Beulque 2014):

- Poor or lacking measurement data of the cavity width
- Ignorance towards the minimum cavity width (< 5 cm)
- Insufficient number of bore holes
- Incorrect adjustment of the glue flow rate of the EPS pearls

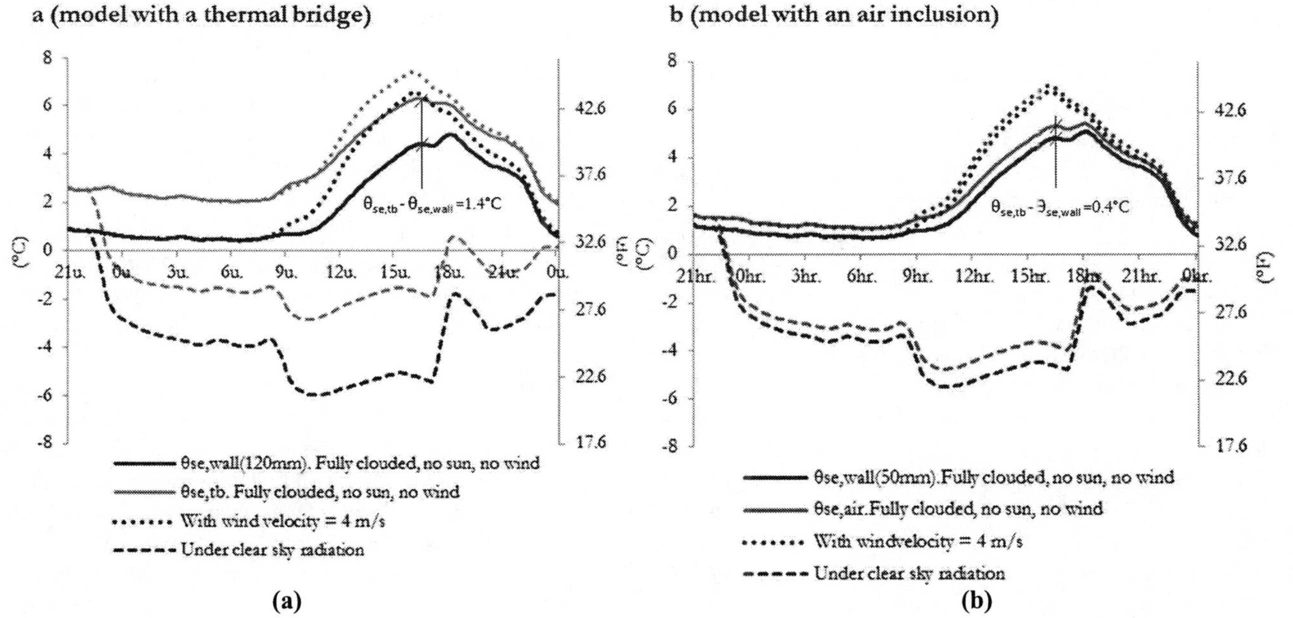

Figure 8 *Effect of clear-sky radiation and wind on the surface temperature difference between (a) a thermal bridge and an intact part of the insulated cavity wall and (b) an air inclusion and an intact part of the insulated cavity wall. See Figure 6 for the locations of the output points.*

Table 1. Waiting Times (h) after Sunset for the Inspection of Cavity Walls on Thermal Deficiencies

Global Solar Radiation	Timber Frame	Filled Cavity Wall	Nonfilled Cavity Wall	Partially Filled Cavity Wall
11400 kJ/m^2	0–1.5 h	3.5–7 h	2–6.5 h	3.5–7 h
8400 kJ/m^2	0–1 h	2–6 h	0.2–5 h	2–6 h
5600 kJ/m^2	0 h	0–4.5 h	0–3.5 h	0–4.5 h
2750 kJ/m^2	0 h	0–2 h	0–0.5 h	0–2 h
0 kJ/m^2	0–0.2 h	0–1 h	0–0.2 h	0–1 h

Figure 9 *Infrared picture of a North/East-orientated facade with cavity wall insulation during sunrise (30 minutes after sunrise). The thermal bridge above the windows and doors are still clearly visible (EPS pearls).*

- Incomplete filling at corners and cavity interruptions (drainage membrane, thermal bridges, etc.)
- Missing bore holes at surfaces with vegetation or cables
- Open parpens at the top of the insulated cavity wall, which causes extra heat loss
- Parpens and openings in the inner cavity leaf that were left open during the execution

These flaws mainly cause insufficient filling and airflows in the insulation.

Air Inclusions in the Insulation Layer. Lacking cavity filling was detected in several case studies, despite the limited surface temperature difference found in the simulations. Typical locations for lacking cavity filling are in corners, at window and door junctions, above membranes for rain drainage in the cavity, and at vegetation and cables (e.g., Figure 10) (Beulque 2014). On Figure 10, there are no specific surface deteriorations visible on the analogue picture. By adjusting the temperature scale of the IR picture, the surface temperature difference caused by the lacking insulation is visualized more clearly.

However, as illustrated in the introduction (Figure 3), it is necessary to document these measurements extensively in order to draw right conclusions. In some cases, the surface temperature pattern can be caused by a slightly different condition of the surface (e.g., local vegetation or a different humidity level). In the case of Figure 10, there were fewer bore holes made during the execution of the cavity filling to prevent damage at the electricity cables in the cavity. This made it impossible to bore holes above the window, at the edge of the roof. Next to that, the analogue picture on Figure 10 shows that the facade was not humid or polluted during the IR measurement. With this information, the thermographer can more likely assume that the warm irregular spot above the window is caused by insufficient insulation in the cavity (Beulque 2014).

Airflows in the Insulation Layer. A second consequence of the execution flaws are airflows in the insulation layer, caused by the combined effect of open parpens at the top of the insulated cavity wall and incomplete filling. In one of the case studies, this effect was clearly visible on an infrared picture (Figure 11). A possible explanation for this kind of pattern is the "chimney effect" in small cavities. The warm air from the inner leaf of the cavity wall rises into the cavity, which causes extra heat losses. On an infrared image, the area at the top of the cavity wall appears as a warmer spot. Airflows in the cavity occur in cases were the insulation is not well attached to the inner leaf of the cavity wall. These

Figure 10 *Analog and infrared pictures with adjusted temperature scale of a filled cavity wall. Above the window and at the edge of the roof, warm patterns indicate an insufficient filling. In the execution report, it is stated that the zone at the rooftop was hard to reach because of vegetation. The cavity wall was insulated with EPS pearls; consequently, there is a possible risk that the glue stacked at the corner of the window interface. With adjusted temperature scale, the small temperature differences are more clearly visualized.*

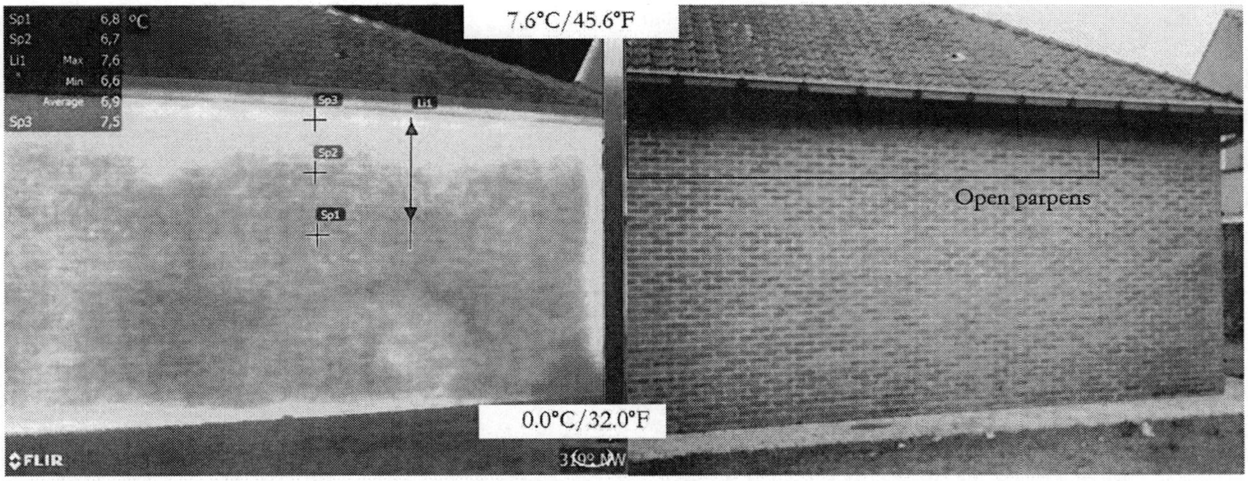

Figure 11 *Two phenomena are visualized with infrared thermography: insufficient filling and open parpens in the upper half. By consequence, there are airflows to the top of the cavity (Beulque 2014). There is a slight temperature gradient visible along the line Li1 on the infrared picture.*

airflows can be induced by natural convection or are forced by the wind entering the cavity (Janssens 2015). A gap of 15 mm can already cause an increase of 50% of the thermal conductivity of the entire layer (Hens et al. 2007).

In the case of Figure 11, the cavity insulation was executed with EPS pearls. Unfortunately, the filling was not sufficient, as the EPS pearls could escape through openings in the inner cavity wall. The parpens at the bottom were also left open during the filling works. The U-factor calculated from flux measurements on this wall indicated a thermal transmittance of 0.476 $W/m^2 \cdot K$, while the thermal transmittance expected from the material characteristics was 0.377 $W/m^2 \cdot K$. The measured values indicate, indeed, a lower thermal performance than expected. On the infrared image, there is a warmer area at the top over the length of the external wall. There is even a slight temperature gradient visible towards the top (Beulque 2014).

CONCLUSIONS

Infrared thermography has the potential to signal thermal deficiencies of a building component. In the case of cavity filling works, this measurement tool can be used to generate an overview of the situation before and after the insulation works. It is therefore a useful supplement for endoscopy and heat flux measurements, which are instruments that are only capable of evaluating the cavity wall pointwise.

However, it is not recommended to draw direct conclusions from a thermographic image alone. For example, a filled cavity wall that shows a deviating surface temperature pattern on an infrared image can have several explanations. The ease by which thermal deficiencies are recognized depends on the following:

- The type of thermal deficiency (thermal bridge or insufficient insulation)
- Shape of the pattern
- The geometry of the building
- The condition of the surface
- Location of the pattern
- Prior weather conditions

The simulations and the case studies in this paper showed that a thermal bridge is easier to recognize on an IR image than air inclusions. For example, thermal bridges in a brick cavity wall typically occur above windows and doors, foundation/floor interface, and floor/wall interface. They generally show large surface temperature differences and have a rectangular shape. Thermal bridges are most detectable when direct sun radiation is avoided during and 1 day prior to the measurement (Table 1).

Air inclusions, on the other hand, cause minor surface temperature changes due to the smaller difference in thermal conductivity compared with the insulation material (see Figure 6). A deviating pattern with a small temperature difference can therefore easily have other causes, apart from insufficient filling (see, for example, Figure 3). With help of an analysis of the condition of the surface, weather conditions, and building geometry, possible causes can be further determined. Additionally, an execution report of the cavity filling works can help to localize potential flaws. However, final conclusions can only be drawn with an endoscopy. Next to that, special attention must be drawn to the use of the camera. The measurement distance and the field of view of the lens is determining for the visibility of small surface temperature differences and thus for the detection of insufficient cavity insulation.

ACKNOWLEDGMENTS

The authors would like to acknowledge the financial support of this work by IWT (Institute for the Promotion of Innovation by Science and Technology in Flanders), Belgium.

NOMENCLATURE

W	=	emitted radiation, W/m^2
ε	=	emissivity of the surface
τ	=	transmittance through a surface
h_c	=	convective heat transfer coefficient, W/m^2·K
θ	=	temperature, °C (°F)

Subscripts

obj	=	the measured object
atm	=	the atmosphere
amb	=	the environment
se	=	external surface

APPENDIX: SIMULATION INPUT DATA

This appendix includes the climate data used as input in the simulation model on Figure 6 (Figure A1 and Figure A2). On the fourth day (marked on Figure A1 and A2), the input weather data for wind velocity, global radiation, and sky temperature varied according to the values in Table A1. By

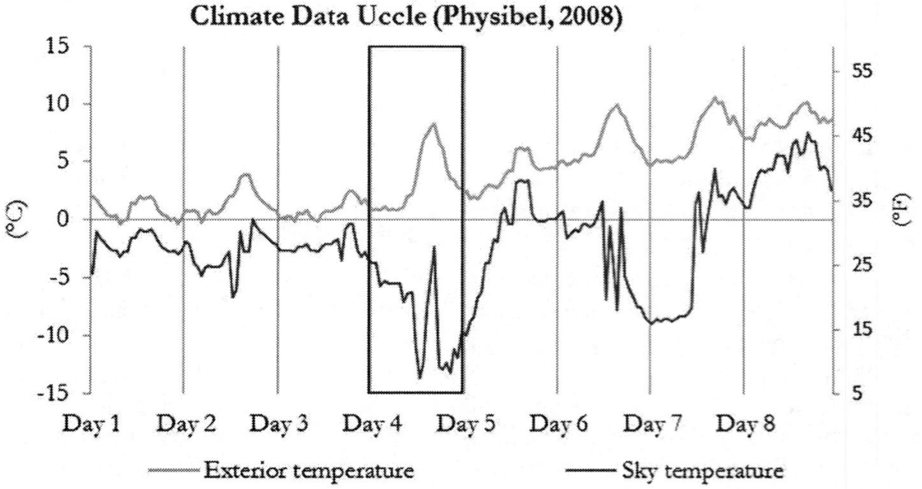

Figure A1 *Outside air temperature and sky temperature used over 1 week of simulation (°C and °F). On Day 4, the sky temperature was adjusted to the values in Table 1. The inside air temperature was minimum 10°C (18°F) higher than the outside temperature in this simulation.*

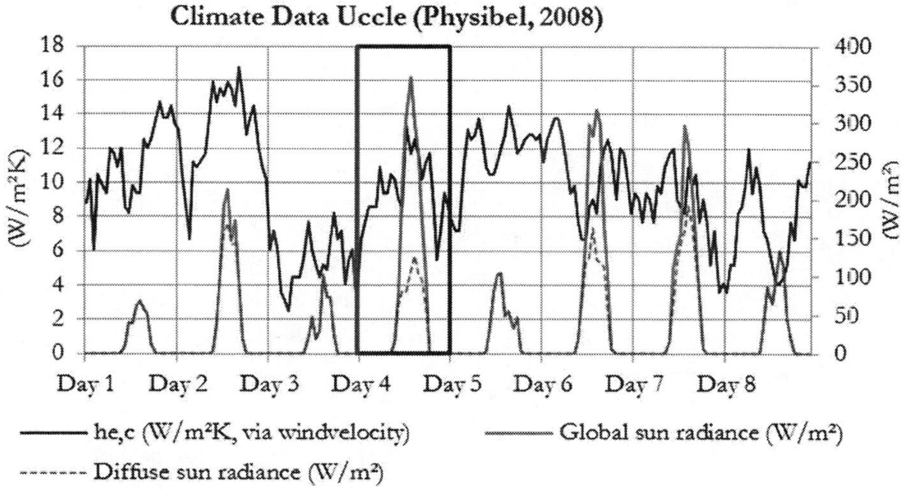

Figure A2 *Convective heat transfer coefficient for the outside surface ($h_{e,c}$) and the global and diffuse sun radiation used over 1 week of simulation. On Day 4, the convective heat transfer coefficient ($h_{e,c}$) and the global sun radiance was adjusted to the values in Table 1. The inside air temperature was minimum 10°C higher than the outside temperature in this simulation.*

Table A1. Climate Data Used as Input Data for the Simulation

Wind Velocity →	$h_{e,c}$	Solar Radiation	Sky Radiation Temperature	
0 m/s →	1.8 W/m²·K	0* ($I_{globalsun}$)	$\Delta T_{sky\text{-}air}$ = 2,5°C (4.5°F)	→ clouded
1 m/s →	5.9 W/m²·K	0.25* ($I_{globalsun}$)	$\Delta T_{sky\text{-}air}$ = 5°C (9°F)	
2 m/s →	8.4 W/m²·K	0.50* ($I_{globalsun}$)	$\Delta T_{sky\text{-}air}$ = 10°C (18°F)	
3 m/s →	10.4 W/m²·K	0.75* ($I_{globalsun}$)	$\Delta T_{sky\text{-}air}$ = 15°C (27°F)	
4 m/s →	12.0 W/m²·K	1* ($I_{globalsun}$)	$\Delta T_{sky\text{-}air}$ = 20°C (36°F)	→ clear sky
5 m/s →	13.5 W/m²·K		$\Delta T_{sky\text{-}air}$ = 20°C (36°F)	
6 m/s →	14.9 W/m²·K		$\Delta T_{sky\text{-}air}$ = 20°C (36°F)	
7 m/s →	16.1 W/m²·K			

*$I_{globalsun}$ is the global sun radiance in the climate file (see Figures A1 and A2).

this, the influence of different values of these weather conditions could be examined. The inside air temperature was at minimum 10°C higher than the outside temperature in this simulation.

The possible weather conditions (Table A1) were combined, resulting in total in 200 possible combinations to simulate. For each weather condition, the surface temperature difference ($\theta_{se,tb} - \theta_{se,wall}$ and $\theta_{se,air} - \theta_{se,wall}$, see Figure 6) was calculated. In the paper, the maximum values of surface temperature differences are displayed.

The convective heat transfer coefficient ($h_{e,c}$) at the outside surface was dependent on the wind velocity from the climate file (Equation A1). The correlation was derived by linking wind velocity measurements to heat flux and surface temperature measurements of a brick surface at 20 m height in an urban environment (Taki and Loveday 1996).

$$\text{Windward side: } h_{e,c} = 16.15 V_s^{0.397}$$

$$\text{Leeward side: } h_{e,c} = 16\ 25 V_s^{0.503} \tag{A.1}$$

The sky radiation temperature (T_{sky}) in the model was found in the climate data of the weather station in Uccle (Physibel 2008). Only on the fourth day T_{sky} was calculated from the outside air temperature T_{air} to model the variation in cloudiness (Table 1). A difference of 20°C between T_{sky} and T_{air} symbolizes the clear-sky conditions, whereas a difference of 2.5°C indicates cloudy conditions.

REFERENCES

Albatici, R, and A. Tonelli. 2010. Infrared thermovision technique for the assessment of thermal transmittance value of opaque building elements on site. *Energy and Buildings* 42: 2177–83.

ASTM. 2011. ASTM Standard C1060-11a, *Standard practice for thermographic inspection of insulation installations in envelope cavities of frame buildings*. West Conshokocken, PA: ASTM.

ASTM. 2002. ASTM Standard E1862-97, *Standard test method for measuring and compensating for reflected temperature using infrared imaging radiometers*. West Conshokocken, PA: ASTM.

ASTM. 2005. ASTM Standard E1993-99a, *Standard test method for measuring and compensating for emissivity using infrared imaging radiometers*. West Conshokocken, PA: ASTM.

Behaneck, M. 2011. Bauthermografie-Praxis: Tipps & Tricks für Einsteiger. no. Gebäude energie berater 2: 44–49.

Beulque, Sarah. 2014. Uitvoeringskwaliteit van na-isolatie van spouwmuren. Gent: Universiteit Gent.

Charlier, Laurence. 2007. Utilisation de la thermographie infrarouge pour la determination des déperditions thermique d'un bâtiment. Liège: Université de Liège.

Dall'O', G., L. Sarto, and A. Panza. 2013. Infrared screening of residential buildings for energy audit purposes: results of a field test. *Energies* 6: 3859–78.

Dall'O, G. 2013. Infrared audit. *Green energy audit of buildings*. London: Springer-Verlag. 111–25.

Eads, L., R. Epperly, and J. Snell. 2000. Thermography: Practical guide. *ASHRAE Journal* (March 2000):51–55.

European Commission. 2011. Routekaart naar een concurrerende koolstofarme economie in 2050. Europe 2020 targets. Brussels.

FLIR. 2015. Gebruikershandleiding FLIR Tools/Flir Tools+. Boston: FLIR.

FLIR. 2012. Thermal camera guide for research professionals. FLIR Systems, Inc. Boston: FLIR.

Fokaides, P.A., and S.A. Kalogirou. 2011. Application of infrared thermography for the determination of the overall heat transfer coefficient (U-value) in building envelopes. *Applied Energy* 88: 4358–65.

Griffith, B.T., D. Türler, and D. Arasteh. 1995. Surface temperatures of insulated glazing units: Infrared thermography laboratory measurements. *ASHRAE Transactions*.

Hens, H., A. Janssens, W. Depreatere, J. Carmeliet, and J. Lecompte. 2007. Brick cavity walls: A performance analysis based on measurements and simulations. *Journal of Building Physics* 31(2).

Janssens, Arnold. 2015. Bouwsfysica 1 (Cursus 3de bachelor ingenieur-architect). Gent.

Jewett, J, and R Serway. 2008. *Physics for scientists and engineers with modern physics*. Belmont: Thomson Higher Education.

Lehmann, B., and K. Ghazi Wakili. 2013. Effects of individual climatic parameters on the infrared thermography of building. *Applied Energy* 110:29–43.

Maroy, Katrien, Kim Carbonez, Marijke Steeman, Nathan Van Den Bossche, and Sven Van De Vijver. 2015. Thermografie als graadmeter van de gebouwschil. Praktische gids voor de thermograaf. Ghent, Belgium: University Press.

NBN. 1999. NBN EN 13187, Thermal performance of buildings—Qualitative detection of thermal irregularities in building envelopes-infrared method. Brussels, Belgium: Bureau voor Normalisatie.

Orlove, G. 2012. What does sensitivity (NETD) mean when applied to a thermal imager? Infrared Training Center (ITC). http://irinformir.blogspot.be/2012/05/what-does-sensitivity-netd-mean-when.html. (Accessed October 28, 2015).

Pearson, C. 2011. Thermal Imaging Of Building Fabric. *BSRIA Guide* 39. BSRIA. 30.

Physibel. Voltra. 2008. *3-dimensional dynamic simulation tool using the finite element method*, Version 7.0. Maldegem, Belgimum.

Taki, A.H., and D.L. Loveday. 1996. Convective heat transfer coefficients at a plane surface on a full-scale building facade. *Heat Mass Transfer* 39(8): 1729–42.

Tanner, C., B. Lehmann, F. Thomas, and K. Ghazi Wakili. 2011. Vorschlag zur standardisierten darstellung von wärmebildern mit QualiThermo. *Bauphysic Heft* 6.

Taylor, T., J. Counsell, and S. Gill. 2013. Energy efficiency is more than skin deep: Improving construction quality control in new-build housing using thermography. *Energy and Buildings* 66:222–31.

Van De Vijver, Sven, Marijke Steeman, Nathan Van Den Bossche, Kim Carbonez, and Arnold Jansenss. 2014. The influence of environmental parameters on the thermographic analysis of the building envelope. QIRT. Bordeaux, Frankrijk. 10.

VATh. 2011. Bauthermografie: zur Planung, Durchführung und Dokumentation infrarotthermografischer Messungen an Bauwerken oder Bauteilen von Gebäuden. no. Bundesverband für Angewandte thermografie e.V. 2011. 15.

WTCB. 2012. Na-isolatie van spouwmuren door het opvullen van de luchtspouw (Dutch). Brussels, Belgium, June 2012.

BIBILIOGRAPHY

Asdrubali, F., G. Baldinelli, and F. Bianchi. 2012. A quantitative methodology to evaluate thermal bridges in buildings. *Applied Energy* 35:365–73.

Balaras, C., and A. Argiriou. 2002. Infrared thermography for building diagnostics. *Energy and Building* 34: 171–83.

Barreira, E., S. de Freitas, V. de Freitas, and J. Delgado. 2013. Infrared thermography application in building diagnosis: A proposal for test procedures. In *Infrared thermography application in building diagnosis*. 91–117. Springer-Verlag Berlin Heidelberg.

BSI. 1999. BS EN 13187, *Thermal performance of buildings—Qualitative detection of thermal irregularities in building envelopes-infrared method*. London: BSI.

Grinzato, E., V. Vavilov, and T. Kauppinen. 1998. Quantitative infrared thermography in buildings. *Energy and Building* 29:1–9.

Hens, Hugo. 2012. *Building physics—Heat, air and moisture*. Berlijn: Ernst & Sohn.

ISO. 2007. EN ISO 6946, Building components and building elements—Thermal resistance and thermal transmittance-calculation method. Geneva: International Organization for Standardization.

Loveday, D.L., and A.H. Taki. 2001. Section 3: Heat transfer. In Reference Data, CIBSE Guide C, edited by P.D. Compton, W.P. Jones, D.L. Loveday, M.R.I. Purvis and A.C. Watson, 84–92. London: Butterworth-Heinermann.

NBN. 2008. NBN B62-002, Thermische prestaties van gebouwen-Berekening van de warmtedoorgangscoëfficiënten (U-waarden) van gebouwcomponenten en gebouwelementen-Berekening van de wartmeoverdrachtscoëfficiënten door transmissie (Ht-waarde) en ventilatie (Hv-waarde). Brussel, België: Belgisch Bureau voor Normalisatie.

RESNET. 2012. Interim guidelines for thermographic inspections of buildings. RESNET Residential energy services network.

Thermal Analysis of Curtain Wall Systems—A Parametric Study

Nathan Van Den Bossche, PhD

Wahid Maref, PhD
Member ASHRAE

ABSTRACT

The thermal performance of curtain wall systems can easily be calculated using 2-D numerical simulations. Several commercial software packages are available, and international standards provide a methodology to calculate the thermal transmittance. However, even though these methods are well known in academia and the research community in general, thermal optimization has not reached its full potential in the building industry yet, and there is considerable margin for improvement. Based on a market survey in North America and Europe, generic curtain wall sections were developed, and typical approaches for improving the thermal performance of curtain wall systems have been identified and described. Subsequently, the impact of separate improvements, as well as combined effects, have been studied using both standardized and advanced calculation methods. For this, the heat transfer phenomena and the way these are modeled according to standard calculation procedures are discussed. Moreover, for a number of important aspects, the current calculation standards do not provide adequate guidelines, which results in diverging interpretations with a large impact on the overall performance. Next to that, a number of secondary effect originating from standards are discussed (e.g., the thickness of the IGU, the depth of the window rebate, equivalent thermal conductivities, and the impact of reduced heat transfer coefficients). Finally, the relative impact of the thermal performance of a curtain walls system on typical commercial buildings is analyzed and evaluated.

INTRODUCTION

Throughout the last few decades there has been a growing interest in reducing the energy use in buildings. There are many aspects that contribute to the overall energy efficiency of a building (e.g., orientation; compactness; heating, ventilation, and air conditioning [HVAC] systems; airtightness; and, of course, an appropriate insulation level). Particularly, the performance of insulation has been investigated thoroughly in past studies as it may be perhaps the first aspect that needs to be tackled in reducing the energy use of buildings. Moreover, countries throughout the world are now implementing specific standards and guidelines to take the impact of thermal bridges and 2-D-effects into consideration in the overall heat loss coefficient calculation. However, one building component remains a source of concern, specifically, window frames and curtain wall systems. These are rather complex components due to specific boundary conditions relating to mechanical performance, operability, acoustics, etc. For walls, roofs, and floors, typical guidelines for maximum thermal transmittance in central to North European countries are situated between 0.1 and 0.3 $W/m^2{\cdot}K$, which can easily be realized with common construction types. For windows, the guidelines are less strict and typically vary between 0.8 and 2.4 $W/m^2{\cdot}K$. In some countries there are specific requirements on the IGUs as well. Commercial double glazing with low-e coating and argon gas filling has a thermal transmittance of 1.1 $W/m^2{\cdot}K$, whereas triple glazing and vacuum glazing can go as low as 0.5 $W/m^2{\cdot}K$. To the knowledge of the authors, there are no specific restrictions in different countries on the thermal conductivity of the window frames or curtain wall systems. Note that imposing specific restrictions would render it impossible to construct some specific window configurations. Moreover, a performance-based approach towards energy use should be preferred over a prescriptive regulation.

Nathan Van Den Bossche is a professor in the Department of Architecture and Urban Planning, Ghent University, Ghent, Belgium. **Wahid Maref** is a professor at École de technologie supérieure, Départment de génie de la construction, Université du Québec, Montreal, QC, formerly senior research officer at National Research Council Canada, Ottawa, ON.

The information on thermal optimization of curtain wall systems in scientific literature, and literature in general, is rather scarce. The research community has moved away from the rather simple heat transfer calculations of components towards more dynamic and complex effects (e.g., impact of shading devices, computational fluid dynamics of cavities in double skin facades). Not a single scientific publication was found that provides clear guidelines on the thermal optimization of mullions and transoms. Alternatively, a few publications can be found on the thermal performance and optimization of window frames. Gustavsen et al. (2008, 2010) studied the impact of thermal conductivity of frame material and thermal breaks to define material performance targets for current window designs. Note that a U-factor of 0.5 $W/m^2 \cdot K$ was chosen as required performance level simply based on the fact that the best commercial insulating glass units (IGUs) on the market now have a U-factor of 0.5 $W/m^2 \cdot K$. Based on that approach, it was concluded that thermal breaks should have a thermal conductivity below 0.02 W/m·K (or 0.005 W/m·K if "new" materials are developed), structural insulating materials for wood composite profiles should have a thermal conductivity below 0.03 W/m·K, and ideally aluminum and PVC frames should comprise cavities with an emissivity below 0.05 for the surrounding materials. No design guidelines for window geometry were presented, and no pathways were specified on how to acquire the specified conductivities. Similarly, Byars and Arasteh (1992) also focused on the impact of thermal conductivity on the U-factor of the frame. Research by Gustavsen et al. (2008) indicated that although convection is modeled in EN ISO 10077-2 by adopting a simplified approach with equivalent thermal conductivities, results compare well with the fluid flow simulations. ISO 10077-2 prescribes that cavities with an interconnection not exceeding 2 mm are to be considered as separate. Any reference to papers or research is lacking for that assumption, and, by means of CFD simulations, it was shown by Gustavson that 7 mm would be a more realistic criterion.

SIMULATION METHOD

In this paper an analysis is presented on the thermal performance of typical North American and European curtain wall systems. In order to assess and evaluate that performance, a uniform calculation methodology was adopted. Note that the North American and European approach are not completely identical. This has two consequences: first of all, small discrepancies may arise between the results in the analysis presented in this report and results published by manufacturers using a different calculation method. Different calculation rules lead to different heat transfer results. However, in this report, only a relative comparison is made to evaluate the impact of specific parameters. Consequently, small differences in absolute value are not as important, and the specific choice of calculation method should not be overrated. Secondly, the sheer existence of very specific calculation rules inevitably leads to adapted industrial designs. For example, in North America two cavities are considered as independent and connected with a "throat" region when the shortest distance between two sides is smaller than 5 mm (see ISO 15099 [2003]), whereas in Europe a value of 2 mm is used as criterion (ISO 10077-2 [2012]). So, when two parts of one cavity are separated by a throat region that is only 2 mm wide, the two zones of the cavity can be considered as two separate zones. By consequence, window frame and curtain wall designs might slightly differ in different regions, as profiles are most likely optimized according to local calculation methods. For the example described above relating to cavity subdivision, one will find a lot of 4.9 mm throats in North America and 1.9 mm throats in Europe. By consequence, simulating frame sections according to a calculation sequence for which it was not optimized might lead to a suboptimal design.

Only the U-factor of the curtain wall is considered in this analysis, which excludes the impact of the IGU (center of glazing), the spacer (edge of glazing), and the window/wall interface. The specific geometry of the IGU (i.e., thickness and width) does have an impact on the U-factor of the frame and will thus be considered. The spacer and window/wall interface do not have an impact on the U-factor of the frame and, by consequence, is not comprised in the analysis. This approach isolates the impact of the curtain wall itself on the overall performance, and the results of this analysis can be combined with separate simulations on IGU and IGU spacers. Furthermore, there is a big discrepancy between the European and North American approach. In Europe, only the center-of-panel U-factor of the IGU is considered (and multiplied with the total IGU surface), whereas the impact of the IGU edge spacer is accounted for by means of a linear thermal transmittance. In contrast, in North America, the IGU area is separated in two distinct areas: the center-of-panel and a band of 63.5 mm wide with a U-factor that incorporated the edge effect of the spacer (i.e., the alternative approach described in Section 4.1.4 of ISO 15099 [2003]).

NFRC 100-2014

In North America, the NFRC 100-2014, *Procedure for Determining Fenestration Product U-factors* by the National Fenestration Rating Council is typically used to evaluate the thermal performance of window and curtain wall systems. The document provides guidelines on how to determine the U-factor of product lines (how to cope with different sizes, how to select a test matrix, rules for extrapolation, etc.) and custom products. For the simulation and calculation individual frames, the document refers to ISO 15099. Next to that, it is specified that only area-weighed methods are allowed. Furthermore, a number of technical aspects are listed. Only the most important ones are summarized below:

- Boundary conditions: T_{in} 21°C, T_{out} −18°C, wind speed 5.5 m/s.
- Indoor convective heat transfer coefficient is based on center-of-panel IGU surface temperature.

- On the interior side a detailed gray body radiation model shall be used.
- The exterior convective heat transfer coefficient is based on the exterior wind speed (26 $W/m^2 \cdot K$).
- On the exterior side a detailed black body radiation model shall be used.
- A cross-section should include at least 150 mm of glazing section.

Note that NFRC 100-2014 does not specify what correlation should be used to calculate the convective heat transfer coefficient on the inside and outside. For both, several correlations can be found in literature. Next to that, it does not include specific calculation procedures for curtain wall systems. ISO 15099 provides guidelines on the calculation of the thermal performance of windows and doors. Two different thermal indices are presented to normalize absolute heat loss through a component: the linear thermal transmittance ψ, $W/m \cdot K$, and the frame thermal transmittance U, $W/m^2 \cdot K$. ISO 15099 refers to ISO 10077-2 and ISO 10211 for the details regarding the required two-dimensional numerical analysis and specifies following boundary conditions:

- T_{in} 20°C
- T_{out} 0°C
- Interior convective heat transfer coefficient $h_{cv,in}$ is 3.6 $W/m^2 \cdot K$
- Exterior convective heat transfer coefficient $h_{cv,ex}$ is 20 $W/m^2 \cdot K$

ISO 15099 also provides correlations for the exterior convective heat transfer coefficient: $h_{cv,ex} = 4.7 + 7.6 \cdot v$.

Next to that, the treatment of cavities according to ISO 15099 depends on the orientation. In this report, only horizontal heat flows are considered (the curtain wall is assumed to be in a vertical position). Please refer to the standard for the detailed calculation of the Nusselt number. Finally, the Therm 6.3/Window 6.3 NFRC simulation manual comprises an approach to take the effect of screws into account. Basically, the impact of thermal bridges is accounted for by simulating the cross section at the thermal bridge itself, adopting a surface-area-based "effective" thermal conductivity for the component that introduces the thermal bridge.

ISO 12631

The only ISO standard that specifically deals with calculating the thermal performance of curtain wall systems is ISO 12631, *Thermal Performance of Curtain Walling – Calculation of Thermal Transmittance*. Two different calculation approaches are presented: the single assessment method and the component assessment method. The component assessment method is more suited for evaluating product lines and considers the frame, IGU, and spacer of the IGU as separate components, which allows an easy approach towards calculating the performance of different combinations of frames, IGUs, and spacers. The energy transmission is normalized by dividing the heat loss by the surface area of the frame. This approach is often used in both Europe and North America (where it is the only accepted method). The single assessment method was developed to calculate the thermal performance of systems that were developed for a specific project and allows to combine the effect of the frame and spacer in one calculation. But every time a single parameter is changed, the whole simulation needs to be repeated. The heat loss is expressed as a linear additional heat loss, similar to linear thermal transmittance that is used in thermal bridge calculation. NFRC 100-2014 does not support the use of this approach. For determining the thermal transmittance of frames, mullions, and transoms, ISO 12631 refers to the numerical approach in ISO 10077-2, or the experimental assessment with a hot box. ISO 10077-2 is the reference standard on the numerical aspects for determining the thermal performance of window frames. According to this approach, the following boundary conditions should be adopted:

- T_{in} = 20°C
- T_{out} = 0°C
- Interior surface resistance is as follows:
  - 0.13 $m^2 \cdot K/W$ (~ heat transfer coefficient of 7.7 $W/m^2 \cdot K$) in plane surfaces
  - 0.20 $m^2 \cdot K/W$ (~ heat transfer coefficient of 5 $W/m^2 \cdot K$) at edges or corners with reduced heat transfer
- Interior surface resistance is 0.04 $m^2 \cdot K/W$ (~ heat transfer coefficient of 25 $W/m^2 \cdot K$)
- A cross-section should include at least 190 mm of glazing section

The radiation heat transfer is calculated in a similar way in NFRC 100-2014. Please refer to the ISO 10077-2 standard for the detailed calculation of the Nusselt number. Note that this calculation procedure does differ significantly from the one described in ISO 15099. Mind that the ISO 10077-2 was developed specifically for the calculation of heat transport in window frames, which arguably would result in a narrower range of Nusselt correlations, in accordance with the typical geometry of the cavities found in window frames. The only specific aspect of curtain wall systems that is not addressed in the ISO 10077-2 is the impact of screws connecting the internal to the external sections of mullions and transoms. ISO 12631 provides two approaches to take this into account: either a default ΔU is added to the U-factor of the frame without screws, or it can be calculated according to the approach described in Annex C of ISO 12631. The only default value for ΔU that is listed is 0.3$W/m^2 \cdot K$. This can be used for screws in stainless steel, with a diameter $\leq$ 6 mm, and a distance between the connectors between 200 and 300 mm. Annex C provides an approach using an equivalent thermal conductivity for the screw. The simulation is done assuming the screw is a continuous metal plate in the profile, but by changing the thermal conductivity, the actual performance can be determined.

Simulation Approach

The NFRC 100-2014 refers to the ISO 15099 for the calculation procedures but specifies a number of small changes with respect to boundary conditions and modeling the surface heat transfer coefficients. In turn, the ISO 15099 standard refers to ISO 10077-2. The ISO 12631 standard also refers to ISO 10077-2 to determine the heat transmittance but adds specific guidelines to account for screws (similar to the approach in the Therm 6.3/Window 6.3 NFRC Simulation Manual). Consequently, the ISO 10077-2 is the basis for numerical simulations in both North America and Europe. The most important differences refer to the boundary conditions and convection in cavities. Next to that, the calculation of thermal bridges is also of importance. As stated in the introduction, in this report the focus lies on a comparative analysis, and, as such, the relative difference between different systems is more important than the absolute value of an individual calculation.

A comparison of convection correlations can be found in (Gustavsen 2001). Considering a range of aspect ratios, he concluded that the ISO 10077-2 overpredicts the convection for aspect ratios below 0.4 compared to measurements reported in literature, whereas it underpredicts the convection for aspect ratios between 0.4 and 10 and shows a good correlation for higher aspect ratios. The ISO 15099 seems to work well for Rayleigh numbers up to 105. However, the impact of the chosen correlation should not be overestimated. Simulations by Blanusa et al. (2007) indicate that the difference in the overall thermal performance is never larger than 3%. Mind that this only refers to the difference in convection correlations in the cavities. By consequence, the ISO 12631 (and ISO 10077-2, which both ISO 12631 and NFRC 100-2014 refer to) is adopted here. Furthermore, this is the only ISO standard that was specifically designed for curtain walls, and is used on several continents. Next to that, as European optimization parameters and approaches will be considered for thermal optimization, the European approach is best fit to evaluate the effectiveness of the thermal improvements. Similarly, fixed heat transfer coefficients are adopted that take into account reduced heat transfer at corners according to ISO 10077-2. Again, the impact of different heat transfer coefficients is of limited importance for a comparative analysis, and the impact reduces for thermally improved systems.

Simulation of Mechanical Fixings

Although heat conduction through mechanical fixings is a 3-D phenomenon, a 2-D simulation method can provide a relatively precise estimation. According to Annex C of NBN EN ISO 12631:2012, the screw has to be modeled as a "smoothed" screw with its thickness equal to the real diameter with an equivalent thermal conductivity. The airspace surrounding the screw needs to be treated as a single air space to avoid inaccurate results.

Table 1 provides an overview of some results on simulating the screws in curtain wall systems. For these simulations, software programs Bisco, Trisco, and Solido were used (please refer to the manual for a more detailed description of the calculation procedure, Bisco [2012]). These programs have been developed by Physibel and are generally considered state-of-the-art models in the research community. The three programs are rather similar, but Bisco uses a triangular grid, Trisco has an orthogonal grid, and Solido also uses a rectangular grid with node fitting so it can almost perfectly replicate the original geometry (Bisco 2012). The results

Table 1. Simulation Results on Simulation of Screws in Bisco, Trisco, and Solido

	Q, W/m	Q, W	U_f, W/m²·K	
		Frame without Screw		
Bisco	8.3127		2.256	
Trisco	8.3205		2.259	
Solido	8.3205		2.259	
		Frame with Screw		
Solido		1.1049	2.526	Reference
Trisco		1.1141	2.564	+2%
Bisco[1]	8.4918		2.363	–6%
Bisco[2]	8.9182		2.585	+2%
Trisco[2]	8.9260		2.571	+2%

1 Smoothed lambda method (EN 13947).
2 Smoothed thickness method (EN 13947).

show that the difference between those three programs is limited (0.1%). When the actual geometry of the screw is modeled accurately, the U-factor is 2.526 W/m^2·K. When the smoothed λ value is applied, as adopted in Annex C of NBN EN ISO 12631:2012 and Therm 6.3/Window 6.3 NFRC Simulation Manual, the heat loss through the screws is underestimated by 6% (or 0.16 W/m^2·K), compared to the correct simulations. The simulations point out it is better to adopt the smoothed thickness approach: the screw is replaced by a continuous plate that has the same overall cross section as the screws. For example, if each 250 mm a screw (simplified, 5 × 5 mm) connects both aluminum parts, the smoothed thickness of the thin plate is 0.1 mm (5 × 5 mm = 25 mm^2 = 250 mm × 0.1 mm). In that case, the heat loss through the screws is overestimated by 2%. However, based on the Solido simulations, the impact of the screw itself is 0.267 W/m^2·K, 0.107 W/m^2·K based on the smoothed lambda approach in Bisco, and 0.329 W/m^2·K according to the smoothed thickness method. Based on these results, it was decided to use the smoothed thickness method. All simulations presented in this paper were done with the software program Bisco, and the ISO 10077-2 standard calculation method is adopted.

MODELS

In order to design generic curtain walls systems, an overview of typical sections of North American and European commercial curtain wall systems is listed. Technical information on typical aluminum curtain wall systems is gathered from company websites, industry guidelines, public test reports, CE-marking databases, and local technical approval documents.

Commercial European Curtain Wall Systems

In Europe, a variety of curtain wall systems can be found. In general, the width of the frame varies between 50 and 60 mm, depending on the specific application. The length of the total frame varies as well, but only the interior frame will have a major influence on the thermal performance. In Table 2, three exemplary frames of 60 mm width, 200 mm length, and a glass thickness of 28 mm are simulated and compared. As required by NBN ISO EN 10077-2, the IGU length in simulations is 190 mm, measured from the outer surface of the aluminum frame.

Commercial North American Curtain Wall Systems

In general, North American curtain wall systems do not differ much from European frames, except for the shape of the aluminum frame itself. In most cases, the frame width is equal to 60 mm. In both Europe and North America, the mullions and transoms of curtain wall frames consist of an aluminum structural member on the inside, an aluminum pressure plate on the outside, and, for reasons of aesthetics, an aluminum cap to finish the exterior. The IGUs are installed with dry seals (typically EPDM). The most important difference between systems is the way the pressure plate is mechanically fixed to the structural member. To optimize the thermal performance of the system, the continuity of the insulation should be maintained in the space between the IGUs by avoiding materials with high conductivity and reducing the section of these components. In Table 3, three exemplary frames of 60 mm width, 200 mm length, and a glass thickness of 28 mm are simulated and compared.

Development of a Generic Curtain Wall System

The development of a generic curtain wall system allows for an analysis that is independent of specific products and manufacturers. A generic model is representative for current building practice and allows for the implementation of all relevant thermal enhancements. In a first analysis, common techniques in curtain wall design are investigated and evaluated. The reference profile should act as a basis for further development and should leave sufficient room for improvement. Design parameters were chosen to be rather poor, creating possibilities for optimization.

- Profile lengths usually vary from 100 up to 300 mm. A representative length of 200 mm was chosen as a starting point, leaving room for increasing and decreasing the profile length.
- The width of the curtain wall frame is commonly 50 or 60 mm. A representative width of 60 mm was chosen as a starting point, leaving room for increasing or decreasing the profile width.
- The glazing panel was chosen to be double glazed, resulting in an IGU width of 28 mm.
- The outer point of the IGU was positioned just below the inner side of the gaskets. In this way, the effect of positioning the IGU towards the inside and the outside of the frame can be investigated.
- After a first analysis, it was clear that the length of the screw fins could have a major impact on the overall performance of the frame. The length of the screw fins was chosen to be on the poor side (30 mm), as a result of which the influence of a decrease in length could be investigated.

After analyzing both European and North American curtain wall systems, two generic curtain wall frames are proposed in Figure 1.

European Versus North American Generic Curtain Wall Frames

Two systematic differences that are evident from analyzing a series of European and North American curtain wall systems are the location of the interior gasket and the more or less solid screw connection. In European systems there is an additional member in the interior aluminum profile that holds the gasket, whereas in North America the gasket is often installed directly into the aluminum structural member. Note that this difference in design does not originate from a thermal

Table 2. Selection of Typical European Curtain Wall System Sections

Company A	Simulation Model	Results
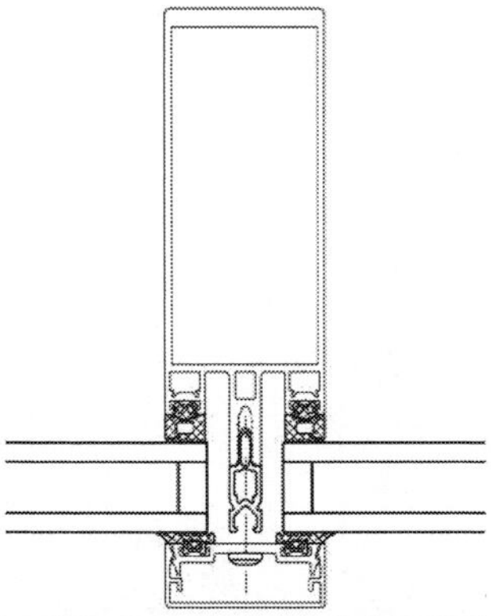	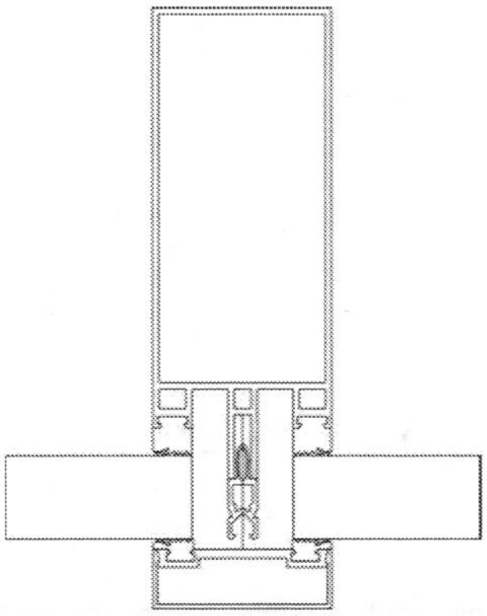	$U_f = 4.203$ W/m^2·K $\Psi = 0.581$ W/m·K
Company B	**Simulation Model**	**Results**
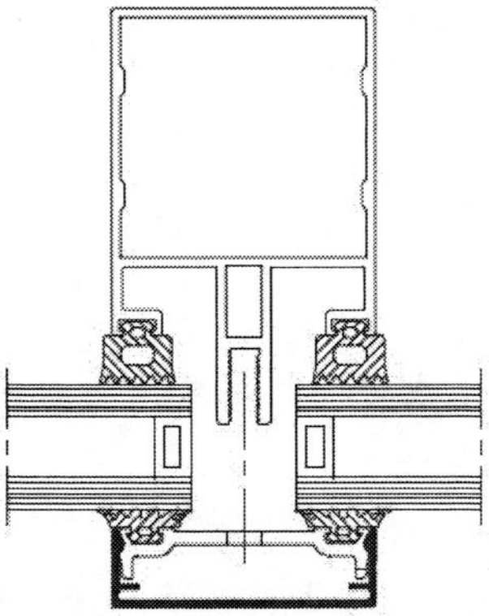	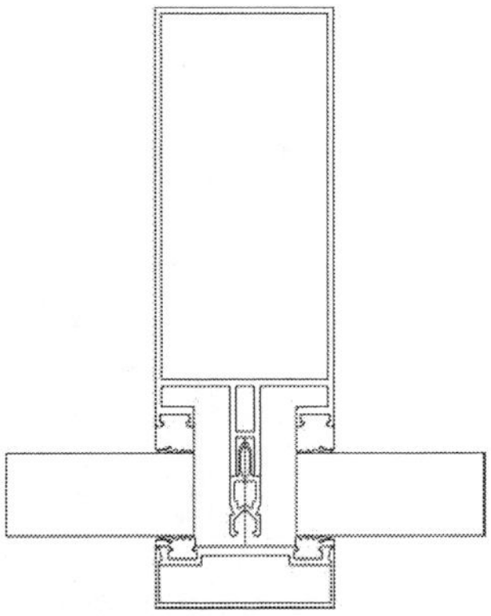	$U_f = 3.961$ W/m^2·K $\Psi = 0.567$ W/m·K
Company C	**Simulation Model**	**Results**
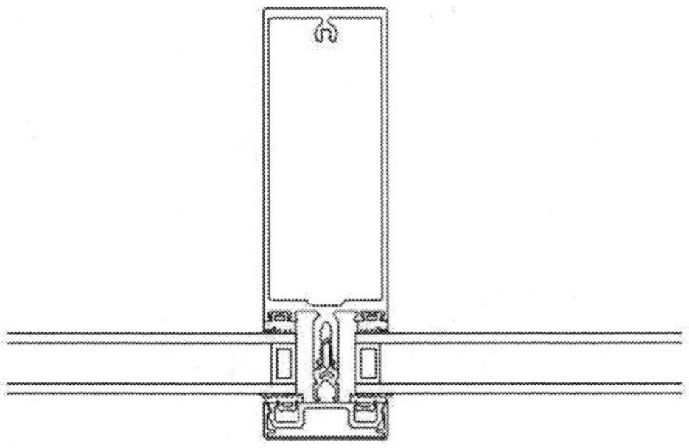	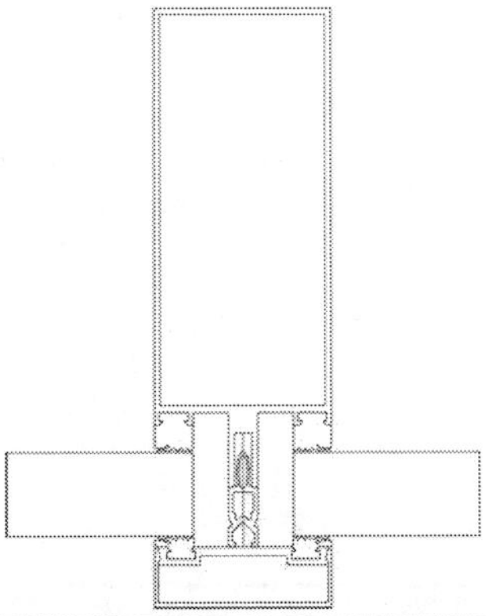	$U_f = 4.352$ W/m^2·K $\Psi = 0.590$ W/m·K

point of view, but from watertightness. As the curtain walls are designed as drained and pressure-equalized systems, the drainage plane is located at the back of the cavity between the IGUs. During rain events, water will reach that plane and drain at the bottom to the exterior by means of weep holes. Due to the use of pressure equalization, the larger part of the wind gusts will be transferred to the interior gaskets. Consequently, if some water would in fact infiltrate into the system and reach the back of the cavity, it will be subjected to significant driving forces to infiltrate into the interior. In European systems the additional member that holds the gasket ensures that the gaskets are not located at the drainage plane. Moreover, in North American systems, the water will run down along the gasket of the mullion, and, due to surface tension, the water

Table 3. Selection of Typical North-American Curtain Wall System Sections

Company D	Simulation Model	Results
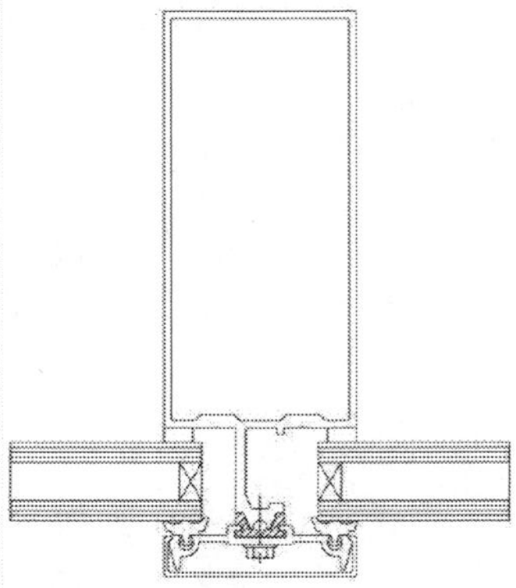	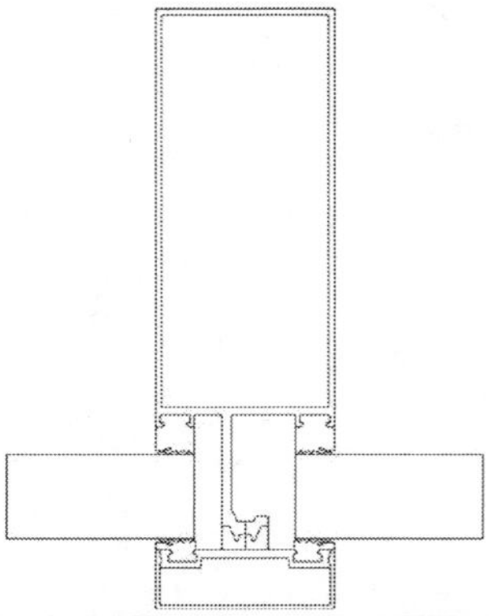	$U_f = 7.344\ W/m^2 \cdot K$ $\Psi = 0.770\ W/m \cdot K$
Company E	**Simulation Model**	**Results**
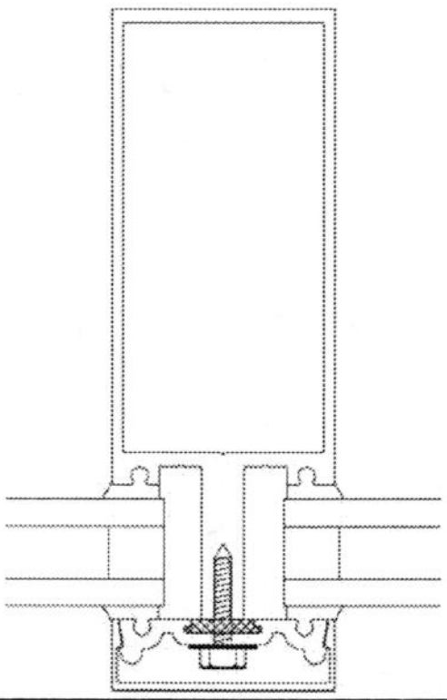	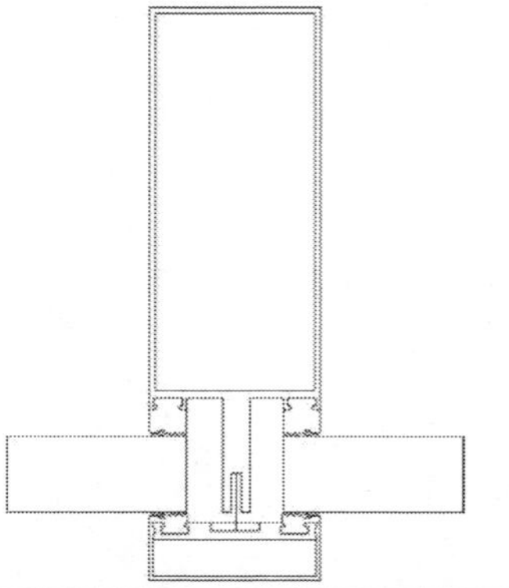	$U_f = 6.648\ W/m^2 \cdot K$ $\Psi = 0.728\ W/m \cdot K$
Company F	**Simulation Model**	**Results**
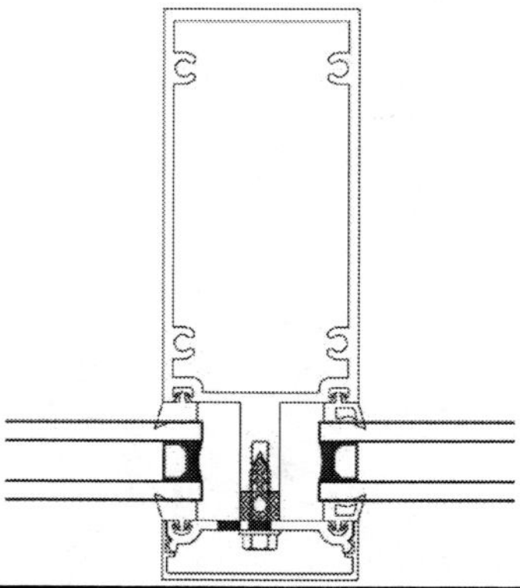	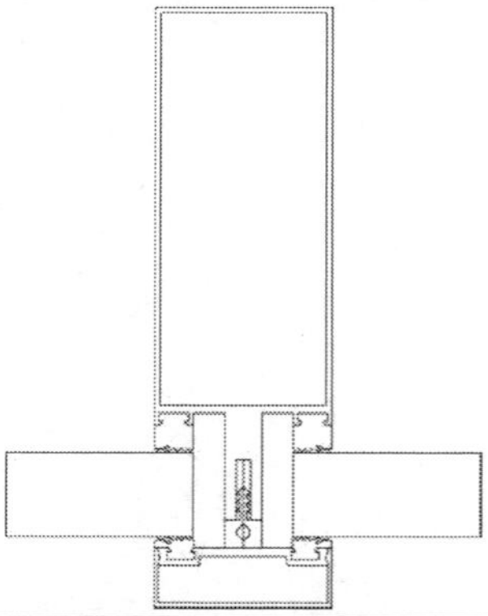	$U_f = 7.453\ W/m^2 \cdot K$ $\Psi = 0.777\ W/m \cdot K$

will be partly redirected horizontally where the vertical gasket connects with the horizontal gasket of the transom. It is well known in the industry that specific location is the primary source of water leaks in lab conditions as well as in practice.

To compare the generic curtain wall frames, simulations were performed to assess the (linear) thermal transmittance. Table 4 shows that little difference between both frames can be discerned; the North American frame performs slightly better. Similar results are evident when simulating a generic European frame with solid screw fin connection, where the U-factor only increased with 0.021 $W/m^2 \cdot K$. The lower U_f-value of the North-American frame can be linked to the shorter length of the screw fins. The absence of the profile member holding the gaskets in the North American reference frame, leads to different heat flux

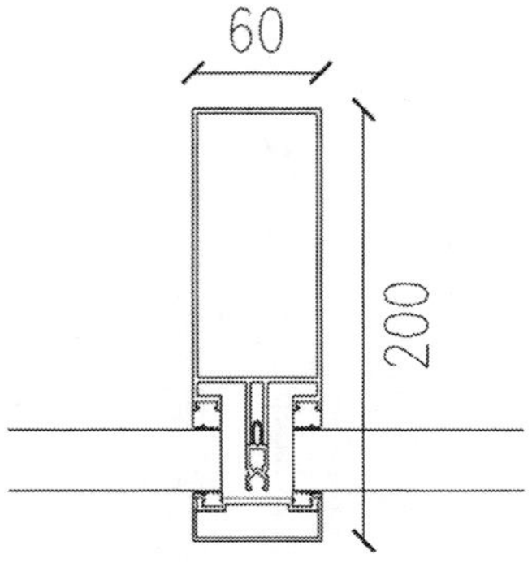

(a)

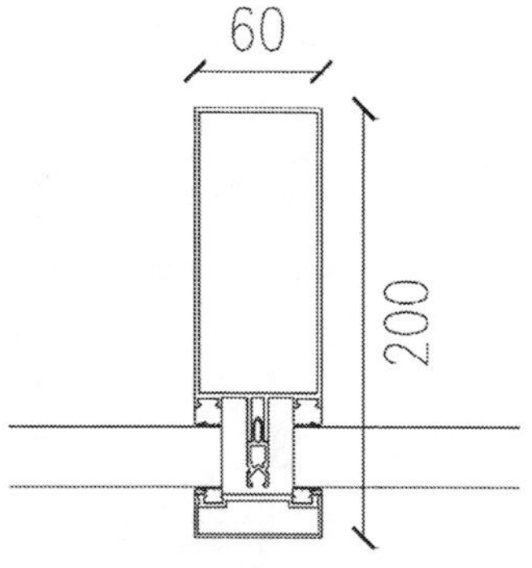

(b)

Figure 1 *(a) European and (b) North American generic curtain wall frame.*

Table 4. Performance of Generic Curtain Wall Frames

	U_f, W/m²·K	Ψ, W/m·K
Europe	3.947	0.566
North America	3.792	0.557

Table 5. λ Value Materials

Material	λ, W/m·K
Aluminum frame	160
Insulation panel (glass)	0.035
EPDM gaskets	0.25
Fiber reinforced polyamid	0.3
Stainless steel (screw)	17

pathways in both frames. A shorter screw fin length results in less heat loss through these fins, although more heat will pass through the pressure plate below the outer gaskets. Similarly, the solid screw connection does not have a significant impact on the heat fluxes through the curtain wall frame.

When executing 1-D calculations, the combined interior and exterior surface resistance would be equal to 0.04 + 0.13 = 0.17 m^2·K/W, resulting in a U-factor of 5.88 W/m^2·K of a thin aluminum plate (disregarding the thermal resistance in conduction of the plate). One might be temped to consider that value as a theoretical and practical upper limit for U-factors. Nevertheless, the 2-D simulations show results of approximately 7 W/m^2·K. This difference is generated by discarding the material influence in 1-D simulations. In 2-D simulations, the cooling fin effect results in an increase of thermal transmittance, as the most influential parameter for heat loss in highly conducting components is $\Sigma A_i \cdot R_{si}$. A larger interior surface results in a higher U-factor, because the interior heat transfer coefficient is, in fact, the bottleneck for the overall heat flux, whereas in typical building constructions this is the thermal resistance of the component itself. When discarding the interior aluminum frame, the U-factor decreases from 7.344 W/m^2·K to 5.274 W/m^2·K.

PARAMETER ANALYSIS

A series of simulations were executed to analyze the influence on thermal performance of the design parameters shown in Figure 2, which is the same as the European reference in Figure 1. The parameters that will be simulated are as follow: (1) profile width and (2) depth, (3) length of the screw fins, (4) position of the insulating glass units, (5) influence of intermediate distance between mechanical fixings, (6) glass unit thickness, and (7) length of the snap cover. Furthermore, the influence of the λ value of screws and compartmentalization profiles is investigated. To achieve a thermally optimized curtain wall frame, several techniques to insulate the inner gasket are compared and simulated as well.

Table 5 lists the materials used in the simulations and their respective λ value (thermal conductivity). If this value is varied for a specific part of the parameter study in the simulations, the specific λ values are given further in the report. The values reported in Table 5 originate from Annex 1 of EN ISO 10077-2. Output results from the simulations are the thermal transmittance of the frame (U_f) and the linear thermal transmittance (ψ). Results of the simulations are discussed below.

Profile Width and Length

First, the dimensions of the curtain wall frames affect its thermal performance. Figure 2 shows the varying parameters that were simulated; (1) indicates a change in profile width and (2) indicates a varying profile length. Profile widths of 50 and 60 mm and profile lengths of 100, 150, 200, 250 and 300 mm are compared and analyzed in Figure 3. From the results it is

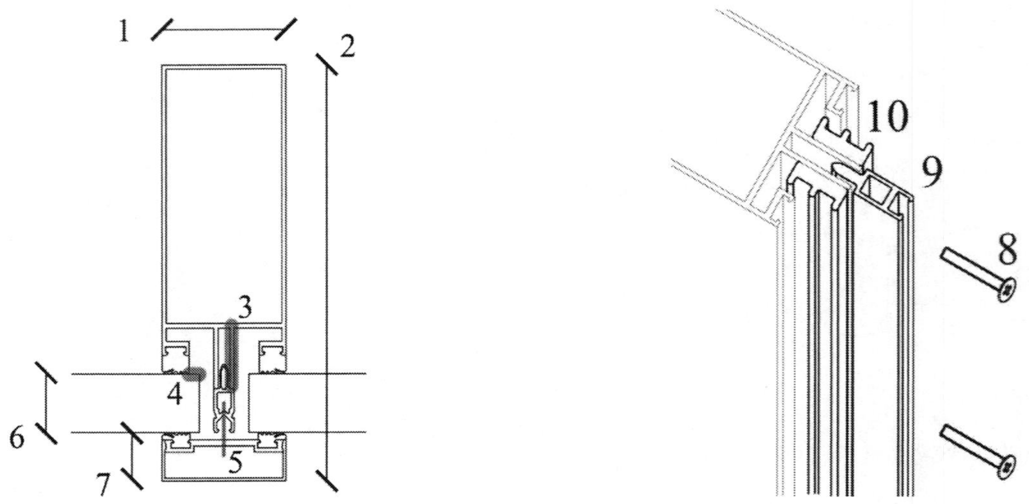

Figure 2 *Visualization of sensitive parameters: (1) profile width, (2) profile length, (3) length of the screw fins, (4) position of the IGU, (5) intermediate distance between mechanical fixings, (6) IGU thickness, (7) snap cover length, (8) λ value of the screws, (9) λ value of compartmentalization profile, and (10) insulating interior gasket.*

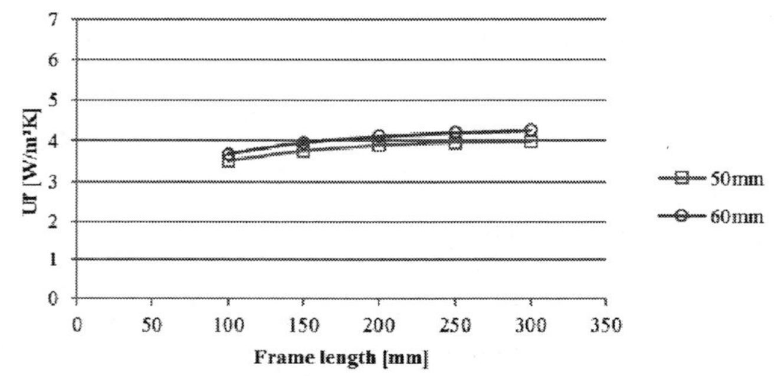

Figure 3 *Impact of profile width and length on thermal performance.*

evident that the 50 mm profiles are accompanied by a slightly lower heat flux as compared to the 60 mm profiles. Similarly, if the depth of a profile increases, the thermal performance decreases. Adding more material to a specific configuration typically reduces the heat flux and as such increases the thermal performance by increasing the thermal resistance. However, in the case of aluminum profiles, the thermal conductivity of the aluminum is excessively high, which leads to the situation that the material itself is not the most important bottleneck for heat transport but the heat transfer coefficient at the interior surface. By increasing the size of the interior aluminum profile, a larger surface area is generated, and the heat transfer from the component to the air is facilitated. The effect is rather similar to the design of cooling fins to facilitate heat transfer. The values vary between 3.5 and 4.5 $W/m^2 \cdot K$, and the lowest U_f-value is found for a curtain wall frame of 50 by 100 mm (the smallest one that was evaluated). Next to that, it can be seen that the influence of the profile depth reduces for increased size. For larger profiles, an equilibrium is becoming apparent for heat flux through the component and maximum heat loss at the interior side (defined by the convective and radiation heat transfer coefficient).

The IGUs are replaced by solid insulation panels in the simulations (please refer to EN ISO 10077-2 for the simulation procedure). Another important fact is the absence of a temperature gradient in the interior and exterior aluminum profiles. The thermal conductivity of the aluminum is easily 1000 times higher than the conductivity of the air inside the cavities. As a result, any thermal resistance that may be found in the air in the cavities is simply thermally bypassed by the adjoining aluminum frame members. To obtain a better understanding of the thermal behavior of the complex geometry of the curtain wall systems, the heat flux density throughout the section is visualized as well. Figure 4 shows the most important pathways that conduct heat from the interior to the exterior (normalized value for visual reference). Given the symmetrical layout of the section, this symmetry is reflected in the reduced heat flux at the central axis of the profile. The high thermal conductivity of the aluminum is likewise evident: the highest heat fluxes can be found at the aluminum parts connecting the interior to the exterior. The screw that is used to fix the pressure plate to the frame is modeled by means of very slender stainless steel continuous plate located in the center of the frame. Even though the screws do not necessarily make contact with the back of the tubular frame in practice, this is superseded by the conductivity through the aluminum. Even if the screws do make contact with the back of the frame, it hardly affects the overall thermal performance. Throughout the parameter study the heat flux density is used to evaluate the effectiveness of specific optimization strategies.

Length of the Screw Fins

From Figure 4 it is evident that the two aluminum fins in the central space between the IGUs are a very important heat flow pathway. It allows the fixation of a screw that holds the pressure plate, and the width and depth are designed to accommodate the easy installation of screws. The thickness of the profile is determined by mechanical strength. Given that the interior and exterior aluminum profiles do not contribute to the overall thermal resistance, the thermal performance of the system basically depends on the configuration of the area between the IGUs and between the pressure plate and the onset of the aluminum fins. Consequently, an important step to improve the thermal performance consists of decreasing the length of the aluminum fins, shown by parameter 3 in Figure 2. Results of the simulations on decreasing screw fin length are shown in Figure 5. It can be seen that a smaller screw fin results in a lower U_f-value. Screw fin were simulated with lengths of 15, 20, 25, 30 and 35 mm. It can be seen that from a certain point, the impact of reducing the screw fins will stabilize and the influence on the thermal optimization will be negligible.

Position of the Glass Units

As discussed in the previous section, the thermal performance mainly depends on the configuration of the frame between the IGUs, and between the pressure plate and the aluminum fins. Given that the thermal resistance of that zone is typically significantly lower than that of the IGUs, one might consider increasing the height and width of the IGU panels. The distance between the glass unit and the screw fins should be at least 5 mm in order to ensure a proper drainage of the system and allow for a practical installation and adjustment on site. On the other hand, there is a minimum contact length between the IGU and gasket as well for a reliable mechanical contact with the frame to transfer wind loads to the secondary and primary structure. Next to that, from an aesthetic point of view, the nontransparent edge seal is typically completely covered by the pressure plate. Note that when the IGU is inserted too much into the curtain wall system, the difference in solar exposure of the IGU at the edge increases the risk for thermal breaking of the glass panels. The effect of the IGU spacer is taken into account separately in the normative documents. In the standard configuration, the IGU is inserted 12 mm into the profile, which corresponds to the width of the gasket (i.e., the gasket is completely covered by

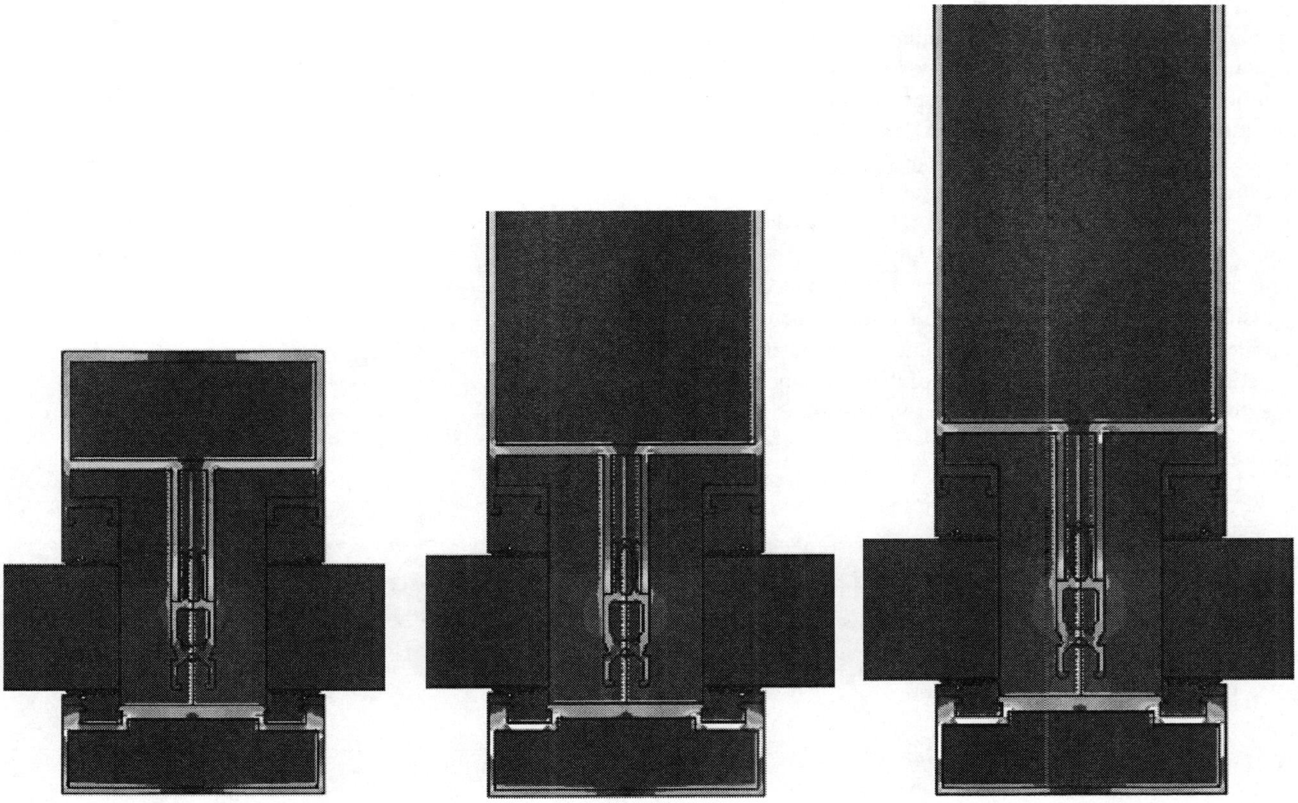

(a) **(b)** **(c)**

Figure 4 *Heat flux profile for (a) 100 m, (b) 200 mm, and (c) 300 mm profile length and 60 mm width.*

the IGU, see position in Figure 1). This corresponds to a U-factor of 3.958W/m^2. When we increase the insertion by 5 mm, the U-factor decreases to 3.434 W/m^2 (this is the situation shown in Figure 2). In contrast, when the IGU is only inserted 7 mm (recessed 5 mm in comparison with the original position), the U-factor increases up to 4.481 W/m^2·K. The increased IGU size narrows the area of the curtain wall where excess heat loss may be found and results in a lower U_f-value of the frame. Hence, this parameter should be considered when comparing different systems with their respective thermal performance. A slight change of IGU size by 5 mm can easily change the thermal performance by 15%.

Spacing between Screws

In practice the pressure plate is fixed to the structural member by means of screws every 150 mm up to 300 mm. The distance between the screws is set by the maximum loads that the screws need to transfer. In turn, this is determined by the size of the IGU, the aspect ratio (height/width of the IGU), and maximum wind loads (suction). Peak negative pressures of building facades are typically higher the peak positive pressures. The mechanical fixation of the pressure plate also generates a number of secondary effects. First of all, by increasing the pressure, a specific effect may arise similar to structural glazing: stress in the glass panes can get transferred to the aluminum profiles. Second, due to the elastic behavior of the gasket and low elastic modulus, it is very difficult to acquire a uniform stress on the gasket over the length of the mullion or transom. Most European system manufacturers mention that a specific torque should be applied for the installation of the screws. In contrast, on construction sites, cordless electrical screwdrivers are used instead of a torque wrench. Even when a torque wrench is used, it typically takes three or four iterations of adjusting the individual screws to acquire a uniform torque and pressure within a 10% variance limit. As a result of these considerations, the number of screws incorporates major safety factors to account for the uncertainties in practice. To evaluate the influence of the mechanical fixings, the distance in between the screws is varied from 150 mm up to 300 mm. Assuming a spacing of 250 mm and a λ value of 17 W/m·K as a reference case, the equivalent λ value of the screw in the simulations is changed in proportion to the specific intermediate distance (see results in Figure 6). For an increasing spacing between the screws, less material will be available for heat to pass through, resulting in a decreasing λ value of the screw and a decreasing U_f-value for an increasing intermediate distance.

Glass Unit Thickness

As discussed in previous sections, only the space between the IGUs, pressure plate, and aluminum fins contributes significantly to the thermal resistance of the frame. This might be optimized by increasing the thickness of the IGU panels, which will lengthen the pathway and reduce the thermal transmittance (see Figure 2, number 6). Figure 7 shows the influence of the glass unit thickness on the thermal transmittance of the frame. Increasing the thickness (20, 28, 36, 44, 52, and 60 mm) has a major influence on the thermal performance of the frame, showing that triple glazing can definitely contribute to the thermal optimization in curtain wall design. Due to the large difference in thermal transmittance of the IGU and the frame, the geometry of the IGU relative to the frame has a very large impact. Similar

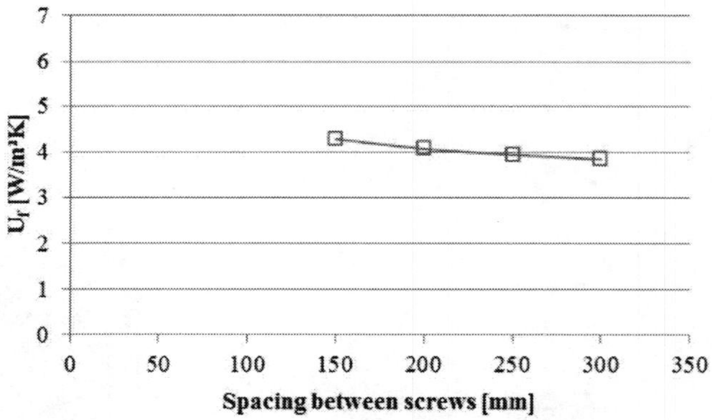

Figure 6 *Spacing between screws.*

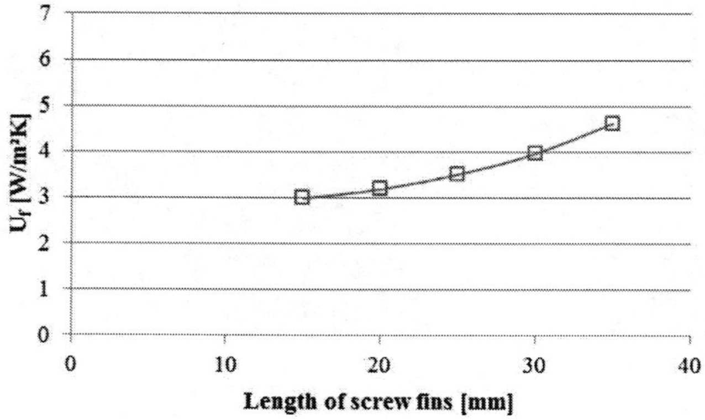

Figure 5 *Impact of length screw fins.*

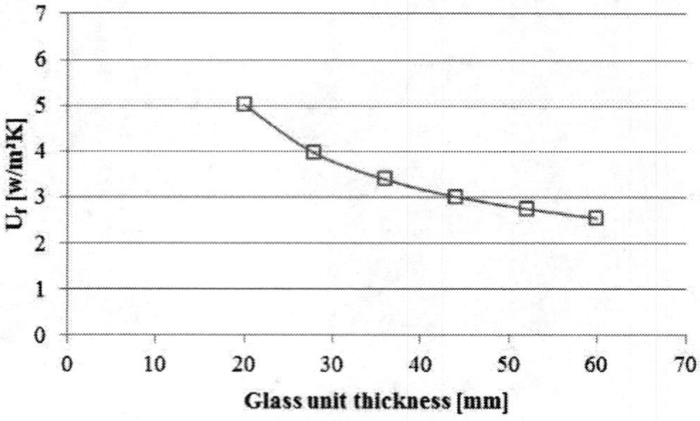

Figure 7 *Glass unit thickness.*

to the width and height of the IGU, the thickness of the IGU should always be considered when different systems are compared. In the simulations, no modifications were made to the frame when altering the thickness of the glazing units. In Figure 7, it can be seen that just by replacing a thin double glazing by thick triple glazing, the thermal transmittance of the frame can be reduced by half. This only accounts for the frame; the U-factor of the IGU (center or edge) is not considered.

λ-Value of Screws and Compartmentalization Profile

Figure 2 (number 8 and 9) shows the influencing construction parts of which the λ value will be varied during the simulations described below. Previous sections highlighted the impact of the screws and spacing of the screws on the overall thermal performance of the system. In practice the screws are stainless steel or galvanized steel. Figure 8(a) shows that this choice in screw material already changes the thermal transmittance by 1 $W/m^2 \cdot K$. As a calculation example, following materials are considered: rigid polymers (λ value typically 0.3 W/m·K), different types of stainless steel (λ value 14, 15, 16, 17, and 18 W/m·K), steel (λ value 50 W/m·K), and aluminum (λ value 160 W/m·K). Figure 8(a) indicates that once the thermal conductivity approaches that of the adjoining materials and air cavities, the relative impact decreases. Although the stainless steel screws are now considered state-of-the-art, the adoption of, for example, fiber-reinforced screws would further reduce the transmittance. Note that carbon fiber reinforced anchors are increasingly used in precast industrial concrete sandwich panels in Europe, which already indicates that the material is starting to become more popular in the building industry. Similarly, the thermal conductivity of the compartmentalization profile has an impact on the overall performance (Figure 8[b]). The λ values of the compartmentalization profile that are considered are 0.1, 0.2, 0.28, 0.3, and 0.4 W/m·K. In this case, a decrease of λ value only induces a more moderate decrease in thermal transmittance (as the considered range is more limited as well).

Snap Cover Length

Previous simulations reveal that the pressure plate and snap cover only have a minor influence on the overall thermal performance of the curtain wall frame. To investigate whether or not a variation in length of the snap cover has an effect on the thermal transmittance, simulations are executed for a snap cover length of 13.5, 18.5, 23.5, 28.5, and 33.5 mm (please refer to Figure 2, number 7). The results confirm the aforementioned statement and show that the snap cover length has a minor influence on the thermal performance of the frame. An increase of 20 mm increases the thermal transmittance by 0.1 $W/m^2 \cdot K$. The minor impact of the snap cover on the overall thermal performance is due to the higher exterior heat transfer coefficient (although the snap cover has a limited surface, this is compensated for by the high convective heat transfer coefficient).

Materialization of the Curtain Wall Frame

Since most of the heat loss can be assigned to the aluminum frame, a more insulating material could improve the thermal performance drastically. A first option is to tackle the insulating capacity of the overall frame (see Figure 9). Glass fiber reinforced plastic (GFRP, λ = 0.40 W/m·K) is one option found in the industry to match the strength of aluminum and at the same time improve the thermal performance of the curtain wall system. Note that, although the thermal performance of the curtain wall frame increases drastically, the mechanical connection of the pressure plate to the inner frame can be questioned. Secondly, in traditional aluminum systems a significant amount of heat is directed outwards through the pressure plate. Another approach is to replace the aluminum pressure plate by a polymer pressure plate (λ = 0.30 W/m·K), creating an insulating barrier at the outside of the curtain wall frame.

In Table 6, it can be seen that these changes in materialization have a major impact on the thermal performance of the overall frame. Both options result in a thermal optimization as compared to the reference frame. Note that, although the

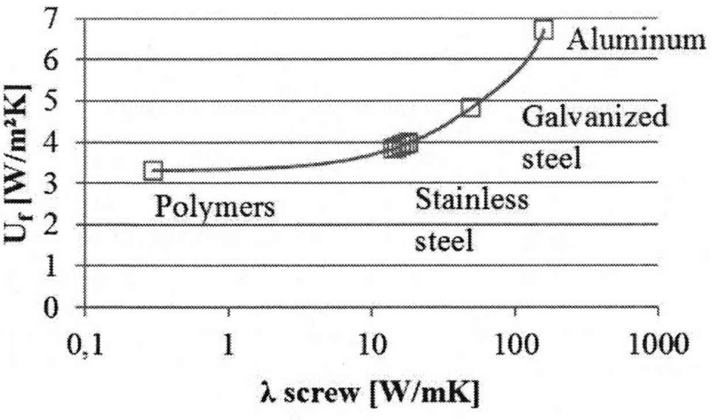

(a)

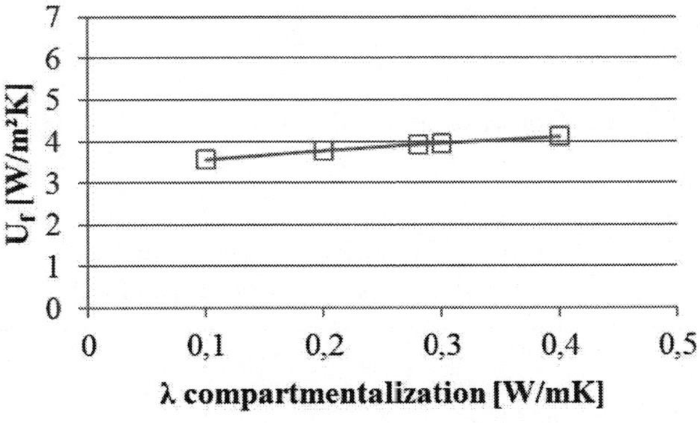

(b)

Figure 8 *λ values of (a) different screw and (b) compartmentalization profile materials.*

GFRP frame drastically decreases the heat loss through the frame, the mechanical connection can be questioned. Therefore, this option will not be taken into account in further optimization strategies.

Additional Insulation in the Cavity

As a final step in thermal optimization of curtain wall frames, additional insulation can be installed in the inner cavity. Figure 10 shows three different options for insulation of the inner frame; the insulation material has a λ-value of 0.035 W/m·K.

Table 7 shows the thermal transmittance for the insulated curtain wall frames as compared to the European reference frame. It is evident that the thermal transmittance decreases with an increasing amount of insulation in the inner cavity. Note that a complete filling of the inner cavity will prevent drainage of infiltrating water, increasing the risk for damage to the curtain wall. Because in this research only an analysis of existing techniques was performed, there is still room for optimization of the shape of the insulation.

CONCLUSIONS

In this paper, a comparative analysis of current European and North American curtain wall systems is performed to evaluate the thermal performance of existing curtain wall systems. When comparing existing standard systems (i.e., not thermally improved) for both continents, it can be seen that in general the European frames perform better and are more thermally efficient. A generic curtain wall frame was developed to investigate the influence of several design parameters on the thermal performance of the system.

Two systematic differences that are evident from analyzing a series of European and North American curtain wall systems are the location of the interior gasket and the more or less solid screw fin connection. In European systems there is an additional member in the interior aluminum profile that holds the gasket, whereas in North America the gasket is often installed directly into the aluminum structural member. Nonetheless, simulations show that the profile member that holds the gasket, which is only present in the European frame, does not have a major influence on the heat loss. The higher U_f-

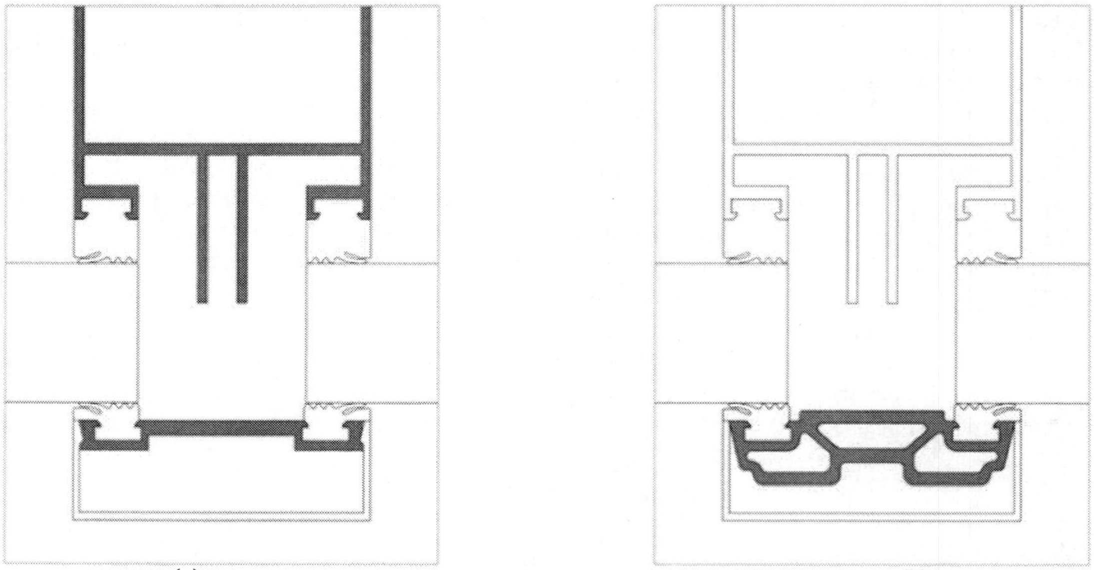

Figure 9 *Improved materialization of (a) GFRP frame and (b) polymer pressure plate.*

Table 6. Thermal Performance due to Improved Materialization of the Frame

	U_f, W/m²·K	Ψ, W/m·K
Reference frame	3.947	0.566
GFRP frame	2.105	0.455
Polymer pressure plate	2.788	0.496

Table 6. Effect of Additional Insulation

	U_f, W/m²·K	Ψ, W/m·K	I, dimensionless
Reference frame	3.947	0.566	0.818
Option A	3.046	0.512	0.821
Option B	2.634	0.487	0.819
Option C	2.497	0.479	0.819

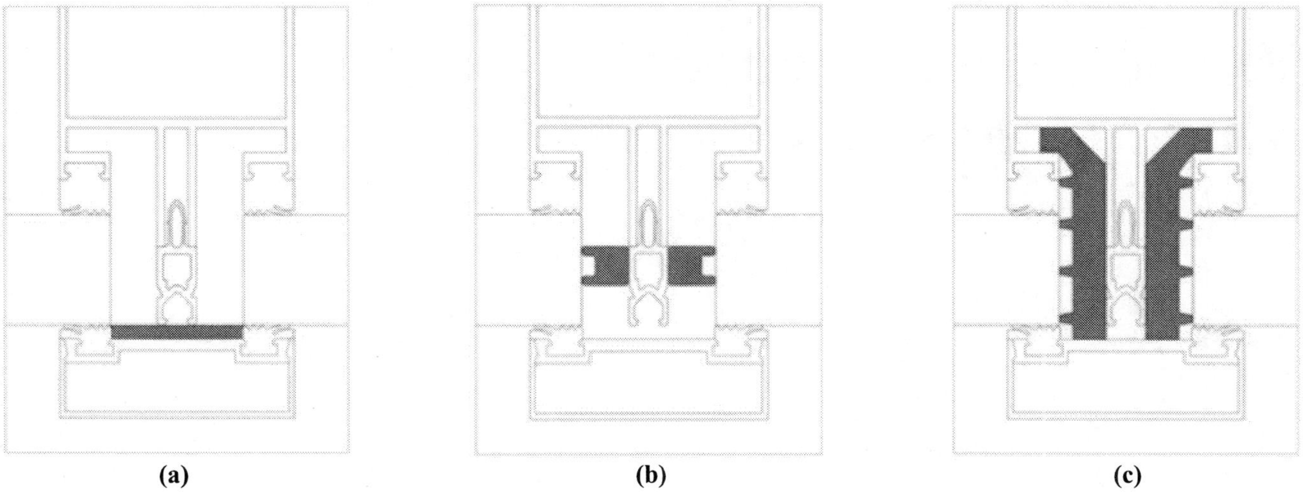

Figure 10 *Additional insulation in the inner cavity: (a) Option A, (b) Option B, and (c) Option C.*

value of the North American frame can be linked to the increased length of the screw fins.

A series of simulations was executed to evaluate the influence on thermal performance of several design parameters, such as profile width and length, position of the insulating glass units, glass unit thickness, λ values of screws, and compartmentalization profile. Results of the simulations show that the most influential parameters on the thermal performance of the curtain wall systems are the following:

- Length of the screw fin
- Glass unit thickness
- Additional insulation in the cavity

To design effectively thermal optimized curtain wall frames, the parameters mentioned should be combined into one optimal system.

REFERENCES

Bisco. 2012. *Bisco 10.0w, User Manual*, Physibel, Adegem, Belgium.

Blanusa, P., W.P. Goss, H. Roth, P. Weizmannn, C.F. Jensen, S. Svendsen, and H. Elmahdy. 2007. Comparison between ASHRAE and ISO thermal transmittance calculation methods. *Energy and Buildings* 39(3):374–84.

Byars, N., and D. Arasteh. 1992. Design options for low-conductivity window frames. *Solar Energy Materials and Solar Cells* 25(1992):143–48.

Gustavsen, A. 2001. Heat transfer in window frames with internal cavities, PhD thesis, 2001, Trondheim, Norway.

Gustavsen, A., D. Arasteh, B.P. Jelle, C. Curcija, and C. Kohler. 2008. Developing low-conductance window frames: capabilities and limitations of current window heat transfer design tools—State-of-the-art review. *Journal of Building Physics*, 32(2):131–53.

Gustavsen, A., S. Grynning, D. Arasteh, B.P. Jelle, and H. Goudey. 2010. Key elements of and material performance targets for highly insulating window frames. *Energy and Buildings* 43(10):2583–94.

ISO. 2003. ISO 15099:2003, *Thermal performance of windows, doors and shading devices—Detailed calculations*. Geneva: International Organization for Standardization.

ISO. 2007. EN ISO 10211:2007, *Thermal bridges in building construction—Heat flows and surface temperatures —Detailed calculations*. Geneva: International Organization for Standardization.

ISO. 2012. EN ISO 10077-2:2012, *Thermal performance of windows, doors and shutters—Calculation of thermal transmittance Part 2: Numerical method for frames*. Geneva: International Organization for Standardization.

ISO. 2012. ISO 12631:2012. *Thermal performance of curtain walling—Calculation of thermal transmittance*. Geneva: International Organization for Standardization.

NFRC. 2013. *Therm 6.3/Window 6.3, NFRC Simulation manual 2013*. Lawrence Berkeley National Laboratory, California.

NFRC. 2014. ANSI/NFRC 100-2014, *Procedure for determining fenestration product U-factors*. Greenbelt, Maryland: National Fenestration Rating Council Incorporated.

Window Condensation—Theory into Practice

Alex McGowan, PEng
Member ASHRAE

ABSTRACT

Thermal performance of fenestration systems often refers to U-factors or solar heat gain coefficients, but condensation in windows is an even more direct aspect of thermal performance. Sweat on windows does not require expert interpretation: building occupants will quickly and loudly identify acceptable levels of performance without reference to consensus-based standards or building codes.

Many years ago, the author developed a computer model to predict window condensation and this paper describes the use of that model, as well as tools and techniques for in-field assessment and measurement, to either prevent condensation at the design stage or—in some cases—to address condensation-related problems that developed after construction. Specific examples are used to show how even high-performance window designs can be compromised as a result of poor installation practice—and what to do about it.

INTRODUCTION

Thermal performance of fenestration systems often refers to U-factors or solar heat gain coefficients, but condensation in windows is an even more direct aspect of thermal performance. U-factors and solar heat gain affect homeowners when they pay their utility bills; condensation is evident when they step out of the shower. And, when condensation leads to mold growth, damage to interior finishes, and structural damage, the cost of condensation can dwarf the cost of heating and cooling a building.

This paper describes the causes of window condensation, identifies some useful tools to assess problems in consulting practice, and presents three case studies intended to provide insight into some aspects of condensation and potential solutions, as these occur in the field. Simulation is presented as one such tool—not in its usual capacity of rating standardized product performance or conducting research, but as a way to evaluate design solutions for real window condensation problems.

CAUSES OF WINDOW CONDENSATION

Although condensation may manifest itself in many ways in windows, it always occurs because the local surface temperature is lower than the dew point temperature of the air in contact with the surface. The original development of the psychrometric chart (Figure 1 [Gatley 2004]) identified two conditions when this would happen: at constant humidity, the local temperature is reduced until the state point is on the saturation curve (arrow A, Figure 1), or at constant dry-bulb temperature, the humidity is increased until the state point is on the saturation curve (arrows B and C, Figure 1).

Figure 2 indicates a variety of ways that the room-side temperature of a window assembly can be reduced, resulting in potential condensation problems:

- Convection between the panes of glass in the glazing unit develops a flow pattern down surface No. 2 (as counted from the exterior) as the surface cools the gas in the sealed unit, and up surface No. 3 (as the gas is heated by the warmer glass surface). The result is a ther-

Alex McGowan is an vice president of technical services for TSP Canada, Inc., Victoria, BC, Canada.

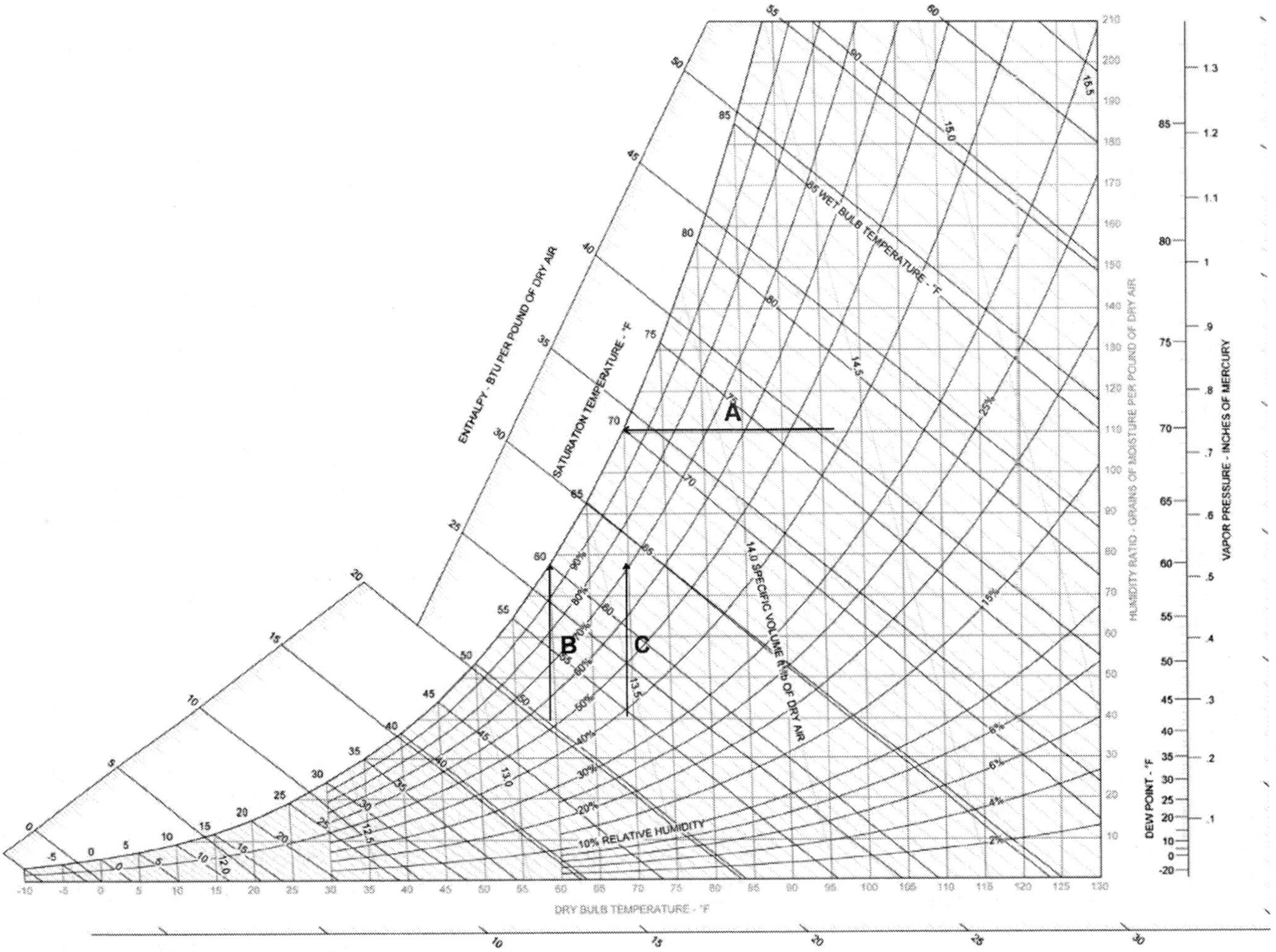

Figure 1 *Typical psychrometric chart.*

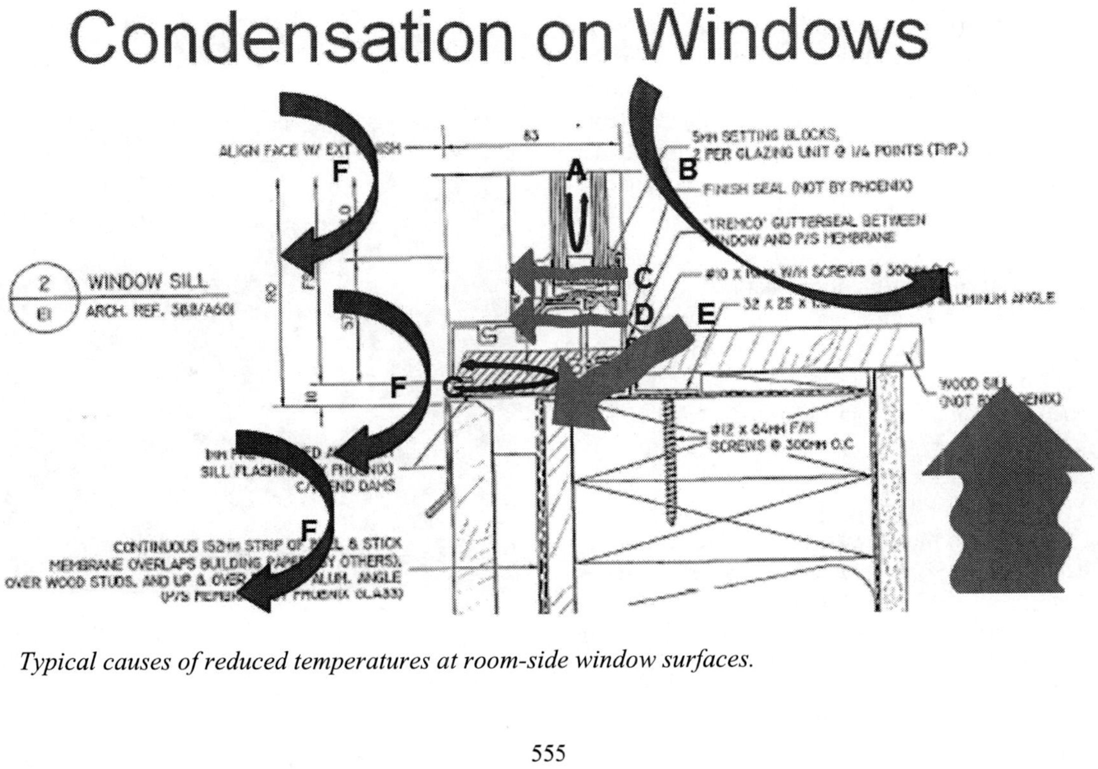

Figure 2 *Typical causes of reduced temperatures at room-side window surfaces.*

mal transport phenomenon that brings cooler temperatures to the lower sightline of the window, which is why condensation typically shows up there first.

- A similar convective pattern develops in the room when the glass cools the room air next to the window. The cooled air becomes denser and flows down the window, resulting in an induced draft that can create occupant discomfort. This is why heat is often delivered to the conditioned space with baseboard heaters or forced-air registers below that window. The increased velocity of air tends to promote local evaporation of condensation, but this effect is reduced at inside corners where dead-air pockets are formed. The geometry of inside corners also reduce radiative heat transfer between room surfaces and the window, as the view factor reduces to 0.5 at an inside corner because of self-viewing (i.e., what the window surface "sees" at inside corners comprises 50% of room surfaces and 50% of the window itself). Thus, the thermal contact between the room and the window is reduced, so inside corners tend to be colder than adjacent window surfaces.
- Increased conduction through the edge-of-glass spacer assembly that keeps the panes of glass at a constant distance will reduce the room-side surface temperature.
- Increased conduction through the window sash and frame also reduces the room-side surface temperature.
- As does increased conduction through the wall framing or other components around the rough opening.
- High wind conditions increase exterior heat loss, reducing interior surface temperature.
- This also increases wind washing in the window assembly, thus cooling room-side surfaces.

However, as shown in Figure 1, the problem is not just caused by colder interior surfaces; high humidity (arrows B and C) also contributes to condensation potential. Therefore, diagnostic procedures should consider both interior surface temperature and interior air humidity as clues to the potential source of the problem. Proper diagnosis of window condensation will require tools that can measure these parameters.

TOOLS

Figure 3 shows some of the tools that may be useful in diagnosis of window condensation. Some advantages and disadvantages of each tool as tabulated below Figure 3 (Table 1).

Thermocouples are used in measuring condensation potential in the laboratory, especially when rating the fenestration system performance to national standards such as NFRC 500 (NFRC 2010), but are not always appropriate for *in situ* diagnostics. Instrumenting a specimen with thermocouples can be quite complex, and if not done properly, the thermocouple leads can interfere with convective patterns (thus affecting the parameters that were to be measured). For the case studies for which this author has experience, the conglomeration of thermocouples, leads, and securing tape would not have been acceptable to the owner/occupants of the subject properties.

SIMULATION

Computer simulation techniques developed to predict window condensation potential—and shown to be as accurate as thermocouple tests (McGowan 1995)—are also useful in diagnosing *in situ* problems. This method uses FRAME (Enermodal Engineering 1996) and VISION (University of Waterloo 1996), which model the convective fill gas motion in the glazing cavity, as described by Wright (1998), and allows local variation of the room-side heat transfer coefficient to more accurately capture mechanisms A through E as described in Figure 1. If the problem is shown to be specific to thermal bridging in the frame of the window, then THERM (LBNL 2013) could be used to assess mechanisms C through E in Figure 1, but that method does not capture interglazing convection and is cumbersome when modifying the room-side film coefficient.

Figure 4 shows the results of this method: Figure 4a shows the conventional model that assigns an effective conductivity to the fill gas inside the sealed glazing unit, with the resulting room-side surface temperature shown at the cross-hair cursor, in this case placed at 13 mm (1/2 in.) above the sightline, which is the standard location for thermocouple measurement of condensation potential. Figure 4b shows the same window cross section, with the VISION model for interglazing convection included: the difference in resulting temperatures is displayed in the upper left of each screen capture. In this case, including inter-glazing convection in the model reduced the predicted sightline temperature from 10.9°C to 7.5°C (51.6°F to 45.5°F). In a typical interior condition (20°C or 70°F dry-bulb temperature), this reduces the maximum relative humidity from 55% to 45% for this example.

Figure 3 *Typical tools used in field diagnoses of window condensation problems.*

Table 1. Advantages and Disadvantages of Select Window Condensation Detection Tools

Tool (Figure 3)	Advantages	Disadvantages*
Sling psychrometer (upper left)	No batteries Easy to use Closest to a direct measure	Only precise to ±0.5°C Requires a psychrometric chart Breakable (= mercury leak) Single-point reading
Electronic psychrometer (middle)	Easy to use text No psychrometric chart required Impresses the client	Relies on batteries Precise (but accuracy depends on calibration) Single-point reading Delayed readings†
Data logger (upper right)	Long-term readings Provides ability to identify trends Impresses the client	Relies on batteries (but these last longer) Influences occupants‡
Infrared thermometer (lower right)	Noncontact measurement Impresses the client	Relies on batteries Indirect measurement
Anemometer (lower right)	More accurate measurement of ventilation effectiveness	

* Another disadvantage of infrared thermography is that reflective surfaces such as glass and metal may reflect the image of the equipment operator or other surfaces in the room, producing erroneous readings.

† It can take several minutes for the measuring chamber of the electronic device to reach equilibrium with the environs; premature (inaccurate) readings are possible.

‡ Occupants can sometimes change their behavior when they know the building is being monitored. It may be advisable to ignore the first week of data until the occupants forget about the data loggers.

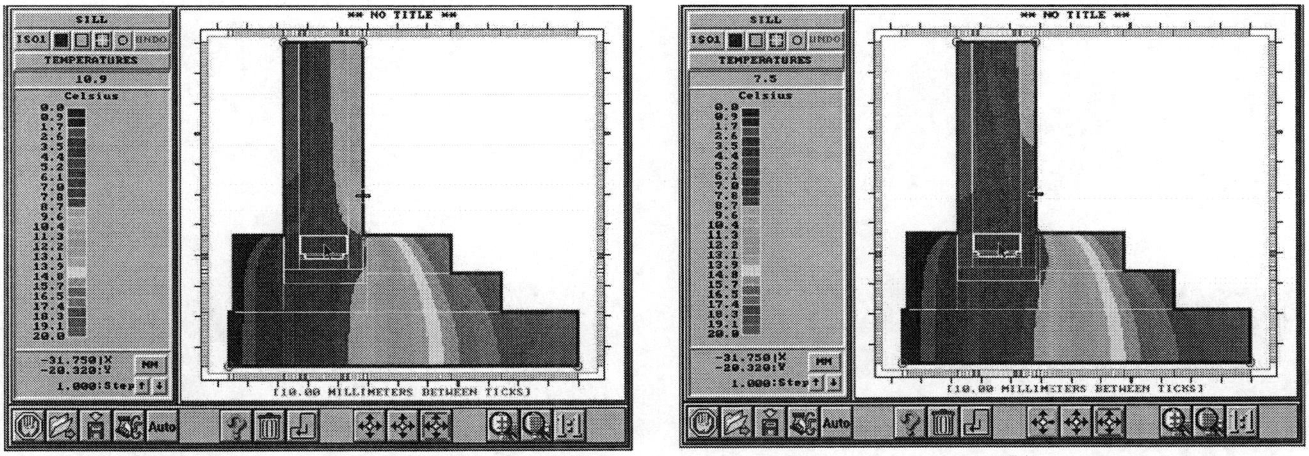

(a) (b)

Figure 4 *(a) Window cross section showing simulation with no convective model in the glazing unit and (b) with convective model, both under standard boundary conditions. Predicted temperature at the crosshair is shown on the upper right (in °C in this example).*

Computer simulation of the thermal performance of windows was originally developed as a design tool, and has since been proven a useful tool for rating the thermal performance of windows. It is also an effective diagnostic tool for evaluating the effects of various conditions in the field and predicting the results of remedial measures, as shown in Case Study 2. The simulation can be done using standard boundary conditions to identify the relative changes in interior surface temperature predicted as a result of suggested design changes. Predicting actual surface temperature is more difficult because of the challenge in determining *in situ* boundary conditions.

CASE STUDIES

Case Study 1: At the Design Stage

The easiest and least expensive way to correct a potential window condensation problem (apart from avoiding it in the first place) is to address it at the design stage. Figure 5 shows a design drawing of a window sill detail, provided for the sub-consultant's review during a new construction project. In a coastal climate, the architect was rightly concerned about wind-driven rain and sought to maximize drainage from the sub-sill area. In doing so, however, the slope of the sub-sill would have exposed the interior half of the window frame to the exterior air temperature, bypassing the thermal break. In this case, the sub-consultant was able to persuade the design team that the slope of the sub-sill should start from the outside edge of the thermal break, which would provide adequate drainage without short-circuiting the thermal performance of the window frame. The design was revised accordingly, at no additional cost to the project.

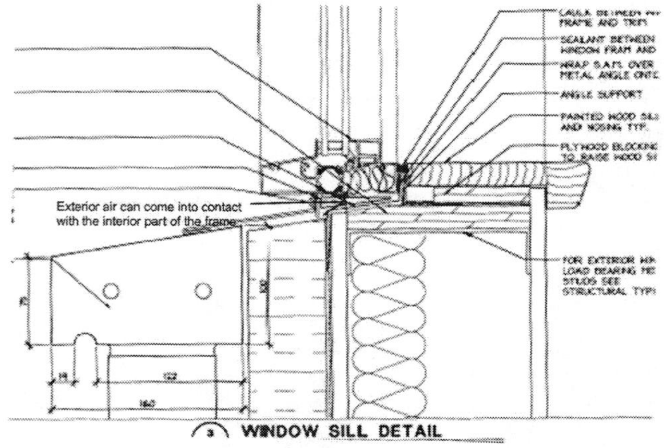

Figure 5 *Design detail showing bypass of the thermal break.*

Case Study 2: *In Situ* Retrofit

The consultant was called in to diagnose condensation observed on windows in a newly constructed multiresidential building (Figure 6a). Knowing the window system that was installed, the manufacturing shop drawings of the window were used to create a computer model of the assembly. The base-case simulation indicated that the window surface temperature as designed should have been approximately 5°C (9°F) warmer than the amount of condensation observed would suggest. It was apparent that the thermal break was being bypassed, which was confirmed upon review of photographs taken during construction (Figure 6b). A well-meaning window installer set the window on a metal sill flashing that spanned the full depth of the window rough opening. This bypassed the thermal break of the window frame by bringing the outdoor temperature into contact with the interior portion of the window frame and created a separate thermal bridge of its own by bringing the outdoor temperature into contact with the room-side air. The magnitude of the effect was confirmed with simulation, which then used to predict surface temperatures for of several alternative solutions, presented in Table 2, with most options illustrated in Figure 7.

The owner rejected the "heat sink" options in Table 2 (i.e., attaching an extended surface area to increase heat transfer to the frame, thus passively warming the surface). We proposed a heat-trace (electrically heating the frame), and the owner rejected that option as well. The most cost-effective approach of those the owner would accept was predicted to be Option 8: lowering the sill covering to expose more of the window frame to the room temperature, partially closing the exterior cavity between the sash

(a) (b)

Figure 6 *(a) Window condensation at the sill (on the frame but not on the glass) and (b) a solid metal sill pan, photographed during the installation of the windows during the new construction.*

and frame to reduce wind washing, and (especially) removing the back leg of the sill flashing shown in Figure 6b and physically separating it from the interior portion of the metal window frame. With this done, the room-side sill temperature (as measured by handheld infrared thermometer) increased by 6°C(10°F), as predicted by simulation.

The owner subsequently reported that the condensation was not completely eliminated, which led to a return visit to the site. Humidity levels were measured to be high: the only way to remove moisture from the room air was two small exhaust fans, in the bathroom and above the kitchen range. The space was rather small, so excess humidity in one area diffused rapidly into adjacent areas and appeared quickly on the windows. This was demonstrated by slightly opening a balcony door near the windows at the start of the site visit: with no other changes during the time we took additional measurements and observations, the indoor relative humidity dropped from 73% to 61% in approximately one hour, although it had been raining outside. The recommended remedy to recurring condensation was to increase exhaust-fan operation, or install larger capacity fans. One should always be careful, however, not to recommend that the fans be direct-wired for continuous operation, as low-cost fans often

Table 2. Effect of Various Proposed Retrofits in Case Study 2

Option	Configuration	Predicted Sill Temperature, °C (°F)	Difference from Base Case, °C (°F)
0	Theoretical performance (thermal break intact)	17.0 (62.6)	+5°C (+9°F)
1	Actual performance (as discovered on site)	12.0 (53.6)	—
2	Remove backleg of sill flashing, reduce thermal bridge	12.1 (53.8)	0.1 (0.2)
3	Lower interior sill liner to expose sill frame	12.4 (54.3)	0.4 (0.7)
4	Close in exterior cavity to reduce wind washing (+ Option 3)	13.4 (56.1)	1.4 (2.5)
5	Separate sill flashing from frame (+ Option 3)	16.3 (61.3)	4.3 (7.7)
6	Separate sill flashing from frame (+ Options 3 and 4)	16.6 (61.9)	4.6 (8.3)
7	Remove sill flashing from interior frame (+ Option 3)	17.4 (63.3)	5.4 (9.7)
8	Remove sill flashing from interior frame (+ Options 3 and 4)	17.6 (63.7)	5.6 (10.1)
9	Heat sink on the horizontal face of the sill liner	15.5 (59.9)	3.5 (6.3)
10	Heat sink on horizontal and vertical faces of the sill liner	15.8 (60.4)	3.8 (6.8)
11	Base case, but with interior blinds closed	10.6 (51.1)	–1.4 (–2.5)

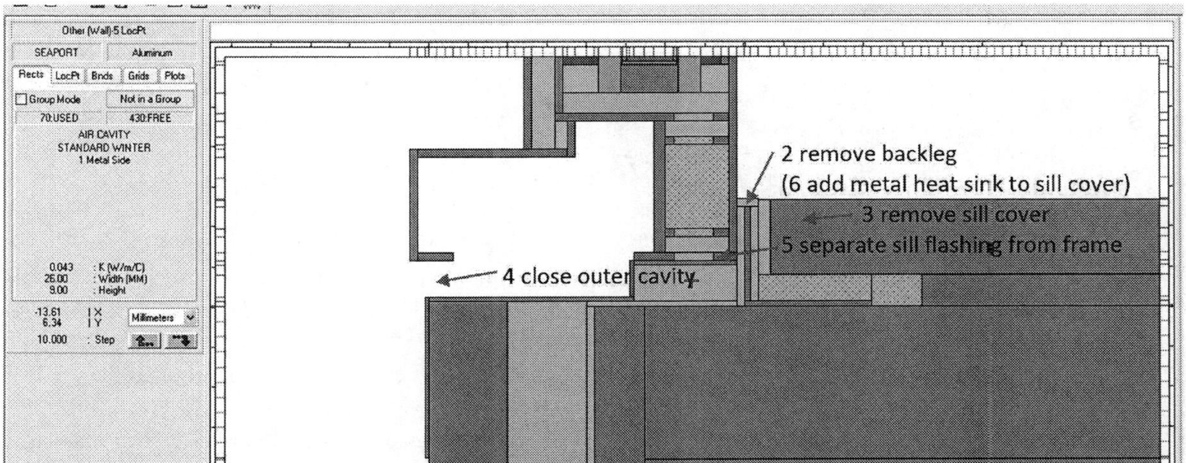

Figure 7 *Simulated cross section, showing several proposed alternative solutions for window condensation.*

installed in new construction are not intended for such heavy usage and will provide no ventilation at all when the motor burns out.

Case Study 3: Remediation via Litigation

Undoubtedly the most expensive and least effective way to address condensation related problems is via legal avenues. This case study provides an unusual variation on the discussion of window condensation: the author was called upon to provide an expert witness report on a particular type of condensation that is becoming more prevalent, especially in high-performance windows. The subject windows were wood-framed triple-glazed picture windows in a residential heating-dominated application, and the homeowners were assured that condensation would not be a problem. They were quite surprised to therefore find condensation obscuring most of the vision area on the outside of the window, similar to the effect shown in Figure 8.

Exterior condensation is not unusual in a hot, humid climate: the occupied space is cooled and, because windows typically represent the lowest thermal resistance in the building enclosure, the outer glazing is a relatively cold surface that can easily be colder than the outdoor dew-point temperature. In a heating-dominated situation, the mechanism for exterior condensation is quite different. The high-performance window assembly is the key: heat transfer through the assembly (especially the glazing) is reduced to the point that the outer surface of the glass is not significantly warmer than the ambient outdoor condition. On a cloudless night, radiative heat loss from the glass surface to the night sky reduces the glass surface temperature below the ambient dew-point temperature, resulting in condensation. The amount of condensate accumulated depends on the exterior humidity level and the amount of time the cloudless condition persists. Convincing the occupants that this represents exceptional thermal performance, and is therefore a desirable attribute, is

Figure 8 *Exterior condensation on high-performance windows.*

a difficult proposition—they typically want a solution that restores the view to the outdoors.

The solution can be determined by examining the Stefan-Boltzmann equation (Equation 1), which defines radiative heat transfer between two surfaces (in this case, between the exterior glass surface and deep space). The only parameters that can be modified to reduce the radiative loss to the night sky are the combined emittance E and the geometric shape factor F. The solution, therefore, would involve either a low-emissivity coating on the exterior glass surface, or a reduction in the shape factor, which could be achieved by installing a canopy or overhang above the window.

$$Q = EFA\sigma(T_H^4 - T_C^4) \quad (1)$$

CONCLUSION

Window condensation can carry significant costs, in consideration of mold growth, damage to interior finishes, and structural damage. Fortunately, there are several tools and techniques available to assess the magnitude of the problem, and to develop solutions. Success in this process depends upon moving away from considering windows as merely part of the building enclosure, and toward a true appreciation for building science; that is, the way in which the enclosure interacts with the other systems in the building, including heating, cooling, and ventilation systems.

REFERENCES

Enermodal Engineering. 1996. *FRAME, a computer program to evaluate the thermal performance of window frame systems*. Kitchener, ON: Enermodal Engineering Limited.

Gatley, D.P. 2004. Psychrometric chart celebrates 100th anniversary. *ASHRAE Journal*, November.

LBNL. 2103. *THERM 6.3/WINDOW 6.3 reference manual*. Berkeley, CA: Regents of the University of California.

McGowan, A.G. 1995. Numerical prediction of window condensation potential. *ASHRAE Transactions* 101(1). Adapted as an article in the July 1995 issue of *ASHRAE Journal*.

McGowan, A.G., and J.L. Wright. 1998. Computer simulation of window condensation potential. Proceedings of the Thermal Performance of the Exterior Envelopes of Buildings VII Conference, Clearwater Beach, FL.

NFRC. 2010. *Procedure for determining fenestration product condensation resistance values*. Greenbelt, MD: National Fenestration Rating Council, Inc.

University of Waterloo. 1996. VISION 4.0 glazing system thermal analysis: user's and reference manual. University of Waterloo Advance Glazing System Laboratory, 1998.

Wright, J.L. 1998. A simplified numerical method for assessing the condensation resistance of a window. *ASHRAE Transactions* 104(1).

Thermal Performance Evaluation of Walls with Gas-Filled Panel Insulation

Som Shrestha, PhD, BEMP
Member ASHRAE

Andre Desjarlais

Dahtzen Chu

Axy Pagan-Vazquez, PE

ABSTRACT

Gas-filled insulation panels (GFP), which were initially developed with a grant from the U.S. Department of Energy, are very lightweight and compact (when uninflated) advanced insulation products. GFPs consist of multiple layers of thin, low-emittance (low-e) metalized aluminum. When expanded, the internal, low-e aluminum layers form a honeycomb structure. These baffled polymer chambers are enveloped by a sealed barrier and filled with either air or a low-conductivity gas such as argon, krypton, or xenon. The sealed exterior aluminum foil barrier films provide thermal resistance, flammability protection, and properties to contain air or a low-conductivity inert gas.

The unexpanded product is nearly flat for easy storage and transport. Therefore, the transportation volume and weight of the GFP needed to fill the volume of wall cavity is much smaller than that of conventional insulation products. This feature makes this product appealing to use at U.S. Army contingency and forward operating bases, where transportation costs are much higher than the cost of materials. This paper presents lab test results conducted in a rotatable guarded hotbox to evaluate the thermal performance of walls, similar to those used in typical barrack hut (B-hut) hard shelters, with GFPs in cavities. The test result show that while the effective surface-to surface thermal resistance (R-value) of the wall with two layers of GFP decreases sharply as the temperature of the GFPs increases, the R-value of the wall with only one layer of GFP remains fairly steady. That is because the R-value of 0.5 in. air space decreases and that of 1.5 in. air space increases as the mean temperature increases.

INTRODUCTION

Gas-filled insulation panels (GFP) are very compact (when uninflated) and lightweight (weight per unit volume when inflated for use) advanced insulation products. GFPs consist of multiple layers of thin, low emittance (low-e) metalized aluminum. When expanded, the internal, low-e aluminum layers form a honeycomb structure (Figure 1). These baffled polymer chambers are enveloped by a sealed barrier and filled with either air or a low-conductivity gas. The sealed exterior aluminum foil barrier films provide thermal resistance, flammability protection, and properties to contain air or a low-conductivity inert gas. This product was initially developed with a grant from the U.S. Department of Energy (DOE). The unexpanded product is nearly flat for easy storage and transport. Therefore, transportation volume and weight of the GFP to fill unit volume of wall cavity is much smaller compared to that of other conventional insulation products. This feature makes this product appealing to use at U.S. Army contingency bases (COBs) where fuel supplies are highly sensitive and

Figure 1 *Honeycomb structure of the GFP.*

Som Shrestha is research and development staff and **Andre Desjarlais** is the program manager for building envelope systems at th Building Technologies Research and Integration Center, Oak Ridge National Laboratory, Oak Ridge, TN. **Dahtzen Chu** and **Axy Pagan-Vazquez** are researchers at the Construction Engineering Research Laboratory (CERL), Champaign, IL.

expensive commodities, energy efficiency is mission critical, and transportation cost is significant compared to the cost of materials. The objective of this study is to evaluate thermal performance of walls, similar to those used at typical U.S. Army temporary hard shelters known as barracks huts (B-huts), when GFPs are used in the wall cavities. Oak Ridge National Laboratory (ORNL) evaluated performance of the wall in the rotatable guarded hotbox (RGHB) following the ASTM C1363 (ASTM 2011) standard test method.

The U.S. Army builds and uses temporary shelters, either hard- or soft-sided, at COBs. These B-huts are typically wood-framed structures constructed by soldiers or contracted out to local nationals. They have an expected lifetime of 3 to 4 years, and are demolished at the end of their service life. Usually, the huts are not insulated and lack airtightness. Therefore, they are inherently very energy inefficient and have relatively high heating and cooling requirements. A typical 300-person camp uses more than 400,000 gallons of fuel per year, almost 50% of it to heat and cool shelters. Improvements to the building envelope can therefore potentially save energy and consequently reduce the amount of fuel that the buildings use. This research studied potential use of GFPs in B-huts to reduce the fuel requirements. Further details on the test apparatus and results are described by Shrestha et al. (2014).

PREVIOUS STUDIES

Griffith et al. (1991) and Griffith and Arasteh (1992) described the potential use of low thermal conductivity gases such as argon or krypton to produce insulating panels. Griffith and Arasteh (1992) estimated GFP R-values as 5.2 ft^2·h·°F/Btu·in. for air, 7.1 ft^2·h·°F/Btu·in. for argon, and 12.5 ft^2·h·°F/Btu·in. for krypton. However, measurements on handmade specimens resulted in 4.3 ft^2·h·°F/Btu·in. for air, 6.3 ft^2·h·°F/Btu·in. for argon, and 10.1 ft^2·h·°F/Btu·in. for krypton. While the measured values were lower than the estimates, the results indicated the possibility of a new type of building insulation with thermal resistance values greater than conventional air-filled insulations (Yarbrough et al. (2007).

Yarbrough et al. (2007) studied the thermal performance of nominally 1.5 in. thick GFPs, with internal and external reflective surfaces, filled with argon and with air, in the Large-Scale Climate Simulator at Oak Ridge National Laboratory. The result shows total contribution of the GFP layer installed above fiberglass batt insulation as 5 to 6 ft^2·h·°F/Btu for winter conditions with an outside temperature of 25°F and an inside temperature of 70°F. The results from the Yarbrough et al. (2007) study show significantly lower R-values of GFPs compared to the results from Griffith and Arasteh (1992).

Baetensa et al. (2010) reviewed the state-of-the-art on GFPs for building applications, which shows the R-values of 3.14 to 3.61 ft^2·h·°F/Btu·in. have so far been achieved for air-filled and argon-filled GFPs, respectively. The R-values are slightly lower than currently available traditional building insulation materials.

None of these studies explored the impact of mean temperature of GFPs R-value.

TEST WALL DETAILS

The test wall construction resembles that of typical B-hut walls, except that GFPs were used in cavities of the test wall whereas no insulation is used in typical B-Hut wall cavities. The test wall was constructed of nominal 2 × 4 in. lumber studs 16 in. on center (OC) with 0.5 in. plywood on both sides of the wall. Figure 2 shows the test wall built for RGHB test.

The GFPs were inflated with compressed air with a pressure regulator set at 7 psig. Some of the GFPs were bulged during inflation. Severely bulged GFPs were not used in this test. One GFP was punctured by a splinter while installing the GFP on the test wall and replaced. The honeycomb structure of the GFP was visually inspected (Figure 1) by cutting one panel. The thickness of the inflated GFP was about 1.5 in.; thus, two layers of GFPs fill the cavities between 2 × 4 in. studs.

EXPERIMENTAL APPARATUS AND INSTRUMENTATION

ORNL operates and maintains a guarded hotbox that is used to measure the effective thermal resistance (R-value) and thermal transmittance (U-factor) of full-size wall and window assemblies. The box operates under the requirements of ASTM C1363 (ASTM 2011). Test assemblies are installed in a specimen frame mounted on a movable dolly. The specimen frame has an aperture of 13 ft. long by 10 ft. high. The specimen frame/test assembly is inserted between two "clamshell" chambers of identical cross-section. The placement of the test wall assembly between the chambers allows the chamber temperatures to be independently controlled, thus creating a temperature difference across the specimen. The chambers are designated as the climate (cold) and metering/guard (hot) chambers. Pictures of the RGHB in Figures 3 and 4 show a

Figure 2 *Test wall with GFPs in the cavity.*

typical wall specimen installed in the test frame. The central 8 × 8 ft wall section, which aligns with the metering chamber boundary, is used for the actual test.

CALCULATION METHODOLGY

The meter and climate-side average surface temperatures were calculated in an appropriate area-weighted manner. The surface area of the wall used for the test was 64 ft^2. The percentage of the total wall surface area that each individual wall component comprised was determined. The average temperatures were then computed by area-weighing the average cavity, stud, and track surface temperatures. The energy exchange rate from the metering box to the guard chamber Q_{mb} is calculated by Equation 1.

$$Q_{mb} = \frac{A_{mb} \cdot \Delta t_{mb}}{R_{mb}} \quad (1)$$

where

Q_{mb} = heat flow rate through metering box walls, Btu/h

A_{mb} = surface area of the metering box, 100.08 ft^2

Figure 3 *Rotatable guarded hot box.*

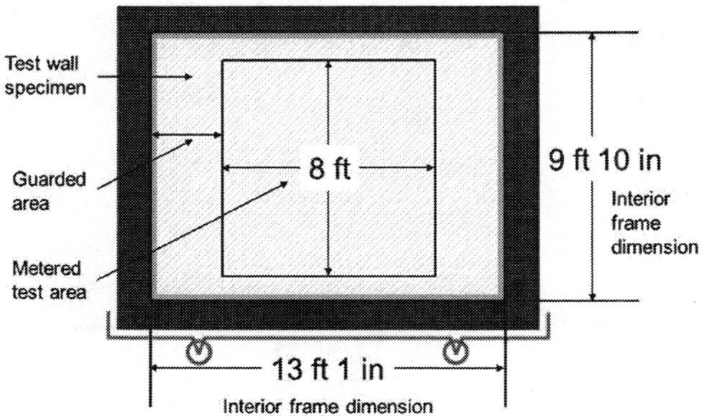

Figure 4 *Schematic of a typical test wall within the hotbox test frame.*

Δt_{mb} = temperature difference between the guard side and meter side of the metering box walls, °F

R_{mb} = thermal resistance of the metering box walls, 15 h·ft^2·°F/Btu

The total energy flow through the wall assembly Q_{wall} is calculated from

$$Q_{wall} = Q_h + Q_{fan} + Q_{mb} + Q_{vp} \quad (2)$$

where

Q_{wall} = total energy flow rate through the wall assembly, Btu/h

Q_h = energy input to the resistance heaters in the metering chamber, Btu/h

Q_{fan} = energy input to the fans in the metering chamber, Btu/h

Q_{vp} = energy input to the velocity probes in the metering chamber, Btu/h

The surface-to-surface thermal resistance of the wall assembly R_{wall} is calculated from

$$R_{wall} = \frac{A_{wall} \cdot (t_{ms} - t_{cs})}{Q_{wall}} \quad (3)$$

where

R_{wall} = surface-to-surface thermal resistance of the wall assembly, h·ft^2·°F/Btu

A_{wall} = area of wall, ft^2

t_{ms} = area weighted average metering-side surface temperature, °F

t_{cs} = area weighted average climate-side surface temperature, °F

TEST CONDITIONS

The tests were conducted with the following three configurations:

1. Two layers of GFPs on cavities between studs
2. One layer of GFP on climate side
3. Without GFPs

The length of the GFPs was 4 ft, so two GFPs were used to cover the 8 ft height of the test wall. The motivation for conducting the test with one layer of GFP was to evaluate if only one layer of GFP can be used without compromising the effective R-value significantly. Each configuration was tested at three temperature conditions in order to evaluate change in R-value as a function of temperature. The following are the test conditions used:

- Meter side surface temperature 80°F and climate side surface temperature 40°F
- Meter side surface temperature 100°F and climate side surface temperature 60°F

- Meter side surface temperature 120°F and climate side surface temperature 80°F

The meter side and climate side surface temperatures were determined by area-weighted averaging of the thermocouples attached to the individual components. On each side of the wall, 32 thermocouples were used to measure surface temperature. The perimeter of the test wall and the joints were caulked and taped to prevent air leakage. The data from the final 15 hours after achieving stable temperature and heat flow conditions were used for the analysis.

TEST RESULTS

Temperatures, heat flows and R-values are presented in Table 1. Table 1 also summarizes the calculated surface-to-surface R-values. Figure 5 shows R-value as a function of temperature for walls with two layers of GFP and one layer of GFP. It is interesting to note that while the R-value of the wall with two layers of GFP decreases sharply as the temperature increases, the R-value of the wall with only one layer of GFP remains fairly steady. This is, a change from 8.662 h·ft²·°F/Btu at 80°F/40°F (hot side and cold side surface temperature) to 7.762 h·ft²·°F/Btu at 120°F/80°F, resulting 10.4% decrease in R-value for the wall with two layers of GFP and a change from 7.241 h·ft²·°F/Btu at 80°F /40°F to 7.183 h·ft²·°F/Btu at 120°F /80°F, resulting only 0.8% decrease in R-value for the wall with only one layer of GFP. At 120°F hot side surface temperature and 80°F cold side surface temperature, the R-value of the wall with one layer of GFP is only 7.5% less than that of the wall with two layers of GFP. This phenomenon can be explained with the data available in Chapter 26 of *ASHRAE Handbook—Fundamentals* (ASHRAE 2013).

Figure 6 presents the relevant data from Chapter 26 (ASHRAE 2013): effective R-value of plain air spaces as a

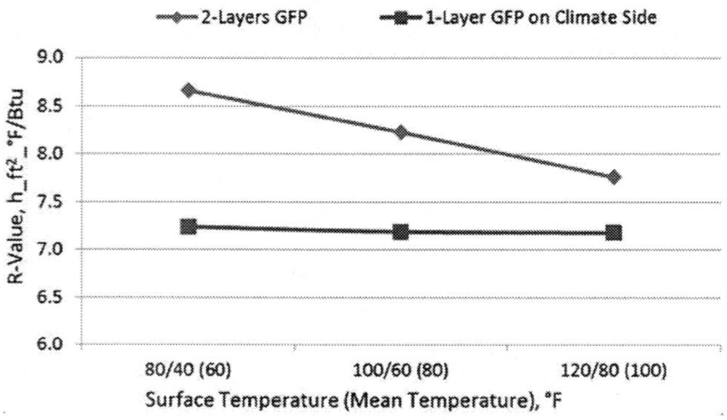

Figure 5 *R-value as a function of temperature.*

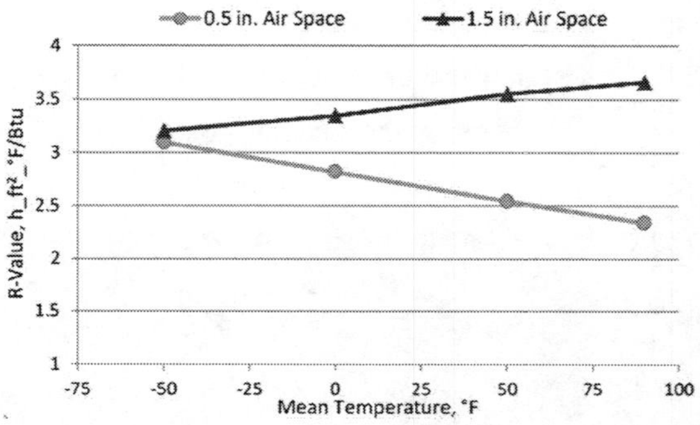

Figure 6 *Effective thermal resistance of plane air spaces (ASHRAE 2013).*

Table 1. Summary Test Results

Wall Configuration	Two layers GFP			One layer GFP on climate side			No GFP		
Surface Temperatures, °F	80/40	100/60	120/80	80/40	100/60	120/80	80/40	100/60	120/80
M-FAN-PWR*	3.9	3.9	3.8	3.9	3.8	3.8	3.9	3.8	3.8
M-HTR-PWR†	271.4	287.5	305.3	329.8	331.9	332.7	1132.6	1013.2	1052.3
M-VEL-PWR‡	20.2	20.0	19.7	20.0	19.6	19.3	19.7	9.9	9.9
Q_{mb}**	–0.1	–0.1	–0.1	–0.1	–0.1	–0.1	–0.1	–0.1	–0.1
TMP_DC††	295.4	311.3	328.7	353.6	355.3	355.8	1156.0	1026.9	1065.9
T-DIFF‡‡	40.0	40.1	39.9	40.0	39.9	40.0	40.1	40.1	39.9
R-value***	8.662	8.232	7.762	7.241	7.192	7.183	2.218	2.500	2.396

* M-FAN-PWR = meter fan power, Btu/h
† M-HTR-PWR = meter heater power, Btu/h
‡ M-VEL-PWR = meter velocity probe power, Btu/h
** Q_{mb}= heat flow rate through metering box walls, Btu/h
†† TMP-DC= total heat flow rate into meter box, Btu/h
‡‡ T-DIFF= Average temperature difference across wall surfaces, °F
***R-value= Surface to surface R-value of the wall, h•ft2•°F/Btu

function of mean temperature for horizontal heat flow at a 10°F temperature difference and effective thermal emittance 0.05. Note that while R-value of 0.5 in. air space decreases as the mean temperature increases, it is just opposite for 1.5 in. air space. Even though the conditions in GFP test are not exactly the same as in the ASHRAE *Handbook* (2013), an analogy can be made between test data and the ASHRAE (2013) data. The test wall with two layers of GFP contains six layers of approximately 0.5 in. air spaces. Therefore, the R-value decreases as the mean temperature increases. For the test wall with one layer of GFP, the cavity between studs are filled with three layers of approximately 0.5 in. air spaces and one layer of approximately 1.5 in. air space. The R-value of 0.5 in. air space decreases and that of 1.5 in. air space increases as the mean temperature increases. Thus, the resulting R-value for the test wall with one layer of GFP remains fairly steady over the range of temperature used in the tests.

Figure 7 shows the increase in effective surface-to-surface R-value due to the use of GFPs in wall cavities. While the increase in R-value is significant, several factors (such as cost of materials, transportation, energy, weather conditions, application etc.) would need to be considered to determine whether it is cost effective to use one layer or two layers of GFP in the B-hut cavity.

SUMMARY

Thermal performance of wall assembly with two layers of GFP, one layer of GFP, and without GFP was evaluated at Oak Ridge National Laboratory. The test wall was constructed of nominal 2 × 4 in. lumber studs 16 in. on center with 0.5 in. plywood on both sides of the wall. The walls were tested in the RGHB according to the ASTM C1363 (ASTM 2011) standard test method. The test results show that while the effective surface-to surface thermal resistance (R-value) of the wall with two layers of GFP decreases sharply as the temperature of the GFPs increases, the R-value of the wall with only one layer of GFP remains fairly steady. That is because the R-value of 0.5 in. air space decreases and that of 1.5 in. air space increases as the mean temperature increases.

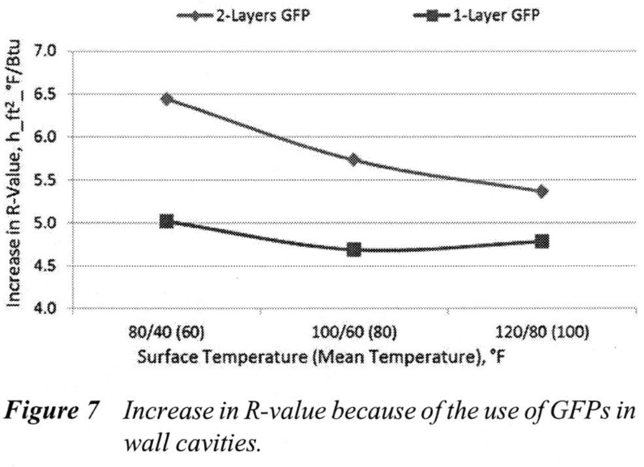

Figure 7 *Increase in R-value because of the use of GFPs in wall cavities.*

For the range of test conditions (80°F, 100°F, and 120°F hot side surface temperature and 40°F, 60°F, 80°F cold side surface temperature), the R-value range from 8.662 to 7.762 h·ft^2·°F/Btu for the wall with two layers of GFP, 7.241 to 7.183 h·ft^2·°F/Btu for the wall with one layer of GFP, and 2.218 to 2.396 h·ft^2·°F/Btu for the wall without GFP. Thus, the increase in R-value was 6.444 to 5.366 h·ft^2·°F/Btu for the wall with two layers of GFP and 5.024 to 4.787 h·ft^2·°F/Btu for the wall with one layer of GFP compared to the wall without GFP. One interesting finding was that while the R-value of the wall with two layers of GFP decreases sharply as the temperature increases, the R-value of the wall with only one layer of GFP remains fairly steady.

Factors such as cost of materials, transportation, energy, weather conditions, and application would need to be considered to determine whether it is cost effective to use one layer or two layers of GFP in the B-hut cavity. A simulation study is recommended to quantify potential energy savings and associated cost savings potential of GFPs while used in B-huts. A follow-up paper will present EnergyPlus simulation results for B-huts with GFP insulation.

ACKNOWLEDGMENTS

This study was conducted for Headquarters, U.S. Army Corps of Engineers (HQUSACE) and the Office of the Assistant Secretary of the Army for Acquisition, Logistics, and Technology.

Thanks to Jerry Atchley of ORNL for his technical contributions to this research.

REFERENCES

ASHRAE. 2013. *ASHRAE Handbook—Fundamentals*. Atlanta: ASHRAE.

ASTM. 2011. ASTM Standard C1363-11. *Standard test method for thermal performance of building materials and envelope assemblies by means of a hot box apparatus*. West Conshohocken, PA: ASTM International.

Baetensa, R., B.P. Jellea, A. Gustavsend, and S. Grynninga. 2010. Gas-filled panels for building applications: A state-of-the-art review. *Energy and Buildings* 42(11):1969–75.

Griffith, B.T., D. Arasteh, and S. Selkowitz. 1991. Gas-filled panels high performance thermal insulation. *Insulation Material: Testing and Applications*. ASTM STP 1116 2: 441–54. R.S. Graves and D.C. Wysocki, eds. Philadelphia, PA: American Society for Testing and Materials.

Griffith, B.T., and D. Arasteh. 1992. Gas-filled panels: A thermally improved building insulation. *Thermal Performance of the Exterior Envelopes of Buildings* V, December 7–10.

Shrestha, S., A. Desjarlais, and J. Atchley. 2014. *Thermal performance evaluation of walls with gas filled panel insulation*. Oak Ridge, TN: Oak Ridge National Laboratory. http://info.ornl.gov/sites/publications/Files/Pub52624.pdf.

Yarbrough, D.W., T.W. Petrie, D. Kinninger, and R. Graves. 2007. Thermal performance of gas-filled panels with reflective surfaces installed in an attic. Thermal Performance of the Exterior Envelopes of Whole Buildings X International Conference, Dec 2007. http://web.ornl.gov/sci/buildings/2016/2007%20B10%20papers/049_Yarbrough.pdf.

Mind the Gap: Summary of Window Residential Retrofit Solutions

Joseph Petersen **Katherine Cort** **Sarah Widder**
Associate Member ASHRAE

Thomas Culp
Member ASHRAE

Greg Sullivan
Associate Member ASHRAE

ABSTRACT

Improving the insulation, solar heat gain, and infiltration characteristics of windows in a home has the potential to significantly improve the overall thermal performance by reducing heat transfer through the window and also by decreasing air leakage into and out of the home. As approximately 43% of existing homes still have single-pane clear windows (~50 million houses) and millions of other homes have only double-pane clear windows (Cort 2013), improving window performance also presents a significant opportunity for energy savings in the residential sector. Today, various energy-saving window retrofit opportunities are available to homeowners, ranging from window coverings and storm panels to highly insulating triple-pane R-5 window replacements. Many of these technologies have been evaluated in the field in the Lab Homes at Pacific Northwest National Laboratory, and through modeling to prove their cost-effectiveness and performance in different climate regions. Recently, the Pacific Northwest's Regional Technical Forum approved a utility measure for low-emissivity storm windows based on such data. This action represents a watershed moment for increasing the variety and prevalence of fenestration options in utility programs, especially for the low-income demographic. This paper will review various window retrofit options, the most recent field test and modeling data regarding their performance and cost-effectiveness, and discuss future rating efforts. This information is useful for utilities and energy-efficiency program managers to help effectively implement incentive measures for these technologies.

INTRODUCTION

Residential buildings in the United States currently require energy amounting to approximately 8 quadrillion Btu/y for heating and cooling, which accounts for more than 40% of the primary energy consumed by homes (EIA 2015). Windows are a major source of heating losses and gains in residential buildings because of their heat transfer and infiltration properties, especially relative to other building shell components. For example, it has been estimated that windows account for approximately 25% of energy use in a typical residential building (Huang et al. 1999). Retrofitting and renovating existing homes to save energy has become an increasingly important component of United States energy strategy, and window replacement and energy-efficient window attachments, such as high-efficiency shades, blinds, storm windows, or other window coverings, can significantly improve the thermal performance of a window for minimal financial cost.

Improving the insulation and solar heat gain characteristics of a home's windows has the potential to significantly improve the overall thermal performance by reducing heat loss (in the winter) and cooling loss and solar heat gain (in the summer) through windows. A high-quality retrofit will also minimize or reduce air leakage through the building envelope, decreasing infiltration and thereby contributing to reduce heat loss in the winter and cooling loss in the summer. These improvements all contribute to decreasing overall annual home energy use.

Over the past 15 years, new window replacement and retrofit technologies have been developed that significantly increase the options available to homeowners and utilities when considering upgrading existing windows, including

Joseph Petersen is an associate research engineer, **Katherine Cort** is a project manager and economist, and **Sarah Widder** is a research engineer at Pacific Northwest National Laboratory (PNNL), Richland, WA. **Thomas Culp** is a senior research engineer at Birch Point Consulting, La Crosse, WI. **Greg Sullivan** is a senior research engineer at Efficiency Solutions, Richland, WA.

low-emissivity (low-e) storm windows, insulating blinds, and other window attachments. These technologies all have various performance levels in terms of U-factor reduction, solar heat gain and air infiltration impacts, and cost. This article reviews previous research conducted on window retrofit technologies, including experimental evaluation at the Lab Homes located at Pacific Northwest National Laboratory (PNNL), and discusses how these data are currently being used to support utility incentive and market transformation programs. The performance data are compared to performance data from baseline existing windows as well as highly-insulating replacement windows. While there are a number of window retrofit technologies, including low-e storm windows, insulating blinds, awnings, window films, and other innovative products, this article focuses on low-e storm windows and high-efficiency cellular shades, as most experimental data to date have been collected on these two technologies.

REVIEW OF TECHNOLOGIES

With a thermal conductance of approximately 1 watt per meter Kelvin (W/m·K), glass alone is a poor insulator. As a result, single-pane windows often have an overall heat transfer coefficient (U-factor) around 5.5 $W/m^2 \cdot K$ or 1 $Btu/h \cdot ft^2 \cdot °F$, which is only R-1. Clear glass also transmits between approximately 70% and 90% of light/heat at all wavelengths, including infrared (IR) radiation (LBNL 2016). Many window attachment technologies look to improve the thermal and optical properties of the window via coatings or coverings that, depending on the climate, decrease solar heat gain (by decreasing transmission of light through the window), increase the reflectance of the window in the IR wavelengths, and decrease the amount of thermal conduction or convection between indoor spaces and the outdoors.

Low-e Storm Windows

Traditional storm windows consisted of a single piece of clear glass (or plastic) mounted in a wood or aluminum frame, and were installed on the outside of an existing window. This window retrofit design focused on reducing thermal conduction and convection. Modern storm windows can be operable or fixed in place and can significantly reduce air leakage through the window opening. As modern storm windows are intended to be permanently mounted, they typically feature tighter seals and gasketed or caulked window frames.

However, in addition to conduction and convection, radiation is an important mechanism for heat gain and heat loss through windows. All materials radiate heat in the form of longwave IR energy depending on the emissivity and temperature of their surfaces, which contributes to heat loss from buildings in addition to conductive, convective, and air leakage heat loss. Typical low-e storm windows include a low-e pyrolytic coating that lowers the emissivity of glass for certain wavelengths, effectively reducing heat transmission through the storm window (Culp et al. 2015). Specifically, low-e coatings are microscopic coatings consisting of very thin, electrically conductive material. Uncoated glass typically has an emissivity of around 0.84, while low-e coated glass can have an emissivity of 0.16 or lower. In general, terrestrial radiation is generated by most appliances and occupants of the home. This generated radiation is absorbed by the interior window pane and radiated across the air gap between the primary and storm window or reflected back into the interior space. The addition of a low-e coating reduces the reradiation by reducing the emissivity of the interior window pane. When the interior heat energy tries to escape to the colder outside during the winter, the low-e coating reflects the radiative heat back to the inside, reducing the overall heat loss through the glass. The reverse transfer of heat occurs during the summer if the low-e coating is designed to reflect solar energy. This is not the case for all low-e coatings (Culp et al. 2015).

For exterior low-e storm windows, the coating resides on the window pane side that faces the interior of the conditioned space. This placement removes any possibility of damage to the low-e coating by the elements throughout the life of the window. In contrast, the orientation of the low-e coating on the interior storm window does not alter the performance, and it can be effective in reducing heat loss when facing either direction, although it is still most common to have the low-e coating face toward the existing primary window.

Insulating Blinds

Window attachments, such as window shades and blinds, have provided privacy for as long as windows have been in use. Only recently, however, has attention been focused on the energy savings potential achieved by increasing the insulting values of window coverings and reducing or optimizing the solar gains added to the space. The type and selection of differing attachment technology has greatly expanded in previous years; however, there is limited information on the differing energy-saving characteristics of these products, and currently no comprehensive rating system exists to help distinguish the energy-saving features of one window covering from another (Curcija et al. 2013). High-efficiency shading devices, such as cellular shades (i.e., honeycomb shades), are one such window attachment technology that can provide significant energy savings and thermal comfort improvement.

Within the window coverings category, honeycomb cellular shades are typically considered to have the highest R-values. Introduced in the 1980s, cellular shades are designed to trap air inside pockets that act as insulators, and this design can increase the R-value of the window covering and reduce the conduction of heat through the window that it covers (Ariosto and Memari 2013). Insulating shades also impact solar heat gains if managed properly.

Along with the added insulating properties of the shades, many of these attachments have a built in automation feature to assist in optimizing management of solar gains throughout the year. This allows the blinds to open and close via predefined schedule or as the homeowner desires. The automation process and scheduling can be optimized based on the solar calendar and geographical location. The schedule reduces the HVAC load while ensuring that adequate light and thermal comfort is achieved within the conditioned space. During the heating season, the schedule can optimize visible light and solar heat gain to the space during the daylight hours.

PREVIOUS RESEARCH

As utility and government programs and regulations continue to drive reduced energy use in new and existing site-built and manufactured homes, new energy-efficient technologies and measures are needed to cost-effectively achieve energy-reduction goals. The U.S. Department of Energy (DOE) has identified highly-insulating windows with U-factors around 0.2 Btu/h·ft^2·°F (W/m^2·°C) as a key technology that could play an important role in the next phase of energy-efficiency improvements in the residential sector. Improving the insulation and solar heat gain characteristics of windows in a home has the potential to significantly improve the home's overall thermal performance by reducing heat loss (in the winter), and cooling loss and solar heat gain (in the summer) through the windows. In 2012, research conducted in the PNNL Lab Homes showed the added benefit in thermal and HVAC system performance by replacing the installed window technology with R-5 windows or other window replacements. Window replacement has been shown as a first step at envelope improvement and energy reduction. The results are also relevant to residential retrofit programs being implemented across the nation. Windows are a key aspect of the building envelope, accounting for approximately 10% to 13% of an average home's energy use (EIA 2015).

Low-e storm windows and other window attachments have been evaluated via modeling and at the laboratory scale over the past 10 to 15 years. Beginning in 2007, DOE conducted a series of laboratory tests at the component level to validate the savings applying a low-e coating to storm windows. The performance improvements were validated with field tests and case studies supported by DOE's Buildings Technology Emerging Technology team. The approaches and results of these field tests and case studies are described and summarized by Cort (2013). A high-level summary of these activities is provided in Table 1.

In 2013, DOE sponsored a comprehensive energy modeling study led by Lawrence Berkeley National Laboratory (LBNL) that focused on a range of window attachments, including products such as shades, blinds, storm window panels, and surface applied films simulated in four types of "typical" houses, located in 12 characteristic climate zones. The simulations captured the optical and thermal complexities of these products (Curcija et al. 2013) and also considered typical operation and usage patterns based on a separate study focusing on user behavior with respect to operable window coverings (Bickel et al. 2013). The study found that many of the window attachments examined can yield significant energy savings when installed over windows; however, the degree of savings depends on the attachment type, baseline window conditions, seasonal and climate factors, and how the attachment is operated, when applicable.

Calculations of energy savings and the cost-effectiveness of low-e storm windows have been conducted with two software platforms: 1) the National Energy Audit Tool, used by weatherization programs, and 2) RESFEN (RESidential FENestration), used to compare the annual energy performance of different window options in single-family homes (Culp and Cort 2015).

In addition to DOE's research focusing on window coverings, a number of research institutions, energy-efficiency programs, and utilities have completed characterization and meta-analyses (Ariosto and Memari 2013) and energy simulation analyses (CEE 2014; Garber-Slaght and Craven 2011; Zirnhelt et al. 2015) validating energy savings from cellular shades and other window attachments in multiple climate zones and prototype residential buildings.

Field data and case studies provide valuable insights related to the savings potential of window attachments in specific applications or climate zones, but the variability that occurs because of home type and occupancy behavior can make it difficult to isolate the savings from the window attachment and project these savings to alternative circumstances. Controlled side-by-side experiments, such as those conducted in the PNNL Lab Homes, provide a platform for more detailed and comprehensive data collection on HVAC system energy performance. The PNNL Lab Homes provide controlled experimental HVAC data can be used to appropriately tailor and calibrate building simulation models to account for relevant interactions, occupancy, climate zones, and baseline characterizations.

PNNL LAB HOMES EXPERIMENTS

The experiments described below were conducted in the PNNL side by side Lab Homes, which form a platform for precisely evaluating energy-saving and grid-responsive technologies in a controlled environment. The PNNL Lab Homes are two factory-built homes installed on the PNNL campus in Richland, WA. Each Lab Home has seven windows and two sliding glass doors, for a total of 196 ft^2 of window area. To be representative of the Pacific Northwest climate, clear double-pane windows were installed as the baseline technology. For the primary experiments examined in this study, the "experimental home" was retrofitted with the fenestration technology under evaluation, while a matching "baseline home" remained unaltered. The floor plan of the Lab Homes as constructed is shown in Figure 1. The specific construction characteristics of the Lab Homes have been documented previously (Widder et al. 2012).

The approach to the metering includes metering and system-control activities taking place at both the electrical panel and at the end-use. For all experiments, metering was completed using data loggers and matching sensors. Two data loggers were installed in each home, one allocated to electrical measurements and one to temperature and other data collection. Data from all sensors were

Table 1. Summarized Case Studies Focused on Low-e Storm Windows

Study	Sponsor	Baseline Description	Findings
Chicago case study (2007)	DOE, HUD, NAHB Research Center, LBNL	Six low-income homes; single-pane wood-framed windows	21% reduction in overall home heating load 7% reduction in overall home air infiltration Simple payback of 4 to 5 years
IR camera imaging	DOE, LBNL, Building Green	Single-pane wood-framed windows	Images showed that interior low-e storm windows performed equivalently or better than new double-pane replacement windows with low-e glass and argon fill
Atlanta case study (2-year study)	DOE, Quanta,* Larson,†	Ten occupied homes; single-pane wood-framed windows	High variability, but approximately: • ~15% heating energy reduction • ~2 to 30% cooling reduction (highly variable) • 17% reduction in overall home air infiltration
Philadelphia multifamily case study	DOE, Quanta, Larson, NAHB, AGC Flat Glass, NSG-Pilkington	Two large multifamily buildings; single-pane metal-framed windows	Replacing old clear glass storm windows with new low-e storm windows provided: • 18%–22% reduction in heating energy use • 9% reduction in cooling energy use • 10% reduction in overall apartment air leakage
Field air-leakage testing (Bronx, NY, 2013)	Steven Winter Associates, Quanta	Multifamily dwellings in the Bronx	Interior low-e panels reduced the effective leakage area by: • 77% for windows without air-conditioning units • 95% for windows with air-conditioning units
Pennsylvania weatherization technical support (2010)	DOE, Birch Point Consulting	37 model homes with range of window types	Modeled results for 7 climate zones: • 12%–33% overall HVAC savings
Energy savings from window shades (Zirnhelt et al. 2015)	Hunter Douglas and Rocky Mountain Institute	EnergyPlus modeling of DOE residential buildings	Denver max cooling savings—25% Denver max heating savings—10% Peak electrical demand reduction of 9% for new homes Increased thermal comfort
Evaluating window insulation for cold climates (Garber-Slaght and Craven 2011)	Cold Climate Housing Research Center	Modeled interior insulated blinds	U-factor reduction of 15%

HUD = U.S. Department of Housing and Urban Development; LBNL = Lawrence Berkeley National Laboratory; NAHB = National Association of Home Builders.

* Quanta Technologies, Inc., Lancaster, PA.
† Larson Manufacturing Company, Brookings, SD.

collected via cellular modems that were individually connected to each of the loggers. All data were captured at 1-minute intervals by the data loggers. The 1-minute data were averaged over hourly and daily time intervals to afford different analyses. Occupancy in the homes was simulated via a programmable commercial lighting breaker panel (one panel per home) using motorized breakers. These breakers were programmed to activate connected loads on schedules to simulate human occupancy by introducing heat to the space. The analysis by Widder et al. (2012) contains more details on the metering and data collection capabilities of the Lab Homes.

R-5 and Low-e Storm Windows

In March 2013 through March 2014, PNNL evaluated commercially-available low-e storm windows in the PNNL Lab Homes for their thermal performance, including energy savings and impact on interior temperature distributions. During these experiments, IR images were taken of both the primary and interior low-e storm windows (see Figure 2). These images help visualize the temperature differential between the conditioned space and outdoor air and the interior photos help characterize the impact of the storm windows on the thermal comfort. The pictures were taken on February 2, 2015. During this day, the average temperature was 34°F with a low of 17°F and a high of 40°F.

Reviewing the exterior images (Figure 2) shows the effect of the storm windows on the envelope. The external surface temperature of the baseline home was measured to be 43°F. Compared to the 39.4°F of the experimental home, the change in temperature between the two surfaces is 3.6°F. The insulating values gained by the interior storm windows reduce the amount of heat that is transferred though the window (Petersen et al. 2015b).

Similarly, the internal temperature master-bedroom glass surface temperature in the baseline home was 59.9°F (Figure 3). Compared to the internal glass temperature of the experimental home (66.6°F), there is a differential of 6.7°F between the two window panes. These temperatures were motored throughout the full duration of the experiment. This differential can be attributed to the insulating properties of the interstitial space, but also to the low-e coating on the windows. The low-e surface was more effective at trapping and reflecting the internal heat back into the space.

Both interior and exterior storm windows were evaluated in the PNNL Lab Homes throughout differing heating and cooling seasons. During the heating season, the heat pump was disabled, and a forced-air electric resistance furnace supplied the required heating to the Lab Homes. In general, measured HVAC savings because of the exterior storm windows averaged 10.5% for the heating season and 8.0% for the cooling season for identical occupancy conditions (Knox and Widder 2014). Because of limitations on the manufactured size of interior storm windows, only an estimated 74% of the window

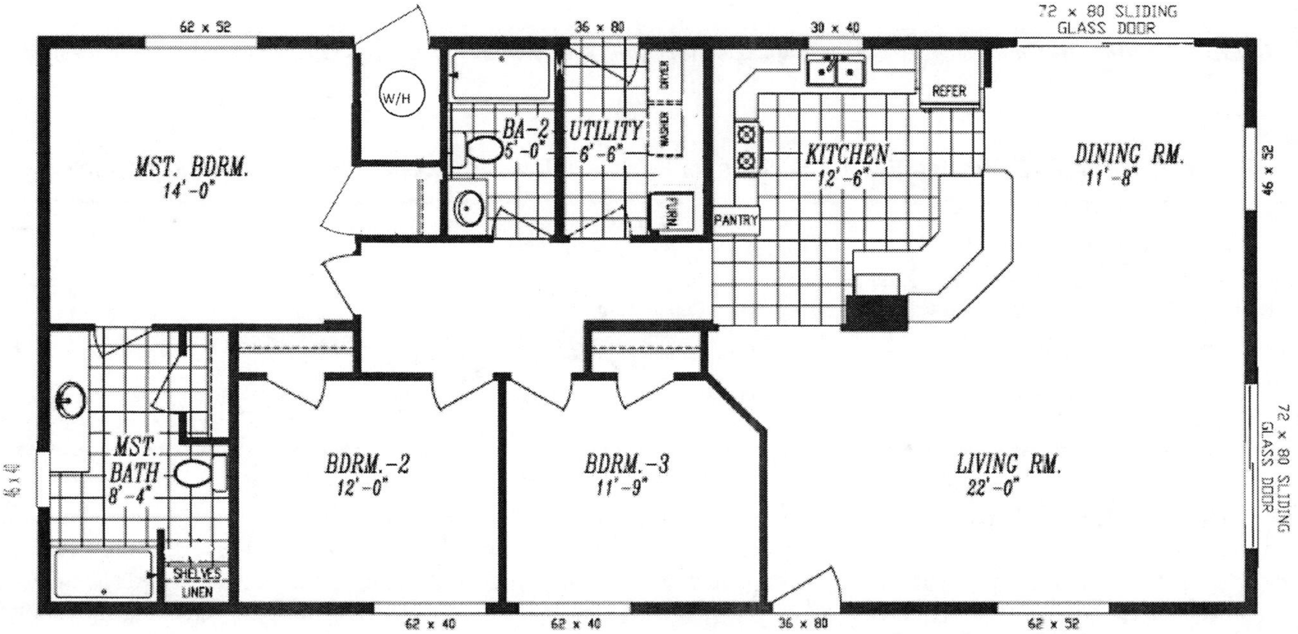

Figure 1 *Floor plan to the PNNL Lab Homes.*

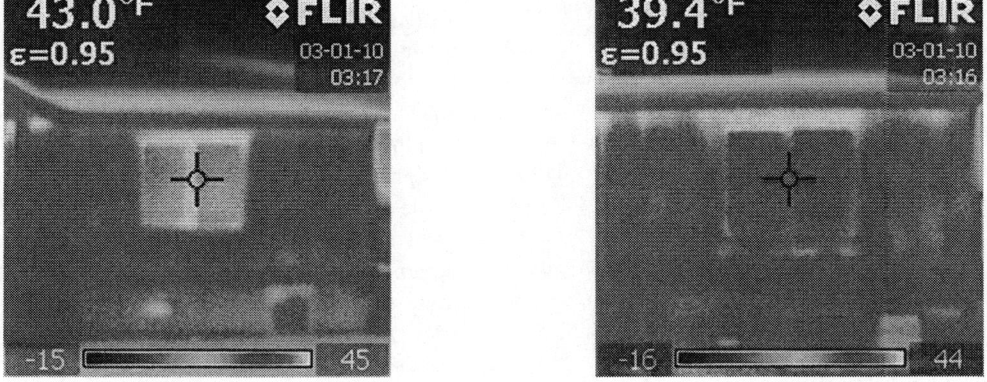

(a) **(b)**

Figure 2 *(a) Baseline home master bedroom exterior and (b) experimental home master bedroom exterior with interior storm windows.*

(a)

(b)

Figure 3 *(a) Baseline home master-bedroom interior and (b) experimental home master-bedroom interior with interior storm windows.*

Table 2. Annual Estimated Energy Savings for Each Window Replacement or Attachment Technology

Technology (Experiment)	Baseline	Average Annual Energy Savings, %
Highly insulating window replacements (R-5)	Double-pane aluminum frame clear glass	12.2 ±1.3
Exterior low-e storm windows	Double-pane aluminum frame clear glass	10.1 ±1.4
Interior low-e storm windows	Double-pane aluminum frame clear glass covering 74% of window area	7.8 ±1.5

area could be covered during the test. Testing was conducted during the winter heating and summer cooling seasons, and the collected data showed that interior low-e storm windows installed over 74% of the window area in experimental home resulted in an 8.1 ±1.9% and 4.2 ± 0.7% reductions in HVAC energy use during the heating and cooling seasons, respectively (Petersen et al. 2015b). The energy savings results from the interior and exterior low-e storm window evaluations are summarized in Table 2, and are compared to the energy savings achieved from primary window replacement with highly-insulating triple-pane windows (i.e., R-5). The data from the highly insulating triple-pane windows were gathered using the same baseline window and represent window replacement, another option that is often considered when retrofitting windows. It can be seen that the highly insulating R-5 windows offer the largest annual HVAC energy savings. These do require a complete replacement of the primary windows. In comparison, the interior and exterior storm windows offer substantial annual HVAC energy savings without the need to replace the primary window.

High-Efficiency Cellular Shades

PNNL evaluated cellular shades during the 2015-2016 heating and cooling seasons. The specific technology examined as part of this study was a set honeycomb fabric cellular shades, which are made with six layers of fabric including two opaque layers and five insulating air pockets (Figure 4). The inclusion of insulating air pockets as well as the layer of metalized biaxially-oriented polyethylene terephthalate (boPET) that lines the air pockets, minimizes radiant heat transfer and effectively increases the R-value of the fabric.

These cellular shades were also equipped with the HD Green automated scheduling technology, which allowed researchers to evaluate both the thermal improvement of the shades and the impact of automated shading devices on optimal management of solar gains. Developed by Hunter Douglas, the HD Green Mode operation schedule is based on the solar calendar and the latitude of the location at which the window attachments are installed. The schedule is specifically designed to optimize HVAC operation and solar heat gain while allowing adequate light into the conditioned space. A local area network (LAN) is created within the home to communicate with individual shades. Equipped with built-in motors, the shades open and close per the HD Green schedules (Petersen 2015a). To evaluate the full functionality of the blinds and associated automation schedules, PNNL has tested the cellular shades when compared to a residential building without window attachments and with standard vinyl blinds. The specific experiments and the results obtained are described in Table 3.

Because data presented in Table 3 are preliminary results, each operational mode has not been annualized over both the heating and cooling seasons. Insulating values associated with cellular shades adds a reduction in total HVAC system load by a significant amount when compared to standard vinyl blinds. The maximized savings realized though the implementation of the HD green mode schedule can be seen in the optimum operation experiment.

UTILITY ACCEPTANCE FOR WINDOW ATTACHMENTS

Until recently, energy-efficient window attachments have received only modest support and recognition by utility energy-efficiency programs. Despite the demonstrated energy savings and cost-effectiveness, the Database of State Incentives for Renewables and Efficiency (DOE 2016) indicates only 18 states have about 26 utility-sponsored incentive programs between them that explicitly identify storm windows as qualified measures (Cort 2013). Recently, however, there have been some major efforts by utility programs in Vermont and the Pacific Northwest focused on integrating storm windows into utility incentive programs. In the Pacific Northwest, the Regional Technical Forum (RTF)[1] conducted a modeling study based on its Simplified Energy Enthalpy Model (SEEM)[2] and the RTF Operating Guidelines to evaluate the cost-effectiveness of low-e storm windows in the region (RTF 2015). Supported by the existing field and experimental data discussed above, SEEM results demonstrated that low-e storm windows met the criteria to be considered a "proven" and cost-effective energy-saving measure in the Bonneville Power Administration region (includes Oregon, Washington, Idaho, and part of Montana),[3] which led to approval by the RTF board for the measure in the Pacific Northwest.[4] In January, 2016, the RTF expanded the study to include manufactured homes, and the board's approval followed.[5] In 2015, Efficiency Vermont launched a successful education and awareness campaign about low-e storm windows in conjunction local retailers. The promotion resulted in a 37% increase in overall storm window sales, and a 337% increase in low-e storm window sales (Bonn et al. 2015). These cases demonstrate the feasibility of low-e storm windows as a utility incentive measure. In addition, RTF's conclusion regarding the proven status of the measure demonstrates the sufficiency and depth of the existing data on interior and exterior low-e storm windows.

Utility programs have not yet been established for other window attachment technologies, as measured field and/or

Figure 4 *Cellular shades.*

1. The RTF is an advisory committee for the Northwest Power and Conservation Council established to develop standards to verify and evaluate energy savings from technologies, approaches, systems, and measures for the Bonneville Power Administration.
2. The SEEM program is designed to model small-scale residential building energy use and consists of hourly thermal simulation and humidity simulation that interacts with duct specifications, equipment, and water parameters to calculate the annual heating and cooling energy requirements of the home.
3. Meeting minutes and RTF staff presentations are available online at http://rtf.nwcouncil.org/meetings/2015/07/minutes20150721 .pdf (accessed September 2015). [[cite this instead of footnote?]]
4. Meeting minutes and RTF staff presentations are available online at http://rtf.nwcouncil.org/meetings/2015/07/minutes20150721 .pdf (accessed September 2015). [[cite this instead of footnote?]]
5. See meeting agenda and minutes online: http://rtf.nwcouncil.org/ meetings/2016/01/. [[cite this instead of footnote?]]

Table 3. Results from Testing High-Efficiency Cellular Shades

Experiment	Description	Season	Estimated Savings*
Optimum operation	Blinds operated per the HD Green schedule. Compared to no window attachments on the baseline home.	Cooling	14.4 ±2.0%
		Heating	17.6 ±8.1%
Optimum operation comparison	Blinds operated per the HD Green schedule. Compared to standard vinyl blinds operated per the HD Green mode on the baseline home.	Cooling	10.4 ±6.5%
		Heating	16.6 ±5.3%
Static operation	Blinds remain closed for the duration of the experiment. Compared to standard vinyl blinds remaining closed for the full experiment.	Cooling	13.3 ±2.8%
		Heating	10.5 ±3.0%

* Note: Because these are preliminary results, the savings have not been annualized.

laboratory performance data is just now becoming available. Based on the experience with low-e storm windows and the tested performance data from studies, such as the PNNL Lab Homes evaluation, window attachments may also be considered for utility incentive programs. However, confirming the persistence of savings from window attachments and user-based schedules will present additional hurdles for such technologies. However, just as incentives for programmable thermostats began to be offered, it is possible that programmable shade operation and highly-insulating shades could also be considered for incentive programs, provided there is sufficient evidence that supports consistent savings over time.

WINDOW ATTACHMENT ENERGY RATING COUNCIL

One market barrier to achieving more widespread penetration of energy-efficient window attachment programs and utility incentives is the lack of standardized ratings for such products. Currently, there are not standardized metrics, test procedures, or performance criteria available for evaluating and describing the performance of window attachment technologies. Standards and ratings can help inform consumers and drive the market for energy-efficient products. Because no energy-related rating or standards existed for window attachments, in 2014 the DOE helped establish a voluntary rating council for energy-efficient window attachments, which has become known as the Attachments Energy Rating Council[6]. The purpose of the council is to develop a comprehensive energy rating, labeling, and certification program for window attachments. The AERC is run as an independent, public interest, nonprofit organization that serves the public interest by providing accurate and credible information about the energy performance of window attachments. AERC's goals include establishing a rating for two window attachments by the end of 2016.

CONCLUSION

Research conducted at the PNNL Lab Homes has begun to highlight the energy savings potential of window attachment products such as low-e storm windows and cellular blinds with automating scheduling. For residential customers, the highlighted window attachment products offer a reduction in HVAC system energy consumption without sacrificing utility or comfort. On a broader scale, high efficiency window fenestration products have great potential for reducing energy consumption in the residential sector and offer a cost-effective incentive option for utilities. The addition of energy ratings and labels for window attachment and additional support from energy-efficiency programs should help consumers, designers, and home performance contractors make informed decisions about window retrofit options, helping to save some of the vast amounts of energy lost through windows each year.

6. http://aercnet.org/

REFERENCES

Ariosto, T. and A. Memari. 2013. Evaluation of residential window retrofit solutions for energy efficiency. No. 111. University Park, PA: Pennsylvania Housing Research Center (PHRC).

Bickel, S., E. Phan-Gruber, and S. Christie. 2013. *Residential windows and window coverings: A detailed view of the installed base and user behavior*. Silver Spring, MD: U.S. Department of Energy's Office of Energy Efficiency and Renewable Energy, D&R International. http://energy.gov/sites/prod/files/2013/11/f5/residential_windows_coverings.pdf.

Bonn, L., J. Rivest, E. Phan-Gruber, J. Winer, S. Bickel, T. Stratmoen, and D. Bailey. 2015. *Lo(w-e) and Behold: Low-e storm windows provide a new way to solve the window conundrum*. Efficiency Vermont, D+R International, and Larson Manufacturing. www.efficiencyvermont.com/Media/Default/docs/white-papers/efficiency-vermont-low-e-and-behold-white-paper.pdf

CEE. 2014. Window retrofit technologies. Center for Energy and Environment. Minnesota Department of Commerce, Division of Energy Resources. COMM-0519012-53155.

Cort, K.A. 2013. Low-e storm windows: Market assessment and pathways to market transformation. PNNL-22565, Richland, WA: Pacific Northwest National Laboratory.

Culp, T.D., and K.A. Cort. 2015. Energy savings of low-e storm windows and panels across U.S. climate zones. PNNL-24826. Richland, WA: Pacific Northwest National Laboratory.

Culp, T.D., S.H. Widder, and K.A. Cort. 2015. Thermal and optical properties of low-e storm windows and panels. PNNL-24444. Richland, WA: Pacific Northwest National Laboratory.

Curcija, D.C., M. Yazdanuan, C. Kohler, R. Hart, R. Mitchell, and S. Vidanovic. 2013. *Energy savings from window attachments*. DOE EERE award #DE-FOA-0001000. U.S. Department of Energy. Berkeley, CA: Lawrence Berkeley National Laboratory.

DOE. 2016. DSIRE database. Raleigh, NC: N.C. Clean Energy Technology Center. http://www.dsireusa.org/.

EIA. 2015. Annual energy outlook 2015 with projection to 2040. U.S. Energy Information Administration. Washington, DC: U.S. Department of Energy. www.eia.gov/forecasts/aeo/.

Garber-Slaght, R. and C. Craven. 2011. *Evaluating window insulation curtains, blinds, shutters, and more*. Cold Climate Research Institute. www.cchrc.org/sites/default/files/docs/JGB_windows_article.pdf.

Huang, J., J. Hanford, and F. Yang.1999. *Residential heating and cooling loads component analysis*. LBNL-44636, Berkeley, CA: Lawrence Berkeley National Laboratory.

Knox, J.R., and S.H. Widder. 2014. *Evaluation of low-e storm windows in the PNNL Lab Homes*. PNNL-23355, Richland, WA: Pacific Northwest National Laboratory.

LBNL. 2016. *International glazing database* v. 45.0. Berkeley, CA: Lawrence Berkeley National Laboratory. https://windows.lbl.gov/materials/igdb/.

Petersen, J.M., K.A. Cort, M.B. Merzouk, G. Sullivan, and J.M. Weber. 2015a. *Evaluation of cellular shades in the PNNL Lab Homes*. PNNL-24857 Rev 1. Richland, WA: Pacific Northwest National Laboratory.

Petersen, J.M., K.A. Cort, M.B. Merzouk, G. Sullivan, and V. Srivastava. 2015b. *Evaluation of interior low-e storm windows in the PNNL Lab Homes*. PNNL-24827. Richland, WA: Pacific Northwest National Laboratory.

RTF. 2015. Regional Technical Form. *Residential: Weatherization—Manufactured home (electric resistance heat)*. Portland: Bonneville Power Administration.

Widder, S.H., G.B. Parker, M.C. Baechler, and N.N. Bauman. 2012. Side-by-side field evaluation of highly insulating windows in the PNNL Lab Homes. PNNL-21678. Richland, WA: Pacific Northwest National Laboratory.

Zirnhelt, H., B. Bridgeland, and P. Keuhn. 2015. *Energy savings from window shades*. Hunter Douglas. Boulder, CO: Rocky Mountain Institute.

Application of Glass Bubbles for Vacuum Insulation

Boyce Chang, MS **Mufit Akinc, PhD**

ABSTRACT

Glass bubbles (GB) are hollow silica glass microspheres. These materials have very high effective porosity (>90%) due to their unique structure. Composites of GB and fumed silica (FS) were evaluated for their potential application in vacuum insulation panels. Thermal conductivity measurements were performed using the transient plane source technique. Experimental results show that as much as 20% (by volume) GB can be added to FS in order to produce a composite that is similar to pure FS in terms of thermal conductivity up to approximately 1000 Pa ($1.45 \cdot 10^{-1}$ psi). This result is associated with the ability of FS to form a matrix that completely envelops the GB particles, which eliminates the large interparticle pores that are responsible for gaseous conduction at low vacuum (>1000 Pa [$>1.45 \cdot 10^{-1}$ psi]), in accordance with the Kaganer relation. Above 20% GB, macropores begin to form within the composite, leading to a more pronounced increase in thermal conductivity induced by the pressure rise. The formation of macropores occurred much earlier than the theoretical maximum amount of GB estimated from the product data sheet. This was associated with inhomogeneity in the composite originating from the method of sample preparation.

INTRODUCTION

Approximately half the total energy consumed by the average household in the United States is used for heating and cooling (EERE 2011). The rising cost of energy and strict government policies for greener technology have boosted the initiative for commercializing vacuum insulation panels (VIPs), which have 10 times lower thermal conductivity, *k*, at center of panel (edges are prone to thermal bridging across the laminate) than traditional insulation materials (Ghazi Wakili et al. 2004). This can be achieved because heat transfer under vacuum is limited to conduction at the particle or fiber contact points and radiative contribution (however, this is greatly affected by operating temperature). Yet there is a major disparity in terms of cost between traditional insulating materials and VIPs. The cost of VIPs is approximately $54/m^2 ($5/ft^2), which is 5–10 times higher comparatively due to the use of fumed silica as its core material and the need for a barrier laminate to secure the vacuum. Low-cost alternatives such as glass fibers and expanded polystyrene exist, but these materials are not suitable for long-term applications because of their large pore size distribution (Di et al. 2013; Wong and Hung 2008). Such VIPs rely on added getter to improve their otherwise short service life (Alotaibi and Riffat 2014).

Traditionally, porosity is the main parameter that dictates *k* for a given material, owing to the fact that air has a very low *k* (0.0263 W/m·K [0.0152 Btu/h·ft·°F]). It is known that in highly porous materials (>90%) heat is primarily transmitted by gas conduction (Incropera et al. 2007). However, for VIPs the pore size distribution also significantly affects *k*. Materials with smaller pore sizes are desired because they are capable of maintaining low *k* as pressure inside the panel rises. A relationship between pore size, gas pressure, and thermal conductivity developed by Kaganer (1969) is widely used to model gaseous *k* for a single pore size.

Naturally occurring materials such as diatomaceous earth and perlite have been studied for their potential as low-cost core materials for VIPs (Chang et al. 2016; Alam et al. 2014; Caps and Fricke 2000). Composites of these materials were

Boyce Chang is a doctoral candidate and **Mufit Akinc** is a professor in the Department of Materials Science and Engineering, Iowa State University, Ames, IA.

made with fumed silica in order to reduce pore size distribution. Other studies on core materials include pumice, polymer foams, and organic fibers (Mukhopadhyaya et al. 2008; Wong and Hung 2008; Yang et al. 2007; Nemanic and Žumer 2015). An alternative is to introduce glass bubbles (GB) into the composites (see Figure 1b). GB consist of hollow borosilicate microspheres, which result in a highly porous material (Beck and O'Brien 1968). Although the insulation performance of GB has been studied in a vacuum (Allen et al. 2003), dry-pressed composites (for vacuum insulation) that incorporate GB have not been investigated. In this work, we measured the thermal conductivity, k, of varying compositions of GB and fumed silica. The results were then correlated with their corresponding microstructures.

THEORY

Thermal conductivity of the gas phase is typically described using the Kaganer relation, which describes the effective pore size, gas pressure, and gaseous thermal conductivity (Kaganer 1969). The equation utilizes the Knudsen number, which is the fraction between the mean free path of particles (molecules) with the characteristic length of the system (pore size) (Kennard 1938):

$$k_{gas} = \frac{k_o}{1 + 2\beta\left(\frac{P_o l_p}{PD}\right)} \quad (1)$$

where

k_{gas} = effective gaseous thermal conductivity, written as a function of k_o

k_o = gaseous thermal conductivity at ambient pressure

β = constant that represents the efficiency of energy transfer from gas molecules to solid surface; the value of this constant ranges from 1.5 to 2

l_p = mean free path of gas particles at ambient pressure

D = effective pore diameter

P = pressure

P_o = atmospheric pressure

Variation of gaseous thermal conductivity, k_{gas}, as a function of pressure for several pore sizes is illustrated in Figure 2. Clearly, thermal conductivity is less sensitive to pressure increase in smaller pores compared to larger pores.

EXPERIMENT

Instruments

The transient plane source technique was used for k measurements. In this technique, a disk-shaped nickel thin film sensor is sandwiched between two samples and acts as both a resistance thermometer and a heat source. Power supply into the sensor is set and the change in temperature is recorded as a function of time; k is then obtained from Equation 2 (Gustafsson 1991):

$$\Delta T(\tau) = W_0(\pi^{3/2}\alpha k)^{-1}D_s(\tau) \quad (2)$$

where

W_0 = heating power supplied into sensor

k = thermal conductivity

π = sensor radius

α = dimensionless time variable

$D_s(\tau)$ = complex time-dependent function

Thermal conductivity is calculated from the slope when ΔT is plotted with $D_s(\tau)$.

This technique has been reported to produce different k values compared to steady-state techniques such as the guarded hot plate and heat flow methods. However, it has been

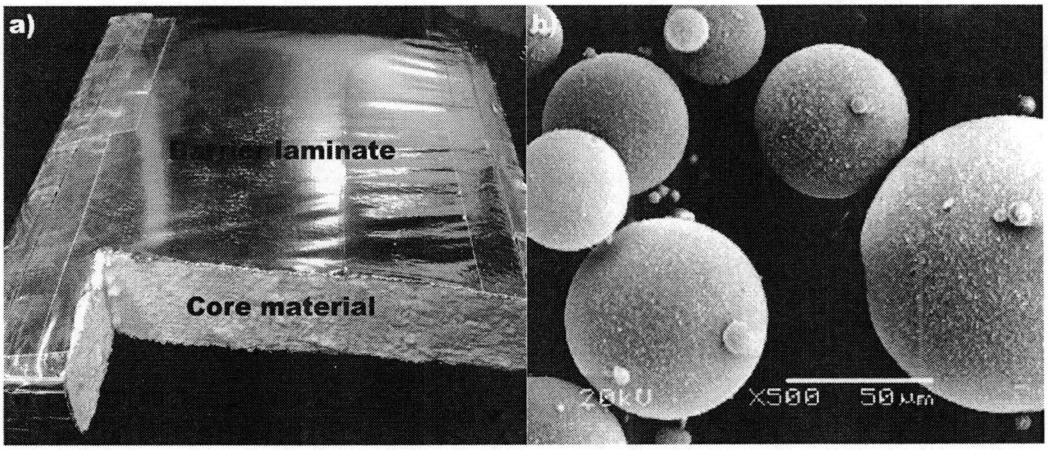

(a) (b)

Figure 1 a) Cross section of a VIP and b) scanning electron microscope image of glass bubbles.

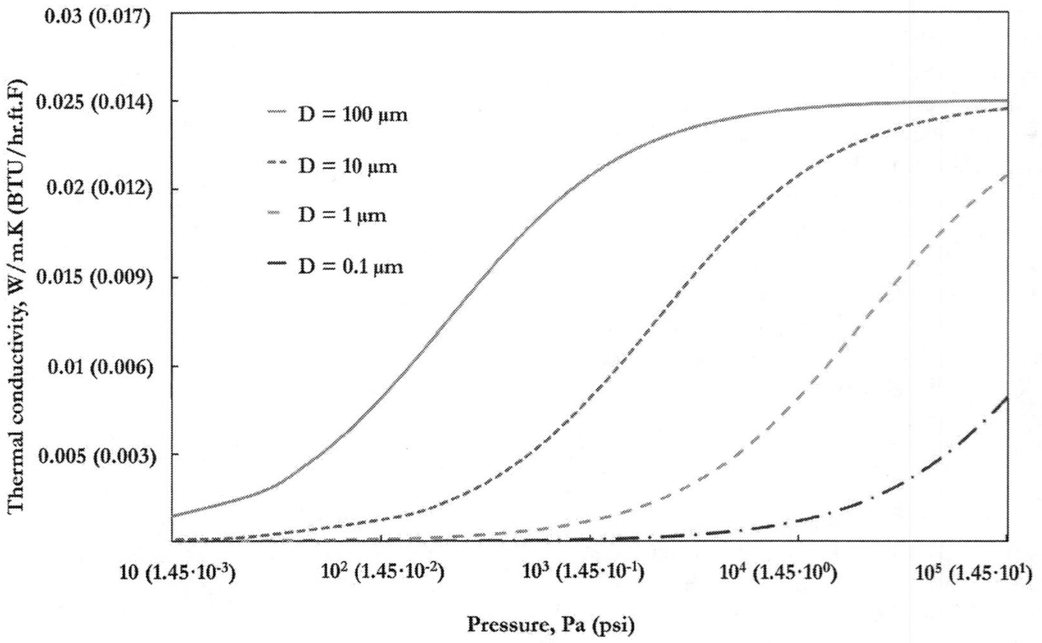

Figure 2 *Kaganer relation for several effective pore sizes,* D *ranging from 0.1 to 100 μm (4 × 10^{-6} to 4 × 10^{-3} in.).*

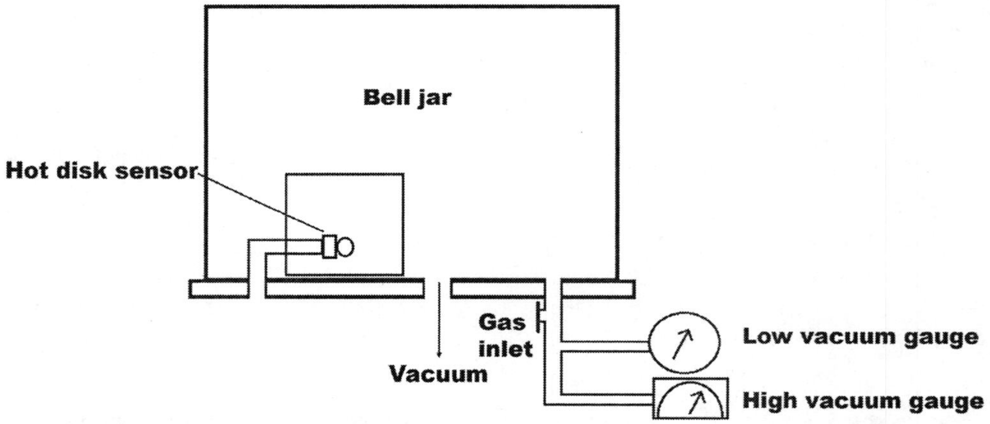

Figure 3 *Schematic of experimental setup.*

shown that the trends between both techniques are equivalent (Karami et al. 2014; Almanza 2004). A bell jar vacuum chamber was used to make measurements under variable pressure; a schematic of the setup is shown in Figure 3. The morphology of the selected samples was analyzed using a scanning electron microscope (JEOL JSM-6060LV).

Sample Preparation

The powders used in this study are glass bubbles (GB) (3M™ K20) and fumed silica (FS) (Sigma-Aldrich S5505). The characterization of these powders was done in a previous study (Chang et al. 2016). The composites of these powders were mixed using a vibratory mill (Sweco GM18L) for 12 hours and then dry-pressed into compacts at 1.76 MPa (255 psi). The compacts were 5 cm in diameter and approximately 1.5–2.0 cm in thickness. All samples were analyzed immediately after preparation in ambient air. Table 1 lists all the samples prepared. The volume fractions of the constituents were calculated using the following equation:

$$V = \frac{1}{\rho_{GB}}(m_{GB}) + \frac{1}{\rho_{FS}}(m_{FS}) \quad (3)$$

where

ρ_{GB} = average particle density of K20 GB (0.2 g/cm³)

ρ_{FS} = density of FS compacted at 1.76 MPa (255 psi)

m_{GB} = mass of GB added in the mixture

m_{FS} = mass of FS added in the mixture

V = measured volume of pressed composite

Porosity of the compact was determined using Equation 4, whereby the density of the porous sample is taken as a fraction of the density of a completely solid material:

$$\varphi = 1 - \frac{\rho_{compact}}{\rho_{SiO_2}} \quad (4)$$

where

φ = porosity of composite

$\rho_{compact}$ = density of composite

ρ_{SiO_2} = density of amorphous silica (2.2 g/cm^3), which constitutes the solid portion of fumed silica and glass bubbles

RESULTS AND DISCUSSIONS

The thermal conductivity, k, of different grades of GB (K15, K20, and K25) were measured from 26 to 101325 Pa ($4{\cdot}10^{-3}$ to 14.7 psi) (Figure 4). The three types of GB have nearly identical particle size distribution and mainly differ in terms of particle density; in other words, the wall thicknesses are greater for dense particles (3M 2009). Thus, the resulting porosities also vary with different grades. As expected, the denser the particles, the higher the k. Given that the particle size distributions of the particles are very similar, the effective pore size, which in this case corresponds to the size of the interstitial spaces between stacked particles (Figure 5), are also similar. Thus, as predicted by the Kaganer relation, the thermal conductivity versus pressure (k/P) curves overlap each other and only diverge at high pressure.

The k/P curves of composites listed in Table 1 are plotted in Figure 6. Sample FS (pure fumed silica) was included as a benchmark for comparing the performance of the composites. As mentioned in the Experiment section, the absolute values of k reported in this work should not be compared with those found in the literature due to the nonstandard method of measurement. Instead, the relative performance between composites and fumed silica, which is the standard core mate-

Table 1. Composition, Density, and Porosity of GB/FS Composites

Composite	Composition by Volume %[a]		Density, g/cm^3	Porosity[b]
	GB (K20)	FS		
47% GB	47	53	0.133	0.94
43% GB	43	57	0.142	0.94
40% GB	40	60	0.144	0.93
37% GB	37	63	0.136	0.94
34% GB	34	66	0.140	0.94
30% GB	30	70	0.154	0.93
20% GB	20	80	0.137	0.94
10% GB	10	90	0.145	0.93
FS	0	100	0.130	0.94

[a] Calculated from Equation 2.
[b] Calculated from Equation 3.

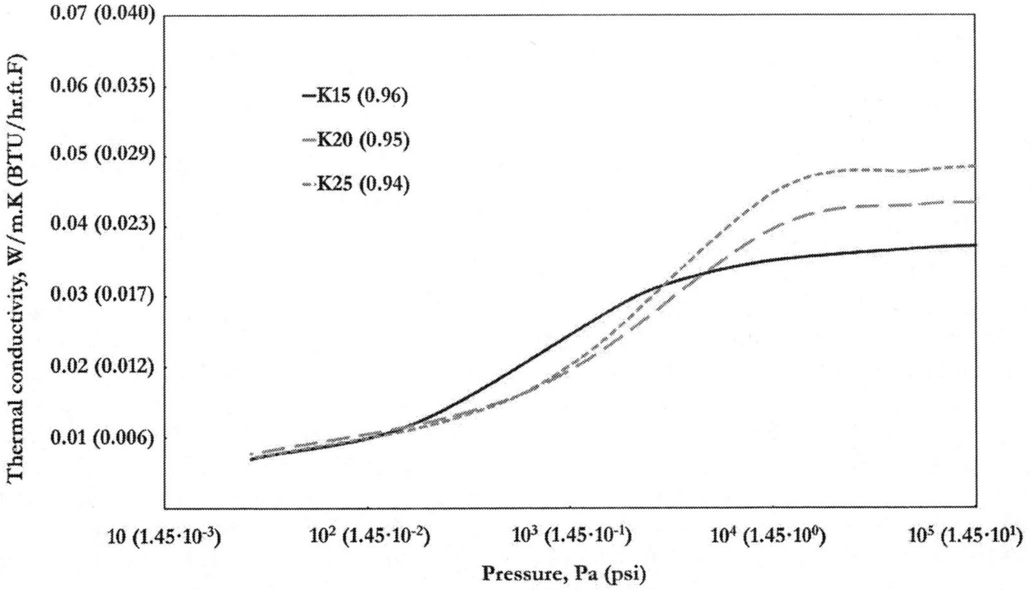

Figure 4 *Thermal conductivity versus air pressure of various GB. Parentheses in legend represent porosity of each GB grade.*

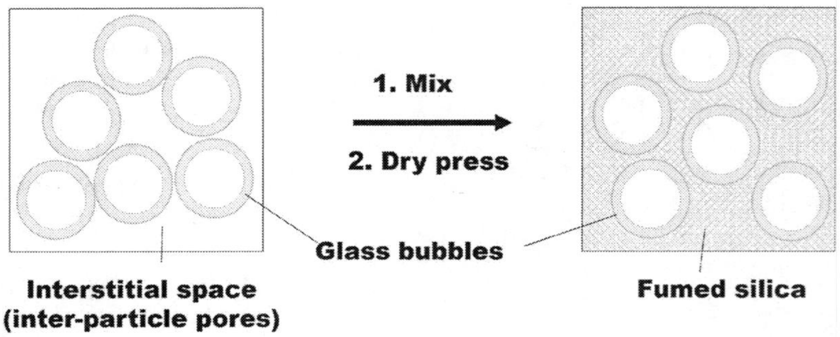

Figure 5 *Pore structure before and after forming composite. Fumed silica fills the interstitial spaces, significantly lowering the effective pore size of the composite.*

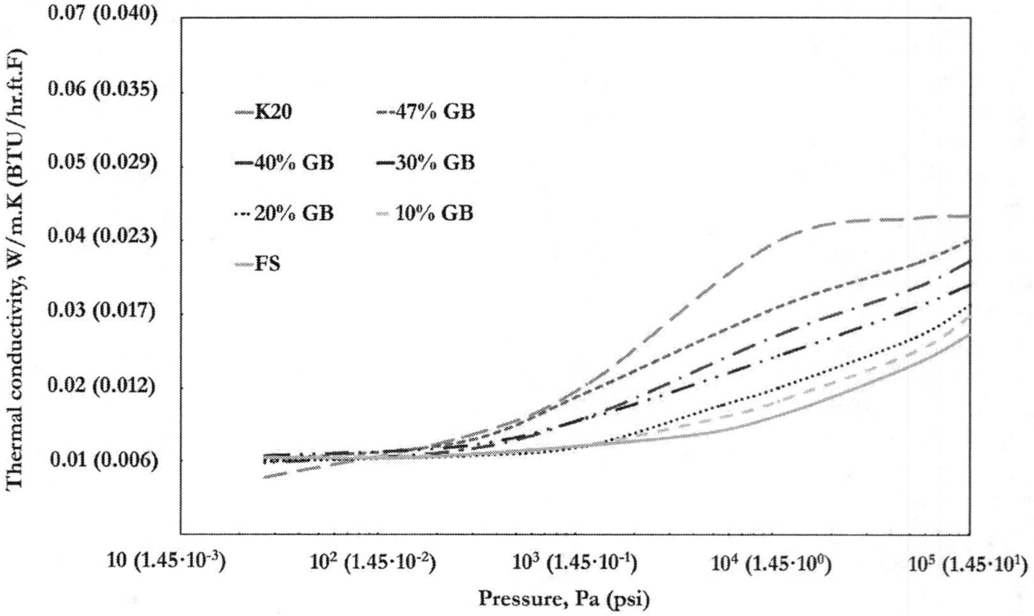

Figure 6 *Thermal conductivity versus air pressure of K20 GB/FS composites.*

rial for VIPs, provide valuable information. Because all measurements were performed in ambient air (60% rh), the effect of moisture was treated as a controlled variable.

The results indicate that up to 20% GB containing composite has similar *k* to that of pure FS up to 1100 Pa ($1.6 \cdot 10^{-1}$ psi). Beyond that composition, the thermal conductivity curves quickly deviate from that of FS, whereby the *k* of such composites are much more sensitive to the change in pressure. This behavior is highly unfavorable because it entails a lower service life, given that the level of vacuum inside the panel degrades over time due to diffusion of gases across the barrier laminate as well as the seals at the edges. Adoption of VIPs for building insulation will require long-term superinsulative performance in order for this technology to be cost-competitive. Aging studies by Simmler and Brunner (2005) have shown that for a typical multilayer metalized polymer barrier laminate, the pressure within the panel increases by 100 Pa ($1.45 \cdot 10^{-2}$ psi) per year. Assuming that the building insulation service life requirement will be on the order of decades, the critical pressure whereby composites must maintain superinsulation lies on the order of 10^3 Pa ($1.45 \cdot 10^{-1}$ psi). Here, concentrating on data collected at 1100 Pa ($1.6 \cdot 10^{-1}$ psi), we plotted *k* as a function of volume percent GB (Figure 7).

Above 20% GB, the *k* at 1100 Pa ($1.6 \cdot 10^{-1}$ psi) significantly increases until 47% GB, which is approximately equal to the *k* of pure GB at this pressure (Figure 7). Considering that the chemical compositions (silicon dioxide, SiO_2) of the composites are mostly unchanged and the porosities (Table 1) are approximately equal, the only parameter that is significantly altered in this study is pore structure—more specifically, effective pore size. The *k* of uncompressed fumed silica was measured and it appears to align with composites from 30% to 40% GB. It has been shown that the *k* of low-density

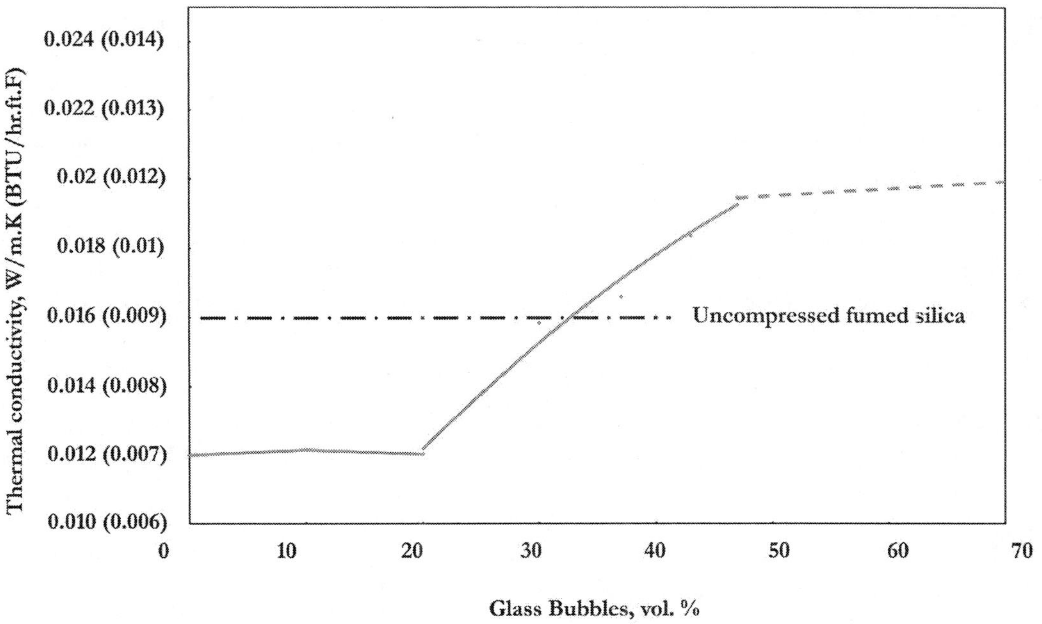

Figure 7 *Thermal conductivity of K20 GB/FS composites at 1100 Pa ($1.6 \cdot 10^{-1}$ psi). The black dotted line represents the thermal conductivity of uncompressed fumed silica.*

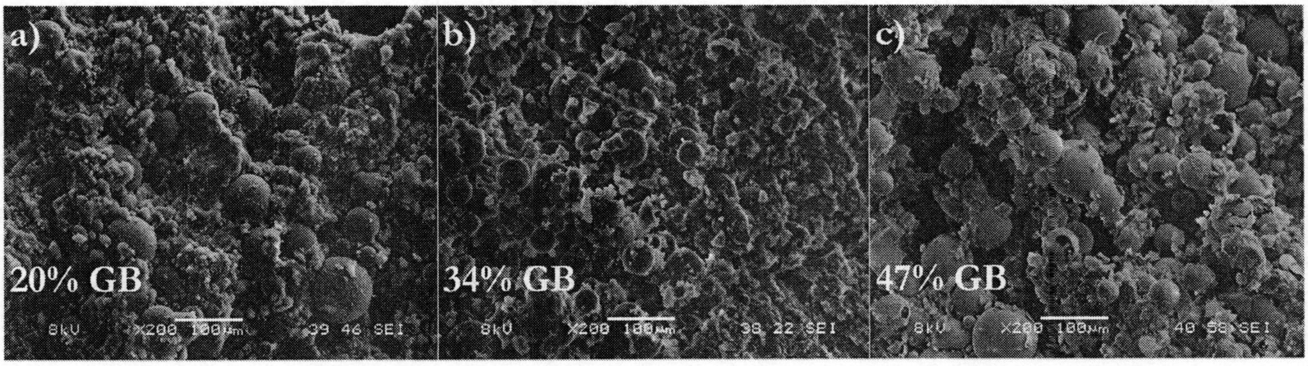

Figure 8 *SEM micrographs of selected GB/FS composites: (a) 20% GB—FS completely fills the interparticle pores, (b) 34% GB—composite contains pores that are partially filled with FS, and (c) 47% GB—interparticle pores can by clearly observed throughout the composite.*

(lightly compressed) FS is higher at low pressure due to the presence of large pores (macropores) (Kim and Song 2013). Thus, the microstructure of the composites at 20%, 34%, and 47% GB were analyzed by scanning electron microscopy (SEM). Figure 8a illustrates 20% GB, which appears to be well packed compared to 34% and 47% GB (Figures 8b and 8c). This finding verifies that the increase in k at 1100 Pa ($1.6 \cdot 10^{-1}$ psi) above 20% GB is due to the introduction of macropores into the composite.

The packing factor of GB was given by 3M (2009) to be 60%. Hence, this also amounts to the theoretical maximum volume percent of GB that can be added into the composite without introducing interstitial space (macropores). However, experimental results show that beyond 20% GB, macropores begin to develop. The discrepancy observed may be associated with the unidirectional pressing of the powders during sample preparation, which presumably promoted local inhomogeneity in the composite leading to the formation of macropores. Thus, the amount of GB could potentially be increased without sacrificing performance by introducing a uniform compression.

CONCLUSION

Three grades of glass bubbles (GB) were measured for their thermal conductivity as a function of gas pressure. As expected, thermal conductivity increased with decreasing porosity. The applicability of GB composites for vacuum insulation panels (VIPs) was demonstrated by measuring composites of GB and fumed silica (FS) under ambient conditions. It was found that up to 20% (volume) of GB can be added into FS to obtain a composite with similar thermal conductivity as pure FS up to 1100 Pa ($1.6 \cdot 10^{-1}$ psi). This result was linked to the microstructure of the composite whereby FS formed a matrix surrounding the GB particles, which eliminated large interstitial spaces (macropores) between GB particles. Above 20%, macropores begin to surface, presumably due to unidirectional compression applied during sample preparation, which leads to inhomogenous sections in the composite.

ACKNOWLEDGMENTS

This project was supported by Iowa Energy Center Grant #13-04.

REFERENCES

3M. 2009. Product information: 3M™ glass bubbles. St. Paul, MN: 3M.

Alam, M., H. Singh, S. Brunner, and C. Naziris. 2014. Experimental characterisation and evaluation of the thermo-physical properties of expanded perlite-fumed silica composite for effective vacuum insulation panel (VIP) core. *Energy and Buildings* 69:442–50.

Allen, M.S., R.G. Baumgartner, J.E. Fesmire, and S.D. Augustynowicz. 2003. Advances in microsphere insulation systems. *Advances in Cryogenic Engineering* 49 A 710:619–26.

Almanza, O., M.A. Rodriguez-Perez, and J.A. De Saja. 2004. Applicability of the transient plane source method to measure the thermal conductivity of low-density polyethylene foams. *Journal of Polymer Science Part B: Polymer Physics* 42(7):1226–34. DOI: 10.1002/polb.20005.

Alotaibi, S.S., and S. Riffat. 2014. Vacuum insulated panels for sustainable buildings: A review of research and applications. *International Journal of Energy Research* 38(1):1–19. DOI: 10.1002/er.3101.

Beck, W.R., and D.L. O'Brien. 1968. *Glass bubbles prepared by reheating solid glass particles*. U.S. Patent 3,365,315.

Caps, R., and J. Fricke. 2000. Thermal conductivity of opacified powder filler materials for vacuum insulations. *International Journal of Thermophysics* 21 (2):445–52.

Chang, B.S., L. Zhong, and M. Akinc. 2016 Low cost composites for vacuum insulation core material. *Vacuum* 131:120–26.

Di, X., Y. Gao, C. Bao, Y. Hu, and Z. Xie. 2013. Optimization of glass fiber based core materials for vacuum insulation panels with laminated aluminum foils as envelopes. *Vacuum* 97:55–59.

EERE. 2011. Buildings Energy Data Book. Washington, DC: U.S. Department of Energy, Office of Energy Efficiency and Renewable Energy, Building Technologies Program. buildingsdatabook.eere.energy.gov.

Ghazi Wakili, K., R. Bundi, and B. Binder. 2004. Effective thermal conductivity of vacuum insulation panels. *Building Research and Information* 32(4):293–99.

Gustafsson, S.E. 1991. Transient plane source techniques for thermal conductivity and thermal diffusivity measurements of solid materials. *Review of Scientific Instruments* 62(3):797–804.

Incropera, F.P., D.P. DeWitt, T.L. Bergmann, and A.S. Lavine. 2007. *Fundamentals of Heat and Mass Transfer*, 6th ed. New York: John Wiley and Sons.

Kaganer, M.G. 1969. *Thermal insulation in cryogenic engineering*. Jerusalem: Israel Program for Scientific Translations.

Karami, P., E.T. Afriyie, P. Norberg, and K. Gudmundsson. 2014. A study of the thermal conductivity of granular silica materials for VIPs at different levels of gaseous pressure and external loads. *Energy and Buildings* 85:199–211.

Kennard, E.H. 1938. *Kinetic theory of gases, with an introduction to statistical mechanics*. New York: McGraw Hill.

Kim, J., and T. Song. 2013. Vacuum insulation properties of glass wool and opacified fumed silica under variable pressing load and vacuum level. *International Journal of Heat and Mass Transfer* 64:783–91.

Mukhopadhyaya, P., K. Kumaran, N. Normandin, D. van Reenen, and J. Lackey. 2008. High-performance vacuum insulation panel: Development of alternative core materials. *Journal of Cold Regions Engineering* 22(4):103–23.

Nemanic, V., and M. Žumer. 2015. New organic fiber-based core material for vacuum thermal insulation. *Energy and Buildings* 90:137–41.

Simmler, H., and S. Brunner. 2005. Vacuum insulation panels for building application: Basic properties, aging mechanisms and service life. *Energy and Buildings* 37(11):1122–31.

Wong, C.-M., and M.-L. Hung. 2008. Polystyrene foams as core materials used in vacuum insulation panel. *Journal of Cellular Plastics* 44(3):239–59.

Yang, C.G., L. Xu, J. Wang, and F.L. Shi. 2007. Outgassing of rigid open-celled polyurethane foam used in vacuum insulation panels under vacuum condition. *Journal of Cellular Plastics* 43(1):17–30.

Big Area Additive Manufacturing Applied to Buildings

Kaushik Biswas, PhD **Randall Lind, PhD** **Brian Post, PhD**
Roderick Jackson, PhD **Lonnie Love, PhD**
Associate Member ASHRAE
Johney Green, Jr., PhD **Lucas Tryggestad, AIA** **Andrew Obendorf, AIA**
Leif Eikevik, AIA **Maged Guerguis**

ABSTRACT

Developments at Oak Ridge National Laboratory (ORNL) in Big Area Additive Manufacturing (BAAM) have enabled acceleration of new technology into the marketplace and can revolutionize the way products are designed and built. In January 2015, ORNL demonstrated how additive manufacturing can be used to drive rapid innovation in vehicles and has now applied BAAM in building envelopes. In partnership with Skidmore, Owings, & Merrill (SOM), an innovative building design was developed to incorporate low-cost vacuum-insulated panels into a three-dimensional (3-D) printed 20 × 13 × 10 ft structure. These panels use a new vacuum insulation technology, called modified atmosphere insulation *(MAI), which has been shown to achieve the same performance but at significantly reduced cost than traditional vacuum insulation. The radically new approach to integrating new technologies such as vacuum panels into building enclosure design and final assembly is described in this paper. This paper also details the design process and the final 3-D printed design.*

INTRODUCTION

Business-as-usual approaches to increase the energy efficiency of homes and businesses make it difficult to achieve the target energy efficiency and energy savings stated by the U.S. Department of Energy (DOE). Even though national laboratories, industry, and academia are developing new approaches to energy efficient solutions, continuing to push innovation into age-old construction practices that haven't changed during the last century can only yield incremental improvement. The first step to break that paradigm was the ZEBRAlliance project that brought several government and industry stakeholders together to evaluate next-generation technologies and strategies that might generate cost-effective energy savings for residential buildings (Jackson et al. 2013). After the successful ZEBRAlliance collaboration, we took a step back and asked, "How can we take the next step to not only develop innovative building technologies, but also augment the building construction practices to facilitate successful integration and market deployment?" Furthermore, because the building is part of a larger energy ecosystem that includes power generation, consumption, and storage for homes and vehicles, disruptive thinking is needed in how homes are built and designed, how vehicles are powered, and how the two consume, generate, and store energy synergistically. New industry partners and relationships working together to rapidly develop and deploy building energy-efficiency solutions are needed.

The Additive Manufacturing Integrated Energy (AMIE) demonstration project, shown in Figure 1, leverages rapid innovation through additive manufacturing to connect a natural-gas-powered hybrid electric vehicle to a high-performance building designed to produce, consume, and store renewable energy. AMIE is the realization of rapid innovation, given that all components of the demonstration were designed, tested, built, and assembled in nine months. Components of the building and vehicle were additively manufactured (3-D printed). To offset the uncertainty of power, the vehicle's natural-gas engine provides complementary power to the building. When coupled with

Kaushik Biswas is research and development staff and **Roderick Jackson** is the group leader of building envelopes research, **Randal Lind** and **Brian Post** are research and development staff and **Lonnie Love** is the group leader of manufacturing systems research, and **Johney Green, Jr.** is the director of the energy and transportation science division at Oak Ridge National Laboratory, Oak Ridge, TN. **Lucas Tryggestad** and **Andrew Obendorf** are senior architects and **Lief Eikevik** and **Maged Guerguis** are designers at Skidmore, Owings, & Merrill, Chicago, IL.

integrated demand-side controls to enable responsive loads, and then scaled up, this concept can enhance the resiliency, cost-effectiveness, and reliability of buildings worldwide.

There were five key components to AMIE: the house or enclosure, printed vehicle, secondary-use battery storage system, bidirectional wireless power transfer system, and building control and power management. By incorporating these five components, AMIE is able to produce, consume, and store renewable and conventional energy. The technologies and concepts incorporated into AMIE crosscut numerous research divisions at Oak Ridge National Laboratory (ORNL), so a team of experts was formed to cover the following vital specialties: advanced manufacturing, vehicle technologies, building technologies and sustainable electricity. Numerous industry partnerships were also leveraged to ensure market-appropriate solutions were developed and enabled. The remainder of this article focuses on the additive manufacturing and building-related aspects of AMIE.

ENCLOSURE AND ASSEMBLY

Additive manufacturing or 3-D printing has been investigated for building applications using concrete (Valkenaers et al. 2014; Gosselin et al. 2016). For the AMIE demonstration, in collaboration with ORNL research staff, Chicago-based Skidmore, Owings, & Merrill (SOM) designed an innovative single-room building module to demonstrate new manufacturing and building technology pathways to high-performance buildings. The design incorporates lower-cost vacuum-insulated panels into a 3-D printed shell. The structure was created via the Big Area Additive Manufacturing (BAAM) technology at ORNL's Manufacturing Demonstration Facility (Holshouser et al. 2013). BAAM technology enables rapid production of strong, lightweight composite parts, and it is up to 1000 times faster and capable of printing components 10 times larger than other industrial additive machines. The accelerated creation and printing of the house further demonstrated the program's function as an applied science tool to get products to market more quickly than traditional manufacturing.

The AMIE house, designed by SOM, is a single-room building module with 11 major segments (9 standard rings, 2 end walls); if desired, the length can be extended by adding standard rings to the midsection. SOM designed the house to have a footprint of 19.5 m^2 (210 ft^2) with 2.8 m (9 ft 3 in.) of headroom. AMIE's design explored the future potential for a 3-D printed panel system to condense the structure, insulation, air and moisture barriers, and exterior cladding into one vertically integrated building shell. This innovative approach allowed for material use efficiency as complex geometries with rounded and curved surfaces can reduce localized stresses. This could also lead to a future of zero-waste building construction and retrofit.

The printing was done for half rings that were combined to form full rings and then joined along the length. Approximately 80% of the house, including the segments, was 3-D printed with carbon fiber-reinforced acrylonitrile butadiene styrene (ABS) plastic composite material using ORNL's BAAM system. The printed half rings were designed to incorporate all the building systems into a repeatable module. Each C-shaped section incorporates not only the exterior wall layers, but also glazing, flooring, electrical runs, lighting and tension cables (Figure 2).

The design leverages BAAM's capacity for complex geometry, bending the plane of the exterior ring to incorporate glazing directly into the structural wall. One edge of each ring pushes outboard of the wall plane, widening the interior space and creating a glazed reveal that brings in natural light. At the same time, this small design move allows the majority of the exterior wall to remain opaque and insulated, achieving a total overall window/wall ratio of under 20% (Figure 3).

Despite the potential for complex form-making, two major constraints of the BAAM printer significantly shaped the building form. The first constraint concerned the inherent structural weakness in the z-direction of the printed structures. The plane of the printing bed formed the x and y plane. The rings were printed on their side, starting with the truss form and moving up in the z direction along the outer wall of the shell. As consecutive layers of ABS beads were deposited, they stacked in the z direction, creating the 3-D form of the print. The structure was always strongest in the x-y plane of the print and weakest in the z direction, partly because of the partial cooling of the material between layers of printing. The

Figure 1 *AMIE demonstration.*

Figure 2 *AMIE ring components and assembly.*

team designed for the stronger plane of the print by tilting the rings 90 degrees from their printed orientation, thus utilizing the *x-y* direction for vertical and gravitational forces. To mitigate any separation of beads in this *z* direction, now longitudinal to the building, it was determined that AMIE would be post-tensioned with steel rods running the full length of the building, keeping the stacked layers of ABS print in constant compression.

The second major constraint of printing governed the angle at which subsequent layers of print could be cantilevered. As ABS beads were deposited in the *z* direction they could not cantilever at more than a 40 degree angle over the bead below. Angles above 40 degrees prevented adequate support and caused slump or delamination of the ABS strands. This constraint dictated the angle of elements such as the exterior shell shown in Figure 4.

The building was rapidly prototyped, direct design to print, which enabled flexibility/power in design by removing constraints of traditional methods. The total weight of the house is approximately 13,500 lbs. At a rate of 60 lb/hr, the total print time was estimated to be four weeks: 13,500 lbs. * 60 lb/h = 225 h., not including machine setup time, etc. Figure 5 shows the printing of the ABS rings and Figure 6 shows the assembled exterior rings.

PROJECT DELIVERY

Through the process of completing AMIE, an alternative process to constructing buildings was demonstrated. Traditionally, architects design a building and provide drawings to the contractor. The contractor then prepares shop drawings for review by the architect. The integrity of the design can be affected by multiple layers of drawing review and communication exchanges.

In AMIE, the designers worked hand-in-hand with ORNL researchers and the builders to share building models in a manner that fostered real-time innovation, as the project was progressing. Rather than issuing the typical 2-D representational drawings, the AMIE deliverable was the actual 3-D digital design model, shared directly with the printing team as an IGES data file. These data were processed and fed directly into the 3-D printer. The built result was a physical manifestation of the actual design model. Any errors in the build could be immediately relayed to the design team for revision in the design model. By this delivery method, problems were quickly reviewed and resolved with input from all members of the team, greatly reducing the project timeline and eliminating chances for the design to be lost in translation. This integrated design approach allowed the AMIE team to reduce the production time by 40% and perform innovative troubleshooting in real time.

INSULATION

Lower-cost vacuum-insulated panels (VIPs), called *modified atmospheric insulation* (MAI) (Biswas 2014), were incorporated into the AMIE house walls for high thermal resistance or R-value across a small thickness (2.54 cm). A VIP offers very low conductivity (or high R/in.) compared to conventional insulation materials and has attracted interest for building applications (Alam et al. 2011; Alotaibi and Riffat, 2014; Cho et al. 2014).

MAI panels were evaluated at ORNL using ASTM C518 (ASTM 2015) and were observed to have the same thermal performance as regular VIPs. Figure 7 shows the measured steady-state thermal conductivity of MAI panels. The standard internal core pressure of MAI panels is about 10 mbar; at that pressure the thermal conductivity ranged from 0.0038 to 0.0048 W/m·K. For comparison, current foam insulation materials have steady-state conductivities in the range 0.024 to 0.032 W/m·K. Further, to evaluate the impact on the thermal performance if the internal vacuum was partially or fully lost, tests were done by varying the internal pressure of the MAI panels. The results from 0.0041 W/m·K at 12 mbar internal pressure to 0.019 W/m·K at atmospheric pressure internally. It should be noted that even at atmospheric pres-

(a) (b)

Figure 3 *(a) Joined rings from exterior and (b) joined rings from interior.*

sure, the thermal conductivity of MAI was lower than foam insulation. All thermal conductivity measurements of the MAI panels represent center-of-panel values. Like all insulation materials, thermal bridging around the insulation (due to studs, edge-effects in vacuum insulation, etc.) in the building envelope assemblies need to be considered when evaluating the effective thermal performance of MAI panels.

As a lower cost alternative to the expensive traditional vacuum insulation panels, MAI panels offer greater market deployment potential. However, unlike traditional insulation,

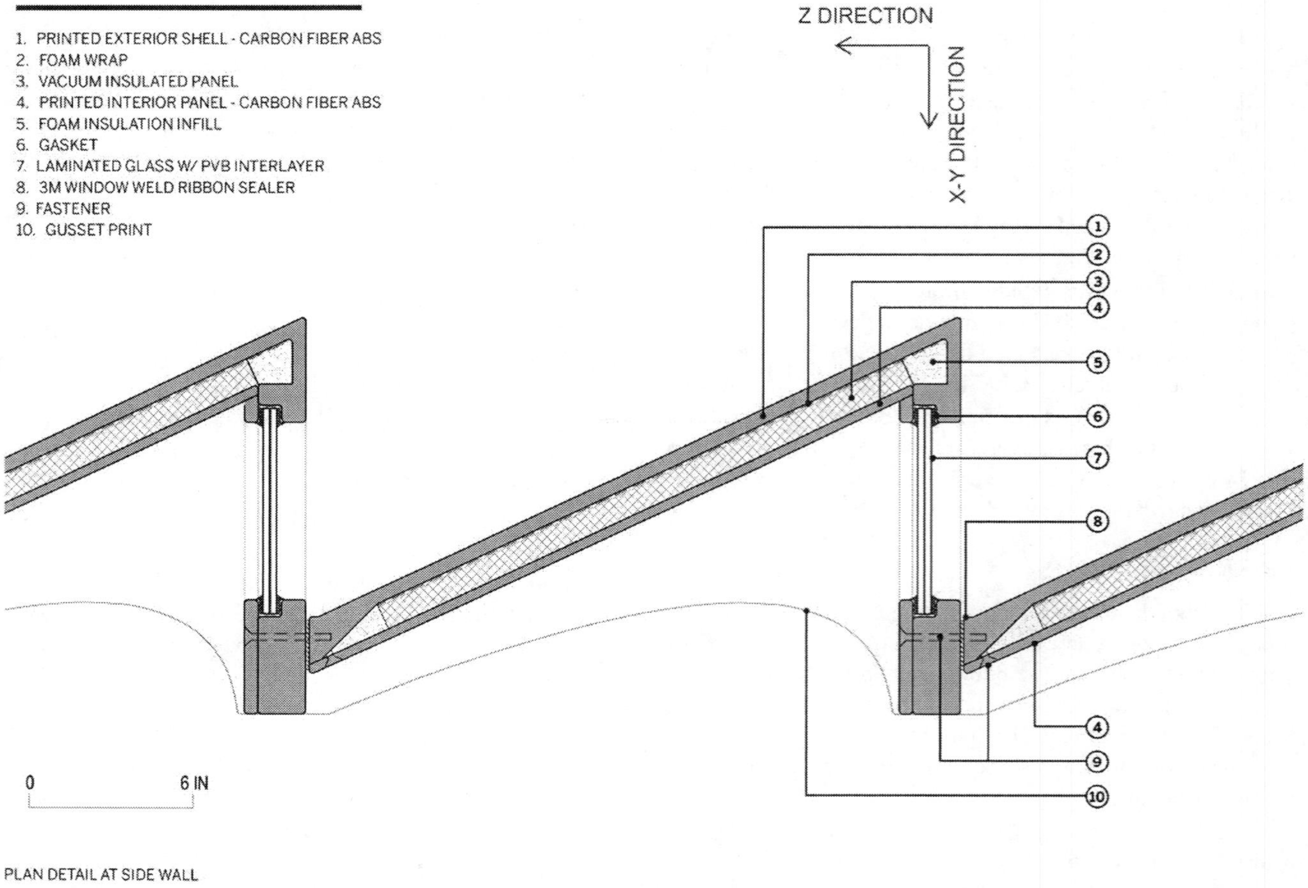

Figure 4 *Plan detail of ring connection and glazing.*

(a) (b)

Figure 5 *(a) Printing ABS layers and (b) printing AMIE ring on the BAAM.*

MAI requires additional protection to retain the internal vacuum. Although other protective schemes can be employed, the structure design in AMIE addressed this issue without the need for additional material layers incorporated either in-plant or on-site whose only role is protection. As shown in Figures 8 and 9, the 3-D printed enclosure was designed to accept the MAI panels with carbon fiber reinforced polymers 3-D printed material protection from both the interior and exterior of the building.

MAI was developed as a lower-cost alternative to the expensive traditional VIPs. One of the requisites for lower-cost, volumetric production of MAI panels is standardization of dimensions. The AMIE structure was designed such that the dimensions of the rings were constantly changing from one end to the other. Each half ring was designed to be insulated with ten MAI panels, with the insulation layout being symmetric for the half rings facing each other. During the initial design iteration, the dimensions of the 10 MAI panels were chosen to maximize MAI coverage without any gaps between the adjacent MAI panels. However, following discussion between ORNL, the insulation manufacturer, and SOM, the insulation layouts were revisited and redesigned. The MAI panels that differed in size in only one edge dimension and had edge dimensions within 1 to 2 in. were consolidated. The trade-off considered an MAI coverage that yielded an adequate overall R-value of the opaque sections while limiting the total number of different MAI panel types. The gaps between the MAI panels were filled with foam insulation boards. Table 1 lists the final dimensions and numbers of the different MAI panels.

OTHER FEATURED TECHNOLOGIES

AMIE has the potential for adoption of advanced optimized building control and power management strategies to integrate the various energy systems (photovoltaics, batteries, power grid) for routing energy while also leveraging the build-

Figure 6 *The assembled AMIE house.*

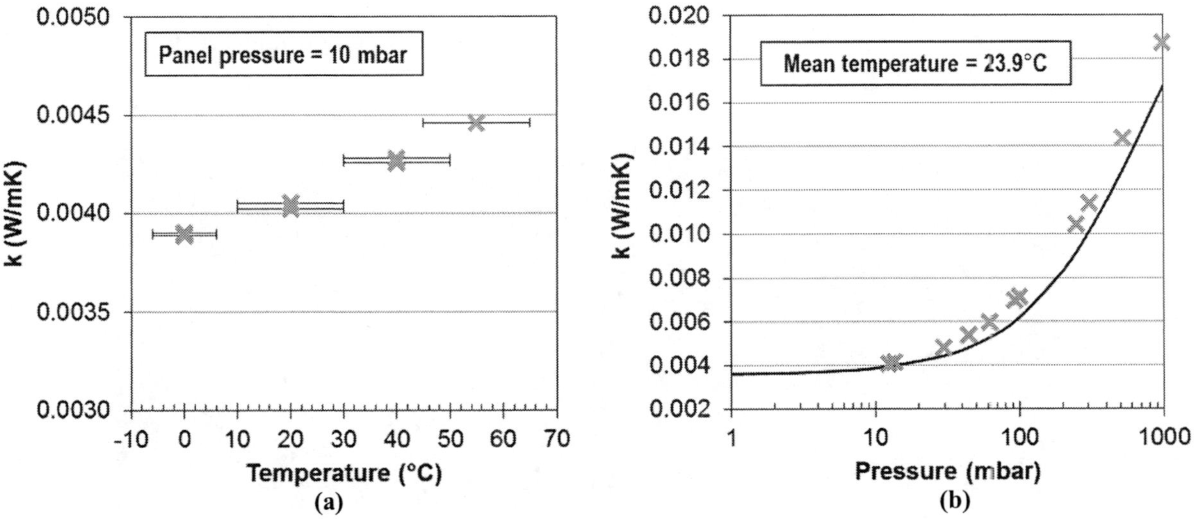

Figure 7 *Thermal conductivity of MAI panels at (a) variable temperatures and (b) internal pressures.*

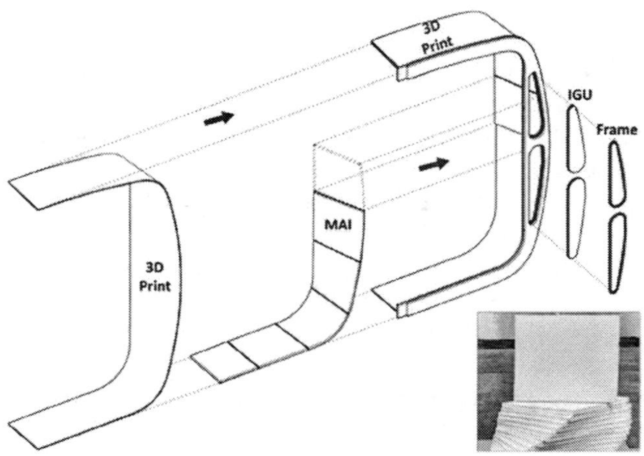

Figure 8 *MAI incorporated in the AMIE design.*

Table 1. Final Dimensions and Number of Different MAI Panels

Side lengths (cm)				Panel Number	Count
Top	Bottom	Left	Right		
50.8	50.8	50.8	53.3	1	4
50.8	50.8	50.8	55.9	2	4
50.8	50.8	53.3	55.9	3	4
50.8	50.8	53.3	58.4	4	4
50.8	50.8	53.3	61.0	5	24
50.8	50.8	61.0	61.0	6	80
50.8	53.3	58.4	61.0	7	20
53.3	50.8	61.0	61.0	8	20
53.3	55.9	61.0	61.0	9	20
55.9	53.3	61.0	61.0	10	20
				Total	200

ing secondary use battery as an energy storage system for demand-side load management (Figure 10). AMIE can serve as a flexible platform to integrate the various energy systems while also using the building as a virtual battery through demand-side load management.

Hybrid Electric/Natural-Gas Driven Vehicle

Most car engines run less than three hours each day. However, the use factor could be increased by using the vehicle as a distributed power generation resource for peak shaving or other power management scenarios in buildings. The vehicle uses an extended range, hybrid electric powertrain, and energy is generated onboard using a very small (5.5 kW) internal combustion engine natural gas generator to demonstrate the extreme downsizing potential with range-extended electric vehicle configurations. The natural gas engine is also capable of providing power to house via wireless power transfer. The vehicle can also accept and store renewable electricity generated by the house.

Renewable Power Generation

A flexible 3.2 kW solar photovoltaic system is integrated into the roof surface to generate renewable power and supplement the vehicle energy source. On sunny days, the system can meet most of the power needs; at night or on cloudy days, the occupant could rely more heavily on previously stored electricity, electricity generated by the vehicle, or pull from the grid. When electricity is plentiful, the vehicle's battery can be recharged from the photovoltaic system.

Secondary Use Battery Storage System

A battery storage system, consisting of a 24 kWh lithium-ion battery pack from an electric vehicle, is stored under the porch of the AMIE house. The system can provide self-power or grid-interactive control. Inverter control is obtained through a Labview-based (National Instruments2016) system design/control software, using multiple control modes for different types of interconnections.

Figure 9 *MAI inserted into the 3-D printed rings during assembly.*

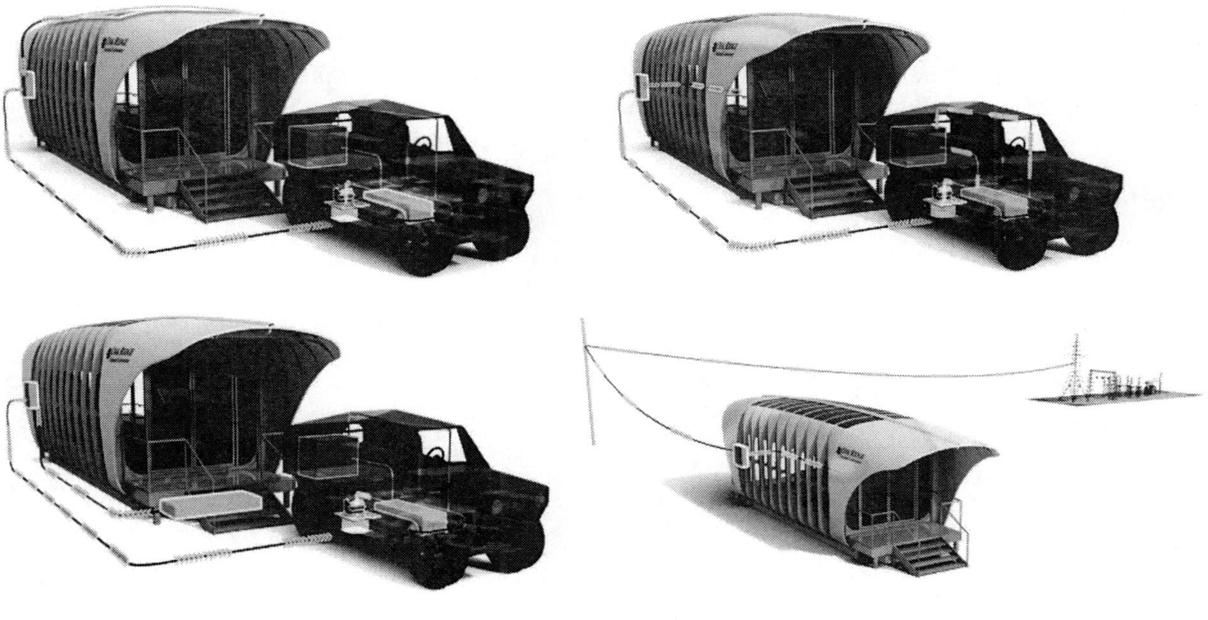

Figure 10 *Building control and power management strategies directing electrical energy flow.*

Bidirectional Wireless Power Transfer System

AMIE housed the world's first level 2 (6.6 kW) bidirectional wireless power transfer system, which was designed, built, and pioneered by ORNL researchers. The system initiates when the driver aligns the vehicle's wireless charging plates over the charging pad during parking. The intelligent control system is designed to decide which way the wireless power transfer should send energy and the origin of the energy (e.g., solar, wind, battery storage, the grid). This method offers a convenient and safer alternative to plugging the electric vehicles into wall outlets.

CONCLUSION

AMIE is the world's largest 3-D printed polymer structure that can share power with a vehicle. Equally importantly, completed in a period of nine months, AMIE is an examination of new ways and technologies to design and build energy-efficient homes that can also efficiently integrate with the grid and vehicles. For example, low-cost vacuum-insulated panels (MAI) have traditionally presented challenges for integration into traditional construction practices. However, the design of the AMIE structure offers the flexibility to incorporate the MAI panels and offer needed protection. In addition, AMIE's design explores the future potential for 3-D printed systems to condense the structure, insulation, air and moisture barriers, and exterior cladding into one vertically integrated building shell.

AMIE is a step toward exploring an integrated and efficient energy future, while also providing a flexible research platform to investigate the challenges and opportunities of integrated energy research with physical components. Printing vehicles and houses with polymers may not be part of this future, but the need to explore outside-of-the-box avenues will continue to be valuable in exploring potentially disruptive scientific achievements and inspiring creativity in multidisciplinary research and development for a cleaner future of integrated energy systems.

The engineering and science generated in AMIE also provides momentum for other examinations of how large-scale additive manufacturing can be used in buildings. For example, 3-D printing can be leveraged in precast concrete construction to provide forms for cast in architectural detail with demoldability and durability for repeated concrete casting. The high geometric control of additive manufacturing allows precise openings for fenestration, doors, and other penetrations. With industry partners, the AMIE team will continue to investigate this and other opportunities to leverage the success of AMIE to achieve energy efficient building technologies solutions.

ACKNOWLEDGMENTS

The Additive Manufacturing Integrated Energy (AMIE) demonstration project is supported by the U.S. Department of Energy's Office of Energy Efficiency and Renewable Energy (EERE) under Contract No. DE-AC05-00OR22725 with UT-Battelle, LLC. We acknowledge Dr. Martin Keller for his leadership from the original conception of this project through final completion. We also acknowledge Dr. Karma Sawyer for her support, feedback, and direction provided. We acknowledge the many contributors to the success of this demonstration project who range from industry, academia, and government research laboratories. These organizations include: Alcoa/Kawneer, Cincinnati Incorporated, Clayton Homes, DowAksa, EPB of Chattanooga, General Electric

(GE), Hexagon Lincoln, IACMI the Composites Institute, Johnson Controls, Knoxville Utilities Board, Liberty Utilities, Line-X, Mach Fuels, NanoPore, Spiers New Technologies, Techmer ES, Tru-Design, and the University of Tennessee (UT).

REFERENCES

Alam, M., H. Singh, and M.C. Limbachiya. 2011. Vacuum insulation panels (VIPs) for building construction industry—A review of the contemporary developments and future directions. *Applied Energy* 88:3592–602.

Alotaibi, S.S., and S. Riffat. 2014. Vacuum insulated panels for sustainable buildings: a review of research and applications. *International Journal of Energy Research* 38:1–19.

ASTM. 2015. ASTM C518-15. *Standard test method for steady-state thermal transmission properties by means of the heat flow meter apparatus.* Philadelphia: ASTM International.

Biswas. 2014. *Modified atmosphere insulation—Scoping study for next generation building envelope materials.* Oak Ridge, TN: Oak Ridge National Laboratory. http://energy.gov/sites/prod/files/2014/07/f17/emt60_Biswas_042314.pdf.

Cho, K., Y. Hong, and J. Seo. 2014. Assessment of the economic performance of vacuum insulation panels for housing projects. *Energy and Buildings* 70:45–51.

Gosselin, C., R. Duballet, P. Roux, N. Gaudillière, J. Dirrenberger, and P. Morel. 2016. Large-scale 3D printing of ultra-high performance concrete—A new processing route for architects and builders. *Materials Design* 100:102–9.

Holshouser C., C. Newell, S. Palas, L. Love, V. Kune, R. Lind, P. Lloyd, J. Rowe, C. Blue, and C. Duty. 2013. Out of bounds manufacturing. *Advanced Materials and Processes* 171(3):18–20.

Jackson, R., P. Hughes, G. Khowailed, and J. Green Jr. 2013. Lab takes on residential energy efficiency in the Southeast. *Home Energy* May/June.

National Instruments. 2016. Labview System Deisgn Software. Austin: National Instruments www.ni.com/labview/.

Valkenaers, H., D. Jansen, A. Voet, A. Van Gysel, and E. Ferraris. 2014. Additive manufacturing for concrete: a 3D printing principle. Proceedings of the 14th European Society for Precision Engineering and Nanotechnology (EUSPEN) International Conference, Dubrovnik 1:139–42. https://lirias.kuleuven.be/handle/123456789/459412.

Economic Efficiency of Thermal Insulation Aimed at Saving Energy: A Critical Assessment

Andreas Holm, PhD, PE **Christoph Sprengard**

ABSTRACT

Repair work, regardless of energetic improvement, costs money. Assuming that owners or investors of buildings don't see the necessity or can't afford to pay for the implementation, money becomes a severe obstacle for any energy-oriented refurbishment. The crucial question concerning all energy-saving measurements is the following: how realistic is the chance that additional costs for refurbishment, from day one of the construction period, can be amortized by a reduction of heating costs during the utilization period of the building?

Assessments of the economic viability of various measurements given by industry, the housing sector, and owners, but also by science, show significant differences. Besides climatic, building physics, geometrical, and economic parameters, the results are also influenced by the particular calculation method. Depending on the calculation approach, different statements concerning the amortization of energy-saving methods will be reached. This leads to fierce public discussions.

Very often, "pseudo" calculation methods are being used. This means inaccurate, incomplete, or only estimated calculation methods. However, one has to admit that simple approaches cannot be avoided because exact cost ratios are missing or cannot be achieved within a reasonable cost framework. Important in this case would be an open statement of the simplifications and a discussion of the resulting consequences.

In general, type, volume and execution of restoration measures have to be exactly optimized to the building. For this purpose, a competent, individual expert analysis by a qualified energy-efficiency specialist is needed. A careful survey will enable the expert to develop a restoration schedule, *which describes the chronological order of measurements that will make good sense and define saving potentials. One also has to be aware that insulation measures are especially profitable when repair measures, for example, of the roof or the renovation of the plaster of a facade, are due. If energetic improvement measures are being integrated into a general modernization, one-off costs such as scaffolding costs, site setup, and rubble trough are due only once. Therefore the costs for the original energetic renovation are reduced.*

This study clearly shows that the evaluation of the economic efficiency of a restructuring measure with different methods and specifications is possible. Based on the examples and variations, the strong impact of different parameters becomes clear. To make a general statement concerning economy and profitability of a restoration, for example, additional insulation of an outer wall, there are a lot of unknown results. First, the boundary conditions that determine the energetic savings potential such as U-factor in the building stock, internal and external climatic conditions, and the energetic state of the other building parts are decisive. Besides the project specifications, universal but unsteady parameters such as energy prices and rise in energy prices are of importance. A further factor to be noted is that, in the end, it the inhabitant and their behavior that mostly influences the efficiency of the building. They determine the energy consumption with regulation of heating, room temperature, and ventilation.

All of these parameters are necessary to come to a conclusion concerning economic evaluation. Still, it is difficult to generalize them only to come to a universal statement concerning the economic efficiency of insulation measurements. The margin of fluctuation of the results, assuming various positive and negative scenarios, that shows the limits of plausible and realistic parameter variations is relatively large.

Andreas Holm is director of FIW Munich and professor of the University of Applied Science, Munich, Germany. **Christoph Sprengard** is department leader at FIW Munich.

INTRODUCTION

By increasingly negative reports in the media, the economical benefit of thermal protection measures is currently discussed controversial. Therefore, this study will examine the conditions under which thermal insulation measures becomes economical. It is important to remember that these investments may not immediately lead to an economic success. Usually their profitability can only be seen in longer-term cost savings. This means that the calculations to be carried out and temporal changes in the calculation parameters must be included (for example, the price of energy or the interest rate), which naturally have development uncertainties in their future. A deterministic calculation with fixed parameters would not consider these effects. The impact of these uncertainties on the accuracy of the calculation result can be estimated by variation of calculation parameters by dynamic time series and special stochastic methods.

To perform return on investment (ROI) calculations, a number of (time-dependent) calculation parameters such as the development of energy price, the real interest rate, or the desired service life must be known. Further information concerning the construction costs and especially to energy-related charges are needed, too. These data are provided in the present work first. Then the different calculation methods have to be explained, because these—depending on investor interest—can lead to different results. Afterwards the effects of the individual calculation parameters and their influence on the economy result must be checked. An example of application completes this study.

TIME-DEPENDANT CALCULATION PARAMETERS

Energy price

If energy-saving targets are met, heating costs are lowered for building users and residents by means of renovation measures and a reduction in the heating requirement. To take these costs into account over the entire useful life of the building, the trend in energy prices over time is of great importance. Thermal energy savings can be converted into cost savings with the help of estimates and assumptions relating to the annual energy price increase. It is important to note that long assessment periods lead to a relatively high level of uncertainty with regard to energy price trends. Figure 1 shows the annual expenditure on energy (light, process heat and central heating with hot water) of a private household in Germany for the period from 1996 to 2012. The graph initially shows a clear upward trend. If the current energy costs for central heating and hot water from 2012 are compared with those from 1996, we can see a price increase of just under 50%.

Figure 2 plots the domestic consumer prices per kilowatt of energy since 1990. The prices for electricity and gas have more than doubled over this period, and in fact have more than tripled in the case of fuel oil. If we look at the price increase curve, it is noticeable that the consumer prices were relatively stable up to the year 2000. Initially, the energy consumption price remained relatively constant, the energy prices then began to show a significant increase. The oil price is currently showing a trend towards lower prices again. In 2014, the price per 100 L remained virtually constant at €80. Since September 2014, however, the price has fallen and was €55 per 100 L in January 2015. This, of course, has an impact on the profitability of renovation work.

Real Interest Rate

Energy-saving investments must be seen in relation to the expected interest yields that an investor would receive if they placed their money in the bank rather than investing in thermal

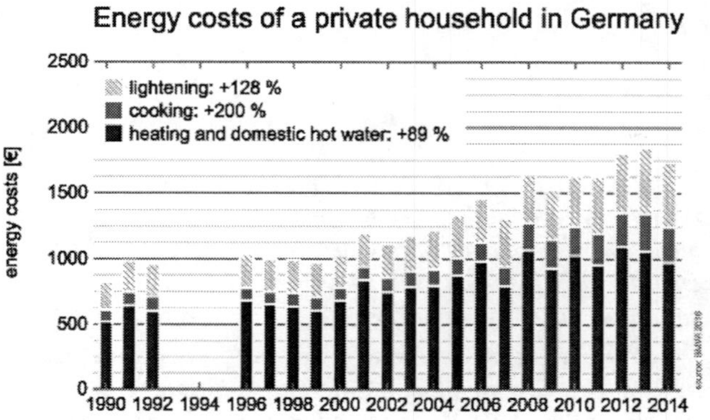

Figure 1 Trend in energy costs of a private household in Germany according to BMWi (2016). The percentage increase is related to the year 1990.

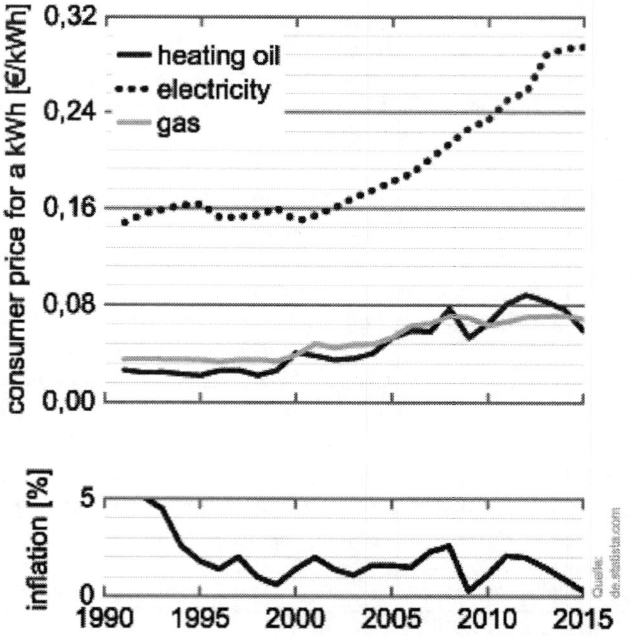

Figure 2 Trend in consumer prices for energy in private households in Germany compared with the trend in price inflation in the same period, according to Statistisches Bundesamt (2016).

insulation. The most cost-efficient option can be assessed using real interest rates and what is known as *current yield*. The current yield is the average yield of all fixed-interest domestic federal bonds in circulation with a duration of at least 3 years and a maximum of 30 years. The real interest rate is a frequently used concept. Here, the inflation rate (Figure 2) is subtracted from the nominal interest rate to allow inflation-adjusted analysis. The real interest rate trend is particularly well-suited to illustrating fluctuations in the cur-rent yield. For this interest rate, there are long-term time series which show the interest rate cycles with upswing and downturn phases. During the 2008 financial crisis, real interest rates in Germany fell to below 2% per annum. Real interest rates were in fact negative for several months. In this case, banks do not offer interest on cash deposits, but charge interest instead.

Useful Life

Because evidence of the trend in energy prices and interest rates becomes increasingly unreliable as time periods extend, lifespans and/or renovation cycles and consequently the useful life of building parts play a key role in long-term cost-efficiency studies. Normal lifespans and/or renovation cycles, which are approximately 30 to 60 years, are well above the typical mortgage term of 10 to 20 years. Based on information gathered over many years, it is generally assumed that insulation materials have a long lifespan and that if insulation material is properly fitted, it will not measurably change for more than 20 years. Therefore, lifespans of 50 years are used in standardization and when issuing technical European approval certification.

ADDITIONAL COSTS OF THERMAL ENERGY SAVING

Energy-saving measures cost money. Costs become a barrier to energy-related renovation if owners or investors are not conclusively convinced that the financial investment is profitable. The key question in all energy-saving steps is whether the additional costs to be paid during the renovation can be recouped by reducing fuel costs in the course of the useful life of the building.

As with cost-efficiency calculations in general, the cost groups to be considered must also be defined. These form the basis of the calculation of an economic key indicator, which describes the cost-efficiency of an energy-saving measure according to the calculation method. In the course of the useful life, the investment constantly generates incidental income and expenditure flows, which should lead to an amortization of the investment made. Building and material costs are key to a cost-efficiency analysis of an investment in insulation measures. These can be estimated by obtaining a variety of quotes for specific projects.

DETERMINING THE COST-EFFICIENCY OF ENERGY-SAVINGS MEASURES

Basic Principles

In principle, cost-efficiency analysis methods can be subdivided into static and dynamic methods. Static methods are characterized by the fact that the time factor is not factored into the analysis. In the case of dynamic methods, however, the changes in interest and compound interest over the analysis period are factored in. Based on the findings of Werner and Gertis (Werner and Gertis 1976a, 1976b, 1976c; Gertis1977, 1978), the following analysis illustrates key calculation methods for instances where running costs are constant every year or change over time as a result of constantly rising energy prices. The various calculation methods may of course vary still further. Time-varying factors (e.g., different lifespans of some building sections) could also be considered or the various methods could be combined. However, these calculation versions are not considered here, because this study is concerned with illustrating the central differences between the individual methods.

The methods of calculating the cost-efficiency of energy-saving measures have already been presented in detail through the work of Werner and Gertis (Gertis 1977). Five methods are considered especially important; these are as follows:

I. Static amortization (simple amortization)
II. Dynamic amortization without changes in energy prices
III. Dynamic amortization with changes in energy prices
IV. Net present value and annuity
V. Internal rate of return

The individual methods stem from the different mindsets of the investor. It is relatively difficult to settle on a specific method, because investors have adopted different approaches which have also proven successful for them in practice. Different amortization models have been implemented and changed in public and/or publicly funded residential construction projects over the years, the effects of which are reflected in time series. The investor's decision to use a specific method also therefore depends on the tax treatment of his asset and is therefore influenced to some degree by tax advisors. These widely varying and sometimes extremely subtle factors are a key reason why the cost-efficiency of thermal insulation measures sometimes fluctuates and why they are therefore considered to be controversial.

Selected Calculation Methods

Figure 3 shows the calculated amortization periods for calculation methods I to V as a function of the additional cost benefit ratio. The assumptions on which this example is based are a current energy price of 0.08 €/kWh, an annual energy price rise of 5%, as well as imputed real interest rates of 1%, 3%, and 5%.

The calculated amortization periods increase as the additional cost benefit ratio rises. The change in the real interest rate has no effect on the result produced by method I (static amortization). While growth is constantly linear using method I, method II produces a progressive curve, the function of which is strongly influenced by the assumed real interest rate. The amortization periods in method II increase if the additional cost benefit ratio is poor and the real interest rate is higher. This is because the energy price rise, which generally has a positive impact on the amortization period, has been neglected. Methods III, IV, and V are dependent on the cost benefit ratio and show a slightly declining curve. Method IV is also relatively strongly influenced by the real interest rate. It is striking that the amortization periods calculated according to all five methods do not differ greatly up to an additional cost benefit ratio of approximately 0.5.

ADDITIONAL COST BENEFIT RATIO (MNV)

The payback period may also be expressed as *additional cost benefit ratio* (MNV). This is defined as the ratio of the additional cost of improved thermal insulation (energy-related costs) and the annual heating energy savings. The MNV therefore describes the energy-related costs per saved kWh of heat energy. Accordingly, the following applies: the smaller the MNV, the more effective and economical a retrofitting measure is.

EFFECT OF THE CALCULATION PARAMETERS ON COST EFFICIENCY

Effect of the Energy Price

Method IV (net present value and annuities methods) and/ or method V (internal rate of return) are used in the following diagram (Figures 4–5), because they come closest to the mind-sets of today's investors. Figure 4 shows the effect of the energy price rise and the energy price at the time of the investment on the amortization period. Three different investments for insulation measures, corresponding to three different additional cost benefit ratios, have been examined. The effect of the energy price rise on the amortization period can be clearly seen from the diagrams. We can also see that the effect of the energy price rise is reduced by higher initial energy prices. Conversely, clear differences in amortization periods as a function of the energy price rise can be seen if the initial energy prices are low. If the additional cost benefit ratio is rising, e.g., if the investment costs for a measure are high, the energy price rise has a greater impact than if the additional cost benefit ratio and/or the investment costs are low. If the additional cost benefit ratio is, for example, 0.3 €/kWh/a and the current energy price is 0.08 €/kWh, the energy price rise has barely any effect on the calculated amortization period. In the case of a high additional cost benefit ratio of 0.9 €/kWh/a and the same current energy price, an energy price rise of 7.5% per annum reduces the amortization period by approx. one half compared to the calculation without a price rise.

Effect of the Insulation Thickness and U-Factor

Figure 5 shows the additional cost benefit ratio curve for both renovation variants as a function of the achieved U-factor when insulating a facade in a thermal insulation composite system with mineral wool and expanded polystyrene (EPS). Both rise relatively sharply from a U-factor of approx. 0.3 $W/m^2{\cdot}K$.

DIFFERENTIAL SENSITIVITY ANALYSIS

Using a differential sensitivity analysis, the effect of the various parameters on the amortization period of insulation is to be examined initially. This is to show how sensitively the result of a cost-efficiency calculation made on the basis of quasi-reliable data reacts when some of the input data is varied. Table 1 shows the influencing parameters with the examined minimum and maximum values.

Figure 6 shows the effect of the minimum and maximum values of individual parameters on the amortization period as a function of the additional cost benefit ratio. The blue curve

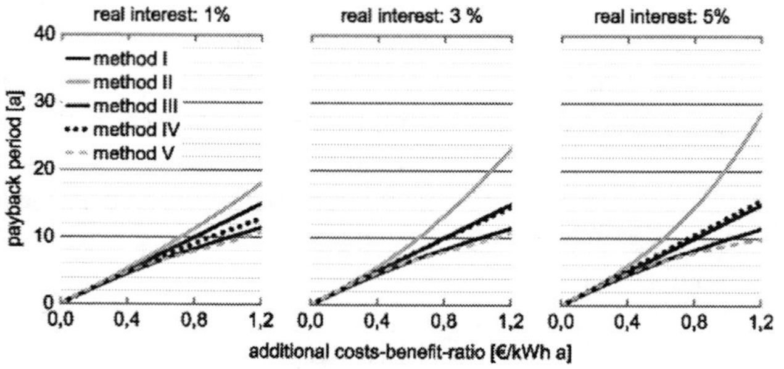

Figure 3 *Amortization period, calculated using various methods, as a function of the additional cost benefit ratio. Assumptions on which this example is based: energy price 0.08 €/kWh, annual energy price rise of 5%, as well as imputed real interest rates of 1%, 3%, and 5%.*

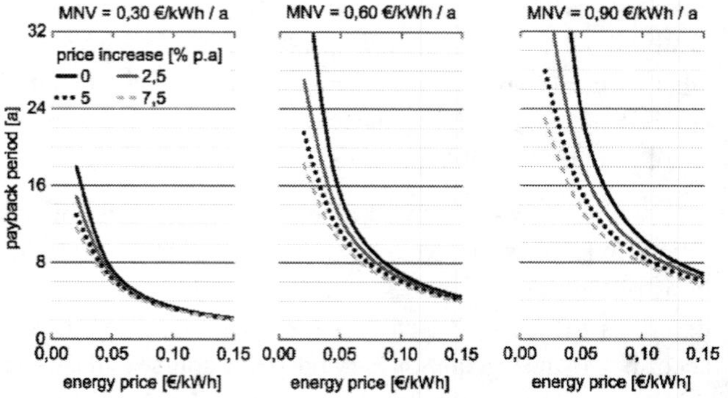

Figure 4 *Amortization period over time as a function of energy price in €/kWh and rise in energy prices for various levels of investment. Assumption: U-factor of existing housing stock = 1.4 $W/m^2{\cdot}K$; U-factor after renovation = 0.24 $W/m^2{\cdot}K$.*

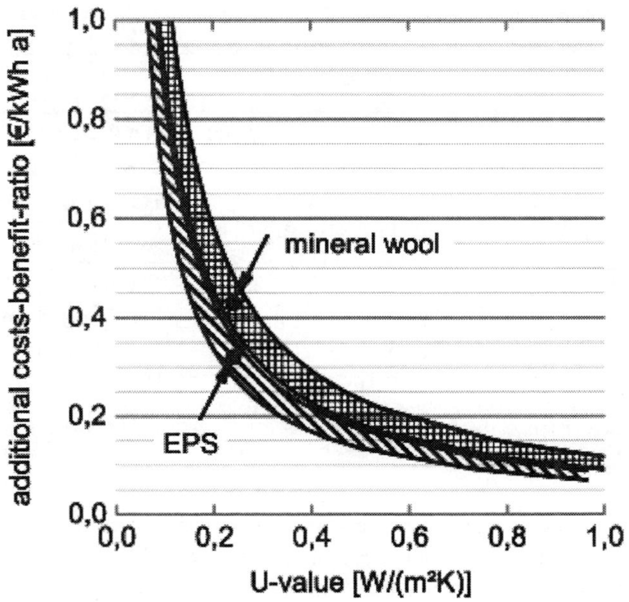

Figure 5 *Additional cost benefit ratio for a thermal insulation composite system with mineral wool or expanded polystyrene as a function of the achieved U-factor. U-factor of existing housing stock = 1.4 $W/m^2{\cdot}K$, U-factor after renovation = 0.24 $W/m^2{\cdot}K$.*

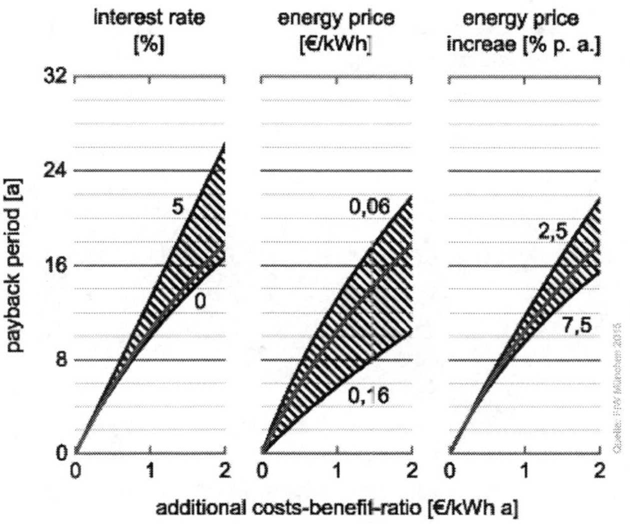

Figure 6 *Effect of the additional cost benefit analysis (investment costs) on the amortization period, taking into account the fluctuation ranges of various parameters as a function of the additional cost benefit analysis.*

Table 1. Ranges of Various Influencing Parameters Examined for the Sensitivity Analysis

Influencing Parameter	Unit	Minimum	Mean	Maximum
Annual thermal energy-saving by insulating the building section	kWh/m^2	68	85	102
Interest used for calculation	%	6.0	2.0	1.0
Inflation rate	%	1.0	1.0	1.0
Real interest rate	%	5.0	1.0	0.0
Initial energy price	€/kWh	0.06	0.08	0.16
Energy price rise	%	2.5	5	7.5

represents the result with the mean of all parameters. Only the studied parameter was changed for individual curves; the defined means were applied to the other parameters. It is evident that the real interest rate and energy price rise have a major effect on the amortization period of an investment. The average value applied to these parameters also lies in the upper and/or lower range of results. These parameters should be selected extremely carefully in order to obtain a realistic result.

MONTE CARLO ANALYSIS

The approaches addressed thus far essentially allow us to form a picture of the type and extent of the uncertainties relating to them. This type of sensitivity analysis however does not provide conclusions regarding the probability of a change occurring to the examined parameter changes. The question of what conclusions can be drawn from sensitivity with respect to the cost-efficiency analysis of energy-saving renovations, also remains open.

Stochastic simulation methods can provide assistance here. This technique is referred to as *Monte Carlo Analysis* (MCA) or *probability-based analysis*. This involves generating possible combinations of uncertain parameter values through artificial random experiments. The reliability of the analysis rises with the number of repeated calculation runs. To consider the effect of the parameters listed in Table 1 and their range on the amortization period, a Monte Carlo simulation was performed for the amortization period as a function of the investment costs. The simulation was repeated several thousand times, the characteristic value of each parameter being selected at random from the predefined range during each simulation.

The result of this was a certain range, as shown in Figure 7, which was determined by using a confidence interval of 95%. This means that 95% of all simulation results lie within this range. Obvious outliers are excluded. The effect of investment costs on the amortization period under the minimum values described in Table 1 was also determined and shown as a worst-case scenario. This curve is clearly steeper,

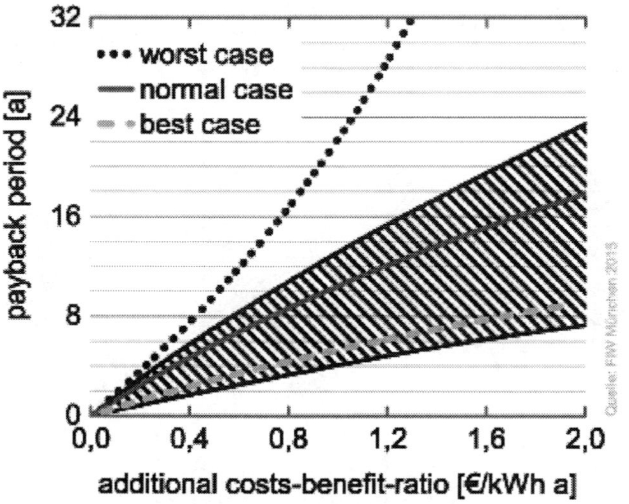

Figure 7 *Effect of the additional cost benefit ratio (investment costs) on the amortization period, stating the confidence interval for the MCA and showing the best- and worst-case scenarios as a function of investment costs in €/m^2. The mean amortization period is shown as a solid bold line.*

which can be attributed to the fact that selecting all worst-case scenarios by randomly selecting from the parameter range during the Monte Carlo simulation is relatively improbable. By contrast, a best-case scenario was created by applying all maximum values, its curve clearly lying below the confidence interval of the MCA.

Table 2 shows the 95% probability range of the possible amortization period for external wall insulation with a thermal insulation composite system. The mean is also given for comparison. Table 6 shows that the amortization period may vary from around 2 to 7 years. The maximum amortization period of 7.2 years involves investment costs of €150/m².

The Monte Carlo calculation of other cases produced the minimum and maximum values shown in Table 2. The wide range of values show that sweeping statements cannot be made about the amortization periods due to the uncertain parameter trends. If such statements are made, reference must also be made to the calculation parameters on which the statements are based. It must be accepted that the calculation parameters are generally associated with uncertainties.

SUMMARY

The cost-efficiency of energy-saving thermal insulation has become a much discussed and, in part, controversial topic. Although increasingly negative reports in the media have repeatedly cast doubt on the cost-efficiency of thermal insulation measures, observations of this type are often based on “mock” calculation methods. This means that the calculation methods are imprecise, incomplete, or only approximate. However, it must be admitted that straightforward approaches are often essential in instances where no precise cost relationships and correlations are available, or if these cannot be obtained at reasonable expense. In this instance, it is important to clearly represent the simplifications applied and to hold a critical discussion of the resulting consequences.

Table 2. Calculated Additional cost benefit Ratio and the Resulting Amortization Periods According to Investment Costs*

Investment Costs, €/m^2	Additional cost benefit Ratio, €/kWh	Amortization Period			
		Deterministic Approach	Range with 95% Probability		
			Mean	From	To
90	0.32	3.6	3.0	1.9	4.5
120	0.42	4.8	3.9	2.5	5.9
150	0.53	6.2	4.8	3.2	7.2

* Assuming an energy-related cost share of 30% based on the results in Figure 7. The calculation is based on a renovation with an initial U-factor of 1.4 W/m^2·K, which is improved to meet the 0.24 W/m^2·K currently required by EnEV.

This study shows the strong influence of different key indicators on an analysis of the cost-efficiency and profitability of renovation measures such as retrofitting insulation to exterior walls. In order to make a general statement, a wide range of parameters need to be taken into account. Firstly, calculation parameters and constraints are key in determining the potential energy saving, such as the U-factor in existing housing stock, outdoor and indoor climate, and the energy-efficiency of the other parts of the building. Secondly, and equally importantly, financial constraints are also required, such as actual renovation costs, borrowing costs, runtimes, and intended useful life. Alongside this project-specific information, parameters that change over time, such as the energy price trend and increase and real interest rate trend, also carry weight. All of these parameters are essential before a statement can be made on the cost-efficiency analysis; it is not only extremely difficult but also inappropriate to generalize these parameters to make generally valid statements about the cost-efficiency of insulation measures. This means that an accurate knowledge of the input data is required in order to make sufficiently accurate statements on the cost-efficiency and sense of an energy-efficiency-oriented renovation measure.

Of the currently 18.2 million residential properties in Germany, just under 9 million have had no improvements made to their thermal insulations, or only slight improvements (BMWi 2016). This means that virtually every second period property in Germany has insufficient thermal insulation. If the current annual rate of renovation of less than 1% (to be more precise, around 0.9%) is maintained, it would take more than 100 years to complete the renovation of the existing housing stock of period properties. This seems ludicrous. Instead, it is possible to demand that each generation performs one-off renovation (with each generation's active period being 30 years). However, this would require the annual renovation rate to increase to at least 3%.

A series of (time-related) calculation parameters such as energy price and real interest rate trends, as well as the target useful life, is required for cost-efficiency calculations. Details of the building costs and especially the energy-related costs are also needed. These calculation parameters are presented in this study. Although energy prices initially increased significantly in the past, they have fallen again recently. During a 30-year useful life, they will probably be subject to vast changes. The real interest rate (i.e., current yield) changes significantly over time. Recently, it has even become negative, meaning that no revenue is achieved if money is placed in a bank account, and that it may even be necessary to pay interest on that deposit.

With respect to renovation costs, a distinction must be made between business-as-usual and energy-related costs. If a specific action is required regardless, such as fixing a crumbling facade or leaking roof etc., thermal insulation should also be improved alongside this business-as-usual renovation. The coupling of these two should be a matter of course for every investor, even if investor decisions are often based on different cost-efficiency calculation models. Depending on their monetary attitude, an investor has five calculation methods available, namely:

- static amortization
- dynamic amortization without changes in energy prices
- dynamic amortization with changes in energy prices
- net present value and annuity
- internal rate of return

Questions relating to amortization also play a role for brokers when brokering purchases or lettings; they should—at least for clients who take a long-term approach—use the cost-efficiency of thermal insulation as a form of advertising. To enable clear statements to be made and, with otherwise equivalent propositions, to avoid diverging results, agreement must be reached on a specific calculation method, with potential government aid and fiscal capital allowances being taken into account where applicable. In this respect, as is also the case in the present study, the net present value and internal interest rate methods are favored in most cases.

Given that cost-efficiency results depend on a very wide range of, in part, conflicting calculation parameters, and given that these parameters change over time through the course of useful life, a sensitivity analysis has been applied. This analysis has shown how sensitively the cost-efficiency statement reacts to parameter variation. Although differential analysis provides the fluctuation margin of a statement, it does not provide the probability of the selected parameters occurring. Stochastic simulations (e.g., using the MCA) are required to include the probability of occurrence. These use artificial random experiments to show which statements generate uncertain parameter combinations. The reliability of the statement rises with the number of repeated calculation runs. Several thousand calculation repetitions were completed in this study. Condensed into confidence intervals of 95%, amortization periods of between 2 and 30 years were created here for renovations with thermal insulation measures, and the high value of 30 years corresponds to a single-generation period. Essentially, it can be stated that it is possible to renovate the thermal insulation of the existing, period property housing stock under these conditions if the annual renovation rate increases to 3%. If this does not take place, period property renovation—and with it large portion of the energy turnaround—will fall by the wayside.

REFERENCES

BMWi. 2016. Bundesministerium für Wirtschaft und Energie. Zahlen und Fakten. Energiedaten. Berlin: Nationale und internationale Entwicklung.

Gertis, K. 1977. Was bedeutet wirtschaftlich optimaler Wärmeschutz? *WKSB* 22(4):1–4.

Gertis, K. 1978. Kosten für die Verbesserung des Wärmeschutzes von Gebäuden. Kosten-Nutzen Analysen. BW 32(31):1238–0.

Statistisches Bundesamt. 2016. Data on energy price trends—Long-time series. Wiesbaden: Statistisches Bundesamt.

Werner, H., and K. Gertis. 1976a. Zur Wahl von Kalkulationsmethoden bei der Er-mittlung der Wirtschaftlichkeit von Energiesparmaßnahmen. Baumasch. u. Bautechnik 26(2):65–72.

Werner, H., and K. Gertis. 1976b. Wirtschaftlich optimaler Wärmeschutz von Einfa-milienhäusern. *Kritische Gedanken zu Optimierungsrechnungen.* Ges.-Ing. 1/2:27–31.

Werner, H., and K. Gertis. 1976c. Wirtschaftlich optimaler Wärmeschutz von Einfa-milienhäusern. *Kritische Gedanken zu Optimierungsrechnungen*. Ges.-Ing. 5:97–103

Evaluation of a Modified Co-Heating Test for In-Situ Measurements of Thermal Transmittance of Single-Family Houses

Angela Sasic Kalagasidis, PhD **Emma Brycke, PE**
Jannicke Nilssen, PE **Pär Johansson, PhD**

ABSTRACT

Within two years after commissioning, the energy use for heating and operation of new buildings in Sweden should be verified by measurements. These have to be corrected for energy usage deviating from what has been defined as normal during the building design, e.g., excessive venting and hot-water use. This is practically difficult because the transmission losses of the building in use cannot be verified due to lack of a standard practical method for their evaluation. Designers and producers of low-energy houses would benefit from such a method, as the design of well-insulated envelopes is an essential quality of these buildings. A recently reintroduced method, the so-called co-heating test, *could be used for the verification of the overall thermal transmittance of buildings. To test the applicability of the co-heating test in situ, measurements were performed on two test objects: a two-year-old low-energy house and a new summer cottage. During the measurements, the latter was placed in a laboratory environment with a stable climate. Airtightness was measured on both houses. The overall average heat transfer coefficients were obtained and compared to theoretical values. This paper describes how the co-heating test has been modified to be used in in-situ conditions. Findings from both the measurements and following analyses are presented. The results obtained indicate that there is a clear potential for further simplifications of the co-heating test in the future.*

INTRODUCTION

The co-heating test is an experimental method for in-situ evaluation of a building's overall thermal conductance (W/K [Btu/h·°F]), also called the overall heat loss coefficient (HLC). This value describes the insulation level of a building envelope, which is one of the most influential parameters in the energy use for heating of a building. A measured value of HLC is needed when it is not possible to make reliable predictions of the building's energy use by simulations or to verify the building simulation results. Aging and deterioration of buildings in use may cause permanent changes in building envelopes that cannot be quantified by available building simulation software due to their inherent limitations. The same applies for varying workmanship standards and on-site invented solutions during the production of buildings. The co-heating test, therefore, can be used both as a diagnostic tool for buildings in use and as a quality control method for the building's design and construction processes. The latter could be of particular interest for designers and constructors of low-energy houses, where such a test could be a certificate of their particular expertise. A similarity can be drawn with a blower door test thanks to which the airtightness of a building envelope is reported as evidence of a building's particular quality.

After the co-heating test was introduced in the late 1970s (Sonderegger and Modera 1979), a vast number of modified co-heating tests have been developed with the aim of achieving better accuracy and repeatability of the test. The interested reader may refer to Bauwens and Roels (2014) for an overview of existing co-heating methods. Yet, the co-heating test still remains a challenging task for practical applications. In this work we would like to further clarify some issues that should be taken into account when preparing and executing the test. Based on two field studies, we show how the test can be simplified and shortened.

Angela Sasic Kalagasidis is a professor and **Pär Johansson** is an assistant professor in the Department of Civil and Environmental Engineering, Chalmers University of Technology, Gothenburg, Sweden. **Emma Brycke** is a design engineer at Skanska, Gothenburg, Sweden. **Jannicke Nilssen** is a design engineer at Sweco, Trondheim, Norway.

Basics of the Co-Heating Test

The co-heating test involves heating of the indoor environment to a constant temperature by an electrical heater over a period of time. Continuous recording of the heating output, the ambient conditions, and processing of data reveal the HLC. Besides electrical heaters with sufficient heating power and data loggers (for electrical power input and indoor and outdoor temperatures), fans may be needed to mix the indoor air to a uniform temperature. While the test is in progress, the building should be unoccupied and intentional ventilation, domestic appliances, and other heat sources inside the building should be shut down, and solar radiation through windows should be prevented.

The stationary heat transfer through the building envelope establishes the theoretical basis of the test. The heating temperature should be high enough to ensure that the heat flow through the building envelope during the test is transferred from the inside to the outside, to avoid the need for a cooler. From the data collected, the HLC of the building can be found by regression analysis of daily mean heating power input, Q_h (W), and the daily mean inside-to-outside temperature difference, ΔT (K):

$$Q_h = \text{HLC} \cdot \Delta T \quad (1)$$

With relatively simple equipment it is possible to conduct a co-heating test in any type of building. However, difficulties associated with ensuring stationary ambient conditions, separation of transmission heat losses from other heat transfer processes through the building envelope, as well as the need to have an unoccupied building can make the test both time consuming and expensive. These challenges are discussed briefly in the following section.

Conditions for a Successful Co-Heating Test

The reliability of the co-heating test depends largely on the stationarity of ambient conditions and on how well the transmission heat losses can be separated from other heat losses and gains through the building envelope. It is important to subtract infiltration losses from the measured power supply from the heater so that Q_h in Equation 1 encloses only the transmission losses through the building envelope. The air leakage characteristics of the building should be measured by a blower door or tracer gas technique (Bauwens and Roels 2014). This also implies that an adequate air infiltration model of the building should be established in order to utilize the results of the air leakages tests. Therefore, wind data (wind direction and speed) during the co-heating test should be recorded.

Varying outdoor conditions, particularly sunny and windy periods, disturb the co-heating test by forcing a building into nonstationary thermal balance with the surroundings. To illustrate this, the heat demand in a model building has been calculated by a building simulation tool and plotted against both outdoor air temperature and solar-air temperature in Figure 1. For this purpose, a transient energy balance model of the building has been constructed in MATLAB®/Simulink® (MathWorks 2015a, 2015b) using a building physics toolbox (ibpt 2015), and the building's thermal characteristics have been simplified to a one-capacity model as described in ISO 13790 (ISO 2008). As can be seen in Figure 1, there is no unique linear correlation between the heat supply and the

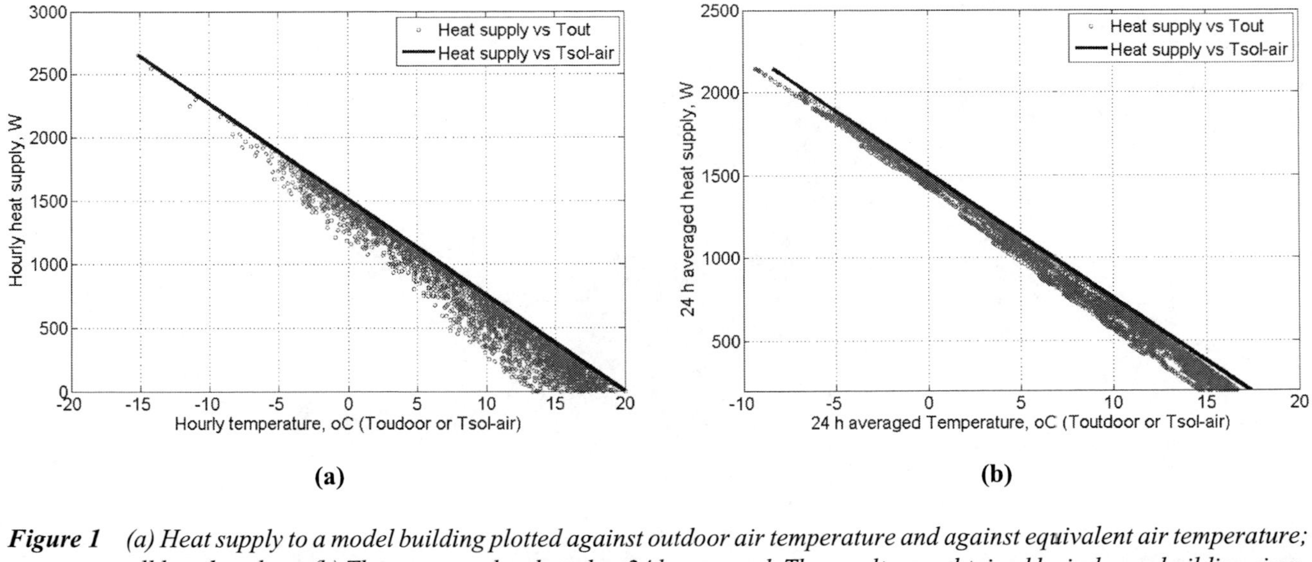

Figure 1 *(a) Heat supply to a model building plotted against outdoor air temperature and against equivalent air temperature; all hourly values. (b) The same results plotted as 24 h averaged. The results are obtained by in-house building simulation software (ibpt 2016), where the building's thermal characteristics have been simplified in accordance with recommendations in SS-EN ISO 13789 (SIS 2007). The indoor temperature in the building is kept constant at 20°C (68°F).*

outdoor air temperature, regardless of the time step (hourly or 24 h average). On the other hand, a clear and unique linear correlation exists between the heat supply and the solar-air temperature, which takes into account solar radiation on the building, sky radiation, wind, etc. In order to measure the solar-air temperature, additional measuring equipment should be added to the basic measuring kit (electrical heater, indoor and outdoor temperature loggers, and wind data loggers) such as pyranometers and pyrgeometers.

The importance of excluding tenants from the measurements is illustrated in Figure 2, which shows the measured heating power supply during a heating season in 16 identical and occupied single-family houses. These houses were designed according to the Swedish low-energy house standard FEBY09 (SCN 2009) and built in 2012 in the neighborhood of Vallda Heberg, outside Gothenburg, in the southwest of Sweden. To give readers a reference, FEBY09 limits the total annual energy demand for heating, cooling, and ventilation to 65 kWh/m^2 (20.6 kBtu/ft^2) heated floor area (60% of the national requirements at that time) at an indoor temperature of 21°C (70°F). To fulfill these requirements, the thermal transmittance (U-factor) of the exterior walls and windows in the tested houses is designed to 0.11 and 0.7 W/m^2·K (R-52 and R-8 ft^2·°F·h/Btu), respectively, while the air leakage characteristic is 0.25 ach at 50 Pa (2394 lb$_f$/ft) pressure difference. Although all 16 houses in Figure 2 have the same design value of the heat loss coefficient, HLC = 112.7 W/K (692.2 Btu/h·°F), the heating power supply varied largely between them during the heating season as a consequence of different users and their preferences for indoor environment (Jimmefors and Östberg 2014).

TWO REASONS TO SHORTEN THE CO-HEATING TEST

Based on the literature, the length of the co-heating experiment varies between different methods from days to weeks (Bauwens and Roels 2014). Weekly periods are too long for practical applications because adequate solutions for the relocation of the tenants may not be available. It would be easier if the test could be performed over a course of a day or two, as there is a fair chance for tenants to spend that time at another location. The results from Figure 1 suggest that there is no reason for prolonging the test more than a day or two. More measured data will not improve the accuracy of the test unless they are correlated to the correct temperature, solar-air instead of air temperature, and there is a proper air leakage model.

Another way to shorten the co-heating test is to keep the indoor temperature at the same (average) level as in normal use of the building. While raising the indoor temperature may be needed to ensure a positive temperature difference between the indoor and outdoor environment, it should be definitely avoided whenever possible. Each change of the indoor temperature leads a building into nonstationary thermal balance with the environment. Thermal inertia of the building will determine the time that is basically wasted on waiting until the building reaches a new steady state. Theoretically, assuming a lumped model of the building, the difference between the steady state and the current average temperature

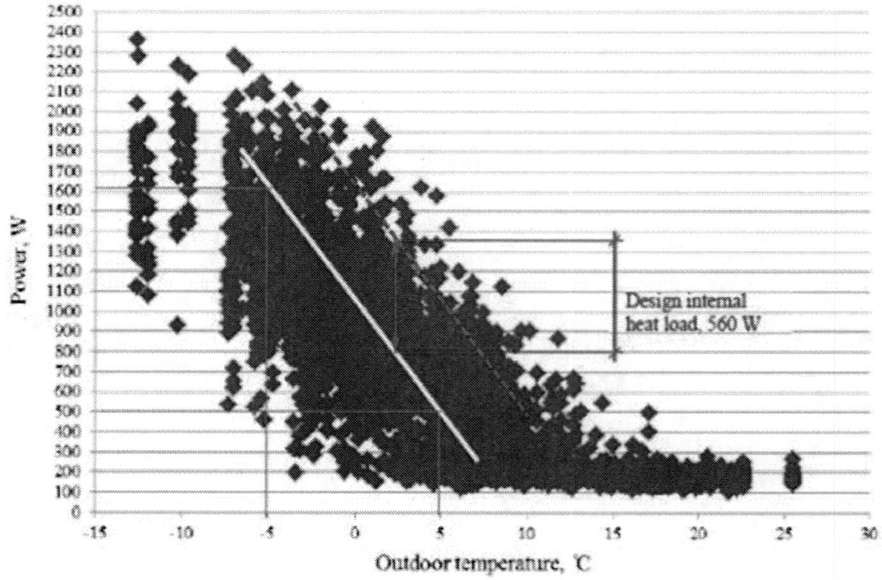

Figure 2 *Daily averaged heating power supply measured in 16 identical and occupied houses during 2013. All houses fulfill the passive house standard FEBY09 (SCN 2009). The white line represents a designed HLC = 112.7 W/K (692.2 Btu/h·°F) of the building envelopes. The distance between the white and the red line represents the designed internal heat gains (Jimmefors and Östberg 2014).*

in the building is about 13.5%, if the waiting time is equivalent to two time constants of the building (Hagentoft 2001).

FIELD TESTS

The major hypothesis in this work is that the heat loss coefficient of a house can be obtained from short-term co-heating tests without changing the indoor air temperature. To test this hypothesis, two field tests were performed. The first test was conducted on a single-zone building, a summer cottage, and under stable ambient conditions. The second test was conducted in a community building in the Vallda Heberg neighborhood. This is a multizone building in real climate conditions. The testing procedures and major findings are summarized in the following section.

First Test Object: Summer Cottage

The co-heating test under almost ideally stable ambient conditions was conducted in a simple building during February 2015 in an experimental hall at Chalmers University of Technology. The ambient temperature was around 19°C (66°F) throughout the test, without solar radiation and wind (see Figure 3). The building, a summer cottage (*Friggebod* in Swedish), was constructed by students of architecture as a part of their spring-term course, after which it was sold and profits were given to charity. It was a single-zone building, with the floor area approximately 10 m^2 (108 ft^2), with one window and one door. The building envelope was timber-framed and insulated. The overall thermal transmittance of the building, including thermal bridges, is estimated at 0.47 W/m^2·K (R-12 ft^2·°F·h/Btu) by detailed calculations (Brycke and Nilssen 2015).

All ventilation openings were sealed to exclude the ventilation heat losses during the test. A blower door measurement was also performed. Sensors for logging the temperature and relative humidity were placed both inside the building and outside, at different heights. Inside the object, four devices were placed on tripods with a vertical distance of 0.5 m (1.64 ft) between them. They were placed in a suitable location where the influence of direct heat from the radiator should be as low as possible. Outside the building, four sensors were placed along a vertical line with approximately 0.8 m (2.63 ft) distance between them. Further details can be found in the work by Brycke and Nilssen (2015).

The house was not heated before the test. Due to a relatively high temperature in the test hall, the indoor temperature in the house was raised to about 30°C (86°F) to ensure sufficient temperature difference during the co-heating test in respect to the ambient temperature. An electrical radiator was placed in the middle of the room in order to spread the heat uniformly. The data from the power/energy meter and the temperature sensors were logged hourly. The measurements lasted for about a week. A fan was used to mix the indoor air only during the last 7 h of measurements, to evaluate possible effects on the results.

Results from the test are presented in Figure 4. The ambient temperature during the test was rather stable, and the small variations that are visible in the figure were caused by occasional openings of the exterior doors at the test hall. The indoor temperature during the test was somewhat higher than intended, i.e., 31.7°C (89.1°F) on average, due to a coarse regulation of the thermostat at the radiator. There was clear temperature stratification that did not decrease after the fan was turned on (the last 7 h).

From the measured data, the average thermal transmittance of the building envelope U_m (W/m^2·K [Btu/h·ft^2·°F]) was found as

$$U_m = \frac{Q_h}{A \cdot (T_{in} - T_{out})} \tag{2}$$

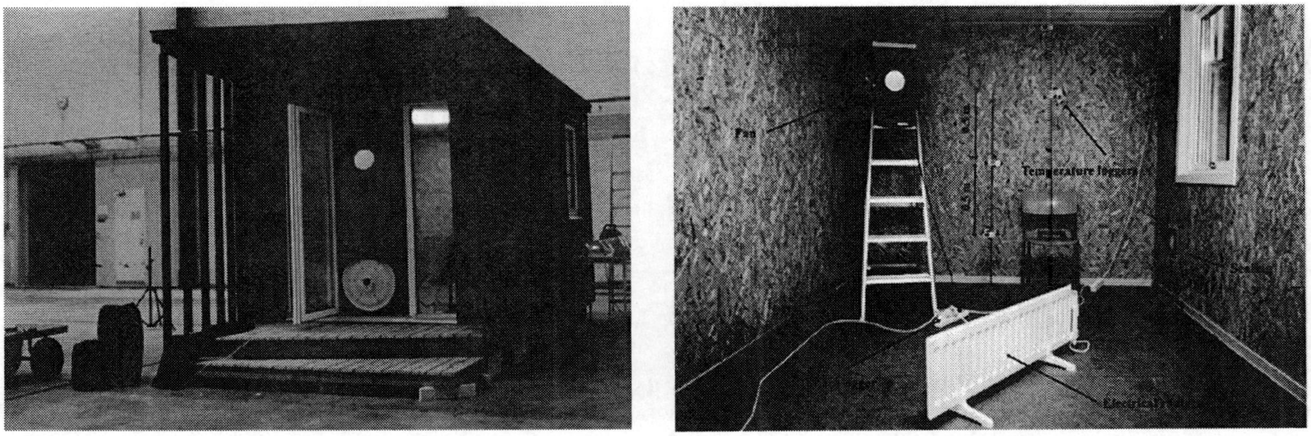

(a) (b)

Figure 3 (a) The summer cottage from the outside, during a blower door test, and (b) the interior of the summer cottage during the co-heating test.

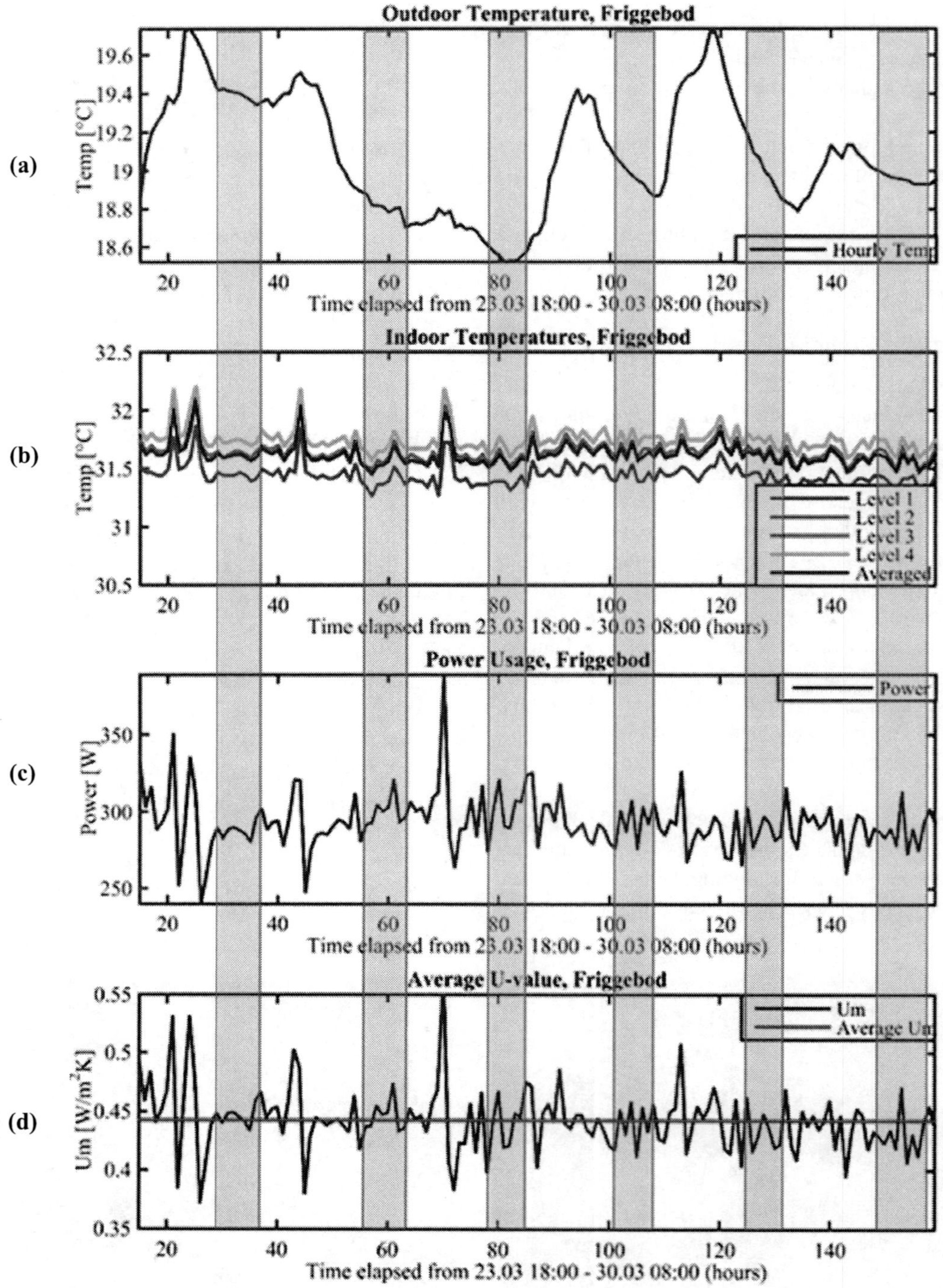

Figure 4 *Results of the co-heating test in the summer cottage (Friggebod): (a) ambient temperature in the test hall; (b) indoor temperature—level 1 at 0.5 m (1.6 ft) from the floor, all other levels offset by 0.5 m (1.6 ft); (c) power usage by the electrical radiator; and (d) estimated thermal transmittance of the building envelope. Shaded areas indicate six short-time periods for the estimation of the average thermal transmittance of the building envelope (Equation 2 and Table 2).*

where

A = total area of the building envelope, m^2 (ft^2)

T_{in} = average indoor temperature over a period of interest, °C (°F)

T_{out} = average outdoor temperature over a period of interest, °C (°F)

Q_h = average electric power supply from the radiator, W (Btu/h)

The air infiltration losses were deducted from Q_h.

The average test-based U_m values were found for both the whole measuring period (7 days) and for the six selected shorter periods (5–10 h), as summarized in Table 1. The average test-based U_m value from all test periods is 0.44±0.008 W/m^2·K (12.9±0.23 ft^2·°F·h/Btu). The results from the short periods diverge for not more than 2.3% from the U_m value found over the whole test period. We can conclude that a stable outdoor environment and indoor temperature provide good conditions for successful use of the co-heating test, regardless the length of the test. One can notice a slight increment of U_m during the periods 1–3 and then decrement during the periods 4–6, which coincide with the increment and decrement of the relative humidity inside the house (see Figures 4 and 5). The house was finished just before the measurements, so the drying of the built-in moisture very likely occurred during the test. Finally, the difference between the test-based U_m values and the theoretical one are in the range of 4% to 9%, which is considered acceptable. However, it should be kept in mind that the accuracy of the theoretical value is not known.

Table 1. Measured Average Heat Transfer Coefficient in the Summer Cottage

Interval	Period of Logging: Start	Period of Logging: Stop	Average U_m, W/m^2·K (ft^2·°F·h/Btu)	Deviation from the Theoretical Value, %	Deviation from the Whole-Period Averaged Value, %
Whole period	2015-03-24 09:00	2015-03-30 08:00	0.44 (R-12.9)	–6.4	—
1	2015-03-24 23:00	2015-03-25 09:00	0.45 (R-12.6)	–4.3	2.3
2	2015-03-26 02:00	2015-03-26 09:00	0.45 (R-12.6)	–4.3	2.3
3	2015-03-27 01:00	2015-03-27 06:00	0.44 (R-12.9)	–6.4	0.0
4	2015-03-27 23:00	2015-03-28 05:00	0.44 (R-12.9)	–6.4	0.0
5	2015-03-28 23:00	2015-03-29 05:00	0.43 (R-13.2)	–8.5	–2.3
6	2015-03-29 23:00	2015-03-30 07:00	0.43 (with fan) (R-13.2)	–8.5 (with fan)	–2.3 (with fan)

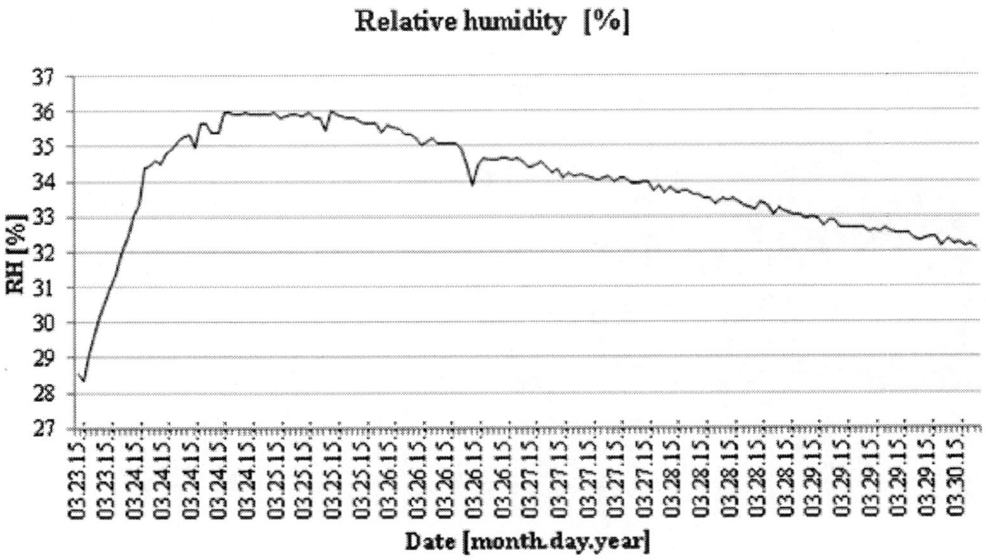

Figure 5 *Relative humidity inside the summer cottage during the co-heating test.*

Second Test Object: The Community Building in Vallda Heberg

The second co-heating test was performed in a multizone building located in the Vallda Heberg neighborhood residential area outside Gothenburg under normal weather conditions. This is a rentable house for the residents in the neighborhood (the same as for Figure 2), and as such it was found suitable because it was possible to keep it both uninhabited and unventilated during the test, with all domestic appliances turned off. The building has a floor area of approximately 200 m^2 (2100 ft^2) divided on two floors, as shown in Figure 6.

The community building is much larger than an average single-family house and would require several electric heaters for the co-heating test. Due to financial reasons, it was decided to use the existing hydronic heating system, i.e., radiators placed below windows, since it was already equipped with energy and power meters. This decision was further motivated by the lack of reported results from co-heating tests where the existing heating system in the house was used.

The co-heating test was performed on two different occasions, in March and April 2015, which were chosen based on the building's availability (i.e., the building was fully booked on other occasions). The ventilation system was turned off and all visible openings in the building envelope were closed or sealed. The air leakage characteristics was determined from a blower door test. Temperature loggers were distributed evenly in the building, mainly at a height of 1.5 m (4.9 ft) above the floor level and typically in the centerline of the non-exterior walls. Rooms with a higher ceiling height, such as those on the second floor and the kitchen, were equipped with temperature loggers at two levels to account for possible temperature gradients. All windows were covered by aluminum foil to block the solar irradiation through windows. It was estimated that the solar radiation would be of modest intensity during this period of the year, so only the outdoor air temperature was logged during the test. Further details about the test can be found in the work by Brycke and Nilssen (2015).

Results from the co-heating tests from both test periods are presented in Figures 7 and 8 and summarized further in Tables 2 and 3. As a basis for comparison, a theoretical thermal transmittance of the entire building envelope was estimated to 0.19 $W/m^2{\cdot}K$ (R-30 $ft^2{\cdot}°F{\cdot}h/Btu$) (HLC = 94.2 W/K [578.6 Btu/h·°F], total area of the building envelope 496 m^2 [5300 ft^2]) by detailed calculations, where both thermal bridges and air infiltration losses were taken into account. Further details are provided by Brycke and Nilssen (2015). The test-based average U_m values were found in the same manner as presented in the test on the summer cottage. Besides the average values over the entire measuring periods of 4 and 5 days, respectively, the values from shorter periods, mostly nighttimes, were also calculated.

The average test-based U_m value from all shorter periods in test 1 is 0.19±0.02 $W/m^2{\cdot}K$ (R-29.9±0.31 $ft^2{\cdot}°F{\cdot}h/Btu$) (Table 2) and in test 2 is 0.17±0.02 $W/m^2{\cdot}K$ (R-33.4±0.39 $ft^2{\cdot}°F{\cdot}h/Btu$) (Table 3). These differences are in the range of 10% ± 10%. Approximately the same range of difference is found when the short-term test-based U_m values are compared to the theoretical one. The average test-based U_m values for the entire periods are in a similar absolute range. It should be emphasized again that the accuracy of the theoretical U_m value cannot be estimated, although standard procedures have been followed.

There were some notable differences between the conditions during these tests in the community building and the one performed on the summer cottage. Namely, the temperature

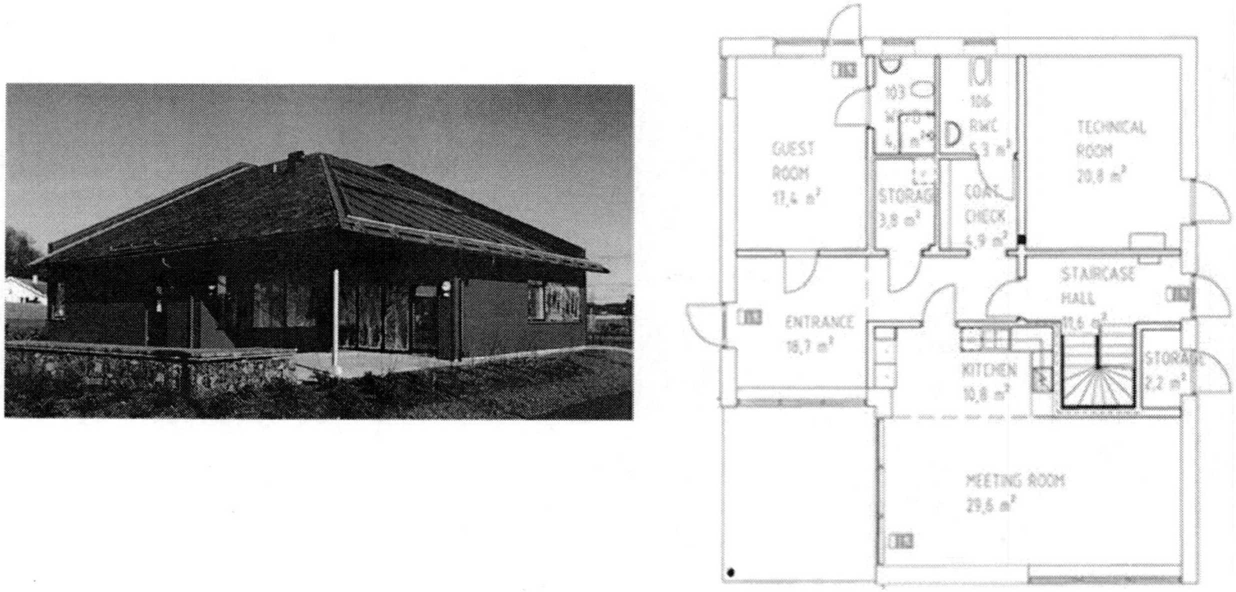

Figure 6 *The community building in the Vallda Heberg neighborhood and the layout of the ground floor.*

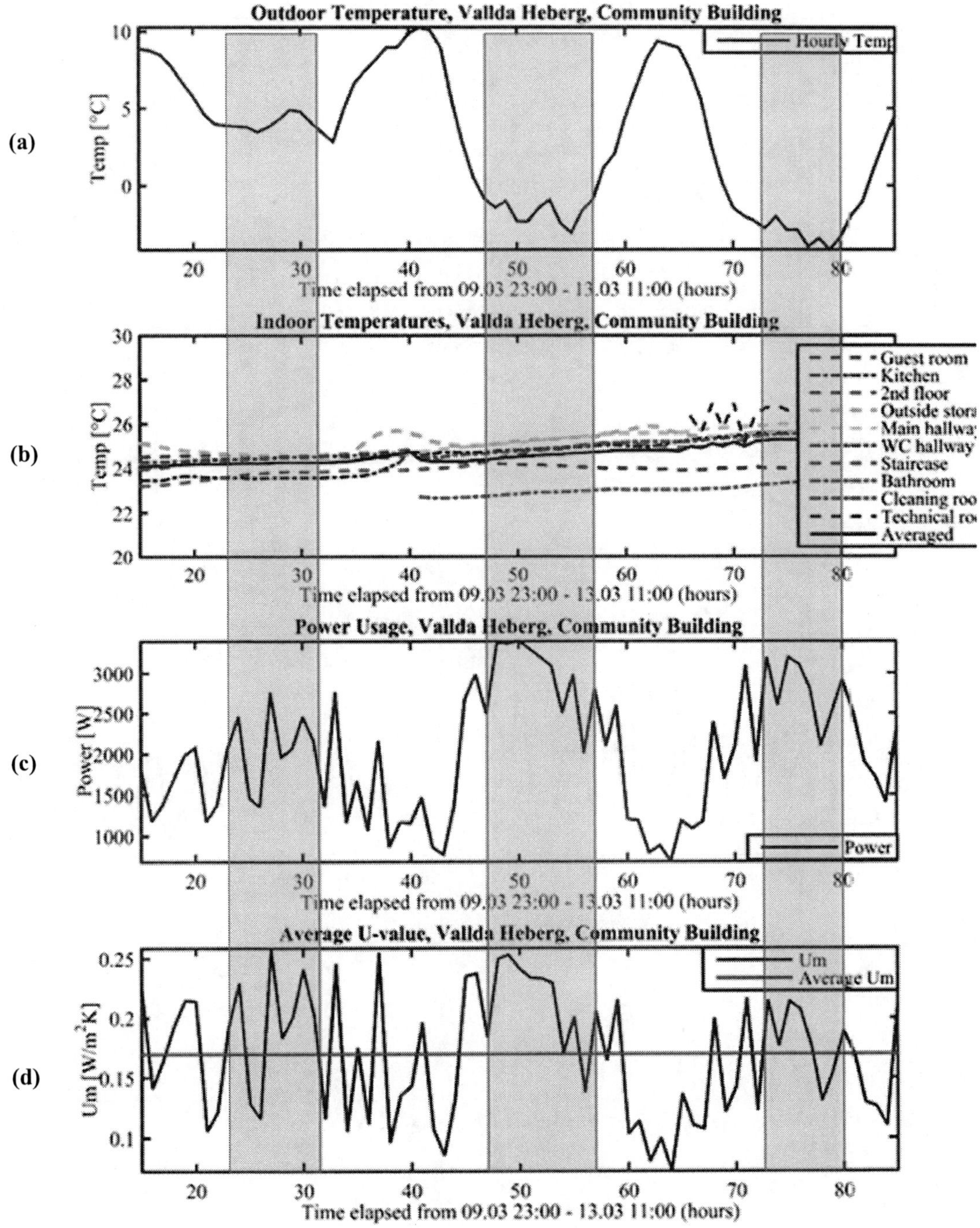

Figure 7 *Results of co-heating test 1 in the community building: (a) ambient temperature, (b) indoor temperature, (c) power usage by the electrical radiator, and (d) estimated thermal transmittance of the building envelope. Shaded areas indicate three short-time periods for the estimation of the average thermal transmittance of the building envelope (Equation 2 and Table 2).*

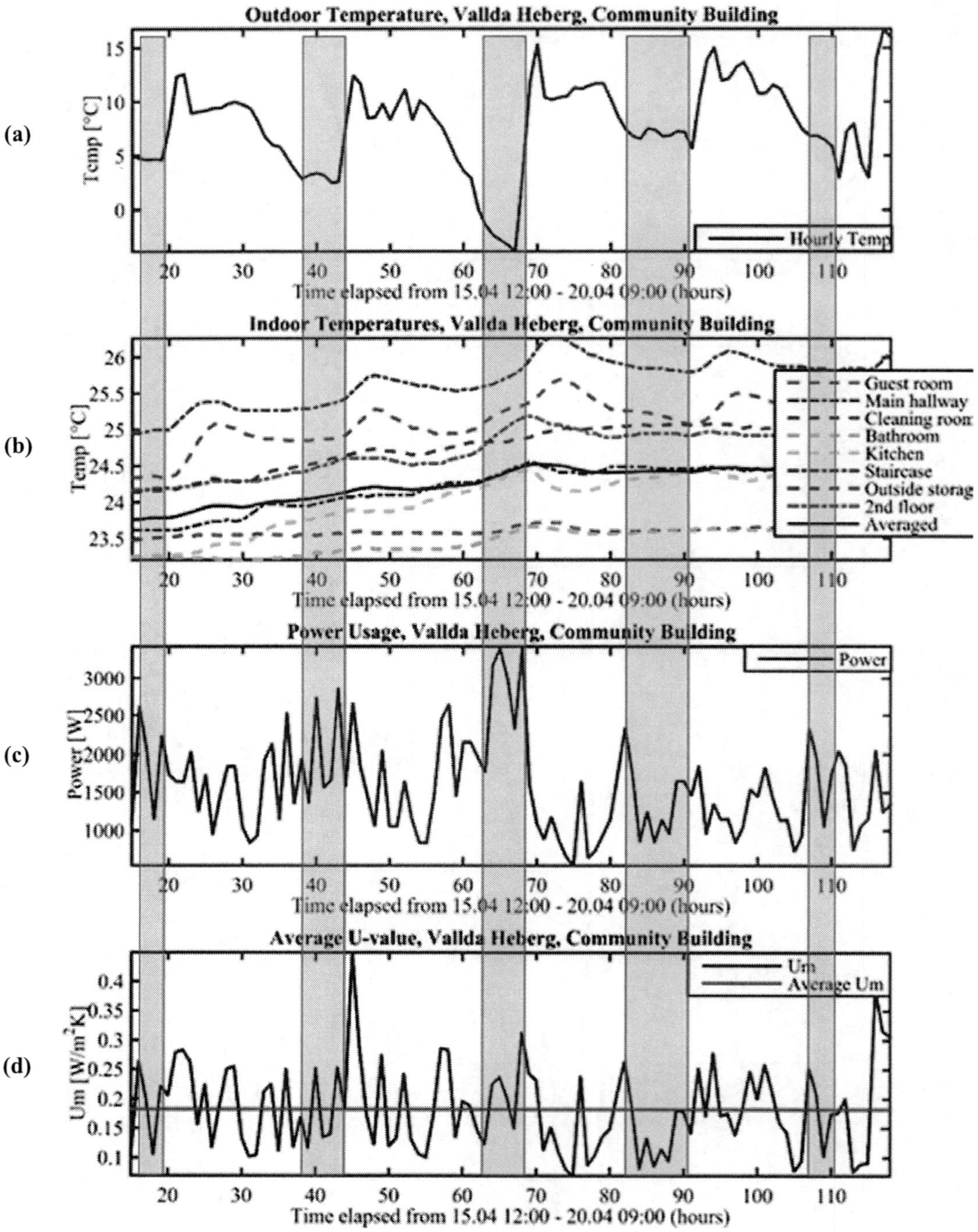

Figure 8 *Results of co-heating test 2 in the community building: (a) ambient temperature, (b) indoor temperature, (c) power usage by the electrical radiator, and (d) estimated thermal transmittance of the building envelope. Shaded areas indicate five short-time periods for the estimation of the average thermal transmittance of the building envelope (Equation 2 and Table 3).*

Table 2. Measured Average Heat Transfer Coefficient in the Community Building—Test 1

Interval	Period of Logging		Average U_m, $W/m^2 \cdot K$ ($ft^2 \cdot °F \cdot h/Btu$)	Deviation from the Theoretical Value, %	Deviation from the Whole-Period Averaged Value, %
	Start	Stop			
Whole period	2015-03-10 23:00	2015-03-13 11:00	0.17 (R-33.4)	–10.5	—
1	2015-03-10 23:00	2015-03-11 06:00	0.19 (R-29.9)	–1.6	25.3
2	2015-03-11 22:00	2015-03-12 08:00	0.21 (R-27.0)	12.1	3.5
3	2015-03-12 23:00	2015-03-13 07:00	0.18 (R-31.5)	–7.4	10.0
Periods 1, 2, and 3			0.19±0.02 (R-29.9±0.31)	1.0±10.0	12.9±11.2

Table 3. Measured Average Heat Transfer Coefficient in the Community Building—Test 2

Interval	Period of Logging		Average U_m, $W/m^2 \cdot K$ ($ft^2 \cdot °F \cdot h/Btu$)	Deviation from the Theoretical Value, %	Deviation from the Whole-Period Averaged Value, %
	Start	Stop			
Whole period	2015-04-15 12:00	2015-04-20 09:00	0.18 (R-31.5)	–4.2	—
1	2015-04-16 02:00	2015-04-16 07:00	0.20 (R-28.3)	4.2	8.8
2	2015-04-17 02:00	2015-04-17 07:00	0.18 (R-31.5)	–6.3	–2.2
3	2015-04-18 02:00	2015-04-18 07:00	0.18 (R-31.5)	–5.3	–1.1
4	2015-04-18 22:00	2015-04-19 06:00	0.15 (R-37.9	–23.2	–19.8
5	2015-04-19 22:00	2015-04-20 02:00	0.17 (R-33.4)	–12.6	–8.8
Periods 1–5			0.17±0.02 (R-33.4±0.39)	–8.6±10.1	–4.6±10.6

difference inside the community building differed between the zones and changed over time (see Figures 7 and 8). The reason for the latter could be found in the unprecise control of the indoor temperature as provided with the hydronic heating system, but also because of the sunny periods during the measurements. Besides, the outdoor temperature varied both on average and in amplitude. Nevertheless, the accuracy of the short-term test-based U_m values is found acceptable, as it is in the range of accuracy of the theoretically estimated values.

CONCLUSIONS

This study investigated the usability and reliability of a modified co-heating test, which is both shortened to a couple of hours (nighttime) and conducted without changing the indoor air temperature. Both simplifications are motivated by theoretical analyses that suggest that a large set of data does not necessarily improve the accuracy of the test. On the contrary, substantial time is wasted if the indoor air temperature is changed during the measurements.

These hypotheses have been tested in two field tests, the first conducted under stable ambient conditions in a simple single-zone building and the second under realistic ambient conditions in a complex multizone building. The results from the first test differed 2% to 3% from the theoretically estimated overall thermal transmittance of the building envelope. The difference between the measured and theoretical values in the second test was about 10% ± 10%, which is considered acceptable since the uncertainties in estimating the theoretical values could be in the same range.

The results from the field tests indicate that the suggested modifications of the co-heating test are reasonable. The test could be performed with basic equipment composed of a heater, energy and electric power meters, and temperature loggers. Besides, the airflow characteristics of the building should be measured and a building air leakage model should be established in order to differentiate between infiltration and transmission losses. Moreover, it was shown that the test could be performed using the existing hydronic heating system in a building. Finally, the results indicate that it is sufficient to conduct the test under several nighttime periods during cold months.

ACKNOWLEDGMENTS

The authors would like to thank Martin Sandberg at NCC Construction Sweden for giving access to one of the test objects; Peter Lindblom in the Department of Architecture, Chalmers University of Technology, for practical assistance; and Niklas Christensson at Eksta Bostads for helping with equipment and measurement data.

REFERENCES

Bauwens, G., and S. Roels. 2014. Co-heating test: A state-of-the-art. *Energy and Buildings* 82:163–72.

Brycke, E., and J. Nilssen. 2015. Assessment of co-heating test: A practical method to evaluate the in-situ heat transfer coefficient in dwellings. Master's thesis, Department of Civil and Environmental Engineering, Chalmers University of Technology, Gothenburg, Sweden. http://studentarbeten.chalmers.se/publication/228073.

Hagentoft, C.-E. 2001. *Introduction to building physics*. Lund: Studentlitteratur AB.

ibpt. 2015. International Building Physics Toolbox in Simulink. www.ibpt.org.

ISO. 2008. ISO 13790:2008, *Energy performance of buildings—Calculation of energy use for space heating and cooling*. Geneva: International Organization for Standardization.

Jimmefors, H., and J. Östberg. 2014. Energy performance and indoor climate investigations in the passive house residential area Vallda Heberg. Master's thesis, Department of Civil and Environmental Engineering, Chalmers University of Technology, Gothenburg, Sweden. http://studentarbeten.chalmers.se/publication/215321.

MathWorks. 2015a. MATLAB®. Natick, MA: The Mathworks, Inc. http://www.mathworks.com/products/matlab/?s_tid=hp_ff_p_matlab.

MathWorks. 2015b. Simulink®. Natick, MA: The Mathworks, Inc. http://www.mathworks.com/products/simulink/?s_tid=hp_ff_p_simulink.

Sonderegger, R., and M. Modera. 1979. Electric co-heating: A method for evaluating sesonal heating efficiencies and heat loss rates in dwellings. Berkeley, CA: Lawrence Berkeley Laboratory, University of California–Berkeley.

SCN. 2009. FEBY Kravspecifikation för Passivhus (FEBY09). (Specification of requirements for passive houses, version 2009) [in Swedish]. Sveriges Centrum för Nollenergihus (Swedish Centre for Zero Energy Houses. http://www.nollhus.se/rapporter.

SIS. 2007. SS-EN ISO 13789:2007, *Thermal performance of buildings—Transmission and ventilation heat transfer coefficients—Calculation method.* Swedish Standards Institute.

A Simplified Energy Savings Calculation Tool for the Lifetime Energy and Environmental Impact Assessment of Insulation Materials

Mahabir Bhandari, PhD
Member ASHRAE

Som Shrestha, PhD
Member ASHRAE

Kaushik Biswas, PhD

Andre Desjarlais

ABSTRACT

Application of insulation materials to building envelopes is a relatively easy and effective way of reducing energy use for space conditioning and, consequently, limiting the negative environmental impacts from the buildings sector. While insulation materials have a positive impact on the environment by reducing energy consumption in buildings, they also have some negative environmental impacts associated with their embodied energy. The total lifetime environmental impacts of insulation materials are a summation of the following: (1) direct impacts due to their embodied energy and (2) indirect or environmental impacts avoided due to the reduced building energy consumption. The indirect impacts are generally much higher than direct impacts of insulation materials and are often calculated by means of numerical modeling. However, whole-building energy modeling tools are usually fairly complex, and building managers may require significant training to be able to appropriately use those tools. To aid in this effort, a simplified regression algorithm-based spreadsheet was developed to quantify energy savings potential without requiring whole-building energy simulations. The objective was to develop a simple, Excel-based tool that can be used to estimate the electricity and gas savings due to insulation upgrades by providing only a few parameters (i.e., the building type, floor area, heating and cooling degree days, and solar radiation). The regression tool was based on the simulation study carried out to estimate the energy savings potential of insulation upgrades of 16 different types of commercial buildings in 15 climatic locations in the United States.

INTRODUCTION

U.S. commercial buildings consumed 17.6 quadrillion Btu (quad) of primary energy, which was 19% of the total U.S. primary energy consumption, and is projected to increase by 3.3 quads from 2012 to 2040, the second largest increase after the industrial sector (DOE 2014). These buildings consume 29% of the delivered (or site) energy for heating and cooling the buildings. Application of insulation in building envelopes is an effective method of reducing the heating and cooling energy consumption of these buildings. In the U.S., the adoption of building insulation has been largely driven by building codes and standards, with little attention paid to the environmental benefits of more advanced insulation products. However, the concept of lifetime environmental impact assessment is relatively new in the area of envelope insulations applications. The lifetime environmental impacts of insulation materials can broadly be divided into two categories: (1) direct impacts due to the embodied energy of the insulation materials and (2) indirect or environmental impacts avoided as a result of reduced operational energy consumption of the buildings due to the addition of insulation. It is important to identify insulation materials for buildings that will lead to minimum negative environmental impacts over the insulation lifetime. However, current lifetime environmental impacts assessment methods mainly focus on the direct impacts due to the embodied energy and generally ignore the indirect impacts due to the complexity of using the tools for estimating the indirect impacts.

Mahabir Bhandari, Som Shrestha, Kaushik Biswas, and **Andre Desjarlais** are research and development staff at Oak Ridge National Laboratory, Oak Ridge, TN.

The primary energy saved, and hence CO_2 emissions avoided, over the building service life are the indirect environmental impacts of the insulation materials, which usually dominate the total lifetime environmental impacts, often exceeding the direct impacts by orders of magnitude (Satori and Hestnes 2007; SPFA 2012; Kosny et al. 1998). However, the current life-cycle assessment (LCA) methods lack the standardization in how the indirect impacts are calculated. To reduce variability in energy savings calculations due to the subjective selection of input parameters, standardized and appropriate sets of building models are required. For commercial buildings, Shrestha et al. (2014) recommended the use of a standardized set of building models derived from DOE's commercial reference and prototype building models (Deru et al. 2011; Goel et al. 2014) and typical meteorological year (TMY)[1] weather files for the locations of interest.

The indirect environmental impact factors used in LCA are usually calculated using correlations, analytical models, or whole-building energy models. Different LCAs use customized building models, and there appears to be no standardized set of modeling parameters. Furthermore, the energy saved and environmental impact avoided over the lifetime of the insulation material depends on geographical location, climate, building characteristics, and use. A large number of whole-building energy modeling tools are available (BEST 2016). However, use of these tools requires a certain level of expertise. Simplified tools such as Roof Savings Calculator (RSC 2016), ConstructIGU (2016), Energy Calc (2016), and Insulation Savings Calculator (CIMA 2016) exist, but most of theses tools are focused on a specific building component, specific category of products, and/or a particular manufacturer's product. To the best of authors' knowledge, a generic simplified tools for energy savings calculation related to added building envelope insulation, specifically used for LCA analysis, does not exist. A simplified spreadsheet-based tool was developed in this project to calculate the site-energy savings with applications of ASHRAE Standard 90.1-2013 compliance-level insulation for different commercial building types in different climate locations. The indirect impact factors of avoided primary energy generation and CO_2 emissions can be estimated from the site energy savings using appropriate regional or national conversion factors

SIMULATION STUDY

The simulation study was carried out to estimate the energy savings potential of insulation upgrades of 16 different types of commercial buildings in 15 climatic locations in the United States. These buildings represent the majority of commercial buildings types in the United States.

The 16 prototype building models considered were based on the DOE commercial prototype building models for post-1980 and new construction[2] (Deru et al. 2011; Goel et al. 2014). Two sets of buildings were selected for this study. The first set consisted of existing buildings built according to ASHRAE Standard 90.1-1989 compliance and the second set represented the new construction based on ASHRAE Standard 90.1-2013 compliance. The models used for new construction were significantly enhanced from the existing prototype models by PNNL (Goel et al. 2014). The following sets of building models for each of the 16 building type were created in this study to represent the baseline case and cases with insulation upgrades for 15 climatic locations:

Existing construction (retrofit application):

- Baseline case: ASHRAE 90.1-1989 compliance compliant
- Upgraded wall insulation: modified wall insulation to ASHRAE Standard 90.1-2013 compliant wall insulation
- Upgraded roof insulation: modified roof insulation to ASHRAE 90.1-2013 compliant roof insulation
- Upgraded wall and roof insulation: modified wall and roof insulation to ASHRAE 90.1-2013 compliant wall and roof insulation

New construction:

- Baseline case: no wall or roof insulation
- Upgraded wall insulation: modified wall insulation to ASHRAE Standard 90.1-2013 compliant wall insulation
- Upgraded roof insulation: modified roof insulation to ASHRAE Standard 90.1-2013 compliant roof insulation
- Upgraded wall and roof insulation: modified wall and roof insulation to ASHRAE Standard 90.1-2013 compliant wall and roof insulation

The reason for dividing the simulation study into two separate categories (i.e., existing and new construction) was to cover the cases with no insulation and with insulation as prescribed by ASHRAE Standard 90.1-1989. The existing construction case pertains to retrofitting an existing building constructed after 1980 (ASHRAE Standard 90.1-1989 compliant building) and adding insulation to walls and roofs as per the requirements of ASHRAE Standard 90.1-2013. The existing construction cases assumed that the building received only an envelope retrofit and all other building characteristics and equipment (lighting and HVAC, etc.) were not changed/upgraded.

New construction pertains to adding insulation to a building with no wall or roof insulation but with all other buildings characteristics (equipment, lighting, HVAC, etc.) as per the ASHRAE Standard 90.1-2013. The added wall and roof insulation were as per the requirements of ASHRAE Standard 90.1-2013. These two categories/scenarios are assumed to cover the extreme cases of retrofitting scenarios. This was done to avoid having to simulate all the possible permutations and combinations of various retrofitting measures in buildings. Various cases include scenarios, such

1. https://energyplus.net/weather

2. Commercial Prototype Building Models: http://www.energycodes.gov/commercial-prototype-building-models.

as "what if only the wall is retrofitted or just the roof or both?" It was assumed that if an existing building with no insulation is being retrofitted, it would be retrofitted to current compliance level and not in increments; for example, first to 1989, then 2004 and finally to 2013 compliant level. Therefore, only no insulation is considered as a base case for new construction, and intermediate compliance levels were considered under existing construction. In future, the calculation could be extended to cover the above/beyond the current compliant scenario, such as ASHRAE Standard 189.1.

In both existing and new construction models, all parameters except the envelope were maintained at compliant levels, as per the 1989 and 2013 versions of ASHRAE Standard 90.1 for existing and new construction, for the baseline and upgraded insulation cases. The wall and roof insulation values varied depending on the building type and climatic location.

Prototype building models for post-1980 and new construction were used for the existing and new buildings, respectively. Besides 240 baseline prototype models of existing and new construction each (16 building types × 15 climatic locations), overall modified EnergyPlus model sets for this simulation study consisted of 720 (16 building types × 15 climatic locations × 3 upgraded insulation case) models for existing construction and 720 models for new construction (i.e., a total of 1920 (2 × 720 + 2 × 240) EnergyPlus models.

For new construction, the EnergyPlus models, except the base case that assumed no wall or roof insulation, assumed the minimum wall and roof insulation values required as per ASHRAE Standard 90.1-2013. For existing construction, the EnergyPlus models assumed the minimum wall and roof insulation values required as per ASHRAE Standard 90.1-1989. The simulations were carried out for 15 locations in different ASHRAE climate zones.

Construction of Basic Walls and Roofs

Basic constructions for the walls and roofs were taken from ASHRAE Standard 90.1-1989. There were three basic constructions for roof assemblies. The attic construction was composed of, from the interior side of the assembly, an interior air film, 0.5 inch thick gypsum board, an insulation layer of fiberglass loose fill, another 0.5 inch thick gypsum board, and an outside air film. For the insulation entirely above deck (IEAD) construction, an interior air film, metal deck, an insulation layer of polyisocyanurate foam, a roof membrane, and an outside air film was used. For the metal construction, an interior air film, an insulation layer of fiberglass batt, a metal roof, and an outside air film were employed.

There were five basic wall constructions. The mass wall with interior insulation was composed of an interior air film, 0.5 inch thick gypsum board, a 3.5 inch fiberglass batt insulation layer thermally shorted with steel framing, an 8 inch heavyweight concrete layer, 1 inch of stucco, and an outside air film. The mass wall with exterior insulation was composed of an interior air film, 0.5 inch thick gypsum board, an 8 inch heavyweight concrete layer, polyisocyanurate foam insulation, 1 inch of stucco, and an outside air film. The steel wall included an interior air film, 0.5 inch thick gypsum board, a 3.5 inch fiberglass batt insulation layer thermally shorted with steel framing, wood siding, and an outside air film. The wood wall was comprised of an interior air film, 0.5 inch thick gypsum board, a 3.5 inch fiberglass batt insulation layer thermally shorted with wood framing, wood siding, and an outside air film. The metal wall was made up of an interior air film, 0.5 inch thick gypsum board, fiberglass batt insulation layer thermally shorted with steel purlins, metal siding, and an outside air film.

SIMULATION RESULTS

The prototype EnergyPlus models were modified for various wall and roof insulation levels for all 16 building types in 15 climatic locations. Overall 1440 models were created based on the modifications. All the 1920 models (prototype models and modified models) were simulated and the results for further regression analysis were collected in terms of electricity and gas savings for cooling and heating energy consumption.

As the main aim of this analysis is to come up with the simplified equations, it was decided to use the commonly used units and types by the industry stakeholders (i.e., the savings were presented in terms of electricity and natural gas savings and the units kWh and kBtu, respectively were used for the data presentation).

Regression Analysis

Once the simulation results were collected for all the cases, the next step was to determine if a simplified method of estimating the impact of insulation can be obtained based on the simulation results for all the building types. The simplified equations for determining heating and cooling energy savings based on the annual cooling degree days (CDD), heating degree days (HDD), and diffuse and direct solar radiation as the independent variable were developed for each building type considered in this analysis. Engineering Equation Solver (EES) and Excel were used for the regression analysis. Several combinations of input variables were considered during the simplified equation development process. The main purpose of development of the regression-based equations was to come up with a simplified set of equations that can predict the electricity and gas savings with a small number of input variables. The number of independent (input) variables was also limited by the total number of data points available. For any insulation upgrade for each building type, only 15 data points, corresponding to 15 climatic locations, were available. As expected, CDD and HDD were most influential in electricity and gas savings equations, respectively. After analyzing several combinations of independent input variables, it was determined that CDD or HDD, direct normal solar (DS) and diffuse solar (DIFF) would produce the best set of equations for most for most of the building types. Final set of equations selected were of the following form:

$$\text{Savings} = a_1 \cdot \text{CDD} + a_2 \cdot CDD^2 + a_3 \cdot \text{DS} + a_4 \cdot \text{DS}^2 + a_5 \cdot \text{DIFF} + a_6 \cdot \text{DIFF}^2 + a_7 \cdot \text{CDD} \cdot \text{DS} + a_8 \cdot \text{CDD} \cdot \text{DIFF} + a_9 \cdot \text{DS} \cdot \text{DIFF} \quad (1)$$

where

CDD = cooling degree days, base 18.3°C (65°F)

DS = annual daily average direct normal solar radiation, $Wh/m^2/day$

DIFF = annual daily average diffuse solar radiation, $Wh/m^2/day$

The independent input variables are generally available in SI units (e.g., all the typical meteorological year [TMY] data, which are available for a large number of locations in the U.S. are in SI units). The ".stat" file available with TMY data, which can be downloaded from EnergyPlus website,[3] contains all the independent variables needed for using these equations for predicting the electricity and gas savings.

The goal of the final analysis was to come up with the equations that utilized simplified and easily available input data. These equations can then be used for calculating the approximate heating and cooling energy savings due to the insulation upgrade in a typical building without using the detailed whole-building energy simulation tools. Because the energy savings calculations presented here are based on the typical buildings with a specific set of building parameters, such as building geometry, floor area, number of stories, window to wall ratio, internal loads, lighting, and HVAC systems, it was difficult to generalize the results with a generic set of equations that can be used for any building with different parameters in a particular building category. One of the objectives of this study was to develop a simplified set of equations with a limited number of input variables for the ease of use. Therefore, the simulated electricity and gas energy savings for each building type were normalized with respect to building floor area (i.e., the electricity and gas savings results were presented in terms of kWh/ft^2 and $kBtu/ft^2$, respectively). Separate sets of equations were developed for electricity and gas savings. Several iterations were made so that the input variables, such as CDD, HDD, WS, DS, and Diff were considered during the analysis to determine the best set that can be used for the majority of building types with a high coefficient of determination, R^2. It is to be noted that, as with any simplified approach, this analysis was based on a small dataset (i.e., 15 data points representing 15 climatic locations for each type of building). Therefore, the large degree of freedom for input variables did not exist for increasing the accuracy of the equations. For example, just using four input variable with cross terms can give a perfect set of equations ($R^2 = 1$) for any building type but would not be a representative equation for the data points outside this range. In general, a set of three input variables with cross term variables resulted in a good fit for most of the building types. The following sets of regression equations were developed for new and existing construction:

- Electricity savings: upgraded wall insulation
- Gas savings: upgraded wall insulation
- Electricity savings: upgraded roof insulation
- Gas savings: upgraded roof insulation
- Electricity savings: upgraded wall and roof insulation
- Gas savings: upgraded wall and roof insulation

Overall, a total of 192 sets of coefficients (a_1 through a_9), 96 each for new and existing construction, were developed for all 16 building types considered in this analysis. For the majority of the buildings, input variables of CDD, DS, and DIFF for electricity savings and HDD, DS, and DIFF for gas savings were used to derive the equations. The derived equations were of the following form:

$$\text{Electricity Savings} = a_1 \cdot \text{CDD} + a_2 \cdot \text{CDD}^2 + a_3 \cdot \text{DS} + a_4 \cdot \text{DS}^2 + a_5 \cdot \text{DIFF} + a_6 \cdot \text{DIFF}^2 + a_7 \cdot \text{CDD} \cdot \text{DS} + a_8 \cdot \text{CDD} \cdot \text{DIFF} + a_9 \cdot \text{DS} \cdot \text{DIFF} \quad (2)$$

For gas savings, a similar equation is used, with CDD replaced by HDD.

Table 1 summarizes the constants of the equations for new constructions for electricity savings. Similar tables were created for the rest of the cases.

As seen in Table 1, the equations showed a strong fit ($R^2 > 0.95$) for the majority of the building types for new construction. However, for a small set of existing construction buildings, equations for gas savings for upgraded insulation roof and wall insulation did not show a strong fit. A different set of equations was attempted for the cases not showing a strong fit. The new set of equations did improve the fitness in the cases considered. For example, for an existing large office building, the gas savings fit improved to $R^2 = 0.86$ from R^2 of 0.53.

The modified equation for an existing large office building has the following form:

$$\text{Gas Savings} = a_0 + a_1 \cdot \text{DS} \cdot \text{DIFF} + a_2 \cdot \text{HDD} \cdot \text{DIFF} + a_3 \cdot \text{HDD}^2 + a_4 \cdot \text{DS}^3 + a_5 \cdot \text{DIFF} \cdot \text{HDD}^2 + a_6 \cdot \text{DIFF} \cdot \text{DS}^2 \quad (3)$$

In general, electricity and gas savings for the majority of the building types can be expressed by simplified sets of equations. For the majority of building types, the correlation was quite good. For example, in new construction, the difference in insulation levels between the base case (no insulation) and target case (ASHRAE Standard 90.1-2013 compliant insulation levels) was consistent, which was reflected in a very good correlation ($R^2 > 0.80$) between the simulation and equation predicted savings results. Similar results were obtained for existing construction. However, the correlation was poor in some cases due to the inconsistency in insulation level in base cases in the existing buildings.

3. https://www.energyplus.net/weather.

Table 1. Summary of Coefficient for Upgraded Roof Insulation Electricity Savings for New Construction

Building Type/ Coefficients	a1	a2	a3	a4	a5	a6	a7	a8	a9	R^2
Small office	−7.90E-03	−8.50E-07	5.72E-03	−3.04E-07	−9.45E-03	5.16E-06	1.11E-06	2.25E-06	−2.29E-06	0.98
Medium office	−1.01E-02	−6.70E-07	1.15E-03	−5.29E-08	7.97E-03	−2.74E-06	8.60E-07	4.22E-06	−1.13E-06	0.96
Large office	9.02E-04	3.41E-08	−4.04E-04	1.73E-08	1.03E-03	−5.18E-07	−6.21E-08	−2.77E-07	1.46E-07	0.97
Mid-rise apartment	5.75E-04	−1.42E-07	1.05E-03	−7.63E-08	−7.50E-04	3.51E-07	1.53E-07	−2.47E-07	−2.94E-07	1.00
High-rise apartment	−6.95E-03	−5.25E-07	7.32E-04	−5.94E-08	6.31E-03	−2.38E-06	7.04E-07	2.39E-06	−6.19E-07	0.98
Quick service restaurant	−3.58E-03	−7.17E-07	5.14E-03	−3.59E-07	−1.13E-02	4.84E-06	7.89E-07	8.72E-07	−1.33E-06	0.99
Full service restaurant	−3.96E-03	−4.36E-07	2.19E-03	−1.36E-07	−4.45E-03	2.17E-06	6.10E-07	1.14E-06	−7.26E-07	0.99
Hospital	1.09E-03	6.60E-08	−5.72E-04	3.03E-08	1.61E-03	−7.77E-07	−1.02E-07	−3.06E-07	1.84E-07	0.96
Outpatient healthcare	−1.44E-02	−1.24E-06	3.86E-03	−2.01E-07	5.95E-03	−1.27E-06	1.48E-06	5.55E-06	−2.30E-06	0.98
Small hotel	−1.77E-02	−1.18E-06	3.19E-03	−1.50E-07	6.94E-03	−1.71E-06	1.60E-06	6.93E-06	−2.22E-06	0.95
Large hotel	1.75E-03	3.30E-08	−4.55E-04	1.50E-08	1.49E-03	−7.23E-07	−5.94E-08	−6.24E-07	1.51E-07	0.98
Primary school	8.81E-04	8.38E-08	−8.31E-04	3.60E-08	3.33E-03	−1.55E-06	−9.45E-08	−1.30E-07	2.75E-07	0.99
Secondary school	8.19E-04	7.67E-08	−4.48E-04	1.73E-08	2.38E-03	−1.13E-06	−7.46E-08	−1.50E-07	1.49E-07	0.99
Stand-alone retail	5.10E-03	5.39E-07	−5.16E-03	2.77E-07	1.59E-02	−7.55E-06	−7.48E-07	−8.24E-07	1.60E-06	0.89
Strip mall	−2.15E-03	−4.54E-08	−2.63E-04	4.85E-09	3.57E-03	−1.67E-06	1.21E-07	1.11E-06	4.83E-08	0.99
Warehouse	6.71E-04	1.12E-07	−2.72E-04	2.34E-08	1.25E-03	−5.39E-07	−9.49E-08	−2.03E-07	4.87E-08	0.98

Based on the regression equations developed (e.g., Equation 1 through 3), a simple Excel-based tool was developed that could be used to estimate the electricity and gas savings due to insulation upgrades by providing only a few parameters (i.e., the building type, floor area, heating degree days, cooling degrees days, direct normal and diffuse solar radiation). Figure 1 shows a screenshot of the calculator. It is to be noted that the purpose of the calculator is to estimate the energy savings using only a very limited number of parameters. The green highlighted parameters are the only input required for estimating electricity and gas savings due to the insulation upgrades.

To check the validity of the developed equations, the medium office building was simulated for Oak Ridge, TN (ASHRAE Climate Zone 4A), and the results were compared with the regression-equations-based electricity and gas consumption using HDD,CDD, DS, and DIFF values for Oak Ridge.

Table 2 shows the EnergyPlus simulated and regression-equation-predicted electricity and gas savings results of a medium office retail building.

For most cases, the regression-equation-predicted electricity and gas savings results for both the building types and insulation upgrade cases agree with the EnergyPlus simulated results. As mentioned earlier, for some locations the electricity and/or savings results did not follow the increasing or decreasing CDD or HDD trends. Based on the regression equations developed above, a simple Excel-based tool was developed that could be used to estimate the electricity and gas savings due to insulation upgrades by providing only a few parameters (i.e., the building type, floor area, heating degree days, cooling degrees days, direct normal and diffuse solar radiation). It should be noted that this tool has several limitations and should be used for specific purposes such as LCA calculations for building insulation materials. The main limitation is that the tool is based on prototype buildings and will not be a true representative of a real building. For example, the geometry, window-to-wall ratio, plug loads, lighting and occupancy schedule, HVAC systems, and set point temperature of actual buildings may not match with the corresponding parameters of the prototype building used in developing the tool.

CONCLUSION

The objective of this study was to develop a simple Excel-based tool that can be used to estimate the electricity and gas savings due to insulation upgrades by providing only a few parameters (i.e. the building type, floor area, heating and cooling degree days, direct normal and diffuse solar radiation). The regression tool was based on the simulation study carried out

Insulation Upgrade	Building Vintage	
Wall & Roof	Existing (ASHRAE 1989)	
Building Type	**State**	**Location***
Medium Office	Maryland_MD	Baltimore-Washington.Intl.AP, MD
Weather Data		
CDD	683	
HDD	2,538	
Direct solar	4,020	
Diffuse Solar	1,737	
Floorspace (ft^2)**	53,628	
Energy savings		
Electricity (kWh)	**7,211**	
Natural gas (kBtu)	**27,120**	
Notes:		
Choose the building, insulation upgrade and location-related options from the drop-down menus in the highlighted cells (***all highlighted cells should be updated***).		
* Choose location of closest weather station		
** Total conditioned floor area of the building (accounting for all storeys)		

Figure 1 *Simplified spreadsheet based calculator for electricity and gas savings due to insulation upgrades.*

to estimate the energy savings potential of insulation upgrades of 16 different commercial buildings in 15 climatic locations in the United States. To check the validity of the developed equations, the medium office building was simulated for Oak Ridge, TN (ASHRAE Climate Zone 4A), and the results were compared with the regression-equations-based electricity and gas consumption using HDD,CDD, DS, and DIFF values for Oak Ridge.

For most cases, the regression-equation-predicted electricity and gas savings results for both the building types and insulation upgrade cases agree with the EnergyPlus simulated results. As mentioned earlier, for some locations the electricity and/or savings results did not follow the increasing or decreasing CDD or HDD trends. Based on the regression equations developed above, a simple Excel-based tool was developed that could be used to estimate the electricity and gas savings due to insulation upgrades by providing only a few parameters (i.e., the building type, floor area, heating degree days, cooling degrees days, direct normal and diffuse solar radiation). It is to be noted that the calculator is based on a very limited set of prototype building simulations in 15 climatic zones of the U.S. The purpose of this tool is to use the results of the tools for estimating the indirect environmental impact in the tools used for the life-cycle impact assessment of insulating materials. The results should not be used for a real building for estimating the accurate electricity and gas savings for insulation upgrades.

ACKNOWLEDGMENTS

This work was supported by the Building Technologies Office of the U. S. Department of Energy under Contract No. DE-AC05-00OR22725 with UT-Battelle, LLC. The authors would like to acknowledge the support from Ms. Kristen Taddonio, Better Buildings Alliance technology development manager at the U.S. Department of Energy.

REFERENCES

ASHRAE. 1989. ASHRAE Standard 90.1-1989, *Energy standard for buildings except low-rise residential buildings*. Atlanta: ASHRAE.

ASHRAE. 2013. ASHRAE Standard 90.1-2013, *Energy standard for buildings except low-rise residential buildings*. Atlanta: ASHRAE.

Table 2. EnergyPlus Simulated and Regression Equation Predicted Electricity and Gas Savings for a Medium Office and a Standalone Retail Building in Oak Ridge, TN

Case Description	Electricity Savings, kWh/ft^2		Gas savings, kBtu/ft^2	
	E+ Simulated	Equation Predicted	E+ Simulated	Equation Predicted
Existing construction				
Baseline case—ASHRAE Standard 90.1-1989 compliant				
Upgraded wall insulation (ASHRAE Standard 90.1-2013 compliant wall)	0.27	0.07	0.03	0.02
Upgraded roof insulation (ASHRAE Standard 90.1-2013 compliant roof)	0.06	0.07	0.05	0.05
Upgraded wall and roof insulation (ASHRAE Standard 90.1-2013 compliant wall and roof)	0.33	0.13	0.09	0.07
New Construction				
Baseline Case—No Wall and Roof Insulation				
Upgraded wall insulation (ASHRAE Standard 90.1-2013 compliant wall)	2.12	2.20	1.11	1.57
Upgraded roof insulation (ASHRAE Standard 90.1-2013 compliant roof)	1.13	1.22	1.06	0.65
Upgraded wall and roof insulation (ASHRAE Standard 90.1-2013 compliant wall and roof)	3.29	3.39	2.52	2.24

ASHRAE. 2014. ASHRAE Standard 189.1-2014, *Standard for the design of high-performance green buildings*. Atlanta: ASHRAE.

BEST. 2016. BEST Directory, Building Energy Simulation Tools, formerly hosted by U.S. Department of Energy, http://www.buildingenergysoftwaretools.com/ (last accessed May 10, 2016).

CIMA. 2016. Insulation Savings Calculator. Cellulose Insulation Manufacturer Association. http://www.cellulose.org/HomeOwners/CalculateSavings.php (last accessed May 10, 2016).

ConstructIGU. 2016. IdeaScapes. PPG Industries, http://construct.ppg.com/(last accessed May 10, 2016).

Deru, M, K. Field, D. Studer, K. Benne, B. Griffith, P. Torcellini, B. Liu, M. Halverson, D. Winiarski, M. Rosenberg, M. Yazdanian, J. Huang, and D. Crawley. 2011. *U.S. Department of Energy commercial reference building models of the national building stock*, Technical Report, NREL/TP-5500-46861. Available online at http://www.nrel.gov/docs/fy11osti/46861.pdf (last accessed on September 14, 2015).

DOE. 2014. EnergyPlus (Energy Simulation Software and Supplemental Documentation). Department of Energy. https://www.energyplus.net/.

EIA. 2014. U.S. Energy Information Administration, Annual Energy Outlook 2014. Available online at http://www.eia.gov/forecasts/aeo/ (last accessed Dec 27, 2014).

EnergyCalc. 2016. Energy Calc: How much can you save using energy efficient windows? http://www.cardinalcorp.com/technology/applications/energy-calculator/ (last accessed May 10, 2016).

Goel, S., R. Athalye, W. Wang, J. Zhang, M. Rosenberg, Y. Xie, R. Hart, and V. Mendon. Enhancements to ASHRAE Standard 90.1 Prototype Building Models. Pacific Northwest National Laboratory, PNNL-23269, April 2014.

Kosny, J, A. Desjarlais, and D.W. Yarbrough. 1998. A procedure for analyzing energy and global warming impacts of foam insulation in U.S. commercial buildings. 1998 ACEEE Summer Study on Energy Efficiency in Buildings, Pacific Grove, CA.

RSC. 2016. Roof Savings Calculator (RSC). http://rsc.ornl.gov/ (last accessed May 10, 2016).

Sartori, I., and A.G. Hestnes. 2007. Energy use in the life cycle of conventional and low-energy buildings: A review article. *Energy Build* 39 249–57.

Shrestha, S.S., K. Biswas, and A.O. Desjarlais. 2014. A protocol for lifetime energy and environmental impact assessment of building insulation materials. *Environmental impact assessment review* 46:25–31.

SPFA. 2012. Life Cycle Assessment of Spray Polyurethane Foam Insulation for Residential & Commercial Building Applications. Spray Polyurethane Foam Alliance. Available online at http://www.certainteed.com/resources/SPFALCALongSummary.pdf. (Last accessed Dec 17, 2012).

Moisture-Related Durability of In-Service High-R Wall Assemblies in Pacific Northwest Climates

Jonathan Smegal, MASc **Robert Lepage, PEng, MASc** **Chris Schumacher, MASc**
Member ASHRAE

ABSTRACT

The Passive House program is becoming increasingly well-known in North America. This program aims to produce a superior thermal enclosure such that the space conditioning systems may be minimized, if not appreciably eliminated. Many Passive House designs use double-stud or thick wall enclosures. By adding significant thicknesses of cavity insulation, this approach provides a high level of thermal performance (R-value). However, the decreased heat flow through these walls reduces drying and can therefore place colder building components at risk of moisture damage. Constructing walls with exterior continuous insulation is a method that helps minimize these effects by warming critical layers, reducing potential condensation and moisture concerns.

This paper reports on two ongoing projects to assess the durability of these assemblies by comparing a 2 × 10 wood-framed wall with nominal exterior insulation in Portland, Oregon to a 2 × 10 wood-framed wall with an interior insulated service cavity in Victoria, British Columbia. Both walls are instrumented to measure temperature and moisture profiles through the assembly, for both the north and south orientations. Historical data from previous research programs will be discussed and compared to measurement data from the current ongoing project. The threshold of performance for these assemblies will be discussed along with solutions to address potential concerns. Analysis of preliminary data shows that it is consistent with past research, and supports the conclusion that exterior insulation can improve durability relative to double-stud walls.

INTRODUCTION

Passive House construction is growing in popularity in North America. The Passive House standard includes an ultra-low energy consumption target that is achieved through measures such as a high-R-value enclosure, passive solar heating, heat recovery ventilation, and seasonal shading (Passive House Institute 2014). More specifically, enclosures are specified and constructed to be well-insulated and airtight, with minimal thermal bridging. To supplement the limited maximum R-value of cavity insulation between dimensional lumber, additional insulation is sometimes required. This additional insulation is commonly added by building and insulating a thick wood-framed wall (e.g., using a double-stud wall, wood I-joists, or similar) or by adding exterior insulation (i.e., continuous insulation on the exterior of a more conventional, dimensional lumber wall). The positioning of this additional insulation will affect the hygrothermal behavior of the wall assembly by affecting the structural sheathing temperature, and thus the potential for durability concerns.

This paper reports on two Passive House projects that use different high R-value wall enclosure strategies to help achieve their low energy consumption targets. The two projects are in the Pacific Northwest, International Energy Conservation Code (IECC) Climate Zone 4C: one in Victoria, British Columbia (using a double-stud approach) and the other in Portland, Oregon (using exterior insulation). Both locations have approximately 2500 heating degree days at 18°C (4500 heating degree days at 65°F).

The Pacific Northwest was chosen for this comparison and analysis to address a gap in the research of high-R wall performance. Exterior-insulated and thick wood-framed wall assemblies have been studied and compared in IECC Climate

Jonathan Smegal is an associate and **Chris Schumacher** is a principal, senior building science specialist at RDH Building Science Laboratories, Waterloo, ON. **Robert Lepage** is a building science research engineer at RDH Building Science Inc., Victoria, BC.

Zones 5 and 6, and builders continue to gain experience with these systems in dry climates. However, there remains a need to study and compare the performance of these systems in more humid coastal climates like those of the Pacific Northwest.

To meet the energy requirements of the Passive House program, the building in Victoria was constructed with a thick stud wall with an interior insulated service wall, and the building in Portland was constructed with a thick stud wall with continuous exterior insulation. Both of these strategies increase the overall effective R-value from standard construction, and decrease thermal bridging through the clear-wall portion of the enclosure. Enclosure monitoring sensor packages were installed in both projects to monitor the conditions of the assembly and compare durability performance. Detailed wall assembly drawings and measured monitoring data from both of these assemblies are presented further in the report. Preliminary data is analyzed and compared to previous research on double-stud and exterior insulation wall assemblies.

BUILDING ENCLOSURE

The building enclosure is designed to separate the interior space from the exterior on every surface of the building both above and below ground. The most important function of the enclosure, with respect to long-term durability, is the control of liquid water, including rainwater, surface water, and ground water. For the purposes of this study, it is important to note that failure to adequately control liquid water will eventually result in the failure of almost any wall system, regardless of the design. This means that monitored enclosure walls that show good measured performance are not immune from failure should the critical water management details be missed. Air control across the enclosure is also important in terms of energy consumption and the movement of water vapor into and out of the enclosure. Failure to control air leakage and subsequent vapor movement can result in long-term durability problems and moisture-related failures as a result of condensation on surfaces within the enclosure. The Passive House program requires airtightness testing to ensure that air leakage is below a specified limit. This reduces the risk of cold weather air leakage condensation but does not eliminate it.

An air barrier must be continuous. Its components (i.e., materials and systems) should be air impermeable, strong, stiff, and durable. When dealing with airtightness, it is also important to consider both windwashing and re-entrant looping. If the air barrier is located near the interior of the enclosure, it is possible for wind-driven air to flow from the exterior into any cavities and back to the exterior. This can reduce the effective R-value of the insulation and may add moisture to the enclosure. Similarly, if the air barrier is located near the exterior of the enclosure it is possible for interior air to move into the enclosure and then return to the interior. This also results in increased energy demands, and in colder climates, will add moisture to the enclosure that could potentially be detrimental. It is important to note that neither windwashing nor re-entrant looping can be measured by an air leakage test. It is very difficult to measure the effects of re-entrant looping or windwashing within wall constructions being monitored. Controlling vapor movement by diffusion is also important, but does not move as much water vapor as quickly as air leakage.

Air and vapor control can be achieved with a single material (such as polyethylene sheeting), but it is not necessary to combine these functions in one layer. Some insulation materials such as spray foam may be part of an air barrier system, whereas some materials such as dense-packed cellulose retard air flow but are too air permeable to be considered part of an air barrier system. With respect to this study, the air barrier in Portland was the plywood sheathing, taped at the joints, using transition membranes at penetrations such as windows and doors and also at the transition to the foundation. The interior vapor control was a continuous layer of a smart vapor retarder material (not taped at the edges, since the vapor control layer was not also the air control layer). A smart vapor retarder has a relative humidity (RH) dependent vapor permeance which means that at the low relative humidities (e.g., ~50% rh) that would be expected in the interior space, the vapor permeance is low (≤1 Perm), but at the higher relative humidities (e.g., ~80% rh) that could be expected if there was water in the wall cavity, or strong inward vapor drives, the permeance increases significantly (>10 Perms) to allow moisture to pass. In Victoria, the vapor and air barrier were the same material, a trilaminate polyolefin/polypropylene membrane installed at the interstitial layer between an exterior and interior framing layer so that services could be run on the interior without penetrating the membrane and allowing air leakage. The membrane was well sealed to all transitions and any penetrations.

LITERATURE REVIEW

Neil Hutcheon was one of the first to address the benefits of exterior insulation. He explained that moving the insulation from the interior of a masonry wall to the exterior surface increases the temperature of the structure during cold weather and can increase the minimum temperature of the condensation plane above the dew point of the interior air, which minimizes the risk of condensation and improves the durability of the wall assembly (Hutcheon 1964). Others (Handegord 1985, Straube 2011) built on Hutcheon's work, adding simple numerical analysis to determine the ratio of insulation R-value required on each side of the condensation plane to minimize the risk of condensation based on the interior dew point and exterior temperature.

Measured performance bears out these analyses. Past published research on high-R wall assemblies in cold climates, including measurement of moisture content, has been conducted in a variety of locations, on many different types of wall assemblies. This review highlights several studies with particular relevance to double-stud wall assemblies and exterior-insulated wall assemblies. All of the reviewed studies were conducted in IECC Climate Zones 5–7.

Tsongas (1991) investigated the performance of wall assemblies in 86 homes in Montana and Washington State in Climate Zones 5 and 6. There was a range of construction types, vapor control layers, and insulation types (with varying vapor permeance) through the monitored regions. The study houses included some exterior insulating sheathing construction and some double-stud wall construction. The exterior insulating sheathing was either foil-faced polyisocyanurate, extruded polystyrene (XPS) or expanded polystyrene (EPS). It was found that building walls with more cavity insulation led to increased moisture levels in the wall assemblies, and that walls with exterior insulating sheathing were significantly drier than walls without exterior insulating sheathing. These walls were believed to be drier because the insulating sheathing kept the wall cavity wood members warmer and also because the insulated sheathing was an excellent exterior moisture barrier that prevented wet siding from transmitting or wicking moisture into the wall cavity.

Arena et al. (2013) monitored the performance of a double-stud wall in Boston, MA, in Climate Zone 5A. A hygrothermal model was used to correlate the measured data. The results showed that the hygrothermal simulation predicted lower sheathing moisture contents than were measured, even though the simulated RH in the model was 20%–30% higher than the measured RH. This means that the measured performance of the double-stud wall assembly was worse than predicted, even with much lower risk factors such as significantly decreased interior relative humidity. Ueno (2015) also found that in Devens, MA (Climate Zone 5), monitoring of a dense-packed cellulose wall assembly showed measured moisture contents greater than 30% with elevated interior relative humidity levels, and measured moisture contents greater than 20% with controlled interior relative humidity levels.

In 2014, at the University of Waterloo in Waterloo, Ontario, Canada (Climate Zone 5), a research program was conducted on a field test hut with several different wood-framed high-R wall assemblies side by side (Fox et al. 2015). Test walls consisted of thick cavity walls such as double-stud walls, I-joist walls with dense-packed cellulose, walls with three different types of exterior continuous insulation (with varying vapor permeance), and a typical 2 × 6 wall assembly with R-22 fiberglass batt and no exterior insulation. All of the test walls were challenged with intentional controlled water and air leakage at different times. It was concluded that the oriented strand board (OSB) sheathing in the exterior-insulated walls was warmer and reduced the risk of moisture-related issues, especially near the location of the controlled air leak. The sheathing in the exterior-insulated walls was consistently below 16% moisture content. The deep-cavity cellulose-insulated walls had the highest moisture content levels in both the framing plates and OSB sheathing throughout the duration of the study period. These walls also had the highest number of condensation hours and were slower to dry in comparison with the exterior-insulated walls.

In 1987, Robert Kane and Guy Titley designed a research program to address the concern that the use of an exterior low vapor permeance foam sheathing allows an accumulation of moisture between the vapor barrier and the exterior sheathing. Walls in occupied buildings in four Canadian cold-climate cities were chosen. The primary method of analysis was moisture content readings of the wood framing in all cases, in combination with temperature measurement and moisture measurements at the sheathing/insulation interface. It was concluded that XPS used as exterior sheathing does not cause moisture accumulation in the wood studs and that XPS sheathing will warm the batt/sheathing interface, thereby controlling the amount of possible air leakage or vapor-driven condensation. It was also concluded that regardless of the design and materials chosen, the importance of installing a proper air barrier to control the moisture cannot be overemphasized.

From the reviewed papers on the measurement of both thick and exterior-insulated wall assemblies, it is clear that there is a general pattern of findings. Monitoring of exterior-insulated walls indicates lower measured moisture contents and higher condensing plane temperatures. The exterior-insulated wall results do not appear to depend on the vapor permeance of the exterior insulation. These walls have a reduced risk for moisture problems, whether caused by vapor diffusion or air leakage. Monitored thick walls with dense-packed cellulose often showed higher measured moisture contents, and a greater risk of moisture-related durability issues, although it was noted in at least one paper that all of the elevated measured moisture contents dried to safe levels in the summer months (Ueno 2015).

MONITORING

The buildings in the current study are wood-framed multi-unit residential buildings (MURBs), 3 to 4 stories in height. Instrumentation was installed on both the north and south orientations. It is the authors' experience that the north and south orientations can have very different experimental outcomes because of the solar influence on the south orientation. Hence, both the north and south orientations are instrumented whenever possible during monitoring. Both of the monitoring locations were instrumented with a series of temperature, relative humidity, and wood moisture content sensors at strategic locations in the enclosure to measure and record the hygrothermal conditions on an hourly basis. The monitoring techniques are based on those described by Straube et al. (2002). The interior and exterior temperature and RH (boundary conditions) are critical to understanding the wall performance and temperature and moisture gradients across the enclosure. The exterior conditions were derived from nearby meteorological station data.

Wood moisture content was determined by measuring the electrical resistance across a sample of wood and correlating the electrical resistance with a wood moisture content according to the Garrahan equation (Onysko et al. 2010).

Dryability of water in the enclosure is often used as a comparison criterion for different wall assemblies, but this is difficult to do in real building enclosures due to the risk of moisture damage by intentionally injecting water. Using dryability is only practical in test huts, where walls can be stressed with moisture without affecting a building. Analysis of enclosure dryability is beyond the scope of this study because real buildings are used for comparison.

Exterior-Insulated Assemblies and Instrumentation (Portland)

In Portland, four walls in four different suites were instrumented, two on the north orientation, and two on the south orientation, as shown in Figure 1.

The interior conditions in every suite were measured immediately inside the continuously-operated heat recovery ventilator (HRV) exhaust fan. A detailed cross section of the Portland wall assembly, showing the typical monitoring locations, is shown in Figure 2. The sensors were installed at the mid-point between framing members, at the mid-height of the wall. The data acquisition system was installed above the ceiling in the corridor between the suites.

Thick Double-Stud Wall Assemblies and Instrumentation (Victoria)

In Victoria, two walls in two different suites were instrumented, for a total of four locations, two on the north orientation, and two on the south orientation (Figure 3). The units, and thus the walls, are stacked one above the other. The interior conditions in every suite were measured in both the common area and the master bedroom, at ceiling height. A detailed cross section of the Victoria wall assembly, showing the typical monitoring locations, is shown in Figure 4.

Evaluation Criteria

In colder climates, accumulation of moisture in the sheathing or elevated sheathing surface RH are used as key performance criteria because the sheathing is typically the first location where water vapor condensation would occur in a cold climate during the heating season as a result of air leakage

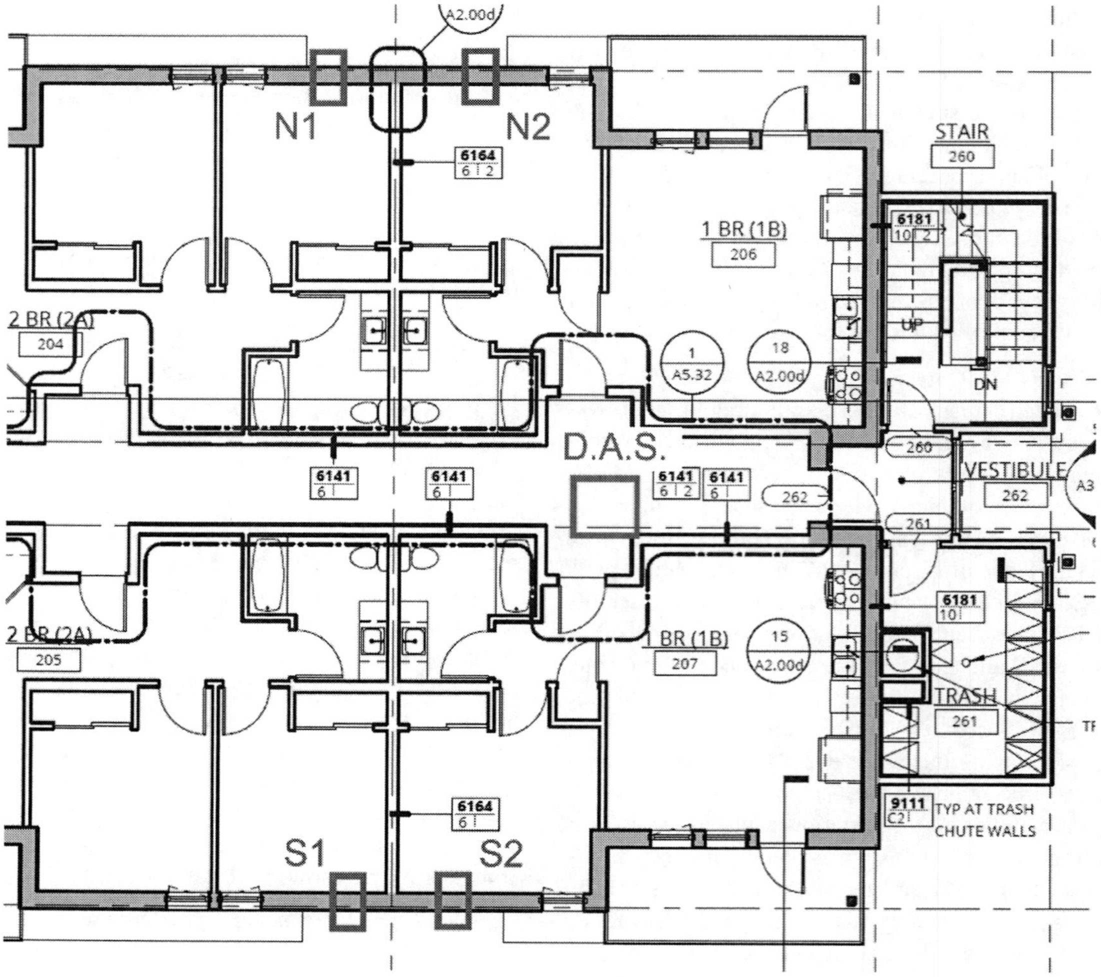

Figure 1 Four monitoring locations and the location of the data acquisition system on the third floor (Portland).

or vapor diffusion from the interior. The sheathing may also experience moisture accumulation from the exterior in wet climates.

There are several ways to assess moisture accumulation and the moisture-related durability of structural wood sheathing. ASHRAE Standard 160, *Criteria for Moisture-Control Design Analysis in Buildings*, Addendum A (ASHRAE 2009), uses a pass-fail evaluation based on the IEA Annex 14 standard (IEA 1991) to minimize mold growth. Specifically, the criteria for low mold risk is a 30-day running average surface RH <80% when the 30-day running average surface temperature is between 5°C and 40°C (41°F and 104°F)

However, the ASHRAE 160 criteria have been found to poorly correlate to actual biological growth risk (Lstiburek et al. 2015). To help provide more accurate analysis, the results will be compared using the Finnish VTT Technical Research

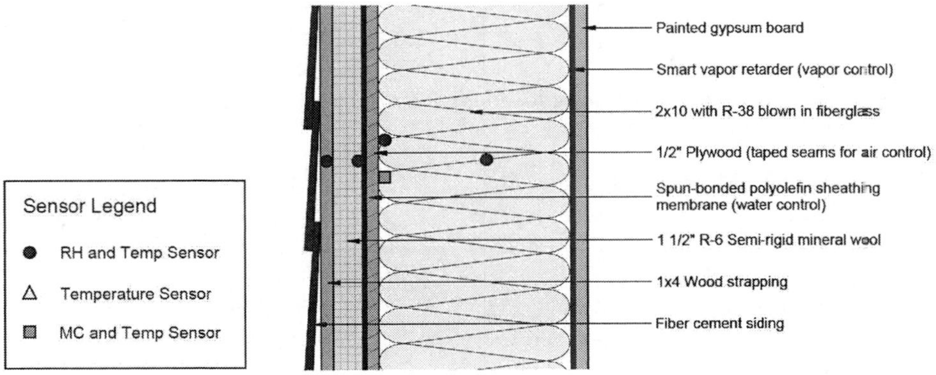

Figure 2 *Detailed assembly and typical monitoring locations (Portland).*

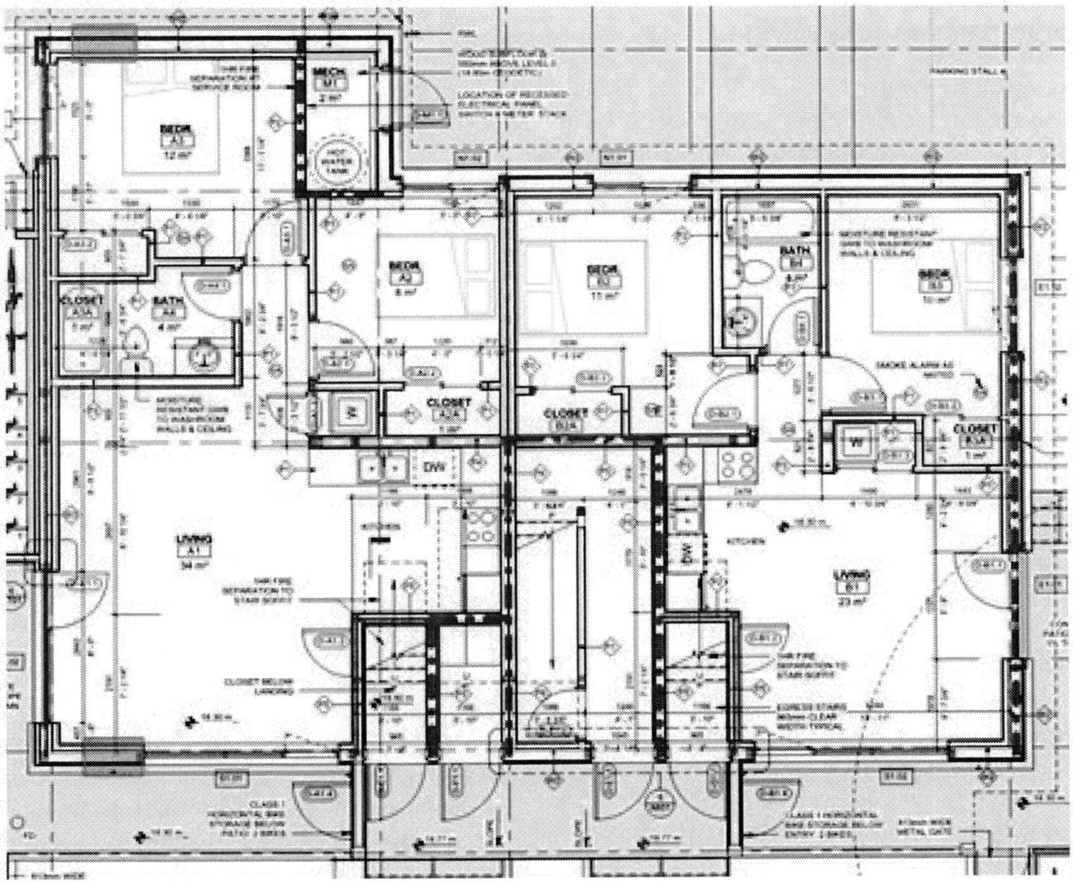

Figure 3 *Plan view showing monitoring locations (Victoria).*

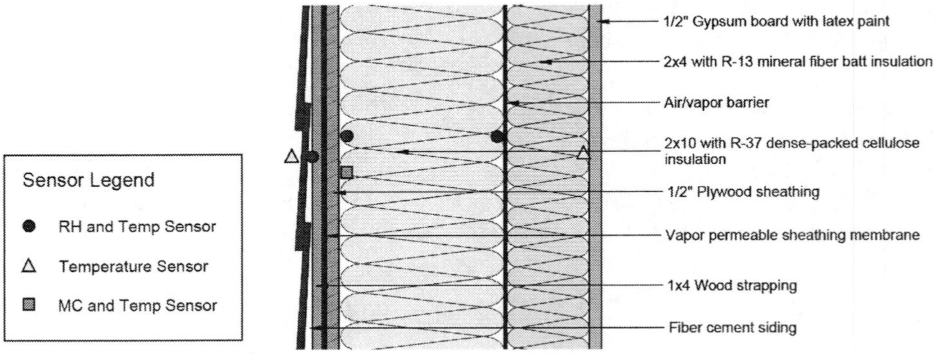

Figure 4 *Detailed assembly and typical monitoring locations (Victoria).*

Centres' Improved Model to Predict Mold Growth in Building Materials (Viitanen and Ojanen 2007). This model is based on empirical regressions of actual mold growth on building materials in varying climatic conditions. While the VTT model results do not necessarily guarantee the presence of mold, they do provide a greater degree of reliability than categorical limits. The VTT model output is a mold index, summarized in Table 1. Mold index values less than 3 are generally not visible to the naked eye, and therefore mold indices greater than 3 are considered a fail.

MONITORING DATA AND ANALYSIS

Monitoring at the Portland location (exterior-insulated assemblies) began during construction in November 2014, but the suites were not occupied until June of 2015. The data for this analysis begins on June 4, 2015 and finishes in June of 2016, for one full year of data. Monitoring will continue until at least 2017.

The interior and exterior relative humidity and temperature boundary conditions are shown in Figure 5. The interior conditions are an average of all four suites, and all four suites were quite similar in their conditions. The interior average temperature was relatively constant at 20°C–22°C (68°F–72°F), and the interior relative humidity fluctuated between 45% and 65% but had a trend to the lower threshold throughout the coldest months, because cool humid exterior fresh air was brought into the building and heated, reducing the relative humidity of the ventilation air. The interior relative humidity trend is correlated to changes in the exterior temperature, as well as occupancy behavior. There was no elevated interior relative humidity at the start of the research program that would indicate a source of construction moisture in the suites.

Monitoring of the Victoria location (double-stud assemblies) was delayed and began in December 2015, four months after the suites were instrumented and occupied in September of 2015. The data for this analysis begins on December 19, 2015 and will continue until 2017.

The interior and exterior relative humidity and temperature are shown in Figure 6. The interior average temperature

Table 1. Mold Index for the VTT Model (Viitanen and Ojanen 2007)

Index	Growth Rate	Description
0	No growth	Spores not activated
1	Small amounts of mold on surface (microscope)	Initial stages of growth
2	<10% coverage of mold on surface (microscope)	—
3	10%–30% coverage of mold on surface (visual)	New spores produced
4	30%–70% coverage of mold on surface (visual)	Moderate growth
5	>70% coverage on mold on surface (visual)	Plenty of growth
6	Very heavy and tight growth	Coverage around 100%

was relatively constant at 21°C (71°F), and the interior relative humidity fluctuated between approximately 35% and 55%.

The sheathing temperature and relative humidity used to calculate the mold index results (see the Conclusions section) are shown in Figures 7 and 8. Results are provided for both monitored buildings with a runninf 24 h average. The start of monitoring was delayed in Victoria, and only overlapping data is shown in this comparison between December 2015 and June 2016. [[this comes well before reference to Figures 9 and 10. Renumber the figures so they are mentioned in order in the text?]]. Portland results are colored in blue, whereas Victoria results are colored red, with the north elevation in a lighter shade for both buildings. All of the average temperatures are similar, but there are peaks on the south orientation in Portland (dark blue) that have higher measured temperatures through the monitoring period as a result of solar heating in Portland.

The measured relative humidity at the sheathing in Victoria (Figure 8) is higher on both the north and south in comparison to the Portland measurements even though the exterior ambient relative humidity was very similar in both Portland

and Victoria. The measured differences in sheathing RH from Figure 8 is further analyzed later in the paper.

Figure 9 compares the exterior temperature and sheathing temperature of each building and Figure 10 compares the exterior relative humidity and sheathing relative humidity for each building. These comparisons were conducted between December 22, 2015, and February 29, 2016. Only the coldest months of the year were used because the effect of the exterior

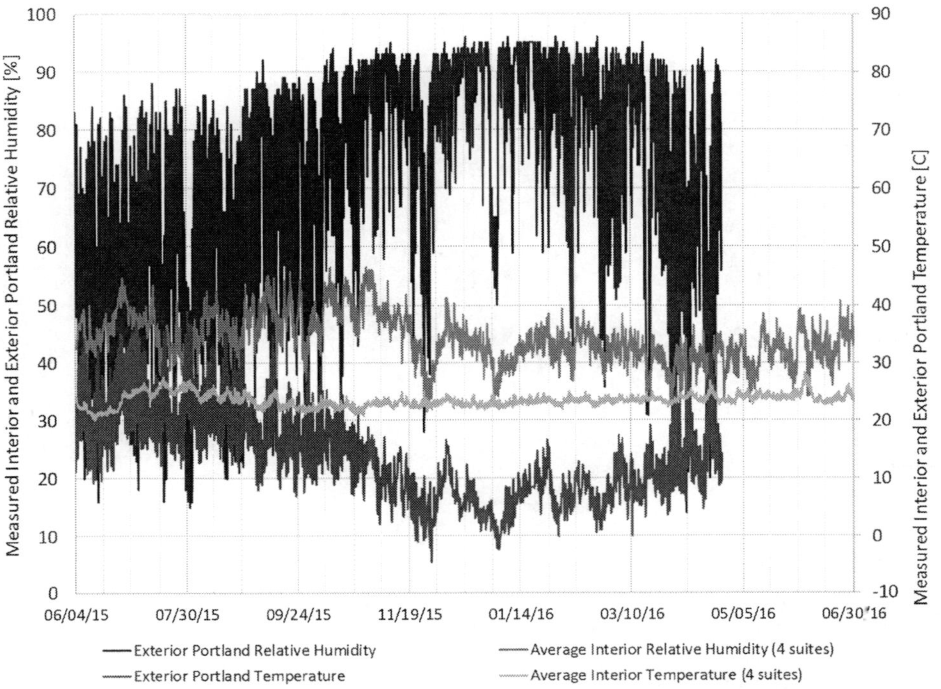

Figure 5 *Interior and exterior temperature and relative humidity (Portland).*

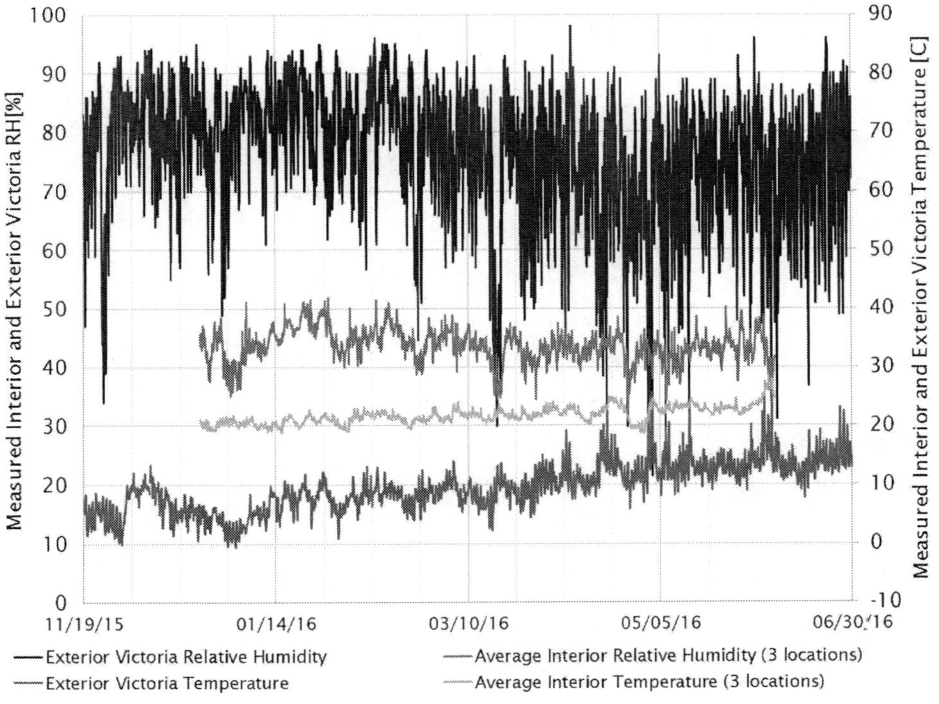

Figure 6 *Interior and exterior temperature and relative humidity (Victoria).*

insulation on wall performance is more significant to the durability of the enclosure during the coldest months when there is a risk of moisture accumulation. Because the two cities have similar climates, these figures can be used to demonstrate the effect on the enclosure of the two wall construction strategies. For example, the blue dashed lines of Figure 9 show that, at an

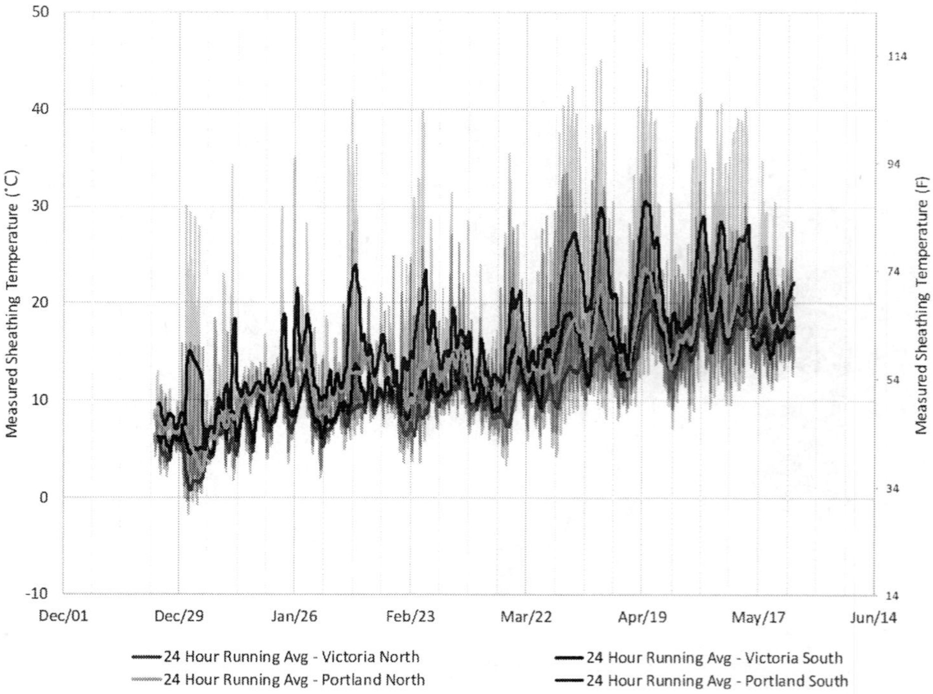

Figure 7 *Sheathing temperature and 24 h running average for Portland (blue) and Victoria (red), north and south orientations.*

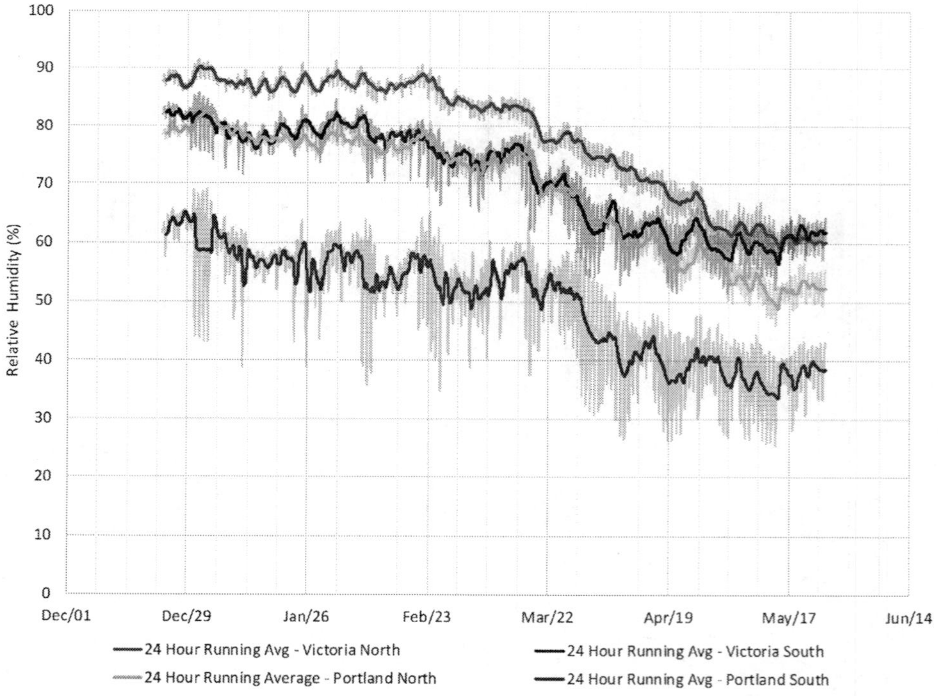

Figure 8 *Sheathing relative humidity and 24 h running average for Portland (blue) and Victoria (red), north and south orientations.*

exterior temperature of 3°C (37.5°F) in either city, the sheathing without exterior insulation (Victoria) will be approximately 4.5°C (40°F) on average and the sheathing with exterior insulation (Portland) will be approximately 7.5°C (45.5°F) on average. The exterior insulation is clearly shown to have a warming effect on the sheathing as indicated by warmer average sheathing temperatures in Portland for the same exterior temperature. This effect would be more

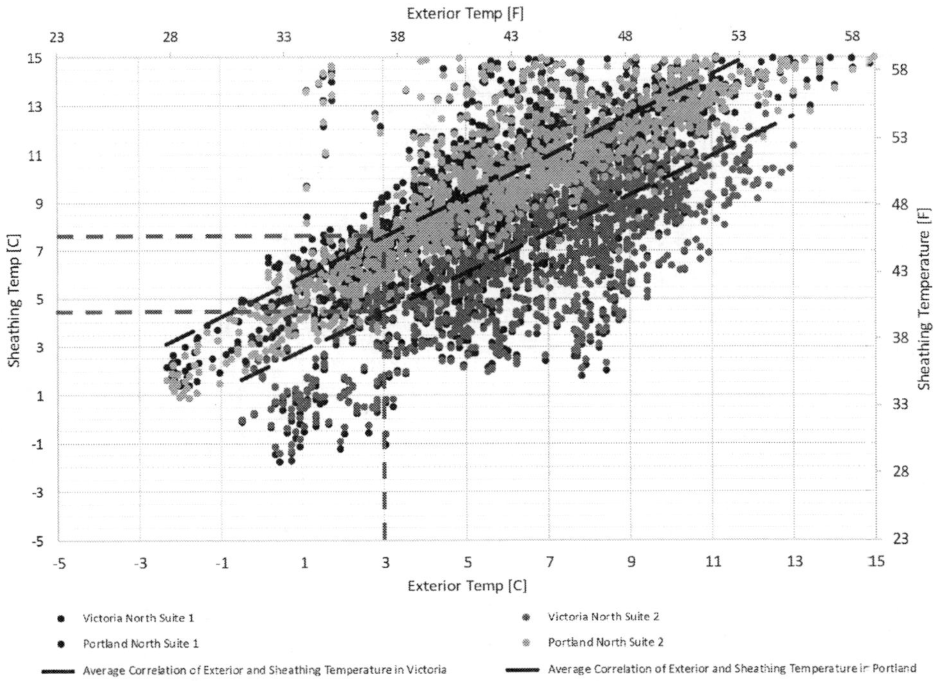

Figure 9 *Distribution of exterior temperatures and sheathing temperatures for Victoria (orange) and Portland (green), on the north orientation.*

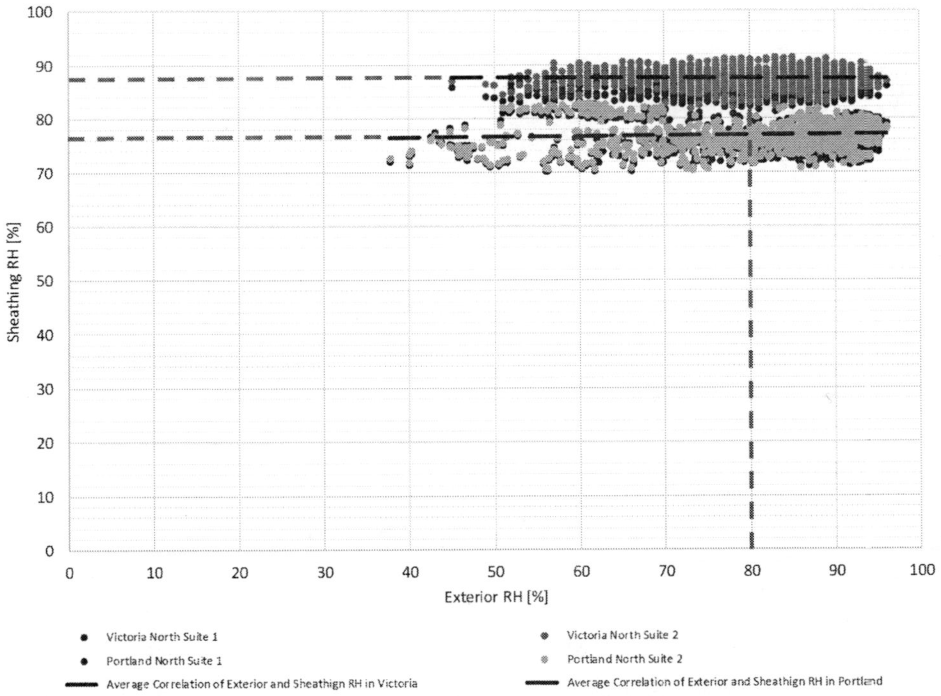

Figure 10 *Distribution of sheathing relative humidity for Victoria (orange) and Portland (green), on the north orientation.*

pronounced in colder climates, or with a more significant portion of the insulation R-value on the exterior of the sheathing relative to the total R-value of the wall.

As a result of the increased sheathing temperatures in the exterior-insulated (Portland) building demonstrated in Figure 9, Figure 10 shows that the relative humidity is decreased at the sheathing layer in Portland. For example, at a measured exterior RH of 80%, the average measured RH at the interior surface of the sheathing of the double-stud walls (Victoria) is approximately 88%, while in the exterior-insulated walls (Portland), the average measured sheathing RH is 76%.

Plotting the measured sheathing surface temperature and relative humidity and applying the VTT model under a sensitive moisture class (typical for wood board products), with a decline coefficient of 0.3, the mold indices for all eight evaluated walls are shown in Figure 11.

Using a mold index criterion of 3 as explained earlier, the results indicate no appreciable risk of mold growth on the sheathing based on measured temperature and relative humidity in either of the monitored buildings, despite the north walls of the Victoria building experiencing 30-day running average relative humidities and temperatures sufficient for concerns of mold growth. Most importantly, none of the exterior-insulated walls suggest the ability to harbor even microscopic levels of mold growth.

CONCLUSIONS

A literature review of field research results found that there were greater moisture-related durability risks with thick double-stud walls compared to exterior-insulated wall assemblies. It was concluded there were no increases to the moisture durability risks as a result of the addition of exterior insulation. This finding is supported by decades of field experience.

In this research program, analysis of the measured sheathing temperature and RH compared to exterior temperature and RH showed that the sheathing of the exterior-insulated wall assembly (Portland) had higher sheathing temperatures and lower sheathing RH for a given exterior temperature and RH, compared to the double-stud assembly (Victoria). Increasing the R-value ratio of exterior insulation to cavity insulation would further increase these effects.

Analysis using the VTT mold index model showed that neither wall assembly experienced measured conditions that show a risk of mold growth on the sheathing during the monitored period. The mold index was 0 for all four monitored locations on both orientations in Portland and on the south orientation in Victoria. On the north orientation in Victoria, the mold index reached approximately 0.4, which is very low. Initial stages of microscopic mold start at a mold index of 1.

Both of the monitored buildings in the current study meet very strict airtightness levels to achieve the Passive House certification, so the workmanship and attention to detail of the air barrier system was higher than in standard construction practices. As well, it is important to note that failure to

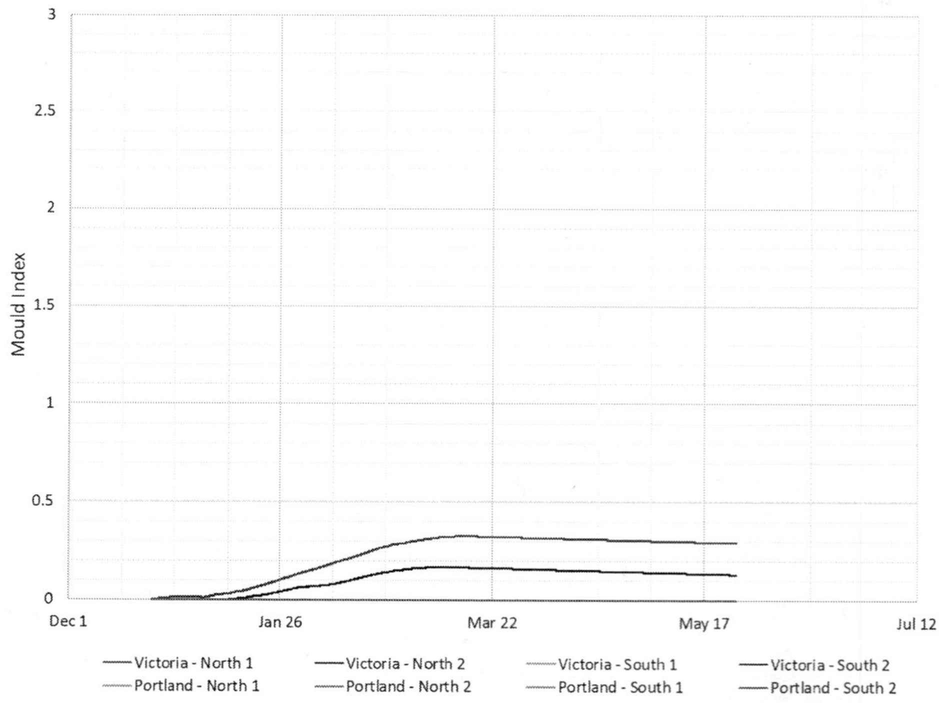

Figure 11 *Mold index results for Victoria and Portland, modeled from December 2015 to May 2016. The mold index was 0 for all four monitored locations in Portland and on the south orientation in Victoria.*

adequately control precipitation will eventually result in the failure of almost any wall system, regardless of the design. This means that monitored enclosure walls that show good performance are not immune from failure should important rainwater management details be missed.

Moisture-related durability also relates to the drying ability of the assembly, which was not addressed in this research program because determining drying ability requires intentionally wetting the wall systems. Intentional wetting of wall systems is typically done on test hut walls, rather than real buildings, because of the potential for damage.

The results of this study support the substantial body of research suggesting that exterior-insulated assemblies can be used without increasing moisture-related durability risks.

ACKNOWLEDGMENTS

The authors would like to thank ROXUL Inc. and Canadian Mortgage and Housing Corporation for providing the monitoring data in Portland, OR, and Victoria, BC, for this research study.

REFERENCES

Arena, L., P. Owens, and P. Mantha. 2013. *Measured performance of an R-40 double-stud wall in climate zone 5A*. Thermal Performance of the Exterior Envelopes of Whole Buildings XII International Conference, Clearwater, FL.

Fox, M., J. Straube, H. Ge, and T. Trainor, T. 2014 F*ield test of hygrothermal performance of highly insulated wall assemblies*. Presented at the Canadian Conference on Building Science and Technology, Toronto, ON. October 28–30.

Handegord, G.O.P. 1985. Prediction of the moisture performance of walls. *ASHRAE Transactions* 91(2).

Hutcheon, N.B. 1964. CBD-50 principles applied to a masonry wall. *Canadian Building Digest*. Ottawa: National Research Council of Canada.

IEA. 1991. Sourcebook. Report of Annex XIV, Volume 1. Leuven, Belgium: International Energy Agency

Kane, R., and G. Titley. 1987. Case histories of moisture monitoring in residential walls. *Thermal Insulation: Materials and Systems* ASTMSTP 922, 615–29.

Lstiburek, J, K. Ueno, and S. Musunuru. 2015. *Modeling enclosure design in above grade walls* Building America Report. Gaithersburg, MD: National Renewable Energy Laboratory.

Onysko, D., C. Schumacher, and F. Garrahan. 2010. *Field measurement of moisture in building materials and assemblies: Pitfalls and error assessment*. Proceedings of the Building Enclosure Science and Technology Conference (BEST2), Portland, OR.

Passive House Institute. 2014. *Passive house requirements*. Darmstadt, Germany: Passive House Institute. www.passiv.de/en/02_informations/02_passive-house-requirements/02_passive-house-requirements.htm.

Straube, J.F., D. Onysko, and C. Schumacher. 2002. Methodology and design of field experiments for monitoring the hygrothermal performance of wood frame enclosures. *Journal Of Thermal Insulation and Building Envelopes* 26(2):123–51.

Straube, J. 2011. BSD-163: Controlling cold-weather condensation using insulation. *Building Science Digests*. Westford, MA: Building Science Corporation.

Tsongas, G.A.,1991. The effect of exterior insulating sheathing on wall moisture. Insulation Materials: Testing and Applications, vol. 2. ASTM STP 1116.

Ueno, K. 2015. Monitoring of double-stud wall moisture conditions in the northeast. *Building America Report*. Gaithersburg, MD: National Renewable Energy Laboratory.

Viitanen, H., and T. Ojanen; 2007. *Improved model to predict mold growth in building materials*. Proceedings: Thermal Performance of the Exterior Envelopes of Buildings X. Atlanta: ASHRAE.

Using the Thermal Mass of a Building to Reduce the Magnitude of the Peak Power Demand of the Primary Heating System—A Whole-Building Simulation with Parametric Analysis

Victor Fransson

Dennis Johansson
Associate Member ASHRAE

Hans Bagge, PhD
Associate Member ASHRAE

ABSTRACT

The idea of using the thermal mass of a building to maintain a more constant indoor climate is not new. This particular property of a building, including its exterior envelope, plays an important part in the reduction of peak power demands, which is becoming an increasingly urgent issue for the heating energy suppliers. The latest smart-grid technology creates new opportunities where communication between customers and heating energy suppliers, and the resulting invoicing on a shorter time-basis, is concerned. Smart-grid technology can be used to lower the primary energy use as the magnitude of the power peaks and, consequently, their impacts can now be reduced. Reducing peak power demand will, however, impact the indoor temperature. At present, there is a lack of knowledge regarding how different factors determine the magnitude and rate of temperature drop when power levels are reduced. These factors include a combination of the heat transmittance, airtightness, and thermal mass of the exterior envelope, as well as the internal thermal mass of the building, the building services, and occupant behavior over time. In this study, the IDA-ICE *building simulation software has been used to perform whole-year simulations of an existing apartment block with 15 apartments. Different power reduction schemes were tested and the impact of variables, such as household electricity usage, occupancy levels, and outdoor climate, were analyzed. The influence of different building envelope parameters, including thermal capacity, insulation levels, and airtightness, were also analyzed. The purpose of this study was to investigate how combinations of these variables and parameters affected the indoor temperature drop during the different power reduction schemes.*

INTRODUCTION

Reducing energy usage is one of the many challenges the building industry has to face. However, a more recently recognized problem is not the energy use itself but how much power is required and the point in time when it is needed. If the simultaneous production and demand levels for power are examined, there are always occasions when these levels will be mismatched. However, this mismatch is not caused by an actual shortage of power (i.e., when the maximum capacity on the supply side is reached) but by the way in which the power is produced or distributed. Mismatches occur during the heating season when demand is high: when cheap fuel is limited there are economic issues, and when fossil fuels are needed to meet demand there are environmental issues. In the case of district heating, which is a major source of heating for multi-dwelling buildings in Sweden, there are other factors to consider, such as the efficiencies of boilers, the starting up and shutting down of boilers, and the capacity of the pipes supplying different parts of the city. There is also the question of significant investments being required to lay new pipes in order to add new customers. To summarize, this paper is based on the fact that utility companies want to reschedule when their power production takes place, as it is sometimes either unnecessarily expensive or "dirty" in terms of CO_2 emissions.

One way to address this issue it to use demand-side management, which has been made possible thanks to the new smart-grid technology, whereby the demand side modifies its heating requirements in order to lessen the demand on the heating supplier. Extensive studies where carried out in the 1980s concerning the benefits of thermal mass during the heat-

Victor Fransson is a doctoral student and **Dennis Johansson** is head of Building Services Division of the Department of Building and Environmental Technology, Faculty of Engineering, Lund University, Lund, Sweden. **Hans Bagge** is an associate professor in the Building Physics Division of the Department of Building and Environmental Technology, Lund University, Lund, Sweden.

ing season. Burch et al. (1984) carried out measurements that were verified by simulations. They showed that having a high thermal mass was advantageous when temperatures were allowed to deviate from the thermostat set points, a consequence of the building structure being charged by internal heat loads. A test to use the thermal mass of buildings in a smart-grid system, by offsetting the supply temperature to the hydronic heating system (Kärkkäinen et al., 2003), was performed and resulted in a reduction of the heating load by up to 25%, with an indoor temperature drop of only 2°C. Werner and Olsson Ingvarsson (2008) carried out measurements in apartments while offsetting the outdoor sensor by up to 10°C, causing the control system to turn the heat off. This, in turn, caused the indoor temperature to decline by up to 2°C. Johansson et al. (2012) had another approach when investigating a system in which an oil-fired heating system handled the peak loads: a scheme to avoid using the heating system as much as possible, by using the properties of the heavy structure of the building, was devised and good opportunities for reducing CO_2 emissions were observed without compromising the indoor climate. Kensby et al. (2015) had a similar approach to that of Werner et al, and offset the outdoor temperature sensor in order to discharge and charge the building in their pilot test. The test was performed in five multi-dwelling buildings, constructed between 1920 and 1940, and the temperature was measured in two apartments in each building. The impact on the indoor temperature was, in the worst case, 0.5°C as a weekly average. Reynders et al. (2013) showed, using simulations of a single-family building, the gains achieved when using its thermal mass and varying the heating loads over time by using a heat pump for which the price of electricity varied. An underfloor heating and radiator system was investigated, and it was found that the peak power demand reductions were between 75% and 95% if a 4°C drop in the design indoor temperature was accepted.

Aim

As stated above, many field tests have been carried out to study this issue, and these have shown promising results. Field tests are usually subject to limited time, money, or access, and although they are authentic, they are usually performed over a rather short period and in a limited number of cases. This study aims to show the impact of reducing the power load during each day of whole heating seasons, of different heating seasons, for different building envelope properties, and for randomized internal loads in 15 apartments. The simulation variables chosen to focus on in this paper are weather, occupancy, and household electricity loads combined with the building envelope properties' U-factor, thermal mass, and air leakage. A whole-building simulation covers differences between apartments due to geometry. For example, corner apartments can have a much larger surface area exposed to the outdoor conditions, and facades with different orientations can be subject to different wind effects. A complete shutdown of the heating system is used to get a clear picture of its impact on the result and to avoid having to deal with the problem of turning off the system in practice. One proposed way of achieving this could be that suggested by Kensby et al. (2015) in which the supply temperature of the water that feeds the hydronic heating system is reduced. However, here it is of greater importance to see how the building reacts with regard to internal loads and the outdoor climate. The reason for not using a more sophisticated model for determining the point in time to reschedule the loads is because, as mentioned in the introduction, it varies depending on the specific location and grid layout, fuel prices, boiler types, etc. This paper thus focuses on what happens in the different apartments in terms of temperature and, at a whole-building level, in terms of the power demand that could be reduced.

METHODS

The simulated building is an existing building in Malmö, in the south of Sweden, built in 2013. It has an apartment floor area of 1077 m^2 divided into 15 apartments, with sizes varying from 60 to 90 m^2. There is an underfloor heating system in each apartment embedded in the concrete floors. For the first and second simulation tests the parameters of the actual constructed building were used, and these can be found in Table 1. The ventilation rate for every simulation was 0.35 $L(s{\cdot}m^2)$, the rate required to fulfill the regulations stated by the Swedish Building Agency, with a heat exchanger with a temperature transfer efficiency of 78%. The design temperature in all the apartments was 21°C.

Table 1. Building Envelope Areas and Thermal Properties

Building Envelope	Area, m^2	U/W, m^2K	U·A, W/K	% of Total
Walls	630	0.16	100.5	16.2
Roof	330	0.17	56.2	9.1
Windows	199	1.36	270.5	43.6
Ground	330	0.07	25.0	4.0
Thermal bridges	—	—	168	27.1

The *IDA-ICE* 4.6.2 (EQUA 2014) building simulation software was used for this study. The model was created in *IDA-ICE* and run through its batch mode using MATLAB (The MathWorks 2015) for automation and changing variables and parameters.

Simulation Setups

Three different simulation setups were used. For each of them, the household electricity load and occupancy load were the profiles shown in Figure 1. The household electricity load was derived from hourly measurements in over 1000 apartments in Sweden during a two-year period (Bagge et al. 2012). The occupancy profiles were derived from measurements using electronic diaries in 86 apartments over a period of 12 to 13 days (Bagge and Johansson 2015). In order to randomly select these variables, the percentiles were assigned different probabilities covering the intervals around the existing five percentiles. For example, the probability of choosing the 50% percentile is, in this case, 25%, corresponding to covering the interval between 37.5% and 62.5% percentiles. The same probability was assigned for the 25% and 75% percentiles, whereas the 5% and 95% percentiles cover only half that interval and are thus assigned only 12.5% probability. This random assignment of internal loads was made prior to each simulation for each of the 15 apartments. The chosen household electricity profile was applied for all the days of the year, and the occupancy profile was randomized every two weeks, corresponding to the measured period from which the data was derived.

Due to the reasons discussed earlier, it is not known when it is most favorable for the heating supply company to desire a lower heating demand. Therefore, three different durations without supplying heat were investigated: two, four, and six hours. These three durations were then investigated using two different starting times, midnight (00:00) and six o'clock in the morning (06:00), resulting in six different ways in which the heating system could be turned off. These cutoff parameters were applied to each day of the simulation.

First Setup. The building envelope was as shown in Table 1. This setup was run through 60 heating seasons with randomized internal loads for each of the six different periods without heat.

Second Setup. The building envelope was as shown in Table 1. Twelve different climate years for the same location were investigated. The weather data was collected from the Swedish Meteorological and Hydrological Institute database (SMHI 2015a, 2015b). The internal load variables were randomized in the same manner as in the first study. Five heating seasons for each year were investigated for each of the six cutoff periods without heat.

Third Setup. In this study, the impact of the building envelope was investigated. The climate year was kept the same and the variables were changed in the same way as before. Building envelopes from three different periods were investigated: building properties were obtained from Bagge et al. (2012) and are presented in Table 2. The three building envelope types are passive house (PH), a standard Building Code building (Std), and a typical building from the 1960s. Changes to the exterior envelope where made, while keeping the heat transfer coefficient

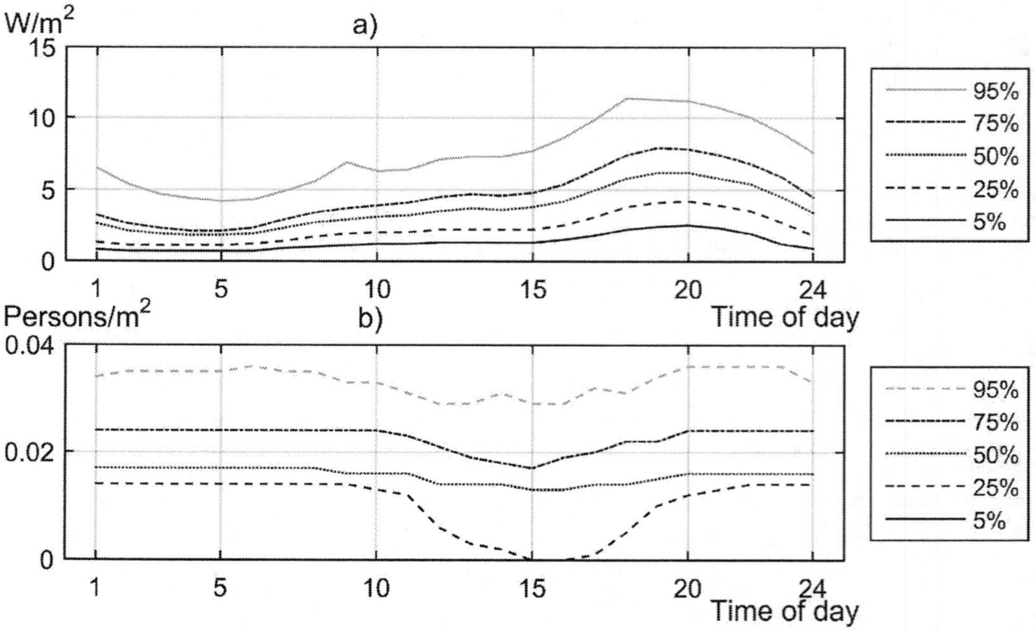

Figure 1 *(a) Household electricity load in W/m^2 apartment area for each hour of the day, shown in percentiles from measured data. 1 W/m^2 = 0.32 Btu/h ft². (b) Number of persons per square meter of apartment area present during each hour of the day, shown in percentiles from measured data. 1 m^2 = 10.76 ft².*

constant. Two cases of thermal mass were investigated for each of the building types, heavy and light. In the heavy case, concrete was added on to the interior sides, 10 cm to exterior walls and 20 cm to the roof. In the light case, there were curtain walls (non-load-bearing facades) with insulation, wooden studs, and a 13 mm gypsum board on the interior side. Three different areas of internal walls were also added (in addition to the walls separating the apartments): 20, 40, and 60 m^2 respectively. The interior walls were simulated as 10 cm concrete walls or as curtain walls in the same way as the exterior walls but with double gypsum boards on both sides. Each of the 18 cases were investigated during five heating seasons for each of the six periods without heat.

RESULTS

The results are presented as distribution functions for the first two setups and summarized in tables; the results from the third study are shown only in table form. Each data point regarding temperature corresponds to the minimum operative temperature in one apartment during the period without power. The power demand is for the whole building and each data point corresponds to the heating demand for all 15 apartments just before the heating system is turned off each day. The percentage results of each setup should not be compared to one another, as they are derived from different numbers of simulations.

First Setup

The results from the first setup are shown in Figure 2 and Table 3. Notable, but reasonable, is that most power can be saved during the night. As discussed previously, the power demands on the heating supply companies can be due to a number of different factors. Not only the outdoor temperature but also behavioral aspects (such as high use of domestic hot water, which is prominent during the morning hours) affect the demands. Turning off the heating during the morning hours almost never causes the temperature to drop more than 0.5°C from the design value, thus "saving" up to 30 kW in power demand.

Second Setup

Three years that differed visibly were chosen and are shown in Figure 3. Here, all the six cutoff periods are combined into a single graph for each year. The results from all twelve heating seasons are summarized in Table 4. For most of the years, the results are similar. However, they can differ. Looking at the two extremes, in the 0.5°C temperature drop there is a twofold spread, which increases to a fivefold spread in the 1°C temperature drop.

Table 2. Building Envelope Properties for the Different Periods

Building Component	PH Values	Std Values	1960s Values
Exterior wall	0.1 W/(m^2·K)	0.18 W/(m^2·K)	0.4 W/(m^2·K)
Roof	0.08 W/(m^2·K)	0.13 W/(m^2·K)	0.25 W/(m^2·K)
Window	0.8 W/(m^2·K)	1.3 W/(m^2·K)	2 W/(m^2·K)
Ground	0.1 W/(m^2·K)	0.18 W/(m^2·K)	0.4 W/(m^2·K)
Air leakage	0.3 l/(s m^2) ext. area at 50 Pa	0.6 l/(s m^2) ext. area at 50 Pa	1.2 l/(s m^2) ext. area at 50 Pa
Thermal bridges	20% (U·A)	20% of U·A	10% of U·A

Table 3. Summary of the Distribution Functions

Time	Lowest Temp.	Probability of <0.5°C	Probability of <1°C	Highest Demand, kW	50% Percentile, kW	90% Percentile, kW
00–02	19.9	14	0	37.6	14.7	23.0
00–04	19.3	43	3	37.7	15.0	23.3
00–06	19.0	60	10	38.6	15.9	24.2
06–08	20.3	1	0	29.4	11.8	18.8
06–10	20.0	5	0	29.3	12.3	19.1
06–12	19.8	10	0	29.4	12.3	19.2

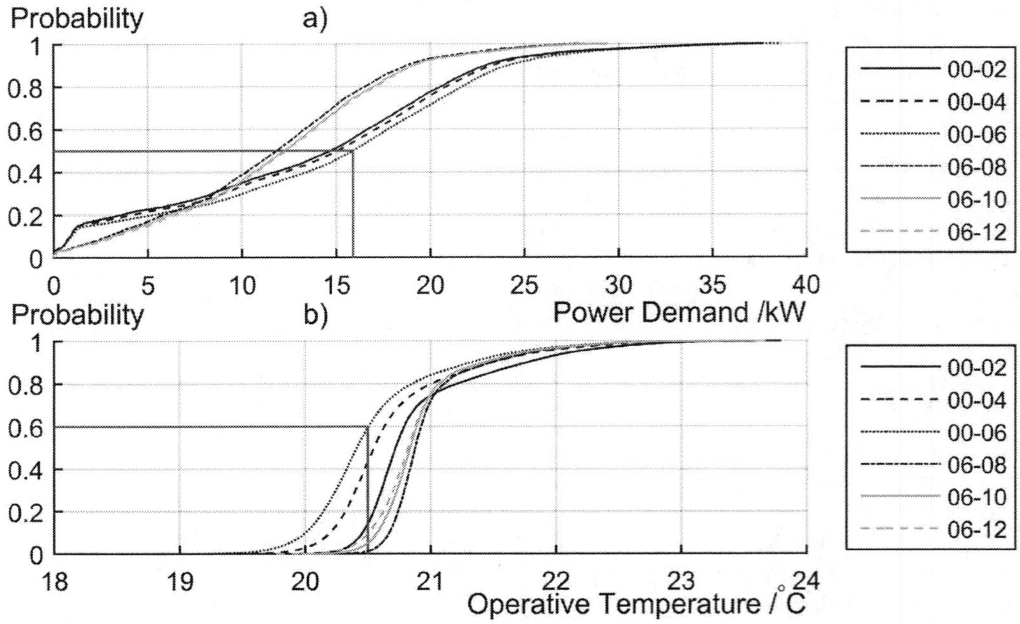

Figure 2 *(a) Distribution of power demand shown for each of the six different durations without power. 1 kW = 3412 Btu/h. (b) Distribution of minimum operative temperature for each of the six different durations without power. 20°C = 68°F. In (a) each curve contains the power demand of each day of the heating season for the whole building just before the power cutoff. The red line shows median power demand for the cutoff period between 00–06 (15.9 kW). In (b) each curve shows the temperature after the cutoff each day of the heating season for the 15 apartments. The red line illustrates the probability of having a lower temperature than 20.5°C for the same cutoff period (00–06).*

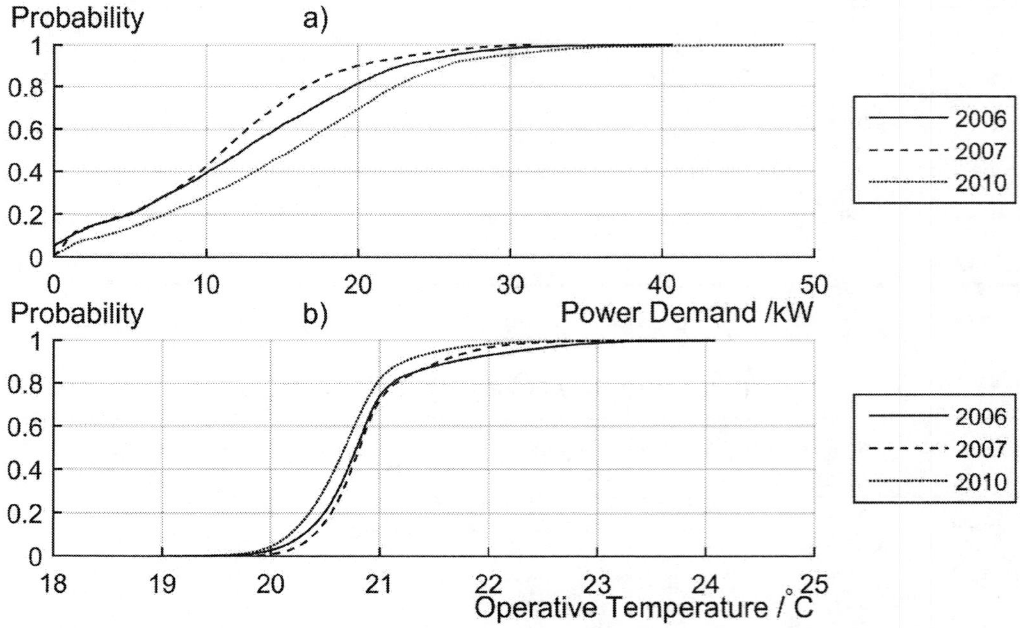

Figure 3 *(a) Distribution of power demand shown for three different heating seasons. The different durations without heat are combined into a single graph. 1 kW = 3412 Btu/h. (b) Distribution of minimum operative temperature. The different durations without heat are combined into a single graph. 20°C = 68°F.*

Table 4. Distributions during All the Investigated Years

Year	Lowest Temp.	Probability of <0.5°C	Probability of <1°C	Demand Highest, kW	50% Percentile, kW	90% Percentile kW
2002	19.2	21	2	43.5	13.2	21.6
2003	18.7	23	3	43.1	13.3	23.1
2004	19.0	21	2	34.9	13.1	21.9
2005	18.8	21	2	37.7	12.6	23.5
2006	18.8	21	3	40.6	12.4	22.8
2007	19.2	16	1	31.4	11.1	20.0
2008	19.2	18	1	33.8	11.6	20.5
2009	19.1	22	2	37.6	13.4	22.0
2010	18.6	32	5	48.0	15.6	25.8
2011	19.1	20	2	35.4	12.3	22.0
2012	18.9	21	2	39.7	12.3	23.9
2013	18.9	23	3	40.6	13.1	23.8

Third Setup

The results from the third setup are summarized in Table 5. Many combinations were investigated, and it is clear that the heavy, more massive buildings offer more advantages when rescheduling power loads; looking at the 1960s case there is twelve-fold difference for the 1°C drop. It can also be seen that the passive house construction has a very low power demand in absolute terms, rendering it a poor subject for heating suppliers and to warrant power load rescheduling. Adding more mass indoors also clearly lowers the risk of unwanted low temperatures when subjecting apartments to a heating power cutoff.

CONCLUSION AND DISCUSSION

Simulation software is a very powerful tool when there is a need to handle many variables and different parameters. It makes it possible to investigate factors that would be impossible to apply in real cases. However, as it only provides simulations, the results should be viewed as indicators. It is not the investigated temperatures themselves that are the major advantage of using a simulation software but being able to compare different simulation setups with one another.

Randomizing occupancy and household power demand is one way of imitating reality. The profiles used might sometimes seem imprecise, as the same profiles are applied for long periods of time. But, as two of the variables are randomized, namely household power demand and occupancy, with the latter over a shorter time period, two weeks, numerous different combinations of these will arise. Repeating them over many heating seasons allows these combinations to occur at different times of the year. This is one way of trying to handle such stochastic variables as occupancy and appliances. However, it should be noted that the values used are averaged percentiles and, for a more realistic simulation, real hourly data of these two variables on a yearly basis would be needed.

It is clear that the most critical case is the six-hour cutoff from midnight, as this would take place during the coldest hours of the day and also the hours with low internal loads. A complete cutoff of power would be quite drastic and would probably not be accepted in reality. However, reducing the power instead of cutting it off completely would result in less impact on the indoor temperature. This paper shows the worst case in terms of temperature drop and maximum reduction of power.

The majority of the buildings in Sweden that use a lot of energy (power) have similar properties to the one representing a typical 1960s building in this study. As they have poor building envelopes and consequently high power demands, they are also the buildings with the best prospects of reducing unfavorable power demands. The results show that the heavy 1960s buildings hardly ever experience a greater drop than 1°C, as shown in previous field tests. However, many of the buildings of this period do not have heat recovery in their ventilation systems, which is the case in this study. On the other hand, these buildings usually have natural ventilation, turning air leakage into ventilation losses instead of having both ventilation and leakage losses. Another factor is the use of underfloor heating, which can be presumed to be favorable in comparison to radiators or airborne heating systems, as the mass of the building structure is directly connected to the heating system.

Table 5. Distributions of All the Investigated Envelope Combinations

House Type	Mass	Added Interior Wall Area, m^2	Lowest Temp.	Probability of <0.5°C	Probability of <1°C	Max. Demand, kW	50% Percentile, kW	90% Percentile, kW
PH	Light	20	19.5	9	2	3.2	1.5	2.1
		40	19.7	6	1	3.2	1.4	2.1
		60	19.7	4	0	3.2	1.4	2.1
	Heavy	20	20.2	1	0	3.2	1.5	2.1
		40	20.3	1	0	3.2	1.5	2.1
		60	20.3	0	0	3.2	1.4	2.1
Std	Light	20	18.7	22	2	38.6	13.5	23.3
		40	18.9	17	1	39.0	13.4	23.3
		60	19.0	14	1	40.9	13.4	23.5
	Heavy	20	20.0	5	0	40.7	13.5	23.8
		40	20.1	3	0	40.7	13.8	24.0
		60	20.1	2	0	40.9	13.3	23.3
1960s	Light	20	17.5	38	12	56.6	23.0	35.3
		40	17.8	32	8	56.9	22.6	35.5
		60	18.0	27	6	59.3	22.5	36.3
	Heavy	20	19.3	18	1	56.4	21.6	34.6
		40	19.4	15	0	57.1	21.6	35.1
		60	19.4	13	0	57.0	21.5	35.1

Finally, we used the instantaneous power demand just before the cutoff, which, of course, is what the heating supply companies would save at that moment. However, what would happen in the hours to come is harder to say. The power demand is dependent on the outdoor conditions as well as the internal loads and, if these remain unchanged, it can be presumed that a constant and reduced level of power demand would be experienced throughout the cutoff period.

REFERENCES

Bagge, H., and D. Johansson. 2015. Measured occupancy levels in apartments and the consequential simulated energy benefit from demand controlled ventilation. *ASHRAE Transactions* 121.

Bagge, H., L. Lindstrii, and D. Johansson. 2012. Brukarrelaterad energianvändning Resultat från mätningar i 1300 lägenheter. (In Swedish).

Burch, D.M., D.F. Krintz, and R.S. Spain. 1984. Effect of wall mass on winter heating loads and indoor comfort—An experimental study. *ASHRAE Transactions* 94–121.

EQUA. 2014. *IDA ICE*, 4.6 ed. Stockholm: EQUA Simulation AB.

Johansson, C., F. Wernstedt, and P. Davidsson, P. 2012. Distributed thermal storage using multi-agent systems. 1st International Conference on Agreement Technologies, AT. Dubrovnik.

Kensby, J., A. Truschel, and J.O. Dalenback. 2015. Potential of residential buildings as thermal energy storage in district heating systems—Results from a pilot test. *Applied Energy* 137:773–81.

Kärkkäinen, S., K. Sipilä, L. Pirvola, J. Esterinen, E. Eriksson, S. Soikkeli, M. Nuutinen, H. Aarnio, F. Schmitt, and C. Eisgruber. 2003. Demand side management of the district heating systems. VTT. VTT Research Notes 2247.

Reynders, G., T. Nuytten, and D. Saelens. 2013. Potential of structural thermal mass for demand-side management in dwellings. *Building and Environment* 64:187–99.

SMHI. 2015a. SMHI Öppna data. http://opendata-catalog.smhi.se/explore/.

SMHI. 2015b. STRÅNG—A mesoscale model for solar radiation. http://strang.smhi.se/.

The Mathworks, I. 2015. *MATLAB*. R2015a ed. Natick, MA: The MathWorks, Inc.

Werner, S., and L. Olsson Ingvarsson. 2008. Building mass used as short term heat storage. The 11th International Symposium on District Heating and Cooling. Reykjavik, Iceland.

Implementation and Validation of a Long-Wave Heat Exchange Model

Matthias Winkler **Florian Antretter** *Member ASHRAE* **Jan Radon, PhD**

ABSTRACT

The hygrothermal whole-building simulation software WUFI® Plus was extended with a new model: the calculation of long-wave radiative heat exchange between surfaces inside a zone. This allows a more detailed calculation of the interior radiative and surface temperatures, which also results in a better simulation of indoor climate conditions. The model also takes into account the optical properties of indoor surfaces, such as low-e coatings. Further benefits are a more comprehensive comfort assessment, for example, for asymmetric radiant temperature. Radiant temperatures can also be calculated for every position inside a zone. The model uses Gebhart factors for the calculations of long-wave heat exchange between surfaces inside a zone. These factors describe the fraction of emitted energy from one surface that is absorbed at another. They are determined from view factors, which only describe geometrical relationships between surfaces. WUFI® Plus uses a numerical triangulation method to calculate the view factors. As the Gebhart factors also consider optical properties of surfaces, all possible radiation paths, including multiple and own reflections, are taken into account for the calculations. Single parts of the calculation process are successfully validated with analytical solutions: the view factor calculation of a nontrivial room geometry and the long-wave heat exchange between two parallel plates show only minimal deviations from the analytical results. The German standard DIN EN 13791 provides tests for the validation of interior long-wave heat exchange. The model was successfully validated with them. In an exemplary application case, simulations with the new model are used as a preliminary study to determine the required accuracy and positioning of measurement equipment of a future test room. The simulated surface temperatures were successfully used as a plausibility control. The measurements will later be used as an experimental validation test for the new model.

INTRODUCTION

WUFI® Plus (Fraunhofer IBP 2016) is a hygrothermal whole-building simulation software developed at the Fraunhofer Institute for Building Physics IBP in Germany. It couples hygrothermal component simulation with a simulation of the indoor environment and additional modules such as dynamic and detailed HVAC systems, dynamic simulations of three-dimensional objects such as thermal bridges, and an airflow model calculating the air exchange between zones and the environment. WUFI® Plus also allows simulation of surface heating or cooling systems, for example thermo-active building systems (TABS). Up to now, the radiative heat exchange on surfaces was calculated by a constant radiative heat exchange coefficient for each surface. This is a legitimate assumption, as the surface temperatures in buildings are often close to each other. If they differ more, for example with surface heating or cooling systems, a more in-depth calculation is necessary. This can be accomplished with a detailed calculation of the inner long-wave heat exchange between the zone's surfaces. It allows a more detailed calculation of the interior radiative and surface temperatures, which also takes into account the optical surface properties, such as low-e coatings. Further benefits are a more comprehensive comfort assessment, for example, for asymmetric radiant temperature. Radiant temperatures can also be calculated for every position inside a zone.

Matthias Winkler is a scientist and **Florian Antretter** is the group manager in the Hygrothermal Building analysis group in the Hygrothermics department at the Fraunhofer Institute for Building Physics, Valley, Germany. **Jan Radon** is a professor at the University of Krakow, Krakow, Poland.

Other software tools also calculate the long-wave radiation exchange between surfaces inside a zone. EnergyPlus (LBNL 2015) uses the so-called "ScriptF"-concept (Hottel and Sarofim 1967), where all radiative exchange paths between two surfaces are part of an exchange coefficient, including reflection, absorption, and re-emission. These coefficients are based on the calculation of direct view factors, which are approximated by the software. TRNSYS (Aschhaber et al. 2008) uses so-called Gebhart factors to determine the radiative exchange between surfaces. These are also based on the calculation of view factors, which are defined as the diffuse radiation emitted by a surface that directly hits another surface. View factors describe only a geometrical relationship and take no optical properties of surfaces, such as emissivity or reflectivity, into account. The Gebhart factor describes the fraction of emitted energy from one surface that is absorbed at another. All possible radiation paths, including multiple and own reflections, are considered, as the optical properties of surfaces are part of the calculation. Both calculation methods are mathematically equivalent to each other (Clark and Korybalski 1974).

Considering the interior long-wave heat exchange in building simulation tools offers many benefits and possibilities, but it has to be used with care: the software tool's methods rely on the correct calculation of view factors, which is a computationally very extensive task. Thus, approximations and internal fixes are used, and the calculation can still be erroneous. Also, new inputs are added, such as the optical properties of interior surfaces, which can have a heavy impact on the simulation results.

CALCULATION OF INTERIOR LONG-WAVE HEAT EXCHANGE IN WUFI® PLUS

The determination of interior long-wave heat exchange in WUFI® Plus consists of multiple steps. First, view factors are calculated for all surfaces inside a zone. The view factor F_{ij} is defined as the fraction of long-wave radiation emitted from a general surface i with a direct incidence on another general surface j (Siegel and Howell 2002). Thus, it can be calculated with Equation 1:

$$F_{ij} = \frac{1}{A_i}\int\int_{A_iA_j}\frac{\cos\theta_i \cdot \cos\theta_j}{\pi \cdot r^2}dA_i dA_j \qquad (1)$$

The geometry used in Equation 1 is explained in Figure 1. The angle θ_i is defined between the line r, which connects the midpoints of both surfaces A_i and A_j, and the normal vector n_i on pane A_i. Thus, θ_j is the angle between r and n_j normal to A_j.

The analytical calculation of the integral of Equation 1 is only possible for a limited number of shapes and mutual positioning. A collection of analytical solutions can be found in the work by Howell (2015). If the connecting line r partially crosses another pane or polygon, the analytical integration becomes more complicated or in most cases even impossible. For that reason, surfaces i and j are triangulated into n triangles on surface i and m triangles on surface j. The triangulation process is described in the next paragraph. View factors are then calculated numerically for each pair of triangles according to Equation 2:

$$F_{ij} = \frac{1}{A_i}\sum_{i=1}^{n}\sum_{j=1}^{m}\frac{\cos\theta_i \cdot \cos\theta_j}{\pi \cdot r^2}dA_i dA_j \qquad (2)$$

WUFI® Plus uses OpenGL for its three-dimensional visualization. The building envelope consists of outer and inner partitions that are represented by flat polygons. They can be edited directly in WUFI® Plus or imported from SketchUp or computer-aided design (CAD) software. Because only convex shapes can be rendered correctly, the polygons are divided into triangles using OpenGL tessellation WUFI® Plus divides the obtained triangles into smaller ones by division of the longest side of particular triangles into two equal sections and connecting with opposite vertexes. This process is repeated until the longest side of each triangle is smaller than a value defined by an assumed accuracy. Thus, a set of small patches is obtained, which tends to be equilateral triangles if the division is dense enough. Figure 2 shows an example of this triangulation process.

After the triangulation process is finished, the view factors are calculated according to Equation 2. For every pair of triangles in surfaces A_i and A_j, the connecting line r is defined. Then it is checked by using vector analysis if the connecting line r crosses any other partition or polygon. Both inner and outer partitions can overshadow surfaces (see Figure 3). Thus, all partitions are taken into account when overshadowing is calculated. This can make the calculation very time and processor intensive when high accuracy is assumed. WUFI® Plus uses parallel processing for this task to reduce computing time. For each surface the sum of view factors should be 1.0. The numerical calculation will hardly

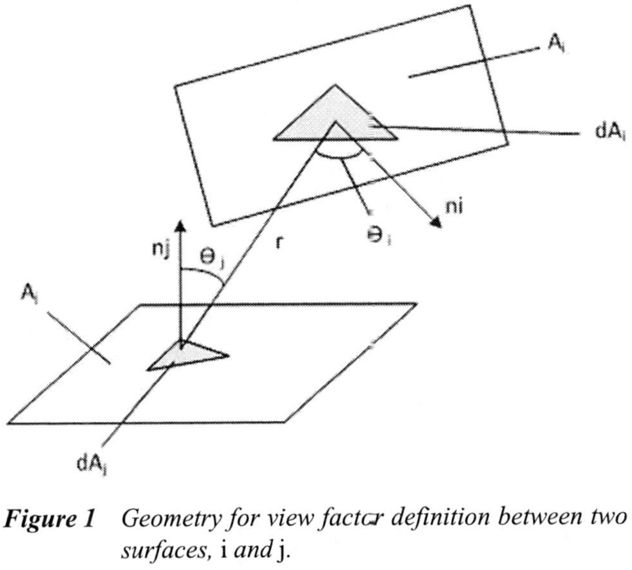

Figure 1 *Geometry for view factor definition between two surfaces,* i *and* j.

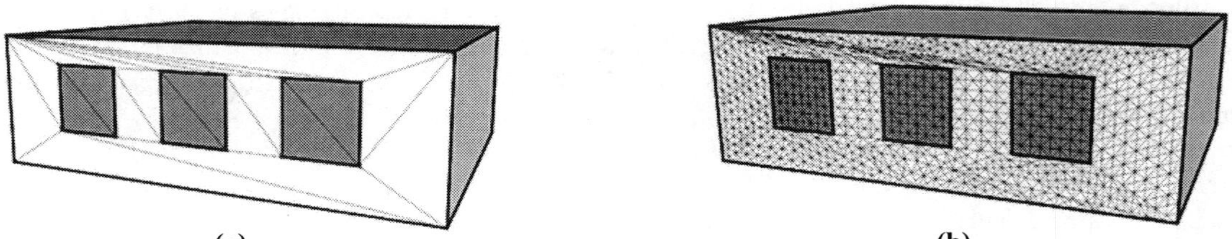

Figure 2 *A façade with three windows (a) tessellated by OpenGL and (b) divided into small triangles by WUFI® Plus.*

give the exact solution, so the view factors are adjusted automatically: the remaining difference to the exact solution of 1.0 is distributed according to the interior surfaces' area.

The next step in the calculation process is the Gebhart-factor G_{ij}. It describes the fraction of emitted energy from one surface *i* that is absorbed at another surface *j* by considering all radiation paths, including multiple and own reflections. The Gebhart factors *G* are calculated from the view factors *F* by considering the surfaces' long-wave reflectivity ρ and emissivity ε (Siegel and Howell 2002). To ease the calculation process, the Gebhart factors for *N* surfaces inside a zone are written as the *N*×*N* matrix *G*:

$$G = (I - F \cdot \rho)^{-1} \cdot F \cdot \varepsilon \tag{3}$$

This matrix equation is solved for all *N* surfaces inside a zone. Its individual parts are the Gebhart matrix *G* (Equation 4), an identity matrix *I* (Equation 5), the view factor matrix *F* (Equation 6), the reflectivity matrix ρ (Equation 7), and the emissivity matrix ε (Equation 8):

$$G = \begin{bmatrix} G_{11} & \dots & G_{1N} \\ \vdots & \ddots & \vdots \\ G_{N1} & \dots & G_{NN} \end{bmatrix} \tag{4}$$

$$I = \begin{bmatrix} 1 & \dots & 0 \\ \vdots & \ddots & \vdots \\ 0 & \dots & 1 \end{bmatrix} \tag{5}$$

$$F = \begin{bmatrix} F_{11} & \dots & F_{1N} \\ \vdots & \ddots & \vdots \\ F_{N1} & \dots & F_{NN} \end{bmatrix} \tag{6}$$

$$\rho = \begin{bmatrix} \rho_1 & \dots & 0 \\ \vdots & \ddots & \vdots \\ 0 & \dots & \rho_N \end{bmatrix} \tag{7}$$

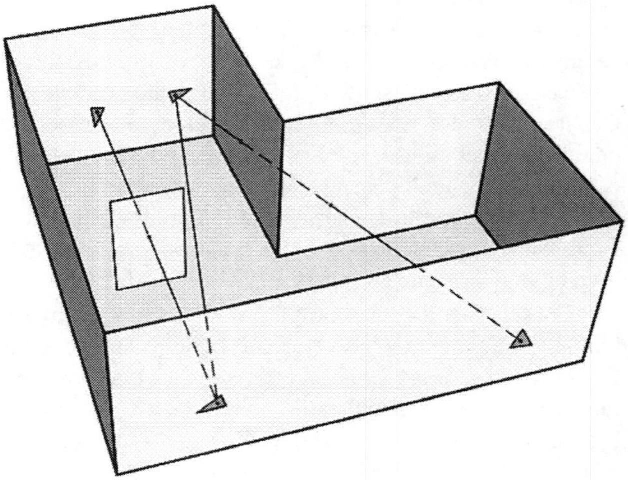

Figure 3 *Calculation of view factors with overshadowing by inner and outer partitions.*

$$\varepsilon = \begin{bmatrix} \varepsilon_1 & \dots & 0 \\ \vdots & \ddots & \vdots \\ 0 & \dots & \varepsilon_N \end{bmatrix} \tag{8}$$

The long-wave radiation absorbed on an interior surface $I_{l,i}$ can be calculated from its emissivity ε, the Gebhart factors *G*, the Stefan-Boltzmann constant σ, and the temperatures of interior surfaces *T* (Siegel and Howell 2002):

$$I_{l,i} = \varepsilon_i \cdot \sum_{j=1}^{N} G_{ij} \cdot \sigma \cdot (T_i^4 - T_j^4) \tag{9}$$

VALIDATION

Before the model can be used in further studies it must be validated. Therefore, the simulation output was compared with analytical solutions and reference values from standards.

View Factors

The precise determination of view factors is of central importance for the calculation of the inner long-wave heat exchange. For many standard geometrical cases, analytical solutions are available that allow calculation of the view factors (Howell 2015). Thus the first validation step is comparing analytically calculated view factors with solutions of WUFI® Plus. Therefore, a nontrivial geometry of a rectangular room with an interior wall was used (see Figure 4). The view factors from the interior wall to the surrounding surfaces were used for the comparison. As WUFI® Plus supports calculations with different levels of accuracy, the view factors were calculated with each level. These levels differ in the density of the triangulation network: a higher number of triangles should result in more accurate view factors. The calculation results of the view factors in WUFI® Plus were not adjusted by surface area, allowing a more significant comparison. The results show good accordance of analytical and simulated data even in the low-accuracy levels (see Table 1).

Long-Wave Heat Exchange between Two Surfaces

The long-wave heat exchange between two surfaces can be calculated analytically for selected geometries. One of these is two identical parallel plates, where the distance between them is very small compared to their length and width. The heat exchange between these two surfaces can be calculated with Equation 10 (Glück 1990):

$$q_{1,2} = \frac{1}{\frac{1}{\varepsilon_1} + \frac{1}{\varepsilon_2} - 1} \cdot \sigma \cdot (T_1^4 - T_2^4) \qquad (10)$$

Emission, reflection and absorption are included in this term. Long-wave radiative exchange to the sides is neglected, as this influence is minimal (Glück 1990).

This type of geometry is used to compare the analytical calculation with the output of WUFI® Plus. The two plates are modelled as identical squares with edge length of 12.00 m. The distance between them is 0.20 m. Four tests were performed with varying surface temperatures and emissivity. Table 2 summarizes their boundary conditions and the results of long-wave heat exchange between the two surfaces for both solutions. An adjustment of view factors wasn't considered in the simulations, resulting in long-wave heat exchange to side

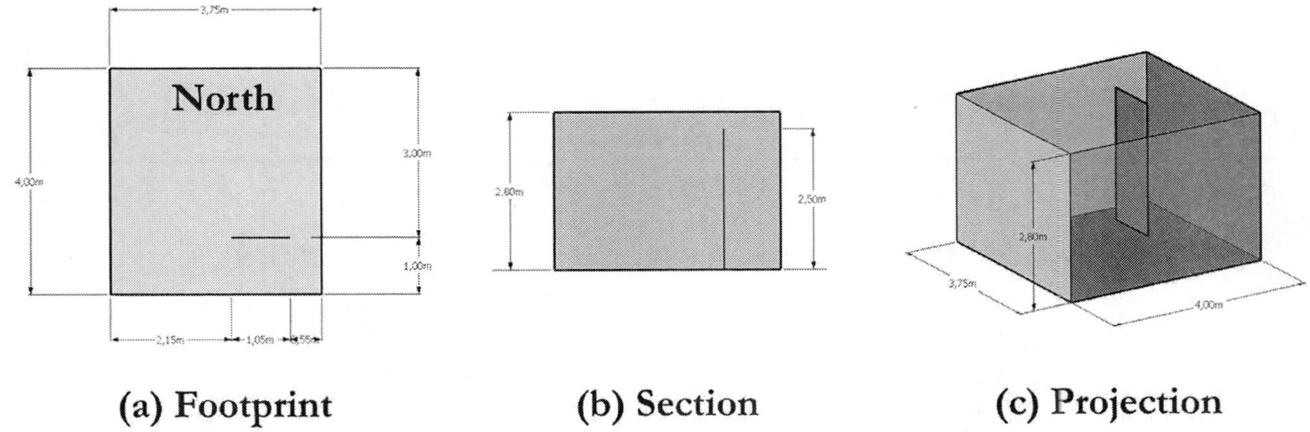

Figure 4 *Visualizations of the test room's geometry for testing of WUFI® Plus view factor calculation.*

Table 1. Comparison of View Factors from Interior Surface to Ambient Surfaces: Analytical Calculation and WUFI® Plus Results with Varying Calculation Accuracy

	Analytical Solution	WUFI® Plus: Low Accuracy	WUFI® Plus: Medium Accuracy	WUFI® Plus: High Accuracy	WUFI® Plus: Very High Accuracy
$F_{1,ceiling}$	0,2104	0,2172	0,2127	0,2109	0,2105
$F_{1,wall\ west}$	0,0787	0,0795	0,0788	0,0787	0,0786
$F_{1,floor}$	0,2538	0,2619	0,2618	0,2565	0,2549
$F_{1,wall\ north}$	0,2308	0,2324	0,2312	0,2309	0,2308
$F_{1,wall\ east}$	0,2262	0,2353	0,2285	0,2268	0,2264
Sum (control)	0,9999	1,0263	1,0129	1,0038	1,0012

surfaces being part of the simulation results. Thus, they were corrected by hand to represent the same assumptions as the analytical solution. The maximum deviation between analytical calculation and simulation is in all tests 0.30%. Thus, the validation of long-wave heat exchange for the test geometry of two identical parallel plates is rated as successful.

DIN EN 13791

The German standard DIN EN 13791 (ISO 2012) describes general criteria and validation procedures for thermal building simulation software that calculates interior temperature. The standard includes test cases for relevant heat transfer processes, including inner long-wave heat exchange. The standard describes four test rooms with only opaque surfaces with varying outdoor climate conditions. To comply with the standard, the indoor temperatures of these rooms have to be calculated for static conditions and meet given reference values. Figure 5 shows the room geometries.

One surface in each room differs from the others: it represents an exterior wall, with an exterior temperature of 30°C on its outer side and absorption of short-wave radiation of 100 W/m² on its interior side. The boundary conditions of all surfaces are summarized in Table 3.

Table 2. Long-Wave Heat Exchange between Two Parallel Surfaces: Comparison of Analytical Calculation and WUFI® Plus Results

	Test 1	Test 2	Test 3	Test 4
Surface temperature 1	50°C	50°C	30°C	30°C
Surface temperature 2	10°C	10°C	20°C	20°C
Emissivity surface 1	0.9	0.9	0.9	0.9
Emissivity surface 2	0.9	0.1	0.9	0.1
Long-wave heat exchange: Analytical solution	207.69 W/m²	25.11 W/m²	49.19 W/m²	5.95 W/m²
Long-wave heat exchange: WUFI® Plus	208.29 W/m²	25.11 W/m²	49.34 W/m²	5.95 W/m²
Deviation	+ 0.29%	+ 0.04%	+ 0.30%	+ 0.03%

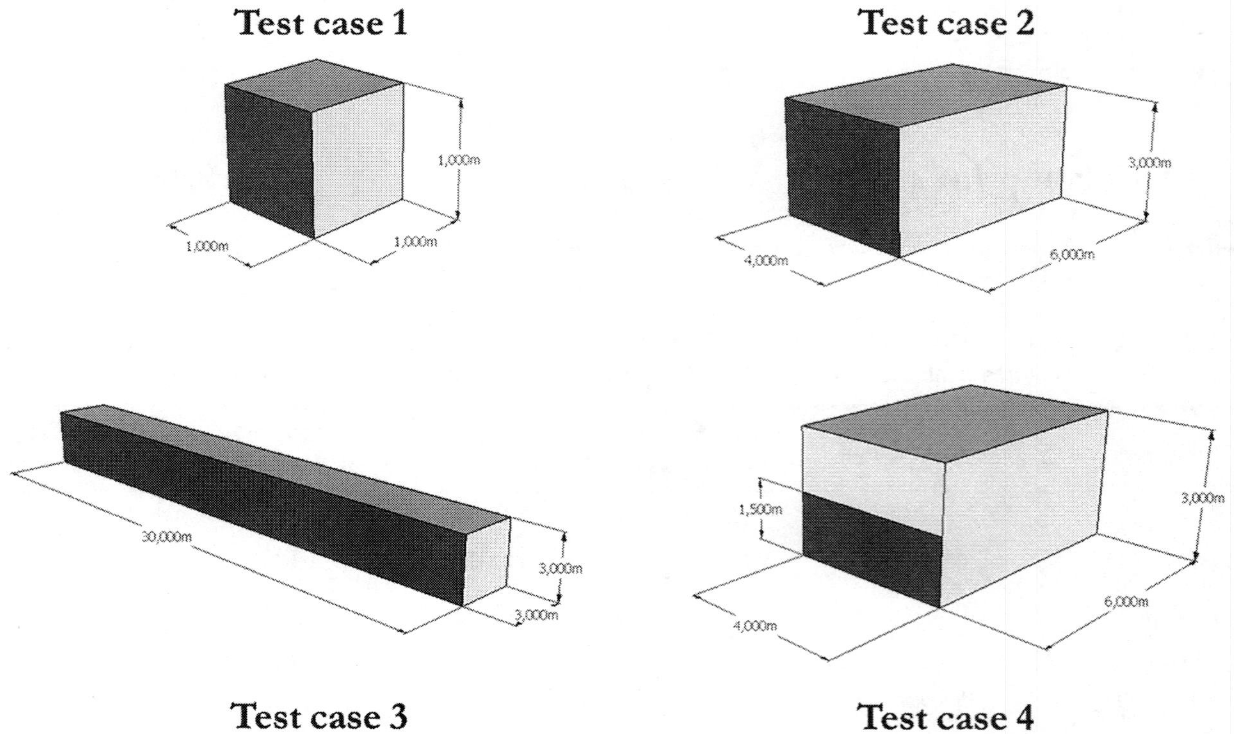

Figure 5 Geometry of DIN EN 13791 test cases 1 to 4. Exterior walls are highlighted in red.

Table 3. Boundary Conditions for the Validation According to DIN EN 13791

Surface	Exterior Temperature, °C	Heat Transfer Coefficient, $W/m^2 \cdot K$	Emissivity of Interior Surface	Absorption of Shortwave Radiation, W/m^2
Exterior wall	30	2.50	0.90	100
Partition walls	20	0.83	0.90	—
Ceiling	20	0.83	0.90	—
Floor	20	0.83	0.90	—

Table 4. Comparing Simulation Results for Interior Temperature with Reference Values of DIN EN 13791

	Test Case 1	Test Case 2	Test Case 3	Test Case 4
DIN EN 13791	34.4°C	30.4°C	38.5°C	25.5°C
WUFI® Plus	34.3°C	30.1°C	38.3°C	25.0°C

To comply with the validation criteria of DIN EN 13791, the simulated indoor temperatures are allowed to differ 0.5 K from the reference values. Table 4 compares the simulation results and the standard's reference values. The maximum deviation is 0.5 K; thus, the validation with inner long-wave heat exchange of DIN EN 13791 is regarded as successful.

EXEMPLARY APPLICATION

The new model for the calculation of interior long-wave heat exchange in WUFI® Plus was described and its output was validated in previous sections. The plausibility of the simulation output is shown in an exemplary application case: a room is conditioned with floor heating and the simulated surface temperatures are evaluated on several measurement positions on the surrounding walls. Depending on their distance to the heated floor, the surface temperatures should decline slightly. In future research, the room will be built up in a climate chamber and used for measurements. Thus, the simulations are used as a preliminary study to determine the required accuracy and positioning of measurement equipment. The measurements will later be used as an additional validation test for the new model.

The room has a cuboid geometry (see Figure 6). Its inner dimensions are 4.50 × 3.50 × 2.80 m. The building components and their boundary conditions are described in Table 5. Two of the walls are exposed to interior climate conditions and two are exposed to exterior climate conditions. Five measurement positions are placed on each wall: one in its center and one in every corner (see Figure 6). The points are labelled to simplify further assessments. The room is heated through a TABS system in the floor construction with a design water supply temperature of 45°C. The heating system is set to maintain a minimum air temperature of 16°C in the night hours. During 08:00 am and 08:00 pm, the minimum temperature is increased to 21°C.

For the simulations, a time step of five minutes is used and the results are assessed for a period of two days. Figure 7 shows the mean inner surface temperatures for all surfaces of two days of the simulation period. The floor heating system increases its heating power as the set point changes at 08:00 am, thus increasing the floor's surface temperature and its long-wave heat emission. As soon as the air temperature set point is reached, the heating system maintains this temperature level, resulting in decreasing heating power and surface temperatures. The surrounding surfaces react fast to these changes. Thereby, the ceiling's response is significantly faster and higher than the walls'. This can be explained with the increased long-wave heat exchange between floor and ceiling, as the latter is the surface that has the highest long-wave heat exchange with the floor. The increase in surface temperatures on the walls is similar for exterior and interior walls, although they are at different temperature levels because of their exterior ambient temperatures. Minimal differences in the mean surface temperatures between large and small walls can be discovered: the temperature on large walls is slightly higher. This effect can also be correlated with the long-wave heat exchange, as the larger walls see more radiation from the floor than the smaller walls. After 08:00 pm, with reduced floor heating, the ceiling temperature again shows the fastest response.

As mentioned previously, five measurement points are included on every wall. The surface temperatures of the points highlighted in Figure 6 are examined here. The small wall beside points 1 and 2 is an interior one. Next to points 4 and 5 is an exterior wall. If the calculation of long-wave heat exchange weren't considered, all measurement points would show the same surface temperature, as the model assumes a uniform temperature on a component. If the calculation of interior long-wave heat exchange is activated, the radiant heat exchange considers the geometric position inside a zone, lead-

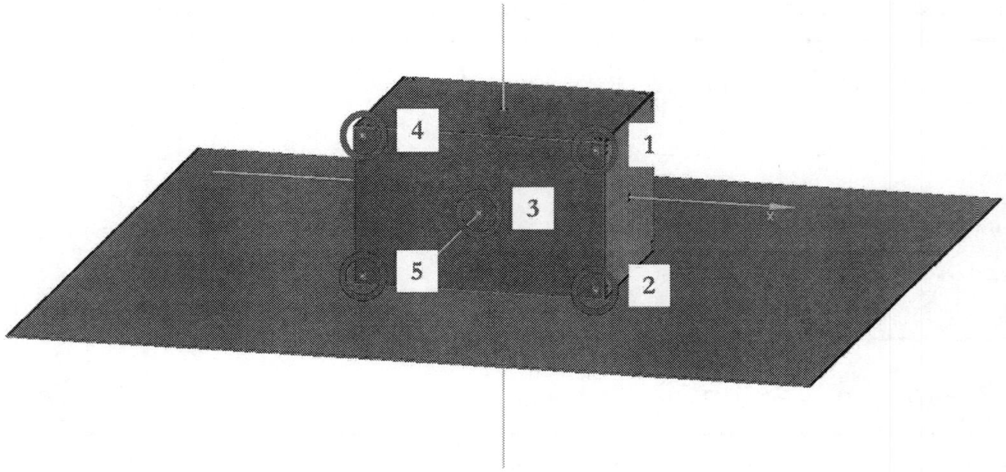

Figure 6 *Visualization of the simulated room. Interior walls are colored red, exteriors walls blue. The measurement positions are labelled and highlighted in color.*

Table 5. Boundary Conditions and Dimensions of the Exemplary Application Test Room

Surface	Exterior Ambient Temperature, °C	Heat Transfer Coefficient, $W/m^2 \cdot K$	Emissivity of Interior Surface	Area, m^2
Floor (heated)	20	0.22	0.90	15.75
Exterior wall small	0	1.55	0.90	9.80
Exterior wall large	0	1.55	0.90	12.60
Interior wall small	20	2.25	0.90	9.80
Interior wall large	20	2.25	0.90	12.60
Ceiling	0	0.19	0.90	15.75

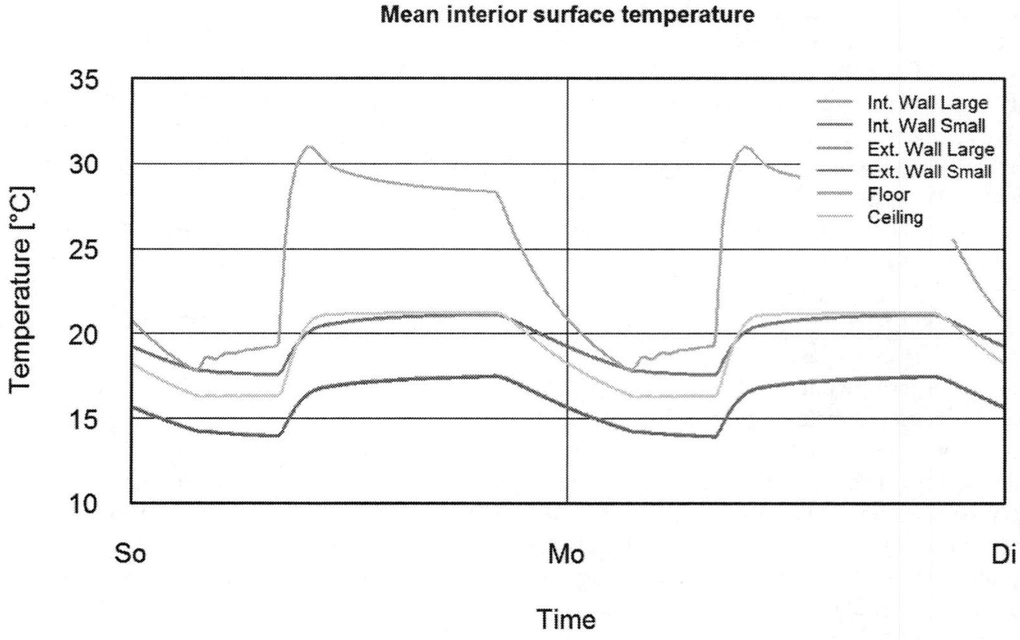

Figure 7 *Mean surface temperatures.*

ing to varying surface temperatures. Figure 8 shows this effect. As expected, the surface temperatures at the bottom points 2 and 5 are higher than their counterparts 1 and 4 on top of the wall. A difference between the adjoining walls can also be seen. Points 1 and 2, which are beside an interior wall, show higher surface temperatures than points 4 and 5. Measurement point 3 is positioned in the middle of the wall and shows a mean temperature curve.

The described simulation results show that the model for the calculation of interior long-wave heat exchange in WUFI® Plus is able to give plausible results. Further variants of this room will be made in future work, including simulations with varying surface properties, heating strategies, and thermal insulation levels of the room's surfaces.

CONCLUSION

A model for interior long-wave heat exchange has been successfully implemented in WUFI® Plus. It uses Gebhart factors for the calculations of long-wave heat exchange between surfaces inside a zone. These factors describe the fraction of emitted energy from one surface that is absorbed at another. This calculation method includes the optical properties of surfaces. Therefore, all possible radiation paths, including multiple and own reflections, are considered in the calculations. The model output was compared with analytical calculations and special tests from an appropriate German standard. All validation tests were completed successfully.

OUTLOOK

A first-use case for WUFI® Plus's new interior long-wave heat exchange model will be studying the impact of interior thermal insulation and intermittent heating on interior climate and its energy savings potential. Today's buildings often are not permanently in use and so no constant heating or cooling is necessary. The heating and cooling control mechanisms can be adjusted to the user's attendance and needs, which leads to an energy savings potential compared to constant conditioning. Different heating systems will be investigated in this future work, including radiant systems coupled to components such as wall heating. In addition, the heat inertia in the attached components also must be considered, as the heating system is part of them, thus conditioning the indoor air indirectly. Energy losses can be reduced during intermittent usage through interior insulation of thermal mass. Especially for radiant heating systems, surface coatings with low long-wave absorption allow faster room heating when applied to nonheated surfaces. Because of the new model for interior long-wave heat exchange discussed in this paper, WUFI® Plus is suitable for this task. The exemplary application case showed the influence of long-wave heat exchange on interior surface temperatures and that the model is capable of calculating these. The room described as an exemplary application will be used for measurements in a climate chamber in future work. These will contribute to studying the effects coming from interior thermal insulation and intermittent heating. Varying insulation types and heating systems will be applied to this room. The measurements will also be used to further validate the interior long-wave heat exchange model in WUFI® Plus. All mentioned effects cannot only be assessed through measurements. Therefore, a parametric simulation study will be made that also allows for transfer of the insights to other buildings. It includes different operating modes for the

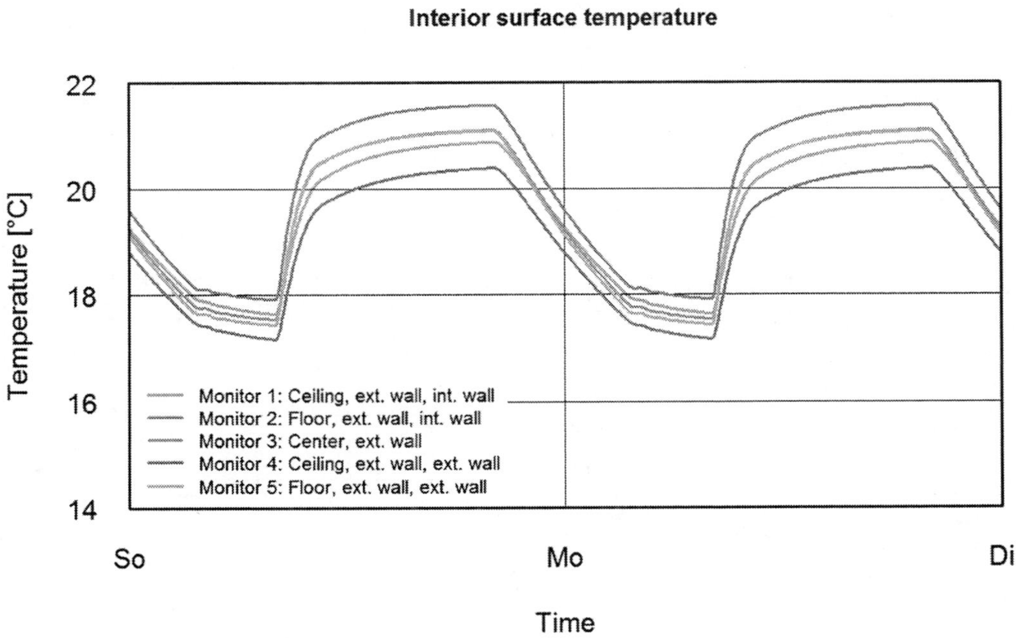

Figure 8 *Large interior wall: surface temperatures at the measurement points.*

rooms, different heating systems, and various types of interior thermal insulation and surface coatings. The resulting indoor climate, its comfort conditions, and the energy consumption will be compared.

NOMENCLATURE

A	=	area of a surface, m^2
$F_{i,j}$	=	view factor from surface i to surface j, dimensionless
$G_{i,j}$	=	Gebhart factor from surface i to surface j, dimensionless
n	=	normal vector on a surface
N	=	number of surfaces inside a zone, dimensionless
$q_{1,2}$	=	long-wave heat exchange between surface 1 and surface 2, W/m^2
r	=	direct connecting line between midpoints of two surfaces
T	=	absolute surface temperature, K
ε	=	emissivity of a surface, dimensionless
ρ	=	reflectivity of a surface, dimensionless
σ	=	Stefan-Bolzmann constant, 5.67×10^{-8} $W/(m^2 \cdot K)$
θ	=	angle between the line r, which connects the midpoints of two surfaces A_i and A_j, and the normal vector n_i on surface A_i, degrees

Subscripts

i	=	surface i
j	=	surface j

REFERENCES

Aschhaber, J., M. Hiller, and R. Weber. 2008. TRNSYS 17: Das 3D-Strahlungsmodell. Presented at the BauSim Conference.

Clark, J.A., and M.E. Korybalski. 1974. Algebraic methods for the calculation of radiation exchange in an enclosure. *Heat and Mass Transfer* 7(1):31–44.

Fraunhofer IBP. 2016. *WUFI® Plus*, ver. 3.0.3. Stuttgart, Germany: Fraunhofer-Institut für Bauphysik.

Glück, B. 1990. *Wärmeübertragung—Wärmeabgabe von Raumheizflächen und Rohren*. Berlin: Verlag für Bauwesen.

Hottel, H.C., and A.F. Sarofim. 1967. Chapter 3. In *Radiative transfer*. New York: McGraw Hill.

Howell, J.R. 2015. *A catalog of radiation heat transfer configuration factors*, 3rd ed. http://www.thermalradiation.net/indexCat.html.

ISO. 2012. DIN EN 13791, *Thermal performance of buildings—Calculation of internal temperatures of a room in summer without mechanical cooling—General criteria and validation procedures* [German]. Geneva: International Organization for Standardization.

LBNL. 2015. *EnergyPlus™ Documentation, v. 8.4.0—Engineering reference: The Reference to EnergyPlus calculations*. The Board of Trustees of the University of Illinois and the Regents of the University of California through the Ernest Orlando Lawrence Berkeley National Laboratory. https://energyplus.net/sites/all/modules/custom/nrel_custom/pdfs/pdfs_v8.4.0/EngineeringReference.pdf.

Siegel, R., and J.R. Howell. 2002. *Thermal radiation heat transfer*, 4th ed. New York: Taylor & Francis.

A New Approach for Analysis of Complex Building Envelopes in Whole-Building Energy Simulations

Jan Kosny
Member ASHRAE

Handan Liu

Charlie Curcija
Member ASHRAE

Elizabeth Kossecka

Anthony D. Fontanini
Student Member ASHRAE

ABSTRACT

The ability for reduction of whole-building energy consumption depends, in large scales, on correct predictions of building thermal loads with the building's envelope characteristics being one of the most important factors. Because most of today's building envelopes are complex, three-dimensional networks of structural, insulation, and finish materials, the potential for correct predictions of their thermal performance depends on the availability of acceptable, scientifically valid, consensus procedures for accurately implementing a building's envelope thermal characteristics into whole-building energy simulation programs.

This paper is discusses a joint LBNL and Fraunhofer CSE project focused on the upgrade of the already existing THERM program and its integration with EnergyPlus, a whole-building energy simulation tool. It is expected that these two programs, combined together, will eliminate typical analytical limitations of most existing whole building energy tools, capable of simulating only simplified, one-dimensional envelopes. The main research challenge is the design of an easy-to-implement upgrade of the THERM numerical tool to allow analysis of complex building envelope structures. The new version of THERM needs to be able to modify thermal characteristics of the complex three-dimensional (3-D) wall assemblies in a way to enable their use in whole-building energy simulation programs. It will be achieved through an application of the unique theoretical procedure, which will allow a generation of the simplified one-dimensional (1-D) wall geometry and material characteristics to fully and accurately capture the dynamic effects of thermal bridges.

At this stage of the project, the research team focuses on development of theoretical bases for necessary changes in the THERM framework. This paper explains the theoretical methodology that is used and presents some results from the series of steady-state and dynamic heat transfer simulations performed on building envelopes' architectural components to illustrate the accuracy limitations associated with thermal calculation methods recommended by building energy codes worldwide.

BACKGROUND AND MOTIVATION

It is widely expected that the exterior building envelopes should perform a variety of different roles, including keeping the exterior climate outside by limiting the transport of heat, moisture, and air, which helps with maintaining the comfortable interior conditions. Additional functions are associated with providing a visual and daylight connection to the outdoors, limiting noise transmission, supporting structural loads, and providing an aesthetically pleasing appearance on both sides. Although building envelopes are usually effective in meeting these requirements, there are situations where they do not perform adequately. Shortcomings in thermal performance are often exhibited by excessive heat transfer (in and out) and unwelcome air or moisture transport across the building envelope. These can lead to durability problems and increased energy consumption, as well as to poor thermal comfort within the occupied space.

During the last half of the century, numerous wall technologies have been introduced to the building marketplace. Some of them represent the complex three-dimensional networks of structural components and thermal insulation (Totten et al. 2008; NCMA 2010; Egelmann et al. 2011).

Jan Kosny is a director, Building Enclosures and Materials, **Anthony D. Fontanini** is a member of technical staff and **Handan Liu** is a postdoctoral fellow at Fraunhofer Center for Sustainable Energy Systems (CSE), Boston, MA. **Charlie Curcija** is a deputy group leader, Windows and Envelope Materials Group, at Lawrence Berkeley National Laboratory (LBNL), Berkeley, CA. **Elizabeth Kossecka** is a professor emeritus at the Polish Academy of Sciences, Warsaw, Poland.

Moreover, building structural systems are getting more advanced, and every year variety of new materials are introduced by builders. Consequently, buildings are often becoming unforgiving for design errors or assembly imperfections. At the same time, the overall thermal efficiency of a building is a function of the thermal performance of the planar exterior envelope elements (e.g., walls, roofs, windows). Large local heat losses, caused by the heat conducting components of the building's envelope, can occur around these planar elements at many different locations. These areas of intense local heat flow, commonly known as *thermal shorts* or *thermal bridges*, can have a significant impact on the thermal performance of the building envelope and the overall building energy consumption. Thermal bridging and construction details in the building envelope change thermal performance of envelope components. These features have both steady-state (Kosny and Desjarlais 1994; Totten et al. 2008) and dynamic repercussions (Stephenson and Mitalas 1971; Burch et al. 1992; Kosny and Kossecka 2000). Capability for reduction of building thermal loads and a successful whole-building integration are dependent on the ability to use either accurate predictions of the building envelope thermal characteristics or actual test-generated data.

While generating actual test data can be expensive and detailed analysis is needed at the design stage of an envelope component, this project addresses some of the shortcomings of current software used to predict building envelope thermal characteristics. A joint research team from Lawrence Berkeley National Laboratory (LBNL) and Fraunhofer CSE is working on the advancement of analytical capabilities of EnergyPlus[1]—a whole-building energy simulation tool, from simple one-dimensional geometries to complex multidimensional features commonly found in modern, highly efficient building envelopes. We plan to achieve this goal through the addition of the multifunctional and dynamic modeling capabilities to the THERM[2] and WINDOW[3] tools, which will allow accurate analysis of complex building envelop components and convert these components into the one-dimensional (1-D) computational format utilized by EnergyPlus. We expect that the results of the current work on a new version of the THERM framework will allow detailed steady-state thermal evaluations of building envelope components (R-value and U-factor computations). It will also allow dynamic thermal performance analysis of complex building envelope assemblies.

However, as required by EnergyPlus, the input data needs to be one dimensional. That is why the essential part of the current work is focused on development of the conversion method of the dynamic performance characteristics of complex multidimensional building assemblies into accurate, equivalent 1-D dynamic thermophysical properties (such as, for example, specific heat and density). It is anticipated that this new feature of the THERM framework will be also very useful in design work and code approvals. Imperfections in the plane wall do not only change the overall thermal resistance, but also modify its dynamic characteristics. To take this effect into account, one should consider the general conditions between the structural and dynamic thermal characteristics of a building envelope object. That is why the main challenge of the current work is how to modify thermal characteristics of the complex multidimensional wall assembly to be able a use the simplified 1-D wall geometry and, at the same time, account for the dynamic effects of thermal bridges. For this purpose, we utilize the theoretical approach based on the equivalent wall theory (Kossecka and Kosny 1999), which was validated during ASHRAE RP-1145 (2001).

This paper presents the results from the first phase of the LBNL-Fraunhofer CSE project focused on development of the conversion methodology allowing modification of thermal characteristics of complex 3-D assemblies into a simple 1-D format. We discuss current theoretical analysis of complex thermal bridges, which is necessary to start development work on the new THERM framework and, later on, integration with EnergyPlus.

THERM AND WINDOW FRAMEWORK INTRODUCTION

THERM is a computer program for calculating two-dimensional (2-D) heat transfer in building components such as windows, walls, foundations, roofs, and doors, as well as other building shell components where thermal bridges are of concern. THERM is based on the finite element numerical method (FEM) and incorporates sophisticated automated meshing and error estimation for rapid model generation and ensuring accurate results. The THERM program has been developed and is maintained by LBNL, with funding from U.S. DOE. The current released version is 7.2 and the current version under development is 7.3. The development of THERM started in 1992, with the Version 1.0 released in 1995. In addition to English version, there are also Russian and Chinese versions of the program. THERM is used by building component manufacturers, engineers, educators, students, architects, and others, but most prominently it is used as an official tool in National Fenestration Rating Council's (NFRC) fenestration rating and certification. THERM's heat transfer analysis allows evaluation of a product's energy efficiency and local temperature patterns, which may relate directly to problems with condensation, moisture damage, and structural integrity. THERM includes a material database, which contains about 100 building materials. THERM has two-way interface to the WINDOW program and allows WINDOW to determine total window product U-factors, solar heat gain coefficients, visible transmittance, and condensation resistance. THERM and WINDOW have large user bases, both nationally and internationally. The latest user and usage statistics show that there are over 50,000 unique downloads of the program in the last year and a half with over 1 million starts (see Figure 1).

1. http://apps1.eere.energy.gov/buildings/energyplus/.
2. https://windows.lbl.gov/software/therm/therm.html.
3. https://windows.lbl.gov/software/window/window.html.

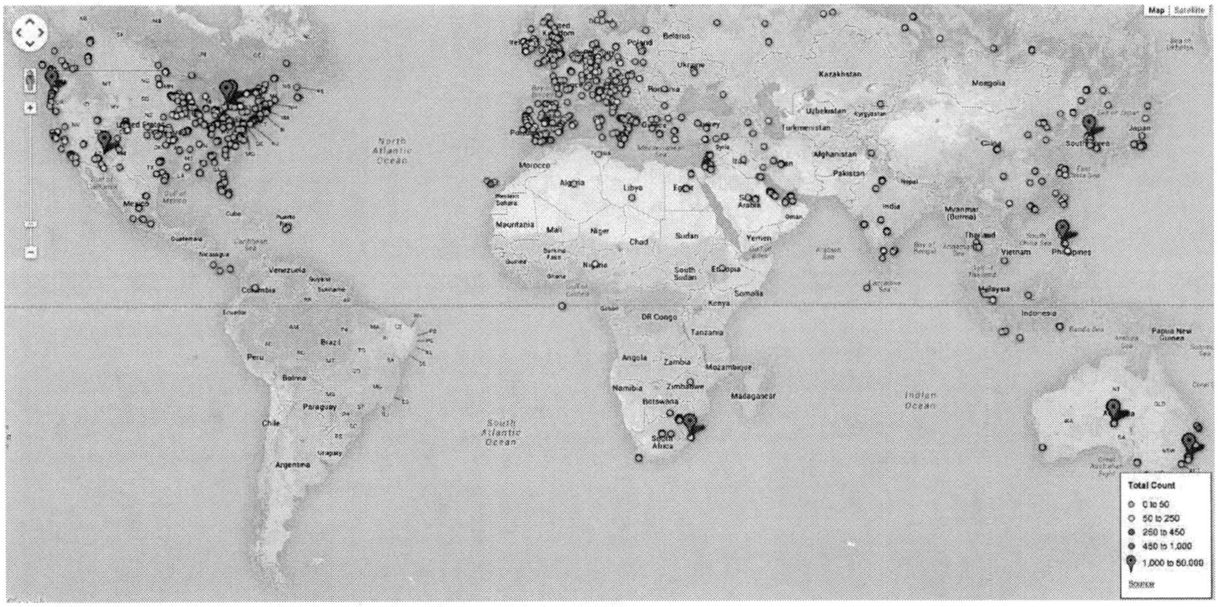

Figure 1 *THERM and WINDOW worldwide usage map.*

Currently, EnergyPlus does not have an easy-to-use interface supporting analysis of complex building envelopes. In addition, the development of numerical descriptions for building envelope components and architectural interfaces is an expensive and time-consuming task for energy modelers. That is why we expect that new numerical capabilities in the THERM tool will provide a relatively inexpensive way of performing the energy analysis, building envelope design work, and code approvals for almost all buildings containing complex building envelope assemblies. We expect that this new development will eliminate several existing EnergyPlus barriers and, at the same time, increase accuracy and credibility of energy predictions with the use of EnergyPlus.[4]

DEFINITION OF THERMAL BRIDGES AND COMMON CONSEQUENCES

Building exterior skin provides a separation between the interior space and the exterior environment. In a perfect theoretical scenario, this separation should be as continuous as possible. This obligation applies specifically to the building thermal insulation as well as to the air and moisture barriers. At the same time, thermal bridges provide a path of higher heat conduction through the insulation, allowing for more heat to bypass the building thermal barrier. Considering that in building components exposed to a temperature gradient, heat always travels across any higher conducting element, and taking into account complexity of today's building envelopes, these thermal pathways often have a strong multidimensional character. They are most often caused by discontinuities in any thermal barrier and are more pronounced when the material creating the bridge is highly conductive.

From the envelope designer's perspective, a thermal bridge is an element or location with missing thermal insulation, less insulation, or reduced insulation performance relative to the adjacent areas of the thermal envelope. The intensity of the thermal bridge is a function of the local area of the heat flow path and thermal conductivity proportions between neighboring materials. In this light, even thermal insulation material can generate thermal bridge effect. The plastic foam can generate the thermal bridge effect (Figure 2) in the case of a significantly less conductive vacuum insulation panel (VIP) packed in to the foam casing (see Kosny et al. 2014). In this case, the plastic foam is about ten times less thermally resistive than the core of the VIP.

In building envelopes, interface details and architectural components make a difference. The consequences of poorly selected connections between envelope components are severe. Taking into account the interface details can easily impact as much as 50% of the overall elevation area. For some conventional wall systems, the whole-wall R-value can be as much as 40% less than what is measured for the clear wall section (Tye et al. 1986; Kosny and Desjarlais 1994; Kosny et al. 1995). In addition, local heat losses through some wall interface details may be twice that estimated by simplified design calculation procedures that focus only on the clear wall, or clear field roof areas (Totten et al. 2008). Furthermore, thermal bridging may affect dynamic thermal performance of a building (Kossecka and Kosny 1998, 1999). Finally, poor interface details also may cause excessive moisture condensation and lead to stains and dust markings on the interior and

4. EnergyPlus will utilize well-documented input data based on the existing experimental data for building envelopes.

Figure 2 *Thermal bridge effect generated by plastic foam in case of the vacuum insulation panel (VIP) packed in the protecting foam covering.*

exterior finishes, which reveal envelope thermal shorts in an unsightly manner. This moist surface area can encourage the propagation of molds and mildews, which can lead to poor indoor air quality (Totten and Pazera 2009; Trechsel and Bomberg 2009).

It is good to remember that in the case of discontinuities in the thermal insulation layer, thermal bridges provide unwanted paths of least resistance for heat to move across the building envelope. The following unwanted effects of thermal bridging can be listed:

- Higher energy consumption. Due to the thermal outflow at the thermally bridged areas, heat is drawn from every room (in heating dominated climates) or unwanted heat gains (in cooling dominated climates). This usually results in a significant rise in whole-building energy consumption.
- Mold formation. In cold climates, interior temperatures of the adjacent rooms can drop well below the dew point. This leads to condensation, deteriorates plaster and paintwork, and is an ideal condition for harmful mold formation!
- If there is sustained exposure to condensation, the building can be subject to a serious fungi grow, leading unavoidably to health hazard.
- Compromised living space thermal comfort. Cold surface temperatures cause uncomfortable living space for occupants.

Nevertheless, in most of buildings, even high-performance buildings, the existence of limited thermal bridging is unavoidable, mostly for structural reasons. That is why it is good to remember that a proper thermal design of building envelope components, combined with good construction execution, can reduce the number of thermal bridges and notably mitigate their effects (Tye et al. 1986; Erhorn et al. 2010).

VARIETY AND COMPLEXITY OF COMMON TYPES OF THERMAL BRIDGES

In building structures, there are three very distinct types of thermal bridges: (1) geometric and architectural detail thermal bridges, (2) material and subsystem thermal bridges, and (3) thermal bridges generated by structural components and intersections. In addition, there is a group of hybrid thermal bridges representing mixed characteristics of the above-listed basic types. Keep in mind that thermal bridges are always associated with relatively higher conductive building materials or elements that penetrate the envelope thermal insulation, thereby leading to increased heat flow rates.

Thermal Bridges Generated by Building Geometry and Architectural Details

These thermal bridges are typical for building geometric details and architectural components (corners, overhangs, wall/floor, wall/roof, wall/internal partition interfaces, window and door perimeters, etc.) In these locations, increased heat flow can be observed. Component details are high strength, high conductivity element assemblies that serve to hold or connect building structural or architectural components within the envelope (balconies, wall connections with terraces, perimeters of wall openings, window and curtain wall mullions, etc.) Geometric thermal bridges do not form literal thermal shorts in the way, for example, construction thermal bridges do. More intense heat transfer can occur in the areas where nominal insulation thickness and insulation continuity are maintained. Usually, geometric thermal bridges can be found in locations where the area of the external building surface is significantly different from the corresponding internal surface area. Best identifiable geometric thermal bridge effects refer to constructions at corners that accentuate two-dimensional heat flow paths that exist at corners (Figure 3). It can be also seen that some corner constructions may also lead to discontinuities in the envelope insulation and/or the air barrier. Geometric thermal bridges are unavoidable. However, the energy impact of geometric thermal bridging generally increases with the complexity of the building form. Therefore, from the thermal and energy standpoint, as well from the long-term durability perspective, it is good to keep the building form as simple as possible.

Thermal Bridges Associated with Building Materials and Structural Subsystems

It is common that construction thermal bridges are highly localized and repetitive along the length of the structure. Figure 4 is a very good example of such thermal bridges. The figure shows metal connectors in concrete sandwich panels (two layers of concrete divided by a layer of foam). Another example are small metal connectors or hangers used in curtain walls. In high-performance buildings, very often in thick foam insulation applications (high R-value walls or roofs), the foam sheathing is connected with the structural substrate using long and relatively thick metal connectors. Beyond their high prices, these connectors often seriously compromise the overall system thermal and hygrothermal performance (Kosny et al. 2013).

A different illustration of these kind of thermal bridges can be found in concrete masonry blocks. As depicted in Figure 5, concrete masonry units have different designs of the central core parts. However, they usually have two or three concrete webs joining the external and internal concrete skins. In situations where the interior of the concrete block is either empty or filled with thermal insulation, these webs generate thermal bridge effects. In some cases, thermal bridge effect can be minimized by an application of the interlocking insulation inserts using a continuous layer of foam.

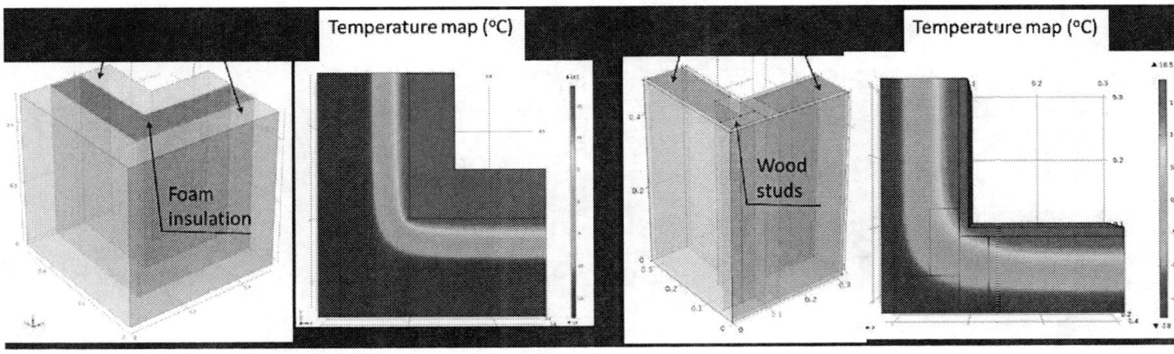

Figure 3 *Example temperature map for the concrete-foam-concrete sandwich corner and thermal insulation discontinuity in the corner configuration using wood studs.*

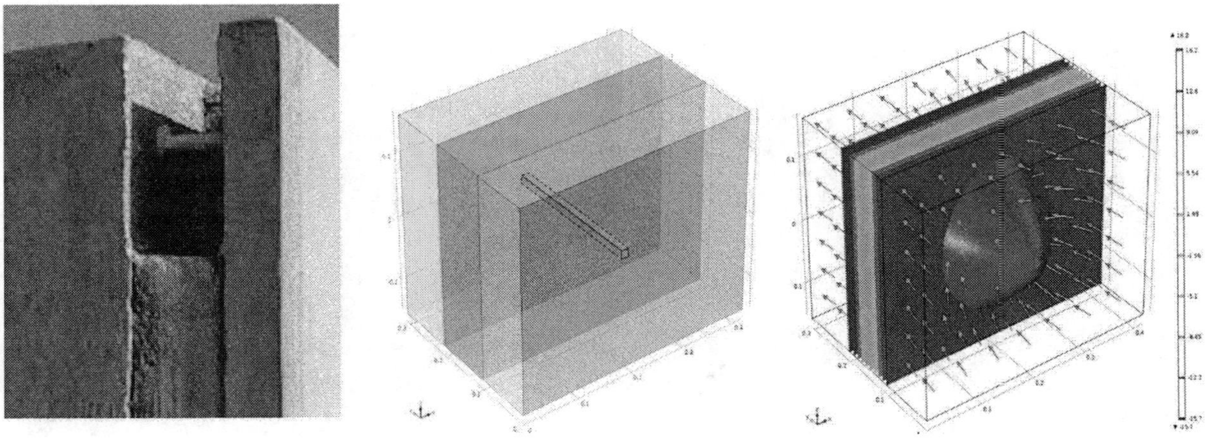

Figure 4 *Thermal effect of metal connector used in concrete-foam-concrete sandwich.*

Construction Thermal Bridges

Construction thermal bridges are probably the largest and most commonly seen in buildings. These thermal bridge configurations are not only the easiest type to recognize in real buildings, but also relatively easy to comprehend and later mitigate through corrections in design. Structural thermal bridging is usually associated with the existence of structural components combined with discontinuities in thermal insulation. They can be easily identified in places where highly conducting components are passing through the exterior thermal insulation barrier.

Some examples of construction thermal bridges include the following:

- Reinforced concrete or steel frame beams and columns
- Roof rafters or floor joists that pass through the external thermal envelope
- Wood, concrete, or metal framing members within the insulation zone
- Cantilevered structures passing through or anchored in the thermal envelope
- Lintels and headers that interrupt envelope thermal insulation
- Structural connectors between metal sandwich panels

Figure 6 shows the wall/floor intersections in the concrete-foam-concrete sandwich wall for two cases of connectors between concrete skins: 0.1 × 0. 1 × 0.1 m. (4 × 4 × 4 in.) concrete connectors and the metal ties with 12 mm (0.4 in.) diameter. They are both very common point thermal bridge configurations, occurring in many commercial and residential buildings. In both detail configurations, the concrete floor slab does not penetrate the vertical wall insulation. In both cases, significant disturbances in temperature fields can be observed in locations of connectors. The floor/wall connections are often the site of significant thermal bridging when the metal floor joists or concrete floor slabs pass through the wall insulation layer (SEI/AISC 2012).

Hybrid Thermal Bridges

In many cases, geometric thermal bridges also include an element of construction thermal bridging. For example, an external wall corner, while being a geometric thermal bridge, will also tend to have additional structure creating construction thermal bridging. Similarly, the ground floor and external wall junction often involve a degree of construction thermal bridging. Very good examples of combined thermal bridges are corners with highly conducting metal studs or wall inter-

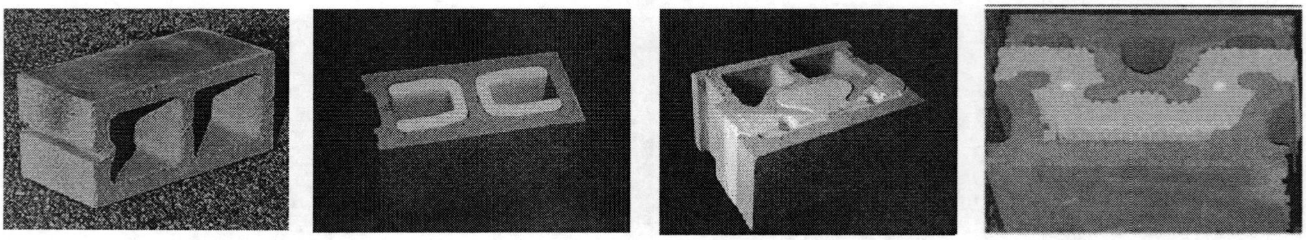

Figure 5 *Concrete masonry units (CMUs) with concrete webs acting as thermal shorts and new types of CMUs using continues insulation inserts.*

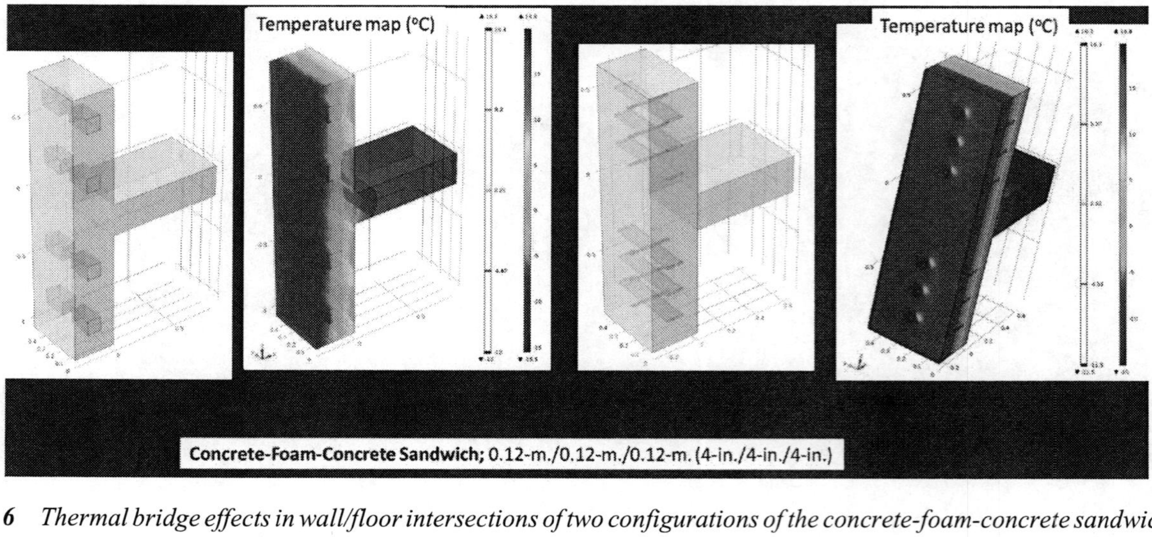

Figure 6 *Thermal bridge effects in wall/floor intersections of two configurations of the concrete-foam-concrete sandwich wall.*

sections with internal partitions. Figure 7 shows construction of the corner using light gage steel studs as structural members. In this case, increased heat flow through this architectural detail is caused by the intense heat transfer through the metal studs, as well as by the corner shape effect (different heat exchange areas on the interior and exterior surfaces).

MAJOR ACCURACY CHALLENGES ASSOCIATED WITH THERMAL BRIDGE ANALYSIS FOR BUILDINGS CONSTRUCTED WITH USE OF NORTH AMERICAN CONSTRUCTION METHODS

A variety of thermal and whole-building energy modeling tools are used today in North America for determining the local thermal performance as well as the overall energy efficiency of buildings. The energy efficiency for the building enclosure is typically based on individual R-values or U-factors of various building envelope components. The effects of thermal bridges are approximated in these calculations. However, the currently available whole-building energy modeling tools do not always accurately account for the real R-values of various interface configurations, such as, for example, window-to-wall, wall-to-partition, wall-to-foundation, and wall-to-roof interfaces (see Kosny and Desjarlais 1994; Kosny and Kossecka 2000). It is even more difficult to consider thermal bridges formed by steel framing. In these cases, two- and three-dimensional heat transfer software needs to be used to better determine local R-values; the overall energy analysis can then be carefully modified to better consider these effects and arrive at a more accurate picture of the overall energy efficiency of the building enclosure.

In 2013, *ASHRAE Handbook—Fundamentals* introduced the linear thermal transmittance method (LTTM), which has been officially used in Europe since 2007 for thermal analysis of building envelopes. The methodology for determining the linear thermal bridging coefficient is specified in several ISO standards including ISO 14683-2007, which deals with methods for determining the heat loss through linear thermal bridges at junctions of building elements (see ISO 2007a), which contains specifications for developing thermal models of thermal bridges, which can be utilized for determination of heat losses and minimum surface temperatures (ISO 2007b).

In the LTTM, each construction detail containing a thermal bridge represents a heat flow through that a detail. The thermal effect of this thermal bridge is characterized by its linear thermal transmittance, which is expressed by the ψ-value ($W/m^2{\cdot}K$).The ψ-value represents the additional heat flow through the area of the linear thermal bridge, which is above the heat flow for the adjacent, thermally undisturbed area. Two types of thermal bridges are specified in this method: linear and point thermal bridges. A linear thermal bridge is one with a uniform cross-section along one of the three orthogonal axes (see ISO 10211-2007). This type of thermal bridge is most commonly found in architectural intersections and structural junctions. An example of point thermal bridges can be visualized by a metal fastener penetrating an envelope assembly, shelf angles, slab edges, balconies, corner framing, parapets, or window/wall interfaces.

In ISO 10211-2007, a linear thermal bridge is defined as follows:

$$\psi = L^{2D} - \sum_{j-1}^{N_j}(U_j L_j) \qquad (1)$$

where,

L^{2D} = thermal coupling coefficient obtained from 2-D simulations (see ISO 10211-2007)

U_j = thermal transmittance of the 1-D component j separating the two environments being considered

L_j = length over which the ψ-value applies

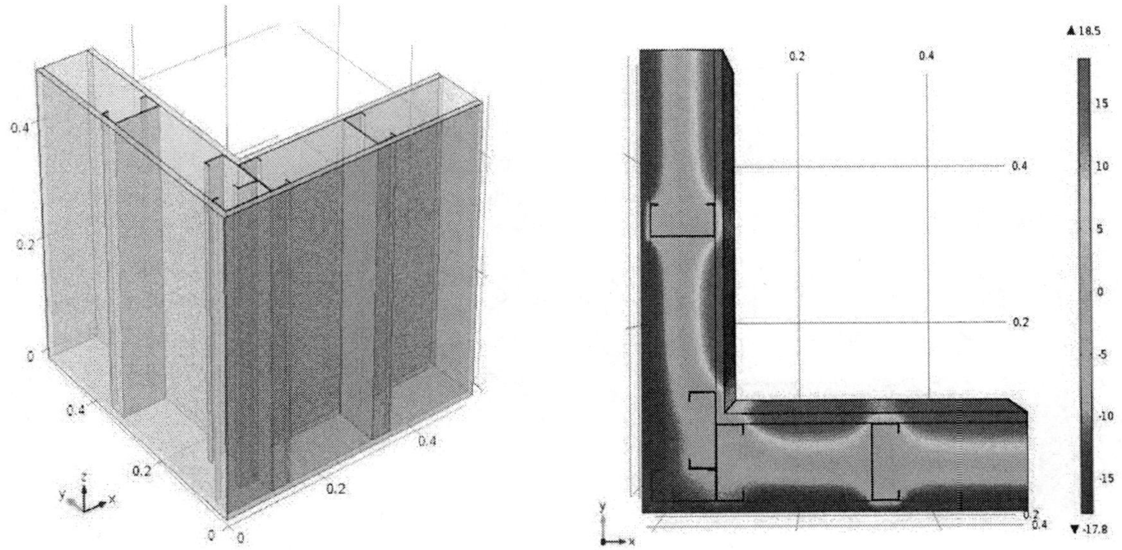

Figure 7 *Thermal bridge effects in the case of the corner assembly using light-gage steel framing.*

Keep in mind that the LTTM was developed in Europe, with consideration for major European construction methods and European architectural customs in designing of building elevations (the great majority of European buildings are using concrete or masonry technologies). That is why the theoretical base for thermal calculations with use of the LTTM is the assumption that thermal effects generated by adjacent thermal bridges do not overlap each other (see ISO 10211-2007 and 2013 *ASHRAE Handbook—Fundamentals* assumption of 1 m [3 ft] thermal zone of influence for thermal bridges). This theoretical assumption might be correct in case of majority of European buildings. However, it is difficult to justify this assumption for the North American buildings, often, with very dense highly conductive framing and busy elevations.

To illustrate this concept, let us consider an example. Assuming that the range of the thermal zone of influence of the window header is 1 m (40 in.) and that the range of the located above wall/floor intersection is also 1 m (40 in.), the edges of these two details will have to be spaced by a minimum of 2 m (over 6 ft) to follow the above theoretical assumption. Keep in mind that in the situation of overlapping thermal zones generated by the adjacent details, the estimation of the overall thermal performance of building elevation can be quite inaccurate when the LTTM is used. The following section of this paper will try to evaluate potential inaccuracies associated with use of the LTTM.

Figure 8 describes, for the case of the corner, the Fraunhofer's criteria for estimation of the detail's zone of thermal influence. This criteria limits the maximum error in local R-value calculation to 5%. It can be seen that for the case of the wood-framed three-stud corner, the European assumption of the 1 m (40 in.) range of the detail thermal zone of influence works very well. The cutoff 5% error range for this detail is at about 0.66 m (26 in.) Keep in mind that often the thermal zone of influence can be significantly larger than 1 m (40 in.). That is why the baseline calculations presented in the Figure 8 need to be sometimes performed for larger areas.

In LTTM analysis, the overall thermal resistance $R_{overall}$ ($m^2 \cdot K/W$) of any type of building envelope can be computed with the use of the following equation (assuming that there is only one type of the clear field envelope assembly within the area considered for calculations):

$$R_{overall} = \frac{A}{\frac{A}{R} + \sum_{j=1}^{J} L_j \psi_j + \sum_{k=1}^{K} n_k \chi_k} \quad (2)$$

where

A = total area of considered assembly, m^2

R = thermal resistance of the assembly without thermal bridge effects, $m^2 \cdot K/W$

L_j = length of linear thermal bridge, m

ψ_j = linear thermal transmittance of linear thermal bridge, W/(m·K)]

n_k = number of the point thermal bridge components

χ_k = point thermal transmittance of the point thermal bridge *k*, W/K

For building envelope assemblies containing multiple clear field configurations, *i*, Equation 2 will have the following form:

$$R_{overall} = \frac{A}{\sum_{i=1}^{I} \frac{A_i}{R_i} + \sum_{j=1}^{J} L_j \psi_j + \sum_{k=1}^{K} n_k \chi_k} \quad (3)$$

where

A_i = area of considered part of the assembly without thermal bridge effects, m^2

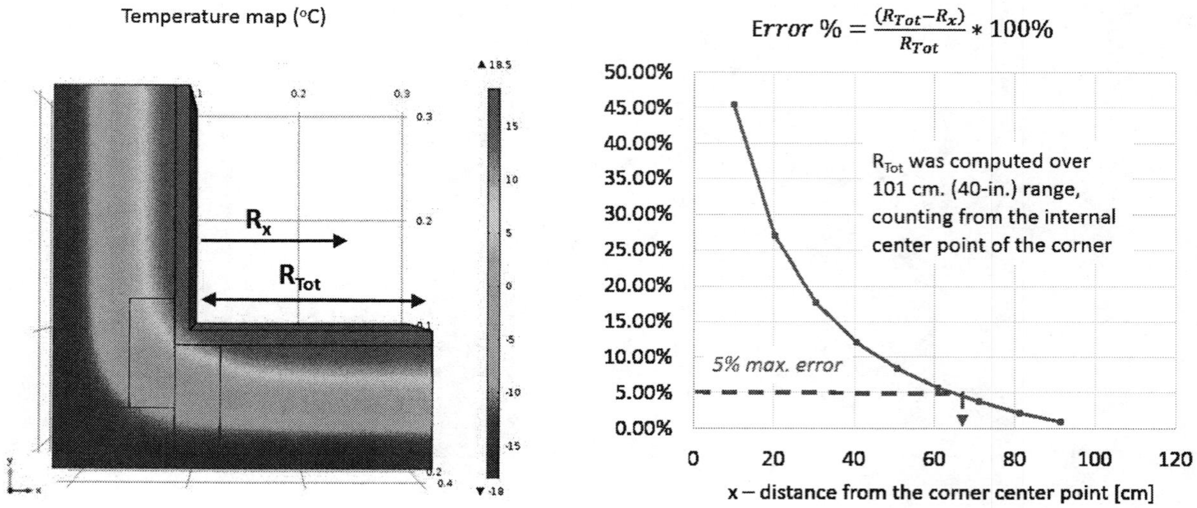

Figure 8 *Methodology used for estimation of the detail zone of thermal influence.*

R_i = thermal resistance of considered part of the assembly without thermal bridge effects, $m^2 \cdot K/W$

Figure 9 shows basic types of thermal bridges used in the LTTM and describes the principles of heat transfer calculations with use of the linear transmission method. The same calculation methodology is used in EN 14683 for calculation of the transmission heat transfer coefficients H (W/K), where the contribution due to the thermal bridges is included according to the following formula:

$$H = \sum_{i=1}^{I} \frac{A_i}{R_i} + \sum_{j=1}^{J} L_j \psi_j + \sum_{k=1}^{K} n_k \chi_k = \sum_{i=1}^{I} A_i U_i + \sum_{j=1}^{J} L_j \psi_j + \sum_{k=1}^{K} n_k \chi_k \quad (4)$$

where

H = object's total heat transfer coefficient, W/K

U_i = thermal transmittance for each considered part of the assembly without thermal bridge effects, $W/m^2 \cdot K$

It can be clearly seen in Figure 9 that for busy elevations, after the addition of all details' zones of thermal influence, the remaining area of so-called clear wall can be either very small or even negative (the combined area of all zones of influence can be larger from the total wall area). This result has two consequences: (1) it is difficult to analyze buildings with complex elevations and highly conductive materials (steel or concrete) using the LTTM method calculations, and (2) this is a very good argument for development of an easy-to-use numerical tool supporting the thermal bridge analysis in building envelopes.

In order to illustrate the sizes of thermal influence range for the basic point and linear thermal bridges, Table 1 shows the results of finite difference thermal simulations performed on concrete-foam-concrete sandwich technology. Simple concrete and metal tie connectors and a corner detail without any structural connectors were considered for the point and linear thermal bridge cases. For each analyzed case, ranges of zone of thermal influence were estimated for maximum errors levels in R-value calculations: 20%, 10%, and 5%, taking as a baseline the results of simulation with assumed 1m (40 in.) maximum size of the detail. Error in R-value calculation was estimated in the same way as explained in Figure 7. It can be seen in the case of these simple thermal bridge details that in order to maintain maximum error of R-value estimations below 5% and to avoid situations where thermal zones of influence generated by neighboring details will overlap each other, these types of details need to be separated by about 2× (0.70 to 0.90 m [28 to 30 in.]). In case of a more "busy" elevation, the error in R-value estimates made with use of the LTTM can be more significant than 5%, which may impact whole building energy performance. Keep in mind that in some cases, thermal zones of influence may have larger range, and in such situations the baseline configurations need to be simulated over the larger areas.

METHODOLOGY FOR THEORETICAL CONVERSIONS OF THREE-DIMENSIONAL THERMAL BRIDGING EFFECTS INTO ONE-DIMENSIONAL EQUIVALENTS

The main question considered in the LBNL-Fraunhofer CSE research program, focused on the upgrade of the THERM numerical tool, is how to modify thermal characteristics of the complex 3-D wall assembly, to be enable the use of the simplified 1-D wall geometry and account for both the steady-state and the dynamic effects of thermal bridges.

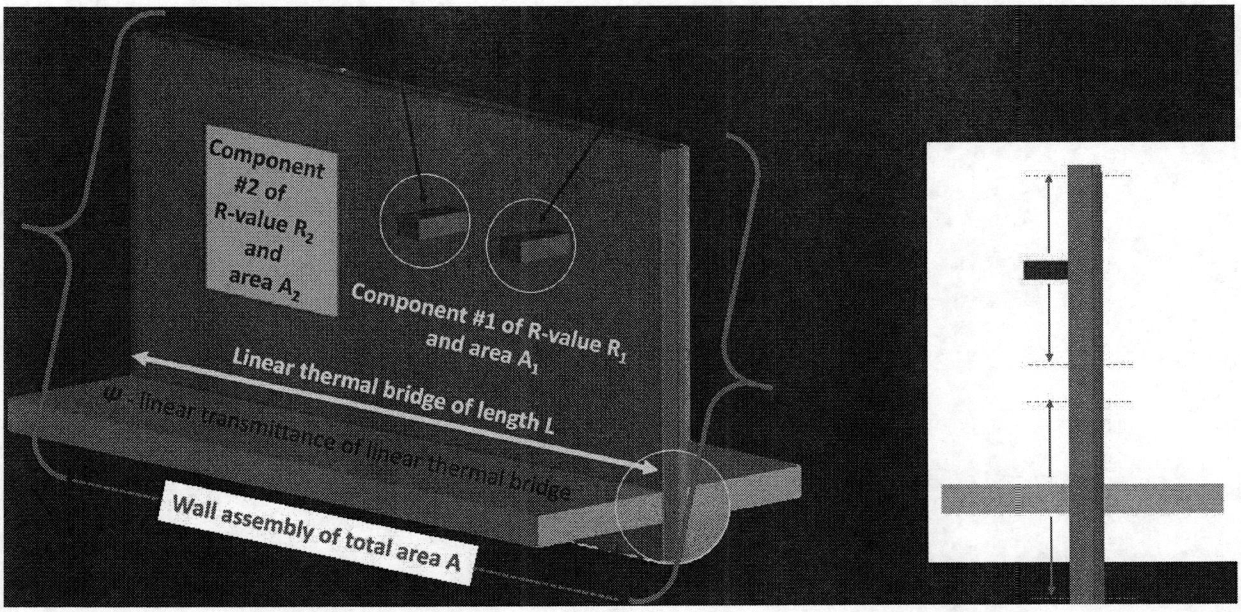

Figure 9 *Basic types of thermal bridges and calculation principles for the linear transmission method.*

Imperfections in the plane wall do not only change the overall thermal resistance but also modify its dynamic characteristics (Kosny and Kossecka 2000). To take this effect into account, one should look for the general conditions between the structural and dynamic thermal characteristics of an object. Such conditions follow from the examination of the integral formulas for the heat flow across the surfaces of the separated wall element due to the temperature difference on both its sides (see Kossecka and Kosny 1999 and ASHRAE RP-1145 2001). These formulas express the role of the heat storage effects in heat flow through an element. They lead to the notion of structure factors, the dimensionless quantities representing the fractions of heat stored in its volume, in transition between two different states of steady heat flow, which are transferred across each of its surfaces. As described in Kossecka and Kosny (1999), these quantities, together with total transmittance and capacity, are the basic thermal characteristics of a structure. But, in a three-dimensional case, through formal analogy with the case of a one-dimensional plane wall (see Carslaw and Jaeger 1959 and Clarke 1985), structure factors are not determined directly by the capacity and resistance distribution in its volume; to calculate them effectively, one has to solve the steady-state heat transfer problem. Thermal structure factors represent, together with the total thermal resistance and heat capacity, the basic thermal characteristics of walls. Thermal structure factors have their counterparts in structures where three-dimensional heat transfer occurs (see Kossecka and Kosny 1999). The products of structure factors and heat capacity are also identified in the ISO Standard 9869 (1994).

Structure Factors for Multilayer Wall Assemblies

The thermal structure of a wall is understood to be the thermal resistance and capacity distribution in its volume. Formal relationships, which describe in a quantitative way the effect of structure on dynamic thermal behavior of walls, follow from the integral formula for the heat flow across the surfaces of a wall in a finite time interval (Kossecka 1992). They include quantities called the *thermal structure factors*. Relationships between the structure factors, response factors, and z-transfer function coefficients have been derived and analyzed by Kossecka and Kosny (1999).

Thermal structure factors appear in expressions for the asymptotic heat flow across the surfaces of a wall, for boundary conditions independent of time. Consider the heat transfer in the exterior building wall of thickness L separating the room at temperature T_i from the environment at temperature T_e (see Figure 10).

Thermophysical properties of this wall include the following: thermal conductivity, k, specific heat, c_p, and density, ρ, as well surface film resistances, R_i and R_e; it is assumed that they do not change with time. Let θ be the dimensionless temperature for the steady-state heat transfer through a wall with boundary temperatures $T_i = 0$ and $T_e = 1$. For a plane wall, for which one-dimensional heat conduction conditions are satisfied, the function $\theta(x)$ is given by

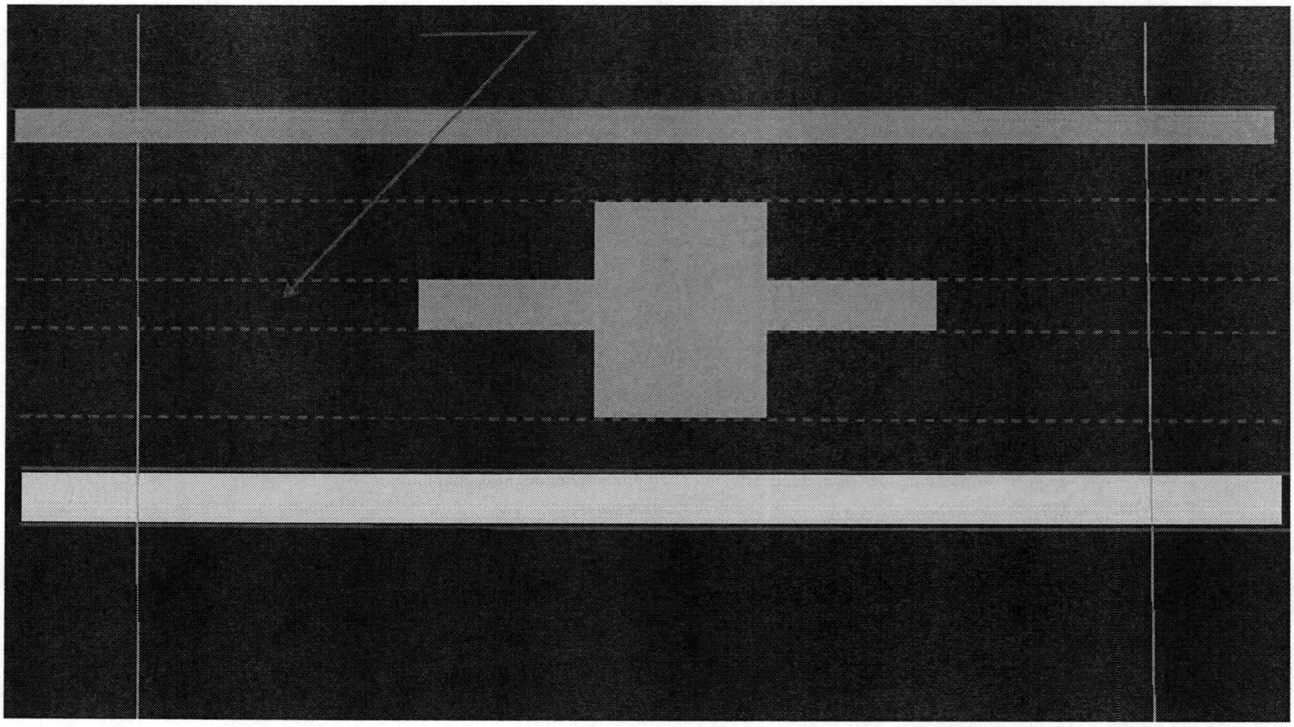

Figure 10 *Cross section of the wall element with thermal bridge.*

Table 1. Summary of the Results of Finite Difference Thermal Simulations, Performed on Concrete-Foam-Concrete Sandwich Technology for Simple Point and Linear Thermal Ridges

Type of Envelope Technology	Structural Detail	Simulated Temperature Map	Target Maximum Error in R-Value Estimates	Range of the Zone of Thermal Influence, m (in.)
Concrete-Foam-Concrete Sandwich	**Concrete Bridge**	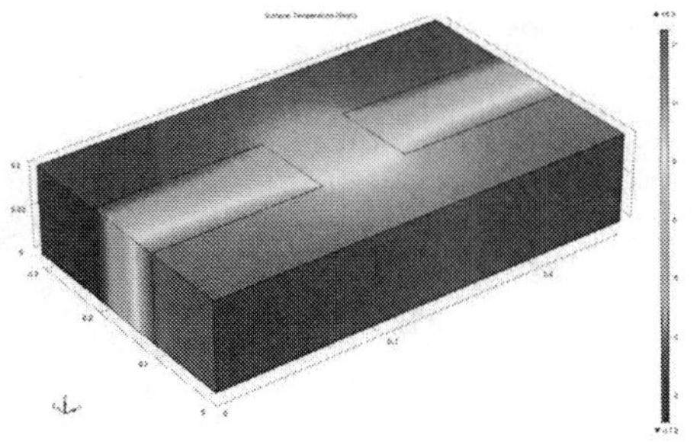	20%	0.61 (24)
0.12 m/0.12 m/0.12 m	0.12 × 0.12 m		10%	0.79 (31)
4 in./4 in./4 in.	4 × 4 in.		5%	0.89 (35)
Concrete-Foam-Concrete Sandwich	**Steel Tie**	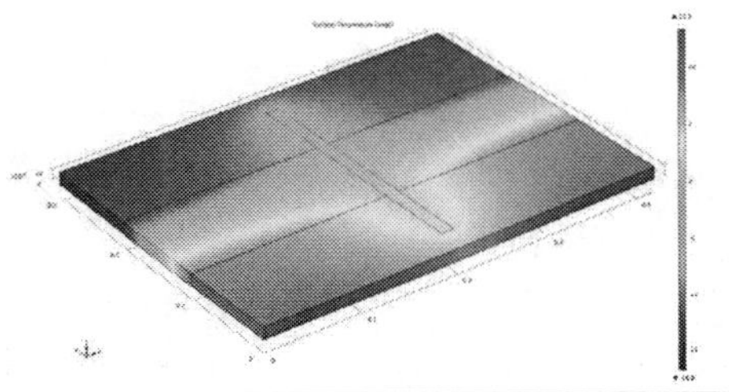	20%	0.36 (14)
	$l = 0.25$ m.; $d = 10$ mm		10%	0.56 (22)
	$l = 10$ in.; $d = 0.4$ in.		5%	0.74 (29)
Concrete-Foam-Concrete Sandwich	**Corner**	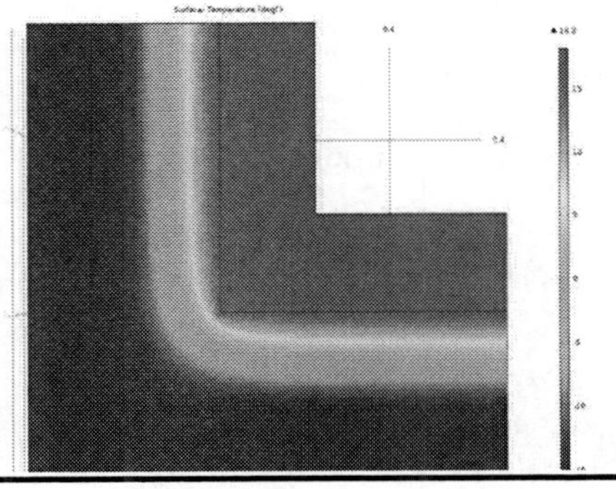	20%	0.30 (12)
	Interior surface calculation		10%	0.51 (20)
	No connectors		5%	0.71 (28)

$$\theta(x) = \frac{R_{i-x}}{R_T},\ 1-\theta = \frac{R_{x-e}}{R_T} \tag{5}$$

where

R_{i-x} and R_{x-e} = resistances for heat transmission from point x in a wall to the internal and external environment, respectively

R_T = total resistance for heat transmission through a wall

R_{i-x} and R_{x-e} can be expressed by the following integrals:

$$R_{i-x} = R_i + \int_0^x \frac{dx'}{k(x')},\ R_{x-e} = \int_x^L \frac{dx'}{k(x')} + R_e \tag{6}$$

Consider now the transient heat transfer process for the ambient temperatures held constant for $t > 0$ and initial temperature in a wall equal to zero. For sufficiently large t, the asymptotic expressions for the total heat flow in time interval $[0, t]$ across the internal and external surface of a wall in the direction from the room to the environment, $Q_i(t)$ and $Q_e(t)$, have the following simple form:

$$Q_i(t) = \frac{t}{R_T}[T_i - T_e] + C\varphi_{ii} + C\varphi_{ie} \tag{7}$$

$$Q_e(t) = \frac{t}{R_T}[T_i - T_e] - C\varphi_{ie} - C\varphi_{ii} \tag{8}$$

where C is the total thermal capacity of the wall element of the unit's cross-sectional area described in Equation 9:

$$C = \int_0^L \rho c_p dx \tag{9}$$

and the quantities φ_{ii}, φ_{ie}, and φ_{ee} are given by

$$\varphi_{ii} = \frac{1}{C}\int_0^L \rho c_p (1-\theta)^2 dx = \frac{1}{C}\int_0^L \rho c_p \frac{R_{x-e}^2}{R_T^2} dx \tag{10}$$

$$\varphi_{ie} = \frac{1}{C}\int_0^L \rho c_p \theta(1-\theta) dx = \frac{1}{C}\int_0^L \rho c_p \frac{R_{i-x}R_{x-e}}{R_T^2} dx \tag{11}$$

$$\varphi_{ee} = \frac{1}{C}\int_0^L \rho c_p \theta^2 dx = \frac{1}{C}\int_0^L \rho c_p \frac{R_{i-x}^2}{R_T^2} dx \tag{12}$$

Dimensionless quantities φ_{ii}, φ_{ie}, and φ_{ee} are called the *thermal structure factors* for a wall. For a plane wall they are determined directly by its structure, represented by the thermal capacity and resistance distribution across its thickness. In transitions between two different states of steady heat flow, they represent fractions of the total variation of heat stored in the wall volume that are transferred across each of its surfaces, due to the ambient temperature variations. The following identity is a consequence of Equations 10 through 12:

$$\varphi_{ii} + 2\varphi_{ie} + \varphi_{ee} = 1 \tag{13}$$

Structure factors can be determined experimentally in processes with steady-state initial and final states of heat flow. As expressed by Equations 7 and 8, they are related to measurable quantities, such as the heat flow, temperature, thermal resistance, and heat capacity. Together with the total thermal resistance R_T and total heat capacity C, they constitute the basic thermal characteristics of a wall and have their counterparts for structures in which three-dimensional heat flow occurs.

Structure factors for a wall composed of n plane homogeneous layers, numbered from 1 to n with layer 1 at the interior surface, are given as follows:

$$\varphi_{ii} = \frac{1}{R_T^2 C}\sum_{m=1}^{n} C_m\left[\frac{R_m^2}{3} + R_m R_{m-e} + R_{m-e}^2\right] \tag{14}$$

$$\varphi_{ie} = \frac{1}{R_T^2 C}\sum_{m=1}^{n} C_m\left[-\frac{R_m^2}{3} + \frac{R_m R_T}{2} + R_{i-m}R_{m-e}\right] \tag{15}$$

$$\varphi_{ee} = \frac{1}{R_T^2 C}\sum_{m=1}^{n} C_m\left[\frac{R_m^2}{3} + R_m R_{i-m} + R_{i-m}^2\right] \tag{16}$$

where

R_m and C_m = thermal resistance and capacity of the mth layer, respectively

R_{i-m} and R_{m-e} = resistances for heat transfer from surfaces of the mth layer to inner and outer surroundings, respectively

The structure factor φ_{ii} is comparatively large when most of the total thermal capacity is located near the interior surface $x = 0$ and most of the insulating materials (resistances) reside in the outer part of the wall, located near the surface $x = L$. The opposite holds for φ_{ee}. The following relations are straightforward: $0 < \varphi_{ii} < 1$, $0 < \varphi_{ee} < 1$. Structure factor φ_{ie} is comparatively large if most of the thermal mass is located in the center of the wall and the resistance is symmetrically distributed on both sides of it. The following limitations on φ_{ie} result from Equation 11: for a two-layer wall $0 < \varphi_{ie} < 3/16$; for an n-layer wall, with $n \geq 3$, $0 < \varphi_{ie} < 1/4$. For a homogeneous wall, with negligibly small film resistances R_i and R_e, $\varphi_{ii} = \varphi_{ee} = 1/3$, $\varphi_{ie} = 1/6$. The products $C\varphi_{ii}$, $C\varphi_{ee}$, and $C\varphi_{ie}$ for a multilayer wall are identified as thermal mass factors, introduced by the ISO Standard 9869 (1994).

Relationships between the Structure Factors and the Response Factors

Quantities $C\varphi_{ii}$, $C\varphi_{ie}$, and $C\varphi_{ee}$, which in Equations 7and 8 determine the role of storage effects in transitions between different states of steady heat flow, affect particular modes of dynamic heat flux responses of a wall. They appear in the constraint conditions on dynamic thermal characteristics of walls such as the response factors, z-transfer function coefficients, and also residues and poles of the Laplace transfer functions (see Clarke 1985; Kossecka 1992, 1996).

Let $X(m\delta)$, $Y(m\delta)$, and $Z(m\delta)$ denote the response factors for a wall, corresponding to different heat flux response modes. Response factor for number n represents the heat flux due to the unit, triangular temperature pulse of base width 2δ, at the time instant $n\delta$ (Kusuda 1969; Clarke 1985). Relationships between the response factors $X(m\delta)$, $Y(m\delta)$, $Z(m\delta)$ and structure factors φ_{ii}, φ_{ie}, φ_{ee} have the following form:

$$\delta \sum_{n=1}^{\infty} nX(n\delta) = -C\varphi_{ii} \tag{17}$$

$$\delta \sum_{n=1}^{\infty} nY(n\delta) = -C\varphi_{ie} \tag{18}$$

$$\delta \sum_{n=1}^{\infty} nZ(n\delta) = -C\varphi_{ee} \tag{19}$$

Similar conditions are satisfied by the response factors for wall elements of complex structure in which three-dimensional heat flow occurs (Kossecka and Kosny 1999). Equations 14 through 16 refer to the response factors with number $n \geq 1$, which represent the storage effects in form of surface heat fluxes after the duration of the temperature pulse. They indicate that, for given total thermal capacity C, the sums of products of response factors of a particular kind and their numbers increase with the appropriate structure factors. This means that the role of response factors having large numbers n increases with values of the appropriate structure factors.

PROCEDURE FOR GENERATION OF THE EQUIVALENT WALL

There are several ways the equivalent wall technique may generate a one-dimensional n-layer structure with similar dynamic thermal properties as the actual wall. At first, a real 3-D or 2-D structure will have to be simulated to generate steady-state and transient thermal characteristics. Next, a simplified, multilayer, 1-D model of the wall will have to be generated using the equivalent wall theory (Kossecka and Kosny 1999). In general, the number of layers in the equivalent wall should be three or more. Three-layer structures are the simplest way of achieving a model that gives good results. This methodology was validated by ASHRAE during the RP-1145 project (see ASHRAE 2001).

The procedure of numerical generation of the equivalent wall is based on distribution of thermally massive and heat resistive components across the cross section of the wall assembly. It is utilizing theoretical formulas for structural factors (see Equations 15 through 16). Performance of the generated equivalent wall can be validated through comparisons with the original 3-D structure. Response factors can be utilized for this purpose (Kusuda 1969; Stephenson and Mitalas 1971; Clarke 1985). Please bear in mind that Equations 14 through 16 are not all independent because of the identity (Equation 13). They are linear in C_n and quadratic in R_n. Summing up Equation 14 with 15 or Equation 15 with 16 yields equations linear in R_n. As the basis of an equivalent wall model one may use the following system of independent equations:

$$\varphi_{ii} + \varphi_{ie} = \frac{1}{RC} \sum_{m=1}^{n} C_m \left[\frac{R_m}{2} + R_{m-e}\right] \tag{20}$$

$$\varphi_{ie} = \frac{1}{R^2 C} \sum_{m=1}^{n} C_m \left[-\frac{R_m^2}{3} + \frac{R_m R}{2} + R_{i-m} R_{m-e}\right] \tag{21}$$

$$R = \sum_{m=1}^{n} R_m \tag{22}$$

$$C = \sum_{m=1}^{n} C_m \tag{23}$$

The response factors can be generated using any dynamic heat transfer code, including a new version of THERM with transient capabilities. The determination of structure factors follows from Equations 10 and 11.

Assuming that number of layers $n = 3$, and given R, C, φ_{ii}, φ_{ie}, and a set of first approximations for R_1, R_2, and R_3, one may easily solve the system of linear Equations 20, 21, and 22 for C_1, C_2, and C_3. However, the solution obtained in this way may not be correct in the sense that some C_n may be negative. It is thus necessary to generate, randomly or with some logic (taking, for example, R_n and C_n proportions between individual layers from the analyzed multilayer assembly), many combinations of R_n values, to find admissible combinations of C_n values. One of the easiest ways leading to the generation of the equivalent wall is to assume first a set of capacitances C_n and then seek for the resistances R_n, to satisfy Equations 20, 21, and 22. For $n > 3$ more R_n and C_n values must be treated as "parameters;" they may be either generated or fixed.

Many equivalent 1-D wall models can be obtained in this way for the same original 3-D wall assembly. When assuming the initial R_n or C_n values, it should be taken into account that material configuration of a "real 3-D wall." This can be often described, as "thermal mass inside" or "material of low conductivity" outside. The new-generated theoretical configuration should be similar to that of the original wall assembly. It is recommended that internal and external finish layers of the wall (for example gypsum board and siding, respectively) are entered to this calculation procedure using the real thickness, heat capacity, density, and thermal conductivity values. The other internal core layers of the equivalent wall may have factitious properties. The equivalent wall model is not unique, even for a given n number of layers. However, different models lead, in general, to very similar dynamic thermal performances.

At the end of the generation process, the new THERM tool will examine generated equivalent models to choose the best one. This step is to calculate response factors for the newly generated equivalent wall and compare them with 3-D response factors (using finite differences or, for simpler configurations, with the Laplace transform method) (see Kusuda 1969; Stephenson and Mitalas 1971; Clarke 1985).

Some equivalent wall models, for given thickness of the layers and specific heat, may produce values of conductivity and density that are outside of the range accepted for building materials. While this effect is of no importance for the response factors calculations, which depend on resistance and capacitance only, it should be avoided when using whole building energy simulation tools.

WHAT NEW FEATURES WILL BE AVAILABLE IN UPGRADED VERSION OF THERM TO ENABLE GENERATION OF THE EQUIVALENT WALL AND INTEGRATION WITH ENERGYPLUS?

As stated earlier, we anticipate that an addition of the new capabilities to the LBNL THERM and WINDOW tools may, in an economical way, remove several existing EnergyPlus barriers and increase accuracy and credibility of energy predictions with the use of EnergyPlus. Currently, Energy Plus does not have an easy-to-use interface supporting analysis of complex building envelopes. In addition, interface development is an expensive and time-consuming task for energy modelers. On the other hand, an addition of the new numerical capabilities in the THERM framework is relatively inexpensive compared to expected cost and energy saving achievable during energy analysis, building envelope design work, and code approvals.

At the present time, the THERM software tool includes the following numerical modules:

- Graphical user interface (GUI)
- CAD drawing translator (DXF CAD file format) and underline capability for both DXF and BMP file formats
- Native drawing module
- Steady-state, 2-D finite element method conduction and gray body radiation simulation engine
- Correlation-based convective modeling
- Automated meshing engine
- Error estimation and automated re-meshing module
- Material, boundary conditions, gas, glazing, and U-factor libraries
- Reporting module
- Data storage module, including new, partially implemented XML file format

THERM incorporates steady-state FEM simulation engine for conduction and gray body radiation modeling, called "conrad." FEM can naturally model nonrectangular geometries, and it currently includes linear finite elements, allowing it to represent linear geometrical segments. Quadratic finite elements are also available but are currently not used, primarily because of additional complications with representing curved geometry. When geometry includes curved segments, current THERM approximates such geometry with the series of smaller linear segments. This approximation is minimal and it does not result in any appreciable loss of accuracy. However, it is prudent to consider implementation of a quadratic element-based engine, which will increase efficacy and speed of calculation, even if linear geometrical segments are used.

Conduction and radiation heat transfer are solved explicitly, with local numerical errors calculated after each simulation is completed; and if errors are higher than prescribed tolerance, the geometry is re-meshed only in regions with numerical errors higher than the tolerance, and simulation is repeated. This process is repeated until final error estimate returns a solution that is within tolerance in all areas of the model. Radiation heat transfer is solved for both internal gaps (frame cavities) and boundary conditions (exposed portions of the model). Radiation heat transfer includes both black body and gray body view-factor based models.

Convective heat transfer is not modeled explicitly (i.e., no CFD numerical modeling); so, to account for convection heat transfer, THERM includes an extensive set of convective heat transfer correlations based on prior experimental and CFD numerical work. Whether the convection is modeled in internal gaps (e.g., frame cavities) or on the boundary, heat transfer correlations are applied as an average value. In the case of glazing systems, when condensation resistance is calculated, locally varying convection heat transfer correlations are applied.

While the calculations in THERM are largely based on ISO Standard 15099, EN/ISO Standard 10077 is also supported, allowing types of calculations widely used in European Union (EU). Currently, only the area-weighted method of calculating overall U-factor is supported, however the linear thermal transmittance method, favored by ISO and EN standards can also be implemented with minimal modifications. Both methods are also applicable to opaque components, just as they are to fenestration products.

THERM has a mature drawing package, widely adopted by users domestically and internationally. In addition to being able to convert properly drawn DXF CAD files into THERM models, the drawing module also allows user to either trace-over DXF or BMP image files (underlay concept) or simply draw from dimensioned drawing using a set of tools for precise movement of cursor. The sophisticated but user-friendly drawing module allows experienced modelers to complete even the most complex and detailed fenestration geometries in couple of hours time frame.

The planned development of the opaque envelope modeling capability and calculation of the equivalent 1-D dynamic thermo-physical properties calculations for EnergyPlus would not require any significant internal changes within Energy Plus, and it will significantly simplify development of the building envelope associated input files for the Energy Plus simulations, because THERM and WINDOW would serve as preprocessors.

From the midterm perspective, the development effort will include the following:

- User-friendly interface GUI elements from modifications in THERM
- Addition of time domain to the THERM's Finite Element Method numerical simulation engine
- Modifications to the THERM's error estimator and automated mesh refinement
- GUI modifications in WINDOW
- An addition of opaque envelope configurator in WINDOW (addition of R-value and U-factor generator for the clear-wall-scale building envelopes and for architectural details)
- Development of R-value, U-factor report generator for design work and code approvals
- Addition of the equivalent dynamic performance calculation engine for E+
- Upgrade of file format to XML based format. THERM already includes partial implementation of XML file format (.thmx), so this upgrade would complete full

implementation of thermal model and equivalent dynamic performance calculation

- Upgrade of CAD import functionality. Current CAD import in THERM includes DXF and bitmap trace-over functionality.

The upgraded model would fix some of the weaknesses of the current DXF import, such as support for newer DXF formats and increased robustness. Also, it would add functionality to automatically fix incorrectly formed models in CAD programs, such as open polygons, overlapping regions, and voids.

By adding transient simulations to THERM it will allow preparing accurate thermal performance characteristics of complex building envelope components for EnergyPlus whole-building energy simulations. Namely, currently walls with thermal bridges (the vast majority of wall constructions) are analyzed using the simplified model where 2-D and 3-D effects of thermal bridging are accounted using idealized, 1-D construction without thermal bridging through modified layer-by-layer data. The most common way is to modify thickness of wall layers and/or their density, ρ and specific heat, C_p. This methodology is not accurately reflecting steady-state characteristics but, what is even more important, is it is not accurately reflecting the dynamic thermal behavior of building assemblies (see Kosny and Kossecka 2000). These simplified calculations are usually done for a predefined set of known wall assemblies and cannot easily be extended to new constructions. This normally means that thermal bridging within wall construction is not modeled correctly. The new transient simulation engine in THERM will be able to model any thermal bridging without approximation compromising accuracy and will allow building simulation practitioners to accurately model expending selection of wall, roof, and foundation constructions. The new upgraded THERM will generate information required for the equivalent 1-D layer-by-layer construction from the real 2-D construction, using the earlier described equivalent wall methodology (see Kossecka and Kosny 1999; ASHRAE RP-1145 2001; Kossecka and Kosny 2004).

Similar numerical capability will be added for complex window/wall interfaces and wall framing. This new feature will be very useful in both whole-building energy analysis as well in code approval work. Later on, THERM's 2-D model of conduction and radiation heat transfer could be used directly in EnergyPlus, without the need for 1-D simplification. Clearly, for this to happen computers will need to be much faster, as transient 2-D numerical simulation would take substantially longer for large building layouts.

The current meshing engine used in THERM is somewhat limited when extremely large domains are being modeled (i.e., approximately over 20,000 finite elements). In the process of developing a unified program, it might be practical to upgrade this engine to handle those large geometry domains. This ability to handle very large domains would be very useful when direct 2-D envelope modeling is integrated directly in EnergyPlus.

SUMMARY

Work is on the way focused on upgrading the numerical capabilities of the THERM framework to allow accurate simulations of complex building envelope assemblies. The main objective of the above described project is to extend analytical capabilities of EnergyPlus from simple one-dimensional geometries to complex multidimensional features commonly found in modern, highly efficient building envelopes. This goal will be achieved by an addition of the multifunctional and dynamic modeling capabilities to the THERM tool, which will allow accurate analysis of complex building envelop components and convert them into the one-dimensional computational format utilized by EnergyPlus. This work will utilize the equivalent wall methodology developed almost two decades ago and validated by ASHRAE. The new upgraded THERM framework will not only allow detailed steady-state thermal evaluations of building envelope components (R-value and U-value computations—required for designing and code approval purposes), but it will also allow an accurate dynamic thermal simulation, supporting energy analysis and optimization of complex building envelopes and whole buildings.

REFERENCES

ASHRAE. 2001. *Modeling two and three-dimensional heat transfer through composite wall and roof assemblies in hourly simulation programs*. ASHRAE RP-1145. Atlanta: ASHRAE.

ASHRAE. 2013. *ASHRAE Handbook—Fundamentals*. Atlanta: ASHRAE.

Barbour, E., J. Goodrow, J.Kosny, and J.E. Christian. 1994. Thermal Performance of Steel-Framed Walls. Prepared for The American Iron and Steel Institute by NAHB Research Center, November 21, 1994.

Burch, D.M., J.E. Seem, G.N. Walton, and B.A. Licitra. 1992. Dynamic evaluation of thermal bridges in a typical office building. *ASHRAE Transactions*.

Carslaw, H.S., and J.C. Jaeger. 1959. *Conduction of heat in solids*. Oxford: Oxford University Press.

Clarke, J.A. 1985. *Energy simulation in building design*. Bristol, UK: Adam Hilger Ltd.

Curcija C., H. Goudey, R. Mitchell, and E. Dickerhoff. 2013. *Highly insulating window panel attachment retrofit*. Report prepared for the General Services Administration by Lawrence Berkeley National Laboratory. Windows and Envelope Materials Group. March 2013.

Erhorn H., H. Erhorn-Kluttig, M. Citterio, M. Cocco, D. van Orshoven, A. Tilmans, P. Schild, P. Bloem, K.E. Thomsen, and J. Rose. 2010. *An effective handling of thermal bridges in the EPBD context*. Final Report of the IEE ASIEPI Work on Thermal Bridges.

Engelmann P., J. Kosny, and B. Urban. 2011. new developments in mitigation of thermal bridges generated by light gauge steel framing components. 9th Nordic Symposium on Building Physics, Tampere, Finland, May 29th, 2011.

EPA. 2013. Moisture Control Guidance for Building Design, Construction and Maintenance. U.S. Environmental Protection Agency, December 2013.

ISO. 1994. ISO 9869, *International Standard, Thermal Insulation—Building elements—In situ measurement of thermal resistance and thermal transmittance*. Geneva: International Organization for Standardization.

ISO. 2007a. *Thermal bridges in building construction—Heat flows and surface temperatures—Detailed calculations*. Geneva: International Organization for Standardization. Retrieved April 30, 2014, from iso.org: http://www.iso.org/iso/home/store/catalogue_tc/catalogue_detail.htm?csnumber=40967.

ISO. 2007b. *Thermal bridges in building construction—Linear thermal transmittance—Simplified Methods and Default Values*. Geneva: International Organization for Standardization. Retrieved April 30, 2014.

Isover Saint Gobain. (2008). “What is a thermal bridge?” Retrieved April 30, 2014, from http://www.isover.com/Q-A/Implementation/What-is-a-thermal-bridge.

Kosny, J., and A.O. Desjarlais. 1994. Influence of architectural details on the overall thermal performance of residential wall systems. *Journal of Thermal Insulation and Building Envelopes*, July.

Kosny, J., A.O. Desjarlais, and J.E. Christian. 1995. Thermal performance of “energy efficient” metal stud wall systems. ASHRAE, BETEC, U.S. DOE VI Thermal Envelope Conference, Clearwater, Florida, December 1995.

Kosny, J., and E. Kossecka. 2000. Computer modeling of complex wall assemblies—Some accuracy problems. The International Building Physics Conference, Eindhoven, The Netherlands 2000.

Kosny, J., A. Fallahi, and N. Shukla. 2013. *Cold climate building enclosure solutions*. U.S. DOE Building America Program Report, NREL/SR-5500-55875. January 2013. http://www.nrel.gov/docs/fy13osti/55875.pdf.

Kosny, J., C. Misiopecki, A. Fallahi, N. Shukla, W.C. DuPont, and L.D. Carbary. 2014. Thermal design of window-wall interface in wall energy retrofits using high-performance vacuum insulation. *ASHRAE Transactions* 120:90.

Kossecka, E., and J. Kosny. 1998. Effect of insulation and mass distribution in exterior walls on the dynamic thermal performance of a whole buildings. DOE, ASHRAE, ORNL Conference—Thermal Envelopes VII, Clearwater, Florida, December 1998.

Kossecka, E., and J. Kosny. 1999. Equivalent wall as a dynamic model of the complex thermal structure. *Journal of Thermal Insulation and Building Envelopes*, 249–68.

Kossecka E., and J. Kosny. 2004. Two-step procedure for determining three-dimensional conduction Z-transfer function coefficients for complex building envelope assemblies. IX Conference—Thermal Performance of the Exterior Envelopes of Buildings, December 2004 Clearwater, Florida.

Kosny, J., C. Misiopecki, A. Fallahi, N. Shukla, W.C. DuPont, and L.D. Carbary. 2014. Thermal design of window-wall interface in wall energy retrofits using high-performance vacuum insulation. *ASHRAE Conference Proceedings*.

Kossecka, E. 1992. Heat transfer through building wall elements of complex structure. *Arch. Civ. Eng.* 38(1–2): 117–26.

Kossecka, E. 1996. The effect of structure on dynamic thermal characteristics of multilayer walls. *Arch. Civ. Eng.* 42(3):351–69.

Kusuda, T. 1969. Thermal response factors for multi-layer structures of various heat conduction systems. ASHRAE *Transactions* 75(1):241–71.

NCMA. 2010. Thermal bridges in wall construction. TEK 6-13B Energy & IAQ (2010). Herndon, VA: National Concrete Masonry Association.

SEI/AISC. 2012. Thermal bridging solutions: Minimizing structural steel’s impact on building envelope energy transfer. SEI/AISC Thermal Steel Bridging Task Committee, Supplement to Modern Steel Construction (March 2012).

Sedlbauer, K. 2001. Prediction of mold fungus formation on the surface of and inside building components. Stuttgart: Fraunhofer Institute for Building Physics.

Stephenson D.G., and G.P. Mitalas. 1971. Calculation of heat conduction transfer functions for multiplayer slabs. *ASHRAE Transactions* 77(2):117–26.

Totten P.E., S.M. O’Brien, and M. Pazera. 2008. The effects of thermal bridging at interface conditions. *Proceedings of the Building Enclosure Science and Technology (BEST 1) Conference*. June, 2008, Minneapolis, MN.

Totten, P.E., and M. Pazera. 2009. Thermal inefficiencies in building enclosures—Causes of moisture related performance problems. ASCE Forensics Engineering Congress (2009).

Trechsel H.R., and M. Bomberg. 2009. Moisture control in buildings: The key factor in mold prevention. ASTM International 2009.

Tye, R.P, J.P. Silvers, D.L. Brownell, and S.E. Smith. 1986. New materials and concepts to reduce energy losses through structural thermal bridges. ASHRAE/DOC/BETEC Thermal Performance of the Exterior Envelopes of Buildings III, ASHRAE SP 49, 1986.

Ziegel. 2010. Thermal Bridges. Retrieved April 30, 2014, from www.ziegel.at: http://www.ziegel.at/gbc-ziegel handbuch/eng/ressourcen/energie/prim/waermebrue cken.htm.

Evaluation of Energy Efficiency of U.S. Army Hard Shelters

Som Shrestha, PhD
Member ASHRAE
Axy Pagan-Vazquez, PE
Dahtzen Chu
Megan Kreiger
Kaushik Biswas, PhD

ABSTRACT

The Army uses a variety of soft shelters and semipermanent structures at Contingency Operating Bases as barracks, dining halls, administrative offices, and maintenance shops. These structures, which are commonly built by hand by soldiers or local nationals, often show performance problems. The use of an improved barracks hut (B-hut) can offer significant benefits such as enhancing the performance of building envelopes, reducing energy use, and improving comfort. Nevertheless, the effectiveness of these technologies must first be proven in experimental building stock to enable integration of this technology into Department of Defense (DoD) contingency bases. This paper presents field data that demonstrates performance of an improved B-hut compared to a co-located typical Army B-hut.

Detailed EnergyPlus models of the huts were developed and simulation results were compared against the field measured data. Validated EnergyPlus models were used to simulate the performance of huts at Champaign, IL and Kabul, Afghanistan climate conditions. Results show that the improved B-hut requires only one-fifth of the shelter conditioning energy compared to that for the baseline B-hut.

INTRODUCTION

The U.S. Army builds and uses temporary shelters, either hard-sided or soft-sided, at Contingency Operating Bases (COBs). Significant opportunities exist to improve the energy performance of temporary shelters, such as huts, used at COBs where fuel supplies are highly sensitive and expensive commodities and energy efficiency is mission critical. Hard-sided shelters, commonly known as *barracks huts* or *B-huts*, are typically wood-framed structures with exterior and interior plywood walls and a metal- or plywood-covered truss roof. Although their primary use is to house military personnel, B-huts can also serve in a variety of other capacities, e.g., as dining, administration, and maintenance facilities.

The huts are constructed by soldiers or contracted out to local nationals, have an expected lifetime of 3 to 4 years, and are demolished at the end of their service life. Typically these huts are not insulated and lack airtightness. Therefore, they are inherently very energy inefficient and have relatively high heating and cooling requirements. A typical 300-person camp uses more than 400,000 gal of fuel per year, almost 50% of it to heat and cool shelters. Improvements to the building envelope can therefore potentially save energy and consequently reduce the amount of fuel the buildings use. This research studied potential upgrades to such hard-sided shelters by using commercially available materials and construction technologies to dramatically improve energy performance that will reduce the fuel requirements at COBs and increase occupants' comfort.

Extensive field studies were conducted to evaluate performance of a typical Army hut (referred to as *baseline B-hut* in this paper), an improved B-hut, and a structurally insulated panel (SIP) hut. All three huts were co-located at Engineer Research and Development Center, Construction Engineering Research Laboratory (ERDC-CERL) Forward Operating Base Laboratory (EFOB-L) in Champaign, Illinois. However,

Som Shrestha and **Kaushik Biswas** are research and development staff at the Building Technologies Research and Integration Center at Oak Ridge National Laboratory, Oak Ridge, TN. **Dahtzen Chu**, **Axy Pagan-Vazquez**, and **Megan Kreiger** are researchers at the Construction Engineering Research Laboratory (CERL), Champaign, IL.

performance of the SIP hut is not discussed in this paper. Construction of the B-huts was completed in July 2013 and performance data was collected until September 2015.

FEATURES OF THE B-HUTS

Two shelters were monitored and studied for this energy analysis. The layouts used for the shelters were derived from the typical Army huts. Each B-hut is 16 ft wide by 32 ft long with an interior height of 8 ft (Figure 1). In addition to a conditioned living space, each hut contained an unconditioned attic and a crawlspace. The slope of the corrugated galvanized steel roof was 4:12 and its overall height to the roof peak (including crawlspace) was ~15 ft. Doors were located at both gable ends. Stairs, with guardrails and a landing, were included to provide access to the shelter.

The huts were constructed with nominal 2 × 4 (1.5 × 3.5 in.) wood framing walls, 16 in. on center, and an open-web roof truss. The interior and exterior sides of walls and the interior ceiling plane were enclosed with 0.5 in. plywood. The flooring system was assembled with 2 × 6 (1.5 × 5.5 in.) floor joists, covered with 0.75 in. plywood flooring. The roofs were composed of #15 felt attached to 0.5 in. plywood-sheathed roof trusses, roofed with 24-gauge corrugated metal roofing panels. The roof extended 2 ft beyond the walls to create a continuous eave around each structure.

Baseline B-Hut

The Army-typical B-hut shelter baseline structure building envelope consisted of exterior plywood, wood studs, and interior plywood with no insulation in the cavities. The entire structure was attached to 18 to 24 in. high "floating" wood piers set on a gravel base to create a crawlspace. The crawlspace was enclosed by 0.5 in. plywood skirting panels.

Improved B-Hut

The improved B-hut is identical to the baseline B-hut except it had air barriers, insulation, and all joints were sealed using tape.

Figure 1 The baseline B-hut (left) and improved B-hut (right).

Air Barrier. A continuous air barrier of #15 asphalt saturated organic felt was installed behind the exterior plywood and under the plywood floor assembly. Felt was selected as an air barrier because of its availability, low cost as compared to typical air barriers, and ease of installation in the field.

Insulation. Kraft-faced fiberglass batt insulation was applied in the wall cavities and between the floor joists. Unfaced fiberglass batt insulation was applied on the attic floor. Two layers of 1.5 in. extruded polystyrene (XPS) insulation were used on the doors and at the crawlspace perimeter skirting. Table 1 lists the materials, specifications, and locations of the insulation.

Air Sealing the Joints Using Tape. The B-hut was further improved with the application of air-sealing tape at all unintended openings (such as all openings around the electrical wiring, outlets, and switches) and the seams between the plywood panels on the interior. Performance of various types of air-sealing tapes tested at the improved B-hut is discussed in the works by Kreiger et al. (2015a, 2015b) and Pagan-Vazquez et al. (2015).

Blower door tests were conducted following ASTM E779-10 (ASTM 2010) to measure airtightness of each hut and the test results were used to model infiltration in EnergyPlus.[1] The test results clearly show that the improved B-hut is significantly tighter than the baseline B-hut (Table 2).

Table 1. Insulation Materials Properties

Location	Material	Thickness, in. (cm)	Nominal R-Value, h·ft²·°F/Btu (m²·K/W)
Attic	Fiberglass	12 (30.5)	39 (6.9)
Wall	Fiberglass batt	3.5 (8.9)	13 (2.3)
Floor	Fiberglass	8 (20.3)	25 (4.4)
Skirting	XPS	3 (7.6)	15 (2.6)
Door	XPS	3 (7.6)	15 (2.6)

Table 2. Blower Door Test Results

Test Results	Baseline B-hut	Improved B-hut
Date tested	8 July 14	2 September 14
Leakage rate (cfm/sf) at 50 Pa	1.909	0.232
Flow coefficient (cfm/Pa^n)	318.12	69.94
Pressure exponent	0.549	0.52

1. EnergyPlus is a whole-building energy simulation program that engineers, architects, and researchers use to model both energy consumption—for heating, cooling, ventilation, lighting, and plug and process loads—and water use in buildings. Its development is funded by the U.S. Department of Energy Building Technologies Office. http://apps1.eere.energy.gov/buildings/energyplus/.

Air-Conditioning Unit Specifications

Initially, both B-huts were equipped with mini split air conditioner units which include an interior fan coil unit and an exterior condensing unit. Chigo mini split units are commonly used in COBs in Afghanistan and can be acquired at minimal cost. Figure 2 shows the unit as installed and Table 3 provides the manufacturer's details of the system. In addition to the mini split units, electric resistance heaters, and improved environmental control units (IECUs) were also tested at a later time. However, this paper describes only the test with mini split units.

ENERGYPLUS MODELING AND SIMULATION

Detailed EnergyPlus models of the huts were developed and simulation results were compared against the field data. Agreement between the two gives the assurance that the models are accurate and that the simulation results are meaningful. The calibrated models were then used to predict energy savings potential of improved huts compared to the baseline B-hut at other climate conditions.

Thermal and physical properties such as thermal conductivity, specific heat, thickness, density, solar reflectance (SR), and thermal emittance (TE) of building materials were either measured on site, determined by conducting laboratory tests at Oak Ridge National Laboratory, gathered from the ASHRAE Handbook (ASHRAE 2013), or obtained from manufacturers' data sheets. Each type of envelope system was then assigned one or more layers of materials based on the actual construction, and each surface was assigned its respective construction, outside boundary condition, and relative geometry.

Figure 2 Mini split unit.

Table 3. Mini Split Unit Specifications

Description	Outdoor Unit	Indoor Unit
Model	CDS-35H1A-E2	CDS-35H1A-P94A
Cooling capacity	12 kBtu/h	
Heating capacity	12 kBtu/h	
Fan motor	45W	16W
SEER	13.0 (Btu/h)/W	
COP (calculated from SEER)	3.43	

Building geometry was set up using architectural drawings, while important parameters such as the exact location of the heat flux transducers (HFTs), and shading surfaces were verified with field measurements.

Figure 3 shows the renderings of the EnergyPlus model of the B-hut. Note that the sections with HFTs were modeled as small sections of the wall at exact locations. When a wall is partially shaded by overhangs, EnergyPlus calculates area-weighted solar radiation on the exterior surface. However, the incident solar radiation on the small section of the wall where the HFT is mounted could be different than the area-weighted value. Therefore, it is important to model small wall sections at exact locations for meaningful comparison between the model-predicted and measured heat flux.

Unpainted exterior plywood surfaces that are exposed to weather conditions discolored over time, which changed their SR values. To evaluate the change, SR of exterior surfaces was measured three times at 10-month intervals, using a solar spectrum reflectometer and used in the model.

The huts were monitored unoccupied, with no internal load. Thus, heat flow through the building envelope and air leakage are the two major sources of space conditioning load. Therefore, validations of EnergyPlus simulation results are focused mainly on those two parameters and the mini split unit energy use. Weather data files for model validation were created using the actual weather data collected at on-site weather station.

The models were validated using graphical and statistical methods at hourly resolution by comparing simulation results against the measured data, both in heating and cooling mode in a wide range of operating conditions. While the graphical methods are useful to assist in this process, the ultimate determination of acceptable calibration is the statistical method (ASHRAE 2002). The statistical indexes allow the modeler and any other interested parties to judge how closely

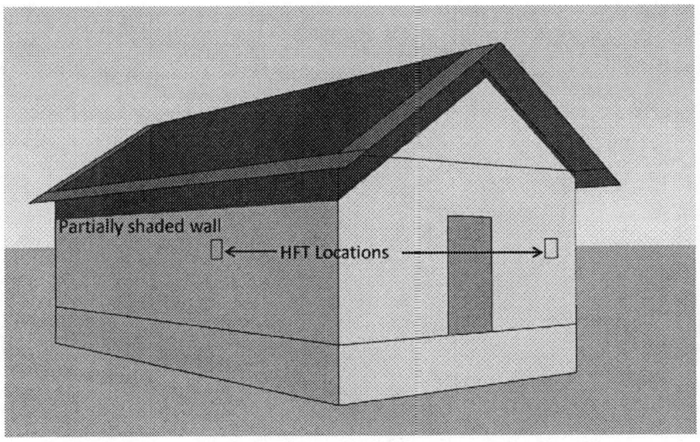

Figure 3 Renderings of EnergyPlus model of the B-hut.

the simulation output matches the measured data. The graphical techniques provide information regarding the time periods during which the two data sets diverge. Graphical comparisons for exterior and interior surface temperatures for one week in August 2014 are presented in Figure 4 and Figure 5, respectively. When EnergyPlus-calculated hourly average values for each side of the walls are compared against measured values, exterior surface temperatures were within 0.4°F and interior surface temperatures were within 1.0°F (Table 4).

The common method to validate simulation predicted heat flux through an opaque building envelope is to compare simulation results against field-measured heat flux. HFTs are used to measure heat flux through the wall. Typically, HFTs are either attached on the interior surface of the wall and then covered with a grooved plate to fit the heat flux sensor (Figure 6, left) or attached on interior drywall (or plywood) surface facing toward the insulation (Figure 6, right). When a HFT is mounted on an interior surface exposed to the indoor environment, it is essential to cover the heat flux sensor to avoid surface temperature disruption problems. At the baseline B-hut walls, HFTs were attached on interior plywood surfaces facing toward the air cavity. At the improved B-hut walls, HFTs were attached on interior plywood surfaces facing toward the insulation (Figure 6, right). In both cases HFTs were installed halfway between studs to measure the heat flux through the wall section with minimal effect from the studs.

EnergyPlus Version 8.4.0 and older calculates heat flux at an exterior surface (facing the outdoor weather) or an interior surface (facing the conditioned space) of a wall, but not at the interface between layers where HFTs were installed. Therefore, there was no direct way to compare simulation-predicted heat flux with measured heat flux. EnergyPlus was modified to calculate and to report heat flux at each node for this study. The new output variable "CondFD Surface Heat Flux" was used to calculate heat flux at the HFT location. The new output variable will be available in the next version of EnergyPlus. Graphical comparisons for heat fluxes are presented in Figure 7. When EnergyPlus-calculated hourly average values for each side of the walls are compared against measured values, heat fluxes were within 0.35 Btu/h·ft^2 (Table 4).

As a temporary solution for the heat flux measurement location problem discussed earlier, simulations were performed using COMSOL Multiphysics® (finite element analysis, solver and simulation software [COMSOL 2016]) in combination with field-measured surface temperature and materials properties to calculate heat flux at the HFT location.

Figure 8 compares the COMSOL Multiphysics®-calculated and EnergyPlus-calculated heat flux at the HFT location of the baseline B-hut south wall. The average values calculated by COMSOL Multiphysics® and EnergyPlus were 2.77 and

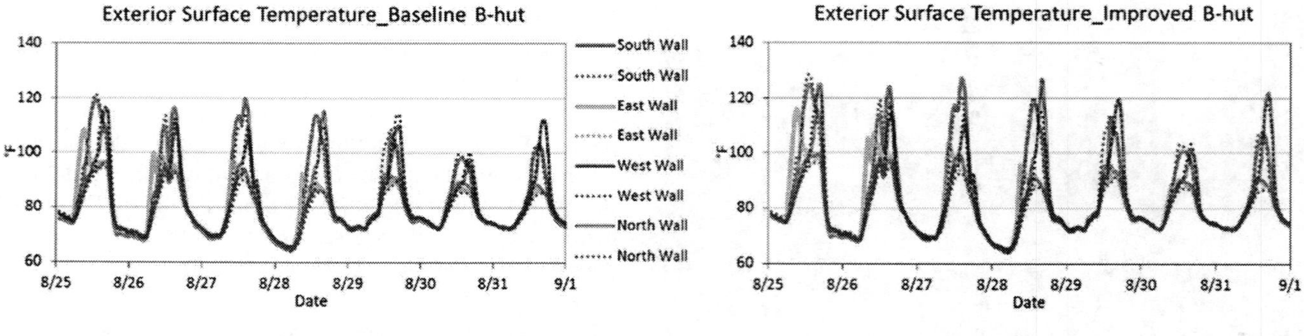

Figure 4 *Comparison between measured and EnergyPlus-calculated exterior surface temperatures. Solid lines represent the measured data and dotted lines represent the EnergyPlus results.*

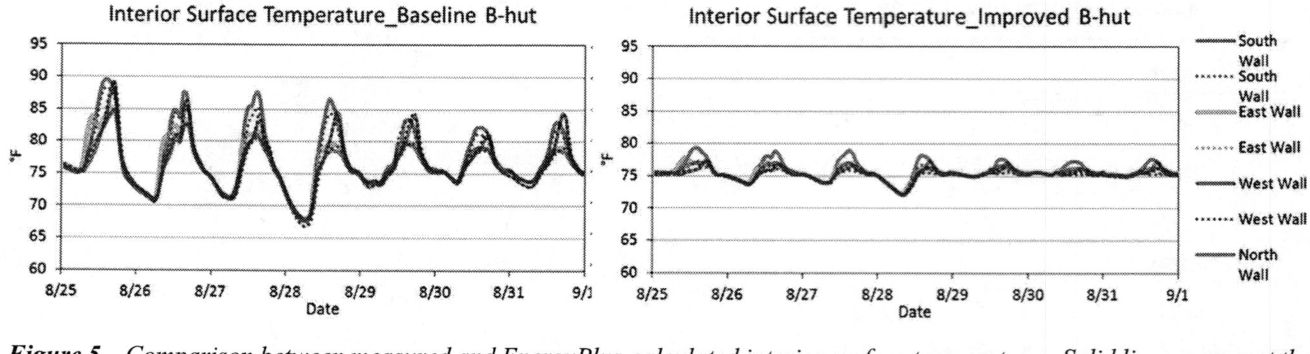

Figure 5 *Comparison between measured and EnergyPlus-calculated interior surface temperatures. Solid lines represent the measured data and dotted lines represent the EnergyPlus results.*

Table 4. Average Surface Temperature and Heat Flux Comparison

Parameters and Locations		Baseline B-Hut			Improved B-Hut		
		Exterior Surface Temperature, °F	Exterior Surface Temperature, °F	Heat Flux, Btu/h·ft^2	Interior Surface Temperature, °F	Interior Surface Temperature, °F	Heat Flux, Btu/h·ft^2
	Measured	83.6	77.6	3.13	84.8	75.9	0.70
South Wall	EnergyPlus	83.6	76.6	3.02	85.1	75.3	0.68
	Difference	0.0	–1.0	–0.11	0.2	–0.6	–0.02
	Measured	80.1	76.6	1.61	81.3	75.6	0.40
East Wall	EnergyPlus	80.2	76.9	1.80	81.1	75.2	0.42
	Difference	0.1	0.4	0.19	–0.2	–0.4	0.01
	Measured	81.9	76.6	2.80	83.3	75.3	0.62
West Wall	EnergyPlus	82.0	75.8	2.45	83.3	75.2	0.54
	Difference	0.1	–0.7	–0.35	0.0	-0.1	–0.07
	Measured	78.7	76.0	1.21	79.6	75.4	0.28
North Wall	EnergyPlus	78.3	75.8	1.12	79.6	75.1	0.27
	Difference	–0.4	–0.2	–0.09	0.0	–0.4	–0.01

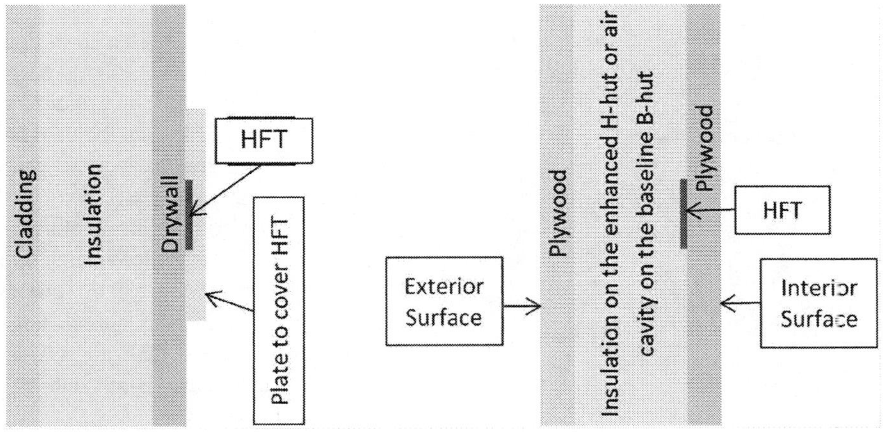

Figure 6 *Typical methods used to mount HFTs to measure heat flux through building envelope.*

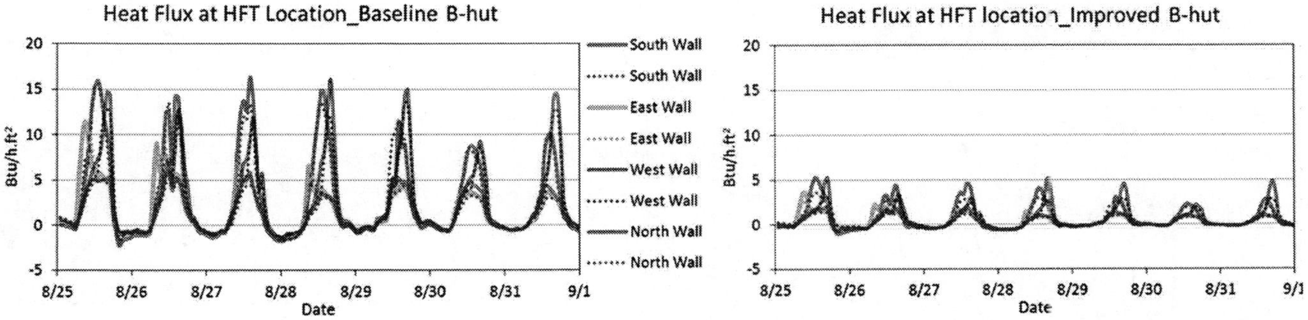

Figure 7 *Comparison between measured and EnergyPlus-calculated heat flux. Solid lines represent the measured data and dotted lines represent the EnergyPlus results.*

3.02 Btu·h·ft^2, respectively. The difference could have been smaller if EnergyPlus-calculated surface temperatures were used to run the COMSOL Multiphysics® simulation, rather than the measured surface temperatures.

The mini split unit was modeled as *package terminal heat pump* (PTHP) in the EnergyPlus model. The rated efficiency of the unit was available from the manufacturer (Table 3). However, other performance curves, such as cooling capacity, heating capacity, and energy efficiency as a function of temperature and flow fraction, that are needed to model the performance of PTHP were not available. Therefore, EnergyPlus default performance curves for PTHP were used to calculate mini split unit energy use.

Figure 8 compares simulation-predicted and measured energy use. Measured and EnergyPlus-calculated average cooling electricity use was 336 and 325 W (in agreement within 3%) for the baseline B-hut and 106 and 98 W (in agreement within 8%) for the improved B-hut, respectively. This indicates that the improved B-hut required less than one-third of the energy as compared to the baseline B-hut during the measurement period.

Typical statistical parameters to compare simulation predicted energy use against measured energy use include the daily mean, max, and min of hourly data, normalized mean bias error (NMBE), root mean square error (RMSE), and coefficient of variance RMSE (CV-RMSE) from the hourly data. ASHRAE Guideline 14 (ASHRAE 2002) requires using

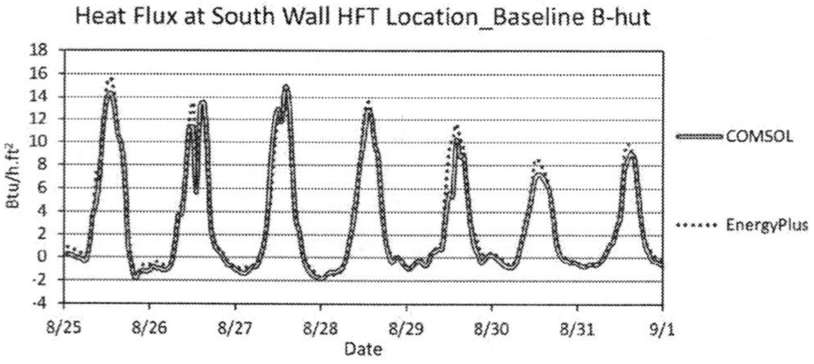

Figure 8 *Comparison between measured, EnergyPlus-calculated, and COMSOL Multiphysics®-calculated heat flux at the HFT location.*

NMBE and CV-RMSE to determine compliance. Acceptable tolerances for this comparison range from ±10% for NMBE to ± 30% for CV-RMSE when using hourly data and ±5% for NMBE to ±15% for CV-RMSE when using monthly data. The NMBE measures how closely the model-predicted values correspond to the measured data. The RMSE is typically referred to as a measure of variability, or how much spread exists in the data. The CV-RMSE (%) is calculated by dividing the RMSE by the measured mean of the data, which is often used in calibration studies.

These variables are calculated as follows:

$$\text{NMBE} = \frac{\sum_{i=1}^{N} \text{Meas}_i - \text{Sim}_i}{(N-1) * \overline{\text{Meas}}} * 100 \quad (1)$$

$$\text{CV-RMSE} = \frac{\sqrt{\frac{1}{N-1}\sum_{i=1}^{N}(\text{Meas}_i - \text{Sim}_i)^2}}{\overline{\text{Meas}}} * 100 \quad (2)$$

Where

Meas_i = measured value at hour i (for i from 1 to N hours)

Sim_i = simulation predicted value at hour i

N = number of observation points

$\overline{\text{Meas}}$ = arithmetic mean of measured values

Based on the data presented in Figure 9, NMBE and CV-RMSE were 0.5% and 22.4% for the baseline B-hut and 7.7% and 24.7% for the improved B-hut. Similar analysis was done for a week in October 2013 when the mini split unit was running on heating mode. During that period, measured and EnergyPlus-calculated average heating electricity use was 442 and 456 W (in agreement within 3%) for the baseline B-hut and 112 and 114 W (in agreement within 2%) for the improved B-hut. The improved B-hut needed one-fourth of the energy as compared to the baseline B-hut during this period. For combined heating and cooling periods, the NMBE and CV-RMSE were –0.7% and 27.1% for the baseline B-hut and 5.7% and 21.3% for the improved B-hut.

Validated models were then used to estimate the performance of these huts at other operating conditions, internal load, occupancy schedule, and geographical locations. PTHP was

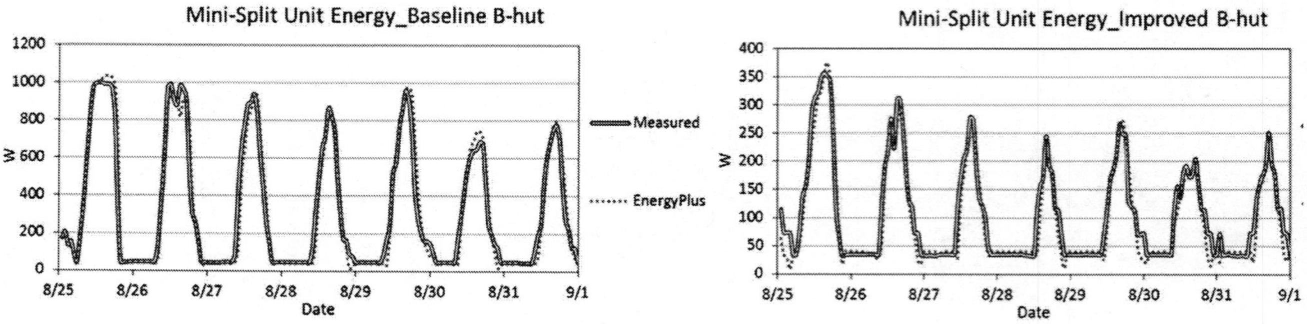

Figure 9 *Comparison between measured and EnergyPlus-calculated mini split unit energy use.*

sized based on cooling load, and therefore supplemental electrical heating was considered to meet the heating load. Annual simulations were performed for Champaign, IL and Kabul, Afghanistan. Eight occupants were assumed in each hut; 100 W/person plug load, 70°F heating set point, and 75°F cooling set point along with TMY3 weather files were used for annual simulations.

Figure 10 compares the air-conditioning energy use for two huts at two locations. EnergyPlus estimated annual air conditioning energy for the baseline and improved B-huts as 19123 and 3636 kWh for Champaign, IL and 13792 and 2553 kWh for Kabul, Afghanistan, respectively. Therefore, the improved B-hut would need only one-fifth of air conditioning at both locations.

CONCLUSIONS

Significant opportunities exist to improve the energy performance of U.S. Army temporary shelters, where fuel supplies are highly sensitive and expensive commodities, and energy efficiency is mission critical. This research studied potential construction upgrades by using commercially available materials and construction technologies to dramatically improve energy performance that will reduce fuel requirements.

Extensive field studies were conducted to evaluate performance of a typical Army hut (referred to as *baseline B-hut* in this paper) and an improved B-hut. Both huts were co-located at EFOB-L in Champaign, Illinois. Performances of the well-instrumented huts were monitored for almost two years.

Detailed EnergyPlus models of the field-tested huts were developed and simulation results were benchmarked against measured data. The validated models were used to estimate energy performance of the huts at other climate and operating conditions. The result show that the improved hut needs only one-fifth of the annual air conditioning energy compared to the baseline huts in Champaign, IL and in Kabul, Afghanistan.

ACKNOWLEDGMENTS

This study was conducted for Headquarters, U.S. Army Corps of Engineers (HQUSACE) and the Office of the Assistant Secretary of the Army for Acquisition, Logistics, and Technology, ASA (ALT).

Thanks to Lake Lattimore, Jedediah Alvey, and Sam Hunter of ERDC-CERL, and to Phillip Childs and Jerry Atchley of Oak Ridge National Laboratory (ORNL) for their technical contributions to this research. Special thanks to Michael J Witte of Gard Analytics for modifying EnergyPlus to calculate and output heat flux at each conduction finite difference heat transfer nodes for this project.

REFERENCES

ASHRAE. 2002. Guideline 14—*Measurement of energy and demand savings*. Atlanta: ASHRAE.

ASHRAE. 2013. *ASHRAE Handbook—Fundamentals*. Atlanta: ASHRAE.

ASTM. 2010. ASTM E779-10, Standard test method for determining air leakage rate by fan pressurization. West Conshohocken, PA: ASTM International.

COMSOL. 2016. COMSOL Multiphysics. Burlington, MA: COMSOL. www.comsol.com/.

Kreiger, M., J. Alvey, A. Pagan-Vazquez, and D. Chu. 2015a. Environmental degradation effect on airtightness of pressure-sensitive adhesive exterior housing tapes on plywood. *ASHRAE Transactions* 121:2.

Kreiger, M., A. Alvey, and D. Chu. 2015b. Reducing Energy Loss in B-Huts. *The Military Engineer* 694:71–2. http://themilitaryengineer.com/index.php/tme-articles/tme-magazine-online/item/443-reducing-energy-loss-in-b-huts.

Pagan-Vazquez, A., D. Chu, M. Kreiger, S. Shrestha, A. Latino, C.T. Decker, D.J. Lawrence, and A. Kumar. 2015. Comparison and analysis of energy performance of baseline and enhanced temporary army shelters. ERDC/CERL TR-15-26. http://acwc.sdp.sirsi.net/client/en_US/search/asset/1045948.

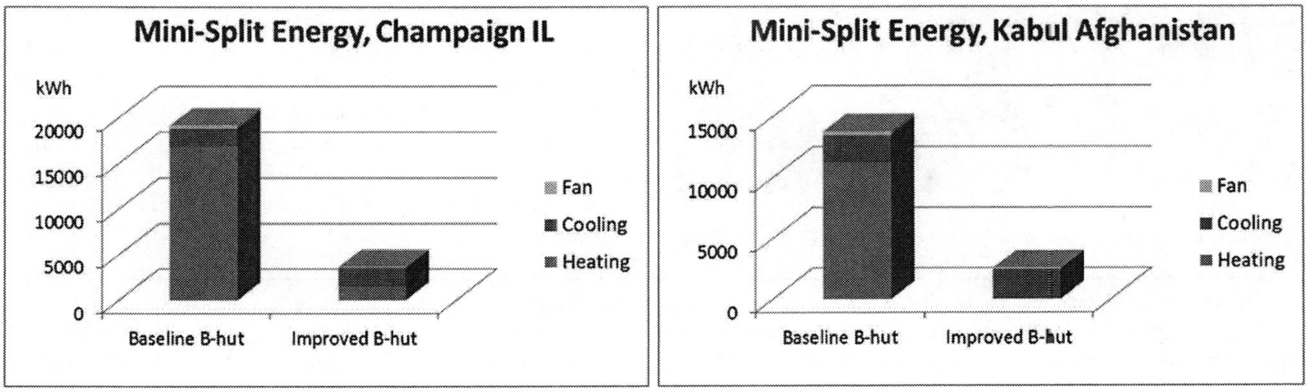

***Figure 10** Annual air conditioning energy comparisons.*

Building Energy and Envelope Performance of a Near Net Zero Energy Building

Stanley D. Gatland II
Member ASHRAE

Julie M. Dinh-Ngoc

Remi Goulet, PhD

Parhum Delgoshaei, PhD

David Riley, PhD

Diana Elliott

ABSTRACT

Pennsylvania State University's GridSTAR Experience Center, located at the Philadelphia Navy Yard, features a near net zero energy building. A unique combination of building materials, construction methods, energy management strategies, and renewable energy sources were employed to approach net zero. The building is heated using the combination of a solar thermal hot-water system, a high-efficiency electric hot-water heater, and high-performance mini split heat pumps.

Research was conducted to evaluate building energy and envelope performance. The results of the measured building energy performance were compared to a whole-building energy simulation. One-dimensional hygrothermal modeling results were compared to measured data for the above-grade wall assemblies. Measured temperatures and relative humidity were found to agree with simulated results within the uncertainty of the thermocouples and relative humidity sensors. Total measured building energy consumption compared well to simulated results.

INTRODUCTION

Pennsylvania State University's GridSTAR Experience Center was created to be an education and research resource for "smart grid" technologies. The building complex illustrated in Figure 1 includes a microgrid demonstration structure and research facility—the near net zero energy building (NZEB) (1), a solar training center (2), a commercial-grade battery system (3), and an electric vehicle charging station (4).

Figure 1 *GridSTAR Experience Center building complex.*

The near NZEB features energy-efficient, residential modular construction using integrated solar photovoltaic (PV) and thermal systems, battery systems for security/demand response load control, and grid-interactive control lighting, electrical, and mechanical systems. Passive solar design techniques combined with high thermal performance building envelope assemblies enclose 2600 ft^2 (242 m^2) of conditioned space (Figure 2).

BUILDING ENVELOPE DESIGN

The building's foundation used a 4 ft (1.22 m) high, conditioned crawlspace design. Concrete footings and walls were created using a form-in-place foundation drainage system and traditional wooden forms, respectively. The foundation floor was isolated from the soil using a water proofing membrane, insulated with 2 to 2.5 in. (51 to 64 mm) of closed-cell spray polyurethane foam, and finished with 2 to 3 in. (51 to 76 mm) of coarse gravel. Crawlspace walls were insulated with 2.5 to 3 in. (64 to 76 mm) of closed-cell spray polyurethane foam. The foundation's design features are illustrated in Figure 3.

Stanley D. Gatland II is manager of building sciences and comfort at CertainTeed Corporation, Malvern, PA. **Julie M. Dinh-Ngoc** is a senior research engineer I and **Remi Goulet** is a principal scientist at the Saint Gobain Research & Development Center, Northborough, MA. **Parhum Delgoshaei** is a research associate at The Pennsylvania State University, The Navy Yard, Philadelphia, PA. **David Riley** is an associate professor of architectural engineering at The Pennsylvania State University, University Park, PA. Diana Elliott is a research and development engineer at Isover Sant-Gobain Centre de Recherche Isolation, Rantigny, France.

A modular building manufacturer constructed the above-grade walls using nominal 2 by 8 in. (38 by 184 mm) wood studs spaced 24 in. (610 mm) on center. Advanced framing techniques were used to allow for greater stud spacing. This differs from traditional construction where wall studs are more commonly nominal 2 by 4 in. (38 by 89 mm) or 2 by 6 in. (38 by 140 mm) wood studs spaced 16 in. (406 mm) on center. The 7.25 in. (184 mm) deep wall cavity allowed for an increased level of insulation.

Above-grade walls are traditionally constructed from the outside to the inside. Modular construction methods require that materials be installed in reverse order from the inside-out to expedite the assembly process at the off-site manufacturing facility (Figure 4).

The building's roofing systems consist of two different assemblies, as illustrated in Figure 5. An architectural-style, steep-sloped (9:12 pitch) asphalt roofing shingle system with ridge and soffit ventilation was installed over a full-height, unconditioned open attic space. The attic floor was constructed using nominal 2 by 10 in. (51 by 254 mm) wood joists spaced 16 in. (406 mm) on center and insulated with two layers of fiberglass batt insulation. The clearance between the top plate of the exterior wall and the

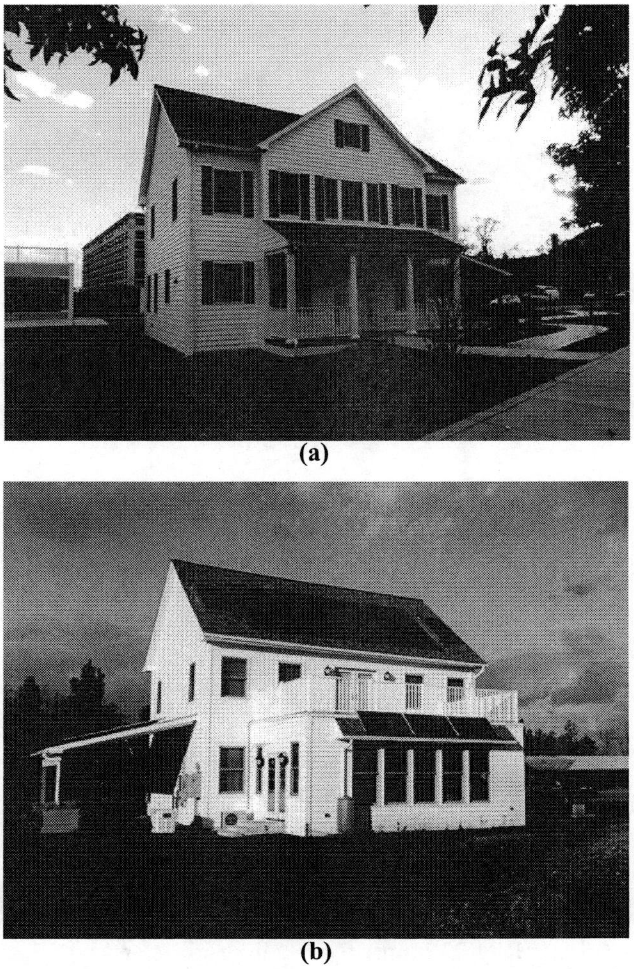

Figure 2 *(a) NZEB north (front) and (b) south (back) facing views.*

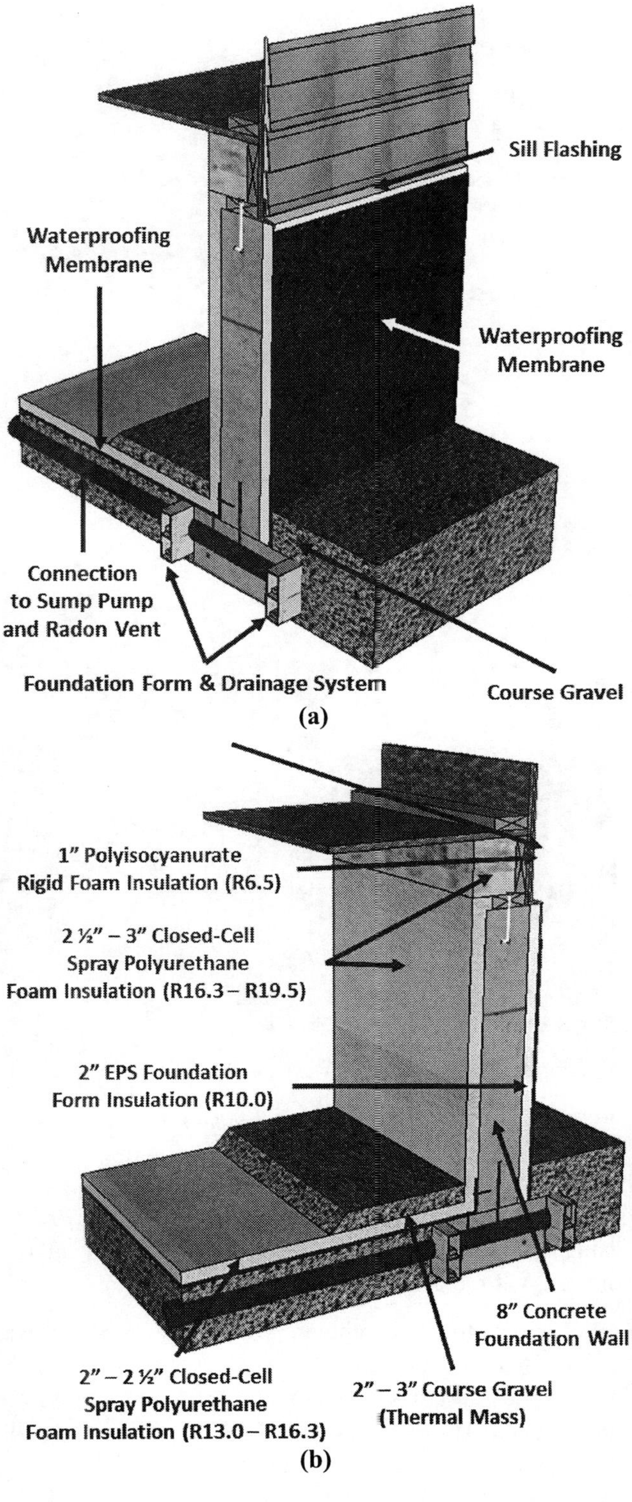

Figure 3 *(a) Foundation moisture management and (b) thermal insulation design features.*

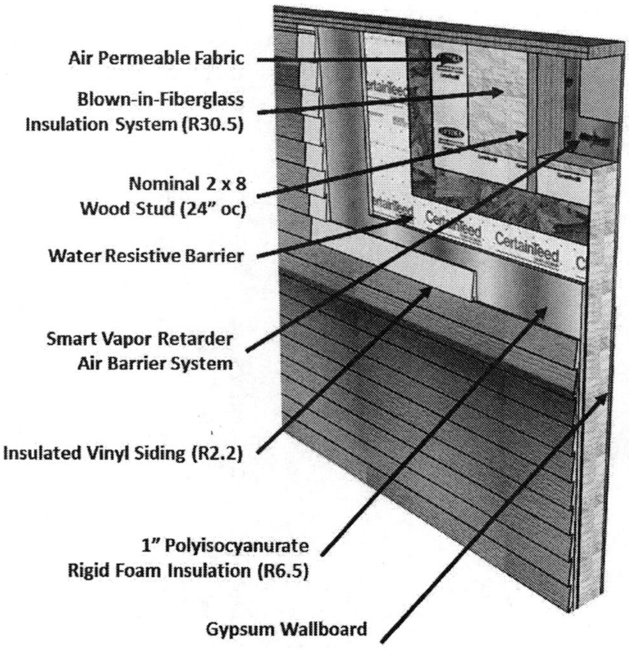

Figure 4 *Above-grade wall assembly details.*

Figure 5 *NZEB roof height, south-facing view.*

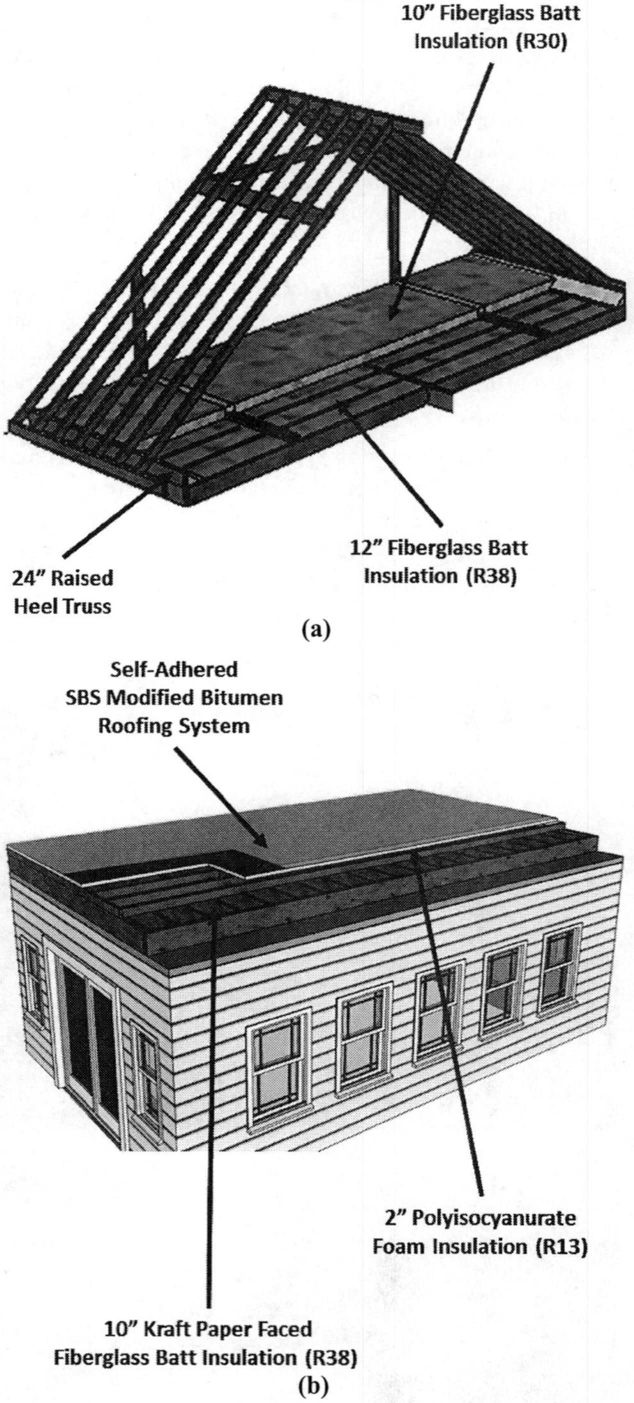

Figure 6 *(a) Steep-slope open attic and (b) low-slope roof assembly framing and insulation details.*

roof joists bottom edge was increased by 24 in. (610 mm) using a raised-heal roof truss to allow for greater insulation thickness at the perimeter, as shown in Figure 6.

A low-sloped, commercial roofing system was constructed over the conference room at the back of the building to support a deck connected to the second-floor meeting and office space. The self-adhered styrene-butadiene-styrene (SBS) modified bitumen roof system was insulated with 2 in. (51 mm) polyisocyanurate foam insulation board. The perimeter edge was insulated with 6 in. (152 mm) of extruded polystyrene foam and sealed with slow expanding closed-cell spray polyurethane foam. The floor-ceiling structure was constructed using nominal 2 by 10 in. (51 by 235 mm) floor-ceiling joists spaced 16 in. (406 mm) on center, insulated with 10 in. (254 mm) fiberglass batts (Figure 6).

BUILDING MECHANICAL SYSTEMS

Passive Solar Collectors

The solar thermal system contains the following elements: two solar thermal collectors (premium low profile flat plate solar collector); circulation system unit, which includes the pump to control hot water temperature; controller; storage tank (80 gallon [302.8 L]) with dual heat exchanger coil); and piping.

The collector system runs according to a program based on the temperature difference between the collection tank and the roof-mounted collection panels. If the switch-on temperature difference (12°F [6.7°C]) is exceeded, the circulation pump will engage. If the switch-off temperature difference is reached, circulation will stop. The solar collection loop runs into the solar storage tank.

Domestic Hot Water

The domestic hot water (DHW) system consists of a heat pump water heater boosted by a solar thermal system (Figure 7). The air-source heat pump water heater has a tank capacity of 80 gal (302.8 L). The unit has a reported coefficient of performance (COP) of 3.18 at testing conditions of 59°F (15°C) air temperature and 70% rh, while heating up the water from 59°F to 140°F (15°C to 60°C). The COP varies between 2 and 6 between cold (41°F [5°C]) and hot (95°F [35°C]) temperatures, respectively. The heat pump storage tank is filled with preheated water from the solar storage tank when the water level in the tank gets low.

Space Heating

Heating needs of the space are satisfied by two different heating systems: in-floor radiant heating and mini split heat pumps. The radiant system uses an electric boiler supplemented by a passive solar thermal hot-water heating system. Hot water circulates through radiant tube panels installed under the building's flooring. The first and second floors are zoned separately. Also, each floor has two mini split heat pumps that supply conditioned forced air through ceiling-mounted evaporator coil units. The different systems are controlled by separate thermostats and operated independently of each other.

Figure 7 Domestic hot-water heating system.

One mini split heat pump has a heating seasonal performance factor (HSPF) of 10.2 Btu/W·h and a heating capacity of 12,000 Btu/h (3.5 kW). An additional split-system heat pump (with three indoor units) has a HSPF of 10.0 Btu/W·h and rated heating capacity of 14,400 Btu/h (4.1 kW).

The radiant floor heating system contains the following: thermostat, two zone switching relay (zone controller), circulation pumps (1/25 hp [29.8 W]) for the two zones, supply and return radiant floor, micro boiler, mixing valve, motorized diverting valve, and associated temperature controller. The radiant return mixes with the radiant supply in the mixing valve before going through the modulating micro boiler. The micro boiler is rated at 40,946 Btu (12 kW).

Zone thermostats control circulation pumps that supply hot water to the radiant floor panels. If the temperature setpoint is not met the micro boiler starts. After passing through the micro boiler the water recirculates through the supply manifolds to the two zones (Figure 8).

Space Cooling

The building's cooling load is served by two systems: a 12,000 Btu/hr (3.52 kW) split-system heat pump (with three indoor units) with a SEER value of 20.5 Btu/W·h and a 11,000 Btu/h (3.2 kW) mini split heat pump with a seasonal energy-efficiency ratio (SEER) value of 19.3 Btu/W·h. When solar power is available, the second system receives power from three 230 W (785 Btu/h) polycrystalline solar modules that increases the SEER value of the unit to 35 Btu/W·h by driving the direct-current (DC) indoor and outdoor fans, as well as the brushless DC (BLDC) rotary compressor (Figure 9).

Roof-Integrated Photovoltaic System

Electricity demand for the building is supplemented by a 4.2 kW (14,331 Btu/h) array, roof-integrated photovoltaic (RIPV) system installed on the south-facing steep-sloped roof of the building. The original specification of 5.0 kW (17,061 Btu/h) RIPV system was reduced to accommodate the installation of two flat-plate solar thermal collectors (Figure 10). An additional twelve RIPV panels rated for 0.7 kW (2388 Btu/h) were removed from service due to voltage output conversion limitations of the NZEB inverter. A 2.1 kW (7165 Btu/h) solar PV array was placed over the carport for the electric vehicle charging station (Figure 10). Excess electricity can be stored in a 10 kW (34,121 Btu/h) commercial battery storage facility located on site. In addition, a 12 kW (40,945 Btu/h) backup gasoline-powered generator is available to provide emergency electricity to the building if the power goes out (Figure 11).

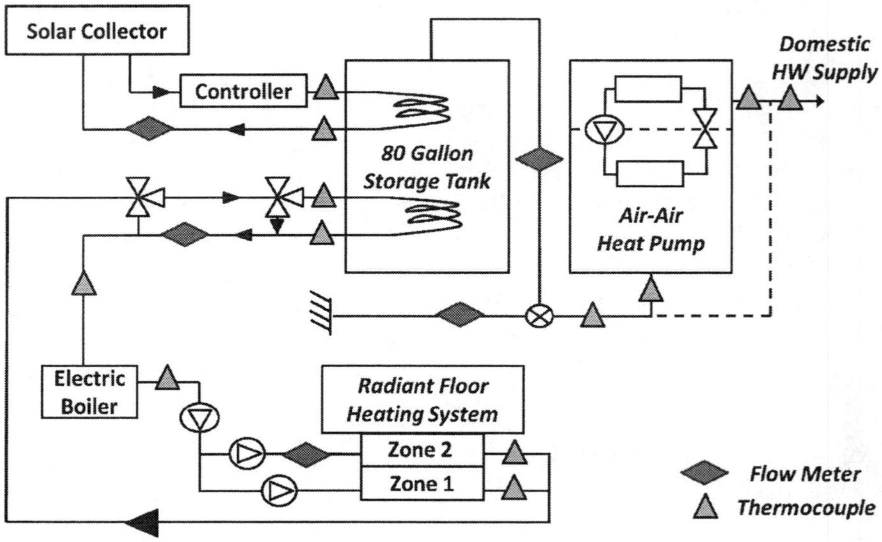

Figure 8 *Space heating flow diagram.*

(a)

(b)

Figure 9 *(a) Solar PV-assisted and (b) high-efficiency mini split heat pump systems.*

Light Management System

The light management system consists of a connection bridge, main repeater, keypads (some wireless), dimmers and switches (including fan controls), lighting fixtures, sensors, window treatment (remote-operated shades), and the associated programming. A mode from a preprogrammed list can be selected from the main control node in the building.

Energy Recovery Ventilation System

An energy recovery ventilator (ERV) was installed in the conditioned crawlspace having a 96% efficiency recovery rate (Figure 12). The system runs on the basic principle that air entering the building is preconditioned by the air exiting the building. This is accomplished by means of an exchange manifold in which intake air and exhaust air cross paths through separate channels that are interconnected. During the cooling season, intake air is precooled, and in the heating season, the air is preheated. The ERV enables fresh air exchange while capturing the energy that would normally be lost with ventilation.

ENERGY CONSUMPTION: PREDICTED VERSUS ACTUAL

The near NZEB's energy performance was verified through the U.S. Environmental Protection Agency (EPA) ENERGY STAR® v2.0 Home Verification Program as part of the LEED® for Homes certification process. A whole-building

(a) (b)

Figure 10 *(a) RIPV and solar thermal panels and (b) 2.1 kW (1705 Btu/h) carport PV array over electric vehicle charging station.*

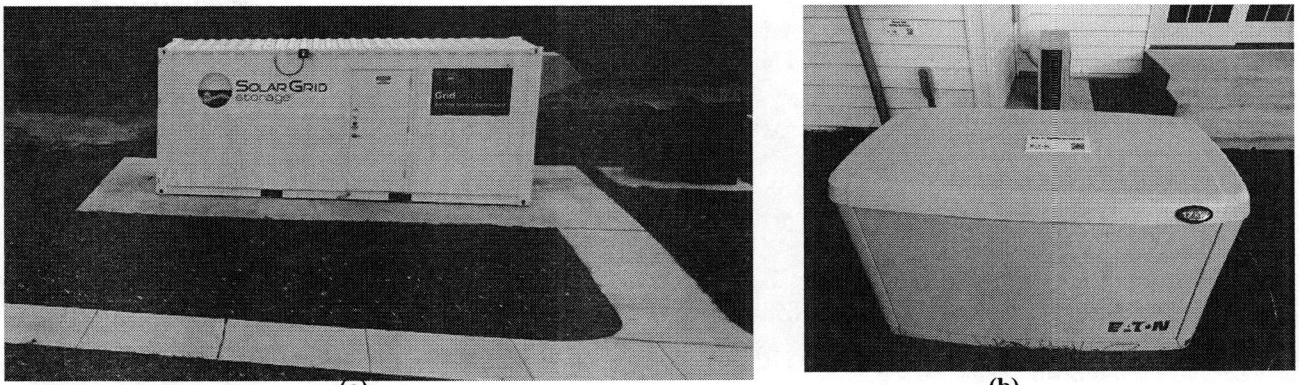

(a) (b)

Figure 11 *(a) 10 kW (34,121 Btu/h) commercial battery storage facility and (b) 12 kW (40,946 Btu/h) backup gasoline-powered generator.*

(a) (b)

Figure 12 *(a) Energy recovery ventilator control dial and system and (b) energy recovery ventilator installed in crawlspace.*

energy analysis was performed using REM/Rate™ software[1]. A Home Energy Rating System (HERS) index (Residential Energy Services Network 2013) of 21 was achieved that greatly contributed to receiving LEED-H platinum status for the building.

Three additional energy simulations were performed using the changing requirements of the International Energy Conservation Code (IECC) for 2006, 2009 and 2012 (ICC 2006, 2009, and 2012[2]). The energy models were compared to the NZEB results. Each simulation accounts for the actual geometric parameters of the building (exterior wall area, floor surface area, and window area) and predicts the energy consumption, according to the insulation levels, heating/cooling setpoints, and mechanical equipment specifications for each code year.

Table 1 shows a comparison of the heating and cooling setpoint temperatures and insulation levels for the different code years as imposed in the simulation software. The climate data used as a boundary condition were obtained as a typical meteorological year (TMY) from the Philadelphia International Airport, approximately 6 mi (10 km) from the Philadelphia Navy Yard location. The consumption of NZEB mechanical and electrical systems are monitored through several different interfaces (Figure 13). A summary of monitored data is shown in Table 2.

Both the predicted and actual energy consumption were broken down into the categories space heating, space cooling, DHW, and lights and appliances. The cooling systems are combined mini split heating and cooling units, which are controlled by manual switching between the two modes. The cooling season was defined as May 1 through September 30. PV production is presented separately.

Figure 14 compares the predicted energy breakdown for different IECC code year configurations of the NZEB results versus the actual energy breakdown. The actual consumption is presented for the period between September 1, 2014 and August 31, 2015, and represents the building's response and energy consumption for a complete year directly.

The initial comparison between the NZEB simulation and the measured results shows that the overall energy consumption level is in very good agreement with the predicted value differing by less than 3%, see Table 3. However, the breakdown profile is quite different from what was predicted. Two primary differences are seen in the lights and appliances, in which the actual usage is below that of the predicted values, and in the space heating in which the actual usage is almost double the predicted value.

One of the main assumptions for the different simulated cases is that the building is a single family dwelling, with associated occupancy schedules and heating / cooling loads. This assumption is part of the HERS rating process required for the LEED certification. In reality, the building is used as a small office building, typically housing between three and six people during the hours of 9 am to 5 pm. In some instances, the building is used as a classroom, where as many as 20 people may be present for several hours at a time. Based on the reduced occupancy of the building during nighttime hours versus a single family home, the energy consumption because of lights and appliances is reduced. The increased space heating load is possibly caused by the fact that the heating thermostat programs are not set to decrease heating during the nighttime hours, when the occupancy is low. In addition, the programmable thermostats allow the occupant to interact with the system, allowing for adjustments, which fall outside the programmed schedule, and possibly increase the heating consumption.

The PV production for the time period between September 1, 2014 and August 31, 2015 was found to be 11.2 kWh/m²/y (8.4 mm Btu/y). This accounts for 21.7% of the total consump-

Table 1. Building Energy Simulation—Insulation Levels and Temperature Setpoints for Code-Compliant Design Years 2006, 2009, 2012, and As-Built NZEB

	2006 IECC	2009 IECC	2012 IECC	NZEB
Wall insulation, $m^2 \cdot K/W$ ($ft^2 \cdot °F \cdot h/Btu$)	2.3 [C]* (13 [C])	2.3 [C] (13 [C])	3.5 [C] (20 – [C]) or 2.3 [C] + 0.9 [CI] (13 [C] + 5 [CI])	5.4 [C] + 1.5 [CI] (30.5 [C] + 8.5 [CI])
Attic insulation, $m^2 \cdot K/W$ ($ft^2 \cdot °F \cdot h/Btu$)	6.7 (38)	6.7 (38)	8.6 (49)	12 (68)
Window U-factor, $m^2 \cdot K/W$	0.40	0.35	0.35	0.33
Heating setpoint, °C (°F)	20.0 (68)	22.2 (72)	21.1 (70)	N/A†
Cooling setpoint, °C (°F)	25.6 (78)	23.9 (75)	25.6 (78)	

* [C] indicates cavity and [CI] continuous insulation.
† Smart thermostat with adaptive technology occupant interaction.

Table 2. Building Monitoring Summary

Monitored Parameters		
• PV production • Battery status • Critical loads • EV charging station	• Lights • Appliances • Outlets • Electric boiler • Heat pump water heater	• Solar assist AC • PV inverter • Solar storage temperatures • Critical load panel demand • Energy to grid

1. www.remrate.com
2. iccsafe.org

tion of the building. The initial analysis predicted a PV production of 24.5 kWh/m^2/y (18.4 mm Btu/y), which is above the measured value. There are two possible reasons for this overestimate. First, the designed PV array was sized at 5 kW (17,061 Btu/h); however the system was reduced to 4.2 kW (14,331 Btu/h) to accommodate the solar thermal collectors. The second reason is that the RIPV system was further reduced in size by an additional 0.7 kW (2388 Btu/h) to accommodate the voltage limitations of the building's inverter system, reducing the seasonal output further to 3.5 kW (11,950 Btu/h). Additional electric resources exist with the PV array on the carport (2.1 kW [7165 Btu/h]), the 10 kW (34,144 Btu/h) battery storage system, and the 16 kW (54,631 Btu/h) gas-powered electric generator. However all additional resources were for microgrid demonstration purposes and not available to supplement the building's energy consumption.

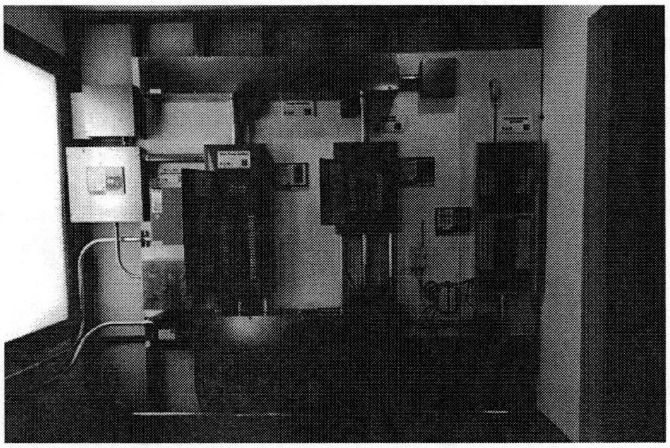

Figure 13 Full electric panel and critical load panel.

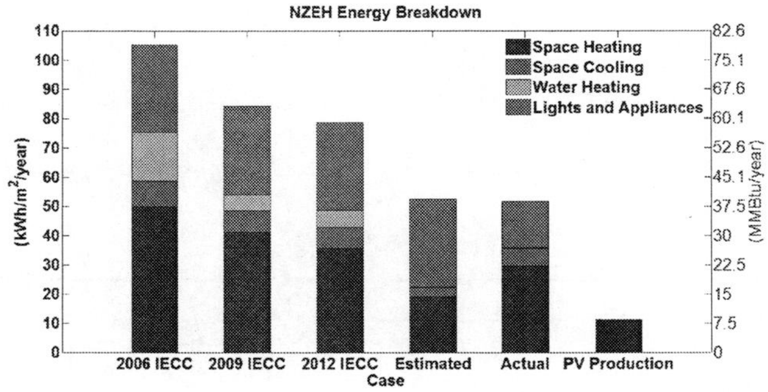

Figure 14 NZEB energy consumption breakdown—predicted versus actual.

Table 3. Total Energy Consumption, kw/m^2/y (MMBtu/y)—Case Comparisons

2006 IECC	2009 IECC	2010 IECC	NZEB Model	NZEB Actual
105.3 (79.1)	84.1 (63.1)	78.7 (59.1)	52.5 (39.4)	51.6 (38.7)

HYGROTHERMAL ANALYSIS COMPARED TO *IN SITU* MEASUREMENTS

In the interest of verifying that the unique wall assembly used at the NZEB performs as designed, a series of sensors were instrumented into the East and West walls, according to Figure 15. A hygrothermal analysis of the above-grade wall assembly, detailed in the section, was completed using WUFI® Pro (2008) transient heat and moisture analysis software while following ASHRAE Standard 160 (2009). Measured data were compared to simulated results.

The climate data used as the exterior boundary conditions were obtained from the historical weather data for the Philadelphia International Airport. The distance is 6 mi (10 km) from the NZEB. The east and west walls measured interior gypsum surface temperatures, and gypsum cavity side surface relative humidity data were averaged and used as the interior boundary conditions. A total time period of three consecutive years was simulated and results of the second and third years were found to be similar. Results from the second--year simulation were compared to *in situ* measurements.

Temperature

Measured surface temperatures at the oriented strand board (OSB)/fiberglass cavity insulation interface were compared to simulation results shown in Figure 16 for the east- and west-oriented walls, respectively. A comparison of measurements and simulated data shows temperature differences that are within the accuracy of the thermocouples (± 3.6°F [±2°C]).

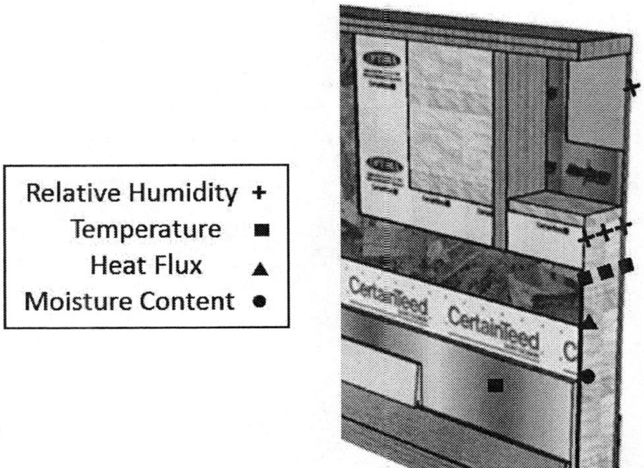

Figure 15 Schematic view of sensor locations and types.

Relative Humidity

At the same wall location, relative humidity was compared as shown in Figure 17 for the east and west walls, respectively. More deviation between the simulated results and measured data was observed, however the results were within the measurement accuracy of the relative humidity sensors, reported as ±3%.

Heat Flux

The heat flux passing through the east and west wall's was calculated using the temperature gradient measured through the fiberglass insulated wall cavity. The accuracy of the estimated value was compared with the results of the hygrothermal simulation. The Fourier number (F_0), shown in Equation1, compares the effects of thermal conduction to the heat storage of the fiberglass insulation, where t in seconds (s) is the characteristic time, α in m^2/s (ft^2/s) is the thermal diffusivity of the fiberglass insulation, and L in meters (m) (inches [in.]) is the cavity depth or insulation thickness. The thermal diffusivity, shown in Equation 2, requires knowledge of the fiberglass insulation's material properties, where k in W/m·K (Btu·in./h·ft^2·°F) is thermal conductivity, ρ in kg/m^3 (lb/in.^{3}) is density, and c_p in J/kg·K (Btu/lb·°F) is specific heat capacity (Bird et al. 2007).

$$F_0 = \frac{\alpha t}{L^2} \tag{1}$$

$$\alpha = \frac{k}{\rho c_p} \tag{2}$$

Since the Fourier number was estimated to be 4 ($F_0 \approx 4$), the heat transfer through the wall cavity is not dominated by the inertia effects of the fiberglass insulation and the heat flux in W/m^2 (Btu/h·ft^2) through the wall cavity is calculated according to Equation 3:

$$Q = \frac{k\Delta T}{L} \tag{3}$$

Temperature measurements were taken from the thermocouples located on the exterior (OSB–fiberglass insulation interface) and interior (fiberglass insulation–gypsum board interface) limits of the cavity. Results were compared for the east wall as shown in Figure 18 for a three-month and two-

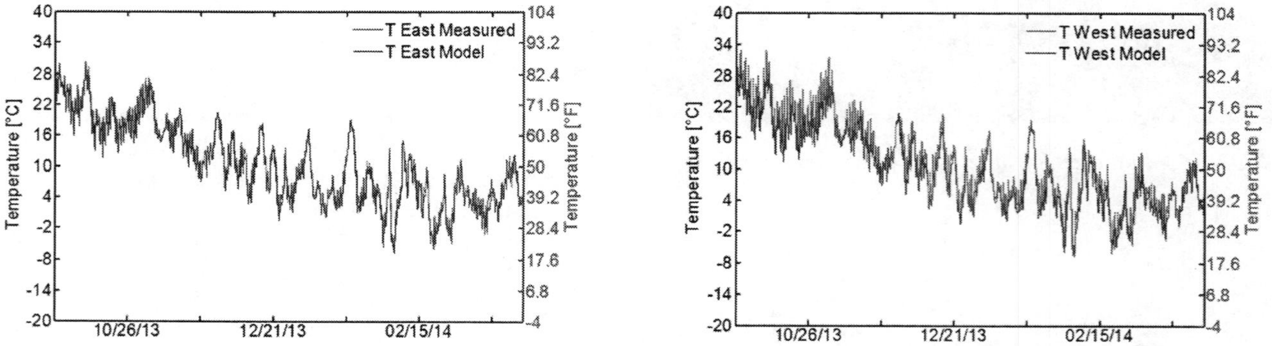

Figure 16 Comparison of east and west wall-measured and simulated OSB/fiberglass cavity interface surface temperatures.

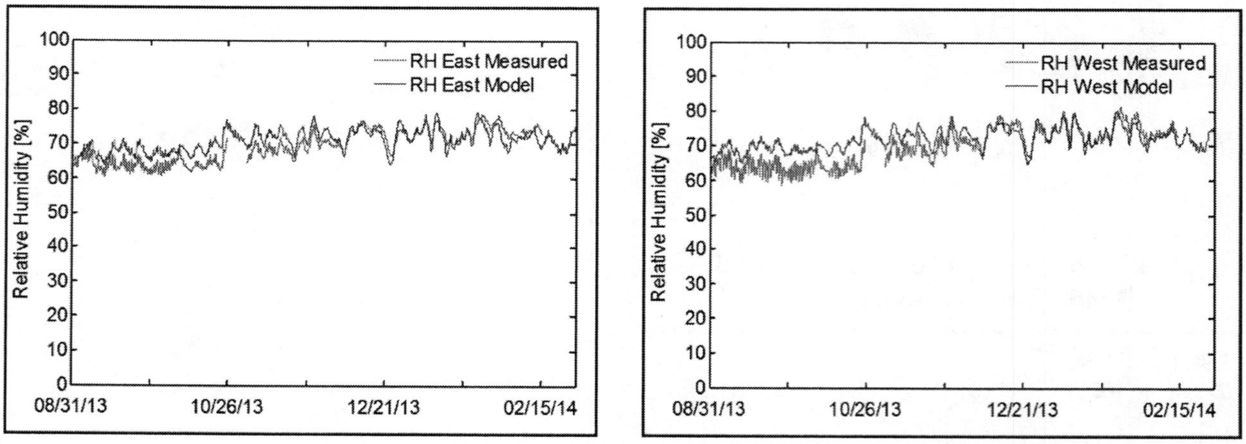

Figure 17 Comparison of east and west wall measured and simulated OSB/fiberglass cavity interface surface relative humidity.

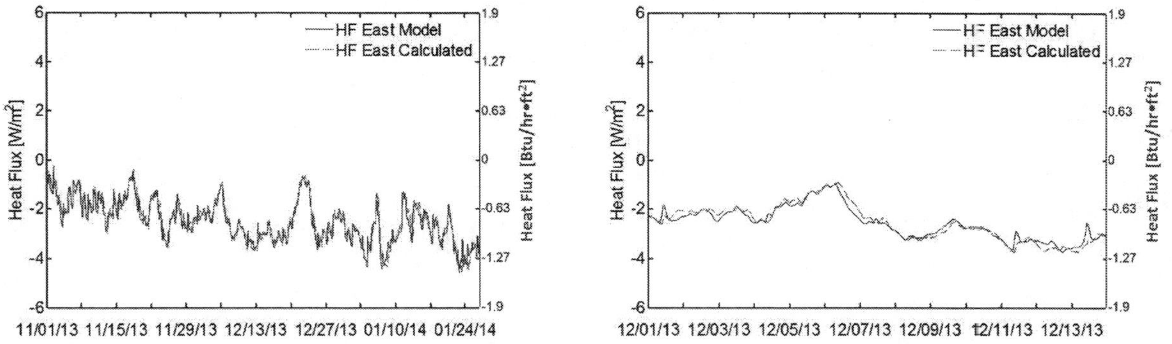

(a) (b)

Figure 18 *(a) Three-month comparison of calculated to simulated east wall heat flux and (b) two-week comparison.*

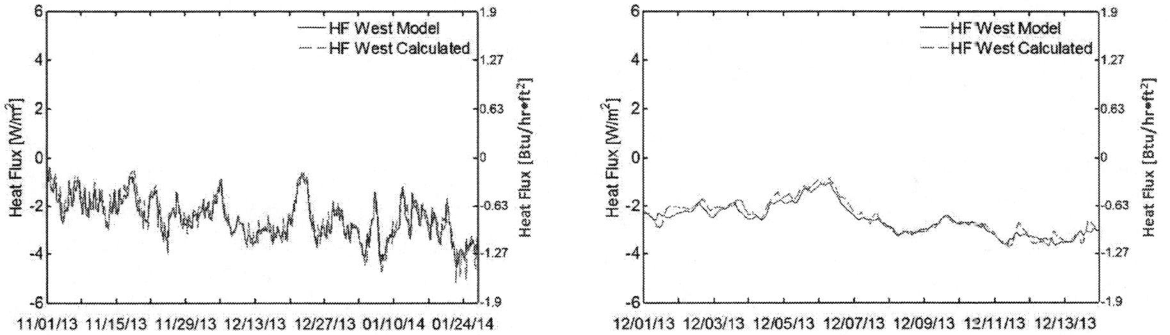

(a) (b)

Figure 19 *(a) Three month comparison of calculated to simulated west wall heat flux and (b) two-week comparison.*

week time period, respectively. In both instances, good agreement between results was observed. Comparisons for the west wall are shown in Figure 19, showing good agreement between calculated and measured results.

Moisture Content

Moisture content of the OSB sheathing was monitored for both wall orientations. Wood moisture content is an indicator of mold growth potential and subsequent material durability. For wood-based materials, the widely accepted threshold for mold growth is 20% moisture content (MC) by weight.

The MC was determined through electrical resistance measurements correlated to an electrical resistance and MC calibration curve. Point Moisture Measurement (PMM) Sensors from SMT Research – Canada, were used to collect data. A generic correlation for OSB provided by the sensor's manufacturer was used to convert resistance measurements to MC (Maref et al. 2010). While the correlation for OSB reports an accuracy of ±2%, variations in grade including wood chip and resin type, as well as density, produce an inherent inhomogeneity. The grade of OSB used in construction may differ in composition and properties from the grade used to determine the correlation parameters. Therefore, the actual error may be higher than reported by the calibration. A comparison of the measurements versus simulation data for the east and west walls is shown in Figure 20.

The time period for the MC comparison followed the construction schedule of the building. The simulated MC values fall below the measured values. One possible reason for the difference between measured and predicted values may be due to the wood-based materials having higher initial MCs than estimated in the simulation.

Further investigation is required to determine if the MC discrepancy lies with the measured or simulated data. However, the measured and simulated OSB MC results for both wall cavities were well below the 20% threshold required for mold growth.

CONCLUSION

Approaching a net zero energy building was achievable using a unique combination of building materials, construction methods, energy management strategies, and renewable

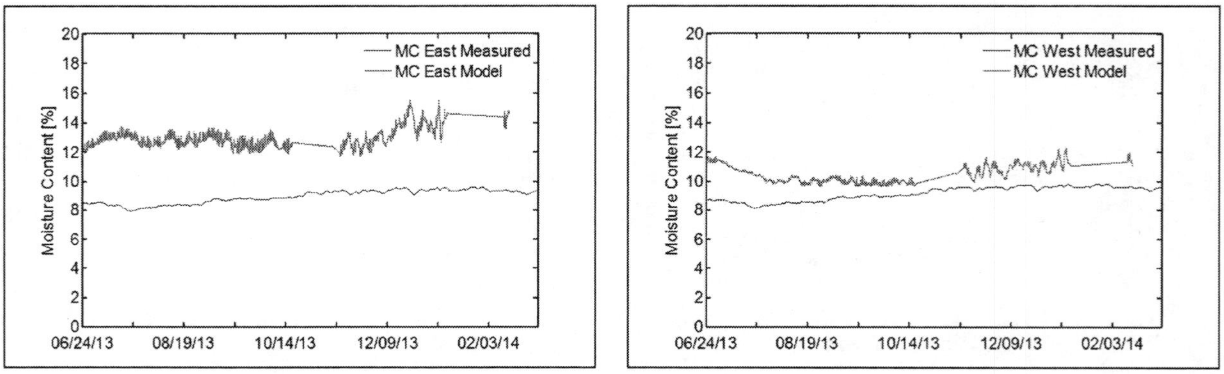

Figure 20 *Comparison of east and west wall OSB measured and simulated MC.*

energy sources. High-performance building envelopes and space-conditioning systems are readily available today to reduce energy consumption and create sustainable built environments that meet the multi-comfort (thermal, visual, acoustic, and indoor air quality) needs of the building's occupants.

A comparison of measured data for both the east and west walls for the NZEB building shows good agreement with hygrothermal simulations. The results confirm that the walls are performing as designed. Temperature and relative humidity sensors are operating as intended. Heat flux calculations, which neglect inertia effects, are in good agreement with simulated values. A discrepancy between MC measurements and simulation results requires further investigation. It is not clear if the observed differences are a result of incorrect correlation factors or model assumptions.

In regards to annual energy use, the total energy consumption of the building was in good agreement with the predicted value, reaching slightly more than 50 $kWh/m^2/y$ (38.7 mmBtu/y). The breakdown showed large deviations, specifically in the areas of lighting, appliances, and space heating. The primary reason for the difference was because the original analysis assumed the building was operated as a single-family residence, not a small office building. The analysis assumed a specific heating and cooling schedule for the conditioned space. In reality, the heating and cooling systems followed a programmable schedule that was changed intermittently by the building's occupants based on personal comfort preference.

REFERENCES

ASHRAE. 2009. Standard 160—*Criteria for moisture-control design analysis in buildings*. Atlanta: ASHRAE.

Bird, R.B., W.E. Stewart, and E.N. Lightfoot. 2007. *Transport phenomena*. John Wiley & Sons.

ICC. 2006. International Energy Conservation Code. Washington, D.C.: International Code Council.

ICC. 2009. International Energy Conservation Code. Washington, D.C.: International Code Council.

ICC. 2012. International Energy Conservation Code. Washington, D.C.: International Code Council.

Maref, W., M.A. Lacasse, and D.G. Booth. 2010. Experimental assessment of hygrothermal properties of wood-frame wall assemblies moisture content calibration curve for OSB using moisture pins. NRC NRCC-51172.

Residential Energy Services Network, Inc. 2013. Mortgage industry national home energy rating systems standards. http://www.resnet.us/hers-index.

WUFI® Pro 5.2. 2008. *A PC program for analyzing the 1-dimensional heat and moisture transport in building components*. Stuttgart: Fraunhofer Institute.

High-Rise Wood Building Enclosures

Graham Finch, MASc, PEng
Member ASHRAE

ABSTRACT

Interest in the construction of taller mid- to high-rise wood structure buildings is growing in Canada and the United States, building off of the success of dozens of projects in Europe and Australia. High-rise wood buildings utilize cross-laminated timber (CLT) and nail-laminated timber (NLT) panels, glulam beams and columns, engineered timber components, as well as steel and concrete to meet the design and construction challenges for these greater heights. The potential for increased time of exposure to weather and heavy timber construction methods pose new challenges for design and construction teams. The design of the building enclosure and the associated details requires careful attention to maintain acceptable durability, fire protection characteristics, and thermal efficiency. Prefabrication of whole floor, wall, or other modular components is utilized to speed up the time of construction and protect the wood assemblies from exposure to moisture.

This paper shares lessons learned through experience in the field and provides guidance on building enclosure design and detailing for mid- to high-rise wood buildings with a focus on control of heat, air, and moisture. Structural, fire, and acoustic considerations are not the focus of the paper nor covered here in detail. Case studies from the building enclosure designs of selected leading new, tall wood buildings are presented, including the Bullitt Center in Seattle, Washington, USA (6 storeys, ~76 ft), the Wood Innovation Design Centre (6 tall storeys, ~97 ft) in Prince George, British Columbia, Canada, and the Brock Commons UBC Student Residence building (18 storeys, ~190 ft) in Vancouver, British Columbia, Canada.

INTRODUCTION

The notion of high-rise wood buildings is gaining traction in many parts of the developed world. With advancements in engineered wood components, along with an industry push to demonstrate the safety, feasibility, and benefits of mid-rise (4- to 9-storey) and high-rise (10-storey and taller) wood buildings, several projects have taken form in Europe, Australia, and now Canada and the United States. This has not been without challenge in demonstrating the safety, durability, and economics of these buildings and moulding public perception of wood at taller heights (Perkins + Will 2014; Green 2012).

Within North America, the height of stick-built wood buildings has been restricted to three or four storeys (low-rise) depending on the jurisdiction since the introduction of the first Canadian and American building and fire codes. It has only been within the last decade that stick-built mid-rise wood buildings have been allowed to be constructed up to five and six storeys in height in some parts of Canada and the United States.

To accommodate code restrictions, the first taller wood buildings (beyond six stories) are being constructed as pilot projects, allowed by the local authority having jurisdiction by utilizing acceptable Alternate Solutions (aka Alternative Means and Methods) to meet fire, structural, life safety, and other requirements. As a result, extensive research, testing, monitoring, and engineering is going into this first cohort of taller wood buildings to demonstrate their suitability. Manufacturers of engineered wood products and the wide array of fasteners necessary to construct these buildings are also starting to innovate in Canada and the United States, joined by

Graham Finch is a principal and building science research specialist with RDH Building Science, Vancouver, BC, Canada. He is the co-author of several publications related to taller wood buildings and has designed the building enclosures for several of the first tall wood buildings in North America.

many European companies who are actively exporting materials into North America.

The North American industry has also been constructing many relatively large, but not necessarily tall, wood buildings incorporating mass timber panel components with cross-laminated timber (CLT), laminated veneer lumber (LVL), laminated strand lumber (LSL), nail-laminated timber (NLT), and glulam columns, beams, and panels. While not tall, these mass timber buildings are providing excellent experience with the durable use of these materials and lessons regarding the appropriate wall and roof systems that would be incorporated into the high-rise buildings that will follow.

A tall wood building is not necessarily constructed entirely out of wood products (see Figure 1). In fact it is likely impossible to do so. Steel and concrete components are generally incorporated into tall wood buildings where it makes practical sense to do so. Examples include concrete footing and floor slabs and concrete toppings on wood floor panels along with the wide array of steel connectors, screws, and nails necessary to connect the wood components together. Building structures that incorporate heavy timber elements along with large amounts of steel and concrete are often referred to as *hybrid structures* within the industry. Above-grade wall and roof enclosures have been wood-based and there is a common theme of prefabrication. Cross-laminated timber wall, floor, and roof panels are relatively common.

This paper describes the current state of the art of the tall wood building industry in North America with a specific focus on the design of durable and efficient building enclosures. The paper provides a discussion of the current practices for mass timber building enclosure design, a review of several lessons learned from a few mass timber buildings that experienced moisture issues during construction, and a case study review of the building enclosure designs from three of the tallest wood buildings in Canada and the United States.

BUILDING ENCLOSURE BEST PRACTICES FOR TALLER WOOD

The building science and physics for a high-rise wood building enclosure are no different than for a low-rise wood-frame building. There are, however, differences in the selection of appropriate materials along with the design considerations to accommodate the increased height and exposure to weather.

Above-grade walls and roof systems may be constructed out of wood, steel, or concrete depending on the building design, economics, and other goals of the project. Local fire codes will also dictate where certain wood products can and cannot be used or left exposed, which means that a variety of structural systems and enclosure assemblies will inevitably be used for taller wood buildings across North America. A few examples of potential high-rise wood structural systems and wall assemblies are shown in Figure 2.

Whether the exterior walls are load bearing or infill will dictate the specific design and materials that can be used. In exterior load-bearing wall designs it will be more common to see CLT or other engineered wood panels designed to accommodate the gravity and lateral loads for the building. If the walls are non-load-bearing infill, subject to wind loads only (as is common for most concrete and steel high-rise buildings), then a wide variety of well-established building enclosure options are available, including wood stud and steel-stud infill walls, curtain wall systems, prefabricated insulated panels such as structurally insulated panels (SIPs) or insulated metal panels, insulated precast concrete, and a number of other wall systems.

The fundamentals of a best-practice building enclosure design for a taller wood building is the same as any other building of similar height and exposure, but with additional consideration for how the wood structure or wood enclosure elements may behave. Design considerations such as cumulative vertical wood movement of the structure (including shrinkage, compression, gap closure, and creep) are poten-

Figure 1 *Three large wood building projects showing newer wood construction components (CLT, NLT, and glulams along with steel and concrete). CLT is shown in the left and center photographs and NLT is shown in the right photograph.*

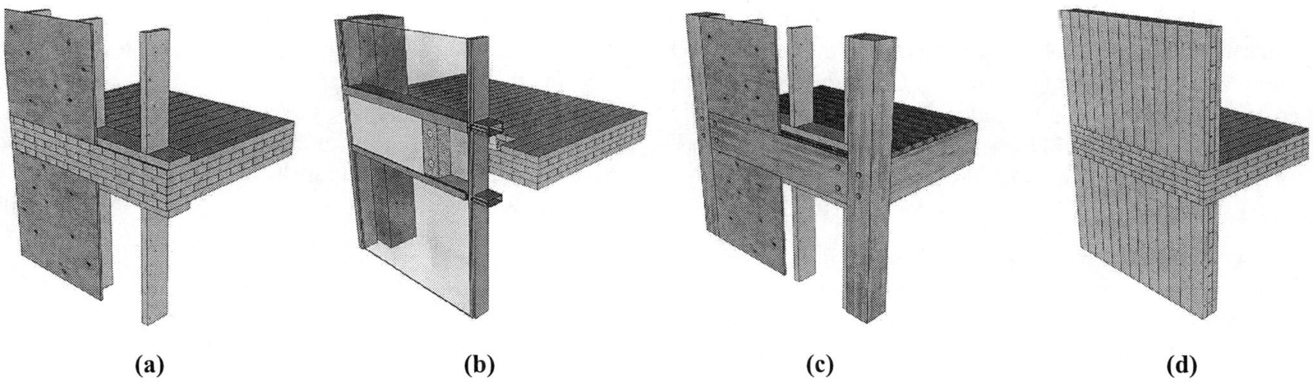

Figure 2 *Examples of some of the different types of exterior wall enclosure systems utilizing mass wood components: (a) platform wood frame infill with CLT floor and interior wall column/panel structure, (b) curtain-wall framing over post and beam structure with CLT floor, (c) wood-stud infill with post and beam structure and heavy timber floor structure, and (d) CLT exterior wall structure (load bearing) with CLT floors.*

Figure 3 *Examples of differential movement considerations for the Wood Innovation and Design Centre in Prince George (covered in more detail in the Wood Innovation and Design Centre section of this paper). The primary consideration here is the change in seasonal thickness of the CLT floor panels and glulam beams in relation to the infill curtain wall and SIPs wall systems. The design of a flexible air and watertight joint that could accommodate up to 1 in. of differential movement was integral to the design of this enclosure at each floor line. Note that the columns run continuous from foundation to roof to limit the amount of "shrinkable wood" within the building.*

tially emphasized in taller wood buildings and need to be accommodated by the building enclosure. It is not so much the overall change in height of the building that is the primary concern but the differential movement between the structure and enclosure and allowing for anticipated initial and seasonal movement. Figure 3 provides an example of how different movement is considered during the design of the curtain wall and infill wall panels at Wood Innovation and Design Centre (WIDC). It is not so much that this strategy differs from steel or concrete buildings in terms of absolute movement but the types of available connectors and materials that need to be adapted to this purpose.

Wood has proven long-lasting performance in properly designed and constructed buildings around the world. The major threat to long-term service life durability comes from decay fungi and mold growth. The key to achieving durability is to prevent excessive moisture accumulation and to allow wood to dry should it get wet during construction or while in service. The risks will be greater in a tall wood building due to greater exposure to weather and potentially longer construction schedules spanning rainy seasons. Addressing this greater risk will typically mean applying appropriate materials to control water, air, and vapor and thermally insulating wood building enclosure elements such as CLTs, NLTs, glulams, etc., on the exterior side. In addition, wood must be properly protected with the use of preservatives or barrier membranes against termites in affected areas of Canada and the United States.

A number of building enclosure design publications, building science textbooks, and other building science references are available and provide general guidance regarding the design and construction of durable and energy-efficient wood-based building enclosures, such as *Guide for Designing Energy-Efficient Building Enclosures* (RDH 2013), *Building Enclosure Design Guide* (RDH 2011), *CLT Handbook* (FPI 2011, 2013), and *High Performance Enclosures* (Straube 2012). The building enclosure design fundamentals described within these publications are typically only for wood-frame buildings up to six storeys; however, they can be applied with caution and consideration for taller wood buildings. The use of CLT and other heavy timber components are covered in detail. It should be noted that many assemblies, details, and materials appropriate for use in low-rise wood-frame buildings may not be suitable for taller buildings due to the increased environmental loads and increased durability expectations.

Technical Guide for the Design and Construction of Tall Wood Buildings in Canada (FPI 2014) provides specific guidance for the construction of taller mid- to high-rise wood building enclosures. Some examples of more robust wood-based building enclosure assemblies suggested for taller wood buildings are provided in Figure 4. The following guidance is suggested for designers of mass timber wall systems in tall wood buildings:

- Use noncombustible drained and ventilated exterior insulated rainscreen claddings. Avoid the use of open rainscreen claddings where large volumes of water may bypass the cladding and excessively wet the cavity materials and/or wood structure.
- Design the cladding with properly flashed movement joints (i.e., cross-cavity flashings at each floor) to accommodate initial and seasonal wood and building movement and provide drainage and ventilation behind the cladding.
- Use noncombustible exterior insulation materials with thermally improved clip and rail or long-screw cladding support systems to minimize thermal bridging through the insulation. Wind loads will typically govern the design of these connections.

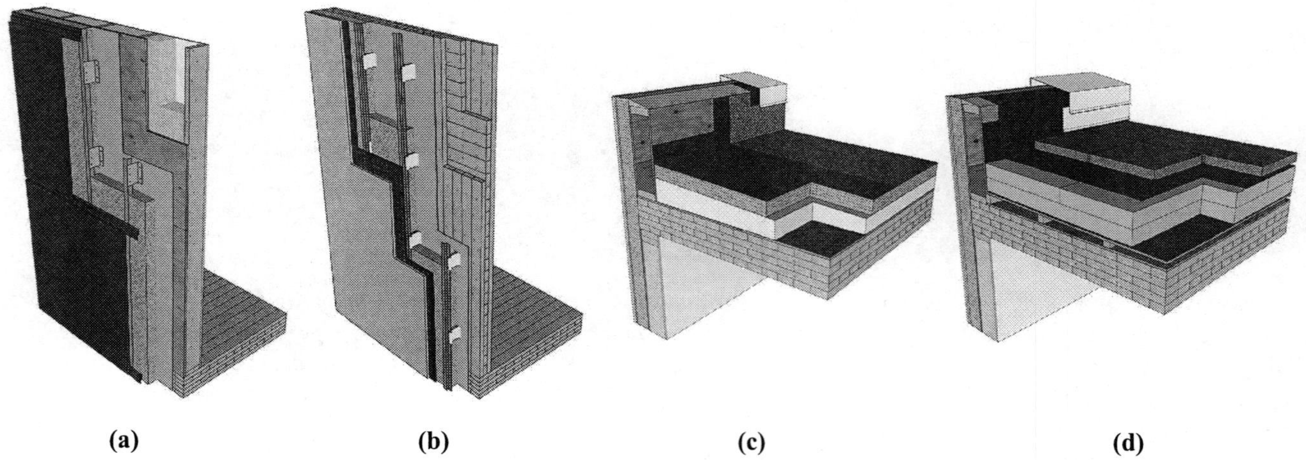

(a) **(b)** **(c)** **(d)**

Figure 4 *Examples of some of the different wood-based building enclosure assemblies suggested as more durable and robust for taller wood buildings in North America (FPI 2014). Note that in all of the assemblies, the wood is well protected by waterproof or water-resistant membranes and that, where possible, the majority of insulation is placed on the exterior of the wood structure. (a) Split insulated wood-frame infill wall with thermally efficient intermittent clip supported metal cladding. Note the use of a self-adhered sheet or liquid-applied vapor-permeable membrane for the air barrier and water-resistive barrier over plywood sheathing. (b) Exterior insulated CLT load-bearing wall with thermally efficient intermittent clip supported stucco cladding. Note the use of a self-adhered sheet vapor permeable membrane for the air barrier and water-resistive barrier over the CLT panels. (c) Conventional roof assembly over sloped CLT roof panels. Note the use of a robust roofing membrane and separated fully adhered air barrier/vapor barrier and temporary roof membrane preapplied to the CLT panels prior to site delivery. (d) Protected membrane roof assembly over tapered wood structure over CLT roof panels. Note the use of a robust fully adhered two-ply SBS waterproofing membrane over the plywood sheathing and venting of the tapered structure to the interior of the building.*

- At minimum, provide cladding and insulation systems that have demonstrated acceptable fire performance in line with NFPA 285 (NFPA 2012) or equivalent.
- Use vapor-permeable exterior insulation and vapor-permeable air barrier (AB)/water-resistive barrier (WRB) materials to facilitate drying from heavy timber wood panels. Where impermeable membranes or insulations are used, care must be taken to ensure the wood is factory dry before applying, and it is preferred that preapplied membranes be used in lieu of site-applied membranes in this application.
- Use self-adhered AB/WRB materials in lieu of mechanically attached AB/WRB materials. Liquid-applied membranes can be considered but are not suitable over the dimensionally unstable faces of CLT or NLT.
- Avoid insulating on the interior of heavy timber panels with more than one-third of the total insulation value unless a more detailed climate-specific hygrothermal analysis is performed. Where stud insulated infill is used, it is suggested to use more exterior insulation to improve durability of the sheathing in this more-exposed assembly.
- Consider prefabrication of panels complete with AB/WRB membranes and potentially cladding and insulation preinstalled.
- Not all tall wood buildings will incorporate wood-based exterior wall assemblies. Steel-stud, precast concrete or aluminum curtain wall systems may also be used. Follow best practices for the design of these materials with consideration for unique connections to wood floor slabs, beams, and columns (instead of concrete) and potentially greater seasonal and long-term building movement.

The following guidance is suggested for heavy timber roof assemblies in taller wood buildings:

- Insulate heavy timber roofs entirely on the exterior side with conventional or protected membrane/inverted membrane assemblies. Avoid use of interior insulation wherever possible with mass timber as to keep the wood warm and allow drying to the interior.
- Strongly consider the application of appropriate temporary roofing and air/vapor barriers to the panels prior to installation on site to protect from construction wetting. These materials tend to be impermeable, so ensure that the panels are factory dry at the time of application.
- Consider roof assembly designs that allow for faster drying of mass timber components such as CLT panels in the event of a roof leak. This could include provisions for venting of interior air above the mass timber components so that leaks can be detected early and may be able to be dried out in both directions. This could take the form of a sloped wood taper package above the CLT deck provided local fire code requirements are also met (Figure 4d).
- Take care when using NLT within roof assemblies as construction wetting will cause the panels to swell significantly and potentially cause damage to the building and/or applied materials. Typically plywood is installed on top of NLT to provide a more dimensionally stable surface for roofing and membrane application as well as to structurally connect panels together.
- Consider sloping the structure of the roof panels in lieu of using tapered insulation or wood-framing systems.
- Consider the use of noncombustible insulation materials and/or gypsum protection boards as additional protection against fire even where fire-rated roof assemblies are not required.

LARGE WOOD BUILDING ENCLOSURES—INDUSTRY LESSONS LEARNED

The building industry is constantly learning from the successes and failures of buildings. For example, the industry is aware of the many hard lessons learned with inappropriately designed and constructed low-rise wood-frame buildings in North America in the past couple of decades. However, as there are very few tall wood buildings to directly learn from (as very few have been built to date), we need to learn from the construction of larger wood buildings incorporating similar wood structural and enclosure systems.

Experience with these recent large and taller wood buildings has identified a few issues and risky practices with mass timber components that could lead to durability problems. Many of the issues can be addressed with construction practices that limit wetting and better accommodate wetting when it does occur, including diligent site protection, changes to sequencing to preprotect wood elements from rainwater, or the selection of more robust materials. These practices are covered in the following section, using unnamed buildings to examine how the industry has learned and is continuing to learn about the construction of larger mass timber projects—particularly in wet environments.

It is important to limit the amount of wetting that occurs as wood assemblies are shipped, as they are stored on-site, and through various construction stages, to the point where the building is closed in or at least protected from exposure to rain. This becomes even more critical where mass timber wood products such as CLT, NLT, LVL, LSL, and glulam panels, beams, and columns are used because these thicker wood elements may absorb and store more moisture than plywood or dimensional lumber and swell when exposed to liquid water sources. In addition, they also dry slowly due to their mass. Large wood members suffer volumetric changes, causing bowing or splitting during severe wetting and drying cycles. Moreover, it can be difficult for enclosure assemblies that are wetted to dry out after insulation and air/vapor/water control membranes are installed. A standard target for wood moisture content (MC) is to keep the wood below 20% MC during storage and construction to provide a margin of safety and reduce long-term differential shrinkage after installation. Evidence from recent North American projects is also showing that the moisture content of mass timber elements should ideally be kept below a lower threshold of 15% MC (closer to its manufactured moisture level of 11%–14%) to reduce dimensional movement and allow the elements to dry out faster in service.

Factory-finishing with special coatings and sealers can provide temporary rainwater protection for wood products such as glulam and CLT, but depending on their properties these products may also trap moisture and slow down drying if water gets into the wood. Good construction sequencing can be used to protect wood and minimize wetting, such as applying wall and roof membranes as soon as the construction allows. In cases of roofs or even floors, this would mean potentially applying temporary or permanent waterproof membranes prior to erection so as to provide a temporary construction roof and protect the underlying wood. Experience has also revealed that some self-adhered or liquid-applied waterproofing or air barrier membranes do not adhere well to damp wood—additional reasons to preprotect wood elements in a tall wood building. The prefabrication of partially or completely finished enclosure wall and roof assemblies is also suggested to reduce the exposure of wood enclosure and structures to weather.

Lesson 1—Protecting Large Prefabricated Roofs from Wetting

This first lesson comes from a large recreational wood structure building constructed in the Pacific Northwest within the past five years. The wood roof consists of a long-span prefabricated curved roof truss with a double layer of plywood as the finished deck. During the tendering and construction of the roof panels, a fully adhered temporary roof membrane that was specified by the designers was value-engineered out of the assembly as a cost-saving measure, leaving the wood panels exposed to weather during storage and until the roofing membrane was installed. It should be noted that the roof was also too large to effectively protect with a scaffold structure or tarps and was exposed to rain for several months in early fall to mid-winter as a result.

Once the panels were erected, plywood sheets were installed along the joints between the trusses to complete the structural diaphragm. The roof was erected in the late summer—later than initially scheduled—and roofing commenced in the fall during the rainy season. As a result, large areas of the double plywood-skinned roof were left exposed to rain and night-sky condensation for several months until the roofing could be completed. Water was readily able to be absorbed into the upper plywood layer as well as reach down between the layers of plywood to wet the lower layer and leak into the interior of the building.

Roofing proceeded with the addition of two layers of paper-faced polyiso insulation and a welded white PVC roof membrane. About halfway through the roofing process, a setback occurred when fungal growth was observed on the surface of the paper-faced polyiso insulation and the surface of plywood decking in an area that had been completed a month or so prior. An extensive investigation and eventual full tear-off of the partially completed roof uncovered that the problem was widespread. It was observed that water from the plywood deck was not drying out to the interior as quickly as anticipated by the designers and was also being readily adsorbed into the polyiso insulation. This was unique in that it is fairly common to see roofing occur over single thinner layers of plywood in this market without the same issue. Figure 5 provides photographs taken during the investigation performed by RDH Building Science.

The repair consisted of attempting to dry the roof, treating the wood with a fungicide, and applying a fully adhered waterproofing membrane to the plywood deck (the same one that was removed earlier as a cost savings) prior to re-roofing with non-paper-faced polyiso and new reflective white PVC.

The primary lesson learned from this project was seeing firsthand how difficult it is to dry out a roof to the inside through two layers of plywood under an insulated and light-colored roof in service. The project also demonstrated how sensitive paper-faced polyiso is to fungal growth when in contact with damp to wet wood and reinforced the need for temporary/construction roof membranes for large panelized roof structures.

Lesson 2—Careful Selection of Preapplied Temporary Roof Membranes over Mass Timber

It should be obvious that preprotecting mass timber roof panels with water-resistant membranes should help protect

Figure 5 *Photographs taken during removal of the roofing showing wetting patterns and absorption of water from the wet double layer of saturated plywood into the polyiso, particularly at the plywood joints.*

them from wetting during shipping and erection and may even speed up construction. The challenge, however, is deciding what type of membrane to use. There are two primary categories of water-resistant membranes: those that are permeable and those that are not. Permeable waterproofing/water-resistant membranes include products such as synthetic house wraps and allow the wood to dry out from beneath the membrane should it get wet. These types of membranes are typically used on walls but are used less so on roofs, as they are not that resistant to standing water. Impermeable waterproofing membranes, such as roofing membranes or asphaltic peel-and-sticks, while they do a good job at protecting from external moisture, do not allow for drying to the exterior should the wood get wet or be wet prior to application of the membrane.

In the second lesson here, the project team used an asphaltic peel-and-stick waterproofing membrane to preprotect a unique exposed LSL roof panel system during construction. Once the panels were installed, the laps were stuck together. Unfortunately, on this project, the peel-and-stick laps were not adequately sealed and penetrations through the roof were not well sealed. Ponding water on the relatively flat surface for several months led to leakage into the LSL panels at many of the peel-and-stick laps and penetrations. Once water was able to get beneath the peel-and-stick, it became essentially stuck within the LSL and unable to dry out fast enough for what the project team deemed safe to the durability of the roof. As an expensive consequence, the membrane had to be stripped off in wetted areas and the panels dried by erecting a temporary roof scaffold and providing a combination of heat and desiccated air on top and below the panels. Figure 6 provides photographs of the roof showing the erection, investigation, and later drying of the roof.

The primary lesson here is discovering that the use of a peel-and-stick membrane with pressure-adhered laps may not have been the best choice for a temporary roof membrane for this type of structure. The tolerance for defects in application is very low. A more robust choice would be the use of a peel-and-stick membrane with torch-welded laps. Care must be taken when using torch-applied roofing around wood and the use of protective elements such as protection boards or self-adhered or cold-applied base/protection sheets applied before torching. Several commercially available peel-and-sticks are available that incorporate cold adhesives with torched laps. An example of such application is shown in Figure 7.

Lesson 3—Caring for NLT and Protecting Wood Before it is Too Late

As NLT consists of 2× nominal lumber stacked together, it is prone to dimensional movement (unlike CLT, which does not have all of the wood grain aligned) and it is able to store a substantial amount of water within the panels. NLT is more sensitive to construction wetting than CLT and requires additional design considerations to accommodate dimensional movement. The author has observed and anecdotally heard of a number of NLT panel projects where moisture movement was not adequately designed for, resulting in the walls being pushed out of plumb by the swelling of panels wetted during construction. Considering that wood will expand/contract by approximately 0.20% to 0.25% in dimension in the radial/tangential grain for every 1% change in moisture content, keeping the wood dry should be paramount.

In the third lesson here, NLT panels were utilized for the floor and roof structure of a four-storey office building in the Pacific Northwest. During construction, the NLT panels of the roof became completely saturated as very little site protection from wetting was provided (Figure 8). It appears that the site was treated like that of a concrete frame building. While expansion of the NLT was allowed by large gaps left between the panels, it became very wet to the point where fungal growth was initiated and a moisture monitoring system was installed to track its drying. To dry the NLT panels and be able to apply the roofing materials within the construction schedule, an expensive scaffold system needed to be erected to cover the entire building. This expense arguably could have been avoided by properly protecting the NLT from wetting in the first place. If the scaffold tent had been erected initially then the construction of the roof would not have been delayed or led to damage to the NLT.

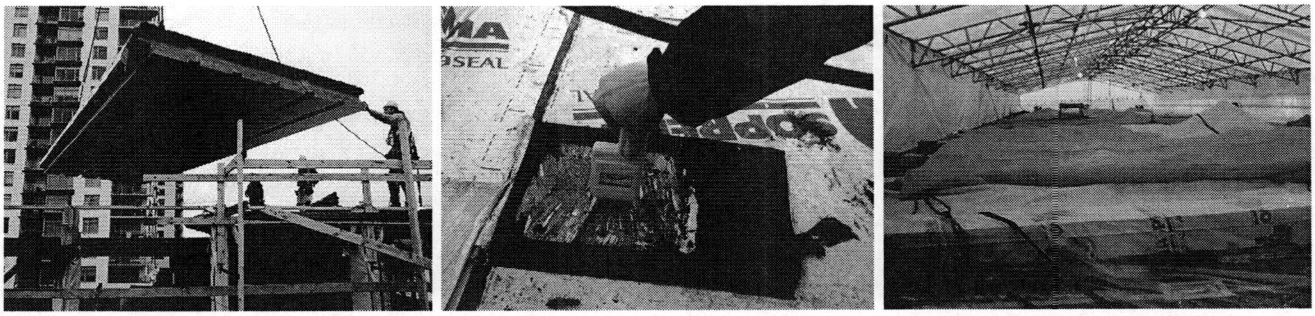

Figure 6 *Photographs taken during the construction, investigation, and later drying process of an LSL panel roof wetted by defects in the laps and damage to the temporary roof membrane.*

Figure 7 *Prefabricated mass timber truss system incorporating a temporary roof membrane consisting of a single-ply SBS peel-and-stick membrane with torch-welded laps immediately sealed after connection is made. Note how the membrane laps protect the sides and edges of the panels during shipping and erection. (Photos courtesy of StructureCraft.)*

Figure 8 *An NLT project where the roof panels were not protected during construction and became saturated, leading to swelling and fungal contamination coupled with a very long drying-out period. As a result of this wetting, a scaffold tent system needed to be erected over the entire building to protect and allow complete drying before roofing could be applied. (Photos courtesy of SMT Research.)*

TALL WOOD BUILDING ENCLOSURE DESIGN CASE STUDIES

Three case studies are included here to discuss the variety of building enclosure designs and assemblies that are currently being incorporated into taller wood buildings in North America. This includes The Bullitt Center in Seattle, Washington, USA (six storeys, ~76 ft), the Wood Innovation Design Centre in Prince George, British Columbia, Canada (six tall storeys, ~97 ft), and the Brock Commons UBC Student Residence building in Vancouver, British Columbia, Canada (18 storeys, ~190 ft).

Bullitt Center

The Bullitt Center is a six-storey, ~76 ft tall wood hybrid office building in Seattle, Washington, USA. The building was designed by The Miller Hull Partnership of Seattle, and RDH Building Science was responsible for the building enclosure design, field review during construction, and air barrier commissioning testing. The Bullitt Center is not only unique in that it was one of the first modern taller wood structures in the United States, but also in that it was the first Living Building Challenge office building in the world (bullittcenter.org). Wood was chosen as one of the primary structural materials for its resource efficiency, weight, and environmental impact. Reinforced concrete was used for the below-grade and podium levels as well as topping for the wood floors; structural steel was utilized for the elevator core and lateral load resisting system.

The roof consists of an R-40 effective insulated conventional roof assembly over wood deck with exposed white TPO roof membrane to harvest rainwater and is not particularly unique for a wood building. The exterior wall assembly consists of a steel-framed wall with studs supported off the exterior of the slab edge to improve thermal performance. The assembly is split-insulated with mineral fiber insulation and incorporates a thermally efficient cladding support system to achieve an effective thermal performance of R-26.6 overall, exceeding the local energy code requirements and living building challenge design target (Figure 9). The wall assembly is unique in its overall thermal performance for a steel-framed wall and demonstrates that a tall wood building does not necessarily need to have wood-framed exterior walls. Windows for the building consist of a high-performance triple-glazed unitized curtain wall system that is air- and water-sealed to the liquid-applied WRB membrane with silicone sealant and EPDM rubber transition strips (Figure 10).

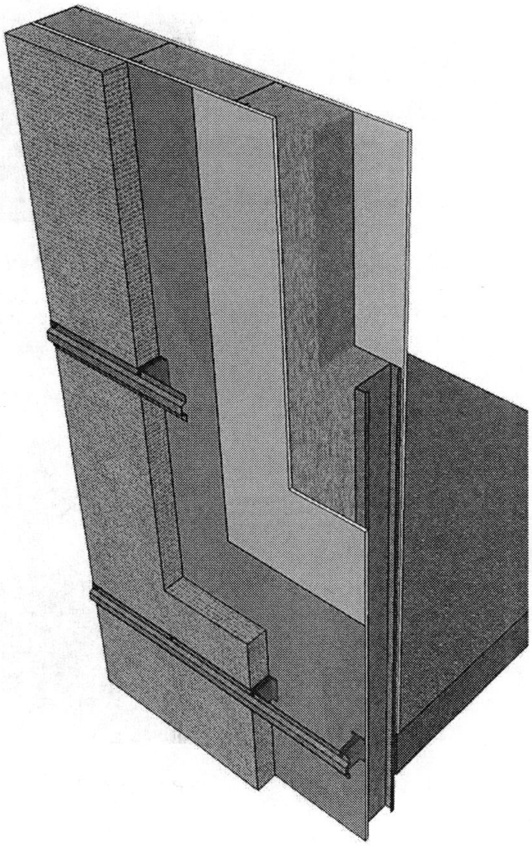

Figure 9 *High R-value steel-stud framed wall assembly utilized at the Bullitt Center.*

The Bullitt Center demonstrates the use of mass timber in a taller building, mainly NLT floor panels and glulam columns and beams in an extremely environmentally sustainable building. The wood elements were protected from excessive wetting during construction and the NLT panels incorporated gaps to facilitate movement due to moisture. Steel and concrete materials were also utilized by the design team, creating a hybrid structure and building enclosure.

Wood Innovation and Design Centre

The Wood Innovation and Design Centre (WIDC) is a six-storey, ~97 ft tall institutional building in Prince George, British Columbia, Canada. WIDC was the first tall wood building in North America and was seen as a pilot project to demonstrate the potential for high-rise wood buildings. Envisioned and designed by Michael Green Architecture (MGA) of Vancouver, the building incorporates a mass timber structure utilizing CLT floor and wall panels along with glulam columns and beams (Figure 9). The columns run continuously the height of the building to minimize dimensional movement, and the beams are hung from the columns to support the floor system. A staggered CLT floor system was innovated by MGA and incorporates spaces for HVAC, plumbing, and electrical services above and below alternate panels (Figure 11).

The author was directly involved with the design and detailing of the building enclosure, which consists of a unique conventional roof assembly over intermittent CLT panels, an exterior wall system consisting of infill prefabricated SIP panels, and a triple-glazed wood-veneer curtain wall system

Figure 10 *Photographs from the construction of the Bullitt Center exterior walls showing fluid-applied AB/WRB and joint sealants being applied to the sheathing, installed curtain wall system, and exterior insulation and cladding. (Photographs by John Stamets.)*

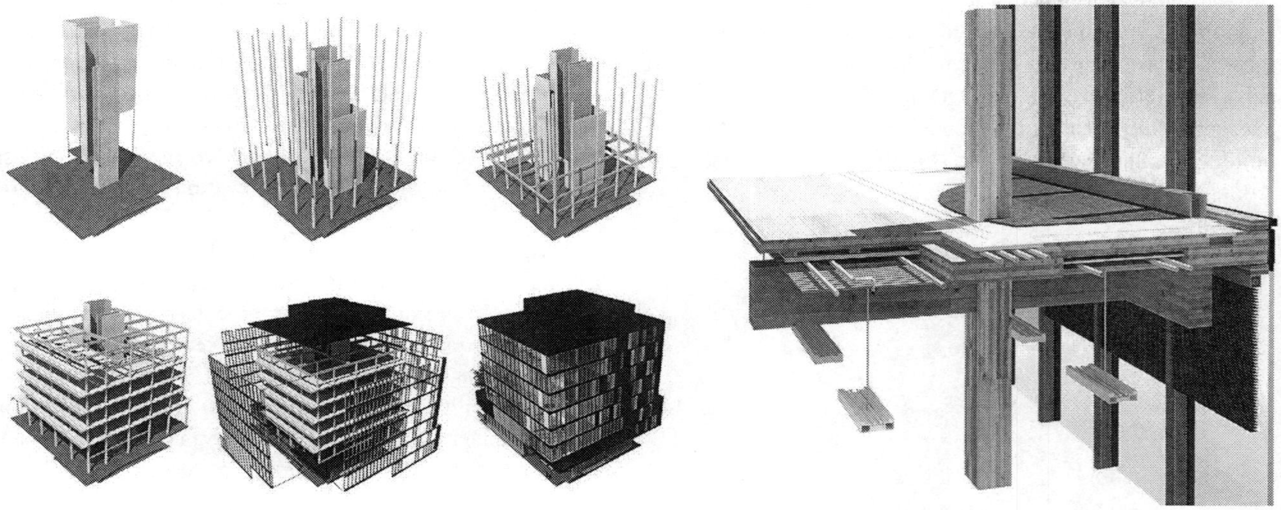

Figure 11 *Architectural renderings showing the unique wood structure and building enclosure elements of WIDC. The same structure and enclosure design can be readily adapted for 20+ storey high-rise wood buildings. (Renders courtesy of MGA.)*

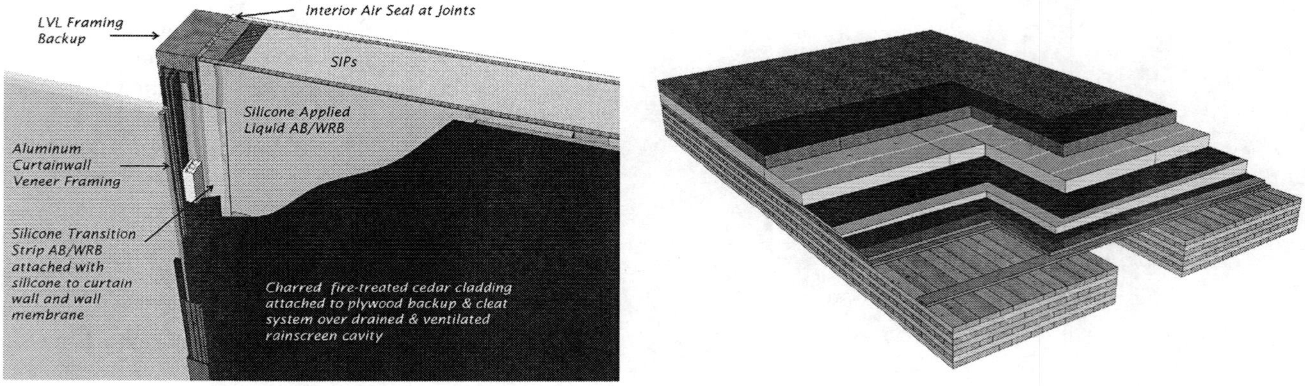

(a) **(b)**

Figure 12 *Primary building enclosure assembly designs showing (a) SIPs wall panel and wood veneer curtain wall details and (b) interior vented conventional roof over intermittent CLT roof panels.*

that was stick built and site glazed (Figure 12). A fluid-applied silicone AB/WRB was preapplied to the oriented strand board (OSB) sheathing of the SIPs in the factory and all joints were sealed in the field. Flexible silicone transition strips were used to span movement joints and to seal the curtain wall and other penetrations in the wall. Because of the simplicity of a single-component fluid-applied system that could be applied in conditions below freezing, detailing of the building enclosure was simple to implement in the field. The use of two colors of silicone (black and grey) were used as contrast between the field-applied wall membrane and joints, which facilitated easier quality assurance and field review. Photographs showing construction of the exterior walls and roof assembly are shown in Figures 13 and 14.

As part of this project, field monitoring was performed by FPInnovations to measure the long-term movement and moisture levels of the heavy timber structure (Wang et al. 2016). Monitoring found initial moisture contents of glulam and CLT components was on average 13% during construction and dropped as low as 4% during the wintertime due to very low indoor relative humidity (below 20%). The total shortening of the building columns was measured to be approximately 11 mm in 24,500 mm (0.044%). Shrinkage in thickness of the CLT floor slabs was measured to be approximately 5 mm in 169 mm (3.0%) on average, which was taken up by the flexible movement joints at the curtain wall and SIPs wall panels (Figure 3). Moisture content monitoring of the CLT panels of the roof assembly demonstrated dry initial and in-service

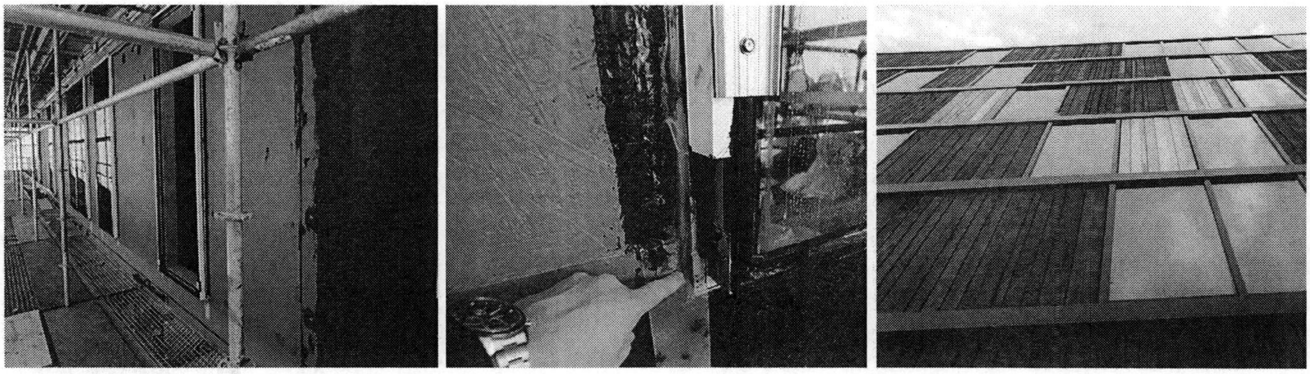

Figure 13 *Exterior walls of WIDC under construction. Note the use of silicone-applied AB/WRB to walls with flexible silicone transition strip movement joints at the curtain wall interfaces. A combination of both charred and bare-stained western red cedar was preapplied to treated plywood panels and the mounted-on French-clip connections and screwed in place to the rainscreen strapping in the field.*

Figure 14 *Conventional roof assembly of WIDC under construction. Note the use of a torch-applied SBS membrane over asphalt protection board functioning as a temporary roof and later air/vapor barrier over the interior vented wood roof deck over the CLT panels.*

conditions at all locations with the exception of one location that was initially wetted during construction but dried out prior to occupation due to the vented design.

The WIDC project demonstrates the use of an all-wood structure over a concrete foundation utilizing CLT walls and floors and glulam beams and columns. The walls incorporated SIPs panels and a wood veneer curtain wall within an infill non-load-bearing exterior wall application with simple connection and sealing details. The roof system demonstrated how the inclusion of a simple vent space above the mass CLT panels can improve the durability of this type of roof system.

Due to the arid inland environment and winter construction season, wetting of the wood components from rain was less of a factor to the construction team. Scaffolding and hoarding was used to facilitate wintertime construction in freezing weather and did protect the walls and curtain wall during erection. The roof was protected by a temporary full-torched roof membrane installed as soon as possible after installation, and the timing avoided heavy rain events (snow was swept off instead). Tarps were used in a few instances by the contractor to protect upper roof areas from wetting in the wetter spring months. All being said, these moisture-management practices would have likely allowed for excessive wetting and potentially delays and drying costs if the building had been constructed during the wet season of a coastal environment, similar to the examples discussed in the Large Wood Building Enclosures—Industry Lessons Learned section.

Brock Commons UBC Student Residence Building

The UBC Brock Commons project is an 18-storey, 180 ft tall hybrid wood residence building currently being constructed in Vancouver, British Columbia, Canada. Completion of the structure and building enclosure is anticipated by early 2017. The building structure consists of a reinforced concrete foundation and first-level and twin elevator cores extending up through the building. The wood structure sits on top of the first level (17 storeys tall) and consists of continuous glulam columns and beams with a five-ply CLT floor plate. A cementitious screed is placed on top of the CLT floor panels, which are encapsulated with gypsum on the underside for fire and acoustic separation. Related to the floor system, research was performed by RDH Building Science to evaluate the impact of construction wetting from rainwater and cementitious screeds on the wetting and drying of CLT panels during construction and to evaluate the benefits of different coatings and membranes applied to the CLT in the factory. Testing of CLT panels indicates that hydrophobic water-resistant coatings are beneficial to reduce rainwater uptake into CLT panels and that some barrier coatings will also reduce the uptake of moisture from the screed into the CLT. The testing found that uncoated CLT panels would be subject to excessively wet conditions if exposed to rain or from the application of the concrete topping and therefore coatings are recommended for CLT floors. Of particular concern is the penetration of water into gaps and cracks between boards past the surface of the panels both within panels and between panels where water is able to bypass taped joints. Based on our research and field experience to date, further research is needed to find optimal factory-applied coatings to protect CLT and other mass timber panels from construction wetting, to allow floors to be made watertight during construction, and to protect the wood from concrete topping moisture.

The building enclosure design process for this building was unique in that four different prefabricated wall systems were designed and prototyped for the project. Prefabrication of wall assemblies including exterior cladding with windows installed was chosen by the design and construction team to speed up site erection and allow a floor to be enclosed within a week. Different panel sizes were considered in early design stages, and it was determined that a module of approximately 30 ft in length by the floor height would be ideal for shipping and erection using the site cranes. Exterior scaffolding is not being utilized for the project, so the preclad panels will need to be air and moisture sealed using a combination of gaskets and sealants incorporated with the panel anchors.

Three of the four prefabricated wall options were prototyped, mocked up (Figure 15), and priced by the construction team and wall subcontractors: 1) steel-stud exterior insulated panel, 2) lightweight insulated precast concrete panel, and 3) wood-stud panel. The fourth option, a solid CLT panel, was eliminated early on and not mocked up, as the CLT panels were determined to be too costly for an infill wall application on the project.

At the end of the mock-up phase, the steel-stud panel option was selected by the owner due to the fact that the wall prefabricator had experience with panelizing similar-size projects and quoted a competitive price on budget for the project. A full laboratory mock-up was then constructed to test the panels and connections against air/water/structural/thermal loads (Figure 16) and tweak the design prior to fabrication of panels for the project.

It should be noted that the lightweight precast concrete panels were also a viable option and the wood structure could easily accommodate the weight of the concrete panels. The wood-stud option was also viable and fully prototyped; however, it ended up being more expensive than the other options and, more critically, there were no local wood panel fabricators capable of taking on such a large and complex project. With time and projects it is hoped that industry capacity will grow to improve the cost-competitiveness of wood options.

The concrete foundation and core, coupled with the wood primary structure and steel-stud prefabricated walls, truly demonstrates the hybrid all-inclusiveness of future high-rise wood buildings (Figure 17). Construction is ongoing and will be completed in 2017; some early construction photos are provided in Figure 18. The intent is to erect the wood structure and install the wall panels during the dry summer months;

(a)

(b)

Figure 15 *Partial two-storey structure and wall panel mock-up constructed to evaluate different wall prefabrication options and to test field exposure wetting, concrete toppings, and structural connections: (a) lightweight precast concrete wall system and (b) wood stud system with metal cladding.*

Figure 16 *Construction of mock-up of the final prototyped preclad steel-stud wall assembly for air/water/thermal/structural load testing at a testing laboratory prior to fabrication and installation on site. Cladding consists of a fire-resistant high-pressure laminate board attached to galvanized steel girts over a fiberglass clip support. The exterior fiberglass-faced gypsum sheathing attached to the steel studs is coated with a liquid silicone AB/WRB that returns around the edges of the panels for ease of sealing at the interior.*

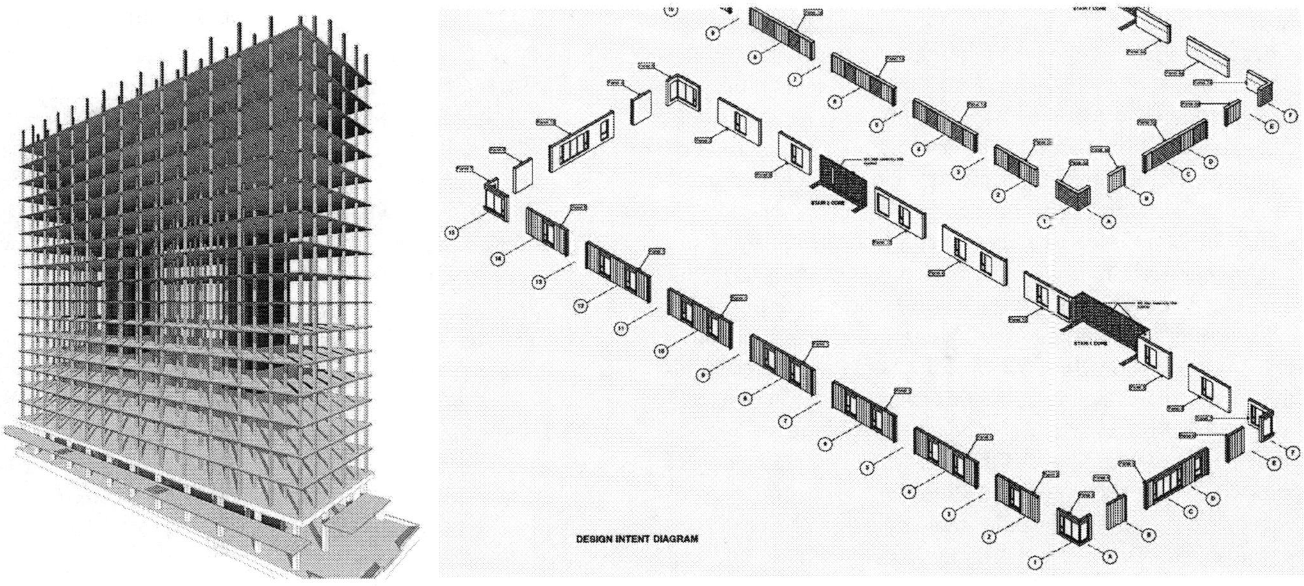

(a) **(b)**

Figure 17 *Architectural renderings of (a) the wood and concrete hybrid structure for the Brock Commons UBC Student Residence building and (b) the prefabricated wall panels.*

Figure 18 *Concrete cores, mass timber structure, and prefabricated steel-stud wall system with integrated aluminum windows.*

however, provisions have been made to address construction wetting (use of moisture-resistant materials within walls and a protective hydrophobic coating on the interior CLT columns and floors). The roof system consists of a traditional steel deck with insulation and exposed membrane in a conventional assembly. Through the summer of 2016, construction demonstrated that one to two floors of structure and enclosure panels can be erected per week, highlighting some of the benefits of both CLT and prefabricated wall construction.

CONCLUSION

Taller mid- and high-rise mass timber buildings are on the near horizon in North America. Through experiences in Europe and the construction of demonstration pilot projects and large mass timber buildings throughout Canada and the United States, the building industry is observing and learning how to design and construct these taller buildings and gaining a better appreciation for which building enclosure assemblies and details work and which are more risky. Best practices for design and construction become even more critical for high-rise wood buildings due to the use of more massive wood components, increased exposure to weather and wetting, and longer construction periods. Recommendations include the design of assemblies with redundancy and the ability to dry out, preapplication of waterproof elements prior to transport and erection, and more diligent protection of large wood components from wetting. The increased use of prefabrication is being observed within many of the larger and taller buildings built to date to allow for more efficient construction techniques and to protect wood components from long exposure to inclement weather. Many of these projects include wood and non-wood based structural and building enclosure components based on economic and technical factors, thereby creating truly hybrid wood building structures and building enclosures.

REFERENCES

FPI. 2011. *Cross-laminated timber (CLT) handbook* (Canadian ed.). Pointe-Claire, QC, Canada: FPInnovations.

FPI. 2013. *Cross-laminated timber (CLT) handbook* (U.S. ed.). FPInnovations, Forest Products Laboratory, BSLC, American Wood Council, APA, WoodWorks.

FPI. 2014. *Technical guide for the design and construction of tall wood buildings in Canada*. Pointe-Claire, QC, Canada: FPInnovations.

Green, M. 2012. *The case for tall wood buildings—How mass timber offers a safe, economical, and environmentally friendly alternative for tall building structures*. Ottawa, ON: Canadian Wood Council.

NFPA. 2012. NFPA 285: *Standard fire test method for evaluation of fire propagation characteristics of exterior non-load-bearing wall assemblies containing combustible components*. Quincy, MA: National Fire Protection Assocation.

Perkins + Will. 2014. *Summary report: Survey of international tall wood buildings*. Forestry Innovation Investment (FII) and Binational Softwood Lumber Council (BSLC). http://www.rethinkwood.com/sites/default/files/Tall%20Wood/Survey%20Tall%20Wood_REPORT%20WITHOUT%20APPENDICES_web.pdf.

RDH. 2011. *Building enclosure design guide—Wood-frame multi-unit residential buildings*. Authored by RDH Building Engineering Ltd. and published by Homeowner Protection Office, Branch of BC Housing.

RDH. 2013. *Guide for designing energy-efficient building enclosures—For wood-frame multi-unit residential buildings in marine to cold climate zones in North America*. Authored by RDH Building Engineering Ltd. and published by FP Innovations.

Straube, J. 2012. *High performance enclosures: Design guide for institutional, commercial and industrial buildings in cold climates*. Sommerville, MA: Building Science Press.

Wang, J., E. Karsh, G. Finch, and M. Chen. 2016. Field measurement of vertical movement and roof moisture performance of the wood innovation and design centre. Presented at the World Conference of Timber Engineering (WCTE), August 22–25, Vienna, Austria.

Guiding Design Teams by Hygrothermal, Energy, and Thermal Comfort Analysis While Managing Uncertainty

Ivan Lee, MASc **Patrick Roppel, PEng, MASc**
Randy Van Straaten, PEng, MASc **Jamie McKay, PEng, LEED Fellow**
Student Member ASHRAE

ABSTRACT

Ideally moisture analysis of buildings is based on statistical analysis where the probability of failure, loads and design parameters are stochastic variables, similar to limit state design for structural engineering. However, many industry professionals and standards, such as ASHRAE Standard 160, acknowledge that sufficient information is rarely available to make a full statistical analysis practical. Informed design decisions must balance the best available information as well as professional judgment. Even so, informed decisions require assumptions that will closely parallel reality, and not be too stringent or optimistic, so that durability, energy efficiency, occupant comfort, and architectural requirements are not unnecessarily sacrificed. This paper presents a case study of systematically evaluating the feasibility of insulating existing uninsulated painted brick masonry walls. Specifically, how to manage uncertainty through computer simulations that incorporate findings from the investigation of study buildings. Critical assessments of the risks are combined with an analysis of the benefits to energy consumption and interior thermal comfort. This project included field investigations, lab testing of brick samples for hygrothermal properties and freeze-thaw damage criteria, and computer simulations. Simulations included 1-D/2-D hygrothermal, 3-D thermal, and whole-building energy analysis. The analysis and investigation informed possible retrofit options that were presented using data visualization techniques so a clear picture of the potential energy savings and improvements to thermal comfort could be effectively presented to the design team.

Uncertainty can never be eliminated and information is not readily available to apply safety factors based on statistical probabilities, notwithstanding all of the effort to ensure that the assumptions were appropriate for the specific project conditions. Design decisions are made based on well-informed findings, sensitivity analysis, and engineering judgment. The presence of painted brick surfaces presented additional challenges from an analysis stand-point but also provided opportunities to correlate observed freeze-thaw damage to simulated damage. Ultimately, uncertainty was managed by systematically evaluating how critical assumptions affected the findings and related these findings to the architectural and owners' vision. The objective of this paper is to highlight how to manage uncertainty in practice when computer simulations are guiding design, so that moisture concerns can be addressed while balancing energy efficiency and thermal comfort objectives.

INTRODUCTION

Managing uncertainty is core to all engineering analysis of building performance. Effective solutions are established when managing the risks associated with designs are balanced with the benefits to the clients and end users. Many engineering disciplines address uncertainty with statistical analysis, such as limit state design in structural engineering, where the probability of failure, loads, and other design parameters are stochastic variables. The same approach cannot currently be readily applied to moisture analysis of buildings, as there is often not enough information to make full statistical analysis practical. Design decisions must be made based on the available information and professional judgment and experience. Most building science problems involve numerous interrelated variables that must be considered when evaluating possible solutions. Assessing or quantifying the impact of each

Ivan Lee is a building science consultant and **Patrick Roppel** is a building science specialist at Morrison Hershfield, Vancouver, BC. **Jamie McKay** is a building sustainability and durability specialist at Morrison Hershfield, Ottawa, ON. **Randy Van Straaten** is a senior engineer at RDH, Toronto, ON.

variable can be challenging and there is often no clear answer. As a result, feasible solutions and opportunities to optimize gains in performance are often discounted in practice because of inadequate information and perceived risk. Such is the case for retrofitting older load-bearing multiple wythe clay brick masonry buildings.

Theory tells us that adding insulation inboard of the brick will reduce heat flow, reduce drying, and increase the risk of freeze-thaw damage in cold climates. Designers have rightly asked the question if there are safe insulation levels from a durability perspective and how to effectively achieve both energy efficiency and thermal comfort goals. Hygrothermal and whole-building energy analysis can help determine optimal insulation levels, but the value of such analysis can be limited if the uncertainty related to the existing brick properties and as-built construction are not adequately managed. This uncertainty can be managed by supplementing the analysis with information gained by field investigation and lab testing of brick samples.

This paper presents case studies where hygrothermal and building energy analysis were used to evaluate the risks and benefits of insulating existing multiple wythe clay brick masonry buildings using an interior insulation strategy. The case study buildings are located in Ottawa and Gatineau, Canada (ASHRAE Climate Zone 6). The buildings formed part of a new sustainable development neighborhood project (One Planned Community), including the adaptive reuse of industrial buildings into new mix-use buildings (commercial, retail and office). For these buildings, the developer was concerned about long-term durability and wished to explore the risks and benefits of improving thermal comfort and energy efficiency through an interior insulation retrofit.

Field investigation combined with lab testing of brick samples informed hygrothermal analysis, 3-D thermal, and building energy analysis to evaluate feasible levels of insulation where energy savings and thermal comfort objectives could be met without increasing the risk of freeze-thaw damage and compromising the structural integrity of the existing buildings.

BACKGROUND

The case study buildings comprised of four 2-story load-bearing clay brick masonry buildings, constructed in the early 1900s, with 3-wythe and 5-wythe load-bearing walls, single-glazed wood framed windows, and an uninsulated roof supported by embedded wood and steel joists. Wood and steel floor and roof joists were embedded within one wythe of the exterior masonry surface. Remedial work in the form of spot repairs and overcladding (metal, brick, and concrete block) has occurred in the past in varying degrees. The existing brick has also been painted numerous times on both the inside and outside faces.

The brick is generally in good condition, with the exception of localized areas highly exposed to water, snow, salt, or poor maintenance. Deterioration is found primarily in details such as window openings, roof parapets, and bases of walls. The paint finish has been worn away or flaked off at many locations and has exacerbated freeze-thaw damage at other locations. Figure 1 shows the overall condition of the existing masonry.

Early meetings with the developer highlighted their interest in better understanding the risks associated with interior insulation; however, this was not their sole interest in commissioning a feasibility study. The team was very responsive to knowing how decisions they were making at the concept design stage could affect energy-efficiency measures, occupant thermal comfort, and the future development of a district heating plant and similar industrial buildings within the development. Through our conversations, we discussed the opportunities for adding insulation to the walls and roofs, improving fenestration design, and HVAC systems and controls. The authors also discussed how poor thermal resistance of the building envelope not only leads to occupant complaints, but buildings with high heat loss can also take longer to warm up in the morning when night setback temperatures are implemented, creating uncomfortable spaces during morning operation. Considering all the possible options for reinsulating the

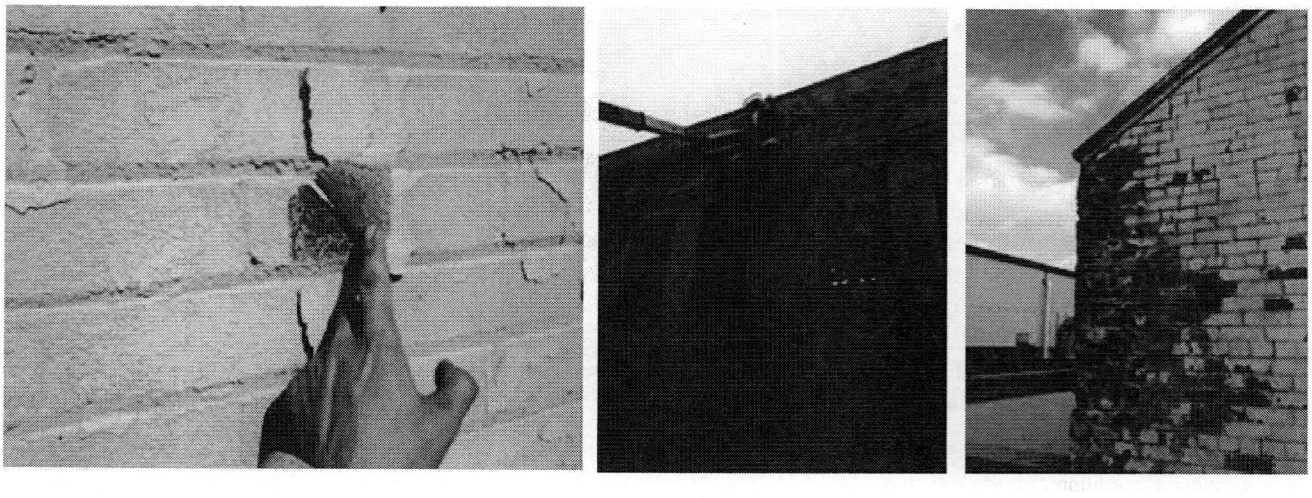

***Figure 1** Overall condition of building and brick masonry.*

walls, an interior insulation strategy was chosen to retain the buildings' industrial exterior appearance of an aged brick facade.

The goal of our early-stage analysis was to influence decisions before the design became too static and to avoid introducing constraints to performance without considering the full implications. However, we had to overcome the uncertainty and endless possibilities that can sometimes paralyze this process. The key to realizing the full value of early stage analysis was to manage the uncertainty and possibilities, with enough detail and insight, so that the design team was empowered to make the right decisions tailored to the priorities of the project. The following discussion uses case study buildings to highlight how computer simulations, with adequate supplemental information, can guide decisions at the early design stage and ensure that durability concerns can be balanced with energy efficiency and thermal comfort objectives.

ASSESSMENT OF DURABILITY CONCERNS

Freeze-thaw damage to bricks occurs when the pores in a brick are saturated with water to a critical point upon freezing and subsequent thawing. Freeze-thaw mechanics are more complicated than water held in the brick expanding upon freezing at 0°C. Past research has provided somewhat different explanations to the contributing forces that lead to damage, such as volume expansion of the water trapped in the brick pores dependent on pore size, hydraulic pressures, and/or ice lenses (Fagerlund 1996; Litvan 1989; Penttala 1998; Prout and Hoff 1991; Roppel 2003; Wessman 1997). Notwithstanding the forces that will lead to freeze-thaw damage, bricks can be sampled and tested in a lab to determine the moisture content at which permanent irreversible expansion (damage) will likely occur when subject to freeze-thaw (Litvan 1973; 1975; Mensinga et al. 2010). This moisture content is called the critical degree of saturation S_{crit}. An indication that a brick may experience freeze-thaw damage in the field can be made by comparing the critical degree of saturation to the free water saturation[1]. There is a possible risk of freeze-thaw damage if the free water saturation moisture content of the brick is higher than the critical degree of saturation of the same brick. However, this risk assessment is conservative and is only useful as a screening tool for assessing whether more analysis is required. Hygrothermal analysis can help evaluate the risk of freeze-thaw damage by comparing simulated brick temperatures and moisture content to the critical degree of saturation S_{crit} for natural weather conditions expected for the building location and exposure.

Hygrothermal simulations informed by testing of brick samples allows the analysis to be targeted to the site specific conditions. However, this higher level of due diligence does not eliminate the uncertainty embedded within this approach. There are several complications to the ascribed methodology that makes a strict performance based analysis unfeasible.

One complication is that the critical saturation is typically measured for only a freeze-thaw cycling scenario to a static temperature (in this case –15°C) and freezing rate. The sensitivity to other freezing temperatures and rates was not done for the case studies. The complication is that the freezing temperature of water held within brick pores decreases exponentially as a function of the pore size, below 0°C, and clay brick has a broad pore size distribution (Arnott 1990; Straube and Schumacher 2006). Moreover, freezing may occur in the larger pores but not the smaller ones. This fact makes a failure criterion that incorporates a 0°C temperature reference too simplistic and conservative for interpreting the results of hygrothermal simulations. However, theory also tells us that freeze-thaw damage can also occur without expansion of water caused by freezing (Litvan 1989, Wessman 1997, Taber 1930). Therefore, there is no obvious "yes/no" function that can be applied to the assessment of risk of freeze-thaw and no universally accepted acceptance criterion. This is where experience and judgment is needed to make the most of hygrothermal simulations to help with making informed and timely decisions. With the lack of better information, the risk of a freeze-thaw cycle where damage might occur can be made using three benchmarks:

1. The instances where the moisture content of the brick is above S_{crit} and the temperature dips below –0.1°C.
2. The instances where the moisture content of the brick is above S_{crit} and then the temperature dips below –2°C and then is warmed above 0°C.
3. The instances where the moisture content of the brick is above S_{crit} and then the temperature dips below –5°C and then is warmed above 0°C.

The first criterion is a conservative acceptance criterion that oversimplifies the risk of freeze-thaw damage and will count cycles that will not likely result in damage in the field. The second criterion filters out many false positive freeze-thaw cycles where the brick may briefly dip below 0°C. This is because some of the pores in clay brick will not likely freeze until below –2°C. The third criterion is the least conservative as it recognizes freezing in some of the smaller pores may not occur until –5°C. Utilizing the three criteria together allows us to evaluate the level of risk: low, medium, and high. This concept is illustrated by comparing an uninsulated wall assembly to the same assembly with R-12 medium-density closed-cell polyurethane foam insulation in Figure 2. The moisture content for both scenarios are briefly above the S_{crit} moisture content, but only the insulated assembly dips below 0°C. One freeze-thaw cycle would be counted for the insulated wall under acceptance criterion 1 but not for criteria 2 and 3. Therefore, the freeze-thaw risk has marginally increased but the authors would conclude that the risk is acceptable.

1. *Free water saturation* is the moisture content when a brick is in contact with liquid water for a prolonged period of time, under atmospheric conditions. This value is the realistic maximum moisture content for bricks in the field subject to precipitation.

An additional consideration is that dissolved salts within the masonry can affect moisture transport and salt crystallization can also damage brick similar to freeze-thaw. The analysis presented in this study looks specifically at freeze-thaw risk and does not directly consider the effect of salts on moisture transport. The assessment of the impact of salt in hygrothermal models and the impact on brick damage is a study area worthy of further investigation, particularly for climates not prone to freeze-thaw damage.

Finally, the critical saturation was determined for brick samples chilled on all sides in a chilled bath (multidirectional freezing). In contrast, the freezing plane for masonry in the field moves inward and outward (unidirectional freezing) because of varying outdoor temperature and solar heating through diurnal cycles. There is some uncertainty related to the use of S_{crit} values measured using multidirectional freezing of brick samples to access the risk of freeze-thaw for brick walls exposed to natural conditions. Nevertheless, some validation has been performed on walls exposed to natural conditions by comparing simulated results to observed deterioration on several projects.

Testing the Brick Properties

Testing of a representative sample of bricks allows for a more accurate hygrothermal analysis of existing brick walls and facilitates a less conservative assessment of risks. The brick sampling can be done in conjunction with making exploratory openings for the investigation of the existing walls.

The approach used at the study buildings was to sample a large quantity of bricks, in the order of magnitude of 50 bricks, from several locations (including interior, mid, and exterior wythes) and representative types: face versus common, painted versus nonpainted. The water absorption (A-value) was measured for all the brick samples. From the large sample size, a smaller sample size was used for freeze-thaw testing using frost dilometry (Mensinga et al. 2010) and detailed measurements of hygrothermal properties. Other properties that were measured included dry bulk density, water uptake (A-value), reference water saturation W_{ref}, free water saturation W_f, vacuum saturation W_{max}, and vapor permeability. The measured and derived brick hygrothermal properties are listed in Table 1. The bricks were also compared to brick properties from other projects and fits well with other bricks from this vintage (Building Science Labs 2014). Photographs of the bricks from site and lab samples are shown in Figure 3.

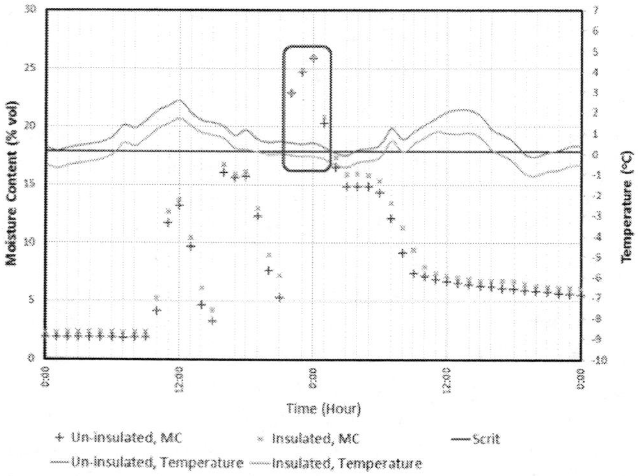

Figure 2 *Brick moisture content and temperature at exterior for uninsulated and insulated wall (red line indicates acceptance criterion 1 for both moisture content and temperature).*

There is a range of hygrothermal properties represented by the brick samples with varying critical degree of saturation and water uptake. From inspection of the brick properties, brick sample 2 might be deemed to be the most likely susceptible to freeze-thaw damage since this sample had the lowest S_{crit}. However, how well the bricks dry out also impacts the risk of freeze-thaw, which is a function of liquid transport and diffusion. Brick sample 3 appears to have the worst properties for the ability to dry out by inspection of the A-value and vapor permeability. Therefore, brick samples 2 and 3 were the focus of the hygrothermal analysis for the case study buildings.

ASSESSMENT OF THE RISKS OF INSULATING USING HYGROTHERMAL ANALYSIS

The risk of freeze-thaw damage on the existing masonry wall assemblies by insulating at the interior was evaluated using hygrothermal model called DELPHIN (IBK 2015).

The models were evaluated with historical weather data for the Ottawa/Gatineau Region over a 20-year period, varying building orientation and rain exposure, and insulation retrofit scenarios with and without paint. All four elevations: north, east, south, and west were evaluated to determine how different levels of rain and sun exposure can affect freeze-thaw cycles. Rain exposure was varied by adjusting the rain deposition factor between 0.5 and 1.0 based on rain deposition ranges determined by Straube and Burnett (2005). The presence of paint was evaluated by reducing the water uptake (A-value) and vapor permeability of an outer layer of the brick using data from the brick samples.

Interior conditions for the simulations were set to 20°C and at a vapor pressure difference of 540 Pa from the indoor to outdoor air (ISO 2000), representative of a medium to high moisture load. This vapor pressure difference accounts for the interior moisture generated by the occupants and activities, such as kitchens, cafes, etc. The indoor humidity is derived from the vapor pressure difference and results in a simulated average winter time relative humidity of 30 to 40% rh.

A total of 564 scenarios were modeled to determine whether interior insulation will increase the risk of freeze-thaw damage.

Existing Wall Reality Check

The findings of the hygrothermal analysis was compared to observed freeze-thaw damage on the building for both heated walls (with high and low exposure to sun and rain) and parapets (unheated wall with high exposure to sun and rain). Field observations noted more freeze-thaw damage at locations with higher rain exposure. Significant damage was also noted at locations where snowmelt and/or runoff are likely, such as window sills and bases of walls. The hygrothermal models correlated well with field observations, which predicted more freeze-thaw cycles for assemblies with high rain exposure and little damage for heated walls with low rain exposure. This correlation is shown in Figure 4, which compares the simulated freeze-thaw cycles with conditions observed on site.

Findings of Hygrothermal Analysis

Freeze-thaw analysis for the case study buildings show that there is little additional risk of freeze-thaw damage with the addition of closed cell spray foam insulation to the interior of the walls with low rain exposure as summarized in Table 2. Freeze-thaw damage is likely to occur for bricks with a low S_{crit}, such as brick sample 2 which has a S_{crit} of 10% moisture content by weight (MC wt), on the north and west elevations when the brick is painted. Brick samples with higher S_{crit}, such as brick sample 3, showed no risk of freeze thaw damage for cases with and without insulation and paint for all three freeze-thaw criteria.

The addition of interior insulation has little effect on the risk of freeze-thaw since it does not significantly alter the moisture content of the bricks for the outer and inner wythes and differences in temperature is not enough to change the number of cycles at S_{crit}. Figure 5 shows the moisture content and temperature of the outer 5/64 in. (2mm) of the brick for a painted west-facing wall with high rain exposure. In a typical rain event, the exterior face of the outer brick reaches the same level of saturation, and the moisture content of the brick spikes due to driving rain and quickly drops as the water dries out. The moisture content levels only differ when the bricks have

Table 1. Hygrothermal Properties of Brick Samples

Sample ID	Dry Bulk Density, kg/m^3	Thermal Conductivity, W/m·K	Heat Capacity, kJ/kg·K	A-Value, kg/m^2·s$^{0.5}$)	W_f, MC wt	Vapor Permeability, ng/s·Pa·m	S_{crit}, MC wt
1	1822	0.82	0.79	0.187	18.2%	16.0	18%
2	1785	0.79	0.79	0.264	15.9%	13.7	10%
3	1864	0.85	0.79	0.128	15.8%	10.0	14%
4	1810	0.81	0.79	0.221	17.3%	15.2	14%
5	1781	0.79	0.79	0.179	16.9%	17.7	14%
6	1809	0.81	0.79	0.166	18.7%	16.8	17%
7	1867	0.85	0.79	0.125	16.7%	13.7	16%
8	1806	0.81	0.79	0.134	17.1%	13.1	15%

(a)

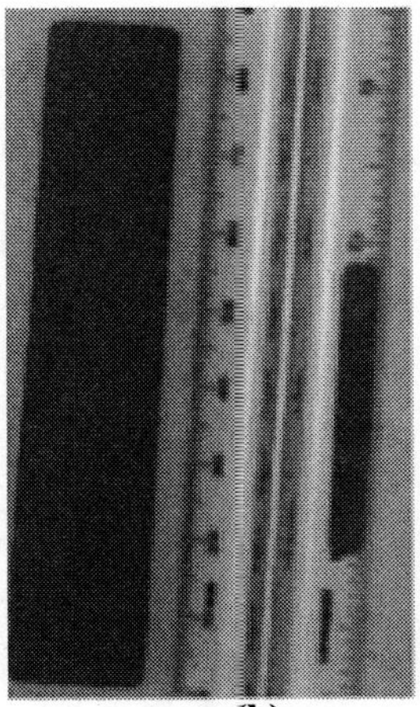

(b)

Figure 3 *(a) Full brick sample and (b) slice for freeze-thaw testing for sample 3.*

dried below S_{crit}. Insulation also did not significantly change the temperature distribution of the bricks at locations where the moisture levels are above S_{crit}. From this analysis, it appears the risk of freeze-thaw is more dependent on hygrothermal properties of the brick, particularly those that affect the higher moisture contents in the liquid phase, than the presence of insulation for the study buildings.

The presence of interior insulation did not increase the risk of freeze-thaw damage deeper within the outer brick and in the inner wythes of the wall assembly. The moisture content deeper into the brick remains below S_{crit} despite experiencing lower temperatures. Freeze-thaw in the inner wythes might occur with bricks with lower S_{crit} levels, higher water absorption rates, and/or wetting from the interior due to condensation. The sensitivity of S_{crit} and water absorption rates is discussed further below.

In addition to the brick samples from the case study buildings, additional scenarios with different brick properties were simulated to determine under which conditions and material properties freeze-thaw damage might occur. The additional brick properties were based on a brick survey completed by Building Science Labs (2014). Results of the additional scenarios were incorporated into a parallel coordinate chart, a tool to help visualize how interrelated variables affect the findings of the hygrothermal analysis. Each line represents one simulation, and the location where the line crosses the axes corresponds to

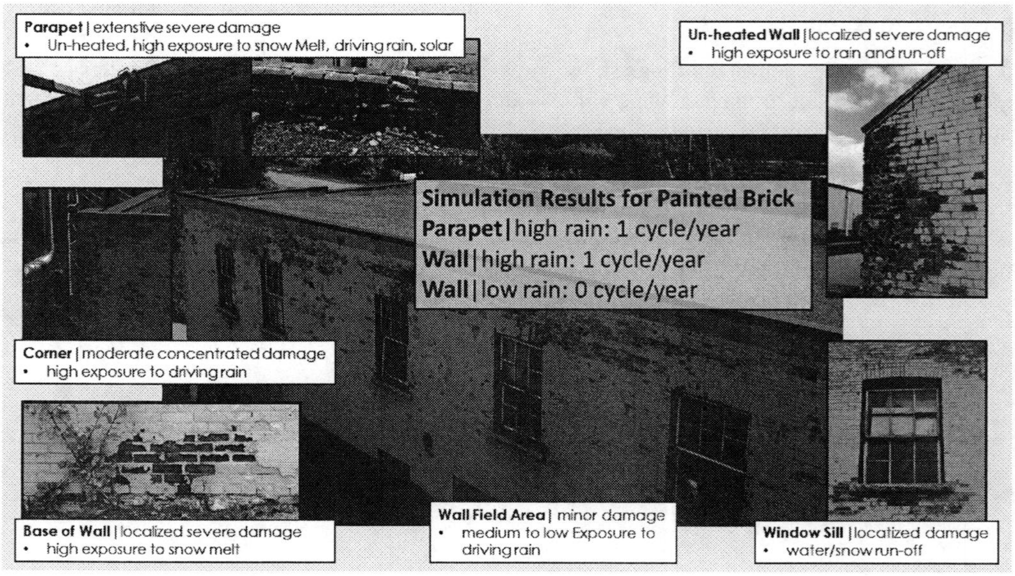

Figure 4 *Simulation results (freeze-thaw cycles below 0°c) for various driving rain exposures compared to observed freeze-thaw damage and likely exposure to moisture (photographs are a combination of elevations).*

Table 2. Estimation of Freeze-Thaw Cycles Where Brick is Above S_{crit}

Sample ID	S_{crit}, MC wt	Paint	Insulation Level	Orientation	Freeze-Thaw Cycles/Year Criterion 1, 0°C	Criterion 2, −2°C	Criterion 3, −5°C
2	10%	Painted	None	North and West	1	0	0
			R-6	North and West	1	0	0
			R-12	North and West	1	0	0
			All other cases had no simulated freeze-thaw cycles				
		Unpainted	None	All cases had no freeze-thaw cycles			
			R-6				
			R-12				
3	14%	All cases had no simulated freeze-thaw cycles					

the value of the parameter set or the output from the simulation. Results can be colored by assigned groups or by the value of an axis. This offers a quick and interactive way to quickly evaluate how different factors including A-value, vapor permeability, W_f, S_{crit}, temperature, interior insulation, and orientation affect the freeze-thaw cycles where the brick is above S_{crit}. The objective of this analysis is not to evaluate specific data sets on individual sampled bricks, but how relative combination of measured properties might affect the results. This analysis helps overcome the uncertainty related to evaluating only small sample sizes and checks the sensitivity of the conclusions reached using the sampled bricks, using an order of magnitude of the measured properties. Figure 6 shows a snapshot of some of the results of 2592 simulations. The right column freeze-thaw cycles per year, lists the maximum number of cycles that occurs during a winter season over the simulated 20-year period most prone to freeze-thaw cycles.

This sensitivity analysis shows the risk of freeze-thaw damage is dependent on hygrothermal properties that describe

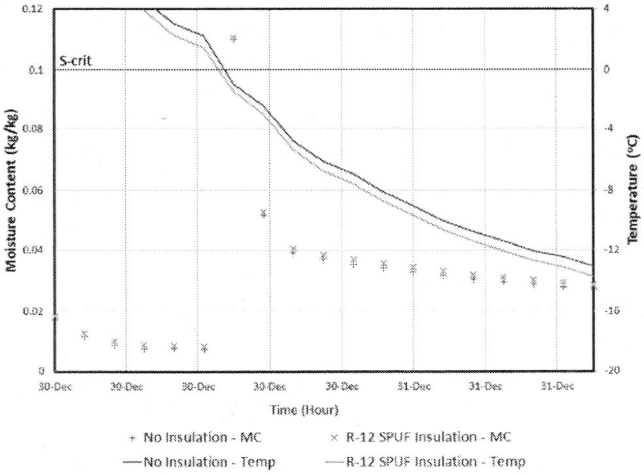

Figure 5 *Moisture content and temperature of painted brick sample 2 for west-facing wall assembly with high rain exposure with and without R-12 closed-cell polyurethane spray-applied insulation.*

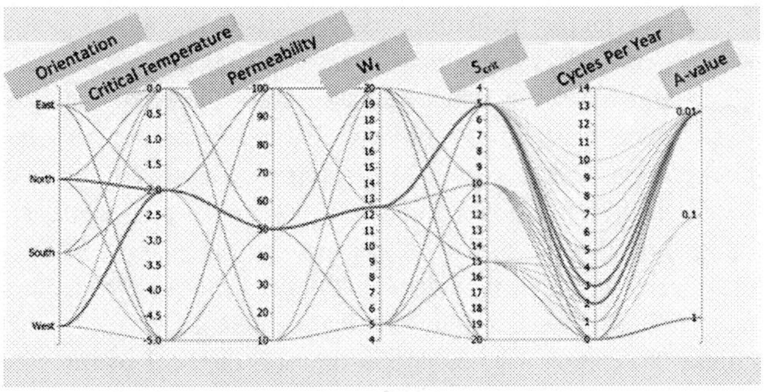

Figure 6 *Freeze-thaw analysis using a parallel coordinates visualization tool with low rain exposure.*

the wetting and drying in the liquid phase rather than in the vapor phase. The critical factors are: W_f, A-value, and S_{crit}. S_{crit} describes the moisture level above which freeze-thaw damage will occur, while W_f describes the maximum water content experienced in the field, and the A-value describes the liquid water uptake. Bricks with lower S_{crit} will experience more freeze-thaw cycles because moisture levels can easily reach and exceed a low S_{crit} during typical rain events.

The relative difference between S_{crit} and W_f can impact the number of freeze-thaw cycles, but not significantly. For example, bricks with a W_f of 20% moisture content (MC) are estimated to only experience one additional freeze-thaw cycle per year than bricks with W_f of 12.5% MC for the same S_{crit}, as shown in Figure 7.

Elevated moisture content is determined by how fast the brick is able to dry out. The moisture mechanisms to create conditions favorable to freeze-thaw is analogous to a bottle. The W_f describes the size or capacity of the bottle, while the A-value describes the size of the spout or opening. During a rain event the brick will be saturated close to the W_f if exposed to water for a sufficient duration for bricks within the range of A-values considered for this analysis (0.01 to 0.1 kg/m^{2}s$^{0.5}$). Bricks with less drying ability, as determined by ability to remove moisture by liquid transport (characterized by A-value), have elevated moisture contents for longer periods and have a higher risk of freeze-thaw damage. This is summarized in Figure 8, where the moisture content is plotted of the bricks with A-values between 0.01 to 1 kg/m^{2}s$^{0.5}$ and high exposure to rain on the west elevation. The analysis shows that bricks with lower A-values have significantly higher moisture levels than bricks with high A-values. This might seem counterintuitive because higher A-values means more rain water absorption and resultant higher moisture contents, which should relate to more opportunity for the bricks to cycle through freeze-thaw conditions when above the critical degree of satu-

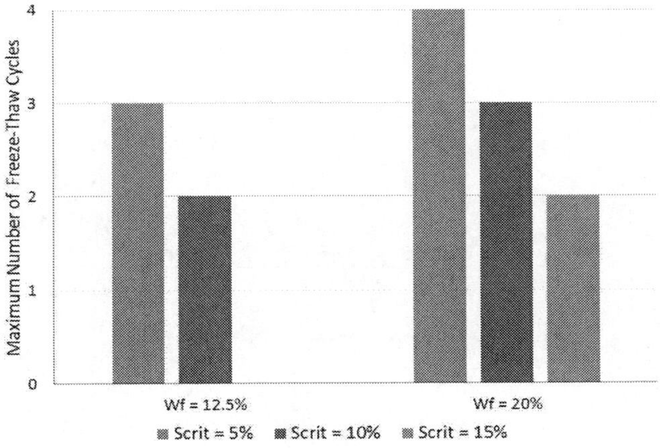

Figure 7 *Maximum number of scenarios with freeze-thaw cycles at S_{crit} for uninsulated wall assembly with low rain exposure and critical freeze-thaw temperature of –2°C.*

ration. However, the hygrothermal models used for this study used the A-value as a proxy for the liquid transport function for desorption as well. Therefore, larger A-values also relate to higher drying characteristics when the A-value is used to compare between tested material properties.[2] The critical A-value which freeze-thaw is likely to occur for the Ottawa climate and the case study buildings is below 0.01 $kg/m^2s^{0.5}$, which is around the levels of painted bricks. Not surprisingly, exterior paint has a more significant effect on increasing the risk of freeze-thaw damage than interior insulation. This finding puts the analysis into perspective and helps relate the analysis to what is already happening at the buildings. Damage has already occurred and the risk of freeze-thaw damage can be reduced by removing the paint, regardless of the insulation retrofit.

The impact of mortar joints, 3 versus 5 wythe bricks, and embedded wood beams were also evaluated as part of the case study. The methodology of this analysis is not detailed in this paper, but the findings are as follows.

Impact of Mortar Joints. Unsatisfactory mortar joints (deteriorated, cracked, or debonded) can introduce additional moisture into the bricks. However, depending on the type of mortar, the mortar can also help dry out the bricks quicker.

Mortar joints appear to have a net reduction in the risk of freeze-thaw damage and the relative number of freeze-thaw cycles can be determined by the 1-D hygrothermal analysis focused on a section through the bricks.

Number of Wythes. Adding insulation has even less impact on freeze-thaw risk for the thicker wythe walls because the relative change in the brick temperatures is small due to the extra thermal resistance that the thicker wythe walls provide.

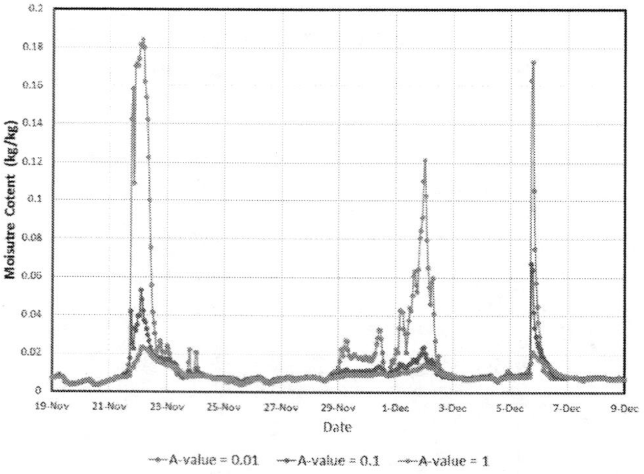

Figure 8 *Moisture content for west-facing uninsulated wall assembly with high rain exposure and 20%* W_f.

2. This is an assumption that warrants further investigation than presented in this paper, due to the relative importance of the liquid transport function for desorption

Embedded Wood Beams. Irrespective of insulation, wood beams that are in a pocket where there is only one wythe outboard the beam end will likely be exposed to moisture levels greater than 28% wt[3] at the outer fibers for the painted brick scenarios where the beams are in direct contact with the outer wythe. The duration above critical levels depends on how wet the bricks get from driving rain. Nevertheless, the wood beam will be exposed to higher moisture contents for the insulated scenarios compared to the uninsulated scenarios.

The difference in hours spent about 28% wt for the beam end is moderate, 85 versus 283 hours over a 2-year period. Nevertheless, the difference in time is significant in terms of the risk of decay.

The wood beam spends no hours about 28% wt for the unpainted brick scenarios, with or without insulation.

BUILDING ENERGY AND COMFORT ANALYSIS

Thermal and whole-building energy simulations were utilized to quantify the benefits of insulating the existing brick walls, adding roof insulation, and glazing improvements with regard to energy consumption and occupant comfort. A building energy model of a representative building was developed in EnergyPlus (DOE 2015) which consisted of a two-story building with a total floor area of 1560 m^2 developed from existing drawings, assumptions based on experience, and industry standards. It was assumed that the window-to-wall ratio of the existing building will be unchanged as no new glazing openings will be added. A variety of likely options for building envelope improvements were evaluated to determine the extent of energy consumption reductions and impact on thermal comfort. Table 3 shows the modeling parameters considered.

Since the objective of the building energy analysis was to quantify the impact of building envelope improvements on energy demand and thermal comfort, the primary HVAC (heating and cooling plant) was modeled as ideal equipment distributing hot water and chilled water at 100% to secondary HVAC equipment. Therefore, the heating and cooling energy, represents annual heating and cooling load delivered to the building. Different mechanical plant options could be considered after an ultimate design is chosen, but the energy demand is dependent on the building envelope performance.

Occupant thermal comfort was evaluated using the Fanger static thermal comfort model (Fanger 1970) to determine the percentage of people who are uncomfortable based on the occupants' expected level of activity, clothing, zone mean radiant temperature, and air speed. The occupants were assumed to be sedentary with the indoor air speed at 0.15 m/s at all times, based on well-designed mechanical systems and placement of diffusers. Occupant clothing level was based on the ASHRAE Standard 55 dynamic clothing model (ASHRAE 2013), which adjusts for clothing values depending on the time of the year. Thermal comfort in a zone is considered to be acceptable when

3. A critical threshold for decay of wood.

fewer than 20% of occupants indicate they are too hot or cold. An annual metric of "number of instances" that greater than 20% of occupants were dissatisfied was recorded for each simulation.

Realistic Building Envelope Performance

Detailed U-factor calculations that fully account for thermal bridging are necessary to realistically determine the energy demand and thermal comfort associated with various building envelope retrofit options. Mitigating thermal bridges for an interior insulation retrofit is challenging. The structure must bypass the thermal insulation at the floor and parapets; aligning the glazing within the plane of the thermal insulation is not likely to happen.[4]

Floor slab and parapet linear transmittances were estimated using the catalog contained in the Building Envelope Thermal Bridging Guide (Morrison Hershfield 2014) as 0.125 and 0.5 W/m·K respectively. The linear transmittance of glazing transitions were determined using 3-D heat transfer software from Siemens called Nx (Siemens 2015), based on field observations of existing window detail geometry. Several details were evaluated to assess the impact on different detailing on the wall U-factor, including insulating the sill and jamb returns with aerogel and spray foam insulation, as shown in Figure 9.

The effective R-values considered for various retrofit scenarios are summarized in Table 4.

Table 3. Building Energy Model Parameters

Building Envelope Assembly	Thermal Performance Range
Clear wall	Uninsulated brick, R-2 to R-30 (RSI 0.35 to RSI[*] 5.28)
Roof	Uninsulated wood joist, R-2.5 to R-40 (RSI 0.42 to RSI 7.04)
Glazing	Single-glazed wood frame, U-0.97 (USI[†] 5.5), SHGC[‡] 0.84
	ASHRAE 90.1-2007 maximum, U-0.55 (USI 3.12), SHGC 0.4
	Double-glazed, U-0.39 (USI 2.2), SHGC 0.4
	Triple -glazed, U-0.28 (USI 1.6), SHGC 0.4

* Thermal resistance (SI).
† Thermal transmittance (SI).
‡ Solar heat gain coefficient.

Building Energy Analysis

The whole building energy simulations also utilized a parallel coordinates map with dynamic sorting capability by energy or comfort outcome, or design variable input. The Building Energy Performance Map (Figure 10) displays outputs from multi-variable parametric simulations and allows the team to quickly analyze the results by different variables. A total of 4400 simulations was performed for this project. The results presented in the Building Energy Performance Map include total energy use, heating and cooling energy, and the south zone predicted percent dissatisfied (PPD) over 20% (thermal comfort metric representing percent of people dissatisfied). Additional metrics were analyzed but not shown for simplicity.

4. A theoretical thermal-bridging-free scenario.

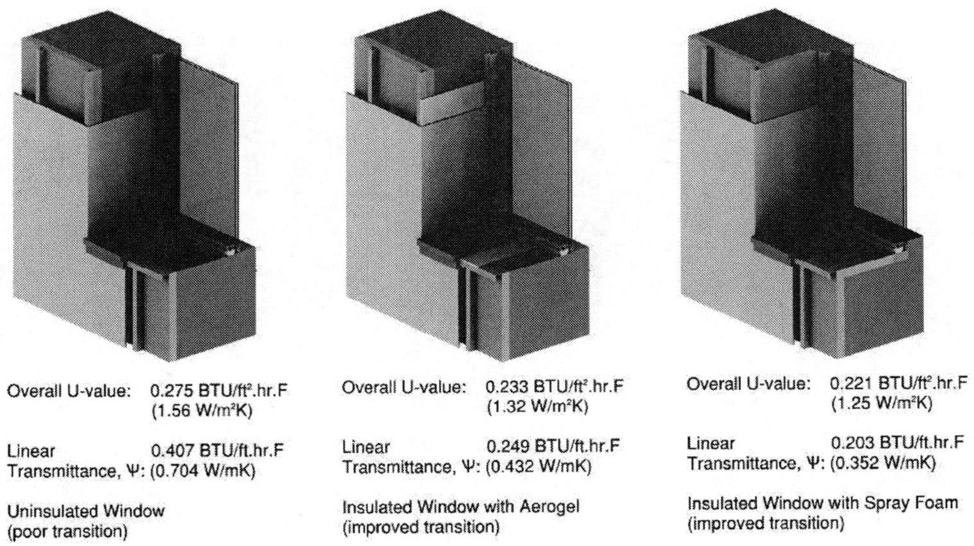

***Figure 9** Window interface details with R-12 interior wall insulation.*

Results from the building energy analysis shows there are multiple ways of decreasing the annual heating and cooling demand of the existing buildings. The most effective way to improve the thermal efficiency of the buildings is to insulate the roof to an effective R-21. Insulating the roof to R-21 results in 36% energy savings compared to the existing roof. Providing insulation beyond this point results in less significant returns with an effective R-40 roof resulting in 38% energy savings over the existing roof or an extra 2% savings beyond the R-21 scenario. Adding insulation to the existing walls was also effective. Adding R-6 insulation results in 36% reduction in energy use, while insulating the walls to an effective R-10 results in 41% energy savings over the existing walls. Figure 11 summarizes the effect of wall R-value on the total energy use and ther-

Table 4. Glazing U-Factors and Effective Wall R-Values

Glazing U-Factor, $W/m^2{\cdot}K$	Roof R-Value, $m^2{\cdot}K/W$	Wall Interior Insulation R-Value, $m^2{\cdot}K/W$	Glazing Transition Linear Transmittance, $W/m{\cdot}K$	Wall Effective R-Value, $m^2{\cdot}K/W$
Single 5.5	0.42	0	Poor 0.297	0.47
Double 2.2	3.67	0	Poor 0.552	0.80
Double 2.2	3.67	1.06	Poor 0.668	1.17
Double 2.2	3.67	1.06	Improved 0.352	1.38
Double 2.2	3.67	1.06	Poor 0.668	1.19
Double 2.2	2.28	1.06	Improved 0.352	1.38
Double 2.2	2.28	2.11	Poor 0.704	1.42
Double 2.2	2.28	2.11	Improved 0.352	1.79
Triple 1.6	2.28	2.11	Poor 0.704	1.42
Triple 1.6	2.28	2.11	Improved 0.352	1.79

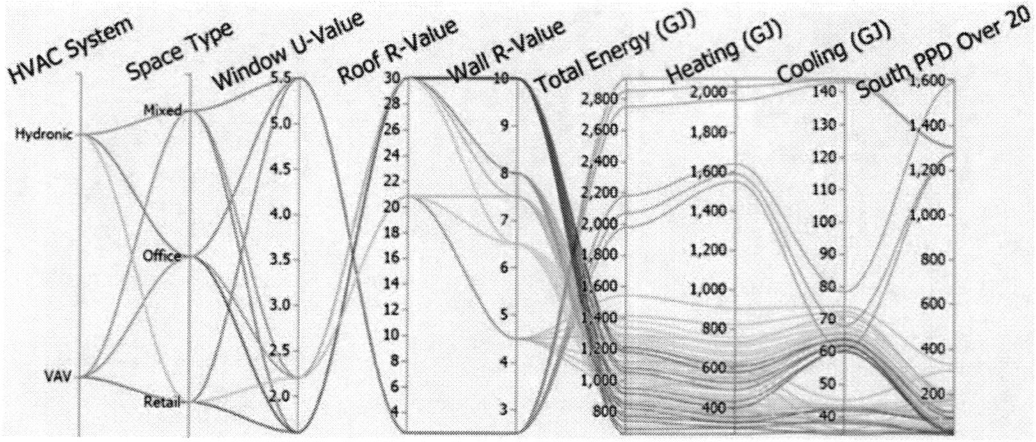

Figure 10 *Building energy performance map.*

mal comfort parameters for a building with R-21 roof and double glazed windows. Conversely, the window U-factor has relatively low impact on both energy savings and occupant comfort. Using windows with a U-factor of 0.55 Btu/ft^2·hr·°F results in only 8.5% energy savings compared to the existing single glazed windows. Upgrading window performance to a double glazed window with U-factor 0.39 Btu/ft^2·hr·°F results in 12% savings, with further improvements to triple glazed windows only increases savings to 13.7%. This is due to the low window-to-wall ratio of the building, which the windows only represent a smaller portion of the building envelope. The annual number of hours occupants are dissatisfied is reduced below 200 at effective wall R-values above R-9.5 with double glazing and an R-21 roof, attaining a significant improvement relative to the original building.

Based on these findings, different retrofit options were evaluated and are presented in Table 5. These options demonstrate that the majority of improvements in energy use are achieved with double glazing, R-21 roof, and R-6 walls. Increasing the roof thermal performance to R-30 and using improved transitions reduces the number of hours when occupants are dissatisfied to 200. Building envelope improvements are effective in reducing the amount of time occupants are dissatisfied to less than 200 hours in a year. The majority of the remaining dissatisfied hours occur in the winter, when occupants feel too cold as a result of morning start-up of HVAC equipment after night setback. Additional gains in occupant comfort can be realized in practice by either a change in HVAC controls (less night setback, earlier start-up of HVAC equipment, etc.), or oversizing HVAC equipment to more quickly warm up air and surface temperatures. Alternatively, substantial improvements in building envelope would also reduce dissatisfied hours due to significant reductions in heat loss which would maintain space and surface temperatures for longer periods, even when setbacks are implemented. However, these further improvements in the building envelope are not cost-effective in terms of their additional contribution to energy efficiency. As a result, adding an additional R-12 insulation to the interior appears optimal for achieving both energy savings and thermal comfort.

CONCLUSIONS

The hygrothermal and whole-building energy analysis completed for these case studies strove for a balanced performance base assessment of an interior insulation retrofit to existing load bearing multiple wythe brick masonry buildings. Despite employing best-practice investigation and testing to help make appropriate assumptions for the project conditions, there was still significant uncertainty to overcome in assessing the capacity of the bricks to resist freeze-thaw damage using hygrothermal simulations. However, with the application of comprehensive sensitivity analysis and engineering judgment, the authors were able to provide the design team with technical and clear guidance moving forward with the schematic design. More importantly, the hygrothermal analysis removed the designer's questions and developer's concerns and presented the optimal retrofits that will reduce energy demand, improve thermal comfort, and best meet the objectives of the design team.

While the decisions for this project cannot be unilaterally applied to other projects, the same approach can be applied. For example, if the existing conditions illustrated no freeze-thaw damage and no paint, then it could be more difficult to interpret the findings of the hygrothermal analysis that showed a small increase in the probability of the brick freezing while above S_{crit}. Other projects could also theoretically consider less discreet retrofits than the scenarios considered for these case study buildings, such as exterior insulation and cladding,

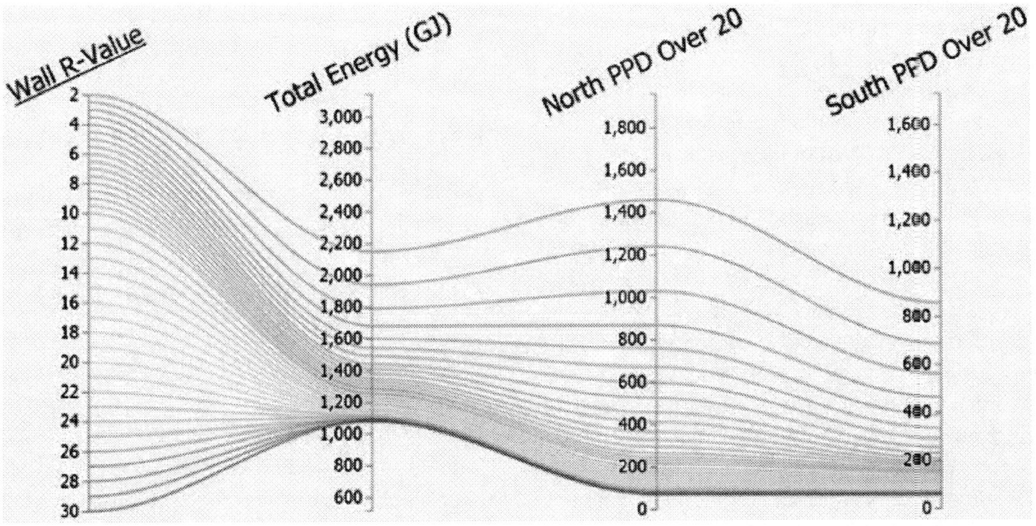

Figure 11 *Impact of wall R-value on total energy use and percent of people dissatisfied with R-21 roof and double glazed windows.*

Table 5. Energy Demand and Thermal Comfort Performance of Retrofit Options

Retrofit	Window U-Factor	Interior Insulation	Roof R-Value	Transitions	Effective Wall R-Value	Total Energy Demand, GJ	Thermal Comfort: PPD Over 20%	
							North	South
1	R-5.5 (single glazed)	None	R-2.4	Poor	R-2.5	2180	1841	1590
2	R-2.2 (double glazed)	None	R-20.8	Poor	R-4.5	1120	639	302
3	R-2.2 (double glazed)	R-6	R-20.8	Poor	R-6.5	986	372	176
4	R-2.2 (double glazed)	R-6	R-20.8	Improved	R-7.5	944	290	152
5	R-2.2 (double glazed)	R-6	R-30	Poor	R-6.5	958	296	147
6	R-2.2 (double glazed)	R-6	R-30	Improved	R-7.5	917	215	126
7	R-2.2 (double glazed)	R-12	R-30	Poor	R-8.0	900	194	116
8	R-2.2 (double glazed)	R-12	R-30	Improved	R-10.0	848	136	89
9	R-1.6 (triple glazed)	R-12	R-30	Poor	R-8.0	869	151	95
10	R-1.6 (triple glazed)	R-12	R-30	Improved	R-10.0	818	106	72

and may have more lofty goals for the reduction of energy demand. Regardless of the objectives, the same parallel coordinates approach at the early stage of design can help guide the design towards the most practical solutions and help the design team make informed decisions. The only difference might be the range of parameters considered in the analysis and the performance metrics considered.

Further research that can support using the performance-based assessment of interior insulation retrofits is considerable. Some ideas, not all covered in this paper, include better defining the freeze-thaw criteria damage functions (rate of cooling, coldest temperature attained, and number of cycles), better methods for defining the desorption curve of brick on projects, better understanding of how salts affect freeze-thaw risk and how to assess the impact of salt in hygrothermal models, and correlations of the loss in compressive strength for each successive brick freeze-thaw cycle at S_{crit}.

More research may allow for increased statistical analysis and/or standardization to be incorporated into the ascribed methodology for assessing freeze-thaw durability using hygrothermal simulations. Nevertheless, judgment can never be replaced and there is an open question as to whether increased statistical analysis and standardization will help industry make better decisions.

ACKNOWLEDGMENTS

Morrison Hershfield collaborated with several internal and external experts in completing this work. We would like to thank these individuals and companies that provided support for the work that went into this paper. A special thanks to Windmill Development Group in Ottawa (client) and Building Science Labs in Waterloo that completed all the brick testing.

REFERENCES

ASHRAE. 2013. ANSI/ASHRAE Standard 55. *Thermal environmental conditions for human occupancy*. Atlanta: ASHRAE.

Arnott, M. 1990. *Investigation of freeze thaw durability*. IRC Report No. cr 5680, IRC/NRCC.

Building Science Labs. 2014. *Masonry testing data set*. Waterloo, ON: Building Science Labs.

DOE. 2015. EnergyPlus. https://energyplus.net. Washington, D.C.: U.S. Department of Energy.

Fagerlund, G. 1996. Assessment of the Service Life of Materials Exposed to Frost. *Durability of Building Materials and Components* 7(1). Prediction, Degradation, and Materials Conference, London.

Fanger, P.O. 1970. *Thermal comfort.* Copenhagen: Danish Technical Press.

IBK. 2015. *DELPHIN.* http://bauklimatik-dresden.de/delphin/index.php. Dresden: Institut fur Bauklimatik.

ISO. 2000. Final draft international standards ISO/FDIS 13788:2000 (E). *Hygrothermal performance of building components and building elements—Internal surface temperature to avoid critical surface humidity and interstitial condensation—Calculation method.* Geneva: International Standards Organization.

Litvan, G. 1973. Pore structure and frost susceptibility of building materials. Research Paper 584, National Research Council Canada.

Litvan, G. 1975. Testing the frost susceptibility of bricks. In Masonry: Past and Present pp. 123–132. ASTM STP 589.

Litvan, G. 1989. Freeze-Thaw Durability of Porous Building Materials. Durability of Building Materials and Components pp. 455-463. ASTM STP 691, P.J. Serada and G.G. Litvan, eds., American Society for Testing and Materials.

Morrison Hershfield Limited. 2014. Building envelope thermal bridging guide—Analysis, applications, & insights. BC Hydro Power Smart, Vancouver, BC.

Mensinga, P., J. Straube, and C. Schumacher. 2010. Assessing the freeze-thaw resistance of clay brick for interior insulation retrofit projects. *Buildings* XI.

Penttala, V. 1998. Freezing-induced strains and pressures in wet porous materials and especially in concrete mortars. *Advanced Cement Based Materials* 7(1): 8–19

Prout, W., and W.D. Hoff. 1991, Fundamental Studies of Frost Damage in Clay Brick. Proceedings of the Fifth International Conference on Durability of Building Materials and Components 5(1) New York.

Roppel, P. 2003. *Masonry walls treated with water-repellents.* Master of Applied Science Thesis, University of Waterloo, Waterloo, Canada.

Siemens. 2015. *Nx.* Plano, TX: Siemens Product Lifecycle Management Software.

Straube, J., and E. Burnett. 2005. *Building science for building enclosures.* Westford, Massachusetts: Building Science Press.

Straube, J., and C. Schumacher. 2006. *Assessing the durability impacts of energy efficient enclosure upgrades using hygrothermal modeling.* Westford, Massachusetts: Building Science Press.

Taber, S. 1930. The mechanics of frost heaving. *Journal of Geology* 38:303–17.

Overcladding for Thermal Performance and Building Resiliency

Daniel McKelvey, AIA
LEED AP BD+C, NIBS

David Copley, AIA
LEED AP, CSI

Kendal Foster, AIA
LEED AP

Allison Wilson, AIA
LEED AP BD+C

ABSTRACT

This case study documents the complete thermal overcladding of an existing nine-story critical-care patient tower in an urban hospital complex. Overcladding was completed while maintaining critical hospital operations and retaining most of the existing construction's exterior masonry envelope dating from the 1970s.

The comprehensive approach used by this project to develop a high-performance enclosure for an aging structure began with developing strategies for hybrid thermal overcladding and "recycling in-place" to achieve passive temperature regulation, air compartmentalization, and increased thermal mass. Overcladding with a thermal high-performance envelope improved energy performance through improved insulation, moisture mitigation strategies, and resilient materials, in addition to providing an aesthetic improvement that successfully positioned the aging building as a viable part of the hospital's master plan for decades to come.

INTRODUCTION

Building owners and facility managers are increasingly confronted by an aging inventory of building stock. These structures generally retain sound structural frames but have failing, poorly insulated, and antiquated building enclosures. The dilemma owners are confronted with is the decision to either remove these structures and design and construct new facilities, a fairly disruptive and costly path, or to attempt to retain some or all of the aging structures and embark on a renovation to repurpose these facilities. Both choices involve careful planning and scheduling and a significant investment of capital funds.

Hospital administrators particularly are reluctant to remove their aging existing structures due to the often limited availability of open areas to erect new buildings and because in many instances the existing facility must remain fully or partially operational during renovation. Selecting a strategy that retains the existing structure in part or in whole and plans for repurposing that existing structure can begin to address these concerns. One such repurposing approach is termed *thermal overcladding* and essentially involves encasing the aging existing structure with a new exterior insulating envelope system. This strategy can allow aging, existing structures to meet contemporary energy conservation and indoor environmental quality concerns, overcome schedule and budget hurdles, and can create opportunities for older healthcare buildings to have new life in hospital master plans.

This case study focuses on the implementation of such an overcladding strategy for the 118,500 ft^2 Nelson Harvey Building, a 33-year-old, nine-story patient care building that occupies a dense urban site in Baltimore, Maryland. The building is situated in the heart of the Johns Hopkins Medical campus and serves as a nexus of circulation and clinical/service support for the adjacent and adjoining medical buildings (see Figures 1 and 2). Hospital administrators elected to renovate the existing structure in lieu of building new to reduce initial capital costs and because the building houses vital shared facilities used by the eight adjoining buildings as part of daily hospital operations.

Daniel McKelvey, **David Copley**, **Kendal Foster**, and **Allison Wilson** are registered architects at Ayers Saint Gross, Inc., Baltimore, MD.

Figure 1 *Johns Hopkins Medical Campus view from the west with the Nelson Harvey Building in the background.*

Figure 2 *Johns Hopkins Medical Campus aerial view with the Nelson Harvey Building project scope highlighted.*

The interior portion of the renovation program encompassed the complete demolition of the interiors on eight patient floors and the first-floor lobby and public waiting areas. First-floor renovations were limited to a new first floor entry vestibule and lobby, public corridors, and a plaza restaurant. The basement, first, and second floors of the Nelson Harvey Building were required to remain occupied during the extensive interior and exterior renovation. New HVAC, electrical, fire protection and plumbing systems were upgraded and/or replaced throughout.

The exterior renovation program focused on the failing exterior envelope, including walls, roofs, and glazing, and targeted those features to be replaced and/or repurposed if possible. The existing entrance canopy structure was intended to be removed and replaced by a more gracious, open, and welcoming entry that would also provide better thermal isolation between the public lobby and the primary entry. The project scope also included landscape and hardscape additions to the entry forecourt and an insulated plaza deck over the hospital's imaging center.

EXISTING CONDITIONS

The existing construction of the Nelson Harvey Building prior to the renovation reflects an all too common assembly of systems for late 1970s buildings—a steel-framed superstructure with cast-in-place concrete over metal deck floor plates and, for the vast majority of the building, a simple, uninsulated, masonry cavity wall cladding. The exterior wall cladding includes a nominal 2 1/4 × 4 × 8 in. face brick, a 2 in. airspace, and cement parged coating over nominal 4 × 8 × 16 in. cored concrete masonry units (CMUs) (see Figure 3). Horizontal ladder reinforcing incorporating eye and pintle masonry ties were used to support the brick veneer. The interior framed walls consisted of 5/8 in. gypsum board on 3 5/8 in. 20 ga. metal studs to provide a finished interior surface over the CMUs of the exterior wall assembly. The only insulation provided in the wall assembly was unfaced fiberglass batts between the metal studs, which is subject to R-value reduction due to the thermal break at each exposed metal stud.

Selected areas on the first and second floors featured an insulated metal panel wall assembly attached to miscellaneous structural steel angles and channels with interior framed walls of 5/8 in. gypsum board on 3 5/8 in. metal studs to provide a finished surface to the interior.

The roof assembly over the composite construction of 6 in. concrete and metal deck consisted of white polyvinyl chloride (PVC) roofing membrane fully adhered to the concrete deck. The roofing membrane was protected with 2 × 4 ft tongue-and-groove insulating roof pavers composed of 2 in. extruded polystyrene topped with a 1/2 in. layer of concrete as a wearing surface.

The Nelson Harvey Building's existing fenestration was characterized by long expanses of aluminum ribbon windows with the masonry spandrel above carried on a system of structural steel lintels, with the structural framing braced to the concrete floor slab. Large expanses of aluminum curtain wall were used at public lobby spaces. The window frames and curtain wall were not thermally improved and lacked a water management system and/or a pressure equalization chamber within the framing system. The vision glazing was single-pane bronze glass (see Figures 4 and 5).

Investigation

Prior to the design team's involvement in the project, Johns Hopkins Hospital commissioned a building envelope analysis of the existing exterior walls to assess existing conditions. Through this analysis it was determined that structural stabilization as well as repairs to the exterior envelope would be required.

Structural Analysis. The structural analysis included a visual inspection of the exterior walls, a lateral analysis of the walls based on the current building code, and a review of the existing building drawings. The analysis revealed lateral loading and thermal failures in the building envelope, including the following:

- Cracking in the brick veneer
- Horizontal displacement of the brick veneer
- Horizontal bowing at the parapets and various other masonry locations
- Vertical bowing at some of the full-height masonry walls

It was suspected, based upon the existing drawing details and visual inspection of the exterior walls, that a soft joint was

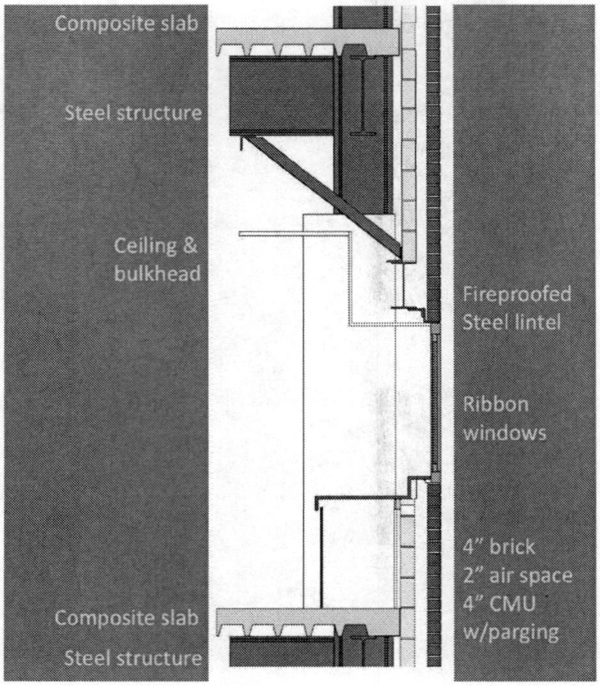

Figure 3 *Existing wall section.*

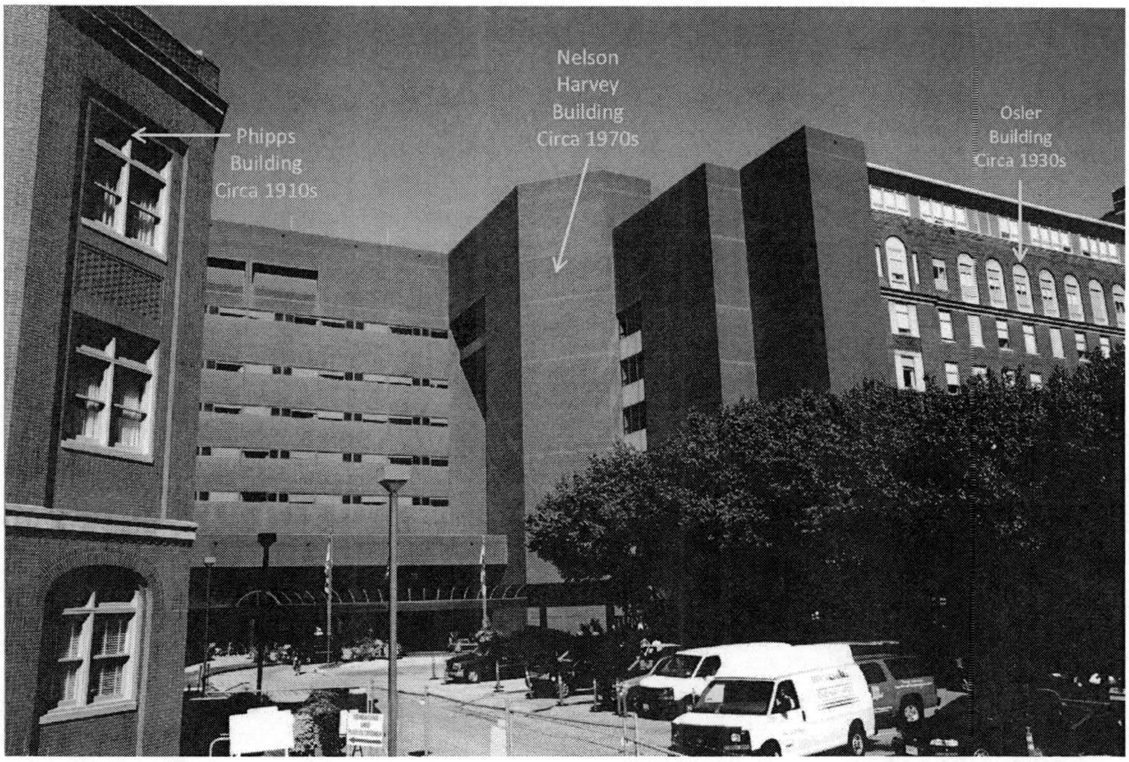

Figure 4 *Existing east elevation of the Nelson Harvey Building and adjacent adjoining buildings.*

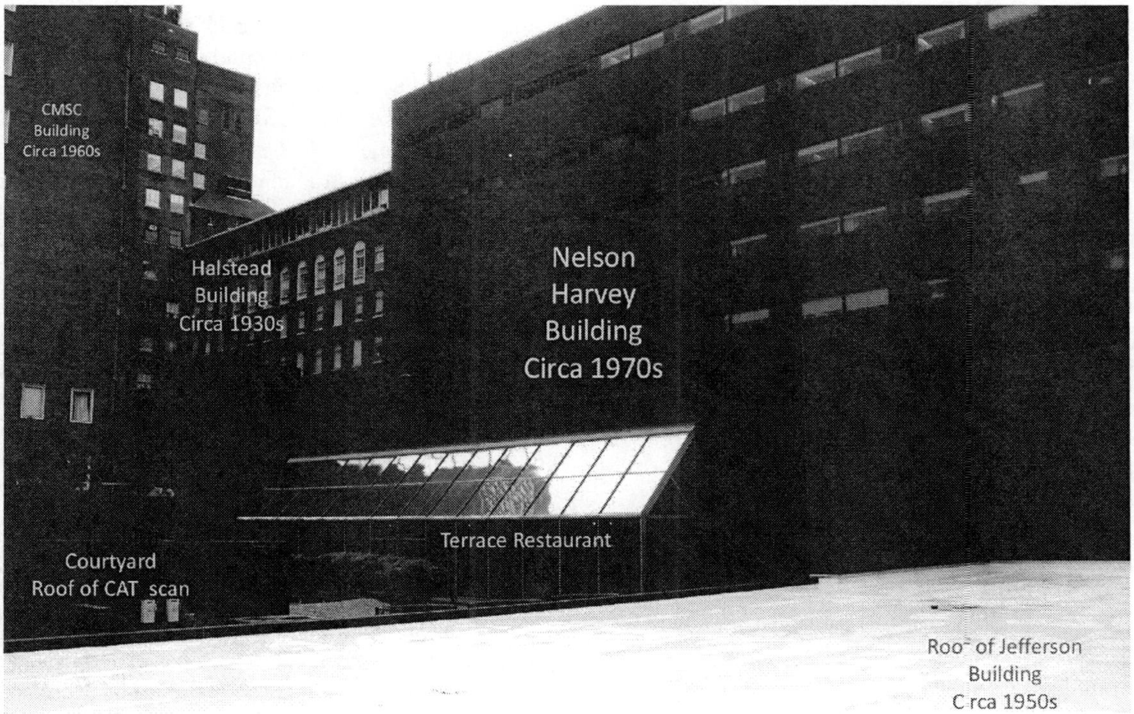

Figure 5 *Existing west elevation of the Nelson Harvey Building and adjacent adjoining buildings.*

not installed below the steel shelf angles or along vertical control joints. This was confirmed through the forensic investigation. At the locations investigated, the control joints at the brick were mortared solid. The walls were also structurally analyzed based on the current adopted code requirements for wind load. It was determined that the exterior walls were significantly overstressed and that in order to restore the structural integrity of the exterior envelope, lateral reinforcing would be required.

Architectural Analysis. Through forensic examination and test cuts, several major construction issues with the envelope were revealed. These conditions had contributed to the envelope's deterioration over time and it was evident that some form of remediation would be required. The following conditions were observed:

- The insulation was inadequate. With no insulation in the cavity, the only existing thermal barrier was located between internal studs. Without a continuous insulation system in the masonry cavity, the CMUs of the backup wall, metal studs, and the edge of the concrete floor slabs became major contributors to heat loss due to thermal bridging in the exterior wall system.
- Control joints were too few, mortared solid, and incorrectly located at the locations investigated. Therefore, areas of cladding were structurally failing, causing the bowing and displacement to occur.
- Air infiltration was observed in the cavity walls due to the lack of an air barrier and gaps in the walls. There were broken and missing portions of the parge coating on the concrete masonry backup wall that were noted in the test cut locations. Air leakage in the building contributed to significant energy loss.
- Nonthermal, single-pane glazing with deteriorating gaskets in the existing ribbon windows were contributing to significant energy loss.
- Wall flashings were not constructed to industry standards, which can contribute to moisture accumulation, resulting in moisture issues such as leaks and material deterioration. The following flashing installation defects were discovered:
  - Improper location of sealant
  - Improper lapping and sealing of adjoining flashing pieces
  - Lack of end dams
  - Flashing edge ending within the veneer
  - Buildup of mortar in wall cavities
  - Lack of weeps

These flashing defects may have contributed to many moisture problems within the building, but there was no record of any reported. It was suspected that any moisture issues were mitigated due to the high volume of internal air exfiltration through the cavity wall, which would have inhibited its accumulation.

Digital Laser Survey. As a direct result of the structural and architectural analyses, a critical component of the project during the design process was the completion of an extensive digital laser survey of the exterior of the building. The digital laser survey data were collected from 37 different locations over a two-day period allowing for the construction of an accurate three-dimensional model of the exterior building surfaces. This made it possible to map in detail the actual physical dimensions of the building, including all dimensional deformities due to thermal movement. The high degree of accuracy conveyed by this in-depth digital survey was necessary to properly construct a new enclosing envelope over the existing building envelope with minimal removal of that existing envelope.

THE OVERCLADDING DESIGN APPROACH

A number of design approaches were evaluated in the concept phase for rectifying the envelope problems. All the options proposed new insulated roof and glazing systems but differed in how to deal with the failing masonry envelope. An initial option proposed removal of all existing brick veneer and CMU backup walls in order to construct new thermally improved masonry wall assemblies on the existing structure. Another option suggested repairing and stabilizing the exterior brick masonry and reinforcing the existing CMU backup wall with 16 ga. cold-formed metal framing (CFMF) to satisfy code requirements for lateral loading to the building envelope. The third idea was to overclad the existing masonry with an insulated metal panel system supported on a cold-formed metal stud frame connected back to the building's main structural steel frame. This last option would require first repairing and stabilizing the existing masonry prior to any overcladding. All of these early proposed options were rejected for reasons of cost, schedule, and aesthetic. The removal of all masonry walls and the construction of new masonry infill walls was deemed too costly and logistically challenging and would result in vast amounts of debris and airborne dust as a result of the demolition process to remove all of the existing masonry. The approach to internally reinforce the CMUs in the exterior walls with CFMF would result in a net loss of valuable square footage for patient rooms already deemed underscaled by industry standards. Insulated metal panel was undesirable as a primary exterior material aesthetic for the prominent building as voiced by the hospital's board of directors, who desired to maintain a campus composed of predominantly masonry buildings.

After thoughtful consideration of a number of different design approaches, the design team decided that the most promising strategy was to overclad the building, as illustrated in Figure 6, provided that the proper materials and structural systems could be found. The selection of overclad materials and assemblies would have to address several important requirements as set forth by the owner, architect, and structural engineer. The owner stipulated that the predominant material expression be brick to complement the medical campus in general and the adjacent historical buildings in particular. The structural engineer required that any overclad assembly be self-supporting, transfer lateral and gravity loads to the building's existing structural frame only, and not exceed the overall building's structural lateral dead load capacity by 10% or an individual column's structural dead load capacity by more than 5%. The architect desired to develop a high-performance enclosure that met the former criteria while being aesthetically modern, minimizing ongoing energy use of the building post-occupancy, using new materials sustainably, and maintaining high indoor and outdoor environmental quality through demolition and material reuse. It was also desirable to enhance the building's resiliency given the vital mission of the hospital and the operation of this building as a center for the treatment of highly infectious patients.

The design approach formulated by the design team is a concept termed *recycle-in-place*. Key to this concept is the notion to encapsulate the existing building envelope and incorporate the existing materials together with the new overcladding materials into a hybrid thermal overcladding that will achieve passive temperature regulation, exterior wall convective air compartmentalization, and increased resistance to fire and sound transmittance through tapping into the potential of thermal mass. The new overcladding envelope meets contemporary energy conservation and indoor environmental quality concerns through improved insulation and moisture mitigation with a double-sealed barrier wall system. The minimal demolition required to implement this strategy limited airborne dust and debris on site and decreased the volume of construction materials hauled and dumped miles away. This approach reduced, and in some cases eliminated, airborne pollutants in both the hospital's sensitive environment and in the community at large. Applying this strategy will have a significant impact on both the short-term and long-term sustainability of the Nelson Harvey Building.

As previously described, forensic and observational investigations indicated the existing Nelson Harvey Building had adversely aged and begun failing to a point that the exterior construction conditions were not accurately represented by the project's as-built drawings. Establishing an accurate three-dimensional model of the existing building in order to design an overcladding system that would essentially slip over the existing building envelope was critical but presented numerous challenges. Prior to the three-dimensional scan of the building, as-built documents from the original 1970s construction documents were used to construct a three-dimensional model using Autodesk's Revit software (Autodesk 2012) that represented at a minimum how the Nelson Harvey Building was originally intended to be constructed. Field surveys were conducted by the design team to ascertain the general completeness of the as-builts as well as to document evident deviations that may have resulted during the building's construction or any subsequent additions and alterations in the years that followed. The deviations were numerous and included the bowing and displacement of the masonry veneer resulting from the lack of provision for material thermal expansion in the original design and construction. The back face of the overcladding system, as envisioned by the design team, would be held off the face of the existing envelope walls by 1 in. in order to limit the cantilever of the new structural system and to reduce the size of the resultant cavity between the existing wall and the new overcladding. This meant that the design team needed to establish the outmost control points of the existing envelope, including any thermally displaced veneers, in order to slip the new overcladding over the existing envelope.

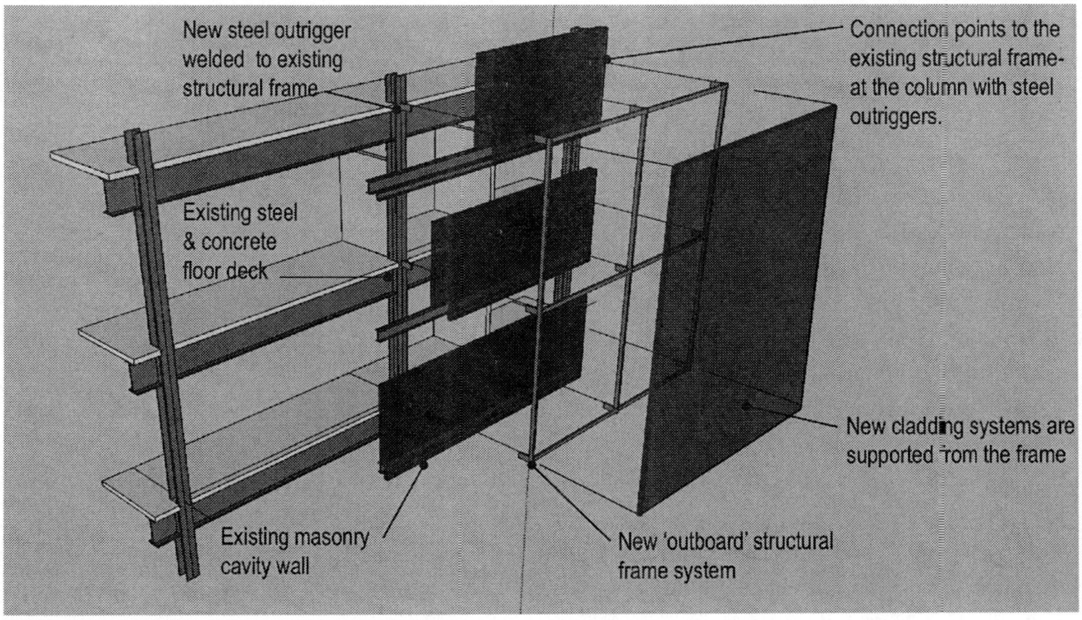

Figure 6 *Design team's selected conceptual design approach following feasibility study.*

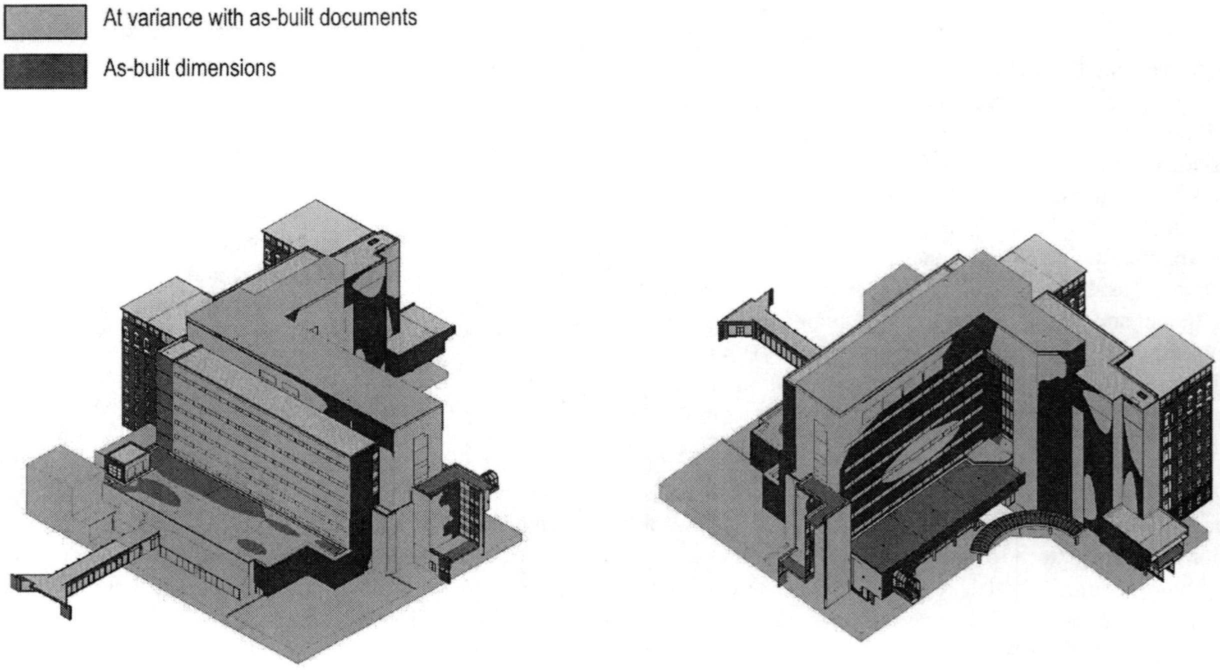

Figure 7 *Point cloud model overlaid onto the Revit model of the existing building.*

A conventional surveying program to establish the existing control points would have been daunting for the size and complexity of the existing building. Instead, digital laser surveying was employed to scan the entire exterior envelope in order to create a digital point cloud file that provided an extremely accurate digital model that could be compared with the Revit model created from the less-than-accurate as-built info. The accuracy of the digital point cloud is within ±1/16 in. The variances between the two models were several inches in most cases, with some areas identified as being several feet off (see Figures 7 and 8). Once the point cloud file and the Revit model were overlaid, the Revit model was adjusted to reflect the true field conditions of the building. Having this very accurate digital model enabled the design team to create the new overcladding envelope that would slip over the existing envelope with confidence that the new envelope would both fit and absorb the building's deflection anomalies. In addition to the irregularities of the existing building envelope that the digital survey accurately recorded, the digital survey was also crucial in modeling the physical building connections that the existing Nelson Harvey Building had with no less than eight adjoining buildings. Any proposed envelope overcladding required close coordination with the adjoining buildings to properly create the necessary wall and roof expansion details between the buildings. The Nelson Harvey Building's renovation could not succeed at the expense of compromising any of its adjoining neighbors.

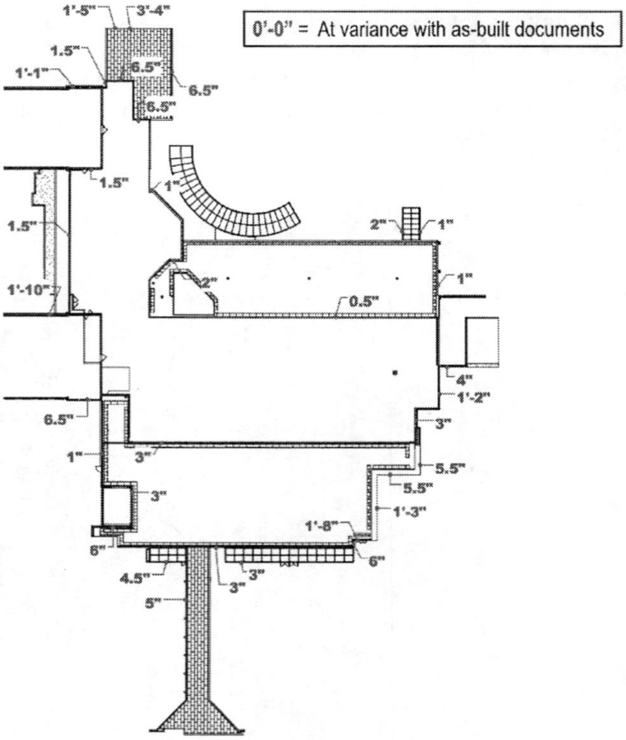

Figure 8 *Variations between the point cloud model and the existing building as-built documents.*

Design and Execution

With reliable data regarding the Nelson Harvey Building's on-site condition, the design team was able to move forward with designing and documenting the new overcladding systems that would revitalize the building. In evaluating the available systems for the overcladding materials, the following systems were identified as most capable of fulfilling the project's established design criteria:

- Over existing masonry portions of the existing envelope:
  - Wall panels of thin-brick cast into thin-precast concrete panels, reinforced by thermally broken, cold-formed metal framing with sprayed polyurethane insulation between the concrete and framing members, between the framing members, and between adjoining precast panels.
  - Wall panels of prefinished, insulated, interlocking aluminum panels, with insulated joints attached to cold-formed metal framing. This system was also used at exterior soffits.
  - Thermally improved, aluminum-framed, and glazed curtain wall using insulating and fritted low-e coated spandrel glazing, backed with mineral board insulation.
- At existing vision-glazed portions of the existing envelope:
  - Thermally improved, aluminum-framed, and glazed curtain wall using insulating and low-e-coated low-iron glass.
- At existing roofs of the existing envelope:
  - Fully adhered PVC roof membrane over multilayered rigid insulation and protected with concrete-topped insulated roof pavers.
  - At roof areas exposed to public view, an extensive green roof system was installed over a fully adhered PVC roof membrane over rigid insulation.

The new envelope system required careful planning and detailing for incorporation into an overcladding system. Each system was selected based on aesthetic appeal, structural integrity, light weight, thermal capability, and moisture resistance.

Precast Panel System as Overcladding

The thin-precast panel system selected as the primary cladding system for the Nelson Harvey Building embodies many favorable attributes for envelope design: aesthetic expression (brick), good weight to size ratio, weather and hazard resistance, continuous insulation, limited thermal bridging, air/vapor barrier, moisture resistance and management, and thermal mass. The precast panel system permits shop fabrication with better quality control, an accelerated construction schedule, and good life-cycle costing.

The thin-precast panel system at 32 lb/ft^2 conformed to the structural engineer's weight limit of a maximum of 35 lb/ft^2 for any overcladding system in order to remain within the code-dictated 5% maximum column dead load and 10% lateral load addition to the building's overall structural frame (see Figure 9). Each panel spans from column to column (22 ft on

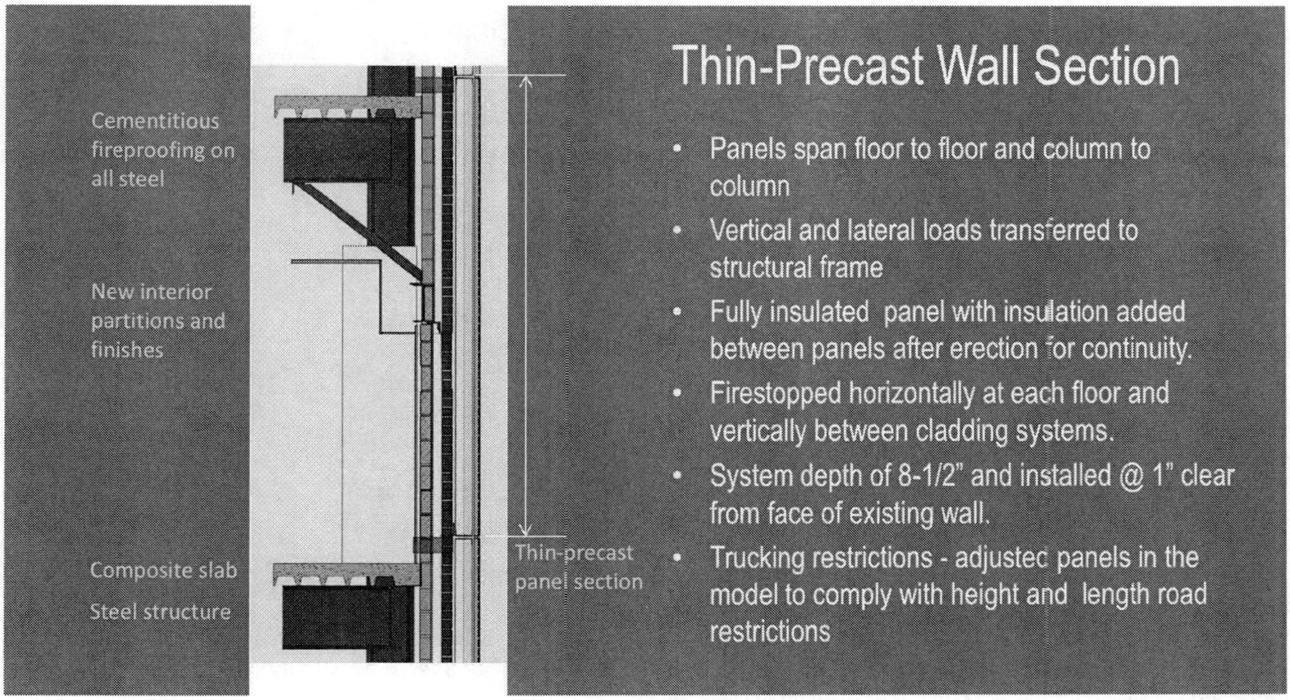

Figure 9 *Thin-brick precast wall section.*

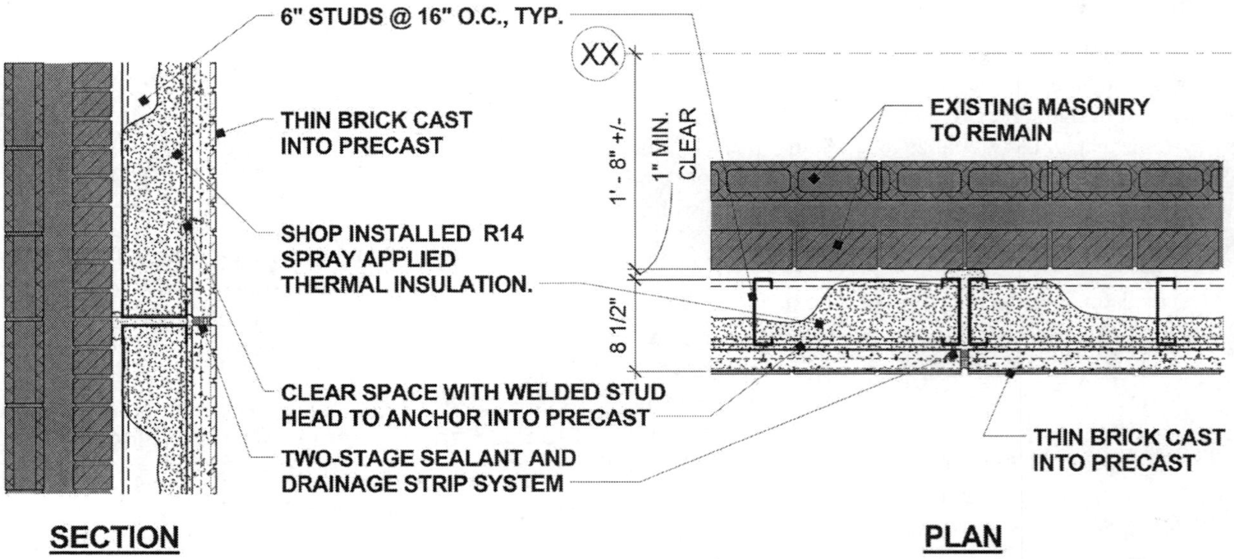

Figure 10 *Thin-brick precast assembly detail.*

average) and has a maximum height limit of 12 ft. Each individual panel is supported at four points on steel tube outriggers "needled" through the existing masonry envelope and welded to the existing steel columns. Two mid-panel connections are used for in and out adjustment of the panel for optimal alignment. The six connection points limit the surface area of thermal bridging between the precast panel and the internal construction. Once the connections were made, the masonry removed to allow access to make the steel connections was infilled with new masonry to provide the desired fire and acoustic resistance. The precast panel as designed and fabricated incorporates a thermal break between the concrete and the cold-formed metal framing via the use of intermittent steel Nelson anchors that provide a 1/2 in. standoff between the two. Sprayed-on polyurethane insulation (SPF) of 4 in. minimum thickness, medium density, and closed-cell foam between and around the steel framing provides a high degree of insulation and also performs as an air/vapor barrier (see Figure 10).

An average of eight precast panels were erected per day during construction, as shown in Figure 11, with each one being craned into place and adjusted using the six anchoring points provided for each panel. Once the panels were installed, the 3/4 in. gaps between panels were completely filled from the exterior with expanding closed-cell polyurethane foam to provide for continuous insulation of the thermal envelope. The joint is ultimately capped with a double sealant barrier encasing a preformed pressure equalizing drain strip completing the barrier wall overcladding system. The 1 in. minimum air space between the back of the precast panel framing and the existing masonry wall was horizontally firestopped at every floor. The firestopping afforded a means of compartmentalizing the vertical chase between the new and existing envelopes to not only limit the spread of fire and smoke in the event of a fire but also provide a means to limit the stack effect of convective air

Figure 11 *Thin-brick precast wall panel installation—southeast elevation.*

movement in the cavity that can have an adverse performance effect on the envelope's ability to limit air infiltration and exfiltration. The 1 in. void space was also periodically firestopped vertically for identical reasons. See Figures 12 and 13.

Prefinished Insulated Aluminum Wall Panel System as Overcladding

The lightweight insulated aluminum wall panel system was selected as a secondary overcladding system to be installed over the existing masonry wall in select areas of the building where construction access by the crane was restricted and the weight of precast panels would have exceeded the safe loading reach distance for the crane's boom (see Figure 14). In these areas (see Figure 15), cold-formed metal framing was installed 1 in. off of the existing face of masonry, spanning floor to floor and connected back to the building's structure along each floor line by means of a structural angle connected to steel tube outriggers "needled" through the existing

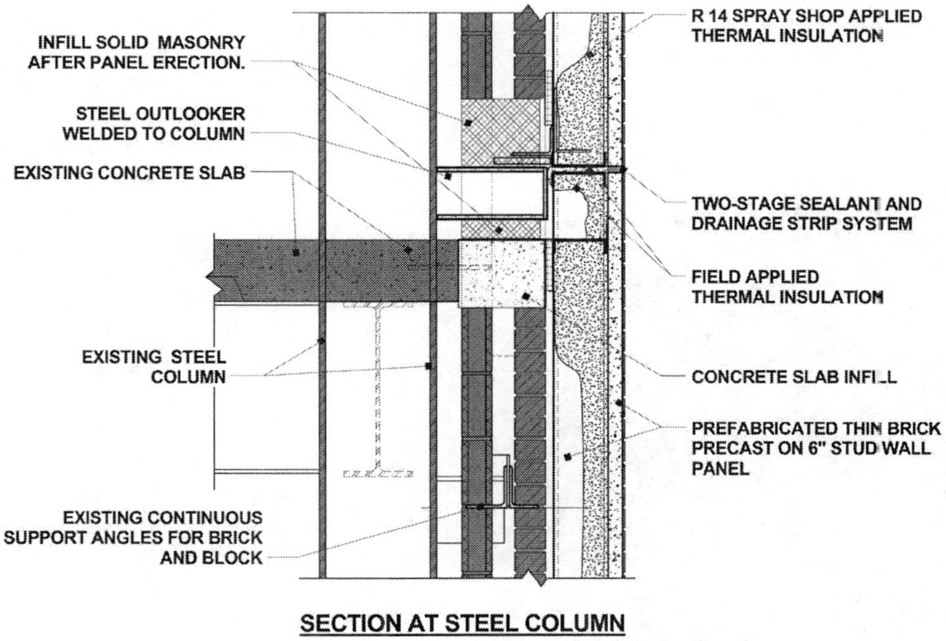

Figure 12 *Thin-brick precast—column support section.*

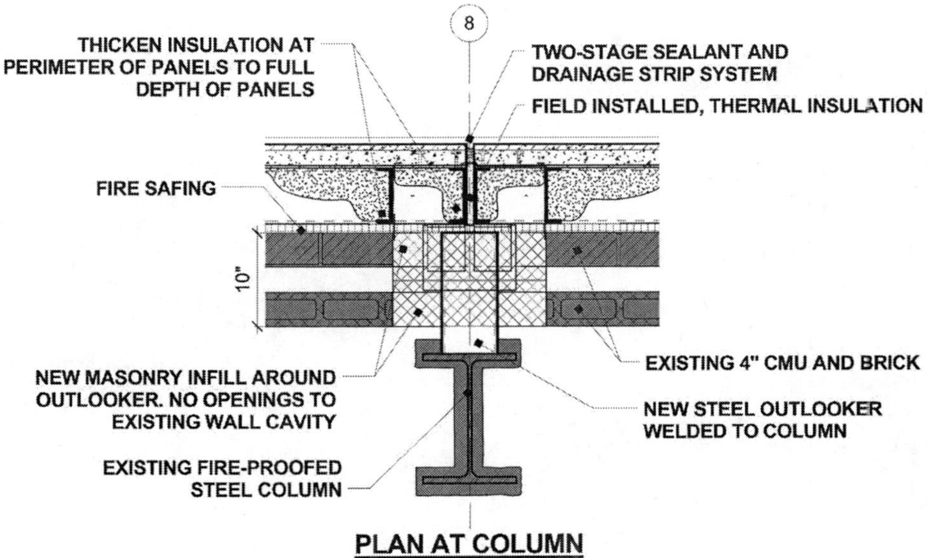

Figure 13 *Thin-brick precast—column support plan.*

masonry to make connections at the columns. The aluminum panel system has an excellent weight to size ratio, is weather resistant, provides continuous insulation when installed in conjunction with insulated joints between panels, limits thermal bridging, acts as an air/vapor barrier, and provides moisture resistance and management. The aluminum panel system permits shop fabrication with better quality control, an accelerated construction schedule, and good life-cycle costing.

This lightweight cladding system aided in keeping the aggregate overcladding within the code-defined limits for added dead load to the building structural frame. The 1 in. minimum air space between the back of the aluminum panel framing supports and the existing masonry wall is horizontally firestopped at every floor. The firestopping affords a means to compartmentalize the vertical chase between the new and existing envelopes to limit the spread of fire and smoke in the event of a fire and provides a means to limit the stack effect of convective air movement and the resulting impacts on the envelope's ability to prevent infiltration and exfiltration. The 1 in. void space was also periodically firestopped vertically to complete the compartmentalization. The insulated aluminum panel system was also used in a limited capacity in soffit areas.

Figure 14 *Metal wall panel and curtain wall installation—west elevation.*

Aluminum Curtain Wall System as Overcladding

The thermally improved aluminum curtain wall system was selected as the primary cladding system for the patient tower portion of the Nelson Harvey Building to provide a viable vision glazing strategy coupled with an aesthetic expression to reduce the visual mass of the building, as originally designed. With its alternating horizontal bands of heavy masonry and narrow windows, the existing building did not project a welcoming sight to patients. The new curtain wall system permitted shop fabrication of some components with better quality control, accelerated the construction schedule, and provided good life-cycle costing.

The portion of the existing masonry wall located from the existing window sill to the floor line was removed to aid in the reduction of the overall dead load of the building, allowed the installation of the new steel structure to support the curtain wall system, and provided additional floor space for the patient rooms (see Figures 16 and 17).

A new concrete floor extension was added to the existing concrete floor slab, bringing the face of the new slab to the exterior face of the existing brick. This created a two-hour rated floor slab to the back face of the curtain wall and effectively capped the former wall cavity. A new structural steel tube spans between existing structural steel columns and is held 1 ft above the existing floor slab to provide a connection point for the vertical curtain wall members (see Figures 17 and 18). The structural tube is supported at the column

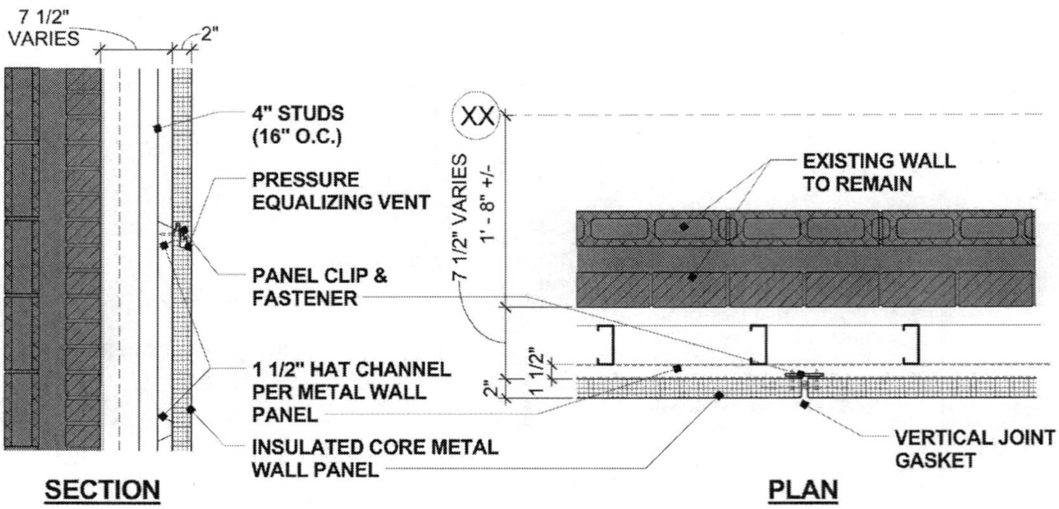

Figure 15 *Insulated aluminum wall panel assembly details.*

connections, in addition to third points along its length by stub-up tubes from the existing concrete floor, designed to limit vertical and horizontal tube deflection. This structural arrangement of a horizontal steel tube to provide curtain wall support has the added benefit of limiting the thermal conductive path between the building frame and the curtain wall mullions that is the bane of more traditionally framed curtain wall systems.

Where the curtain wall system is installed as overcladding of the existing masonry wall, the resultant void is insulated with 4 in. of mineral board insulation and firestopped at the head and sill of the vision-glazed areas of the curtain wall system, as shown in Figure 19. The portions of the aluminum curtain wall framing mullions that are concealed from view are fully wrapped in mineral board insulation. The steel support structure is fully insulated against thermal conductance in the event of fire and also serves to enhance the wall's R-value. The curtain wall system is vertically firestopped where the system interfaces with the precast and aluminum panel systems.

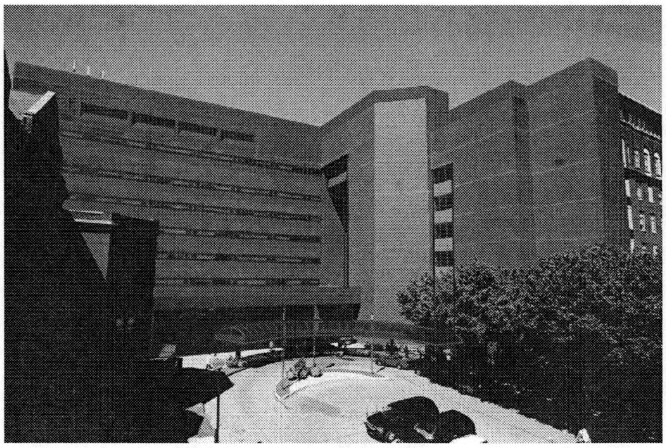

Figure 16 *Location of masonry (red) and glazing (blue) demolition—east elevation.*

The insulating glazing in the spandrel portion of the curtain wall system consists of an inner and outer unit of low-iron glass. The outside glass unit has a white silk-screened dot frit pattern in 20% and 60% densities on the #2 surface, in addition to a low-e coating on the same #2 surface, thus improving the unit's heat gain resistance and shading coefficient. The #4 surface has a gray uniform opacified coating (see Figure 20 inset illustration).

Roofs

Based on the point cloud data and on-site observation, existing parapets were removed, as these exhibited the greatest displacement and bowing. All new roofs and parapets have continuous thermal insulation with continuous air/water barriers. The white PVC roofing membrane with heat-welded seams is fully adhered to a glass-mat faced coverboard over two staggered layers of polyisocyanurate insulation. The PVC membrane is protected from above by both new and salvaged tongue-and-groove concrete-topped insulating roof pavers that serve to enhance the insulating value of the roofing system in addition to providing protection of the roofing system. A highly reflective surface coating was added to the pavers to enhance heat reflectivity (see Figure 21).

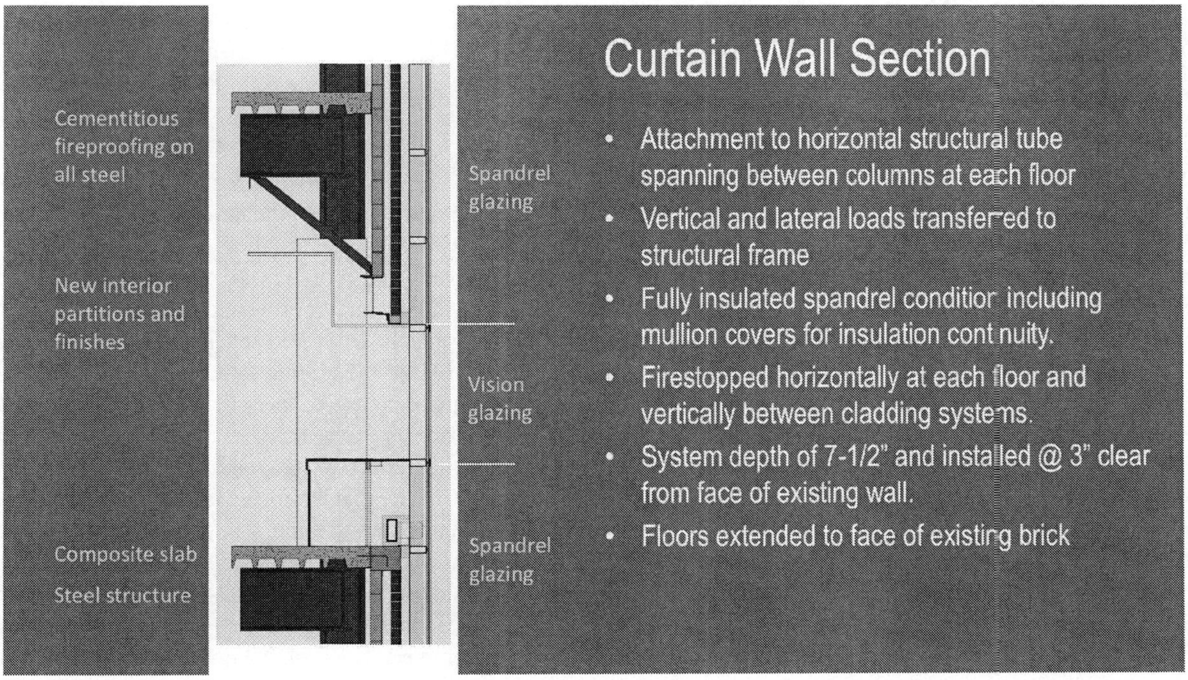

Figure 17 *Curtain wall section.*

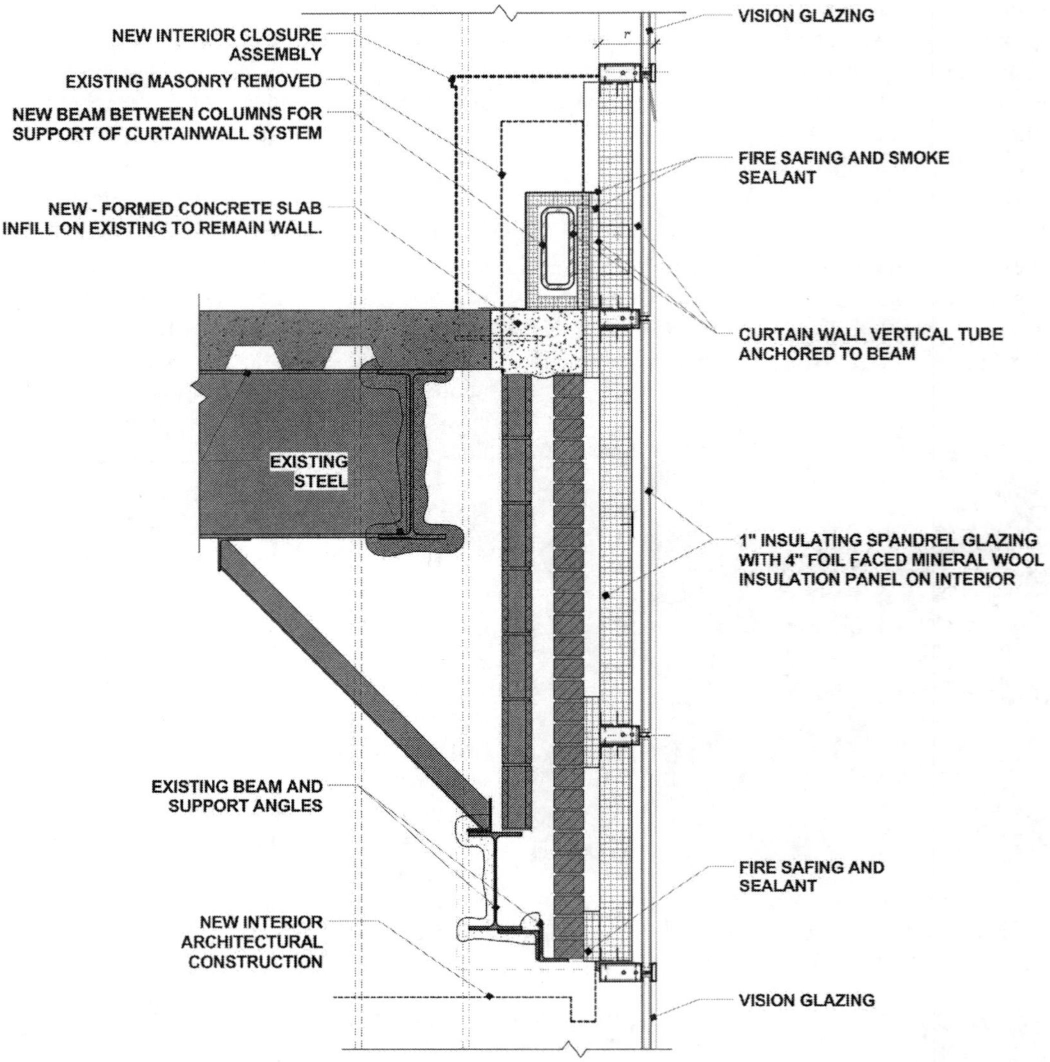

Figure 18 *Curtain wall support section.*

Extensive green roof systems consisting of drought-tolerant sedums are used where patient rooms and public spaces have direct views to the roof areas. The green roofs serve as a means of lowering reflective heat while providing a natural setting that is conducive for patient and staff well-being. A large skylight installed within the green roof area, located above the main lobby, augments the natural light permeating the lobby area through the glazing along the forecourt entry (see Figures 22 and 23).

Installation of Overcladding Systems

Selective cutouts of the existing brick masonry cavity wall occurred over much of the building exterior in order to install new structural outriggers designed to support the overcladding from the existing steel columns. The new steel framing was "needled" through the existing masonry façade through these small access ports cut into the masonry to afford anchorage points for the new façade systems. Once the existing façade was stabilized, the new insulating façade systems and roof systems were installed to encapsulate the existing masonry walls and roof areas. Demolition also included the removal of all existing roofing and glazing on the building, in addition to attached, covered pedestrian walkways connecting the Nelson Harvey Building to adjoining buildings. Most of the existing concrete-topped insulating roof pavers were salvaged for reuse over the new roofs. The existing exterior canopy on the building was removed and replaced by a new entrance structure that includes a large metal canopy roof over a glazed entrance vestibule. The existing masonry parapets were removed due to pronounced bowing and deflection from thermal expansion. Masonry walls below the ribbon windows on patient floors 3 through 8 were removed for reasons previously mentioned. The removal of the existing masonry aided in shedding dead-load weight from the building frame that permitted the extensive overcladding by the precast panels.

Figure 19 Installation of curtain wall and thin-brick precast wall systems over the existing facade—east elevation.

Maintaining the vast majority of the exterior wall system had the additional effect of enabling interior construction work to proceed without the necessity of erecting extensive temporary walls. This strategy allowed for faster dry-in time and improved the overall duration for scheduled construction activities.

PERFORMANCE AND METRICS

Energy Modeling

To measure the efficacy of revising the Nelson Harvey Building's envelope, the baseline and proposed conditions were analyzed using Sefaira, a SketchUp plug-in (Sefaria 2016; Wilson 2016). In both cases, the HVAC and hot-water heating components were modeled identically to isolate the impact of the building envelope changes on energy usage. Both models were run as a healthcare occupancy and located in Baltimore, Maryland, to accurately represent the impacts use group and location have on energy usage.

In the baseline case, a glazing U-factor of 0.61 Btu/h·ft^2·°F and a solar heat gain coefficient (SHGC) of 0.83 were applied to both façade and skylight glazing in all orientations.

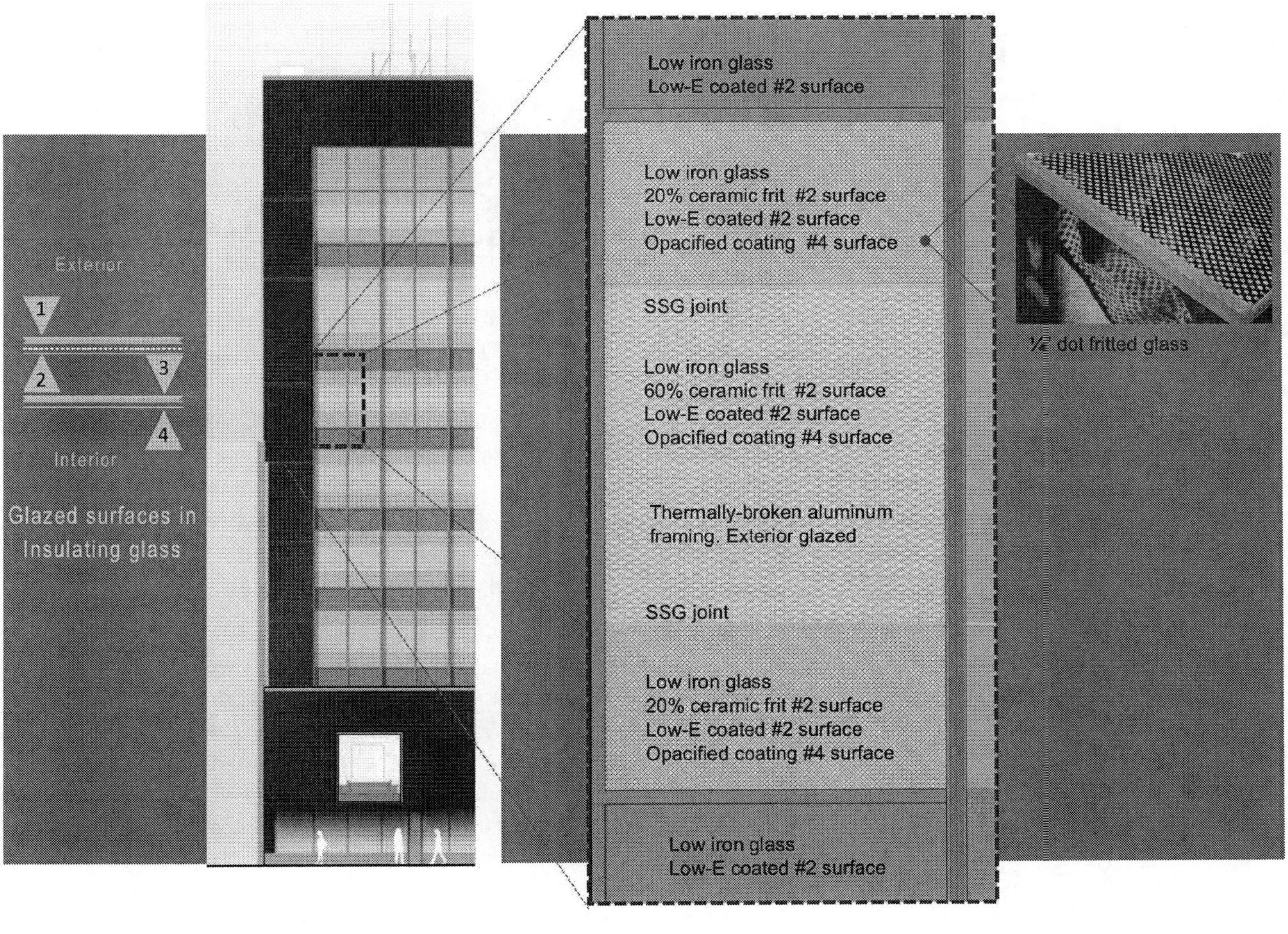

Figure 20 Glazing types at the curtain wall system overcladding.

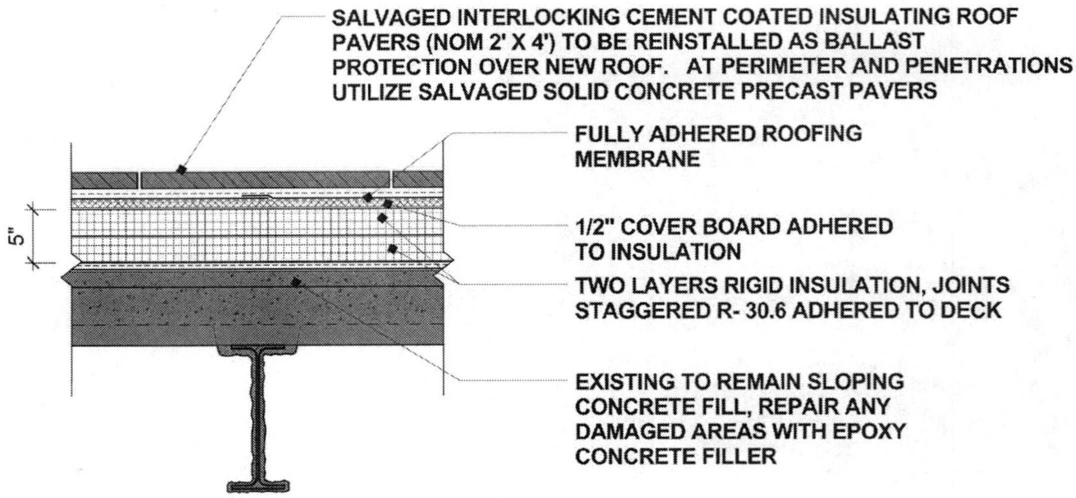

Figure 21 *Salvaged paver roof.*

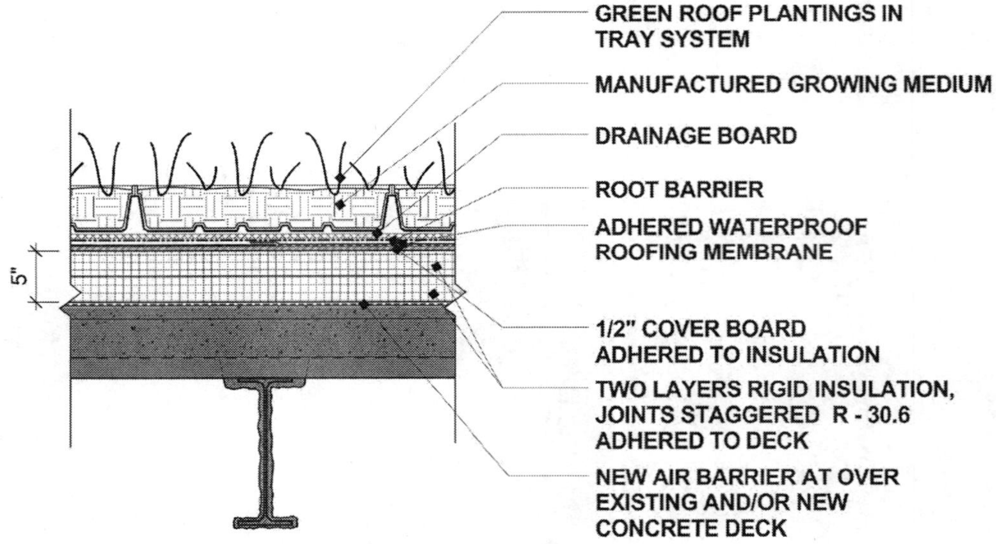

Figure 22 *Green roof.*

The exterior wall was uniformly defined to have an R-value of 9.20 h·ft^2·°F/Btu. The structure was assumed to have a leakage rate of 0.54 cfm/ft^2 at 75 Pa per ASHRAE/IES Standard 90.1 (ASHRAE 2013) and a surface reflectance of 0.5. Because testing data for leakage rate and surface reflectance were not available, these values were maintained in the proposed model as well so that their impact in comparing the performance of each model would be similar. It should be noted that the existing building likely had a higher air leakage rate than modeled owing to its age and exterior envelope failures. The hybrid envelope, conversely, likely has less air leakage than was modeled. ASHRAE/IES Standard 90.1 now includes guidelines about how to account for variable leakage rates between baseline and proposed conditions. In the future, similar work should reference the most recent ASHRAE guidelines and incorporate air leakage rates accordingly. The baseline roof was defined to have an R-value of 11.17 h·ft^2·°F/Btu.

Because Sefaira offers a limited number of opportunities to define materials, approximations were made in the representation of the proposed case that introduces uncertainty in the results. The modeling program allows only one glass and one exterior wall R-value to be defined per cardinal direction and required our team's best judgment to determine how to most accurately represent a materially complex exterior envelope that features multiple assemblies on each exposure. The exterior assemblies in the proposed design are: thin-precast panel, vision glazing, spandrel glazing, metal panel, stucco, green roof, and non-green roof (see Figure 24).

To generate the most accurate proposed model, each elevation's glass and exterior wall R-value were defined independently. The north elevation, featuring stucco and vision glass, was modeled using an exterior wall R-value of 16.8 and the vision glazing properties for the high-performance glazing installed on the project, which carried a U-factor of 0.29 Btu/h·ft^2·°F and an SHGC of 0.29. The south elevation, featuring metal panel as well as thin-precast in addition to vision and spandrel glazing, was defined using the exterior wall R-value of the thin-precast assembly and the vision glazing properties. All spandrel glazing was assumed to have the properties of the vision glazing because of the small quantity of spandrel glazing on this façade, and all metal panels were assumed to have the same R-value as the thin-precast because of the minimal variation in R-value between these two assemblies. The thin-precast assembly has an R-value of 33.17 while the metal panel assembly has an R-value of 31.66. The south elevation's two materials were therefore defined using the vision glazing properties, a U-factor of 0.29 Btu/h·ft^2·°F, an SHGC of 0.29, and the exterior wall R-value as 33.17.

Figure 23 *West elevation of the main patient tower (metal panels, curtain wall, and green roof).*

Wall and Roof System Insulation Comparison

Existing Cladding

Material	R-value	% of cladding	Weighted Average
Brick/CMU Cavity Wall	9.20	88	8.10
Brick/CMU Parapet	4.00	2	0.08
Metal Panels	14.34	10	1.43
		100	9.61 AVG R-value

There was a 227% increase in the insulation value of the building walls with the over-cladding.

New Hybrid Cladding

Material	R-value	% of cladding	Weighted Average
Thin Precast Panel	33.17	35	11.61
Thin Precast Panel Parapet	24.00	2	0.48
Type C1-Spandrel	29.15	6	1.75
Type C2-Spandrel	29.60	6.5	1.92
Type D-Spandrel	29.60	15	4.44
Metal Wall Panels	31.66	35.5	11.24
		100	31.44 AVG R-value

Existing Roofing

Material	R-value	% of Roofing	Weighted Average
PVC w/ insulated pavers	11.17	100	11.17 R-value

New Roofing

Material	R-value	% of cladding	Weighted Average
PVC adhered insulated	43.04	60	25.80
Green Roof	35.54	40	14.22
		100	40.00 AVG R-value

There was a 258 % increase in the insulation value of the building roofs with the new roofing.

Window to Wall Ratio (WWR)

Vision glazing total = 14,281 SF 17% of total wall SF

Wall area (including "glazed" wall) = 82,808 SF

WWR = 17% vision glazing

Figure 24 *Comparison of insulation values of new and existing wall and roof systems.*

The east elevation features vision and spandrel glazing as well as thin-precast panels on the tower and metal panels on the first two levels. Because the quantity of metal panels was minimal compared to the quantity of thin-precast panels, the east exterior wall was defined with an R-value of 33.17 to match the thin-precast assembly. Vision glazing was defined with a U-factor of 0.29 Btu/h·ft^2·°F and an SHGC of 0.29. The spandrel glazing was defined as the thin-precast exterior wall assembly because of the 4 in. of mineral wool insulation provided behind the spandrel glazing and the existing brick cavity wall that was preserved in place. Given the components, the spandrel assembly had an R-value of 29.15, which is far closer to the performance values of the thin-precast assembly than it is to the vision glazing.

The west elevation features vision and spandrel glazing as well as metal panels on the tower and thin-precast panels on the mechanical penthouse. Because the quantity of metal panels is significantly greater than the quantity of thin-precast panels, the west exterior wall was defined using the metal panel assembly and modeled using an R-value of 31.66. Similar to the east elevation, spandrel glazing was treated as having the same performance characteristics as the metal panel assembly because of the similarity in their performance. Vision glazing was defined with a U-factor of 0.29 Btu/h·ft^2·°F and an SHGC of 0.29.

The roof assembly was uniformly modeled without a green roof because of the overwhelming quantity of non-green roof area compared to the area that included a green roof assembly. The roof's R-value was defined as 43.04.

Because the proposed building has a multilayered envelope that must be approximated in the energy analysis tool, the predicted energy use intensity (pEUI) of the proposed building is of limited usefulness. The comparison between the proposed building's pEUI and the baseline building's pEUI, however, is a reasonable estimation of the proposed project's improvement over baseline. The proposed model resulted in a pEUI 17% lower than that of the baseline case, suggesting the improvements made to Nelson Harvey's building envelope resulted in significant energy savings.

Post-Occupancy Observations

The Nelson Harvey Building has been in operation for over two years since construction activities were completed (see Figures 25–27). The new hybrid envelope of insulating overcladding systems combined with the attributes of retaining the existing masonry wall system has greatly improved the building's resistance to heat transfer and moisture transmission. As a general rule, heat and moisture travel with infiltrating and/or exfiltrating air, and as materials get wet, the insulating properties are less effective. As a result, the relative humidity of the air moving in the wall system increases and the temperature of the materials at which the dew point is reached are lowered. The new envelope as a whole, with its improved thermal insulation and moisture-resistant overcladding, manages heat, air, and moisture much more effectively than

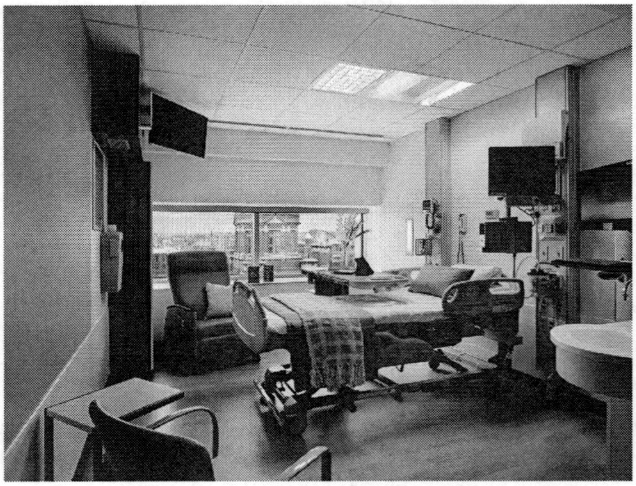

Figure 25 *Patient room in the main tower portion of the building, facing west (courtesy Halkin Mason Photography).*

Figure 26 *"Before" view of the Nelson Harvey Building from the east.*

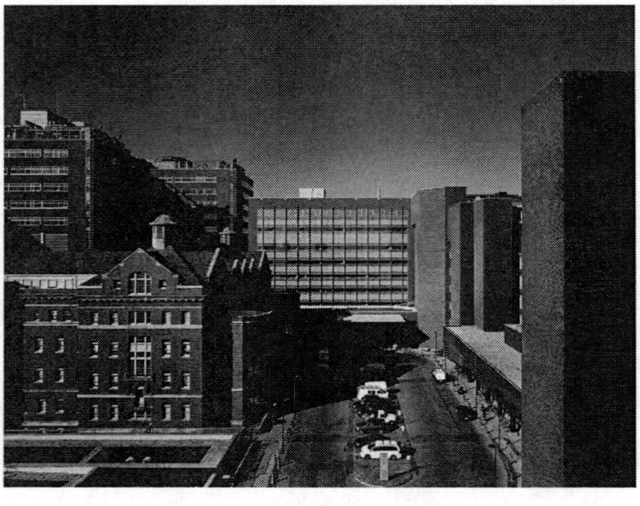

Figure 27 *"After" view of the Nelson Harvey Building from the east (courtesy Halkin Mason Photography).*

the previous construction. The mass walls at the exterior, and especially the mass wall at the interior, have greater inherent capacity to absorb excess moisture within the wall system and to release that moisture slowly over a period of time, either to the exterior or to the interior, depending on the location within the wall assembly.

An inherent property of precast concrete and masonry is the thermal mass. A wall system with high thermal mass has high heat capacity and will absorb and release heat slowly. The exterior overcladding precast panel system stores heat energy during the hot and humid summers in Maryland. Combined with the proper insulation and with minimal thermal bridging, the overcladding precast system slows down heat transfer, which reduces the energy needed to heat and cool the structure. When combined with the thermal mass of the retained existing masonry wall inboard of the insulation, a portion of the internal thermal energy is stored and made available passively to help maintain the desired indoor temperature. This further reduces the amount of energy needed to heat or cool the structure, thus improving energy efficiency even more. This translates to reduced operating costs for the hospital and more comfort for patients and staff. Johns Hopkins Hospital Facilities staff report significantly fewer complaints from patients and staff regarding thermal comfort conditions in the building than what occurred prior to the renovation. In addition to reducing operating costs of the building itself, this also translates into fewer maintenance visits to adjust HVAC equipment to meet occupant needs.

The thermal mass also aids in delaying when the inside peak heating and cooling times occur. This is the result of thermal lag, or the time it takes for heat to transfer through the materials of the wall assembly. As temperatures cycle throughout a 24-hour period, there is a point when the outdoor temperature reaches the peak high and low. The inside of the building generally has a temperature variance with the outdoor conditions and is always trying to equalize with the outdoor temperature. The greater the variance in temperatures, the more energy is required to maintain the indoor temperature at the desired level. If the interior building peak times can be altered to coincide when the variance in temperatures is less, then less energy will be required to maintain the desired interior temperature.

For the Nelson Harvey Building, the thermal mass effect of the combination of the exterior overcladding precast with the interior thermal mass of the existing masonry adjusts the inside temperature peaks by several hours relative to when the outdoor temperature peak occurs. This effect of the double thermal mass wall on the precast portions of the overcladding reduces the variance in temperature and the amount of energy required to maintain the interior temperature at the desired level. The benefit of the thermal mass effect holds true for both the aluminum panel and the glazed curtain wall overcladding where those systems are installed over the existing masonry wall of the building, albeit not to the same degree that the precast concrete overcladding areas over the existing masonry have, due to the lower mass of those systems. The insulating value of the hybrid building wall system has been improved in excess of 200% over the existing building envelope. The overall result is that the HVAC system for the Nelson Harvey Building should be able to perform more efficiently and economically to maintain the desired indoor temperature.

At roof areas, the thermal mass of the existing concrete slab serves to reduce the transfer of heat much as the precast panels and existing masonry do for the wall systems. In addition, the reuse of the concrete-topped insulated roof pavers over the new PVC roof membrane and primary roof insulation further limits heat transfer through the roof. Mass walls and roof systems have a greater effective R-value than the sum of their material R-values and contribute significantly to the building's ability to resist temperature transfer through the envelope. The insulating value of the new building roof system has improved by as much as 250% over the existing building roofing insulation value.

Thermal mass combined with limited thermal bridging can aid in the sizing of HVAC systems and thus aid in reducing the first costs of a project. Having reduced exfiltration and infiltration of air by employing a newer, tighter envelope over the existing building will certainly contribute to improving HVAC performance.

Field Thermal Imaging Analysis

The design team conducted a thermal imaging analysis of the exterior of the Nelson Harvey Building to ascertain the actual field performance of the envelope. The thermal images were taken in the month of January when the outdoor temperature was at the lowest and the differential between outdoor and indoor temperatures exceeded 30°F. The testing occurred at dawn to allow stored heat in the thermal mass of the building from the previous day to release in the overnight so as to not compromise the thermal readings. The result of this testing confirmed that the thermally improved envelope is performing as expected.

The thermal images of the east and west elevations of the curtain wall indicate heat transmission from conductance through the aluminum mullions as would be expected; however, the images also indicate the reduction in heat loss through conductance at the spandrel locations due to the added insulation between and around the aluminum mullions (see Figure 28). It is interesting to note in the thermal images how the vertical aluminum fins on the curtain wall increase thermal transmission due to their increased surface area.

The insulated precast panels are performing well with very limited thermal conductance through the actual panel (see Figure 29). The cold-formed metal framing of the precast panels are held off from the concrete via spaced Nelson studs, and these appear to be quite successful at limiting thermal conductance between the framing and the concrete facing. There are limited locations where a structural connection is evident behind the panel, as revealed by the thermal images.

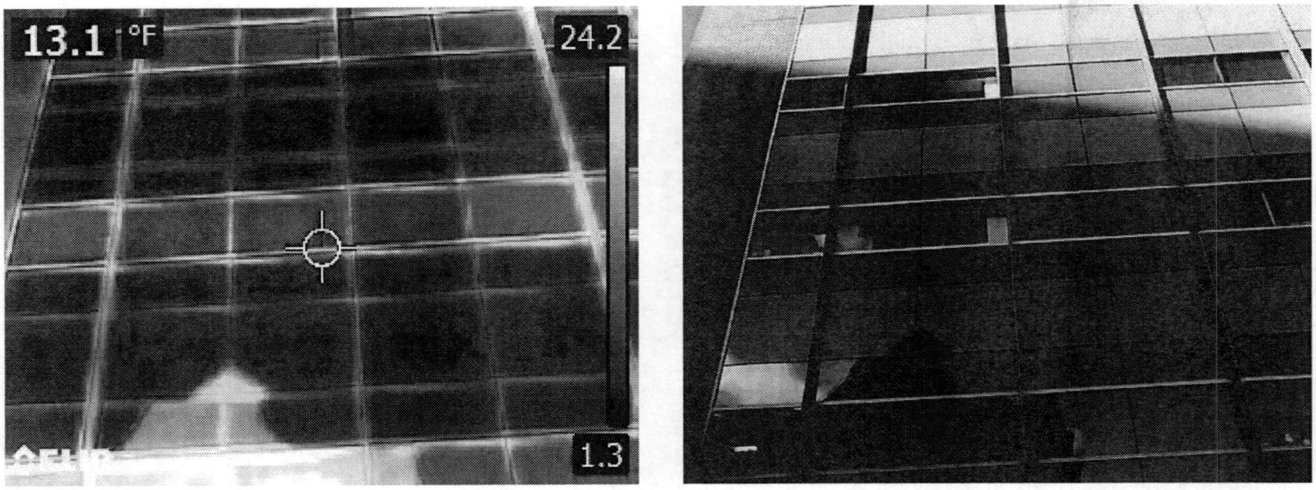

(a) (b)

Figure 28 *Thermal and visible imaging of the new east curtain wall system.*

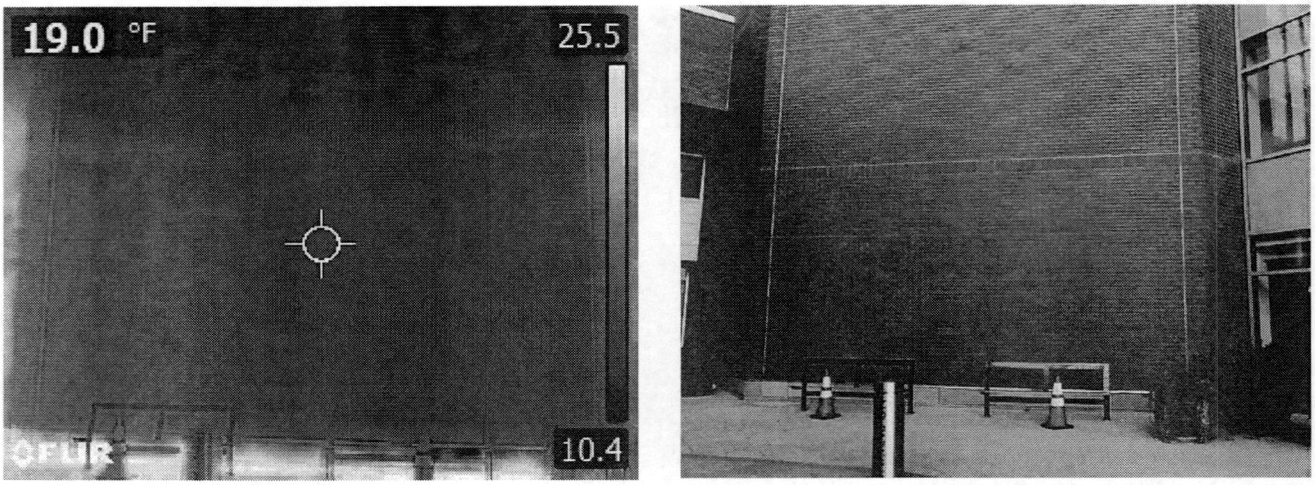

(a) (b)

Figure 29 *Thermal and visible imaging photos of the thin-brick precast wall panel system.*

There is limited thermal conductance through the metal panel system overcladding, as illustrated in Figure 30. As expected, the greatest thermal heat transfer occurs at the panel joints where there is limited insulation, in the case of vertical joints, and no insulation in the gasket or sealant joints.

The original masonry parapets of the building that were the source of so much thermal heat loss and movement were removed in the demolition phase of the project and replaced in the new construction by extending the vertical height of the precast panel overcladding to form a parapet. The new parapet is fully insulated on both sides and horizontally through the panel at the roof. The thermal images show little heat transmittance at the parapets.

Air exfiltration was evident in the thermal images at building entrances where door seals were being compromised. The thermal images of the east elevation exhibit substantial air exfiltration around the intake louvers at the mechanical penthouse level at the top of the building. The interiors of the plenums are fully insulated at that level, so the exfiltration of warmer air from around the louver is not fully understood at this time, but it may be attributed to loss of air around the annular space between the ducts and the insulated plenum panels. The thermal images show a uniform heat loss around the tops of each louver despite the plenum chambers being compartmentalized from each other. Another area of exfiltration that is evident is the horizontal connection between the curtain wall

(a) (b)

Figure 30 *Thermal and visible imaging photos of the metal panel system.*

and the metal panel system on the west elevation at the top of the building. This is likely attributable to an improperly sealed joint that is creating a breach in the insulation envelope. The thermography image revealed that the joint between the curtain wall and the metal panel was not correctly installed at this isolated location. These types of testing support the principle that a combination of on-site visual inspections during and after construction and post-installation field testing is paramount for quality control and quality assurance in validating envelope performance.

CONCLUSIONS

Design began in March 2011 and was completed in June 2012. Construction began October 2012 and was completed October 2014 (see Figures 31 and 32). The final project construction cost was substantially below the hospital's initial projected construction budget estimate; this enabled the hospital to expand upon the original project scope to include a new forecourt and plaza deck for the Nelson Harvey Building. The project achieved a Green Star 2 Level under the Baltimore City Green Building Certification Program (LEED® Silver certification level equivalent).

The design team's and owner team's intent at the outset of this project was to deliver a project that improved the aesthetic, thermal, and structural elements of the existing building through the execution of a design approach that would minimize capital costs, have minimal impact to the daily operations of the hospital during the renovation, and employ sustainable practices in both the building design and construction. The design concept of installing a new thermal envelope overcladding system that incorporates the existing masonry envelope into a new hybrid building envelope exploits the potential of the existing building's infrastructure. Resiliency, sustainability, thermal control, and moisture resistance are all enhanced through the process of recycling-in-place. Beyond the apparent benefits of improved building performance, the retention of a significant portion of the existing masonry envelope aided the construction process by contributing to shorter construction schedules and the limiting of demolition debris. Retaining the existing building's walls provided a more secure construction environment that shielded the interior materials and mechanical equipment from exposure to moisture and dust from the exterior construction site activities. To underscore the contribution that recycle-in-place has on sustainability is the realization that an estimated total of over 2130 tons of debris that would have been hauled by an estimated 71 dump trucks to landfill sites was avoided. Most importantly, critical-care patients within the Nelson Harvey Building and adjacent buildings were not subjected to the construction noise or airborne pollutants that extensive demolition would have wrought.

The comprehensive approach towards developing high-performance enclosures for aging structures that this case study illustrates began with developing alternative strategies for a hybrid thermal overcladding and recycling-in-place that enabled the Nelson Harvey Building to successfully be positioned as a viable part of the Medical Center's master plan for decades to come.

REFERENCES

ASHRAE. 2013. ANSI/ASHRAE/IES Standard 90.1-2013, *Energy standard for buildings except low-rise residential buildings*. Atlanta: ASHRAE.

Autodesk. 2012. AutoDesk Revit. Building information modeling software. San Rafael, CA: Autodesk.

Sefaria. 2016. Sefaira Architecture. Real-time analysis plug-in for SketchUp software program for high-performance building analysis. New York: Trimble Navigation, Ltd.

Figure 31 *New envelope—east elevation (courtesy Halkin Mason Photography).*

Figure 32 *New envelope—entrance canopy and vestibule (courtesy Halkin Mason Photography).*

Wilson, A. 2016. SketchUp 3-D building model and energy modeling with Sefaira plug-in. Baltimore: Ayers Saint Gross Architects, Inc.

BIBLIOGRAPHY

CE. 2010. Nelson Harvey Building structural analysis and condition survey of the exterior walls. Columbia, MD: Columbia Engineering, Inc.

DD. 2012. Digital laser survey: Point cloud 3-D model. Owings Mills, MD: Direct Dimensions, Inc.

Gale. 2011. Exterior evaluation report: Johns Hopkins Hospital Nelson Harvey Building. Towson, MD: Gale Associates, Inc.

Harburger, N. 2012. Revit 3-D building model and integration with point cloud 3-D model. Baltimore: Ayers Saint Gross Architects, Inc.

WS. 2011. Architectural programming and design of clinical areas for the Johns Hopkins Hospital Nelson Harvey Building. Gaithersburg, MD: Wilmot Sanz, Inc.

Oak Ridge National Laboratory
P.O. Box 2008
Oak Ridge, TN 37831-6070

ISBN 978-1-5108-6001-8

9 781510 860018

CURRAN ASSOCIATES INC.
proceedings.com